2013 天线与传播国际会议论文集
Proceedings of the 2013 International Symposium on Antennas and Propagation

（卷一）

东南大学毫米波国家重点实验室　编

东南大学出版社
·南京·

图书在版编目(CIP)数据

2013天线与传播国际会议论文集:Proceedings of the 2013 International Symposium on Antennas and Propagation:英文/东南大学毫米波国家重点实验室编.—南京:东南大学出版社,2013.10

ISBN 978-7-5641-4279-7

Ⅰ.①2… Ⅱ.①东… Ⅲ.①天线-国际学术会议-文集-英文 ②电波传播-国际学术会议-文集-英文 Ⅳ.①TN82-53 ②TN011-53

中国版本图书馆CIP数据核字(2013)第108576号

2013天线与传播国际会议论文集

Proceedings of the 2013 International Symposium on Antennas and Propagation

出版发行	东南大学出版社
社　　址	南京市四牌楼2号(邮编:210096)
出 版 人	江建中
经　　销	全国各地新华书店
印　　刷	常州市武进第三印刷有限公司
开　　本	787 mm×1092 mm 1/16
印　　张	89.75
字　　数	2304千字
版　　次	2013年10月第1版
印　　次	2013年10月第1次印刷
书　　号	ISBN 978-7-5641-4279-7
定　　价	360.00元(共两卷)

本社图书若有印装质量问题,请直接与营销部联系,电话:025-83791830。

Message from the General Chairmen

The ISAP2013 will be held in Nanjing, China on October 23-25, 2013. On behalf of the conference committees, it is our pleasure to welcome all of you to attend the 2013 International Symposium on Antennas and Propagation (ISAP2013) in Nanjing, one of the most beautiful and ancient cities in China.

The 2013 International Symposium on Antennas and Propagation (ISAP2013) provides an international forum for exchanging information on, and updating progresses of, the most recent research and development in antennas, propagation, electromagnetic wave theory, and other related fields. It is also an important objective of this meeting to promote professional networking among conference participants.

Nanjing was awarded the title of Famous Historic and Culture City because she had been Capitals of China for ten times. Today, Nanjing is the Capital City of Jiangsu Province and one of the important wireless communication hubs in China. The ISAP2013 is sponsored and organized by Southeast University, co-sponsored by University of Electronic Sci. & Tech. of China, and technically co-sponsored by CIE Antenna Society, the IEICE Communications Society, IEEE Antennas and Propagation Society, the European Association on Antennas and Propagation (EurAAP), Laboratory of Science and Technology on Antenna and Microwave, IEEE AP/MTT/EMC Nanjing Joint Chapter, the Jiangsu Institute of Electronics, Journal of Microwaves, etc.

ISAP2013 totally received 420 submissions, and finally 359 papers are accepted after its rigorous peer reviews by TPC members based on their technical merits and interests to the antennas and propagation communities. Among a large number of students' papers, 15 papers were selected by the TPC into the final-list for the best student paper contest.

Finally, please enjoy technical sessions of the conference, and also the Chinese culture in, and the beautiful modern city scenery of, the ancient Nanjing.

Prof. Wei Hong, Prof. Joshua Le-Wei Li, and Prof. Shuxi Gong
General Co-Chairs

October 23, 2013

ISAP2013 Committee Officers

General Co-chairs

Wei Hong (Southeast University)

*Le-Wei Li (*University of Electronic Science and Technology of China & *Monash University)*

Shu-Xi Gong (STAML， NRIET & XIDIAN)

IAC Co-chairs

Wen-Xun Zhang (Southeast University)

Koichi Ito (Chiba University)

Zhi-Ning Chen (National University of Singapore)

TPC Co-chairs

Xiao-Wei Zhu (Southeast University)

Dong-Lin Su (Beijing University of Aeronautics and Astronautics)

Yi-Jun Feng (Nanjing University)

Zhi-Peng Zhou (STAML， NRIET & XIDIAN)

Local Arrangement Co-chairs

Ji-Xin Chen (Southeast University)

Ya-Ming Bo (Nanjing University of Posts Telecommunications)

Bing Liu (Nanjing University of Aeronautics and Astronautics)

Exhibition Comm. Co-chairs

Ju-Lin He (CIE-AS)

Zhe Song (Southeast University)

Tie Gao (STAML， NRIET & XIDIAN)

Publication Comm. Co-chairs

Zhang-Cheng Hao (Southeast University)

Wen-Quan Che (Nanjing University of Science and Technology)

Zu-Ping Qian (PLA-UTS)

Can Lin (Nanjing Research Institute of Electronics Technology)

Finance Co-chairs

Guang-Qi Yang (Southeast University)

Chen Yu (Southeast University)

You-Cai Lin (STAML， NRIET & XIDIAN)

ISAP2013 IAC Members

Ajay Chakraborty (Indian Institute of Technology Kharagpur)
Chi Hou Chan (City University of Hong Kong)
Dau-Chyrh Chang (Oriental Institute of Technology)
Jin Pan (Univ. of Electronic Sci. and Tech.)
Juan Mosig (EurAAP)
Kam Weng Tam (University of Macau)
Ke Wu (Polytechnique Montreal)
Makoto Ando (Tokyo Institute of Technology)
Mazlina Esa (Universiti Teknologi Malaysia)
Prayoot Akkaraekthalin (King Mongkut's Inst. of Tech. North Bangkok)
W. Ross Stone (Stoneware Limited)
Wen Bin Dou (Southeast University)
Yang Hao (Queen Mary, University of London)
Yilong Lu (Nanyang Technological University)
Yingjie Jay Guo (CISRO)

ISAP2013 TPC Members

Bingzhong Wang (University of Electronic Science and Technology of China)
Cheng Liao (Southwest Jiaotong University)
Dau-Chyrh Chang (Oriental Institute of Technology)
Derek Gray (The University of Nottingham Ningbo)
Dhaval Pujara (Nirma University)
Fan Yang (Tsinghua University)
Feng Xu (Nanjing University of Posts Telecommunications)
Guohua Zhai (East China Normal University)
Guoqing Luo (Hangzhou Dianzi University)
Hao Xin (The University of Arizona)
Jian Yang (Chalmers Univ. of Technology)
Jiang Zhu (Apple Co.)
Julien Le Kernec (The University of Nottingham Ningbo)
Jun Hu (University of Electronic Science and Technology of China)
Kam Weng Tam (University of Macau)
Keisuke Konno (Tohoku University)
Kin-Lu Wong (Sun Yat-Sen University)
Kwok Wa Leung (City University of Hong Kong)
Lezhu Zhou (Peking University)
Lixin Guo (Xidian University)
Luyi Liu (Antenova Ltd, Cambridge)
Min Zhang (Shanghai Jiaotong University)
Qingxin Chu (South China University of Technology)
Qun Wu (Harbin Institute of Technology)
Ronghong Jin (Shanghai Jiaotong University)
Ruixin Wu (Nanjing University)
Tiejun Cui (Southeast University)
Weixing Sheng (Nanjing University of Science and Technology)
Wen Bin Dou (Southeast University)
Xiangyu Cao (Air Force Engineering University)
Xiaodong Chen (Queen Mary, University of London)
Xiaoxing Yin (Southeast University)
Xiuping Li (Beijing University of Posts and Telecommunications)
Xueguan Liu (Soochow University)
Xuexia Yang (Shanghai University)
Xun Gong (The University of Central Florida)
Ya-Ming Bo (Nanjing University of Posts Telecommunications)

Yan Zhang (Southeast University)
Yang Hao (Queen Mary, University of London)
Yi Huang (University of Liverpool)
Yi-Jun Feng (Nanjing University)
Yongchang Jiao (STAML， NRIET & XIDIAN)
Yongjun Xie (Beijing University of Aeronautics and Astronautics)
Yueping Zhang (Nanyang Technological University)
Yujian Cheng (University of Electronic Science and Technology of China)
Zhengwei Du (Tsinghua University)
Zhenqi Kuai (Southeast University)
Zhijun Zhang (Tsinghua University)
Zhongxiang Shen (Nanyang Technological Univ. Singapore)

ISAP2013 Secretariat Staff

Chuan Ge (Southeast University)
Fan Meng (Southeast University)
Wencui Zhu (Nanjing Normal University)
Yinjin Sun (Southeast University)
Mei Jiang (Southeast University)
Jun Chen (Southeast University)
Tao Zhang (Southeast University)
Lina Cao (Southeast University)
Yao Li (Southeast University)
Maomao Xia (Southeast University)
Zhihao Tang (Southeast University)

ISAP2013 Online Support

Guangqi Yang (Southeast University)
Kaihua Gu (Southeast University)

2013-10-23 AM

Room F, 2nd Floor

Keynote Speech

2013-10-23 PM
Room A

ORAL Session: WP-1(A)
Adv Ant for Radio-Astr. -1
Session Chair: Jian Yang, Bo Peng

ORAL Session: WP-2(A)
Adv Ant for Radio-Astr. -2
Session Chair: Biao Du, Jian Yang

2013-10-23 PM
Room B

ORAL Session: WP-1(B)
New Strategies of CEM-1
Session Chair: Chao-Fu Wang, Haogang Wang

ORAL Session: WP-2(B)
New Strategies of CEM-2
Session Chair: Wen-Yan Yin, Lianyou Sun

2013-10-23 PM

Room C

ORAL Session: WP-1(C)

UWB Antennas

Session Chair: Zhongxiang Shen, Qiang Chen

ORAL Session: WP-2(C)

Broadband Antennas

Session Chair: Derek Gray, Jin Shi

2013-10-23 PM

Room D

ORAL Session: WP-1(D)

Compact Antennas

Session Chair: Kin-Lu Wong, Yuehe Ge

ORAL Session: WP-2(D)

Small Antennas

Session Chair: Seong-Ook Park, Wen-Shan Chen

2013-10-23 PM
Room E

15:10 - 16:00

POSTER Session: WP-C
Best Student Papers Contest

2013-10-23 PM

Room E

15:10 - 16:00	**POSTER Session: WP-P**

2013-10-24 AM

Room A

ORAL Session: TA-1(A)

EurAAP/COST

Session Chair: Per-Simon Kildal

ORAL Session: TA-2(A)

Wireless Power Transmis.

Session Chair: Le-Wei Li, Qiang Chen

2013-10-24 AM
Room B

ORAL Session: TA-1(B)
Computational EM
Session Chair: André Barka, Yaming Bo

ORAL Session: TA-2(B)
EM Scattering
Session Chair: Kiyotoshi Yasumoto, Zhenhai Shao

ORAL Session: TA-1(C)
WLAN Antennas
Session Chair: Jui-Han Lu, Yuan Yao

ORAL Session: TA-2(C)
Patch Antennas
Session Chair: Dhaval Pujara, Xinmi Yang

ORAL Session: TA-1(D)
Measurements
Session Chair: Hiroyoshi Yamada, Ji Yang

ORAL Session: TA-2(D)
Radio Propagation
Session Chair: Mazlina Esa, Dongya Shen

2013-10-24 AM

Room E

09:40 - 10:30	**POSTER Session: TA-P**	

2013-10-24 PM
Room A

ORAL Session: TP-1(A)
Body-central Antennas
Session Chair: Koichi ITO, Zhao Wang

ORAL Session: TP-2(A)
Body-central Propagation
Session Chair: Yang Hao, Yoshinobu Okano

2013-10-24 PM

Room B

ORAL Session: TP-1(B)

SIW Antennas & Devices

Session Chair: Jian Yang, Jiro Hirokawa

ORAL Session: TP-2(B)

Integrated MMW Antennas

Session Chair: Yueping Zhang, Bing Zhang

2013-10-24 PM
Room C

ORAL Session: TP-1(C)

Reflector & Air-fed Array

Session Chair: Hisamatsu Nakano, Zhi-Hang Wu

ORAL Session: TP-2(C)

Array for Radar Systems

Session Chair: Cornelis G. van 't Klooster, Fan Yang

2013-10-24 PM
Room D

ORAL Session: TP-1(D)
Mobile & Indoor Propag.
Session Chair: Zhizhang Chen, Nadir Hakem

ORAL Session: TP-2(D)
Wire Antennas
Session Chair: Dau-Chyrh Chang, Hiroyuki Arai

2013-10-24 PM

Room E

15:10 - 16:00 **POSTER Session: TP-P**

2013-10-25 AM

Room A

ORAL Session: FA-1(A)

A &P for Mobile Comm.

Session Chair: J. W. Modelski, Eko T Rahardjo

ORAL Session: FA-2(A)

A &P for MIMO Comm.

Session Chair: Richard W Ziolkowski, Jiaying Zhang

2013-10-25 AM

Room B

ORAL Session: FA-1(B)

MMW & THz Antennas

Session Chair: Xiaodong Chen, Makoto Ando

ORAL Session: FA-2(B)

MMW Antennas

Session Chair: Takeshi Manabe, Yan Zhang

ORAL Session: FA-1(C)

Ant. Analysis & Synthesis

Session Chair: Riccardo E Zich, Toru Uno

ORAL Session: FA-2(C)

Freq. Selective Surface

Session Chair: Toshikazu Hori, Zhenguo Liu

2013-10-25 AM

Room D

ORAL Session: FA-1(D)

EM in Circuits-1

Session Chair: Dhaval Pujara, Qunsheng Cao

08:00 - 08:20 Simplified Modeling of Ring Resonator (RR) and Thin Wire Using Magnetization and Polarization with Loss Analysis **973**
Dongho Jeon, Bomson Lee (South Korea)

08:20 - 08:40 Transmission Characteristics of Via Holes in High-Speed PCB **977**
He Xiangyang, Lei Zhenya, Wang Qing (China)

08:40 - 09:00 Novel W-slot DGS for Band-stop Filter **981**
Chen Lin, Minquan Li, Wei Wang, Jiaquan He, Wei Huang (China)

09:00 - 09:20 Coupled-Mode Analysis of Two-Parallel Post-Wall Waveguides **984**
Kiyotoshi Yasumoto, Hiroshi Maeda, Vakhtang Jandieri (Japan)

09:20 - 09:40 Systematic Microwave Network Analysis for Arbitrary Shape Printed Circuit Boards With a Large Number of Vias **988**
Xinzhen Hu, Liguo Sun (China)

ORAL Session: FA-2(D)

EM in Circuits-2

Session Chair: Trevor S. Bird, Wenmei Zhang

10:30 - 10:50 A Novel Phase Shifter Based on Reconfigurable Defected Microstrip Structure (RDMS) for Beam-Steering Antennas (Invited Paper) **993**
Can Ding, Jay Y.Guo, Pei-Yuan Qin, Trevor S.Bird, Yintang Yang (Australia)

10:50 - 11:10 Transient Response Analysis of a MESFET Amplifier Illuminated by an Intentional EMI Source **997**
Qifeng Liu, Jingwei Liu, Chonghua Fang (China)

11:10 - 11:30 Crosstalk Analysis of Through Silicon Vias With Low Pitch-to-diameter ratio in 3D-IC **1001**
Sheng Liu, Jianping Zhu, Yongrong Shi, Xing Hu, Wanchun Tang (China)

11:30 - 11:50 Design of a Feed Network for Cosecant Squared Beam based on Suspended Stripline **1005**
Huiying Qi, Fei Zhao, Lei Qiu, Ke Xiao, Shunlian Chai (China)

11:50 - 12:10 The study on Crosstalk of Single Wire and Twisted-Wire Pair **1008**
Lijuan Tang, Zhihong Ye, Linglu Chen, Zheng Xiang, Cheng Liao (China)

2013-10-25 AM
Room E

09:40 - 10:30 **POSTER Session: FA-P**

2013-10-25 PM
Room A

ORAL Session: FP-1(A)
Antennas for RFID
Session Chair: Kyeong-Sik Min, Wen Wu

ORAL Session: FP-1(B)
Inversed Scattering
Session Chair: Yisok Oh, Xincheng Ren

13:30 - 13:50 Inversion of the dielectric constant from the co-polarized ratio and the co-polarized discrimination ratio of the scattering coefficient **1167**
Yuanyuan Zhang, Yaqing Li, Zhensen Wu, Xiaobing Wang (China)

13:50 - 14:10 Microwave Radiation Image Reconstruction Based on Combined TV and Haar Basis **1171**
Lu Zhu, Jiangfeng Liu, Yuanyuan Liu (China)

14:10 - 14:30 Imaging of object in the presence of rough surface using scattered electromagnetic field data **1175**
Pengju Yang, Lixin Guo, Chungang Jia (China)

14:30 - 14:50 A Simple and Accurate Model for Radar Backscattering from Vegetation-covered Surfaces **1179**
Yisok Oh, Soon-Gu Kweon (South Korea)

14:50 - 15:10 The Analysis of Sea Clutter Statistics Characteristics Based On the Observed Sea Clutter of Ku-Band Radar **1183**
Zhuo Chen, Xianzu Liu, Zhensen Wu, Xiaobing Wang (China)

ORAL Session: FP-1(C)
Slot Antennas

Session Chair: Tsenchieh Chiu, Peng Chen

2013-10-25 PM
Room D

ORAL Session: FP-1(D)

A&P in Mata-structures

Session Chair: Youngjoong Yoon, Yijun Feng

2013-10-25 PM
Room E

15:10 - 16:00	**POSTER Session: FP-P**	

Keynote Speech

October 23 (WED) AM

Room F

A teleco's view for better and better customer expectations in multi-band, multi-network, multi-device and multi-demand smart society

Shinichi Nomoto, Ph.D
KDDI R&D Laboratories Inc.
2-1-15 Ohara, Fujimino-shi, Saitama, 356-8502 Japan
sh-nomoto@kddi.com

Abstract:
Since people use smart phones in daily life, the mobile traffic over the network is changing. The rich content such as video streaming with high quality becomes popular and popular, resulting in huge traffic. The 4G (LTE) system with high capacity, launched in 2012 and now under rapid deployment, may not be sufficient to cope with the explosion. KDDI, the second-largest telco in Japan, is accelerating R&D activities towards LTE-Advanced, including Multi-User MIMO, Small Cell Enhancement, and Small-sized Active Antenna for Multi-band Basestations. Also, KDDI has a broader view under the name of "3M strategy" which comprises "Multi- Network", "Multi-Device" and "Multi-Use." We believe that further network enhancement from "Dumb Pipe" to "Smart Pipe" is the key for user-centric smarter life. Related R&D activities backed by Big Data will be introduced.

Biography

Shinichi NOMOTO received B.E., M.E., and Ph.D degrees, all in electrical engineering, from Waseda University, Tokyo, Japan, in 1980, 1982, and 1993, respectively. He joined Kokusai Denshin Denwa Co., Ltd. (now KDDI Corp.), in 1982. Since 1983, he has been engaged in research and development of radio transmission systems. As a professional assignee at Inmarsat HQ's from 1992 to 1995, he has contributed to the "Inmarsat-P (ICO)" project, which includes development of a global personal communications system using a number of non-geostationary satellites.

He is a Vice President, Managing Director, of KDDI R&D Laboratories, Inc., an R&D fellow of KDDI, a fellow of IEICE, a senior member of IEEE, and a Chairman of the Standardization Council in the Telecommunication Technology Committee (TTC). He has also been a visiting professor of Waseda University, Tokyo University of Agriculture and Technology, University of Electro-Communications, Tokyo Institute of Technology, Keio University, and Doshisha University. He received the Shinohara Memorial Young Researchers' Award from IEICE in 1988, the Piero Fanti International Prize from INTELSAT/Telespazio in 1988, and the Radio Distinguished Award from RCR (now ARIB) in 1991. In 2004, two of his published papers received the Best Paper Awards from IEICE, one of which was the recipient of the 10th Inose Award (the very best paper of the year) too. In 2010, he received the Prize for Science and Technology (Development Category) in the Commendation for Science and Technology by the Minister of Education, Culture, Sports, Science and Technology. He also received the 58th Maejima Hisoka Prize from Tsushinbunka Association in 2013.

978-7-5641-4279-7

4G/Multiband Handheld Device Antennas and Their Antenna Systems

Kin-Lu Wong
Department of Electrical Engineering
Sun Yat-sen University, Kaohsiung, Taiwan
http://antenna.ee.nsysu.edu.tw

Abstract

Promising 4G/multiband antennas for handheld devices will be presented. Some low-profile, small-size and wideband techniques for LTE/WWAN antennas will be addressed. The ground antenna design concept and promising ground antenna structure will be introduced, which is especially suitable for slim, flexible handheld devices. The promising antenna systems using the same for achieving wideband high-isolation antenna systems for MIMO, diversity or dual WWAN operation will be discussed.

Future trends for the handheld device antennas including the reconfigurable and tunable antennas that can be adaptive to environmental changes or tunable to cover different bands or switched to have multi-beams or suitable for antenna systems will also be discussed.

Biography

Prof. Kin-Lu Wong is Sun Yat-sen Chair Professor of Sun Yat-sen University, Kaohsiung, Taiwan. He has published more than 500 refereed journal papers and 250 conference articles. He holds over 200 patents and is the author of three books including Compact and Broadband Microstrip Antennas (Wiley, 2002) and Planar Antennas for Wireless Communications (Wiley, 2003). Dr. Wong's published works have been cited over 14,000 times in Google Scholar.

Dr. Wong is an IEEE Fellow and received many awards including Taiwan Science Council Outstanding Distinguished Researcher in 2013, top 50 NSC scientific achievements in past 50 years (1959~2009) in Taiwan, and the Academic Award from Ministry of Education of Taiwan, in 2012. He was selected as top 100 honor of Taiwan by Global Views Monthly in August 2010 for his contribution in mobile antenna researches. Dr. Wong received the 2008 APMC Best Paper Award (APMC Prize), and is an IEEE AP-S Awards Committee member (2011~2013). Dr. Wong was General Chair of 2012 APMC and will also serve as General Chair of 2014 ISAP at Kaohsiung, Taiwan.

978-7-5641-4279-7

Rethinking the Wireless Channel for OTA testing and Network Optimization by Including User Statistics: RIMP, Pure-LOS, Throughput and Detection Probability

Per-Simon Kildal, Professor
Chalmers University of Technology

Abstract

The reverberation chamber has through the last thirteen years been used to emulate a rich isotropic multipath (RIMP) environment, and it has successfully been demonstrated that it can be used to test performance of multiport antennas and complete wireless terminals with MIMO and OFDM. The measured throughputs of practical LTE devices have been shown to be in excellent agreement with basic theoretical algorithms.

Now is the time to use this concept and complete the picture so that also real-life environments can be covered. This is done by introducing the pure-LOS as another limiting environment, and by introducing the statistics of the user. The latter plays a major role in pure-LOS that thereby becomes a random-LOS. The two limit-environments are linked together with a real-life hypothesis, and work has started to test this by simulations.

It will be shown that the major characterizing quantity becomes the detection probability of the single or multiple bit streams (for diversity and multiplexing cases, respectively) over an ensemble of users. This detection probability becomes equal to throughput in a multipath environment, readily seen through a simple threshold receiver model representing an ideal digital receiver.

The new approach represents a way to start optimizing the wireless networks by taking the statistics of the user into account.

Biography

Professor Per-Simon Kildal, Distinguished Lecturer of IEEE Antennas and Propagation Society 2011-2013

Per-Simon Kildal is professor in antennas at Chalmers University of Technology in Gothenburg, Sweden since 1989. He is teaching antennas and heading a group doing research on antenna systems. Until now, 19 persons have received a Ph.D. from him.

Kildal received two doctoral degrees from the Norwegian Institute of Technology in Trondheim. He is a Fellow of IEEE since 1995, and in 2011 he was awarded the prestigious Distinguished Achievements Award from the IEEE Antennas and Propagation Society. Kildal has authored more than 120 articles in scientific journals; concerning antenna theory, analysis, design and measurements, two of which was awarded best paper awards by IEEE (1985 R.W.P. King Award and 1991 Schelkunoff Prize Paper Award).

Kildal's research is innovative and industrially oriented, and has resulted in several patents and related spinoff companies. He has done the electrical design of the 40m x 120 m cylindrical reflector antenna and line feed of the EISCAT scientific organization, and the dual-reflector Gregorian feed of the 300 m Ø radio telescope in Arecibo. He is the inventor

978-7-5641-4279-7

behind technologies such as dipole with beam forming ring, the hat antenna, and the eleven feed. Kildal's hat-fed reflectors have till now been manufactured in more than 930 000 copies for use in radio links.

Kildal was the first to introduce the reverberation chamber as an accurate measurement instrument for Over-The-Air (OTA) characterization of small antennas and wireless terminals for use in multipath environments with fading, commercialized in Bluetest AB.

Kildal is also the originator of the concept of soft and hard surfaces from 1988, today being regarded as the first metamaterials concept. This concept is the basis of his newest and most fundamental invention, the gap waveguide technology.

Kildal organizes and lectures in courses within the European School of Antenna (ESoA, www.antennasvce.org). His textbook Foundations of Antennas - A Unified Approach (Lund, Sweden: Studentlitteratur, 2000) was well received, and is now in the process of being revised.

Rethinking the Wireless Channel for OTA testing and Network Optimization by Including User Statistics:

RIMP, pure-LOS, Throughput and Detection Probability

Per-Simon Kildal, Distinguished Lecturer of IEEE Antennas and Propagation Society
Department of Signals and Systems
Chalmers University of Technology
41296 Gothenburg, Sweden
per-simon.kildal@chalmers.se

***Abstract*-The reverberation chamber has through the last thirteen years been used to emulate a rich isotropic multipath (RIMP) environment, and it has successfully been demonstrated that it can be used to test performance of multiport antennas and complete wireless terminals with MIMO and OFDM. The measured throughputs of practical LTE devices have been shown to be in excellent agreement with basic theoretical algorithms.**

Now is the time to use this concept and complete the picture so that also real-life environments can be covered. This is done by introducing the pure-LOS as another limiting edge environment, and by introducing the statistics of the user. The latter plays a major role in pure-LOS that thereby becomes a random-LOS. The two limit-environments are linked together with a real-life hypothesis, and work has started to test this by simulations.

It will be shown that the major characterizing quantity becomes the detection probability of the single or multiple bit streams (for diversity and multiplexing cases, respectively) over an ensemble of users. This detection probability becomes equal to throughput in a multipath environment, readily seen through a simple threshold receiver model representing an ideal digital receiver.

The new approach represents a way to start optimizing the wireless networks by taking the statistics of the user into account.

I. Introduction

Previously all tests of antenna systems were made in anechoic chambers, i.e. with a pure Line-Of-Sight (pure-LOS) between the wireless device (terminal) and the base station. Therefore, traditionally all accurate and repeatable characterization methods were adapted to this method. However, the wireless systems of today are designed for use in multipath i.e. a field environment with many incident waves. These are due to reflections and scattering from objects in the environment, such as houses, trees, vegetation, humans, cars and so on. Naturally, the pure-LOS environment cannot be used to test the algorithms and hardware that handle the multipath.

The reverberation chambers emulate a rich isotropic multipath environment (RIMP). It was originally, for three decades, used for EMC measurements [1]. If it is well designed and large enough, it represents an ideal RIMP environment [2]. Its performance is based on well-accepted theories [3, 4] and it has during the last decade shown its ability to accurately measure efficiency, diversity gain and maximum available MIMO capacity [2, 5] of passive antenna systems as well as radiated power, receiver sensitivity [6], diversity gain and throughput data rate [7] of active mobile devices. The early basic works [2, 5] were later updated to control the time and frequency domain characteristics of the reverberation chamber i.e. delay spread and coherence bandwidth, respectively [8] as well as fading speed and Doppler spread. The most ground breaking of the recent developments is the introduction of a simple threshold model for an advanced digital receiver. Using this, it is possible to predict curves showing throughput data rate versus signal power that are within fractions of a dB from the same measured curves in a reverberation chamber [7], both for single [9] and multiple bit streams [10]. The accuracy of the prediction is of course also due to a good understanding and accurate calibration of the reverberation chamber [11]. The threshold model is therefore a very useful representation of an advanced digital receiver. The threshold function is observed in the stationary case (i.e. with no fading) with added white Gaussian noise and is a result of advanced error correction coding.

The present paper will link the pure-LOS and RIMP environments to real-life environments by a hypothesis (initially presented in [12]), and then show how we can handle the statistics of the user in the pure-LOS environment by CDFs and detection probabilities. The latter is readily achieved via the threshold receiver described above [7].

II. Two Edge Envirenoments and a Real-life Hypothesis

A real-life multipath environment will have characteristics between the two extreme or rather limiting pure-LOS and RIMP environments, also referred to as edge environments. The real-life environments may have both LOS and multipath, and the latter is not necessarily rich. Therefore, we propose to systematize the research on wireless systems and components by using a "real-life hypothesis" that quite reasonably states that: *wireless systems and devices that are optimized for best performance subject to the statistics of the user, in both the limiting pure-LOS and RIMP environments, will also have*

978-7-5641-4279-7

optimum performance in real-life. The table in Figure 1 summarizes all this.

Thus, the systematic research approach will be to: i) treat the environment as two separate edge environments i.e. the statistical pure LOS and the RIMP, both subject to the statistics of the user, ii) determine technical solutions, algorithms and new systems that are optimum in each of the two reference environments under the statistical variation of different users and user practices (by experimental and theoretical research), and finally iii) prove (mainly by numerical simulation) under which conditions the real-life hypothesis holds, and if needed re-optimize taking some special real-life situations into account with specific weights on the two edge environments.

This systematic approach is new, because the RIMP environment was not defined and used before for system analysis. The RIMP environment has the advantage of providing unique repeatable results from a performance assessment point of view, which thereby enables optimization of the system or the devices. Real-life environments cannot give unique results because the results will depend on the time when they were done, where the device was located, and even how both the user and the device were oriented with respect to both each other and the environment. However, the real-life hypothesis indicates that if performance is evaluated in a large number of real-life environments and scenarios, the overall performance will be optimum if we have ensured optimized performance in the two edge environments.

III. Random pure-LOS environment due to User Variations

Actually, the pure-LOS is also a statistical field environment, due to the arbitrariness of a mobile user both in orientation and in the way the user holds the mobile device. Such user variations have been studied before, but actually not the fact that they will introduce a statistical variation of the received signal also in a pure-LOS environment. This statistical variation is present even for stationary users, within an ensemble of users. This represents the rethinking. The LOS must be considered as being a statistical environment due to the arbitrariness of the orientation of the single user, due to variations between users, and due to different user practices related to e.g. how the mobile device is held. Thus, this also creates a new scenario for wireless communication systems, i.e. *how to deal with the angular variations among users when optimizing systems performance, and in particular in a non-rich multipath environments with a LOS component.*

Probability has previously been taken into consideration when optimizing system performance in fading, but the statistical orientation of the user has been forgotten. In particular, it has been forgotten that we need to ensure good performance for the majority in an ensemble of users. Thus, statistical optimizations must be done over an ensemble of users, not only over a single user, in order to draw fair conclusions about reception quality and data rate. We must ensure that as many users as possible have sufficient communication quality. The optimization criteria should be to minimize the power consumption needed for a certain quality of reception over an ensemble of statistically oriented different users.

The wireless channel is defined as the voltage received on the receiving terminal relative to that transmitted from the base station. Thus, this includes the antennas on both sides and the environment between them, and the user is of course part of this environment. *We claim that the statistical variation among the stationary users is equally important for the performance as the statistics of the fading of a single moving user.*

Environment	Equivalent measurement method	Antenna quality measure	MIMO and diversity capability
Free space (pure-LOS)	Anechoic chamber	Deterministic case: Realized gain User-random case: Not known	To some degree
Real-life environments		No unique quality measure	Yes
Rich isotropic multipath (RIMP)	Reverberation chamber	Total radiation efficiency e_{rad}	Yes

Hypothesis: If wireless terminals work well in RIMP and random pure-LOS environments, they will work well also in real-life environments.
It remains to be proven.

Figure 1. Two limiting reference environments (edge environments) and definition of a real-life hypothesis.

IV. Characterizing Metrics in RIMP and random-LOS

We will now discuss how to characterize the statistics of the user in RIMP and in random pure-LOS. We will also refer to the random pure-LOS case simply as random-LOS.

A. Cumulative random-LOS- and RIMP-Diversity Gains of the 1% Worst Users in dB-Rayleigh (dBR)

We have already defined and calculated some diversity gains in pure-LOS environments [13, 14]; similar to what we did some years ago for RIMP [2]. This is for the random-LOS case defined as the cumulative diversity gain of the 1% worst users when the terminal has a randomized orientation in pure-LOS. Initially, we have chosen a three-dimensionally (3D) random orientation, valid for smartphones that can be used with any orientation of the screen. For this case the distribution of received signal amplitudes over many arbitrary users actually turns out in practice to be very close to the Rayleigh shape obtained during normal fading. Still, this pure-LOS case does not represent fading because each terminal is approximately stationary. *This closeness to the Rayleigh distribution is found to be valid for practical small antennas when located on or close to objects of unknown complex shape like the chassis of a mobile phone and a human hand or head or body.* We have also introduced a unit for quantifying diversity gain. This is dBR meaning dB *relative to Rayleigh distribution*, see [14].

B. Ideal threshold receiver and detection probability

The ideal threshold receiver is a fundament of the present approach. This definition and the resulting throughput model was already published in [7]. The results (also shown in Figure 2 herein) are valid for advanced digital communication systems like LTE, and can be interpreted like this:

The throughput in RIMP shows for the stationary case a step slope, similar to the conducted AWGN case shown in Figure 2. However, when we plot the throughput versus the transmit power (or preferably versus the average received reference power P_{av} at the multi-port antenna like in this paper) in RIMP, the threshold appears to be different from conducted case due to the interference between all incoming waves. When we move to another position, the incoming waves interfere differently, so the threshold appears at a different level of P_{av}. Therefore, for a given P_{av} we will in some places in the RIMP environment be above the threshold, and in other places below it, so the throughput evaluated over many positions in the RIMP (or for a moving user) will correspond to a detection probability: In each position the user either has a channel or he doesn't. Thus, by the above explanation, we have been able to interpret the throughput of a moving user as a detection probability over an ensemble of stationary users in RIMP. This metric of detection probability is directly extendable to the random-LOS case.

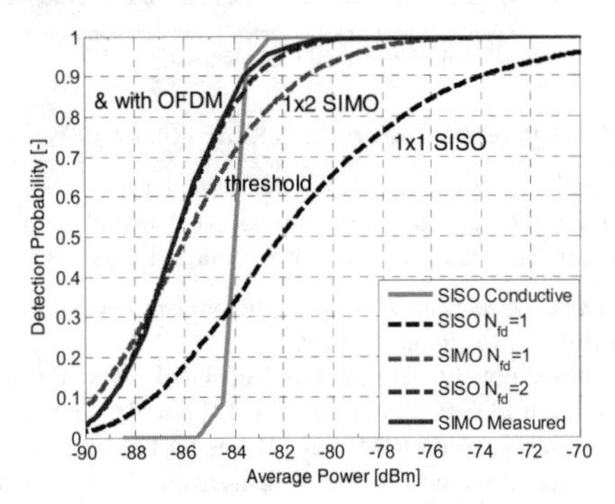

Figure 2. Detection probability of single bit stream in RIMP (i.e. normalized throughput) for 1x1, 1x2 and 1x2 MIMO system with OFDM for a commercial LTE device. The threshold is measured conducted. The dashed curves are simulated from measured threshold values.

C. Detection probability for single bit stream in RIMP

We have done some further studies of OFDM frequency diversity by loading the reverberation chamber for different coherence bandwidths (inverse of RMS delay spread) [9]. We have here in Figure 3 presented the throughputs in [9] as detection probabilities of 8, 16, 24 and 32 Mbps bit streams for the 5, 10, 15 and 20 MHz system bandwidths, respectively. We see that the detection probabilities are larger the lower the coherence bandwidths in the environment is (i.e. more frequency selective fading) The reason is that then the OFDM algorithms improve performance, representing diversity gain in frequency domain. (Note that it is only the large detection probabilities that are of interest, and at large values the small coherent bandwidth is better.) It can be seen that the theoretical and measured detection probabilities are very similar. The theoretical curves are based on thresholds measured for the conducted case with a cable connected to the wireless device [9], but the values have been adjusted by 1.0 - 1.3 dB to match the measurements better, meaning that the threshold is about 1 dB lower (i.e. better) with the antennas connected to the device than what we measure conducted.

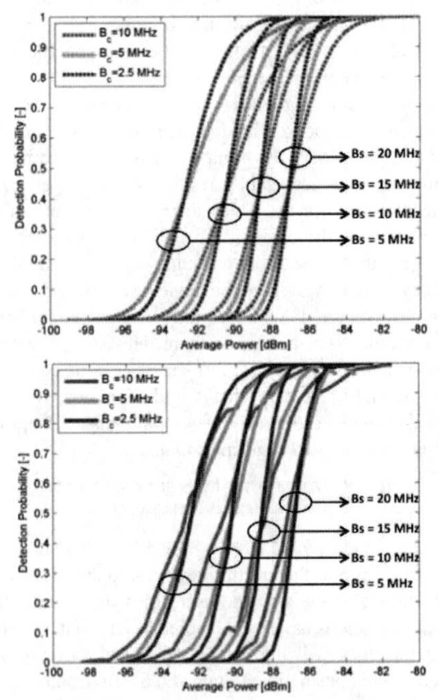

Figure 3. OTA Simulations (top) & OTA measurements (bottom) of 2×2 MIMO LTE throughput in RIMP for system bandwidths of 20, 15, 10, and 5 MHz, and for coherence bandwidths of 10, 5, and 2.5 MHz. The thresholds used in the theoretical model are adjusted by about 1 dB relative to the thresholds in conducted case as explained in the text.

D. Detection probability of multiple bit streams in RIMP

For the purpose of illustration we have for a theoretical i.i.d. channel case (independent and identically distributed) produced detection probabilities of 1 and 2 bit streams in a 2x2 MIMO system, and up to 4 bit streams in a 4x4 MIMO system, as a function of the average received power P_{av} over the threshold power level P_t, see Figure 4. The multiple bit streams are produced with a zero-forcing algorithm [10], i.e. open loop measurements with no CSI available at the transmitter. These i.i.d. curves are reference curves for both RIMP and random-LOS cases. For practical systems the results will degrade compared to these theoretical curves, and we can measure the degradation in dBiid (dB relative to the appropriate i.i.d curve) at a certain probability level. Such curves can be used both for RIMP and random-LOS. We showed in [13, 14] that the i.i.d. makes sense as a reference also in pure-LOS environments, because a channel measured on a practical terminal with a small antenna in 3D-random-LOS will have very close to Rayleigh shape, which is also described in Section IV.A in the present paper.

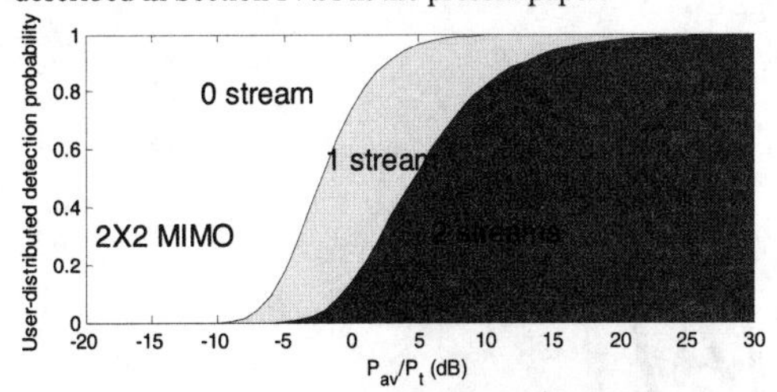

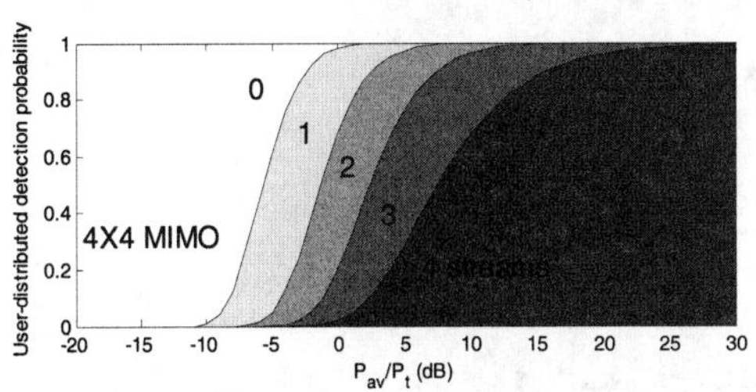

Figure 4. User-distributed detection probability of 2×2 (upper) and 4×4 (lower) open-loop MIMO systems for the i.i.d. cases. The different colors represent the probability of the maximum number of streams supported.

V. Detection Probabilities in random-LOS

The oral presentation will also contain detection probabilities in random pure-LOS. Then, we can detect a maximum of two bit streams if the multiport antenna can provide independently two orthogonally polarized beam patterns in all directions in space.

VI. Conclusion

We have introduced the detection probability as a characterizing metric that can be used in RIMP and random-LOS environments, as well as real-life environments. The detection probability takes account of both the statistics of the environment and the user, and it characterizes a moving user as well as a distribution of stationary users. The simple form of the detection probability is possible due to modern digital receivers that have a detection threshold (i.e. on/off function), so that for each position or time moment the user either has a channel or he doesn't have one.

In a practical design or measurement situation, we would need to define a minimum required detection probability, and then compare devices or solutions with respect to the average power levels required to meet this. Then, quality can be expressed in relative dB values, or we can simply compare the dBiid values of each solution, for each MIMO configuration. The chosen detection probability can be 99% corresponding to the 1% level at which we normally specify diversity gains. Then, the diversity gain improvements will be directly seen as the same improvement in detection probability for the single bit stream case. However, it may be more practical to use 90% or 95% probabilities, because then the characterization becomes less vulnerable to error sources in the measurement setups.

The new approach should enable a new level of system optimization of the complete wireless network, based on collected statistics about how the users use their phones.

VII. References

[1] M. Bäckström, O. Lundén, and P.-S. Kildal, "Reverberation chambers for EMC susceptibility and emission analyses," *Review of Radio Science*, pp. 429-452, 1999.

[2] P. S. Kildal and K. Rosengren, "Correlation and capacity of MIMO systems and mutual coupling, radiation efficiency, and diversity gain of their antennas: simulations and measurements in a reverberation chamber," *IEEE Communications Magazine*, vol. 42, pp. 104-112, 2004.

[3] D. A. Hill, M. T. Ma, A. R. Ondrejka, B. F. Riddle, M. L. Crawford, and R. T. Johnk, "Aperture excitation of electrically large, lossy cavities," *IEEE Transactions on Electromagnetic Compatibility*, vol. 36, pp. 169-178, 1994.

[4] J. G. Kostas and B. Boverie, "Statistical model for a mode-stirred chamber," *IEEE Transactions on Electromagnetic Compatibility*, vol. 33, pp. 366-370, 1991.

[5] K. Rosengren and P. S. Kildal, "Radiation efficiency, correlation, diversity gain and capacity of a six-monopole antenna array for a MIMO system: theory, simulation and measurement in reverberation chamber," *IEE Proceedings - Microwaves, Antennas and Propagation*, vol. 152, pp. 7-16, 2005.

[6] C. Orlenius, P. S. Kildal, and G. Poilasne, "Measurements of total isotropic sensitivity and average fading sensitivity of CDMA phones in reverberation chamber," *2005 IEEE Antennas and Propagation Society International Symposium, 3-8 July 2005*, vol. Vol. 1A, pp. 409-12, 2005.

[7] P. S. Kildal, A. Hussain, X. Chen, C. Orlenius, A. Skårbratt, J. Åsberg, T. Svensson, and T. Eriksson, "Threshold Receiver Model for Throughput of Wireless Devices with MIMO and Frequency Diversity Measured in Reverberation Chamber," *IEEE Antennas and Propagation Wireless Letters*, vol. 10, pp. 1201-1204, October 2011.

[8] X. Chen, P. S. Kildal, C. Orlenius, and J. Carlsson, "Channel Sounding of Loaded Reverberation Chamber for Over-the-Air Testing of Wireless Devices: Coherence Bandwidth Versus Average Mode Bandwidth and Delay Spread," *IEEE Antennas and Wireless Propagation Letters*, vol. 8, pp. 678-681, 2009.

[9] A. Hussain and P.-S. Kildal, "Study of OTA Throughput of 4G LTE Wireless Terminals for Different System Bandwidths and Coherence Bandwidths in Rich Isotropic Multipath," in *EuCAP 2013*, Gothenburg, Sweden, 2013.

[10] X. Chen, P.-S. Kildal, and M. Gustafsson, "Simple Models for Multiplexing Throughputs in Open- and Closed-Loop MIMO Systems with Fixed Modulation and Coding for OTA applications," in *ISAP 2013*, Nanjing, China, 2013.

[11] P.-S. Kildal, X. Chen, C. Orlenius, M. Franzén, and C. Lötbäck Patané, "Characterization of Reverberation Chambers for OTA Measurements of Wireless Devices: Physical Formulations of Channel Matrix and New Uncertainty Formula," *IEEE Transactions on Antennas and Propagation*, vol. 60, pp. 3875-3891, August 2012.

[12] P.-S. Kildal and J. Carlsson, "New Approach to OTA Testing: RIMP and pure-LOS as Extreme Environments & a Hypothesis," in *EuCAP 2013*, Gothenburg, Sweden, 2013.

[13] P.-S. Kildal, U. Carlberg, and J. Carlsson, "Definition of Antenna Diversity Gain in User-Distributed 3D-Random Line-Of-Sight," *Journal of Electromagnetic Engineering and Science (JEES)*, vol. 13, Autumn 2013 2013.

[14] P.-S. Kildal, C. Orlenius, and J. Carlsson, "OTA Testing in Multipath of Antennas and Wireless Devices with MIMO and OFDM," *Proceedings of the IEEE*, vol. 100, pp. 2145-2157, July 2012.

WP-1(A)

October 23 (WED) PM

Room A

Adv Ant for Radio-Astr. -1

Design of Antenna Array for the L-band Phased Array Feed for FAST

Yang Wu, Xiaoyi Zhang, Biao Du, Chengjin Jin, Limin Zhang, Kai Zhu
Joint Laboratory for Radio Astronomy Technology
589 West Zhongshan Road
Shijiazhuang, Hebei 050081 China

Abstract-**The Five-hundred-meter Aperture Spherical radio Telescope (FAST), which is currently under construction, is proposed to be equipped with phased array feed (PAF) to improve survey speed and compensate for the feed cabin vibration, and a study is underway for this. In this paper, the L-band FAST PAF plan is introduced, and the design of antenna array element is presented in detail. The measured results of the fabricated element validate the element design.**

I. INTRODUCTION

The concept of phased array feeds (PAFs), implementing a small phased array with digital beamforming network as the feed of radio telescope, is an emerging technology in radio astronomy to realize fast sky survey. Compared with traditional cluster horn feeds, PAFs are capable of continuous field-of-view (FoV), much more simultaneous beams of similar performance, and the capability of RFI mitigation [1-3]. Currently, a series of development of PAF are underway and many progresses have been reported [3-7]. For the abuilding Five-hundred-meter Aperture Spherical radio Telescope (FAST), PAF will provide a great enhancement in L-band observation than the schemed 19-beam cluster feed receiver.

Key preliminary performance requirements for a PAF on FAST are shown in Table I. The goal is a PAF with more than 100 simultaneous beams overlapped at 3dB points, covering a continuous 0.6 × 0.6 deg^2 FoV at L-band. All LNAs will be cryogenically cooled to realize low system temperature and highest sensitivity possibly. Such a system will be more efficient for wide-field sky surveys.

By the early of 2013, preliminary design of the L-band antenna array is completed and the fabrication for the FAST PAF prototype of smaller scale is ongoing. The prototype will be tested in 2014, and an array element has been manufactured and tested. The results are presented in this paper.

TABLE I
PERFORMANCE SPECIFICATION FOR L-BAND FAST PAF

Parameter	Quantity
Operating frequency range	1.05 ~1.45GHz
Instantaneous bandwidth	≥ 500 MHz
Effective f/D	0.4611
Field of view	0.6 *0.6 deg^2
Simultaneous beams	≥ 100
System noise temperature	25 K
Aperture efficiency	≥ 55% (in 300m diameter)
Low noise amplifier	Cryogenic

II. ARRAY CONFIGURATION

Based on the focal field analysis in conjugate with FAST optics [8], the reflected incident waves converge in a circular region of 2.6m in diameter at the focal plane at L-band. To effectively capture incident wave from different directions, a dense array is required to fill this region. For FAST PAF, a hexagonally arranged array of 217 dual-polarization elements is going to be adopted, with 0.66λ element spacing through optimization.

III. ELEMENT DESIGN

For phased array antenna, mutual coupling effect plays an important role in design. Ideas either utilizing or pressing this effect has been implemented in antenna design. Since the bandwidth of FAST PAF is not very challenging and the element spacing is larger than 0.5λ, mutual coupling reduction scheme is preferred.

The term cavity-backed dipole, which is a combination of dipole and cavity, has better wideband property than dipole and symmetrical radiation pattern, which contributes to improve the illumination efficiency as a feed. Besides above merits, the cavity is introduced to increase isolation between elements in FAST PAF design.

The shape of the cavity is one of the key parameter in element design. Diagram of arrays of hexagonal arrangement consist of distinct cavity-backed dipoles are shown in Fig.1. Due to the physical confliction at each corner, interspaces between square cavities in the same row are the largest among the three candidates. Thus this kind of cavity accommodates better to rectangular arranged arrays. Circular cavity suits both rectangular and hexagonal arrangement, however, gaps between elements are inevitable. Hexagonal cavity makes full use of the array aperture and the increment in cavity aperture slightly improves the gain. Fig.2 gives a pattern comparison between the circular and hexagonal cavity-backed dipoles satisfying the same element spacing.

978-7-5641-4279-7

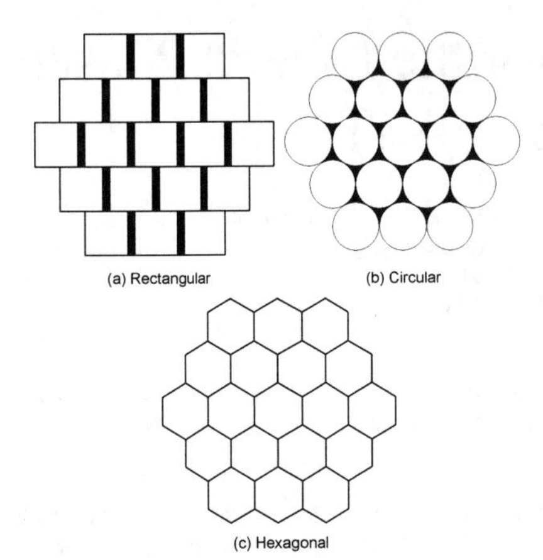

Figure 1. Interspaces between elements caused by different cavities.

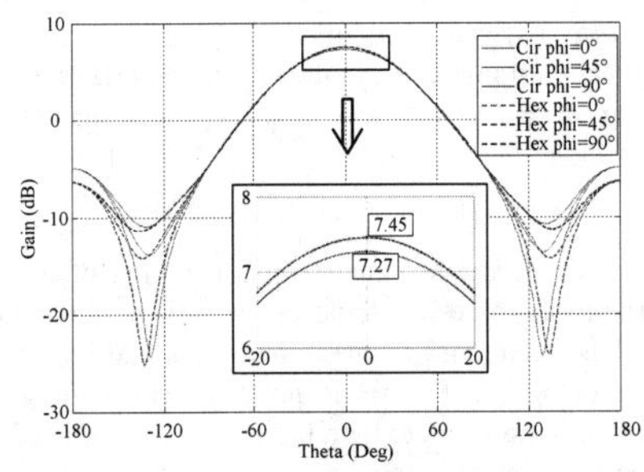

Figure 2. Comparison of patterns in free space between circular and hexagonal cavity-backed dipoles.

As Fig.2 shows, the hexagonal cavity dose not distort the symmetry of pattern and the corresponding gain is about 0.2dB higher than circular one. Since the gain relates to the effective receiving area, the hexagonal cavity-backed dipole intuitively capture more incident wave with the same element spacing.

A view of the hexagonal cavity-backed dipole from the top side is given in Fig. 5.

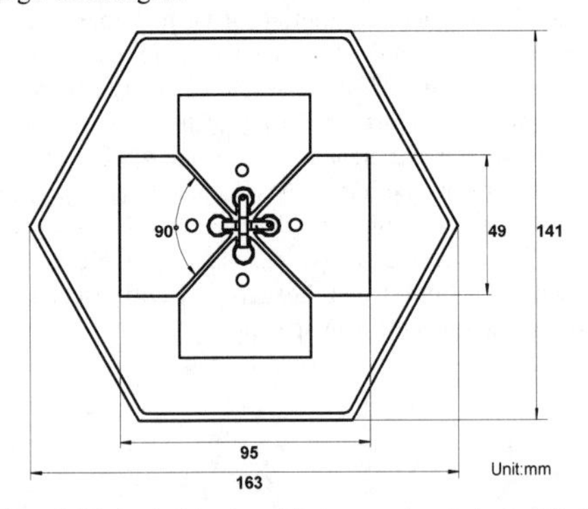

Figure 5. Mechanical drawing of the hexagonal cavity-backed dipole.

The hexagonal cavity is 50mm in height, and 141mm in diameter of the inscribed circle. Two orthogonally placed bowtie dipoles are employed as the radiator on the aperture of the cavity. The arms are design with a 90 degree opening angle and 49mm wide at the end point, 37.5mm long, and 1mm thick.

IV. Measurement Results

Fig. 4 gives a photograph of the fabricated hexagonal cavity-backed dipole. A match plate of 70mm in diameter is employed at the top for lower VSWR, supported by four Teflon posts.

The measured results are shown in Fig.5 and Fig. 6. The measured reflection coefficients are below -15dB over the operating band.

Fig. 6 shows the normalized modeled and measured patterns at 1.25GHz. It is observed that the symmetry of measured pattern is not as excellent as simulation. This is caused by an imperfect balun. To make space for cables and LNAs, the feed gap is design to be 20mm long, resulting in balun degradation and modest pattern distortion. And a reduction in feed gap has been made to mitigate this effect in further design.

Figure 4. Photograph of the hexagonal cavity-backed dipole

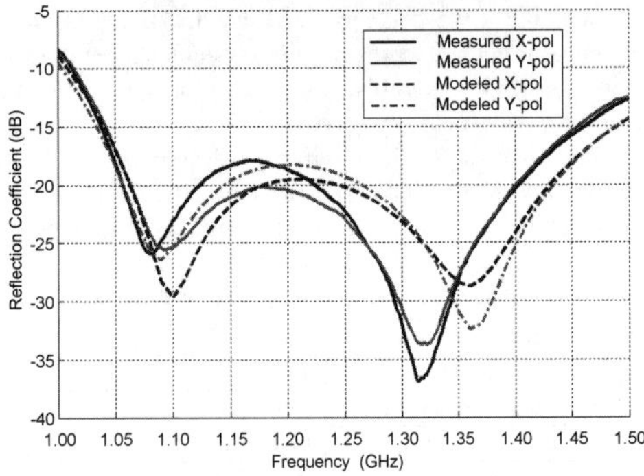

Figure 5. Measured and modeled reflection coefficient of the hexagonal cavity-backed dipole.

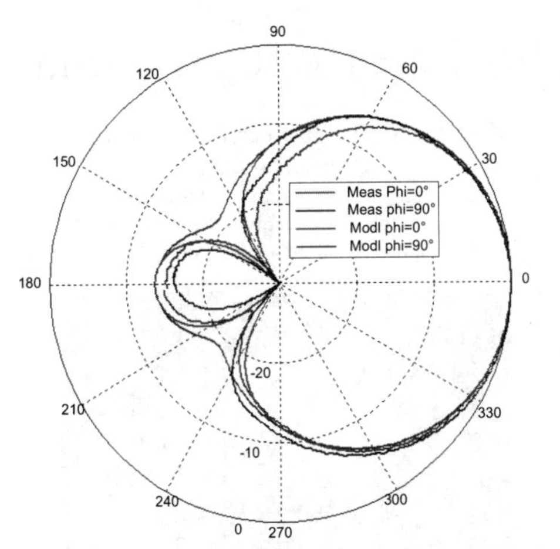

Figure 6. Measured and modeled pattern of the hexagonal cavity-backed dipole.

V. CONCLUSION

A study on PAFs for the FAST is underway. Consideration and design of the antenna array is present in this paper. The measured results of the fabricated element validate the design and a 19-element array is being manufactured.

ACKNOWLEDGMENT

This work is partly supported by China Ministry of Science and Technology under the State Key Development Program for Basic Research (2012CB821800, 2013CB837900), and the key research program of the Chinese academy of sciences (KJZD-EW-T01).

REFERENCES

[1] G. Eason, B. Noble, and I.N. Sneddon, "On certain integrals of Lipschitz-Hankel type involving products of Bessel functions," *Phil. Trans. Roy. Soc. London,* vol. A247, pp. 529-551, April 1955.

[2] J. Clerk Maxwell, *A Treatise on Electricity and Magnetism,* 3rd ed., vol. 2. Oxford: Clarendon, 1892, pp.68-73.

[3] I.S. Jacobs and C.P. Bean, "Fine particles, thin films and exchange anisotropy," in *Magnetism,* vol. III, G.T. Rado and H. Suhl, Eds. New York: Academic, 1963, pp. 271-350.

[1] Van Cappellen W, de Vaate J G B, Warnick K, et al. "Phased array feeds for the Square Kilometre Array". *General Assembly and Scientific Symposium. Istanbul, 2011*, pp. 1-4.

[2] Cortes-Medellin G, Rajagopalan G, Perillat P, et al.. "Field of view characterization of Arecibo radio telescope with a phased array feed". *IEEE International Symposium on Antennas and Propagation (APSURSI). Spokane, 2011*, pp. 847-850.

[3] Hansen C, Warnick K F, Jeffs B D. "Interference cancellation using an array feed design for radio telescopes". *International Symposium on Antennas and Propagation Society, 2004.* Vol. 1, pp. 539–542.

[4] Van Cappellen W A, Bakker L, Oosterloo T A. "APERTIF: phased array feeds for the westerbork synthesis radio telescope". IEEE International Symposium on Phased Array Systems and Technology (ARRAY), Boston, 2010: 640-647.

[5] Veidt B, Hovey G J, Burgess T, et al. "Demonstration of a dual-polarized phased-array feed". *IEEE Trans on Antennas and Propagation, 2011*, Vol. 59(6): pp. 2047-2057.

[6] Hampson, G A, Macleod A, Beresford R J, et al. "ASKAP PAF ADE - Advancing an L-band PAF design towards SKA". *International Conference on Electromagnetics in Advanced Applications. Cape Town, 2012*, pp. 807-809.

[7] Warnick K F, Webb T, Adhikari M, et al. "Progress in high sensitivity phased array feed for single-dish radio telescope". *International Conf. on Electromagnetics in Advanced Applications. Cape Town,* 2012, pp. 199-201.

[8] Wu Y, Warnick K F, Jin C J. "Design study of an L-band phased array feed for wide-field surveys and vibration compensation on FAST". *IEEE Trans on Antennas and Propagation*, Accepted for publication.

Progress in SHAO 65m Radio Telescope Antenna

Biao Du, Yuanpeng Zheng, Yifan Zhang, Wancai Zhang
The 54th Research Institute of CETC
589 West Zhongshan Road
Shijiazhuang, Hebei 050081 China

Zhiqiang Shen, Qingyuan Fan, Qinghui Liu, Quanbao Ling
Shanghai Astronomical Observatory
Chinese Academy of Sciences
80 Nandan Road, Shanghai 200030, China

***Abstract*-The SHAO 65m radio telescope is the largest fully steerable radio telescope antenna in China, operating at decimeter to millimeter wavelength. This paper describes the design, main technical feature and progress in construction of the telescope antenna.**

I. INTRODUCTION

In accordance with the demands of the radio astronomy research as well as the needs of Chinese Lunar exploration and deep space exploration missions, a fully steerable 65m radio telescope with the advanced state of art is being built. It will be located at Sheshan Station, Shanghai Astronomical Observatory (SHAO), in Songjiang County, about 30km far away from Shanghai center.

The SHAO 65m radio telescope antenna is a shaped Cassegrain antenna, operating in the frequency range from 1 to 50GHz. To realize such frequency coverage, seven receivers are adopted, including two dual band ones (S/X and X/Ka). For Phase I construction, the telescope antenna will work below X band using homology method[1]. And in Phase II, active surface will be employed to enhance performance at higher frequency. After over 3-year construction, the Phase I is almost finished. Table I gives a summary of basic parameters of the SHAO 65m radio telescope antenna.

II. TECHNICAL FEATURES

A. Optics

The SHAO 65m radio telescope antenna has a sub-reflector of 6.5m in diameter, illuminated by feeds of 13 degree half opening angle. The reflectors are shaped[2] to achieve higher aperture efficiency (~70%) and control the first sidelobe level below -20dB, minimize the VSWR via reduce center illumination, and decrease the spillover from the edge of the main-reflector, which contributes low side lobes and low antenna noise temperature. An additional merit of shaped configuration is that it is less sensitive to illumination variance from the feed, which is nearly inevitable for wideband feed.

TABLE I
BASIC PROPERTY OF THE SHAO 65M RADIO TELESCOPE ANTENNA.

Aperture	65m
Frequency band	L, S, C, X, Ku, K, Ka and Q dual band observation for S/X and X/Ka band)
Optics	Shaped Cassegrain, f/D(Main-reflector)=0.32, f/D(effective)=2.2
Antenna efficiency (at best elevation angle)	≥55~65% (L-K Band); ≥50% (Ka Band) ≥45% (Q Band)
Main reflector	Single panel surface accuracy ≤ 0.10 mm rms Total accuracy ≤0.6 mm rms (without active surface, best elevation angle)
Sub-reflector	Single panel surface accuracy ≤ 0.05 mm rms Total surface accuracy ≤ 0.1 mm rms.
Maximum slew rates and acceleration	Azimuth: 0.5 (°)/s, 0.25 (°)/s^2 Elevation: 0.3 (°)/s, 0.15 (°)/s^2
Travel range	Azimuth: ≥270°, Elevation: +5°~ +88°
Tracking& Pointing accuracy	High accuracy: ≤ 3 arc-seconds, rms (wind speed≤ 4m/s, temperature variation≤2°C/h)

Figure 1. The SHAO 65m radio telescope antenna.

The corrugated horns are used for all feeds, including two dual band ones for S/X and X/Ka bands.

B. Mechanical Structure

The antenna design is based on the classical wheel-and-track configuration (Fig. 1 and 2), consisting of reflector (main and sub- reflector), mount and feed cabin.

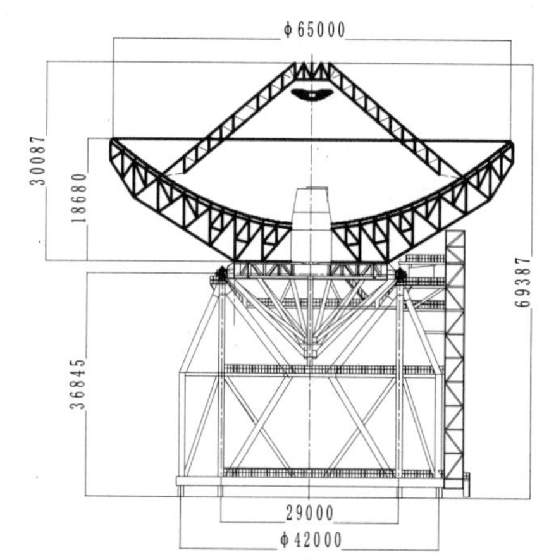

Figure 2 Mechanical drawing of the SHAO 65m radio telescope antenna.

978-7-5641-4279-7

The main reflector consists of 1008 solid aluminum panels and a backstructure. 1104 actuators are mounted on the backstructure to support the 1008 panels with 4-point support for each panel. The high accuracy panels with slot ribs are adopted to realize the surface accuracy ≤ 0.1 mm rms.

The sub-reflector surface consists of 25 solid panels, adopting double layer honeycomb sandwich structure to realize a surface accuracy ≤ 0.05 mm rms. The sub-reflector adopts 6-UPS parallel mechanism to accomplish 5 freedom degrees adjustment in real time for the sub-reflector when the antenna rotates in elevation.

An elevation over azimuth mount with wheel track configuration is adopted in the mount design. The antenna mount is space struss composing rectangular beam. Azimuth structure is supported by 6 groups of wheel. Combination structure of double layer octagonal beam and conical space struss is introduced in antenna support design. The mount structure is symmetrical, and the azimuth and elevation axes intersecting at one point in space.

8 motors are used for azimuth driving and 4 motors for elevation. The elevation driver suspends on the elevation gear. The azimuth and elevation driving devices adopt electric antibacklash to eliminate backlash, all the driving boxes adopt planetary gear reducer. Encoder connection adopts leading axis from the centre of rotation to improve connection precision.

32 sections of track, made of high quality alloy steel, are assembled by welding to form the whole track on the site, with plane accuracy less than 0.45mm rms.

C. *Servo System*

Servo system is a high performance fully digital control system. It can be remote controlled by main controller and automatic working without any local operator. Servo system adopts multi-motor pre-tension anti-backlash drive mode, a classical three feedback loops (position, rate and torque) control architecture. It also has high accuracy encoder (32bit resolution and 1 arcsec system accuracy), high performance safety inter-lock logic, satisfied in EMC, and site environment flexibility. Blended control methods (CPP+FF+PID) are used now and a robust controller (LQG) will be applied for improving tracking accuracy (the aim is 2 arcsec) in near future.

D. *Feed Switching*

The SHAO 65m redio telescope antenna operates at L, S, C, X, Ku, K, Ka and Q bands, and all feeds locate at the second

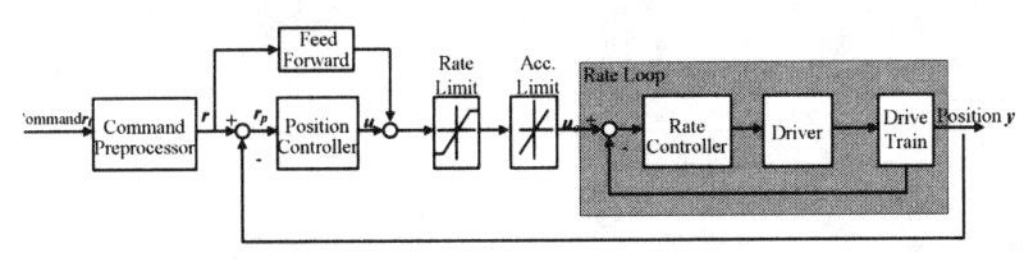

* CPP=Command Pre-processor
FF=feedforward
LQG=Linear Quadratic Gauss

Figure 3. Diagram of servo system.

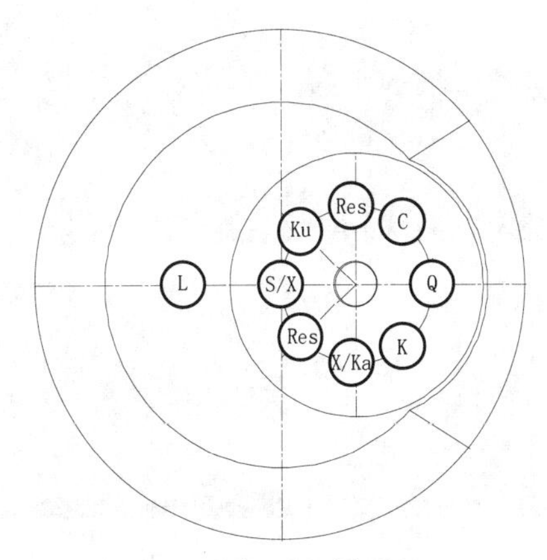

Figure.4 Top View of Feed Placement.

focus. To realize feed switching effectively, a combination scheme is adopted. For the L-band feed, because of its huge physical size, it is placed offset the second focus. And when L-band observation is required, the sub-reflector will point to it. All other feeds are placed at a turret of 2m diameter and frequency switching is achieved by rotating the feed turret. The structure schematic diagrams are shown in Fig. 4. As shown in Fig. 4, two vacant positions are reserved on the turret for further update.

III. Progress in Construction

The milestone of SHAO 65m radio telescope antenna are listed in Table II.

Table II

MILESTONES OF SHAO 65M RADIO TELESCOPE ANTENNA.

No.	Time	Description
1	January 2009	Kicking off
2	September 2009	Passed the international review
3	October 2010	Started antenna assembling
4	July 2012	Completed antenna assembling
5	October 2012	A series of measurements were made and the L, S/X, C band feeds debugging completed.
6	June 2013	Phase I construction acceptance

Fig. 5 shows the pictures of the whole assembling process.

(a) 2011.05 (b) 2011.10

(c) 2011.10 (d) 2012.03

(e) 2012.04 (f) 2012.10

Figure 5 Assembly of the SHAO 65m radio telescope antenna.

After the dish assembly completion, the measurement of the total surface accuracy was made and the result showed that, accuracies of 0.571mm rms and 1.13mm rms were achieved at 50degree and 10 degree elevation respectively. Fig. 6 gives the comparison between measured pattern and simulation at X band.

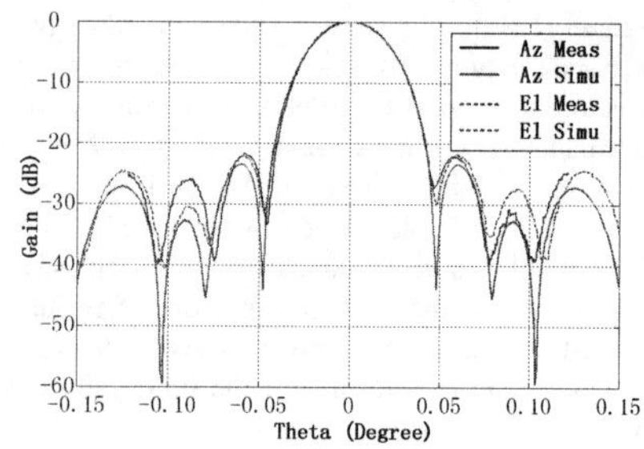

Figure 6. Measured X-band pattern

IV. Conclusion

After four-year design, manufacture, assembling, integration, and debugging, the Phase I construction of the SHAO 65m radio telescope antenna is nearly completed.

References

[1] F. Yang, B. Du. "A Calculation Method of the Best-fit Surface for Shaped Cassegrain Antenna", Radio Engineering, vol. 41, No. 3, pp.38-40, March 2011.

[2] L. Zhang, Z. Shi, "Study of Antenna Aperture Field Distribution Function with High Efficiency and Low Sidelobe Level," *Radio Communications Technology,* vol. 36, No.4, pp. 36-38, April 2010.

[3] "Shanghai 65m Radiotelescope Antenna System Design Document," report, 2009.

Promoting the Planetary Radio Science in the Lunar and Deep Space Explorations of China

J. Ping, M. Wang, Y. Lu, W. Li, Z. Yang
Key Laboratory of Lunar and Deep Space, National Astronomical Observatories, CAS, 20A Datun Rd., Beijing, 10012, China
N. Jan, L. Fung, S. Zhang, K. Shang
Shanghai Astronomical Observatory, CAS, 80 Nandan Rd., Shanghai, 200030, China
Z. Wang
Xinjiang Astronomical Observatory, CAS, Urumuqi, China
Q. Meng, C. Chen, S. Qiu
Faculty of Space Science, South-East University, Sipailou, Nanjing, 210018, China
J.G. Yan
RISE Project, National Astronomical Observatory,NINS,2-12,Hoshigaoka,Mizusawa,Oshu,Japan

***Abstract*-Radio science experiments using the radio link between s/c and ground tracking stations have been playing important roles in lunar and deep space exploration since 1960s, which study the planetary atmosphere and ionosphere, lunar and planetary gravity fields and rotations, carry out precise orbit determination and relevant tests of general relativities. In Chinese lunar and planetary missions since 2004, based on very poor or limited s/c radio tracking conditions, research team improved the lunar gravity field and found new mascons, developed open-loop only OD method for Martian missions, and accomplished 1st round atmospheric occultation experiments for Mars and Venus. In future lunar landing mission and deep space mission of China, radio science experiments will also measure the lunar physical liberation and planetary rotations with very high precisions.**

I. INTRODUCTION

The radio science experiment makes a virtue out of a necessity by using the radio propagation techniques that convey data and instructions between the spacecraft and Earth to investigate the planet's atmosphere, ionosphere, rings, and magnetic fields, surface, gravity field, GM and interior, as well as testing the theories of relativity. The experiments have been conducted by most planetary missions and are planned for many future ones. This is almost like getting science for free some way, and has been used also in Chinese lunar orbiting missions to carry out POD and estimate the lunar gravity field successfully.

Since 2004, lunar and planetary exploration missions have been started in China. Lunar orbiters of Chang'E-1/2 were launched in 2007 and 2010 separately[1]. A failed join Martian mission with Russia, Yinghuo-1(YH-1) and Phobos-Grunt, was launched in 2011. Lunar landing mission Chang'E-3 will be launched in 2013. New Martian missions and asteroid missions are been promoted now.

Besides the payload mission times, a radio science (RS) experiments research team are also involved in above missions. This RS team using Chang'E-1 tracking data improved the lunar gravity field successfully. They also developed radio science receiver system, and connected to the IF and H-Maser system of Chinese VLBI stations. A series of radio occultation experiments have been carried out. In the techniques, radio frequency transmission from MEX and VEX spacecraft, occulted by center planets, and received on Earth probe the extended atmosphere of the planets. The radio link is perturbed in phase and amplitude, the perturbation is converted into an appropriate refractivity profile, from which, information is derived about the electron distribution in the ionosphere, temperature-pressure profile in the neutral atmosphere, or particle size distribution of the ring material surrounding the center planet, in the case of a ring occultation. In Chang'e-3/4 and Luna-Glob landing missions, the two radio downlinks, S/X-bands or X/Ka-bands, will be used for different types of investigation about the physical mechanisms that brought the changes. We will use the signal to measure the lunar physical liberations, so as to improve the lunar interior studies together with LLR data. Above method may be used in Chinese Martian mission.

II. RADIO SCIENCES IN CHANG'E-1

Chang'E-1 was the first Chinese lunar orbiter mission, and its main scientific objectives were to obtain three dimensional images of the lunar surface, and to obtain the distribution of lunar elements[1]. Chang'E-1 was launched on October 24, 2007, and after multiple orbit adjustments, such as phase change, Earth-Moon transfer and lunar capture, it became a lunar satellite in a near-circular orbit with 200 km altitude and inclination ranging from 87.58° to 91.40, which is sensitive to low degree coefficients of lunar gravity field. During nominal mission phase of Chang'E-1 a large amount of range and Doppler tracking data were acquired, which could be used to develop lunar gravity field solutions. A lunar gravity field model labeled CEGM01 using only these data has been obtained by Yan et al[2]. It shows that Chang'E-1 orbital tracking data can contribute more to determining gravity field than Clementine, and they can be used to estimate the gravity field coefficients to degree and order 50 independently.

Using Chang'E-1 orbital tracking data, in combination with orbital tracking data of SELENE, Lunar Prospector, and historical spacecraft, a lunar gravity field model denoted

978-7-5641-4279-7

CEGM02 is developed, see Figure 1. Analyses show that due to its higher orbit altitude (200 km), tracking data of Chang'E-1 contribute to the long wavelengths of the lunar gravity field. When compared to SGM100h, formal error of CEGM02 coefficients below degree 5 is reduced by a factor of about 2~3.

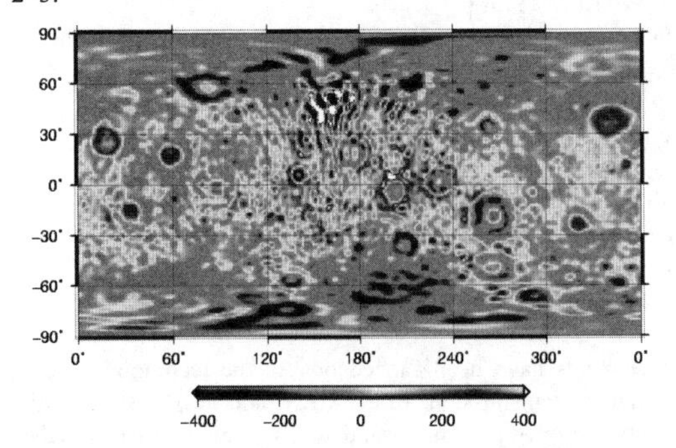

Figure 1. Free air gravity of the moon on the surface of reference radii 1738km (Unit: mgal).

Using the lunar gravity model CEGM02 and the topography model CETM-s01 of Chang'E-1 mission, we calculated the terrain correction for lunar free-air gravity anomaly (FAGA). The obtained lunar Bouguer gravity anomaly (BGA) reveals density irregularities of the interior mass, where the South Pole-Aitken (SPA) basin was found to be the largest mascon basin on the Moon. Another 8 middle size hidden basins are newly identified of showing strong BGA mascon feature. An example is shown in Figure 2. CETM-s01 and CEGM02 models also show some area with strong features of FAGA, BGA and topographic swells of hundreds kilometers in extent, where we identified 3 shield volcanoes with proposed underground magma chambers at the west area of Oceanus Procellarum[3].

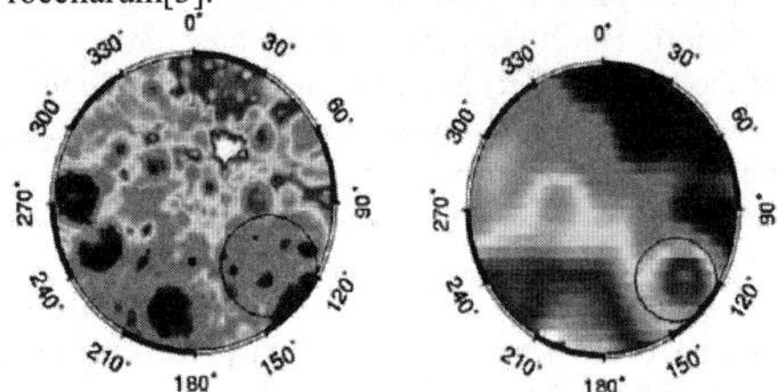

Figure 2. Newly identified Lunar Hidden BGA basins and Amundsen-Ganswindt at South pole area.

III. RADIO SCIENCES IN CHANG'E 3/4

China will launch the 1st lunar landing mission with a rover this year. A backup sister landing and rover mission, Chang'E-4, with identical payloads of Chang'E-3 will be launch and landed at different area of the Moon. They compose the main part of the 2nd phase of Chinese lunar scientific exploration projects. Together with the various in-situ optical observations around the landing sites, missions will also carry out 4 kinds of radio science experiments, cover the various lunar scientific disciplines as well as lunar surface radio astronomy studies.

4 kinds of radio science experiments have been planned in Chang'E 3/4 landing missions: 1) HF and VHF duel-band penetrator radar on the rovers; 2) very low frequency through HF radio astronomy on the surface of Moon; 3) same-beam X-band Very Long Baseline Interferometry (VLBI) for precise positioning of rover; 4) precise radio phase ranging for lunar rotation and dynamics.

HF and VHF duel-band penetrating radar: the radar has center frequencies at 30MHz and 50MHz with bandwidth of 15MHz for each, linear polarized antenna to study the subsurface structure of landing area. In the mission LRS of SELENE/KAGUYA project, Japanese researchers obtained the lunar subsurface structure of 5~10 KM deep with resolution of dozens meters, where the igneous lunar basalt filling at mare area was firstly measured with the maximum thickness of 500~600 meters. Lunar regolith and lunar crust subsurface of shallower than 3km will be firstly studied by using the duel-band GPR on the rover with very high resolution. The thickness of regolith of the soft landing area will be measured [3].

Very low frequency radio astronomy: frequency from several KHz through 10MHz, single polarized dipole antenna, with a spectra analyzer. Due to the frequency truncate by ionosphere of the earth, cosmic radio signal with frequency below 10MHz will be absorb and reflect back to the space. We cannot receive the natural cosmic signal of this band on the ground. To overcome this problem, two sets of antenna and spectra receiver system are installed on the two landing missions separately. The spectra analyzers have frequency resolution better than 1KHz. This instrument is a kind of proto type or path finder for lunar surface low frequency radio network array, or for lunar far side low frequency array in the future. On the surface of the Moon, this payload will be first time to carry the studies of extra-terrestrial solar space VLF radio observations for solar radio burst, space particle flow, kilometric wave radiation, coronal mass ejections and planetary low frequency noise.

Same beam VLBI tracking: Two VLBI beacon transmitters with high stable oscillators are installed on the Chang'E 3/4 landers and the rovers separately. Beacons will transmit X band DOR wave or single carrier wave back to the Earth. The Chinese VLBI network and the new developed Chinese deep space tracking station will observe the DOR signal from two beacons simultaneously by the main lobe of each antenna. In the differential DOR observables, the effects due to the tracking station, the atmosphere and the ionosphere of the earth, as well as the effects due to the lunar rotation, tides and liberation can be cancelled dramatically. Then, the high precise relative position of the rover to the lander of the same mission will be obtained at each observation point of the rover.

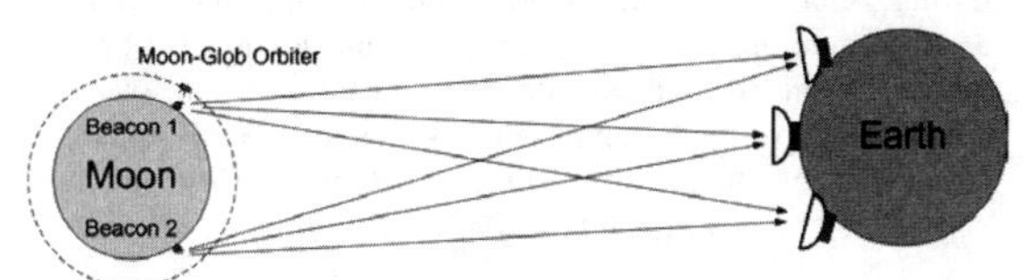

Figure 3. Lunar radio science by radio beacons in landing missions.

Lunar Radio Phase Ranging: Since early Luna and Apollo missions set 5 optical corner reflector systems on the surface of the Moon, Lunar laser ranging have been played key role on measuring the lunar rotation, Physical liberation and surface solid tides. However, the bad weather on laser site, the full Moon and new Moon phase may block the optical observation. Similar to Luna-Glob landers, together with the VLBI radio beacons, the radio transponders are also set on the Chang'E-3/4. Transponder will receive the uplink S/X band radio wave transmitted from the two newly constructed Chinese deep space stations, where the high quality hydrogen maser atomic clocks have been used as local time and frequency standard. The clocks between VLBI stations and deep space stations can be synchronized to UTC standard within 20 nanoseconds using satellite common view methods. In the near future there will be a plan to improve this accuracy to 5 nanoseconds or better, as the level of other deep space network around world. This experiment will improve the study of lunar dynamics [5], by means of measuring the lunar physical liberations precisely together with LLR data.

IV. Radio Sciences in Chinese Martian Mission

Following the successful lunar mission of Chang'E-1, the first Chinese Mars Probe, Yinghuo-1, had been planned to launch in October 2009, delayed in October 2001, together with the Russia Phobos-Grunt landing mission. YH-1 was planned to explore the space weather of the Mars, and to test the deep space communication and navigation techniques. Difference from common deep space probe, the astronomical VLBI, open loop tracking method, like DOR/DOR, seam beam VLBI and 1-way Doppler, were used to determine the s/c orbit and position.

The RS team was responsible for developing the open loop tracking method in YH-1 mission Since 2007. The RS team developed a prototype radio science receiver successfully, based on digital radio technology with associated open-loop Doppler signal processing techniques to measure a spacecraft's line-of-sight velocity [4]. The prototype was tested in Chang'E-1 lunar mission relying on S-band telemetry signals transmitted by the satellite, with results showing that the residuals had a RMS value of ~3 mm/s (1σ) using 1-sec integration, which is consistent with the Chinese conventional USB (Unified S-Band) tracking system. Such precision is mainly limited by the short-term stability of the atomic (e.g. rubidium) clock at the uplink ground station. It can also be improved with proper calibration to remove some effects of the transmission media (such as solar plasma, troposphere and ionosphere), and a longer integration time (e.g. down to 0.56 mm/s at 34 seconds) allowed by the spacecraft dynamics. The tracking accuracy can also be increased with differential methods that may effectively remove most of the long-term drifts and some of the short-term uncertainties of the uplink atomic clock, thereby further reducing the residuals to the 1 mm/s level. The experimental tracking data have been used in orbit determination for Chang'E-1. Successful application of the prototype to the Chang'E-1 mission in 2008 is valuable for the upcoming Chinese YH-1 Mars exploration project.

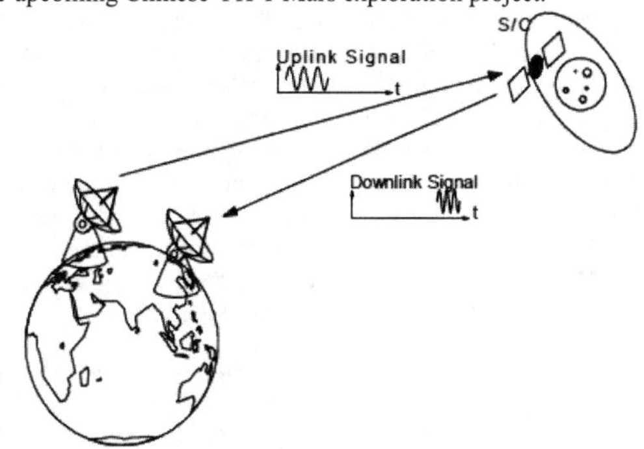

Figure 4. Radio link between ground stations and S/C.

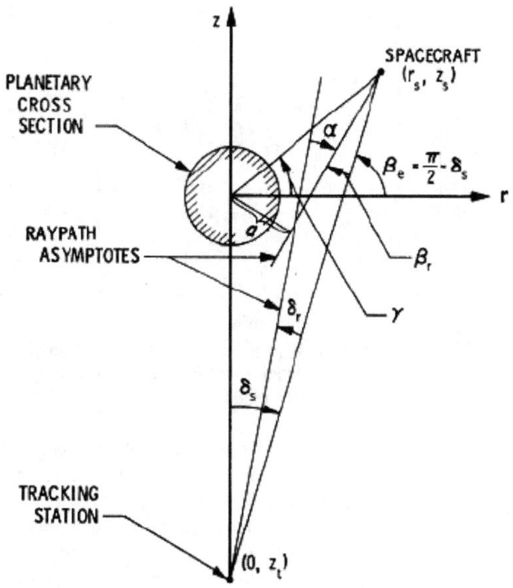

Figure 5. Raypath geometry of the radio occultation experiment following Fjeldbo et al. 1971 [6]

In YH-1 mission the Earth-based planetary radio occultation experiment had been planned [5]. In planetary radio occultation techniques, a radio signal transmitted from a spacecraft is occulted by a planet (or one of its satellites) and received on Earth. See Figure 5. The perturbation of the radio link in phase and amplitude can be converted into an appropriate refractivity profile of the atmosphere by an inversion method in both occultation immersion and emersion. From the refractivity profile, information can be derived about the electron distribution in the ionosphere, temperature, pressure and molecular number density profiles in the neutral atmosphere, or particle size distribution of the ring material surrounding a planet, in the case of a ring occultation.

Additionally, RS team designed and developed Earth-based Planetary Occultation observation Processing system (SPOPs) for the joint Martian exploration project. Utilizing the open-loop and closed-loop Doppler residual data of the Mars Express radio occultation experiment provided by the ESA Planetary Science Archive (PSA) and the NASA Planetary Data System (PDS), the temperature, pressure, number density profiles of the Martian atmosphere and electron density profiles of the ionosphere are successfully retrieved. The results are validated by the released level 04 radio science products of the ESA MaRS group. See Figure 6.

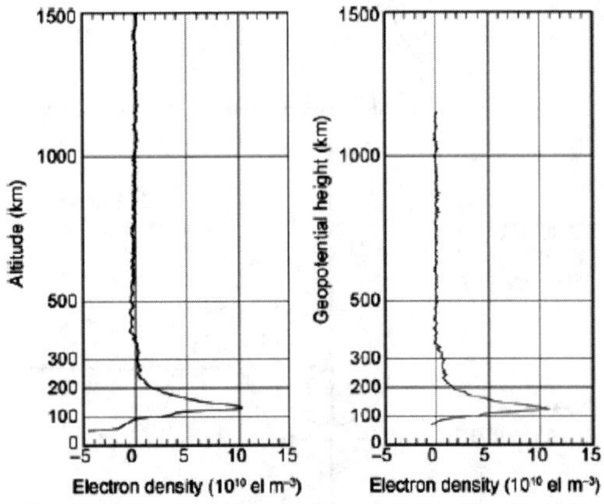

Figure 6. Example of Martian ionspheric profile: Derived ionosphere electron density profiles from the MEXMRS_0046 X-band observation (left panel) and the level 04 product given by MaRS (right panel).

The prototype receiver[7] was updated by the RS team [8]. Finally the multi-channel open loop Doppler receiver was developed for VLBI and Doppler tracking in Yinghuo-1 and Phobos-Glob Martian missions. Although the YH-1 mission failed together with Phobos-Grunt, the radio science technique for ground tracking and VLBI stations has been developed and tested using lunar and planetary missions.

A new Martian mission of China is under promoted and developed for launching during the window of 2018. 4 kinds of radio science experiments have been planned in this mission: 1) HF and VHF duel-band penetrator radar on the orbiter; 2) very low frequency through HF radio astronomy during the cruise transferring orbit from the Earth to the Mars; 3) open-loop and close-loop radio tracking for precise orbit determination, so as to monitor the variations of long wave length gravity field components; 4) precise radio phase ranging for Martian atmospheric occultation. Above methods can be used in the next Chinese Martian mission.

REFERENCES

[1] Z.Y. Ouyang, C.L. Li, Y.L. Zou, et al., 2010. Chang'E-1 lunar mission: an overview and primary science results. *Chinese Journal of Space Science*, vol. 30(5), 2010, pp. 392-403.

[2] J.G. Yan, J. S. Ping, K. Matsumoto, et al. *Sci China Phys Mech Astron*, vol. 41, 2011, pp.870–878, doi: 10.1360/132010-868.

[3] J.S. Ping, X.L. Su, Q. Huang, J.G. Yan, *Sci China Phys Mech Astron*, The Chang'E-1 orbiter plays a distinctive role in China's first successful selenodetic lunar mission, vol.54(12), 2011, pp.2130–2144, doi: 10.1007/s11433-011-4561-0.

[4] N.C. Jian, K. Shang, S.J. Zhang, et al., *Sci China Ser G,* A Digital Open-Loop Doppler Processing Prototype for Deep-Space Navigation, 2009, vol. 52(12), pp. 1849-1857, doi: 10.1007/s11433-009-0283-y

[5] S.J. Zhang, J.S. Ping, T.T. Han, et al., *Sci China Phys Mech Astron*, Implementation of the Earth-based planetary radio occultation inversion technique, vol.54(7), 2011, pp. 1359–1366, doi: 10.1007/s11433-011-4247-7.

[6] G. Fjeldbo, A. J. Kliore, V. R. Eshleman. *Astron J.*, The neutral atmosphere of Venus as studied with the Mariner V radio occultation experiments, 1971, vol.76, pp.123–140.

[7] J.S. Ping, Z.H. Qian, S.H. Ye, et al., IEEE Xplore, APCC2009, Open loop tracking in 1st Chinese Mars exploration mission: Yinghuo-1 Martian orbiter. 2009, pp. 446-449, doi: 10.1109/APCC.2009.5375597.

[8] J.S. Ping, M.Y. Wang, Q. Meng, et al., Developing RSR for Chinese Astronomical Antenna and Deep Space Exploration, in this proceeding.

Initial Considerations of the 5 Meter Dome A Terahertz Explorer (DATE5) for Antarctica

Ji Yang, Zheng Lou, Yingxi Zuo, Jingquan Cheng*
Purple Mountain Observatory, Chinese Academy of Sciences
Key Laboratory of Radio Astronomy, Chinese Academy of Sciences
Nanjing 210008, China
*jqcheng@pmo.ac.cn

Abstract- **Terahertz radiation comes from cold objects in universe. It provides important information of star and galaxy formation. Within these bands, thousands of molecular lines are located, which may reveal important processes in cosmology. In Antarctica, the Dome A area has the best observing windows in the terahertz waveband. In this paper, a 5 meters Dome A Terahertz Explorer (DATE5) telescope concept is provided, which is a fully steerable telescope, working under the harsh Dome A polar environments. It requires a higher surface and pointing accuracies. This paper also discusses major aspects to be taken into consideration in its initial conceptual study.**

I. INTRODUCTION

In the electromagnetic wave spectrum, terahertz radiation, which lies between very far end of infrared and just before the start of microwave radiations, is a unique band for astronomer to observe the cold universe between 4K and 100K[1].

If one wish to know the formations of heavy elements, stars, planets, and galaxies, it is very important to make observations in the terahertz frequencies. In astrophysics, one key issue in the big bang theory was the early life of galaxies. The galaxies we see today are in optical wavebands. However, there must have been a time the galaxies first formed in the universe, radiation from these early galaxies would be in the infrared band and had highly redshifted blackbody spectrum. These radiations would be redshifted into the terahertz band as we observe them today.

The cosmic microwave background (CMB) radiation peaks in the range of 100 to 300 GHz, which lies at the bottom of terahertz frequency range. However, the measured anisotropy has revealed the acoustic modes in the early universe. In the new century, to measure the predicted complicated state of polarization of CMB would reveal the physical mechanism when the radiation became decoupled from matter, and the rest being formed as gravity waves in the earliest moments of creation.

The terahertz region of spectrum is also essential for astronomers in decoding the secrets of star formation. Many thousands lines are now known in the THz bands. By mapping and studying these lines, it is possible to build complete models of objects; these models include temperature, density, movement of material, magnetic field, etc. All these make the terahertz observation and terahertz telescope construction being extremely important.

Before building a successful terahertz telescope, two important issues are the receiver technology and the site selection. Terahertz receiver technology is mainly driven by applications in astronomy. In China, a design team at Purple Mountain Observatory had finished a 500GHz SIS (superconductor-insulator-superconductor) receiver in 2005[2]. The team is now working on the terahertz receiver study. On the site selection issue, a team of Chinese scientists had reached the Dome A plateau in Antarctica also in 2005. The Dome A area is 4,093 m above sea level. The water vapor contents of the site are very low with a year average number of 0.21 mm[3],[4],[5]. Therefore, Dome A provides a rare atmosphere windows between 0.7 and 1.6 terahertz. The transparency of the Dome A is not only higher than that of the Acatama (5105 m above sea level) and Staircabar (5525 m above sea level) sites in Chile, but also higher than Dome B, C, and F in Antarctica[6]. At Dome A, a high quality terahertz telescope can conduct scientific research that would otherwise only be possible from space.

Based on these listed conditions, in 2010 astronomers at Purple Mountain Observatory proposed a two-step plan: first to build a 5 m terahertz telescope and, then, to build a follow-up 15 m terahertz telescope. The 5 m proposed telescope is now named as 5 Meter Dome A Terahertz Explorer, abbreviated as DATE5 telescope.

II. ENVIORMENT CONDITIONS AT DOME A

The recorded annual temperature variations at Dome A, Antarctica, are ranged from -10°C to -80°C with a diurnal variation smaller than 15°C. In summer, the temperature is about -30°C, so that ordinary power machines could work reliably. However, after about a month, the temperature will fall down to very cold region, where no snowmobiles could even start. The year round average temperature is -58.4°C. The vertical temperature gradient is generally smaller than 2°C/m above the inversion layer. Dome A also has a smaller inversion layer and smaller seeing size.

The average wind speed at Dome A is only about 2.5 m/s with the maximum speed of about 14 m/s[7]. Since the site is 4000 m above sea level, the air pressure is lower. The variation of air pressure in the area is between 540 and 610 hPa with an average value of 575 hPa. Low air density will

978-7-5641-4279-7

make convention cooling less efficient. Lower air density brings lower oxygen level. This brings difficulties in human activities.

Air relative humidity is ranged between 10% and 80% with an average relative humidity of only 40% at Dome A. These numbers do not give direct indication of the absolute water contents in the air as they are relative to the maximum allowable water contents at the temperature. At a low temperature the maximum water contents allowable are very small. Snowing is not often at the site and the average snow accumulation is about 23 mm equivalent water every year. Frosting and icing have been seen on top of some over-winter observing instruments.

International shipping of the telescope parts will be provided by XueLong ice breaker, which arrives at Zhongshan Station, a year round seashore station, and the cargo will be subsequently transferred to surface transportations by snowmobiles. A one way surface shipping takes 18-20 days from Zhongshan to Kunlun Station over a 1,200 km journey. The trip can be only performed in summer time.

Telescope assembly will be finally carried out at Kunlun Station of Dome A. As a summer supporting station, Kunlun station could provide accommodations of about 20 people up to one month. There is no winter over stuff in this mountain station. Therefore, only small part of the observation data cumulated can be transferred to observer's home in real time while bulks of data are retrieved by the summer traversers.

III. DESIGN CHALLENGES OF DATE5 TELESCOPE

The proposed DATE5 telescope will include two important bands: Band 1 of 0.78~0.95 THz using SIS receiver and Band 2 of 1.25~1.55 THz using SIS/HEB (hot electron bolometer) technology[8],[9]. Both receivers are in the developing stage.

For antennas working in this very high terahertz frequency, the reflector surface root mean square error has to be smaller than one twentieth of the wavelength operated. The pointing accuracy should be about one tenth of the telescope's main beamwidth[1]. For cancelation of noise from the atmosphere, small fast chopping mirror or chopping secondary mirror may be needed for this telescope. By considering the wide temperature variation, very low temperature, and very low air density of the Dome A site, the above accuracy and fast chopping requirements produce real challenges to the telescope designers.

Stable surface shape under a wide temperature range is usually a problem when materials with high coefficient of thermal expansion (CTE) and low heat conductivity are used. Panels with stretched aluminum skin and stiffening ribs glued with epoxy are not suitable for this telescope as the epoxy has lower heat conductivity. Machined aluminum panels may be a suitable dish panel candidate for this telescope. However, the size of the aluminum panels may be limited when a carbon fiber reinforced plastics (CFRP) backup structure (BUS) is used. Other suitable panel candidates include CFRP replicated sandwich panels with aluminum honeycomb. These panels have been used in submillimeter wavelength telescopes in the past.

Generally, low CTE CFRP material has to be used for the BUS structure of this telescope as the site temperature varies from -10 to -80°C. The BUS can be either in truss or boxed forms. To avoid temperature gradient inside the telescope BUS structure, forced ventilation can be applied to keep the temperature uniform inside. Air ventilation can also be used inside the receiver cabin and the mounting structures. The low air density of the site will reduce the efficiency of all power equipments of the telescope[10]. However, the lower site air temperature will help the heat exchange for power equipments. These effects may cancel each other.

Other design difficulties of the DATE5 telescopes include the anti-frost requirement, stable working temperature for receiver and telescope drive system, limited installation time, and long period remote operation of the telescope.

Frost is one important issue in the telescope design as the frost may block terahertz wave path, damage telescope structure, and stop movement of the drive train. When a solid surface is chilled below the dew point of the surrounding air, frost will be formed on the surface.

Frost is a form of ice crystals growing directly on the structure surface. The crystals size may be different depending on time, temperature, and the amount of water vapor available. Cooling of structure surface is the cause of the frost forming. Radiation and wind are two main reasons in structure cooling. When the structure has sharp edges, holes, or gaps, the structure relative surface area per structure weight is larger than that in other regions. These surfaces will cool down faster due to radiation or air convection. Holes and gaps also attract ice particles and make ice and snow build up.

Structure material and surface coating influences frost forming. Water is a polar molecule with a surface energy of 72 mJ/m^2. It attracts molecules with the same polar property or with higher surface energy. Metals have surface energy of about 500 mJ/m^2 while Teflon or wax have low surface energy. Surface with high surface energy is contaminated easily with low surface energy material. Modern anti-frost coating uses polypyrrole to avoid ice to form on structure surface.

One can also apply heating elements to prevent frost. Profiling the structure to avoid any holes, gaps, and sharp edges is effective in keeping the structure surface temperature stable. Icing may be also caused by ice particles carried by the wind or by interaction of strong temperature changes caused by weather.

To maintain a stable temperature for receiver and drive system, local heating, internal ventilation, and wall insulation are necessary, so that the telescope can be divided into different temperature zones to meet individual zone temperature requirement for different systems.

The DATE5 telescope should have limited number of pre-assembled components before it is shipped to Dome A. The connection and adjustment between components should be simple and easy so that the telescope site assembly can be finished within short period.

IV. DESIGN FEATURES OF THE DATE5 TELESCOPE

A. Optics of the antenna

A classic Cassegrain or an R-C optical systems can be used for DATE5 telescope. An R-C system has a slightly larger field of view, but requires fixed mirror surface shape and mirror separation. Therefore, traditional Cassegrain design is still preferred. In the optics design, the size of subreflector should be slightly larger than it should be, so that the beam pattern will be better when a mirror chopping is applied.

It is preferred to use a two mirror system, resulting in higher efficiency and lower noise level than those of three- or four-mirror systems. However, chopping of the subreflector increases telescope cost because all the subreflector drive and encoder systems are on top of the dish, suffering from Antarctica weather changes. For eliminating the reaction force from subreflector chopping, complex reactionless drive system has to be used. To avoid subreflector chopping and the motion of receiver Dewar, Coude optical system with two additional small mirrors may be used. This system can be easily arranged inside a slant axis mounting structure, resulting in a smoother telescope profile without sharp edges, holes, and gaps.

B. Mounting and structure design

In ground layer, air turbulence at Dome A is serious; the antenna dish should be high above the ground. Therefore, the telescope has a tall base. The distance between the ground level and the elevation axis is about 5 meters. Ice has a small compression strength. The average pressure on the ice surface have to be kept lower than 403 Pa and the maximum pressure on the ice surface under survive force condition should be lower than 470 Pa[11]. In order to reduce forces from telescope steel structure on the ground ice surface, timber footing may be used under the steel structure surface and on top of ice surface. Timber has a Young's modulus of about 6.9×10^9 Pa. For long term stability, the footing of the telescope has to be buried under ice surface.

Both altazimuth and slant-axis mountings can be used for the DATE5 telescope. In an altazimuth mounting, receiver is located at Cassegrain focus. The weight of the receiver cabin is used to balance the dish structure weight. The elevation axis usually uses self-adjusted spherical roller bearings. The azimuth axis will use a double row roller bearing. However, there are gaps existed between yoke and cabin. The ice particles could be trapped in these gaps, blocking the required telescope movement. Therefore, anti-frost heating has to be applied in these areas.

Slant axis design has a much smoother profile without any sharp edges, holes, or gaps for frost to stay. Slant axis mounting includes two axes: one is slant axis and the other is slant-muth axis. The slant-muth axis is the same as azimuth axis in an altazimuth mounting. The name here is specially for slant axis mounting system. All the telescope drive system is within one structure volume. Two identical bearings and drive gears can be used for both the slant-muth and slant axes. If the receiver is at the Coude focus, all the cables and pipes connected to delicate Dewar device are fixed with slant-muth platform. No relative motions are required for the receiver parts cables, and pipes. The whole antenna forms a closed volume which makes telescope thermal control and frost prevention easier. The space inside the telescope volume can be divided into different temperature zones: a warm zone in the main body and a cold zone in the BUS area. With each zone, the temperature is the same and, between different zones, the temperatures are different. The waste heat generated from electronic or electrical parts is used for internal heating.

C. Dish and surface panels

For terahertz observation, high precision surface panels and CFRP backup structure may be necessary. The surface panels can be either computer control machined aluminum ones or replicated CFRP honeycomb sandwiched ones. The machined aluminum panels are weight reduced from solid thick aluminum alloy plates. The replicated CFRP sandwiched panels are formed on top of a precision glass mould. The backup structure can be either CFRP trusses or CFRP box structures.

Reflecting surface error of the DATE5 telescope comes from a number of sources: panel, BUS, and secondary mirror. These error sources also include manufacture, gravity, thermal, wind, and even the measurement itself. The thermal induced panel or BUS error includes two different temperature distributions: 1) absolute temperature error and 2) differential thermal error. Absolute temperature error, also named as temperature soaking error, is caused by the CTE difference between panel and BUS materials. The error is produced as the surface setting temperature is different from the operation temperature. Differential thermal error is caused by temperature difference between different parts of the same panel or BUS structure. Reduced stiffness of the panel adjusters in one or more directions can take care of part of the absolute temperature surface error, while panel material conductivity can affect the panel differential thermal error. To reduce the BUS differential thermal error, low CTE CFRP material has to be used in this structure.

The main component of CFRP is carbon fibers. Within carbon fibers, carbon atoms are in a form of hexagonal layers with very strong bonds between atoms. These layers form rings around fiber axis. In the fiber axial direction, Young's modulus is very high up to many hundred GPa, few times of that of steel. In the perpendicular plane, the modulus is few times smaller. Carbon fibers also have very low, even negative, CTE along the fiber direction. Carbon fibers are held together by polymer resins. Resin usually is poorer in stiffness as well as in CTE, resulting in less stiffness and thermal unstability in the resin direction.

CFRP BUS structure can be truss style made of tubes or box style made of plates. When using truss structure, joints are usually of stainless steel, adding more weight and more thermal expansion, or Invar, adding more cost. All the panels and CFRP BUS structure has to be tested in a low temperature chamber before being used in the DATE5 telescope.

D. Thermal control of the telescope

At Dome A, the temperature is very low (between -20 to -80°C), too cold for operation of receiver, drive and control systems. Under this condition, some components of the systems may be not functioning. In general, receiver cabin requires temperature higher than that at Dome A site. The drive system can have a slightly lower operation temperature. The BUS area can be even lower. Most telescope volume should keep a constant temperature to avoid structure distortion.

Entire telescope body should be insulated using forms and reflecting plates. Heating elements may be needed on panels for frost prevention.

E. Drive and control systems

To maintain an accurate pointing of the DATE5 telescope, backlash free twin gear drive system will be used for both telescope axes. All the motors, gear boxes, and encoders have to be pre-tested in cold chambers before been assembled into the telescope. Closed loop control is used for good pointing and tracking. Before the telescope operation, the pointing correction is carried out using radio as well as optical stars with known coordinate positions.

F. Test operation and site assembly

When the DATE5 telescope is finished, it will be transported to Delingha observing station in west China for the system pre-assembly and testing. All assembly procedure has to be checked to meet the site assembly requirement. This requires using standard interfaces for connections between components. Therefore, the parts can be quickly assembled and detached. Dish surface holographic measurement in Delingha is one important task.

ACKNOWLEDGMENT

This work is supported by Chinese National Science Foundation under grant No. 11190014. The DATE5 work was initiated from discussions of international Cosmic star formation research team. The group includes Qizhou Zhang, Shencai Shi, Jiasheng Huang, Qijun Yao, Zhong Wang, Yu Gao, Lin Yan, Edward Tong, Ruiqing Mao, Longlong Feng, Hongchi Wang, Zhibo Jiang, Ye Xu, and Xianzhong Zheng.

REFERENCES

[1] S. Withington, "Terahertz astronomical telescopes and instrumentation," phil. Trans. R. Soc Lond. A, 362, 395-402, 2004.

[2] S. P. Huang, J.Li, et. Al., "A 500 GHz superconducting SIS reciever for the portable submillimeter telescope," IRMMW-THHz 2006, China National Astronomy Observatory, Nanjing, 2006.

[3] G. Sims, et al., "Parcipitable water vapor above Dome A, Antarctica, determined from diffuse optical sky spectra," PASP, Vol. 124, No. 911, 2012, pp. 74-83

[4] Jingquan Cheng, The principles of astronomical telescope design, Springer, 2009.

[5] Sanunders W. et al., "Where is the best site on earth? Domes A, B, C and F, and Ridges A and B," PASP, 121,2009, pp. 976-992.

[6] D. P. Marrone, et al., "Observations in the 1.3 and 1.5 THz atmospheric windows with the Reciever Lab Telescope," arXiv:astro-ph/0505273, 2005.

[7] http://www.antarctica.gov.au/science/ice-ocean-atmosphere-and-climate/glaciology-research/antarctic-weather/dome-a-details

[8] Zheng Lou and Shengcai Shi, "Design of a multi-channel quasi-optical frontend at terahertz bands," Proc. SPIE 7849, 78490W, 2010.

[9] Jingquan Cheng, Ferromagnetism and its applications, Chinese Science and Technology Press, 2006, Beijing.

[10] Jingquan Cheng, "Forced air cooling at high altitude," ALMA memo 203, NRAO, 1998.

[11] S. Padin, "SPT footing," SPT memo, 2004.

Design Study on Near-Field Radio Holography of the 5-Meter Dome A Terahertz Explorer

Yingxi Zuo[#,*,1], Zheng Lou[#,*], Ji Yang[#,*], and Jingquan Cheng[#,*]
[#]Purple Mountain Observatory, Chinese Academy of Sciences
[*]Key Laboratory of Radio Astronomy, Chinese Academy of Sciences
2 West Beijing Road, Nanjing, 210008, China
[1]yxzuo@pmo.ac.cn

***Abstract*- The 5m Dome A Terahertz Explorer (DATE5) is a proposed terahertz telescope to be deployed in Dome A, Antarctica, to exploit one of the best observing conditions at terahertz wavelengths on earth. In this paper, a design configuration for the near-field holography of the DATE5 surface measurement is presented. Important factors, such as measurement distance, operating frequency, and signal source location, are discussed. Special efforts have been given to the reduction of truncation errors. Simulation results under typical signal-to-noise ratios and design parameters are also provided.**

I. INTRODUCTION

Dome A, Antarctica is one of the best sites on earth for terahertz astronomical observations due to its extremely low water vapor contents in the atmosphere [1]. To exploit this unique site condition, a 5-m Dome A Terahertz Explorer (DATE5) telescope, which is a fully steerable 5-m Cassegrain telescope operating at dual bands of 350μm (Band1) and 200μm (Band2) under remote control [2][3], is proposed for this site. The antenna has to have an overall equivalent surface accuracy of 10 μm rms [3]. The error budget for surface measurement of the main reflector is only ~3 μm rms, which is a challenging target for on-site measurement. Moreover, the extreme site environment and limited assembly and testing time produce further difficulties in achieving such accuracy.

In the past, a number of techniques have been used in the field of antenna reflector surface measurement, including the laser tracker method, digital photogrammetry, and radio holography, etc. Typical accuracy of a laser tracker is 10μm + 5μm/m, and that for the photogrammetric measurement is 5μm + 5μm/m, depending on the measurement distance and the scale of the measured object. These methods will not meet the accuracy requirement of the DATE5 antenna. Radio holography is considered to be more accurate for the antenna reflector measurement and it has been applied to a number of millimeter and submillimeter wavelength telescopes. The ACA (Atacama Compact Array) 12-m antennas have achieved a best surface accuracy of approximately 8 μm, while the ACA 7-m antennas achieved an accuracy of approximately 5.7μm, both using near-field holography [4][5].

Theoretically, radio holography is based on the integral transform relationship between the radiation pattern and the aperture field distribution of a reflector antenna [6]. To carry out holography measurement, it is desired to have a strong source in the farfield region, e.g., a beacon emitter from a satellite on orbit, since in this case the radiation pattern is simply the Fourier transform of the aperture field. If there is no far-field beacon emitter, as in the case of the DATE5 antenna at the Dome A site, a ground transmitter in the vicinity of the telescope has to be used.

In this paper, the system configuration for a near-field holography of the DATE5 is presented. Several important factors such as the measurement distance, the operating frequency, and the signal feed horn location, are discussed. Methods for reducing the truncating error are proposed. Simulation results under a typical signal-to-noise ratio and design parameters are also provided.

II. SYSTEM CONFIGURATION FUNDAMENTALS

A. *Hardware Configuration*

At Dome A, Antarctic, there exist no practicable far-field sources, so that a local transmitter mounted on a tower, as shown in Fig. 1, will be used. Fig. 2 shows the block diagram of a dual-channel holography receiver. In this figure, a signal feed horn illuminates the antenna reflector, while a reference feed horn looks directly to the transmitter. At the backend, the dual-channel output signals are digitized and correlated, obtaining a complex correlation function.

An antenna is normally designed for observing far-field sources. When the signal comes from a transmitter in the near-field, there will be a rapid phase variation over the antenna aperture. The phase variation can be compensated, to a large degree, by an axial displacement of the feed in the direction away from the main dish [6]. By using this method, the antenna is refocused to the transmitter and has a relatively narrow beam. Thus, a small region of the beam pattern can be scanned without much truncation errors.

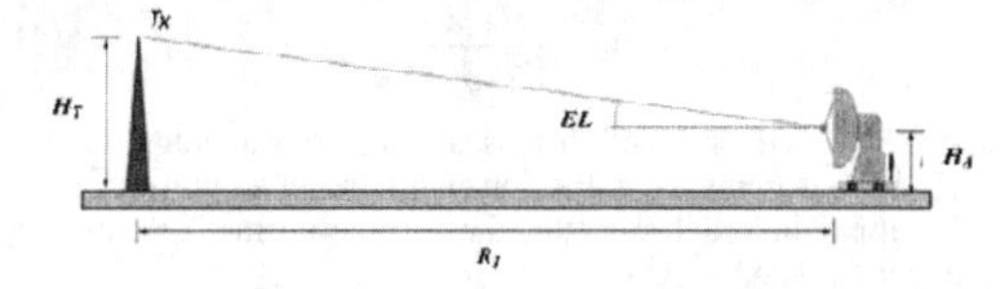

Figure 1. Configuration for near-field holography of DATE5.

978-7-5641-4279-7

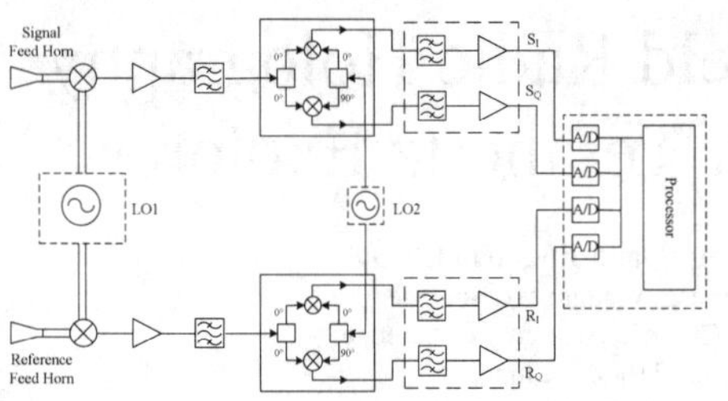

Figure 2. Holography receiver block diagram.

B. Measurement Procedures

The first step in holography measurement is to sample the beam pattern. This can be done by raster scanning the antenna beam around the transmitter. The sampling interval should be no larger than the Nyquist interval of λ/D, where D is the aperture diameter and λ the wavelength. With an oversampling factor k_s ($0.5 < k_s < 1$), the sampling interval Δu and Δv in the beam coordinates can be written as:

$$\Delta u = \Delta v = k_s\lambda/D. \tag{1}$$

The minimum scanning angle of the beam is determined by the required spatial resolution over the aperture. If the measured beam angle is $N\cdot\Delta u$, then the aperture field resolution obtained is:

$$\delta = D/(N\cdot k_s). \tag{2}$$

When the beam sampling is done, the measurement data are calibrated to remove the drifts in amplitude and phase, based on the boresight measurements at the beginning and the end of each row. The data are then fitted into a square grid.

The next step is to obtain the aperture field distribution through integral transform, resulting in the phase distribution over the aperture plane as in formulas of [6]. Phase corrections have to be made for the geometrical phase deviation and for the measured feed phase pattern. Furthermore, several phase terms accounting for phase offset, constant antenna pointing error, and small vector displacement of the signal feed relative to its nominal position have to be fitted and to be removed. Finally, the half pathlength deviation over the aperture plane, Δz, is derived as:

$$\Delta z(x,y)=\frac{\lambda}{4\pi}\cdot\varphi(x,y), \tag{3}$$

where $\varphi(x,y)$ is the phase deviation over the aperture plane of (x,y). The rms error of Δz, in the presence of white noise, can be written as [8]:

$$\sigma_z=\frac{D\cdot\lambda}{16\sqrt{2}\cdot\delta\cdot SNR}, \tag{4}$$

where SNR is the ratio of on-axis signal to rms average noise over all measurements, at the correlation receiver output.

For a paraboloidal reflector, the surface normal deviation ε_s is directly related to Δz by:

$$\varepsilon_s(x,y)=\sqrt{1+\frac{x^2+y^2}{4f^2}}\cdot\Delta z(x,y), \tag{5}$$

where f is the focal length of the primary reflector. The normal deviations are then used for adjusting the reflector panels.

III. Design Considerations

A. Measurement Distance

In order to suppress the ground reflection, doing the holography measurement at a higher elevation is favorable. This is realized by constructing a higher beacon tower, or by reducing the measurement distance (the distance between the transmitter and the antenna under test). Under the extreme conditions of Dome A, it will be very difficult to build a very high tower. Therefore, a short measuring distance is desired. However, when the transmitter is closer to the antenna, the higher order phase terms (refer to [6], these are terms with their orders higher than the Fresnel term) will become significant and, therefore, more measurement error is produced [6][7]. Fig. 3 depicts a simulation result for the DATE5 antenna with N=64 (64×64 map), showing the relationship between the surface rms error increase and the measurement distance (R) decrease. In the simulation, the reduction of the truncation error is not considered. Later, one will find that the truncation error is also one of the major terms for the surface error in the near-field holography measurement. The truncation error can also be reduced significantly. In this paper, a measurement distance of 100m and a tower of 20 meters high are selected, which are thought to be practical numbers at Dome A site. If the antenna elevation axis height is 5m, the measurement elevation angle is about 8.5 degrees.

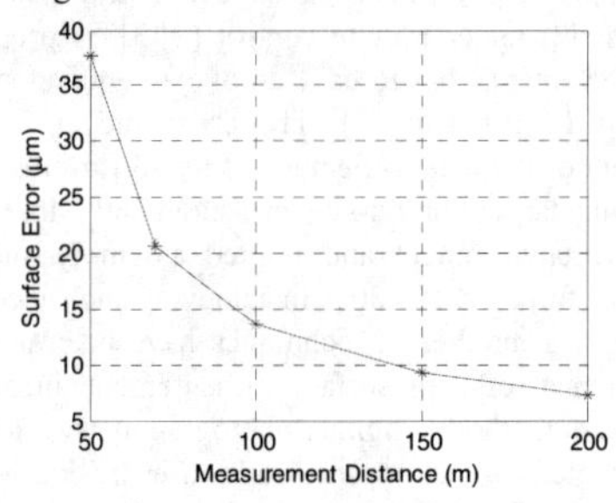

Figure 3. Surface RMS error vs measurement distance R.

B. Operating Frequency

According to Equation (4), for a given accuracy, the required SNR and dynamic range increase inversely with the frequency. Therefore, a higher frequency is preferred, since large dynamic range will increase the difficulty in the receiver development. On the other hand, as the frequency increases, the power of the transmitter will decrease and the receiver noise will increase, making the measurement SNR decrease. At the same time, the cost will increase as frequency increases. For these reasons, the preferred holography system will operate in the W-band.

Another issue in the measurement is to suppress the frequency-dependent effects, such as multipath reflection from the ground and antenna structures, and diffraction from the subreflector support structures. By assuming a path difference between the reflected and the direct signal being dx, the phase difference will change by one cycle for a frequency change of $dF = c/dx$, where c is the velocity of light. Thus, if the holography measurements were taken at several frequencies and averaged for the resulting aperture distributions, the multipath effect would be largely suppressed. For ground reflection, the path difference is usually within several meters, a frequency variation of several hundred MHz is quite enough. For suppressing the diffraction and multipath reflection with short path difference, measurements at more than one frequency with large frequency separations is required. Two nominal operating frequencies of 79GHz and 106GHz with a variation range of several hundred MHz at each frequency will be used for the future DATE5 holography measurement.

C. *Signal Feed Horn Location*

For holographic measurement, there are two options for the location of the signal feed, one is at the primary focus and the other is at secondary focus of the antenna. Both options have pros and cons.

Major advantage for the primary-focus configuration is that the reference and signal feed horns can be arranged back-to-back with the reference horn looking directly to the transmitter, so that a dual-channel receiver frontends can be used with the local oscillator (LO) signal being equal in length. The phase stability of the system is therefore improved. Another advantage for this configuration is that the reflector surface deviations can be obtained directly since the displacement of the signal horn could be easily separated from the measured aperture distribution during data post-processing. A major disadvantage for this configuration is that the subreflector assembly has to be removed from the antenna so that the holography receiver frontend can be mounted.

For the Cassegrain focus configuration, one advantage is that an overall surface deviation is obtained through the measurement, including subreflector surface error and errors from the reflector misalignment. In this configuration, there is no receiver installation problem. However, several disadvantages exist. Firstly, a separate reference frontend has to be installed away from the signal horn, resulting in phase instability of the system. Secondly, it is difficult to isolate the exact contributions from individual reflector and feed misalignment, which include the effects of displacements and tilts as cross correlations may exist between them. Based on such a measurement, the main reflector may be adjusted away from a true paraboloid. In addition, as the DATE5 antenna has a small half extending angle of subreflector from the Cassegrain focus (< 2 deg, refer to [2]), there may be a risk that the signal feed sees the transmitter directly during the beam scanning.

In summary, the primary-focus configuration has less error in measurement. It will be our first choice, especially for the first-time antenna installation on site. Even so, the second option still provides an alternative way, as long as the error contributions are at an acceptable level.

IV. Truncation Error Reduction

Although the radiation field of an antenna extends over the whole 4π space, only a small beam region around the boresight is sampled in holography measurement with a sampling interval stated in (1). This results in a systematic truncation error. Simulations for the DATE5 antenna have been carried out under different beam map size in the absence of noise, and the corresponding residual errors are shown in Table I. The simulation parameters used are: D=5m, f=2m, k_s=0.75, F=100GHz, R=100m, axial refocusing displacement δf=53mm (at which the antenna is refocused on the transmitter), and edge taper ET=6dB. The error reduces as the sampling map becomes larger, showing that truncation effect being one of the major contributions to the systematic error.

TABLE I

Residual Error for Different Size of Beam Map

Map size N×N	64×64	128×128	256×256	512×512
Surface Error (μm rms)	12.9	4.78	2.36	1.13
Aperture spatial resolution (mm)	104	52	26	13

To minimize the truncation error, a larger beam map should be used. However, a larger beam map means longer sampling time for the measurement, resulting in an instability risk. For the DATE5 antenna, a fast holography measurement is of special importance because the limited assembly and testing time is set by the site condition. In fact, an aperture spatial resolution of ~100mm is sufficient for the reflector adjustment. Such a resolution corresponds to N=64. However, from Table I, the truncation error is not acceptable. Therefore, two methods are proposed for further reduction of the truncation error. These two methods are discussed in following paragraphs.

A. *Refocus Displacement Optimization*

Simulations have been performed for different refocus displacement δf, and an optimum value has been found where the residual error reaches its minimum, as shown in Fig. 4. The optimum focus displacement is slightly larger than the value for the antenna being refocused on the transmitter. Simulation parameters are the same as for Table I except N=64. At the optimum refocus displacement, the residual surface rms error is only 1.4μm when δf=76mm instead of 12.9μm when δf=53mm. A simulation for a 128×128 map size is also performed. The error can be further reduced from 4.8μm when δf=53mm to 0.70μm when δf=76mm.

B. *Map Size Extension and Iteration*

The iteration procedure is as follows:

(a) The measured N×N beam map size is extended to 2N×2N by using zero-padding technique.

(b) A 2N×2N aperture field distribution is computed using the extended beam map.

(c) A new 2N×2N beam map is calculated based on the aperture distribution obtained in the previous step.

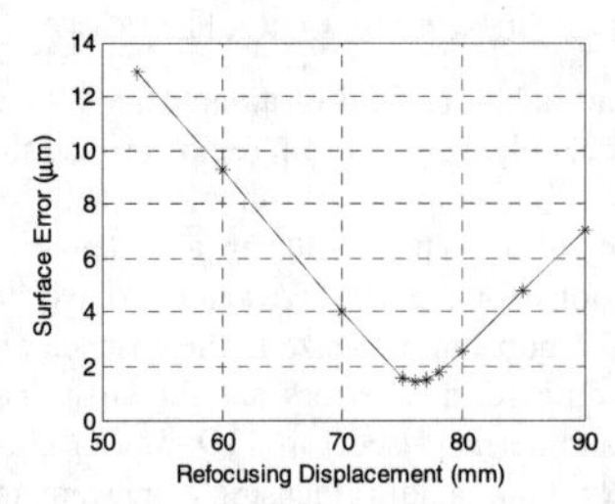

Figure 4. Residual surface error vs refocusing displacement δf.

(d) The central N×N part of the beam map obtained in step (c) is replaced with the original measured beam map.

(e) A new aperture field distribution is computed.

(f) Repeat steps (c) through (e).

Finally, the interpolations are made to obtain an N×N aperture distribution from the derived 2N×2N aperture map.

Fig. 5 shows the iteration procedure for a 64×64 measured beam map (other parameters are the same as for Table I). The map is extended to a 128×128 beam map first. After 9 iterations, the surface rms error is reduced from 12.2μm to 5.85μm. Finally the aperture field distribution is interpolated back to a new 64×64 map, and the error is further reduced to 2.84μm rms. This method reduces the truncation error significantly with the aperture spatial resolution unchanged for the measured beam map size.

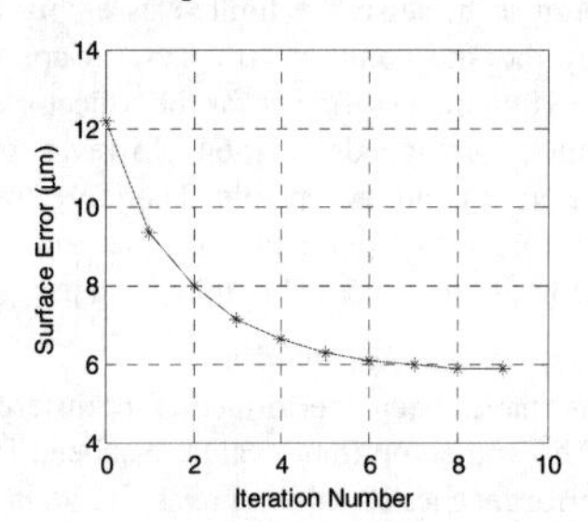

Figure 5. Error reduction iteration procedure.

V. Simulation Results

Simulation is carried out under the following design parameters: $R = 100$m, $F = 106$GHz, $k_s = 0.75$, N=64, $\delta f =$ 76mm (at which the truncation error is a minimum), $ET = 6$dB, $P_t = 1\mu$W (transmitter power), $G_t = G_r = 29$dB (gain of transmitter and reference feed respectively), $T_{sys} = 3000$K (SSB), and B=10kHz, and $\tau = 0.04$s. The system noise temperature T_{sys}, bandwidth B and integration time τ are assumed to be identical for both receiver channels. The resulted SNR for signal channel is SNR_s=84.5dB, and for reference channel SNR_r=46.6dB. The truncation error reduction methods mentioned above have been used in the simulation. The surface deviation distribution is shown in Fig. 6, and a surface rms error of 1.1μm is achieved.

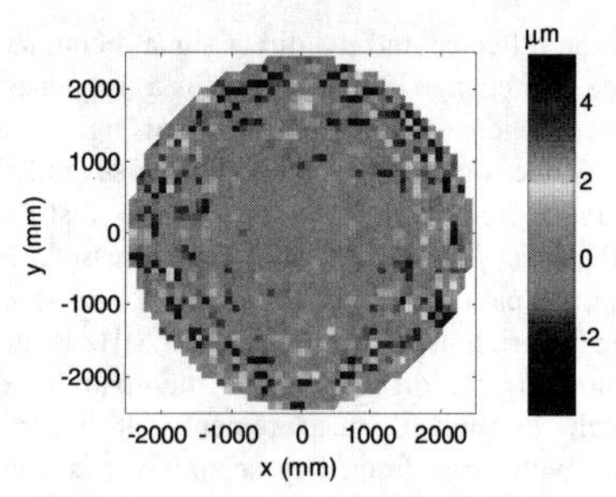

Figure 6. Simulated surface deviation distribution.

VI. Conclusion

Near-field holography is the most practical method for the proposed DATE5 antenna which requires a 10μm rms surface accuracy. The proposed measurement system includes a transmitter on top of a 20m high tower 100m away from the antenna. The transmitter will operate at two frequencies, 79 GHz and 106 GHz, each with a variation of several hundred MHz, to suppress those frequency-dependent errors. The signal feed can be located either at the primary focus or at the secondary focus. The truncation error, which is significant for near-field holography, can be greatly reduced by the proposed methods. Simulation shows that surface rms error of 1.1μm can be achieved when a 64×64 beam map size is used under typical SNR and design parameters.

Acknowledgment

This work was supported by National Science Foundation of China (NSFC) through grants No. 11190014 and 11003050.

References

[1] H. Yang, C. Kulesa, C. K. Walker, et al., "Exceptional Terahertz Transparency and Stability above Dome A, Antarctica," *PASP*, Vol.122, pp. 490-494, April 2010.

[2] Z. Lou, Y. X. Zuo, J. Q. Cheng, J. Yang, S. C. Shi, "Study on the Optics of the 5 Meter Dome A Terahertz Explorer (DATE5) for Antarctica,Antennas," *Propagation & EM Theory (ISAPE), 2012 10th International Symposium on,* 2012, ():47-50.

[3] J. Yang, Y.X. Zuo, Z. Lou, J.Q. Cheng, et al, "Conceptual Design Studies of the 5 Meter Terahertz Antenna for Dome A, Antarctica," submitted to *RAA*.

[4] M. Saito, J. Inatani, K. Nakanishi, et al., "Atacama Compact Array Antennas," *Proc. of SPIE* Vol. 8444, 84443H (Sep. 17, 2012).

[5] S. Asayama, L. B. G. Knee, P. G. Calisse, P. C. Cortés, R. Jager, et al. "ALMA array element astronomical verification", *Proc. SPIE* Vol. 8444, 84443F (Sep. 17, 2012).

[6] J. W. M. Baars, R. Lucas, J. G. Mangum, and J. A. Lopez-Perez, "Near-Field Radio Holography of Large Reflector Antennas," *IEEE Antennas and Propagation Magazine*, vol. 49, no. 5, Oct. 2007.

[7] D. Morris, "Errors in near-field radioholography," *IET Microwaves, Antennas & Propagation*, 2007, 1, (3), pp. 586–591.

[8] D'Addario, L., "Holographic antenna measurements: further technical considerations", *12m Telescope Memo No.202, NRAO, Charlottesville,* Nov. 1982.

WP-2(A)

October 23 (WED) PM

Room A

Adv Ant for Radio-Astr. -2

Calculation of the Phase Center of an Ultra-wideband Feed for Reflector Antennas

Jian Yang

Dept. of Signals and Systems, Chalmers University of Technology, S-41296 Gothenburg, Sweden
Email: jian.yang@chalmers.se

***Abstract*—Next generation ultra-wideband (UWB) radio telescopes require UWB feeds for reflector antennas. Different from narrow band feeds, how to determine the optimal phase center location for a UWB feed over a wide operating frequency band has not been investigated much and therefore analysis and discussions on this issue are needed. In this paper, a method for calculating the optimal phase center of a UWB feed is presented. Examples of the Eleven feeds, a decade bandwidth feed, are used to demonstrate the applications of the new method.**

***Index Terms*—Phase center, ultra-wideband antenna, feed, reflector antenna, Eleven antenna**

I. Introduction

The ongoing developments of the next generation ultra-wideband (UWB) radio telescopes, such as the Square Kilometer Array (SKA) [1] and VLBI2010 [2], have pushed technologies forward on a broad front. For example, UWB feed technologies for reflector antennas have made substantial progress [3], such as the Eleven feed [4]–[6], the quadruple-ridged flared horn [7], the sinuous feed [8] and the quasi self-complementary antenna [9].

The issue of the phase center of a feed for reflectors has been discussed in [10]–[14], mainly for narrow band, and the results have been applied in designs of different feeds, for example in hat feeds [15]–[18].

In this paper, we discuss how to determine the optimal phase center for a wideband feed.

II. Definition

The aperture efficiency e_{ap} of a feed for parabolic reflectors can be expressed by its sub-efficiencies as [19]–[21]

$$e_{\text{ap}} = e_{\text{sp}} e_{\text{BOR1}} e_{\text{ill}} e_{\text{pol}} e_{\phi}, \quad (1)$$

where e_{sp}, e_{BOR1}, e_{ill}, e_{pol} and e_{ϕ} are the spillover efficiency, the BOR_1 efficiency, the illumination efficiency, the polarization efficiency and the phase efficiency, respectively. The phase efficiency can be expressed by

$$e_{\phi}(f,z) = \frac{\left|\int_0^{\theta_0} G_{\text{co45\_BOR1}}(\theta,f) tan(\theta/2) d\theta\right|^2}{\left[\int_0^{\theta_0} |G_{\text{co45\_BOR1}}(\theta,f)|\, tan(\theta/2) d\theta\right]^2}, \quad (2)$$

where $G_{\text{co45\_BOR1}}(\theta,f) = |G_{\text{co45\_BOR1}}(\theta,f)|\, e^{j\phi_z(\theta,f)}$ is the co-polar radiation function of the BOR_1 component in $\varphi = 45^\circ$ plane of the feed. The phase function is $\phi_z(\theta,f) = \phi(\theta,f) - kzcos(\theta)$ when the phase reference point is moved from the origin to a point z in the coordinate system of the feed, where $k = 2\pi/\lambda$ is the wave number.

In order to find an optimal phase center over an ultra-wide frequency band, we define first a new characterization - the optimal frequency-weighted phase efficiency as

$$e_{\phi,\text{opt}}(z) = \frac{\int_{f_1}^{f_2} w(f) e_{\phi}(f,z) df}{\int_{f_1}^{f_2} w(f) df} = \int_{f_1}^{f_2} \bar{w}(f) e_{\phi}(f,z) df, \quad (3)$$

where $w(f)$ is an optimal frequency weighting function according to the application, and $\bar{w}(f)$ the normalized optimal weighting function defined by

$$\bar{w}(f) = \frac{w(f)}{\int_{f_1}^{f_2} w(f) df}. \quad (4)$$

The choice of $w(f)$ is very much depending on applications. For example, in radio astronomy, the phase efficiency may not be important at low frequency where the sky noise is very high but it is very critical to have high phase efficiency at high frequencies. So the weighting function can be chosen to weight on high frequencies.

Then, the optimal phase center over an ultra-wide band is defined as the phase reference point which maximizes the optimal frequency-weighted phase efficiency, i.e.

$$Z_{\text{pc,opt}} = \arg\max_z e_{\phi,\text{opt}}(z). \quad (5)$$

III. Formulation

Then, the optimal phase center can be determined efficiently by the formulas and procedure as follows.

A. Initial optimal UWB phase center

As it is known from [10], the phase center for a single frequency point can be determined preliminarily (or approximately) by

$$Z_{0\text{pc}}(f) = \frac{\phi(\theta_0,f) - \phi(0,f)}{(cos(\theta_0) - 1)k}, \quad (6)$$

where θ_0 is the half subtended angle of the reflector, which makes $\phi_z(\theta_0,f) = \phi_z(0,f)$.

Then, we define the preliminay optimal phase center location $Z_{0\text{pc,opt}}$ as

$$Z_{0\text{pc,opt}} = \int_{f_1}^{f_2} \bar{w}(f) \frac{\phi(\theta_0,f) - \phi(0,f)}{(cos(\theta_0) - 1)k} df. \quad (7)$$

978-7-5641-4279-7

B. Determining the optimal UWB phase center

We assume then that the optimal UWB phase center $Z_{\text{pc,opt}}$ is not far from the preliminary optimal UWB phase center $Z_{0\text{pc,opt}}$, i.e.,

$$Z_{\text{pc,opt}} = Z_{0\text{pc,opt}} + \Delta z. \tag{8}$$

and $|k\Delta z| \ll 1$ over the operating frequency band of (f_1, f_2), which means that $|k_2\Delta z| = |2\pi/\lambda_2\Delta z| \ll 1$, where λ_2 is the wavelength at f_2. Now the phase function can be expressed as

$$\phi_z(\theta, f) = \phi(\theta, f) - kZ_{0\text{pc,opt}}\,cos(\theta) - k\Delta z\,cos(\theta)\,. \tag{9}$$

Then, by using the Taylor expansion

$$e^{-jx} = 1 - jx - \frac{x^2}{2}, \quad when\ |x| \ll 1\,,$$

the optimal phase efficiency in (3) can be expressed to second order in Δz as

$$e_{\phi,\text{opt}}(z) = I_0 + I_1(\Delta z) + I_2(\Delta z)^2\,, \tag{10}$$

where I_0, I_1 and I_2 can have analytic expressions in a format similar (but much more complicated) to those in [11], [12], [14]. Actually, there is no advantage to use the analytic but very complicated formulas to find the optimal phase center. Instead, we can use the following simple method to determine the phase center location.

Choose three points of Δz around $\Delta z = 0$ randomly, say Δz_1, Δz_2 and Δz_3. Calculate the optimal phase efficiencies $e_{\phi,\text{opt}}(\Delta z_1)$, $e_{\phi,\text{opt}}(\Delta z_2)$ and $e_{\phi,\text{opt}}(\Delta z_3)$ so we can obtain the values of I_1, I_2 and I_3 from (10)

$$\begin{bmatrix} I_0 \\ I_1 \\ I_2 \end{bmatrix} = \begin{bmatrix} 1 & \Delta z_1 & (\Delta z_1)^2 \\ 1 & \Delta z_2 & (\Delta z_2)^2 \\ 1 & \Delta z_3 & (\Delta z_3)^2 \end{bmatrix}^{-1} \begin{bmatrix} e_{\phi,\text{opt}}(\Delta z_1) \\ e_{\phi,\text{opt}}(\Delta z_2) \\ e_{\phi,\text{opt}}(\Delta z_3) \end{bmatrix} \tag{11}$$

Thus, the optimal phase center of a feed for paraboloids over a ultra-wide band can be determined by

$$Z_{\text{pc,opt}} = Z_{0\text{pc,opt}} - \frac{I_1}{2I_2}\,, \tag{12}$$

and the maximum phase efficiency is

$$e_{\text{pc,opt}} = I_0 - \frac{I_1^2}{4I_2}\,. \tag{13}$$

IV. Example

The Eleven feed is a decade bandwidth feed [22], [23]. Fig. 1 shows a 1.5–14GHz Eleven feed for reflectors [24]. This feed has been measured and we have the measured complex $G_{\text{co45\_BOR1}}(\theta, f)$. Applying the above formulas, we can obtain the phase center as shown in Fig.2. First, we calculate $Z_{0\text{pc,opt}}$ by (7), which is $Z_{0pc,opt} = -62.0$ mm in this case. Then, choose three values of Δz, as shown by the red circles in Fig. 2. Using (11)-(13), the phase center can be calculated ($\Delta z = -10.86$ mm), depicted by the black square in Fig. 2. As a verification, we calculate the phase efficiency $e_{\phi,\text{opt}}(z)$ over $Z_{0\text{pc,opt}} + \Delta z$ where $k_2\Delta z$ is from -0.6 to 0.6, shown by the blue line. From this curve, we see that the phase center is

Fig. 1. Photo of the 1.5–14GHz Eleven feed.

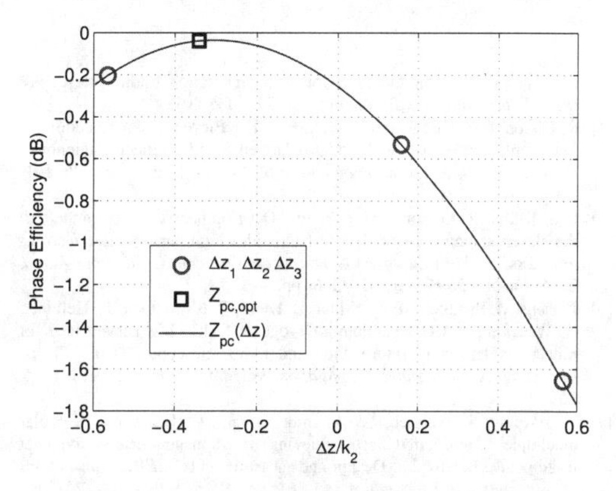

Fig. 2. Phase center calculation.

located at the point marked by the black square. Fig. 3 shows the aperture efficiency and its sub-efficiencies, including the phase efficiency, calculated based on the measured radiation patterns, when the feed illuminates a primary-focused reflector with a subtended angle of $2 \times 60°$. From this figure, the phase efficiency e_ϕ is very high (close to 1) over the whole frequency band of 1–14 GHz.

V. Conclusion

In the paper, a simple and fast formula for calculation of the phase center of a wideband feed for reflectors is presented and applied to an example of 1.5–14GHz Eleven feed.

References

[1] P. Hall, "The square kilometre array: An international engineering perspective," *The Square Kilometre Array: An Engineering Perspective*, pp. 5–16, 2005.

[2] A. Niell, A. Whitney, B. Petrachenko, W. Schluter, N. Vandenberg, H. Hase, Y. Koyama, C. Ma, H. Schuh, and G. Tuccari, "VLBI2010: current and future requirements for geodetic VLBI systems," *Report of Working Group*, vol. 3, 2005.

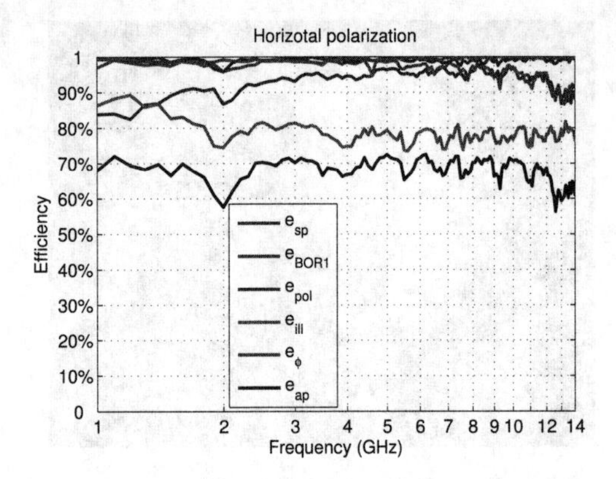

Fig. 3. Efficiency of the 1.5–14GHz feed based on the measured data, with a subtended angle of $2 \times 60°$.

[3] The Special Issue on Antennas for Next Generation Radio Telescopes, *IEEE Trans. Antennas Propagat.*, vol. 59, June 2011.

[4] R. Olsson, P.-S. Kildal, and S. Weinreb, "The Eleven antenna: a compact low-profile decade bandwidth dual polarized feed for reflector antennas," *IEEE Trans. on Antennas Propagat.*, vol. 54, no. 2, pp. 368–375, Feb. 2006.

[5] P.-S. Kildal, R. Olsson, and J. Yang, "Development of three models of the eleven antenna: a new decade bandwidth high performance feed for reflectors," in *First European Conference on Antennas and Propagation, 2006. EuCAP 2006.* IEEE, 2006, pp. 1–6.

[6] J. Yang, M. Pantaleev, P.-S. Kildal, B. Klein, Y. Karandikar, L. Helldner, N. Wadefalk, and C. Beaudoin, "Cryogenic 2–13 GHz Eleven feed for reflector antennas in future wideband radio telescopes," *IEEE Transactions on Antennas and Propagation*, vol. 59, no. 6, pp. 1918–1934, 2011.

[7] A. Akgiray, S. Weinreb, W. A. Imbriale, and C. Beaudoin, "Circular quadruple-ridged flared horn achieving near-constant beamwidth over multi-octave bandwidth: Design and measurements," *IEEE Transactions on Antennas and Propagation*, vol. 61, no. 3, pp. 1099–1108, 2013.

[8] R. Gawande and R. Bradley, "Towards an ultra wideband low noise active sinuous feed for next generation radio telescopes," *IEEE Trans. Antennas Propag.*, no. 99, pp. 1945–1953, June 2011.

[9] G. Cortes-Medellin, "Non-planar quasi-self-complementary ultra-wideband feed antenna," *IEEE Trans. Antennas Propagat.*, vol. 59, no. 6, pp. 1935–1944, June 2011.

[10] Y. Hu, "A method of determining phase centers and its application to electromagnetic horns," *Journal of the Franklin Institute*, vol. 271, no. 1, pp. 31–39, 1961.

[11] P.-S. Kildal, "Combined E- and H-plane phase centers of antenna feeds," *IEEE Trans. Antennas Propogat.*, vol. 31, no. 1, pp. 199–202, 1983.

[12] J. Yang and P.-S. Kildal, "Calculation of phase centers of feeds for reflectors when the phase variations are large," in *IEEE 1998 Antennas and Propagation Society International Symposium*, vol. 4. IEEE, 1998, pp. 2050–2053.

[13] K. Rao and L. Shafai, "Phase center calculations of reflector antenna feeds," *IEEE Transactions on Antennas and Propagation*, vol. 32, no. 7, pp. 740–742, 1984.

[14] J. Yang and P.-S. Kildal, "Calculation of ring-shaped phase centers of feeds for ring-focus paraboloids," *IEEE Trans on Antennas and Propagat.*, vol. 48, no. 4, pp. 524–528, April 2000.

[15] M. Denstedt, T. Ostling, J. Yang, and P.-S. Kildal, "Tripling bandwidth of hat feed by genetic algorithm optimization," in *2007 IEEE Antennas and Propagation Society International Symposium.* IEEE, 2007, pp. 2197–2200.

[16] J. Yang and P.-S. Kildal, "FDTD design of a Chinese hat feed for shallow mm-wave reflector antennas." Atlanta, Georgia: Proc. 1998 IEEE AP-S International Symposium, 21-26 June 1998, pp. 2046–2049.

[17] W. Wei, J. Yang, T. Östling, and T. Schafer, "New hat feed for reflector antennas realised without dielectrics for reducing manufacturing cost and improving reflection coefficient," *IET microwaves, antennas & propagation*, vol. 5, no. 7, pp. 837–843, 2011.

[18] E. G. Geterud, J. Yang, T. Ostling, and P. Bergmark, "Design and optimization of a compact wideband hat-fed reflector antenna for satellite communications," vol. 61, no. 1, pp. 125–133, 2013.

[19] P.-S. Kildal, "Factorization of the feed efficiency of paraboloids and cassegrain antennas," *IEEE Trans on Antennas and Propagat.*, vol. 33, no. 8, pp. 903–908, Feb. 1985.

[20] P.-S. Kildal and Z. Sipus, "Classification of rotationally symmetric antennas as types BOR_0 and BOR_1," *IEEE Antennas Propaga. Mag.*, Dec. 1995.

[21] J. Yang, S. Pivnenko, and P.-S. Kildal, "Comparison of two decade-bandwidth feeds for reflector antennas: the eleven antenna and quadridge horn," in *Antennas and Propagation (EuCAP), 2010 Proceedings of the Fourth European Conference on.* IEEE, 2010, pp. 1–5.

[22] J. Yang, M. Pantaleev, P.-S. Kildal, and L. Helldner, "Design of compact dual-polarized 1.2–10 GHz Eleven feed for decade bandwidth radio telescopes," *IEEE Transactions on Antennas and Propagation*, vol. 60, no. 5, pp. 2210–2218, 2012.

[23] J. Yin, J. Yang, M. Panteleev, and L. Helldner, "A circular eleven feed with significantly improved aperture efficiency over 1.3–14 ghz," in *Antennas and Propagation (EUCAP), 2012 6th European Conference on.* IEEE, 2012, pp. 2353–2356.

[24] J. Yin, J. Yang, M. Pantaleev, and L. Helldner, "The circular eleven antenna: A new decade-bandwidth feed for reflector antennas with high aperture efficiency," *IEEE Transactions on Antennas and Propagation*, vol. 61, no. 8, 2013.

Dish Verification Antenna China for SKA

Xiaoming Chai[1,3], Biao Du[2,3], Yuanpeng Zheng[2,3], Lanchuan Zhou[1,3], Xiang Zhang[1,3], Bo Peng[1,3]
[1]National Astronomical Observatories, Chinese Academy of Sciences, [2]the 54th Research Institute of the China Electronics Technology Group Corporation, [3]Joint Laboratory for Radio Astronomy Technology, NAOC & CETC54
20A Datun Road, Chaoyang District
Beijing 100012 China

Abstract- **The SKA (Square Kilometre Array), being the next-generation largest radio telescope, has now stepped into the pre-construction phase. China has joined this long march since the early 1990s and made significant contributions during this process. In this paper, a brief review of the SKA concept and detailed description of the challenges for the SKA dishes are given. Then two concept designs for the SKA dishes made by Chinese, the Dish Verification Antenna China (DVAC), are presented. The designs including microwave optics, antenna structure and servo system are carefully carried out to meet the intended SKA requirements. Comparison of these two designs is made in the end and one of them is selected as a candidate for the SKA dishes.**

I. Overview of SKA Development

In 1993, astronomers from ten countries including China proposed to build the next-generation Large Telescope (LT) for in-depth exploration of the universe, and the LT was renamed as the Square Kilometre Array (SKA) later. Once being constructed successfully, the SKA will become the world's largest and most sensitive radio telescope, and will remain in that position for 20~30 years.

Through earlier research and trade-off, two technology roadmaps were developed, that is LDSN (Large Diameter Small Number) and LNSD (Large Number Small Diameter). Five potential engineering concepts for the SKA were proposed including KARST (Kilometre Square Area Radio Synthesis Telescope, by Chinese), LAR (Large Adaptive Reflector, by Canadian), ATA (Allen Telescope Array, by American), AAT (Aperture Array Tile, by Netherlander) and Luneburg lens antenna (by Australian). Considering the technical issues and science requirements, the LNSD roadmap was finally adopted for the SKA.

No single technology has been identified to cover the frequency range from several tens of MHz up to 20GHz. The SKA has been divided into three arrays: the low frequency array that covers the range of 70-300MHz, the mid-frequency array that covers the range of 300 MHz -1.8GHz and the high frequency array that covers the range of 300MHz - 20GHz [1]. In the high frequency array, about 3300 dishes with 15m in diameter for each is adopted as shown in Fig. 1.

Unprecedented challenges need to be overcome in the SKA dishes design[2,3], that are: very high imaging dynamic range, mass production, ease of transportation, rapid installation with minimum manpower and equipment at remote sites, low operating costs, maximum sensitivity, etc. None of the present dish design will meet all these requirements in combination.

Figure 1. Artist's impression of SKA dishes.

Chinese has presented two concept designs for the SKA dishes DVAC-1 and DVAC-2 [4-6], which are offset Gregorian dish and prime focus symmetric dish, respectively.

II. Offset Gregorian Dish (DVAC-1)

This section presents the concept design and main specificaitons of DVAC-1, a 15 meter offset Gregorian antenna. The main attractions of this design are as follows.

(a) Offset-Gregorian optics is adopted to enable high aperture efficiency and low system noise temperature.

(b) Wideband single-pixel feed (WBSPF) is used to cover the entire frequency range with less number of feeds.

(c) One-piece main reflector enables fast installation with low man power, and the ajustment is almost free for the shape of the primary.

(d) Sealed and lubricated driving devices are used for high reliability and low maintenance cost.

(e) Wherever possible, mature technology is used for low cost, high reliability and convenient maintenance.

The design includes microwave optics, antenna structure, servo control, antenna radiation pattern calculation, etc. Key technological issues lying in the design study are outside the normal experience, and need to be dealt with creativly.

The schematic diagram of the 15 meter antenna is shown in Fig. 2.

This work is supported by China Ministry of Science and Technology under grant No. 2013CB837900, the National Science Foundation of China under grant No. 11261140641, and the Chinese Academy of Sciences under grant No. KJZD-EW-T01

978-7-5641-4279-7

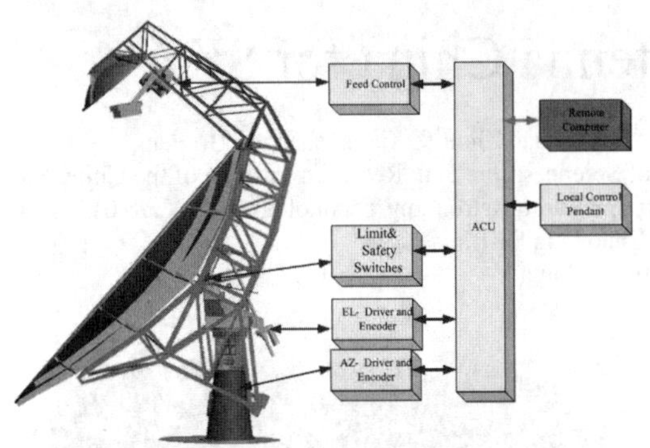

Figure 2. Schematic diagram of DVAC-1.

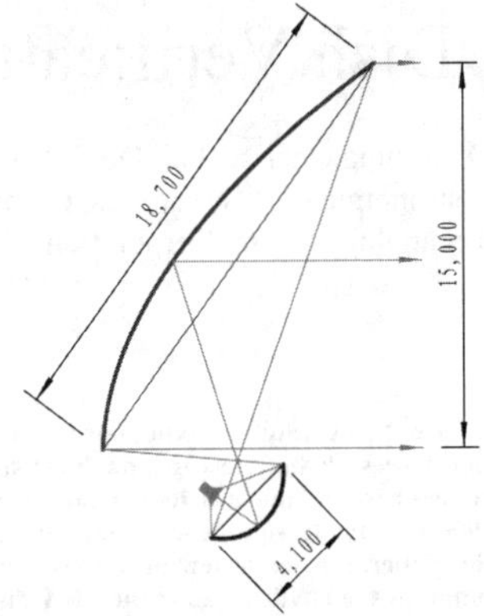

Figure 4. Shaped Offset-Gregorian Antenna

A. *Microwave optics design*

The microwave optics design includes the design of feed, the shape of main reflector and subreflector.

To cover the frequency range of 0.3-10GHz, two WBSPF feeds are planned. Feed 1 and feed 2 are for ranges of 0.3-1.5GHz and 1.5-10GHz, respectively. The radiation pattern of the feed, and far field pattern of the antenna are simulated [4, 7]. The prototype of feed 1 is shown in Fig. 3.

Dual reflector is shaped to increase the aperture efficiency and lower the first and far-out sidelobes. The shaped optics may also help to obtain a compact structure and minimize the main reflector area. Design parameters are optimized and the results [4, 7] are shown in Fig. 4.

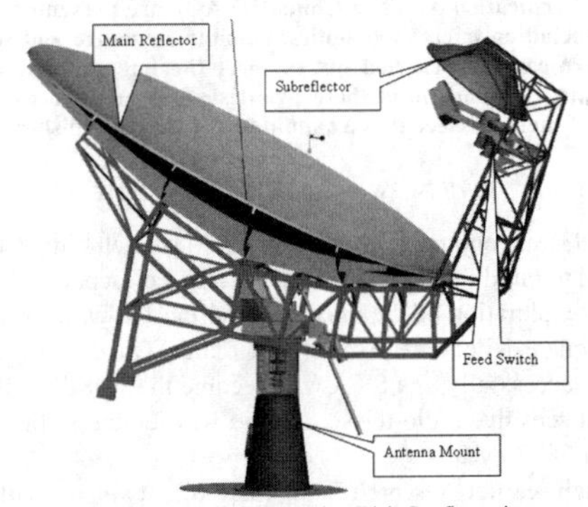

Figure 5. Offset-Gregorian Dish Configuration

B. *Antenna structure design*

As shown in Fig. 5 the antenna structure consists of the reflector and the mount.

The reflector includes the main reflector, subreflector, backing structure and a feed switch mechanism.

The main reflector adopts an integrated one-piece surface approach, which is benificial for fast installation with less man power. The one-piece panel is partitioned into panel elements as in Fig. 6. The sandwich panel element is made by skin (made of carbon fiber) and foam (made of hard polyurethane), and is shaped by vacuum negative pressure on the mould. The surface accuracy of sandwich panel element is $\sigma \leq 0.2$mm rms [4].

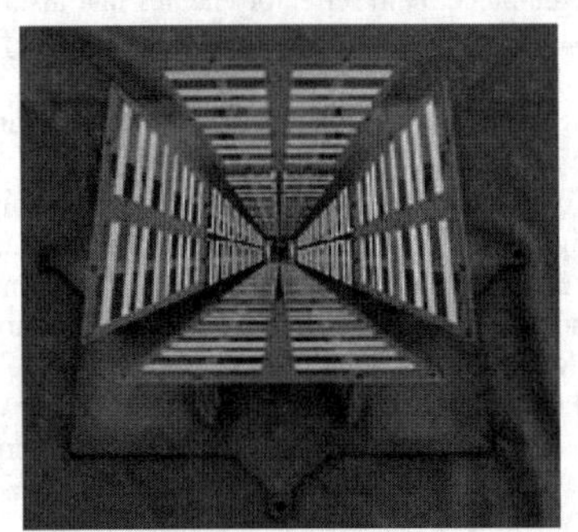
Figure 3. Prototype of WBSPF (feed 1).

On the site, the main reflector is formed by integrating these sandwich panel elements and ribs on a complete mould. Joint methods between different panel elements are shown in Fig. 7. The surface accuracy of the main reflector is $\sigma \leq 0.6$mm rms [4].

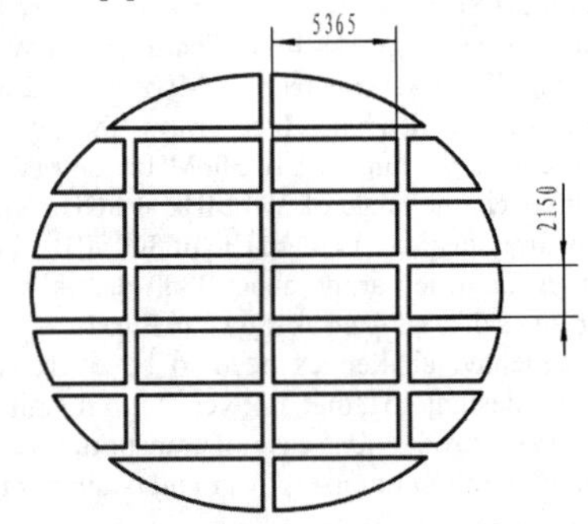

Figure 6. The One-piece Panel Partition.

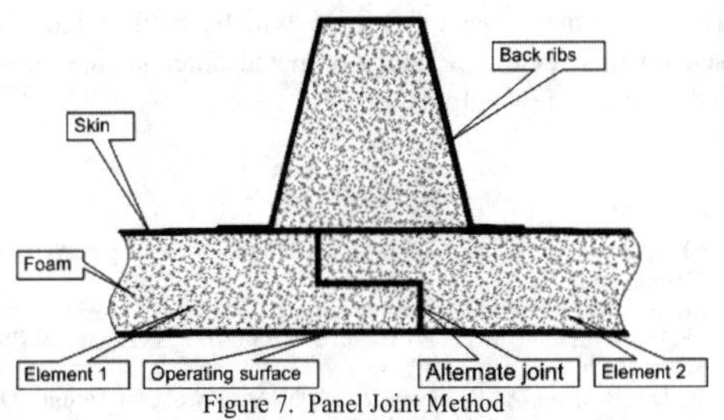

Figure 7. Panel Joint Method

Subreflector is also a one-piece composite foam sandwich structure with metalizing surface, and its reflectivity is higher than 99.5%. The material of skin, sandwich foam and back ribs is as same as that of the main reflector.

The antenna mount is an AZ (Azimuth)-EL (Elevation) type structure, with a gear drive in Azimuth and screw drive in Elevation. The mount has a strong bearing capacity, compact structure and is easy to manufacture and transport.

The total weight of DVAC-1 antenna structure is estimated as 18 300kg, with the reflector 7 050kg and the mount 11 250kg.

C. Antenna servo control system design

The antenna servo control system consists of an Antenna Control Unit (ACU), drivers and motors, power distribution devices, encoders, local control pendant, and a limit and safety protection device, etc. The ACU, drivers and power distribution devices are placed in a RFI-tight cabinet. The block diagram of the antenna control system is shown in Fig. 8.

III. PRIME FOCUS DISH (DVAC-2)

This section presents the conceptual design of DVAC-2, a 15 meter prime focus symmetric antenna with three rotational axis. Its main attractions are similar with DVAC-1, such as the use of WBSPF, integrated modular, one-piece relfector and so on [6].

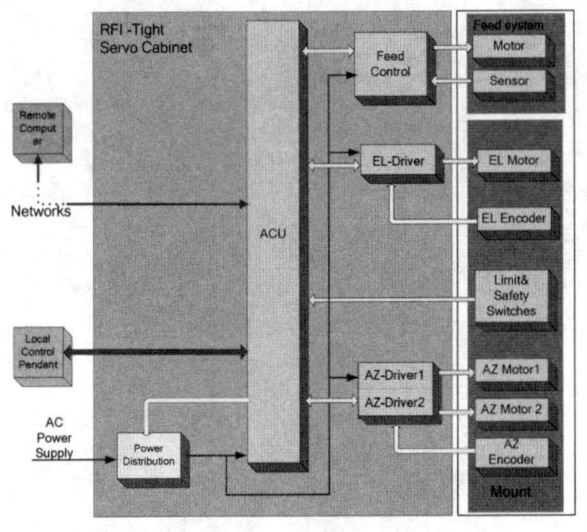

Figure 8. Block diagram of antenna control system.

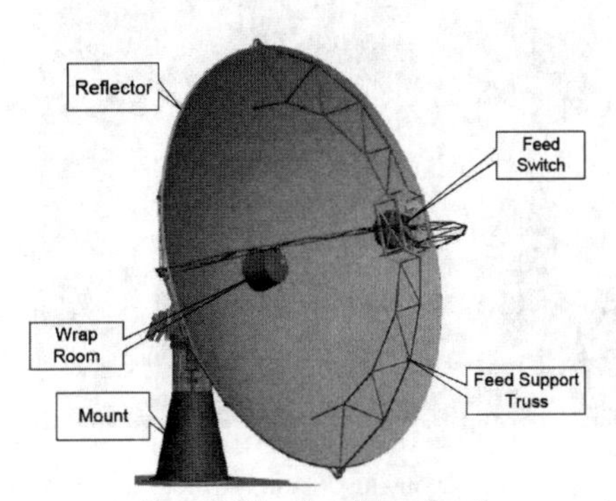

Figure 9. Prime foucs dish configuration.

Configuration of the prime focus dish [6, 8] is shown in Fig. 9. A one-piece reflector is supported by a simple spar structure. The mount has three freedom degrees with an AZ-EL-POL(Plorization) type structure, including a turning head for azimuth, a lead screw for elevation, and a plorization axis as shown in Fig. 10.. A support and interchange mechanism is designed for feeds combination of a PAF and 3 SPFs or 2 WBSPFs.

The reflector also adopts sandwich structure. The skin material can either adopt aluminum or carbon fiber. Total surface accuracy $\sigma \leq 0.8$mm rms [6].

The total weight of DVAC-2 antenna structure is estimated as 19 050kg, with the reflector 6 550kg and the mount 12 500kg.The antenna servo control is similar with that of DVAC-1.

This design has been used on Australian SKA Pathfinder (ASKAP), as shown in Fig. 11. ASKAP is comprised of 36 identical dishes, three-axis prime focus symetric antennas, each 12 metres in diameter. The construction and assembly of the dishes was completed in June 2012. All 36 antennas have been designed and manufactured by Chinese.

Figure 10. Antenna mount structure.

Figure 11. Antenna for ASKAP.

IV. COMPARISON AND FUTURE

Both antenna designs for DVAC-1 and DVAC-2 have their advantages and disadvantages.

The primary focus reflector antenna has the advantage of simpler optics and mechanical structure which are easier to manufacture and at low cost etc. But its shortcomings may include:

(1) It is difficult to simultaneously optimize aperture efficiency and noise temperature;

(2) It has lower efficiency and higher sidelobes due to blockage;

(3) It is mechanically more difficult to accommodate multiple feeds and a PAF at the primary focus.

The advantages of the offset Gregorian reflector antenna are:

(1) Shaped optics may enable high aperture efficiency and low system noise temperature;

(2) No blockage design helps to enhance aperture efficiency and reduce far-out sidelobes;

(3) Feed spillover pointing at the sky can further reduce the system noise temperature;

(4) It is mechanically easier to accommodate multiple feeds at the secondary focus and a PAF at the primary focus.

But its disadvantage is that the asymmetry increases the complexity of the mechanical design and the cost for one single dish. As for mass production of the antennas, the total cost may not increase.

Now offset- Gregorian DVAC-1 is identified as a candidate dish by the SKA and is renamed as DVA-C (Dish Verification Antenna - China). Another two SKA dish candidates are DVA-1 (by Canadan) and MeerKAT (by South African). Studies on these three options are undertaken at the same time. SKADC (SKA Dish Consortium) is formed. SKADC is responsible for the design and verification of dish structure, optics, feed suites, receivers and all supporting systems and infrastructure suitable for SKA in the pre-construction.

Folding in results of the research on the three candidates, the SKADC dish prototype (SKA-P) will be designed and realized by the SKA. Specially points out, no one group or institute will own the SKA-P dish antenna design. SKA-P will integrate advantageous parts of the three candidates from the perspective of system engineering.

The Preliminary Design Review will be held in late 2014. Based on this review, a final optimized antenna may emerge based on the review output.

REFERENCES

[1] SKA Organisation, SKA-TEL-SKO-DD-001, SKA1 System Baseline Design, 2013
[2] SKA website http://www.skatelescope.org
[3] SKA Organisation, MGT-001.005.010-WBS-001-1_SKAStage1WBSand SOW, 2012
[4] B. Du, B. Peng, Y. P. Zheng, et al. DVAC-1 Concept Design: Offset Gregorian Antenna version F, 2012
[5] B. Peng, Y. P. Zheng, Y. Lu, et al. DVAC-1 Concept Description: Offset Gregorian Dish Version E, 2011
[6] B. Peng, Y. P. Zheng, Y. Lu, et al. DVAC-2 Concept Description: Prime Focus Dish Version D, 2011
[7] JLRAT, DVAC-1 Antenna Design and Analysis, 2011
[8] JLRAT, DVAC-2 Antenna Design and Analysis, 2011

Telescopes for IPS Observations

Li-Jia Liu, Lan-Chuan Zhou, Bin Liu, Cheng-Jin Jin, Bo Peng
National Astronomical Observatories, Chinese Academy of Sciences
20A Datun Road, Chaoyang District, Beijing, China

Abstract- **The sun especially the solar wind have influenced our Earth in many ways, and observing the solar wind is an important method to study the solar-earth environment. Interplanetary Scintillation (IPS) observations are an effective method of monitoring solar wind, forecasting the solar-terrestrial space weather and studying the structures of the long distance compact radio sources. Since the discovery of Interplanetary Scintillation, many countries began to use this method to study the solar wind and the interplanetary plasma, so some appropriative radio telescopes were built to achieve these goals. In this paper, we will review some typical IPS radio telescopes and also the current status of IPS study in China.**

I. INTRODUCTION

The solar-terrestrial environment is filled with solar wind plasma. The solar wind is the material in interplanetary space that comes from coronal expansion in the form of inhomogeneous plasma flow. The solar wind is the main source of interplanetary medium, which connects the variations in the Sun with terrestrial physical phenomena.

Radiation from a distant compact radio source is scattered by the density irregularities in the solar wind plasma, which produces a random diffraction pattern on the ground. The motion of these irregularities converts this pattern into temporal intensity fluctuations that are observed as interplanetary scintillation (IPS). There are two ways to observe the solar wind: spacecraft measurement and ground-based measurement. Using spacecraft is a direct method to observe the solar wind, but it has one limitation. It can only travel near the ecliptic plane and sample from a stable orbit, while the solar wind phenomenon is very stochastic [1]. Ground-based IPS observations can measure the solar wind at any distance and also for a long period [2], so it is an effective method to study on the Sun-Earth system [3].

Since the discovery of IPS in 1964, many countries like Britain, former Soviet Union , America, India and Japan have begun to observe this phenomenon. China began IPS studies in the 1990's with the phased array mode of the Miyun Synthesis Radio Telescope (MSRT) at 232 MHz [4]. It used the Single-Station Single-Frequency (SSSF) mode. Recently a new IPS observation system using the 50 m parabolic radio telescope, which is based on the Single-Station Dual-Frequency (SSDF) mode at S/X and UHF bands, is built to serve the National Meridian Project of China .

The theory of IPS, SSSF mode and SSDF mode are discussed in section 2, section 3 reviews the IPS radio telescopes and the conclusions are presented in section *4*.

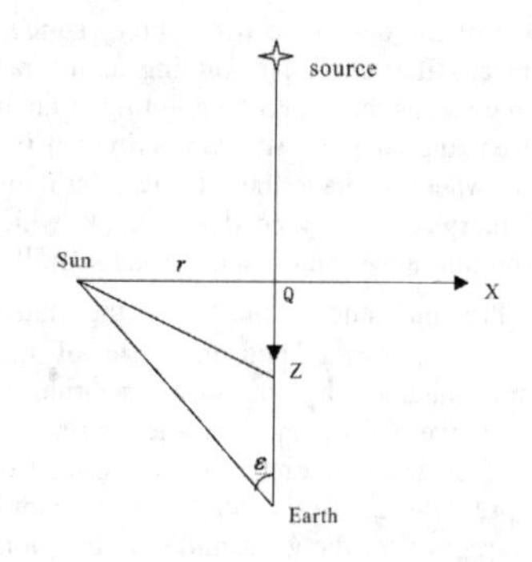

Figure 1 Geometry of IPS Concept

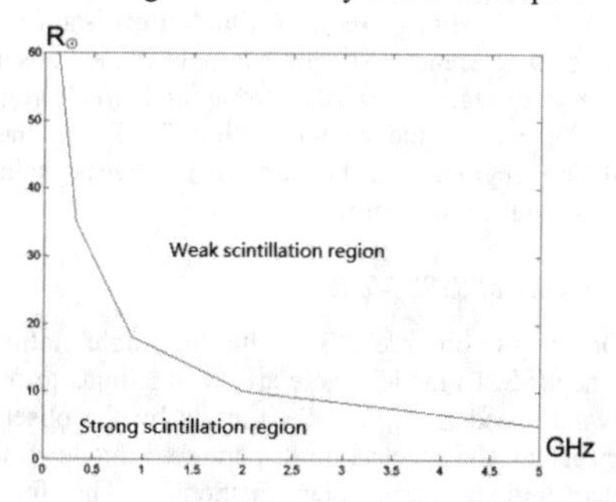

Figure 2 Frequency vs. distance for weak and strong scintillation regions

II. THEORY

A The Theory of IPS

Figure 1 shows the geometry of IPS. The *z*-axis is along the line-of-sight, and the *x*-axis is in the direction perpendicular to the *z*-axis pointing away from the Sun, with the *y*-axis being normal to the paper. Q is the point closest to the Sun along the line-of-sight, and ε is the elongation angle, Sun-Earth-source. r is the distance between the Sun and Q. z is the distance between Q and the Earth.

The degree of scintillation is characterized by the scintillation index *m*, which increases with decreasing distance r. The expression of *m* is shown below [5]:

This work is supported by China Ministry of Science and Technology under grant No. 2013CB837900, the National Science Foundation of China under grant No. 11261140641, and the Chinese Academy of Sciences under grant No. KJZD-EW-T01

978-7-5641-4279-7

$$m = \frac{\sqrt{\sigma_{on}^2 - \sigma_{off}^2}}{C_{on} - C_{off}} \quad (1)$$

Where C_{on} (C_{off}) is the average intensity of the on-source (off-source) signal, and σ_{on}^2 (σ_{off}^2) is the intensity-squared error of the on-source (off-source) signal. Here the on-source means the telescope pointing at the radio source, and off-source means the telescope pointing at the background sky away from the source. IPS is maximized in the region near the Sun, where we have the "strong scintillation region". In most of interplanetary space IPS is weak, which is called the "weak scintillation region". In the weak scintillation region $m^2 \ll 1$. Previous studies show that the statistics of the scintillation are simply related to those of the turbulent interplanetary medium by a linear relationship, if the scintillation is weak [6] [7]. In the "strong scintillation region", however, the relationship is not straightforward, and most of the present study deals with the weak scintillation case. The distance regime for the weak and strong regions is related to observing frequencies. Figure 2 shows the relationship between the observing frequency and the distance regimes according to reference [8]. For example, when observing at 327MHz, the regime for the strong and weak regions is: putting the Sun at the center, within 35 $R_{\odot}$ is the strong scintillation region, and beyond is the weak scintillation region ($R_{\odot}$ is the radius of the Sun).

B The Theory of SSSF Mode

One way to observe IPS is with the Single Station Single Frequency (SSSF) mode. There are two methods to obtain the solar wind speed from the SSSF mode by the observed IPS power spectra: the spectral multi-parameter model-fitting, and the characteristic frequencies methods. The former can measure the speed by adjusting the main parameters of the solar wind to fit the observed scintillation power spectra. The parameters are: α -power law index of the spatial spectrum of electron density, AR-axial ratio of solar wind irregularities, and V-solar wind speed. The latter can measure the speed by calculating two characteristic frequencies of the spectra: the Fresnel knee frequency f_F corresponding to Fresnel diffraction theory, and f_{min} the first minimum of the spectra. Then the solar wind speed can be calculated by either of the formulae shown below [1]:

$$V = f_F \sqrt{Z\pi\lambda} \quad (2)$$

$$V = f_{\min} \sqrt{Z\lambda} \quad (3)$$

where λ is the observing wavelength, and z is the distance between Q and the Earth according to Figure 1.

C The Theory of SSDF Mode

Being different from the SSSF mode, the SSDF mode adopts simultaneous Single Station Dual-Frequency method to measure the solar wind speed, by deriving the first zero point of the normalized cross-spectrum (NCS) at two different frequencies, which wavelengths are λ_1 and λ_2, and $\lambda_1 > \lambda_2$. The expression for solar wind speed using the SSDF mode is shown below [8] [9]:

$$V = Af_{zero} \sqrt{Z\lambda_1} \quad (4)$$

where A is a correction factor which varies slightly with the solar wind parameters and is usually set to be 1.1. In most cases, taking A=1 will cause no more than a 10 % error in the measured solar wind speed, which is acceptable [8] [9]. SSDF mode can measure the solar wind speed more precisely, for f_{zero} is only sensitive to the solar wind speed V.

III. REVIEW OF IPS TELESCOPES

A Some Typical IPS Telescopes

IPS is a fast changing phenomenon; it needs the observing instrument to have a short integration time. That is why the Interplanetary Scintillation was first discovered by the 3.6-ha array in Cambridge [10], which is shown in Figure 3.

From the 1970s, some countries like India and Japan began to observe IPS on regular basis. Figure 4 and Figure 5 show the Ooty radio telescope in India and the four stations IPS system in Japan. The two systems are both parabolic cylinder telescopes, which can have a large collecting area in order to observe more radio sources every day.

Figure 3 The Cambridge 3.6-ha array

Figure 4 The Ooty radio telescope in India

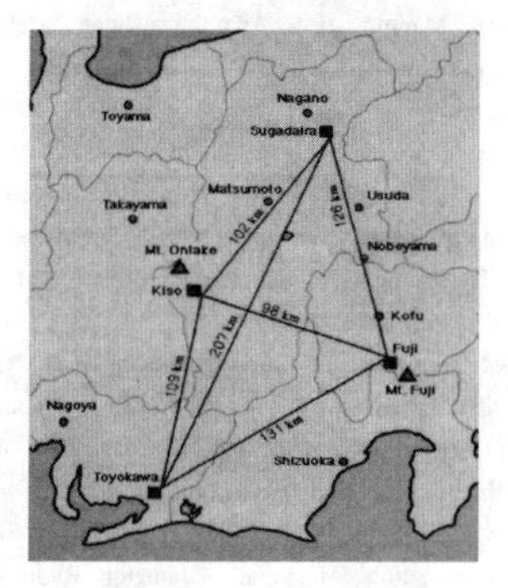

Figure 5 The four stations of STELab IPS radio telescopes in Japan

Recently, some new types of IPS radio telescopes were built along with the development of new data processing method. Like the Murchison Widefield Array (MWA) (shown in Figure 6) in Western Australia [11] and the Mexican Array Radio Telescope (MEXART) in Mexico (shown in Figure 7) [12]. These two telescopes are both constructed with dipole antennas, and have simple front end and complicated data processing back end.

B IPS Telescopes in China

IPS observations have been carried out in China for decades, and some achievements have been made [4][13][14].

China began IPS studies from the 1990s first with the phased array mode of the Miyun Synthesis Radio Telescope (MSRT), which observed at 232MHz. It can investigate the region $R>60\ R_{\odot}$. The equivalent diameter of MSRT is about 47m. Figure 8 shows the distribution of the 28 radio telescopes of MSRT.

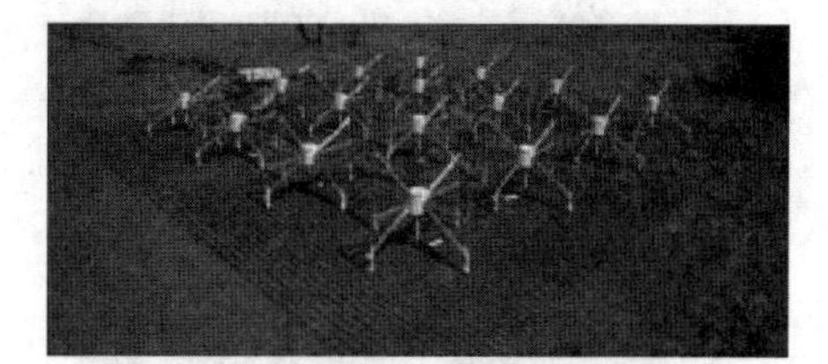

Figure 6 The picture of MWA

Figure 7 The picture of MEXART

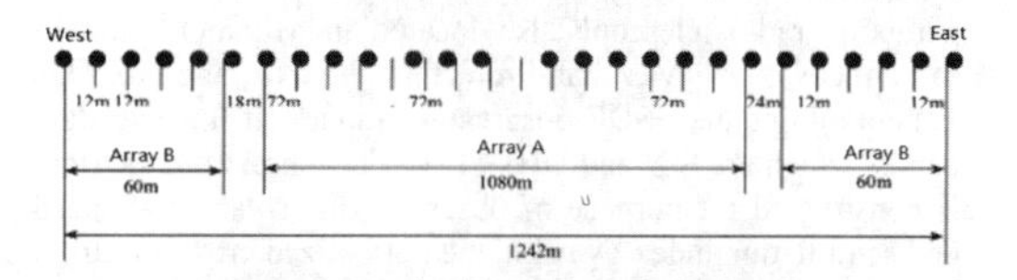

Figure 8 The distribution of the 28 radio telescopes of MSRT

Figure 9 The 25m radio telescope in Urumqi

Figure 10 The 50m radio telescope in Miyun

From 2008, a new SSSF mode IPS observation system was built in Urumqi with the 25m radio telescope in Xinjiang Province China (Figure 9). After a series of experiments, the 18 cm dual-polarization receiver was chosen for the observations, and a data acquisition/receiving system was also established. This is the first SSSF mode IPS observation system in China to carry out quasi regular IPS observations [13][14][15].

In order to explore the IPS ability of domestic radio telescopes, since 2010, some experimental observations also carried out with the 40m radio telescope in Kuming [16].

Recently a new IPS observation system using the 50m parabolic radio telescope also located in Miyun is built to serve the National Meridian Project of China (Figure 10). This system adopts the SSDF observation mode. It has two dual bands, which are S/X and UHF [17]. This observation system is constructed for purpose of observing the solar wind speed and scintillation index by using the normalized cross-spectrum of simultaneous dual-frequency IPS measurement. The system consists of a universal dual-frequency front-end and a dual-channel multi-function back-end specially designed for IPS. This radio telescope is now the only SSDF IPS telescope around the world.

IV. CONCLUSIONS

Nowadays human activity in the outer space is becoming more and more frequently. So the forecasting of space weather becomes more and more important than in the past. IPS observations provide us an effective way to monitor the Sun and solar wind, so it is necessary to build particular instrument to carry out IPS observations. The IPS observation in China is still under developing, and the observation will try to be a routine observation in the future.

REFERENCES

[1] G. Y. Ma. "Reach and application of IPS", PHD thesis. BAO, 1993, pp. 1-3.

[2] A. Hewish, M. D. Symonds, "Radio investigation of the solar plasma", Planetary and Space Science, Vol. 17, 1969, pp.313.

[3] L. Scott, B.J. Rickett, J.W. Armstrong, "The velocity and the density spectrum of the solar wind from simultaneous three-frequency IPS observations", Astronomy and Astrophysics, vol. 123, no. 2, July 1983, p. 191-206.

[4] J.H. Wu, X.Z. Zhang, Y.J. Zheng, "IPS Observations at Miyun Station, BAO", Astrophysics and Space Science, v. 278, Issue 1/2, 2001, pp. 189-192.

[5] M.H. Cohen, E.J. Gundermann, H.E Hardebeck, et al., "Interplanetary Scintillations. II Observations", Astrophysical Journal, vol. 147, 1967, pp.449.

[6] W. A. Coles, J.K. Harmon, A.J. Lazarus, et al., "Comparison of 74-MHz interplanetary scintillation and IMP 7 observations of the solar wind during 1973", Journal of Geophysical Research, vol. 83, July 1, 1978, pp. 3337-3341.

[7] P.K. Manoharan, S. Ananthakrishnan, "Determination of solar-wind velocities using single-station measurements of interplanetary scintillation", Monthly Notices of the Royal Astronomical Society, vol. 244, June 15, 1990, pp. 691-695.

[8] X.Z. Zhang, "A Study on the Technique of Observing Interplanetary Scintillation with Simultaneous Dual-Frequency Measurements", Chinese Journal of Astronomy and Astrophysics, Volume 7, Issue 5, 2007, pp. 712-720.

[9] M. Tokumaru, H. Mori, T. Tanaka, et al., "Solar Wind Velocity Near the Sun: Results from Interplanetary Scintillation Observations in 1989-1992", Proceedings of Kofu Symposium, Kofu, Japan, Sept. 6-10, 1993, pp.401-404.

[10] A. Hewish, P. F. Scott, and D. Wills, "Interplanetary Scintillation of Small Diameter Radio Sources", Nature, Volume 203, Issue 4951, 1964, pp. 1214-1217.

[11] C. J. Lonsdale, R. J. Cappllo, M. F. Morales, et al., "The Murchison Widefield Array: Design Overview", Proceedings of the IEEE, Vol. 97, Issue 8, 2009, pp.1497-1506.

[12] J. C. Mejia-Ambriz, P. Villanueva-Hernandez, J. A. Gonzalez-Esparza, et al., "Observations of Interplanetary Scintillation (IPS) Using the Mexican Array Radio Telescope (MEXART)", Solar Physics, Volume 265, Issue 1-2, 2010, pp. 309-320.

[13] L.J. Liu, B. Peng, "Simulation of interplanetary scintillation with SSSF and SSDF mode", Science China Physics, Mechanics and Astronomy, Volume 53, Issue 1, 2010, pp.187-192.

[14] L.J. Liu, X.Z. Zhang, JJ.B. Li, et al., "Observations of interplanetary scintillation with a single-station mode at Urumqi", Research in Astronomy and Astrophysics, Volume 10, Issue 6, 2010, pp. 577-586.

[15] L.J. Liu, B. Peng, "Data Reduction for Single-Station Single-Frequency Interplanetary Scintillation Observation", Astronomical Research & Technology, Volume 7, Issue 1, 2010, pp. 21-26.

[16] L.J. Liu, B. Liu, L. Dong, and Bo Peng, "Development of Observational System at S band for the 40 m Radio Telescope of Yunnan Observatory", Astronomical Research & Technology, Volume 10, Issue 2, 2013, pp. 134-141.

[17] X.Y. Zhu, X.Z. Zhang , D.Q. Kong, et al. "IPS observation system for the Miyun 50 m radio telescope and its commissioning observation", Research in Astronomy and Astrophysics, Volume 12, Issue 7, 2012, pp. 857-864.

Design of the 4.5m Polar Axis Antenna for China Spectral Radio Heliograph Array

NIU Chuanfeng[1,3], GENG Jingchao[1,3], YANG Guodong[1,3], ZHAO Donghe[1,3], LIU Chao[1,3], YAN Yi-hua[2,3], CHEN Zhijun[2,3], LI Sha[2,3], DU Biao[1,3], WU Yang[3]
[1]The 54th Research Institute of CETC, Shijiazhuang 050081, China
[2]Key Laboratory of Solar Activity, National Astronomical Observatories, CAS, Beijing 100012, China
[3]Joint Laboratory for Radio Astronomy Technology, Shijiazhuang 050081, China

***Abstract-The 4.5*m polar axis antenna with polar Axis mount and dual circular polarization ultra-wideband feed over 0.4-2GHz for China Spectral Radio Heliograph (CSRH) array is presented in the paper. The CSRH array and specifications of 4.5m polar axis antenna are introduced. The detailed design procedure about antenna system together with the structure of polar Axis mount, ultra-wideband feed, parabolic reflector and consideration for practical installation are described. The voltage standing wave ratio (VSWR) and radiation patterns of the feed for 4.5-m antenna have been tested, and the measured efficiency of the 4.5-m reflector antenna is better than 40%, The performance fulfills application requirements.**

I. INTRODUCTION

The sun is very important for human. She supplies light and heat constantly for life on earth just like the earth's mother. But the sun's activity has a significant impact on the earth and human. Such as the rainfall at an area has some correlation with the variation cycle of macula, and also the electromagnetic wave produced when solar flare breaking out can result in the discontinuous of radio-communication. The radio wave radiated by sun has different characteristics, so it's a very important method for the sun research by the radio observations with the radio-wave[1].

Chinese Solar Radio heliograph project（CSRH） observes the activity of the sun and explores the coronal atmosphere with high resolution over 0.4-15GHz[2]. The scientific goals include the instantaneous high-energy phenomenon, the Coronal magnetic field and the structure of solar atmosphere[3]. By confirming the source area features of solar flare and mass ejection of the coronal materials, we can understand the solar dynamic transition zone and the Coronal.

Construction of CSRH are planned into two stages: the phase I of CSRH is comprised of 40 reflector antennas with 4.5-m dishes from 0.4-2GHz and the phase II of CSRH is comprised of 60 reflector antennas with 2-m dishes from 2 - 15GHz. CSRH is located in Inner Mongolia about 400km away from Beijing. All the 100 antennas are mounted equatorially and spread in $3\times3km^2$ areas, a radio quiet zone surrounded by hills [4]. The 4.5m antenna array is shown in Figure.1[5]. The reflector antennas are arranged in three spiral beams in the plane with same height to get good UV coverage. The electrical specifications of the dual polarization feed are summarized in Table 1.

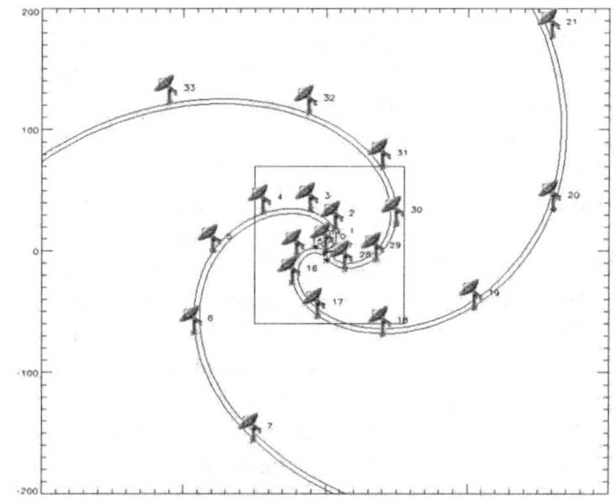

Figure.1. 4.5-m Radio Telescope Antenna Array.

Tab.1. Specifications of the 4.5m antenna

Items	Parameters
Antenna type	Polar Axis Antenna
Diameter	4.5m
Diameter ratio (f/d)	0.4
Frequency band	0.4~2.0 GHz
Antenna efficiency	40%
First side-lobe level	≤-14dB
Polarization	Dual-CP
Surface accuracy	≤ 3mm
Pointing accuracy	≤9′ (R.M.S)
Travel range (Equatorial)	-95°~95°
Travel range (Latitude)	30°~ 48°

In this paper, detailed design of the constituted antenna is presented, including ultra-wideband dual circular polarized feed, parabolic reflector, polar mount, together with the measured results of the antenna.

II. ANTENNA DESIGN

A. Antenna System

The 4.5m antenna consists of the antenna feed, the reflector, mount, pedestal, and servo control system. The reflector is paraboliod, assembled with 12 identical fan-shaped reflectors and the framework. And the mount consists of latitude support saddle, latitude bearing, latitude gear, declination gear, equatorial driving device, latitude driving device, and a counter weight[6].

978-7-5641-4279-7

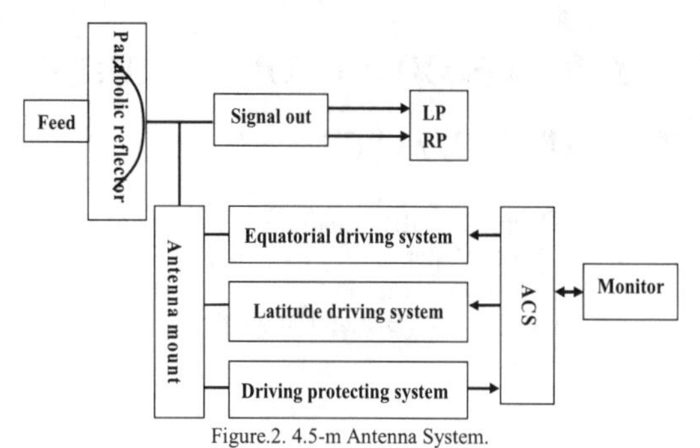

Figure.2. 4.5-m Antenna System.

The servo control system consists of the control circuit, the equatorial power amplifier, the latitude power amplifier, the outward interface and so on. Figure.2 shows the system composition block diagram.

B. Polar Axis Mount

The polar axis mount is also called the equator mount, which consists of two axes. The down axis is called the latitude axis because it parallel with the earth's rotational axis. And the upper axis is called the equatorial axis[6]. This mount is widely used in the radio telescopes because of its ease to tracking a fixed star. The antenna can be kept aiming at the object star just by rotating the latitude axis.

The 4.5m antenna uses the equator mount, which consists of the latitude axis device, the equatorial axis device, and the foundation. Figure.3 shows the structure form of the mount. The latitude axis and the equatorial axis cross at a point. The merit of this kind mount is that only the equatorial axis balance is needed.

C. Parabolic Reflector

The CSRH antenna uses a front-fed parabolic reflector of 4.5m diameter, and 0.4 f/D ratio, corresponding to an illumination angle of 128°.

Figure.4 gives the structure of the parabolic reflector. It consists of reflector, radial beam unit, and centrosome.

The reflector is assembled with 12 identical fan-shaped mesh panels and the framework. The merit of the mesh panel is the good consistency and surface accuracy[7].

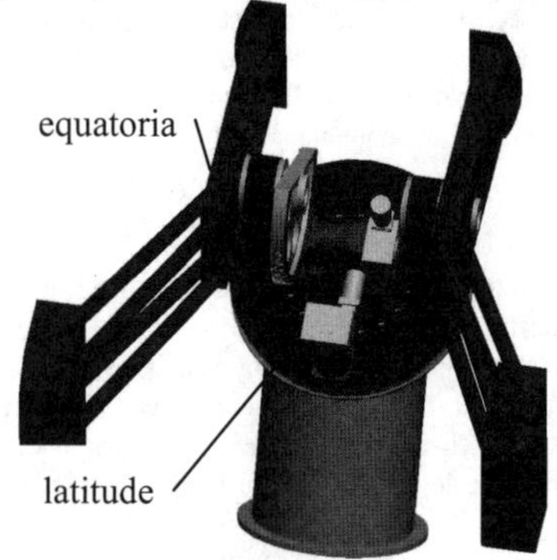

Figure.3. Structure of polar Mount.

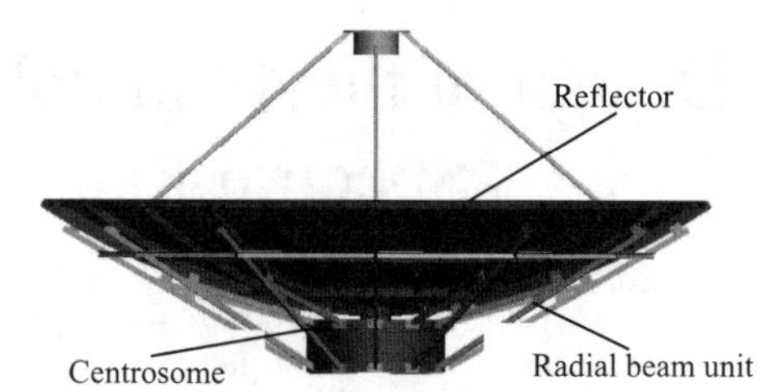

Figure.4. Structure of polar Mount.

To make sure that the panel has a good manufacture precision, uses the high-precision mould and frock are adopted in every manufacture procedure. All of the above ensure that the antenna has good rebuilt precision and a low air drag coefficient.

D. Ultra-wideband Feed

As a wideband feed, both the impedance and radiation requirement should be fulfilled over the operating band, including return loss, illumination pattern and phase center[8]. Frequency independent antennas are good candidates to satisfy the first two requirements, and the conception that using a set of Log-Periodic antennas for the feed has been implemented in a few projects. However, the inconstant phase center results in pattern degradation and efficiency loss, which becomes unacceptable when the operating band is very wide. The idea inspired by the Eleven feed that implementing a ground plane to stabilize the phase center was introduced in CSRH array feed design[8][9][10].

Photograph of the ultra-wideband feed is shown in Figure.5. The dimensions of the feed are 465mm (l) ×465mm (w) ×185mm (h), including 20mm feed radome and bridge height. The radiation part of the feed is machined from metal plate with 3mm thickness to enhance mechanical performance and only supported by Teflon posts at certain points to reduce loss.

III. MEASUREMENT RESULTS

Photograph of the fixed 4.5m polar axis antenna is shown in Figure.6. The measured VSWR with 90° hybrid of the UWB feed are shown in Figure.7, and the measured patterns of the 4.5m polar axis antenna over operating band are shown in Figure.8[11]. It can be seen that a VSWR below 1.7 was achieved, and the first side-lobe level is better than -14dB.

The measured antenna efficiency on 4.5-m reflector antenna is shown in Figure.9 and it is better than 40% over the operating band.

Figure.5. Photograph of the 0.4-2GHz UWB feed.

Figure.6. Photograph of the 4.5m polar axis antenna.

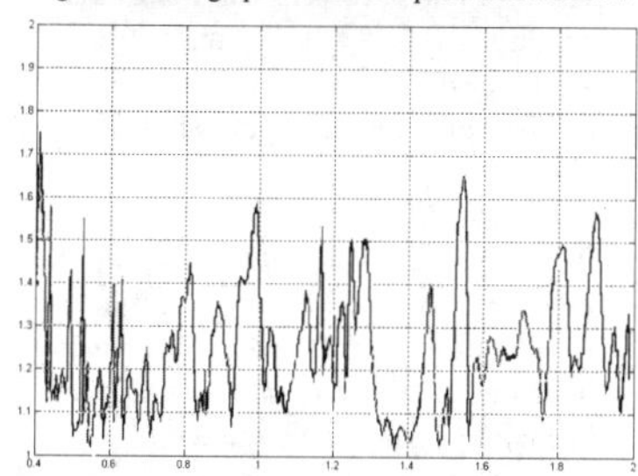

Figure.7. Measured VSWR of the UWB feed.

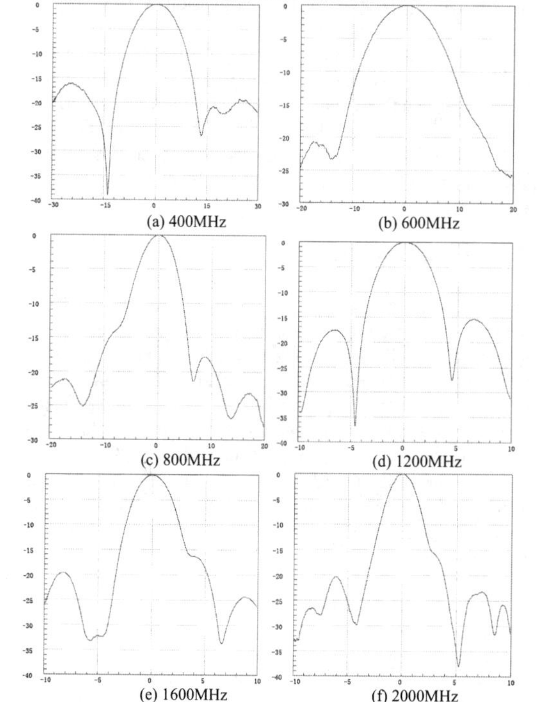

Figure.8. Measured patterns of the 4.5m antenna

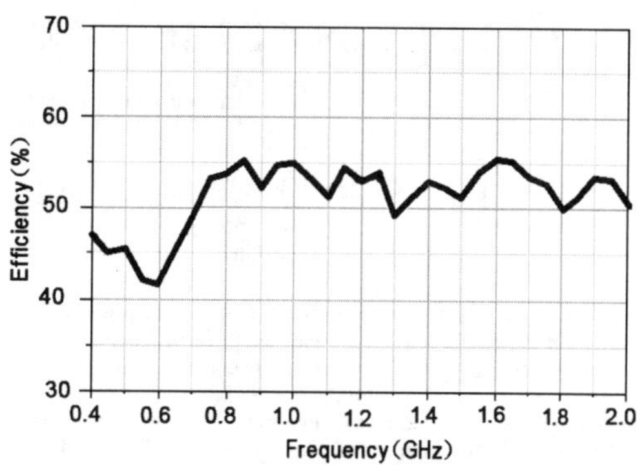

Figure.9. Measured antenna efficiency on 4.5-m antenna.

IV. SUMMARY

In this paper, the 4.5-m polar axis antenna with polar Axis mount and dual circular polarization ultra-wideband feed over 0.4-2GHz for CSRH array is presented. Several key technologies are solved, such as dual circular polarization ultra-wideband and high efficiency feed technology[12], low loss feed network, new concept design on the polar axis mount. The measured results show that the performance fulfills application requirements.

REFERENCES

[1] Yihua Yan, Jian Zhang, Guangli Huang, "On the Chinese spectral radioheliograph (CSRH) project in cm- and dm-wave range." Radio Science Conference, 2004 Asia-Pacific, pp. 391-392, Aug. 2004.

[2] Wang Wei, Yan Yihua, Liu Donghao, Chen Zhijun, Liu Fei, "Data processing for 5-element experiments of Chinese Spectral Radioheliograph." General Assembly and Scientific Symposium, 2011.

[3] Y.Yan , J.Zhang, W. Wang, F.Liu, Z.Chen, G. Ji, "The Chinese Spectral Radioheliograph—CSRH, "*Earth Moon Planet*, 104, pp. 97–100,2009.

[4] CHEN Zhi-jun, YAN Yi-hua, LIU Yu-ying, ZHANG Jian, WANG Wei, "Site Survey and RFI Test for CSRH." Astronomical Research & Technology, Vol.3, No.2, pp.168-175, Jun.2006

[5] Yan Yihua, Zhang Jian, Chen Zhijun, Wang Wei, Liu Fei, Geng Lihong, "Progress on Chinese Spectral Radioheliograph - CSRH Construction." General Assembly and Scientific Symposium, 2011.

[6] Zhang Yalin, "Design of the Pedestal of Polar Mount for Solar Radio Observing System ", Radio Communication Technology,Vol. 35, No 5, pp.40-43, May. 2009.

[7] Rong Jianfeng, Rong Hongfei, "Structure Design And Analysis of an Antenna Frame." Radio Engineering, Vol. 42, No 8, pp.54-55, Aug 2012.

[8] J.Yang, X.Chen, N.Wadefalk, and P.-S. kildal, "Design and realization of a linearly polarized Eleven feed for 1-10GHz," IEEE Antennas Wireless Propag. Letter, vol. 8, pp.64-68,2009.

[9] Olsson, R., Kildal, P.S., and Weinreb, S, "The Eleven Antenna: A compact low-profile decade bandwidth dual polarized feed for reflector antennas." IEEE Trans. Antennas Propag, 54, No.2, pp. 368-374, February 2006.

[10] J.Yang and P.-S. kildal, "Calculation of ring-shaped phased centers of feeds for ring-focus paraboloids." *IEEE Trans. Antennas Propag*, vol. 48, no.4, pp. 524-528, Apr.2000.

[11] Richard C. Johnnson, "Antenna Engineering Handbook," McGraw-Hill, inc, 1984.

[12] K.M.P. Aghdam, R. Faraji-Dana, J.Rashed-Mohassel. "Compact dual-polarisaton palanar log-periodic antennas with integrated feed circuit.," IEEE Proc.- Microw. Antennas Propag., vol.152,No.5,October 2005.

The optics of the Five-hundred-meter Aperture Spherical radio Telescope

C. Jin, K. Zhu, J. Fan, H. Liu, Y. Zhu, H. Gan, J. Yu, Z. Gao, Y. Cao, Y. Wu, H. Zhang
National Astronomical Observatories, Chinese Academy of Sciences
Chaoyang District, Datun Road, A20, Beijing, China

Abstract- **In this paper, we give a brief description of the evolution of the optics of FAST telescope. Several milestones that lead to the current FAST concept are presented. These include the lightweight focus suspension, active main reflector, backward illumination and new feed technology such as Phase Array Feed, etc. A perspective for future development is given in the end.**

I. INTRODUCTION

FAST is the currently largest single dish radio telescope that has been proposed. The project was approved by the Chinese National Development and Reform Commission in 2007. The construction was officially started in March 2011, and will be completed in September 2016 [1] [2].

There are mainly three innovative features of FAST telescope [3]. The telescope is built in a nearly spherical Karst depression in Guizhou province. The second is the active main reflector. The neutral shape of the main reflector is spherical. The illuminated part is deformed into a paraboloid of revolution [4]. The last is the light focus cabin suspension system, which enables the positioning of the feed with an accuracy of less than 10mm [5].

A brief description of the evolution of the optics of FAST will be presented in this paper. In the next section, several milestones of the formation of the FAST concept are reviewed. The current optics of FAST is described in the third section. A concluding remark is made in the end.

II. MILESTONES OF THE CONCEPT FORMATION

FAST telescope is originated from the Chinese SKA (LT in early days) effort. Chinese astronomer proposed an array of 20-30 Arecibo-type telescopes to obtain a collecting area of one square kilometer. And we did a comprehensive site survey of Karst depressions in the Guizhou province in Southwest part of China. Several hundred depressions have been found to be proper site candidates to accommodate large spherical radio telescope.

This work is partly supported by China Ministry of Science and Technology under State Key Development Program for Basic Research (2012CB821800), China Ministry of Science and Technology under State Key Development Program for Basic Research, Grant NO. 2013CB837900, Projects of International Cooperation and Exchanges NSFC, Grant NO. 11261140641 and the key research program of the Chinese academy of sciences, Grant NO. KJZD-EW-T01.

978-7-5641-4279-7

FAST, as a single dish, was proposed as a pathfinder of the Chinese SKA concept. Feasibility study has been carried out to tackle the technical challenges of building Arecibo-like telescope. In spite of the researches about the structure, measurement/control and receivers, the optics also underwent substantial evolutions. Here we list and describe the most significant milestones of this evolution.

A. Spherical main reflector and light weight feed suspension system

Since it is very difficult to realize a fully steerable telescope with a diameter of 200-300m, an Arecibo-like spherical surface seems was a natural choice. In the early days of Arecibo, long line-feed was used to illuminate part of the main reflector. However, the very narrow band line-feed is not suitable for a general purpose radio telescope.

Hybrid feed was proposed to illuminate a spherical radio telescope[6]. A lightweight feed suspension system was also contemplated [7]. Like Arecibo, there is no rigid connection between the feed and the main reflector. But this suspension mechanism use six cable to position the feed, thus eliminate the very large up-side-down AZ-EL feed platform as appeared at Arecibo.

The early FAST technical scheme was featured by a very large spherical main reflector and a light feed suspension system [5].

B. Active main reflector

An active main reflector was proposed to correct the spherical aberration on the ground [4]. After a proper f/D is chosen, a minimal radial deviation of a spherical surface with a paraboloid of revolution could be obtained. The main reflector is then segmented into some 1100 hexagons of 10-m edge. If a point-feed is put at the focus point, the illuminated aperture of 300m diameter is deformed into a paraboloid of revolution in real-time. This scheme is somewhat different from the active surface adopted at other large radio telescope, such as GBT, LMT and the newly built 65m radio telescope near Shanghai in China, which is used to compensate gravitational and thermal deformations when the telescope is pointing different points on the sky.

This active main reflector scheme allows highly-advanced point feed to be adopted for FAST. And since the elementary panels stays approximately at the same tilting angle during observation, thus the gain remains almost constant when tracking a source.

Two main schemes were proposed to support the thousands of panels. One is rigid structures consist of concrete pillars, and actuators were used to push the panels from underneath. Down scale model experiments has been carried out to demonstrate the feasibility. Self-adaptive mounting of the panels was also proposed during the research. Another scheme is to use cable network and downlink cables to form a virtual "back structure" to support the panels. In this scheme, the elasticity of the cables is used to realize the deformation, and the driving mechanism is on the ground which makes maintenance more easier.

The cable network is chosen to be the final technical scheme for active main reflector for FAST. Studies have been carried out to address key technical issues such as the type of driving mechanism, material selection for the cables and self-adaptive mounting of panels, etc.

C. Focus suspension system (with Stewart platform)

Simulation study of the original focus cabin suspension system for the line feed using six cables shows inadequate accuracy for positioning the point feed at higher frequencies. Stewart platform was introduced to compensate the residual positional errors caused by wind and the inherent vibration of the cable suspension system [8]. A X-Y rotation mechanism was also introduced to get the orientation of the Stewart platform roughly correct.

In order to achieve satisfactory position accuracy of the lower platform, it is suggested that the ratio of the mass between upper and lower platform should be no less than about 10:1. During the same period, a set of receiver was proposed by a joint study of NAOC and JBO in the UK. Though the backend changes dramatically during the last decade, the frontend technology didn't change very much. So the 2-3 tons weight of the feeds and low noise frontend to be mounted on the lower platform remains effective.

D. Zenith angle and "backward illumination"

Compared with the Arecibo telescope, which remained the largest radio telescope since the completion of it's construction, has a zenith angle limit of about 20 degrees. FAST has larger zenith angle coverage. The reason is as follows:

First, the spherical surface of FAST is deeper, which allows the main reflector to collect radiation from lower elevations.

Second, FAST adopted a cable-driven focus cabin suspension system. This system is much flexible compared with the large triangular platform. I.e. the zenith angle is mainly limited by the ability of the focus cabin suspension system. Larger zenith angle could be achieved if more power is available or a lighter focus cabin is used.

When the zenith angle is larger than ~26.4 degree, the illuminated 300m aperture will go beyond the edge of the 500m spherical surface. As the zenith angle continues to become larger, the feed will see more noise from the surroundings. A ground screen made from wire mesh was proposed to block the noise from the ground. In order to achieve zenith angle of 40 degree, this ground screen would need to be some 45m high. This one mile long and 45m high ground screen would then become a noticeable infrastructure by itself.

An offset illumination was considered for the above situation. When the illuminated 300m aperture exceed the edge of the spherical surface, the feed will be rotated about it's phase center backward towards the center of the 500m aperture. This "backward illumination" will eliminate the need of a ground screen since the feed will see the un-deformed surface on the other side instead of the warm ground out of the edge. Simulation has shown that the on-axis gain under the "backward illumination" mode differs very little compared with the normal mode where a ground screen is adopted.

E. New feed technology: Phased Array Feed and line feed to illuminate the whole reflector

New feeding technologies have been investigated during the last decade for FAST telescope. These include Phase Array Feed (PAF) and long line feed. Though this is not a milestone yet, when the technology is ready, it will enhance the ability of FAST telescope enormously.

FAST would be a powerful instrument for pusar and HI survey, the current 19-horn receiver provides 19 simultaneous beams on the sky. Same horns are used for all the beams, and the off axis beam will have lower gain and worse far field pattern. And the beams on the sky have gaps since apertures of the adjacent feeds can not overlap. The PAF occupies the same area of the focal plane may be able to provide continuous sky coverage, and the far field patterns for each beam may essentially be the same, with on-axial gain higher than that obtained by using horns. The PAF may provide more advantages compared with using horns [9].

When joining the current international VLBI network, FAST will greatly increase the baseline sensitivity, thus allow more weaker sources to be observed. The PAF will enable FAST to have comparable FoV as smaller telescope, thus allow specific observation modes (such as in-beam phase referencing) to be carried out.

If we used a 140m long line feed, we may be able to effectively illuminate the whole 500m spherical aperture. By putting wideband elements along the line, and compensating the different relative time delay of the signals from the various elements digitally, a wideband line feed may be realized. Though the sky coverage is limited to a stripe on the sky, it may become a quarter of SKA in terms of collecting area.

III. OPTICS OF FAST

Based on the above description, FAST can be seen as a prime focus paraboloid radio telescope with zenith angle limit of 40 degrees. But FAST has it's own specific characters, e.g. the gain remains almost constant for different elevation angles, the metal panels out of the illuminated area will reduce the noise due to feed spillover.

IV. CONCLUDING REMARKS

FAST is an unusual telescope. Compared with Arecibo, the current largest single dish radio telescope, FAST is featured by its active main reflector and delicate lightweight focus suspension system, but this simplicity and delicacy in structure dictates much more difficulty and challenges in the measurement and control of all the parts to work in a coherent manner (private communication with Donald Campbell, the former director of Arecibo).

The main reflector of FAST could be deformed into any shape between a sphere and a paraboloid, this flexibility may allow various optics to be investigated. The shape of the panel will give a limit of the highest frequency end. Improvement of the focus cabin suspension and dynamic beam forming using PAF may help to reach this high frequency limit.

ACKNOWLEDGMENT

Jin, C. thanks Prof. Nan Rendong, Dr. Peng Bo and other FAST staff on the discussion of the history of FAST development.

REFERENCES

[1] R. Nan, D. Li and C. Jin, et al. "The Five-Hundred Aperture Spherical Radio Telescope(fast) Project," International Journal of Modern Physics D, Volume 20, Issue 06, pp. 989-1024, 2011

[2] L. Di, R. Nan, C. Jin, B. Peng, J. Yan, "The Five-hundred-meter Spherical Aperture Telescope(FAST) Project," EGU General Assembly 2009, P.3733

[3] R. Nan, "Five hundred meter aperture spherical radio telescope (FAST)," Science in China Series G, vol. 49, issue 2, pp. 129-148

[4] Y. Qiu, "A novel design for a giant Arecibo-type spherical radio telescope with an active main reflector," Mon. not. R. Astron. Soc. 301, 827-830, 1998

[5] B. Duan, Y. Qiu, Y. Su, W. Wang, R. Nan, B. Peng, "Modelling, simulation and testing of an optomechatronics design of a large radio telescope," Astrophysics and Space Science, Vol. 278, Issue 1-2, pp. 237-242, 2001

[6] J. Xiong, S. Xie, "Hybrid Feeds for a Spherical Reflector", Proceedings of the 3rd meeting of the Large Telescope Working Group and a Workshop on Spherical Radio Telescopes, ed. R. Strom, B. Peng and R. Nan. 1995

[7] B. Duan, Y, Zhao and J. Wang, "Study of the Feed System for a Large Radio Telescope from the Viewpoint of Mechanical and Structural Engineering," Proc of the LTWG-3, ed. R. Strom, B. Peng and R. Nan, pp 85-102, 1996

[8] B. Peng, C, Jin, Q. Wang, L. Zhu, W. Zhu, H. Zhang and R. Nan, "Preparatory Study for Constructing FAST, the World's Largest Single Dish," Proceedings of the IEEE, Vol. 97, Issue 8, P. 1391-1402

[9] Y. Wu, K. Warnick, C. Jin, "Design Study of an L-Band Phased Array Feed for Wide-Field Surveys and Vibration Compensation on FAST," Antennas and Propagation, IEEE Transactions, Vol. 61, Issue. 6, pp. 3026-3033, 2013

WP-1(B)

October 23 (WED) PM

Room B

New Strategies of CEM-1

A General MoM-PO Hybrid Framework for Modelling Complex Antenna Arrays Mounted on Extremely Large Platform (Invited Paper)

Zi-Liang Liu, Xing Wang, Chao-Fu Wang
Temasek Laboratories, National University of Singapore
5A Engineering Drive 1, Singapore 117411

Abstract- **A general method of moments (MoM) - physical optics (PO) hybrid framework is proposed to fast and efficiently analyze complex antenna arrays installed on extremely large-scale platforms. In this general MoM-PO hybrid framework, MoM is applied to simulate the antenna array and the integral equations can be flexibly set up either on the antenna array or on a Huygens surface enclosing the antenna array. PO is employed to efficiently describe the contribution of the electrically large platform and an iterative process is implemented to take the interaction between the antenna array and the platform into account. To further reduce the peak memory usage and CPU time, the adaptive integral method (AIM) is adopted to speed up the solving of MoM equations. Numerical results show that the proposed general MoM-PO hybrid technique can greatly reduce the CPU time and peak memory usage compared with the conventional MoM-PO method.**

I. INTRODUCTION

Onboard antenna placement is an important problem in EMC study and antenna design, especially for complex antenna arrays on large-scale platforms. An efficient iterative method of moments (MoM) – physical optics (PO) technique has been proposed [1], where the antenna array is modeled by MoM, and the electrically large platform is simulated by PO. Then MoM and PO is hybridized through an iterative process to take the interaction between antenna array and platform into account. The efficient iterative MoM-PO (EI-MoM-PO) can greatly reduce the number of unknowns and CPU time in the analysis of onboard antenna arrays with large-scale platforms, and the ground effects can also be coupled into the iterative framework to solve half-space problems [2]. However, the capability and accuracy of EI-MoM-PO depends on whether the antenna array can be accurately calculated with MoM.

On the other hand, a lot of commercial software and testing facilities can well simulate or measure the performance of complex antenna arrays. Therefore, to enhance the capability of the EI-MoM-PO technique, a general MoM-PO hybrid framework is proposed where the Huygens principle is combined with EI-MoM-PO. In this general MoM-PO framework, a Huygens surface, which encloses the complex antenna array, is introduced to replace the antenna array. On the Huygens surface, the equivalent electric and magnetic currents are defined and MoM equations are established. The initial MoM currents depend on the electromagnetic fields radiated from the antenna array on the Huygens surface, which can be given by measurement or simulating with commercial software. Then the iterative process is implemented between MoM and PO to consider the interaction of the antenna array and the platform.

For large and complex antenna array, the MoM part still consumes a large amount of memory and CPU time. To further improve the efficiency of the proposed general MoM-PO technique, the adaptive integral method (AIM) [3] is adopted to reduce the peak memory usage and speed up the solving of MoM equations. Numerical results show very good capability and accuracy of the general MoM-PO hybrid framework.

II. FORMULATION

A. Efficient Iterative MoM-PO Method

Consider an electromagnetic system consisting of antennas and the PEC platform where the integral equations can be established as

$$\vec{E}^s = L^E\vec{J} - C\vec{K}, \tag{1}$$

$$\vec{H}^s = C\vec{J} + L^H\vec{K}, \tag{2}$$

where the operators $L^E\vec{X}$, $L^H\vec{X}$ and $C\vec{X}$ are given by

$$L^E\vec{X} = -j\omega\mu\int_{S'}\vec{X}G(\vec{r},\vec{r}')dS' + \frac{1}{j\omega\varepsilon}\nabla\int_{S'}\nabla'\cdot\vec{X}G(\vec{r},\vec{r}')dS', \tag{3}$$

$$L^H\vec{X} = -j\omega\varepsilon\int_{S'}\vec{X}G(\vec{r},\vec{r}')dS' + \frac{1}{j\omega\mu}\nabla\int_{S'}\nabla'\cdot\vec{X}G(\vec{r},\vec{r}')dS', \tag{4}$$

$$C\vec{X} = \nabla\times\int_{S'}\vec{X}G(\vec{r},\vec{r}')dS'. \tag{5}$$

$G(\vec{r},\vec{r}')$ is the free space Green's function. $\vec{J}$ and $\vec{K}$ are the induced electric and magnetic currents due to feed of antenna.

Then MoM is applied to the analysis of the antenna array and PO is used to approximate contribution of the electrically large platform. By employing proper boundary conditions, one can set up the MoM matrix equation only for MoM-region

$$\begin{bmatrix} Z_{km}^{EE} & Z_{km}^{EH} \\ Z_{km}^{HE} & Z_{km}^{HH} \end{bmatrix}\begin{bmatrix} I_m^{0,MoM} \\ K_m^{0,MoM} \end{bmatrix} = \begin{bmatrix} V_k^E \\ V_k^H \end{bmatrix}. \tag{6}$$

Solving equation (6) yields the electric and magnetic currents in MoM-region, and these currents are regarded as the sources for PO-region (PEC platform) to calculate the induced PO current

978-7-5641-4279-7

$$\bar{J}^{1,PO} = \sum_{m=1}^{M} [2\delta \hat{n} \times (I_m^{0,MoM} C\bar{f}_m + K_m^{0,MoM} \bar{L}^H \bar{f}_m)], \quad (7)$$

where $\hat{n}$ is the outer unit normal of PO-region. δ is the shadowing effect coefficient, and $\bar{f}_m$ is the MoM basis function.

To consider the effects of the platform on the antenna array, additional exciting voltages due to $\bar{J}^{1,PO}$ are added to the right side of the MoM matrix equation (6), which are expressed as

$$\Delta V_k^{E,1} = -<\bar{f}_k, \bar{L}^E \bar{J}_S^{1,PO}>, \quad (8)$$

$$\Delta V_k^{H,1} = -<\bar{f}_k, C\bar{J}_S^{1,PO}>. \quad (9)$$

Substituting the modified sources $V_k^E + \Delta V_k^{E,1}$ and $V_k^H + \Delta V_k^{H,1}$ into (6) obtains the new MoM currents in the presence of the platform. An iterative process will be implemented from equation (6) to (9) by replacing the superscript "*1*" with the number of iteration "*i*", until the errors of MoM currents $\| [I_m^{i,MoM}] - [I_m^{i-1,MoM}] \| / \| [I_m^{i-1,MoM}] \|$ and $\| [K_m^{i,MoM}] - [K_m^{i-1,MoM}] \| / \| [K_m^{i-1,MoM}] \|$ are not more than the prescribed threshold.

B. General Hybrid MoM-PO Framework

In practice, some types of widely used antenna arrays can not be well simulated with MoM, such as Vivaldi arrays. To enhance the capability of EI-MoM-PO technique, the Huygens principle is introduced into this hybrid framework, and the MoM equations in (6) can just be established on a Huygens surface, which covers the antenna array, instead of on the antenna array. The iterative process form (6) to (9) should also be implemented to consider the interaction between the antenna array and the platform. A difference between this general MoM-PO hybrid framework and the EI-MoM-PO is that the initial MoM currents $I_m^{0,MoM}$ and $K_m^{0,MoM}$ are determined by electric and magnetic fields on the Huygens surface radiated from the antenna array. These fields can be obtained from measurement or simulated results with commercial software. Therefore, most types of antenna arrays can be handled with this proposed general MoM-PO hybrid technique.

C. Acceleration with Adaptive Integral Method

It's very time-consuming to analyze large and complex antenna array with MoM, and a great number of unknowns are required. To overcome this difficulty, AIM can be applied to the MoM part of the proposed general MoM-PO hybrid framework to reduce the memory requirement and speed up the matrix-vector multiplication process $[Z][I]$ when we try to solve (6) with iterative method, such as BiCGStab or GMRES. The matrix vector multiplication $[Z][I]$ can be accelerated as follows:

(i). spliting the matrix-vector multiplication $[Z][I]$ into near interaction term $[Z]^{near}[I]$ and far interaction term $[Z]^{far}[I]$;

(ii). directly computing the near interaction $[Z]^{near}$;

(iii). projecting the MoM basis function to the grid points;

(iv). calculating near interaction term $[Z]^{near}[I]$ directly and the far interaction term $[Z]^{far}[I]$ using FFT;

It should be mentioned that procedures (i) - (iii) just need to be executed once when AIM is implemented for the first time. Within each iterative process between MoM and PO, one just needs to run the iterative solver to get the MoM currents. So the efficiency of the general MoM-PO technique can be greatly improved.

III. NUMERICAL RESULTS

A 24×6 longitude waveguide slot array mounted on an aircraft is simulated with the proposed hybrid technique, as shown in Figure 1. The working frequency of the slot array is 10 GHz and the corresponding wavelength is 0.03 m. The length of the aircraft is 47 m and wingspan is 38.05 m, which are more than 1500λ and 1200λ, respectively. The BJ-100 waveguide is used in the design of the slot array and the details of the array are depicted in Figure 2. Taylor distributions are adopted along the two directions of the slot array aperture. The aperture distributions and the array factor patterns are illustrated in Figure 3 to Figure 6.

This model is simulated serially on a Linux server equipped with an Intel Xeon CPU at 3.1 GHz and 128 GB of RAM. The gain patterns in three cut-planes obtained from the proposed general MoM-PO and the conventional MoM-PO of FEKO are plotted in Figure 7 to Figure 9, and show good agreement. The general MoM-PO only takes 172.5 hours to get the results, while the conventional MoM-PO of FEKO needs 1492.3 hours. The peak memory usage of proposed general MoM-PO and the conventional MoM-PO of FEKO is 1.5 GB and 79.7 GB, respectively. We also try MLFMA of FEKO to handle this model. However, it is too large for MLFMA of which the peak memory usage is 278 GB. So we just show MLFMA results of the waveguide slot array for comparison. From Figure 7 to Figure 9, it can be obviously seen that the gain pattern of the slot array is disturbed by the aircraft. Figure 10 displays the 3D gain pattern of the slot array and Figure 11 depicts the 3D gain pattern of the array installed on the aircraft and the PO current on the platform. By comparing the two 3D gain patterns, the effect from the platform can be more clearly observed.

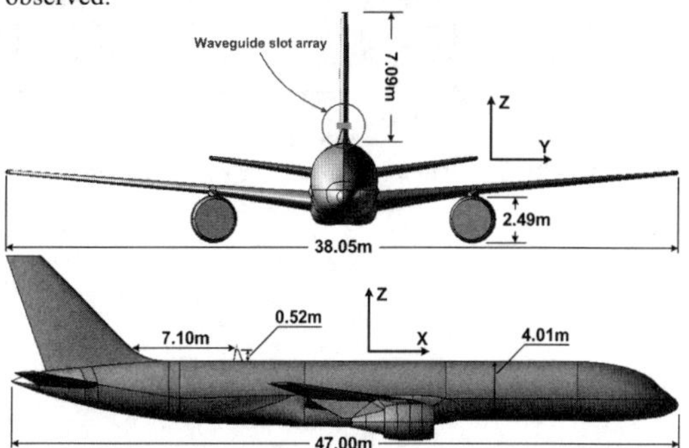

Figure 1. A 24×6 waveguide slot array mounted on an aircraft.

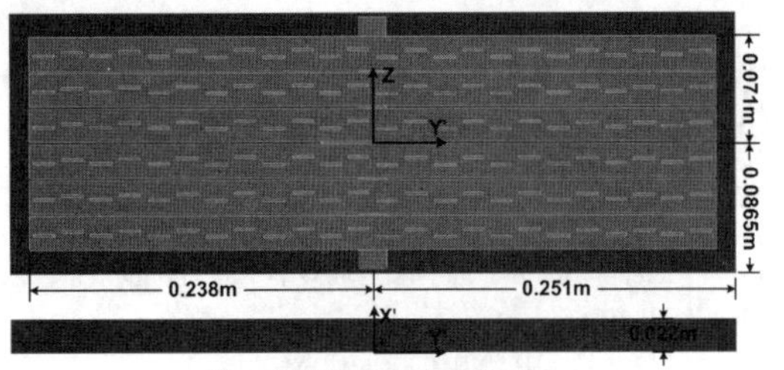

Figure 2. Gain pattern in XOY plane.

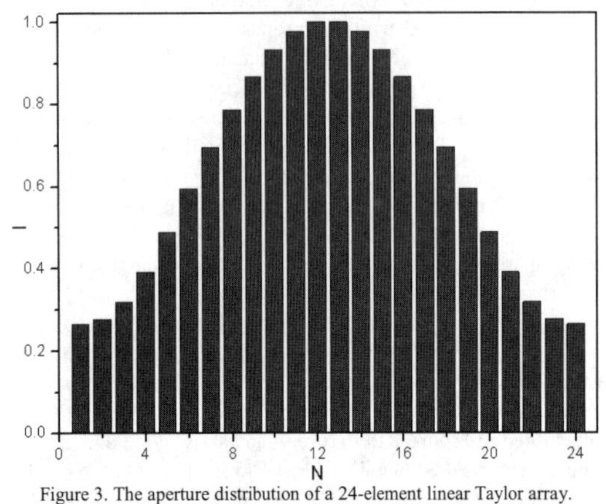
Figure 3. The aperture distribution of a 24-element linear Taylor array.

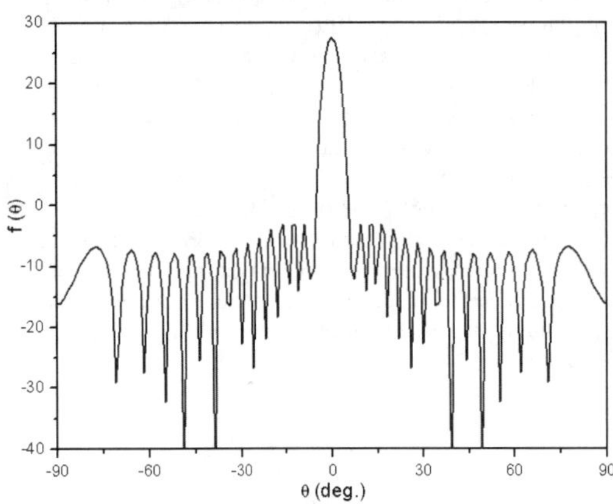
Figure 4. Array factor pattern of a 24-element linear Taylor array.

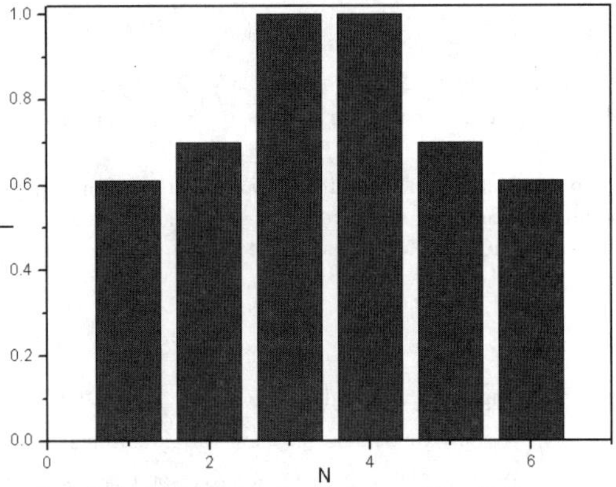
Figure 5. The aperture distribution of a 6-element linear Taylor array.

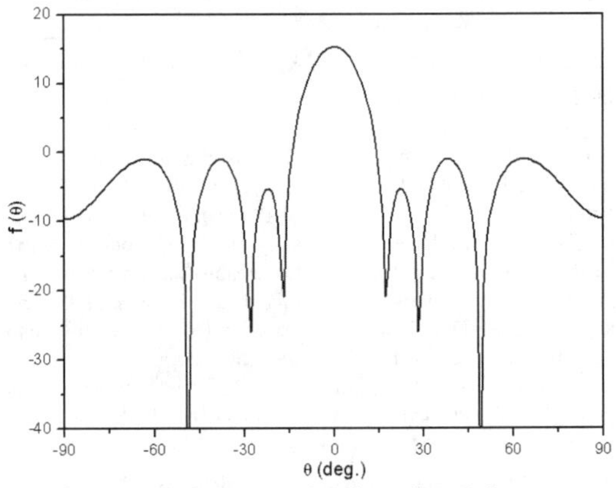
Figure 6. Array factor pattern of a 6-element linear Taylor array.

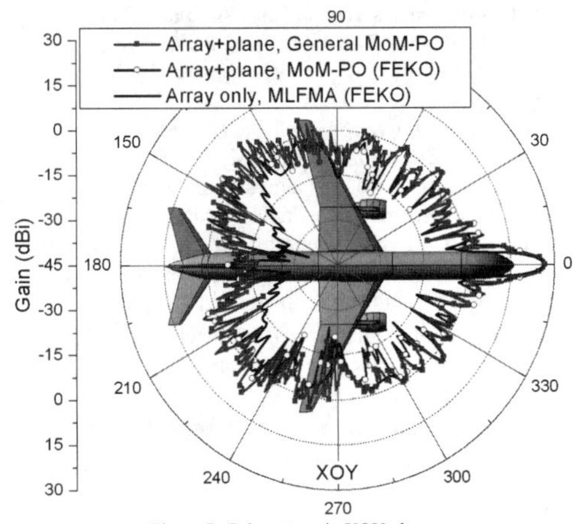

Figure 7. Gain pattern in XOY plane.

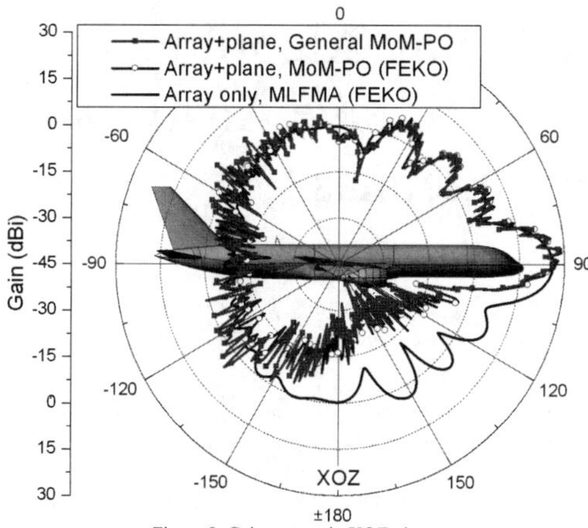

Figure 8. Gain pattern in XOZ plane.

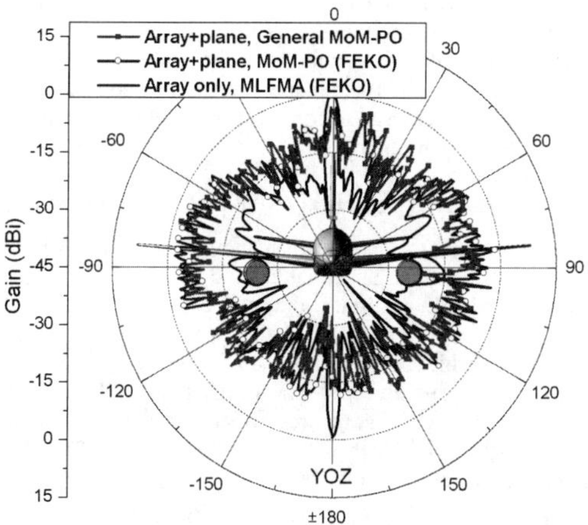

Figure 9. Gain pattern in YOZ plane.

Figure 10. 3D gain pattern of the waveguide slot array

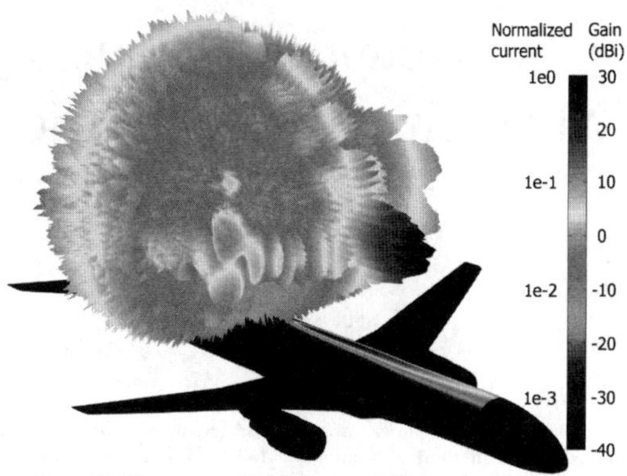

Figure 11. PO current and 3D gain pattern of the waveguide slot array mounted on the aircraft.

IV. CONCLUSION

A general MoM-PO hybrid framework has been presented to fast and efficiently analyze complex antenna arrays mounted on extremely large-scale platforms. In this general hybrid MoM-PO framework, various types of onboard antenna arrays can be well handle by replacing the antenna array with a surrounding Huygens surface where the equivalent electric and magnetic currents are define. MoM equations are established on this Huygens surface and PO is employed to efficiently modelling the contribution of the extremely large platform. An iterative process is implemented to characterize the interaction between MoM-region and PO-region. To further reduce the peak memory usage and CPU time, AIM is applied to MoM-region to accelerate the solving of MoM equations. Numerical results show that the proposed general MoM-PO hybrid technique can dramatically reduce the consumption of computer memory and CPU time compared with the conventional MoM-PO method.

REFERENCES

[1] Z. L. Liu and C. F. Wang, "Efficient iterative method of moments – physical optics hybrid technique for electrically large objects," IEEE Trans. Antennas Propagat., vol. 60, no. 7, July 2012, pp. 3520-3525.

[2] Z. L. Liu and C. F. Wang, "An efficient iterative MoM-PO hybrid method for analysis of onboard wire antenna array with large-scale platform above infinite ground," IEEE Antennas Propagat. Mag., accepted and to be published in Dec. 2013.

[3] C. F. Wang, F. Ling, J. Song, and J. M. Jin, "Adaptive integral solution of combined field integral equation," Microwave Opt. Tech. Lett., vol. 19, no. 5, Dec. 1998, pp. 321-328.

An Adaptive Frequency Sweeping Algorithm of MoM Impedance Matrices in Full-Wave Analysis of Microstrip Patch Antennas

Shi Fei Wu, #Zhe Song and Wei-Dong Li

State Key Laboratory of Millimeter Waves, Southeast University, Nanjing, China
zhe.song@seu.edu.cn

***Abstract*-In this paper, a study on frequency interpolation algorithms of method of moments (MoM) impedance matrices is discussed in detail, which is successfully applied into the full-wave analysis of a microstrip patch antenna in a relatively wide band. By using Lagrange interpolation scheme in an adaptive system, two interpolating rules are realized and their accuracies are defined by Frobenius norms of the impedance matrices in entire frequency band. A microstrip fed patch antenna is considered to verify the algorithm from 1 to 5 GHz. The numerical results have shown that by selecting the Chebyshev zeros in the frequency band for polynomial interpolation is of high accuracy and the simulation efficiency can be highly elevated simultaneously. Besides, a statistical conclusion on the tradeoff between accuracy and efficiency issue has also been made quantitatively.**

I. Introduction

With the fast development in microwave and millimeter wave integrated circuit design and VLSI technology, more and more attention has been paid to the rigorous, accurate and fast modeling and simulation methods of layered circuits. By applying the MoM with layered medium dyadic Green's functions into the mixed potential integral equation (MPIE) has been one of the most popular methods for microstrip structures [1-5]. As is well-known, the spectral-domain Green's functions of layered medium structures can be expressed in closed form [6], and then inversed to spatial domain through the Sommerfeld integrals (SI). Based on our recent works [7, 8], the Green's functions in spatial domain have been fast and accurately obtained by means of the combination of the discrete complex image method (DCIM) and the all modes method in near and non-near region, respectively. With the closed-form spatial domain dyadic Green's functions, the MoM have been constructed for modeling and simulation for planar layered circuits, which is based on the RWG basis functions [9] and Delta-Gap voltage excitation model [10]. Although the computer codes could reach almost the same efficiency as some commercial software at single frequency point, it becomes inefficient for wide-band frequency sweeping situation. Considering the smooth property of MoM impedance matrix elements, it is possible to build interpolating schemes in a relative wide band, therefore, the efficiency of MoM can be highly elevated.

In this paper, a study on frequency interpolation algorithms of MoM impedance matrices is discussed in detail, which has been successfully applied into the full-wave analysis of a microstrip patch antenna in a relatively wide band. By using Lagrange interpolation scheme, two sampling rules are realized and their accuracy are defined by Frobenius norms of the impedance matrices in entire frequency band. A microstrip fed patch antenna is considered to verify the algorithms from 1 to 5 GHz. The numerical results have shown that by selecting the Chebyshev zeros in the frequency band for polynomial interpolation is of high accuracy and the simulation efficiency can be highly elevated simultaneously. Very good agreement on S-parameters between the proposed method and commercial software have been found.

II. MPIE Formulation and Parameter Extraction

By enforcing the boundary condition that revokes the vanishing of the total tangential electric field on the conductor surface, the EFIE governing the total current density can be established. However, to avoid the two-dimensional infinite integrals with highly oscillating, slowly decaying and hyper-singular kernel involved in the EFIE, the MPIE has been widely used in layered structures, which is composed of vector and scalar potentials with weakly singular kernels. The MPIE can be formulated as below [11]:

$$\hat{n}\times\vec{E}^{imp}=\hat{n}\times\begin{bmatrix} j\omega\mu_0\left\langle \underline{\underline{G}}^A(\vec{r}\,|\,\vec{r}\,');\vec{J}(\vec{r}\,')\right\rangle \\ -\dfrac{1}{j\omega\varepsilon_0}\nabla\left(\left\langle G^{\Phi}(\vec{r}\,|\,\vec{r}\,'),\nabla_S'\cdot\vec{J}(\vec{r}\,')\right\rangle\right)\end{bmatrix} \tag{1}$$

where $\vec{E}^{imp}=V_p\delta(\vec{r}-\vec{r}_p)\cdot\hat{n}_p$ stands for the impressed electric field, $\vec{r}_p$ is the location of the port and $\hat{n}_p$ is the outward normal parallel to the feed line. $\underline{\underline{G}}^A(\bullet)$ and $G^{\Phi}(\bullet)$ refer to the dyadic and scalar Green's functions of the vector and scalar potential, respectively. With the spatial Green's functions, the MoM can be applied to converting the MPIE into an matrix equation. For the sake of modeling the arbitrarily shaped geometries, the RWG triangular patches are adopted in this paper. With the Galerkin's procedure, (1) becomes [11]

$$\begin{aligned} &-j\omega\varepsilon_0\int_{T_m}\vec{E}_t^{imp}(\vec{r})\cdot\vec{f}_m(\vec{r})ds \\ &=k_0^2\sum_{n=1}^{N}I_n\int_{T_m}\int_{T_n}\vec{f}_n(\vec{r}\,')\cdot\underline{\underline{G}}^A(\vec{r}\,|\,\vec{r}\,')\cdot\vec{f}_m(\vec{r})ds'ds \\ &+\sum_{n=1}^{N}I_n\cdot\int_{T_m}\nabla\left(\int_{T_n}G^{\Phi}(\vec{r}\,|\,\vec{r}\,')\left(\nabla_S'\cdot\vec{f}_n(\vec{r}\,')\right)ds'\right)\cdot\vec{f}_m(\vec{r})ds \end{aligned} \tag{2}$$

978-7-5641-4279-7

where $\vec{f}_n$ and $\vec{f}_m$ stand for the RWG basis and weighting functions, respectively. T_n and T_m are the triangular pairs containing the source ($\vec{r}'$) and field ($\vec{r}$) point, respectively. By using the Green's identity and the numerical Gaussian integral over triangular meshes, the integral equation (2) can be deduced as an algebraic linear system. As is proposed in [10], the delta-gap voltage model is adopted to excite the physical port, and the matrix element involved can be calculated as [11]:

$$\begin{aligned} Z_{mn} &= k_0^2 \int_{T_m}\int_{T_n} \vec{f}_n(\vec{r}')\cdot \underline{\underline{G}}^A(\vec{r}\,|\,\vec{r}')\cdot \vec{f}_m(\vec{r})ds'ds \\ &-\int_{T_m}\int_{T_n} \left(\nabla'_S\cdot \vec{f}_n(\vec{r}')\right) G^{\Phi}(\vec{r}\,|\,\vec{r}')\left(\nabla_S\cdot \vec{f}_m(\vec{r})\right) ds'ds \end{aligned} \quad (3)$$

For the microstrip planar circuits, the S-parameters are usually extracted, which depend on the incident and reflected wave of the dominant mode. To observe the recognizable standing-wave feature on the feed line, the reference planes should be selected away from not only the discontinuities but also the exciting ports. The generalized pencil-of-function (GPOF) is adopted in this paper. After the curve-fitting operation, the current distribution can be written as [11]:

$$\begin{aligned} I(z) &\approx \sum_{i=1}^{N} p_i \exp(\gamma_i z) \\ &= \sum_{i=1}^{N} p_i \exp\left[(\alpha_i + j\beta_i)z\right], \qquad z>0 \end{aligned} \quad (4)$$

where p_i is the amplitude of the i-th mode. α_i and β_i stand for the propagation constant of the i-th mode. From the physical point of view, the first two terms, namely, (p_1 α_1 β_1) and (p_2 α_2 β_2) are just the incident and reflected wave of the dominant mode. Therefore, the S_{11} can be easily obtained.

III. Impedance Matrix Interpolation Scheme

It is a well-known fact that although most parameters of microstrip structures, such as S-parameters and the induced current distributions, varies rapidly with frequencies, however, the impedance matrix elements appear much smoother behaviors. Therefore, it enlighten us to introduce Lagrange polynomial interpolations to fit the impedance matrix elements.

In this paper, two sampling rules are realized, namely, equally spaced sampling and sampling with Chebyshev zeros in the frequency band of interest, as are shown in (5) and (6), respectively.

$$f_i = f_l + (f_h - f_l)\frac{i}{N} \quad (5)$$

$$f_i = \frac{(f_h - f_l)}{2}\cdot\cos\left(\frac{2i+1}{2N+2}\right) + \frac{(f_h + f_l)}{2} \quad (6)$$

where $[f_l, f_h]$ is the frequency band of interest and the number of sampling points is N+1. With accurate calculate the impedance matrix at these sampling frequencies, the rest can be approximated by Lagrange polynomial interpolation:

$$Z_{mn}^{\text{Inter}}(f) = \sum_{i=0}^{N} Z_{mn}^{\text{sample}}(f_i)\Psi_i(f) \quad (7)$$

$$\Psi_i(f) = \prod_{j=0, j\neq i}^{j=N}\left(\frac{f-f_j}{f_i - f_j}\right) \quad (8)$$

To estimate the relative error of the interpolation scheme, as well as to establish an adaptive algorithm, the Frobenius norms of impedance matrices are adopted [13] and the relative error can be expressed as:

$$\delta_{P+1} = \frac{\left\|[Z]_{P+1} - [Z]_P\right\|_{\text{Frobenius}}}{\left\|[Z]_P\right\|_{\text{Frobenius}}} \quad (9)$$

where the subscripts P stand for the number of sampling points. Considering the imagine parts of impedance matrices are much larger than the real parts, it is more reasonable to calculate δ for real and imagine parts separately.

IV. Numerical Examples

In this paper, a microstrip-fed patch antenna is modeled [12] and analyzed by the proposed method. By using the "PDETOOL" in Matlab, totally 353 triangular elements are generated, corresponding to 486 RWG pairs, as is shown in Fig. 1. According to the definition of relative error in (9), the threshold should be appointed before the adaptive scheme. The behavior of the relative errors is shown in Fig. 2, in which both equally spaced sampling and sampling with Chebyshev zeros in the whole frequency band of interest are calculated. Fig. 3 shows the relative errors between the interpolating impedance matrix and the standard matrix from direct MoM.

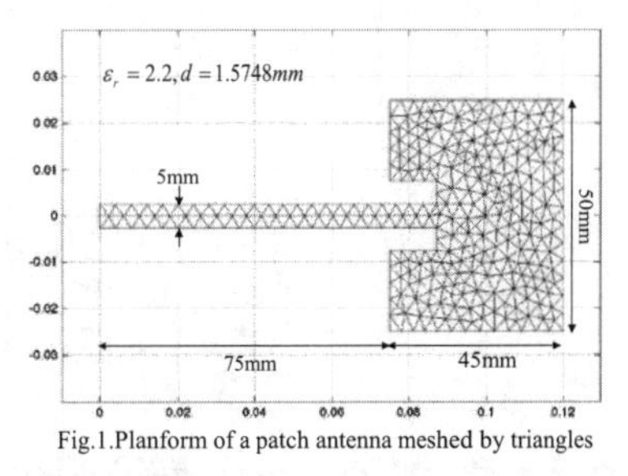

Fig.1.Planform of a patch antenna meshed by triangles

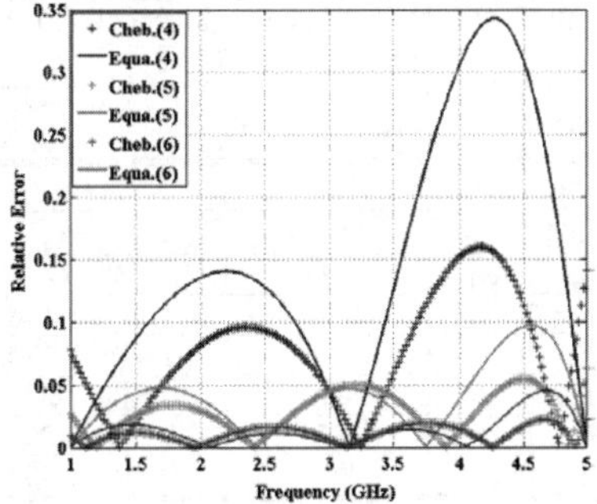

Fig 2. Relative errors of impedance matrix by different interpolation schemes

From these two figures, as the number of sampling nodes N increases, the relative error decreases rapidly. Also, the relative error of Chebyshev zeros interpolation scheme, comparing to that of equally spaced sampling interpolation scheme, tends to be more stable within the whole frequency band. Fig.4 shows the scattering parameter, S_{11}, calculated by the two interpolation schemes and the direct MoM, where 6 sampling points are selected in each interpolation scheme. Very good agreements can be found in Fig.4. However, from the stable point of view, interpolation by sampling with Chebyshev zeros is preferred because of its steady behavior of the relative error in the whole frequency band.

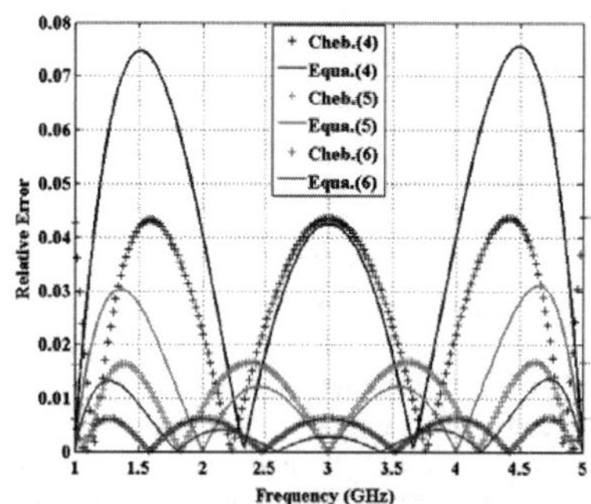

Fig. 3 Relative errors between impedance matrix interpolated by proposed schemes and that by direct MoM

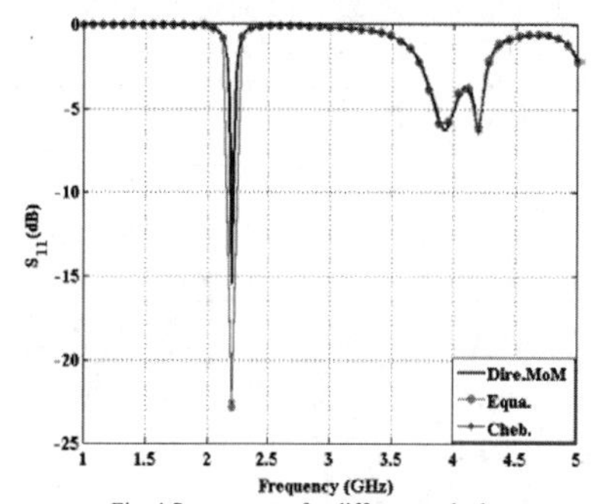

Fig. 4 S_{11} parameter for different methods

Table 1. Accuracy and Efficiency of the proposed method

RWG: 468, [1:0.02:5] GHz || Macbook Pro @ 2.5GHz, 16G RAM		
Scheme	Time Cost (s)	Max Relative Error (Compare to Dire. MoM)
HFSS(discrete)	3,686	—
Dire. MoM	25,037	—
3-Cheb. Interp.	529	11.15% (-10dB)
4-Cheb. Interp.	690	4.37% (-14dB)
5-Cheb. Interp.	836	1.67% (-18dB)
6-Cheb. Interp.	949	0.63% (-22dB)
7-Cheb. Interp.	1071	0.26% (-26dB)
8-Cheb. Interp.	1189	0.11% (-30dB)

Table. 1 shows the time consumptions for different methods performed on a Macbook Pro with 2.5GHz and 16G RAM. From this table, we can find that interpolation by sampling with 6 Chebyshev zeros consumes only one third and one twenty-fifth of time by ANSYS HFSS (discrete model) and direct MoM, respectively, while the relative error is only 0.63%, which verifies the high efficiency and accuracy of the proposed scheme. Besides, a good statistical property can be found from the last two columns in this table, that is, as the sampling nodes increase by one, the time cost for calculation increases 132s in average, while the maximum relative error decreases 4dB.

V. CONCLUSION

An adaptive frequency sweeping algorithm based on Lagrange polynomial is investigated for interpolating impedance matrix of MoM. Both equally spaced sampling and sampling with Chebyshev zeros are realized and discussed. By introducing the Frobenius norms, the relative error of the interpolation matrices are effectively evaluated. From the numerical example, sampling with Chebyshev zeros has the advantage of error control within the whole frequency band. This algorithm yields accurate result in scattering parameters as those from ANSYS HSFF and direct MoM, while enhancing the efficiency more than 3 and 25 times, respectively.

ACKNOWLEDGMENT

This work was supported in part by the National Basic Research Program of China (No. 2009CB320203, and No. 2010CB327400) and in part by the National Nonprofit Industry Specific Research Program of China (No. 200910041-2).

REFERENCES

[1] T. Itoh, Ed., *Numerical Techniques for Microwave and Millimeter-Wave Passive Structures*, New York: Wiley, 1989, ch,3.

[2] R. E. Collin, *Field Theory of Guided Waves*, New York: McGraw-Hill, 1960.

[3] W. C. Chew, *Waves and Fields in Inhomogeneous Media*, ser. Electromagn. Waves Piscataway, NJ: IEEE Press, 1995.

[4] R. F. Harrington, *Field computation by Moment Methods*. Melbourne, FL: Krieger, 1983.

[5] D. G. Fang, *Antenna Theory and Microstrip Antennas*. Beijing: Science Press, 2006.

[6] D. G. Fang, J. J. Yang, and G. Y. Delisle. "Discrete image theory for horizontal electric dipoles in a multilayered medium above a conducting ground plane," *IEE Proc. H*, pp.135 (5): 297-303, 1988.

[7] Z. Song, H. -X. Zhou, J. Hu, W. -D. Li and W. Hong, "Accurate Location of All Surface Wave Modes for Green's Functions of a Layered Medium by Consecutive Perturbations," *SCIENCE CHINA Information Sciences*. 2010;53(11):2363-2376.

[8] Song Z, Zhou H-X, Zheng K-L et. al, "Accurate Evaluation of Green's Functions for Lossy Layered Medium by Fast Locating Surface and Leaky Wave Modes". *IEEE Antennas and Propagation Magazine*. Vol. 55, pp.92-102.

[9] S. M. Rao, D. R. Wilton, A. W. Glisson., "Electromagnetic scattering by surfaces of arbitrary shape," *IEEE Trans. Antennas Propagat.*, vol. 30, pp. 409-418, 1982.

[10] G. V. Eleftheriades and J. R. Mosig, "On the network characterization of planar passive circuits using the method of moments," *IEEE. Trans. Microwave Theory Tech.*, vol. 44, pp. 438-445, 1996.

[11] Z. Song, Fast Algorithms for Spatial Domain Dyadic Green's Functions of Stratified Medium and its Applications in Electromagnetic Simulation of Layered Passive Circuits. Ph. D. Thesis in Southeast Univ. 2011.

[12] Yeo J, Mittra R. An algorithm for interpolating the frequency variations of method-of-moments matrices arising in the analysis of planar microstrip structures [J]. Microwave Theory and Techniques, IEEE Transactions on. 2003;51(3):1018-1025.

[13] W. D. Li, H. X. Zhou, J. Hu, Z. Song and W. Hong, "Accuracy improvement of cubic polynomial inter/extrapolation of MoM matrices by optimizing frequency samples". Antenna and Wireless Propagation Letter, IEEE. 2011;10:888-891.

Combination of Ultra-Wide Band Characteristic Basis Function Method and Improved Adaptive Model-Based Parameter Estimation in MoM Solution

A.-M. Yao[1], W. Wu[1], J. Hu[1,2] and D.-G. Fang[1]
[1] School of Electronic and Optical Engineering, Nanjing University of Science and Technology, China, 210094
[2] State Key Laboratory of Millimeter Waves, China, 210096

***Abstract*-An efficient technique which combines the ultra-wide band characteristic basis function method (UCBFM) and the improved adaptive model-based parameter estimation (IA-MBPE) is introduced for analyzing the wide band electromagnetic scattering problems. The ultra-wide band characteristic basis functions (UCBFs) are generated from characteristic basis functions at the highest frequency in the range of interest. These UCBFs are also available for the entire band. The IA-MBPE is the application of improved adaptive sampling algorithm (IASA) in MBPE. The high sampling efficiency in IASA is realized through extending the regions of searching for the sampling points from one to two simultaneously with the help of parallel computation algorithm. The combination of UCBFM and IA-MBPE results in significant enhancement of computational efficiency reduction in MoM solution. Numerical results from two examples of wide band frequency responses of monostatic RCSs validate the proposed method.**

***Index Term*-ultra-wide band characteristic basis function method; model-based parameter estimation; improved adaptive sampling algorithm; method of moments**

I. INTRODUCTION

The computation of the electromagnetic radar cross section (RCS) by the complex objects with large electrical size, such as ships on the sea and tanks on the ground, is important for military applications. Although the parallel technology of the computer is rapidly developing, it is still an arduous task to create a large RCS database of those objects. Meanwhile the computation by using numerical method, such as method of moments (MoM) [1], is very time consuming. The MoM not only places a heavy burden on the CPU time as well as memory requirements but also requires the impedance matrix to be generated and solved for each frequency point. Hence, if the response over a wide frequency band is of interest, the MoM is more computationally intensive.

Several techniques have been proposed to alleviate this problem. In [2-3], the characteristic basis function method (CBFM), is able to reduce the size of the MoM matrix. In the CBFM, the object is divided into a number of blocks, and high-level basis functions called characteristic basis functions (CBFs) are derived for these blocks, which are discretized by using the conventional triangular patch segmentation and Rao-Wilton-Glisson (RWG) basis functions [4]. In[5-7], an interpolation model known as the model-based parameter estimation (MBPE), which takes into account the physics behind the problem, is proposed to minimize the computational cost with a desired accuracy. The MBPE is based on the rational function interpolation. Since it needs solving MoM matrix equation at sampling points directly, the MBPE can hardly deal with wideband electromagnetic scattering problems from electrically large objects. So in [8], MBPE combined with AMCBFM, is proposed to analyze wide band electromagnetic scattering problems. This method uses the mutual coupling method for generating CBFs, that is time consuming and memory demanding, in CBFM and applies an adaptive sampling algorithm (ASA) [9] for MBPE. In [10], an improved adaptive sampling algorithm (IASA) is presented to obtain the high sampling efficiency. Since the CBFs depend upon the frequency, they need to be generated repeatedly for each frequency. Hence, in [11], the ultra-wide band characteristic basis function method (UCBFM) is developed, without having the generation of CBFs for each frequency repeatedly. The CBFs calculated at the highest, termed UCBFs, entail the electromagnetic behavior at lower frequency range; thus, it follows that they can also be employed at lower frequencies without going through the time consuming step of generating them again. However, in the UCBFM, it is still time consuming for the computation of wideband RCS since it requires repeated solving of the reduced matrix equations at each frequency.

In this paper, the combination of the UCBFM and IA-MBPE is introduced for fast evaluation of wide band scattering problems. In the following sections, the principles of UCBFM and the IA-MBPE are outlined firstly, and then the combination scheme of the UCBFM/IA-MBPE method is developed. Finally, two classical scattering problems are analyzed and the comparisons of the proposed method and traditional methods are provided.

II. FORMULATION

A. UCBF Method[11]

Let us consider a complex 3-D object illuminated by a plane wave. In a conventional MoM, the whole surface is divided into triangles with size ranging from $\lambda/10$ to $\lambda/20$. Applying

[2] This work is supported by State Key Laboratory of Millimeter Waves (K201306).

978-7-5641-4279-7

this to the electric field integral equation, one can obtain a dense and complex system of the form

$$Z(k)I(k)=V(k) \tag{1}$$

In (1), Z is the MoM matrix of dimension $N\times N$, I and V are vectors of dimension $N\times 1$, where N is the number of unknown current coefficients and k is the wave number of the free space. For large and complex problem, the matrix filling and matrix equation solving are quite time consuming.

The CBFM begins by dividing the object to be analyzed into blocks. For the best division scheme and the number of blocks M one may refer to [12]. These blocks are characterized through a set of CBFs, constructed by exciting each block with multiple plane waves (MPW), incident from N_{PW} uniformly spaced θ and φ-angles. To calculate the CBFs on the generic ith block , one must solve the following system

$$Z_{ii}(k)J_i^{CBF}=V_i^{MPW} \tag{2}$$

In (2), Z_{ii} is an $N_i\times N_i$ sub-matrix corresponding to the ith block, J_i^{CBF} is a $N_i\times N_{PW}$ matrix containing original CBFs, and V_i^{MPW} is a $N_i\times N_{PW}$ matrix containing excitation vectors, where N_i is the number of unknowns relative to ith block. In order to extract Z_{ii} from the original MoM matrix, a matrix segmentation procedure can be used. Next, a new set of orthogonal basis functions, which are linear combinations of the original CBFs, are constructed via the singular value decomposition (SVD) approach. Thus, the redundant information because of the overestimation is eliminated. For simplicity, one can assume that the average number of CBFs after SVD is K. Consequently, the solution to the entire problem is expressed as a linear combination of the $M\times K$ CBFs, as follows

$$I(k)=\sum_{m=1}^{M}\sum_{k=1}^{K}\alpha_m^k(k)J_m^{CBF_k}(k) \tag{3}$$

where $J_m^{CBF_n}$ is the nth CBF of the mth lock. By using the above CBFs, the original large MoM matrix can be reduced, and unknowns are changed to weight coefficient vector α whose order is much smaller than that of the original current coefficient vector I. Finally, after solving the reduced system and substituting solution back to (3), one can obtain the solution of single frequency point.

The ultra-wide band characteristic basis functions (UCBFs) is the CBFs generated at the highest frequency. Since the UCBFs can adequately represent the solution in the entire band of interest, they are used for lower frequencies without going through the time consuming step of generating them again. Fig.1 shows the flowchart of the UCBFM.

B. IA-MBPE Method[10]

The rational function in the form of a fractional polynomial function of the ζ-order numerator and the ν-order denominator employed commonly in MBPE is represented as

$$R(f)=\frac{P_\zeta(f)}{Q_\nu(f)}=\frac{p_0+p_1f+\cdots+p_\zeta f^\zeta}{q_0+q_1f+\cdots+q_\nu f^\nu} \tag{4}$$

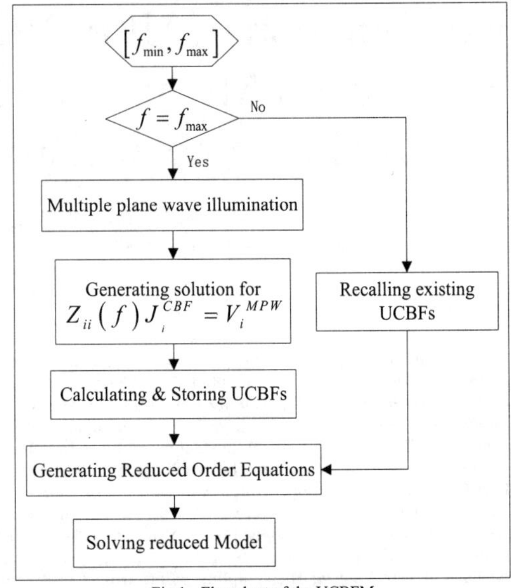

Fig 1. Flowchart of the UCBFM.

where $R(f)$ represents a frequency-domain fitting model and f is the frequency of interest. Since there are n+d+1 unknown coefficients (q_0 being arbitrary), a set of T+1=ζ+ν+1 sample points f_i are required to completely determine $R(f)$. $R(f)$ will then be a curve passing through S_i at f_i for i=0,1,...,T. We assume that $R(f)$ exists and has no unattainable frequency points [13] (frequency at which $R(f)$ has a common zero in the numerator and denominator polynomials) for the scattering parameter model that we are trying to attain.

The interpolation function $R(f)$ can be calculated with the recursion formulas in (6)-(7) initialized with (5) and using the inverse differences defined in (8)-(9).

$$\begin{cases}P_0=S(f_0)\\Q_0=1\\P_1=\varphi_1(f_1,f_0)P_0+(f-f_0)\\Q_1=\varphi_1(f_1,f_0)\end{cases} \tag{5}$$

$$\begin{cases}P_t=\varphi_1(f_t,f_{t-1},\ldots,f_0)P_{t-1}+(f-f_{t-1})P_{t-2}\\Q_t=\varphi_1(f_t,f_{t-1},\ldots,f_0)Q_{t-1}+(f-f_{t-1})Q_{t-2}\\t=2,3,\ldots,T\end{cases} \tag{6}$$

$$R_t(f)=\frac{P_t(f)}{Q_t(f)},\quad t=0,1,\ldots,T \tag{7}$$

$$\varphi_1(f_i,f_0)=\frac{f_i-f_0}{S(f_i)-S(f_i)},\quad i=1,2,\ldots,T \tag{8}$$

$$\varphi_t(f_i, f_{t-1}, \ldots, f_0) = \frac{f_i - f_{t-1}}{\varphi_{t-1}(f_i, f_{t-2}, \ldots, f_0) - \varphi_t(f_{t-1}, f_{t-2}, \ldots, f_0)} \tag{9}$$
$$i = t, t+1, \ldots, T;\ t = 2, 3, \ldots, T$$

The inverse differences in(8)-(9), determined recursively from the sample points, are essentially the polynomial coefficients defining $R(f)$. The rational expressions $R_t(f)$ are partial fractions of (4). Every new sample point increases the order of the rational function by one, until $R(f)= R_T(f)$.

The IASA is defined to work in the interval $[f_0, f_1]$. Define the residual error as

$$E_t(f) = |R_t(f) - R_{t-1}(f)| \tag{10}$$

Suppose the accuracy is δ. As a first step, an arbitrary third frequency point f_2 is selected which lies in the interval $[f_0, f_1]$. The interpolation function $R_1(f)$ is generated from the samples (f_0, S_0) and (f_2, S_2), while $R_2(f)$ is recursively updated using (6)-(9) and the sample (f_1, S_1). S_t is determined from CEM analyzes at f_t. The residual $E_2(f)$ is determined in the interval $[f_0, f_2]$ by evaluating it at a large number of equi-spaced frequency points over that frequency band. At the maximum of the residual, a new sample point (f_3, S_3) is selected.

For iteration t, the algorithm is used to calculate φ_t, P_t and Q_t from (6)-(9) in order to recursively determine $R_t(f)$. Assuming that the sample (f_t, S_t) was selected in the interval $[f_i, f_j]$, the residual $E_t(f)$ is determined in the intervals $[f_0, f_i]$ and $[f_j, f_1]$. Two new sample points (f_{t+1}, S_{t+1}) and (f_{t+2}, S_{t+2}) are chosen at the maximum of the relative residual in the intervals $[f_0, f_i]$ and $[f_j, f_1]$, respectively. The process is repeated until the relative residual becomes less than δ. The IASA automatically selects and minimizes the number of sample points.

C. UCBFM/IA-MBPE Method

The frequency point $f_{\max}$ is calculated first by the UCBFM for storing the UCBFs. The flowchart of UCBFM/IA-MBPE is shown in Fig.2. The frequency point $f_{\max}$, $f_{\min}$ and f_2 are selected firstly and then f_t (t>2) is selected. S_t represents for the RCS, which taking the UCBFM analyzing at f_t. $\psi_t=\{f_0, f_1, \ldots, f_t\}$.

III. Numerical Results

To demonstrate the efficiency and accuracy of the UCBFM/IA-MBPE technique, two numerical examples are investigated. The objects of the numerical simulations are illuminated by a normally incident theta-polarized plane wave from $\theta_{in}=180°$, $\varphi_{in}= 0°$, and the frequency range starts from 0.1GHz and terminates at 0.3GHz. We set δ= 0.01 for the IASA and chose N_{PW} = 800 for UCBFM. All the simulations were run on a notebook equipped with 2 Dual Core at 2.3GHz (only one core was used) and 8GB of RAM.

The first example is the monostatic RCS by a PEC ellipsoid shown in Fig.3. The geometry is automatically divided into four blocks and the discretisation in triangular patches involves

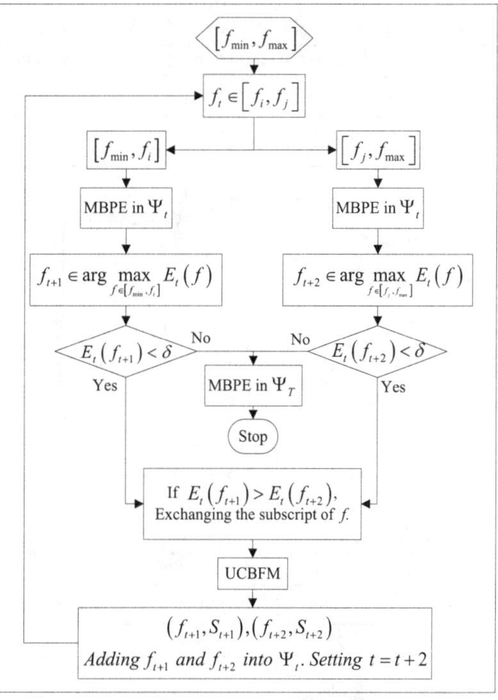

Fig 2. Flowchart of UCBFM/IA-MBPE technique.

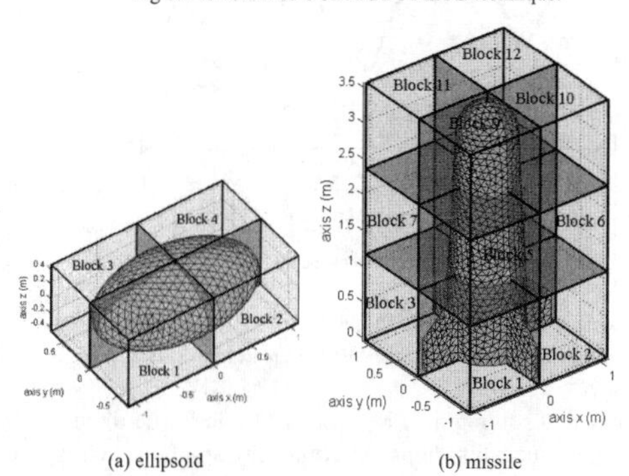

(a) ellipsoid (b) missile

Fig 3. Geometries of two examples.

almost 2115 unknowns. After SVD procedure we totally obtain 283 UCBFs. The results are compared with those derived by using UCBFM and direct MoM, as shown in Fig.4. The monostatic RCS calculated by UCBFM/IA-MBPE coincide very well with direct MoM. Hence, the presented method is accurate in wide band electromagnetic scattering analysis.

With a frequency increment of 1MHz, we obtain 201 sample points for direct MoM, while only 22 sample points are needed for UCBFM/IA-MBPE. To show the efficiency of the presented method, the total simulation time which include matrix fill-

TABLE I
COMPUTATIONAL TIMES FOR THE DIFFERENT METHODS OF AN ELLIPSOID

	MoM	UCBFM	MoM/IA-MBPE	UCBFM/IA-MBPE
Total time	8.008 h	7.764 h	1.137 h	0.846 h
Saving	—	3.05%	85.80%	89.44%

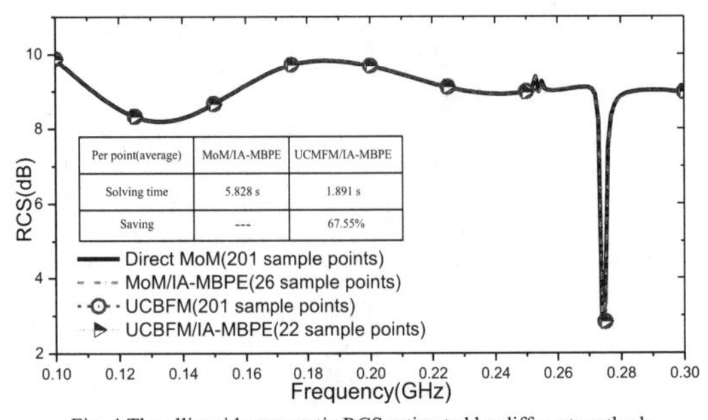

Fig. 4 The ellipsoid monostatic RCS estimated by different methods.

TABLE II
COMPUTATIONAL TIMES FOR THE DIFFERENT METHODS OF A MISSILE

	MoM	UCBFM	MoM/IA-MBPE	UCBFM/IA-MBPE
Total time	35.079 h	27.986 h	11.689 h	8.306 h
Saving	—	20.22%	66.68%	76.33%

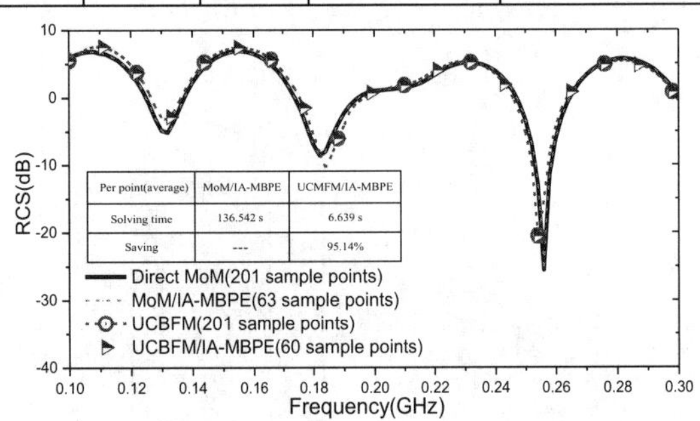

Fig. 5 The missile monostatic RCS estimated by different methods.

ing time and solving time is shown in Table I. The direct MoM requires about 8.008 hours to obtain the solution, While only 0.846 hours are needed for UCBFM/IA-MBPE.

The second example is a PEC missile shown in Fig.3. The geometry is automatically divided into twelve blocks, and the discretisation in triangular patches leads to total unknowns of 4263. After SVD procedure we obtain 765 UCBFs. Fig.5 shows the monostatic RCS by the conductive missile. It can be seen from this figure that the calculated results from UCBFM/IA-MBPE has a good agreement with those of the point to point calculation by direct MoM.

Table II shows the efficiency of the UCBFM/IA-MBPE method. Over the whole frequency band, the total number of the sample points for UCBFM/IA-MBPE is 60, while the direct MoM need 201 sample points. The computational time is reduced from 35.079 hours to 8.306 hours.

The comparison between MoM/IA-MBPE and UCBFM/IA-MBPE shows that the superiority of UCBFM/IA-MBPE over IAMBPE reveals with the increase of the size of the scatterer indicating that UCBFM/IA-MBPE is most suitable for the solving of wide band scattering problems with electrically large objects.

IV. CONCLUSION

In this paper, an efficient approach that combines the UCBFM with IA-MBPE is successfully implemented to fast and efficiently analyze wide band scattering problems. Numerical results demonstrate the high accuracy and efficiency of the proposed hybrid method. The scattering problem of electrically very large and complex objects can be handled, since UCBFM is able to reduce the size of the MoM matrix for fast solving. The IA-MBPE was used in order to further speed up the wide band analysis. Hence, the hybrid method takes both the advantages of UCBFM and IA-MBPE and therefore can reduce considerable total computational time and memory requirements in wide band and electrically large size problems.

REFERENCES

[1] R.H. Harrington, *Field Computation by Moment Methods*, New York: Macmillan, 1968, pp.49-70.

[2] V.V.S. Prakash and R. Mittra, "Characteristic Basis Function Method: A new technique for efficient solution of method of moments matrix equation," *Microwave Opt. Technol. Lett.*, vol.36, no.2, pp.94-100, Feb. 2003.

[3] E. Lucente, A. Monorchio and R. Mittra, "An iteration free MoM approach based on excitation independent characteristic basis functions for solving large multiscale electromagnetic scattering problems," *IEEE Trans. Antennas Propag.*, vol.58, no.7, pp.999-1007, Apr. 2008.

[4] S.M. Rao, D.R. Wilton and A.W. Glisson, "Electromagnetic scattering by surfaces of arbitrary shape," *IEEE Trans. Antennas Propag.*, vol.30, no.3, pp.409-418, Feb. 1982.

[5] E. K. Miller, "Model-based parameter estimation in electromagnetic-I: Background and theoretical development," *Appl. Comput. Electromagn. Soc. Newslett.*, vol.10, no.3, pp.40-63, 1995.

[6] E. K. Miller, "Model-based parameter estimation in electromagnetic-II: Applications to EM observables," *Appl. Comput. Electromagn. Soc. Newslett.*, vol.11, no.1, pp.35-56, 1996.

[7] E. K. Miller, "Model-based parameter estimation in electromagnetic-III: Applications to EM integral equations," *Appl. Comput. Electromagn. Soc. Newslett.*, vol.10, no.3, pp.9-29, 1996.

[8] G.-D. Han, Y.-H. Pan, B.-F. He and C.-Q, Gu, "Fast analysis for 3-D wide band and wide angle electromagnetic scattering characteristic by AMCBFM-MBPE," *Journal of Microwaves(CHINA)*,vol.25,no.6, pp.32-37, Jun. 2009.

[9] R. Lehmensiek and P. Meyer, "An efficient adaptive frequency sampling algorithm for model-based parameter estimation as applied to aggressive space mapping," *Microwave Opt. Technol. Lett.*, vol.24, no.1, pp.71-78, Jan. 2000.

[10] J.-X. Wan, Y. Zhang, T.-M. Xiang and C.-H. Liang, "Rapid solutions of monostatic RCS using FMM with adaptive MBPE technique," *Chinese Journal of Radio Science*, vol.19, no.1, pp.72-76, Feb. 2004.

[11] M.D. Gregorio, G. Tiberi, A. Monorchio and R. Mittra, "Solution of wide band scattering problems using the characteristic basis function method," *IET Microw. Antennas Propag.*, vol.6, no.1, pp.60-66, Jan. 2012.

[12] K. Konno, Q. Chen, K. Sawaya and T. Sezai, "Optimization of block size for CBFM in MoM," *IEEE Trans. Antennas Propag.*, vol.60, no.10, pp.4719-4724, Jan. 2012.

[13] T.J. Rivlin, *An introduction to the approximation of functions*, Dover, New York, 1969.

Analysis of Millimeter-Wave Exposure on Rabbit Eye Using a Hybrid PMCHWT-MoM-FDTD Method

Jerdvisanop Chakarothai [†], Yukihisa Suzuki [†], Masao Taki [†], Kanako Wake [‡], Kensuke Sasaki [‡], Soichi Watanabe [‡] and Masami Kojima [§]
† Tokyo Metropolitan University, 1-1 Minami-Osawa, Hachioji-shi, Tokyo, Japan 192-0397
‡ National Institute of Information and Communication, 4-2-1 Nukui-Kitamachi, Koganei, Tokyo, Japan 184-0015
§ Kanazawa Medical University, 1-1 Daigaku, Uchinada-machi, Ishikawa, Japan 920-0293
Email: † jerd@tmu.ac.jp

***Abstract*-In order to investigate thresholds of biological effects due to millimeter-wave (MMW) exposure, A hybrid method, which combines the Poggio-Miller-Chang-Harrington-Wu-Tsai method of moments (PMCHWT-MoM) and the finite-difference time-domain (FDTD) method, has been developed and applied to analysis of millimeter-wave exposure on rabbit eye using dielectric lens antenna as an electromagnetic source. The PMCHWT MoM has been validated by comparison of calculated and measured electric fields radiated from the dielectric lens antenna. Then, the hybrid PMCHWT-MoM-FDTD method is used to determine specific absorption rate due to MMW exposure on rabbit eye, which was placed at the focal point of the antenna. It was found that the dielectric lens antenna with an input power of 10 mW produces a peak SAR of 86 W/kg in the eye cornea at 35 GHz.**

I. INTRODUCTION

With recent advances in electromagnetic (EM) field technology, the application of millimeter-waves (MMWs) is increasing in daily life, ranging from wireless communications to intelligent transport system such as automobile collision-prevention radar. The rapid development of these new technologies increases public concern about possible biological effects due to EM exposure in millimeter-wave range.

Safety guidelines on EM field exposures published by various organization, including the International Committee on Electromagnetic Safety (ICES) of the Institute of Electrical and Electronics Engineers (IEEE) [1] and the International Commission on Non-Ionizing Radiation Protection (ICNIRP) [2], cover the frequency range of MMWs, i.e., up to 300 GHz. These guidelines are based on established scientific evidence. Few studies, however, address the rationale for the exposure limits in MMW band. The limit values for MMW exposure were derived primarily from extrapolations of experimental data for frequencies up to several GHz, or from the effects of infrared radiation. To confirm the scientific basis for the exposure limit in the MMW region, it is necessary to perform experiments to obtain reliable evidences in this region.

Experimental research in the MMW region, which occupies a broad frequency range from 30 GHz to 300 GHz, however, is difficult from the practical viewpoint due to limit number of animal individuals used in experiments and availability of EM sources in this frequency region. Therefore, the simulation based on actual physical model for calculating power absorption of EM waves into biological bodies is needed.

In this paper, a hybrid method, which combines the Poggio-Miller-Chang-Harrington-Wu-Tsai (PMCHWT) formulation of the method of moments (MoM) [3] and the finite-difference time-domain (FDTD) method [4], is proposed to analyze an exposure system using dielectric lens antenna as an EM source in a MMW range. We are investigating the relationship between injuries for ocular tissues and exposure level of MMW by using rabbit eyes. Since biological effects for MMW exposure is mainly due to EM power absorbed at surface of biological bodies and eyes are considered most vulnerable to MMW [5], [6]. Hence, bare rabbit eye model is employed for our target of numerical simulation. This paper is organized as follows: First, we briefly describe a hybridization of the PMCHWT-MoM and the FDTD method. The PMCHWT MoM is then experimentally validated for a dielectric lens antenna model. Finally, we use the hybrid PMCHWT-MoM-FDTD method to calculate the specific absorption rate (SAR) in rabbit eye by MMW exposure at 35 GHz.

II. HYBRIDIZATION OF PMCHWT-MOM AND FDTD METHOD

The concept of the hybrid PMCHWT-MoM-FDTD method is illustrated in Fig. 1. There are two analysis regions; one is the MoM region including homogeneous dielectrics and metal structures, and the other is the FDTD region including only heterogeneous bodies, e.g. biological bodies. The PMCHWT MoM is applied in the MoM region in order to solve a radiation/scattering problem including dielectric objects. $\mathbf{J}_a$, $\mathbf{J}_d$, and $\mathbf{M}_d$ are electric current density residing on an antenna, electric and magnetic current densities residing on a dielectric structure, respectively. Electric and magnetic current densities on a closed surface S_{JM} surrounding an object in the FDTD region are denoted by $\mathbf{J}_s$ and $\mathbf{M}_s$, respectively. S_{JM} can be chosen arbitrarily as long as it totally encompasses the object. Calculation procedures of the PMCHWT-MoM-FDTD method are as follows:

1) Apply an incident voltage at the antenna terminal, and then $\mathbf{J}_a$, $\mathbf{J}_d$, and $\mathbf{M}_d$ are determined from the MoM procedures. It is assumed that the exposed subject is outside the MoM region and excluded from the calculation.

978-7-5641-4279-7

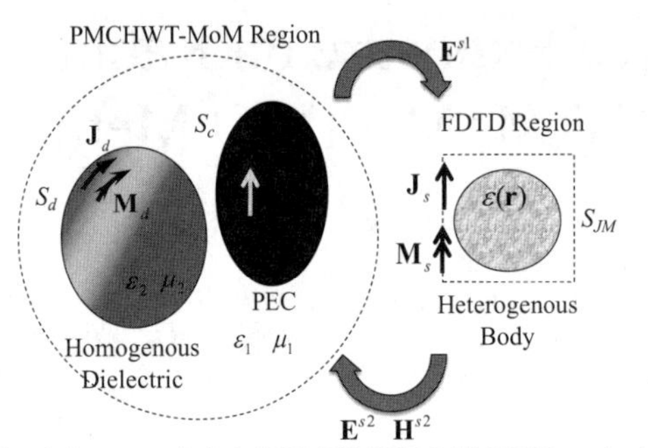

Fig. 1 Concept of a hybrid PMCHWT-MoM-FDTD method.

2) Radiated/scattered electric fields $\mathbf{E}^{s1}$ is calculated for observation points in the FDTD region. Since the solutions of the MoM are found in the frequency domain, we then transform them into the time-domain sinusoidal waveform in order to apply them to the FDTD method as the incident field.

3) Scattered electric and magnetic fields are calculated with the FDTD method. The structures in the MoM region are excluded from the calculation in this step and the perfectly matched layer (PML) is applied at the border of computation space to absorb the outgoing EM fields. The SAR is also calculated from

$$\mathrm{SAR} = \frac{\sigma|\mathbf{E}|^2}{2\rho}, \qquad ...(1)$$

where $\mathbf{E}$ is the peak values of the total electric field (V/m), σ and ρ represent the conductivity (S/m) and the mass density of the tissue (kg•m^{-3}), respectively.

4) $\mathbf{J}_s$ and $\mathbf{M}_s$ on a closed surface surrounding the exposed object are obtained from scattered electric and magnetic fields using the equivalence principle:

$$\mathbf{J}_s = \hat{n} \times \mathbf{H}^s, \mathbf{M}_s = \mathbf{E}^s \times \hat{n}. \qquad ...(2)$$

Since the cell size in the FDTD method is small, actually on the order of one-tenth to -twentieth of wavelength, electric and magnetic dipole moments on a face of the Yee's cell can be considered infinitesimal and can be expressed as

$$\mathbf{j}_p = \mathbf{J}_s(\mathbf{r}_p')\Delta s, \mathbf{m}_p = \mathbf{M}_s(\mathbf{r}_p')\Delta s, \qquad ...(3)$$

where $\mathbf{r}_p'$ is defined as the center coordinate of the small area Δs which corresponds to a face area of the Yee's cell. The scattered electric and magnetic fields at an observation point outside the volume enclosed by S_{JM} can be analytically computed from the following expressions:

$$\mathbf{E}^{s2}(\mathbf{r}) = -jk_0 \sum_{p=1}^{N_d} \frac{e^{-jk_0R_p}}{4\pi R_p} \Big\{ \eta_0(\mathbf{j}_p - \mathbf{p}_p) + \frac{\eta_0 D}{jk_0}(\mathbf{j}_p - 3\mathbf{p}_p) - D(\mathbf{R}_p \times \mathbf{m}_p) \Big\}, \qquad ...(4)$$

$$\mathbf{H}^{s2}(\mathbf{r}) = -jk_0 \sum_{p=1}^{N_d} \frac{e^{-jk_0R_p}}{4\pi R_p} \Big\{ \frac{1}{\eta_0}(\mathbf{m}_p - \mathbf{p}_p') + \frac{D}{jk_0\eta_0}(\mathbf{m}_p - 3\mathbf{p}_p') + D(\mathbf{R}_p \times \mathbf{j}_p) \Big\}, \qquad ...(5)$$

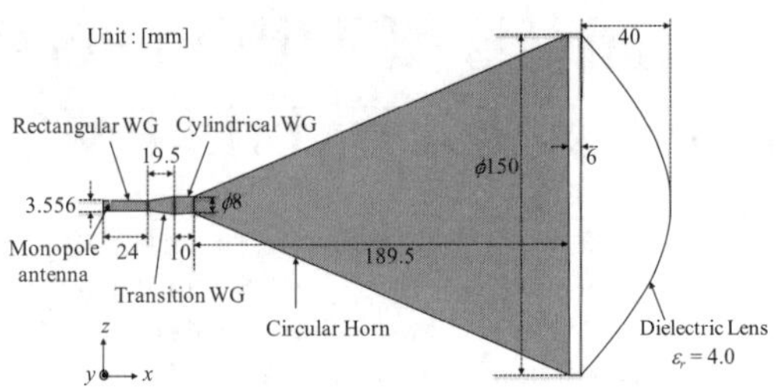

Fig. 2. Structure of the dielectric lens antenna.

$$\mathbf{p}_p = \frac{(\mathbf{R}_p \cdot \mathbf{j}_p)\mathbf{R}_p}{R_p^2}, \mathbf{p}_p' = \frac{(\mathbf{R}_p \cdot \mathbf{m}_p)\mathbf{R}_p}{R_p^2}, D = \frac{1}{R_p}\left(1 - \frac{j}{k_0R_p}\right),$$

$$\mathbf{R}_p = \mathbf{r} - \mathbf{r}_p,$$

where N_d is a total number of dipole moments.

5) Induced voltages on every edge element in the MoM region are calculated using electric and magnetic fields in (4) and (5) as incident fields. Since they are treated as additional sources in the PMCHWT MoM, $\mathbf{J}_a$, $\mathbf{J}_d$, and $\mathbf{M}_d$ obtained in this step must be added to those obtained in the previous iterations. Steps 2-5 are then repeated until the convergence is reached.

III. Experimental Validation of Dielectric Lens Antenna

A. Structure of Dielectric Lens Antenna

Fig. 1 shows the structure of dielectric lens antenna used in numerical analysis. The dielectric lens antenna is composed of rectangular waveguide (WG) with a monopole antenna inside, transition WG, cylindrical WG, circular horn, and dielectric lens. Aperture size of the rectangular WG for Ka band (26.5 GHz to 40 GHz) is 7.112 mm × 3.556 mm. The monopole antenna has a length of quarter wavelength at 35 GHz, i.e., 2.143 mm. An aperture diameter of the cylindrical WG is 8 mm and the tapered part of the horn has a length of 189.5 mm. An aperture diameter of the horn at the open end is 150 mm. Relative permittivity and conductivity of the dielectric lens is 4.0 and 0.001 S/m, respectively. All components are put together in a serial connection as shown in Fig. 1. Analysis frequency is fixed to 35 GHz for all numerical simulations and experiments.

The horn antenna, including the rectangular, transition, and cylindrical waveguides was discretized into 31060 triangular patches while there are 38598 triangular patches for the dielectric lens. Total number of unknowns in the model is 123,659. The PMCHWT MoM code developed in-house was parallelized by adopting the OpenMP interface software platform. The parallelized version of the LU decomposition in the Intel Mathematical Kernel Library (IMKL) was used to accelerate the calculation. All numerical analyses were performed on a high-performance computer with 48 cores (CPU Opteron 6174) in our laboratory. Calculation time for the model was approximately 4 hours and 40 minutes using about 256 GBytes.

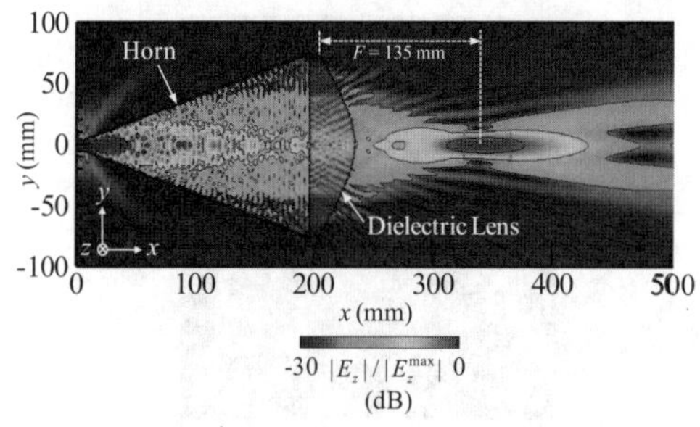

Fig. 3. Near E-field distribution in the xy-plane at 35 GHz.

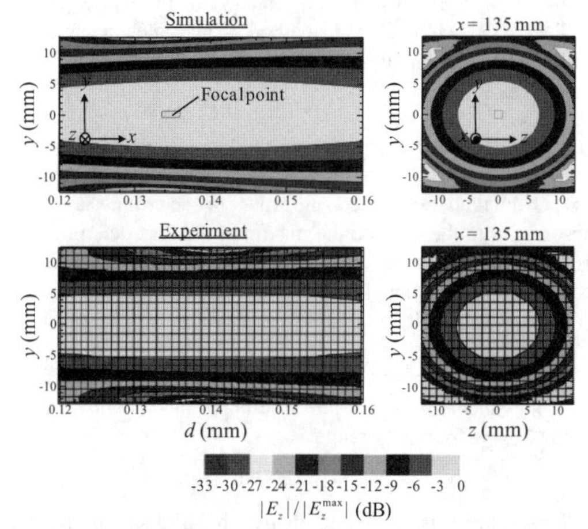

Fig. 4. Comparison of normalized electric field distribution at the focal point of dielectric lens in the xy-plane (left) and yz-plane (right) at 35 GHz. d is the distance from the base of dielectric lens to an observation point.

B. *Numerical Analysis and Experimental Validation*

First, we calculate electric field distribution created by the dielectric lens antenna. Fig. 3 shows the normalized electric field strength in the xy-plane. The focusing behavior of EM waves by dielectric lens can be clearly observed. Focal point of the dielectric lens is then determined as d = 135 mm with a beam width of 10.5 mm × 12.8 mm (half power value), where d is the distance from the base of the dielectric lens. The input impedance of the monopole antenna is 37.12 – j51.06 Ω. The maximum electric field strength, and power density at the focal point are determined as 86.69 V/m and 2.00 mW/cm^2, respectively, for an antenna input power of 4.66 mW. The electric field distribution was also measured by using near-field measurement system (NSI2000, Nearfield System, Inc.). Fig. 4 shows the normalized electric field strength at the focal point in the xy-plane and yz-plane, calculated by the PMCHWT MoM and measured experimentally, respectively. It can be seen that the numerical results are almost identical to those of the experiment. Fig. 5 shows the electric field strength on the

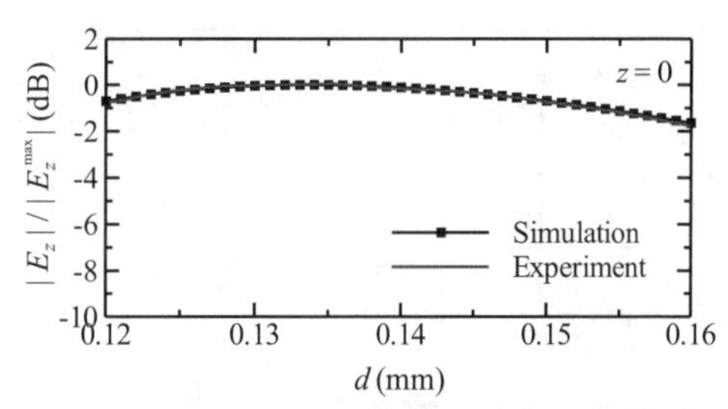

Fig. 5. Comparison of normalized electric field distribution, on the x-axis, calculated by the PMCHWT MoM and that obtained from the experiments.

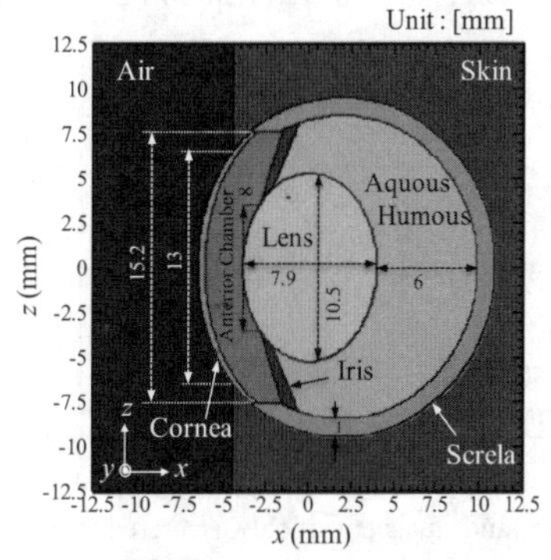

Fig. 6. Structure of eye used in numerical analysis

x-axis when z = 0. Again, the numerical result shows good agreement with the experiment, demonstrating the validity of our PMCHWT MoM.

IV. Numerical Experiment of Millimeter-wave Exposure on Rabbit Eye

The validity of the PMCHWT MoM has been demonstrated in comparison with the measured electric fields. We then combine the PMCHWT MoM with the FDTD method in order to analyze millimeter-wave exposure on rabbit eye. Since eye has a complex and fine structure as shown in Fig. 6, we have modeled the eye in the FDTD region with a spatial grid size of 0.05 mm in each axis. The eye is placed at the focal point of the dielectric lens. There are 7 types of tissues used in the eye model and their dielectric constants are tabulated in Table I. Dimensions of the FDTD region are 25 mm × 25 mm × 25 mm. The 8-layer perfectly matched layer (PML) is utilized to absorb the outgoing wave in order to truncate the FDTD region. The time step size is 96.29 fs and the number of time steps is 2000. Calculation time for the eye model was approximately 2 hours and 40 minutes using about 62 GBytes.

First, the radiated/scattered electric fields calculated from the PMCHWT MoM, when only the dielectric lens antenna exists in the analysis region, are used as incident fields in the FDTD

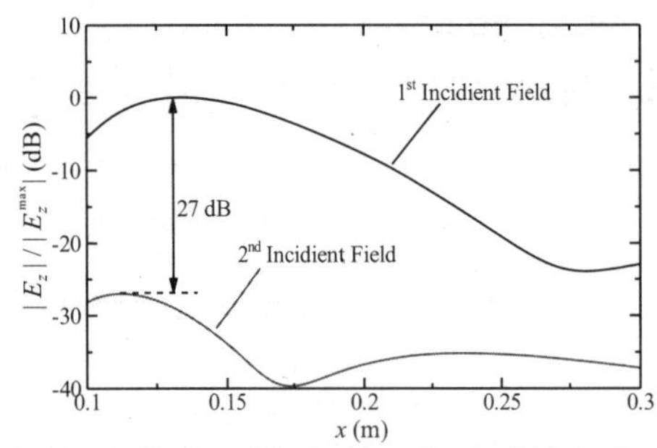

Fig. 7. Distribution of the incident electric fields in the 1st and 2nd iterations of the PMCHWT-MoM-FDTD method.

TABLE I
DIELECTRIC PROPERTIES OF EYE'S TISSUES AT 35 GHZ

	ε_r	σ(S/m)	ρ(kg·m^{-3})
Dry skin	13.34	29.76	1100
Eye sclera	18.87	38.69	1032
Cornea	18.39	37.77	1051
Vitreous humor	22.40	58.53	1005
Lens	16.46	32.34	1076
Anterior chamber	22.41	58.53	1005
Iris	20.42	39.65	1090

region. The SAR is then calculated using the FDTD method. The electric and magnetic fields scattered from the eye, calculated using (4) and (5) are used as a new source for the PMCHWT MoM. We repeated the calculation procedure until the convergence of the SAR is reached. As a result, only two iterations is needed in order to satisfy the convergence condition, since the coupling between the eye and the dielectric lens antenna is quite small. Fig. 7 shows the incident electric field distribution on the x-axis in each iteration step. For the 2nd iteration step, the maximum of the electric field strength is approximately 27 dB less than that in the 1st iteration step. Therefore, only one iteration would be sufficient for simulating our numerical exposure experiment

Finally, the SAR distribution calculated by using the hybrid PMCHWT-MoM-FDTD method after two iterations is shown in Fig. 8. The maximum SAR is approximately 40.1 W/kg for an antenna input power of 4.66 mW. As can be seen from the figure, the power absorption of EM wave occurred only nearby the eye surface where it is exposed to the air region. However, the temperature rise due to the MMW exposure would cause a thermal convection inside the anterior chamber of the eye and this may induce the damages inside the eye [7]. This phenomenon will be taken into account in our future research.

V. CONCLUSIONS

In this paper, a hybrid method combining the PMCHWT-MoM and the FDTD method has been proposed to simulate a situation of the MMW exposure on the eye at 35 GHz. The validity of the PMCHWT-MoM has been confirmed by

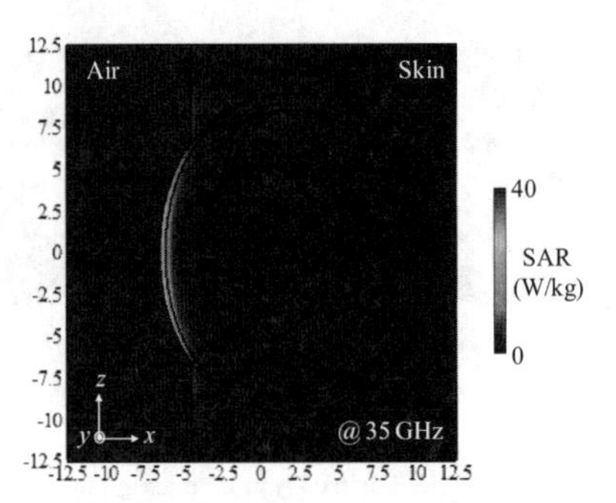

Fig. 8. SAR distribution calculated by using the PMCHWT-MoM-FDTD method. The antenna input impedance is 35.60 – j50.55 Ω and the input power is 4.66 mW.

comparing the calculated electric field distributions with those obtained from the experiment. Then, the PMCHWT-MoM-FDTD method was used in order to determine a power absorbed into the eye and the maximum SAR is determined as 86 W/kg for an antenna input power of 10 mW. It was also confirmed that the coupling between the dielectric lens antenna and the eye is small so that only one iteration will be sufficient for analyses.

A future subject is to include the thermal convection effects into our model and determine thresholds of biological effects in the MMW frequency range.

ACKNOWLEDGMENT

This study was supported in part by Ministry of Internal Affairs and Communications, Japan.

REFERENCES

[1] IEEE International Committee on Electromagnetic Safety, IEEE standard for safety levels with respect to human exposure to radio frequency electromagnetic fields, 3 kHz to 300 GHz, Piscataway, NJ: IEEE; IEEE Std C95.1-2005; 2005.

[2] International Commission on Non-Ionizing Radiation Protection, Guidelines for limiting exposure to time-varying electric, magnetic, and electromagnetic fields (up to 300 GHz), *Health Phys.*, vol. 74, pp. 494-522; 1998.

[3] E. Arvas, A. Rahhal-Arabi, A. Sadigh, and S. M. Rao, "Scattering from multiple conducting and dielectric bodies of arbitrary shape," *IEEE Trans. Antennas Propag. Mag.*, vol. 33, no. 2, pp. 29-36, 1991

[4] A. Taflove, *Computational Electrodynamics: The Finite-Difference Time-Domain Method*, Norwood, MA: Artech House, 1995.

[5] S. W. Rosenthal et. al., *Biological effects of electromagnetic waves vol.1*, Johnson C. C. & Shore M. L., Ed., HEW Publication (FDA) 77-8010, Rockvill, Maryland, 110-128.

[6] H. A. Kues, S. A. D. Anna, R. Osiander, W. R. Green, and J. C. Monahan, "Absence of ocular effects after either single or repeated exposure to 10 W/cm^2 from a 60 GHz CW source," *Bioelectromagnetics*, vol. 20, no. 8, pp. 463-473, 1999.

[7] M. Kojima, M. Hanazawa, Y. Yamashiro, H. Sasaki, S. Watanabe, M. Taki, Y. Suzuki, A. Hirata, Y. Kamimura, and K. Sasaki, "Acute ocular injuries caused by 60 GHz millimeter-wave exposure," *Health Physics*, vol. 97, no. 3, pp. 212-218, 2009.

Time Domain Integral Equation Method Using FFT-Based Marching-On-in-Degree Method for Analyzing PEC Patches On Substrate

Jian-Yao Zhao[1], Wei Luo[2], and Wen-Yan Yin[1,2]
[1]Centre for Optical and EM Research (COER), State Key Lab of MOI, Zhejiang University
Hangzhou 310058, CHINA. zhaojianyao@zju.edu.cn
[2]Key Lab of Ministry of Education of Design and EMC of High-Speed Electronic Systems,
School of Electronic Information and Electrical Engineering, Shanghai Jiao Tong University
Shanghai 200240, CHINA. wyyin@zju.edu.cn and longdi312@sjtu.edu.cn

***Abstract*-One time domain method, based on the TDIE solved by the marching-on-in-degree (MOD) scheme, is presented in this paper for capturing transient electromagnetic responses of dielectric-metallic composite structures. For perfect electrically conducting (PEC) surfaces patched on the dielectric substrate, the time-domain electric field integral equations (TD-EFIE) is used to analyze PEC patch, while PMCHW (Poggio, Miller, Chang, Harrington and Wu) integral equation is utilized for describing the dielectric substrate. The whole set of integral equations is solved simultaneously, and the temporal basis function is chosen to be the Laguerre polynomials (LP), which is called marching-on-in-degree scheme. For the spatial basis function, the classical RWG basis function is usually the suitable choice. Since in the right-hand-side of the iteration equation, during the matrix-vector multiplication, the impedance matrix has the form of Toeplitz matrix, the fast Fourier transform (FFT) can be used to accelerate this multiplication. Numerical results are given to demonstrate our proposed method used for analyzing characteristics of structures containing PEC patch on the substrate.**

I. INTRODUCTION

Time-domain integral equation can be widely used for analyzing transient scattering responses of some classic electromagnetic structures [1-2]. Time domain electric field integral equations (TD-EFIE) are used for analyzing perfect electrically conducting surface, while PMCHW (Poggio, Miller, Chang, Harrington and Wu) integral equation is very suitable for dealing with homogeneous dielectric structures.

As for TDIE formulation, there are usually two schemes to solve it, including marching-on-in-time (MOT) method and marching-on-in-degree (MOD) method [3-5]. Both schemes have their strengths and weaknesses. MOT methods are prone to late time instabilities while MOD schemes appear highly stable due to the high-order limit. Both MOT and MOD methods suffer from a high computational cost. Since MOD is stable even in very late time, it is adopted to solve the integral equation in this paper. In the MOD scheme, the weighted Laguerre polynomials form a set of the temporal basis function, by which the temporal currents are represented together with the spatial Rao-Wilton-Glisson (RWG) basis function.

What is more, the computational cost of MOD scales as $O(N_O^2 N_S^2)$, where N_S is the number of spatial unknowns, and N_O is the highest order of the LP. A fast Fourier transform (FFT)-based blocking scheme [4] can be used for accelerating the temporal convolutions, and the computational complexity is reduced to $O(N_S^2 N_O \log^2(N_O))$.

As we discuss about dielectric-metallic composite structures, the patch structure is one classical part of them [6-7]. The PEC patch is directly attached to the dielectric substrate, so in the analysis using time domain integral equations, the dielectric-metallic junctions must be treated carefully in order to obtain high accuracy. One of the methods is the contact-region modelling (CRM) technique [8], and it supposes the dielectric-metallic surface to be two surfaces infinitesimally close to each other. However, this increases the number of unknowns because one more surface needs to be considered. Another effective way is to use triangular half-basis functions based on the full RWG basis to treat the current on the junction [9].

The organization of this paper is arranged as follows. In Section II, the basic knowledge of the electromagnetic problem and mathematical treatment is presented. The TDEFIE-PMCHW equation is solved using MOD scheme, while the FFT is used to accelerate the order iteration of MOD. In Section III, numerical results of some patch structures are given to demonstrate the capabilities of our proposed method. Then the conclusion is made in Section IV.

II. BASIC THEORY

A. Description of The Geometry

Figure 1 shows a generalized composite structure in free space, including homogeneous dielectric objects and PEC patches, both of arbitrary shape. The structure is illuminated by an electromagnetic pulse (EMP) with an arbitrary magnitude, polarization and incident direction. From Fig.1, it is shown that one PEC patch is directly attached on the dielectric substrate, and there are two metallic surfaces need to be considered, including the one toward the free space and the other one toward the inner region of the dielectric [5]. The dielectric surface is denoted by S^D, while the PEC surfaces toward the free space and the dielectric are denoted by S^P and S^I, respectively.

978-7-5641-4279-7

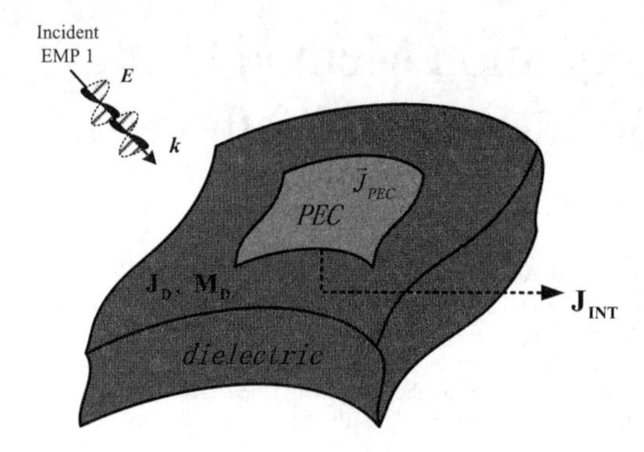

Fig. 1. Description of the simulated dielectric-metallic patch structure.

B. TDIE Formulation

Under the EMP illumination, there are induced electric and magnetic currents on the dielectric surface, which are denoted by $\mathbf{J_D}$ and $\mathbf{M_D}$, respectively. For the PEC object, only induced electric currents $\mathbf{J_{PEC}}$ and $\mathbf{J_{INT}}$ on the surfaces S^P and S^I, respectively.

In the presence of electric and magnetic current sources, the scattered fields of the dielectric object can be described by

$$\mathbf{E}_\mathbf{v}^{\mathbf{sca}}(\mathbf{r},t) = L_v(\mathbf{J_v}) - K_v(\mathbf{M_v}) \quad (1)$$

$$\mathbf{H}_\mathbf{v}^{\mathbf{sca}}(\mathbf{r},t) = L_v(\mathbf{M_v}) + K_v(\mathbf{J_v}) \quad (2)$$

$$L_v(\mathbf{I}) = \frac{-\alpha_v}{4\pi}\int_S \frac{1}{R}\frac{\partial}{\partial t}\mathbf{I}(\mathbf{r},\tau_v)\,dS' + \frac{1}{4\pi\beta_v}\nabla\int_S\int_0^\tau \frac{1}{R}\nabla'\cdot\mathbf{I}(\mathbf{r},t_v)\,dt\,dS' \quad (3)$$

$$K_v(\mathbf{I}) = \frac{1}{4\pi}\nabla\times\int_S \frac{1}{R}\mathbf{I}(\mathbf{r},\tau)\,dS' \quad (4)$$

where $\mathbf{I}$ represents either electric current or magnetic current. v='*1*' and '*2*' represent exterior and interior domains of the dielectric, respectively. So the PEC currents $\mathbf{J_{PEC}}$ and $\mathbf{J_{INT}}$ belong to domain '*1*' and domain '*2*', respectively. $R = |\mathbf{r}-\mathbf{r}'|$ is the distance between an arbitrary observation point $\vec{r}$ and the source one $\vec{r}'$. $\tau_v = t - R/c_v$ is the retarded time. When $\mathbf{I}$ is electric current, $\alpha_v = \mu_v$, $\beta_v = \varepsilon_v$, and for the magnetic current $\mathbf{I}$, they are inverse.

As for dielectric, the current satisfies the property that $\mathbf{J_{D,1}} = -\mathbf{J_{D,2}}$ and $\mathbf{M_{D,1}} = -\mathbf{M_{D,2}}$, so we define $\mathbf{J_{D,1}} = \mathbf{J}_D$ and $\mathbf{M_{D,1}} = \mathbf{M_D}$ for simplicity. Therefore, by applying boundary conditions on the dielectric surface and the two PEC surfaces, we obtain a set of time-domain integral equations for the problem as shown in Fig. 1:

$$-L_2(\mathbf{J_D}) + K_2(\mathbf{M_D}) - L_1(\mathbf{J_D}) + K_1(\mathbf{M_D}) - L_1(\mathbf{J_{PEC}}) + L_2(\mathbf{J_{INT}}) = \mathbf{E^{inc}}(\mathbf{r},t)|_{\tan}, \quad \forall\mathbf{r}\in S^D \quad (5)$$

$$-L_2(\mathbf{M_D}) - K_2(\mathbf{J_D}) - L_1(\mathbf{M_D}) - K_1(\mathbf{J_D}) - K_1(\mathbf{J_{PEC}}) + K_2(\mathbf{J_{INT}}) = \mathbf{H^{inc}}(\mathbf{r},t)|_{\tan}, \quad \forall\mathbf{r}\in S^D \quad (6)$$

While for the metallic surface S_{PEC}, we have

$$-L_1(\mathbf{J_D}) + K_1(\mathbf{M_D}) - L_1(\mathbf{J_{PEC}}) = \mathbf{E^{inc}}(\mathbf{r},t)|_{\tan}, \forall\mathbf{r}\in S^P \quad (7)$$

And for the interface S_{INT}, we have

$$L_2(\mathbf{J_D}) - K_2(\mathbf{M_D}) - L_2(\mathbf{J_{INT}}) = 0, \quad \forall\mathbf{r}\in S^I \quad (8)$$

C. FFT-based MOD Procedure

A set of weighted Laguerre polynomials $L_j(st)$ forms the temporal basis function, as given by $\phi_j(st) = e^{-st/2}L_j(st)$, where j is its order, and s is the scaling factor. Then, the electric and magnetic currents expanded in general as follows

$$\mathbf{I}(\mathbf{r},t) = \sum_{n=1}^{N_S}\frac{\partial}{\partial t}\left[\sum_{j=0}^{\infty}\gamma_{n,j}\phi_j(st)\right]\mathbf{f}_n^S(\mathbf{r}) \quad (9)$$

where $\mathbf{f}_n^S(\mathbf{r})$ is the RWG basis function [10] based on a pair of triangles, *i.e.*

$$\mathbf{f}_n^S(\mathbf{r}) = \mathbf{f}_n^+(\mathbf{r}) + \mathbf{f}_n^-(\mathbf{r}) \quad (10a)$$

$$\mathbf{f}_n^\pm(\mathbf{r}) = \begin{cases} \pm\dfrac{l_n}{2A_n^\pm}\boldsymbol{\rho}_n^\pm, & \mathbf{r}\in T_n^\pm \\ 0, & otherwise \end{cases} \quad (10b)$$

And here N_S denotes the number of edges on the surfaces S^D, S^P and S^I, respectively. $\mathbf{I}(\mathbf{r},t)$ can be $\mathbf{J_D}$, $\mathbf{M_D}$, $\mathbf{J_{PEC}}$ and $\mathbf{J_{INT}}$, corresponding to different current coefficients $\gamma_{n,j}$.

In the final order iteration equation, the temporal convolutions on the right-hand side can be accelerated by the applying FFT, since it satisfies the properties of Toeplitz matrix. Thus the LP order iteration can be accelerated.

III. Numerical Results and Discussion

Here, a temporal modulated Gaussian pulse is used as the incident wave and described by

$$\mathbf{E_{inc}}(\mathbf{r},t) = \mathbf{E_0}(4/\sqrt{\pi}/T)e^{-\gamma^2}\cos(2\pi f_0 t) \quad (30)$$

$$\gamma = 4\left(c_0 t - c_0 t_0 - \mathbf{r}\cdot\hat{\mathbf{k}}\right)/T \quad (31)$$

where E_0 is the amplitude of the incident EMP, $\hat{\mathbf{k}}$ is its wave vector, f_0 is its central frequency, c_0 is its velocity in free space, T is its pulse width, and c_0t_0 is its time delay.

The first structure consists of one dielectric hemisphere and one circular PEC patch just directly attached on the hemisphere as shown in Fig.2. The relative permittivity is ε_r=4.4, and the radius of the composite hemisphere is set to be R=0.2m. It is illuminated by the incident EMP with f_0=800MHz, $\hat{k} = -\hat{z}$ and its polarization chosen along the $+\hat{x}$ axis. The dielectric is meshed by 862 edges of unknowns, and the PEC surface S^P and S^I have both 485 unknowns. Particularly, there are 40 edges at the junction.

Fig.2 shows the calculated bi-static radar cross section (RCS) at f=800 MHz, which is also compared with the results from the commercial software FEKO.

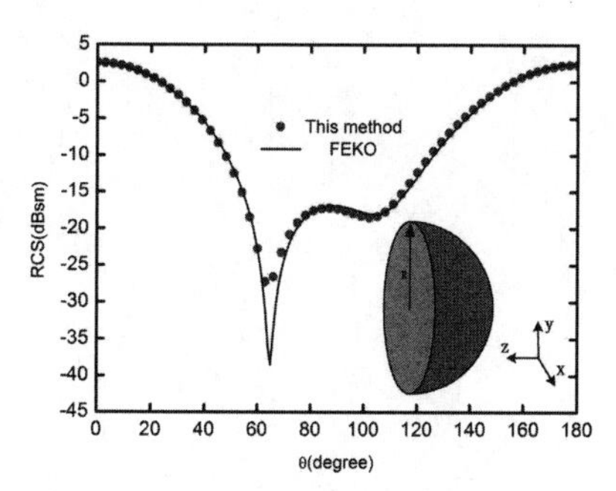

Fig. 2. Bi-static RCS at f=800MHz of the composite hemisphere.

The second example is four square patches mounted on the cylindrical substrate. The radius and the height of the substrate are 0.6m and 0.1m, respectively, with the relative permittivity ε_r=2.0. The four squares are the same, and the dimension of each is 0.3m×0.3m. The centres of these PEC patches are (±0.2, ±0.2) in the x-y plane. The parameters of the incident EMP include f_0=300MHz, $\hat{k}=-\hat{z}$, the polarization chosen along the $+\hat{x}$ axis, and the amplitude E_0=2000V/m.

Fig.3 and Fig.4 show the transient electric current and magnetic current we captured at the point very near the center of the upper surface of the substrate. By our proposed MOO method, we can capture transient electromagnetic characteristics for the composite patch structures.

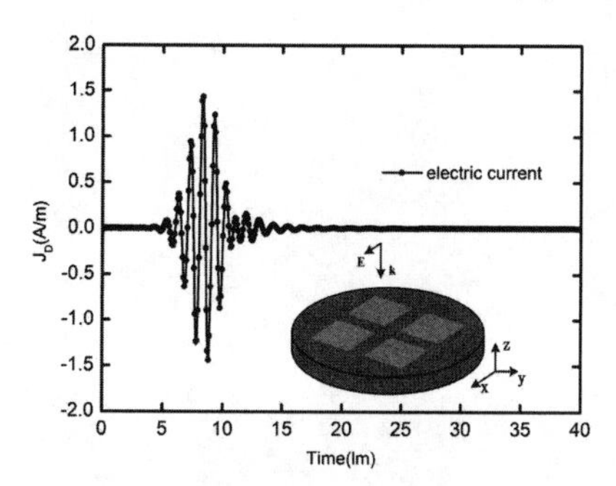

Fig. 3. Transient electric current response on the dielectric.

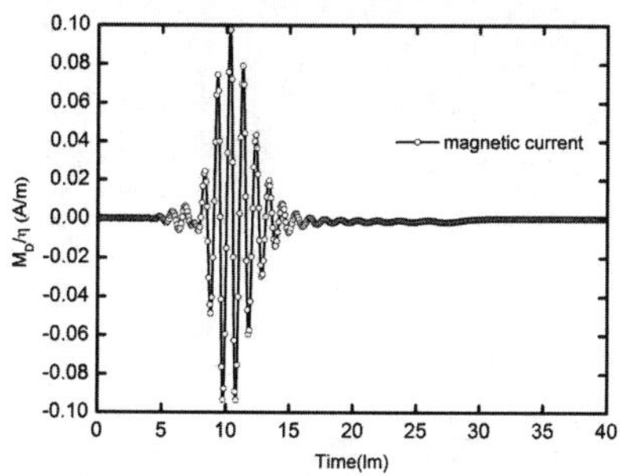

Fig. 4. Transient magnetic current response on the dielectric.

IV. CONCLUSION

In this paper, one time domain method, based on the TDIE solved by the marching-on-in-degree (MOD) scheme is developed for predicting time- and frequency-domain responses of perfect electrically conducting (PEC) surfaces patched on the dielectric substrate. The TDEFIE-PMCHW formulation is derived for the problem, and FFT-based MOD is utilized to solve the iteration. Numerical results show that our proposed method is accurate for handling composite structures, especially PEC patch mounted on the substrate.

ACKNOWLEDGMENT

This work was supported by the NSFC under Grant 60831002 of China.

REFERENCES

[1] D. S. Weile, G. Pisharody, N. W. Chen, B. Shanker, and E. Michielssen, "A novel scheme for the solution of the time-domain integral equations of electromagnetics," *IEEE Trans. Antennas Propagat.*, vol. 52, no. 1, pp. 283–295, Jan. 2004.

[2] W. Luo, W. Y. Yin, M. D. Zhu, J. F. Mao, and J. Y. Zhao, "Investigation on time- and frequency-domain responses of some complex composite structures in the presence of high-power electromagnetic pulses," *IEEE Trans. Electromagn. Compat.*, vol. 54, no. 5, pp. 1006-1016, Oct. 2012.

[3] Z. Ji, T. K. Sarkar, B. H. Jung, M. Yuan, and M. Salazar-Palma, "Solving time domain electric field integral equation without the time variable," *IEEE Trans. Antennas Propagat.*, vol. 54, no. 1, pp. 258–262, Jan. 2006.

[4] M. D. Zhu, X. L. Zhou, and W. Y. Yin, "An adaptive marching-on-in-order method with FFT-based blocking scheme," *IEEE Antennas Wireless Propag. Lett.*, vol. 9, pp. 436–439, 2010.

[5] J. Y. Zhao, W. Y. Yin, M. D. Zhu, and Wei Luo, "Time domain EFIE-PMCHW method combined with adaptive marching-on-in-order procedure for studying on time- and frequency- domain responses of some composite structures," *IEEE Trans. Electromagn. Compat.*, 2013. *(in press)*

[6] B. Shanker, M. Y. Lu, J. Yuan, and E. Michielssen, "Time domain integral equation analysis of scattering from composite bodies via exact evaluation of radiation fields," *IEEE Trans. Antennas Propagat.*, vol. 57, no. 5, pp.1506–1520, May 2009.

[7] S. N. Makarov, S. D. Kulkarni, A. G. Marut, and L. C. kempel, "Method of moments solution for a printed patch/slot antenna on a thin finite dielectric substrate using the volume integral equation," *IEEE Trans. Antennas Propagat.*, vol. 54, no. 4, pp. 1174–1184, Apr. 2006.

[8] Y. Chu, W. C. Chew, J. Zhao, and S. Chen, "A surface integral equation formulation for low-frequency scattering from a composite object," *Trans. Antennas Propagat.*, vol. 51, no. 10, pp. 2837–2844, Oct. 2003

[9] P. Ylä-Oijala, M. Taskinen, and J. Sarvas, "Surface integral equation method for general composite metallic and dielectric structures with junctions," *Progress in Electromagn. Research, PIER 52*, 81–108, 2005.

[10] S. M. Rao, D. R. Wilton, and A. W. Glisson, "Electromagnetic scattering by surfaces of arbitrary shape," *IEEE Trans. Antennas Propagat.*, vol. 30, no. 3, pp. 409–418, May 1982.

WP-2(B)

October 23 (WED) PM

Room B

New Strategies of CEM-2

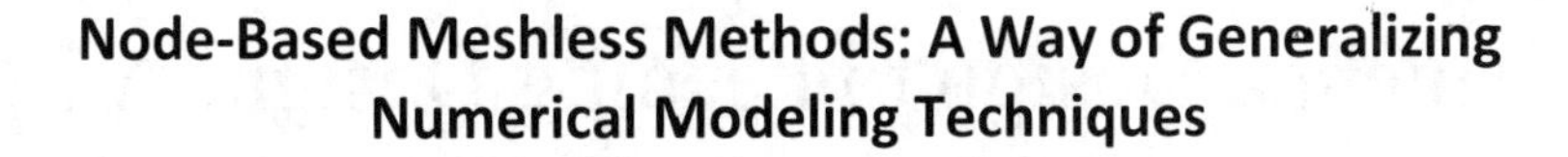

Node-Based Meshless Methods: A Way of Generalizing Numerical Modeling Techniques

Zhizhang (David) Chen
The University of Electronic Science and Technology of China
On leave from Dalhousie University of Canada

Abstract:

Science and Engineering Problems are normally quantitatively described with mathematical models. These mathematical models are often represented by mathematic equations which are often difficult to solve due to irregular boundaries and materials. To address the issue, numerical methods have been developed to find the approximate solutions of the equations. Thanks to the tremendous increase in computing power of modern computers, numerical methods have become widely used techniques nowadays and consequently a large number of numerical methods have been developed. On one hand, the numerical methods have led to solutions of many problems that could not be solved before; on the other hand, they have caused confusions and become challenging to understand and choose. In this talk, we will report our recent finding on the essence of the numerical methods for solving Maxwell's equations: a numerical method is essentially a special case of the method of weighted residuals (minimization). In other words, numerical methods can be unified under the framework of the method of weighted residuals, and a new method can be developed with selection of different basis and testing functions. From there, we will show the recently emerging node-based meshless methods can be used as a way of generalizing or embodying numerical modelling techniques.

Biography:

Zhizhang (David) Chen received the B. Eng. degree from Fuzhou University of China, the M.A.Sc. degree from Southeast University of China, the Ph.D. degree from the University of Ottawa of Canada, and was a NSERC post-doctoral fellow with the ECE Department of McGill University, Montreal, Canada. He is currently with the University of Electronic Science and Technology of China, on leave from the Department of Electrical and Computer Engineering, Dalhousie University, Halifax, Nova Scotia, Canada, where he is a Professor and the Department Head. Dr. Chen has authored and coauthored over 200 journal and conference papers in computational electromagnetics, RF/microwave electronics and wireless systems. He has served on various IEEE society committees, numerous conference committees and guest editors of special issues of professional journals. He was the Chair of IEEE Atlantic Section and founder/chair of its joint Signal Processing and Microwave Theory and Technique Chapter. He received the 2005 Nova Scotia Engineering Award, a 2006 Dalhousie graduate teaching award, the 2007 Dalhousie Faculty of Engineering Research Award, and 2013 IEEE Canada Fessenden Medal. He is a Fellow of the IEEE and the Canadian Academy of Engineering. His current research interests include numerical time-domain modeling and simulation, RF/microwave electronics, smart antennas, ultra- wideband and wireless transceiving technology and applications.

978-7-5641-4279-7

From Antenna Design to Feeding Design: A Review of Characteristic Modes for Radiation Problems

(Invited Paper)

Yikai Chen and Chao-Fu Wang
Temasek Laboratories, National University of Singapore
Emails: tslcheny@nus.edu.sg, cfwang@nus.edu.sg

***Abstract*-This paper briefly reviews the recent progress of characteristic modes (CMs) for radiation problems made in the Temasek Laboratories at National University of Singapore (TL @ NUS). Unique features of the CMs offer great flexibility for the design of high performance antennas. Based on our newly developed in-house software tools for modeling CMs for 3D EM structures and *N*-port networks, the applications we considered include reactively controlled antenna arrays, circularly polarized antennas, platform integrated antennas, and resonant frequency characterization of microstrip antennas. Effective use of CMs can be great help in many antenna designs through proper feeding designs for an existing radiating aperture. This proposed design concept will provide an alternative to change the conventional process of antenna designs, and is also what the CM theory can really benefit the antenna community.**

I. INTRODUCTION

Characteristic mode (CM) theory was initially developed by Garbacz [1] and then refined by Harrington and Mautz [2] in 1971. CMs define a set of orthogonal currents resulted from a generalized eigenvalue system. They can be used to accurately represent the total current induced by any external sources. Such attractive property made CMs helpful for many radiation problems [3]-[11]. Our studies on CMs mainly focus on three topics [11]. The first one is to develop in-house code for the CM computations of *N*-port networks, three-dimensional PEC structures, composite PEC-dielectric structures, and EM problems in multilayered dielectric materials. The second one is to develop techniques to compute the optimal excitations and loadings for certain radiation performances (*e. g.* radiation pattern, polarization, and resonant frequency) from a given radiating aperture. The third one is to design practical feeding structures to excite either mode currents or synthesized currents for the mentioned radiation performances.

This paper briefly reviews our recent progress, made in Temasek Laboratories at National University of Singapore (TL @ NUS), on how to apply the CM theory to diverse radiation problems, including reactively controlled antenna arrays, circularly polarized antennas, platform integrated antennas, and resonant frequency computations for microstrip antennas. Detailed discussion for techniques involved in the analysis and designs are reported in our previous publications [6]-[11]. Throughout these studies, we realized that many antennas can be effectively designed through the proper design of feeding structures for an existing radiating aperture with the help of CMs. The review of these studies also demonstrates CMs are very promising for many practical applications, and the concept of "From Antenna Design to Feeding Design" would change the way for many antenna designs.

II. METRIC QUANTITIES FOR CHARACTERISTIC MODES

For the sake of brevity, the details of the CM theory [2] are omitted here and only two quantities, i.e., characteristic angle and modal significance as the function of eigenvalue λ are respectively defined in (1) and (2) as follows [4]:

$$\alpha = 180^\circ - \tan^{-1}(\lambda) \quad (1)$$

$$MS = 1/|1 + j\lambda| \quad (2)$$

The characteristic angle physically characterizes the phase angle between a characteristic current and the associated characteristic field. Hence, a mode is at resonance when its characteristic angle is close to 180°. The modal significance represents the contribution of a particular mode to the total radiation when an external source is applied.

III. REACTIVELY CONTROLLED ANTENNA ARRAYS

Reactively controlled antenna arrays (RCAA) is an *N*-port radiating system consisting of one element connected to a RF port and a number of surrounding parasitic elements with reactive loads that can be realized by reversely biased varactor diodes. In [6], the RCAA is first analyzed as an *N*-port network using the *N*-port CM theory. Differential evolution (DE) algorithm is then applied to find the optimal reactance loadings using the CMs to avoid demanding a full-wave analysis of each possible candidate. Our study proves that the CMs are important decisions to the final design and demonstrates the effectiveness of the proposed synthesis method based on CMs.

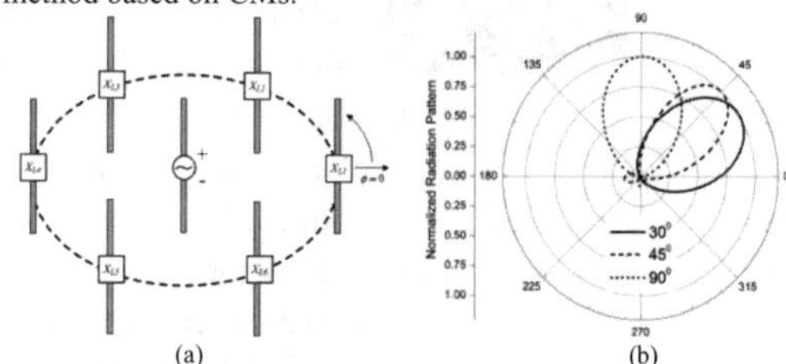

Fig. 1 Reactively controlled steered radiation patterns of a 7-element RCAA.

A 7-element circular array with one quarter wavelength spacing is considered for the synthesis of steered radiation patterns with maximum front-to-back (F/B) ratio. Dipoles of

978-7-5641-4279-7

length $\lambda/2$ and diameter $\lambda/200$ are taken as the antenna elements. Fig. 1(b) shows the steered radiation patterns obtained from the 7-element RCAA. As can be seen, when different loadings are applied, the beam was steered to the directions of $\phi_0 = 30^\circ$, 45°, and 90°, while the F/B ratio have successfully been minimized as low as possible.

IV. Circularly Polarized Antennas

In [8], we focus on how to apply the CM theory as an efficient way to design circularly polarized (CP) antennas. An offset probe feeding for a U-slot CP antenna is determined from the CMs for better axial ratio performance. Fig. 2 shows the characteristic angle and modal significance of the first two modes. It is found that the two modes present exactly the same current amplitude and 90° phase difference at 2.3 GHz. Moreover, Fig. 3 shows that mode $\bar{J}_1$ is characterized as horizontal mode, with intense currents concentrate at the end of U-slot's short arm. It is also evident that mode $\bar{J}_2$ can be identified as the vertical mode, with its intense currents concentrate at the end of U-slot's long arm. To find the optimal probe positions, we subtract the vertical mode from the horizontal mode. Fig. 3(c) shows the current distribution after subtraction, we refer to it as the 'H-V' current $\bar{J}_1 - \bar{J}_2$ for convenience. It is interesting to note that the minimum current area locates at the inner edge of the U-slot's long arm, which indicates that the two modes present almost the same current amplitude in this area. Thus it is possible to get a CP antenna by placing feedings in this area to excite the two orthogonal modes simultaneously. Simulation and measurement results for the axial ratio and CP radiation patterns are presented in [8] to demonstrate the good performance of the U-slot CP antenna with offset feeding.

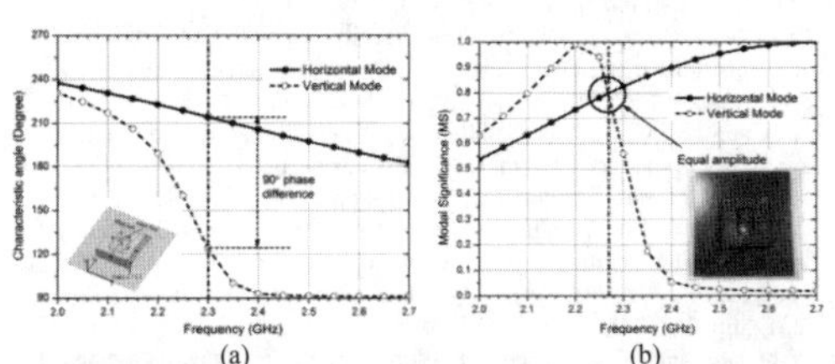

Fig. 2 Horizontal and vertical modes of the U-slot antenna: (a) characteristic angle; (b) modal significance.

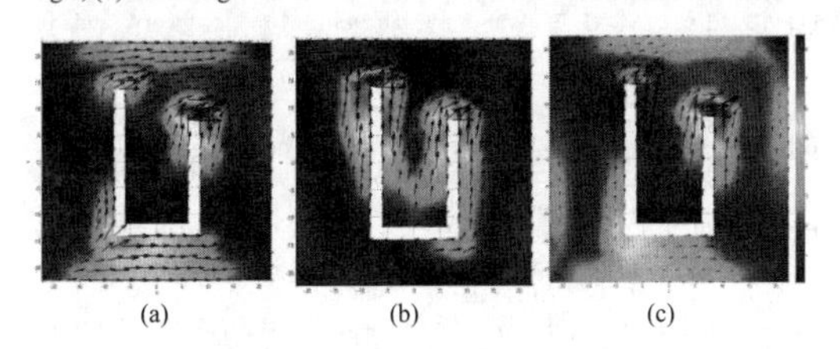

Fig. 3 Current distribution of the U-slot antenna at 2.3 GHz. (a) normalized horizontal mode $\bar{J}_1$; (b) normalized vertical mode $\bar{J}_2$; (c) 'H-V' current $\bar{J}_1$ - $\bar{J}_2$.

V. Platform Integrated Antennas

A novel concept of platform integrated antenna designs for electrically small unmanned aerial vehicle (UAV) using the CM theory is presented in [9], [10]. With the knowledge of the CMs of the UAV body, a multi-objective evolutionary algorithm is implemented to synthesize currents on the UAV body that will radiate desired power patterns. Compact and low-profile feeding structures are then designed to excite the synthesized currents, thus the UAV body serves as the radiating aperture. Reconfigurable radiation patterns are obtained through feeding the probes with proper magnitude and phase excitations. In this method, aperture of the UAV body is fully utilized, practical issues associated with large antennas at low frequency band are eliminated, and knowledge for probe placement on such platforms becomes more explicit.

Fig. 4 shows the synthesized currents and radiation patterns using the CMs of the UAV body. Simulated and measured results are presented in [10] to verify the feeding design approach for such electrically small platform antenna designs.

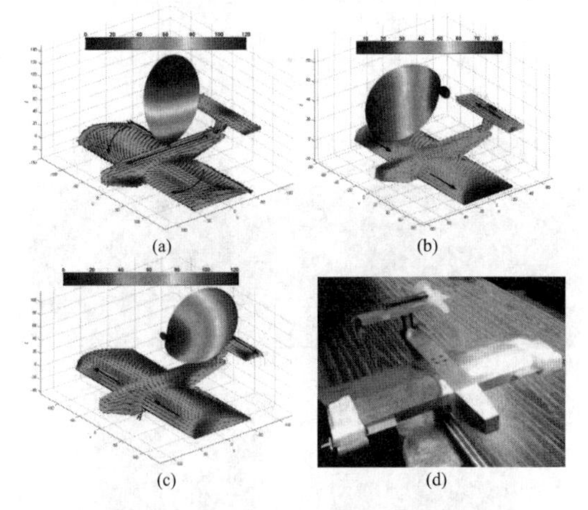

Fig. 4 Synthesized currents and radiation patterns for the UAV antenna systems using CMs. (a) broadside radiation; (b) forward radiation; (c) backward radiation; (d) prototype of the UAV integrated antenna system.

VI. Resonant Frequency Computations for MSAs

To cater to irregular shaped microstrip antenna (MSA) designs using the feeding design concept, we have also addressed the accurate resonant frequency computations for irregular shaped MSAs using the CMs in spectral domain. When the resonant frequency and current distributions of both fundamental and high order radiating modes are accurately described, the rest work for MSA design is simply to design proper feeding structures for certain resonant frequency and radiation patterns.

Fig. 5 shows the modal significances of an equilateral triangular patch with edge length of 10 cm. It is printed on a dielectric substrate with relative permittivity ε_r = 2.32 and thickness h = 1.6 mm. The resonant frequencies marked in Fig. 5 agree well with those obtained from measurement [12], cavity modal [13], and the said plane-wave incident method [14]. Fig. 6 gives the dominant characteristic modes at the resonant frequencies. The mode pairs that resonant at the same frequency have different current distributions. On the other hand, mode currents computed from the plane-wave incident method [14] are dependent on the incident field, and will sometime leads to misunderstanding and confusing mode currents.

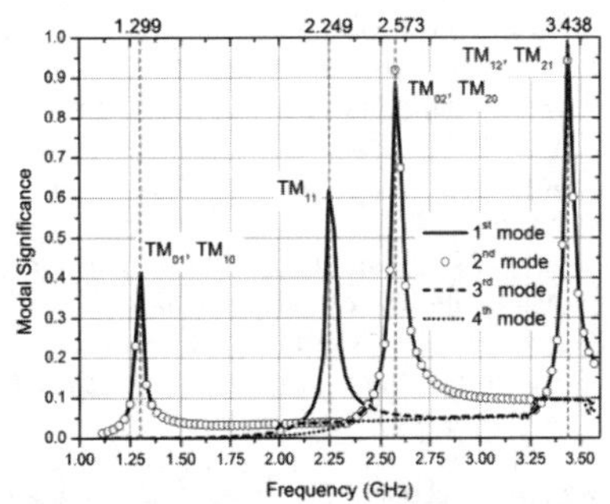

Fig. 5 Modal significance for the equilateral triangular patch antenna.

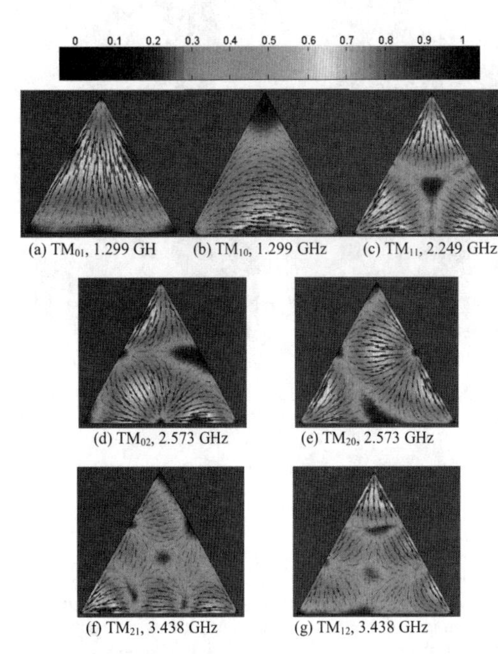

(a) TM_{01}, 1.299 GH (b) TM_{10}, 1.299 GHz (c) TM_{11}, 2.249 GHz

(d) TM_{02}, 2.573 GHz (e) TM_{20}, 2.573 GHz

(f) TM_{21}, 3.438 GHz (g) TM_{12}, 3.438 GHz

Fig. 6. Dominant characteristic modes at the resonant frequencies.

VII. CONCLUSIONS

Recent achievements on the theoretical study, numerical implementation, and practical applications of CMs in various radiation problems are reported. With the help of many unique and attractive features of CMs, physical understandings to the radiating problems can be much clearer, computation burdens in antenna optimization procedure can be greatly alleviated, and designs with favorite features such as compact and low-profile can often be obtained. Throughout this study, we have proposed the CM based feeding design concept that can be great help in simplifying many radiation problems into feeding structure designs for an existing radiating aperture. The proposed concept will provide an alternative to change the conventional process of antenna designs. It is also hoped that the CM theory will benefit the antenna community in more radiation problems in the future.

REFERENCES

[1] R. Garbacz and R. Turpin, "A generalized expansion for radiated and scattered fields," *IEEE Trans. Antennas Propag.*, vol. 19, no. 3, pp. 348-358, May 1971.

[2] R. Harrington and J. Mautz, "Theory of characteristic modes for conducting bodies," *IEEE Trans. Antennas Propag.*, vol. 19, no. 5, pp. 622-628, Sep. 1971.

[3] B. A. Austin and K. P. Murray, "The application of characteristic-mode techniques to vehicle-mounted NVIS antennas," *IEEE Antennas Propag. Mag.*, vol. 40, no. 1, pp. 7-21, Feb. 1998.

[4] M. Cabedo-Fabres, E. Antonio-Daviu, M. Ferrando-Bataller, *et. al.*, "On the use of characteristic modes to describe patch antenna performance," *in IEEE Antennas Propag. Society Int. Symp.*, 2003, pp. 712-715.

[5] J. J. Adams and J. T. Bernhard, "A modal approach to tuning and bandwidth enhancement of an electrically small antenna," *IEEE Trans. Antennas Propag.*, vol. 59, no. 4, pp.1085-1092, Apr. 2011.

[6] Y. Chen and C.-F. Wang, "Synthesis of reactively controlled antenna arrays using characteristic modes and DE algorithm," *IEEE Antennas Wireless Propag. Lett.*,, vol.11, pp.385-388, 2012.

[7] Y. Chen and C.-F. Wang, "Electrically loaded Yagi-Uda antenna optimizations using characteristic modes and differential evolution," *J. Electromagn. Waves Appl.*, vol. 26, no. 8-9, 2012.

[8] Y. Chen and C.-F. Wang, "Characteristic-mode-based improvement of circularly polarized U-slot and E-shaped patch antennas," *IEEE Antennas Wireless Propag. Lett.*, vol.11, pp.1474-1477, 2012.

[9] Y. Chen and C.-F. Wang, "Synthesis of platform integrated antennas for reconfigurable radiation patterns using the theory of characteristic modes," in *10th Int. Symp. Antennas, Propag. EM Theory (ISAPE)*, 2012, pp. 281-285.

[10] Y. Chen and C.-F. Wang, "Platform integrated antenna design for electrically small UAV using characteristic modes," submitted to *IEEE Trans. Antennas Propag.*, April, 2013.

[11] Y. Chen and C.-F. Wang, "Characteristic mode theory and its applications to radiation and scattering problems," *Technical Report No.: TL/EM/13/012, Temasek Laboratories, National University of Singapore*, February, 2013.

[12] M. D. Deshpande, C. R. Cockrell, F. B. Beck, *et al*, "Analysis of electromagnetic scattering from irregularly shaped, thin, metallic flat plates," *NASA Technical paper* 3361, Sep. 1993.

[13] K.-F. Lee, K.-M. Luk, and J. S. Dahele, "Characteristics of the equilateral triangular patch antenna," *IEEE Trans. Antennas Propag.*, vol. 36, no. 11, pp. 1510-1518, Nov. 1988.

[14] M. D. Deshpande, D. G. Shively, and C. R. Cockrell, "Resonant frequencies of irregularly shaped microstrip antennas using method of moments," *NASA Technical paper 3386*, Oct. 1993.

Parallel Higher-Order DG-FETD Simulation of Antennas

(Invited Paper)

Fu-Gang Hu* and Chao-Fu Wang
Temasek Laboratories
National University of Singapore, Singapore
Email: fugang@nus.edu.sg; cfwang@nus.edu.sg

***Abstract*—A discontinuous Galerkin finite-element time-domain (DG-FETD) code is developed to simulate some interesting and challenging antennas. It incorporates several advanced techniques, such as higher-order tetrahedral elements, conformal perfectly matched layer (PML), and local time-stepping scheme. To further speed up the calculation, the DG-FETD method is parallelized by using Message Passing Interface (MPI).**

I. Introduction

The discontinuous Galerkin finite-element time-domain (DG-FETD) method [1]-[5] is one of the most important time-domain methods for solving complex electromagnetic (EM) problems. The DG-FETD method is not only globally explicit, but also capable of dealing with arbitrarily-shaped and inhomogeneously-filled objects. In this paper, the well-developed DG-FETD code is applied to simulate some interesting and challenging antennas. It involves the implementation of the higher-order tetrahedral element technique [6]-[8], conformal perfectly matched layer (PML) technique [1], local time-stepping (LTS) scheme [4], and parallelization scheme using Message Passing Interface (MPI) [3]. Good numerical results demonstrate the validity and capability of the parallel higher-order DG-FETD method.

II. Basic Formulation

The Maxwell's curl equations in the non-PML region are written as

$$\nabla\times\mathbf{E}=-\frac{\mu_r}{c}\frac{\partial\bar{\mathbf{H}}}{\partial t},\quad \nabla\times\bar{\mathbf{H}}=\frac{\epsilon_r}{c}\frac{\partial\mathbf{E}}{\partial t} \tag{1}$$

where $\bar{\mathbf{H}}=\eta_0\mathbf{H}$. The boundary conditions is imposed on the interface of elements [1]-[3]

$$\bar{\mathbf{J}}_s=\hat{n}\times(\bar{\mathbf{H}}^+-\bar{\mathbf{H}}),\qquad \mathbf{M}_s=-\hat{n}\times(\mathbf{E}^+-\mathbf{E}) \tag{2}$$

Applying the Galerkin's approach in each element V_i [1]-[2] and taking advantage of the leap-frog (LF) scheme [2], one can obtain the matrix equations [2]

$$A_{hhv}(h^{n+\frac{1}{2}}-h^{n-\frac{1}{2}})=-(A_{hev}e^n+A^+_{hes}e^{+n})-b_{hs} \tag{3a}$$

$$A_{eev}(e^{n+1}-e^n)=A_{ehv}h^{n+\frac{1}{2}}+A^+_{ehs}h^{+(n+\frac{1}{2})}-b_{es} \tag{3b}$$

In the simulation, the conformal PML is applied to terminate waveguide. The corresponding differential equations of auxiliary variables $\tilde{\mathbf{E}}$, $\tilde{\mathbf{H}}$, $\mathbf{P}$, and $\mathbf{Q}$ are given by [1]-[2]

$$\nabla\times\tilde{\mathbf{E}}=-\frac{\mu_r}{c}\frac{\partial\bar{\bar{A}}_1\cdot\tilde{\mathbf{H}}}{\partial t}-\mu_r\bar{\bar{A}}_2\cdot\tilde{\mathbf{H}}-\mu_r\bar{\bar{A}}_3\cdot\mathbf{P} \tag{4a}$$

$$\nabla\times\tilde{\mathbf{H}}=\frac{\epsilon_r}{c}\frac{\partial\bar{\bar{A}}_1\cdot\tilde{\mathbf{E}}}{\partial t}+\epsilon_r\bar{\bar{A}}_2\cdot\tilde{\mathbf{E}}+\epsilon_r\bar{\bar{A}}_3\cdot\mathbf{Q} \tag{4b}$$

$$\bar{\bar{A}}_5^{-1}\cdot\tilde{\mathbf{H}}-\bar{\bar{A}}_4\cdot\mathbf{P}=\frac{1}{c}\frac{\partial\mathbf{P}}{\partial t} \tag{4c}$$

$$\bar{\bar{A}}_5^{-1}\cdot\tilde{\mathbf{E}}-\bar{\bar{A}}_4\cdot\mathbf{Q}=\frac{1}{c}\frac{\partial\mathbf{Q}}{\partial t} \tag{4d}$$

where $\bar{\bar{A}}_j=\bar{\bar{J}}^T\Lambda_j\bar{\bar{J}},\quad j=1,\cdots,5$. Λ_j are 3×3 diagonal matrices. $\bar{\bar{J}}$ is a tensor related to the local coordinate system on the interface between the PML and non-PML regions. Discretizing the above differential equations yields [2]

$$\begin{aligned}(A_{hha}+A_{hhb})h^{n+\frac{1}{2}}=&(A_{hha}-A_{hhb})h^{n-\frac{1}{2}}-A_{hp}p^n\\&-(A_{hev}e^n+A^+_{hes}e^{+n})-b_{hs}\end{aligned} \tag{5a}$$

$$(A_{qq}+A_{qqd})q^{n+\frac{1}{2}}=(A_{qq}-A_{qqd})q^{n-\frac{1}{2}}+A_{qe}e^n \tag{5b}$$

$$\begin{aligned}&(A_{eea}+A_{eeb})e^{n+1}=(A_{eea}-A_{eeb})e^n-A_{eq}q^{n+\frac{1}{2}}\\&+(A_{ehv}h^{n+\frac{1}{2}}+A^+_{ehs}h^{+(n+\frac{1}{2})})-b_{es}\end{aligned} \tag{5c}$$

$$(A_{pp}+A_{ppd})p^{n+1}=(A_{pp}-A_{ppd})p^n+A_{ph}h^{n+\frac{1}{2}} \tag{5d}$$

The higher-order interpolatory vector basis functions on tetrahedral elements are applied in the above coefficient matrices. The surface magnetic current $\mathbf{M}_s$ is imposed on the excitation port. The incident electric fields can be found in terms of $\mathbf{M}_s$. Hence, pre-simulation of uniform waveguide can be avoided [2].

III. Numerical Results

As an example, the Vivaldi antenna shown in Fig. 1 is simulated. This antenna is fed by a shielded microstrip line that is homogeneously filled with the dielectric $\epsilon_r=2.32$. Figs. 2 and 3 show the incident and reflection coefficients in the time domain, respectively. The S-parameters are shown in Fig. 4. Fig. 5 shows the directivity patterns at 10 GHz. The DG-FETD results agree well with the HFSS results.

978-7-5641-4279-7

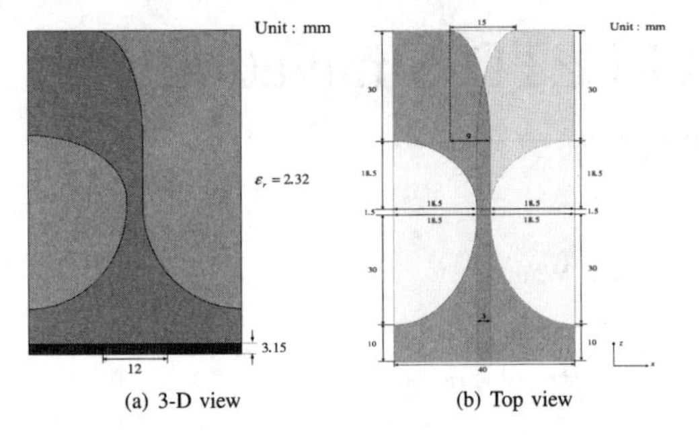

Fig. 1. Configuration of the antipodal Vivaldi antenna.

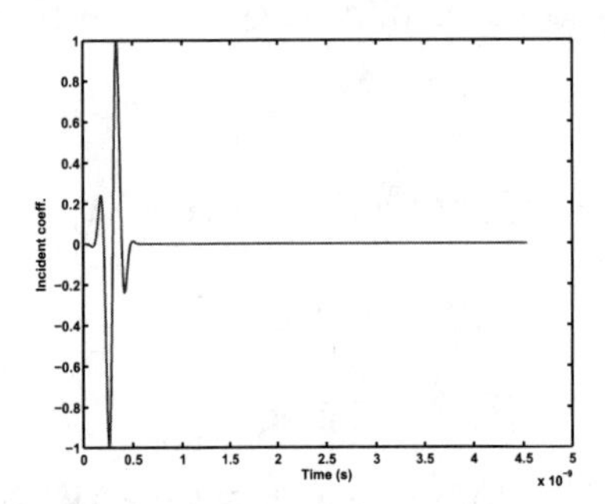

Fig. 2. Time-domain incident coefficients.

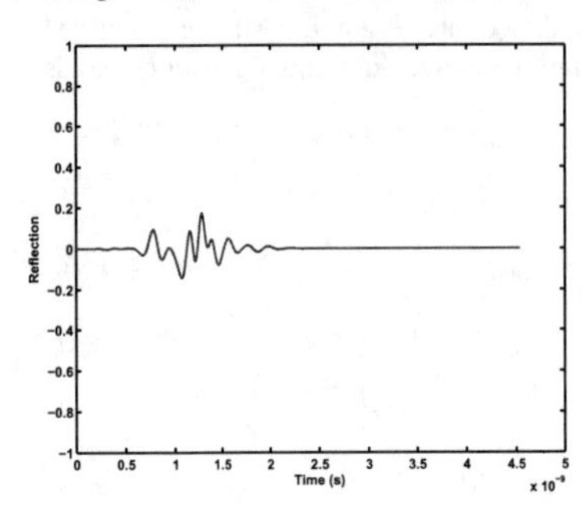

Fig. 3. Time-domain scattered coefficients.

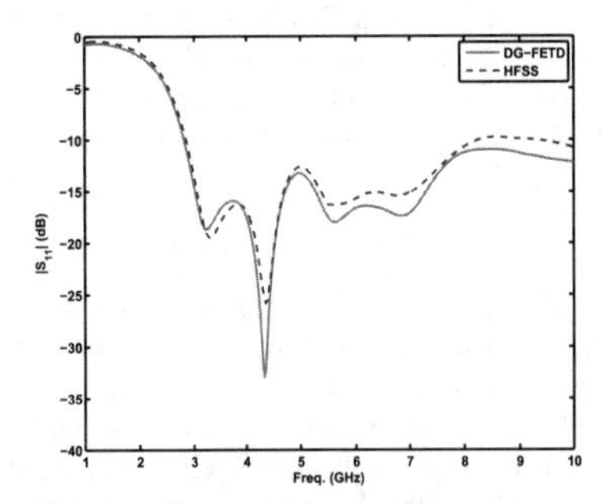

Fig. 4. S-parameters of the Vivaldi antenna.

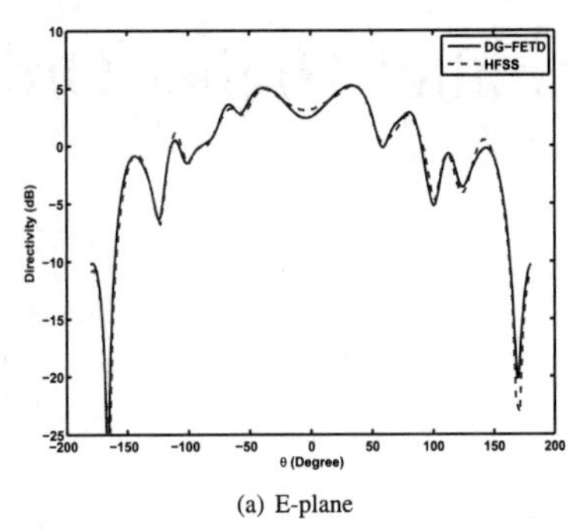

(a) E-plane

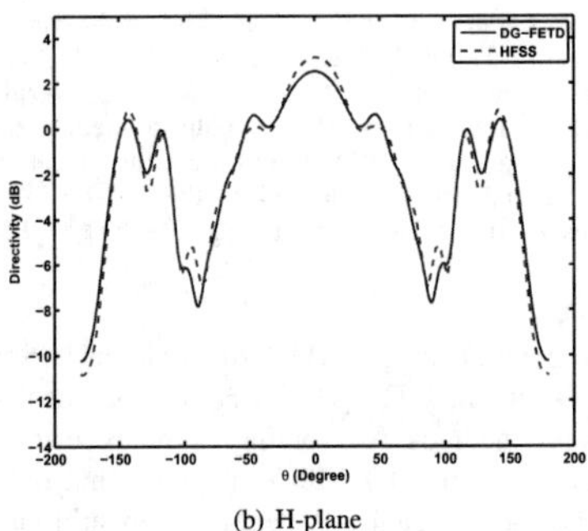

(b) H-plane

Fig. 5. Directivity of the Vivaldi antenna at 10 GHz.

IV. Conclusions

This paper presents the simulation of antennas using the parallel higher-order DG-FETD method. The successful simulation of the challenging antennas demonstrates the capability of the DG-FETD method as an important time-domain technique in the computational electromagnetics.

References

[1] S. D. Gedney, C. Luo, J. A. Roden, R. D. Crawford, B. Guernsey, J. A. Miller, Tyler. Kramer, and E. W. Lucas, "The discontinuous Galerkin finite-element time-domain method solution of Maxwell's equations," *ACES Journal*, vol. 24, pp. 129-141, April 2009.

[2] F. G. Hu and C. F. Wang, "Modeling of waveguide structures using DG-FETD method with higher-order tetrahedral elements," *IEEE Trans. Microwave Theory Tech.*, vol. 60, no. 7, pp. 2046-2054, July 2012.

[3] F. G. Hu and C. F. Wang, "Simulation of EM structures using parallel DG-FETD method," *APCAP*, Aug. 2012.

[4] E. Montseny, S. Pernet, X. Ferrires, and G. Cohen, "Dissipative terms and local time-stepping improvements in a spatial high order discontinuous Galerkin scheme for the time-domain Maxwell's equations," *J. Comput. Phys.*, vol. 227, pp. 6795-6820, 2008.

[5] Z. Lou and J. M. Jin, "A new explicit time-domain finite-element method based on element-level decomposition," *IEEE Trans. Antennas Propag.*, vol. 54, pp. 2990-2999, Oct. 2006.

[6] R. D. Graglia, D. R.Wilton, and A. F. Peterson, "Higher order interpolatory vector bases for computational electromagnetics," *IEEE Trans. Antennas Propag.*, vol. 45, pp. 329-342, March 1997.

[7] J. M. Jin, *The Finite Element Method in Electromagnetics*, 2nd ed. New York: Wiley, 2002.

[8] F. G. Hu, C. F. Wang, and Y. B. Gan, "Efficient calculation of interior scattering from large three-dimensional PEC cavities," *IEEE Trans. Antennas Propag.*, vol. 55, pp. 167-177, Jan. 2007.

A Hybridization of Multi-Level UV with the Hierarchical Fast Far Field Approximations for 3D Rough Surface Scattering

Haogang Wang[1] Biao Wang[1] Tien-Hao Liao[2] Leung Tsang[2]
[1]Department of Information Science and Electronic Engineering at Zhejiang University, Hangzhou, China
[2]Department of Electrical Engineering at University of Washington, Seattle, USA

***Abstract*-This paper presents a hybrid multilevel UV method with the fast far field algorithm for the random rough surface scattering calculations. In the MLUV, a pre-ranked sampling algorithm is used to further compress the UV matrix pairs. To further improve the efficiency of the algorithm, from the second and third coarsest level, a hierarchical FAFFA is employed to further enhance the efficiency of the algorithm. A rough surface with 32-by-32 wavelengths is simulated with 1.5 million unknowns. The results show the efficiency of the algorithm.**

I. INTRODUCTION

Random Rough surface scattering is a very interesting and important phenomenon in surface science. In real world, there exist a lot of examples that can be explained using the random rough surface scattering theorem, e.g., in macroscopic size world, the electromagnetic wave scattering from soil, lake, river and ocean; and in the microscopic size world, the electromagnetic wave scattering from the rough surface of the silicon substrate surface, and the surface plasma excitation on the silver film with rough surface.

The difficulties in study the random rough surface scattering is that the rough surface is always infinite large and the complex multiple scattering by the roughness of the surface. To model the rough surface scattering, one needs to truncate the surface. However, to ensure the accuracy of the modeling of the surface scattering, the truncated surface size should be electric large. To study the rough surface scattering problem, the MoM based fast algorithms i. e., fast integral equation solutions are always employed because of their accuracy and efficiency.

The Multi-Level UV (MLUV) factorization method is a kernel independent fast integral equation method that factorizes and compresses the low rank off-diagonal submatrices in the MoM matrix using a multi-level scheme [1][2]. It has been used for some scalar scattering problems [3]. In [4], MLUV has been employed for object scattering composited with object and rough surface. In it, the surface is the PEC. In [3][5], UV method is combined with Sparse Matrix Canonical Grid (SMCG) for soil scattering problems, the rough surface is soil interface that can be viewed as the rough interface of a dielectric region.

Instead, in this paper, we employ Multi-Level UV directly. A pre-ranked sampling algorithm is employed to further compress the UV matrix pairs. For the second or the third coarsest levels, the fast far field approximation is also used to further enhance the efficiency of the algorithm. Comparing with other fast methods such as MLFMA, MLUV has a great advantage that it is very easy to be parallelized because 1) the calculation of each UV pairs is independent and 2) in a matrix-vector multiplication, each submatrix-vector multiplication using the UV pairs is also independent. A rough surface of 32-by-32 square wavelengths is simulated with 1.5 million unknowns. The results show the efficiency of the algorithm.

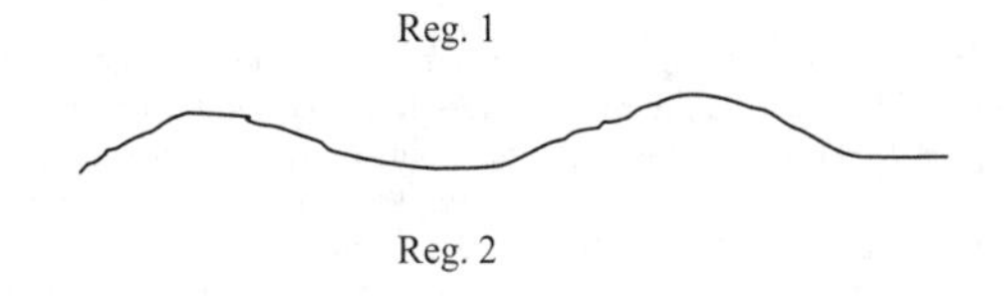

Fig. 1. Two regions rough surface problem.

II. PRE-RANKED MLUV+FAFFA

A. Integral equations

For two regions rough surface problem as shown in Fig. 1, the PMCHWT is used and listed below.

$$\boldsymbol{E}^{inc}(\boldsymbol{r})|_{\tan} = (L_1 + L_2)\boldsymbol{J}'(\boldsymbol{r})|_{\tan} -(K_1 + K_2)\boldsymbol{M}(\boldsymbol{r})|_{\tan}, \tag{1}$$

$$\eta_0 \boldsymbol{H}^{inc}(\boldsymbol{r})|_{\tan} = (K_1 + K_2)\boldsymbol{J}'(\boldsymbol{r})|_{\tan} +(\varepsilon_{r1}L_1 + \varepsilon_{r2}L_2)\boldsymbol{M}(\boldsymbol{r})|_{\tan}. \tag{2}$$

978-7-5641-4279-7

$$L_i = \frac{-i\omega\mu_i}{4\pi\eta_0}\int_S ds'(\overline{\overline{I}} - \frac{1}{k_i^2}\nabla\nabla')G_i(\boldsymbol{r},\boldsymbol{r}')\cdot, \quad (3)$$

$$K_i = -\frac{1}{4\pi}\int_S ds'\nabla G_i(\boldsymbol{r},\boldsymbol{r}')\times, \quad (4)$$

where $\boldsymbol{J}'(\boldsymbol{r}') = \eta_0 \boldsymbol{J}(\boldsymbol{r}')$ and η_0 is the free space wave impedance, and subscript $i = 1, 2$ is the indices of the regions.

Expanding the currents with the Rao-Wilton_Gillison Basis functions and using the Galerkin method, equations (1) and (2) can be discretized into a MoM matrix equation.

$$\begin{bmatrix} \overline{\overline{Z}}^{L_1} + \overline{\overline{Z}}^{L_2} & -(\overline{\overline{Z}}^{K_1} + \overline{\overline{Z}}^{K_2}) \\ \overline{\overline{Z}}^{K_1} + \overline{\overline{Z}}^{K_2} & \varepsilon_{r1}\overline{\overline{Z}}^{L_1} + \varepsilon_{r2}\overline{\overline{Z}}^{L_2} \end{bmatrix} \begin{bmatrix} \overline{I}^E \\ \overline{I}^M \end{bmatrix} = \begin{bmatrix} \overline{V}^E \\ \overline{V}^M \end{bmatrix}. \quad (5)$$

where

$$Z_{p,q}^{L_i} = \frac{-i\omega\mu_i}{4\pi\eta_0}\int_S ds\boldsymbol{f}_p(\boldsymbol{r})\cdot \int_S ds'[\overline{\overline{I}} - \frac{1}{k_i^2}\nabla\nabla']G_i(\boldsymbol{r},\boldsymbol{r}')\cdot \boldsymbol{f}_q(\boldsymbol{r}'), \quad (6)$$

$$Z_{p,q}^{K_i} = -\frac{1}{4\pi}\int_{S_p} ds\boldsymbol{f}_p(\boldsymbol{r})\cdot\int_{S_q} ds'\nabla G_i(\boldsymbol{r},\boldsymbol{r}')\times \boldsymbol{f}_q(\boldsymbol{r}'), \quad (7)$$

(5) can be solved iteratively using method such as GMRES, BICG-stab etc. In this paper we use a hybrid pre-ranked MLUV+FAFFA to accelerate the matrix-vector multiplications.

B. Pre-ranked MLUV

In the original MLUV, the coarse-coarse sampling algorithm is used to sample the subscatterers from the field and source groups. This sampling scheme always gives over sampled subscatterers sets and thus ensures the accuracy of MLUV. However, because both rows and columns are over sampled, the computational load for UV factorizing the submatrices will be unnecessary large. In this paper, we use a pre-ranked approach. In this approach the coarse-coarse sampling is applied first to obtain a submatrix that is greatly smaller than the original submatrix. Subsequently the rank-prevealed QR factorization is employed to obtain the indices sets of the significant rows and columns. According to these two indices sets, we can sample the rows and columns from the original submatrix and then construct the UV pairs following the original UV procedure.

C. Hierarchical Fast Far Field Approximations

When the electric size of the problem becomes very large, the sub-matrix describing the interactions between a node at finer level and its interaction list node can be very large although the sub-matrix is low rank. The complexity of the MLUV thus becomes also very large for solving this kind of problem. In [6], a FAFFA method is employed to solve for the electric large problem. In the FAFFA method, the Green's function of any two points respectively in two well separated groups can be approximated using the following equation [7]

$$e^{ikR}/R \simeq \frac{\eta_0}{-i4\pi\omega\mu_i} e^{i\boldsymbol{k}_{m,n}\cdot(\boldsymbol{r}-\boldsymbol{r}_m)}\alpha_{i,m,n}^{far} e^{-i\boldsymbol{k}_{m,n}\cdot(\boldsymbol{r}'-\boldsymbol{r}_n)}, \quad (8)$$

where $\alpha_{i,m,n}^{far} = \frac{-i\omega\mu_i}{4\pi\eta_0} e^{ik_i R_{m,n}} / R_{m,n}$.

However, to ensure the accuracy of the far field approximation, we should use the weak criterion [7]

$$ratio = R_{m,n} / L_{gp} \geq \beta, \text{ i=1,2} \quad (9)$$

where β is a real constant and k is the wave number. In [8], a multilevel FAFFA is implemented for solving scattering problems. However, in the coarsest level the distances between each far interaction group is different which will definitely affect the efficiency.

In order to further enhance the efficiency, in this paper, an adaptive hierarchical FAFFA (HFAFFA) algorithm is devised. In this method, the FAFFA starts from the second and third coarsest levels. For any two well separate groups at these levels, if the *ratio* satisfies the weak criterion (9), the FAFFA is performed to approximate the interactions between these two groups. Otherwise, divide each group into four sub-groups and perform the same procedure to these sub-groups recursively. Thus an HFAFFA is constructed. The HFAFFA has the advantage that the *ratio* of any two well separate groups between which the interactions are approximated using FAFFA will not change greatly and hence the efficiency enhanced.

III. NUMERICAL RESULTS

All results in this paper are calculated using a server with two 2.60 GHz Intel Xeon CPUs with 16 cores and 32 threads. The shared memory is 128 GB. The program is parallelized employing OpenMP technique with all the 32 threads.

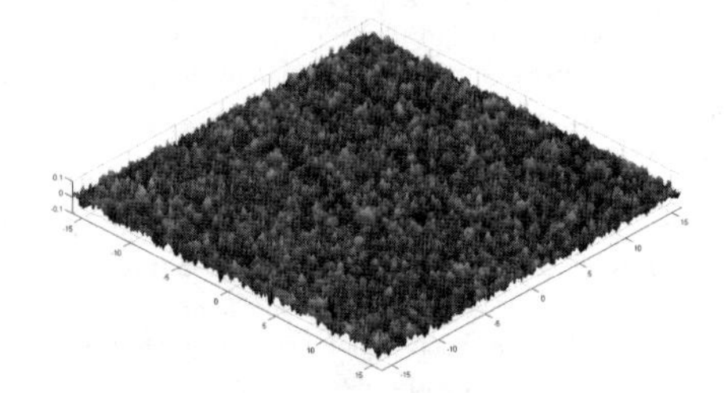

Fig. 2, The profile of Gaussian random rough surface with exponential correlation function. RMSh=0.021; Corre. Length=0.21; Epsr=15.14+1.27i.

The rough surface simulated is ploted in Fig. 2. The root mean square height (RMSh) is 0.021 wavelength. The Coorelation Length is 0.21 wavelength. The dielectric constant of region 2 is 15.14+1.27i. The infinite rough surface is truncated into a 32-by-32 square wavelengths finite surface. If one wavelength is meshed into 16 grids and RWG basis functions are used, the total number of unknowns including both the electric current unknowns and the magnetic current unknowns is 1,572,864.

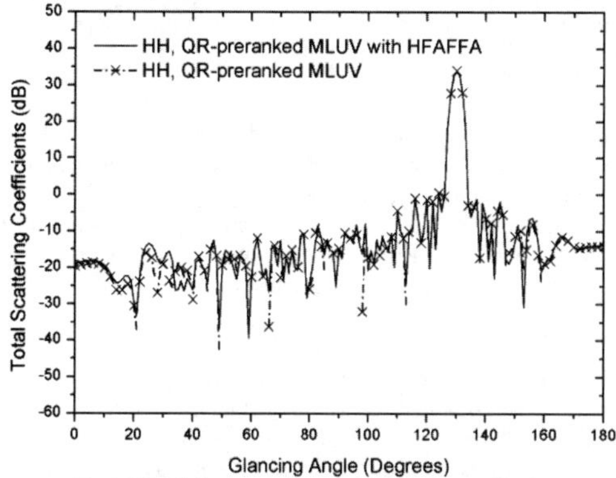

Fig. 3 HH polarized scattering coefficients of one realization.

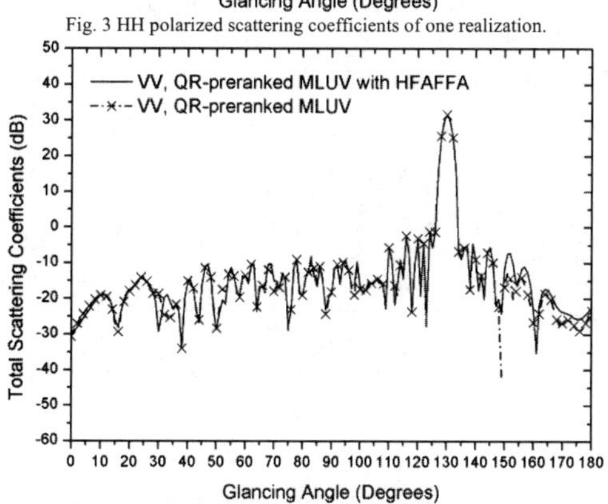

Fig. 3 VV polarized scattering coefficients of one realization

Fig. 3 and Fig. 4 are the calculated HH and VV polarized total scattering coefficients of one realization. We see that the results obtained with using HFAFFA and without using HFAFFA agree very well. Note the incident angle is 40 degree which is just 50 degrees for the glancing angle. Tapered plane wave is used for the rough surface is finite in size.

TABLE I
PERFORMANCE COMPARISON

Methods	Preranked MLUV	Preranked MLUV with HFAFFA
Memory	123 GB	108 GB
Total CPU time	5 hours	2 hours

Table I is a brief comparison of these two methods. We see that with using HFAFFA the CPU time consumption is less than half of that without using HFAFFA, while the memory consumption is also reduced.

IV. CONCLUSION

This paper presents a hybrid multilevel UV method with the fast far field algorithm for the random rough surface scattering calculations. In the MLUV, a pre-ranked sampling algorithm is used to further compress the UV matrix pairs. To further improve the efficiency of the algorithm, from the second and third coarsest level, the FAFFA is employed to further enhance the efficiency of the algorithm. A rough surface with 32-by-32 wavelengths is simulated with 1,572,864 unknowns. The results demonstrate the accuracy and efficiency of the proposed algorithm.

ACKNOWLEDGMENT

The research is supported by the OFSLRSS201310 and the National Natural Science Foundation of China No. 61231001.

REFERENCES

[1] L. Tsang, D. Chen, P. Xu, Q. Li, and V. Jandhyala, "Wave scattering with the UV multilevel partitioning method: 1. Two-dimensional problem of perfect electric conductor surface scattering," *Radio Sci.*, vol. 39, no. 5, p. RS5010, 2004, 10.1029/2003RS003009.

[2] L. Tsang, Q. Li, and P. Xu *et al.*, "Wave scattering with the UV multilevel partitioning method: 2. Three-dimensional problem of nonpenetrable surface scattering," *Radio Sci.*, vol. 39, no. 5, p. RS5011, 2004, 10.1029/2003RS003010.

[3] P. Xu and L. Tsang, ''Scattering by rough surface using a hybrid technique combining the multilevel UV method with the sparse-matrix canonical grid method,''*Radio Sci.*, vol. 40, 2005, RS4012.

[4] S. Y. He and G. Q. Zhu, "A hybrid MM-PO method combining UV technique for scattering from two-dimensional target above a rough surface," *Microw. Opt. Technol. Lett.*, vol. 49, no. 12, pp. 2957–2960, Dec. 2007.

[5] S. Huang and L. Tsang, "Electromagnetic scattering of randomly rough soil surfaces based on numerical solutions of Maxwell equations in 3 dimensional simulations using a hybrid UV/PBTG/SMCG method," *IEEE Trans. Geosci. Remote Sens.*, vol. 50, no. 10, Oct. 2012, pp. 4025-4035.

[6] C. C. Lu and W. C. Chew, "Fast far field approximation for calculating the RCS of large objects," *Micro. Opt. Tech. Lett.*, 1995, 8, pp. 238-241.

[7] W. C. Chew, T. J. Cui, and J. Song, "A FAFFA-MLFMA algorithm for electromagnetic scattering," *IEEE Trans. Antennas Propagat.*, vol. 50, no. 11, Nov. 2002, pp. 1641-1649.

[8] L. Rossi, P. J. Cullen, and C. Brennan, "Implementation of a multilevel fast far-field algorithm for solving electric-field integral equations," *IEE Proc. -Microw. Antennas Propag.* Vol. 147, no. 1, Feb. 2000, pp. 19-24.

Analysis of Electromagnetic Scattering from Complicated Objects Using Nonconformal IE-DDM

X. Wei, J. Hu, M. Tian, R. Zhao, M. Jiang, Z. P. Nie
Dept. of Microwave Engineering,
University of Electronic Science and Technology of China
Cheng du, 610054, China
hujun@uestc.edu.cn

***Abstract*-This paper extends the integral equation domain decomposition method (IE-DDM) with Robin Transmission Condition(Robin TC) to the electromagnetic analysis of general objects of arbitrary shape, even for composite conducting-dielectric structures. Meanwhile, nonconformal mesh processing is formulated for the IE-DDM. The method is based on the CFIE-GCFIE integral equation formulation. First, we present the IE-DDM formulations for composite conducting-dielectric objects. Then, the nonconformal mesh processing is introduced into the IE-DDM. And last, the numerical results are also given to prove the validity of the present method.**

I. INTRODUCTION

The integral equation (IE) method is a popular tool which has many important applications e.g., in the radar technology, antenna design and microwave engineering. An efficient and stable integral equation framework for solving scattering problem of complex and multi-scale platform is a great need, especially for composite conducting-dielectric structures. And lots of studies have been conducted by many researchers [1-5]. Several methods based on IE have been proposed, including surface coupled integral equations method [6] , hybrid volume-surface integral equations method [7] . IE method can analyze conducting-dielectric structures of arbitrary shape by using the combination of electric-field integral equation(EFIE) for PEC part and other kind SIEs for dielectric part. The popular methods such as EFIE-PMCHWT [8-9] or EFIE-JMCFIE still remain challenges on the treatment of junctions between conducting part and dielectric part. The complex combinations of integral equations and conformal mesh processing are uneasy for sophisticated models, even though EFIE-CFIE-PMCHWT [10-11] formulation was present for solving special junction parts. Surface integral equation domain decomposition method (IE-DDM) [12-13] for field problem of composite object with homogeneous material properties performs well while a new generalized combined field integral equation is also employed for IE-DDM, which leads to well-conditioned matrix equations.

In this paper we analyze conducting object and composite conducting-dielectric structures using IE-DDM that makes treatment of junctions uncomplicatedly. Moreover, nonconformal interpolation method [14] is applied in it to make mesh processing more flexible.

The rest of this paper is organized as follows. In Section 2 the formulations about IE-DDM and general combined field integral equation (GCFIE) are given. Nonconformal interpolation method is described in Section 3. In Section 4 numerical results are shown to prove the validity of this method, followed by the conclusions.

II. IE-DDM FORMULATIONS

In this section, firstly we present the basic theorem of integral equation domain decomposition method. Then G-CFIE will be discussed in IE-DDM framework when analyzing the scattering field of a composite conducting-dielectric structure.

A. *Framework of IE-DDM*

We consider that the whole computational domain Ω is divided into some non-overlapping subdomains Ω_i ($i=1,\cdots,N$) (Fig.1), $\Omega=\cup_{i=1}^{N}\Omega_i$.

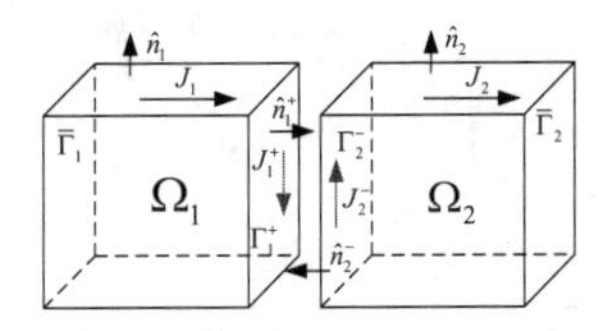

Fig.1. Non-overlapping IE-DDM Scheme

$\overline{\Gamma_i}$ and $\overline{\Gamma_{ij}}$ are the original as well as new interface respectively, that is to say

$$\begin{aligned}\overline{\Gamma_i} &= \partial\Omega_i \cap \partial\Omega \\ \overline{\Gamma_{ij}} &= \partial\Omega_i \cap \partial\Omega_j\end{aligned} \tag{1}$$

While applying the idea above to electromagnetic scattering model, we can get the basic formulations for each subdomain Γ_i, (2.1) for $\overline{\Gamma_i}$ and (2.2) for $\overline{\Gamma_{ij}}$.

978-7-5641-4279-7

$$\left(\overline{E}_i^s\Big|_{\overline{\Gamma}_i}+\overline{E}_i^{inc}\Big|_{\overline{\Gamma}_i}+\sum\nolimits_{j=1,\neq i}^{N}\overline{E}_j^s\Big|_{\overline{\Gamma}_i}\right)\times n_i=0$$
$$n_i\times\left(\overline{H}_i^s\Big|_{\overline{\Gamma}_i}+\overline{H}_i^{inc}\Big|_{\overline{\Gamma}_i}+\sum\nolimits_{j=1,\neq i}^{N}\overline{H}_j^s\Big|_{\overline{\Gamma}_i}\right)=\overline{J}_i\Big|_{\overline{\Gamma}_i} \tag{2.1}$$

$$\left(\overline{E}_i^s\Big|_{\overline{\Gamma}_{ij}}+\sum\nolimits_{j=1,\neq i}^{N}\overline{E}_j^s\Big|_{\overline{\Gamma}_{ij}}\right)\times n_i=0$$
$$n_i\times\left(\overline{H}_i^s\Big|_{\overline{\Gamma}_{ij}}+\sum\nolimits_{j=1,\neq i}^{N}\overline{H}_j^s\Big|_{\overline{\Gamma}_{ij}}\right)=\overline{J}_i\Big|_{\overline{\Gamma}_{ij}} \tag{2.2}$$

Scattering fields in (2.1), $\overline{E}_m^s$ and $\overline{H}_m^s$ $(m=i\ ,\ j)$, arise from the sources on $\overline{\Gamma}_m$, however fields in (2.2) are only the product of sources on interfaces , $\overline{\Gamma}_{ij}$ and $\overline{\Gamma}_{ji}$.

B. *Formulation for Composite Object in G-CFIE*

Quite a few SIE formulations have been proposed for solving EM scattering from homogeneous region. However not all of them perform well when they encounter field problem of composite conducting-dielectric structure. G-CFIE[13] as a stable SIE method will be discussed in IE-DDM framework when analyzes these problems.

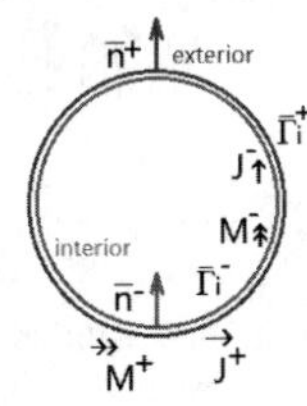

Fig.2. Homogeneous dielectric: G-CFIE Scheme

Unlike other SIEs, the homogeneous region is considered two subdomains, exterior and interior part, Fig.2., each of them has its own SIE formulations (3.1) and (3.2).

$$\left(\overline{E}_i^s\Big|_{\overline{\Gamma}_i^+}+\overline{E}_i^{inc}\Big|_{\overline{\Gamma}_i^+}\right)\times\overset{+}{n_i}=\overline{M}_i^+\Big|_{\overline{\Gamma}_i^+}$$
$$\overset{+}{n_i}\times\left(\overline{H}_i^s\Big|_{\overline{\Gamma}_i^+}+\overline{H}_i^{inc}\Big|_{\overline{\Gamma}_i^+}\right)=\overline{J}_i^+\Big|_{\overline{\Gamma}_i^+} \tag{3.1}$$

$$\left(\overline{E}_i^s\Big|_{\overline{\Gamma}_i^-}\right)\times\overset{-}{n_i}=\overline{M}_i\Big|_{\overline{\Gamma}_i^-}$$
$$\overset{-}{n_i}\times\left(\overline{H}_i^s\Big|_{\overline{\Gamma}_i^-}\right)=\overline{J}_i\Big|_{\overline{\Gamma}_i^-} \tag{3.2}$$

Now we apply G-CFIE into IE-DDM framework then derive the matrix equation of IE-DDM for composite object as (4), assuming Γ_1 is a PEC sub-domain and Γ_2 is homogeneous dielectric region.

$$\begin{bmatrix} A_{\Gamma_1\Gamma_1}^{J^+J^+} & & & & \\ & A_{\Gamma_2^+\Gamma_2^+}^{J^+J^+} & A_{\Gamma_2^+\Gamma_2^+}^{J^+M^+} & & \\ & A_{\Gamma_2^+\Gamma_2^+}^{M^+J^+} & A_{\Gamma_2^+\Gamma_2^+}^{M^+M^+} & & \\ & & & A_{\Gamma_2^-\Gamma_2^-}^{J^-J^-} & A_{\Gamma_2^-\Gamma_2^-}^{J^-M^-} \\ & & & A_{\Gamma_2^-\Gamma_2^-}^{M^-J^-} & A_{\Gamma_2^-\Gamma_2^-}^{M^-M^-} \end{bmatrix}\begin{bmatrix} \overline{J^+}_{\Gamma_1} \\ \overline{J^+}_{\Gamma_2^+} \\ \overline{M^+}_{\Gamma_2^+} \\ \overline{J^-}_{\Gamma_2^-} \\ \overline{M^-}_{\Gamma_2^-} \end{bmatrix}$$
$$=\begin{bmatrix} b_{\Gamma_1}^J \\ b_{\Gamma_2^-}^J \\ b_{\Gamma_2^-}^M \\ 0 \\ 0 \end{bmatrix}+\begin{bmatrix} D_{\overline{\Gamma}_1\overline{\Gamma}_{12}}^{J^+J^+} & B_{\overline{\Gamma}_1\overline{\Gamma}_2}^{J^+J^+} & B_{\overline{\Gamma}_1\overline{\Gamma}_2}^{J^+M^+} & & \\ B_{\overline{\Gamma}_2\overline{\Gamma}_1}^{J^+J^+} & & & C_{\Gamma_2^+\Gamma_2^-}^{J^+J^-} & C_{\Gamma_2^+\Gamma_2^-}^{J^+M^-} \\ B_{\overline{\Gamma}_2\overline{\Gamma}_1}^{M^+J^+} & & & C_{\Gamma_2^+\Gamma_2^-}^{M^+J^-} & C_{\Gamma_2^+\Gamma_2^-}^{M^+M^-} \\ & C_{\Gamma_2^-\Gamma_2^+}^{J^-J^+} & C_{\Gamma_2^-\Gamma_2^+}^{J^-J^+} & & \\ & C_{\Gamma_2^-\Gamma_2^+}^{J^-J^+} & C_{\Gamma_2^-\Gamma_2^+}^{J^-J^+} & & \end{bmatrix}\begin{bmatrix} \overline{J^+}_{\Gamma_1} \\ \overline{J^+}_{\Gamma_2^+} \\ \overline{M^+}_{\Gamma_2^+} \\ \overline{J^-}_{\Gamma_2^-} \\ \overline{M^-}_{\Gamma_2^-} \end{bmatrix} \tag{4}$$

Where, the matrix blocks $A_{\Gamma_i\Gamma_i}^{J^\pm J^\pm}$ are self-coupling and $B_{\overline{\Gamma}_i\overline{\Gamma}_j}^{J^\pm J^\pm}$ stands for mutual coupling in (6). Blocks $C_{\overline{\Gamma}_2^+\overline{\Gamma}_2^+}^{J^+J^+}$ or $B_{\overline{\Gamma}_{ij}\overline{\Gamma}_{ji}}^{J^\pm J^\pm}$ is sparse motar matrices derived from Robin or Neumann transmission conditions , respectively.

III. Computation of Mortar Matrix on Non-conformal Touching Face

In non-overlapping IE-DDM, to enforce field continuity on nonconformal touching face is very important for the accuracy of DDM. Here, a cement technique [14] is used to allow nonconformal interpolation to overcome this difficulty.

Firstly, we assume that all subdomains are have non-conforming meshes with each other. Meanwhile, we could introduce the discrete spaces: each Ω_i is a space provided with its own grid T_i ,1<i<N, such that $\Omega_i=\cup_{\mathrm{T}_i^m\in\mathrm{T}_i}w_i^m$ and for grid $w_i^m\in\mathrm{T}_m$, $1<m<\#\mathrm{T}_i$. We consider that the w_i^m sets belonging to the meshes of simple type (triangles or tetrahedras).

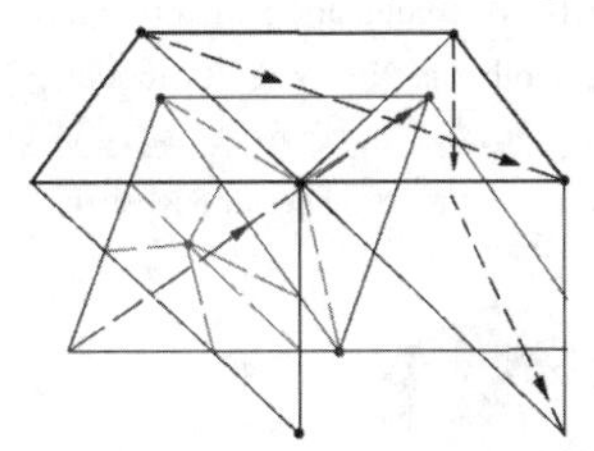

Fig.3. Nonconformal meshes interpolation

Then, we define over each subdomain two non-conforming spaces X_k and Y_k by the space of traces over each Γ_{ij} of elements as follows:

$$X_i = \{v_m | v_m \in T_i, v_i \in \Gamma_{ij}, 1 < m < \#(T_i \cap \Gamma_{ij})\}$$
$$Y_j = \{u_m | u_m \in T_j, u_m \in \Gamma_{ji}, 1 < m < \#(T_j \cap \Gamma_{ji})\} \quad (5)$$

Then , we can get a set of elements P_{ij} by projecting space X_i and Y_j with each other , that $P_{ij} = \{p_m | p_m \in (X_i \cap Y_j)\}$ and $v_p = \sum_{i=1}^{m} p_i$, $v_p \in X_i$.

The projection set P_{ij} between nonconforming space makes the mortar element method easily. The Robin TC formulation in Galerkin testing is as :

$$\begin{aligned}&\alpha\left(\int_{\Gamma_{ij}} \overline{w}_i^k(\overline{r})\cdot \overline{E}_i^s\left(\overline{v}_i^l(\overline{r}')\right)\Big|_{\Gamma_{ij}} d\overline{r} - \sum_{j=1,j\neq i}^{N}\int_{\Gamma_i^+} \overline{w}_i^k(\overline{r})\cdot \overline{E}_j^s\left(\overline{p}_{ij}(\overline{r}')\right)\Big|_{\Gamma_{ij}} d\overline{r}\right)\\ &+(1-\alpha)\left(\int_{\Gamma_i^+} \overline{w}_i^k(\overline{r})\cdot\left(\overset{+}{n_i}\times \overline{H}_i^s\left(\overline{v}_i^l(\overline{r}')\right)\Big|_{\Gamma_{ij}}\right)d\overline{r} - \int_{\Gamma_i^+} \overline{w}_i^k(\overline{r})\cdot \overline{v}_i^l(\overline{r}')\Big|_{\Gamma_{ij}} d\overline{r}\right)\\ &+(1-\alpha)\left(\sum_{j=1,j\neq i}^{N}\int_{\Gamma_i^+} \overline{w}_i^k(\overline{r})\cdot\left(\overset{+}{n_i}\times \overline{H}_j^s\left(\overline{p}_{ij}(\overline{r}')\right)\Big|_{\Gamma_{ij}}\right)d\overline{r}\right) = 0\end{aligned} \quad (6)$$

Where, $\overline{w}_i^k$ is testing function and $\overline{v}_i^l$ is original basic function in $\overline{\Gamma_{ij}}$, $\overline{p}_{ij}$ is basic function which belongs to P_{ij} , projecting space. In (6), double integral terms in scattering fields can be done in original basic functions and surface current terms , single integral term , should be expanded in projecting space.

IV. NUMERICAL RESULTS

In this section the present method is verified and compared with other available data. A coated PEC sphere and two aircrafts models are investigated.

A. Coated PEC sphere model

Consider the case of a perfectly conducting sphere with two different materials, which are permittivity 6.0 and permittivity 4.0 respectively. A combination of three parts that two are dielectric and another is PEC is set for model, λ_0 is wavelength in free space. The bistatic RCS is calculated with 38,835 unknowns and result is correspond to MIE result, in Fig.4.

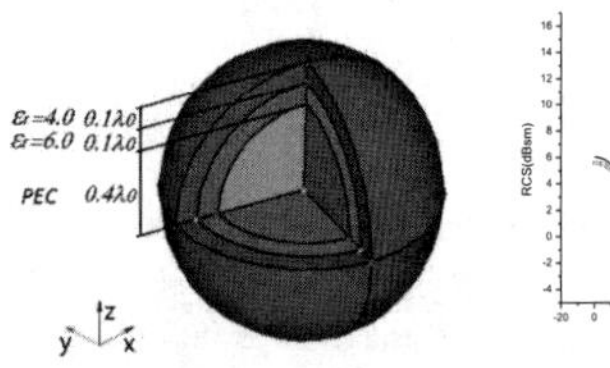

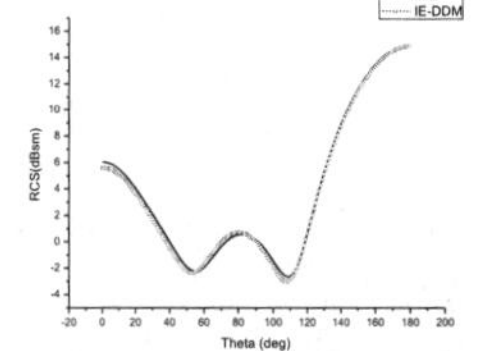

Fig.4.Domain decomposition for coated PEC sphere model

B. EM scattering from simplified aircraft model

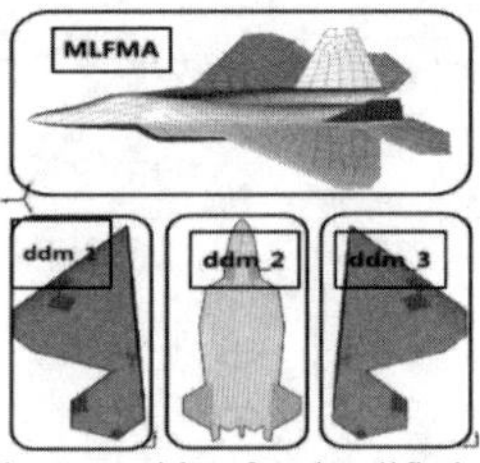

Fig.5.Domain decomposition for simplified aircraft model

In order to show the efficiency and accuracy of nonconformal IE-DDM, the bistatic RCS of a simplified PEC aircraft model is solved. An incident plane wave at 1.0 GHz frequency illuminates from $\theta = 90°$, $\phi = 0°$, the whole model is divided into 3 parts showed in Fig.4. The number of unknowns by the MLFMA and IE-DDM is about 1.6 million and 1.38 million respectively. A good agreement between MLFMA and IE-DDM is also achieved in Fig.5. The SAI preconditioner is used to accelerate the iteration of matrix equation.

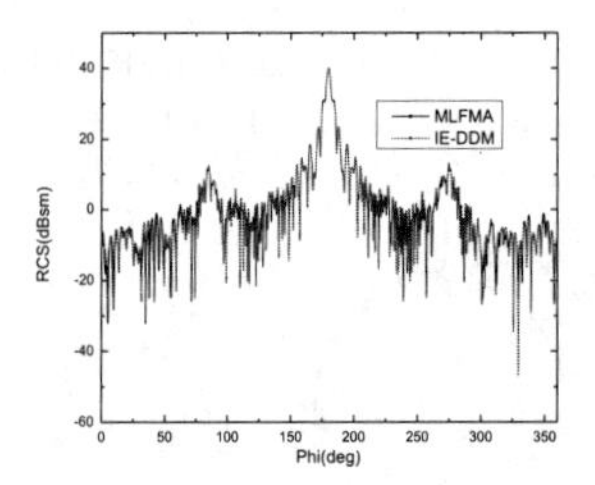

Fig.6.Bistatic RCS for simplified aircraft model

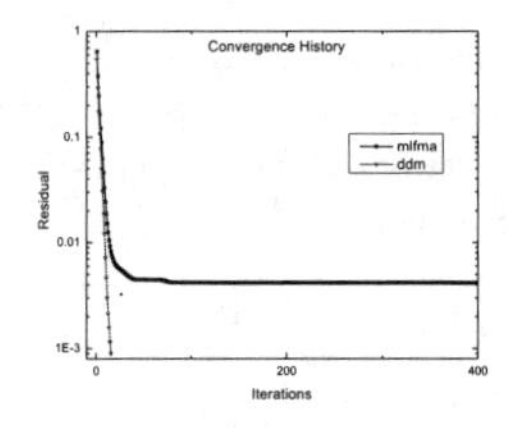

Fig.7. The convergence of IE-DDM and MLFMA for simplified aircraft model

	Memory/GB	CPU Time/h:m
MLFMA	21.6	22:10
IE-DDM	18	15:40

Table 1: Computational Statics of IE-DDM and MLFMA

C. Simplified helicopter model

As shown in Fig. 7, the entire model is divided into 6 closed regions. Each region is meshed independently according to geometry complexities and available computational resources. Due to the non-conformal feature of the proposed IE-DDM, each of the sub-regions can be meshed independently.

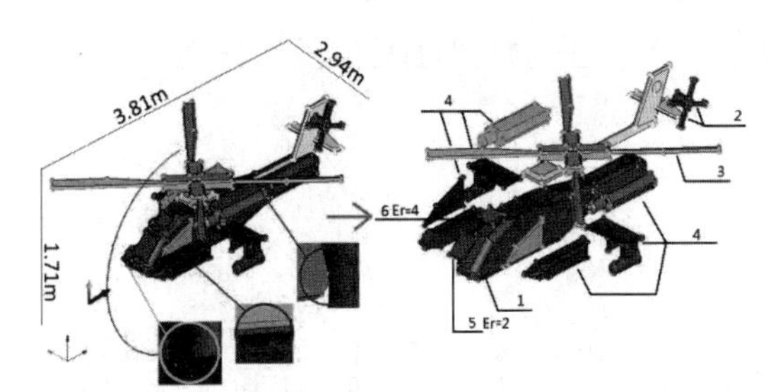

Fig.8.Geometry Model of a simplified helicopter
(Incident wave from θ=90°,φ=0° in 1.5GHz frequency.)

Part 5 and 6 are dielectric structures, their permittivities are 2.0 and 4.0 respectively, while others are all PEC structures. Then the results of scattering fields by different parts of this model are given in Fig.8.

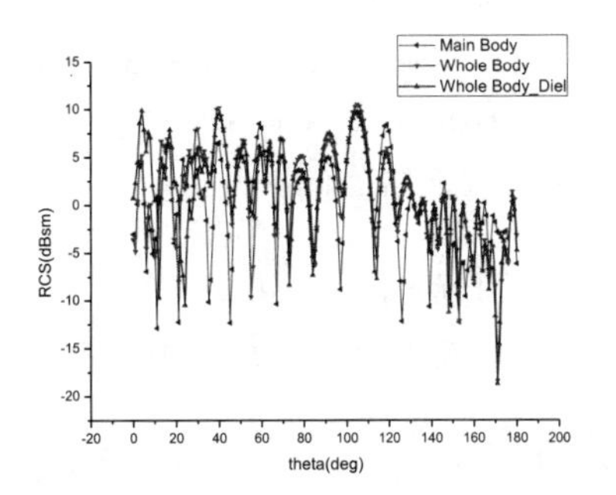

Fig.9.Results comparison of different combined model
(Main and Whole Body are all PEC Parts combination, difference in part 4, while Whole Body_Diel has dielectric structure substituted for PEC, part 5 εr=2.0 and part 6 εr=4.0)

Part	1	2	3	4	5	6	Total
Unknowns	61,389	8,148	28,335	63,510	3,684	5,520	170,586

Table 2: Unknowns of each Part

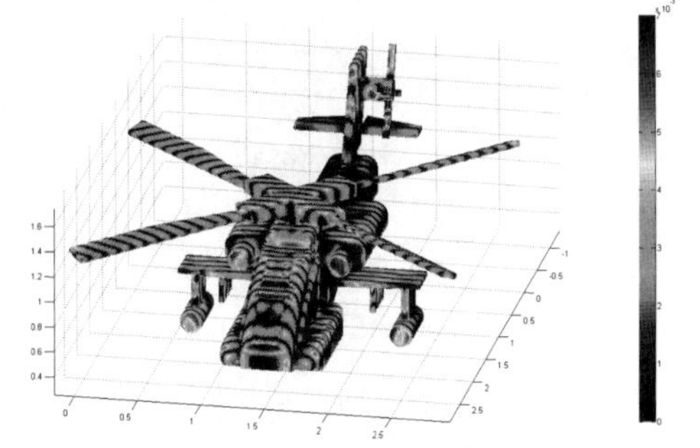

Fig. 10.The current distribution of helicopter model

V. CONCLUSION

In this paper, nonconformal IE-DDM is applied for analysis of electromagnetic scattering from complicated conductor and composite objects. The great benefit of this framework is that it divides the computational domain into subdomains independently, simplifies model processing and reduces the heavy burden of mesh generation. Moreover, it improves greatly the property of matrix, can achieve stable and accurate solution of complicated multi-scale objects in real world.

ACKNOWLEDGEMENT

This work was supported by the National Science Foundation of China (NSFC: No.61271033).

REFERENCES

[1] W. C. Chew, J. M. Jin, E. Michielssen and J. Song, "Fast and Efficient Algorithms in Computational Electromagnetics," *Boston*, MA: Artech House, 2001.

[2] Branko M. Kolund zi ja, ''Electromagnetic Modeling of Composite Metallic and Dielectric Structures,'' *IEEE Trans. Antennas Propag.*, vol. 47, no. 7, pp. 1021–1032, July. 1999.

[3] J. L .Yin, J. Hu, and Z. Nie, "IE-FFT solution of volume-surface integral equation for composite conducting-dielectric objects," in Proceedings of the International Conference on Microwave Technology and Computational Electromagnetics, pp. 414–417, 2009.

[4] F. Vipiana, M. A. Francavilla , and G. Vecchi , ''EFIE modeling of high-definition multiscale structures,'' *IEEE Trans. Antennas Propag.*, vol. 58, no. 7, pp. 2362–2374, Jul. 2010.

[5] J. Liu and J. Jin, ''Scattering analysis of a large body with deep cavities,'' *IEEE Trans. Antennas Propag.*, vol. 51, no. 6, pp. 1157–1167, Jun. 2003.

[6] Putnam, J. M. and L. N. Medgyesi-Mitschang, "Combined field integral equation for inhomogeneous two- and three-dimensional bodies: The junction problem," *IEEE Trans. Antennas Propag.*, vol. 39, No. 5, 667-672, May 1991.

[7] W. B. Ewe, L. W. Li, and M. S. Leong, "Fast solution of mixed dielectric/conducting scattering problem using volume-surface adaptive integral method," *IEEE Trans. Antennas Propag.*, vol. 52, no. 11, pp. 3071-3077, Nov. 2004.

[8] Chen, Q. and D. R. Wilton," Electromagnetic scattering by three dimensional arbitrary complex material/conducting bodies," *IEEE AP Symposium 1990*,V ol. 2,590–593,1990.

[9] Arvas , E. ,A. Rahhal-Arabi,A. Sadigh,and S. M. Rao,"Scattering from multiple conducting and dielectric bodies of arbitrary shape," *IEEE Antennas and Propagation Magazine* ,Vol. 33,No. 2,29–36, April 1991.

[10] Y. Chu. and W. C. Chew, "A multi-level fast multipole algorithm for low-frequency scattering from a composite object, " *IEEE Antenna Pmpagot.*.vol. 1, pp. 43-46. *Columbus, OH, USA* 2003.

[11] Pasi Yla-Oijala and Matti Taskinen, "Electromagnetic Modelling of Composite Structures with Surface Integral Equations of the Second Kind," Computational Electromagnetics Workshop 2007. *Zmir*, pp. 54 - 58, Aug. 2007.

[12] Zhen Peng, Kheng-Hwee Lim, Jin-Fa Lee,''Nonconformal Domain Decomposition Methods for Solving Large Multiscale Electromagnetic Scattering Problems,'' *Proceedings of the IEEE.*, Vol. 101, No. 2, February 2013.

[13] Zhen Peng, Kheng-Hwee Lim ; Jin-Fa Lee , ''Computations of Electromagnetic Wave Scattering from Penetrable Composite Targets using a Surface Integral Equation Method with Multiple Traces,'' *IEEE Trans. Antennas Propag.*, vol. 61, no. 1, pp. 256 - 270, Jan. 2013.

[14] Y. Achdou, C. Japhet, Y. Maday, and F. Nataf. "A new cement to glue nonconforming grids with Robin interface conditions: the finite volume case." *Numer. Math.*, 92(4):593 - 620, 2002.

Validation of FETI-2LM formulation for EBG material prediction and optimal strategy for multiple RHS

André Barka and François-Xavier Roux

ONERA The French Aerospace Lab
BP 4025 – 2 Avenue Edouard Belin – F-31055 Toulouse Cedex 4
andre.barka@onera.fr, francois-xavier.roux@onera.fr

Abstract This paper presents the validation of finite element tearing and interconnecting (FETI) methods for solving electromagnetic frequency domain problems encountered in the design of electromagnetic band gap (EBG) materials. Validation results presented at the EM-ISAE 2012 workshop are discussed.

I. Introduction

The Finite Element Method (FEM) is one of the most successful frequency domain computational methods for electromagnetic simulations . It combines, very efficiently, a geometrical adaptability and ability to handle arbitrary materials for modelling complex geometries and materials of arbitrary composition, including meta-materials that have recently become popular. Finite element approximation of Maxwell's equations leads to a sparse linear system, usually solved by using direct or iterative solvers. However, modern engineering applications dealing with antennas, scatterers or microwave circuits, style require the solution of problems with hundred millions of unknowns.

Domain Decomposition Methods have demonstrated efficiency and accuracy for the solution of Maxwell equations in the frequency domain for both RCS applications and antenna structures interactions [2]. In the domain of finite element methods (FEM), and for the solution of acoustic Helmoltz equations, efficient sub-domain connecting techniques have been applied and called dual-primal finite element tearing and interconnecting [3][4]. These techniques are known under the acronym FETI-DP and have been adapted to electromagnetics (FETI-DPEM) for the calculation of antenna arrays and metamaterial periodic structures [5][6].

This paper presents some recent developments on the Finite Element Tearing and Interconnecting with two lagrange multipliers (FETI-2LM) techniques, for solving large scale FEM problems encountered in electromagnetic applications. Validation results presented at the EM-ISAE 2012 workshop are discussed including cross comparison of the FETI-2LM techniques with FDTD, MLFMM, and Time Domain Discontinuous Galekin methods (TDDG).

II. FETI-2LM Formulation

A. Weak formulation

The general principle of the FETI methods for Maxwell equations is to decompose the global computational domain in non overlapping sub-domains in which local solution fields are calculated by solving the finite element system with a direct method. We then impose the tangent field continuity on the interfaces by using Lagrange multiplier. It results in a reduced problem on interfaces and can be solved by an iterative method. The solution of the interface problem is used as a boundary condition for evaluating the field in each sub-domain. We denote $\Omega = \Omega_1 \cup \Omega_2 \ldots \Omega_N$ a partition of the initial computation domain. In each sub-domain Ω_i (Fig. 1), we are calculating the scattered fields $\vec{E}^i$ verifying on Ω_i:

$$\nabla\times(\mu_{r,i}^{-1}\nabla\times\vec{E}^i)-k_0^2\varepsilon_{r,i}\vec{E}^i = k_0^{\,2}(\varepsilon_{r,i}-\mu_{r,i}^{-1})\vec{E}_{incident} \quad (1)$$

$$\vec{n}_{ext}\times\nabla\times\vec{E}^i + jk_0\vec{n}_{ext}\times(\vec{n}_{ext}\times\vec{E}^i) = 0 \quad (2)$$

The vector $\vec{E}_{incident}$ is representing the electric incident field in the volume Ω_i. Γ_{ABC} represents the boundary of the volume Ω_i, where the field is verifying absorbing boundary conditions (ABC).

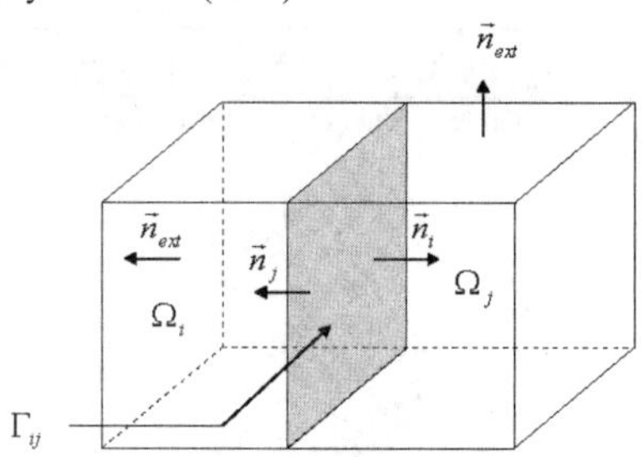

Fig. 1. Interface problem.

In the following, we will denote with $\vec{E}_j^i$ the electric field on the interface of the sub domain Ω_i adjacent to the sub domain Ω_j . On the interfaces Γ_{ij}^{robin} separating two sub-

978-7-5641-4279-7

domains Ω_i and Ω_j, we impose Robin type boundary conditions by using Lagrange multipliers $\vec{\Lambda}^i_j$ and $\vec{\Lambda}^j_i$ which will be the new unknowns:

$$\vec{n}_i \times (\mu_{r,i}^{-1} \nabla \times \vec{E}^i_j) + jk_0 \vec{n}_i \times (\vec{n}_i \times \vec{E}^i_j) = \vec{\Lambda}^i_j \quad (3)$$

$$\vec{n}_j \times (\mu_{r,j}^{-1} \nabla \times \vec{E}^j_i) + jk_0 \vec{n}_j \times (\vec{n}_j \times \vec{E}^j_i) = \vec{\Lambda}^j_i \quad (4)$$

The tangential electric and magnetic field continuity on the interfaces Γ_{ij}^{robin} separating the two sub domains Ω_i and Ω_j leads to the following relations that should be verified by the two Lagrange multipliers $\vec{\Lambda}^i_j$ and $\vec{\Lambda}^j_i$ on Γ_{ij}^{robin}:

$$\begin{aligned} \Lambda^i_j + \Lambda^j_i - 2jk_0 \vec{n}_i \times (\vec{n}_i \times \vec{E}^i_j) = 0 \\ \Lambda^i_j + \Lambda^j_i - 2jk_0 \vec{n}_j \times (\vec{n}_j \times \vec{E}^j_i) = 0 \end{aligned} \quad (5)$$

The weak formulation used for the computation of the scattered fields $\vec{E}^i$ belonging to the space $H(Rot^0, \Omega_i)$'in each volume Ω_i is:

$$\begin{aligned} &\int_{\Omega_i} \left[\mu_{r,i}^{-1} (\nabla \times \vec{E}^i).(\nabla \times \vec{W}) - k_0^2 \varepsilon_{r,i} \vec{E}^i.\vec{W} \right] d\Omega \\ &+ jk_0 \int_{\Gamma_{ext}} (\vec{n} \times \vec{E}^i).(n \times \vec{W}) d\Gamma \\ &+ jk_0 \int_{\Gamma_{ij}^{robin}} (\vec{n} \times \vec{E}^i_j).(n \times \vec{W}) d\Gamma \\ &= k_0^2 \int_{\Omega_i} (\varepsilon_{r,i} - \mu_{r,i}^{-1}) \vec{E}^i.\vec{W} d\Omega \end{aligned} \quad (6)$$

The iterative resolution of the interface problem (5) is based on a Krylov sub-space method. We write equivalently:

$$\lambda^i_j + \lambda^j_i - (M^i_j + M^j_i) E^i_j = 0 \quad \forall \quad i = 1,..,N \quad (7)$$

where j is neighbouring i and

$$M^i_j = jk_0 \int_{\Gamma_{ij}} (\vec{n}_i \times \vec{W}_i).(\vec{n}_i \times \vec{W}_i) d\Gamma \quad (8)$$

The iterative method consists of four steps:

1. Calculation of local solutions in each sub domain with the use of Robin type conditions by solving the problem (5).
2. Exchange fields $\vec{E}$ and Lagrange multipliers $\vec{\Lambda}$ on each interface.
3. $g^i_j = \lambda^i_j + \lambda^j_i - (M^i_j + M^j_i) E^i_j$ computation on each interface.
4. Implementation of ORTHODIR iterations with a stop criterion $\|g\| < \varepsilon$.

B. Optimal strategy for multiple Right Hand Sides

Iteration algorithms, for all FETI methods, are based on Krylov space methods with full orthogonalizations, such as those found in the CG, ORTHODIR or GMRES algorithms. In the event that the same problem must be solved with several right hand sides, one drawback with iterative methods is that they generally need to restart from scratch for each new right hand side. The cost of storing and orthogonalizing a set of interface vectors is small compared to the cost of computing one matrix-vector product for the condensed interface operator.

Let us consider equation $Ax = b$. Once the problem has been solved for the first right hand side, a set of search direction vectors has been built. If the Krylov space method is the conjugate gradient method, these n_c vectors, $(v_j), 1 \le j \le n_c$, as well as their products by the FETI operator, $(Av_j), 1 \le j \le n_c$ have been computed and can be stored in memory. They provide a natural set of vectors to be used to implement a preconditioner based on projection for the next right hand side. Furthermore the $V^t AV$ matrix associated with this set of vectors is diagonal, and it has also been computed during the iterations. Thus, the computation of the projection associated with V is easy.

The new search direction vectors to be computed for the next series of iterations using the preconditioner based on projection would be automatically A-orthogonal to V, and would also be A-orthogonal to one another, thanks to the properties of the conjugate gradient algorithm. So the initial set of vectors can be augmented, for each new right hand side, with the set of newly computed search direction vectors. With this technique, the number of actual new iterations required for each new right hand side tends to decrease dramatically [7].

Of course this method requires us to store all the successive search direction vectors, and to perform a full orthogonalization procedure when computing the projection. But the associated overhead is not so large in the context of the FETI method. Furthermore, the number of stored vectors can be arbitrary limited on demand, for instance according to some memory limits. In practice, the full orthogonalization procedure may also be used for the search direction vectors of the current iterations. It makes the conjugate gradient method even more robust, because it prevents loss of orthogonality due to accumulated round-off errors.

For the case of a non-spd FETI method, if the ORTHODIR algorithm is used, all the search direction vectors as well as their product by the FETI operator have to be stored anyway. They are $A^* A$-orthogonal. They can be straightforwardly used to implement the optimal projection operator with a matrix $V^* A^* AV$ that has already been computed during the ORTHODIR iterations, and is diagonal. The set of vectors can be augmented with the new search direction vectors built for any new right hand side.

The same methodology can be implemented with GMRES, but it is a little bit more technical since the basis built by GMRES is not A^*A-orthogonal, but simply orthogonal. Nevertheless, the advantage of GMRES over ORTHODIR is not so evident, in the context of FETI methods, since the storage of extra interface vectors is not so expensive. Furthermore, GMRES does not compute the value of the approximate solution of the condensed interface problem at each iteration. It is necessary to derive the approximate solution to compute the associated approximate solution of the global problem if the convergence is to be monitored according to the residual of the global problem.

III. EBG MATERIAL

C. EM-ISAE-2012 WORKSHOP

We consider the test of the EBG material designed by ONERA for the EM-ISAE 2012 Workshop.

This test is concerned with the simulation of the fields diffracted by an 8x20 EBG array comprising of alumina dielectric rods (Fig. 2) and centred at the origin (0,0,0). All the dimensions are given in millimetre (mm) in the Table I.

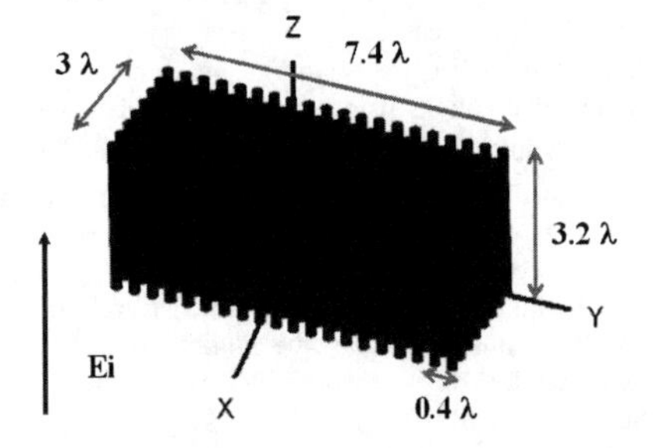

Fig. 2. 8x20 EBG array

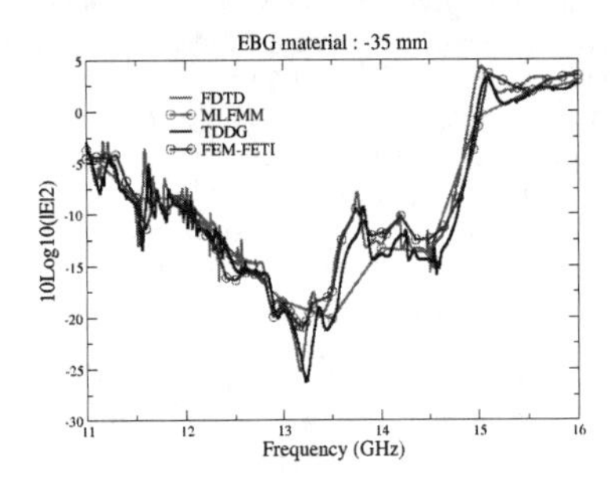

Fig. 3. Near field at point (-35,0,0) mm.

Table I
Geometry of EBG rods.

Rods diameter (mm)	4
Rod length (mm)	60
Array step (mm)	7
Alumina permittivity	$\varepsilon_r = 9.4$

The array is excited by a uniform and unitary plane wave (|Ei|=1), whose electric field is polarized parallel to the axis OZ of the rods and whose incidence direction is collinear to the OX axis (θ=90, φ=0). The parameters to be simulated are the total electric field transmitted behind the array at point (-35 mm,0,0).

The results are obtained with the Finite Element module of the FACTOPO frequency domain code [2] calling the FETI-2LM [1] library. Four empty unit cells are introduced around the 8x20 array to keep the ABC surface sufficiently far away from the scatterer. The size of the simulated array is then 16x28 corresponding to 448 sub-domains. Each domain contains one rod and is meshed with 139,547 edges. The total number of unknowns is 62.52 millions, and the computation parameters are presented in the Table II. The zero-order Whitney edge elements (6 degree of freedom in each tetrahedral) are considered. The local linear system resolution in each sub-domain is performed with the Intel MKL PARDISO solver. The sub domains are connected together iteratively with the ORTHODIR and the simulation results are plotted in pink on Fig. 3. They are in good agreement with the Finite Difference Time Domain simulations (FDTD [2], brown curve), Time Domain Discontinuous Galerkin simulations [1] (TDDG [3], magenta curve) and Multi Leval Fast Multipole Methods simulations (MLFMM [4], green curve). The FEM-FETI results need to be improved for the lowest levels by implementing the PML layers instead of the ABC. The calculation of the electric field in the EBG structure also shows a strong attenuation of the field (Fig. 4) in the band gap around 12 GHz.

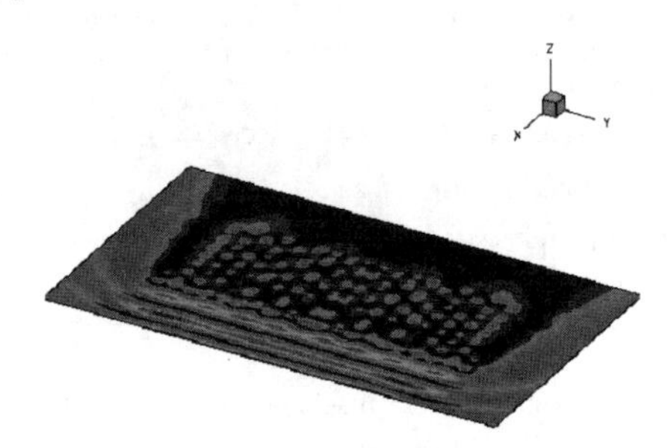

Fig. 4. E field at 12 GHz.

Table II
EBG: computation parameters.

[1] FEM-FETI results provided by the FACTOPO code of ONERA, France
[2] FDTD results provided by the SOPHIE code of CEA, France
[3] TDDG results provided by the SEMBA code of CASSIDIAN/UGR, Spain
[4] MLFMM results provided by the HP-TEST code of CASSIDIAN, Spain

Unknowns	62.52 millions
Longest edge	λ/13
Elapse time per frequency	1,372 s
Computer	448 cores X5560, 2.8 GHz
Memory used per core	1.8 Gb
Iterations par frequency	500
Residual (stop criteria)	1e-3

D. Analysis of scalability

The convergence of the FETI-2LM method is analysed by considering arrays with progressively increasing size (2x3, 4x7, 8x14, 16x28). The unit cell was done, as previously with 139,547 edges, for all of the arrays. The elapse time and the evolution of the number of iterations required for a convergence lower than 10^{-3} is indicated in Table. III. It is observed a linear increase of the number of iterations required for convergence as the array size is increased. This is due mainly to the global distribution of the diffracted field induced by the plane wave in the EBG material, which leads to strong interactions between the cells of the array.

Table III

EBG: statistics versus array size.

Array size	2x3	4x7	8x14	16x28
Cores (X5560 2.8 GHz)	6	28	112	448
Unknowns (millions)	0.837	3.9	15.6	62.52
Elapse time (s)	113	232	785	1,372
Iterations (stop criteria 10^{-3})	70	123	201	500

E. Assessment of the multiple Right Hand Side strategy

Radar Cross Section (RCS) analysis is requiring the computation of the diffracted fields for the vertical and horizontal polarisation of the incident field and a large number of incidence directions. Similarly, in the context of antenna array applications, there is an interest in evaluating the coupling levels between the array elements. This information is usually obtained by calculating the scattering matrix of the array. The procedure consists of sequentially exciting all of the elements of the array and computing the behaviour of the electromagnetic fields at all the feeding ports. The number of right hand sides is potentially prohibitive and one drawback with iterative methods is that they need to restart from scratch for each new right hand side. The optimal strategy developed in subsection II.B is assessed for the EBG material test. We consider a uniform plane wave exciting the arrays 2x3 and 8x14 for 10 incidences angles varying from (ϴ=90, φ=0) to (ϴ=90$, φ=10) with an angular step of 1 degree. We observe from Fig. 5 that the number of iterations required for convergence is significantly reduced by a factor of 8 and 5.6 respectively for the 2x3 and 8x14 arrays. The maximum number of direction vectors stored (subsection II.B) is 150 for the 2x3 array, while it is 656 for the 8x14 array.

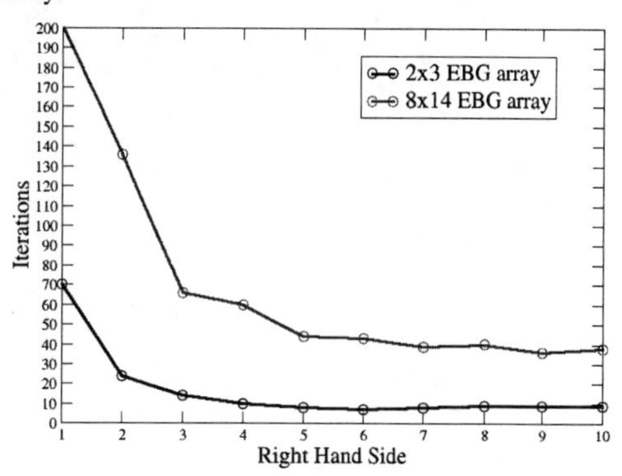

Fig. 5. EBG: iterations versus right hand side

REFERENCES

[1] J. Alvarez, L. D. Angulo, A. R. Bretones, C. de Jong, and S. G. Garcia, "3D discontinuous Galerkin time domain method for anisotropic materials," IEEE Antennas and Wireless Propagation Letters, vol. 11, pp. 1182-1185, 2012.}

[2] A. Barka and P. Caudrillier "Domain Decomposition Method based on Generalized Scattering Matrix for installed performances of Antennas on Aircraft", *IEEE Trans. Antennas and Propagation,* Vol. 55, No 6, June 2007.

[3] C. Farhat, A. Macedo, M. Lesoinne, F-X Roux and F. Magoules, "Two-levels domain decomposition methods with Lagrange multipliers for the fast iterative solution of acoustic scattering problems", *Comput. Methods Appl. Mech. Engrg.,* vol.184, pp.213-239, April 2000.

[4] F-X Roux, F. Magoules, L. Series and Y. Boubendir, "Approximation of the optimal interface boundary condition for two-Lagrange multiplier FETI method", *Lecture Notes Computational Sci. Engrg,* vol. 40, pp. 283-290, 2005.

[5] MN. Vouvakis, Z. Zendes and J-F Lee, "A FEM Domain Decomposition Method for Photonic and Electromagnetic Band Gap Structures", *IEEE Trans. on Antennas and Propagation,* Vol. 54, No 2, pp. 3000-3009, February 2006.

[6] Y. Li, J-M Jin," A Vector Dual-Primal Finite Element Tearing and Interconnecting Method for Solving 3-D Large-Scale Electromagnetic Problems", *IEEE Trans. on Antennas and Propagation,* Vol. 54, No 10, pp. 721-733, October 2006.

[7] C. Farhat, L. Crivelli and F.-X. Roux, "Extending Substructure Based Iterative Solvers to Multiple Load and Repeated Analyses", Compt. Methods Appl. Mech. Engrg., vol. 117, 1994.

WP-1(C)

October 23 (WED) PM

Room C

UWB Antennas

Low VSWR High Efficiency Ultra-Wideband Antenna for Wireless Systems Applications

Li Na* Zhang Haili** Jiang Yi**

*(Institute of Electronics, Chinese Academy of Sciences, Beijing, 100190, China)

**(ZheJiang Textile & Fashion College, Ningbo, 315211, China)

Abstract- **In this paper, a novel and compact ultra-wideband (UWB) antenna with frequency range 800MHz to 6GHz for application in UWB systems and wireless indoor coverage systems is presented. It is composed of a V-cone and a three-quarter sphere geometry, two quarter-wavelength short stubs are also used in the antenna to match the impedance, which can significantly improve the voltage standing wave ratio (VSWR) of the antenna. Details of the antenna design and measurement results are presented in order to demonstrate the performance of the proposed antenna.**

I. Introduction

The past decade has seen a phenomenal growth in wireless communications and antennas with wide bandwidth are in strong demand, so that various application are covered with fewer or preferably with a single antenna. It will be preferred that an antenna has bandwidth in exceed of frequency range from 800MHz to 11GHz or even more [1], to include all the existing wireless communication systems. To satisfy such a requirement, a variety of studies have been explored in the frequency range 3.1-10.6GHz [2-7]. However, the UWB antenna used in frequency band from 800MHz to 6000MHz is rarely reported.

In this paper, a new type antenna with the V-cone and three-quarter sphere geometry is presented. The cone-sphere geometry is eventually developed for a UWB antenna which the ultra antenna can used in frequency band 800-6000MHz; Two quarter-wavelength short stubs are used in the novel UWB antenna to match the impedance, which can significantly achieve omni-directional pattern, and high gain, and improve the voltage standing wave ratio (VSWR) of the antenna. As expected at the beginning of this research, the VSWR is less than 2 in the frequency band 800-6000MHz.

II. ANTENNA DESIGN

Since cutting effect, finite V-cone antenna leads to large reflected current, and the impedance bandwidth is bad, so it's popular that the loading method is used to design UWB antennas. In order to get excellent impedance performance, a V-cone antenna of sphere loaded is designed in this paper. The geometry and dimensions of the UWB antenna is shown in Fig. 1, including ground plane, radial element of cone-sphere, short stubs and feed element. As the radial element, cone-sphere structure is composed of V-cone and one-second of the sphere. The height and angle of V-cone separately are 25.3mm and 90 degree, respectively, and the radius of loaded sphere is 40mm. Quarter of the sphere crown is cut to reduce the profile. Moreover, there are two quarter-wavelength short stubs inside the UWB antenna, which can be used to match the impedance. Consequently, the antenna bandwidth is future broadened. The one end of strip line is fixed in ground plane, the other end is fixed in the edge of spherical crown, and two short stubs are placed symmetrically, the length and width are 90mm and 10mm.

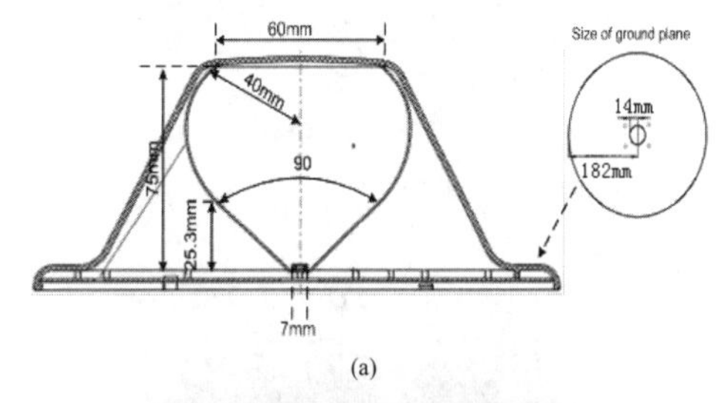

(a)

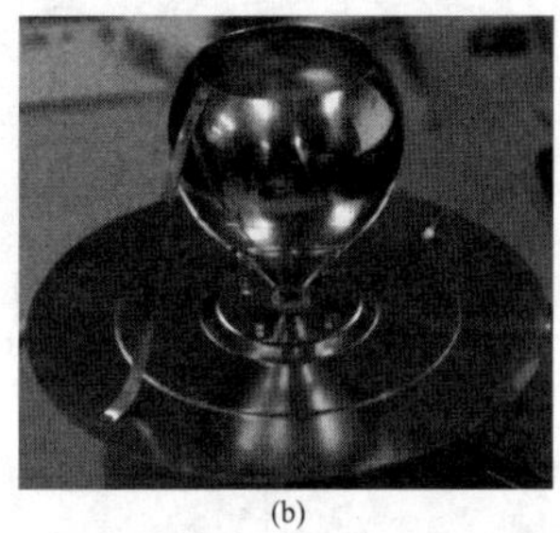

(b)

Fig. 1 The geometry configuration of the proposed antenna
(a) Schematic (b) Prototype front view

III. Experimental Results and Discuss

To verify the performance of the proposed approach, the UWB antenna was measured after fabrication. Fig. 2 shows the measured radiation patterns in E-plane (x-z) and H-plane (y-z) at 890MHz, 2010MHz, 3300MHz and 5900MHz, respectively.

978-7-5641-4279-7

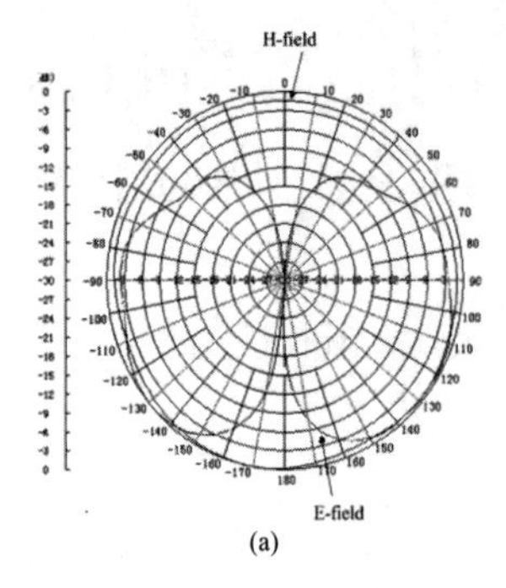

(a)

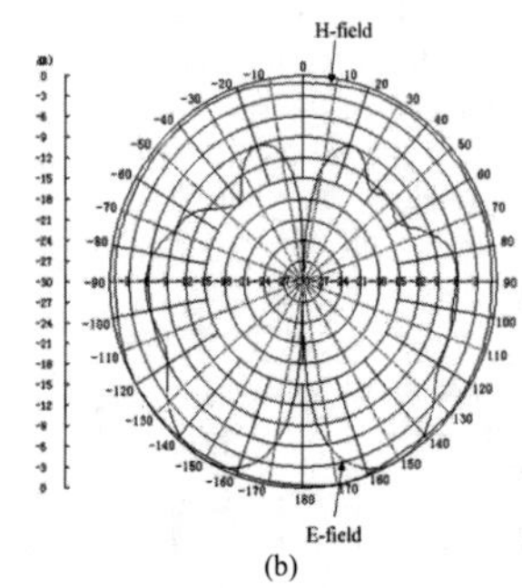

(b)

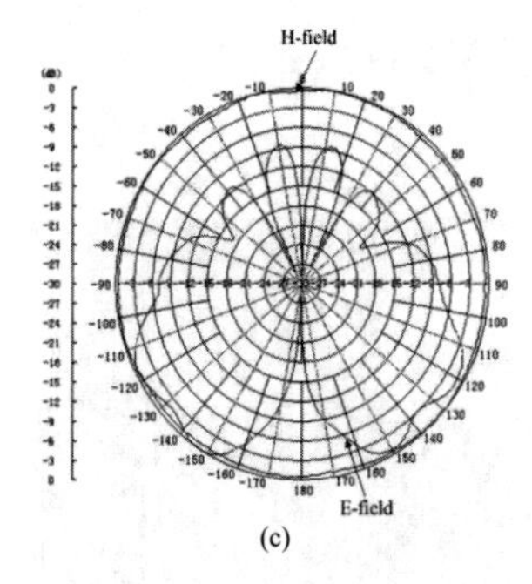

(c)

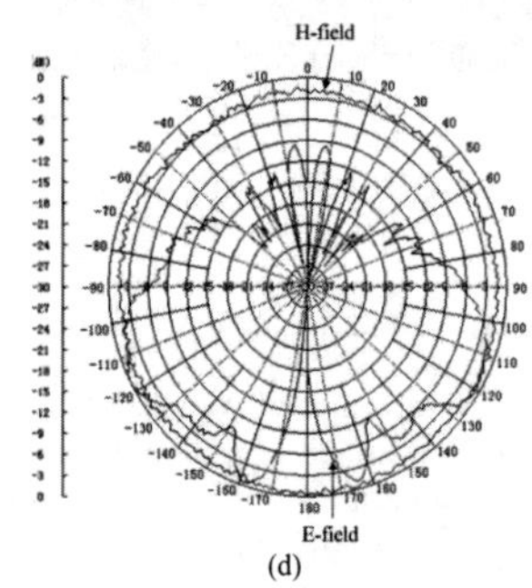

(d)

Fig. 2 Radiation patterns of the E-field and H-field at
(a) 890MHz (b) 2010MHz (c) 3300MHz (d) 5900MHz

As shown in Fig. 2(a) and 2(b), it can be perceived that the omni-directional patterns of E-filed are achieved when the frequency is at relative low value, but when the frequency is above 3GHz, the E-filed radiation patterns show the direction, especially in the near 6GHz. And the reason is contributed that the size of radius ground plane approximately equal or exceed the wavelength, and affect the radiation patterns. However it can be accepted in application of indoor coverage. The gain of the antenna is larger than 2.5dBi in the frequency band and it increases monotonously with the frequency band from 800MHz to 3550MHz. The maximal gain is 6.8dBi at 3550MHz. Therefore, the gain of the UWB antenna can be calculated by Eq.(1).

$$G_x(dBi)=G_s(dBi)+10lg(P_{xr})-10lg(P_{sr}) \qquad (1)$$

where, $G_x(dBi)$ is the gain of the UWB antenna, $G_s(dBi)$ is gain of the standard antenna, P_{xr} is the power of the UWB antenna, P_{sr} is the power of the standard antenna. Table 1 represents the gain of the UWB antenna.

Table 1
Gain of the antenna

Freq.(MHz)	G_s(dBi)	10lg(P_{xr})(dBm)	10lg(P_{sr})(dBm)	G_x(dBi)
824	12.94	-43.07	-32.70	2.57
860	13.30	-44.14	-33.20	2.36
890	13.60	-44.53	-33.21	2.28
960	14.30	-45.26	-33.45	2.49
1880	15.26	-47.16	-37.13	5.23
1990	15.62	-47.39	-37.26	5.49
2170	16.17	-48.47	-37.71	5.41
2700	17.57	-40.86	-28.98	5.69
3300	19.98	-41.79	-27.2	5.39
3550	20.37	-40.57	-27.08	6.88
5100	20.34	-47.41	-32.5	5.43
5900	21.13	-49.23	-32.95	4.85

Fig. 3 shows the comparison of the measured typical VSWR values of the UWB antenna with one short-stub, two short-stubs and no short stubs, respectively. It is apparent that the VSWR<1.5 in the frequency band 800-3800MHz and 4.5-6GHz when using two strips. Obviously, the UWB antenna can be used in indoor coverage system for GSM, CDMA, TD-SCDMA, WCDMA, WiMax, WLAN and CDMA2000 when using two stubs. It's sure that short-stubs main to match the impedance, and improve the poor VSWR performance in 800-1300MHz frequency band.

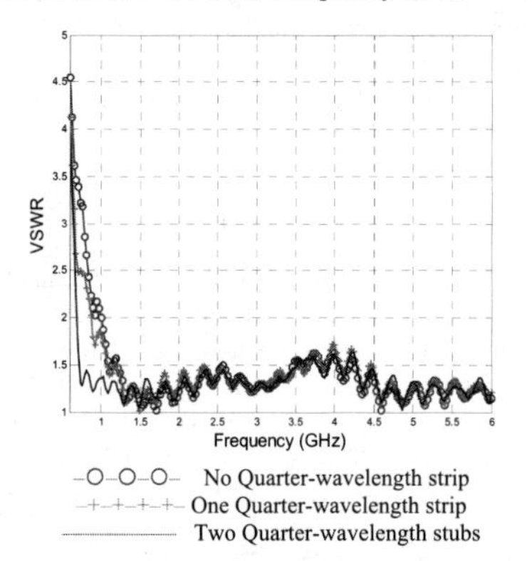

Fig. 3 The comparison of VSWR of the UWB antenna

IV. CONCLUSIONS

From the result of measurement of VSWR, radiation patterns and gain, the good performances are achieved, the VSWR of the UWB antenna is less than 1.7 in frequency band 800-6000MHz. The gain of the UWB antenna is more than 5dBi, when the frequency band is 1500-6000 MHz. Therefore, the UWB antenna can be used in GSM, CDMA, WCDMA, CDMA2000, TD-SCDMA, WiMax and WLAN.

REFERENCE

[1] Abdallah A. A. and Sebak A.R.:'A novel UWB planar patch antenna for wireless communications'. IEEE Transactions on Antennas and Propagation, Vol. 55(2), 2008, pp. 383-388.

[2] Chin Yeng Tan, Selvan K. T: 'The metallic cone-sphere-inserted conical hom as an alternative to $10\lg(P_{sr})$(dBm) the dielectric-loaded conical hom'. IEEE Applied Electriomagnitics Conference (AEMC), 2009.

[3] Gronich I.: 'Omni directional ultra wideband asymmetric biconical antenna'. IEEE International Conference on Microwaves, Communications, Antennas and Electronics Systems (COMCAS), 2009.

[4] Xing Ping Lin and Lingxi Luo: 'A low profile L shape meander circular ultra wideband antenna'. IEEE International Symposium on Antennas and Propagation Society (AP-S), 2008.

[5] Wang Feng, Zhang Kun-fan, Li Dong-hai,: 'An novel design for an ultra-wide band antenna'. The 7th international Symposium on Antennas, Propagation & EM Theory, 2006.

[6] Yang X., Yu Z., Shi Q. and Tao R.: 'Design of novel ultra-wideband antenna with individual SRR'. Electronics Letters', Vol. 44, No. 19, 2008, pp.1109-1110.

[7] Kawakami H., Sato G.: 'Broad-band charactristics of rotationally symmetric antennas and thin wire construct'. IEEE Transactions on Antennas and Propagation, Vol. 35, Issue 1, 1987, pp.26-32.

Printed Slot Antenna for WLAN/WiMAX and UWB Applications

Pichet Moeikham[1] and Prayoot Akkaraekthalin[2]
[1]Department of Electrical Engineering, Faculty of Engineering, Rajamangala University of Technology Lanna Chiang Rai, Thailand
[2]Department of Electrical Engineering, Faculty of Engineering, King Mongkut's University of Technology North Bangkok, Thailand

***Abstract*-A printed slot antenna for WLAN/WiMAX and UWB applications is proposed. The rounded corners of slot and spade-shaped exciting patch are applied to the pentagonal slot. The proposed antenna capably responds to extremely wide impedance bandwidth. For impedance matching enhancement, a tapered CPW transmission line is introduced. The impedance bandwidth over 138 % is obtained. The radiation patterns likely omni-directional is given by the proposed antenna at all frequencies in xz plane. Therefore, the proposed antenna is suitable to use for various WLAN/WiMAX and UWB applications.**

I. INTRODUCTION

An ultrawideband (UWB) occupies frequency range from 3.1 to 10.6 GHz employing for short distance indoor wireless communications, high data rate, and low power consumption. Wireless local area network (WLAN) with IEEE802.11b/g/a (2.4-2.484 GHz, 5.15-5.35 GHz, and 5.725-5.825 GHz) standards is short range communication which prefers to use as indoor backbone. Besides, the wireless broadband internet has emerged recently known as worldwide interoperability for microwave access (WiMAX) supporting the IEEE802.16e (2.5-2.69 GHz, 3.3-3.7 GHz, and 5.25-5.85 GHz) standard. Clearly, the operating frequencies of WLAN and WiMAX have been allocated close together and the frequency bands of 3.5 GHz and 5.5 GHz have been allocated within UWB range. The wide band antenna, which occupies entire frequencies of WLAN/WiMAX and UWB systems, has possessed this vantage. Printed slot antenna is one of most antenna which suits for those modern wireless applications. By designing with various slot shapes and tuning stubs, the printed slot antennas capably operate with multi band or extremely wide band response. Many techniques had been used for the wide bandwidth slot antennas designing. For examples, the compact printed slot antenna [1] is proposed with 169 % impedance bandwidth, the flare angle of bow-tie slot and tapered CPW fed for UWB response is used in [2]. Although, these antennas capably operate with very wide bandwidth, the lower operating frequency is confined and no covered WLAN/WiMAX frequencies. The rounded corner technique of rectangular slot is applied in [3] and the circular slot with arc-shaped tuning stub is used in [4]. However, both antennas have been designed with large size which no proper to use with compact wireless applications. In this article, an alternative printed slot antenna with extremely wide impedance bandwidth is proposed. The prior printed slot antenna [5] is modified. The rounded corners with gradual slop of the slot and spade-shaped exciting patch are applied. The proposed antenna capably affords the impedance bandwidth covering entire WLAN/WiMAX and UWB frequency bands. All antenna designs and discussion are following.

II. ANTENNA DESIGN

Fig. 1 (a) illustrates the geometries of antenna design procedures which lie on *xy* plane. Fig. 1 (b) depicts antenna model simulation results. All models are designed and simulated by the software IE3D^{TM} with one metallic side on FR4 substrate with the thickness of 1.6 mm (permittivity of 4.4 and loss tangent of 0.019). The design procedures are as follow. Firstly, the metallic on an FR4 substrate with *W1* of 31 mm and *S1* of 35.55 mm is determined and cut with trapezoid slot as shown in Fig. 1 (a). The proper spade-shaped exciting patch, which determined *D1* of 11.58 mm, *D2* of 9.3 mm, *P1* of 9.33 mm and *P2* of 5.75 mm, is used. The tapered CPW with 50 ohm characteristic impedance is applied for feeding signal corresponding to the signal line width *FD1* of 3 mm and the gap distance of 0.5 mm. The antenna model 1 capably responds very wide impedance bandwidth. However, the higher frequency is no cover whole UWB range. Secondly, by using sloped lower edge of the slot, the antenna model 2 maintains the lower edge frequency with more steep than the previous model. Although the high frequency has improved impedance matching, the overall $|S_{11}|$ response is not proper. Finally, by making rounded top corners of slot as shown in model 3, the antenna affords appropriate $|S_{11}|$ response with very wide bandwidth from 2.15 to over 12 GHz. This antenna has potential to operate covering for many WLAN/WiMAX and UWB applications. The third model is therefore chosen to propose. The configuration of proposed antenna is shown in Fig. 2. All optimal parameters have got from software IE3D^{TM} including *S2* of 15.9, *S3* of 14, *S4* of 11.5, *F1* of 16.96, *F2* of 6.34, *FD1* of 3, *W2* of 29.4, and *W3* of 26.93, respectively, which are in mm unit.

Clearly, the gradual-sloped lower edge and rounded-corners of the slot are applied for achieving the appropriate frequency response. The *W2*, *W3*, and *F3* are major parameters affecting to the antenna characteristics, which are studied. Each

978-7-5641-4279-7

parameter is varied while the others are fixed. Fig. 3 (a) depicts the return loss simulated results when *W2* is varying. It is found that the lower edge frequency of the antenna is controlled by the *W2* parameter. When varying of *W2* from 27 to 29.4 mm, the lower frequency varies from 2.3 to 2.15 GHz. There is moderate effect at frequency interval from 3 to 5.5 GHz. When W_2 is increased, the impedance matching is slightly improved. A little effect to the higher band (6-10.6 GHz) is also found. The optimal *W2* of 29.4 mm can provides the operating frequency from 2.15 to 12 GHz covering all WLAN/WiMAX and UWB bands. At frequency from 2 to 4 GHz and 7.5 to 10 GHz, the variation of *W3* has major effect to impedance matching, as shown in Fig. 3 (b). It can be seen that the optimal $|S_{11}|$ is obtained when the *W3* is set to be 27 mm. Additionally, almost operating band is evidently achieved the enhancement impedance matching by using the tapered central line of CPW denoted by *F3* except the frequency of 5.5 GHz, as shown in Fig. 3 (c). As mentioned the simulated results, the gradual increment of tapered part has more effect with slightly improve impedance matching. The appropriate impedance matching of entire operating frequency is given by the optimal *F3* of 10.6 mm.

III. Implementation and Measurement Results

All parameters, optimized by software IE3D^{TM}, were used to implement the prototype antenna. The FR4 substrate with the thickness of 1.6 mm, permittivity of 4.4 and loss tangent of 0.019 is used for fabrication the prototype antenna by chemical method. A photograph of the prototype antenna is shown in Fig. 4 (a). The SMA connector is employed for interfacing the prototype antenna and measuring equipment. A comparison of simulated and measured return loss results is illustrated in Fig. 4 (b). It is found that the lowest edge frequency from measured result is slightly shifted to higher frequency. Although the resonant frequencies (higher than 5 GHz) have been slightly shifted to lower frequency, the comparison results are good enough for acceptable. The fabrication tolerance and discontinuous transformation from the SMA to CPW transmission line are causes of those discrepancies. The operating frequency from 2.2 to over 12 GHz is given by the proposed antenna which has potential to operate for all WLAN/WiMAX and UWB applications.

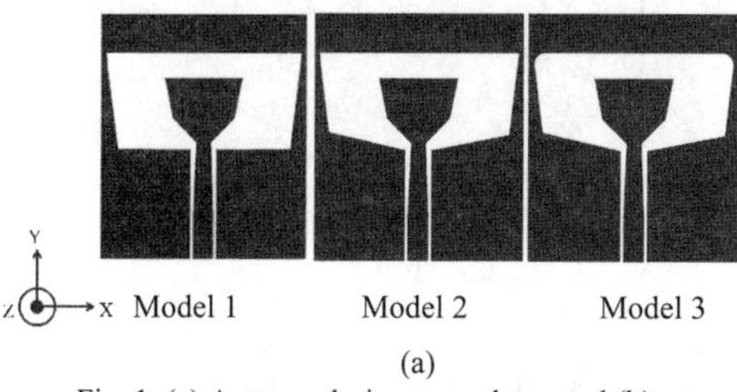

(a)

Fig. 1. (a) Antenna design procedures and (b) return loss simulated results of antenna designs.

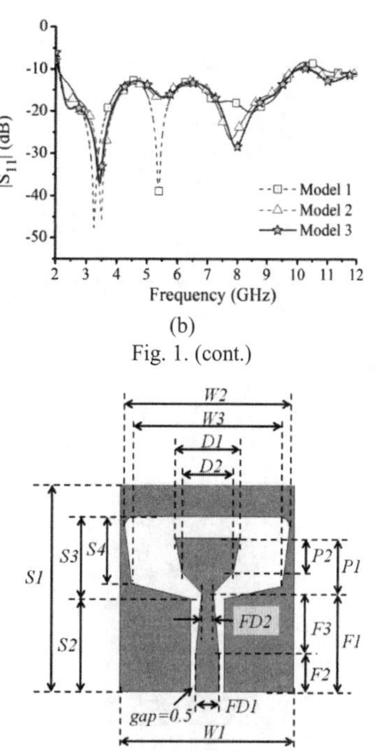

(b)

Fig. 1. (cont.)

Fig. 2. Geometry and configuration of the proposed antenna.

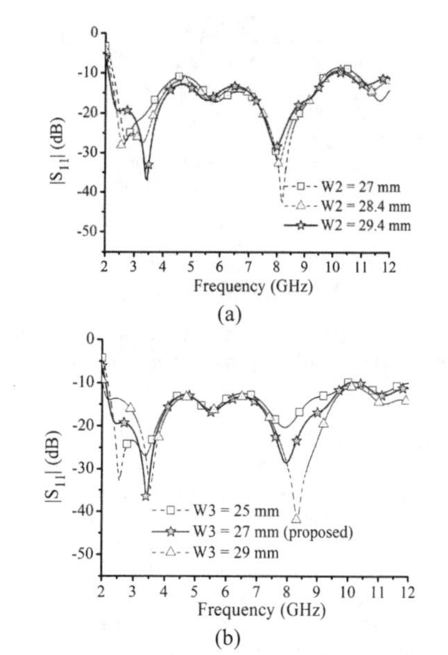

(b)

Fig. 3. Return losses of simulated results when varying of (a) *W2*, (b) *W3*, and (c) *F3*.

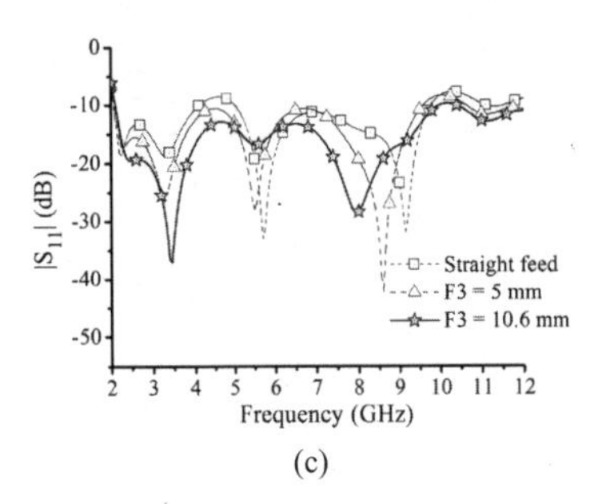

(c)

Fig. 3. (cont.)

(a)

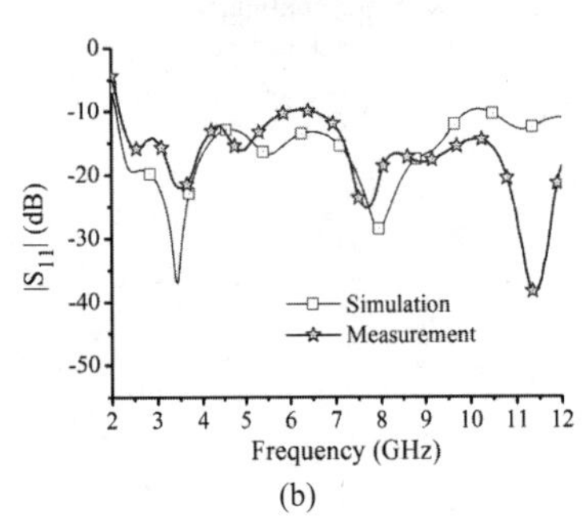

(b)

Fig. 4. (a) photograph of prototype antenna and (b) comparison return loss of simulated and measured results.

The measured results of radiation patterns in *xz* and *yz* planes at the frequencies of 2.5, 3.5, 5.5 and 8.5 GHz are depicted in Figs. 5 (a)-(d), respectively. In *xz* plane, the measured results show that the co-polarization of radiation patterns provide nearly omni-directional at all frequencies, as shown in Figs. 5 (a) and (b). Additionally, high cross polarization is also found at all frequencies, owing to the current distribution flows along the rims around slot and the exciting patch acting as radiating part. In *yz* plane, the co-polarization of radiation patterns shows likely bi-directional at all frequencies. Moreover, some distortions of radiation pattern are found about at 120 to 240 degree at frequency of 8.5 GHz, due to the effect of electromagnetic self-interference at high frequency. Besides, the cross-polarization is still high level at all frequencies. It can be concluded that the proposed antenna operates with linear polarization which suitable for various mobile applications.

Fig. 6 (a) illustrates $|S_{21}|$ and group delay time of the proposed antenna. By setting up two identical antennas with face to face placing distance of 35 mm, time domain measured results are obtained from network analyzer. It can be seen that the $|S_{21}|$ at frequency from 3.5 GHz to 5.5 GHz is fluctuated and slightly decreased at the frequency higher than 8 GHz. The average group delay time shows nearly constant approximate 2 ns with deviation of 8 ns at frequency form 2 GHz to 12 GHz. The measured results proved that the proposed antenna provides adequately linear operation for impulse transmission. Fig. 6 (b) shows measured broadside gain result. The peak level of 3.8 dB at 4.5 GHz is obtained while more fluctuation is occurred at frequency from 8 GHz to 10 GHz. The average broadside gain measured results of entire operating frequencies is about 1.36 dB.

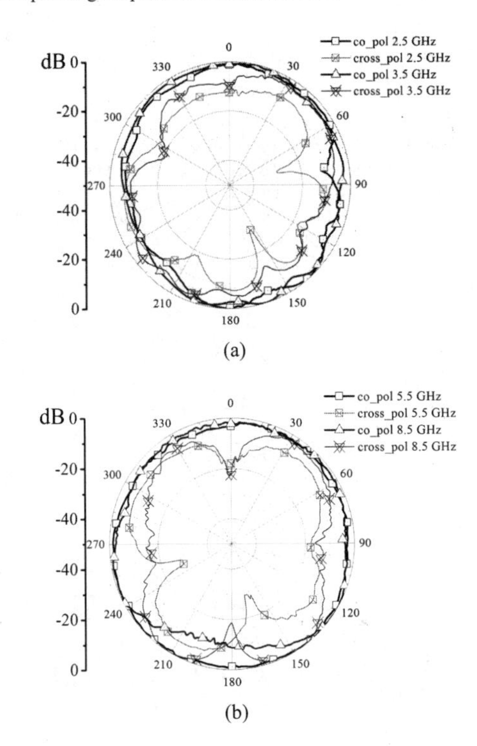

Fig. 5. Measured radiation patterns in *xz* plane (a) at 2.5 GHz, 3.5 GHz, (b) at 5.5 GHz, 8.5 GHz, in *yz* plane (c) at 2.5 GHz, 3.5 GHz and (d) at 5.5 GHz, 8.5 GHz.

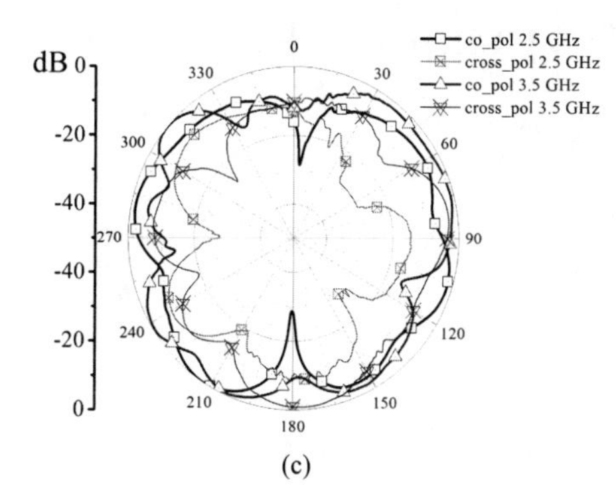

(c)

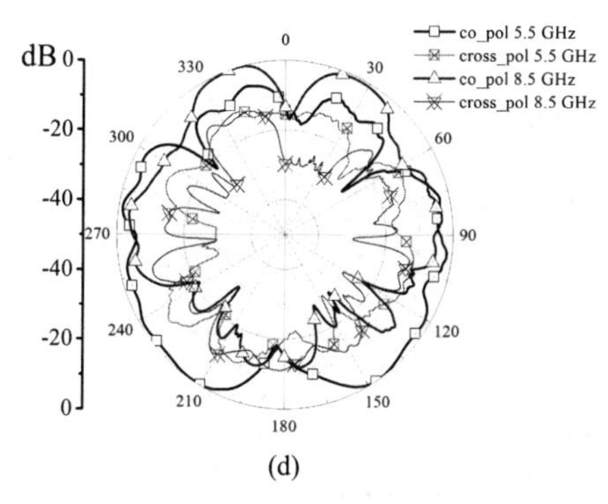

(d)

Fig. 5. (cont.)

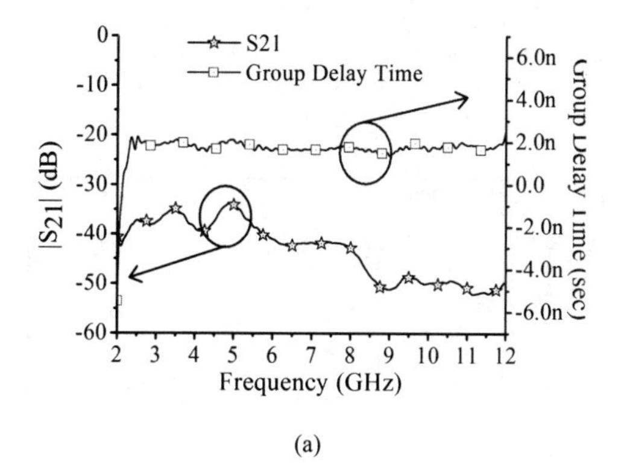

(a)

Fig. 6. (a) Measured time domain and (b) broadside gain of the proposed antenna.

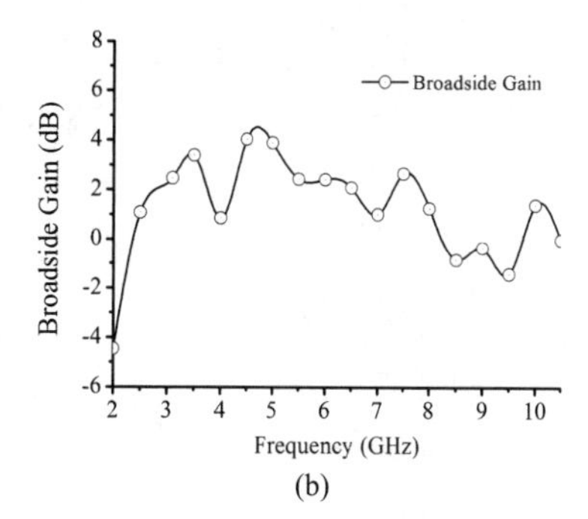

(b)

Fig. 6. (cont.)

IV. CONCLUSION

A compact printed slot antenna for WLAN/WiMAX and UWB applications is designed and presented. The conductor ground plane slot is modified by making gradual slope and round the top corners. The proper spade-shaped exciting patch with tapered CPW-fed is also used. The proposed antenna capably operates with extremely wide impedance bandwidth from 2.2 GHz to over 12 GHz. The dimension of proposed antenna is less than a half guide wavelength of the lowest operating frequency. The measured results proved that the proposed antenna capably operates supporting multi standards of IEEE802.11b/g/a, IEEE802.16e, and IEEE802.15. Additionally, nearly omni-directional radiation patterns are obtained at all frequencies in *xz* plane. Therefore, the proposed antenna is suitable to apply for various wireless applications.

ACKNOWLEDGMENT

Authors would like to thank the Telecommunication Laboratory of Engineering Faculty, Rajamangala University of Technology Krungthep (RMUTK) for antenna measurement.

References

[1] M. Koohestani and M. Golpour, "Very ultra-wideband printed CPW-fed slot antenna," Electron. Lett., vol. 45, no. 21, 8th Oct. 2009.

[2] C.-Y. Huang and D.-Y. Lin, "CPW-fed bow-tie slot antenna for ultra-wideband communications," Electron. Lett., vol. 42, no. 19, 14th Sep. 2006.

[3] X. Chen, W. Zhang, R. Ma, J. Zhang and J. Gao, "Ultra-wideband CPW-fed antenna with round corner rectangular slot and partial circular patch," IET Microw. Antennas Propag., vol. 1, no. 4 pp. 847–851, 2007.

[4] Meng-Ju Chiang, Tian-Fu Hung, Jia-Yi Sze, and Sheau-Shong Bor, "Miniaturized dual-band CPW-fed annular slot antenna design with arc-shaped tuning stub," IEEE Trans. Antennas Propag., vol. 58, no.11, pp. 3710–3715, Nov. 2010.

[5] P. Moeikham and P. Akkaraekthalin, "A pentagonal slot antenna with two-circle stack patch for WLAN/WiMAX," In Proc. ISPACS, Dec. 2011.

A Compact Lower UWB Band PIFA for BAN Applications

Xiong-Ying Liu
School of Electronic and Information Engineering,
South China University of Technology
Guangzhou 510640, China
liuxy@scut.edu.cn

Chun-Ping Deng
MOBI Antenna Technologies (SHENZHEN) Co., Ltd.
Shenzhen 518057, China

Yu-Hao Fu
Samsung Guangzhou Mobile R&D Center
Guangzhou 510640, China

***Abstract*- A compact planar inverted-F antenna (PIFA) with a simple feed structure in lower ultra-wideband (3.1~4.9 GHz) is presented for BAN (Body Area Network) applications. A tapered patch and a folded ground plane are employed to improve the narrow bandwidth characteristics of traditional PIFA, in addition, the proposed antenna is low-profile with the thickness of 7 mm and compact with the area of 60 × 22 mm², making it suitable for the integration in the medical devices. The radiation performance of the modified PIFA preserves well when placed on the human body.**

I. INTRODUCTION

Body area network (BAN) systems have received much more attention due to their potential in several areas [1], such as telemedicine and mobile-health. Low emission power and high data rate make ultra-wideband (UWB) technology a promising candidate for BANs. Besides having the requirements of light weight and low cost, UWB antennas should be with the characteristics of impedance matching and high radiation gain over a broad frequency range, typically, of 3.1~4.9 GHz, which corresponds to the lower UWB band [2].

Several papers with antennas design have been published for BAN applications [3-7]. Although planar antennas, which have advantages of being low profile and easy to be fabricated, are proposed by researchers [3-5], their performance will be deteriorated due to the presence of lossy human tissue. Planar inverted-F antennas (PIFAs) in [6-7] are good candidates with the reduced radiation on the human body, therefore lowering specific absorption rate (SAR). But they are complicated. In this letter, we adopt a kind of modified PIFA with tapered patch and folded ground plane, making it small size, fed conveniently and cover the lower band (3.1~4.9 GHz) in the UWB frequency range.

II. ANTENNA DESIGN STRUCTURE

Figure 1 shows the configuration of proposed PIFA with an SMA connector, having a dimension of 60 × 22 × 7 mm. An air gap of thickness h_f = 1 mm is filled to separate feeding plate from ground plane. Here Air ($\varepsilon_r = 1$; $\tan\delta \approx 0$) is used as substrate so that it could be free from problems relating to loss caused by other dielectric, in addition, the space can be to left for other electronic devices setup. A shorting plate is used

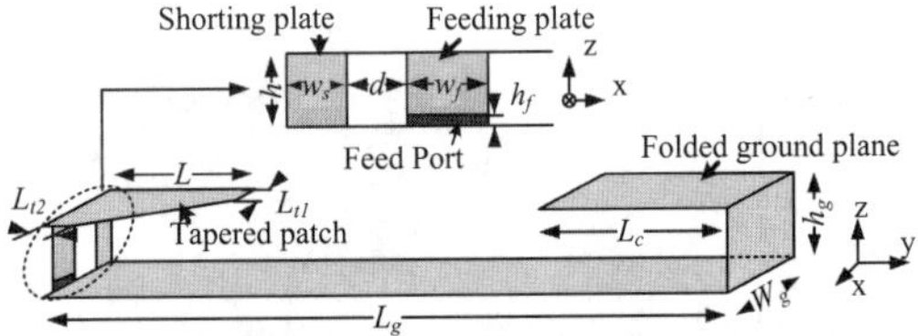

Fig. 1. Geometry of the proposed PIFA.

to connect ground plane with the tapered patch.

The following design procedure is simple and straightforward. Firstly, classical PIFA design techniques were implemented to make a quarter-wavelength resonant patch of PIFA, with the deficiency that higher frequency band is not covered entirely. The corresponding reflection coefficient is plotted with blue dotted line in Figure 2. Subsequently, the top patch was tapered with the characteristic of improved impedance matching, however, the reflection coefficient in 3.1~3.4 GHz band is above -10 dB, as shown with red dashed line in Fig. 2. Finally, the ground plane was lengthened and folded as illustrated in Figure 1, the total lower UWB band (3.1~4.9 GHz) is achieved, as depicted using dark solid line in Figure 2.

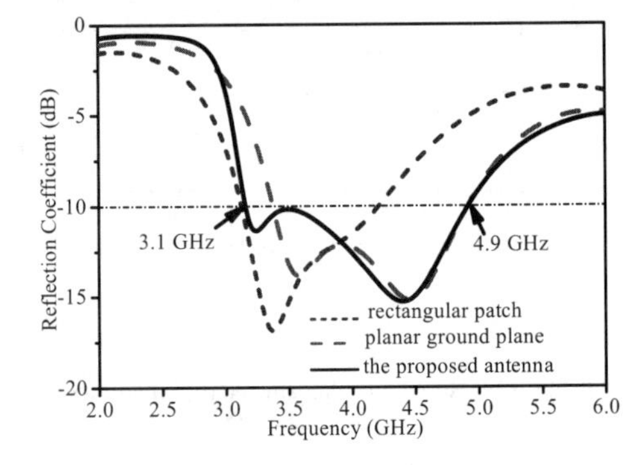

Fig. 2. Simulated reflection coefficients at the different stage of design procedure.

978-7-5641-4279-7

The software package used for simulation is Ansys High Frequency Structure Simulator (HFSS) based on the finite element method. Through adjusting the geometric parameters of the design, we, finally, obtain the optimal dimensions to be: $W_g = 22$ mm; $L_g = 60$ mm; L = 13 mm; h = 6 mm; $L_{t1} = 7$ mm; $L_{t2} = 3$ mm; $w_s = 5$ mm; d = 5 mm; $w_f = 7$ mm; $h_f = 1$ mm; hg = 7 mm; and $L_c = 17$ mm.

III. PRESENCE OF BIOLOGICAL TISSUE

Since the modified PIFA is designed for body area network applications, it is mandatory to study the effect of biological tissue. In this letter, the proposed PIFA was placed on a three-layered phantom: skin ($\varepsilon_r = 36.6$; $\sigma = 2.3$ S/m), fat ($\varepsilon_r = 5.1$; $\sigma = 0.2$ S/m) and muscle ($\varepsilon_r = 50.8$; $\sigma = 3.0$ S/m), with a dimension of $100 \times 60 \times 30$ mm^3, as exhibited in Figure 3. The permittivity and conductivity are given at 4 GHz [8]. The space between the antenna and the human tissue model is set to 0 mm.

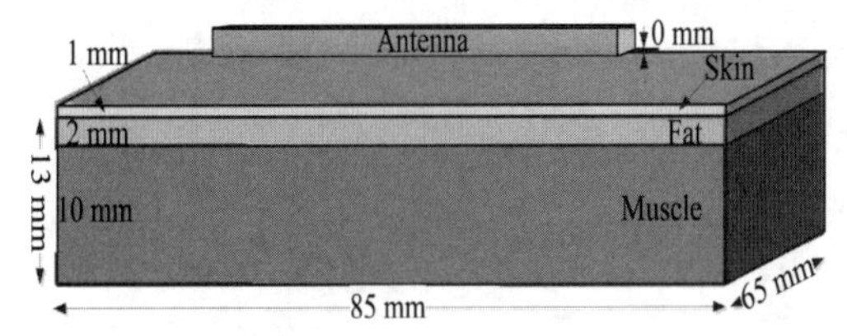

Fig. 3. A three-layered phantom.

IV. RESULTS AND DISCUSSION

The feed port lies in the air gap between feeding plate and ground plane, thus avoiding drilling a hole in the ground plane as other PIFAs [6-7]. The simulated values for reflection coefficients are compared in Figure 4. With reference to the figure, the antenna in free space covers frequency spectrum from 3.1 GHz to 4.9 GHz in simulation. When the proposed antenna is placed directly on the body, the simulated reflection coefficient below -10 dB is met in 3.5~4.7 GHz. Compared to free space, the modified PIFA's performance on the body is slightly affected by the presence of body tissue, but still acceptable.

Figure 5 shows the radiation patterns at 3.2, 4.4 GHz in free space and 4 GHz on the body for two principal planes, i.e. xoz and yoz planes. As can be seen, the simulated radiation patterns are in agreement well and nearly directional with the reduced backward radiation on the human skin.

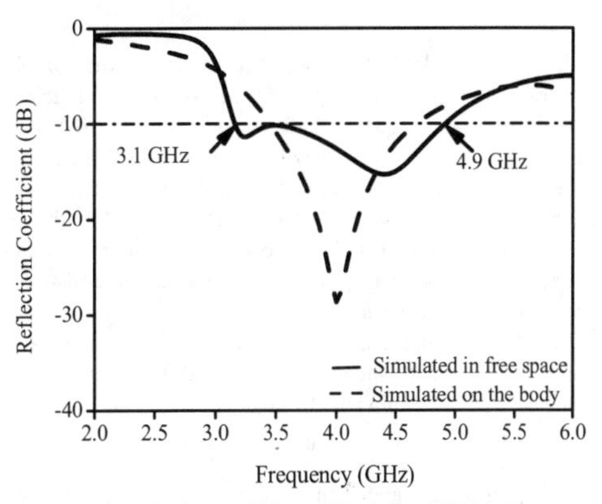

Fig. 4. Simulated reflection coefficients of the designed PIFA in free space and on the body.

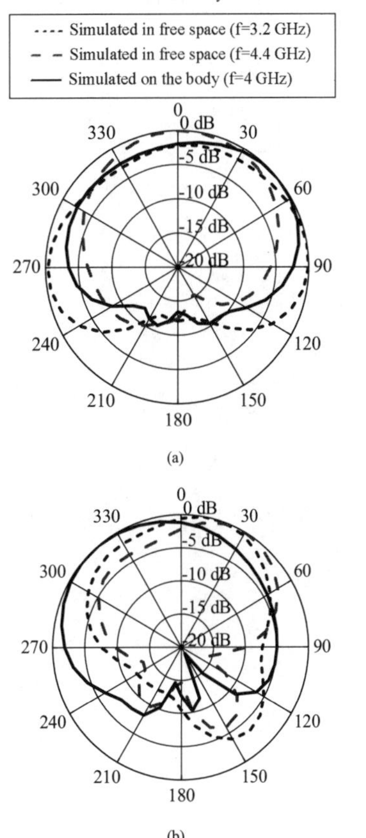

Fig. 5. Simulated and measured radiation patterns of the designed PIFA (a) xoz-plane. (b) yoz-plane.

V. CONCLUSION

A compact UWB PIFA with easily feed port for BAN applications has been proposed. Using tapered patch and folded ground plane, -10 dB bandwidths of 3.1～4.9 GHz in free space and 3.5～4.7 GHz on the body are achieved in simulation, respectively. The modified PIFA's performance on the body degrades a little, due to the effect caused by the lossy human body. The reflection coefficients and directional radiation patterns make the PIFA suitable for body-centric wireless communications.

In order to validate the simulated results, future work will focus on the prototype fabrication and performance measurements, such as gain and efficiency.

ACKNOWLEDGMENT

This work was supported in part by the National Natural Science Foundation of China (60802004), in part by the Guangdong Province Natural Science Foundation (9151064101000090).

REFERENCES

[1] B. Latré, B. Braem, I. Moerman, C. Blondia, and P. Demeester, "A survey on wireless body area networks," *Wireless Netw.*, vol. 17, pp. 1-18, 2011.

[2] W. H. Astrin, B. Li, and R. Kohno, "Standardization for body area networks," *IEICE Trans. Commun.*, vol. E92-B, no. 2, pp. 366-372, Feb. 2009.

[3] J. R. Verbiest, and G. A. E. Vandenbosch, "Small-size planar triangular monopole antenna for UWB WBAN applications," *Electron. Lett.*, vol. 42, no. 10, pp. 566-567, 2006.

[4] T. S. P. See, and Z. N. Chen, "Experimental characterization of UWB antennas for on-body communications," *IEEE Trans. Antennas Propag.*, vol. 57, no. 2, pp. 866-874, Apr. 2009.

[5] C. P. Deng, X. Y. Liu, Z. K. Zhang, and M. M. Tentzeris, "A miniascape-like triple-band monopole antenna for WBAN applications," *IEEE Antennas Wireless Propag. Lett.*, vol. 11, pp. 1330-1333, 2012.

[6] C. M. Lee, T. C. Yo, F. J. Huang, and C. H. Luo, "Dual-resonant π-shape with double L-strips PIFA for implantable biotelemetry," *Electron. Lett.*, vol. 44, no. 14, pp. 837-838, 2008.

[7] C. H. Lin, K. Saito, M. Takahashi, and K. Ito, "A compact planar inverted-F antenna for 2.45 GHz on-body communications," *IEEE Trans. Antennas Propag.*, vol. 60, no. 2, pp. 4422-4425, Apr. 2012.

[8] D. Andreuccetti, R. Fossi, and C. Petrucci, (1997-2012). Dielectric properties of body tissues in the frequency range 10 Hz - 100 GHz. [Online]. Available: http://niremf.ifac.cnr.it/tissprop/.

A TE-shaped Monopole Antenna with Semicircle Etching Technique on ground plane for UWB Applications

Amnoiy Ruengwaree[1], Watcharaphon Naktong[2], and Apirada Namsang[3]
[1]Department of Electronics and Telecommunications Engineering, Faculty of Engineering,
Rajamangala University of Technology Thanyaburi, Klong 6, Phatumtani, 12110, THAILAND
[2]Department of Telecommunications Engineering, Faculty of Engineering and Architecture,
Rajamangala University of Technology Isan, Muang, Nakhon Ratchasima, 30000, THAILAND
[3]Aviation Electronics Division Civil Aviation Training Centre, Chomphon, Jatujak, Bangkok, 10900, THAILAND
amnoiy.r@en.rmutt.ac.th[1]

Abstract- **This paper proposes the TE-shaped monopole antenna with semicircle etchings on the ground plane for Ultra-wideband (UWB) applications of IEEE 802.15.3a standard with the frequency range between 3.1 to 10.6 GHz. The antenna structure was adjusted using the semicircle etching technique at both sides of the ground plane along with the I-shaped stubs. The antenna was fabricated on the FR-4 PCB with size (width x height) of 30 x 30 mm^2, dielectric constant (ε_r) of 4.3 and substrate thickness (*h*) of 0.764. The study of the antenna properties and parameters were accomplished by simulations with the computer program for analyzing important characteristics and structural optimizations. From the simulations and measurements, the results were in good agreement, with the impedance bandwidth of 110.56% (3.09 - 10.73 GHz). Also, the time-domain measurements had been conducted with fairly good characteristics for pulse-radio transceiving applications in the UWB frequency range.**

I. Introduction

In currently the long distance wireless communication, or in specific area to offer that popular systems that is WiMAX or WLAN which has effect for human very much by using trendy wireless communication devices such as mobile phones portable computer and tablet etc [1].

In the beginning of the short communication, the current system for an ultra-wide band (UWB) is high-speed, very wide bandwidth and low-power. Especially UWB which was established in 2002 for FCC supporting has 2 bands. One is the lower the frequency which is below 960 MHz and the higher will be the frequency range from 3.1 to 10.6. GHz [2-3], which is a system that became popular at present. As communication systems mentioned above, antenna is one important device in the system. It can transmit and receive data wirelessly that is many researches and antenna developments in different ways to respond for a particular frequency band. It found that in some researches using larger antenna complexity structures or very tuning configuration points. Most of them use the PCB two-sided [4-21]. On [22] is the monopole antenna structure which is adapted the etching ground technique of [5, 20-21] and adjusting on antenna patch [19, 23] to develop a new antenna structure for reducing the size and decreasing antenna tuning on FR-4 PCB one-page-type in order to material is thin, lightweight, easy to buy. CST Microwave Studio is employed to adjust parameters and analysis features of the proposed antenna structure to make the best effective experiential method.

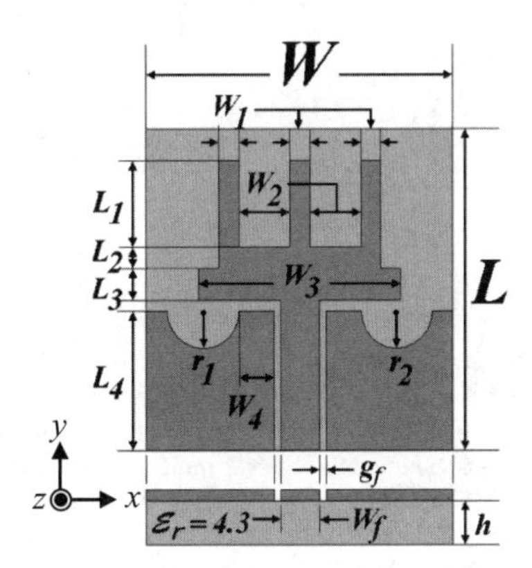

Figure 1. The proposed antenna.

II. Design and Simulation

A. Antenna Structure

As studied the original structural of the prototype antenna, [22] used the etching technique of on the rectangular antenna [19, 23] including the semicircular etching on both sides of the ground plane [5, 20-21]. The prototype antenna structure is built on the FR-4 PCB which has the dielectric constant (ε_r) = 4.3, the material thickness (*h*) = 0.764 mm. Using experiential method and CST-program to scaling with the appropriate structure. After adjusting the parameters and make the most appropriate width W = 30 mm, W_1 = 2 mm, W_2 = 5 mm, W_3 = 20 mm, W_4 = 3.5 mm, the total length, L = 30 mm, L_1 = 7.5 mm, L_2 = 2 mm, L_3 = 3 mm and L_4 = 13 mm, respectively. The feed line of the proposed antenna is CPW structure which is

978-7-5641-4279-7

estimated from the formula which has W_f = 3.8 mm and g_f = 0.6 mm, as shown in Fig. 1

B. Antenna Simulation

CST simulator produces the return loss, radiation pattern and impedance bandwidth for use as input data for the antenna structure size tuning. It found that adjusting the parameters of the proposed antenna, the effective frequency response is the best available three main parts. The first part is the etching on the rectangular antenna patch as shown in Fig. 2 by fix the width of the spacing between strips W_2 of 5 mm and varying its length L_1 which is the stubs length to compromise the return loss and bandwidth in ranging from 2.5, 5.0, 7.5 and 10 mm. It found that the appropriate value of L_1 = 10 mm, which generates the low-frequency response, the frequency range of 51.73% (2.89 - 4.53 GHz) and the high-frequency response 72.51% (from 7.81 to 13.73. GHz), but the operated frequency is not yet covered by the IEEE 802.15.3a standard.

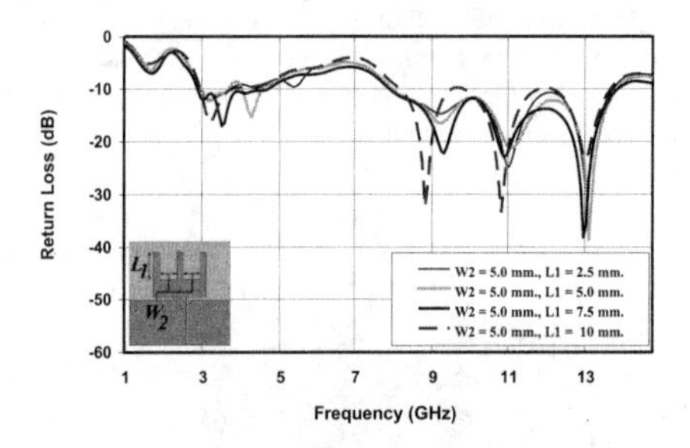

Figure 2. The return loss when L_1 varied.

So that the second section adjusting is adding I-shaped on the presented antenna. It seen that the bandwidth is changed. By resizing the width of the strip W_3 of 17, 18, 19 and 20 mm with the fix I-shaped length (L_3) = 3 mm. was found that the appropriate value is W_3 = 20 mm. The suitable value of L_3 is 3 mm, resulting in a lower frequency response less than -10 dB and its bandwidth of the low-frequency range of 78.70% (2.82 - 6.48 GHz), the middle frequency range of 17.90% (8.24 - 9.86 GHz) and the highest frequency range of 33.93% (11.23 - 15.82 GHz) which is better than the first part as shown in Fig. 3

The final part is the etchings of ground plane into semicircular shape on both ground plane sides. It found that the return loss and the bandwidth are also better changed than Fig. 3 by controlling the radius r_1 and r_2 from 2.0, 2.5, 3.0 and 5.3 mm. The appropriate values are r_1 and r_2 equal 3.5 mm, resulting in a very wide frequency response 132.41% (2.82 - 13.87 GHz), which resulted in increased bandwidth covers the standard of IEEE 802.15.3a, that is the frequency range of 3.1 - 10.6 GHz desired as shown in Fig. 4

In order to the etching of the ground plane into semicircular shape on both ground plane sides, the effective results are to the middle frequency range (about 6 - 8 GHz) more than the other. It can be seen the impedance in the three frequency range including with and without etching as shown in Table 1.

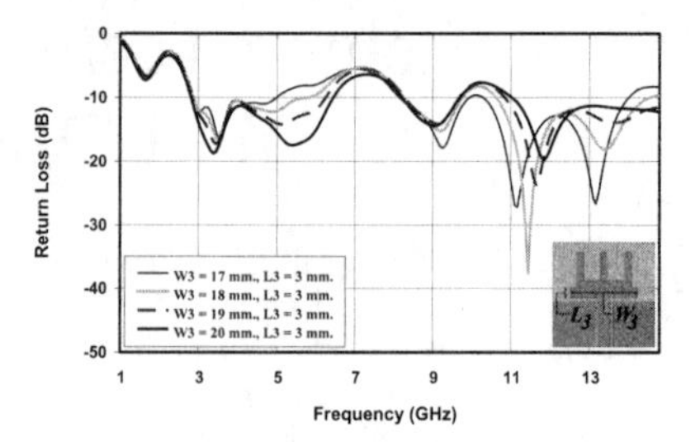

Figure 3. The return loss when W_3 varied

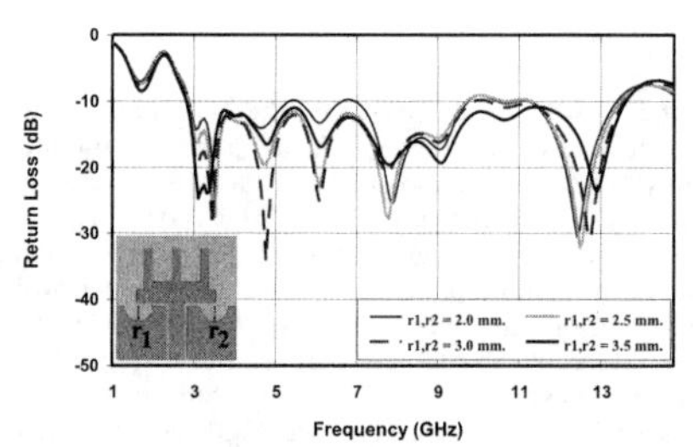

Figure 4. The return loss when r_1 and r_2 varied.

TABLE I
COMPARISON IMPEDANCE ON WITH AND WITHOUT ETCHING

Frequency	Z_{in} without Etching	Z_{in} with Etching
3 GHz	18.4 – j0.5 Ω	27.5 + j6.1 Ω
7 GHz	46.0 – j51.8 Ω	53.1 + j1.0 Ω
10 GHz	32.9 – j18.3 Ω	42.8 – j17.0 Ω

In table I, it found that at the lower frequency (3 GHz) and the high frequency (10 GHz), the semicircular etching will cause the value of reactance drop. In the case of 7 GHz, the value of reactance decrease when is etched. It shown that the semicircular etching on ground plane that help to improve the ground plane impedance as it is added value of inductive load to the antenna at the center frequency.

Fig. 5 (a) - (c) shows the current density on the antenna in three frequency ranges with and without etching. It can see that the semicircular etching on the ground plane will affect the current density near the feed line more than on a dipole. The impact on the current mode of the antenna is not too much. It can be noted that at 7 GHz in case of no etching, the current is more density at the edge of the ground plane that is parallel to the arm of the T-shaped, but when semicircular etching on ground plane, it resulted of the capacitance value of the antenna at the second frequency band width is reduced significantly. This is in similar way with the results in table I.

C. *Fabrication and Measurement*

In the process of fabricating, the structure of propose antenna is on PCB using FR4. The proposed antenna dimension is 30 x 30 mm^2. The measurement system is shown in Fig. 6

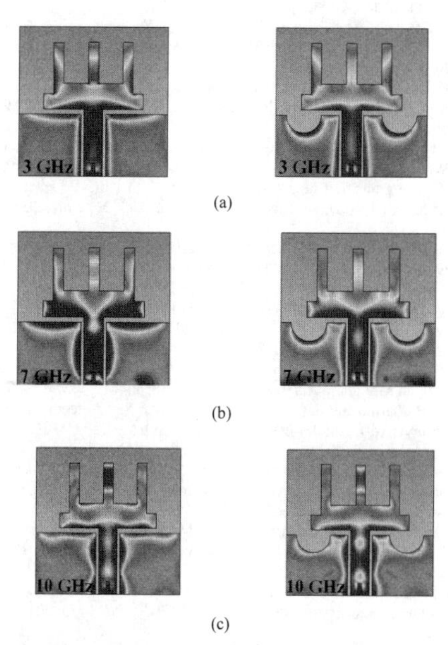

Figure 5. Comparison of the current density on the antenna without etching (left) and with etching (right) at (a) 3 GHz (b) 5 GHz (c) 10 GHz.

Figure 6. Antenna and install a real measurement.

Then measuring the voltage standing wave ratio and comparing the simulation and measurement results. The results are agreed very well. It has impedance bandwidth (at VSWR 2:1) 110.56% (3.09 - 10.73 GHz) as shown in Fig. 7. At 3 GHz, the gain is 3.84 dBi and rising up gain 6.12 dBi at 10 GHz. So the average gain is around 4.85 dBi as shown in Fig. 8. Fig. 9 shows the result of measurement of the time delay group or group delay that shows the phase delay of each frequency. It found that the average is about 6 ns and the swing of the delay does not exceed ±1.5 ns. That the above results are considered eligible, and it is possible for applications to signal pulse. In Fig. 10 and 11 show the comparison of the radiation pattern in the electric field (E-plane) and the magnetic field (H-plane) at a frequency of 3 GHz, 7 GHz and 10 GHz from the simulation and measurement results which is agree very well. The radiation pattern of the presented antenna is bidirectional, but at the same time it is very close to the Omi- directional by observing the plane of the magnetic field as shown in Fig.11

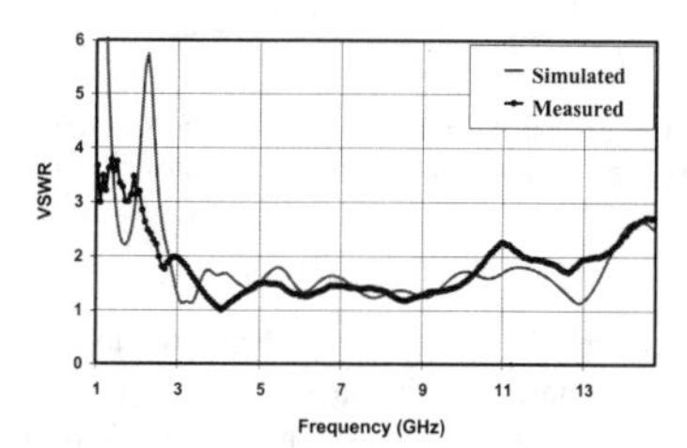

Figure 7. Comparison of simulation and measurement of the voltage standing wave ratio VSWR of the antenna.

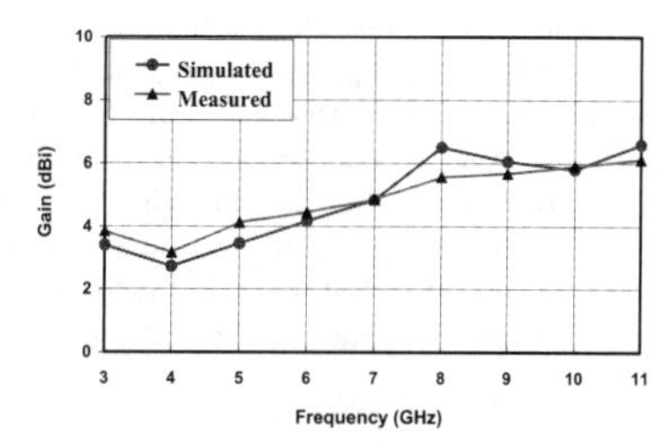

Figure 8. The gain of the antenna relative to the replication master antenna actually created.

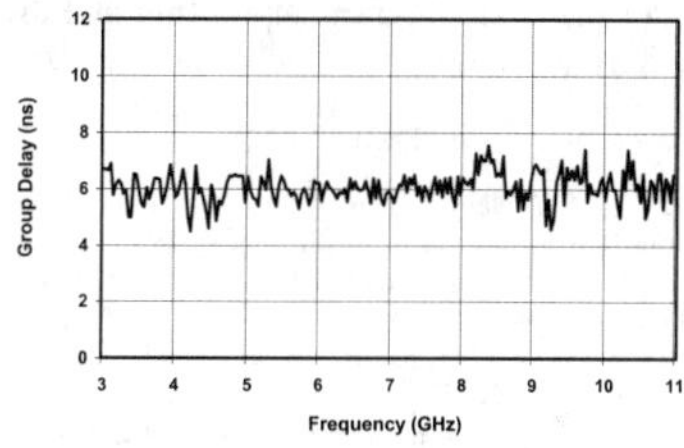

Figure 9. Group Delay.

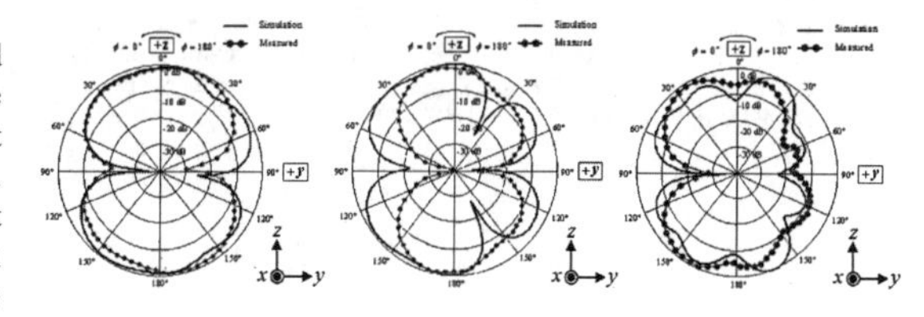

Figure 10. Comparison of simulation results with the measured radiation pattern at frequency 3 GHz, 7 GHz and 10 GHz (E-plane).

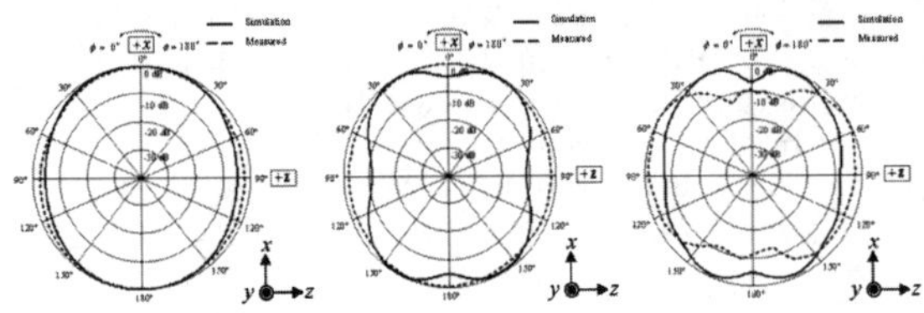

Figure 11. Comparison of simulation results with the measured radiation pattern at frequency 3 GHz, 7 GHz and 10 GHz (H-plane).

III. Conclusion

This paper presents the TE-shaped monopole antenna with semicircular etchings on the ground plane including the I tuning on patch antenna for size reducing when comparing with the previous research, which reduces the complexity of creating the actual antenna and found that the proposed antenna is the impedance bandwidth of 110.56% (3.09 - 10.73 GHz), which is wider than the previous one. The comparison results of the simulation and measurements are agreeing very well and. consistent trend over the frequencies used by IEEE 802.15.3a (3.1 - 10.6 GHz). The overall radiation pattern is bidirectional, but very close to the Omi-directional. The average gain is around 4.85 dBi, It is possibility to use in radio pulse to apply for wireless communication systems.

Acknowledgment

Thank you the Faculty of Technical Education and the Faculty of Engineering of the Rajamangala University of Technology for the CST program supporting and the network analyzer E8363B measuring.

References

[1] IEEE 802.11. "Wireless Access Method and Physical Layer Specifications," *New York, NY, USA,* September 1994.

[2] J FCC, "FCC Report and Order for Part 15 Acceptance of Ultra Wideband (UWB) Systems from 3.1-10.6 GHz," *Washington DC*, 2002.

[3] H. Schantz, "The Art and Science of Ultrawideband Antennas," *Boston, London, Artech House*, 2005.

[4] X. L. Liang, S. S. Zhong and F. W. Yao, "Compact UWB Tapered-CPW-fed Planar Monopole Antenna," *Microwave Conference Proceedings, 2005, APMC2005, Asia-Pacific Conference Proceedings* Vol. 4, 4-7 December, 2005.

[5] J. Y. Jan, J. C. Kao, Y. T. Cheng, W. S. Chen and H. M. Chen, "CPW-Fed Wideband Printed Planar Monopole Antenna for Ultra- Wideband Operation," *Antennas and Propagation Society International Symposium 2006, IEEE,* 9-14 July 2006, pp. 1697-1700.

[6] X. C. Yin, C. L. Ruan, C. Y. Ding, and J. H. Chu, "A Planar U Type Monopole Antenna for UWB Applications," *Progress in Electromagnetics Research Letters,* vol. 2, 2008, pp. 1-10.

[7] V. Shrivastava and Y. Ranga, "Ultra wide band CPW-fed printed pentagonal antenna with modified ground plane for UWB Applications," *Mobile and Multimedia Networks, 2008, IET International Conference on*, 11-12 Jan, 2008, pp.1-2.

[8] X. -C. Yin, C.-L. Ruan, S. -G. Mo, C.-Y. Ding, and J.-H. Chu, "A Compact Ultra-Wideband Microstrip Antenna with Multiple Notches," *PIER 84,* 2008, pp. 321- 332.

[9] T.Hong. S.X.Gong. W.Jiang. Y.X. Xu, and X.Wang. "A Novel Ultra-wide Band Antenna with Reduced Radar Cross Section," *PIER 96*, 2009, pp. 299-308.

[10] D. Chen and C.-H. Cheng, "A Novel Compact Ultra-Wideband (UWB) Wide Slot Antenna with via Holes," *PIER 94*, 2009, pp. 343-349.

[11] S. Hong and J. Choi, "Miniaturization of an Ultra-Wideband Antenna with Two Spiral Elements," *ETRI Journal*, Vol. 31, No.1, 2009, pp. 71-73.

[12] S. Pokapanic and A. Ruengwaree, "CPW–Fed Rectangular Slot Antenna with Mortar Shape Stub Tuning for UWB Application," *32nd Electrical Engineering Conference*) EECON-32), 2009.

[13] C. Deng, Y. Xie, and P. Li. "CPW-Fed Planar Printed Monopole Antenna with Impedance Bandwidth Enhanced," *IEEE Antennas and Wireless Propagation Letters,* Vol. 8, 2009.

[14] H. Zhang , G. Li, J. Wang, X. Yin, "A Novel Coplanar CPW-Fed Square Printed Monopole Antenna for UWB Applications," *ICMMT,* 2010.

[15] S. Barbarino and F. Consoli, "UWB Circular Slot Antenna Provided with an Inverted-L Notch Filter for the 5 GHz WLAN Band," *PIER 104*, 2010, pp. 1-13.

[16] F. Zhu, S.-C. S. Gao, A. T. S. Ho, C. H. See, R. A. Abd-Alhameed, J. Li, and J.-D. Xu, "Design and Analysis of Planar Ultra-Wideband Antenna with Dual Band-Notched Function," *PIER 127*, 2012, pp. 523-536.

[17] M. T. Islam, R. Azim, and A. T. Mobashsher, "Triple Band-Notched Planar UWB Antenna Using Parasitic Strips," *PIER 129*, 2012, pp. 161-179.

[18] R. Azim and M. T. Islam, "Compact Planar UWB Antenna with Band Notch Characteristics for WLAN and DSRC," *PIER 133*, 2013, pp. 391-406.

[19] W. Naktong, B. Kaewchan, A. Namsang, and A. Ruengwaree. "Bidirectional Antenna on Flambeau-Shape," *International Symposium on Antennas Propagation*)ISAP 2010), *Macao, China*, 23-26 November 2010.

[20] W.Naktong and A.Ruengwaree, "Increasing bandwidth of Flambeau-Shape monopole antenna for UWB Application," *8th International Conference on Electrical Engineering / Electronics, Computer, Telecommunications and Information Technology* (ECTI-CON 2011), *Khon Kaen, Thailand*, 17-19 May 2011, pp. 172-175.

[21] W. Naktong and A. Ruengwaree, "Enlargement Bandwidth of slot antenna with tuning-fork stub by triangle grooving on Ground Plane," *34rd Electrical Engineering Conference*)EECON-34), vol. 2, *Chonburi. Thailand,* 30 November - 2 December 2011, pp. 673 -676.

[22] W.S.Chen. Y.C.Chang. H.T.Chen. F.S.Chang and H.C.Su. "Novel Design of Printed Monopole Antenna for WLAN/WiMAX Applications," *Antennas and Propagation Society International Symposium, 2007 IEEE,* 9-15 June 2007, pp. 3281-3284.

[23] P. Jearapraditkul, W. Kueathaweekun, N. Anantrasirichai, 0. Sangaroon and T. Wakabayashit "Bandwidth Enhancement of CPW-Fed Slot Antenna with Inset Tuning Stub," *Communications and Information Technologies, 2008. ISCIT 2008. International Symposium on.* 21-23 October 2008, pp. 14-17.

Design and Analysis of a Modified Sierpinski Carpet Fractal Antenna for UWB Applications

Bin Shi, Zhiming Long, Jili Wang, Lixia Yang
(School of computer science and Communication Engnerring, Jiangsu University,
Zhenjiang,Jiangsu 212013, China)

***Abstract*-This paper presents the design and analysis of a modified Sierpinski Carpet fractal antenna that contributes ultra wideband characteristics. The proposed antenna with coplanar waveguide-fed structure is compact, which has a total size of 40mm×50mm×1.8mm. An experimental prototype has been fabricated and tested .The measured results show the impedance bandwidth of the proposed fractal antenna that the return loss below -10dB is 8.1 GHz ranging from 2.2GHz to 10.3 GHz. Corresponding 129.6% impedance bandwidth. Analysis of antenna is done using Ansoft High Frequency Structure Simulator (HFSS v 10).Very good agreement obtained between simulation and experimental results.**

I INTRODUCTION

The ultra wideband (UWB) system have become emerging research topic in the field of modern wireless communications since Federal Communication Commission (FCC) band 3.1GHz-10.6GHz declared in Feb.2006 [1]. In recent years several printed planar micro-strip antennas are developed for UWB range due to their low profile, small volume, easy to integrate into the communication circuit [2-6]. In a UWB antenna system the construct is designed with the help of rectangular patch consisting of two steps and partial ground plane. This method can be used to achieve wideband characteristics.

The concept of fractal was first introduced by French mathematician B. B. Mandelbrot. Fractals have self similarity properties which are used in achieving multi band characteristics and hence can be optimized for UWB applications. Fractal shaped antennas have already been proven to have some unique characteristics that are linked to the geometrical properties of fractal. Self-similarity of fractal makes them especially suitable for Ultra Wide Band and multi-frequency applications [7-11]. Very recently, several antenna based on the fractal topology has been proposed and investigated. They have been applied to Loops, dipoles, monopoles, patches, slots, arrays and PIFA [12-13].

In this paper, a planar micro-strip antenna with modified Sierpinski Carpet Fractal is presented for UWB characteristics. This antenna exhibits the properties, such as miniature size, wideband phenomenon and Omni-directional radiation pattern. This antenna is developed which covers the UWB range of 2.2 GHz to 10.3 GHz achieving impedance bandwidth of 129.6%. The proposed antennas are simulated using Ansoft High Frequency Structure Simulator (HFSS v 10) to study the performance parameters related to size reduction.

II DESIGN OF FRACTAL PATCH ANTENNA

The initial design of the self similar Sierpinski Carpet Fractal is a rectangle patch. The dimensions of the fractals, which forms the basis of the antenna, are shown for first, second and third iterations. The zero element antenna is called Fractal 0. to form the first iteration Fractal 1, the rectangular in the center of the original patch is one third of its original size and remove the changed small unit from the original patch .The second Fractal 2 and third Fractal 3 iteration are formed by translating their previous iteration, which the dimension of each rectangular with the same size.

Figure 1 illustrates the geometry of the proposed ultra wideband patch antenna. As shown, the designed antenna which is printed on the FR-4 substrate with a thickness of 1.8 mm, relative dielectric constant of 4.4 and a dielectric loss tangent of 0.02, has a compact dimension of 40 mm× 50 mm. As in Figure 1, the proposed antenna consists of two asymmetrical part patches, one patch on the top called radiated element and the other patch on the bottom of the substrate called the limited ground plane. The feeding structure consists of a microstrip line, which is designed to have characteristic impedance of 50 Ohm. There also an impedance convertor for the microstrip structure. The convertor is determined according to optimization to implement 50 Ohm characteristic impedance, and a standard SMA connector is connected to facilitate its connection with other communication devices.

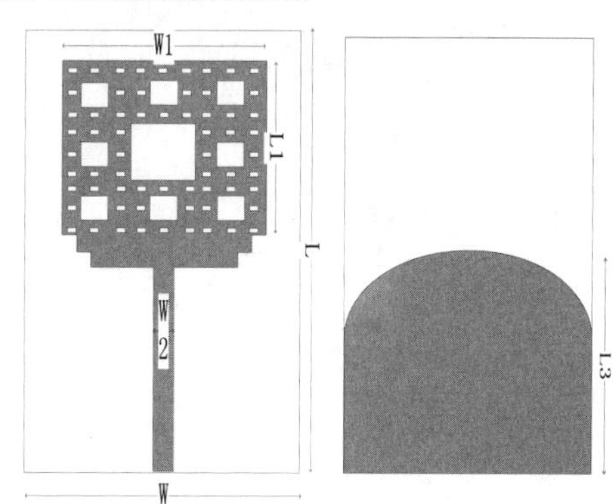

978-7-5641-4279-7

Figure 1 Geometry of the proposed antenna

The proposed antenna parameters are L=50mm, L1=18.9mm, L3=18mm, W=40mm, W1=27mm, W2=4mm. The parameters of a Sierpinski Carpet Fractal Antenna above has been optimized and simulated.

III SIMULATRION AND MEASUREMENT RESULTS

A SIMULATRION RESULTS

Prior to fabrication of the final structure, a parametric study is carried out to pre-determine the cause-effects of fabrication inaccuracies. The study is carried out on proposed structures. Parameters investigated are critical parameters: (1) The times of Fractal on the patch is shown in Figure 2.With the time increasing its effects on the antenna's Return Loss (S11) and bandwidth , it is observed that the third iteration is better than the first and the second iteration in bandwidth: (2) It is seen from Figure 3 that a slight upwards shift and downward shift of the resonant frequency and *S*11 occurred with the increase in width: (3) As we see from Figure 4, with the length increasing, the matching of impedance is not very good in the high frequency range but match in the low range.

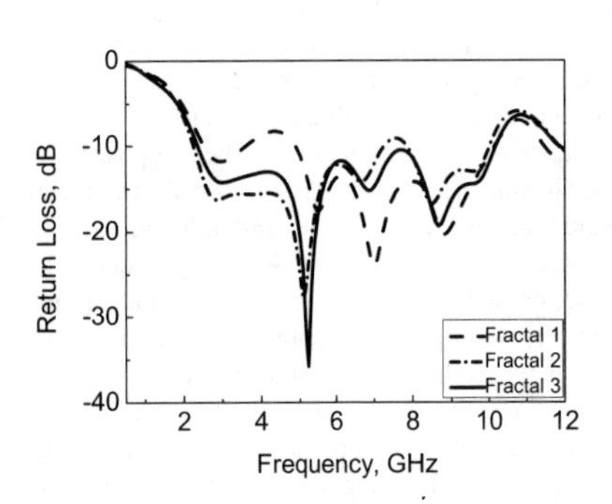

Figure 2 Return Loss for the proposed antenna with fractal

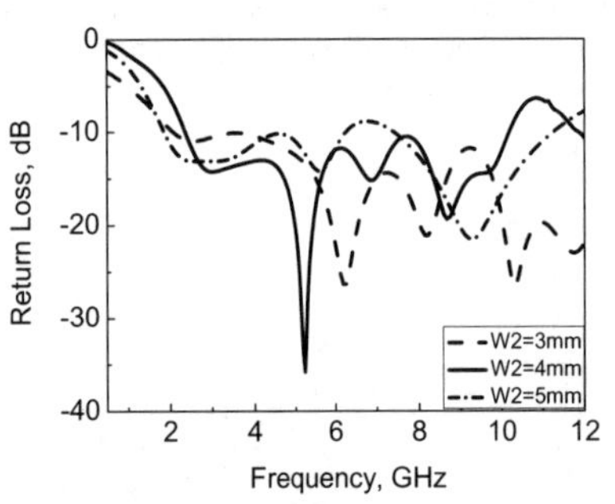

Figure 3 Return Loss for the proposed antenna with W2

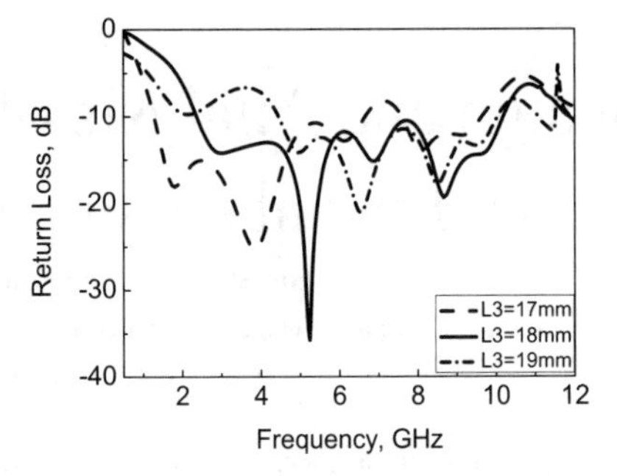

Figure 4 Return Loss for the proposed antenna with L3

B MEASUREMENT RESULTS

Base on the design dimensions shown in above parameters. An experimental prototype of the proposed antenna is fabricated as depicted in Figure 5, Figure 6 and Figure 7 show the measured and simulated the return loss (S11) and Voltage Standing Wave Ratio (VSWR) for the proposed antenna.

Figure 5 The manufactured of the proposed antenna

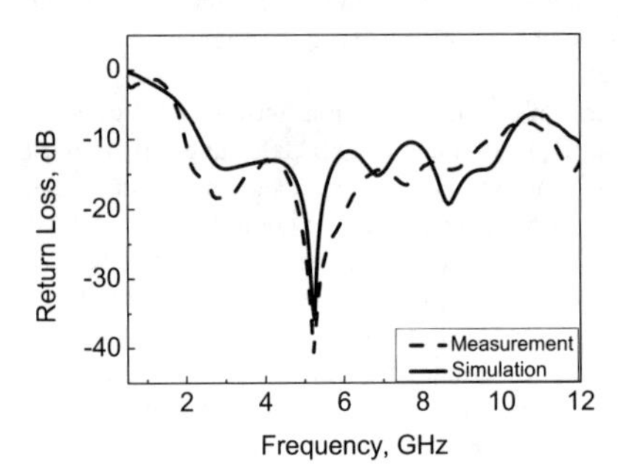

Figure 6 Measurement and simulation results of the antenna return loss

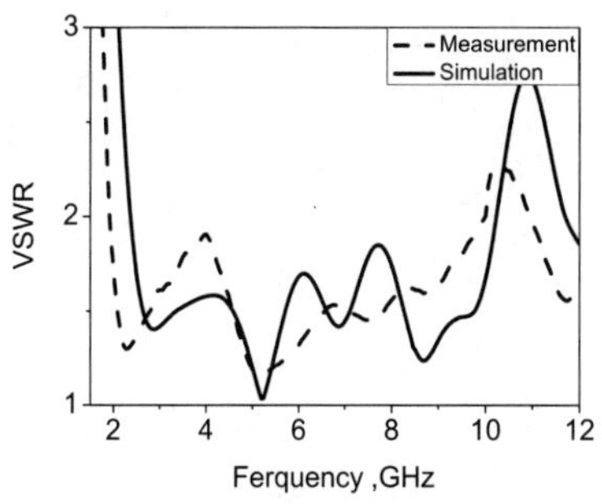

Figure 7 Measurement and simulation results of the antenna

VSWR

The far-field radiation characteristics of the proposed ultra wideband antenna have also been studied. The normalized radiation patterns measured at 2.8 GHz, 5.2 GHz, 6.8 GHz and 8.7 GHz are plotted in Figure 8.

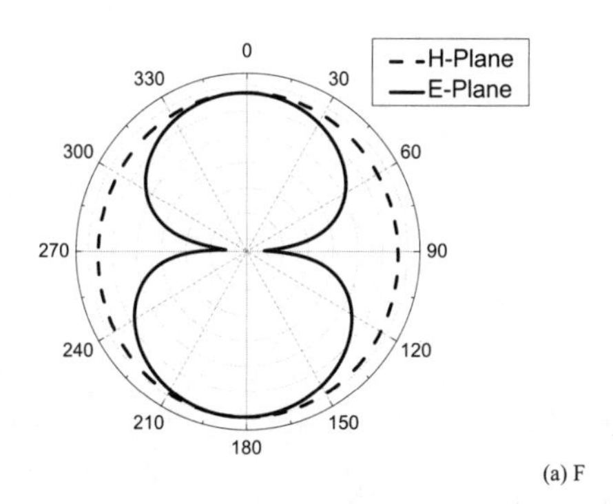

(a) F

=2.8GHz

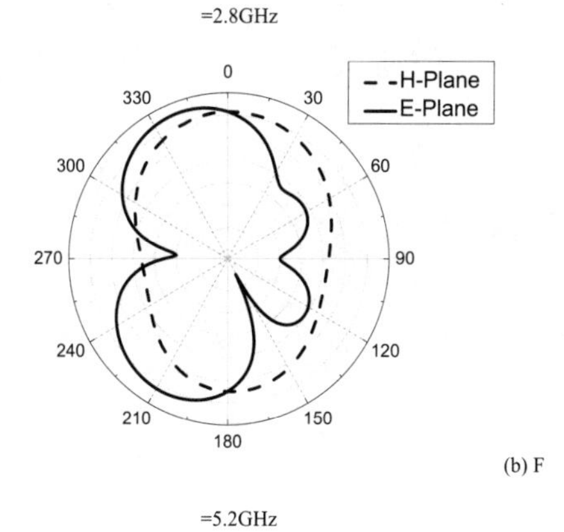

(b) F

=5.2GHz

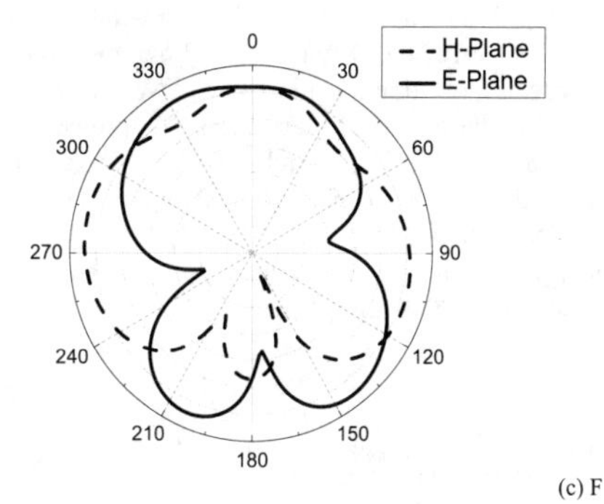

(c) F

=6.8GHz

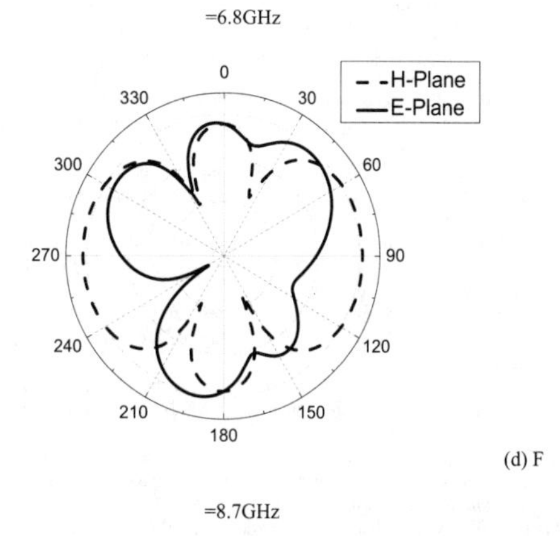

(d) F

=8.7GHz

Figure8 The far-field radiation patterns at resonant frequencies 2.8 GHz, 5.2 GHz, 6.8 GHz and 8.7GHz, respectively

IV CONCLUSION

A new compact ultra-wideband antenna has been proposed for UWB applications. The measured and simulated results conducted by the Ansoft High Frequency Structure Simulator (HFSS v 10) show a good performance. The frequency range obtained for return loss below -10 dB is 2.2 GHz -10.3 GHz. Corresponding 129.6% impedance bandwidth, the simulated radiation patterns at 2.8 GHz, 5.2 GHz, 6.8 GHz and 8.7 GHz is presented. The usage of the Sierpinski Carpet Fractal topology, in particular, has managed to enhance the antenna in terms of additional higher resonance frequencies and broad bandwidth. At the same time, it also contributed to space saving through electrical length shortening. These size compactness and multi-resonant antenna behaviors are particularly significant, in line with the miniaturization of today's electronic devices.

ACKNOWLEDGMENT

Authors would like to acknowledge the financial support

from the National Natural Science Foundation of China (Grand No. 61072002), the Ph.D. Programs Foundation of Ministry of Education of China (Grand No. 20093227120018), the eighth "Elitist of LiuDa Summit" project of Jiangsu Province in 2011 (Grand No.2011-DZXX-031), the Postdoctoral Science Foundation in Jiangsu (GrandNo.1201001A), the Undergraduate Research Foundation of Jiangsu University (Grand Nos.11B009 and 12B050).

REFERENCES

[1] K.-H. Kim, Y. -J. Cho, S. -H. Hwang and S. –O. Park, Band-notched UWB Planar Monopole Antenna With Two Parasitic Patches, *ELECTRONICS LETTERS*, (2005), pp. 783-785

[2] D.Liu, Analysis of a closely-coupled dual band antenna, Proceeding of Wireless Communications Conference, Boulder, Colorado, August 1996.

[3] Rohit K.Raj, Monoj Joseph, C.K.Anandan, K.Vasudevan, P.Mohanan, A New Compact Microstrip-Fed Dual-Band Coplanar Antenna for WALN Applications, *IEEE Trans. Antennas Propagate*, (2006), pp. 3755-3762.

[4] I.Sarkar, P.P.Saekar, S.K.Chowdhhury, A novel compact microstrip antenna with multi frequency operation, *IEEE Trans. Antenna Propagate*, (2009), pp. 147-151.

[5] T.Y.Wu and K.L.Wong, On the impedance bandwidth of a planar inverted-F antenna for mobile handsets, Microstrip Opt Technol Lett. (2002), pp 249-251.

[6] Qu S. –W. Ruan, C. –L. and Wang, B, Bandwidth enhancement of wide-slot antenna fed by CWP and micostrip line, *IEEE Antenna Wire. Propagate*, Lett. (2006), pp. 15-17.

[7] Douglas H W, Suman G An overview of fractal antenna engineering research [J]. *IEEE Antennas and Propagate, Magazine*, (2003), 45 (1): pp 39-57.

[8] C.Puente, J,Romeu, R.Pous, and A.Cardama, On the behavior of the Sierpinski multi frequency antenna, *IEEE Trans. Antenna Propagate* 46, (1998), pp .517-524.

[9] C.Puente, C.Borja, M.Navarro, and J.Romeu, An iterative model for fractal antenna: application to the Sierpinski gasket antenna, *IEEE Trans, Antenna Propagate*, vol. 48, (2000), pp. 713-719.

[10]K. J.Vinoy, K.A.Jose, V.K.Varadan, and V.V.Varadan," Hilbert Curve Fractal Antenna with Reconfigurable Characteristics," *2001 IEEE MTT-S Digest*, 2001, pp. 381-384.

[11] C.P.Baliarda, J.Romeu and A.Cardama,The Koch monopole: A small fractal antenna, *IEEE Trans. Ant. Propagate*, (2000), pp.1773-1781.

[12] N.A.Saidatul, A.A.H, Azremi, R.B.Ahmad, P.J.Soh, and F.Malek,Multiband fractal planar inverted F-antenna (F-PIFA) for mobile phone application, *Progress in Electromagnetic Research*, (2009), pp .127-148.

[13] S.N.Shafie, I.Adam, and P.J.Soh,Design and Simulation of a Modified Minkowski Fractal Antenna for Tri-Band Application, in Mathematical Modeling and Computer Simulation (AMS), 2011 Fourth Asia International Conference on, (2010), pp. 567-570.

[14] A.Ismahayati, P.JSoh, R.Hadibah, and G.A.E Vandenbosch, Design and Analysis of a Multiband Koch Fractal Monopole Antenna, *IEEE International RF and Microwave Conference*, (2011), pp. 58-56.

[15] Salisa Binti Abdul Rahman, P.C.Sharma, Varun Jeoti, and Zainal Arif Burhanuddin, "Influence of Fractal on Resonant Frequencies of a Rectangular Microstrip Antenna", *RF and Microwave Conference*, (2004), pp. 124-127

[16] Mircea Rusu, Mervi Hirvonen, Hashem Rahimi, Peter Enoksson, Cristina Rusu, Nadine Pesonen, Ovidiu Vermesan, and Helge Rustad, "Minkowski Fractal Microstrio Antenna for RFID Tags", *Proceedings of the 38th European Microwave Conference*, (2008), pp. 666- 669.

WP-2(C)

October 23 (WED) PM

Room C

Broadband Antennas

Broadband Loaded Cylindrical Monopole Antenna

Solene Boucher [(1)], Ala Sharaiha [(2)], Patrick Potier [(1]
[(1)]DGA–Maitrise de l'information, France, [(2)] *Institute of Electronic and Telecommunications of Rennes, UEB, UR1, 263 Avenue du général Leclerc, France.*

***Abstract-* A broadband printed monopole antenna based on the variation of the conductivity along its length is proposed.. The result indicates that a non-monotonous repartition provides interesting performances in terms of impedance bandwidth but also concerning antenna gain. The achievement of the method is demonstrated through its application, using the carbon fibers to perform this conductivity variation. Monopole antenna presents a large impedance bandwidth of 123% with an interesting gain. Measurement and simulation present a good agreement.**

I. INTRODUCTION

With the rapid development of communication standards, wide band antennas are needed. Several authors have suggested widening monopole antenna bandwidth by resistive loading continually [1,2]. This kind of resistive distribution leads to a very wide bandwidth but damages significantly antenna gain and its efficiency becomes very low. Antenna characteristics could be improved by modifying the resistive profile.

In this paper, we propose to use a variable conductivity along the antenna length applying an optimized profile which gives a broadband impedance bandwidth, a stable radiation pattern, a stable gain level over a large bandwidth especially in the horizontal plane and a higher radiation efficiency. The values of conductivity used is achieved by using various materials such as copper and specially carbon fibers for which we can easily tune its conductivity by modifying the concentration of carbon nanofibers [3]. The main aim is to find a compromise between a large impedance bandwidth at -5 dB of reflection coefficient and a good efficiency. So, a new broadband monopole antenna is presented working in the VHF-UHF band. Simulation results and measurements are presented in order to illustrate the characteristics of the proposed antenna.

II. ANTENNA DESIGN

The goal of our study is to broaden impedance bandwidth by varying the conductivity along the length of the monopole antenna and maintaining a stable gain level especially in horizontal plane (xOy, $\theta=0°$). For this, we have used a multi-objective optimization software named mode Frontier□, Esteco coupled with a three-dimensional time domain software named Microwave Studio□, CST. The selected method for optimization is based on a non-dominated sorting genetic algorithm [4]. The monopole antenna is composed of five equal segments length of 50 mm and 20 mm of diameter (figure 1).

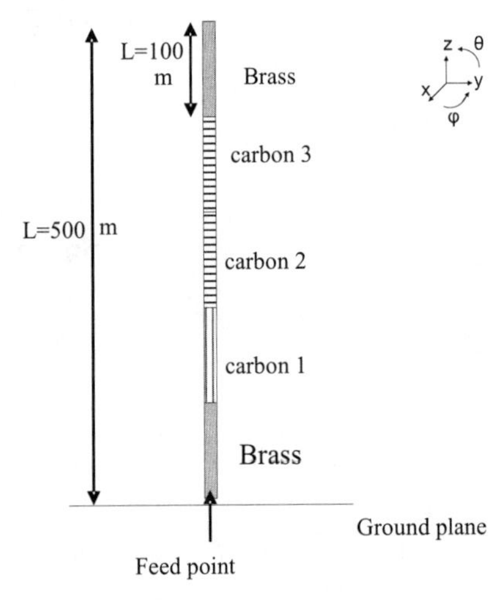

Fig 1. Monopole antenna geometry.

The first segment and last segment is made of Brass and the three other segments was made of carbon tubes.

The optimization solver suggests a distribution conductivity law with a minimum value placed not at the extremity of the monopole antenna as a classical Wu-King profile, but in its centre as shown in Table 1.

TABLE I
CONDUCTIVITY VALUES

Segment	Conductivity (S/m)	Equivalent resistance (Ω)
Carbon 1	2	159
Carbon 2	0.5	6294
Carbon 3	1,5	212

978-7-5641-4279-7

In Figure 2, we present the reflection coefficient of this antenna as well as the gain for θ=0° (horizontal plane).

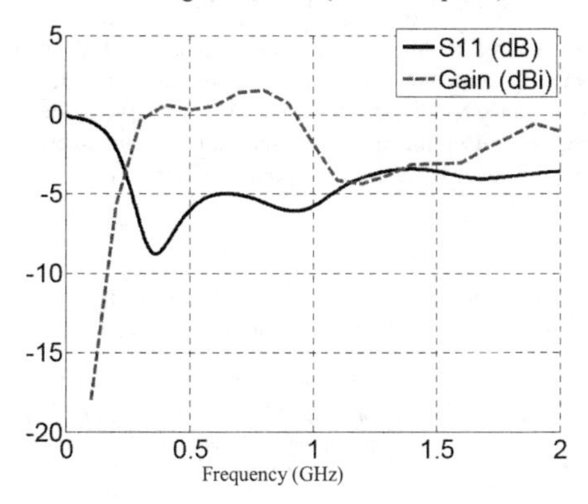

Fig 3. Simulation of monopole antenna reflection coefficient and gain in horizontal plane (θ=0°) vs. frequency.

We can note that the gain level stays quite constant from 300 MHz to 900 MHz in the Oy axis direction. The maximum gain reaches a value of 2 dB. As for the impedance matching, we obtain 121% of bandwidth for -5 dB of reflection coefficient.

III. Measurement results

A prototype has been realized. The five segments are held by a dielectric support and mounted on a circular ground plane of 800 mm of diameter as shown in Figure 3. It is fed by a coaxial cable fixed on a SMA connector soldered on the monopole antenna.
The carbon tubes are formed by winding on a mandrel fabric carbon fibers impregnated with resin, which is horizontally or vertically deposit on the tubular segment. On curing the resin, the reinforcing strip is integrated with the tubular member.

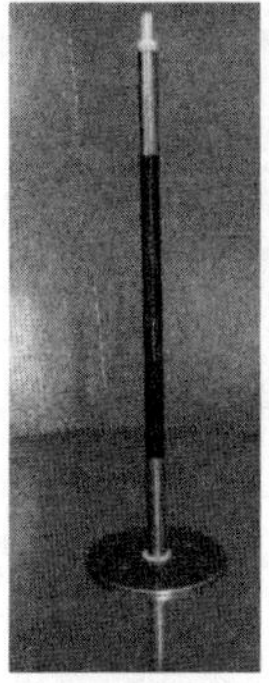

Fig 3. Picture of the realized antenna.

The obtained carbon segments (1-3) conductivity are : 468 SM, 3 S/m and 2 S/m respectively. Figure 4 shows monopole antenna reflection coefficient in simulation and in measurement using the new values. We note the conductivity values are different from the optimized value. We simulate again the prototype with the realized conductivity and compare results to measurements.

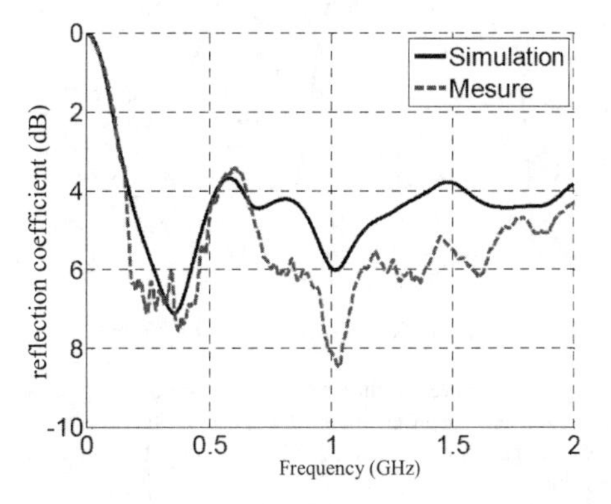

Fig 4. Simulated and measured reflection coefficients vs. frequency.

A double band behaviour at -6 dB of reflection coefficient is observed in Figure 4, starting at 180 MHz MHz to 500 MHz GHz and from 650 MHz to beyond 1 GHz. We note a good agreement between simulation and measurement.
The radiation measurements were performed in an anechoic chamber. In Figures 5, we present at 300 MHz, 600 MHz and 900 MHz, the measured E-plane radiation patterns compared with the simulation.

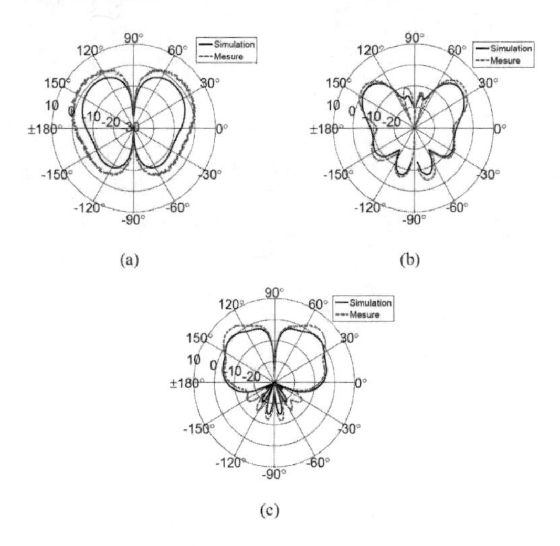

Fig 5. Simulated and measured radiation patterns in E plane at (a) 300 MHz, (b) 600 MHz and (c) 900 MHz

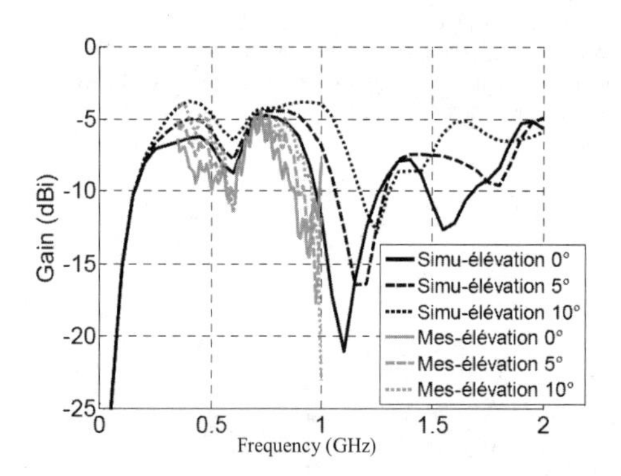

Fig 6. Simulated and measured realized gain vs. frequency for 3 elevations.

In figure 6, we present the measured and simulated gain for various values of elevation angle (θ=0°, θ=5° and θ=10°).. We observe that gain value is quite stable for a more than 100% of bandwidth.

IV. Conclusion

A broadband monopole antenna has been fabricated in varying conductivity values along antenna length. Its impedance bandwidth is about 120% in VHF-UHF band if we realize the optimized value. Its gain in one direction has a correct level because of the introduction of high conductive material at feeding position.

Acknowledgment

The authors would like to thank the DGA (Defense Department) and COMROD France for the financial support.

References

[1] Altshuler, "The travelling-wave linear antenna" , *IRE Transactions on Antennas and Propagation*, Vol. 9, Issue 4, pp. 324-329, July 1961

[2] T. T. Wu and R. W. P. King, "The cylindrical antenna with nonreflecting resistive loading", *IEEE Transactions on Antennas and Propagation,* Vol. 13, Issue 3, pp. 369-373, May 1965

[3] J.-H. Du, Z. Ying, S. Bai, F. Li, C. Sun and H.-M. Cheng, "Microstructure and resistivity of carbon nanotube and nanofiber/epoxy matrix nanocomposite" **INT J NANOSCI**,1 (5-6), 719-723,200

[4] K. Deb, A. Pratap, S. Agarwal and T. Meyarivan, "A fast and elitist multiobjective genetic algorithm: NSGA-II", *IEEE Transactions on Evolutionary Computation*, Vol. 6, N°. 2, pp. 182-197, August 2002

Wideband frangible monopole for radio monitoring

D. Gray†

† The University of Nottingham, Ningbo, Zhejiang, China

***Abstract*- In preparation for an urban environment radio localization demonstration using a Direction of Arrival system carried by a Zeppelin NT airship, some initial flights to check the reference transmitter broadcast power at altitude were conducted. A small handheld monopole antenna was required for 1.8 to 2.2GHz which could safely be used in the cabin of the Zeppelin NT airship in flight. The soft metal – no sharps monopole antenna had 29% bandwidth and was successfully used for 2 flights.**

I. INTRODUCTION

A Direction of Arrival (DOA) system for stratospheric unmanned aircraft to localize radio transmitters on the ground was proposed [1-4]. Potential localization applications were finding cell phones post-disaster, distress beacons, illegal radio transmitters, wildlife tracking, and high speed car positioning, all of which required accuracies of the order of meters. A major complication was that the stratospheric unmanned aircraft were flexible causing the array elements spread along the wingspan to be constantly in relative motion. Placing co-band reference transmitters on the ground in the vicinity of the target transmitter allowed for a simplification of the mathematics and removed the requirement to precisely know the positions of the array elements [1-4], Figure 1. The ground reference transmitters could be purpose laid beacons or as mundane as existing 3G cell phone base stations. Four 3G bands were considered for ground reference stations, Table 1.

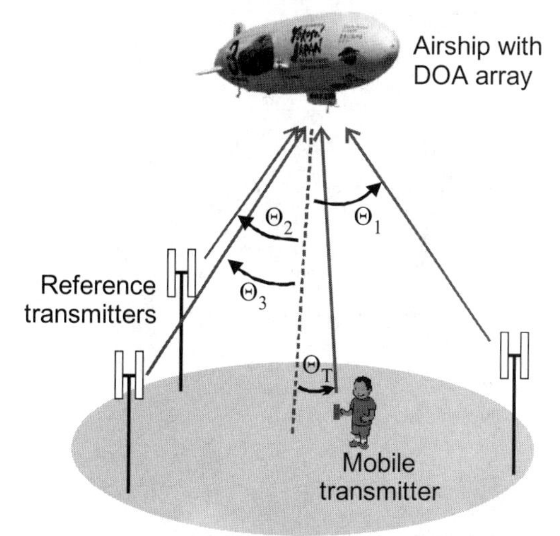

Figure 1. System concept for airborne Direction of Arrival system onboard an airship using 3G cell phone base stations as reference transmitters.

TABLE I
3G FREQUENCY ALLOCATIONS IN JAPAN

Bands	Frequencies (MHz)
VI & XIX	875 - 888
VIII	925-960
IX	1,845 – 1,880
I	2,110 – 2,170

Figure 2. Ground handling of the Zeppelin NT airship at Honda Airport, Okegawa, Saitama Prefecture; note the relative size of the gondola.

A Zeppelin NT airship was available for airborne radio communications demonstrations and field trials during the Japanese 2006/2007 and 2007/2008 financial years, Figure 2. As an airborne communications platform this 75m long airship had excellent characteristics, been capable of station keeping at 1,000m altitude, having a spacious cabin seating 12, and having a toilet enabling full-day field trials. Additionally, the Zeppelin NT airship was of particular interest for post-disaster recovery as it was a stable and quiet platform with no downdraft (contrast to a helicopter), and was perfect for observations of the ground either by human spotters or using high resolution digital camera. However, antennas for specific field trials could not be attached to the outside of the gondola due to flight worthiness certificate restrictions. The project plan is given as Table 2, and required 2 initial flights to confirm the reception of sufficient signal strength at an altitude of 1,000m. A small omni-directional antenna attached to a portable spectrum analyzer was to be used for these 2

978-7-5641-4279-7

flights, which raised a number of safety concerns related to handling a monopole antenna in the cabin of a vehicle in motion.

TABLE II
PROJECT SCHEDULE

Flight	Date	Antenna used
1	August 2006	This work
2	August 2006	This work
3	January 2007	Medium gain [8]
4	January 2008	4x4 patch array [9]

The Zeppelin NT airship in flight rolled and pitched gently in a similar fashion to a yacht. As sudden vehicle movement were thus discounted in the risk analysis, the main safety concern were accidents from human clumsiness working in a confined space such as tripping on improperly stowed equipment. Two possible types of injury while handling a monopole antenna were expected:

- stick-stabbing wounds from been impaled from a straight fall onto a monopole,
- slash-cut from brushing past a thin metal ground plane

The first injury type would be mitigated by using soft and thin metals, and an antenna structure which would readily collapse and crumple if fallen on by an adult. For the second type of injury, the number of sharp corners should be minimized and all edges rounded by filing.

The full bandwidth from the bottom of the 3G VI Band to the top of the I Band was 85.1%, Table 1, which would conventionally be met by a multi-band antenna based on microstrip technology with passive matching elements, such as [5]. However, the rigidity of the microstrip substrate would violate the frangibility requirement. Having separate antennas for the lower VI/XIX and VIII bands and upper IX and I bands was more readily achievable by a wideband monopole design. The bandwidth for the IX and I bands was 16.1%. A flat thin metal square monopole such as [6] would satisfy the $S_{11} \leq -10$dB bandwidth, but fail on the safety requirement due the square monopole been akin to a blade. In contrast, a thin metal folded monopole by the same authors [7] was enclosed and judged to be less likely to inflict slash-cut injuries, Figure 3.

II. WIDE-BAND METAL-PLATE MONOPOLE ANTENNA

The reported VSWR 2:1 bandwidth for a monopole made of folded thin sheet metal was 110%, [7]. Also, the reported lower bound of the VSWR 2:1 bandwidth was 1.92GHz, which only need be moved 100MHz lower to encompass the 3G IX band.

Simulating the antenna in FEKO™ gave a return loss characteristic loosely resembling the reported experimental measurements but significantly having narrower bandwidth and a distinct dual resonance, Figure 4. The currents on petals of the monopole tended to hug the sides of the petals across the entire bandwidth studied, Figure 5. Best return loss occurred when the currents at the bottom of the petals were well below the peak current, with the corners (left side of Figure 5) having low values.

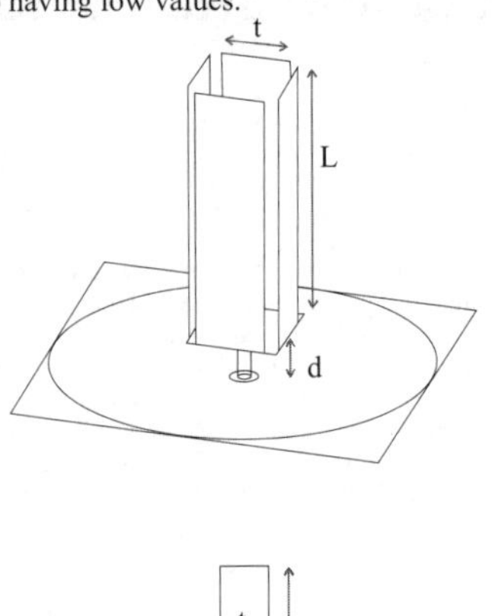

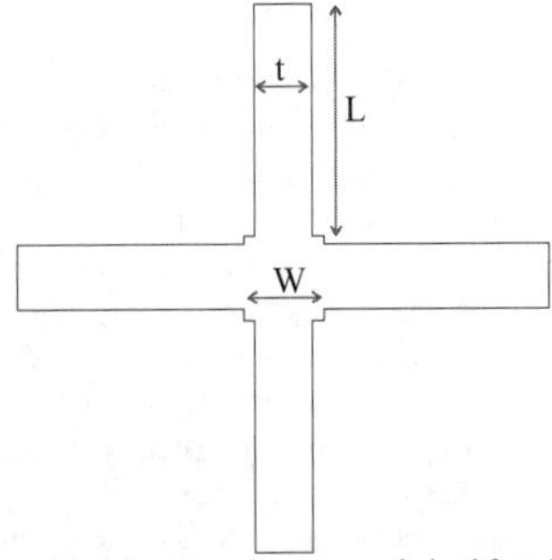

Figure 3. The metal-plate monopole antenna, derived from Figure 1 of [7].

TABLE III
ANTENNA DIMENSIONS

Dimension	Original [7]	This work
L	23	48
W	8	16
t	6	11
d	4.5	5
Ground Plane Side	150mm square	150mm circle

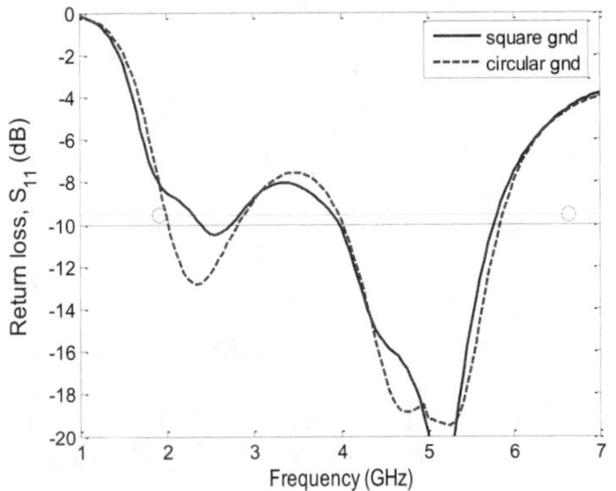

Figure 4. Return loss of metal-plate monopole antenna, from FEKO™.

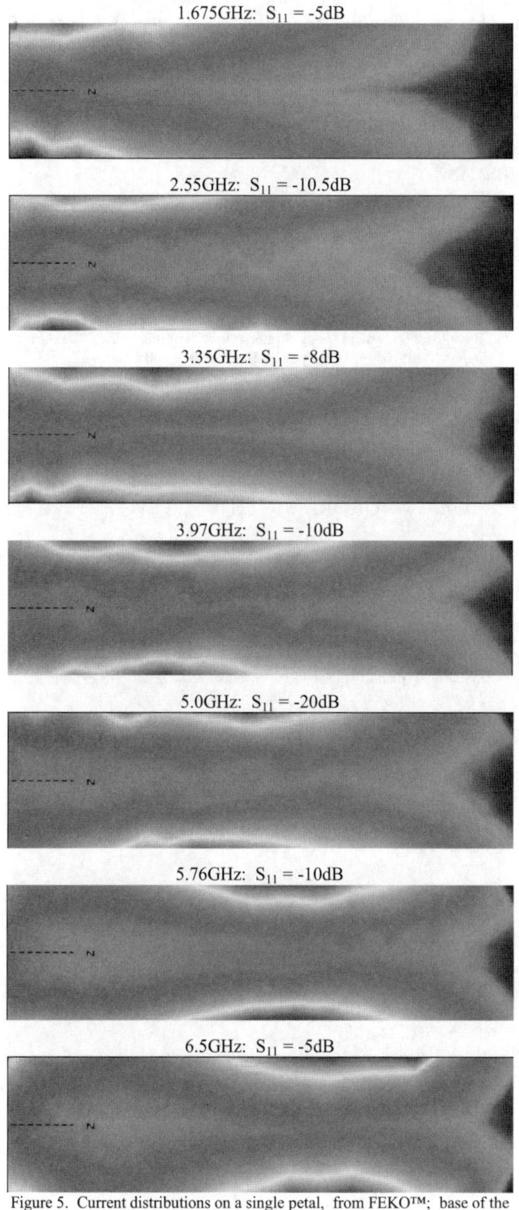

Figure 5. Current distributions on a single petal, from FEKO™; base of the petal on the left, top on the right.

III. Circular Ground Plane Effect

Having achieved a rough repeat of the antenna characteristics reported in [7] and gained some insight into mode of operation, the design was adapted for the application described above.

The first step was to change the ground plane from a 150mm sided square to a 150mm diameter circle, thus eliminating the corners. No radical changes were seen in the return loss, Figure 4. The first resonance around 2.1GHz improved in depth, and moved a little higher in frequency.

The second step was to move the first resonance lower in frequency so that the 3G IX and I bands would be covered. This required a close to doubling of the dimensions of the monopole, Table 3 and Figure 6. This was in no way a disadvantage in that the larger monopole was easier the cut out of 0.3mm thickness brass by hand. The radiation patterns were likewise satisfactory, exhibiting a typical monopole radiation pattern shape at the centers of both IX and I bands, Figure 7.

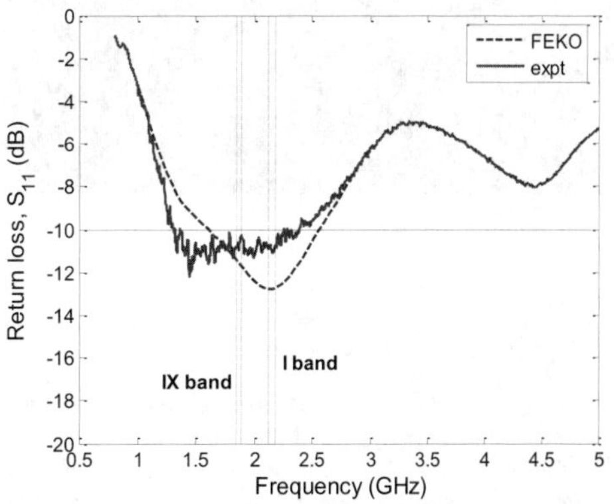

Figure 6. Return loss of the metal-plate antenna with a circular ground plane.

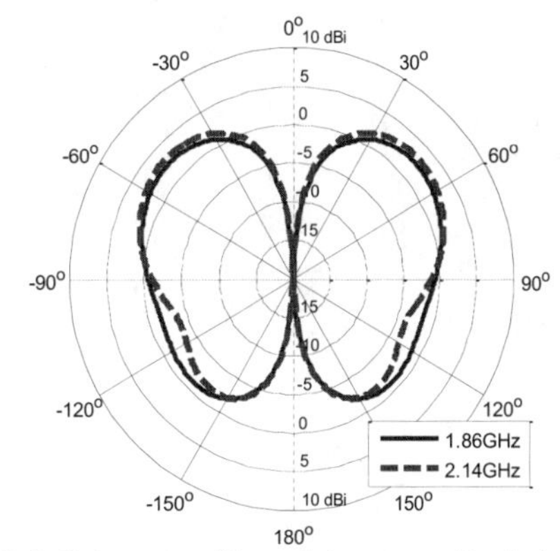

Figure 7. Radiation patterns of the metal-plate antenna with a circular ground plane, from FEKO™.

The ground plane for the metal plate monopole was cut from 0.5mm copper, which was thin and soft enough to crumple if fallen on. This handheld antenna was used for 2 demonstration flights in August 2006.

Figure 8. Antenna in use while flying at 1,000m altitude.

IV. CONCLUSIONS

A wideband frangible monopole antenna was made from thin metal sheeting, and proved its worth as a handheld antenna for receiving 3G cell phone signals during 2 demonstration flights in a Zeppelin NT airship. It was found that the monopole antenna dimensions had to be roughly doubled to effect a small downward shift in resonant frequency.

ACKNOWLEDGMENT

The experimental results reported in this paper were measured at the Yokosuka Research Park during the 2006/2007 financial year, with the airborne DOA project been undertaken and fully sponsored by the Ministry of Infrastructure and Communications of the Government of Japan. The Zeppelin NT airship was owned and operated by Nippon Airship Corporation, flew from Honda Airport, and the author is deeply indebted to the staff of both. The author also thanks EMSS SA for providing an academic license of FEKO™.

REFERENCES

[1] R. Miura, H. Tsuji, M. Suzuki, H. Arai & R. Kohno, "A Study on Radiolocation System Using Stratospheric Platforms," *Proceedings of the 2004 IEICE General Conference*, Tokyo Institute of Technology, March 2004, paper B-1-239.

[2] R. Miura, H. Tsuji, M. Suzuki, H. Arai & R. Kohno, "A Study on Radiolocation System Using Stratospheric Platforms," *IEICE Technical Report*, vol. 104, no.681, pp. 49-54, February 2005.

[3] H. Tsuji, D. Gray, M. Suzuki & R. Miura, "A Mobile Localization Experiment Using Array Antennas for High Altitude Platforms", *13th International Conference on Telecommunications (ICT 2006)*, Funchal, Madeira Island, Portugal, 9-12 May, 2006.

[4] R. Miura, H. Tsuji & D. Gray, "Radiolocation System Using Distributed Sensor Array Onboard a High Altitude Aerial Vehicle," *The 9th International Symposium on Wireless Personal Multimedia Communications (WPMC 2006)*, San Diego, USA, September 2006, paper TP1 322

[5] T. Watanabe & H. Mizutani, "Antenna," Japanese patent JP2003-258527, granted 12th September, 2003.

[6] K.-L. Wong, C.-H. Wu & S.-W. Su, "Ultrawide-band square planar metal-plate monopole antenna with a trident-shaped feeding strip," *IEEE Trans. Antennas Propag.*, vol. 53, no. 4, April 2005, pp. 1262-1269.

[7] K.-L. Wong & C.-H. Wu, "Wide-band omnidirectional square cylindrical metal-plate monopole antenna," *IEEE Trans. Antennas Propag.*, vol. 53, no. 8, Aug. 2005, pp. 2758-2761.

[8] D. Gray & H. Tsuji, "Medium gain antenna for radio monitoring from manned airship", *The Second European Conference on Antennas and Propagation (EuCAP 2007)*, Edinburgh, November 2007, paper Fr-1-6-4.

[9] H. Tsuji, "Radio location estimation experiment using array on airship," *The 11th International Symposium on Wireless Personal Multimedia Communications (WPMC 2008)*, San Diego, USA, September 2006, paper TP1 322

Hybrid Antenna Suitable for Broadband Mobile Phone MIMO System

Minkil Park*, Taeho Son* and Youngmin Jo**

*) Soonchunhyang University, Dept. of IT Eng.,
**) Skycross Korea Co. LTD
POB97 Asan, Chungnam 336-745 KOREA

Abstract- **In this paper, a hybrid antenna for suitable MIMO (Multi Input Multi Output) system for the hexa band mobile phone included LTE700 was designed and implemented. Hybrid consists of a monopole and an IFA (Inverted F Antenna). Monopole is fed, and IFA is simultaneously operated by the capacitive coupling of the monopole. Increase of capacitance by the coupling structure can perform both the low frequency resonance and the wide antenna bandwidth. To achieve small size, antenna is used vlade technique. PCB embedded type was applied to antenna design. For verification of this study, we designed hexa band as LTE700, CDMA, GSM900, DCS1800, PCS and WCDMA, and implemented on the bare board that is same size of real board of the handset. VSWR measurement showed within 3:1 over the whole design band. Average gains and efficiencies were measured as -3.74 ~ -1.31dBi and 42.40 ~ 73.91% for proposed antenna. H-plane radiation patterns were almost omni-directional.**

I. INTRODUCTION

While long term evolution was issued, the implementation of MMO system has been problem especially at LTE700 (class. Because of the low frequency band, the physical length of antenna for LTE700 needs longer than other antennas for the mobile communication. Antenna space trend of modern mobile phone is toward miniaturization with narrow and thin. This trend makes very difficult to design antenna. Currently, there are many studies to achieve small sized antenna [1]–[3]. However, no mobile phone manufacturers realize MIMO system for the LTE700. The reasons are several. First, small size antenna is hardly achieve low frequency resonance as LTE700. Second, bandwidth becomes so narrow due to complex L and C antenna equivalent elements. Antenna radiation efficiency is dropped by the surrounding grounds as LCD, battery, flat connectors and so on. And two and more LTE antenna for the MIMO are not allowed by the space problem. To overcome these, many studies were presented and suggested on the papers [4]-[10].

This paper is focused to have the broad bandwidth with low resonance frequency to be applied MIMO system including LTE700. We suggest a hybrid antenna to achieve both. Hybrid antenna consist of a monopole and IFA [11]-[12]. Vlade (vertical lade) technique is used to realize smaller antenna. Proposed antenna is designed and implemented on the bare board that is same size of real phone. Antenna is tested on the frequency band of LTE700, CDMA, GSM, DCS, PCS, and WCDAM. This paper is organized as follows. Section II describes the proposed antenna design using vlade. In Section III, the implemented antenna are experimented and evaluated in an echoic chamber for the antenna performances. Finally, the Conclusion mentions the technical issues and commercial applications

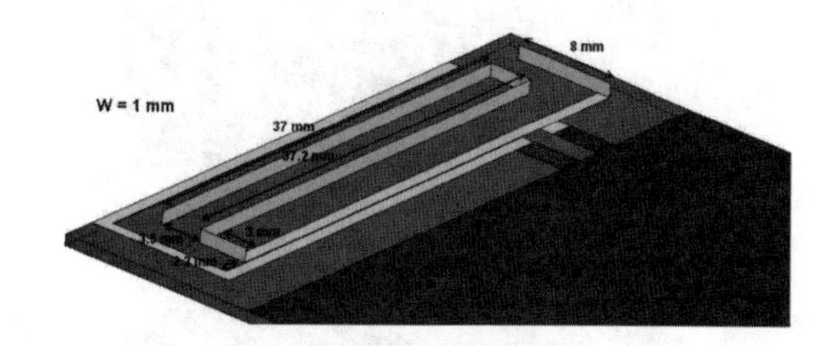

Figure 1. Detailed geometry of hybrid antenna.

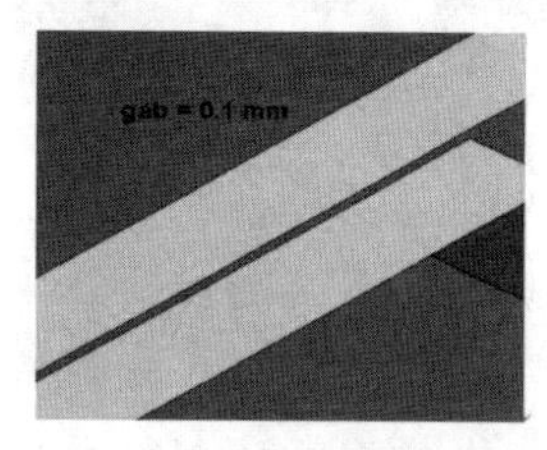

Figure 2. Feeding gap

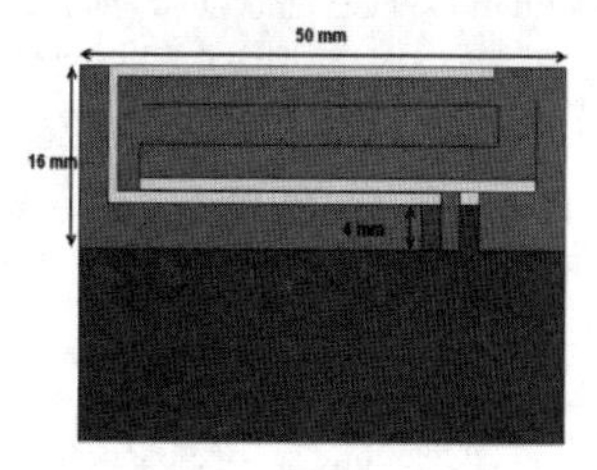

Figure 3. Front view of antenna.

978-7-5641-4279-7

II. HYBRID ANTENNA

A hybrid antenna is designed on the bare board as geometry depicted in Figure 1. Antenna is implemented on 50mm by 96mm FR4 PCB substrate of dielectric constant ε_r =4. Thickness of PCB is 1.0mm. The ground size of PCB is 50mm by 80mm. Antenna is basically fed to monopole. And IFA is fed by the coupling structure of monopole. Coupling gap is 0.1mm as in figure 2. Figure 3 shows front view of antenna with dimension of PCB space. Antenna size is 50mm by 16mm. Metal conductor of IFA is vertically laded to the PCB. By vlade, it can reduce size as much as conductor widths. Simulated current densities of designed antenna are as figure 4. Figure 4(a) is antenna current density at 700MHz frequency for LTE700 band, and 4(b) is at 1850MHz frequency for PCS band. Simulation tool is Ansoft HFSS v13. As shown in figure, current flows on the IFA(a) and both(b) as we predict.

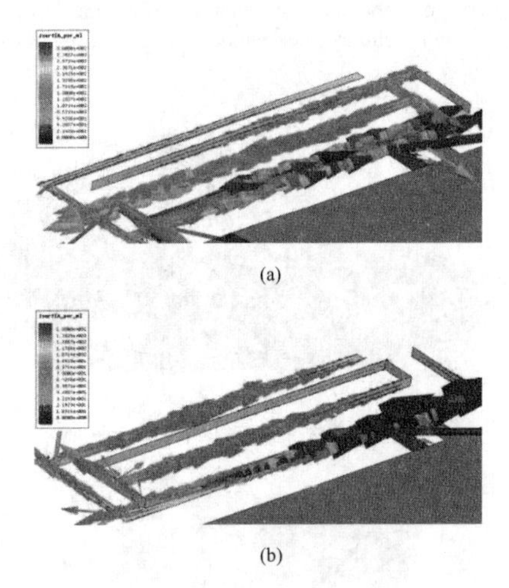

(a)

(b)

Figure 4. Simulated current density of designed antenna (a) LTE700 (b) PCS

III. IMPLEMENTATION AND MEASUREMENT

Implemented antenna is as figure 5. Antenna is directly implemented on the PCB without any carrier. So we can say that this is one kind of the PCB embedded antenna.

Implemented antenna was measured return loss and VSWR by the Agilent network analyzer. Measurement result is as figure 6. By measurement, proposed antenna is satisfied design hexa band within VSWR 3:1 (6dB return loss). Bandwidth of low band (LTE/CDMA/GSM) is 260 MHz that is 35.29% percent bandwidth. It allows antenna to cover the mobile service band of LTE700 (class 13, 14), CDMA, and GSM900. Also high bandwidth is 460 MHz that is 23.17% percent bandwidth. It covers the mobile service band of DCS, PCS and WCDMA.

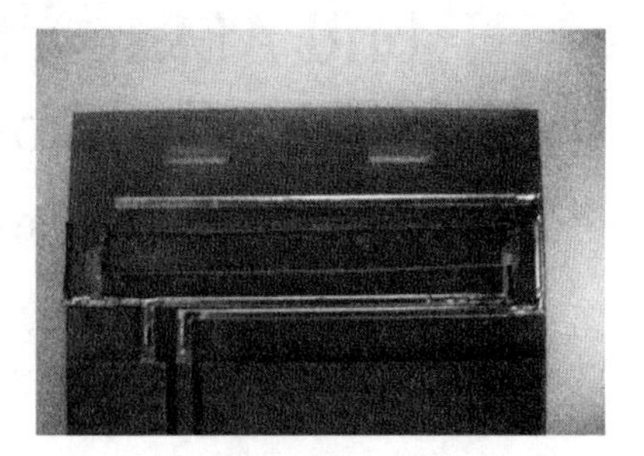

Figure 5. Implemented antenna.

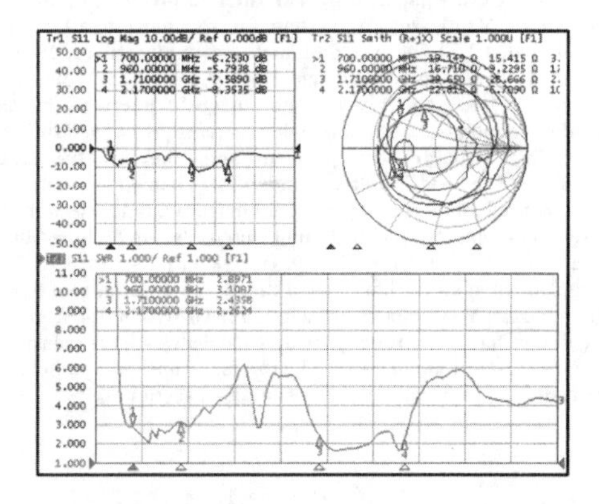

Figure 6. Measured S11 and VSWR

Figure 7 shows measured 2 dimensional radiation patterns. 7(a) shows 2D patterns for the low band, and 7(b) is patterns for the high band. Radiation patterns are measured in MTG far field anechoic chamber. The 2D radiation patterns in the low bands show omni directional pattern at the H–plane. But the radiation patterns of high bands have more variations and nulls. The reason is that the radiation patterns of mobile phone antenna are dependent on the ground plane of the mobile phone set which is also an efficient radiator, especially for frequencies in the low band. Wavelength of high band is shorter than length of mobile phone ground plane that led to nulls observed in radiation patterns at upper band.

Table 1 shows the measured antenna average gain and radiation efficiency for frequencies. Average gains for low band were -3.73 – -2.53dBi, and radiation efficiencies were 42.40 – 55.84%. Average gains for high band were -3.27 – -1.31dBi, and radiation efficiencies were 47.08 – 73.91%.

I. CONCLUSIONS AND DISCUSSIONS

In this study, we designed a hexa band hybrid antenna that can be applied to MIMO system. Vlade technique was applied to realize small antenna. Proposed antenna was implemented on the bare PCB, and measured VSWR and radiation pattern. Measured VSWR was satisfied design hexa band within 3:1. Average gain and radiation efficiency were -3.74 ~ -1.31dBi and 42.40 ~ 73.91%. Proposed antenna had good performance over the hex broadband including LTE700. Therefore, proposed hybrid antenna can be applied to broadband MIMO system.

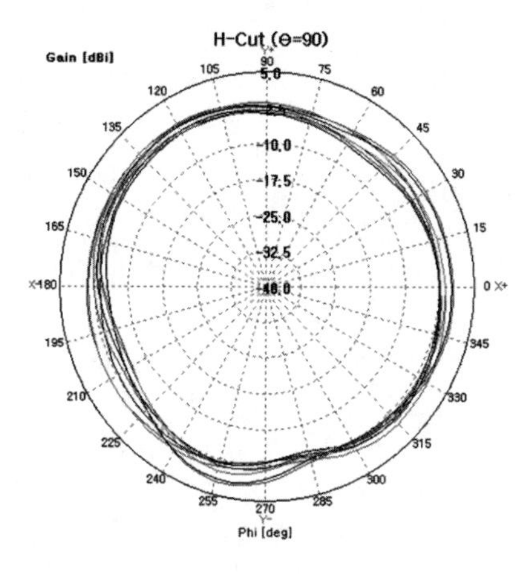

(a)

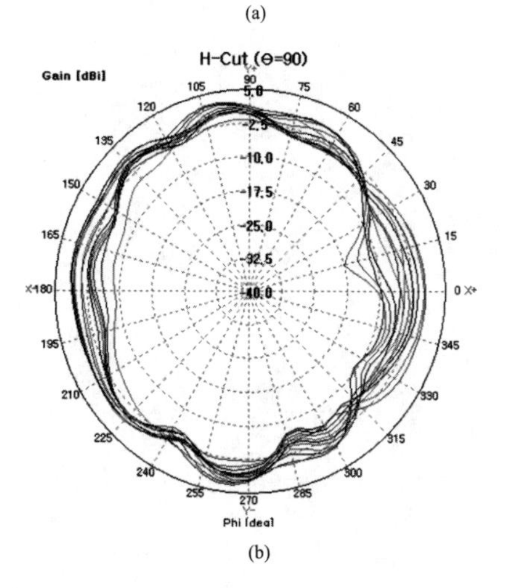

(b)

Figure 7. 2D radiation pattern of the proposed antenna. (a) LTE 700, CDMA, and GSM. (b) DCS, PCS, and WCDMA.

ACKNOWLEDGMENT

This research was supported by Basic Science Research Program through the national Research Foundation of Korea (NRF) funded by the Ministry of Education (2013006254)

TABLE I
THE MEASURED ANTENNA AVERAGE GAIN AND RADIATION EFFICIENCY

	Antenna 1	
Freq.[MHz]	Eff.[%]	Avg.[dBi]
700	42.40	-3.73
740	48.34	-3.16
780	42.06	-3.76
820	43.00	-3.67
860	44.32	-3.53
900	48.94	-3.10
940	55.37	-2.57
960	55.84	-2.53
1710	47.08	-3.27
1750	60.65	-2.17
1790	63.75	-1.95
1830	71.04	-1.48
1870	73.91	-1.31
1910	64.91	-1.88
1950	61.62	-2.10
1990	67.58	-1.70
2030	67.33	-1.72
2070	60.44	-2.19
2110	58.68	-2.31
2170	58.04	-2.36

REFERENCES

[1] T. W. Kang and K. L. Wong, "Chip-inductor-embedded small-size printed strip monopole forWWAN operation in the mobile phone," *Microwave Opt. Technol. Lett.*, vol. 51, pp. 966–971, Apr. 2009.
[2] K. L. Wong and S. C. Chen, "Printed single-strip monopole using a chip inductor for penta-band WWAN operation in the mobile phone," IEEE Trans. Antennas Propag., vol. 58, Jan. 2010.
[3] C. M Peng, I. F. Chen, C. C. Hung, S. M. Shen, C.T Chien, and C. C. Teng, "Bnadwidth enghancement of internal antenna by using reactive loading for penta–band mobile handset application," IEEE trans antenna propag., pp 1728–1733, vol. 59, May. 2011.
[4] F. H. Chu, K. L. Wong , "Planar Printed Strip Monopole With a Closely-Coupled Parasitic Shorted Strip for Eight-Band LTE/GSM/UMTS Mobile Phone," IEEE Trans Antennas Propagat., vol. 5, pp. 3426–3431 October 2010.
[5] S. C. Chen and K. L. Wong, "Bandwidth enhancement of coupled-fed on-board printed PIFA using bypass radiating strip for eight-band LTE/GSM/UMTS slim mobile phone," *Microwave Opt. Technol. Lett.*, vol. 52, pp. 2059-2065, Sep. 2010.
[6] K. L. Wong, M. F. Tu, T. Y. Wu and W. Y. Li, "Small-size coupled-fed printed PIFA for internal eight-band LTE/GSM/UMTS mobile phonen antenna," *Microwave Opt. Technol. Lett.*, vol. 52, pp. 2123–2128, September 2010.
[7] J. Cho and K. Kim, "A frequency-reconfigurable multi-port antenna operating over LTE, GSM, DCS, and PCS bands," *presented at the IEEE Antennas Propasgat.,* Soc. Int. Symp., Charleston, SC, 2009, Session 317

[8] B. N. Kim, S. O. Park, "Wideband Bulit-In Antenna With New Crossed C–Shaped Coupling Feed for Future Mobile Phone Application," *IEEE Antenna Wireless Propag.*, vol. 9, pp. 572–575, 2010.

[9] Minh Zheng, Hanyang Wang, "Hexa–band quad–mode folded monopole/dipole/loop antenna," *IEEE Antenna Tecgnology (iWAT) International Workshop* pp. 48–51 March 2012.

[10] Yonghun Chen, Jungyub Lee, and Joonghee Lee, :"Quad–Band Monopole Antenna Including LTE 700MHz With Magneto - Dielectric Material," IEEE Antenna Wireless Propag., vol. 11 pp.137–140 2012.

[11] Seung Jin Lim, Taeho Son, "Hybrid antenna for the all band mobile phone service including LTE,"KIEES , vol. 22, pp. 737–743, July, 2011

[12] M. K. Chun, T. H. Son "PIFA and IFA hybrid antenna for the data communication terminal," KITS, vol. 10, no. 4 pp. 65–70 August, 2011

Design of a High-Gain Wideband Microstrip Antenna with a Stepped Slot Structure

ZHU Wei-gang ,YU Tong-bing, Ni Wei-min
Institute of Communications Engineering , PLA University of Science and Technology
Nanjing ,Jangsu 210007 China

***Abstract*-This paper focuses on an antenna with a stepped slot structure. The paper analyzes the impedance and radiation characteristics of this type of antenna and designs a high-gain wideband antenna that operates in the L-band and S-band. The simulation results show that the antenna has a comparatively high gain, wide impedance bandwidth and beamwidth in the entire L-band and S-band, and can thus be used in the design of antenna arrays.**

Index Terms-Stepped slot structure, wideband, high-gain antenna.

I. Introduction

As a high-gain wideband traveling-wave slot antenna, the tapered slot antenna has been playing an important role in the research field of wideband antennas since it was put forward. Tapered slot antennas transmit energy through a microstrip slot structure and ultimately couples the energy to antenna patches for radiation.[1],[2]. The patch opening in a gradually widened horn shape constitutes the subject to radiate or receive energy, where different operating frequency points correspond to different 1/2 λ opening width values. Tapered slot antennas can realize highly wide frequency bands, good radiation pattern symmetry, high gains, high beamwidths and pretty low sidelobe levels.

This chapter deals with an antenna with a stepped slot structure, analyzes the impedance and radiation characteristics of the antenna in the S-band, and designs a linear polarized high-gain wideband microstrip antenna[3], [4],[5].

II. Antenna Structure

The structure of the stepped slot antenna described in this paper is shown in Fig. 1 below. The maximum opening width of antenna corresponds to minimum operating frequency of the antenna. The stepped slot structure determines that the antenna can achieve a very wide frequency band, good radiation pattern symmetry, and high gain. It is a linear polarized antenna. In the two main radiation planes, the antenna has a very high linear polarization purity, pretty low cross polarized levels, wide beamwidths and pretty low sidelobe levels.

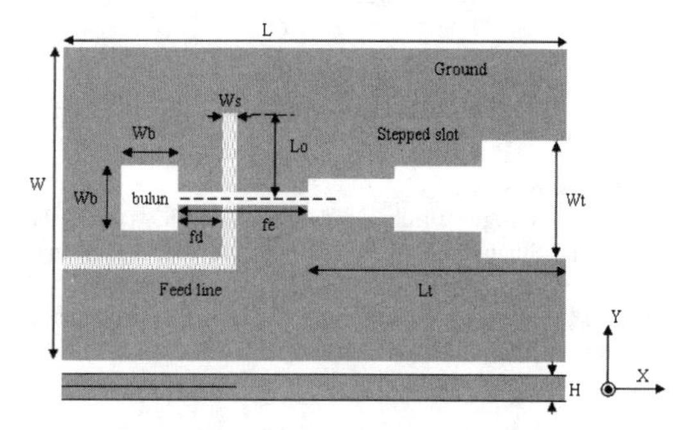

Fig. 1. Antenna with a stepped slot structure.

III. Analysis of Antenna Characteristics

A. Antenna structure

The antenna in Fig. 1 is made up of strip lines. Stepped slots are made in mutually symmetric positions of the grounding plates at the top and base in order to radiate electromagnetic signals. The central conductor is a feed microstip line. The number of stepped slot lines depends on the antenna width. The more the number of stepped slot lines is, the wider the antenna width is. To improve antenna's impedance characteristics, cut out rectangular holes in the mutually symmetric positions at the feed end of both grounding plates at the top and base. The holes are used as Balun for feeding structure adjustment and has an important influence on the optimization of the antenna's impedance characteristics.

B. Antenna performance

The structure of the antenna in this paper is built using the finite element method for tetrahedral mesh generation. With this method, the electrical property of the stepped slot antenna can be worked out efficiently. Conventional slot antennas usually has a three-layer structure: metal radiation patches, dielectric substrate, and feeding structure. The tetrahedral mesh structure can accurately fit multi-layer metal patches and dielectric structure on the plane and fit small electrical structure in a highly accurate way through local uneven segmentation.

978-7-5641-4279-7

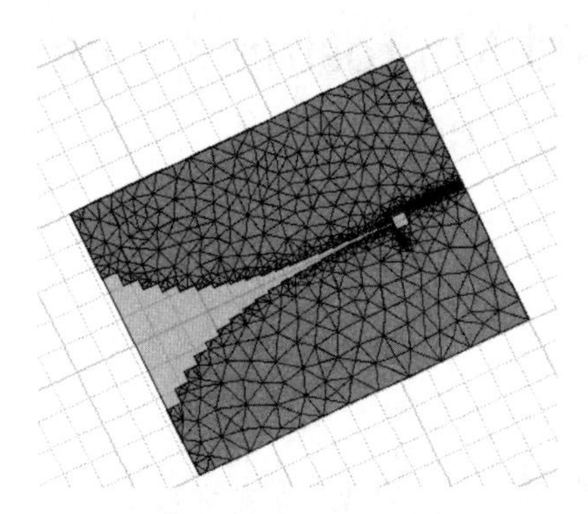

Fig. 2. Antenna segmentation with finite element method.

C. *Antenna simulation*

The paper designs an ultra-wide antenna which covers the L-band and S-band. The dielectric constant of the dielectric substrate of the strip line is 4.4 and the thickness of the substrate is 1.6 mm. Fig. 3 shows a screenshot of the antenna simulation.

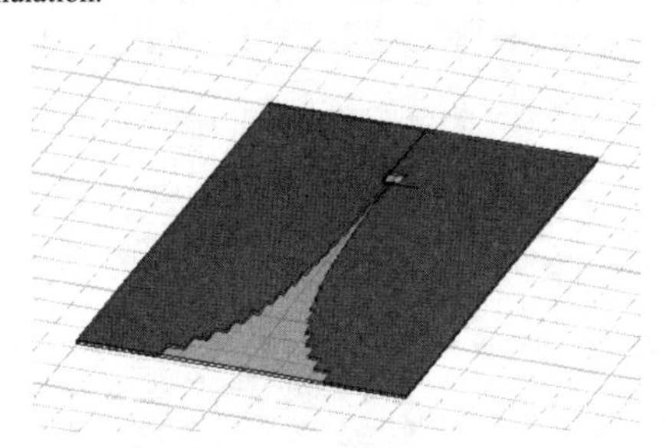

Fig. 3. Screenshot of the antenna electromagnetic characteristics simulation.

The impedance and radiation characteristic curves of the antenna as well as its geometry are given in the table below, which are obtained through the antenna optimization and simulation analyses:

(unit: mm)

W	L	Wb	Ws	fd	fe	Wt	Lt
100	200	10	0.7	5	15.8	100	200

D. *Antenna radiation characteristics*

Fig. 4 shows the 3D beam direction pattern of the antenna. The data column to the left shows antenna gains. The maximum antenna gain is 9.8 dB. The 3D beam direction pattern is to the right of the pattern, where the maximum radiation direction lies in the X axis which has a better radiation symmetry and thus serves as the radiating elements of the antenna array. Thanks to the strip line structure, the antenna has a high linear polarization purity, pretty low cross polarized level, wide beamwidth, and pretty low sidelobe level, all of which makes the antenna ideal for the application of antenna arrays. See Fig. 5 below.

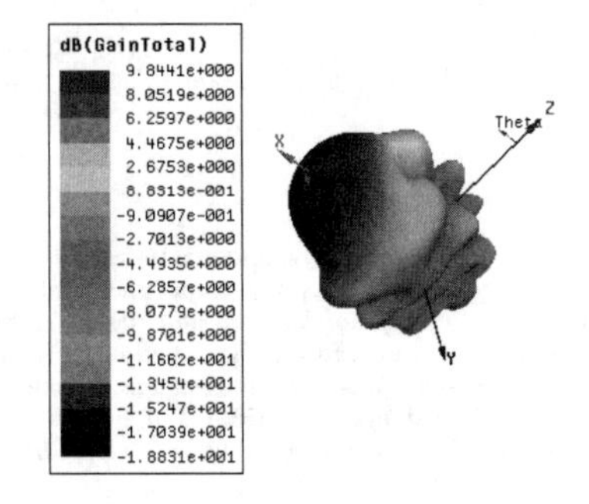

Fig. 4. Antenna 3D direction pattern.

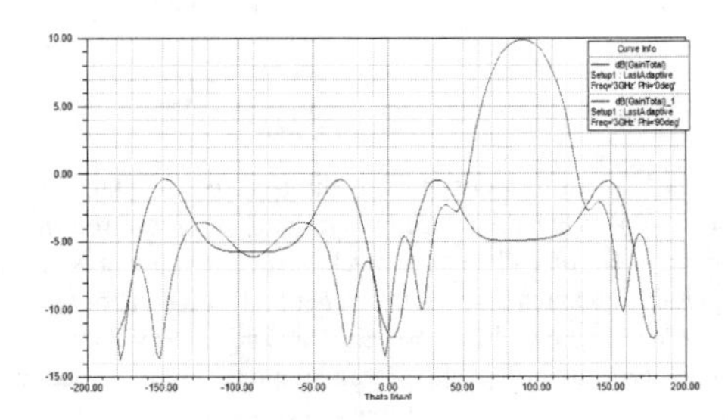

Fig. 5. Antenna 2D direction pattern.

E. *Antenna return loss characteristics*

Fig. 6 shows the return loss characteristics of the antennal. It can be seen that when the antenna operates in the L-band and S-band, if the antenna return loss is less than 10 dB, then the antenna bandwidth can reach 1800 MHz or so, and the relative bandwidth is about 60%. The frequeny band is very wide.

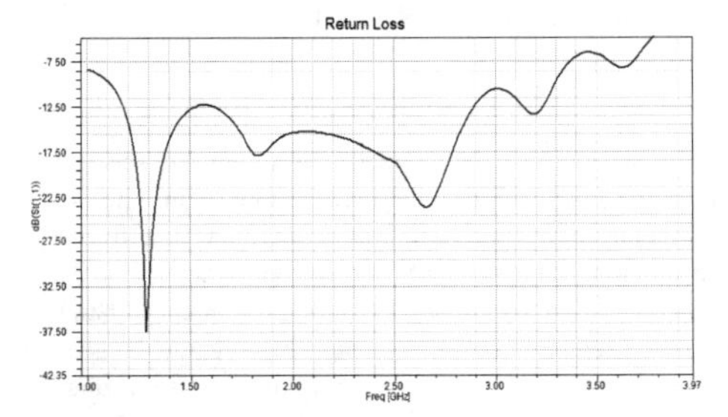

Fig. 6. Antenna return losses.

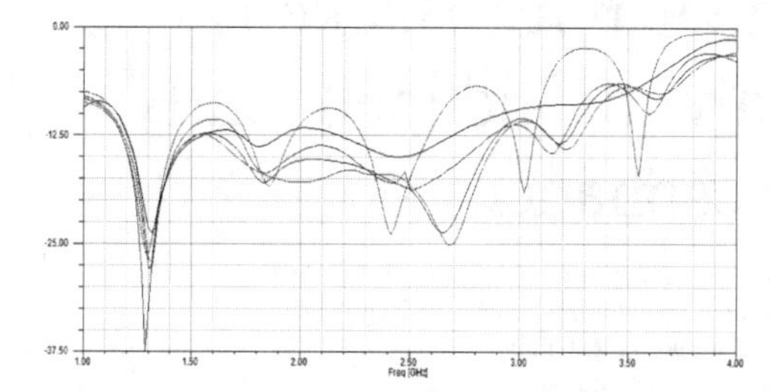

Fig. 7. Influence of the number of stepped slots on return losses.

It can be seen from Fig. 7 that the number of stepped slots of the antenna has a certain influence on the impedance and radiation of the antenna. The impedance width increases with the increase of stepped slots.

IV. CONLUSION

One can make a conclusion from above analyses that the antenna with a stepped slot structure has a higher gain of up to 10 dB, impedance bandwidth of 60%, low sidelobe levels, extremely small cross polarization component, and good radiation characteristics and bandwidth performance, which make the antenna ideal for the applications of antenna arrays. This type of antenna has a marked advantage compared with conventional microstrip antennas; nevertheless, the type of antenna has a comparatively complex structure and more parameters to adjust. In particular, any slight deviation in the feeding structure of the antenna may result in performance deterioration, which may cause certain difficulty in antenna manufacturing.

REFERENCES

[1] .Janaswamy and D. H. Schaubert, "Analysis of the Tapered Slot Antenna," *IEEE Trans . Antennas and Propagation* , vol . 35, pp. 1058-1065, Sep. 1987.

[2] J. Shin and D. H. Schaubert, "A Parameter Study of Stripline. Fed Vivaldi Notch-Antenna Arrays" , *IEEE Trans . Antennas and Propagation*, v01. pp. 879-886, May 1999.

[3] U.Bhobe and Christopher L.Holloway, "Wide-Band Slot Antennas With CPW Feed Line: Hybrid and Log-periodic Designs," *IEEE Trans. Antennas and Propagation*, vol. 52, pp. 2545-2554, Oct.2004.

[4] S . Nikolaou , G E . Ponchak , J . Papapolymerou and M. M. Tentzeris, "Conformal Double Exponentially Tapered Slot Antenna(DETSA)011 LCP for UWB Applications , " *IEEE Trans. Antennas and Propagation*, vol. 54, pp. 1 663-1 669, Jun. 2006.

[5] Shing-Lung steven Yang and Ahmed A.Kishk, "Wideband Circularly Polarized Antenna With L-Shaped Slot," *IEEE Trans. Antennas and Propagation*, vol. 56, pp. 1780-1783, Jun. 2008.

A Q-Band Dual-Mode Cavity-Backed Wideband Patch Antenna with Independently Controllable Resonances

Tao Zhang, Yan Zhang, *Member, IEEE*, Shunhua Yu, Wei Hong, *Fellow, IEEE*, and Ke Wu, *Fellow, IEEE*
State Key Laboratory of Millimeter Waves, School of Information Science and Engineering,
Southeast University, Nanjing, 210096, P. R. China
taozhang@emfield.org, weihong@seu.edu.cn

***Abstract*-In this paper, a dual-mode cavity-backed patch antenna and a 2x2 array is proposed for Q-band wireless applications. In addition to the original TM01 mode of patch antenna, an independently controllable resonance is introduced by loading the antenna with an inductive via and an annular slot. The bandwidth (return loss<-10 dB) of the antenna element is thus expanded to 12.4% while the bandwidth of the 2x2 array is 12.7%, with a peak gain of 6.57 dBi and 10.6 dBi, respectively. The 2-dB gain bandwidth of the array is wider than 13.95%. Besides, in order to provide a deep understanding of the antenna, a fully physically based circuit model is presented. The circuit simulation and full wave simulation agree well with the measurements.**

***Key words*- dual-mode, cavity backed, patch antenna, circuit model**

I. INTRODUCTION

With a rapid development of millimeter-wave wireless communications, many researches are focusing on antennas operating in millimeter-wave bands with a sufficient bandwidth.

Conventional cavity backed patch antennas enjoy the advantages of easy fabrication compatibility with integrated circuits systems, low weight, low mutual coupling when forming an array, and then it is reasonable to adopt this kind of antennas for millimeter-wave applications. However, patch antennas only have one resonance, which implies narrow bandwidth. In order to overcome such shortcoming, numerous efforts have been dedicated to expanding the bandwidth of patch antennas, including: 1) using thick substrate with low permittivity, 2) employing matching stubs on the feeding lines, and in millimeter-wave applications, 3) introduce more resonances. However, 1) and 2) are limited by some disadvantages such as the easy excitation of surface wave on thick substrates, the tedious size introduced by the matching stubs. As for 3), parasitic patches [1], L-shaped probes [2], slots on patches [3], are adopted. Many of these techniques may have their own application contexts and may have some drawbacks: the tedious size introduced, the structural complexities, and high cross polarization, etc.

In this paper, a method to expand the bandwidth of patch antennas is presented for Q-band wireless communications, such as the point-to-point system (40.5-43.5GHz) for backhual application and IEEE 802.11aj etc. By loading an inductive via and a capacitive annular slot, an additional resonance is introduced. The additional resonance, in combination with the original one of the patch mode, expands the -10dB-returnLoss bandwidth of the antenna element to 12.4%, ranging from 38.85 GHz to 43.98 GHz. Moreover, a physically based circuit model is presented to illustrate the mechanism of the antenna.

II. THE ANTENNA ELEMENT

A. Antenna Configuration

The geometry of the antenna element is shown in Fig. 1. The feeding CPW is etched on the ground of the antenna, and at the end of the CPW, a pad is introduced on which a feeding via is used to connect the patch and the pad. The feeding via also serves as an inductor. Under the via is a pad which serves as the positive plate of the introduced capacity while the negative plate is the ground of the antenna. The antenna is designed on a substrate of Rogers 5880 with a thickness of 0.508mm.

B. Circuit Model and Antenna Mechanism

In Fig. 2, it is shown that the antenna has two resonances: one is at f_1=40.45GHz, another is at f_2=43GHz. As is shown in Fig. 3(a),(b), at f_1, the field resonates as the typical TM01 mode of patch antennas while the field resonates around the inductive feeding via and the annular slot on the ground at f_2, which verifies that a new mode is introduced by the above two elements.

In order to deeply explain this phenomenon, a detailed circuit model is presented in Fig. 4(a), with its simplified counterparts in Fig. 4(b). The TM01 model is modeled by *Lh*, *Lpa*, *Cpa*, and *Ri*. The inductive feeding via is modeled by *Lprb* while the annular slot is modeled by *Cgap*. A mutual inductor is adopted to model the distorting effects of the via on the TM01 model as shown in Fig. 3(c). *Cp2p* illustrates the capacity between the pad and the patch while *Cgap* represents the capacity between the upper metal and the patch. *Lcav* is used to model the cavity, which is negligible since the cavity is formed by many inductive vias in parallel, making *Lcav* rather small. Moreover, *Lh*, used to model the effects of higher modes, is also neglected since only the dominant mode is considered in our design. The capacity denoted by Cp2p is ignored for the illustrative circuit model. Finally, *Cgap* is

978-7-5641-4279-7

absorbed into *Cpa*, resulting in *Cpa2* in Fig. 4(b). Since many complex effects are neglected, the circuit model is quite an illustrative one, used only to qualitatively explain the antenna mechanism. As shown in Fig. 2, the simulation between the circuit and HFSS agree well, which demonstrates the validity of the circuit model.

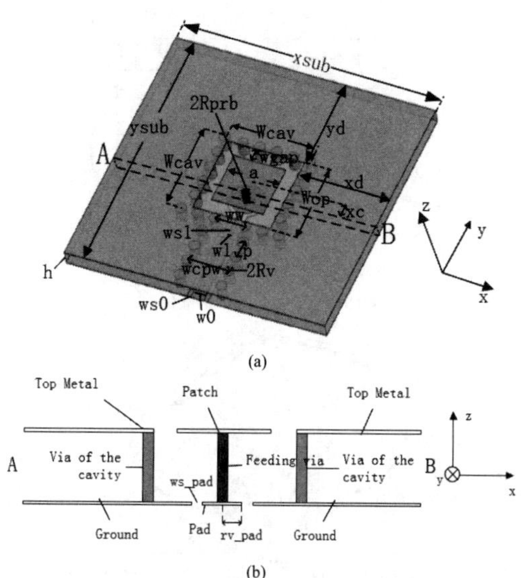

Fig. 1. Configuration of the antenna element with a typical set of parameters: *xsub*=6.7, *ysub*=10, *xd*=0.5, *yd*=3.5, *xc*=0.3, *Wcav*=3.3, *Wop*=3, *ww*=1.2, *wcpw*= 1.6, *w0*=0.5, *w1*=0.15, *ws0*=0.15, *ws1*=0.2, *ws_pad*=0.15, *wgap*=0.5, *rv_pad*=0.53, *rprb*=0.15, *a*=2, *Rv*=0.2, *p*=0.65, all parameters are in "mm". (a) overall view, (b) sectional view.

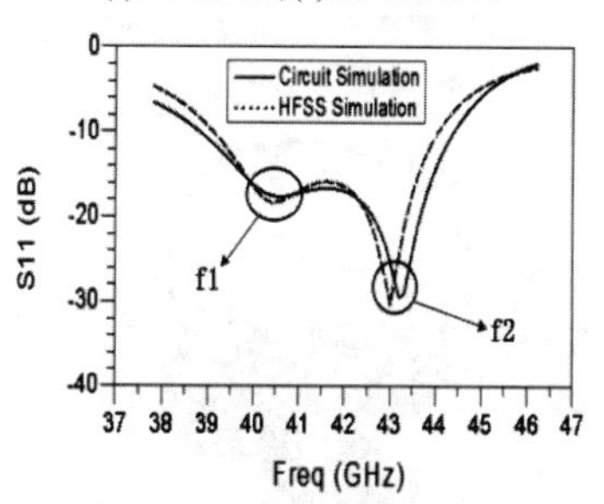

Fig. 2. Simulated return loss (in dB) by HFSS and circuit model.

Fig. 5 presents the behaviors of the two resonances as parameters change. In Fig. 5(a), as "a" (the length of the square patch) increases by even steps, f_1 decreases with equal intervals with a merely slight influence on f_2, as is expected for TM01 mode of patch antennas. In Fig. 5(b), as *rv_pad* increases, f_2 shifts down. This attributes to the increase of *Cgap* in Fig. 4(b) based on formula (1). Similar rules and explanations stand for "*ws_pad*", which is not provided here. In Fig. 5(c), as the radius of the feeding via goes higher, f_2 increases since *Lprb* decreases. As shown in Fig. 5(d), *xc* influences the depth of the S11 curve, that is, the matching of the antenna, while maintaining the resonant frequencies. We have to inform that the analysis by (2) is quiet a rough one, since it is based on single RLC series or parallel circuit. In fact, a more rigorous one, done by simulating the circuit when tuning the relative element in Fig. 4(b), agrees with the above explanation, which is not provided here.

$$f_0 = \frac{1}{2\pi\sqrt{LC}} \tag{1}$$

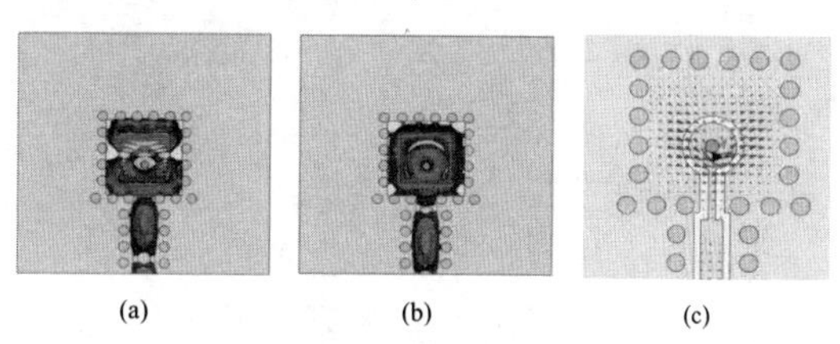

(a) (b) (c)

Fig. 3. Field distribution in the cavity. (a) E-Field at f_1, (b) E-Field at f_2, and (c) H-field distorted by the feeding via.

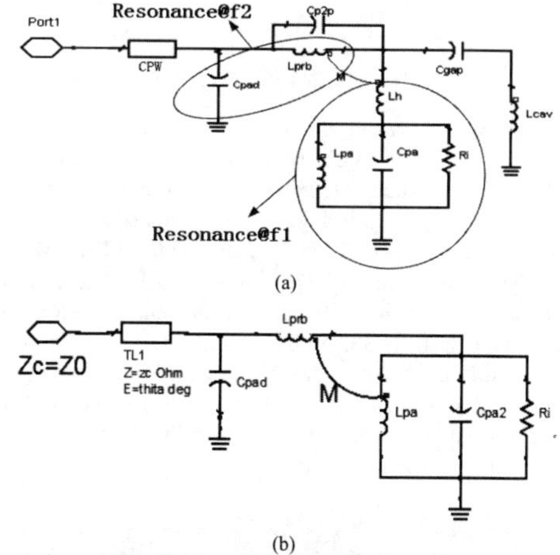

(b)

Fig.4. Circuit model for the proposed antenna.
(a) Detailed circuit model, (b) Simplified circuit model: *Z0*=75Ohm, *Cpad*=0.166pF, *Lprb*=0.117nH, *Lpa*=0.017nH, *M*=0.00627nH, *Cpa2*=0.863pF, *Ri*=300hm, *zc*=2340hm, *thita*=194deg.

C. Design Procedure

First, the size of the square patch is determined based on center frequency of a traditional patch antenna. Second, the size of the annular ring is adjusted until a new resonance appears at a proper frequency. Finally, the position of the feeding point, denoted by *xc*, is adjusted to achieve a good matching.

Determination of other parameters: *wgap* is selected the same as the thickness of the substrate, which in turn determines the size of the cavity; the feeding via is set to be as thin as the fabrication standards permit (this has two benefits: the feeding point would be smaller so that it is closer to a lumped excitation; the inductor introduced is larger so that the annular slot could be smaller based on (2), which will help reduce backward radiation.); the width of the annular slot should be as small as possible because this indicates larger capacity, which in turn will help reduce the annular slot.

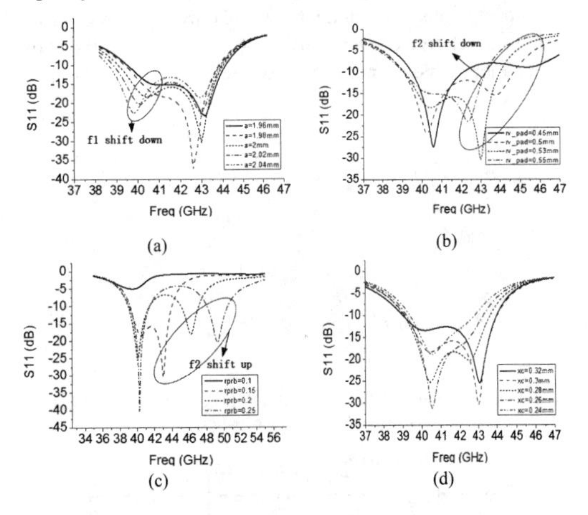

(a) (b)

(c) (d)

Fig. 5. Parameter study of the proposed antenna by HFSS.

D. *Antenna Performance*

In order to test the performance of the antenna, the 75Ohm feeding line is matched to 50Ohm by a quarter-wave CPW. The structure is shown in Fig. 6. The simulated and measured performance is shown in Fig.7. The simulated band width is 14.2%, ranging from 39.3 GHz to 45.3 GHz. The measured bandwidth (return loss<-10 dB) of the antenna element is 12.4%, ranging from 38.85 GHz to 43.98 GHz, with a gain of 6.57dBi at f_1 and 6.2dBi at f_2. The measured return loss shifts lower by 2%, causing a bandwidth sacrifice of 1.8%. This mainly attributes to fabrication errors and improper estimation of dielectric constant at Q bands.

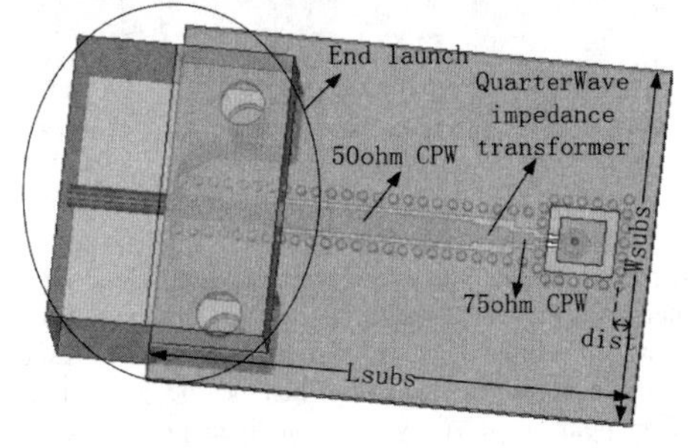

Fig.6. Antenna structure for measurement.
Lsub=21mm, *Wsub*=16mm, *rv_pad*=0.5mm, *dist*=1.5mm.

In Fig. 7(c), the side lobe at 90° is irregularly high. This is due to the influence of the metal end launch which serves as a reflector. As a result, the boresight gain suffers from a drop at higher frequencies as shown in Fig. 7(a). In fact, the simulated gain without the end launch shown in Fig. 1(a) is 7.27dBi at f_1 and 6.52dBi at f_2, which verifies that the gain drop is mostly caused by the end launch rather than the annular slot on the ground. In applications, this effect could be avoided by etching the patch on the ground of the circuit boards when integrated with circuit systems.

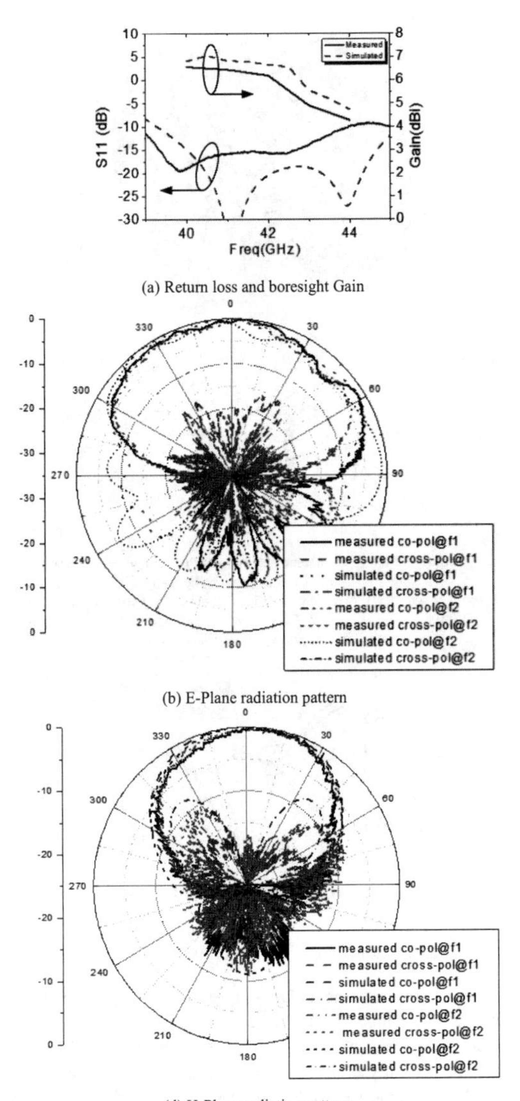

(a) Return loss and boresight Gain

(b) E-Plane radiation pattern

(d) H-Plane radiation pattern

Fig.7 Performance of the antenna element.

III. ANTENNA ARRAY

In addition to the antenna element, we also designed a 2x2 array using parallel-feeding technique. The feeding network consists of a 180° self-compensating phase shifter [6], a divider and a SIW to CPW transition [7]. The overall structure is shown in Fig.8 (the end launch is not shown) with its performance shown in Fig.9. The size of the square patch is optimized to 1.92mm and *rv_pad* to 0.48mm, *xc* to 0.33mm. The simulated impedance bandwidth and the 2-dBi gain bandwidth are 13.95%, both of which ranging from 40 GHz to 46 GHz. The measured return loss shifts higher by 2%, ranging from 41 GHz to 46.7 GHz, with a sacrificed bandwidth of 0.3 GHz. The discrepancy between the simulated and measured results in Fig.9 (a) mainly attributes to fabrication errors and improper estimation of dielectric constant of the substrate at Q-Bands. The overlapping bandwidth is 11.4%, from 41GHz to 46 GHz.

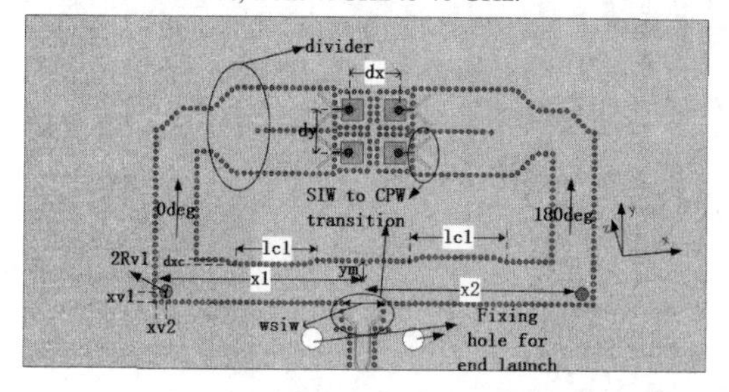

Fig.8. Structure of the 2x2 array. $xv1$=1.05mm, $xv2$=1mm, dxc=0.36mm, $Rv1$=9mm, $lc1$=$lc2$=9mm, dx=4mm, dy=3.9mm, $x1$=19.57mm, $x2$=21.53mm, $wsiw$=3.9mm. The size of the substrate is 64mmx32mm.

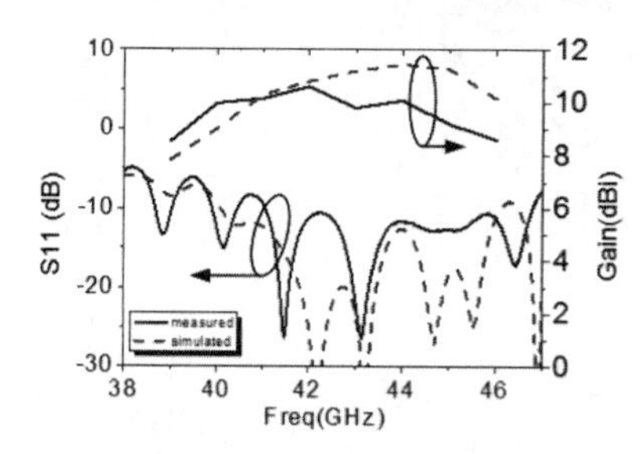

(a) Gain and return Loss

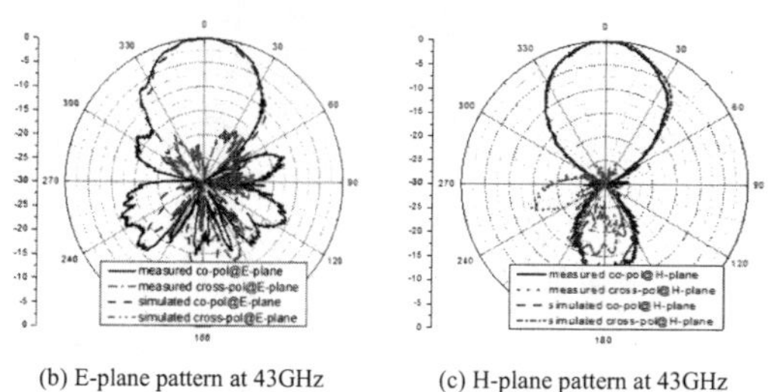

(b) E-plane pattern at 43GHz (c) H-plane pattern at 43GHz

Fig.9. Performance of the 2x2array.

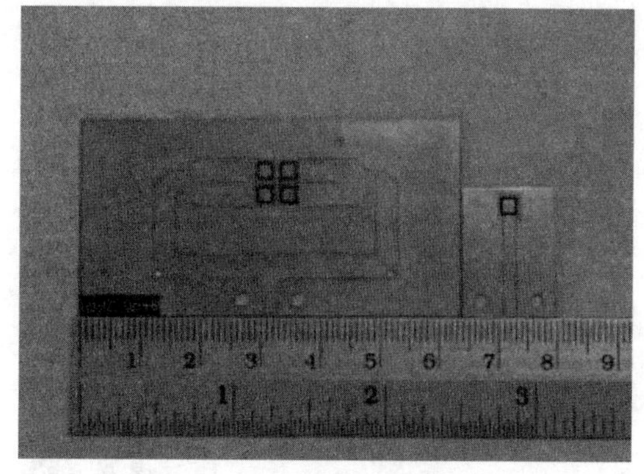

Fig.10. Photograph of the antenna element and array.

IV. CONCLUSION

In this paper, a dual-mode cavity-backed patch antenna with independently controllable resonances is proposed and studied in detail. Moreover, a 2x2 array based on this antenna is presented. The photograph of the antenna element and array are presented in Fig. 10.

ACKNOWLEDGMENT

This work was supported in part by National 973 project 2010CB327400 and in part by NSF of Jiangsu province under Grant SBK201241785.

REFERENCES

[1] Sang-Hyuk Wi, Yong-Shik Lee and Jong-Gwan Yook, "Wideband Microstrip Patch Antenna With U-Shaped Parasitic Elements", *IEEE Trans. Antennas Propagat.*,vol.55,Issue 4, April 2007.

[2] Mak, C.L, Luk,K.M, Lee,K.F "Experimental study of a microstrip patch antenna with an L-shaped probe", *IEEE Trans. Antennas Propagat.*,vol.48, Issue 5, May 2000.

[3] K. F. Lee, K. M. Luk, K. F. Tong, S. M. Shum, T. Huynh, and R. Q. Lee, "Experimental and simulation studies of coaxially fed U-slot rectangular patch antenna," *Inst. Elect. Eng. Proc. Microwave Antennas Propagat,* vol. 144,1997.

[4] Yu Jian Cheng, WeiHong and KeWu, "Broadband Self-Compensating Phase Shifter combining delay line and quual-length unequal-width phaser", *IEEE trans. Microw. Theory Tech.*,vol.58,no.1,Jan.2010.

WP-1(D)

October 23 (WED) PM

Room D

Compact Antennas

Compact Hybrid Dielectric Resonator with Patch Antenna Operating at Ultra-High Frequency Band

M. I. A. Sukur, M. K. A. Rahim, and N. A. Murad
Communication Engineering Department, Faculty of Electrical Engineering, Universiti Teknologi Malaysia, 81310 Skudai, Johor, Malaysia
ishak_1703@yahoo.com.my, mkamal@fke.utm.my, asniza@fke.utm.my

Abstract- **This paper describes the design of the hybrid dielectric resonator with patch antenna operating at Ultra-High Frequency band. At this particular band, the size of the antenna is excessively big, thus consuming bigger space. As a kind of hybrid antennas, which consist of two different antennas integrated on a single body, the design achieved more compact design, especially the patch antenna. The design was designed and simulated using CST Microwave Studio. The patch antenna's size is reduced about 50 % while the dielectric resonator antenna's size is about the same size as the estimated dimension. The hybrid antenna managed to obtain a gain of 2.8 dBi and capable of achieving wide bandwidth, around 49 %.**

I. INTRODUCTION

Microstrip patch and dielectric resonator antennas have often been compared with each other, due to the capability of the dielectric resonator antennas (DRAs) to replace the microstrip patch antennas in various applications. By employing the hybrid antenna design as proposed in [1], both antennas are combined, resulting in single antenna with dual-frequency operation. A single antenna that is small in size and can provide the dual-frequency or multiband operations are highly desired [2] due to cost reduction and less area occupied by that particular antenna. Patch antennas are often used in various applications since it is low-profile, can be mounted on a flat surface and ease of fabrication [3]. However, the size of the patch antennas increased tremendously at low frequencies, below 1 GHz, up to fifteen centimeters.

Several techniques were used in order to reduce the size of the antenna such as meandering [3], using high permittivity substrates [4], and shorting pins [5].

$$W = \frac{c}{2f_c}\sqrt{\frac{2}{\varepsilon_r+1}} \tag{1}$$

$$L = \frac{c}{2f_c\sqrt{\varepsilon_{r_{eff}}}} - 2\Delta L \tag{2}$$

The length and width of the basic rectangular patch antennas are calculated theoretically using (1) and (2), where c is the speed of light, f_c is the operating frequency, ε_{reff} is the effective dielectric constant and ΔL is the length approximation. At 0.85 GHz, the calculated length and width of the antenna are 11.92 and 13.95 cm by using $\varepsilon_r = 2.2$ and substrate thickness of 1.575 mm, which is considered big. Alternatively, a compact patch antenna can be designed by introducing the ground, which reduces the dimension of the antenna from half-wavelength to a quarter-wavelength, which is equivalent to 7.5 cm.

Normally, within the same operating frequency, the size of the DRAs are smaller than microstrip patch antennas due to the influence of the value of the dielectrics used, thus enable the DRA to be placed symmetrically on the top of the patch antenna in this configuration. Identical to the patch antennas, the compact dielectric resonator antennas can also be designed using several approaches such as shape modification [6], using high dielectric constant [7] and metal loading [8]. The dimension of the DRA approximated using (3);

$$F = \frac{2\pi a f_c\sqrt{\varepsilon_r}}{300} \tag{3}$$

Where F is the normalized frequency, a is the length that parallel with y-axis, f_c is the operating frequency, and ε_r is the dielectric constant. At 0.85 GHz, the length and width of the DRA are the same, which is 4 cm, while its height is 1.33 cm.

Even several techniques can be used to create a compact design, both patch antennas and DRAs still experience narrow bandwidth, for example, due to the usage of materials with high dielectric constant, ε_r. The bandwidth degradation can be avoided using slot-feeding techniques, in addition to the size reduction. However, the slot itself will become big; with the length of about 12 cm around 1 GHz. The hybrid antennas often have a dual - frequency operation, one frequency for each antenna. The bandwidth can be further increased by merging both frequencies, resulting in wider bandwidth. In this design,

978-7-5641-4279-7

the DRA will cover higher frequency while lower frequency will be covered by the patch antenna in order to create a compact hybrid antenna design. Two circular slots were used to feed both the patch antenna and the DRA, where upper slot is used to feed the DRA while lower slot is used to feed the patch antenna, instead of normal rectangular slot in order to reduce the space consumed by the slot itself.

In the previous design [9], dual-frequency resonance is recorded in 5.40 and 6.18 GHz which later merged to produce a wide bandwidth of 23.5 %, from 5.14 to 6.51 GHz, where the patch antenna with the dimension of 1.5 x 1.5 cm^2, resonates at lower frequency (5.4 GHz), while the DRA with the dimension of 1.27 x 1.27 x 0.95 cm^3 resonates at a higher frequency (6.18 GHz). Similarly, the patch antenna resonates at lower frequency while the DRA resonates at a higher frequency, achieved a bandwidth of 49.88 %, from 0.8 to 1.2 GHz where the size of the patch antenna is only 5.1 x 5.1 cm^2, while the size of the DRA is only 4.27 x 4.27 x 2.45 cm^3.

II. Antenna Configuration

As mentioned in the previous section, the hybrid antenna consists of two different antennas which are the patch antenna and the DRA, fed by two different slots, the upper and lower slot. In short, this antenna has three layers stacked up together. Starting from the bottom layer, it consists of the feed substrate, followed by the patch antenna with its substrate at the middle layer, and the DRA at the top layer.

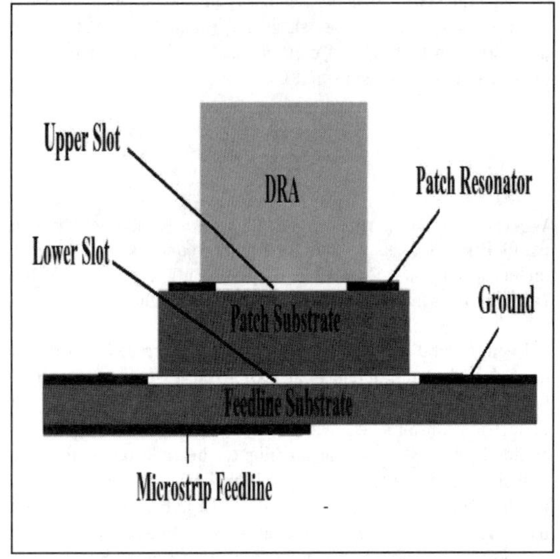

Figure 1: Side View of the Proposed Hybrid Antenna

Figure 1 shows the layout of the hybrid antenna. This kind of excitation is known as electromagnetic coupling (EMC) since there is no physical contact between the feeder and the radiating elements. Both slots are circular, where the length of normal slot is equivalence to its diameter. Lower slot has a radius, R_l of 34 mm while the upper slot has a radius, R_u, of 22 mm. Other related parameters are summarized in Table 1.

Table 1: Summarized Parameters of the Hybrid Antenna

Antenna Parameters	Values
DRA	
• ε_r	9.2
• l x w	42.7 mm x 42.7 mm
• h	24.5 mm
Patch Substrate	
• ε_r	2.2
• l x w	60 mm x 60 mm
• h	1.575 mm
Patch Resonator	
• l x w	51 mm x 51 mm
Ground	
• l x w	80 mm x 80 mm
Feed Substrate	
• ε_r	10.2
• l x w	80 mm x 80 mm
• h	1.27 mm

III. Results and Discussions

The introduction of this hybrid antenna concept provides the solution in obtaining compact antenna design with the capability of achieving wide bandwidth. Figure 2 shows the resonant frequencies of both antennas. Based on Figure 2, it can be seen that the patch antenna covered lower frequency (0.6 GHz), which represented by the solid line, while DRA covered higher frequency (0.95 GHz), which represented by the dotted line. By comparing the size of both antennas with their respective theoretical values, it can be seen that the size of patch antenna in this configuration is smaller, nearly 50 % size reduction while the DRA is almost of the same size. By comparing both antenna's dimensions with the previous compact antennas, it seems that some of the previous antennas have smaller size, but they have a very narrow bandwidth, below 10 %. By using this configuration, both frequencies can merge, resulting in a new frequency response with wide bandwidth, as shown in Figure 3.

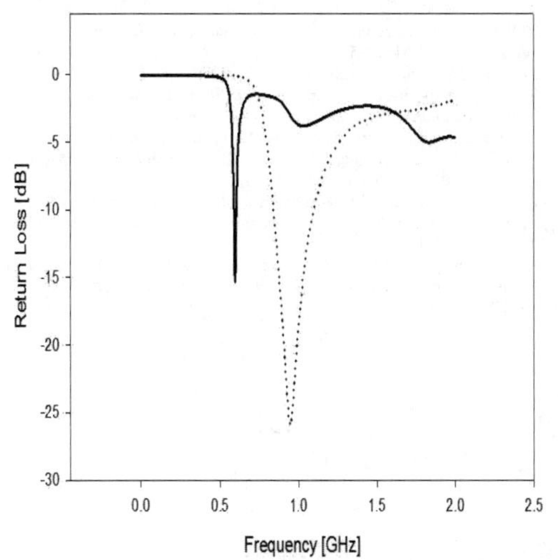

Figure 2: Frequency Response for both Patch Antenna and DRA

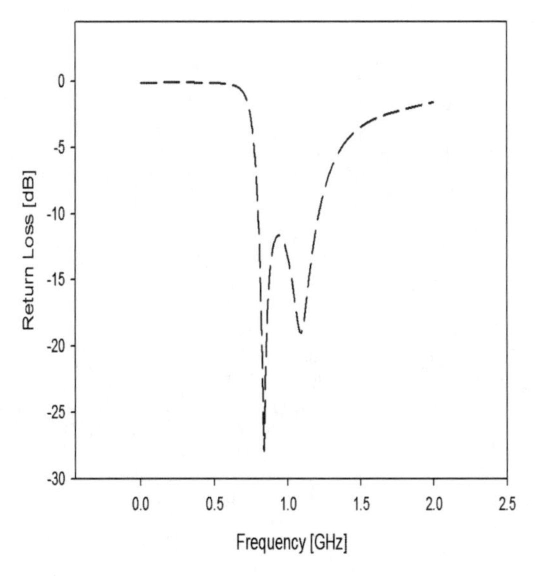

Figure 3: Overall Frequency Response of the Proposed Hybrid Antenna

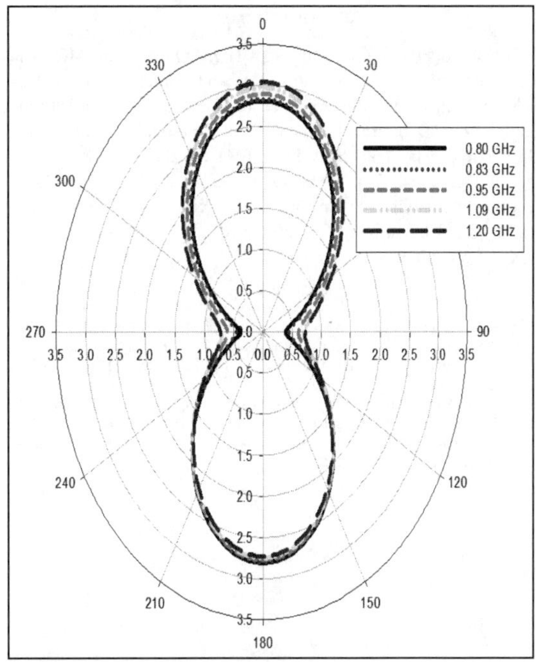

Figure 4: The Simulated Radiation Pattern of the Proposed Hybrid Antenna

The hybrid antenna radiates like a dipole-antenna, due to the presence of the slots, through the whole bandwidth with a gain that varies around 2 to 3 dBi as shown in Figure 4. The size of the patch antenna can be further reduced by using one of the techniques mentioned, especially meandering, but it is crucial to ensure that the bandwidth will not be degraded. Same thing applies to the DRA, which size can be reduced by loading metals on top of it. Using high permittivity materials as the patch substrate can reduce its size, but this is not applicable in this design due to the presence of a dielectric mismatch between the DRA and the feed line substrate with the patch substrate. Dielectric mismatch is also one of the causes of the bandwidth degradation.

IV. Conclusions

A compact hybrid antenna was designed and simulated using CST Microwave Studios. The size of the patch antenna is 50 % smaller compared to theoretical estimation while the size of the DRA almost the same. The hybrid antenna has a wide bandwidth, around 49.88 %, from 0.8 to 1.2 GHz with a gain of 2.8 dBi. The design will be later fabricated in order to measure its performance.

ACKNOWLEDGMENT

The authors thank the Ministry of Higher Education (MOHE) for supporting the research work, Research Management Centre (RMC), School of Postgraduate (SPS) and Communication Engineering Department, Faculty of Electrical Engineering, Universiti Teknologi Malaysia (UTM) for the support of the research grant no. R.J130000.7323.4B070, 4L811 and 04H38, and the International Symposium on Antennas and Propagation (ISAP) 2013 Organizing Committee for all the assistances and services provided.

REFERENCES

[1] K. P Esselle, "A dielectric-resonator-on-patch (DROP) Antenna Wireless Applications: Concept and Results," *Proc. IEEE Antennas and Propagation Society (AP-S) Int. Symp.*, vol. II, Boston, MA, Jul. 8–13, 2001, pp. 22–25.

[2] M. T. Ali, N. Ramli, M. K. M. Salleh and M. N. Md. Tan, " A Design of Reconfigurable Rectangular Microstrip Slot Patch Antennas," *IEEE International Conference on System Engineering and Technology (ICSET)*, 2011, pp. 111-115.

[3] Balanies, C. A., *Antenna Theory: Analysis & Design*, 2nd edition, John Wiley & Sons, Inc., 1997.

[4] Nurul H. Noordin, Ahmet T. Erdogan, Brian Flynn, Tughrul Arslan, "Compact Directional Patch Antenna with Slotted Ring for LTE Frequency Band", *Loughborough Antennas & Propagation Conference*, 2012, pp. 1-4.

[5] R. Waterhouse, "Small microstrip patch antenna," IEE Electronics Letters, vol. 31, pp. 604-605, 1995.

[6] M. T. K. Tam and R. D. Murch, "Compact Circular Sector and Annular Sector Dielectric Resonator Antennas," *IEEE Transactions on Antennas and Propagation*, vol. 47, no. 5, pp. 837-842, 1999.

[7] A. Petosa, A. Ittipiboon. "Dielectric Resonator Antennas: A Historical Review and the Current State of the Art", IEEE Antennas and Propagation Magazine, 2010, vol. 52, no. 5, pp. 91-116.

[8] R. K. Mongia. "Reduced Size Metallized Dielectric Resonator Antennas", *IEEE Antennas and Propagation Society International Symposium*, vol. 4, pp. 2202-2205, 1997.

[9] K. P. Esselle, T. S. Bird. "A Hybrid-Resonator Antenna: Experimental Results", *IEEE Transactions on Antennas and Propagation*, 2005, vol. 53, no. 2, pp. 870-871

A UHF Band Compact Conformal PIFA Array

J. Zhou[1], K. Chen[1], H. Wang[1], Y. Huang[2], J. Wang[2] and X.-Q. Zhu[1]
[1]Millimeter Wave Technique Laboratory, Nanjing University of Science & Technology, Nanjing, 210094, China
[2]Suzhou Bohai Microsystem CO., LTD., Suzhou, 215000, China
haowangmwtl@gmail.com

***Abstract*-A compact conformal planar inverted-F antenna (PIFA) array, working at UHF band, is presented in this paper. This array consists of four PIFA elements. The parallel feeding network has been designed to drive the elements. The size of the proposed array is 100.5 mm × 48 mm. The proposed antenna is fabricated on a very thin substrate with a thickness of 0.254 mm, which is benefit for conformal structure. For the prototype, this array is assembled around a 16 mm diameter cylinder. The antenna has been successfully simulated, fabricated and measured, and the simulated and measured results show a good agreement with each other. The antenna has an omni-directional radiation pattern and a good gain in a very compact structure. The measured peak gain is -0.2 dBi at 432 MHz, and the bandwidth of the conformal antenna is 6 MHz.**

I. INTRODUCTION

Antennas are essential components in communication systems. In recent years, there has been an increased interest in operating at the ultra-high frequency (UHF) band [1]. Conventional antennas can be very large in size when they operate at UHF band. With the development of technology, all the systems demand small footprint and excellent performance, and a minimized antenna which is one of the important parts in the design of such systems. As we know, gain and bandwidth of the antenna are rapidly decreased when the antenna size is reduced to be less than $\lambda_0/8$ (λ_0 is the free-space wavelength), thus it is a challenging task to miniaturize the antenna into a specified form factor while ensuring reasonably good performance [2]. In recent years, there are various types of miniaturized UHF antennas in use, such as miniaturized helix antennas, miniaturized patch antennas, small spirals and so on.

A compact coplanar waveguide (CPW)-fed multi-band monopole antenna loaded with spiral ring resonators is presented in [3]. An ultra-compact antenna using a magneto-dielectric material suited for Digital Video Broadcasting is presented in [4], but its design takes long times and costly. A new miniaturization technique based on an original combination of a modified fractal shape and a sinusoidal profile is proposed in [5]. The bandwidth of a small circularly polarized antenna can be significantly increased by employing four radiating elements and exciting them in equal magnitude and successive phase difference of 90° [6]. When a single or dual radiator is used instead of four, the bandwidth is reduced. The miniaturized version [7] of the classic quadrifilar helix antenna [8] is a familiar example of this approach. In [9] and [10], four straight or meandered inverted-F elements have been employed to realize a circularly polarized antenna.

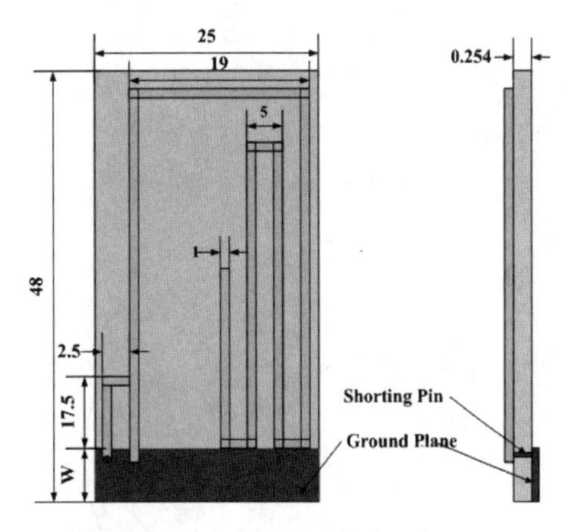

Fig.1 Geometry of antenna element.(unit: mm)

In this paper, we present a compact inverted-F antenna used for radiating elements of a conformal array. A four-port microstrip-line power divider is employed to excite radiating elements. The antenna is constructed in a flexible substrate so that it can stick on a teflon cylinder.

The paper is organized as follows. In section II, a antenna element is studied first, and parameters of the antenna element are discussed; In section III, a planar array structure is simulated and measured; Furthermore, a performance comparison between planar array and conformal array is also given in this section.

II. ANTENNA ELEMENT DESIGN

Fig. 1 shows the proposed PIFA element. The antenna is designed and operated at the 430 MHz frequency band that is used in communication system. A very thin substrate with a thickness of h = 0.254 mm, and dielectric constant of $\varepsilon_r = 2.2$ is used to fabricate the antenna. The PIFA element with dimensions of 25 mm×48 mm, is designed to roughly resonate as a quarter wavelength structure. Moreover, shorting pin with a radius of 0.2 mm and the meander line are used to reduce the size. The ground plane is located at the bottom of the antenna element, which is used to support the feeding microstrip line. More detail parameters have been given in Fig. 1.

978-7-5641-4279-7

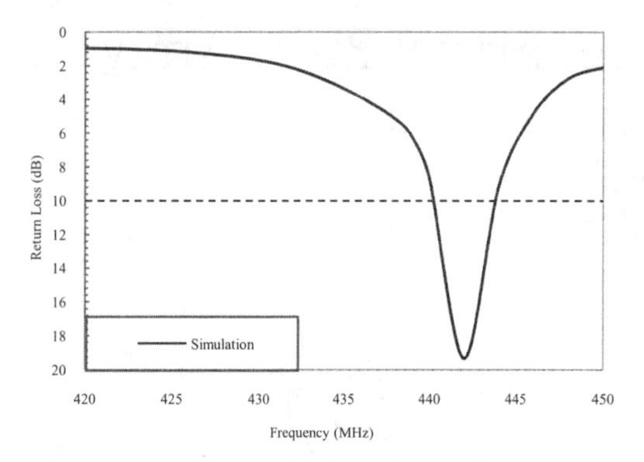

Fig. 2 Simulated return loss of antenna element.

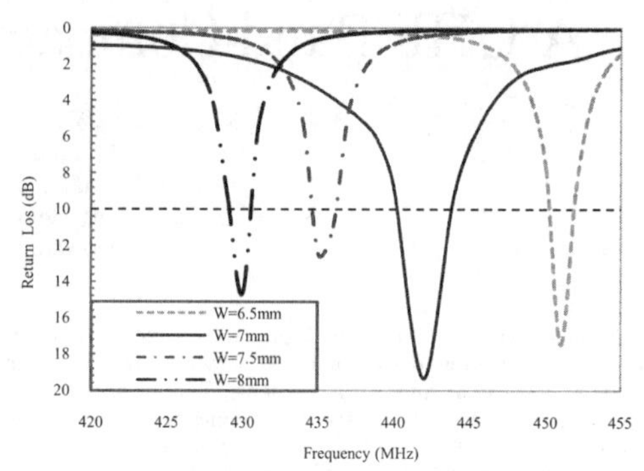

Fig. 4 Return loss of the antenna with four different lengths.

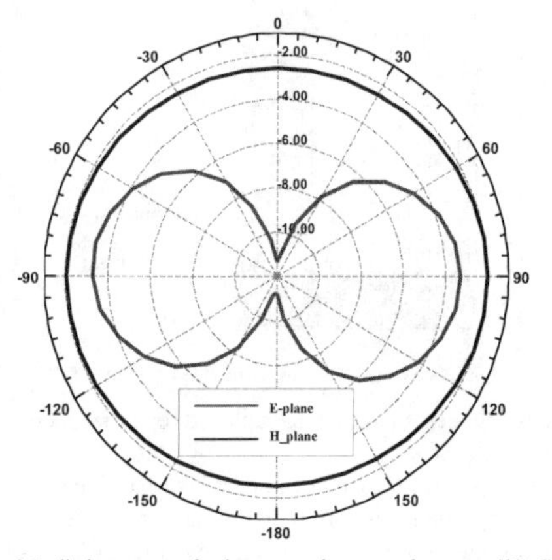

Fig. 3 Radiation patterns for the proposed antenna element at 432MHz.

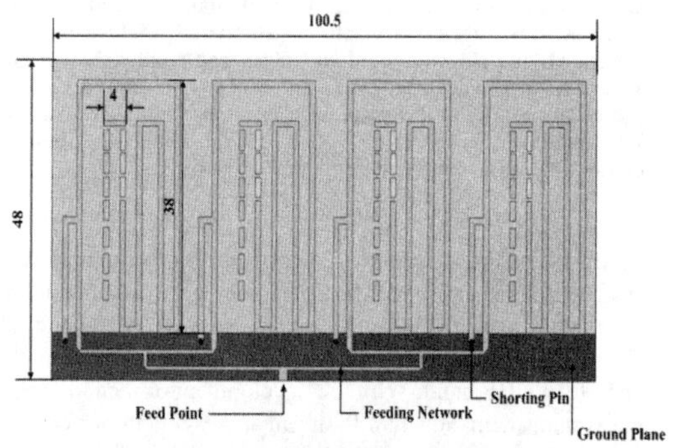

Fig. 5 Structure of the proposed antenna. (unit: mm)

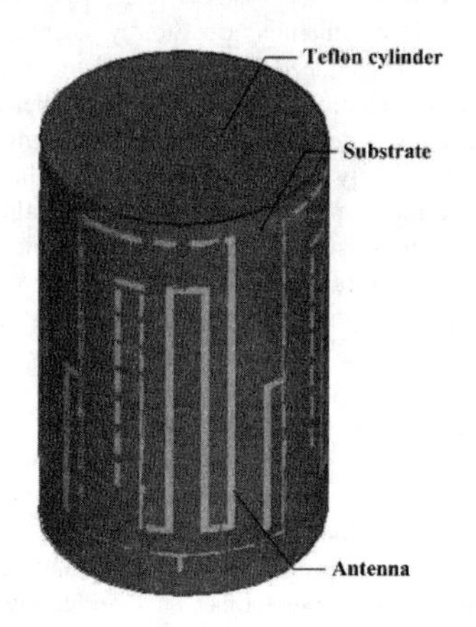

Fig. 6 Geometry of the conformal antenna.

Fig. 2 presents the simulated return loss of the antenna element. The measured 10dB band of the element is from 440 MHz to 444 MHz. The radiation patterns of the antenna element have been given in Fig. 3. From the figure, the antenna has a good omni-directional along horizontal plane. The maximum element gain is -1.9 dBi, which achieves at 432 MHz. In the investigation of the antenna element, besides the length of antenna element, the size of ground is also a key parameter that should be optimized to achieve the good performance. In Fig. 4, it reveals that the height of the ground plane has a significant impact on shifting the working frequence. The resonate frequency moves towards the low frequency with the increment of W. It is seen that the antenna with a length of W = 8 mm has resonating frequency about 430 MHz, while the antenna with a length of W = 6.5 mm has resonating frequency about 450 MHz. This investigation can be served as a guideline to optimize the antenna element. Through the optimization, the final W is chosen as 7 mm, which is considered the effects of the conformal structure.

III. ANTENNA ARRAY DESIGN

Fig. 5 shows the proposed PIFA array, which is parallel fed by a microstrip-line network. It consists of four identical PIFAs and a four-port microstrip-line power divider. The total

(a) Planar case

(b) Conformal case

Fig. 7 Photographs of fabricated antenna.

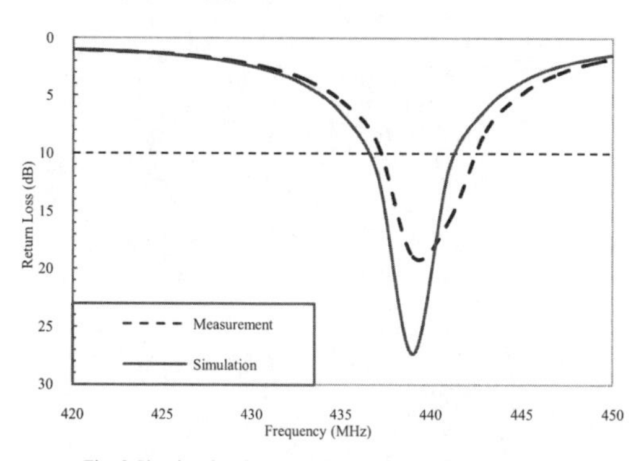

Fig. 8 Simulated and measured return loss of planar array.

size of the PIFA array antenna structure is 100.5 mm×48 mm. This size is fit for the circumference of conformal structure. Geometry of the conformal antenna is shown in Fig. 6. The assembled cylinder with 16 mm diameter is used to support the antenna array. The antenna model was designed and simulated in Ansoft HFSS software.

The proposed antenna array is fabricated. The photos of planar array and conformal array assembled on the teflon cylinder are shown in Fig. 7. The performances of the antenna are measured with a network analyzer. Fig. 8 presents

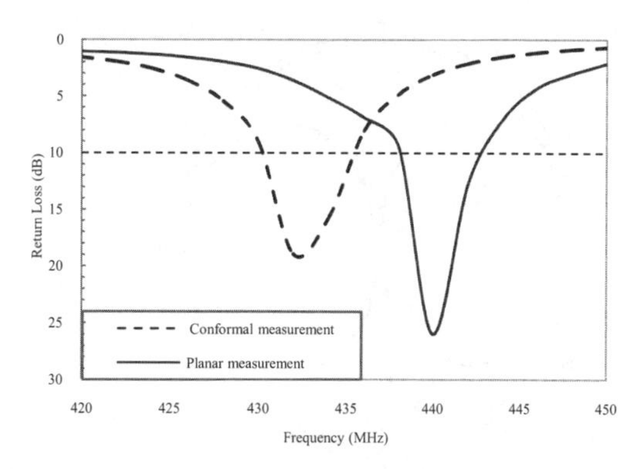

Fig. 9 Measured return loss of planar array and conformal array.

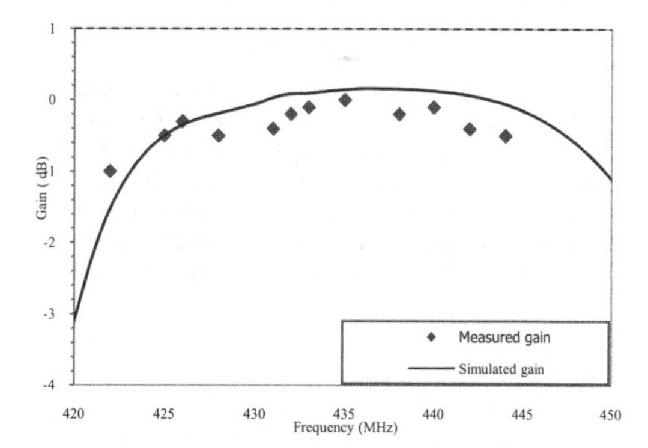

Fig. 10 Simulated and measured gain of conformal array.

simulated and measured return loss of planar array. The agreement is good between the measurement and the simulation, but a 2 MHz shift of the central operating frequency is observed. The 10 dB bandwidth is about 5 MHz (437 – 442 MHz). Fig. 9 shows the measured return loss of planar and conformal case. As shown in Fig. 9, the resonating frequency of the conformal antenna is about 432 MHz with the working band (430 – 436 MHz). The working frequence shifts down by 8 MHz when assembled around the cylinder. It is seen that the resonating frequency is shifted to the lower band with the effect of the conformal structure. The gain of the conformal antenna is given in Fig. 10. The results show that peak simulated gain can reach as high as 0.16 dBi, meanwhile the measured value is about -0.2 dBi. Moreover, the simulated and measured gain curves versus frequency are very stable, and the bandwidths of ± 1 dB gain are more than 30 MHz. Fig. 11 shows the simulated radiation patterns for the conformal array. In Fig. 11, we need to note that the radiation pattern of the array antenna at the frequency of 432 MHz is well omni-direction.

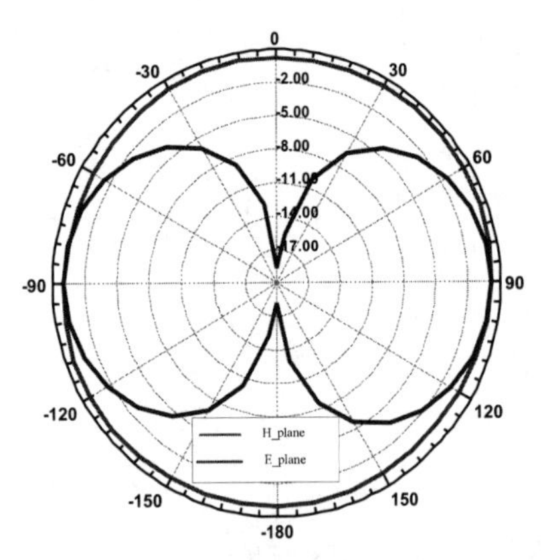

Fig. 11 Radiation patterns of conformal array at 433 MHz.

IV. CONCLUSION

In this paper, a compact conformal PIFA array antenna for UHF band has been presented. It consists of four identical PIFAs and a four-port microstrip-line power divider. The planar antenna and conformal antenna are fabricated and measured. The antenna has an omni-directional radiation pattern and a good gain in such a small volume. The peak gain is -0.2 dBi at 432 MHz, and the bandwidth is nearly 6 MHz.

REFERENCES

[1] H. Li, Zhu, G. X. Zhu, Z. Liang and Y. G. Chen, "A survey on distributed opportunity spectrum access in cognitive network," *6th International Conference*, pp. 1-4, WICOM 2010.

[2] A. J. Compston, J. D. Fluhler, and H. G. Schantz, "A fundamental limit on antenna gain for electrically small antennas," *IEEE Sarnoff Symposium*, pp. 1-5.SARNOF 2008.

[3] Y. T. Wan, D. Yu, F. S. Zhang and F. Zhang, "Miniature multi-band monopole antenna using spiral ring resonators for radiation pattern characteristics improvement," *Electron. Lett*, Vol. 49, 2013.

[4] L. HUITEMA, T. Reveyrand, J. Mattei, E. Arnaud, C. Decroze, T. monediere, "Frequency tunable antenna using a magneto dielectric material for DVB-H application," *IEEE Transaction on Antennas and Propagation*, pp. 1, 2013.

[5] A. Takacs, T. Idda, H. Aubert, H. Diez, "Miniaturization of quadrifilar helix antennas for space applications" *15th International Conference*, pp. 1-3, ANTEM, 2012.

[6] J. Huang, "A technique for an array to generate circular polarization with linearly polarized elements," *IEEE Transaction on Antennas and Propagation*, vol. 34, pp. 1113–1124, September. 1986.

[7] Y. S. Wang and S. J. Chung, "A miniature quadrifilar helix antenna for global positioning satellite reception," *IEEE Transaction on Antennas and Propagation*, vol. 57, pp. 3746–3751, December. 2009.

[8] A. T. Adams, R. K. Greenough, R. F. Wallenberg, A. Mendelovicz, and C. Lumjiak, "The quadrifilar helix antenna," *IEEE Transaction on Antennas and Propagation*, vol. 22, pp. 173–178, March. 1974.

[9] M. Huchard, C. Devalveaud, and S. Tedjini, "Miniature antenna forcircularly polarized quasi isotropic coverage," *2nd European Conference*, pp. 1-5, EuCAP, 2007.

[10] W. I. Son, W. G. Lim, M. Q. Lee, S. B. Min, and J. W. Yu, "Design of compact quadruple inverted-F antenna with circular polarization for GPS receiver," *IEEE Transaction on Antennas and Propagation*, vol. 58, pp. 1503–1510, May 2010.

Small-Size Printed Antenna with Shaped Circuit Board for Slim LTE/WWAN Smartphone Application

Hsuan-Jui Chang[1], Kin-Lu Wong[1], Fang-Hsien Chu[1], and Wei-Yu Li[2]

[1]Department of Electrical Engineering
Sun Yat-sen University, Kaohsiung 80424, Taiwan
[2]Information and Communications Research Laboratories
Industrial Technology Research Institute, Hsinchu 31040, Taiwan

Abstract—**An on-board printed WWAN/LTE antenna of simple structure disposed in a small clearance of 8 × 36 mm^2 in the ground plane of shaped circuit board in a slim handset is presented. The shaped circuit board has a large rectangular notch such that the battery of the handset can be embedded therein to decrease the thickness of the handset, and lead to much enhanced bandwidth of the antenna disposed thereon. This is mainly because stronger surface currents on the ground plane of the shaped circuit board can be excited, which greatly helps improve the operating bandwidth of the antenna disposed thereon. The proposed design makes a simple, small-size printed PIFA capable of providing two wide operating bands to cover the GSM850/900 and GSM1800/1900/UMTS/LTE2300/2500 bands.**

I. INTRODUCTION

For the slim handset, the system circuit board therein may have a large rectangular notch or slot to accommodate the battery to decrease the total required thickness of the handset. In the proposed design, a battery generally having a metal enclosing is disposed in the large notch of the shaped circuit board. A metal midplate, which is used to provide the handset with structural support, is also included in the proposed design. By connecting both the battery and metal midplate at proper locations to the ground plane of the shaped circuit board, large bandwidth enhancement for a small-size antenna can still be obtained.

The antenna used in this study is a printed inverted-F antenna, which can be disposed in a small clearance of 8 × 36 mm^2 in the ground plane of the shaped circuit board. In the proposed design, stronger surface currents excited on the ground plane of the shaped circuit board, compared to those on the ground plane of a corresponding simple circuit board, are observed. This behavior leads to enhanced bandwidth of the small-size antenna. Two wide operating bands are generated to cover the GSM850/900 and GSM1800/1900/UMTS/LTE2300/2500 bands.

II. PROPOSED DESIGN

As shown in Figure 1(a), there are three parts in the proposed design, which include a shaped circuit board with an on-board printed inverted-F antenna thereon, a battery, and a metal midplate. The shaped circuit board is modeled using a 0.8-mm thick FR4 substrate of relative permittivity 4.4 and loss tangent 0.024. The width and length of the circuit board are respectively 60 and 115 mm, which are reasonable dimensions of practical smartphones. A ground plane is printed on the back side of the shaped circuit board. A large notch of size 45 × 50 mm^2 is cut in the circuit board to accommodate the battery of the handset. In this study, the shaped circuit board with a notch size which is reasonable for some practical handset batteries can have stronger surface currents excited on the ground plane thereof, compared to those on the ground plane of a corresponding simple circuit board.

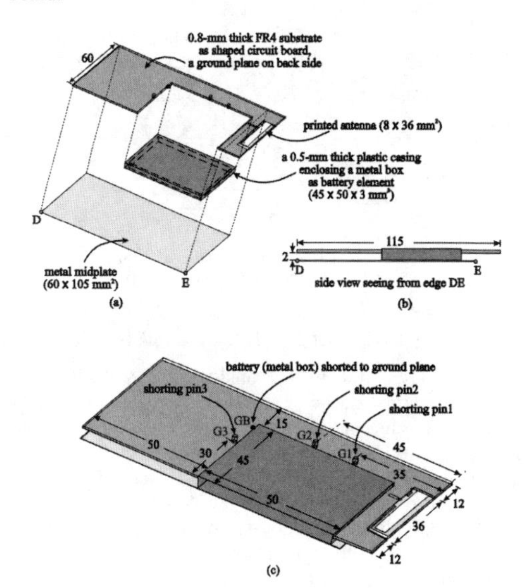

Figure 1. Proposed WWAN/LTE handset antenna with shaped circuit board, battery and metal midplate. (a) Exploded view. (b) Side view. (c) Front view.

978-7-5641-4279-7

As shown in Figure 2, the inverted-F antenna has a folded radiating arm of length 66 mm (section AB) and a shorting strip of length 10 mm (section AS). A chip capacitor of 3.3 pF is added at feeding point A, which mainly compensates for large inductance for frequencies at about 900 MHz to achieve improved impedance matching for the proposed antenna. In this study, two wide operating bands can be provided to cover the WWAN/LTE operation bands, although the antenna has a small size of 8×36 mm^2 only.

The battery is modeled as a metal box enclosed by a 0.5-mm thick plastic casing. The metal box is also short-circuited at the location GB to the ground plane of shaped circuit board. The selected location GB can make the excited surface currents on the ground plane of shaped circuit board slightly affected such that enhanced bandwidth of the antenna can still be obtained, when the battery is embedded in the notch of shaped circuit board. The midplate is usually added to provide the handset with structural support and can be used to support the display of the handset. In this study, the midplate is considered to be a metal plate and has a size of 60×105 mm^2.

The configuration of the proposed design can be seen more clearly in Figure 1(b) and (c). By properly selecting the shorting positions (G1, G2, and G3 in this study) to connect the midplate to the ground plane, the excited surface currents on the region of the midplate facing the battery (that is, facing the notch of shaped circuit board) can be suppressed. Hence, enhanced bandwidth can still be obtained for the small-size antenna in this study.

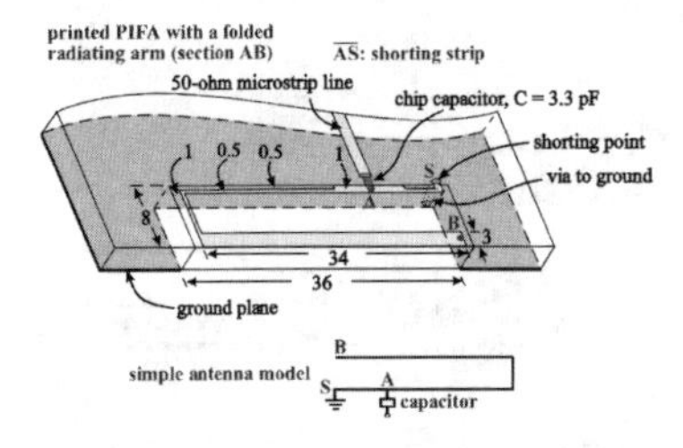

Figure 2. Dimensions of the metal pattern of the antenna (printed PIFA with a folded radiating arm).

III. RESULTS AND DISCUSSION

The proposed design was implemented and shown in Figure 3. Figure 4 shows the measured and simulated return losses of the antenna. The simulated results are obtained using simulator HFSS version 14, and agreement between the simulation and measurement is seen. The obtained lower and upper bands of the antenna cover the WWAN bands and the LTE2300/2500 bands, the impedance matching is better than 3:1 VSWR or 6-dB return loss, which is the design specification widely used for the internal WWAN/LTE handset antennas [1]-[3].

The measured antenna efficiency is shown in Fig. 5. The antenna efficiency includes the mismatching loss. Over the lower band (824~960 MHz), the antenna efficiency is better than about 60%, while that over the upper band (1710~2690 MHz) is better than about 65%. The measured antenna efficiencies are good for practical handset applications. More results on parametric analysis, measured three-dimensional total-power radiation patterns, surface current distributions, and SAR results will be discussed in the presentation.

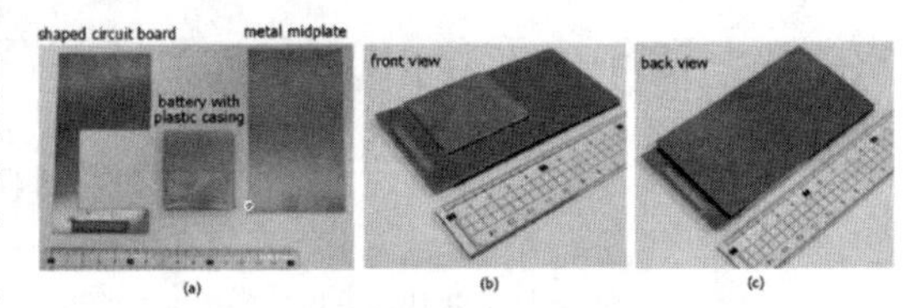

Figure 3. Photos of the fabricated prototype. (a) Shaped circuit board, battery element and metal midplate. (b) Front view seeing from the battery side. (c) Back view seeing from the midplate side.

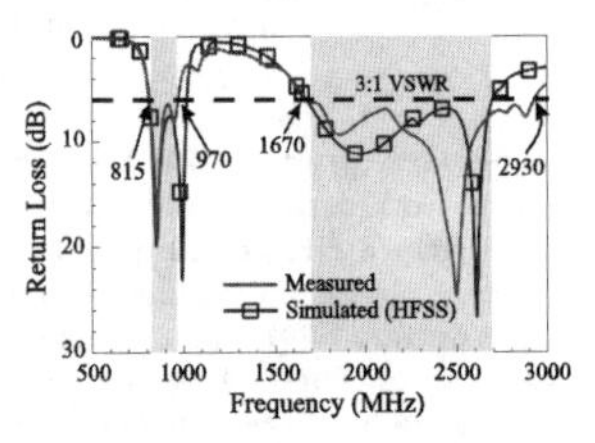

Figure 4. Measured and simulated return losses of the antenna.

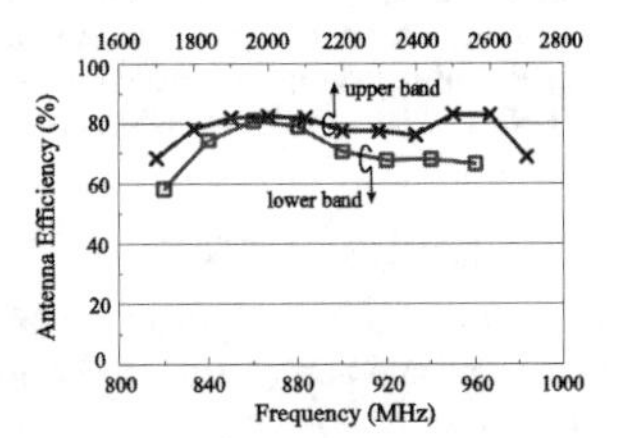

Figure 5. Measured antenna efficiency with mismatching loss.

IV. CONCLUSION

A simple, small-size printed inverted-F antenna has been shown to provide wide operating bands to cover the WWAN and LTE2300/2500 operation bands for the slim handset application. The proposed design can still have the ground plane of shaped circuit board excited as an efficient radiator to greatly enhance the operating bandwidth of the antenna. Good far-field radiation characteristics for frequencies in the operating bands have also been obtained. The proposed design is especially suitable for the slim handset application.

REFERENCES

[1] C. T. Lee and K. L. Wong, "Planar monopole with a coupling feed and an inductive shorting strip for LTE/GSM/UMTS operation in the mobile phone," *IEEE Trans. Antennas Propagat.*, vol. 58, pp. 2479-2483, Jul. 2010.

[2] K. L. Wong, Y. W. Chang and S. C. Chen, "Bandwidth enhancement of small-size WWAN tablet computer antenna using a parallel-resonant spiral slit," *IEEE Trans. Antennas Propagat.*, vol. 60, pp. 1705-1711, Apr. 2012.

[3] C. L. Liu, Y. F. Lin, C. M. Liang, S. C. Pan and H. M. Chen, "Miniature internal penta-band monopole antenna for mobile phones," *IEEE Trans. Antennas Propagat.*, vol. 58, 1008-1011, Mar. 2010.

Compact and Planar Near-field and Far-field Reader Antenna for Handset

Wenjing Li, Yuan Yao, Junsheng Yu, Xiaodong Chen
School of Electronic Engineering, Beijing University of Posts and Telecommunications
No.10 Xitucheng Road, Beijing, China

***Abstract*-In this paper, we proposed a compact near-field and far-field RFID reader antenna with a dimension of 50 mm × 50 mm × 0.8 mm in the UHF band. The fabricated antenna printed on a FR4 substrate operates from 915 to 935 MHz with reflection coefficient less than −10dB, covering the China UHF band. The strong and uniform magnetic field is excited by magnetic dipole source. The measured reading distance are up to 58 mm and 110 mm for near-field and far-field applications, respectively. This antenna can be applied for handset in mobile applications due to its small size.**

I. INTRODUCTION

Radio frequency identification (RFID) in ultra high frequency (UHF) is gaining popularity in a number of practical applications, due to automatic identification for efficiently tracking and managing of objects. Based on types of objects and applications, inductively coupled near-field operation are used to transfer information between reader and tag. Far-field communication is widely used due to its long read range. Near-field reading can be useful for objects having metals and liquids in their vicinity, because normal far-field tags' performance is affected by the presence of these objects [1], [2]. Inductive coupling is conventionally used at low frequency (LF) and high frequency (HF). Due to promising performance at item-level tagging (ITL) of small, expensive, and sensitive objects and different applications such as pharmaceutical logistics and bio-sensing applications, it is considered as a possible solution for ITL in pharmaceutical and retailing industry.

To design a UHF near-field and far-field RFID antenna for handset, the challenge is that the reader antenna must be miniaturized, and at the same time, the antenna have a uniform magnetic field in the near field region. Some structure have been presented. Shrestha et al use a segmented loop and a patch, respectively, to achieve near-field and far field operations, but they have too large size of 184 mm×174 mm [3]. Borja et al's antenna has a dimension of 72.3 mm×72.3 mm [4].

In this paper, a UHF RFID reader antenna was proposed with simple and compact configuration for both near-field and far-field operations. The impedance bandwidth is suitable for China standard (920-925MHz), and it can provide the strong and uniform magnetic field in an adequate interrogation zone. Both simulation and measurements results are provided to illustrate the good performance of the designed antenna.

II. ANTENNA DESIGN AND STRUCTURE

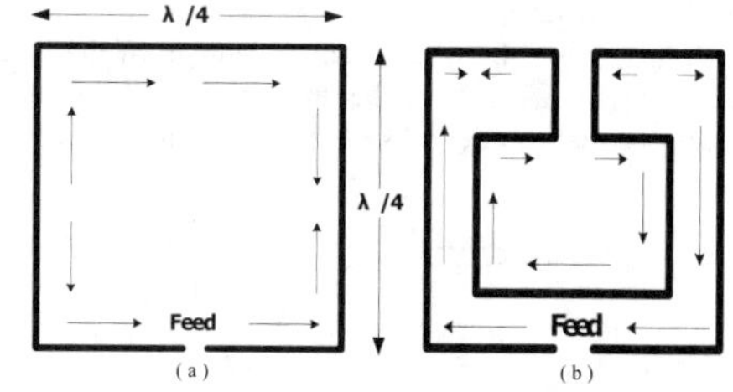

Fig. 1 Current distributions of (a) one-wavelength-perimeter loop and (b) folded-dipole loop.

Loop antennas are commonly used for inductively coupled near-field RFID systems. At UHF band, the optimal size of the loop antenna is electrically large or comparable to the wavelength. The amplitude and phase distribution of the current, in the case, is not uniform, and it reverses at every half-wavelength, which results in a weak and nonuniform magnetic field at the center of the loop [5], as shown in Fig.1.(a). Some ideas were proposed to solve the problem of current reverse. A segmented loop antenna can avoid in-phase of current [6]. Dual-dipoles also can achieve a uniform magnetic field in near-field region for pure near-field operation [7].

A novel folded-dipole loop antenna was proposed as shown in Fig.1.(b). This antenna can achieve a uniform magnetic field easily, and has a good far-field gain. The folded-dipole is fabricated on the FR-4 substrate with a dimension of 50 mm ×50 mm, dielectric constant ε_r=4.4, loss tangent tanδ=0.02, thickness h=0.8 mm. Detailed dimensions are shown in the Fig.2 (a), s=3 mm, s1=8.5 mm, the antenna size is L(45mm)× L(45mm), Line width is 0.75 mm.

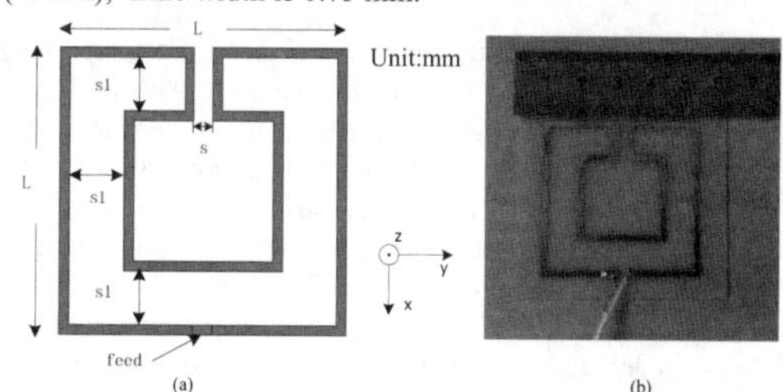

Fig. 2 prototype of the folded-dipole loop antenna

978-7-5641-4279-7

III. Results and Discussion

Simulations were performed using Ansoft High Frequency Structure Simulator (HFSS) software, which uses the finite element method (FEM).

Fig.2 (b) shows the fabricated antenna after parameter optimization. The measured S11 agrees with the simulated one with a slight deviation to the right side by 1 MHz as shown in Fig.3. The measured bandwidth ranges from 916 to 936 MHz, which covers the China UHF band.

The simulated current distribution along this folded-dipole loop is shown in Fig.4. We can see that the current reverses at outer loop, however, which is unidirectional along the inner loop. Fig.5 shows the resulting z-component of the magnetic field on an xy-plane above the antenna at z=0 and z=10mm. We can see that the field distribution is uniform in the center region.

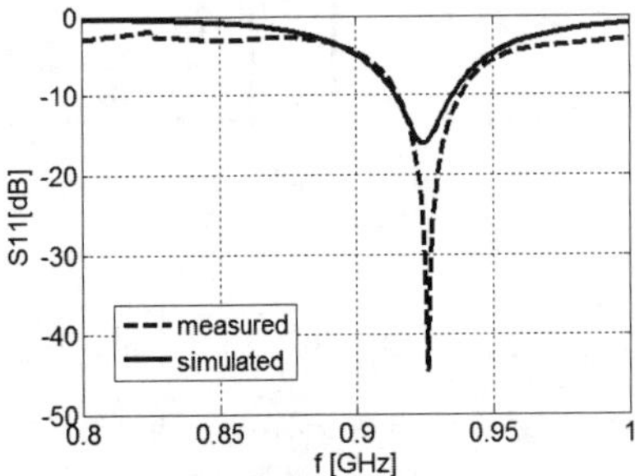

Fig. 3 Simulated and measured S11 of the antenna

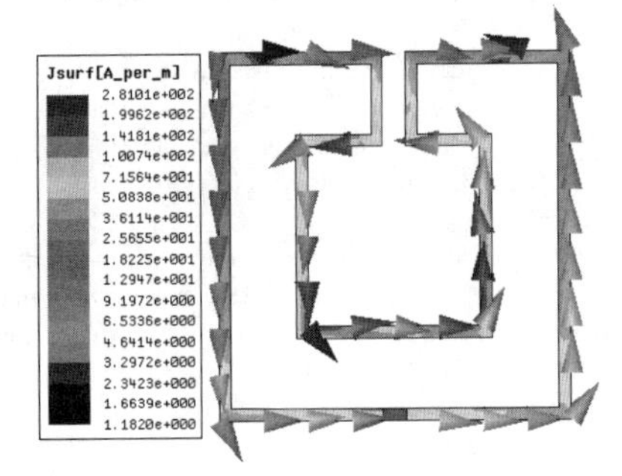

Fig. 4 simulated current distribution on the antenna

Based on the fabricated antenna, we measure the read range and width by using the Impinj UHF button [Fig.6(c)]. The test scene is presented in Fig.6(a), Fig.6(b) shows the prototype of the presented antenna. Under the transmission power level of 15dBm, the measured reading distance is 58 mm. When the tag is attached to a water-item container, the reading range is still the same. Additionally, a far-field tag is also measured, reading range is around 110 mm due to the good far-field gain of the reader antenna. When the far field tag is attached to the bottle of water, the reading range is reduced to 4.5 mm.

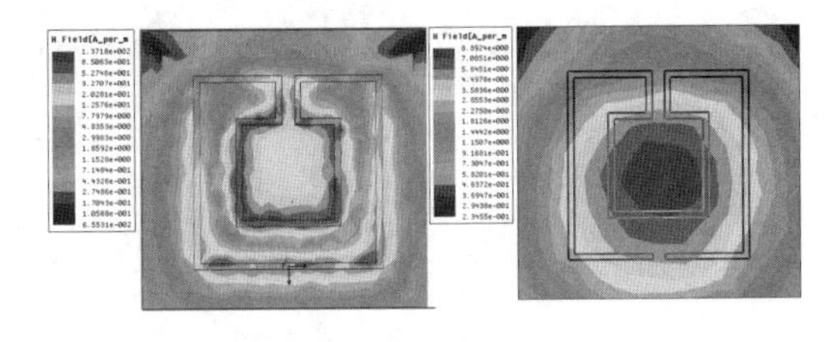

(a) (b)

Fig. 5 Magnetic field distribution Hz at (a)z=0 (b)z=10mm

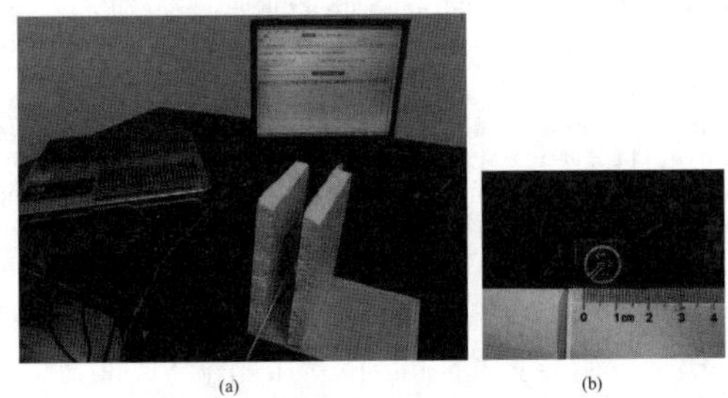

(a) (b)

Fig. 6 Antenna measurement. (a) Test scene of read range measurement. (b) Impinj UHF button near-field tag.

The Impinj UHF button whose diameter is around 1cm measured read width is shown in Fig.7. From Fig.7 (a), we use the square lattice, its size is 1cm×1cm.

The simulated and measured radiation patterns of the proposed antenna are respectively shown in Fig.8 which makes the antenna suitable for far-field application.

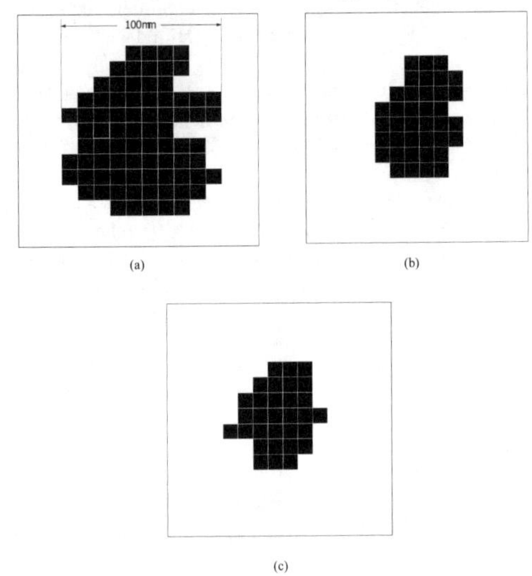

(a) (b) (c)

Fig.7 Measured read width at xy-plane. (a) z=10mm (b) z=20mm (c) z=40mm

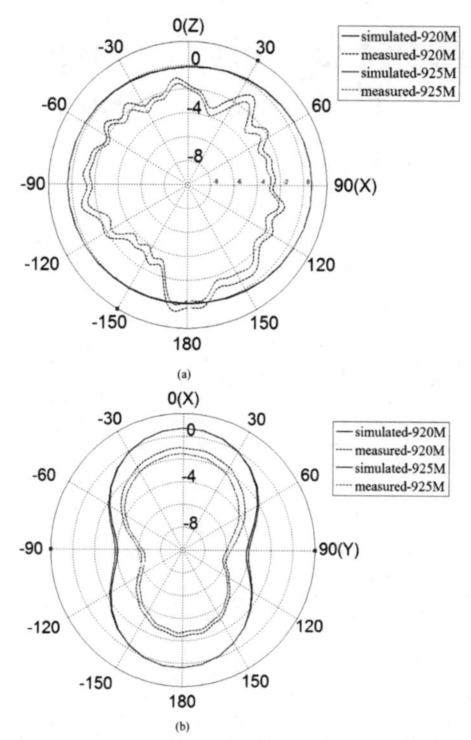

Fig.8 Simulated and measured far-field radiation patterns (a) xz-plane (b) xy-plane

IV. Conclusion

In this paper, a novel UHF reader antenna was proposed for near-field and far-field simultaneous operation. A magnetic dipole was folded to produce a uniform magnetic field distribution at UHF. This RFID reader antenna is designed for the China UHF RFID band. With the Impinj UHF button near-field tag, the maximum read range obtained was 58mm under 15dBm transmission power. The near-field reading performance was not degraded when the tag was attached to a water container. The far-field read range, with a commercial far-field tag, was approximately 11 mm. This novel RFID reader antenna can be applied for near-field and far-field operations.

Acknowledgment

This work is supported by the Fundamental Research Funds for the Central Universities, the National Natural Science Foundation of China under Grant No.61201026, and Beijing Natural Science Foundation (4133091).

References

[1] D. M. Dobkin and S. M. Weigand, "Environmental effects on RFID tag antennas," in IEEE Microw. Symp. Dig. , Jun. 2005, pp. 135–138.

[2] X. Qing, Z. N. Chen, and C. K. Goh, "Platform effect on RFID tag antennas and co-design considerations," inProc. IEEE Microw. Conf. , Dec. 2008, pp. 1–4.

[3] B. Shrestha, A. Elsherbeni, and L. Ukkonen, "UHF RFID Reader Antenna for Near-Field and Far-Field Operations", IEEE Antennas and Wireless Propagation Letters, vol.10, pp. 1274–1277, 2011.

[4] A.L.Borja, A.Belenguer, J.Cascon, and J.R.Kelly, "A Reconfigurable Passive UHF Reader Loop Antenna for Near-Field and Far-Field RFID Applications", IEEE Antennas and Wireless Propagation Letters, vol.11, pp. 580–583, 2012.

[5] A. L. Popov, O. G. Vendik, and N. A. Zubova, "Magnetic field intensityin near-field zone of loop antenna for RFID systems," Tech. Phys. Lett. , vol. 36, no. 10, pp. 882–884, 2010.

[6] X. Qing, C.K. Goh and Z. N. Chen, "Segmented loop antenna for UHF near-field RFID applications," Electron. Lett. , vol. 45, no. 17, pp. 872–873, Aug. 2009.

[7] X. Li and Z. Yang, "Dual-printed-dipoles reader antenna for UHF near field RFID applications," IEEE Antennas Wireless Propag. Lett. , vol. 10, pp. 239–242, 2011.

Printed Loop Antenna with an Inductively Coupled Branch Strip for Small-Size LTE/WWAN Tablet Computer Antenna

Meng-Ting Chen and Kin-Lu Wong
Department of Electrical Engineering, Sun Yat-sen University
Kaohsiung 80424, Taiwan

Abstract-**The technique of using an inductively coupled branch strip for bandwidth enhancement of a simple printed loop antenna to achieve small size yet multiband operation to cover the LTE/WWAN bands (704~960 and 1710~2690 MHz) in the tablet computer is presented. The antenna's metal pattern occupies a small area of 10 × 34.5 mm^2, and is printed on a thin FR4 substrate. The branch strip can cause an additional resonance and thereby greatly widens the antenna's low-band bandwidth to cover the 704~960 MHz band for the LTE700/GSM850/900 operations. While in the antenna's higher band, the chip inductor provides a high inductance and limits the excitation of the branch strip. In this case, with the printed loop antenna excited in the higher band, the antenna can generate two higher-order resonant modes to form a wide operating band for the GSM1800/1900/ UMTS/LTE2300/2500 operations (1710~2690 MHz).**

Keywords: mobile antennas, LTE/WWAN antennas, tablet computer antennas, inductively coupled branch strip, small-size antennas

I. INTRODUCTION

Owing to the very limited space inside the tablet computers, the embedded antennas having planar structure and small size for multiband operation are generally demanded. In this paper, we demonstrate that a quarter-wavelength printed loop antenna [1], [2] can be applied in achieving two wide operating bands for the LTE/WWAN operations. By applying an inductively coupled branch strip for bandwidth enhancement of the antenna's lower band to cover the LTE700/GSM850/900 operations in the 704~960 MHz band, the antenna's metal pattern occupies an area of 10 × 34.5 mm^2 only. A wide higher band contributed by two higher-order resonant modes of the printed loop antenna alone is also obtained to cover the GSM1800/1900/UMTS/LTE2300/2500 operations in the 1710~2690 MHz band. Design considerations and operating principle of the proposed antenna are described. A parametric study on the proposed antenna is also conducted, and results of a fabricated antenna are presented and discussed.

II. PROPOSED ANTENNA

Fig. 1 shows the geometry of the proposed LTE/WWAN tablet computer antenna. The antenna has a uniplanar structure and is printed on an FR4 substrate of thickness 0.8 mm. The antenna is mounted along an edge and at a corner of the device ground plane of size 150 × 200 mm^2, which is a reasonable size for the tablet computer with a 10-inch display panel. The antenna is formed by a loop strip configured into an inverted-L shape and a branch strip inductively coupled to the loop strip through a chip inductor (L_2) of 22 nH. The loop strip (section ADB) has a length of 64 mm (about 0.16λ at 750 MHz) only. Aided by a series chip capacitor (C) of 1.2 pF and a series chip inductor (L_1) of 5.6 nH, the loop antenna can generate a resonance at about 750 MHz, and the 750-MHz mode generation is mainly owing to the series chip capacitor added to compensate for the large inductive reactance around the lowest resonant mode or the quarter-wavelength mode of the printed loop antenna. In addition, it can be seen in Fig. 2 that two higher-order resonant mode are generated with good impedance matching, which is aided by the chip inductor L_1. In the proposed design, the antenna's higher band cover the desired 1710~2690 MHz band.

In order to achieve a wide lower band, a branch strip of length (t) 39 mm and width 0.3 mm is coupled to the loop strip through a chip inductor of 22 nH. A parallel resonance controlled by the additional inductance and capacitance occurs at about 1150 MHz (see the input impedance of the proposed antenna in Fig. 2). This parallel resonance causes the resonant mode owing to the loop strip shifted to be at about 750 MHz and having improved impedance matching. Further, an additional resonance at about 950 MHz is also generated, which greatly widens the bandwidth of the antenna's lower band to cover the desired 704~960 MHz band.

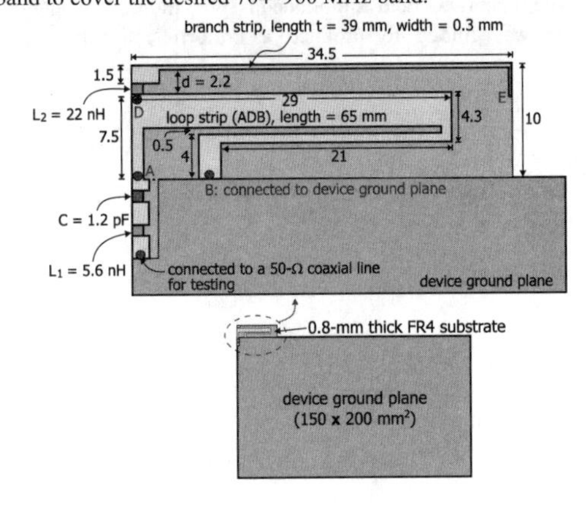

Figure 1. Geometry of the proposed LTE/WWAN antenna.

978-7-5641-4279-7

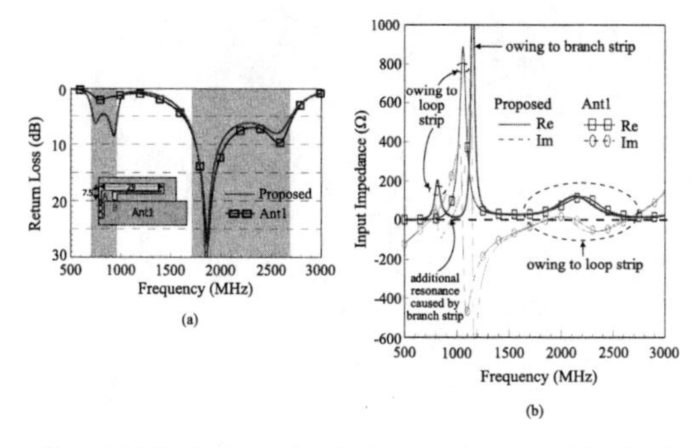

Figure 2. (a) Simulated return loss for the proposed antenna and the printed loop antenna only (Ant1). (b) Simulated input impedance.

III. PARAMETRIC STUDY

To further analyze the effects of the inductively coupled branch strip, Fig. 3 shows the simulated return loss as a function of the chip inductor. Results for the chip inductor with L_2 = 0, 7, 14, 22 nH are shown. For L_2 = 0, the added branch strip makes the proposed antenna operates more like a simple shorted strip antenna having a radiating strip of length about 49 mm (section ADE) and a shorting strip (section DB) of length about 57 mm. This explains the resonant mode excited at about 1.5 GHz, which is mainly owing to the radiating strip being about a quarter-wavelength at 1.5 GHz. In this case, the section ADB can still function as a loop antenna, and a resonant mode at about 750 MHz and two higher-order resonant modes in the higher band are still generated, with the first mode in the higher band being affected and shifted to higher frequencies. The simulated input impedance for the proposed antenna with L_2 = 0, 7 and 14 nH is also shown in Fig. 3 for comparison. It is seen that for L_2 = 7 and 14 nH, the resonances in the antenna's higher band are almost not affected. Furthermore, the lowest resonant mode of the printed loop antenna is also seen to be shifted to lower frequencies with the added L_2 having a larger inductance, and the impedance variations thereof become much smoother, thereby causing good excitation of the resonant mode at about 750 MHz when the chip inductor of 22 nH is used (the proposed antenna).

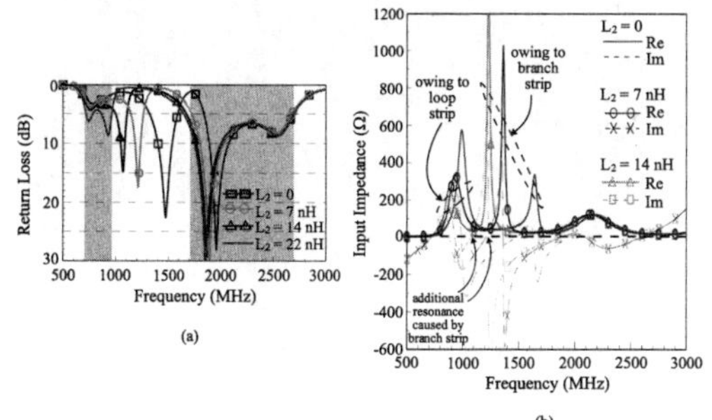

Figure 3. (a) Simulated return loss as function of the chip inductor L_2 = 0, 7, 14, 22 nH. (b) Simulated input impedance.

IV. EXPERIMENTAL RESULTS AND DISCUSSION

The proposed antenna was fabricated and tested. A photo of the fabricated antenna mounted along an edge and at a corner of a 0.2-mm thick copper plate to simulate the device ground plane in Fig. 1 is shown in Fig. 4(a). Measured and HFSS simulated return loss of the proposed antenna is presented in Fig. 4(b). Good agreement between the measurement and simulation is seen. The measured antenna efficiency of the fabricated antenna is shown in Fig. 4(c). The antenna efficiency includes the mismatching loss and was measured in a far-field anechoic chamber. The antenna efficiency is about 40~62% and 64~92% in the lower and higher bands, respectively. The obtained antenna efficiency is acceptable for mobile communications.

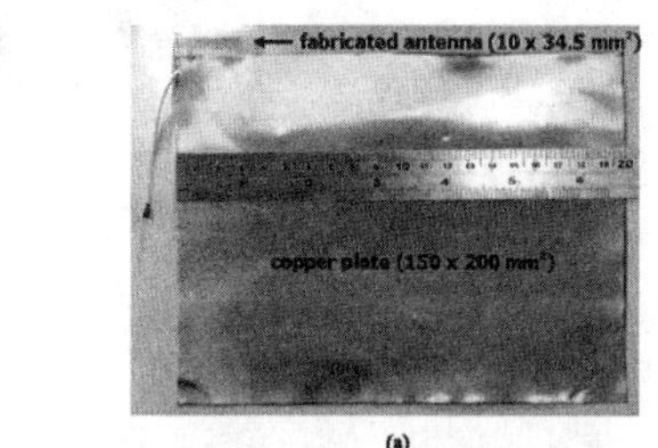

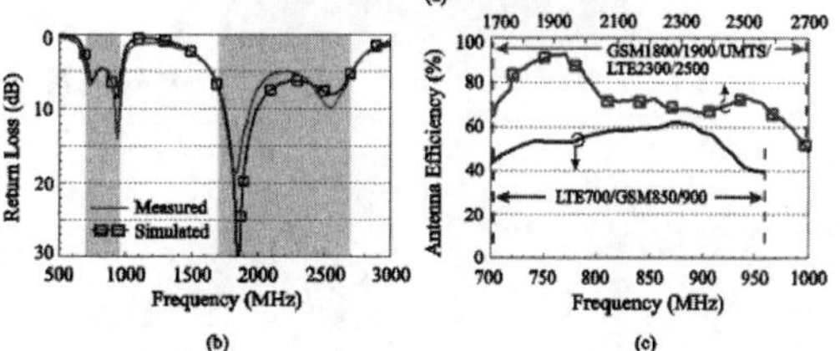

Figure 4. (a) Photo of the fabricated antenna. (b) Measured and simulated return loss. (c) Measured antenna efficiency.

V. CONCLUSION

A small-size, planar LTE/WWAN tablet computer antenna has been proposed. The antenna can provide two wide operating bands for the LTE/WWAN operations. The operating principle of the proposed antenna, especially the effects of the inductively coupled branch strip on the low-band bandwidth enhancement has been described. The proposed antenna is especially suitable for the LTE/WWAN operation in the slim tablet computer applications.

REFERENCES

[1] Y. W. Chi and K. L. Wong, "Very-small-size printed loop antenna for GSM/DCS/PCS/UMTS operation in the mobile phone," *Microwave Opt. Technol. Lett*, vol. 51, pp. 184-192, Jan. 2009.

[2] Y. W. Chi and K. L. Wong, "Quarter-wavelength printed loop antenna with an internal printed matching circuit for GSM/DCS/PCS/UMTS operation in the mobile phone," *IEEE Trans. Antennas Propagat.*, vol. 57, pp. 2541-2547, Sep. 2009.

WP-2(D)

October 23 (WED) PM

Room D

Small Antennas

An Idea for Low-Profile Unidirectional Slot Antennas Based on Its Complementary Dipoles

J. Xiong, X. L. Li, and B. Z. Wang
Computational Electromagnetics Laboratory
Institute of Applied Physics, UESTC
Chengdu, P. R. China

***Abstract*-In this paper, a novel scheme of constructing low-profile unidirectional slot antennas has been proposed. The scheme is based on the complementary dipole concept. The basic principle of the complementary concept has been revisited, with a discussion on the effect of excitation amplitude on the unidirectional radiation performance. The practical antenna configuration has then been given, followed by some preliminary simulated results. Compared to some conventional techniques for designing unidirectional slot antennas, the proposed structure has several advantages such as low profile, low Q, and ease of fabrication.**

I. Introduction

Slot antennas are promising candidates for military, wireless communication, and medical applications, due to its attracting features such as compact size, light weight, low profile, conformability to planar or nonplanar surfaces, and compatibility with MMIC designs [1-3]. Particularly, compared to their planar counterpart, i.e. microstrip patch antennas, they have generally wider bandwidth, less interaction via surface waves and better isolation from feed net-works [1]. However, one salient drawback of a slot radiator cut on the metallic plane is its inherent bi-directional radiation, which means half of the fed power radiates in most cases in the unwanted direction. To mitigate this undesired characteristic, one traditional solution is to place a metallic plane reflector on one side and parallel to the slot surface. However, a minimum distance of a quarter wavelength between the slot and the reflector is required to guarantee an optimal matching [4]. In addition, if one wishes to reduce this distance, the parallel plate TEM mode between the two layers dominates and causes great energy leakage [3], [5], [6]. Another commonly used technique is to replace the reflector by a closed cavity to form a cavity-backed slot (CBS) antenna [7]-[10]. However, the CBS antenna can hardly be low-profile, and due to its closed form, the bandwidth, which is an advantage of a slot antenna, is inevitably affected. Other techniques include using shorting pins around the slot [11], twin slot configuration for phase cancellation [5], [11], and EBG or AMC as the reflector [12], [13].

The concept of complementary electric dipole and magnetic dipole was first studied by Clavin [14], [15], and recently based on this concept Luk et al. have proposed a series of novel antennas with improved characteristics such as wide band, symmetric radiation pattern in the E- and H-planes [15]-[17]. In this paper, we try to discuss the possibility of constructing novel unidirectional radiating slot antennas based on the complementary dipole concept. The basic scheme of the antenna configuration and its working principle is discussed in section II. The practical antenna structure and some initial simulation results are given in section III. Finally the conclusion is given in section IV.

II. Basic Principle and Discussion

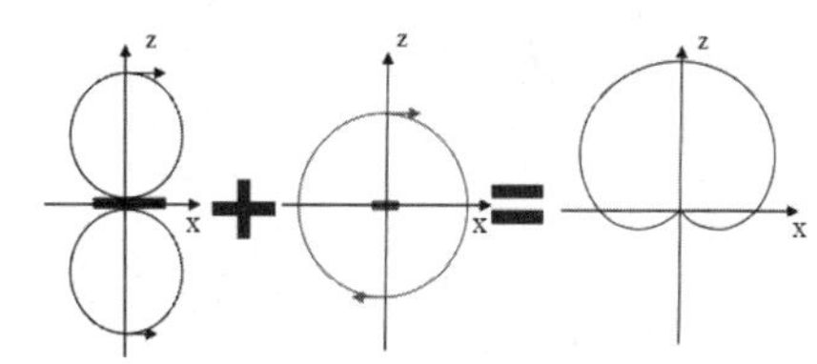

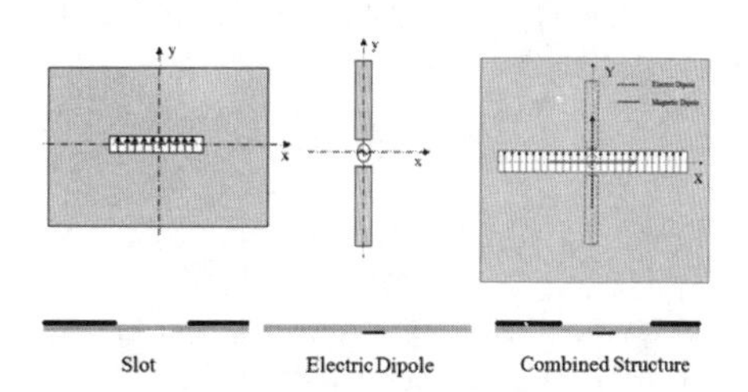

Figure 1. The scheme of combining a slot and its complementary E-dipole to produce a unidirectional radiation.

Previous studies have shown that an electric dipole and a magnetic dipole that are simultaneously excited with proper amplitudes and phases can have a unidirectional beam and symmetric cardiac pattern, where in ideal case the backward radiation is completely suppressed [14], [16]. As a thin slot cut on an infinite ground plane can be viewed as a magnetic dipole [18], it is straightforward to infer that if an electric dipole is placed orthogonally near the slot (e.g. on the opposite side of the very electrically thin substrate to the slot plane, see the right column of Fig. 1) with proper excitation, the backward radiation of the conventional bi-directional slot antenna can also be similarly suppressed and unidirectional pattern can be obtained. This aforementioned scheme, with the antenna configuration and corresponding radiation pattern, is plotted in Fig. 1. Note in Fig. 1 only H-plane pattern is given, and according to the complementary dipole concept the same effect can be observed in the E-plane.

Previous study has already shown that the unidirectional pattern with effective backward radiation suppression largely depends on the excitation of the two complementary sources with proper amplitude and phase. Here a set of complete mathematical derivation is given below, followed by a

978-7-5641-4279-7

discussion on the excitation amplitude. Suppose the finite length slot (i.e. magnetic dipole) and its complementary electric dipole are placed along y and x axis, respectively, and the magnetic and electric current on these dipoles are

$$\vec{I}_e(x',y',z') = \begin{cases} \hat{a}_y I_e \sin[k(\frac{l_e}{2}-x')], 0 \le x' \le \frac{l_e}{2} \\ \hat{a}_y I_e \sin[k(\frac{l_e}{2}+x')], -\frac{l_e}{2} \le x' \le 0 \end{cases} \quad (1)$$

$$\vec{I}_m(x',y',z') = \begin{cases} \hat{a}_x I_m \sin[k(\frac{l_m}{2}-y')], 0 \le y' \le \frac{l_m}{2} \\ \hat{a}_x I_m \sin[k(\frac{l_m}{2}+y')], -\frac{l_m}{2} \le y' \le 0 \end{cases}$$

where $\vec{I}_e$, l_e and $\vec{I}_m$, l_m are the current and length of the electric and magnetic dipole, respectively. The electric field of the electric dipole ($\vec{E_e}$) and magnetic dipole ($\vec{E_m}$) are then

$$\vec{E}_e = j\eta \frac{I_e e^{-jkr}}{2\pi r}(-\vec{a}_\theta \cos\theta\cos\varphi + \vec{a}_\varphi \sin\varphi)\cdot[\frac{\cos(\frac{kl_e}{2}\sin\theta\cos\varphi)-\cos(\frac{kl_e}{2})}{1-\sin^2\theta\cos^2\varphi}]$$

$$\vec{E}_h = j \frac{I_m e^{-jkr}}{2\pi r}(-\vec{a}_\theta \cos\varphi + \vec{a}_\varphi \cos\theta\sin\varphi)\cdot[\frac{\cos(\frac{kl_m}{2}\sin\theta\sin\varphi)-\cos(\frac{kl_m}{2})}{1-\sin^2\theta\sin^2\varphi}]$$

(2)

The total field of the two sources is a combination of the field in (2). Let A_1 and A_2 be the amplitude of $\vec{E_e}$ and $\vec{E_m}$, the total field is thus

$$E_\theta = -\cos\varphi\cdot\{A_1\cos\theta[\frac{\cos(\frac{kl_e}{2}\sin\theta\cos\varphi)-\cos(\frac{kl_e}{2})}{1-\sin^2\theta\cos^2\varphi}] + A_2[\frac{\cos(\frac{kl_m}{2}\sin\theta\sin\varphi)-\cos(\frac{kl_m}{2})}{1-\sin^2\theta\sin^2\varphi}]\}$$

$$E_\varphi = \sin\varphi\cdot\{A_1[\frac{\cos(\frac{kl_e}{2}\sin\theta\cos\varphi)-\cos(\frac{kl_e}{2})}{1-\sin^2\theta\cos^2\varphi}] + A_2\cos\theta[\frac{\cos(\frac{kl_m}{2}\sin\theta\sin\varphi)-\cos(\frac{kl_m}{2})}{1-\sin^2\theta\sin^2\varphi}]\}$$

(3)

It can be readily concluded that when $A_1 = A_2$ and $l_e = l_m$ are satisfied, the backward radiation ($\theta = \pi$) can theoretically be zero. Fig. 2 shows 2D and 3D radiation pattern of such half-wavelength complementary dipoles ($l_e = l_m = \lambda/2$).

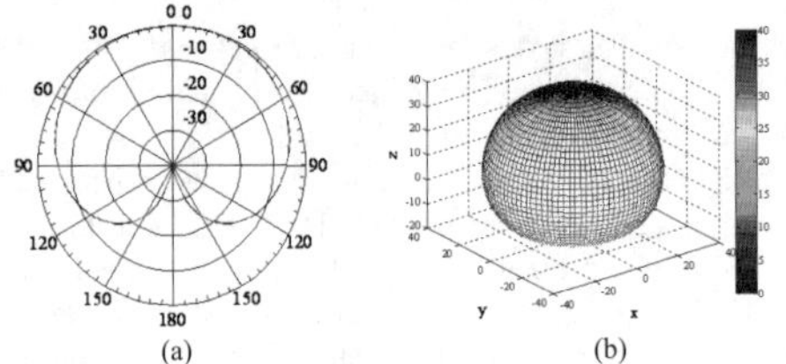

(a) (b)

Figure. 1 The (a) E-plane and (b) 3D pattern of $\lambda/2$ complelemtary E- and M- dipoles

If the two dipoles are not excited with the same amplitude, an increased backward radiation is observed. This is illustrated in Fig.3, and the ratio of the field intensity with different feeding amplitude is listed in Table I. This indicates if one wishes to construct such a slot antenna with good unidirectional radiation, it is vital to apply a proper excitation amplitude ratio of the slot and its complementary electric dipole.

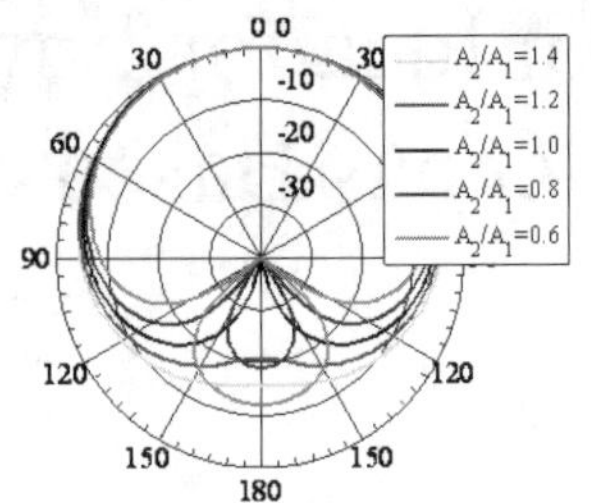

Figure. 2 $\lambda/2$ complelemtary E- and M- dipoles with different excitation amplitude

TABLE I
THE RATIO OF THE FIELD INTENSITY WITH DIFFERENT FEEDING AMPLITUDE

A_1/A_2	1.4	1.2	1.0	0.8	0.6
$\frac{E_\theta\|_{\theta=\pi}}{E_\theta\|_{\theta=0}}(dB)$	15.5630	20.8279	∞	19.0849	12.0412

III. Practical Antenna Configuration and Preliminary Simulated Results

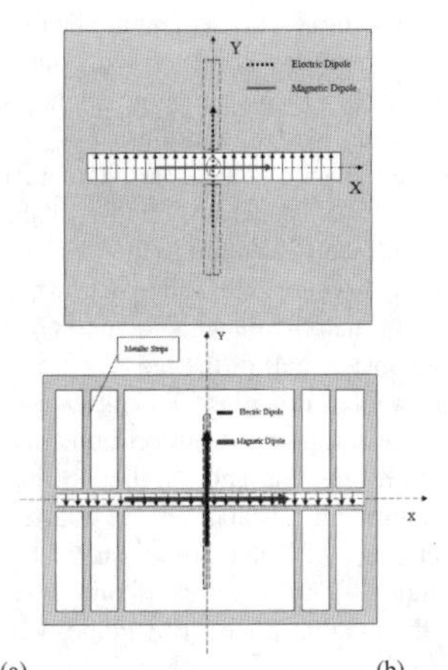

(a) (b)

Figure. 3 The configuration of the (a) idealized and (b) practical slot and its complementary E-dipole.

In section II, the radiation performance of an idealized complementary E- and M-dipoles has been discussed. However, when a practical antenna configuration is considered, the E-dipole can not be directly placed near the surface of the slot plane to form complementary dipoles. This is because, according to the image theory, an E-dipole near an parallel to a conducting plate has an image with opposite direction and thus the actual and image sources together produce very poor radiation. In [2], a conventional slot but with modified ground plane has been studied. It has been demonstrated that when the metallic conducting plate around the slot is replaced by a series of strips, the matching and radiation performance of the slot is not significantly altered. Therefore, when the E-dipole is orthogonally placed near such a slot, as shown in Fig. 3 (b), it is reasonable to believe that its radiation properties, particularly along the ground plane surface normal, will almost keep unchanged.

From the practical configuration point of view, the antenna in Fig. 3 (b) can be taken as a combination of two parts, a slot on a finite ground plane and an E-dipole surrounded by a metallic frame. In this paper we only conduct separate full-wave simulation for these two parts, and investigate the combination of their radiation pattern.

Fig. 4 and Fig. 5 show our model in CST and corresponding 3D radiation pattern of the slot and E-dipole, respectively. In accordance with Fig. 3, the E-dipole and slot are along x and y axis, respectively, and they both are excited by a discrete port (voltage source of 1V). One sees that, with the existence of the surrounding metallic frame, the radiation of the E-dipole on the metallic plane (also the plane of the slot) is zero. As to the slot radiator, the finite plane also brings a radiation null on this plane and introduces ripples on the E-plane as well [4]. However, one sees that the radiation property in $\pm z$ direction is not significantly altered, and this allows us to believe that a unidirectional pattern can still be obtained by a combination of such two patterns (if proper excitation ratio of the two sources is applied).

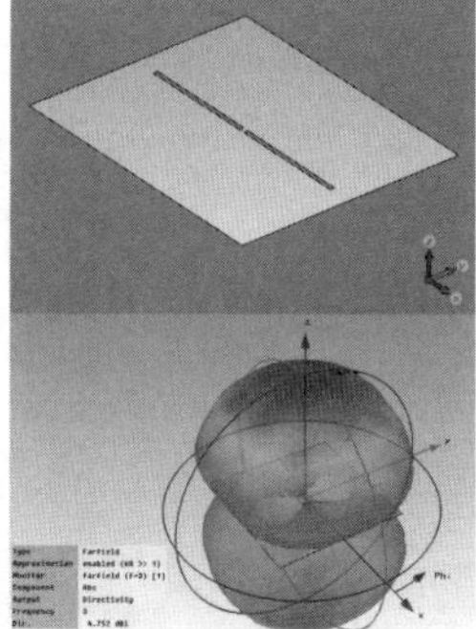

Figure. 4 The (a) CST model and (b) simulated 3D radiation pattern of the E-dipole with a surrounded metallic frame.

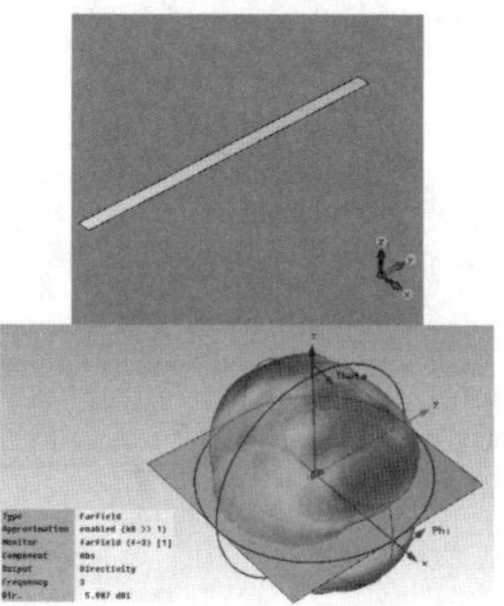

Figure. 5 The (a) CST model and (b) simulated 3D radiation pattern of the slot on a finite ground plane.

As discussed in section II, the backward radiation suppression largely depends on the excitation amplitude of the slot and the E-dipole, which is denoted by k in what follows. Fig. 6 shows the E-plane radiation pattern of the separate E-dipole, slot and their combination when k=0.96. One sees that in this case, the backward radiation is greatly suppressed, and particularly, the radiation in $\theta=\pi$ is theoretically zero. Two tiny side lobes are observed, which comes from the incomplete field cancellation. In fact, the pattern is very alike to the cardiac pattern in Fig. 1, except that due to the ground plane effect, radiation along this plane ($\theta=\pi/2, 3\pi/2$) is also zero. In fact, the magnitude of back lobes can be regulated by k. Fig. 7 plots the combined pattern of the complementary E-dipole and slot on the E-plane when k is set to be three different values, and one sees that an optimized k (around 1.4) can be obtained to keep all three back lobes equally small (Fig. 7 (b)).

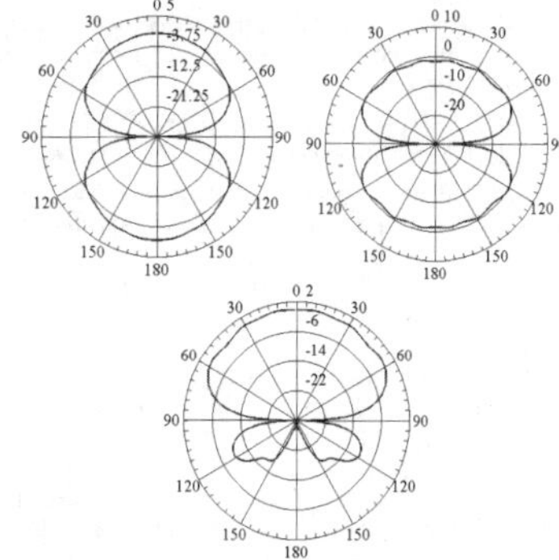

Figure. 6 The E-plane pattern of (a) the E-dipole, (b) the slot, and (c) their combination.

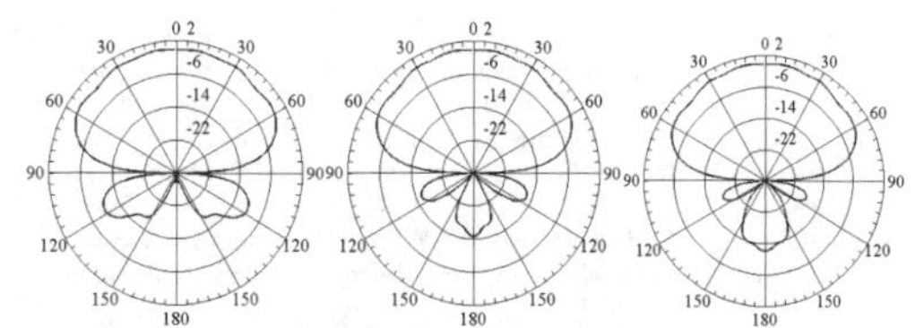

Figure. 7 The E-plane pattern of the complementary E-dipole and slot when k is (a) 0.96, (b) 1.4, and (c) 1.6, respectively.

Here in this paper we only report our idea and some preliminary simulated results. The next step of our work includes the arrangement of the feeding for the practical structure, elimination of mutual coupling of the two sources, and fabrication and measurement, and these are currently under study.

IV. Conclusion

Based on the complementary dipole concept, a novel scheme for constructing low-profile unidirectional slot antennas has been proposed. Preliminary study shows that when the original slot and its complementary E-dipole are excited with proper amplitude, unidirectional radiation with low backward radiation can be achieved. Compared to conventional unidirectional slot techniques, the proposed structure has several advantages such as low profile, low Q, and ease of fabrication.

References

[1] Q. Li and Z. Shen, "Microstrip-fed cavity-backed slot antennas," *Microw.Opt. Technol. Lett.*, vol. 33, no. 4, pp. 229–233, 2002.W. Hong,

[2] N. Behdad, and K. Sarabandi, "Size reduction of cavity-backed slot antennas," *IEEE Trans. Antennas Propag.*, vol. 54, pp. 1461–1466, May. 2006.

[3] J. Joubert, J. C. Vardaxoglou, W. G. Whittow, and J. W. Odendaal, "CPW-fed cavity-backed slot radiator loaded with an AMC reflector," *IEEE Trans. Antennas Propag.*, vol. 60, pp. 735–742, Feb. 2012.

[4] Y. Yoshimura, "A microstripline slot antenna," *IEEE Trans. Microwave Theory Tech.*, pp. 760–762, Nov. 1972.

[5] M. Qiu, M. Simcoe, and G. V. Eleftheriades, "Radiation efficiency of printed slot antennas backed by a ground reflector," in *Proc.2000IEEE AP-S Symp. Dig.*, pp. 1612–1615.

[6] G. V. Eleftheriades and M. Qiu, "Efficiency and gain of slot antennas and arrays on thick dielectric substrates for millimeter-wave applications: a unified approach," *IEEE Trans. Antennas Propag.*, vol. 50, pp. 1088–1098, Aug. 2002.

[7] S. Hashemi-Yeganeh and C. Birtcher, "Theoretical and experimental studies of cavity-backed slot antenna excited by narrow strip," *IEEE Trans. Antennas Propag.*, vol. 41, pp. 236–241, Feb. 1993.

[8] Q. Li and Z. Shen, ."Inverted microstrip-fed cavity-backed slot antennas," *IEEE Trans. Antennas Propag.*, vol. 1, pp. 98–101, 2002.

[9] B. Zheng and Z. Shen, ."Effect of a finite ground plane on microstrip-fed cavity-backed slot antennas," *IEEE Trans. Antennas Propag.*, vol. 53, pp. 862–865, Feb. 2005.

[10] A. Vallecchi and G. B. Gentili, "Microstrip-fed slot antennas backed by a very thin cavity," *Microw. Opt. Tech. Lett.*, vol. 49, no. 1, pp. 247–250, Jan. 2007.

[11] M. Qiu and G. V. Eleftheriades, "Highly efficient unidirectional twin arc-slot antennas on electrically thin substrates," IEEE Trans. Antennas Propagat., vol. 52, no. 1, pp. 53–58, Jan. 2004

[12] C. Löcker, T. Vaupel, and T. F. Eibert, "Radiation efficient unidirectional low-profile slot antenna elements for X-band applications," *IEEE Trans. Antennas Propagat.*, vol. 53, no. 8, pp. 2765–2768, Aug. 2005.

[13] J. Y. Park, C. -C. Chang, Y. Qian, and T. Itoh, "An improved low-profile cavity-backed slot antenna loaded with 2D UC-PBG reflector," in *Proc.2000 IEEE AP-S Symp. Dig.*, pp. 194–197.

[13] F. Elek, R. Abhari, and G. V. Eleftheriades, "A uni-directional ring-slot antenna achieved by using an electromagnetic band-gap surface," *IEEE Trans. Antennas Propag.*, vol. 53, pp. 181–190, Jan. 2005.

[14] A. Clavin, "A new antenna feed having equal E- and H-plane patterns," *IRE Trans. Antennas Propag.*, vol. 2, no. 3, pp. 113–119, Jul. 1954.

[15] A. Clavin, D. A. Huebner, and F. J. Kilburg, "An improved element for use in array antennas," *IEEE Trans. Antennas Propag.*, vol. 22, no. 4, pp. 521–526, Jul. 1974.

[15] H.Wong,K.-M. Mak, and K.-M. Luk, "Wideband shorted bowtie patch antenna with electric dipole," *IEEE Trans. Antennas Propag.*, vol. 56, no. 7, pp. 2098–2101, Jul. 2008.

[16] L. Ge and K.-M. Luk, "A low-profile magneto-electric dipole antenna ," *IEEE Trans. Antennas Propag.*, vol. 60, no. 4, pp. 1684–1689, Apr. 2012.

[17] M. Li and K. -M. Luk, "A differential-fed magneto-electric dipole antenna for UWB applications," *IEEE Trans. Antennas Propag.*, vol. 61, no. 1, pp. 92–99, Jan. 2013.

[18] C. A. Balanis, Antenna Theory: Analysis and Design, 2nd ed. New York: Wiley, 1997, ch. 12, pp. 616–620.

A New Radiation Method for Ground Radiation Antenna

Hongkoo Lee, Jongin Ryu, Jaeseok Lee, *Hyung-Hoon Kim, Hyeongdong Kim
Department of Electronics and Communications Engineering, Hanyang University, 17 Haengdang-Dong, Seongdong-Gu, Seoul 133-791, Korea
*Kwangju Women's University, Korea
E-mail : hdkim@hanyang.ac.kr

***Abstract-* This paper presents a new radiation method for exciting small size Universal Serial Bus (USB) dongle applications. New radiation method is designed on a thin line that fully covers Bluetooth and Wireless Lan (Wi-Fi) services. The simulation of the proposed antenna has a bandwidth of 123 MHz, from 2388 MHz to 2512 MHz under Voltage Standing Wave Ratio (VSWR) = 2 : 1. The proposed antenna consists of a feeding line, capacitor (C) and inductor (L). The basis for radiating USB dongle is to form a resonating structure on a thin line connecting USB dongle and Printed Circuit Board (PCB) ground of an external device, where current is strong. The antenna is designed on a Frame Retardant Type 4 (FR-4) substrate (ε_r = 4.4, tan δ = 0.02) and occupies area of 2.5 X 7 mm^2. New feeding structure was designed on a thin line, which is the USB port of a USB dongle that does not require additional clearance area.**

***Index Terms* – PIFA, Ground Radiation Antenna**

I. Introduction

Mobile devices currently on the market require both a wide bandwidth capability and a high realized efficiency because of their rapidly increasing data usage requiring high data transfer rates. The many functions of today's mobile devices are capable of require many components; for this reason, the components must be compactly arranged within the handsets and mobile antennas must become ever smaller. Expansion of the impedance bandwidth in an electrically small antenna is difficult because the impedance bandwidth of a small antenna is closely related to its physical volume [1].

According to 'Chu-Harrington limit' [2, 3, 4, 5], the radiation quality factor of an ideal small antenna is approximately inversely proportional to the volume of the antenna in wavelengths and thus its impedance bandwidth is limited by the antenna size. From these reasons, the main challenge in the field of antennas is to make the optimal compromise between their size and performance. Therefore, we need to develop a new radiation method with high performance within allowed volume for the antenna.

Antennas are indispensable components in wireless communication devices. They play a pivotal role in determining the performance of a whole communication system. Recently, the concept of ground radiation is extended by previously published papers [6, 7]. This new radiation technique (so called 'Ground Radiation Antennas') is to utilize the ground plane effectively as a radiator for high frequency applications (Wi-Fi and Bluetooth band). Instead of actual antenna structures, this technique uses only reactive components such as commercial chip-capacitors. In other word, two capacitors replace both matching circuits and antennas typically used in mobile handsets. One of reactive components is inserted in the strip line to connect two ground points. Its value is related to a resonant frequency similar to a branch line length of Planar Inverted F Antenna (PIFA). The other component is located on the feeder and its value is involved with input impedance of antennas.

With the wide increase in the use of mobile devices, Bluetooth service has become an essential component in wireless service. All mobile devices require wireless service that enables communication with other devices.

The mobile devices are becoming smaller day by day and as a result, the area allowed for antenna is also becoming smaller. An obvious way to enhance the performance is to increase the antenna size. However, it is impossible to increase the antenna size. It is becoming difficult for antenna designers to fulfill the performance in a limited given space.

In this paper, a new radiation method to excite USB dongle at Wi-Fi frequency (2.4-2.5 GHz) was proposed. The simulation was performed with the Finite Elements Method (FEM) 3D simulator 'ANSYS HFSS v.13' and measurement data was obtained using network analyzer and three dimensional anechoic chamber.

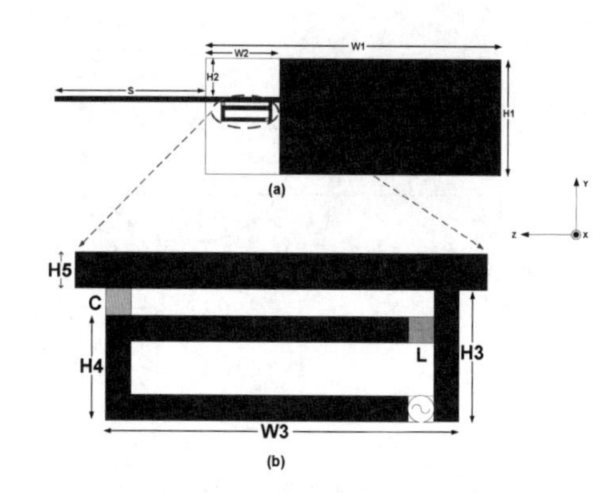

Figure 1. Geometry of the proposed antenna (a) General view; (b) Dimensions of the feeding structure

978-7-5641-4279-7

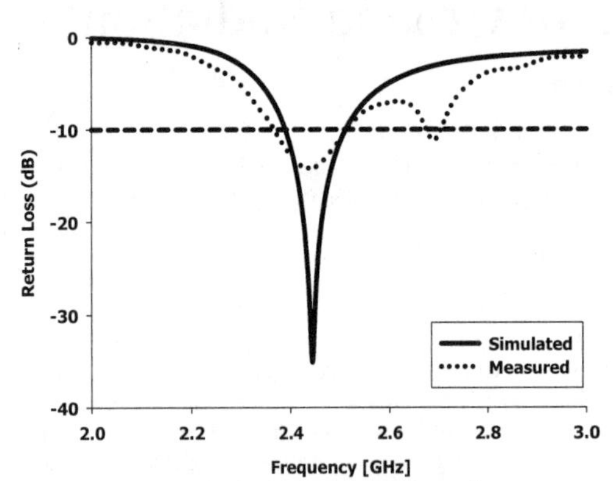

Figure 2. Simulated and measured return loss

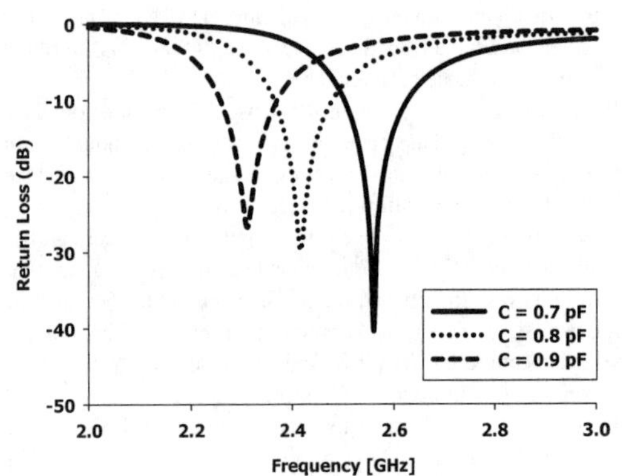

Figure 3. Return loss with C variation

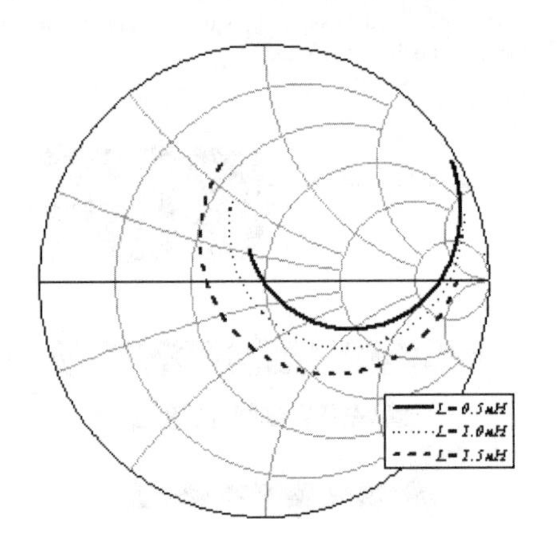

Figure 4. Input impedance with L variation

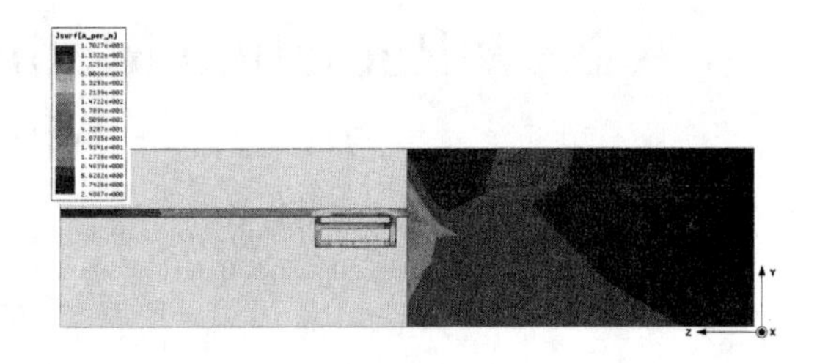

Figure 5. Computed surface current density

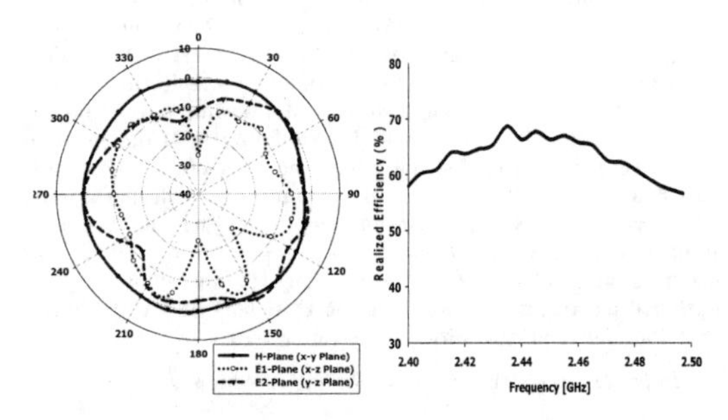

Figure 6. Measured radiation patterns and realized efficiency

II. ANTENNA DESIGNS AND ANALYSIS

The geometry of the proposed antenna is shown in Figure. 1. The optimized design parameters for the proposed antenna are: H1 = 15mm, H2 = 5 mm, H3 = 2.5 mm, H4 = 2 mm, H5 = 0.7 mm W1 = 40 mm, W2 = 10 mm, W3 = 7 mm, S = 20 mm, thickness = 1.2 mm, L = 0.4 nH and C = 0.775 pF.

Technique for controlling resonance frequency is shown in Figure. 3. Resonance frequency can be lowered by increasing the values of C. Figure. 4 shows how the impedance matching can be achieved by controlling the value of L. Figure. 5 is computed surface current density, implying strong current on the thin line of the USB dongle. Figure. 6 is the measured radiation patterns and realized efficiency.

The resonator forms a feeding loop exciting the thin line where current is strong, with magnetic coupling. This coupling can be controlled by variation of L and the resonance frequency can be controlled by variation of C.

III. EXPERIMENTAL RESULT AND DISCUSSION

The proposed antenna has been successfully simulated and constructed. The proposed antenna was simulated and measured. It can be seen from Figure. 2 that the proposed antenna fully satisfies the Wi-Fi bandwidth in both simulation and measured results. Figure. 2 illustrates return loss of simulated and measured results, showing similar pattern at

Bluetooth and Wi-Fi bands. Fig. 2 shows simulation result with bandwidth of 123 MHz (2388 - 2512 MHz) and the measured result with bandwidth of 149 MHz (2365 - 2514 MHz) under VSWR = 2 : 1. It shows overall realized efficiency of 63.21% at 2400 MHz to 2500 MHz and omnidirectional pattern.

IV. CONCLUSION

The proposed antenna was designed and achieves sufficient bandwidth covering Bluetooth and Wi-Fi service. It forms a feeding structure on a thin line, where the current is strong. The proposed feeding method occupies area of 2.5 X 7 mm^2. This makes it possible to apply the proposed feeding structure in various devices that include a connection through a thin line because it does not require additional clearance area on the PCB ground.

ACKNOWLEDGMENT

This research was funded by the MSIP (Ministry of Science, ICT & Future Planning), Korea in the ICT R&D Program 2013.

REFERENCES

[1] McLean, J.S.: 'A re-examination of the fundamental limits on the radiation Q of electrically small antenna', *IEEE Trans.* Antennas Propag., 1996, 44, (5), pp. 672-676.

[2] H.A.Wheeler, 'Fundamental limitations of small antennas,' *Proc. IRE*, Vol. 35, No. 12, pp. 1479-1484, Dec. 1947.

[3] H.A.Wheeler, 'Small antennas,' *IEEE Transactions on Antennas and Propagation*, Vol. AP-23, No.4, pp. 462-469, Jul. 1975.

[4] R.E.Collin, 'Minimum Q of small antennas,' *Journal of Electromagnetic Waves and Applications*, Vol. 12, No. 10, pp. 1369-1393, Dec. 1998

[5] R.F.Harrington, 'Effect of antenna size on gain, bandwidth and efficiency,' *Journal of Research of the National Bureau of Standards- D. Radio Propagation*, 64D, pl-12, Jan-Feb. 1960.

[6] O. Cho, H. Choi, and H. Kim, 'Loop-type ground antenna using capacitor,' *Electronic Letters*, Vol. 47, No. 1, pp. 11-12, Jan. 2011.

[7] Y. Liu, X. Lu, H. Jang, H. Choi, K. –Y. Jung, and H. Kim, 'Loop-type ground antenna using resonated loop feeding, intended for mobile devices,' *Electronics Letters*, Vol. 47, No. 7, pp. 426-427, Mar. 2011.

Design of A Novel Quad-band Circularly Polarized Handset Antenna

Youbo Zhang[#], Yuan Yao[#], Junsheng Yu[#], Xiaodong Chen[*], Yixing Zeng[#], and Naixiao He[#]

[#] School of Electronic Engineering, Beijing University of Posts and Telecommunications, No.10 Xitucheng Road, Beijing, China

[*] School of Electronic Engineering and Computer Science, Queen Mary, University of London, London, UK

bob0313@126.com

***Abstract* — A novel circularly polarized handset antenna which is compatible with multiple Global Navigation Satellite Systems is proposed. The antenna comprises a palm-size slab as the handset test bench, a small antenna chip mounted at the corner of the slab, three branches, a small patch, and a ground plane with rectangle clearance region. The antenna is featured with several characteristics such as quad frequency band operation, circularly polarized reception, low fabrication cost, compact size. Both of the simulated and measured results are shown to illustrate the good performance of the proposed antenna.**

***Index Terms* — Quad-band, Circular polarization, GNSS, Handset antenna.**

I. INTRODUCTION

The fast growth of wireless communication technology in the last decade demands diversified services from wireless communication devices. The Global Navigation Satellite System(GNSS) is the one of the most successful commercial products, and is being used more and more widely in various areas. To overcome the drawbacks of linear polarization signals, such as the problems of polarization alignment and terrestrial mobile communication systems, inclement weather, ionosphere absorption and multipath effects, circularly polarized(CP) signals are needed in these systems.

Nowadays, the American Global Positioning System(GPS), the Russian GLONASS and the Chinese Beidou navigation systems are already in operation. And EU's Galileo positioning system are scheduled to provide full function in the near future. Integrating multiple navigation systems is a good choice for commercial devices to improve the positioning performance. Our goal in this paper is to develop a compact, low cost, multi-band circularly polarized antenna for handheld navigation devices.

Many methods have been taken to realize circularly polarized reception, such as helixes, four-arm spirals, patches bended monopole and crossed dipoles. The curl antenna [1] uses a bended monopole to yield a CP radiation pattern for satellite communication. Reference[2] presents a compact quadrifilar helix antenna with the self-phasing technique. It is installed externally to the portable de-vice with excellent circular polarization performance. However, it works in one band only and requires a wideband balun as well as a matching network, which, in turn, increases the design complexity.

An alternative approach to producing circular polarization is to create two orthogonal resonant modes on the radiating element with a 90 phase difference. For patch antennas, perturbation cuts or strips are used to produce two perpendicular modes. Many studies have been reported in the literature that describe different methods of achieving multiband. A stacked patch technique with a high permittivity dielectric material is used in [3] to achieve a compact triple band antenna design. However, a dual orthogonal feed is required to drive the antenna. Reference [4] proposes a novel CP antenna design based on stacked patches with a single feed for the GPS operation in L1 (1.575GHz), L2 (1.227GHz), and L5(1.176GHz) band. However, these kinds of antenna's CP bandwidth are usually narrow and cannot meet requirements of modern wireless communication systems.

The previously mentioned antenna types may not be suitable for handheld devices for its large sizes or big weights. So a general solution for portable device antennas is linear polarization. But using linearly polarized antennas to receive CP signals from satellite suffers a 3-dB loss in terms of signal to noise ratio. [5] uses a bended monopole to realize dual-band CP radiation and successfully solved the problem mentioned previously. To study new structure of CP antenna for handset devices, an antenna configuration with CP radiation pattern is proposed in this paper. The antenna comprises a palm sized slab and a small chip which are both made of FR4 substrate. The antenna works at L1 and L2 band of GPS. Actually the working band around L1 is quite broad which can cover all the near bands for other navigation systems like B1 (1.561GHz) band for Compass and L1C/A (1.602GHz) band for GLONASS.

II. ANTENNA DESIGN

The structure of the proposed quad-band circularly polarized GNSS antenna design for handheld devices is illustrated as fig 1. Taking into considerations the effect of the big ground on the handset circuit board, a palm-size test bench is included in the simulation model. The small antenna chip with size of 30mm × 6mm × 1.6mm is fabricated

978-7-5641-4279-7

perpendicularly on the corner of the 120mm (L) × 67mm (W) × 1.6mm (h) slab. Both the antenna chip and the slab is made of substrate FR4 with relative permittivity of 4.4 and dielectric loss tangent of 0.02.

TABLE I

Dimensions of the proposed antenna for GNSS handset

Lz	H1	lg	wg	L	W
30	6	40	15	120	67

Unit :mm

The antenna design mainly comprises of a rectangle big ground with a rectangle clearance, three branches (a folded one, a bended one and a meander one), two short pins and a small patch. On the back surface of the slab, the big ground is printed, while on the up surface of the slab and the inner surface of the antenna chip, the three branches are printed. One short pin is placed between the folded branch and the small patch which is printed on the outer surface of the antenna chip, while another shorting pin is used to connect the ground and the end of the meander branch. A 50 ohm coax line is attached to the feed point to connect to the testing instrument. The detailed parameter for the antenna design is given in table 1.

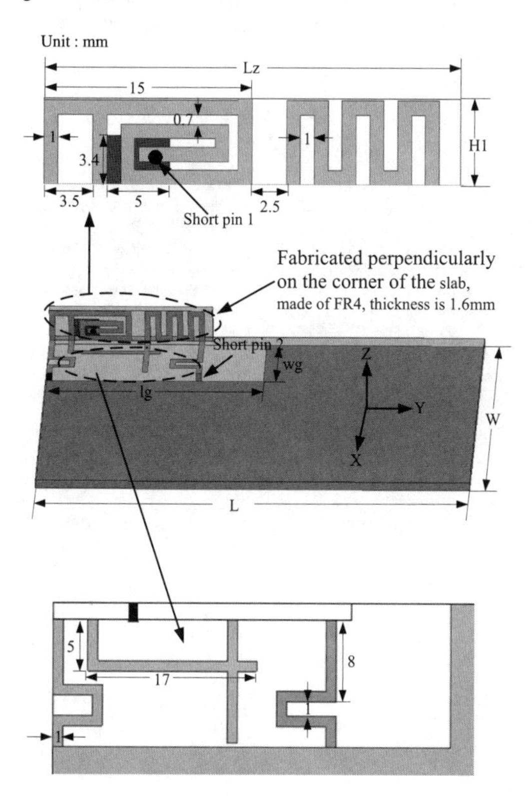

Fig. 1. Detailed geometry of the quad band GNSS reception antenna

In fact, most of the space is occupied by the test bench used to emulate the environment of the hand held devices in which the antenna works, so the space occupied by the antenna is relatively small.

III. Simulation and Measurement Results

The simulation results of the antenna was carried out via HFSS, which is a full wavelength numeric electromagnetic simulation tool. To validate the performance of the antenna design, a prototype was fabricated and the photo of the prototype is shown in fig 2. The measurement was carried out in chamber.

Fig. 2. Fabricated quad band GNSS reception antenna

Fig 3 shows the simulated and the measured return loss. From the figure we can see that the measured and simulated results match well. The -10dB bandwidth is from 1.22GHz to 1.24GHz and from 1.51GHz to 1.66GHz for each operating bands respectively, which cover several frequency bands for navigation systems. Fig 4 compares the measured and simulated axial ratio (AR). At each operating frequency bands the AR is below than 3dB, which means that antenna has a good CP performance. Both Fig.3 and Fig.4 show that the lower band covers L2 band of GPS and the higher frequency band is broad enough to cover all the frequency bands around 1.575GHz for other navigation systems besides GPS L1 band, like B1 band for Compass and L1C/A band for GLONASS.

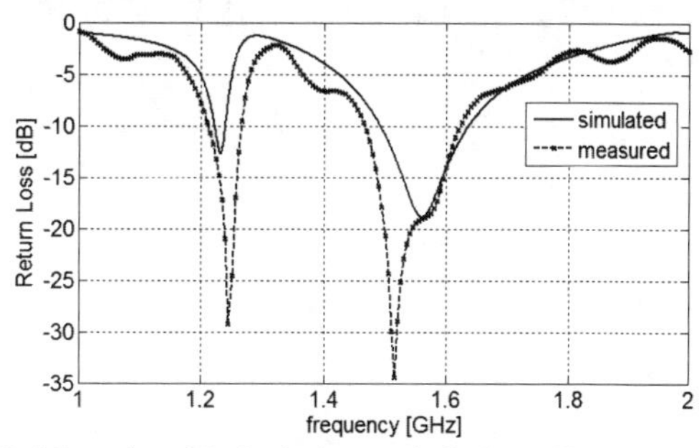

Fig. 3 Comparison of simulated and measured reflection coefficient spectra of the proposed quad-band GNSS reception antenna

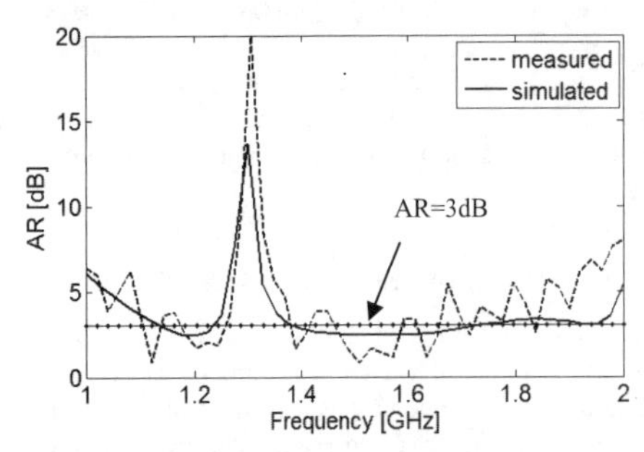

Fig.4 Comparison of simulated and measured axial ratio spectra of the proposed quad band GNSS reception antenna.

Fig.5 shows the measured and simulated radiation patterns of both frequency bands. Both XZ plane and YZ plane patterns are given. Fig.6 is the 3D graph of measured RHCP gain at 1.227GHz and 1.575GHz respectively. The RHCP gain maximum, which is about 1.5dBi, is found about 10 degree away from +z direction, and the 3dB beam width is about 150 degree at 1.227GHz. The measured peak RHCP gain at 1.575GHz is about 3.2dBi, and the beam width is about 120 degree, narrower than the lower band which is caused by the higher peak gain.

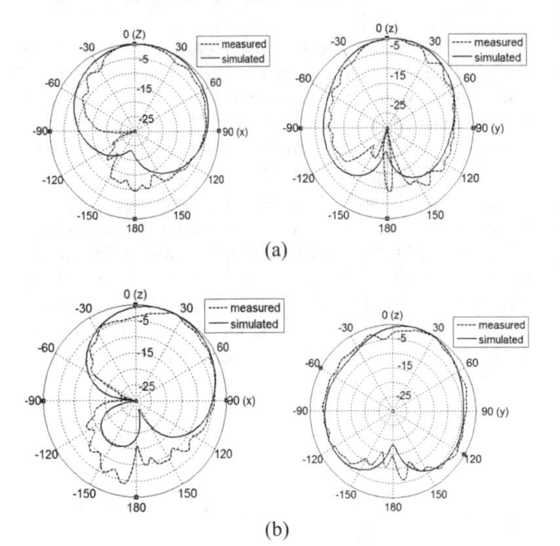

Fig.5 Comparison of simulated and measured radiation patterns of the proposed quad band GNSS reception antenna. (a) at 1.227GHz, (b) at 1.575GHz

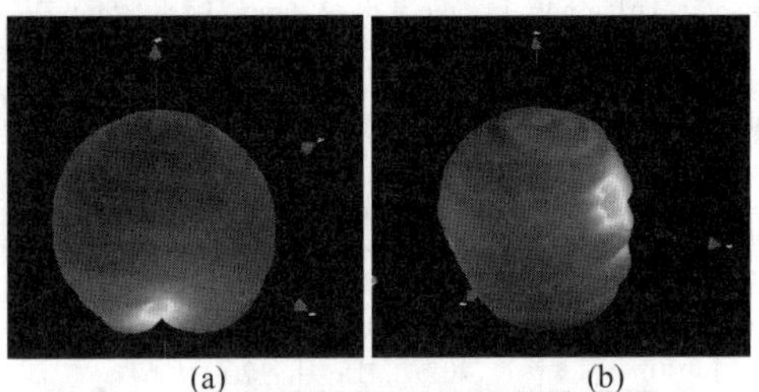

(a) (b)

Fig.6. Measured 3D radiation pattern of the proposed quad band GNSS reception antenna. (a) at 1.227GHz, (b) at 1.575GHz

IV. CONCLUSION

A compact low cost handheld GNSS device antenna for L1 band and L2 band of GPS, B1 band of Compass and L1C/A band of GLONASS was designed and simulated in HFSS. A prototype of the antenna design was fabricated successfully. Both the simulated and measured results demonstrate that the antenna at all operating frequency bands has good circular polarization radiation and impedance matching

ACKNOWLEDGMENT

This work is supported by the Fundamental Research Funds for the Central Universities, the National Natural Science Foundation of China under Grant No.61201026, and Beijing Natural Science Foundation (4133091).

REFERENCES

[1] H. Nakano, S. Okuzawa, K. Ohishi, H. Mimaki, and J. Yamauchi, "A curl antenna," *IEEE Transactions on Antennas and Propagation.*, vol. 41, no. 11, pp. 1570–1575, November 1993.

[2] Y. S. Wang and S. J. Chung, "A miniature quadrifilar helix antenna for global positioning satellite reception" *IEEE Transactions on Antennas and Propagation*, vol. 57, no. 12, pp. 3746–3751, December 2009.

[3] Z. Yijun, C. Chen, and J. L. Volakis, "Proximity-coupled stacked patch antenna for tri-band GPS applications," *IEEE Antennas Propagation. Soc. Int. Symp.*, pp. 2683–2686, July 2006.

[4] Oluyemi P. Falade, Masood Ur Rehman, Yue (Frank) Gao, Xiaodong Chen, Clive G. Parini, "Single Feed Stacked Patch Circular Polarized Antenna for Triple Band GPS Receivers" *IEEE Transactions on Antennas and Propagations*, vol. 60, no. 10, pp. 4479-4484, October 2012.

[5] Shih-Hsun Chang, Wen-JiaoLiao, "A Novel Dual Band Circularly Polarized GNSS Antenna for Handheld Devices" *IEEE Transactions on Antennas and Propagations,*, vol. 61, no. 2, pp. 555-562, February 2013.

Mobile handset antenna with parallel resonance feed structures for wide impedance bandwidth

Yongjun Jo, Kyungnam Park, Jaeseok Lee, *Hyung-Hoon Kim, Hyeongdong Kim
Hanyang University, Korea
*Kwangju Women's University, Korea

Abstract-**In this paper, mobile handset antenna with a novel feed structure is proposed. The feed structure composed of two resonators significantly enhances the impedance bandwidth of reference PIFA. The measured impedance bandwidth of the proposed antenna is 245 MHz in the lower band and 475 MHz in the upper band, covering LTE13, GSM850/900, DCS1800, PCS1900, and WCDMA under a VSWR of 3:1 with good realized efficiency.**

I. Introduction

Mobile handsets currently on the market require both wide bandwidth capability and high realized efficiency because of their rapidly increasing data usage and high data transfer rates. Recently, Long-term Evolution (LTE) services have been provided by wireless communication service providers. The mobile handsets are being constantly miniaturized due to consumer preference while supporting multi and wideband operation like a versatile electric device. The dimension for antenna design is getting limited as the mobile handsets are being more compact. Unfortunately, expansion of the impedance bandwidth of electrically small antennas is very difficult because the most of radiation performance of the mobile antenna are closely related to its physical volume [1]. Thus, it is impossible for antenna engineers to avoid serious challenges for designing a wide impedance bandwidth antenna with good radiation performance within limited dimension.

In order to expand an impedance bandwidth, [2, 3] modified feed structures of conventional planar inverted-F antennas (PIFAs) with lumped elements to form loop-type parallel resonators seen by the source.

We propose the method to obtain multi and wideband operation. The proposed antenna fully covers LTE 13 (745-787 MHz), GSM850 (824-894 MHz), GSM900 (880-960 MHz) and DCS1800/ PCS1900/WCDMA(1710~2170 MHz). Details of the proposed antenna design and operating mechanism will be presented. The proposed antenna was designed and analyzed using commercial HFSS v13 software.

II. Antenna Design and Simulated Result

The geometry of the proposed antenna is shown in Fig. 1. The proposed antenna is based on a simple quarter wavelength (λ /4) PIFA. The proposed antenna consists of a feed structure, an antenna element and a ground plane. The only difference between the proposed antenna and the reference PIFA is to insert two shunt capacitors (C_L and C_H) in the feed structure. The size of the ground plane etched on a PCB (FR4 substrate, ε_r = 4.4, tan δ = 0.02) with 1mm thickness is 60 × 110 mm^2. The dimension assigned for designing the proposed antenna is 60 × 10 × 5 mm^3. Fig. 2 shows that fundamental and high-order mode of the reference PIFA operate at 770 and 1820 MHz, respectively. In general, PIFAs suffer from narrow impedance bandwidth characteristics because an impedance of the loop formed of the shorting and feed line, operating as a loop-type inductor, is very high. In order to enhance the impedance bandwidth of the reference PIFA, two branch capacitors were inserted to form two resonators operating at different frequencies as shown in Fig. 1. The lower resonant frequency of the resonator composed of C_L and shorting line is decided by loop-size and the component value. When C_L is 6.6 pF, the impedance bandwidth in the lower band is significantly expanded as shown in Fig. 2 because the impedance of the feed structure is decreased by C_L. In a similar way, the resonator composed of C_L and C_H, which is operating at frequencies above 1.5 GHz is resonated by loop-size and the component values. When C_H is 1.1 pF, the impedance bandwidth in the higher band is significantly expanded as shown in Fig. 2. Compared with the reference PIFA, the proposed antenna's impedance bandwidth is enhanced approximately 3.5 times in the lower frequency bands and 5.9 times in the higher frequency bands with criteria of voltage standing wave ratio (VSWR) = 3. That is, the proposed antenna can obtain a wider bandwidth characteristic than reference PIFA by using the parallel resonance in feed structure.

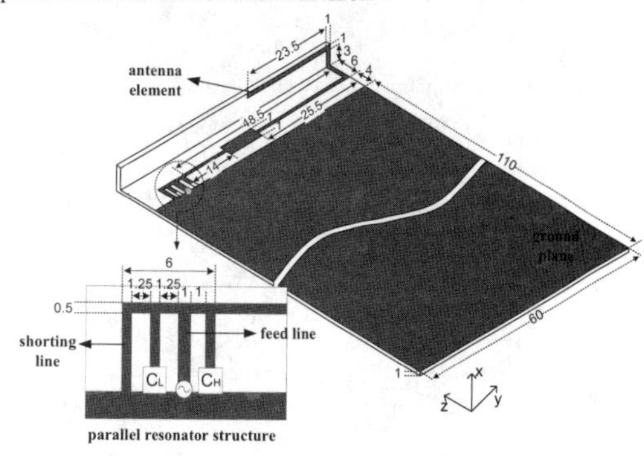

Figure 1. Geometry of proposed antenna.

978-7-5641-4279-7

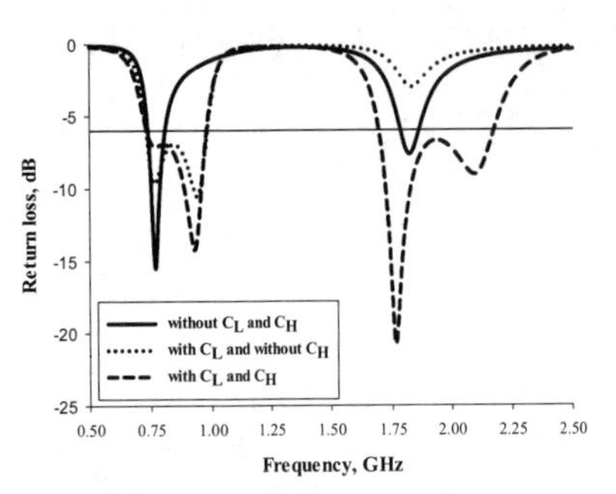

Figure 2. Simulated return loss characteristics with and without C_L and C_H

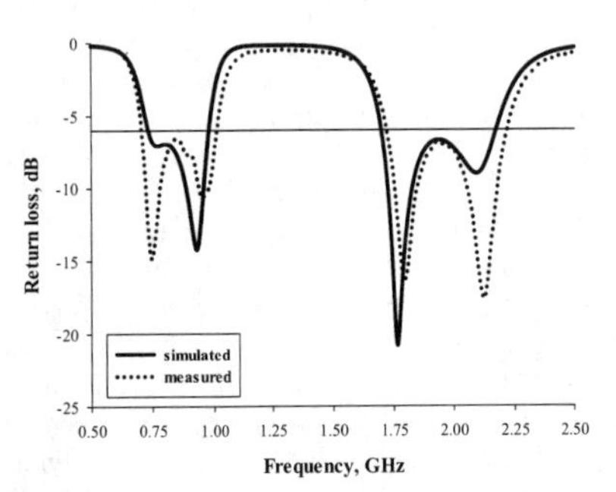

Figure 3. Simulated and measured return loss

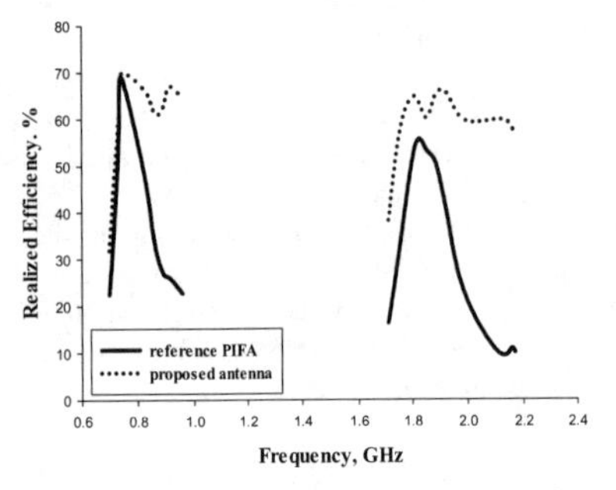

Figure 4. Efficiency of conventional PIFA and proposed antenna

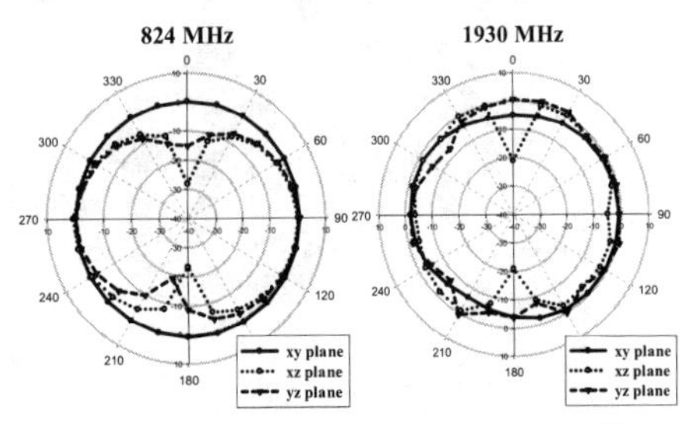

Figure 5. Measured radiation patterns at 824 and 1930 MHz

III. Experimental Result and Discussion

The simulated and measured return loss characteristics are shown in Fig. 3. The impedance bandwidth with VSWR = 3 are 245MHz (735-980MHz) for the lower band and 475MHz (1700-2175MHz) for the higher band. It is obvious that the proposed antenna can cover all of the desired bands, which are LTE 13, GSM850/900, DCS1800, PCS1900, and WCDMA. In Fig. 4, the measured realized efficiency of the proposed antenna is shown. The average realized efficiency of the proposed antenna is 62.85% in lower band and 57.14% in higher band while the reference PIFA's average realized efficiency is much less in all bands because of its narrow impedance bandwidth characteristic. The measured total radiation patterns at 824 MHz and 1930 MHz are shown in Fig. 5. These are the total electric field of the theta and phi components on the xy, xz, and yx plane. The measured radiation patterns in desired bands are omni-directional like monopole antennas, which is good for practical application.

IV. Conclusion

The proposed antenna significantly enhances an impedance bandwidth in the lower and high band by inserting parallel resonance circuit in the feed structure. In addition, the proposed antenna can be easily fabricated because it used only one antenna element without any branch line and parasitic element for wideband operation. Since resonance frequencies of two loop-type resonators can be individually controlled, we easily carry out the enhancement of bandwidth in desired bands. The measured impedance bandwidth with VSWR = 3 are 245MHz (735-980MHz) for the low band and 475MHz (1700-2175MHz) for the high band. The proposed antenna for mobile handsets is competitive for practical applications because of sufficient wide impedance bandwidth, simple antenna pattern and good radiation performance, which covers LTE 13, GSM850/900, DCS1800, PCS1900, and WCDMA applications.

Acknowledgment

This research was funded by the MSIP (Ministry of Science, ICT & Future Planning), Korea in the ICT R&D Program 2013.

References

[1] J.S. McLean, 'A re-examination of the fundamental limits on the radiation Q of electrically small antenna', *IEEE Trans.* Antennas Propag., Val 44 (2002), (5), pp. 672-676

[2] S. Jeon, H. Kim, 'Mobile terminal antenna using a planar inverted-E feed structure for enhanced impedance bandwidth', *Microwave and Opt Technol Lett*, vol. 54, No.9, pp. 2133-2138, Sep. 2012.

[3] S. Jeon, S. Oh, H.H Kim and H. Kim, 'Mobile handset antenna with double planar inverted-E (PIE) feed structure', *Electronics Letters*, vol. 48, No. 11, 24th May 2012.

[4] K.R. Boyle and L.P. Ligthart, 'Radiation and balanced mode analysis of PIFA antennas', *IEEE Trans.* Antennas Propag., vol. 54, No. 1., pp. 231-237, Jan. 2006.

Dual Beam Antenna for 6-Sector Cellular System

#Yoshihiro Kozuki[1], Hiroyuki Arai[1], Huiling Jiang[2], Taisuke Ihara[2]
[1]Graduate School of Engineering, Yokohama National University
79-5, Tokiwadai, Hodogaya-ku, Yokohama-shi, Kanagawa, 240-8501, Japan
e-mail: kozuki-yoshihiro-vk@ynu.jp, arai@ynu.ac.jp
[2]NTT DoCoMo, Inc. Hikarino-oka 3-5, Yokosuka, 239-8536 Japan
e-mail: jiang@nttdocomo.co.jp, iharat@nttdocomo.co.jp

***Abstract*- This paper proposes dual beam antennas for 6-sector cellular systems using multi-element array, suppressing the side lobe level to interfere with adjacent cell area. Dual beam is given by phase controlled feed circuit and parasitic elements.**

I. INTRODUCTION

To increase the channel capacity of current cellular systems, multi-sector antennas are widely introduced. Recent high density base station siting requires 6-sector antenna with the half power beam width less than 60 degree. A narrow beam width is obtained by a wide antenna size, however it is necessary to develop small sized base station antenna due to a lack of place of base station installation.

This paper proposes dual beam antennas for 6-sector base station antenna. The dual beam antenna consists of several antenna elements and a feed circuit to tilt two beams toward ± 30° from the front direction. An array antenna has side lobes and the first side lobe interfere another main beam. A key design issue is to suppress this side lobe level in this dual beam antenna. We examine 2, 3 and 4-elemnt array to obtain dual beam antenna for 6-sector systems.

II. RELATION BETWEEN SIDE LOBE AND ELEMENT NUMBER

First, we show the half power beam width (HPBW) and side lobe level of 2-element array with main beam tilting of ± 30° at 2GHz. This beam tilt is given by exciting antenna elements by 3dB 90° hybrid coupler. As shown in Fig. 1, we set two dipoles with the spacing d=0.5λ placed above a ground plane of 180 × 450mm with the height h=0.25λ. Its radiation pattern is shown in Fig. 2. The pattern has the tilt angle of ± 28°, the HPBW of 54.5° and the first side lobe level of -6.2dB. We should suppress the interference of the adjacent main beam, then we need to suppress the first side lobe level less than -10dB [1]. And, we show the HPBW and the side lobe level of 3-element array with main beam tilting of ± 30°. As shown in Fig. 3, we set three dipoles with the spacing d=0.5λ placed above a ground plane of 180 × 450mm with the height h=0.25λ. This beam tilt is given by exciting antenna elements by the phase difference as in Table 1. Its radiation pattern in Fig. 4 shows that the tilt angle of ± 30°, the HPBW of 42.5° and the first side lobe level of -9.52dB. Both of the HPBW and the side lobe level of 3-element array are suppressed compared with 2-element's ones, however it is difficult to make the feed circuit to provide the excitation phase difference in Table. 1.

TABLE 1
3-ELEMENT ARRAY IN FIG. 3

Tilt angle	Port1	Port2	Port3	Excitation phase difference
+30°	90°	0°	-90°	-90°
-30°	-90°	0°	90°	90°

TABLE 2
4-ELEMENT ARRAY IN FIG. 5

Tilt angle	Port1	Port2	Port3	Port4	Excitation phase difference
+30°	-90°	180°	90°	0°	-90°
-30°	0°	90°	180°	-90°	90°

Next, we show the HPBW and side lobe level of 4-element array with main beam tilting of ± 30°. As shown in Fig. 5, we set four dipoles with the spacing d=0.5λ placed above a ground plane of 280 × 450mm with the height h=0.25λ. This beam tilt is given by exciting antenna elements by the excitation phase difference as in Table 2. Its radiation pattern is shown in Fig. 6. The pattern has the tilt angle of ± 30°, the HPBW of 30° and the first side lobe level of -10.4dB.

These results show that the array to obtain dual beam antenna can suppress the side lobe level by increasing exciting element numbers, and it is better to use even element number for the feed circuit.

III. 4-ELEMENT ARRAY

In preceding section, we show 4-element array suppressing the first level side lobe level less than -10dB, however it is difficult to make the feed circuit as shown in Table. 2. In this section, we use a 4-way butler matrix for the feed circuit. A 4-way butler matrix gives the excitation phase differences ± 45° and ± 135°, respectively [2]. When we incline the antenna by +15° as shown in Fig. 7, a tilted beam toward +15° by the phase difference of -45°, and toward -45° by the phase difference of +135° by the 4-way butler matrix in Table 3, then we obtain tilted main beam toward ± 30°. Its radiation pattern is shown in Fig. 8. The patterns by the phase difference of -45° have the tilt angle of +30°, the HPBW of 27° and the first side lobe level of -11.2dB, and by the phase difference of +135° the tilt angle of -30°, the HPBW of -31° and the first side lobe level of -7.88dB. In [1] and [3], the adequate performances of the antenna are the first level side lobe level to less than -10dB and the HPBW of 30° ~ 45° to suppress the interference with adjacent cell area.

978-7-5641-4279-7

TABLE 3
4-WAY BUTLER IN FIG. 7

Tilt angle	Port1	Port2	Port3	Port4	Excitation phase difference
+30°(+15°)	135°	90°	45°	0°	-45°
-30°(-45°)	45°	180°	-45°	90°	135°

A 4-way butler matrix provides tilted beam toward 4 directions, while we need only two directions for the application of base station antenna, and it is not necessary for symmetric antenna structure. To suppress the side lobe level of -30° we adjust the antenna height as shown in Fig. 9, where we set three dipoles with the spacing d=0.5λ placed above a ground plane of 280 × 450mm with the heights from the left are 0.2425λ, 0.2475λ, 0.2525λ, 0.2575λ, respectively. The radiation pattern shown in Fig. 10 by the phase difference of -45° has the tilt angle of +27°, the HPBW of 27° and the first side lobe level of -10.2dB, and by the phase difference of +135° the tilt angle of -33°, the HPBW of -34° and the first side lobe level of -8.58dB. This shows that inclining the antenna by +15° can suppress the first side lobe level at the -30° direction slightly. Then we use parasitic elements of director and reflector to improve radiation pattern as shown in Fig. 11. We set director in the middle of plane with the height of 0.4λ, and reflector by 0.9λ from the right of the middle of plane with the height of 0.4075λ. The radiation pattern shown in Fig. 12 by the phase difference of -45° has the tilt angle of +27°, the HPBW of 29.5° and the first side lobe level of -10.4dB, and by the phase difference of +135° the tilt angle of -33°, the HPBW of -36° and the first side lobe level of -11.1dB. Its can suppress the first side lobe to less than -10dB.

IV. CONCLUSION

In this paper, we revealed the relation between beam tilting and element number to reduce the HPBW and suppress the first side lobe level by increasing the exciting element number. And we proposed 4-element array with parasitic elements fed by the 4-way butler matrix. Its pattern has the first side lobe level of smaller than -10dB and the HPBW of about 30° with main beam tilting of ± 30°.

REFERENCES

[1] K. Uehara, et al., "Indoor Propagation Delay Characteristics Considering Antenna Sidelobes," General Conf. of IEICE, B-3, May 1995.

[2] S. Yamamoto, et al., "A Beam Switching Slot Array with a 4-Way Butler Matrix Installed in Single Layer Post-Wall Waveguides," IEICE Trans. Commun., Vol. E86-B, No. 5, pp.1653-1659, June 2003.

[3] M. Iwamura, et al., "Optimal Beamwidth of Base Station Antennas for W-CDMA," General Conf. of IEICE, B-5-157, May 1999.

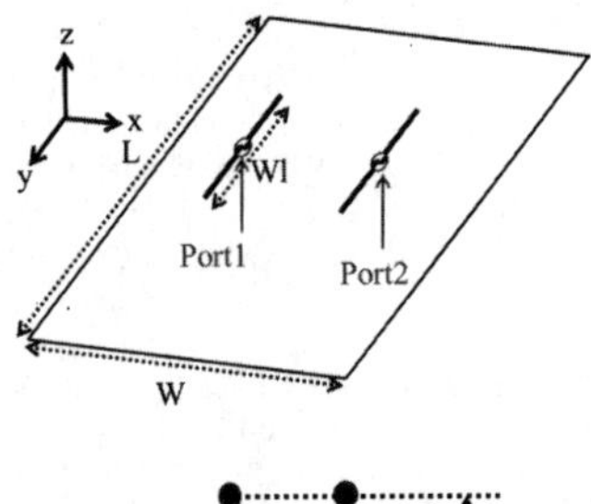

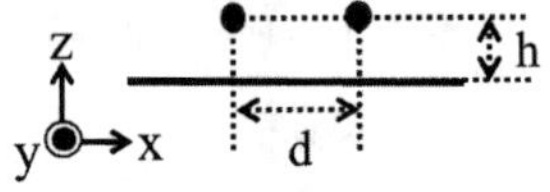

Figure 1. Antenna structure
2-element
W=180,L=450,Wl=64.5,d=75,h=37.5[mm]

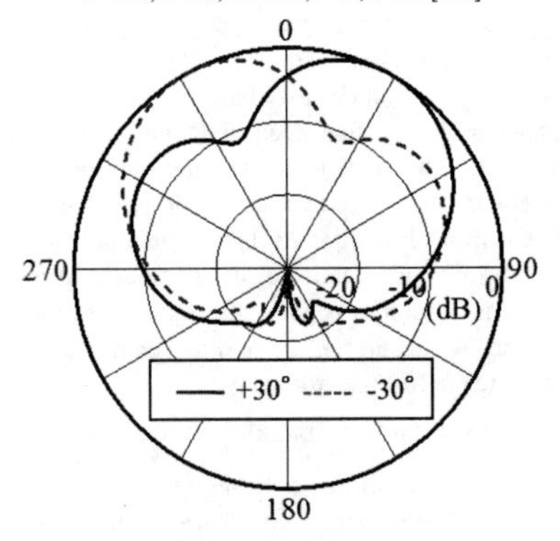

Figure 2. Radiation pattern
zx plane, 2-element

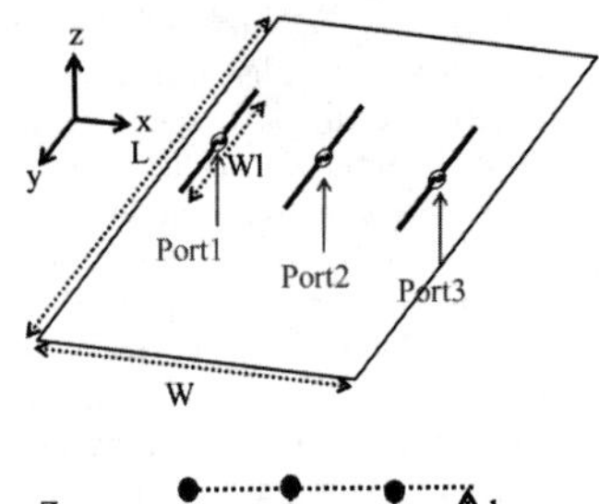

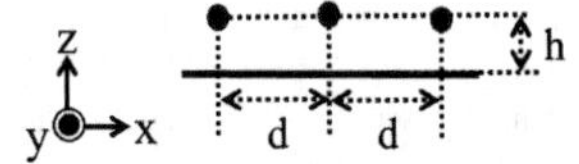

Figure 3. Antenna structure
3-element
W=180,L=450,Wl=64.5,d=75,h=37.5[mm]

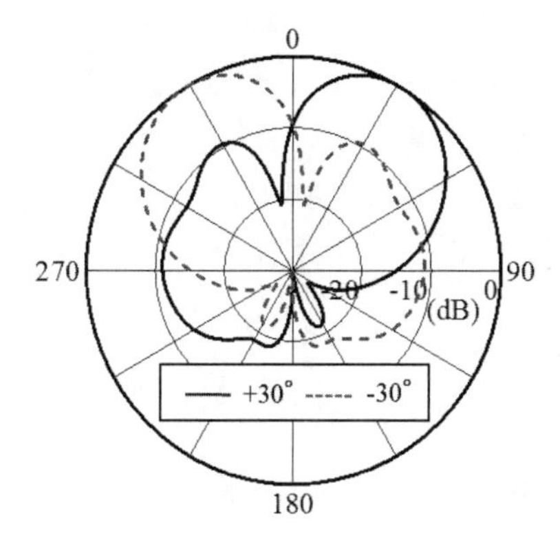

Figure 4. Radiation pattern
zx plane, 3-element

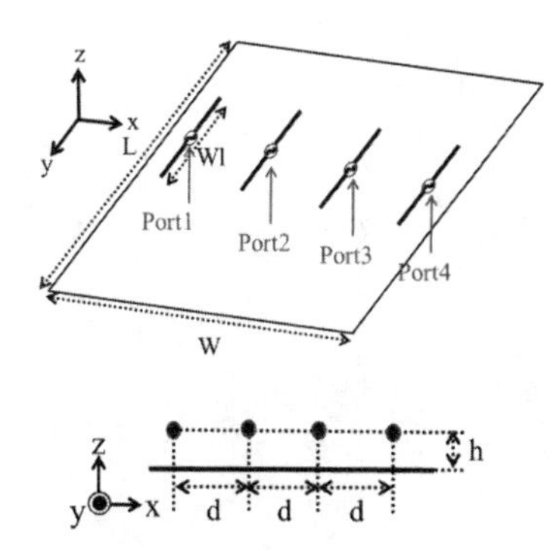

Figure 5. Antenna structure
4-element
W=280,L=450,Wl=64.5,d=75,h=37.5[mm]

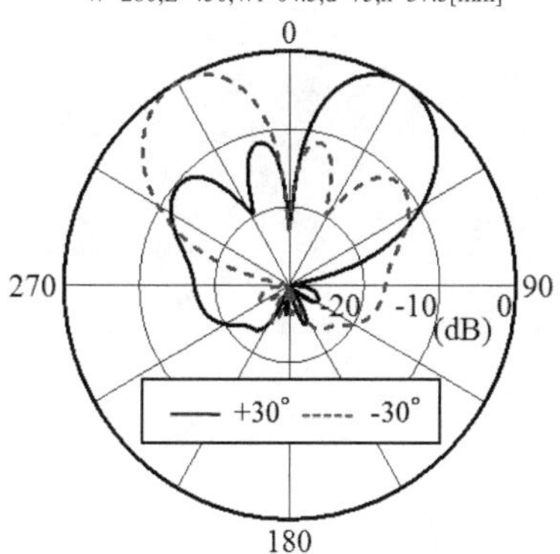

Figure 6. Radiation pattern
zx plane, 4-element

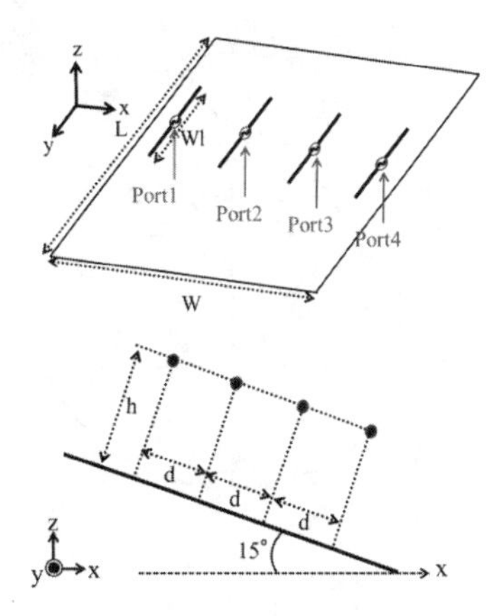

Figure 7. Antenna structure
4-element inclined plane
W=280,L=450,Wl=64.5,d=75,h=37.5,b=7.5[mm]

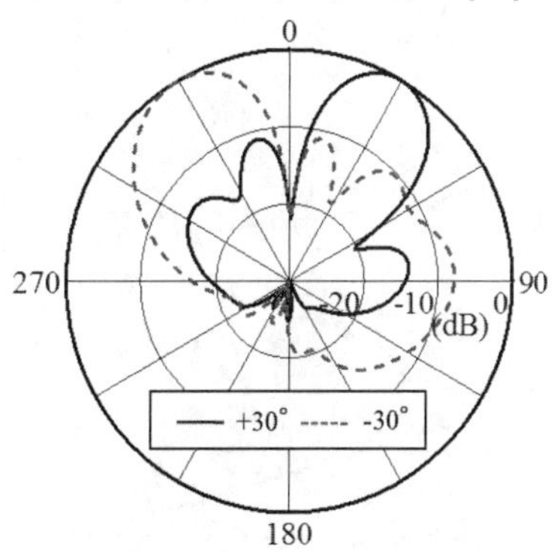

Figure 8. Radiation pattern
zx plane, 4-element inclined plane

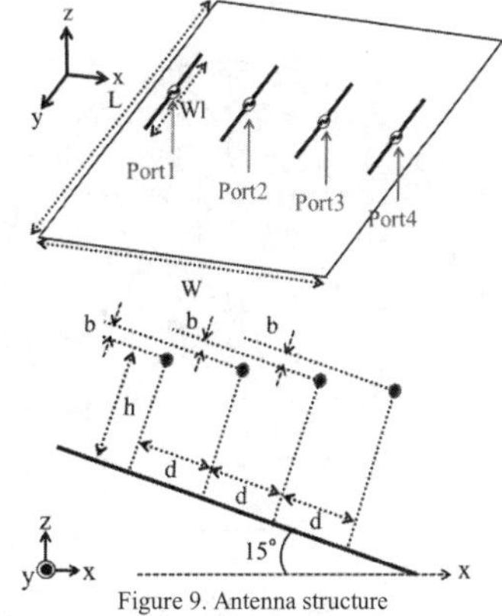

Figure 9. Antenna structure
4-element inclined plane and elements
W=280,L=450,Wl=64.5,d=75,h=26.25,b=7.5[mm]

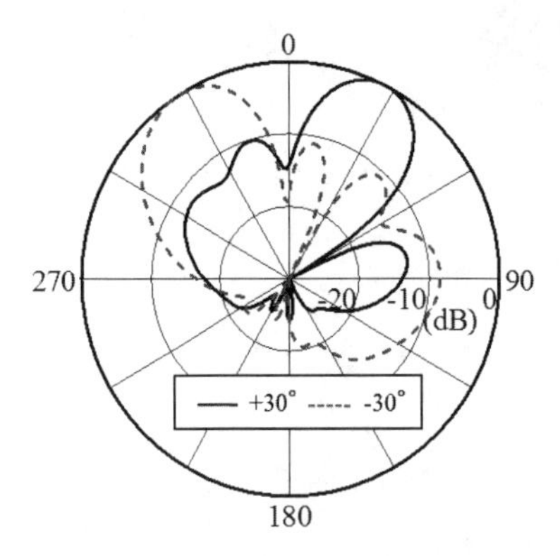

Figure 10. Radiation pattern
zx plane, 4-element inclined plane and elements

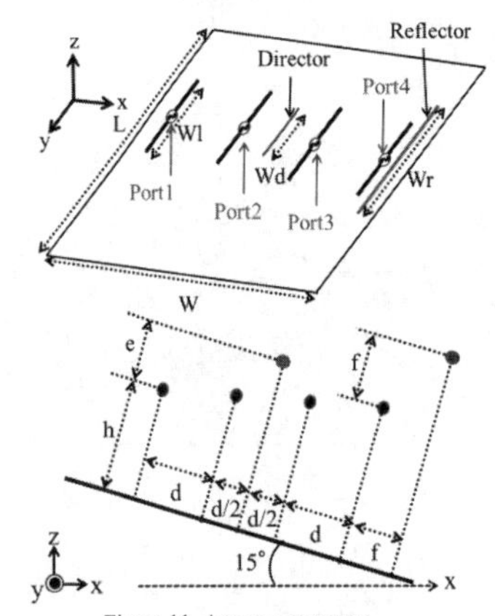

Figure 11. Antenna structure
4-element inclined plane and elements, w/ director and reflector
W=280,L=450,Wl=64.5,Wd=54,Wr=90,d=75,h=26.25,e=33.75,f=22.5[mm]

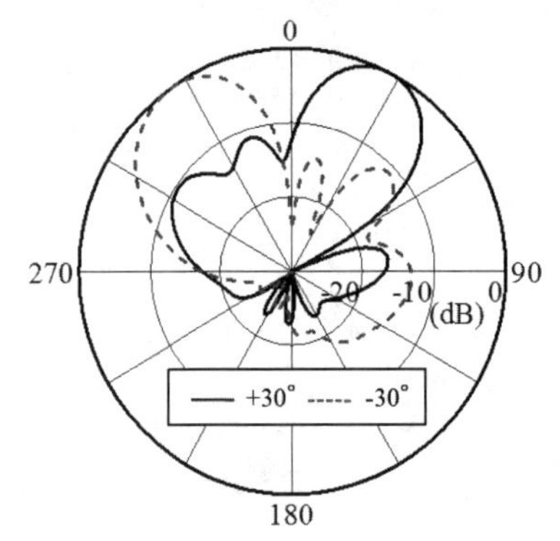

Figure 12. Radiation pattern
zx plane, 4-element inclined plane and elements, w/ director and reflector

WP-C

October 23 (WED) PM

Room E

Best Student Papers Contest

A Method of Moments Analysis and Design of RLSA by Using Only a Dominant Mode Basic Function and Correction Length for Slots

#Nguyen Xuan Tung[1], Jiro Hirokawa[1], Makoto Ando[1], Osamu Amano[2], Takaomi Matsuzaki[2], Shuichi Koreeda[2], Kesato Takahashi[2]

[1] Department of Electrical and Electronic Engineering, Tokyo Institute of Technology, Japan

tungnguyen@antenna.ee.titech.ac.jp

[2] NEC Toshiba Space System, Tokyo, Japan

***Abstract*- In this paper, a well-know Method of Moments (MoM) is introduced together with the definition of a constant slots correction length in order to simulate the behaviors of slots on a rectangular waveguide having narrow wall. Large array radial line slotted antenna (RLSA) is analyzed and designed using this method, and their measured performances suggest the accuracy of the MoM and the importance of the slot correction length.**

I. INTRODUCTION

Recently, as the growing development of computer science and technology, high accurate simulators have been utilized to predict the behaviors of Electromagnetic Field (EMF) based on EM Theory and Analytical Methods. However, when the EM devices are electrically large i.e. 30000 radiating elements [1], those simulators still require a big amount of computational time as well as a large virtue memory and a super high-speed processor. To this end, conventional and well-know MoM turns out to be advantageous.

For slotted array antennas, estimation of coupling characteristics of the slots is very important for an accurate and fast design. Basically, slot coupling can be controlled by geometrical parameters of the slots (length, width and shape). Dependence of slots coupling upon their lengths was thoroughly studied and presented by Hirokawa [2]. On the other hand, an equivalent slot length was introduced to compensate the modeling errors considering the practical slots have round-ended shape [3].

This paper is ordered as follow: the estimation of a newly defined slot correction length to enhance the accuracy in slot coupling analysis of slots is reported at first, and follows by the fast analysis/ design procedure of a large array RLSA in Ka band based on the MoM with consideration of the slot correction length. Finally, measured performances of the fabricated antennas validate the use of the slot correction length as well as the analysis/ design procedure.

II. SLOT CORRECTION LENGTH

In MoM analysis, slot is substituted by a magnetic current M_k (k = 1, 2), and this current distribution is longitudinally expanded by a number of so call basis functions, which are normally in sinusoidal form as in (1), (2).

$$\mathbf{M}_k = \sum_{i=1}^{N_b} A_{ki}\mathbf{m}_i \qquad (1)$$

$$\mathbf{m}_i = \hat{\xi}\sin\{\frac{i\pi}{l}(\xi+\frac{l}{2})\} \qquad (2)$$

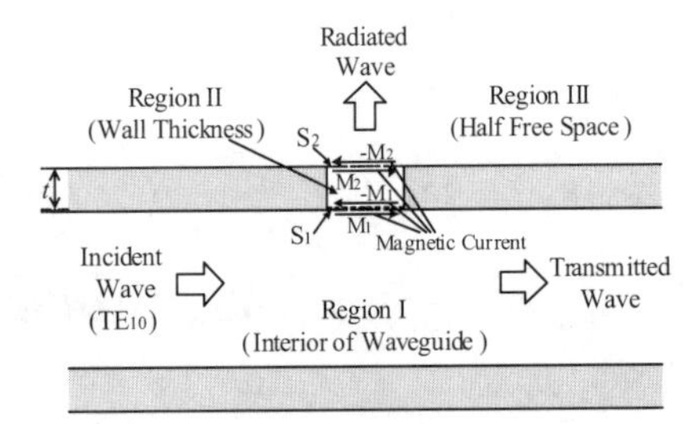

Figure 1: Canonical regions and equivalent slot in MoM

Fig. 1 shows the MoM modeling of a slot cut on a rectangular waveguide. The number of basic functions N_b determines the accuracy of the MoM analysis, the more number is the better accuracy can be achieved. However, in order to reduce the calculation time, only a dominant mode basic function is utilized together with a slot correction length, and this combination can produce the slot coupling just as precise as the use of multi-mode basic functions. This matter is well reported by Ando et al., [4].

Fig. 2 shows the unit slot pair model for design of RLSA. In this particular case, this model consists of a double layer waveguide structure and two orthogonal slots on top. The periodic boundary walls are applied to take into account the mutual coupling between adjacent slots in circumferential direction-ϕ. Fig. 3 presents the coupling factor analysis; the black, red, and dotted lines indicate the results obtained by MoM with 1-mode basic function (N_b = 1) i.e., 1-mode MoM, MoM with multi-mode basic functions (N_b = 15) i.e., multi-mode MoM, and High Frequency Structure Simulator – HFSS, respectively. While the multi-mode MoM and the HFSS lines are in reasonable agreement, it is observed that the 1-mode MoM line differs from those two. Furthermore, in order to produce the same coupling, slots analyzed by 1-mode MoM-

978-7-5641-4279-7

Radiation
slot pair
#3
Periodic boundary
wall
Periodic
boundary
wall
#2
#1
S_ϕ

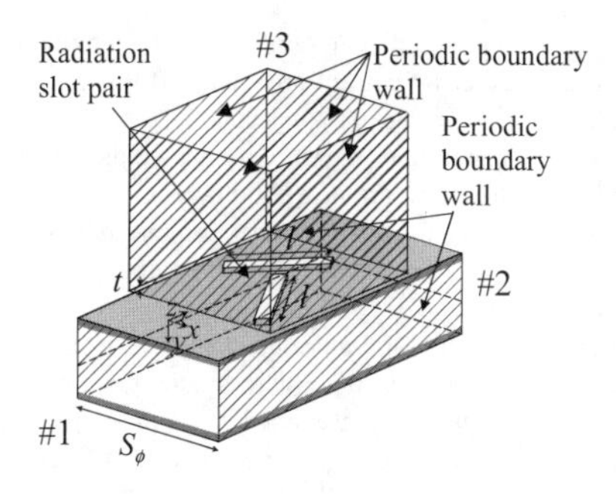

Figure 2: Unit pair model for MoM analysis

Coupling α[1/m]
Slot length [mm]
MoM 1-mode
HFSS
MoM multi-mode
MoM 1-mode w. 90μm shift

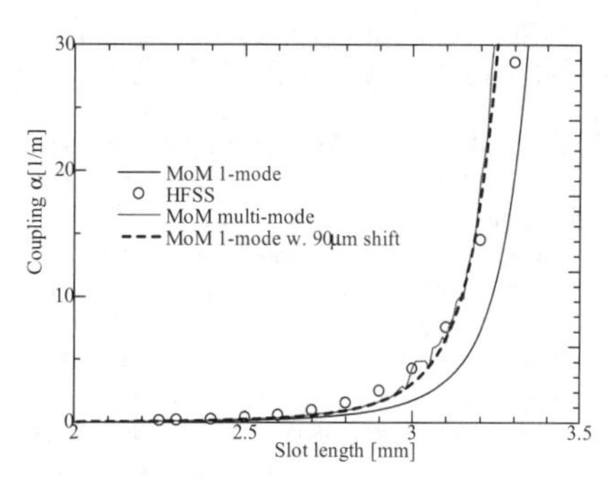

Figure 3: Slot coupling analysis

Slot plate
Skin QFRP
Adhesive
Honeycomb
core
Bottom plate
900mm

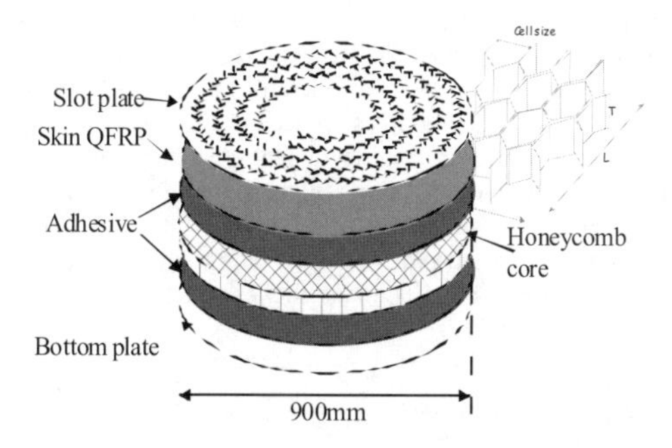

Figure 4: Structure of Multilayer RLSA with honeycomb

Step I: Analyze the unit slot pair model assuming a periodic boundary condition to obtain the slot coupling α and slow wave factor ξ. The equivalent double layer model is used in this analysis [14]-[15].
Step II: Arrange slots spirally on the circular aperture to realize the specified coupling α at respective position ρ and ϕ. The effective dielectric constant εeff is used to define the spiral pitch.
Step III: Extract a 1-dim linear slot array model from the slot array designed in Step II, and then analyze it by MoM. The equivalent double layer model is applied in this analysis.
Step IV: Assign the excitation coefficients for all the slots in the circular aperture of RLSA by extrapolating those analyzed for a 1-dim linear array in Step III.
Step V: Calculate the far-field pattern and the directivity using the slot positions in Step II and the excitation coefficients in Step IV.

Figure 5: Fast Design/ Analysis procedure for large array RLSA

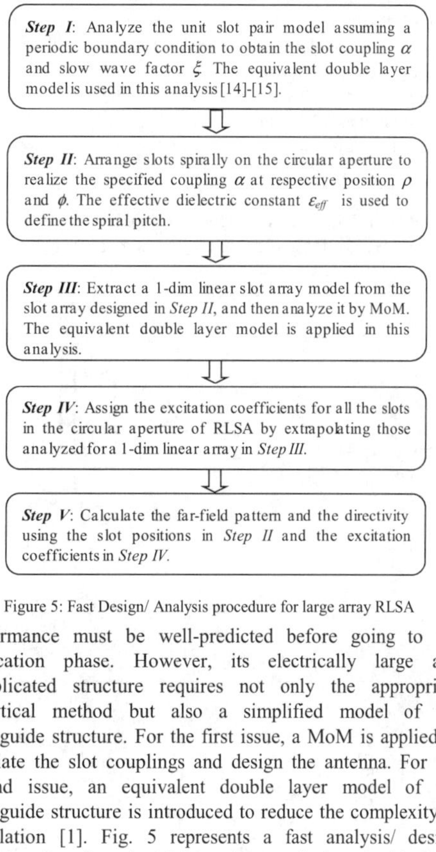

are always longer than that of those analyzed by multi-mode MoM or HFSS. Put differently, couplings produced by 1-mode MoM would be just the same as ones obtained from multi-mode MoM if we just shortened the slots a constant length of Δl. From fig. 3, we can derive a $\Delta l = 90\mu m$ for the slot correction length. The constant shift of $-90\mu m$ (dashed line) from the 1-mode MoM is almost identical with one produced by multi-mode MoM, which reassures our prediction.

III. Fast Design/ Analysis of a High Gain RLSA

A. *Antenna Structure*

A high gain, light weight planar slotted array antenna is required for space application [1]. To enhance the mechanical strength and reduce the antenna's weight, a honeycomb-type structure was utilized. Together with some additional adhesive and Quartz Fibre Reinforced Plastic (QFRP) – made skin layers, it creates a multilayer structure of the oversized radial waveguide. The antenna is fed by a simple coaxial line which produces a cylindrical wave travelling inside the multilayered waveguide. On the top metal plate, thousands of slots are etched to radiate the EM field. Fig. 4 details the antenna's structure and it is also well reported in [1].

B. *Fast Analysis/ Design procedure and Role of the slot correction length*

Since this RLSA is for practical use in outer space and the fabrication cost is expensive, it is important that the antenna's performance must be well-predicted before going to the fabrication phase. However, its electrically large and complicated structure requires not only the appropriate analytical method but also a simplified model of the waveguide structure. For the first issue, a MoM is applied to simulate the slot couplings and design the antenna. For the second issue, an equivalent double layer model of the waveguide structure is introduced to reduce the complexity in calculation [1]. Fig. 5 represents a fast analysis/ design procedure based on the MoM and the equivalent double layer model.

It is noted that in the 1st and 3rd steps, the MoM is used, and so should be the slot correction length in order to obtain high accuracy of slot couplings. The correct use of the slot correction length $\Delta l = 90\mu m$ can be stated as follow:

- Slot couplings are calculated by 1-mode MoM without considering the slot correction length. At this state, we have typical slot with a length of l_{ana}, for example.
- Slots obtained from previous analysis are shortened by a constant Δl=90μm, and sent to the maker for the fabrication. The fabricated length is $l_{fab} = l_{ana} - \Delta l$.
- Extracted slots with typical length of $l_{fab} = l_{ana} - \Delta l$ are used for the 1-dimensional linear array analysis. To obtain high accuracy, slot with a length of l_{fab} should be analyzed by multimode MoM. However, 1-mode MoM can also be utilized by just simply adding Δl to l_{fab}. In other words, at this state, by applying 1-mode MoM to analyze slot with length of $l_{ana} = l_{fab} + \Delta l$, accurate results are still produced.

Fig. 6 explains how the slot correction length was applied in this paper and the correct use of it is left for the future. It suggests that in design/ analysis procedure, the slot correction length can be neglected and the 1-mode MoM still produce accurate results for predicted antenna performance. However, the slot correction length must be taken into account in the fabrication phase. Notations 1, 2 and 3 are used to differentiate the antenna performance results, which will be discussed latter.

IV. Measured Antenna Characteristics

A 900mm RLSA was designed and fabricated. The design frequency is 32GHz and expected gain is more than 44.1dBi. -For the fabrication, the equivalent slot length [3] was introduced but the slot correction length [4] was not taken into account.

Fig. 7 compares the measured and predicted reflection coefficients. These two are in a good agreement. At the design frequency of 32GHz, measured reflection coefficient is about -17dB while it is less than -15dB in the 31.2~32.3GHz frequency band.

Antenna directivity and gain are presented in Fig. 8. Black and red lines account for the directivity and gain, respectively. Directivity and gain are predicted without considering any correction length and are plotted by the dashed lines (noted as no. 1). Peak frequency is 32.1GHz for both directivity and gain. The measured performance is 200MHz shifted to lower frequency since the slot correction length was neglected in fabrication process, as indicated by the solid lines (noted as no. 2). The prediction with consideration of the slot correction length (dashed-dotted lines, noted as no. 3) is in a good agreement with the measurement, which reassures the necessary of slot correction length in the fabrication. As for antenna specification, a gain of 44.6dBi at 32GHz already satisfies the system requirement, even though a considerable loss of 3-4dB still exists due to the lossy honeycomb material [5].

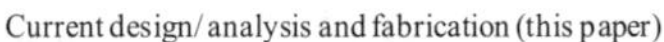

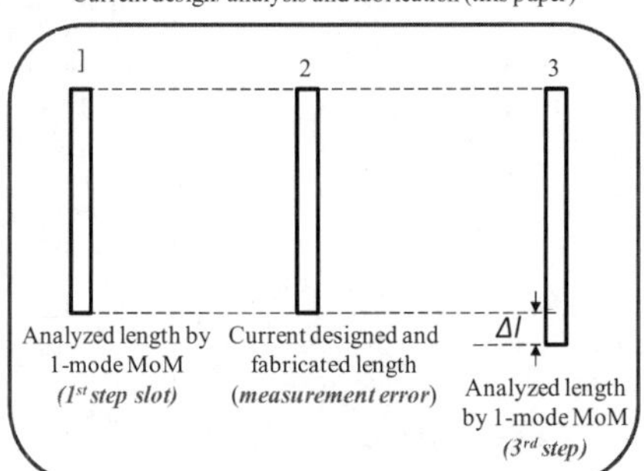

Figure 6: Role of the slot correction length in design/ analysis procedure and fabrication phase

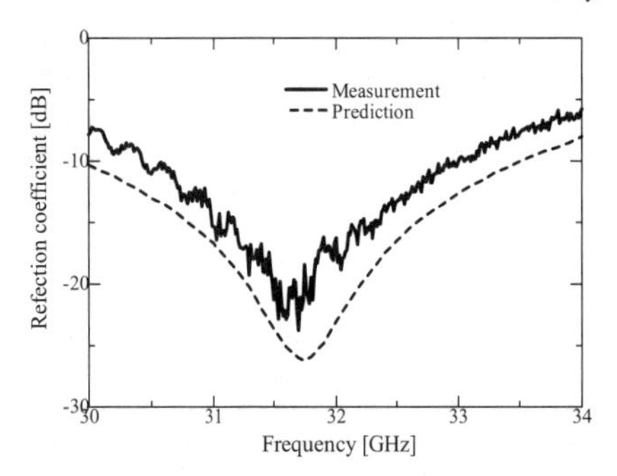

Figure 7: Measured reflection coefficients

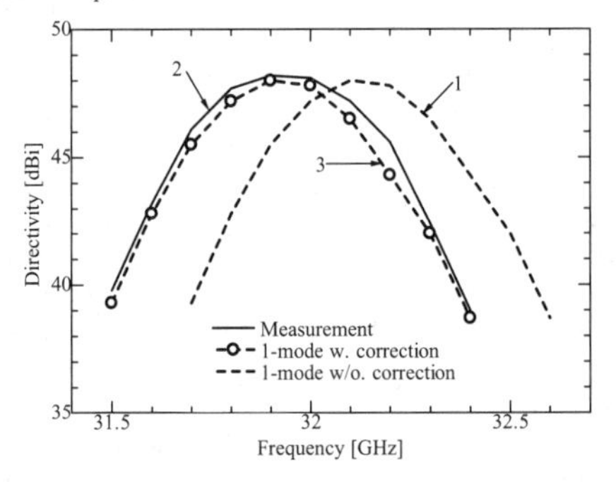

Figure 8: Measured and predicted performance

V. Conclusion

A fast analysis/ design procedure of a large array RLSA was developed based on the MoM and a simplified, equivalent double layer waveguide structure. A constant slot correction length was introduced to enhance the accuracy in MoM analysis using only a dominant mode of the basic function. The measured antenna performance is in a good agreement with the predicted one, which explains our procedure and assures the importance of the use of the correction length. In practical point of view, a RLSA which can produce a very high gain of more than 44dBi and a super light weight of about 1kg is advantageous for space applications.

Acknowledgment

This work was conducted in part as "the Research and Development for Expansion of Radio Wave Resources" under a contract with the Ministry of Internal Affairs and Communications. It was also supported by the JAPAN Aerospace Exploration Agency (JAXA). The authors thank Mr. Harada at NEC Toshiba-Space Systems for his support with the experiments.

References

[1] Tung Nguyen et al., "Characteristics of a High Gain and Light Weight Radial Line Slot Antenna with Honeycomb Structure in 32GHz band for Data Link in Space Exploration". *The 2012 International Symposium on Antennas and Propagations*, A07_1010, Oct 2012, Nagoya, Japan.

[2] J. Hirokawa, "A study of slotted waveguide array antennas," *Doctoral Dissertation*, Dept. Electronics Eng., Tokyo Institute of Technology, 1993.

[3] M. Zhang, T. Hirano, J. Hirokawa and M. Ando, "Analysis of a waveguide with a round-ended wide straight slot by the method of moments using numerical-eigenmode basis funtions," *IEICE Trans. Commun.* vol.E87-B, no.8, Aug. 2004.

[4] J. H. Lee, J. Hirokawa and M. Ando, "Practical slot array design by Method of Moments using one basic function and constant correction length," *IEICE Transactions on Communication*, vol.E94-B, No.1, pp.158-165, Jan. 2011.

[5] R.S. Jayawardene, T. X. Nguyen, Y. Takano, K. Sakurai, T. Hirano, J. Hirokawa, M. Ando, O. Amano, S. Koreeda and T. Matsuzaki, Estimation and Measurement of Cylindrical Wave Propagation in Parallel Plate with Honeycomb Spacer for Space Use in mm-Wave RLSA". *The Asian Pacific Microwave Conference*, Dec 2012, Kaohsiung, Taiwan.

A Compact and Low-profile Antenna with Stacked Shorted Patch Based on LTCC Technology

Yongjiu Li[1], Xiwang Dai[1], Cheng Zhu[1], Gang Dong[3] , Chunsheng Zhao[2] , and Long Li[1], *Senior Member, IEEE*

[1]School of Electronic Engineering Xidian University Xi'an 710071, Shaanxi, China
[2]Shanghai Spaceflight Electronic Communication Equipment Research Institute
[3]School of Microelectronics Xidian University Xi'an 710071, Shaanxi, China

Abstract **- In this letter, a compact stacked shorted patch antenna is proposed using low temperature cofired ceramic (LTCC) technology. The height of the antenna is decreased due to the shorted pin, and this greatly improves the polarization performance of the antenna. At the same time, the height (about λ /25) of the antenna and its polarization state are determined by both the feeding and the shorting pin of antenna integrated. The antenna is mounted on a 44 × 44 mm^2 ground plane to miniaturize the volume of the system. The designed antenna operates at a center frequency of 2.45GHz, and its impedance bandwidth is about 200 MHz, resulting from two neighboring resonant frequencies at 2.41 and 2.51 GHz, respectively. The average gain in the frequency band of interest is about 5.28dBi.**

I. INTRODUCTION

Recent developments in microelectronic technology have created a probability for system-on-package (SoP) antennas [1] in a smaller size. As a result, low temperature cofired ceramic (LTCC) package technology is becoming more and more popular for the production of highly integrated, specifically complicated multilayer modules and antennas. At the same time, this technology is applicable to complex structures and an arbitrary number of layers for its flexibility. However, due to the high dielectric constant of LTCC materials, the size of antenna is decreased. Consequently, it also reduces the impedance bandwidth. The contradiction between the size reduction and bandwidth enhancement of the antenna for a practical wireless communication system is more difficult to be solved, therefore this problem is more challengeable.

In order to reduce the antenna size while maintaining the bandwidth even improving it, some techniques have been proposed and reported by researchers for designing compact broadband antennas over the past several years, e.g., slotting ground plane or embedding narrow slits at the patch's non-radiation edges [2, 3], stacked shorted patches [4-6], and aperture-coupled stacked patch [7]. When slots are embedded in the ground plane or the aperture-coupled feed method is applied for exciting stacked-patch antennas for broadband operation, the backward radiation is increased compared with the traditional patch antenna. This increase in the back-radiation is contributed by the embedded slots in the ground plane and the decreased ground-plane size in wavelength. The backscattered component is detrimental to the chip in the package cavity, that is to say, the back-radiation strengthens electromagnetic coupling between the antenna system and the system on chip. In addition, stacked patch antennas are used for LP or CP, and dual-polarization. However, the total height is still high and not enough compact.

In this letter, a low temperature co-fired ceramic (LTCC) antenna is investigated for a 2.45GHz wireless transceiver module. At first, the size reduction technique using shorting pin is introduced. Note that the position has a significant effect on the polarization state. The bandwidth broadening technology using stacked patches is then described to obtain a broadband microstrip patch antenna. It is noted that the location of the feed point is different from the previous antennas for LP. By placing a shorting pin along the null in the electric field across the patch, the resonant length is reduced and the polarization performance within the frequency band is also greatly improved. Thirdly, the influences of the structure dimensions of the antenna on the performance of the antenna are considered and carefully discussed.

II. ANTENNA STRUCTURE AND OPERATION MECHANISM

The geometry of the proposed LTCC multilayer antenna in this paper is shown in Fig. 1. The shorted stacked-patch antenna is embedded on the top of a RF frond-end module in an LTCC multilayer package. For the sake of simplicity, we consider a probe-fed square patch antenna on a square grounded substrate with thickness h_f=0.79mm. In order to miniaturize the antenna, the shorting technique is applied to the design scheme in this letter. A wide impedance bandwidth is achieved with the stacked patches. Both the upper layer and the lower layer are square patches, but it should be worth noting that the center of the lower patch is offset 2mm from the coordinate center along x and y directions, respectively. The antenna ground plane is under the lower patch. Just for fabrication convenience, all the LTCC substrates use the same material, with the relative dielectric constant $\varepsilon_r = 7.8$ and a dielectric loss tangent $\delta = 0.0015$. In order to reduce patches' size and the height of the antenna, the probe is placed at the corner of the lower patch. The position of the shorting pin depends on the location of the feed, because the antenna polarization state is usually CP due to the phase-angle difference of the two orthogonal components of the electric field when the feed is placed at the corner of the patch without the shorting pin. But the desired polarization in this paper is LP. By adjusting the position of the shorting pin to the

978-7-5641-4279-7

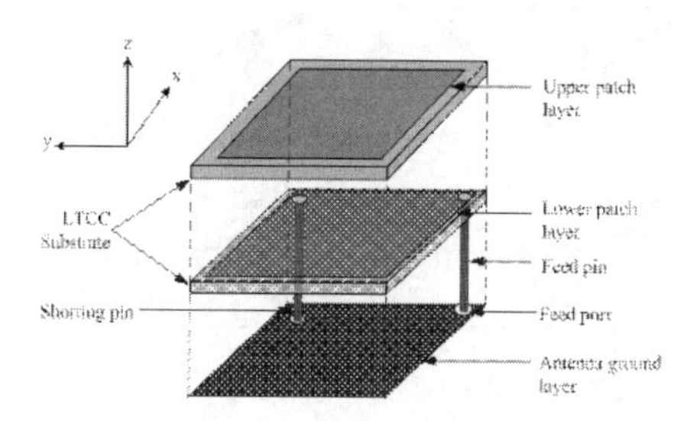

Fig. 1. Expanded view of the proposed antenna integrated in package.

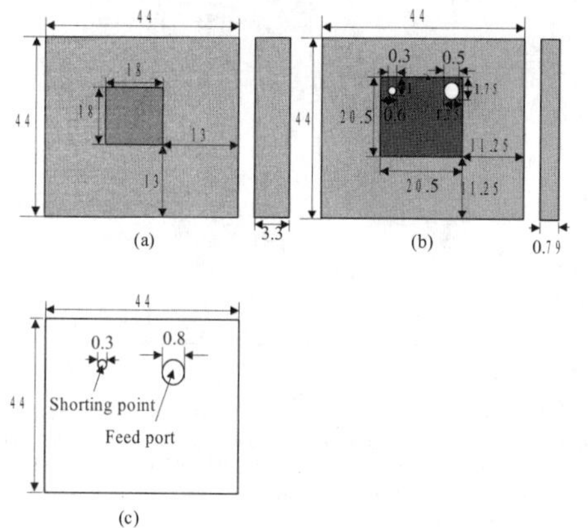

Fig. 2. Design parameters of each layer, (a) Upper patch layer, (b) Lower patch layer, (c) Antenna ground layer. All are in mm.

optimum position in y-direction and x-direction and canceling the phase-shift of the two orthogonal components, good linear polarized radiation over a wide operating bandwidth can be achieved.

The patch sizes were initially estimated by using simple approximations and then patch sizes and positions are optimized by GA algorithm. The upper patch and the lower patch interact with each other, which must be adjusted in length for the proper impedance matching bandwidth. The position of the shorting pin must be also modified appropriately for a wide impedance bandwidth and the desired state of the polarization.

To meet the design requirement of the antenna, the antenna thickness should be less than 5mm without the supersubstrate thickness, while the antenna ground plane's size is assumed to be $44\times44mm^2$. The sizes of the upper patch and the lower patches are determined to be $18\times18mm^2$ and $20.5\times20.5mm^2$, respectively. Due to the different sizes of the two patches and the mutual coupling between each other, these two patches resonate at different modes. At the same time, the patches of variable size have a profound impact on the impedance matching and their influences on the performance of the antenna will be carefully considered and discussed in Section III.

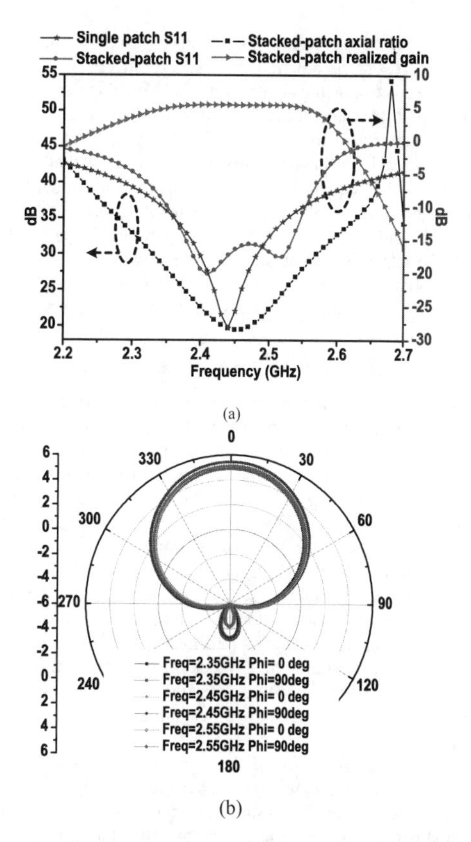

Fig. 3. The calculated characteristics of the proposed (a) S11 including single patch and stacked-patch antenna, stacked-patch antenna realized gain and its axial ratio versus frequency, and (b) Radiation pattern at 2.45GHz of stacked-patch antenna.

The detailed configuration of the proposed antenna as well as the design parameters of each layer is shown in Fig. 2. It is shown that the total height of the antenna is 4.09mm illustrated in Fig. 2. Instead of microstrip line feed, the coaxial feed could be used to excite the lower patch and produce two operating modes. The center frequency is at about 2450MHz, and the simulated S11 is shown in Fig. 3 (a). The substrate thickness of the single patch antenna is 12mm, and the size of the patch is 17.1mm. Comparing the stacked-patch antenna and the single patch antenna, from Fig. 3 (a), we can see that the height of the single patch antenna is much larger than that of the stacked-patch antenna with a same impedance bandwidth. The results are shown in Fig. 3, where the two neighboring resonant frequencies of the designed antenna are 2.41GHz and 2.51GHz for the lower and upper patches and the mutual coupling between them, and -10dB absolute impedance bandwidth of S11 is 200MHz from 2.35GHz to 2.55GHz. The impedance matching characteristic at the low resonant frequency is better than the high one because the current distributing in the lower patch is changed due to the shorting pin. For the present proposed structure, the antenna achieves two resonant frequencies, resulting from the mutual coupling between the upper and lower patches. In this case, the impedance bandwidth, defined by the -10-dB S11, is about 8.16% referenced to the center frequency at 2450 MHz. The minimum axial ratio is 19.3 dB. Fig. 3 (a) shows that the gains across the frequency band are nearly unchanged. The maximum gain of the proposed antenna is 5.48 dB and the average gain is 5.28 dB across the operating frequency band. The broadside radiation patterns are observed in Fig. 3 (b). The broadside patterns are symmetrical because of the modeling structure. The beamwidth within the required gain is greater than 100°.

III. Antenna Parameters Analysis

As described above, the proposed antenna structure is determined owing to a relatively small volume. Considering the mutual coupling between the two patches, the proper sizes of the patches have to be selected. The bandwidth and impedance matching of the antenna are influenced by various structure parameters. That is to say, the performance of the stacked shorted patch antenna is mainly determined by the characteristics of the configuration of the patches including dimensions and heights, and the positions of the feeding pin and the shorting pin. In this Section, the effects of these parameters on the antenna performance are studied in detail. To better understand the effects of the parameters on the performance of the proposed antenna, only one parameter will be varied at a time, while the others are kept unchanged unless especially indicated.

Firstly, the effects of the dimensions of the two patches on the antenna performance are discussed. Fig. 4 (a) compares S11 in different dimensions of the lower patches. The fundamental mode disappears and the impedance bandwidth is decreased as the size of the lower patch increases. For increasing the upper patch L_1 shown in Fig. 4 (b), it is seen that the fundamental resonant frequency is quickly lowered, at the same time, the high mode vanishes. Thus, we know from Fig. 4 (a) and (b) that the upper patch should be increased when the lower patch increases for a good impedance matching in the frequency band of interest.

The thickness of the dielectric substrate under the lower patch of the antenna, h_1, is then considered. Generally speaking, the thicker the substrate, the wider the impendence bandwidth. It is well understood that the inductance of the feeding pin becomes larger and the impedance matching becomes better as the height of the feeding pin increases. With such a large length, the effect of the inductive reactance of the probe pin in the air substrate on the impedance matching is large and makes the antenna match well in the operating frequency band of interest. Fig. 4 (c) gives the effect of substrate thickness under the lower patch on S11 of the proposed antenna simulated using HFSS software. In addition, the interaction between the driven and parasitic patches has a profound impact on the bandwidth of the antenna.

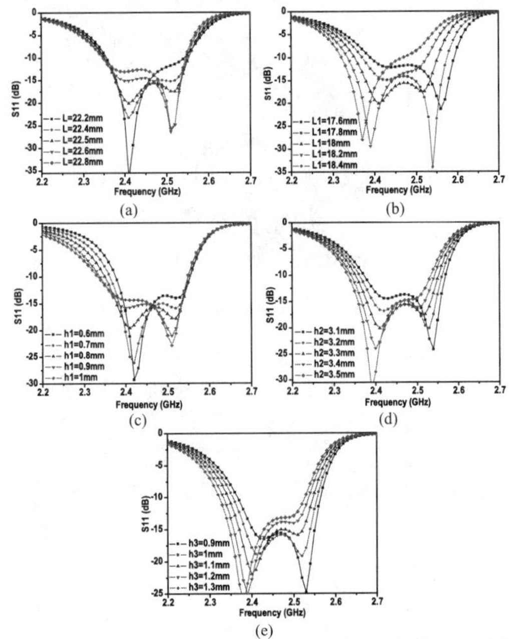

Fig. 4. S11 characteristics, (a) Lower patch L variation, (b) Upper patch L1 variation, (c) Thicknesses of the dielectric substrate under the lower patch of the antenna variation, (d) Separations of the lower and upper patches variation, (e) Thicknesses of the supersubstrate variation

The separation of the lower and upper patches, h_2, has a heavy influence on the antenna performance. The smaller or larger separation strengthens or weakens the coupling of the two patches, which results in a relatively narrow impedance bandwidth. Thus, a proper height should be selected to obtain a wide bandwidth. Simultaneously, a thinner substrate is also necessary to miniaturize the antenna's size. Fig. 4 (d) shows the effect of varying the height of the parasitic patch across the bandwidth of the antenna. The EM coupling between the driven and parasitic patches strengthens as the thickness h_2 increases. In order to enhance the gain and bandwidth of the antenna, a thin supersubstrate, which has a thickness of h_3, is introduced to cover the upper patch. One of its advantages is that the supersubstrate can avoid the abrasion of the antenna. Different thicknesses of the supersubstrate are calculated by HFSS and the results are illustrated in Fig. 4 (e). The characteristics of S11 with different thicknesses of the supersubstrate are similar to that reported simulated S11 for different thicknesses of the supersubstrate show similar results which are obtained with different substrate thicknesses between the lower and upper patches.

In previous description, the thicknesses of the LTCC materials and the patches have an impact on the impedance

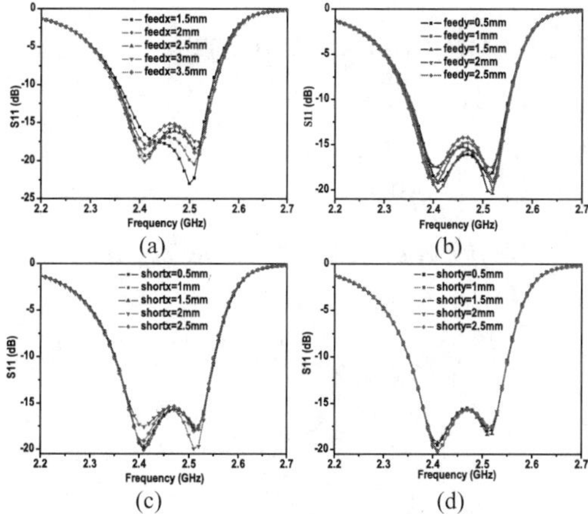

Fig. 5. S11, (a), (b) S11 in different offset of the feed location, (c), (d) S11 with variable position of the shorting pin

match and the bandwidth. Finally, another key factor affecting the antenna performance is the loci of the feeding and shorting pins. Due to the feeding pin placed in the corner of the lower patch without the shorting pin, a 90° phase difference is produced and CP is obtained. Since the desired polarization is LP, a proper position of the shorting pin is necessary and important for achieving the desired polarization state. The results of the proposed antenna with different positions of the shoring and feeding pin are shown in Fig. 5. When the position of the feeding pin is offset in a specific range of a distance *feedx* and *feedy*, from the corner of the patch along x and y direction, respectively, the stacked-patch antenna can excite two resonant modes with good impedance matching shown in Fig. 5 (a) and (b) and S11 across the frequency band is nearly unchanged. The distances, *shortx* and *shorty* along x and y direction, are away from another corner of the lower patch. The simulated results are illustrated in Fig. 5(c) and (d). It is observed that the variations of S11 are not sensitive to the movement of the feed position from 0.5mm to 2.5mm for *shortx* and *shorty*.

IV. CONCLUSION

In this paper, a compact and low-profile stacked-patch antenna using low temperature cofired ceramic (LTCC) package technology is designed. Broadband linear polarization radiation is achieved and the two modes together give a fractional impendence bandwidth of 8.16% with respect to the center frequency at about 2.45GHz. The stable radiation patterns are achieved. The total height of the proposed antenna is 4.09mm and the average gain is 5.28dBi across the operating frequency band. The parameters analyses are carefully conducted in this paper for design guidance. The proposed antenna is suitable for integration into the chip package using LTCC package technology.

ACKNOWLEDGEMENTS

This work is supported by National Natural Science Foundation of China under Contract No. 61072017, and supported partly by the Program for New Century Excellent Talents in University of China, by National Key Laboratory Foundation and Fundamental Research Funds for the Central Universities (K5051202051, K5051302025, K5051302030).

REFERENCES

[1] Sang-Hyuk Wi, Yong-Bin Sun, In-Sang Song, Sung-Hoon Choa, Il-Suek Koh, Yong-Shik Lee,and Jong-Gwan Yook "Package- Level Integrated Antennas Based on LTCC Technology," *IEEE Trans. Antennas Propagat.* **54**, 2190–2197, August 2006

[2] J. S. Kuo and K. L. Wong, "A compact microstrip antenna with meandering slots in theground plane," *Microwave Opt. Technol. Lett.* **29**, 95–97, April 20, 2001S.

[3] S. T. Fang, K. L. Wong, and T. W. Chiou, "Bandwidth enhancement of inset-microstripline-fed equilateral-triangular microstrip antenna," *Electron. Lett.* **34**, 2184–2186, Nov.12, 1998.

[4] J. Ollikainen, M. Fischer, and P. Vainikainen, "Thin dual-resonant stacked shorted patch antenna for mobile communications," *Electron. Lett.***35**, 437–438, March 18, 1999.

[5] L. Zaid, G. Kossiavas, J. Dauvignac, J. Cazajous, and A. Papiernik, "Dual-frequency and broad-band antennas with stacked quarter wavelength elements," *IEEE Trans. Antennas Propagat.* **47**, 654–660, April 1999.

[6] R. Chair, K. M. Luk, and K. F. Lee, "Miniature multiplayer shorted patch antenna," *Electron.Lett.***36**, 3–4, Jan. 6, 2000.

[7] D. M. Pozar and S. M. Duffy, "A dual-band circularly polarized aperture-coupled stacked microstrip antenna for global positioning satellite," *IEEE Trans. Antennas Propagat.* **45**, 1618–1625, Nov. 1997.

An Investigation on the Gain of Folded Reflectarray Antennas with Different *F/D*s

Mei Jiang, Student Member, IEEE, Yan Zhang, Member, IEEE, Wei Hong, Fellow, IEEE, and Shun Hua Yu, Student Member, IEEE

State Key Laboratory of Millimeter Waves, School of Information Science and Engineering, Southeast University, Nanjing, 210096, P. R. China

Abstract — In this paper, an investigation on the effect of *F/D* ratio on gain for folded reflectarray antenna (FRA) is presented. Quantization phase errors which are the major contribution to losses of reflectarray are presented with different *F/D* ratios. In order to reduce the quantization phase error, a folded reflectarray antenna with high *F/D* ratio is preferred. A pair of FRAs with *F/D* of 0.5 and 1 is chosen to fabricated, measured and compared at 42GHz, respectively. A gain improvement of 0.8dB is achieved when the larger *F/D*, namely 1, is adopted to the antenna.

Index Terms — Folded reflectarray antenna, phase shifting surface, parabolic antenna, quantization phase error.

I. Introduction

Recently, millimeter-wave band has attracted more and more interests of wireless communications due to its abundant spectrum. China has proposed Q-LINKPAN as a high speed transmission standard proposal for both short-range and long-range millimeter-wave communications. For developing such short-range and point to point communication systems high gain antennas are most preferred.

Folded Reflectarray antenna (FRA) which was proposed in the 2000s [1]-[4], is attractive high gain antenna due to its low profile, low cross polarization and easily integrated with planar circuits. The antenna system consists of a primary source illuminating a polarizing grid which reflects the linearly polarized spherical wave toward the main reflectarray. The main reflectarray is designed to focus the beam and rotate the incident polarization by 90° to make sure the radiated wave can pass through the polarizer. Due to the folded wave trace, the FRA can reduce its depth by a factor of 2.

According to the reported research results on gain and loss mechanisms of folded reflectarray antenna in [5]-[6], phase errors are the major contributions to losses. For the reflectarray design, the phase distribution on the phase shifting surface (PSS) are often assumed to be continuous, but actually the PSS is compensated by tens of thousands of reflecting cells, so the phase distribution is discretized into many staircase sections, the quantization phase error [7] between the discretized phase cells and the continuous phase distribution on the reflecting surface will cause a lot of losses of the

This work was supported in part by National 973 project 2010CB327400 and in part by Natural Science Foundation of Jiangsu province under Grant SBK201241785.

978-7-5641-4279-7

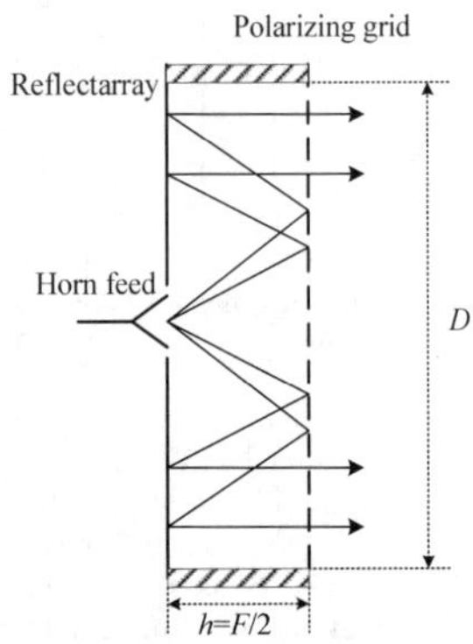

Fig. 1. Principle of FRA and the definition of focus *F* and reflecting surface diameter *D*.

reflectarray antenna. In order to eliminate the quantization phase error, a size reduction of reflector element by choosing high permittivity substrate is adopted in [5], but the high permittivity substrate with high loss tangent also introduces extra dielectric loss, so the gain improvement is not significant. However, from another point of view, the quantization phase error is reduced with the decreased steepness of phase shifting curve. A high *F/D* ratio resulting in a smooth phase shifting curve will causes less phase errors. Thus, an appropriate choose of *F/D* will reduce the loss of folded reflectarray antenna.

In this paper, we present two FRAs with different *F/D*s for Q-LINKPAN application at 42GHz and investigate their gain performance. We firstly definite the quantization phase error and present the relationship between *F/D* and the quantization phase error. And then we study its effect on gain and 3dB beamwidth by simulating two pairs of FRAs and their corresponding parabolic antennas with different *F/D*s. By comparing the gains of these antennas, we study the loss reduction of FRA with the increase of *F/D*. Finally, we fabricate the FRAs with *F/D* of 0.5 and 1, and the experiment results are compared with Ref. 1 and Ref. 2.

II. Quantization Phase Error

The main reflecting surface of the folded reflectarray antenna utilizes many radiation elements with different phase

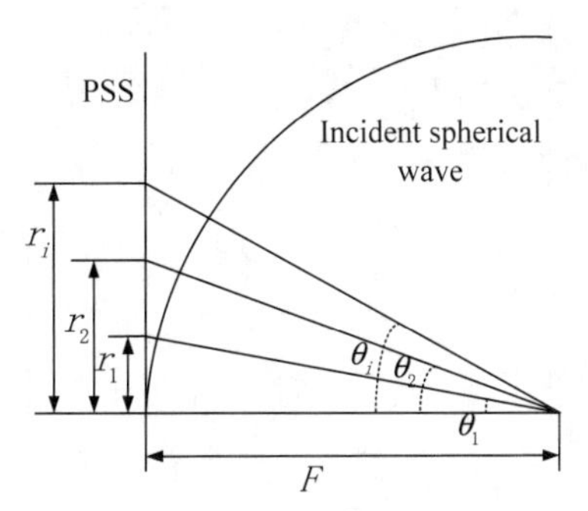

Fig. 2. Wave trace of the classic reflectarray antenna - definition of the phase shifting surface.

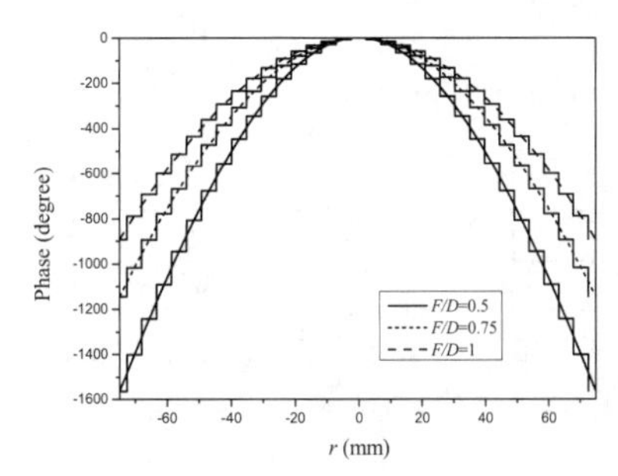

Fig. 3. The compensation phase distributions and the corresponding quantization phase distributions with different F/D.

shifting characteristics to compensate the phase delay from the feed to the reflecting plane by different wave paths. The operation is similar to a parabolic reflector to use its unique curvature to compensate the phase delay and form a planar phase front. The wave trace of the classic reflectarray antenna is shown in Fig. 2. The compensated phase function is determined by [8]:

$$\varphi_i = -k_0 F(1-\cos\theta_i)/\cos\theta_i \quad \text{for } i=1,2\ldots n \tag{1}$$

$$\theta_i = \tan^{-1}(r_i / F) \tag{2}$$

where φ_i is the compensated phase distribution; F is the focus distance; θ_i is the subtend angle from coaxial to the radiation element; r_i is the distance from center of the reflecting plane to the radiation cell; and k_0 is the propagation constant in vacuum.

Fig. 3 shows the required phase shifting on the reflecting surface and the counterpart quantization phase distribution. The root mean square (rms) quantization phase error is defined as:

$$\delta = \left(\frac{\int(\varphi_i - \varphi_0)^2 dS}{\pi(D/2)^2}\right)^{1/2} \tag{3}$$

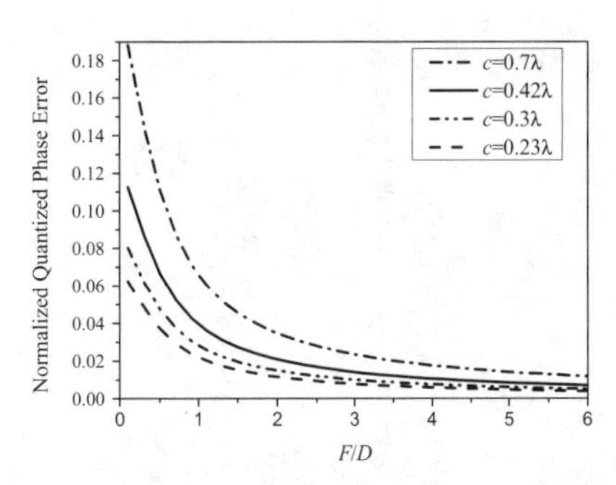

Fig. 4. The normalized phase errors versus F/D with different lengths of square reflecting cells c.

TABLE I
SIMULATED GAIN, BEAMWIDTH AND SIDE LOBE LEVEL OF THE LOSSLESS FRAS AND PARABOLIC ANTENNAS

	Gain (dBi) @42GHz		Beamwidth(°) @42GHz	Sidelobe(dB) @42GHz
	Parabolic antenna	FRA	FRA	FRA
F/D=0.5	34.41	32.86	3.3 ×2.9	-24.2
F/D=1	34.57	33.68	2.7×2.9	-20

where φ_0 is the continuous phase distribution function; and D is the diameter of the reflecting surface. The normalized quantization phase error can be obtain as $\varepsilon=\delta/2\pi$.

Based on the above definition, the normalized phase errors versus F/D with different lengths of square reflecting cells c are illustrated in Fig. 4. The rms quantization phase error reduced with the increasing of F/D due to the decreasing steepness of the phase shifting curve as illustrated in Fig. 3. The quantization phase errors will contribute to the side lobes and further reduce the gains of the reflectarray antennas. Therefore, the folded reflectarray antenna with lower F/D ratio causes more loss due to quantization phase error.

In order to investigate the losses brought by the quantization phase errors with different F/D, the EM simulator of CST Microwave Studio (MWS) is adopted to simulate two pairs of parabolic antennas and FRAs with F/D equaling to 0.5, and 1, respectively. The materials of the simulated antennas are free from losses in order to guarantee the major gain differences between the FRAs and the corresponding parabolic antennas are only contributed by quantization phase errors. The gains, beamwidths and side lobe levels of these antennas are illustrated in table I. The losses caused by quantization phase errors are 1.55dB and 0.89dB, receptively corresponding to F/D of 0.5 and 1. Due to the better compensated phase distribution on the reflecting surface, the FRA with F/D of 1 has narrower beamwidths than the FRA with F/D of 0.5 which resulting in a gain increase of 0.82dB.

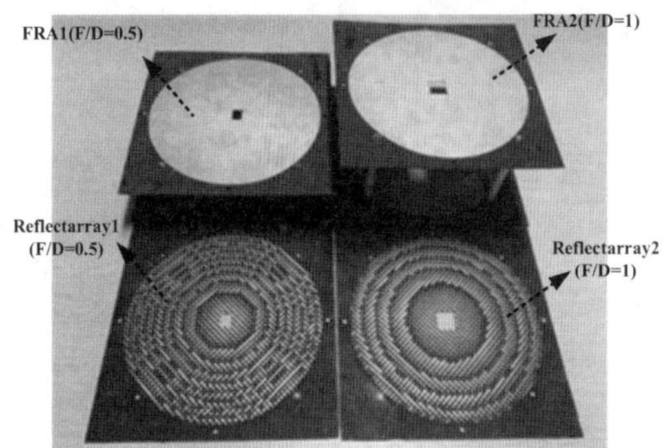

Fig. 5. The photograph of the folded reflectarray antennas.

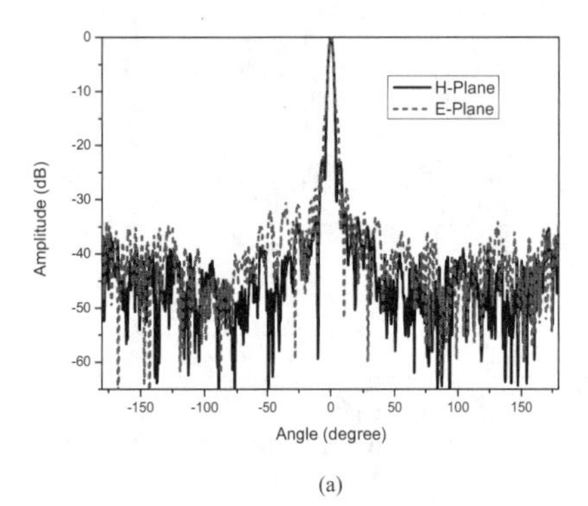

(a)

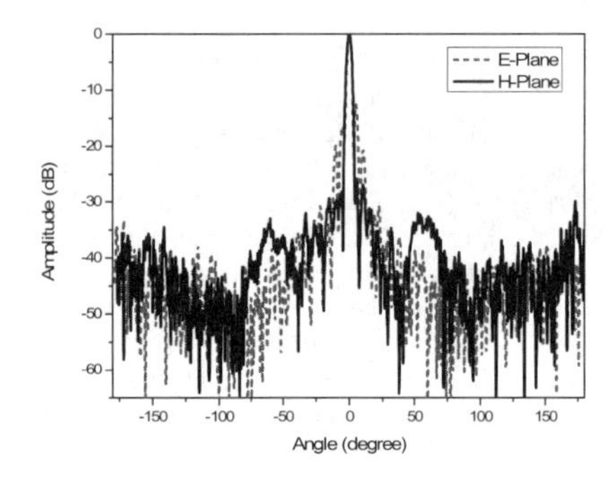

(b)

Fig. 6. The measured radiation patterns in E-plane and H-plane of the FRAs with F/D 0.5 (a) and 1 (b) at 42GHz.

III. Folded Reflectarray Antenna Design and Measurement

The main reflector consists of rectangle patches with various sizes. By varying the lengths and widths of the metal patches, the classic phase compensation as well as rotated

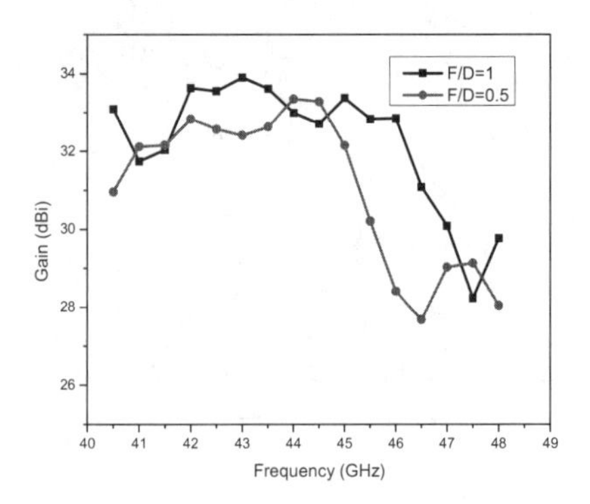

Fig. 7. Measured gains of the FRAs with F/D of 0.5 and 1.

TABLE II
THE COMPARISON OF VARIOUS REFLECTARRAY ANTENNAS

	F/D	Gain(dBi)	Aperture Efficiency
FRA1	0.5	32.8@42GHz	43.8%
FRA2	1	33.6@42GHz	52.6%
Ref. 1 in [6]	0.5	36.5@94GHz	29%
Ref. 2 in [4]	0.36	27.4@38.5GHz	33.9%

electric field polarization by 90° is obtained. All the simulations of reflecting cells by CST are using Floquet boundary condition, which is assuming the unitary cell is surrounded by identical cells. The unit size is 3mm×3mm, and the reflecting surface diameter *D* is 150mm.

Pyramid horn antenna is adopted as the feed for the folded reflectarray antenna. Two sizes of pyramid horns are simulated by CST for the antenna designs of *F/D*=0.5 and *F/D*=1. The average edge amplitude tapers of the pyramid horns which are the average values of E-plane and H-plane are 10.3dB and 10.9dB at 42GHz respectively. Therefore, the folded reflectarry antennas theoretically can achieve amplitude taper loss (ATL) and spillover (SPL) efficiencies [9] of 78% and 80% respectively.

The reflecting planes and polarizing grids are all fabricated on Rogers5880 with the permittivity of 2.2 and loss tangent of 0.0009. Fig. 5 shows the photograph of the proposed FRA1 and FRA2, corresponding to *F/D* of 0.5 and 1 respectively. The measured radiation patterns of both antennas at 42GHz are shown in Fig. 6. From Fig. 6, it can be seen that the measured 3dB beamwidths in E-plane and H-plane are 3.5°×3.4° and 2.9°×3.3° for FRA1 and FRA2 respectively. Fig. 7 shows the measured gains of the FRAs with *F/D* of 0.5 and 1. The gains of the FRAs with *F/D* of 0.5 and 1 are 32.8 dBi and 33.6 dBi respectively at 42 GHz, and the 3dB bandwidths are 11.6% and 13.8% respectively. Table II illustrates the comparison of the proposed FRAs with some reported FRAs.

This comparison shows the both proposed folded reflectarray antennas have good gain and efficiency performances and a gain increase of 0.8dB as well as an efficiency increase of 8.8% is achieved when adopting FRA with *F/D* of 1 to replace the antenna with *F/D* of 0.5.

IV. CONCLUSION

In this paper, quantization phase error of folded reflectarray antenna is proposed and its effect on gain is analyzed and compared with different *F/D*s. The FRA with *F/D* equaling to 1 is preferred to reduce the quantization phase error, and a gain increase of 0.8dB is achieved when adopting FRA with *F/D* of 1 to replace the antenna with *F/D* of 0.5. The good measured results testify the ideas.

REFERENCES

[1] D. Pilz, W. Menzel, "Printed mm-wave folded reflector antennas with high gain, low loss, and low profile," *2000 IEEE Antennas and Propagation Society International Symposium*, vol. 2, pp. 790-793, July 2000.

[2] S. Dieter, C. Fischer, and W. Menzel, "Design of a folded reflectarray antenna using particle swarm optimization," *The 40th European Microwave Conference proceedings*, pp. 731-734, Sept. 2010.

[3] A. Zeitler, J. Lanteri, C. Pichot, C. Migliaccio, P. Feil, and W. Menzel, "Folded reflectarrayswith shaped beam pattern for foreign object debris detection on runways," *IEEE Transactions on Antennas and Propagation,* vol. 58, pp. 3065 - 3068, Sept. 2010.

[4] I. Y. Tam, Y. S. Wang, and S. J. Chung, "A dual-mode millimeter-wave folded microstrip reflectarray antenna," *IEEE Transactions on Antennas and Propagation,* vol. 56, pp. 1510 - 1517, June 2008.

[5] W. Menzel, S. Keyrouz, J. Li, and S. Dieter, "Loss mechanisms of folded reflectarray antennas," *Proceedings of the 7th European Radar Conference*, pp. 180 - 183, Oct. 2010.

[6] B. D. Nguyen, J. Lanteri, J. Y. Dauvignac, and C. Pichot, "94 GHz folded Fresnel reflector using c-patch elements," *IEEE Transactions on Antennas and Propagation,* vol. 56, no. 11, pp. 3373 - 3374, Nov. 2008.

[7] V. Mrstik, "Effects of phase and amplitude quantization errors on hybrid phased-array reflector antennas," *IEEE Transactions on Antennas and Propagation,* vol. 30, no. 6, pp. 1233 - 1236 Nov. 1982.

[8] J. Huang, and J. Encinar, *Reflectarray Antennas,* New York: J. Wiley & Sons, 2007.

[9] D. M. Pozar, S. D.Targonski, and H. D. Syrigos, "Design of millimeter wave microstrip reflectarrays," *IEEE Transactions on Antennas and Propagation,* vol. 45, no. 2, pp. 287 - 296, February 1997.

Planar Multi-band Monopole Antenna for WWAN/LTE Operation in a Mobile Device

Jia-Ling Guo[1], Hai-Ming Chin[2] and Jui-Han Lu[1]
[1]Department of Electronic Communication Engineering
[2]Department of Marine Engineering
Kaohsiung Marine University
Kaohsiung, Taiwan 811

Abstract- **By using the loop parasitic shorted strip, a small-size uniplanar antenna with multi-band WWAN/LTE operation in the mobile device is proposed. The obtained impedance bandwidths across dual operating bands approach 277 MHz and 1176 MHz at the LTE and WWAN bands, respectively. The proposed uniplanar antenna reduces the antenna size by at least 22 % since the overall antenna size is only 35 × 10 × 0.8 mm^3. The measured peak gains and antenna efficiencies are approximately 2.6 / 3.8 dBi and 83 / 81 % for the LTE/WWAN bands, respectively. Moreover, with a compact structure, the proposed planar antenna can be deposited inside the mobile device and complies with the body specific absorption rate (SAR) requirement (< 1.6 W/kg for 1-g body tissue) for practical applications.**

I. Introduction

The long term evolution (LTE) system with three operating bands in the LTE700 (698 ~ 787 MHz), LTE2300 (2300 ~ 2400 MHz) and LTE2500 (2500 ~ 2690 MHz) [1] has attracted considerable attention for use in 4G wireless wide area network (WWAN) systems to incorporate the LTE system with GSM/UMTS operations in mobile devices, because of significantly higher data rate than that of 3G wireless wide area network (WWAN) operations for mobile broadband services. Meanwhile, owing to that an internal LTE/GSM/UMTS antenna occupies a large amount of space, the physical distance from the other embedded antennas (e.g., GPS, WLAN and Bluetooth antennas) decreases, which is a prerequisite for achieving an acceptable isolation between those embedded antennas. Thus, to fulfill the bandwidth specifications of the 4G system and ensure the ability to embed into a limited space, compact multi-band antennas appear to have the potential for providing commercial broadband coverage in the 698–960 / 1710–2690 MHz bands in LTE/WWAN environments. Several planar LTE/WWAN internal monopole antennas (MAs) have been developed for mobile phone [2-13]. However, limitations include a larger antenna size for the above MAs with a greater planar dimension [2-9] or insufficient operating bandwidth [10-13] for LTE/WWAN operations in the mobile phones. To overcome this limitation, this study presents a novel loop shorted strip as the parasitic element, to generate dual 0.25-wavelength resonant modes at approximately 750/940 MHz bands to cover the LTE700/GSM850/900 operating bandwidth. A F-shaped driven monopole strip is devised to excite a resonant mode in the upper (1710–2690 MHz) band of the desired antenna. Meanwhile, in contrast with the schemes developed in [2-13], all monopole strips in this study have the same width (0.6 mm) to simplify the dimensional parameters and minimize manufacturing defects. As for the overall antenna volume, the proposed antenna with a small size of 35 × 10 × 0.8 mm^3 (280 mm^3) has an antenna size at least 22 % less than that of the smallest LTE/WWAN internal antenna with a dimension of 30 × 15 × 0.8 mm^3 [6] for the mobile phone. Moreover, the body SAR [14] for the internal antenna must be tested, and should be less than 1.6 W/kg for 1-g body tissue [15-16]. To comply with this requirement, simulated body SAR results of the proposed antenna are also analyzed.

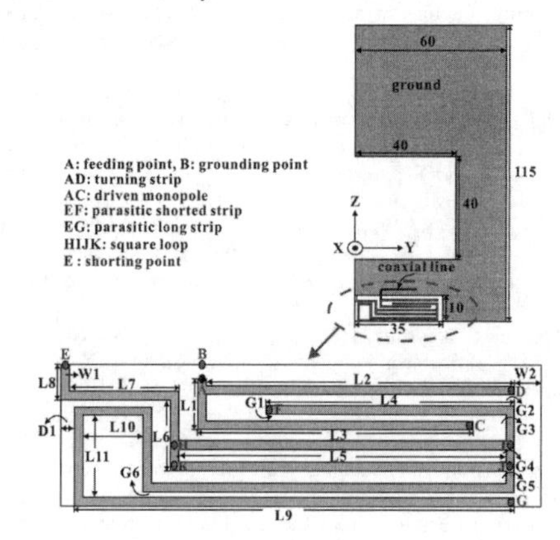

Figure 1. Geometry of the proposed planar multi-band monopole antenna with a F-shaped driven monopole strip for the mobile phone.

II. Antenna Design

Figure 1 displays the geometrical configuration of the proposed planar compact antenna for LTE/WWAN operations in a mobile phone. While printed on the same side of an FR4 substrate with the dimension of 35 × 10× 0.8 mm^3, the antenna is mounted along the top-right edge of the C-shaped system ground with the dimension of 115 × 60× 0.8 mm^3. The proposed antenna consists of a F-shaped driven monopole and a loop parasitic strip shorted at point E. The antenna is fed by a 50-Ω mini coaxial line connected to the feeding point (point A) of the F-shaped driven monopole and the system grounding point (point B). The F-shaped driven strip is first arranged as a quarter-wavelength monopole to generate the fundamental resonant mode at

978-7-5641-4279-7

approximately 2085 MHz. The lower meandered arm of the loop parasitic shorted strip (section EKJG) then contributes to its fundamental (0.25-wavelength) resonant mode at around 793 MHz with a higher-order resonant mode at approximately 1580 MHz. Moreover, the impedance bandwidth for the lower operating band (LTE700/GSM900 MHz) is widened by introducing the upper meandered arm of loop parasitic shorted strip (section EHIF) to generate the fundamental (0.25-wavelength) resonant mode at approximately 956 MHz band with a higher-order resonant mode at approximately 2658 MHz. Moreover, the antenna is optimized based on the above guidelines and by using Ansoft HFSS, a commercially available software package based on the finite element method [17]. Return loss is measured with an Agilent N5230A vector network analyzer.

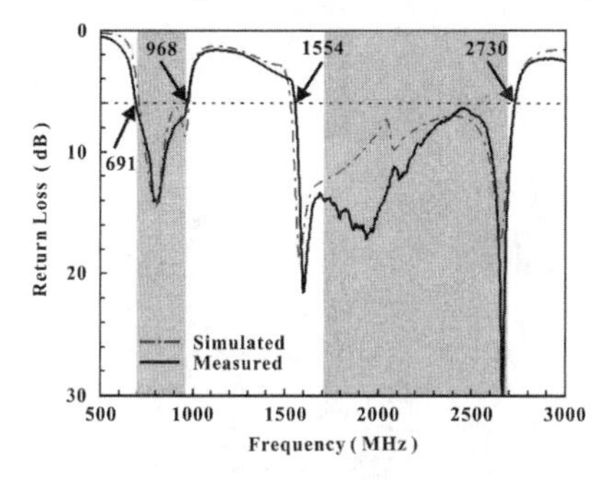

Figure 2. Simulated and measured results against frequency for the proposed planar multi-band monopole antenna. L1 = 3.6 mm, L2 = 22.4 mm, L3 = 20 mm, L4 = 18 mm, L5 = 23.8 mm, L6 = 5 mm, L7 = 8 mm, L8 = 2.5 mm, L9 = 32 mm, L10 = 4.4 mm, L11 = 5.8 mm, W1 = 0.6 mm, G1 = 0.5 mm, G2 = 0.8 mm, G3 = 1.9 mm, G4 = 0.9 mm, G5 = 0.9 mm, G6 = 0.4 mm, D1 = 1 mm

III. Results And Discussion

Figure 2 summarizes the simulation and experimental results for return loss in the proposed monopole antenna. The lower band reveals a measured 3:1 VSWR (6-dB return loss) bandwidth of 277 MHz (691–968 MHz), whereas the upper band has a bandwidth of 1176 MHz (1554–2730 MHz). Dual wide bands can comply with the bandwidth requirements of the desired eight-band LTE/WWAN (LTE700/GSM850/900/ GSM1800/1900/UMTS/LTE2300/2500) operations. Figure 3 presents the measured antenna gain and efficiency (mismatching loss included, [18]) for the proposed compact printed antenna. This figure also shows the simulation results for comparison. For frequencies over the LTE700/GSM850/ 900 bands, the measured antenna gain is approximately 0.8 ~ 2.6 dBi. Meanwhile, that for the GSM1800/1900/UMTS/ LTE2300/2500 bands ranges from approximately 2.4 to 3.8 dBi. The measured antenna efficiency is approximately 54 ~ 83 % over the LTE700/GSM850/900 bands, while that over the GSM1800/1900/UMTS/LTE2300/2500 bands is around 67 ~ 81 %. Figure 4 shows the measured 3-D and 2-D radiation patterns at typical frequencies. At frequencies (740 and 925 MHz) in the antenna's lower band, the radiation patterns are close to the dipole-like patterns. At higher frequencies (1795, 1920 and 2350 MHz) in the antenna's upper band, more dips and more variations are found in the radiation patterns than those at lower frequencies. This is largely owing to that more surface current nulls are excited in the system ground at higher frequencies than at lower ones. Furthermore, the body SAR [14] is tested to verify that the proposed monopole antenna fulfills the requirements for device use. Table 1 lists the SAR results for 1-g body tissue at the central frequencies of the lower and upper operating bands. The SAR results are less than the limit of 1.6 W/kg.

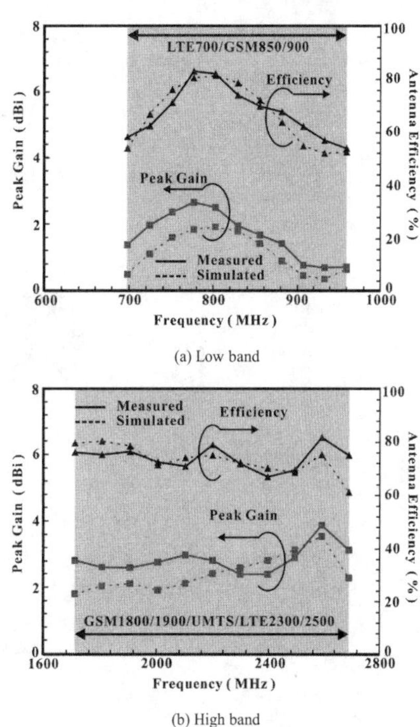

Figure 3. Measured and simulated antenna gain and efficiency for the proposed planar multi-band monopole antenna studied in Fig. 2.

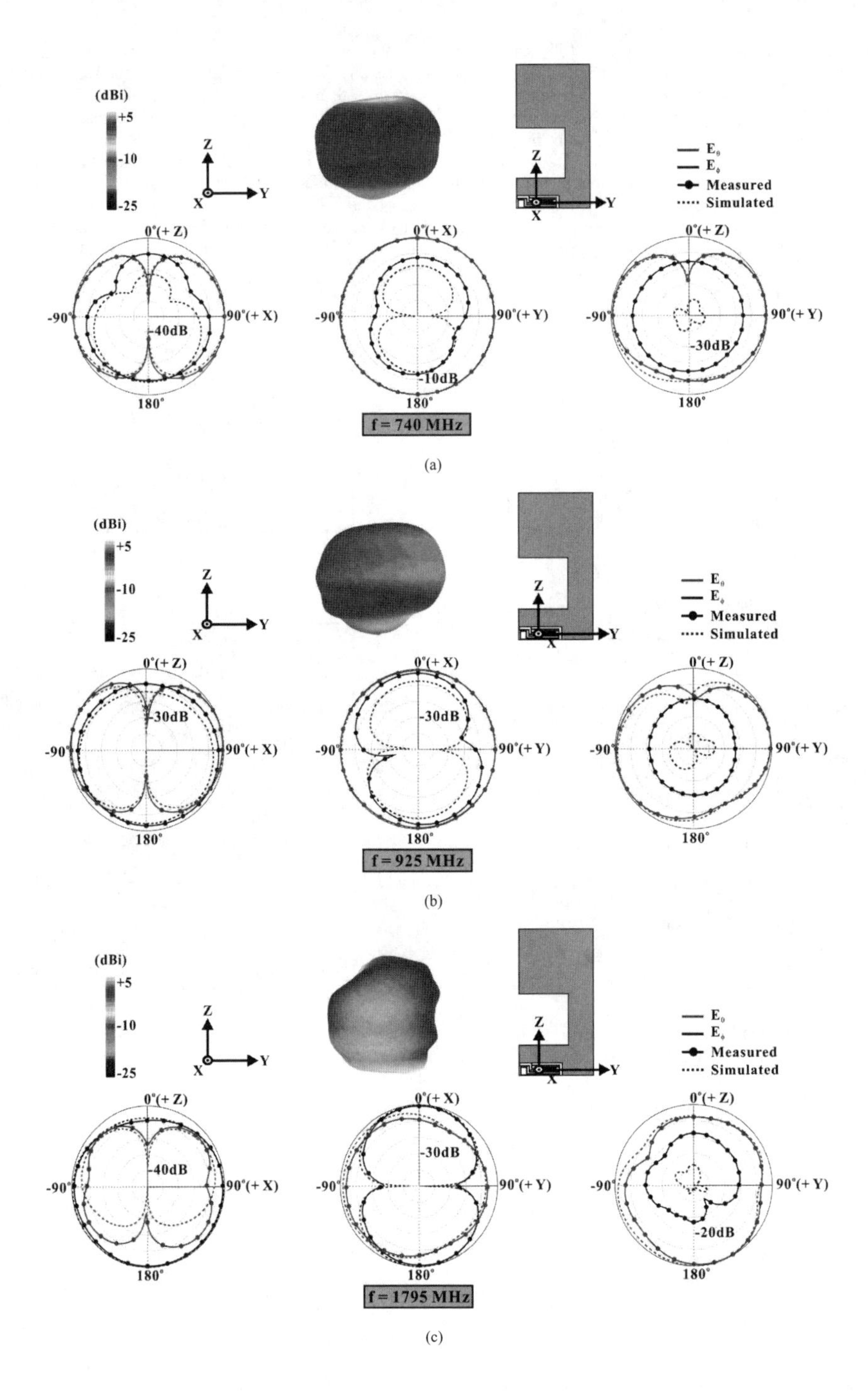

(dBi)
+5
-10
-25
Z
Y
X
E$_\theta$
E$_\phi$
Measured
Simulated
0°(+ Z)
-90°
90°(+ X)
180°
-40dB
0°(+ X)
90°(+ Y)
-10dB
-30dB
f = 740 MHz
f = 925 MHz
f = 1795 MHz
-20dB

(a)

(b)

(c)

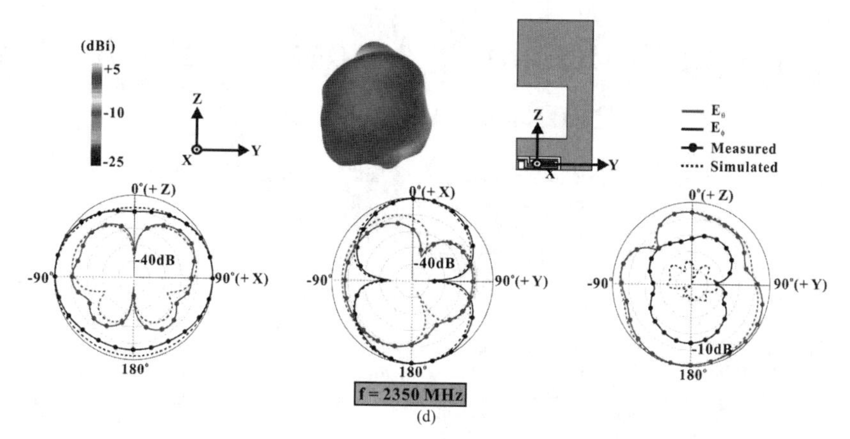

Figure 4. Measured three-dimensional (3-D) total-power and two-dimensional (2-D) radiation patterns for the proposed planar multi-band monopole antenna.

Table 1 : Simulated body SAR results obtained using SEMCAD X [19] for 1-g body tissue.

Frequency (MHz)	740	925	1795	1920	2350
Input power (Watt)	0.125	0.25	0.125	0.125	0.125
1-g SAR (W/kg)	0.96	1.06	0.51	0.47	0.39
Return loss (dB)	9.4	6.3	12.2	10.3	6.7

IV. Conclusions

This work presents a bandwidth enhancement approach to achieve a small-size eight-band LTE/WWAN internal mobile phone antenna. The impedance bandwidth across the operating bands can reach approximately 277 MHz and 1176 MHz at the LTE and WWAN bands, respectively. The measured peak gains and antenna efficiencies are approximately 2.6 / 3.8 dBi and 83 / 81 % for the LTE/WWAN bands, respectively. The proposed uniplanar antenna reduces the antenna size by at least 22% since the overall antenna size is only 35 × 10 × 0.8 mm^3. Moreover, the proposed uniplanar antenna design fulfills the requirement of the body specific absorption rate (SAR) (< 1.6 W/kg for 1-g body tissue) for device use.

References

[1] LTE Advanced The 3rd Generation Partnership Project, 2011. Available: http://www.3gpp.org/LTE-Advanced.

[2] C. H. Ku, H. W. Liu and Y. X. Ding, "Design of planar coupled-fed monopole antenna for eight-band LTE/WWAN mobile handset application", *Progress In Electromagnetics Research C*, vol. 33, pp. 185-198, Oct. 2012.

[3] C. T. Lee and K. L. Wong, "Planar monopole with a coupling feed and an inductive shorting strip for LTE/GSM/UMTS operation in the mobile phone", *IEEE Trans. Antennas Propagat.*, vol. 58, pp. 2479-2483, July 2010.

[4] K. L. Wong and W. Y. Chen "Small size printed loop type antenna integrated with two stacked coupled-fed shorted strip monopoles for eight-band LTE/GSM/UMTS operation the mobile phone", *Micro. Opt. Technol. Lett.*, vol. 52, pp. 1471-1476, July 2010.

[5] K. L. Wong, M. F. Tu, T. Y. Wu and W. Y. Li, "Small size coupled-fed printed PIFA for internal eight band LTE/GSM/UMTS mobile phone antenna", *Micro. Opt. Technol. Lett.*, vol. 52, pp. 2123-2128, Sep. 2010.

[6] F. H. Chu and K. L. Wong, "On-board small-size printed LTE/WWAN mobile handset antenna closely integrated with system ground plane", *Micro. Opt. Technol. Lett.*, vol. 53, pp. 1336-1343, June 2011.

[7] S. C. Chen and K. L. Wong, "Hearing aid-compatible internal LTE/WWAN bar-type mobile phone antenna", *Micro. Opt. Technol. Lett.*, vol. 53, pp. 774-781, April 2011.

[8] Z. Chen, Y. L. Ban, J. H. Chen, J. L.-W. Li and Y.-J. Wu, "Bandwidth enhancement of LTE/WWAN printed mobile phone antenna using slotted ground structure", *Progress In Electromagnetics Research*, vol. 129, pp. 469-483, July 2012.

[9] K. L. Wong and F. H. Chu, "Planar printed strip monopole with a closely-coupled parasitic shorted strip for eight-band LTE/GSM/UMTS mobile phone", *IEEE Trans. Antennas Propagat*, vol. 58, pp. 3426-3431, Oct 2010.

[10] K. L. Wong and C. H. Chang, "Printed λ/8-PIFA for penta-band WWAN operation in the mobile phone", *IEEE Trans. Antennas Propagat.*, vol. 57, pp. 1373-1381, May 2009.

[11] K. L. Wong and Y. W. Chi, "Quarter-wavelength printed loop antenna with an internal printed matching circuit for GSM/DCS/PCS/UMTS operation in the mobile phone", *IEEE Trans. Antennas Propagat.*, vol. 57, pp. 2541-2547, Sep. 2009.

[12] K. L. Wong, W. Y. Chen and T. W. Kang, "On-board printed coupled-fed loop antenna in close proximity to the surrounding ground plan for penta-band WWAN mobile phone", *IEEE Trans. Antennas Propagat*, vol. 59, pp. 751-757, March 2011.

[13] K. L. Wong and W. Y. Li, "Internal penta-band printed loop-type mobile phone antenna", *Micro. Opt. Technol. Lett.*, vol. 49, pp. 2595-2599, Oct. 2007.

[14] Federal Communications Commission, Office of Engineering and Technology, Mobile and Portable Device RF Exposure Equipment Authorization Procedures, OET/Lab Knowledge Database publication number 447498 item 7, Dec. 13, 2007.

[15] American National Standards Institute (ANSI), Safety levels with respect to human exposure to radio frequency electromagnetic fields, 3 kHz to 300 GHz, ANSI/IEEE standard C95.1, April 1999.

[16] IEC 62209–1, Human exposure to radio frequency fields from hand-held and body-mounted wireless communication devices— Human models, instrumentation, and procedures—Part 1: Procedure to determine the specific absorption rate (SAR) for hand-held devices used in close proximity to the ear (frequency range of 300 MHz to 3 GHz), Feb. 2005.

[17] http://www.ansoft.com/products/hf/hfss, Ansoft Corporation HFSS.

[18] Y. Huang and K. Boyle, *Antennas from Theory to Practice*, pp. 124, John Wiley & Sons Ltd, 2008.

[19] http://www.semcad.com, SPEAG SEMCAD, Schmid & Partner Engineering AG.

Development of Rotating Antenna Array for UWB Imaging Application

Min Zhou, Xiaodong Chen and Clive Parini
Electronic Engineering and Computer Science
Queen Mary, University of London
London, United Kingdom
min.zhou@eecs.qmul.ac.uk

***Abstract*-The detection of concealed metallic weapons in bags/suitcase by using UWB (ultra widbeband) imaging is a cheap alternative to the security in the airport, subway or public area, compared to X-ray machines. In this paper, we have proposed and studied a novel UWB imaging system, including a rotating UWB antenna array, RF components and the 2D implementation of the delay-and-sum (DAS) imaging algorithm.**

Keywords—UWB imaging; Antenna array; Time-domain; Delay-and-Sum

I. INTRODUCTION

Imaging is one of the most popular applications of UWB (ultra wideband) technology, for the reasons of low cost and no health hazards. The detection of concealed metallic weapons in bags/suitcases is vital to the security in the airport, subway or public area. To reduce the cost associated with the deployment X-ray machines, UWB imaging technology is proposed to be a cheap alternative. Suitcase imaging is based on the UWB Radar technology to detect and locate the object through the opaque obstacle. This imaging technology works on the contrast of the permittivity of the target and the surrounding material. Due to the strong contrast between metal targets and other objects in the suitcase, it is easier to attain the imaging of metal targets.

In order to achieve better resolution and imaging result, most of the existing UWB imaging systems have a large profile because of the large antenna array structure. For example, the UWB 3D imager presented in [1] is able to image the concealed objects in the suitcase and people over a distance. It uses four antenna array units, operating as two independent bi-static Radars, which are co-polar vertical and co-polar horizontal polarization respectively. Each antenna array unit contains 25 elements, total in 49 cm length. Although this system can be applied in multiple applications, it is not in compact size. The UWB through-wall imaging system in [2] uses 16 Vivaldi antenna array elements for its antenna structure, each of which also consists of 16 single antennas. It is not so complicated in the array structure, but the coupling between so many antennas needs to be considered while designing.

This paper will present a novel UWB imaging system with a spiral rotating antenna array. The rotating antenna array make the imaging system compact and low cost. The paper is structured as follows. The structure and the performance of the rotating antenna array will be described in details in the second section. The two-dimensional reconstructed imaging method based on delay-and-sum (DAS) algorithm will be presented in the third section. Then the simulated results will be discussed in the fourth section. Finally, the conclusion will be given.

II. ANTENNA ARRAY

A. *Corrugated Balanced Antipodal Vivaldi Antenna*

In order to make the UWB imaging system compact and better in imaging performance, the antenna element should be small and high/stable gain over the operating bandwidth. Considering these requirements, the Vivaldi antenna is a good choice for UWB imaging. Among most types of Vivaldi antennas, the balanced antipodal antenna is chosen as the candidate for our UWB imaging system.

A novel corrugated balanced antipodal Vivaldi antenna has been designed for our purpose. This antenna has three metallic layers and two supporting substrates. The structure of the three layers is shown in Fig. 1. Like the sandwich structure, the radiated antenna layer is in the middle of the two supporting FR4 substrates, while the top and bottom metallic layer, which can be treated as the ground layer, are attached on the outermost side of these two substrates. The top and bottom ground layer is in the same shape, as shown in Fig. 1 (a), while the centre radiated one is in the mirrored shape with them, as shown in Fig. 1 (b). The radiated flare is shaped of inner and outer exponential edge, and cut with linear slot lines.

TABLE I. THE DIMENSIONS OF THE CORRUGATED BAVA

Parameter	W	L	wm	L_1	L_2	d_s	d
Value (mm)	35	75	1.5	10	32	0.7	0.9

978-7-5641-4279-7

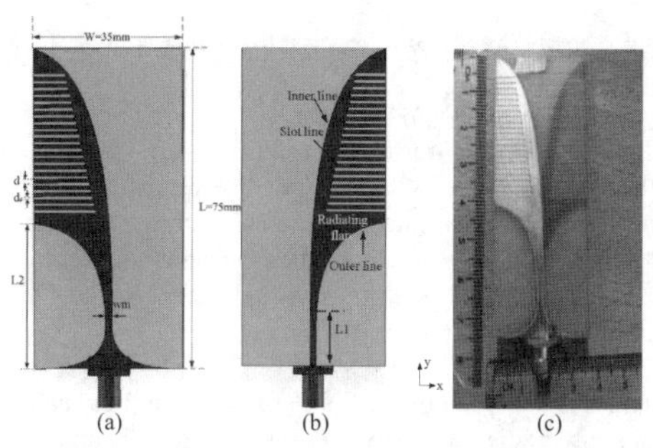

Fig. 1. Physical structure of the corrugated BAVA; (a) Top and Bottom layer view; (b) Middle layer view; (c) The fabricated antenna.

These slot lines are all in the same length ds along y-axis, keeping the same distance d between them. They are reduced in width along x-axis linearly from lower side to upside of the radiating flare. The dimensions of the parameters for this antenna are listed in Table I.

This antenna has an impedance bandwidth from 2.8 GHz up to 12 GHz in simulation and from 3 GHz to 12 GHz in measurement, as shown in Fig. 2. The red line is the measured result while the blue line is the simulated one. The deep attenuation around 5.5 GHz in simulation has been shifted to 4 GHz in measurement, mainly because of the gap between the two substrates at the side of the feeding SMA connector. The fabricated antenna only uses the SMA to connect the two substrates so that there will be a little gap of air between them.

The corrugated structure of this antenna can alter the phase of the currents flowing along the outer edges of the antenna substrate, and change the direction of the electric field at the edge of the substrate. It is evident from the current flow on the surface of the radiating layer at different frequency, shown in Fig. 3 (a). At 3 GHz, the current flow is forced to travel along the edge of the flare. At higher frequency, as shown in Fig. 3 (b), the current is not affected too much due to the slot lines. Thus, it can improve the gain of the antenna in lower frequency, as shown in Fig. 4.

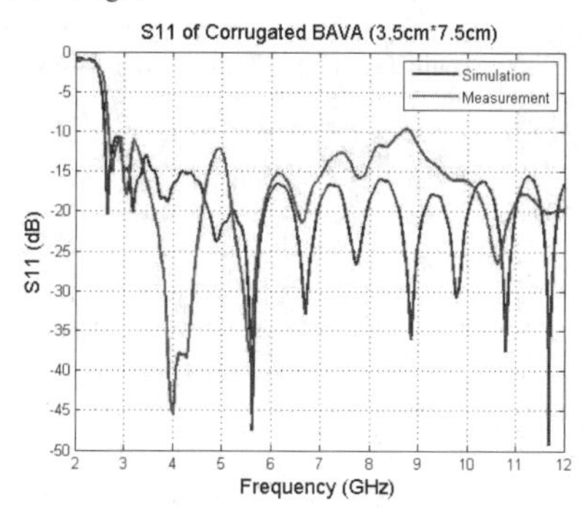

Fig. 2. The simulated and measured S_{11} of corrugated BAVA.

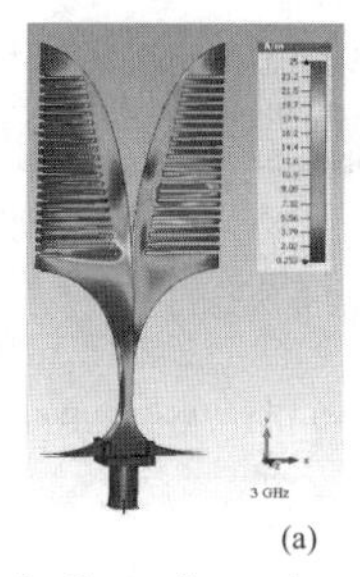

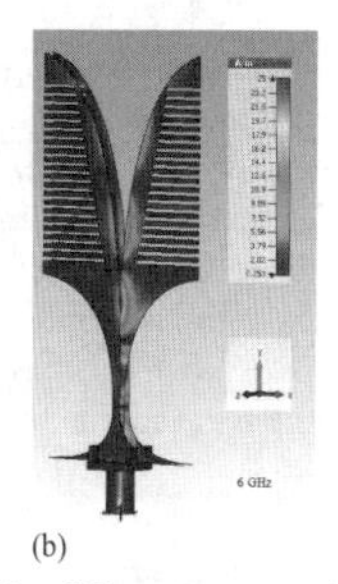

Fig. 3. Current flow on the surface of BAVA at different frequency; (a) 3 GHz; (b) 6 GHz.

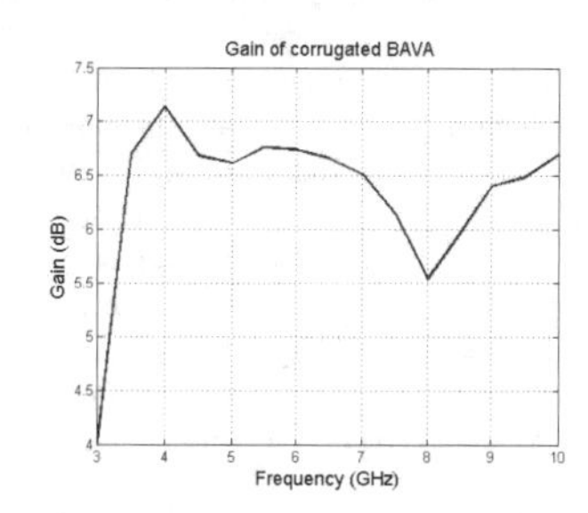

Fig. 4. Simulated gain of the corrugated BAVA.

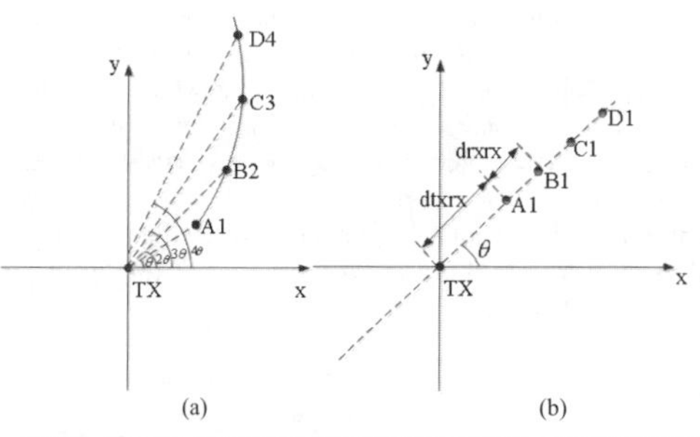

Fig. 5. The structure of the rotating receiving antenna array; (a) Rotating antenna array; (b) Equal arm of the receiving antenna array.

B. Rotating antenna array structure

When used in the UWB imaging system, the rotating receiving antenna array is composed of four elements, placed on a spiral arm, as shown in Fig. 5 (a). The four elements receiving antennas are assumed as A1, B2, C3 and D4. The angle between the neighboring receiving antennas is assumed as θ. After rotating one circle around the centre transmitting antenna, the four receiving antennas are equal to be the one in a straight arm, which are A1, B1, C1 and D1 shown in Fig. 5 (b). In such an arrangement, the distance d_{rxrx} between the neighboring receiving antennas will not be limited by the antenna's width. Also, the mutual coupling between the elements can be reduced.

Although this rotating antenna array is composed of four elements, after rotating a circle, it is equal to have N elements of antennas in space. Here N is followed equation (1). It is obvious that the rotating antenna array structure make it possible to have large numbers of antennas equivalently in space by using fewer antenna elements in reality.

$$N=n\times 360^0/\theta \quad (1)$$

Where n is the number of elements in the receiving antenna array.

C. UWB imaging system

This UWB imaging system used for scanning the concealed target inside the suitcase consists of RF front end, antennas and signal processing parts. The antennas consist of one fixed horn antenna to be the transmitting antenna and one rotating spiral receiving antenna array. An electromagnetic wave transmitting through RF is radiated by transmitting antenna, penetrates through the suitcase, and is reflected by the concealed target. Then it is received back via receiving antenna array. After the signal is received by the receiving antenna, it will travel through RF front end to the signal processing part. Finally through signal processing, the images will be reconstructed on a PC. The whole architecture of the imaging system is shown in Fig. 6.

The receiving antenna array rotates in space with the interval angle of θ. At each position, the receiving antenna elements work one by one controlled by the switch. When the receiving antenna array is rotated over 360 degree, it is equal to have a 2D antenna array with N elements. It is seen that the antenna elements/switches are reduced so that the whole system is cheap and more compact. Another important thing is that reducing the antenna elements, the coupling between antenna elements will be much less, resulting a good SNR.

Therefore, this system has the ability to have better imaging quality when the receiving antenna array is rotating to obtain more measured data.

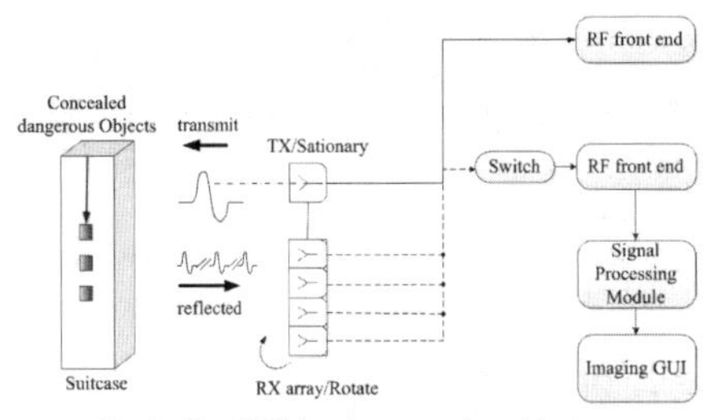

Fig. 6. The UWB imaging system in architecture.

III. TWO-DIMENSIONAL IMAGING METHOD

The imaging scheme is based on DAS algorithm. The fundamental of the method is shown in Fig. 7 (a). Assuming the transmitting antenna is placed at $x_t=(x_t,y_t)$, the reflected signal from the target T is received by m-th receiving antenna located at $x_{rm}=(x_{rm},y_{rm})$. The target is located at $x_j=(x_j,y_j)$. The signal received by the *m-th* receiving antenna is given by [3] in equation (2).

$$b_m(t)=\omega_m a(t-\tau_m) \quad (2)$$

Here, $a(t)$ is the transmitted signal. ω_m is the weighting factor of the *m-th* receiving antenna. It is represented the pulse attenuation and determined by antenna influence, system loss and so on. τ_m is the delay time when the signal travels from the transmitting antenna to the target and then back to the *m-th* receiving antenna, which is given by

$$\tau_m=[d(x_i,x_j)+d(x_j,x_{rm})]/c \quad (3)$$

Here c is the speed of the light in free space (m/s) and $d(x,y)$ is the direct distance between two position x and y. This process is repeated until the *m-th* receiving antenna. The M outputs are processed as follows. The region of space is divided into a finite number of pixels in small square unit. The signal corresponding to the image of the *i-th* pixel, located at (x_i,x_j) is obtained by applying time delays to the outputs of the M receiving antenna, and adding them, which is shown in Fig. 7 (b) and given by

$$f_i(t)=\sum_{m=1}^{M} b_m(t+\tau_m) \quad (4)$$

The imaging value for the *i-th* pixel is obtained by passing through the signal $f_i(t)$. In this case, the pixels in the region will have different values so that the target can be reconstructed.

The 2D reconstructed image is based on combination of each rotation plane of the antenna array. In each plane, the method is based on the delay-and-sum (DAS) imaging method described above. Here, X axis is assumed as the direction along the width of the target while Y axis is along the height of the target and Z axis is along the down range distance.

In XY plane, when receiving antenna array A0, B0, C0 and D0 are at the angle of 0^0, the reconstructed image highlighted in red can be achieved for “image1” in Fig. 8. When rotating the helix receiving antenna array to different angles, the coordinate X-axis and Y-axis will be considered to rotate the same angle with the receiving antenna array so that the reconstructed image will be rotated in terms of the varied coordinate, which is similar to the area “image m”, “image n” and “image q” in Fig. 8. After rotating a circle, the area highlighted in red will be merged into the final reconstructed result.

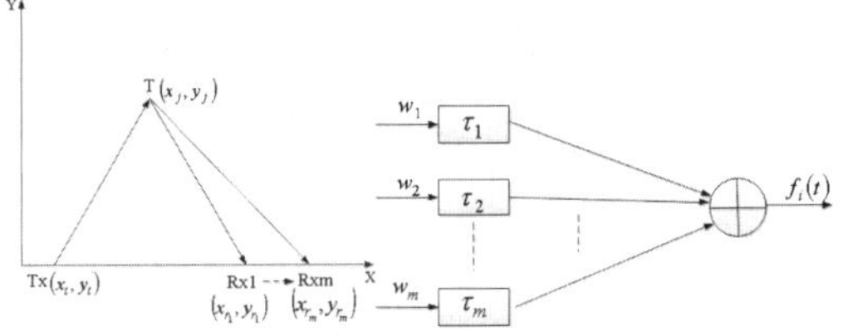

Fig. 7. Delay-and-Sum reconstructed method; (a) Detecting process; (b) Time-domain beam former.

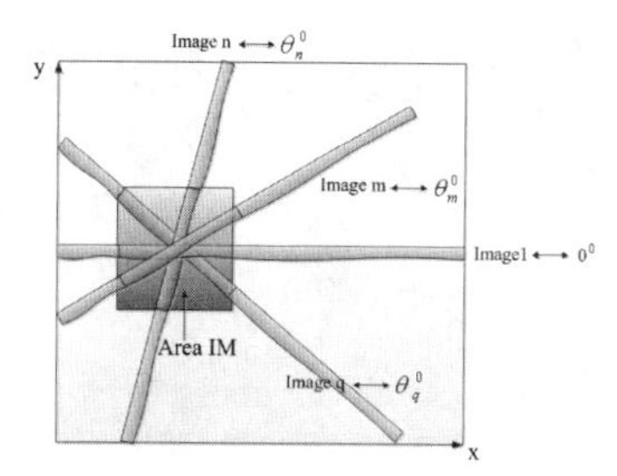

Fig. 8. The 3D reconstructed imaging method.

IV. SIMULATION RESULT

A. The simulation model

The simulated model of this UWB imaging system is shown in Fig. 10, including horn antenna (transmitting antenna), four elements rotating antenna array (receiving antenna) and one metallic target (10 cm × 10 cm). The transmitting antenna and receiving antenna are in the same vertical plane (*XY* plane). The metallic target is placed a distance of dmz at Z direction in front of the antenna array.

In this model, the distance d_{txrx} between TX and RX-A is assumed as 20 cm. The equivalent distance d_{rxrx} between the neighboring receiving antennas is 2.8 cm, while the angle of θ is 10^0. The metallic target is placed at down range distance dmz of 50 cm in front of the system. The truncated sine signal is excited to TX, following the equation (5) as below.

$$f_{TruncatedSine}(t)=\begin{cases}\sin 2\pi f_c t & kT\le t\le kT+\tau\\ 0 & kT+\tau\le t\le (k+1)T\end{cases} \quad (5)$$

Where k is the periodic times; T is the signal repetition period 10 us; τ is the signal duration 1 ns; f_c is the centre frequency 4.5 GHz.

At first, the angle between the receiving antenna A1 and TX is 10^0, thus the angle between receiving antenna B2, C3, D4 and TX are 10^0, 20^0 and 30^0 respectively. Rotate the spiral receiving antenna array three times so that 16 different positions of RX have been obtained. Among them, the equivalent straight arm of receiving antenna array is A4, B4, C4 and D4, which are at the same angle of 30^0 from TX. Repeating the rotation for one circle, there will be 36 groups of equivalent straight antenna array arm covering the whole scanning space.

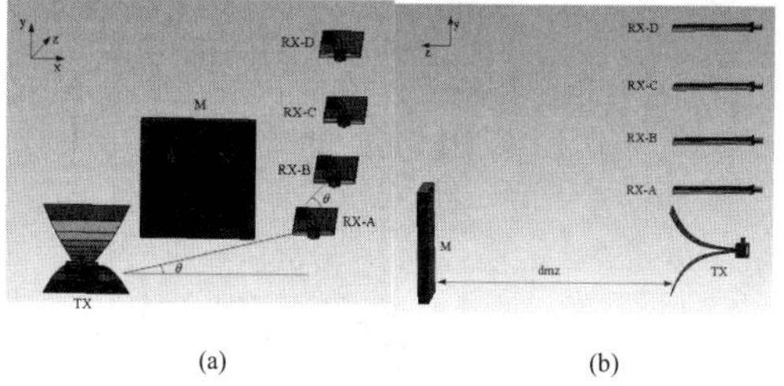

(a) (b)

Fig. 9. The simlatied model of the imaging system; (a) Front view; (b) Side view.

B. The reconstructed image

Based on the reconstructed algorithm described in the third section, the image can be reconstructed via different positions of the receiving antenna. If the receiving antenna array is stable, which means only one group of the received data has been obtained, the UWB imaging system scans only one plane of the target. Therefore, the achievable image result of the target is only to be 1D, as shown in Fig. 10 (a).

If the receiving antenna is rotating, the two-dimensional image can be obtained, just like the one shown in Fig. 10 (b). The part highlight in red is the proposed metallic target in the simulation model. It is located at about 50 cm of down range distance. The size of the target is similar to 10 cm × 10 cm.

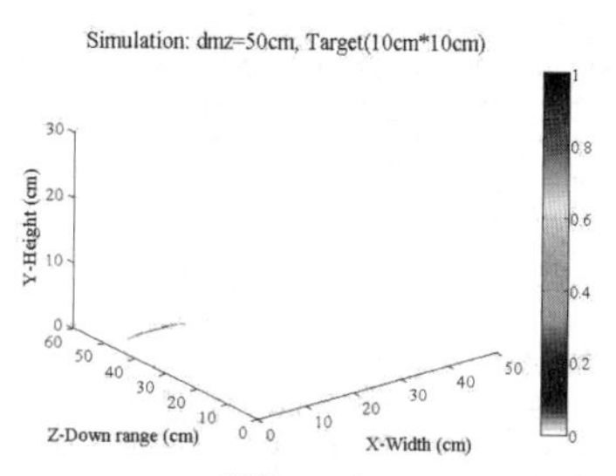

(a) No rotation

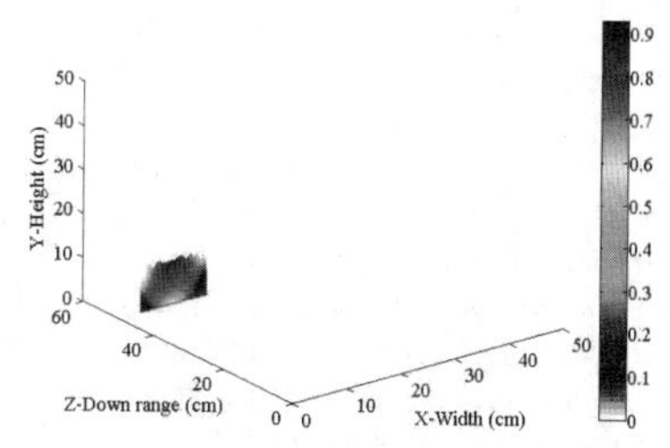

(b) With rotation of the antenna array

Fig. 10. The reconstructed images for the simlation model.

V. CONCLUSION

A spiral rotating antenna array based on corrugated BAVA has been designed for a novel UWB imaging system. The two-dimensional imaging method based on the delay-and-sum algorithm has been proposed and implemented. The 2D imaging result has been achieved in simulation by rotating the receiving antenna array.

REFERENCES

[1] B. J. Harker, A. D. Chadwick, G.L. Harris, "Ultra-wideband 3-Dimensional imaging (UWB 3D Imaging)", *2008 Roke Manor Research Limited*, UK

[2] Y. Wang, A.E. Fathy, "Design of a compact tapered slot Vivaldi antenna array for see through concrete wall UWB applications", *Proceedings of the XXIXth URSI General Assembly in Chicago*, USA, 7-16 August, 2008.

[3] Chen Lei, Shan ouyang, "A Time-domain Beamformer for UWB Through-wall Imaging", *TENCON2007–2007 IEEE Region 10 Conference*, 2007, Page(s): 1-4.

A Singular Value Decomposition Model for MIMO Channels at 2.6GHz

Yang Liu Ye Wang Wen-jun Lu Hong-bo Zhu
Nanjing University of Posts and Telecommunications
Nanjing,Jiangsu Province,China
ly71354@163.com

Abstract- **A novel MIMO model based on singular value decomposition of matrix is proposed in this paper. A corridor to stair MIMO measurement at 2.6GHz is conducted in order to evaluate the performance of the novel model. The joint DoA-DoD-delay power spectrum and the channel capacity of the model are compared with those obtained from the measured channels. The results show clearly that the joint DoA-DoD-delay power spectrum and the channel capacity of the novel model provides a better fit to the measured data than other models.**

I. Introduction

LTE(Long Term Evolution), as one of the key technologies for 4G wireless communication systems, has been put forward to support high rates and large capacity gains. For LTE system, the use of MIMO(Multiple-Input Multiple-Output) combined with OFDM(orthogonal-frequency multiple access) is the most promising strategy[1]. In China, the allocation spectra for indoor LTE wireless communication system is from 2.5 GHz to 2.69 GHz. Accurate MIMO channel model is an important tool to enhance the performance of LTE wireless communication system. In the past years, several MIMO channel models have been proposed. They can be classified into physical models and analytical models [2,3]. Physical models describing electromagnetic wave propagation environment are independent of the antenna configuration. Saleh-Valenzuela model [4,5], Zwick model [6], geometry-based stochastic model(GSCM) [7] are three typical physical models. On the other hand, analytical models considering antenna array configuration are often used to describe the channel impulse response(CIR) matrix.

Among all the analytical models, Kronecker model [8], Weichselberger model [9] and virtual channel representation model [10] are the most typical models. Kronecker model based on the assumption that the spatial correlation coefficient at the transmitter (receiver) is independent of receiver (transmitter). That is to say the one-sided spatial correlation matrix of transmitter (receiver) is independent of receiver (transmitter). It is confirmed that this assumption is in general too strict for channel by realistic channel measurements. Different with Kronecker model, Weichselberger model is based on the assumption that the eigenbases of one-sided spatial correlation matrix at the transmitter (receiver) depend on the environment, i.e., number of antennas, positions, and scatterers. However, it is shown that this assumption holds only approximately and the eigenbases are also influenced by the spatial structure of transmit signals in [11]. In [10], Sayeed develops a virtual channel representation model that the eigenbases at the transmitter and receiver in Weichselberger model are replaced by the steering matrices into the virtual angles at the transmitter and receiver respectively. The virtual channel representation model is only available for uniform linear arrays (ULAs) and spatial resolution of the model deponds on the antenna array size.

In this paper ,a novel MIMO channel model is presented based on the singular value decomposition (SVD) of matrix. Measurement setup and environment are described in Part II. In Part III, a novel MIMO model based on the singular value decomposition of matrix is established. In Part IV, measured results and model results are shown to evaluate the performance of the novel model.

Throughout the paper, we will use the vec(·) operator that stacks the columns of a matrix into one vector. The superscripts $[\cdot]^H$, $[\cdot]^T$, $[\cdot]^*$ stands for complex conjugate matrix transpose operator, matrix transpose operator, and complex conjugate operator, respectively. The symble $\otimes$ stands for the Kronecker product of two matrices. Finally, E(·) denotes the expectation operator.

II. Measurement Setup

A. *Measurement System*

Fig.1 shows a 2×2 MIMO measurement system conducted by Agilent 8753ES vector network analyzer (VNA). The VNA generates a 20MHz swept signal at a center frequency of 2.6GHz. The transmitted power is 10 dBm. The transmit and

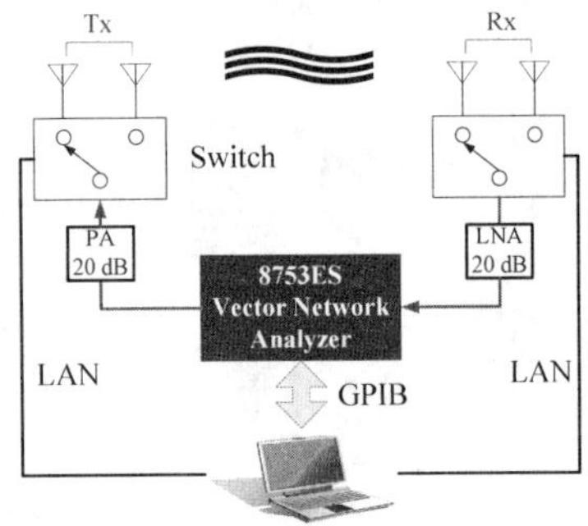

Figure 1: block diagram of MIMO measurement system.

This work was supported by National Science and Technology Major Project under granted no. 2012ZX03001028-005 and 2011ZX03005-004-03 and Jiangsu 973 project (BK2011027).

978-7-5641-4279-7

receive antennas are the same omnidirectional monopoles with a gain of 3 dBi. Transmit and receive antennas are connected to Agilent 34980A switch matrix respectively. The transmit signal is amplified by a 20dB power amplifier. The receive signal is amplified by a 20dB low noise power amplifier and conducted to the VNA via 20 meters coaxial cable. A lap top is used to sample the measured data through GPIB interface and control the switch matrix via LAN. The measurement is carried out without moving persons to make sure of the time-invariant channel behavior during one snapshot, which is about 20 seconds. We conduct measurement according to the transmit antenna spacing and receive antenna spacing as 0.5λ, 1λ, 2λ respectively, where λ is wavelength.

B. *Environment*

As shown in Fig.2, the measurement environment is in an office building that contains a 15m corridor to a stair through a corner. This paper discusses not line of sight(NLOS) case that the transmitter(Tx) is located on the corridor while the receiver(Rx) is moved along the corner and stair. The floor is marble and the ceiling is plaster at a height of 3.2 m. The stair step is made of reinforced concrete with width of 133 cm, length of 26 cm, and height of 16.5 cm. In order to obtain the transfer function of each link(H_{11}, H_{12}, H_{21}, H_{22}), Rx is moved on the prepared six locations(Rx6-Rx11) in the corner and stair as shown in Fig.2.

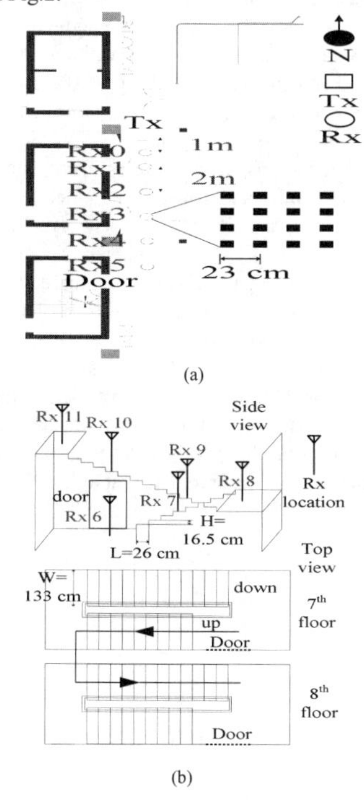

Figure 2: Layout of measurement environment: a) corridor, b) stairwell.

III. A Sigular Value Decomposition MIMO Channel Model

A. *The MIMO Channel Impulse Response Matrix*

For a MIMO system with two transmit antennas and two receive antennas, the relationship between transmitted and received signals can be expressed as

$$y = Hx + n \tag{1}$$

where y is the 2×1 received signal, x is 2×1 the transmitted signal and n is the noise. H is the 2×2 CIR matrix whose entry h_{mn} is the scalar valued frequency response between the nth transmit antenna and the mth receive antenna[12].

$$H = \begin{pmatrix} h_{11} & h_{12} \\ h_{21} & h_{22} \end{pmatrix} \tag{2}$$

For a Rician fading channel, the CIR matrix is given by

$$H = \sqrt{\frac{K}{K+1}} H_{LOS} + \sqrt{\frac{1}{K+1}} H_{res} \tag{3}$$

where H_{LOS} reflects the contribution of line of sight(LOS) componet and H_{res} reflects the residue stochastic part of CIR matrix. K is the Rician factor. This paper discusses the NLOS case only(Rx6-Rx11), then K=0. That is to say $H=H_{res}$ and corresponds to a Rayleigh fading case[12].

B. *Singular Value Decomposition Model*

Based on the singular value decomposition, a novel model is discussed in this part. According to singular value decomposition of matrix, the 2×2 CIR matrix H can be expressed as

$$H = U\Sigma V^H \tag{4}$$

$$\Sigma = \begin{pmatrix} \sigma_1 & 0 \\ 0 & \sigma_2 \end{pmatrix} \tag{5}$$

where σ_1, $\sigma_2(\sigma_1 \geqslant \sigma_2 \geqslant 0)$ are the sigular values of H, both U and V are unitary matrices.

We proposed that the unitary matrix U(V) can be divided into the determined part $U_d(V_d)$ and the stochastic part $U_s(V_s)$. That is to say, U, V can be expressed as

$$U = U_d \times U_s \tag{6}$$

$$V = V_d \times V_s \tag{7}$$

respectively. We difine that $U_d(V_d)$ is the eigen-matrix of recei -ver(transmitter) and $U_d(V_d)$ depends on the environment.

Equation (6) and (7) are substituted into (4) and then H can be modeled as

$$H = U_d U_s \begin{pmatrix} \sigma_1 & 0 \\ 0 & \sigma_2 \end{pmatrix} (V_d V_s)^H \tag{8}$$

C. *Extractions of Model Parameters*

In order to simplified model, we assume that U_s, V_s are any stochastic two dimension unitary matrices.

For any stochastic two dimension unitary, U_s, can be expressed as

$$U_s = \begin{pmatrix} ae^{j\omega_1} & \sqrt{1-a^2}e^{j\omega_2} \\ \sqrt{1-a^2}e^{j\omega_3} & ae^{j(\omega_2+\omega_3-\omega_1+\pi)} \end{pmatrix} \tag{9}$$

where a, ω_1, ω_2, ω_3 are real valued and j is the complex unit. We assume that a is uniform distributed at interval [0,1] and ω_1, ω_2, ω_3 are uniform distributed at interval [0, 2π]. The stochastic two dimension unitary V_s can be assumed as same as U_s

Futhermore, from (8), we can get

$$E_H(HH^H) = U_d \times \Lambda_1 \times U_d^H \tag{10}$$

$$E_H(H^H H) = V_d \times \Lambda_2 \times V_d^H \tag{11}$$

where $\Lambda 1$, $\Lambda 2$ are two diagonal matrices. Through eigen decomposition of matrix $E_H(HH^H)$, the eigen-matrix of receiver U_d can be extracted. The eigen-matrix of transmitter V_d can also be extracted as same as U_d.

As we know, the sigular values σ_1, σ_2 are square root of eigenvalues. In [13,14], Wonsop Kim has discovered that eigenvalues of H follow Gamma distribution and developed the model of eigenvalues about the correlation coefficient. From the model in [13,14], we can easily get the expectation and variance of eigenvalues and then determine the distribution of eigenvalues. That is to say, the sigular values $\sigma 1$, $\sigma 2$ can be determined easily.

IV. Model Validation

In order to evaluate the performance of the novel model presented in this paper, two metrics, the joint DoA-DoD-delay power spectrum and the channel capacity of the model are compared with those obtained from the measured channels.

A. Joint DoA-DoD Power Spectrum

The Joint DoA-DoD Power Spectrum can be expressed using the Bartlett beamformer[12,15]

$$P_{Bart} = (a_{Tx}(\theta_{Tx}) \otimes a_{Rx}(\theta_{Rx}))^H \times R_{WB} \times (a_{Tx}(\theta_{Tx}) \otimes a_{Rx}(\theta_{Rx})) \tag{12}$$

$$R_{WB} = E_H(vec(H)vec(H)^H) \tag{13}$$

where R_{WB} is the full-correlation matrix, $a_{Tx}(\theta_{Tx})$, and $a_{Rx}(\theta_{Rx})$ are normalized steering vectors at θ_{Tx}, θ_{Rx} respectively.

The computed DoA-DoD Power Spectrum can reflect the spatial and delay characteristics of the radio channel. In order to quantify the power spectrum differences, we use the Kullback-Leibler divergence(KLD) to describe the differences[16]. The KLD is defined as follows:

$$\gamma = \int_x \int_y \overline{P_1}(x,y) \log \frac{\overline{P_1}(x,y)}{\overline{P_2}(x,y)} dxdy \tag{14}$$

$$\overline{P_i}(x,y) = P_i(x,y) / \int_x \int_y P_i(x,y) dxdy \tag{15}$$

where $P_i(x,y)$ is the DoA-DoD Power Spectrum.

The KLDs from novel modeled APS to the measured APS can be calculated and the results are shown in Table I. In this table, it can be seen that the KLDs between the measured APS and the novel model are very small.

B. Capacity

The ergodic capacity of MIMO channel with equally allocated transmit power $\overline{C}$ can be determined as[17]

$$C = \log_2 \det(I_2 + \frac{SNR}{2} HH^H) \tag{16}$$

$$\overline{C} = E_H(C) \tag{17}$$

where C is the channel capacity, I_2 is an 2×2 identify matrix,

TABLE I

The KLDs Between Modeled And Measured Joint DoA-DoD Angular Power Spectrums

Loc.	KLD		
	0.5λ Spa.	1λ Spa.	2λ Spa.
Rx6	0.195	0.154	0.08
Rx7	0.01	0.01	0.019
Rx8	0.04	0.11	0.026
Rx9	0.007	0.017	0.013
Rx10	0.02	0.022	0.025
Rx11	0.01	0.014	0.011

SNR is the receive signal power to noise power ratio.

The 10% outage capacity of MIMO channel C_{out} can be calculated as[17]

$$P(C \le C_{out}) = 10\% \tag{18}$$

The capacity error from models to measurement is defined as

$$Error(\%) = \frac{C_{model} - C_{mea}}{C_{mea}} \tag{19}$$

where C_{model} is the modeled capacity and C_{mea} is the measured capacity.

The evaluation SNR is selected as 10 dB.

The ergodic capacity and the 10% outage capacity obtained from the measured channels and the novel model are shown in Fig.3 and Fig.4. For comparision, the ergodic capacity and the 10% outage capacity of Kronecker model and Weichselberger model are also shown in Fig.3 and Fig.4. The ergodic capacity errors and the outage capacity errors from each model to measurement are shown in Table II and Table III. The averaging ergodic capacity errors from novel model, Weichselberger model and Kronecker model to measurement are 0.39%, 5.77%, and 4.15% respectively. The averaging outage capacity errors from novel model, Weichselberger model and Kronecker model to measurement are 2.92%, 12.38%, and 15.07% respectively. From the results, two conclusions can be obtained. First, the capacity error from kronecker model to measurenment in this paper is smaller than other paper (i.e, in[18]) because the rank of 2×2 channel matrix is very small. Finally, the most important conclusion is that analyzing the results, the novel model presented in this paper matches better to the measured data than other models.

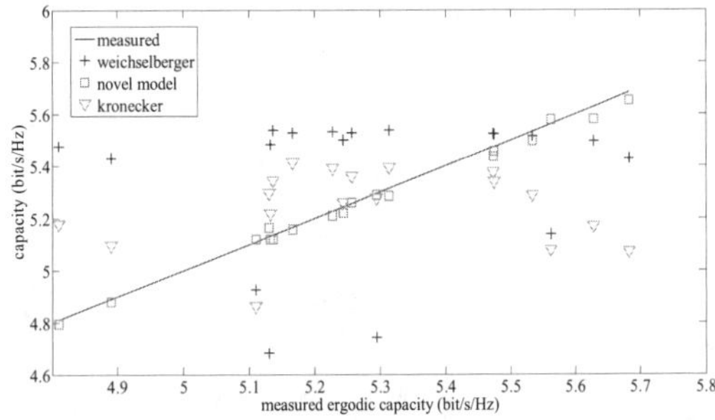

Figure 3: Ergodic capacity of every model and measured channels.

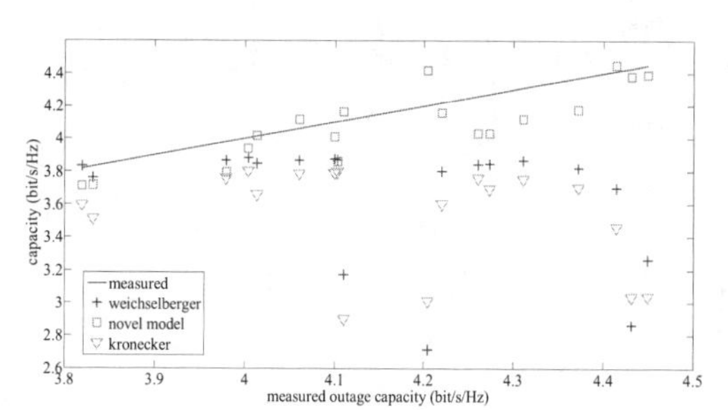

Figure 4: Outage capacity of every model and measured channels.

TABLE II

THE ERGODIC CAPACITY ERRORS FROM EACH MODEL TO MEASUREMENT

Loc. Spa.	Error(%)			Loc. Spa.	Error(%)		
	novel	web	kron		novel	web	kron
Rx6,0.5λ	0.1	10.4	0.3	Rx9,1λ	0.3	5.9	3.1
Rx7,0.5λ	0.7	1.0	1.8	Rx10,1λ	0.2	11.0	4.2
Rx8,0.5λ	0.5	4.5	10.8	Rx11,1λ	0.6	4.2	1.5
Rx9,0.5λ	0.1	5.2	2.0	Rx6,2λ	0.2	3.6	4.8
Rx10,0.5λ	0.3	13.8	7.6	Rx7,2λ	0.7	0.3	4.5
Rx11,0.5λ	0.3	7.8	4.0	Rx8,2λ	0.8	2.3	8.1
Rx6,1λ	0.7	8.7	3.2	Rx9,2λ	0.5	4.9	0.3
Rx7,1λ	0.3	1.0	2.4	Rx10,2λ	0.2	6.9	1.7
Rx8,1λ	0.3	7.63	8.7	Rx11,2λ	0.2	7.0	4.8

TABLE III

THE OUTAGE CAPACITY ERRORS FROM EACH MODEL TO MEASUREMENT

Loc. Spa.	Error(%)			Loc. Spa.	Error(%)		
	novel	web	kron		novel	web	kron
Rx6,0.5λ	5.1	35.5	28.5	Rx9,1λ	1.6	3.0	5.0
Rx7,0.5λ	1.4	4.7	6.8	Rx10,1λ	3.0	1.8	8.3
Rx8,0.5λ	0.8	16.2	21.5	Rx11,1λ	2.1	5.5	7.6
Rx9,0.5λ	5.4	9.8	11.8	Rx6,2λ	1.3	22.8	29.4
Rx10,0.5λ	2.8	0.4	5.8	Rx7,2λ	4.5	12.6	15.3
Rx11,0.5λ	4.6	2.8	5.5	Rx8,2λ	1.5	9.9	14.6
Rx6,1λ	1.1	35.4	31.6	Rx9,2λ	5.6	10.0	13.5
Rx7,1λ	4.4	10.3	13.0	Rx10,2λ	0.1	4.1	8.8
Rx8,1λ	1.3	26.6	31.7	Rx11,2λ	5.9	5.7	7.4

V. CONCLUSION

In this paper, a corridor to stair MIMO measurement at 2.6GHz is carried out. A novel MIMO model based on singular value decomposition of matrix is proposed in this paper. The joint DoA-DoD-delay power spectrum and the channel capacity are selected as two metrics to evaluate the performance of the novel model. The results show that the novel model provides a better fit to the measured data than Kronecker model and Weichselberger model. However, the frequency selectivity of channel are not considered in this novel model. Wider frequency bands are needed to achieve higher throughput in future wireless communication systems. An extension of the novel model adapted to wideband cases may be an important work in the next period.

ACKNOWLEDGEMENT

This work is supported by China Telecommunication Technology Labs(CTTL). Thanks to Yan Qin, Xiao-lei Wang and other employees in CTTL. Thanks to the teachers in Jiangsu Key Laboratory of Wireless Communications

REFERENCES

[1]. S. Sesia, I.Toufik, M.Baker, LTE-The UMTS Long Term Evolution From Theory to Practice. John Wiley & Sons, 2009

[2]. P. Almers, E.Bonek, and A. Burr et al., "Survey of channel and radio propagation models for wireless MIMO systems," EURASIP J. Wireless Commun. New., vol. 2007, 2007, 10.1155/2007/19070, Article ID 19070.

[3]. H. Özcelik, "Indoor MIMO channel models," Ph.D. dissertation, Institut für Nachrichtentechnik und Hochfrequenztechnik, Technische Universität Wien, Vienna, Austria, Dec. 2004

[4]. A. M. Saleh, R. R. Valenzuela, "A statistical model for indoor multipath propagation," IEEE J. Sel. Areas Commun., vol. 52, no. 1, pp. 128-137, Feb. 1987.

[5]. Alighanbari, C. D. Sarris, "Parallel Time-Domain Full-Wave Analysis and System-Level Modeling of Ultrawideband Indoor Communication Systems," IEEE Trans. Antennas Propag., vol. 57, no. 1, pp. 231-240, JAN. 2009

[6]. T. Zwick, C. Fischer, W. Wiesbeck, "A stochastic multipath channel model including path directions for indoor environments," IEEE J. Sel. Areas Commun., vol. 20, no. 6, pp. 1178-1192, Aug.2002.

[7]. P. Petrus, J. H. Reed, T. S. Rappaport, "Geometrical-based statistical mcrocell channel model for mobile environment," IEEE Trans. Commun., vol. 50, no. 3, pp. 495-502, Mar. 2002

[8]. J. P. Kermoal, L. Schumacher, K. I. Pedersen, P. E. Mogensen, F. Frederiksen, "A stochastic MIMO radio channel model with experimental validation," IEEE J. Sel. Areas Commun, vol. 20, pp. 1211–1226, Aug. 2002.

[9]. W. Weichselberger, M. Herdin, H. Özcelik, E. Bonek, "A stochastic MIMO channel model with joint correlation of both link ends," IEEE Trans. Wireless Commun., vol. 5, pp. 90–100, Jan. 2006.

[10]. M. Sayeed, "Deconstructing multiantenna fading channels," IEEE Trans. Signal Process., vol.50, pp. 2563-2579, Oct. 2002.

[11]. R. Vaughan, J. B. Andersen, Channels, Antennas, and Propagation for Mobile Communications. London, U. K. : IEE, 2003.

[12]. Y. Zhang, O. Edfors, P. Hammarberg, et al., "A General Coupling Based Model Framework for Wideband MIMO Channels," IEEE Trans. Antennas Propag., vol. 60, no. 2, pp. 574-586, Feb. 2012

[13]. W. Kim, K. Lee, M. D. Kim, J. J. Park, H. K. Chung, H. Lee, "Performance analysis of spatially correlated MIMO-OFDM beamforming systems with the maximum eigenvalue model from measured MIMO channels," IEEE Trans. Wireless Commun., vol. 11, pp. 3744-3753, Oct. 2012.

[14]. W. Kim, K. Lee, M. D. Kim, J. J. Park, H. K. Chung, "Distribution of Eigenvalues for 2×2 MIMO Channel Capacity Based on Indoor Measurements," IEEE Trans. Wire. Commun., vol. 11, no. 4, pp. 1255-1259, Apr. 2012.

[15]. W. Weichselberger, "Spatial structure of multiple antenna radio channels," Ph.D. dissertation, Vienna Univ. of Technology, Vienna, Austria, 2003

[16]. T. T. Georgiou, "Distances and Riemannian metrics for spectral density functions," IEEE Trans. Signal Process., vol. 55, no. 8, pp. 3995-4003, Aug. 2007

[17]. J. G. Proakis, Digtial Communications, 3rd ed. New York: McGraw-Hill, 1995

[18]. S. Wyne, A. F. Molisch, P. Almers, et al., "Outdoor-to-Indoor Office MIMO Measurements and Analysis at 5.2 GHz," IEEE Trans Veh. Technol., vol. 57, no. 3, pp. 1374-1386, Ma -y 2008.

An Efficient 3D DI-FDTD Method for Anisotropic Magnetized Plasma Medium

Ye Zhou, Yi Wang, Min Zhu and Qunsheng Cao
College of Electronic and Information Engineering
Nanjing University of Aeronautics and Astronautics
Nanjing, 210016 China
Email: zhouye89@gmail.com

Abstract—**An efficient three-dimensional (3D) direct integration (DI) finite-difference time-domain (FDTD) method is introduced to solve electromagnetic wave propagation problems in anisotropic magnetized plasma medium using matrix method. Compared with previous 3D FDTD plasma method, our method is more flexible in adjusting medium parameters thus can deal with varied mediums. Through several simulations, the correctness of the method are proven and the advantages are clarified. Finally, potential applications of this method are discussed, such as the further study of the ionosphere.**

I. Introduction

Accurate simulation of electromagnetic (EM) wave propagation in plasma medium is often a difficult problem for its dispersive and anisotropic properties. However, an time domain approach proposed by Luebbers [1] presented a frequency-dependent formulation for transient propagation in plasma using the finite-difference time-domain (FDTD) method [2], which has become an efficient solution to such problems.

After that, the time domain numerical methods, instead of analytic ones, have become the major methods to solve these problems and many improved techniques were proposed. Some of the techniques [1, 3] are based on a difference approximation of Maxwells equations coupled to an iteration derived from the convolution integral form of the auxiliary differential equation, which are called recursive convolution (RC) methods. Some other techniques [4] are based on direct finite-difference approximations of the complete field equations of the medium, which are commonly referred to as direct integration (DI) methods. A systematic analysis of these techniques was reviewed by S. A. Cummer [5].

Among these techniques, the DI-FDTD method has much simpler forms and the accuracy is close to traditional RC-FDTD method, thus many two-dimensional (2D) and three-dimensional (3D) DI-FDTD methods were developed in the recent years.

Due to the existence of the geomagnetic field, the ionosphere surrounding the Earth becomes a gyrotropic plasma medium. Many plasma algorithms were applied to model the EM wave propagation in Earth-ionosphere system. An initial FDTD scheme that can deal with such an anisotropic medium was presented by Thèvenot [6], allowing the simulation of radiowave propagation in the Earth-ionosphere waveguide using a 2D spherical-coordinate FDTD method. Cummer [7] also reported a 2D FDTD method to simulate EM wave propagation in the Earth-ionosphere waveguide but using the cylindrical-coordinates. A 3D FDTD method was proposed by Lee [8] to study the transformation of an EM wave by a dynamic (time-varying) inhomogeneous magnetized plasma medium. The current density vector of such method is positioned at the center of the Yee cube to accommodate the anisotropy of the plasma medium. Nevertheless, it is only first-order accurate compared to another E-J collocated 3D FDTD method [9, 10] which is more accurate and has less memory-cost. However, the method is still not so perfect regarding the calculating time and computer memory.

In this paper, we have improved Yu and Simpson's 3D E-J collocated FDTD method [10] to a more flexible one, in which the parameter limits to maintain its accuracy are not as restrict as before. The robustness of parameter matrix is discussed to prove the method useful in a wide variety of fields. In Section II, the governing equations for the 3D magnetized cold plasma are described, as well as the the resulting FDTD time-stepping algorithm. Section III illustrate numerical examples of the 3D plasma FDTD method in which both unmagnetized and magnetized plasma cases are provided. Finally, Section IV concludes the paper and forecasts the future applications of the 3D FDTD method.

II. Approach

A. Governing Equations

In this paper, plasma mediums are assumed anisotropic and the method is based on a 3D Cartesian coordinate. Wave propagation effect introduced by electrons, positive ions and negative ions are included for generality. An extra magnetic flux density $\boldsymbol{B}$ is set here to simulate natural geomagnetic field. The governing equations of anisotropic magnetized cold plasma consist of three Lorentz current equations derived from Lorentz equation of motion which modeling the response of each charged particle species to the electric field $\boldsymbol{E}$ and the extra magnetic flux density $\boldsymbol{B}$, as well as two Maxwell's curl equations including total induced current density $\boldsymbol{J}_I$ and source current density $\boldsymbol{J}_S$. The whole governing equation is given by

978-7-5641-4279-7

$$\nabla \times \boldsymbol{E} = -\mu_0 \frac{\partial \boldsymbol{H}}{\partial t} \tag{1}$$

$$\nabla \times \boldsymbol{H} = \varepsilon_0 \frac{\partial \boldsymbol{E}}{\partial t} + \boldsymbol{J}_I + \boldsymbol{J}_S \tag{2}$$

$$\frac{\partial \boldsymbol{J}_e}{\partial t} + v_e \boldsymbol{J}_e = \varepsilon_0 \omega_{Pe}^2 \boldsymbol{E} + \boldsymbol{\omega}_{Ce} \times \boldsymbol{J}_e \tag{3}$$

$$\frac{\partial \boldsymbol{J}_p}{\partial t} + v_p \boldsymbol{J}_p = \varepsilon_0 \omega_{Pp}^2 \boldsymbol{E} - \boldsymbol{\omega}_{Cp} \times \boldsymbol{J}_p \tag{4}$$

$$\frac{\partial \boldsymbol{J}_n}{\partial t} + v_n \boldsymbol{J}_n = \varepsilon_0 \omega_{Pn}^2 \boldsymbol{E} + \boldsymbol{\omega}_{Cn} \times \boldsymbol{J}_n \tag{5}$$

$$\boldsymbol{J}_I = \sum_l \boldsymbol{J}_l = \boldsymbol{J}_e + \boldsymbol{J}_p + \boldsymbol{J}_n \tag{6}$$

Here the subscript l indicates the charged particle species in the plasma (e, p and n as electrons, positive ions and negative ions, respectively). $\boldsymbol{J}_e$, $\boldsymbol{J}_p$ and $\boldsymbol{J}_n$ are the current densities of each species. v_e, v_p and v_n are the collision frequencies of each species. In addition, ω_{Pe}, ω_{Pp} and ω_{Pn} shows the plasma frequencies of each species with detailed structure in (7).

$$\omega_{Pl} = \sqrt{\frac{q_l^2 n_l}{\varepsilon_0 m_l}} \tag{7}$$

Further, $\boldsymbol{\omega}_{Ce}$, $\boldsymbol{\omega}_{Ce}$ and $\boldsymbol{\omega}_{Ce}$ are the cyclotron frequencies of each species given by (8).

$$\boldsymbol{\omega}_{Cl} = \frac{q_l \boldsymbol{B}}{m_l} \tag{8}$$

It is clear that the cyclotron frequency is a function of magnetic flux density $\boldsymbol{B}$ so that if the cross-product terms in (3)–(5) is set to zero, the whole governing equations will reduce to isotropic, in other words, the wave behavior is independent of its propagation direction.

The complete scalar equations derived from equations (1)–(6) are shown in [10].

B. FDTD Discretization Scheme

Take the scalar equations into the FDTD grids. J components locate at the same positions of E ones. Central differencing of the space derivatives is employed to transform them to the update equations which is easy for computing. On time derivatives, J and E components are supposed to be at the integer timesteps, indicated as n, while H components are at the semi-integer timesteps, indicated as $(n + 1/2)$.

Thus, the semi-implicit equations come with a new problem that each of E and J components must be iterated simultaneously. Three parameter matrices are then introduced to solve this problem perfectly. The approach used here is very similar to the matrix method described in [10] considering in detail the spatial averaging but without value scaling of $\tilde{H}_u = (\mu_0 \Delta u / \Delta t) H_u [u = x, y, z]$ and $\tilde{J} = (\Delta t / \varepsilon_0) J$. Matrix A and B represent the parameter coefficients of the E and J components at the present and previous timestep and matrix C represents the coefficients of H and J_S components at previous timestep. However, the matrix C used in Yu's method could be simplified as each two derivatives of H components in third column could be grouped so that the storage of parameter matrix is reduced. The last term of the equation array is described by

$$\begin{bmatrix} \frac{\Delta t}{\varepsilon_0} & & & -\frac{\Delta t}{\varepsilon_0} & & \\ & \frac{\Delta t}{\varepsilon_0} & & & -\frac{\Delta t}{\varepsilon_0} & \\ & & \frac{\Delta t}{\varepsilon_0} & & & -\frac{\Delta t}{\varepsilon_0} \\ \vdots & & & & & \vdots \\ 0 & & \cdots & \cdots & & 0 \end{bmatrix} \cdot \begin{bmatrix} \frac{\Delta H_z}{\Delta y} - \frac{\Delta H_y}{\Delta z} \\ \frac{\Delta H_x}{\Delta z} - \frac{\Delta H_z}{\Delta x} \\ \frac{\Delta H_y}{\Delta x} - \frac{\Delta H_x}{\Delta y} \\ J_{Sx} \\ J_{Sy} \\ J_{Sz} \end{bmatrix} \tag{9}$$

where the left side is the simplified matrix C.

C. Stability and Accuracy Analysis of the Scheme

Since the solutions of the semi-implicit differencing equations have a growth per timestep factor of $(1 - v\Delta t/2)/(1 + v\Delta t/2)$, most of the plasma FDTD methods require a strict criterion $v\Delta t \ll 1$ that limits the efficiency of the FDTD method [5]. However, the matrix method avoids such problem because the elements in diagonal of matrix A equals unit or $(1 + v_l \Delta t/2)$. Whether the criterion $v\Delta t \ll 1$ is fitted or not, the parameters $A^{-1}B$ and $A^{-1}C$ of iteration equation arrays are valid. Even for magnetized case, spatial grid-cell size is not necessary to be chosen carefully to satisfy the stability condition and accuracy requirements. For the same reason, the value scaling could be canceled to simplify the calculation process.

III. Demonstration

In this Section, the modified 3D FDTD plasma method is applied to simulate EM wave propagation in a space with a spheric plasma object inside. The uniaxial perfectly matched layer (UPML) boundary condition [11] is used here to absorb the wave outside the calculation region. All the E, H and J field components are discretized in x, y and z directions indicated i, j and k as their subscripts. $\Delta x = \Delta y = \Delta z = 2$ mm and $\Delta t = 3.3333$ ps. A pair of z-polarized differential Gaussian pulse dipole is generated as the wave source of our numerical demonstrations.

A homogeneous spheric plasma object is set inside the calculation space characterized by
electron plasma frequency:

$$\omega_{Pe} = 1.8 \times 10^{11} rad/s \tag{10}$$

and neutral-electron collision frequency:

$$v_e = 2 \times 10^{11} rad/s \tag{11}$$

Fig. 1 illustrates the snapshot of electric field E_z distribution after the pulsed wave propagating for 100 timesteps. The magnitude of all values is normalized. No extra magnetic flux density is applied here, in other words, $\boldsymbol{B} = 0$. In this case, $v\Delta t < 1$ is achieved to meet the fundamental requirement of traditional plasma DI-FDTD algorithm.

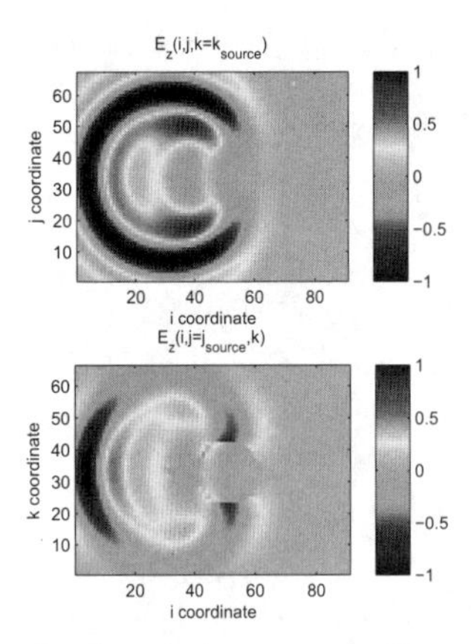

Fig. 1: Snapshot for pulsed wave propagation with a spheric region of unmagnetized plasma (a) upper panel: xOy-plane (b) lower panel: xOz-plane

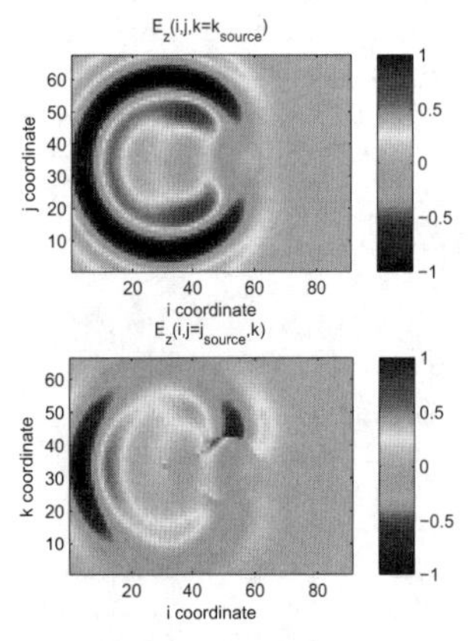

Fig. 2: Snapshot for pulsed wave propagation with a spheric region of magnetized plasma (a) upper panel: xOy-plane (b) lower panel: xOz-plane

We then apply an extra magnetic flux density $\boldsymbol{B}$ as $B_x = B_y = B_z = 2$ T in order to simulate the wave propagation through anisotropic magnetized plasma medium.

Fig. 2 shows the snapshot for the magnetized plasma medium case. The influence of the geomagnetic field is clear compared to Fig. 1 for its effect on wave rotation.

The final demonstration is shown in Fig. 3 as an example of particular case in which $v\Delta t$ is significantly greater than unit. We suppose that $\Delta = 60$ mm and $\Delta t = 100$ ps and other condition keeps the same as the first case so that $v\Delta t = 20$ which is not under the circumstance of $v\Delta t \ll 1$. The result shows that the improved method is still stable and accurate.

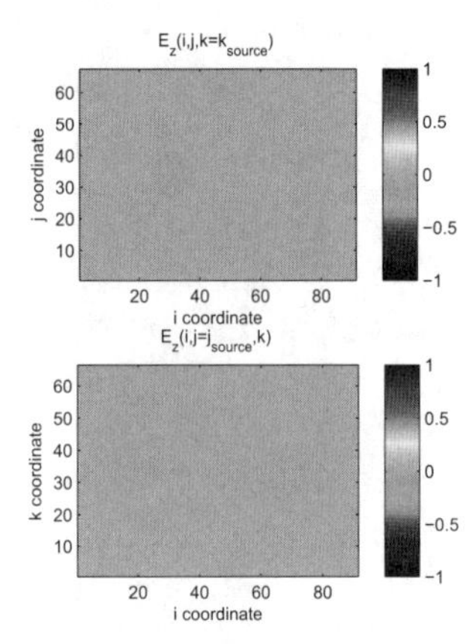

Fig. 3: Snapshot for pulsed wave propagation with a spheric region of unmagnetized plasma under the circumstance of $v\Delta t = 20$ after 10000 timesteps (a) upper panel: xOy-plane (b) lower panel: xOz-plane

IV. Summary and Conclusion

In this work, an efficient DI-FDTD method including anisotropic magnetized plasma medium is presented using the matrix method. The algorithm involves direct integration of current density term in semi-implicit equation arrays. Since the particular structure of the matrix, the parameter of the plasma is more flexible than that of traditional plasma FDTD methods.

Its validity is demonstrated by calculating the wave propagation of pulse generated from a pair of polarized dipole. The results of three simulations under different conditions, even over the limit of tradition algorithm, agree well with the expectation. Therefore, for its nice property in low frequency (Δt is large), the method is supposed to be efficient to a wide variety of applications such as the study of wave propagation in ionosphere.

Since ionosphere is known to be an anisotropic plasma medium to certain extent due to the geomagnetic field, the method presented in this paper may play a role in future study coupling to real models of the Earth-ionosphere system [12].

Acknowledgment

This work was supported by the Natural Science Foundation of China (BK2009368), the Jiangsu Planned Projects for Postdoctoral Research Funds (1201006C) and the Jiangsu Province meteorological detection and information processing Key Laboratory (KDXS1202).

REFERENCES

[1] R. J. Luebbers, F. Hunsberger, and K. S. Kunz, "A frequency-dependent finite-difference time-domain formulation for transient propagation in plasma," *IEEE Trans. Antennas Propag.*, vol. 39, no. 1, pp. 29–34, 1991.

[2] K. Yee, "Numerical solution of initial boundary value problems involving maxwell's equations in isotropic media," *IEEE Trans. Antennas Propag.*, vol. 14, no. 3, pp. 302–307, 1966.

[3] D. F. Kelley and R. J. Luebbers, "Piecewise linear recursive convolution for dispersive media using FDTD," *IEEE Trans. Antennas Propag.*, vol. 44, no. 6, pp. 792–797, 1996.

[4] L. Nickisch and P. Franke, "Finite-difference time-domain solution of maxwell's equations for the dispersive ionosphere," *IEEE Antennas Propag. Mag.*, vol. 34, no. 5, pp. 33–39, 1992.

[5] S. A. Cummer, "An analysis of new and existing FDTD methods for isotropic cold plasma and a method for improving their accuracy," *IEEE Trans. Antennas Propag.*, vol. 45, no. 3, pp. 392–400, 1997.

[6] M. Thèvenot, J. P. Bérenger, T. Monedière, and F. Jecko, "A FDTD scheme for the computation of VLF-LF propagation in the anisotropic earth-ionosphere waveguide," *Ann. Télécommun.*, vol. 54, no. 5-6, pp. 297–310, 1999.

[7] S. A. Cummer, "Modeling electromagnetic propagation in the Earth-ionosphere waveguide," *IEEE Trans. Antennas Propag.*, vol. 48, no. 9, pp. 1420–1429, 2000.

[8] J. H. Lee and D. K. Kalluri, "Three-dimensional FDTD simulation of electromagnetic wave transformation in a dynamic inhomogeneous magnetized plasma," *IEEE Trans. Antennas Propag.*, vol. 47, no. 7, pp. 1146–1151, 1999.

[9] W. Hu and S. A. Cummer, "An fdtd model for low and high altitude lightning-generated EM fields," *IEEE Trans. Antennas Propag.*, vol. 54, no. 5, pp. 1513–1522, 2006.

[10] Y. Yu and J. J. Simpson, "An E-J collocated 3-D FDTD model of electromagnetic wave propagation in magnetized cold plasma," *IEEE Trans. Antennas Propag.*, vol. 58, no. 2, pp. 469–478, 2010.

[11] S. D. Gedney, "An anisotropic perfectly matched layer-absorbing medium for the truncation of FDTD lattices," *IEEE Trans. Antennas Propag.*,vol. 44, no. 12, pp. 1630–1639, 1996.

[12] J. J. Simpson, "Current and future applications of 3-D global earth-ionosphere models based on the full-vector Maxwell's equations FDTD method," *Surveys Geophys.*, vol. 30, no. 2, pp. 105–130, 2009.

Design Principles and Applications of a novel Electromagnetic Spectrum Table

Min Ju, Shi-Lin Xiao
State Key Laboratory of Advanced Optical Communication Systems and Networks,
Department of Electronic Engineering, Shanghai Jiao Tong University,
800 Dongchuan Road, Shanghai 200240, China
E-mail: slxiao@sjtu.edu.cn

Abstract- **Electromagnetic spectrum diagram is used to represent the content of electromagnetic wave. At present different kinds of electromagnetic spectrum representations make people confused. In this paper, design principles and important applications of a novel electromagnetic spectrum table are introduced. In the new table, electromagnetic wave bands are classified in more detail, especially for the first time terahertz wave band is redefined from 0.15 Thz to 6 Thz according to short millimeter wave limit and long infrared wave limit. This novel spectrum table can be used as electromagnetic spectrum representation standard in academic and will bring great convenience for researchers and learners.**

I. INTRODUCTION

The development of electromagnetic wave technology extends scientific research fields and drives many emerging disciplines. However, electromagnetic products make their users under a long time hazards of electromagnetic radiation. The electromagnetic wave application and hazard prevention make it pressing to strengthen the education and popularization of electromagnetic wave knowledge.

Traditionally the content of electromagnetic wave is described by frequency-wavelength diagrams, as shown in figure 1 and figure 2 [1]. In these diagrams, two close parallel lines signify the track of electromagnetic frequency and wavelength, respectively. Based on these two lines, electromagnetic spectrum is commonly divided into radio, microwave, infrared, visible, ultraviolet ray, X-ray and gamma ray. In some diagrams, applications of the waveband are also briefly described in the diagrams. Even though these diagrams can be helpful for people to research on electromagnetic wave, different kinds of electromagnetic wave diagrams make people confused.

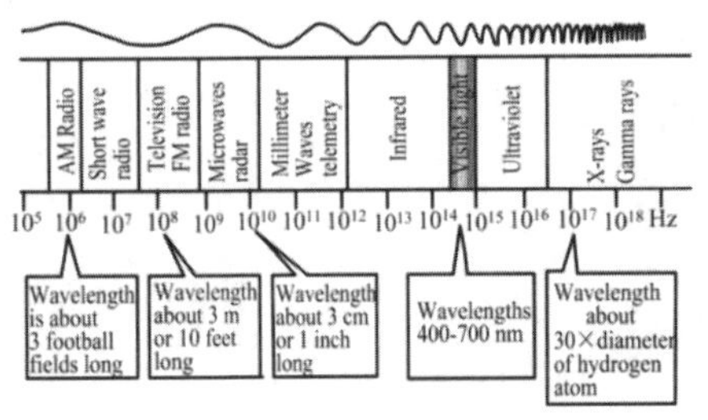

Figure 1. Electromagnetic spectrum

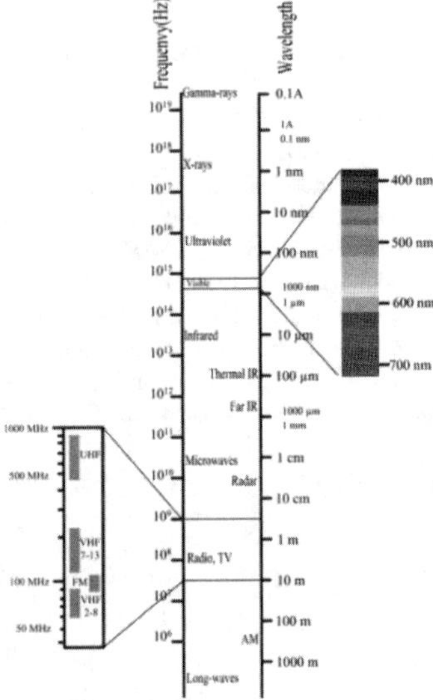

Figure 2. Electromagnetic spectrum

Furthermore, it is limited in these diagrams with one-dimensional space, because electromagnetic wave wavelength and frequency change continuously from zero to infinity. The one-dimensional representation cannot show this nature very well. In this paper, design principles and applications of a novel electromagnetic spectrum table are introduced. In the new table, electromagnetic wave is firstly divided into air wave, terahertz wave and light wave, and finer divisions are made in these three wave bands. In addition, based on short millimeter wave limit and long infrared wave limit, terahertz wave band is redefined from 0.15 Thz to 6 Thz. In terahertz wave band, the concepts and ideas of "near air wave" and "near light wave" are first proposed. "near air wave" band is from 0.15 Thz to 1 Thz and "near light wave" band is from 1 Thz to 6 Thz. These two bands help people to research on terahertz wave band referring to electronics category and photonics category.

II. DESIGN PRINCIPLES OF THE NOVEL "ELECTROMAGNETIC SPECTRUM TABLE"

Electromagnetic wave originates from the vibration of electric charge. This vibration creates a wave which has an electric

978-7-5641-4279-7

component and a magnetic component. The succession of induced fields (electric to magnetic to electric to magnetic, etc.) results in the generation of electromagnetic wave [2], whose propagation velocity c is the product of wavelength λ and frequency ν,

$$c = \lambda \cdot \nu. \quad (1)$$

It is easy to obtain that it is an inverse relationship between electromagnetic wave frequency and wavelength in a specific medium. And electromagnetic wave frequency and wavelength are continuously from zero to infinity, so electromagnetic wave diagram representations in one-dimensional space are limited when comprehensively analyzing the electromagnetic wave properties. As mentioned before, the changes of frequency and wavelength in diagrams are indicated by the two close parallel lines, which cannot clearly describe the properties of electromagnetic wave bands.

In the new table 1 (see the Appendix), two improvements have been done: (1) continuous electromagnetic spectrum is described by discrete wave bands; (2) electromagnetic spectrum description is changed from one-dimensional space to two-dimensional space. In the first column of the new table, electromagnetic wave is firstly divided into air wave, terahertz wave and light wave according to obvious differences of electromagnetic wave nature and characteristics. Different kinds of air wave are listed on the basis of traditional classification (Ultra low frequency, Radio wave, etc.). In addition, light wave is divided into micron wave, nano wave and pico wave. In nano wave band, color symbols indicate the particularity and importance of visible light. Furthermore, for electromagnetic wave, some interesting properties can be learned in quantum domain [3]. The energy of each quantum is indicated at the bottom of the table, which is the product of Planck constant and electromagnetic wave frequency, i.e. E=hv=4.1357×10-15×v, 1 Electron volt (eV) =6.6261×10-34×v(J).

III. New division of terahertz wave band

The first occurrence of the term terahertz is attributed to Fleming in 1974, when the term was used to describe the spectral line frequency coverage of a Michelson interferometer [4]. As early as 1896 and 1897, researches of Rubens and Nichols involved this wave band. Terahertz wave band is defined from 0.1 THz to 10 THz traditionally, which lies between millimeter wave and infrared wave as marked in figure 3.

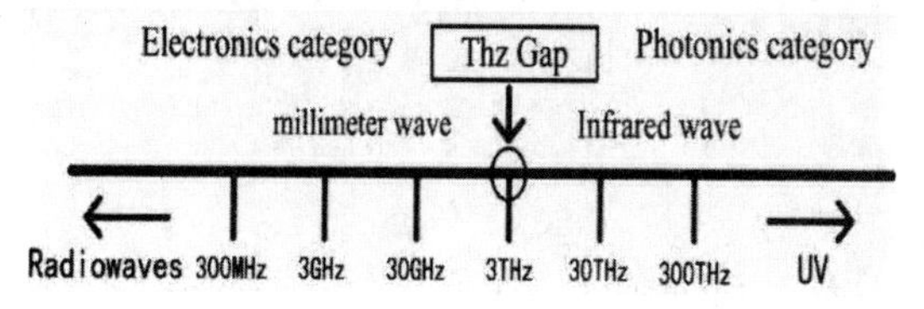

Figure 3 Terahertz wave "gap"

However, the very large THz portion of the spectrum has not been particularly useful because there were neither suitable emitters to send out controlled THz signals nor efficient sensors to collect them and record information. As a result, the THz portion of electromagnetic spectrum was called the THz gap. The long wave of terahertz wave band is overlapped with millimeter wave, and the short wave of this band is overlapped with infrared wave. On the other hand, long terahertz wave mainly belongs to electronics category, and short terahertz wave basically belongs to photonics category, thus terahertz wave band once became the research gap of the electromagnetic spectrum [5].

In the new table, terahertz wave band is redefined from 0.15 Thz to 6 Thz, because short millimeter wave limit is 2 mm (0.15 Thz) [6] and long infrared wave limit is 50 μm (6 Thz) [7]. Furthermore, terahertz wave is finely divided into "near air wave (0.15 Thz -1Thz)" and "near light wave (1Thz -6Thz)". The two names refer to the naming of near infrared wave and near ultraviolet wave. The long light wave limit is 300 μm (1Thz) [8], so it is reasonable that "near air wave" band is from 2 mm (0.15 Thz) to 300 μm (1Thz) and "near light wave" band is from 300 μm (1 Thz) to 50 μm (6 Thz).

With fine division of terahertz wave band, the study of "near air wave" and "near light wave" can refer to electronics technology and photonics technology respectively. Thus, good research ideas and methods in light wave and air wave should be used to solve terahertz wave problems. People will be easier to obtain and understand characteristics of terahertz wave knowledge.

IV. Applications of the novel "Electromagnetic Spectrum Table"

Based on the above, we can clearly find advantages of new electromagnetic spectrum table. From previous electromagnetic spectrum representations, we just roughly know the spectrum division of electromagnetic wave. It is difficult to find the features and applications of more specific electromagnetic wave bands. In addition, the division of wave bands is not very obvious in one-dimensional space, which limits people to get electromagnetic wave information in previous electromagnetic spectrum representations. In the new table, electromagnetic wave bands are arranged scientifically and detailedly, no matter the overall division or the partial division. The frequency range and principal uses of each wave band lies clearly in the new table. For a certain wave band, people can easily find the features and research on this wave band.

Most parts of electromagnetic spectrum are used in science to study and characterize matter. But the research and improvement of terahertz wave band is slower than others in a very long time because of the terahertz wave "gap". Now the applications of terahertz wave have been found in many fields, such as physics, material science, electrical engineering, chemistry and so on. The division of "near air wave" and "near light wave" is significant for terahertz wave study.

Nevertheless, exposure to strong electromagnetic field may cause damage for humans and animals, even though the exposure is of short duration. It can make the long-established physical laws of humanity seriously damaged, and lead to impaired immune function [9]. The new table let us master the electromagnetic wave knowledge better and we should use electromagnetic wave technology carefully to reduce the bad effects as many as possible.

V. CONCLUSION AND DISCUSSION

In the new electromagnetic spectrum table, detailed wave bands and corresponding application descriptions help people research on electromagnetic wave, especially for terahertz wave. Even though the new table has important value in physics, there is still a little imperfection which needs for further study. For example, since the common view of wavelength range in visible light wave band is not clear, the range 730-400 nm in new table needs to be further discussed. However, new electromagnetic spectrum table is very useful for electromagnetic wave researchers and learners. It will play a important role in promoting for electromagnetic wave knowledge popularization and related work in education.

ACKNOWLEDGMENT

Finally, the authors would like to express thanks to professor Wen-Quan Lu for the key recommendation and the careful help in the writing process of this letter.

REFERENCES

[1] Electromagnetic spectrum http://en.wikipedia.org/wiki/Main_Page Jan.8, 2013.

[2] Akira Ishimaru, Electromagnetic wave propagation, radiation, and scattering [M], Prentice Hall, Englewood cliffs, 1991, pp.121-124.

[3] Y. Aharonov, D.Bohm and H.II.Wills, "Significance of electromagnetic potentials in the quantum theory", Phys. Rev. vol. 115, 1959, pp485-487.

[4] Peter H. Siegel, "Terahertz technology,"IEEE Trans. Inf. Microwave theory and techniques. vol.50, 2002, pp910.

[5] E. J. Nichols and J. D. Tear, "Joining the infrared and electric wave spectra," Astrophys. J, 1927, pp61-17.

[6] xueguan Liu and huiping Guo, Microwave and antenna [M],Xidian University Press, xian,2004, pp1-224.

[7] E.M.Sparrow and R,D.Cess,Radiation, et al. Chuanbao Gu and Xuexue Zhang, Heat Transfer[M], Higher Education Press, Beijing,1983.

[8] Jingzhen Li, Handbook of Opties[M], Shaanxi Science Press, xian, 1986, pp64-65.

[9] A.Bhanu Lavanya, "Effects of electromagnetic radiation on biological systems: A short review of case studies", Proceedings of INCEMIC, 2003, pp87-90.

Appendix

Table.1 A novel table of electromagnetic spectrum

Name of wave band					frequency section $\nu=(3\times10^{8}\mathrm{m}/s)/\lambda_0$	Wavelength section $\lambda_0=(3\times10^{8}\mathrm{m}/s)/\nu$	principal use
Air wave	Ultra low frequency (ULF)		ULW		3(Hz)-3 (kHz)	(100000-100) (Km)	Electronics, headphones......
Air wave	Radio wave	Very low frequency (VLF)	VLW		(3-30) (kHz)	(100-10) (Km)	AM radio, walkie-talkie medical, long-haul communication, induction cooker......
Air wave	Radio wave	low frequency (LF)	LW		(30-300) (kHz)	(10-1) (Km)	
Air wave	Radio wave	Middle frequency (MF)	MW		300(kHz) -3 (MHz)	(1000-100) (m)	
Air wave	Radio Frequency	High frequency (HF)	SW		(3-30) (MHz)	(100-10) (m)	TV, AM and FM radio.....
Air wave	Radio Frequency	Very high frequency (VHF)	MW		30(MHz)-1 (GHz)	(10-0.3) (m)	
Air wave	Micro wave	Ultra high frequency (UHF)	Deci meter wave	L	(1-2) (GHz)	(30-15) (cm)(ref.: 22cm)	Mobile communications, microwave oven......
Air wave	Micro wave	Ultra high frequency (UHF)	Deci meter wave	S	(2-4) (GHz)	(15-7.5) (cm) (ref: 10cm)	
Air wave	Micro wave	Super high frequency (SHF)	Centi meter wave	C	(4-8) (GHz)	(7.5-3.75) (cm) (ref: 5cm)	Satellite broadcasting television, medical, communication, radar, telemetry, electronic reconnaissance, detection......
Air wave	Micro wave	Super high frequency (SHF)	Centi meter wave	X	(8-12) (GHz)	(3.75-2.5) (cm) (ref: 3cm)	
Air wave	Micro wave	Super high frequency (SHF)	Centi meter wave	Ku	(12-18) (GHz)	(2.5-1.67) (cm) (ref: 2cm)	
Air wave	Micro wave	Super high frequency (SHF)	Centi meter wave	K	(18-27) (GHz)	(1.67-1.11) (cm) (: ref 1.25cm)	
Air wave	Micro wave	Super high frequency (SHF)	Centi meter wave	Ka	(27-40) (GHz)	(1.11-0.75) (cm)(ref: 0.8cm)	
Air wave	Micro wave	Extremely high frequency (EHF)	Milli meter wave	U	(40-60) (GHz)	(7.5-5) (mm) (ref: 6mm)	Communication, radar, Medical, astronomy, detection.....
Air wave	Micro wave	Extremely high frequency (EHF)	Milli meter wave	V	(60-80) (GHz)	(5-3.75) (mm) (ref: 4mm)	
Air wave	Micro wave	Extremely high frequency (EHF)	Milli meter wave	W	(80-150) (GHz)	(3.75-2) (mm)(ref:3mm)	
Terahertz wave		Near air wave			(0.15-1.0) (THz)	(2-0.3) (mm)	Scientific research tools, Imaging technology, plasma detection......
Terahertz wave		Near light wave			(1/300-1/50)300 (THz)	(300-50) (μm)	
Light wave	Micron wave	infrared wave	Far infrared		(1/50-1/10.6) 300 (THz)	(50-10.6) (μm)	Heating, exploration......
Light wave	Micron wave	infrared wave	Middle infrared		(1/10.6-1/1.675)300 (THz)	(10.6-1.675) (μm)	Medical, convection oven, Laser processing......
Light wave	Micron wave	infrared wave	Near infrared	U	(1/1.675-1/1.625)300 (THz)	(1.675-1.625) (μm)	Optical transmission, wavelength complex, amplification , remote detection, sensing......
Light wave	Micron wave	infrared wave	Near infrared	L	(1/1.625-1/1.566)300 (THz)	(1.625-1.566) (μm)	
Light wave	Micron wave	infrared wave	Near infrared	C	(1/1.566-1/1.53)300 (THz)	(1.566-1.53) (μm)	
Light wave	Micron wave	infrared wave	Near infrared	S	(1/1.53-1/1.46)300 (THz)	(1.53-1.46) (μm)	
Light wave	Micron wave	infrared wave	Near infrared	E	(1/1.46-1/1.36)300 (THz)	(1.46-1.36) (μm)	
Light wave	Micron wave	infrared wave	Near infrared	O	(1/1.36-1/1.26)300 (THz)	(1.36-1.26) (μm)	
Light wave	Micron wave	infrared wave	Near infrared	Short wave	(1/1.26-1/1.06)300 (THz)	(1.26-1.06) (μm)	Laser medical, laser cosmetology, ranging, processing , guidance systems, detection ......
Light wave	Micron wave	infrared wave	Near infrared	Short wave	(1/1.06-1/0.94)300 (THz)	(1.06-0.94) (μm)	
Light wave	Micron wave	infrared wave	Near infrared	Ultra short wave	(1/0.94-1/0.85)300 (THz)	(0.94-0.85) (μm)	
Light wave	Micron wave	infrared wave	Near infrared	Ultra short wave	(1/0.85-1/0.78)300 (THz)	(0.85-0.78) (μm)	
Light wave	Micron wave	infrared wave	Near infrared	Ultra short wave	(1/0.78-1/0.73)300 (THz)	(0.78-0.73) (μm)	
Light wave	Nano wave	Visible light	Red		(1/0.73-1/0.66)300 (THz)	(730-660) (nm)	Lighting, biological photosynthesis, print, copy, scanning, photovoltaic power generation, remote control...
Light wave	Nano wave	Visible light	Orange		(1/0.66-1/0.60)300 (THz)	(660-600) (nm)	
Light wave	Nano wave	Visible light	Yellow		(1/0.60-1/0.54)300 (THz)	(600-540) (nm)	
Light wave	Nano wave	Visible light	Green		(1/0.54-1/0.50)300 (THz)	(540-500) (nm)	
Light wave	Nano wave	Visible light	Cyan		(1/0.50-1/0.46)300 (THz)	(500-460) (nm)	
Light wave	Nano wave	Visible light	Blue		(1/0.46-1/0.44)300 (THz)	(460-440) (nm)	
Light wave	Nano wave	Visible light	Violet		(1/0.44-1/0.40)300 (THz)	(440-400) (nm)	
Light wave	Nano wave	Ultraviolet wave	Near ultraviolet		(1/0.40-1/0.20)300 (THz)	(400-200) (nm)	Disinfection, currency detector, sterilization, fault detection, Communication......
Light wave	Nano wave	Ultraviolet wave	Middle ultraviolet		(1/0.20-1/0.1)300 (THz)	(200-100) (nm)	
Light wave	Nano wave	Ultraviolet wave	Far ultraviolet		(1/0.1-1/0.01) 300 (THz)	(100-10) (nm)	
Light wave	Picometer wave		X-Ray		(100-10000) 300 (THz)	(10-0.1) (nm)	Medical check, material structure analysis ...
Light wave	Picometer wave	Special radiation	γ -Ray		(10000-1000000)300 (THz)	(100-1) (pm)	Fault detection, and strategic weapons......
Light wave	Picometer wave	Special radiation	High energy radiation		>300000000 (THz)	<1 (pm)	Substance processing, strategic weapons......

Note: After quantization the energy of each quantum, E=hv=$4.1357\times10^{-15}\times\nu$, 1 Electron volt (eV) =$6.6261\times10^{-34}\times\nu$(J)

An Integrated Transition of Microstrip to Substrate Integrated Nonradiative Dielectric Waveguide Based on Printed Circuit Boards

Fan Li and Feng Xu
School of Electronic Science and Engineering
Nanjing University of Posts and Telecommunications
Nanjing 210003 China

***Abstract*-Multilayer circuit is essential for hybrid integrated technology. In this paper, a triple-layer transition of microstrip line to substrate integrated nonradiative dielectric waveguide (SINRD) is presented, of which the SINRD waveguide is fabricated directly on PCBs through air holes. Due to this integrated transition, the realization of hybrid integrated systems directly connected with planar circuits is possible. Furthermore, to use SINRD waveguide fabricated on PCBs instead of traditional NRD is because that mechanism and integration of the former are much easier. In addition, a double-layer transition is also proposed to compare with the triple-layer transition and to demonstrate that the leakage loss from the uncovered SINRD waveguide is negligible. Finally, the simulation results of the transition of conventional NRD guides are compared to those of SINRD guides. Good agreements can be found.**

I. INTRODUCTION

The past decade has seen a huge growth in the area of microwave applications, resulting in an ever increasing demand for bandwidths, which pushes the microwave applications in the region of millimeter wave. And nonradiative dielectric waveguide (NRD) has been proved to be a promising candidate of traditional waveguide in microwave and particularly millimeter wave region due to its superior performance.

NRD guide was first proposed by Yoneyama and Nishida [1]. It consists of a dielectric strip sandwiched by two parallel conducting plates, which separates less than $\lambda/2$, λ is the free-space wavelength. In such a case, LSE_{mn} and LSM_{mn} modes of the NRD guide are below cut-off and vanish gradually out of the central dielectric strip. Due to this unique feature, NRD is a waveguide with mostly no radiation loss at bends and discontinuities. Thus it has been used to design a class of active and passive circuits for microwave and millimeter wave systems.

However, NRD, as a nonplanar structure, can hardly meets all the sever requirements independently, e.g., compactness, simple mechanization, low conduction and radiation losses for millimeter wave systems. Resulting from these inevitable problems, planar integrated circuits still play important roles in microwave and millimeter-wave systems. Therefore, it is a vital key to combine planar circuits with the nonplanar structure. To do so, an appropriate transition exploiting both advantages and avoiding both drawbacks is critical [2],[3].

To realize the hybrid planar/NRD integration technology, a series of transitions have been studied, e.g., the transitions from microstrip line to NRD, and slotline to NRD. However, all of these transitions are fabricated in the use of conventional NRD guide, which is not easy to implement and has bad mechanical tolerance, especially in higher frequency spectrum. To overcome these difficulties, a simple type of NRD guide which is formed by drilling via-holes directly on printed circuits boards (PCB) is proposed [4]. As a result, this kind of structure can easily integrate with other planar circuits.

In this paper, a double layers integrated transition of mirostrip line to substrate integrated nonradiative dielectric waveguide (SINRD) is proposed. With the ground of microstrip line covering on one side of the SINRD and none on the other, the performance of the transition is acceptable. It is means that leakage losses derived from the via-air slots are suppressed through carefully designing the dimensions and patterns of the via-air slots. Also, a three layers integrated transition of mirostrip line to SINRD is presented. With both sides of the SINRD are covered by the grounds of the two microstrip lines respectively, leakage loss are completely suppressed. And transitions formed on the top and the bottom plates of the NRD guide means that hybrid integrated transition of planar to NRD guide is flexible, and multilayer systems can be designed as compact as possible and no space is waste. Simulation results have demonstrated that the transition of mirostrip line to SINRD fabricated directly on PCBs offer high coupling efficiency and low power loss. It is means that this new type of transition promises to be useful in the future.

II. EXISTED INTEGRATION TECHNOLOGY BASED ON NRD GUIDE

To throw light on the motivation of proposing a new form of hybrid integrated transition, it is necessary to firstly discuss the shortcomings of the existed integrations based on NRD guide before moving forward to the details of the presented hybrid integrated technology. The traditional active circuits based on NRD guide are usually designed with two-terminal active devices which are mostly inserted into the NRD central strip through a planar surface sheet [5]. However, the active

978-7-5641-4279-7

devices are not involved in the planar circuits to which the signals directly transmit from NRD guide. And impedance mismatch between NRD guide and planar circuits is also a headache problem. To solve it, a thin dielectric sheet with air-gap dimension is used. Therefore, the air-gap (separation between the metallic plates of NRD guide), directly limiting the dimensions of the dielectric sheet which connects the NRD guide and the planar mount, is vitally critical, particularly in mm-wave frequencies even involving with two terminal devices. When three terminal devices are used in practical circuits, it is almost impossible to integrate the NRD guide with the planar mount using the mentioned technology. It is superior to coherently combine these two dissimilar structure, considering the fact that planar circuits is suitable to integrated both two and three active terminal structures but will exhibit bad transmission loss at mm-wave in design of passive devices, and the fact that NRD guide is a counterpart in design of passive devices. Follow this concept, both advantages could be inherited and drawbacks could be eliminated.

III. Description of the Presented Transition

An integrated transition derived from the concept of the aperture coupling is presented in Fig.1 on the basis of the above the background. This geometry consists of a microstrip line deposited on the top plate on NRD guide. The rectangular couple aperture is etched on the ground plane which is also shared by the NRD guide as its metallic plate. And the microstrip line is perpendicular to the rectangular slot and NRD's central dielectric strip simultaneously. As is known, LSM_{11}, as a nonradiative mode in NRD guide, is usually preferred for practical applications. The electrical field of this mode is orthogonal to the air-dielectric interface of the NRD guide, while the magnetic field is parallel to the same interface as is shown in Fig. 2. The fundamental mode of the microstrip line is quasi-TEM, which revolves around the microstrip line and propagates along it. Therefore, both magnetic fields of the quasi-TEM and LSM_{11} in microstrip line and NRD respectively match well through the rectangular coupling slot. Signals could run from the microstrip line via the slot on the common shared ground to the NRD guide system.

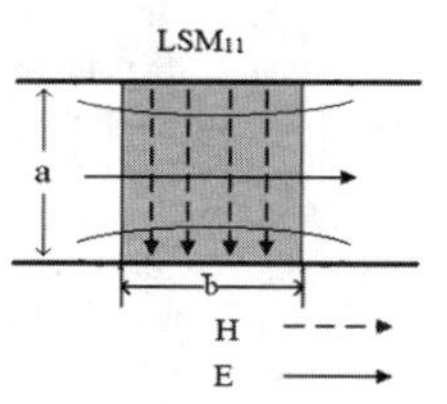

Fig.2 Field lines in a cross-sectional plane of NRD guide

Clearly, the microstrip line can be attached on either side of the NRD guide, which gives rise to both triple-layer transition and double-layer transition in this paper. While double-layer transition is promised to show the acceptable leakage loss from the uncovered SINRD guide, triple-layer transition presents some additional interesting benefits. First, a complete integrated system can be designed as compact as possible if both sides of the NRD guide are used and no space is wasted. On the other hand, by arranging the circuits of possible interference or cross-talk on the opposite sides of the NRD guide, the unwanted effects between them can be partially or completely suppressed. Therefore, this proposed triple-layer hybrid technology is compact, self-packaged and can be developed to reach a higher level of circuit integration.

IV. Topology of the SINRD Based on PCBs

Due to the fact that the substrate thickness should be less than the $\lambda/2$, it is difficult to implement NRD guide circuits in mm-wave frequencies. This is because mechanical tolerances related to the substrate thickness become more and more serious when the wavelength decreases. Following this problem, a lot of attempts have been made to realize NRD guide in planar versions, and so far, have received several effective structures, e.g. nonradiative perforated dielectric waveguide (NRPD)[6] and substrate integrated nonradiative dielectric waveguide (SINRD)[7]. Both cases actually eliminate the alignment problems of conventional NRD guide, however, the potential mechanical difficulty still exists. Fig. 3 presents a new scheme of manufacturing SINRD directly on printed circuit boards (PCBs). It is formed by drilling a series

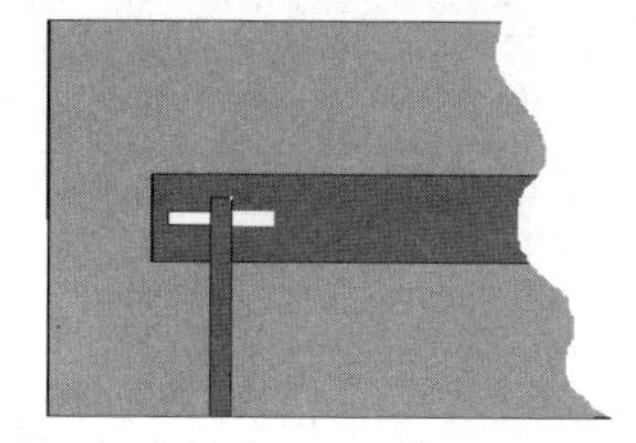

Fig.1 Geometry of the transition from microstrip line to NRD guide

Fig.3 View of the SINRD based on PCBs

of via-slots on PCBs instead of naked dielectric substrates, and with an absence of holes in the central channel of the PCBs. By carefully design of the patterns and dimensions of the via-slots, leakage loss from the uncovered via-slots can be controlled to a minimum value. In Fig. 3, the via-slots are square holes with width d and gap g. In this scheme of SINRD guide, a clear and regular equivalent width of the NRD central dielectric strip W is shown in Fig. 3. Simulation results of a new scheme of transition making use of this SINRD guide technology will be presented in next section.

V. PERFORMANCE OF THE PROPOSED TRANSITION

According to the above mentioned technologies, this section will analysis the electrical performance of the proposed transition of microstrip line to SINRD guide directly formed on PCBs through aperture coupling theory. Fig. 4(a) shows a triple-layer transition of microstrip line to SINRD waveguide. In Fig. 4(a) a pair of transitions are fabricated on the opposite sides of the SINRD guide, and interconnected through SINRD guide with a length of 67.4 mm. The SINRD is made directly on a 67.4×31.6 mm^2 rectangular PCBs with a height of 7.5 mm (Rogers TMM3, ε_r=3.2) by drilling 5 rows of via-slots on each side of the central channel. The width of the hole-absence region is 7 mm. As is shown in Fig. 4(a), the via-slots are square holes with a width of 2.2 mm. In such a structure, the height of 7.5 mm and central width of 7 mm refer to the values of a and b in a conventional NRD guide as is shown in Fig. 2, which decide its operating frequency around 15GHz. The microsrtip line is fabricated on a

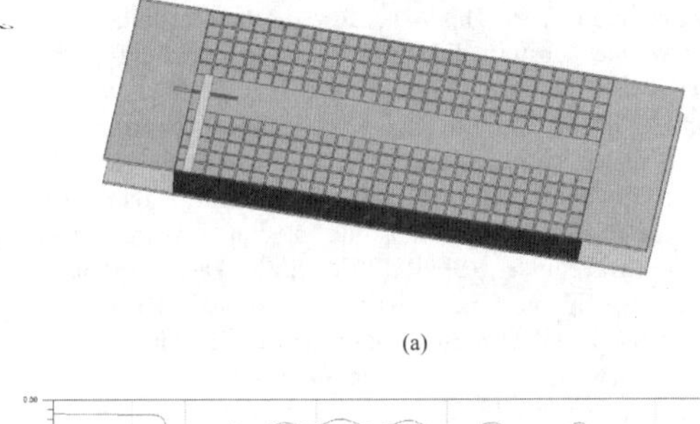

(a)

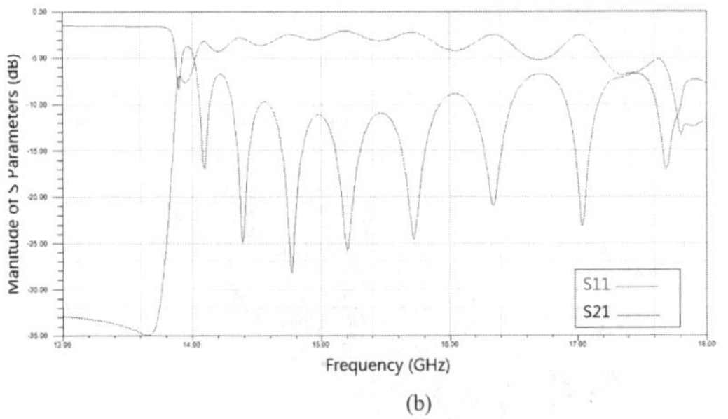

(b)

Fig. 4 (a) View and (b) Simulation results of the triple-layer transition from misrcostrip line to SINRD guide

(a)

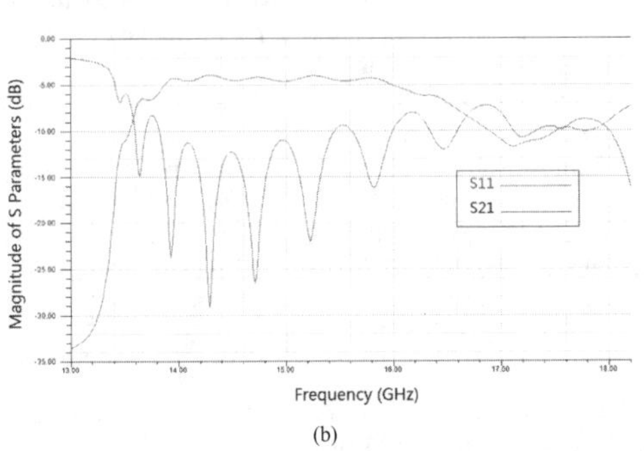

(b)

Fig. 5 (a) View and (b) Simulation results of the double-layer transition from misrcostrip line to SINRD guide

90.6 × 31.6 mm^2 substrate (Rogers 5880, ε_r=2.2) with a thickness of 0.52 mm and designed to have an impedance of 50 Ω with a strip width of 1.56mm. The coupling aperture on the ground plane is a narrow rectangular slot with 10.5×0.55 mm^2. The distance between the open-ended microstrip line and the aperture center is 3.37 mm, while the SINRD guide open-ended position is 2.24 mm apart from the center of the aperture. Different from those of [3], the rectangular slot in this paper scales out of the SINRD waveguide. To do so, a better bandwidth can be acquired. The two identical microstrip lines are designated as input and output ports in this structure, and Fig. 4(b) plots its transmission and reflection coefficients against frequency. Considering the fact that these simulation results have involved the losses of leakage, dielectric, radiation from the open-ended microstrip line and coupling slot, the bandwidth is generally good. It is found that the reflection coefficient is better than -10 dB over 50% of bandwidth and the transmission coefficient is generally better than -5 dB. In this triple-layer calculation, the interference effects between the input and output ports are suppressed. Fig. 5(a) presents a double-layer structure with the two microstrip lines deposited on the same dielectric substrate and leaving one side of the SINRD waveguide uncovered. Fig. 5(b) plots the simulation results. Compared to those of triple-layer structure, S_{11} changes little (negligible), while the S_{21} decreases by about 2dB. According to the analysis of the uncovered SINRD waveguide directly formed on PCBs mentioned above, this decrease is affected by leakage loss. Furthermore, interference between the two

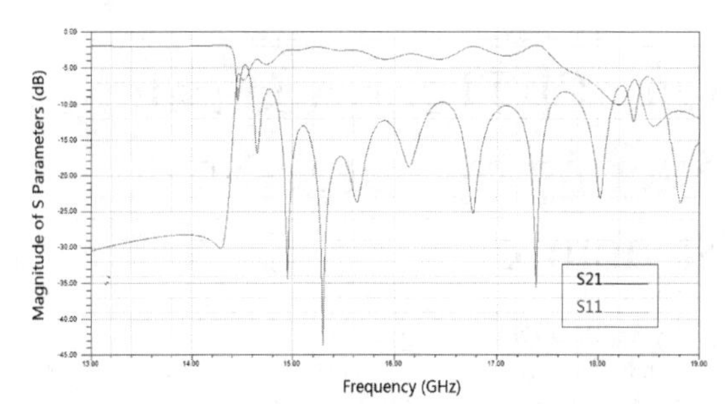

Fig. 6 Simulation results of the transition from misrcostrip line to conventional NRD guide

microstrip lines is complicated and may derogate the electrical performance. Except this, the bandwidths from 14GHz to 16GHz of both double-layer and triple-layer transitions agree quite well. Finally, the electrical performance of a triple-layer transition connected microstrip line with conventional NRD guide is shown in Fig. 6. It is found that in Fig. 6, with a bandwidth of about 3.2GHz, the reflection coefficient is better than -10 dB over 80% of bandwidth and transmission coefficient is generally better than -4 dB. Considering the additional losses initiated by the via-slots in the form of Fig. 4(a), its simulated electrical performance agrees quite well with those in Fig. 4(b). This demonstrates that SINRD guide directly fabricated on PCBs is suitable for hybrid integrated transition structure.

VI. Conclusion

A new type of transition has been proposed, in this paper, to integrate the microstrip line with the planar SINRD guide based on PCBs, at microwave and millimeter-wave frequencies. The shortcomings of traditional hybrid integration techniques have been analyzed. SINRD guide based on PCBs has been proved to be better in mechanism and integration with the planar circuits. Therefore, a new scheme of transition overcoming the drawbacks of the traditional integration techniques and inheriting the benefits of planar SINRD guide based on PCBs has been acquired. It is found that triple-layer transition, with both sides of SINRD guide covered and interference suppressed, agrees quite well with the transition using conventional NRD guide. It is also found that, by carefully designing the dimensions of via-holes, the leakage loss of the double-layer transition can be controlled. Thus, simulation results indicate that this typical transition is flexible to choose either double-layer or triple-layer form. This study proves that, at microwave and millimeter-wave frequencies, this form of transition is potentially suitable for low-cost applications of hybrid integration systems.

References

[1.] T. Yoneyama and S. Nishida, "Nonradiative dielectric waveguide for millimeter-wave integrated circuits," *IEEE Trans. Microwave Theory Tech.*, vol. 29, no. 11, pp. 1182-1192, November 1981.

[2.] K. Wu and L. Han, "Hybrid integrated technology of planar circuits and NRD guide for cost effective microwave and millimeter-wave applications," *IEEE Trans. Trans. Microwave Theory Tech.*, vol. 45, no. 6, pp. 946-954, June 1997

[3.] L. Han, K. Wu, and R. G. Bosisio, "An integrated transition of microstrip to nonradiative dilectirc waveguide for microwave and millimeter-wave circuits," *IEEE Tans. Microwave Theory Tech.*, vol. 44, pp. 1091-1096, July 1996.

[4.] F. Xu and Ke. W, "Substrate integrated nonradiative dielectric waveguide structures directly fabricated on printed circuits and metallized dielectric layers," *IEEE Trans. Microwave Theory Tech.*, vol. 59, no. 12, pp. 3076-3086, Dec. 2011.

[5.] F. Furoki and T. Yoneyama, "Nonradiative dielectric waveguide circuit components using beam-lead diodes," *Electronics and communications in Japan*, vol. 73 , no. 9, pt.2, pp. 35-40, 1990.

[6.] N. Grigoropoulos and P. R. Young, "Low cost nonradiative perforated dielectric waveguides," in *Eur.. Microw. Conf.*, Munich, Germany, 2003, pp. 439-442.

[7.] Y. Cassivi and K. Wu, "Substrate integrated non-radiative dielectric waveguide," *IEEE Microw. Wireless Compon. Lett.*, vol. 14, no. 3, pp. 89-91, Mar. 2004.

FE-BI-MLFMA Combined with FETI for Accurate and Fast Computation of Scattering by Large-Scale Finite Array Structure

Hong-Wei Gao, Li Gong, Ming-Lin Yang, and Xin-Qing Sheng
Center for Electromagnetic Simulation,
School of Information and Electronics, Beijing Institute of Technology, Beijing, 100081 China.

***Abstract*-To effectively solve the electromagnetic scattering problem of large-scale finite array structure, a non-overlapping domain decomposition method(DDM), the dual-primal finite element tearing and interconnecting method(FETI-DP), is applied to the hybrid finite element–boundary integral–multilevel fast multipole algorithm method(FE-BI-MLFMA). The formula of near scattering field is deduced by equivalent electric and magnetic current. Then, the Generalized Radar Cross Section (GRCS) in near scattering region is calculated by the near scattering field. The numerical performance of the proposed method is demonstrated by calculating a patch antenna array's scattering.**

I. INTRODUCTION

The domain Decomposition Method (DDM) has always been recognized as an important scheme to construct highly efficient algorithm. The Finite Element Method (FEM), implemented together with a domain decomposition algorithm in electromagnetic, was first introduced by Després[1]. Subsequently, sorts of advanced finite element domain decomposition methods were proposed [2]-[6]. Among various existing finite element domain decomposition methods, a non-overlapping finite element tearing and interconnection (FETI) method is investigated as a particularly efficient method [2]-[3]. Recently, a dual-primal technology is adopted in FETI denoted as FETI-DPEM1, which is successful applied for the computation of large-scale finite array antenna radiation problem [5]. The convergence of the FETI-DPEM1 algorithm becomes slow at higher frequencies. This disadvantage can be eliminated by using the FETI-DPEM2 algorithm [6], which enhances the dual-primal (DP) idea with two Lagrange multipliers, while the field continuity at the subdomain interfaces is guaranteed by the Robin-type transmission condition. To improve the algorithm [6] accuracy, a second-order absorbing condition is introduced [7], but the approximation for the solution can't be omitted completely.

The hybrid finite element–boundary integral–multilevel fast multipole algorithm method (FE-BI-MLFMA) has been verified as a general and accurate method for inhomogeneous electromagnetic problems [8]. Whereas, the final deduced FE-BI matrix is an ill-conditioned matrix for its partly sparse and partly dense matrix. It becomes difficult to solve the system matrix by general iterative solver and even no convergence for some complex problems. Then, two methods are applied to improve the convergence of traditional FE-BI-MLFMA [9]-[10]. Though the two methods can accelerate the convergence effectively, the process of solving the inverse of a sparse matrix required in each method consumes large memory. This bottleneck limits the application of FE-BI-MLFMA in solving large-scale electromagnetic problems.

To improve the computational accuracy of finite element DDM and reduce the high memory consumption by FE-BI-MLFMA, combining the dual-primal tearing and interconnecting DDM and FE-BI-MLFMA together yields a new hybrid method denoted as FETI-BI-MLFMA. What's more, scattering field anywhere in space is strictly deduced by the current and magnetic sources on the outward surface of object which has been obtained by FETI-BI-MLFMA. Take a patch array antenna as an example of finite array structure and calculate its near and far field by the proposed method.

II. FORMULATION

A. The FETI-BI-MLFMA Method

Considering scattering by an inhomogeneous object, whose surface is denoted as S .According to the conventional FE-BI-MLFMA [8], the computing region is directly divided into interior region and exterior boundary by the surface S .The interior field is formulated as

$$\mathrm{F}(\mathbf{E})=\frac{1}{2}\int_V\left[(\nabla\times\mathbf{E})\cdot\left([\mu_r]^{-1}\nabla\times\mathbf{E}\right)-k_0^2\mathbf{E}\cdot[\varepsilon_r]\mathbf{E}\right]dV+jk_0\int_s(\mathbf{E}\times\bar{\mathbf{H}})\cdot\hat{n}dS \quad (1)$$

Here, $\bar{\mathbf{H}}=Z_0\mathbf{H}$ where $\mathbf{E}$ and $\mathbf{H}$ are the unknown electric and magnetic field respectively. In addition, Z_0 is the free-space impedance and k_0 is the free-space wave number. $\hat{n}$ denotes the outward unit vector normal to S . The field on the exterior surface is formulated into the following combined field integral equation (CFIE):

$$0.5\times \mathrm{EFIE}+0.5\times \mathrm{MFIE} \quad (2)$$

For explicit expressions of the function CFIE in (2), the reader should refer to [8].

In this paper, the FETI-DPEM2 algorithm is applied in the FEM part of FE-BI-MLFMA. To be more specific, the interior FEM domain is decomposed into N non-overlapping sub-domains. Let the sub-domains V_i and V_j be adjacent, we employ $\Gamma_{j,i}$, $\Gamma_{i,j}$ to represent their respective neighboring surfaces, with the corresponding outward unit normal vectors being $\hat{n}_i$ and $\hat{n}_j$ .The edges shared by more than two subdomains

978-7-5641-4279-7

or shared by two sub-domains only for those on exterior boundary are called corner edges denoted as Γ_c.

At the interfaces Γ_i, the continuity of the tangential electric and magnetic field components is guaranteed by imposing Robin-type transmission conditions with extra variable $\mathbf{\Lambda}_i$:

$$\hat{n}_i \times \left(\left[\mu_{r,i}\right]^{-1} \cdot (\nabla \times \mathbf{E}_i)\right) + jk_0(\hat{n}_i \times \hat{n}_i \times \mathbf{E}_i) = \mathbf{\Lambda}_i \quad on\ \Gamma_i \tag{3}$$

Thus the fields in each sub-domain can be independently formulated with combination of (1) and (3), which is given by

$$\begin{aligned} F(\mathbf{E}_i) = & \frac{1}{2}\int_{V_i}\left[(\nabla\times\mathbf{E}_i)\cdot\left(\left[\mu_{r,i}\right]^{-1}\nabla\times\mathbf{E}_i\right) - k_0^2\mathbf{E}_i\cdot\left[\varepsilon_{r,i}\right]\mathbf{E}_i\right]dV + \\ & \alpha\int_{\Gamma_i}(\hat{n}_i\times\mathbf{E}_i)\cdot(\hat{n}_i\times\mathbf{E}_i)dS + \int_{\Gamma_i}\mathbf{E}_i\cdot\mathbf{\Lambda}_i dS + jk_0\int_{S_i}(\mathbf{E}_i\times\bar{\mathbf{H}}_i)\cdot\hat{n}_i dS \end{aligned} \tag{4}$$

The electric field $\mathbf{E}_i$ in the ith sub-domain can be expanded with edge-element vector basis function and (4) is cast in the following form for the ith sub-domain:

$$\mathbf{K}_i E_i = -\mathbf{C}\bar{\mathrm{H}}_{s,i} - \int_{\Gamma_i}\mathbf{W}_i\cdot\mathbf{\Lambda}_i dS \tag{5}$$

where

$$\begin{aligned} \mathbf{K}_i = & \int_{V_i}\left[(\nabla\times\mathbf{W}_i)\cdot\left[\mu_{r,i}\right]^{-1}(\nabla\times\mathbf{W}_i)^{\mathrm{T}} - k_0^2\mathbf{W}_i\cdot\left[\varepsilon_{r,i}\right]\mathbf{W}_i^{\mathrm{T}}\right]dV \\ & + \alpha\int_{\Gamma_i}(\hat{n}_i\times\mathbf{W}_i)\cdot(\hat{n}_i\times\mathbf{W}_i)^{\mathrm{T}}dS \end{aligned} \tag{6}$$

$$\mathbf{C} = jk_0\int_{S_i}(\mathbf{W}_i\times\mathbf{W}_i^{\mathrm{T}})\cdot\hat{n}_i dS \tag{7}$$

with $\mathbf{W}_i$ denotes a column vector containing the edge-element vector basis function in the ith sub-domain.

In each subdomain, the unknown coefficients of the electric field E_i are grouped into three categories $E_{V,i}$, $E_{I,i}$, and $E_{c,i}$:

$$E_i = \left[E_{V,i}^{\mathrm{T}}, E_{I,i}^{\mathrm{T}}, E_{c,i}^{\mathrm{T}}\right]^{\mathrm{T}} = \left[E_{r,i}^{\mathrm{T}}, E_{c,i}^{\mathrm{T}}\right]^{\mathrm{T}} \tag{8}$$

In (8), the superscript T denotes matrix transposition. Furthermore, the subscripts V, I and c denote the degrees of freedom associated with the internal volume, interface, and corner edges, respectively. The unknowns $\mathrm{E}_{r,i}$ associated with the internal volume and interfaces are considered as local variables, whereas the unknowns $\mathrm{E}_{c,i}$ associated with the corner edges are considered as global variables. The global primal variables in interior FEM region can be represented as:

$$\mathrm{E} = \left[\mathrm{E}_{r,1}^{\mathrm{T}}, \cdots, \mathrm{E}_{r,N}^{\mathrm{T}}, \mathrm{E}_c^{\mathrm{T}}\right]^{\mathrm{T}} \tag{9}$$

When the solution vector is constructed in this manner, the tangential electric field at the corner edges is ensured to be continuous. Applying this notation to (5) results in

$$\begin{bmatrix}\mathbf{K}_{rr,i} & \mathbf{K}_{rc,i}\\ \mathbf{K}_{rc,i}^{\mathrm{T}} & \mathbf{K}_{cc,i}\end{bmatrix}\begin{bmatrix}\mathrm{E}_{r,i}\\ \mathrm{E}_{c,i}\end{bmatrix} = \begin{bmatrix}-\mathbf{B}_{r,i}\mathbf{C}_i\bar{H}_{s,i} - \mathbf{B}_{I,i}^{\mathrm{T}}\lambda_i\\ -\mathbf{B}_{c,i}\mathbf{C}_i\bar{H}_{s,i}\end{bmatrix} \tag{10}$$

Here, $\mathbf{B}_{r,i}$, $\mathbf{B}_{I,i}$, and $\mathbf{B}_{c,i}$ are Boolean matrixes and satisfy $\mathbf{B}_{r,i}E_i = E_{r,i}$, $\mathbf{B}_{I,i}E_{r,i} = E_{I,i}$, and $\mathbf{B}_{c,i}E_i = E_{c,i}$. The unknown λ_i is called the dual-variable and defined as $\left[\mathbf{B}_{I,i}^{\mathrm{T}}\lambda_i \quad 0\right]^{\mathrm{T}} = \int_{\Gamma_i}\mathbf{W}_i\cdot\mathbf{\Lambda}_i dS$ whose dimension corresponds to the number of unknowns at interfaces.

From (10), it can be seen that the system matrix characterizing each subdomain decouples, and the interaction of the adjacent subdomains is included in the transmission conditions of the interfaces. Combining the two matrix equations in (10) permits the elimination of the unknowns $\mathrm{E}_{r,i}$. We assemble the subdomains contribution from (10) to obtain a global corner unknowns related system equation written as:

$$\tilde{\mathbf{K}}_{cc}E_c - \tilde{\mathbf{K}}_{cl}\lambda = \tilde{\mathbf{C}}_c\bar{H}_s \tag{11}$$

with

$$\tilde{\mathbf{K}}_{cc} = \sum_{i=1}^{N}\mathbf{O}_{c,i}^{\mathrm{T}}\left(\mathbf{K}_{cc,i} - \mathbf{K}_{rc,i}^{\mathrm{T}}\mathbf{K}_{rr,i}^{-1}\mathbf{K}_{rc,i}\right)\mathbf{O}_{c,i} \tag{12}$$

$$\tilde{\mathbf{C}}_c = \sum_{i=1}^{N}\mathbf{O}_{c,i}^{\mathrm{T}}\left(\mathbf{K}_{rc,i}^{\mathrm{T}}\mathbf{K}_{rr,i}^{-1}\mathbf{B}_{r,i}\mathbf{C}_i - \mathbf{B}_{c,i}\mathbf{C}_i\right)\mathbf{R}_i \tag{13}$$

$$\tilde{\mathbf{K}}_{cl} = \sum_{i=1}^{N}\mathbf{O}_{c,i}^{\mathrm{T}}\mathbf{K}_{rc,i}^{\mathrm{T}}\mathbf{K}_{rr,i}^{-1}\mathbf{B}_{I,i}^{\mathrm{T}}\mathbf{O}_{I,i} \tag{14}$$

where $\mathbf{O}_{c,i}$, $\mathbf{O}_{I,i}$, and $\mathbf{R}_i$ are Boolean matrices satisfying $E_{c,i} = \mathbf{O}_{c,i}E_c, \lambda_i = \mathbf{O}_{I,i}\lambda, \bar{H}_{s,i} = \mathbf{R}_i\bar{H}_s$.

To couple subdomains together, another set of equations representing the tangential electric and magnetic field continuity across the interfaces between different subdomains is required. Based on Robin-type transmission condition given by (3) at interface $\Gamma_{j,i}$, we can get

$$\lambda_{j,i} + \lambda_{i,j} = -2\mathbf{M}_{i,j}E_{i,j} \quad i = 1,2,\cdots N \text{ and } j \in \text{neighbour}(i) \tag{15}$$

Where $E_{i,j}$ denotes the electric field at the interface of the jth subdomain adjacent to the ith subdomain and $\mathbf{M}_{i,j} = jk_0\int_{\Gamma_{i,j}}(\hat{n}_j\times\mathbf{W}_j)\cdot(\hat{n}_j\times\mathbf{W}_j)^{\mathrm{T}}dS$.

Assembling (15) together for all sub-domains to get the following another corner problem equation:

$$\tilde{\mathbf{K}}_{II}\lambda + \tilde{\mathbf{K}}_{Ic}E_c = \tilde{\mathbf{C}}_I\bar{H}_s \tag{16}$$

with

$$\tilde{\mathbf{K}}_{II} = \mathbf{I} + \sum_{i=1}^{N}\mathbf{O}_{I,i}^{\mathrm{T}}\sum_{j\in\text{neighbor(i)}}\mathbf{T}_{j,i}^{\mathrm{T}}(\mathbf{T}_{i,j} - 2\mathbf{M}_{j,i}\mathbf{T}_{i,j}\mathbf{B}_{I,j}\mathbf{K}_{rr,j}^{-1}\mathbf{B}_{I,j}^{\mathrm{T}})\mathbf{O}_{I,j} \tag{17}$$

$$\tilde{\mathbf{K}}_{Ic} = \sum_{i=1}^{N}\mathbf{O}_{I,i}^{\mathrm{T}}\sum_{j\in\text{neighbor(i)}}\mathbf{T}_{j,i}^{\mathrm{T}}(-2\mathbf{M}_{i,j}\mathbf{T}_{i,j}\mathbf{B}_{I,j}\mathbf{K}_{rr,j}^{-1}\mathbf{K}_{rc,j})\mathbf{O}_{c,j} \tag{18}$$

$$\tilde{\mathbf{C}}_I = \sum_{i=1}^{N}\mathbf{O}_{I,i}^{\mathrm{T}}\sum_{j\in\text{neighbor(i)}}\mathbf{T}_{j,i}^{\mathrm{T}}(2\mathbf{M}_{i,j}\mathbf{T}_{i,j}\mathbf{B}_{I,j}\mathbf{K}_{rr,j}^{-1}\mathbf{B}_{r,j}\mathbf{C}_j)\mathbf{R}_j \tag{19}$$

And Boolean matrix $\mathbf{T}_{j,i}$ satisfies $E_{j,i} = \mathbf{T}_{j,i}E_i$ and $\lambda_{j,i} = \mathbf{T}_{j,i}\lambda_i$.

By combining (11) and (16) and eliminating E_c, we can derive the equation for the dual unknowns and the magnetic field on surface S

$$\left(\tilde{\mathbf{K}}_{II} + \tilde{\mathbf{K}}_{Ic}\tilde{\mathbf{K}}_{cc}^{-1}\tilde{\mathbf{K}}_{cl}\right)\lambda + \left(\tilde{\mathbf{K}}_{Ic}\tilde{\mathbf{K}}_{cc}^{-1}\tilde{\mathbf{C}}_c - \tilde{\mathbf{C}}_I\right)\bar{H}_s = 0 \tag{20}$$

For the sake of clarity, let

$$\tilde{\mathbf{D}} = \tilde{\mathbf{K}}_{II} + \tilde{\mathbf{K}}_{Ic}\tilde{\mathbf{K}}_{cc}^{-1}\tilde{\mathbf{K}}_{cl} \tag{21}$$

$$\tilde{\mathbf{F}} = \tilde{\mathbf{K}}_{Ic}\tilde{\mathbf{K}}_{cc}^{-1}\tilde{\mathbf{C}}_c - \tilde{\mathbf{C}}_I \tag{22}$$

Therefore, (20) takes the form

$$\tilde{\mathbf{D}}\lambda + \tilde{\mathbf{F}}\bar{H}_s = 0 \tag{23}$$

Similar to the convenient FE-BI-MLFMA, discretizing (2) by MOM yields:

$$\mathbf{P}\,E_s + \mathbf{Q}\,\bar{H}_s = \mathbf{b} \tag{24}$$

For explicit expressions of the matrix $\mathbf{P}$, $\mathbf{Q}$ and column vector $\mathbf{b}$ in (24), the reader should refer to [8].

Because of (9), the unknown E_s is grouped into $E_{c,s}$ and $E_{r,s}$ .with the aid of (10) and (11), equation (24) is deduced into

$$\tilde{\mathbf{P}}\,\lambda + \tilde{\mathbf{Q}}\,\bar{H}_s = \mathbf{b} \tag{25}$$

where

$$\tilde{\mathbf{P}} = \mathbf{P}\left[\mathbf{O}_{c,s}^{\mathrm{T}}\tilde{\mathbf{K}}_{cc}^{-1}\tilde{\mathbf{K}}_{cl} - \sum_{i=1}^{N}\mathbf{B}_{sr,i}(\mathbf{K}_{rr,i}^{-1}\mathbf{K}_{rc,i}\mathbf{O}_{c,i}\tilde{\mathbf{K}}_{cc}^{-1}\tilde{\mathbf{K}}_{cl} + \mathbf{K}_{rr,i}^{-1}\mathbf{B}_{I,i}^{\mathrm{T}}\mathbf{O}_{I,i})\right] \tag{26}$$

$$\tilde{\mathbf{Q}} = \mathbf{Q} + \mathbf{P}\left[\mathbf{O}_{c,s}^{\mathrm{T}}\tilde{\mathbf{K}}_{cc}^{-1}\tilde{\mathbf{C}}_c - \sum_{i=1}^{N}\mathbf{B}_{sr,i}(\mathbf{K}_{rr,i}^{-1}\mathbf{K}_{rc,i}\mathbf{O}_{c,i}\tilde{\mathbf{K}}_{cc}^{-1}\tilde{\mathbf{C}}_c + \mathbf{K}_{rr,i}^{-1}\mathbf{B}_{r,i}\mathbf{C}_i\mathbf{R}_i)\right] \quad (27)$$

The final system matrix equation of FETI-BI-MLFMA is obtained by combination of (23) and (25) as

$$\begin{bmatrix} \tilde{\mathbf{D}} & \tilde{\mathbf{F}} \\ \tilde{\mathbf{P}} & \tilde{\mathbf{Q}} \end{bmatrix}\begin{bmatrix} \lambda \\ \overline{H}_s \end{bmatrix} = \begin{bmatrix} 0 \\ \mathbf{b} \end{bmatrix} \quad (28)$$

Observing (28), by the introduction of the dual unknown λ, the original 3-D problem is reduced to a problem relating with sub-domain interfaces and exterior boundary surface. It is apparent that the problem always has fewer unknowns which enable us to solve large-scale problems.

B. The Strict Scattering Field Formulation

After solved by FETI-BI-MLFMA, the fields in interior region and on surface S have been obtained. Because the obtained filed is limited in computing region, scattering field anywhere in space out of objects can be strictly deduced by the current and magnetic sources on surface S according to electromagnetic field theory. The scattering electric field $\mathbf{E}^s$ in $\mathbf{r}$ is calculated by

$$\mathbf{E}^s(\mathbf{r}) = Z\,\mathrm{L}(\hat{n}\times\mathbf{H}_S) - \mathrm{K}(\mathbf{E}_S\times\hat{n}) \quad (29)$$

where $Z=\sqrt{\mu/\varepsilon}$. L and K are two operators defined in [11]. Introduce $\mathbf{R}=\mathbf{r}-\mathbf{r}'$, $R=|\mathbf{r}-\mathbf{r}'|$, and $\hat{R}=\mathbf{R}/R$ where $\mathbf{r}'$ denotes any point at surface S . Through a series of mathematical transformation, formula (29) is reduced to

$$\begin{aligned}\mathbf{E}^s(\mathbf{r}) = & \frac{jkZ_0}{4\pi}\int_S \frac{(\hat{n}\times\mathbf{H}_s(\mathbf{r}')\times\mathbf{R})\times\mathbf{R}e^{-jkR}}{R^3}\left(1+\frac{1}{jkR}-\frac{1}{k^2R^2}\right)dS \\ & +\frac{Z_0}{2\pi}\int_S \frac{(\hat{n}\times\mathbf{H}_s(\mathbf{r}')\cdot\mathbf{R})\mathbf{R}e^{-jkR}}{R^4}\left(1+\frac{1}{jkR}\right)dS \\ & -\frac{jk}{4\pi}\int_S \frac{\mathbf{E}_s(\mathbf{r}')\times\hat{n}\times\mathbf{R}e^{-jkR}}{R^2}(1+\frac{1}{jkR})dS\end{aligned} \quad (30)$$

Using (30), the scattering electric filed in near and far region is strictly calculated, we adopt the following general radar cross section (GRCS) [12] to further demonstrate object's near and far field property.

$$GRCS = 4\pi R^2\frac{\left|\mathbf{E}^s\cdot\hat{e}_r\right|^2}{\left|\mathbf{E}^i\right|^2} \quad (31)$$

In formula (31), $\hat{e}_r$ denotes the polarized direction, R denotes the distance between observing point and center of target. When R is big enough, GRCS is consistent with the strict RCS.

III. Numerical Experiments

To verify the validity of FETI-BI-MLFMA and strict scattering formulation, a homogeneous dielectric $4\lambda\times4\lambda\times4\lambda$ cube's GRCS on sphere surface 1000λ away from the center of object is calculated. It is illuminated by plane wave at incidence angle $\theta=0^\circ,\varphi=0^\circ$ and the computing region is divided by cuboid subdomain with dimension of $0.5\lambda\times0.5\lambda\times4\lambda$. A tetrahedral mesh with average edge length of 0.05λ is used. The result from FETI-BI-MLFMA is compared with RCS by MOM in Fig. 1. Investigate Fig. 1 the result from FETI-BI-MLFMA has a good agreement with MOM.

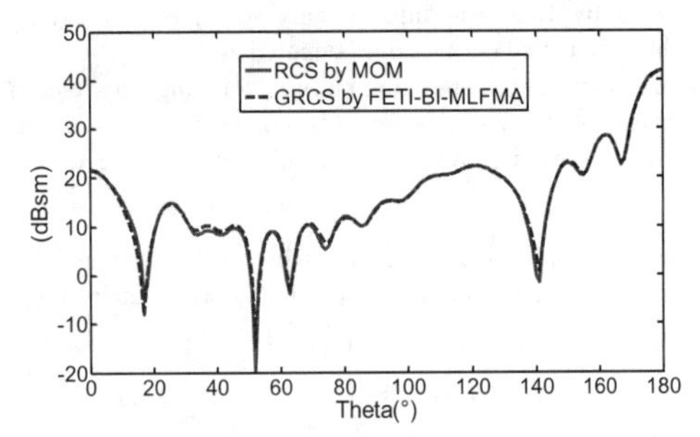

Fig. 1. Bi-static VV-polarized RCS and GRCS of a dielectric cube in the E-plane

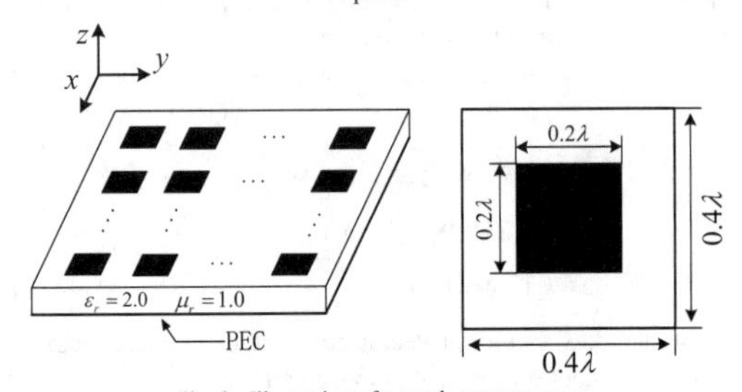

Fig. 2. Illustration of a patch antenna array

1	4	…	7
2	5	…	8
⋮	⋮	x y	⋮
		…	
3	6	…	9

Fig. 3. Illustration of different subdomains needed to be analyzed in $M\times N$ array problem

Next, take a patch array antenna's (Fig. 2) scattering as an example of finite array structure to investigate the numerical scalability of FETI-BI-MLFMA. Each patch element is regarded as a sub-domain and is extended along both x and y directions denoted as $M\times N$ patch array antenna. A tetrahedral mesh with average edge length of 0.05λ is used. To save computer memory and time, the periodicity of structure is sufficiently utilized which is shown by Fig. 3. 8×8 , 16×16 , 32×32 patch array antennas are illuminated by plane wave at incidence angle $\theta=0^\circ,\varphi=0^\circ$ and the GRCS on sphere surface 1000λ away from the center of objects is calculated. Results are showed in Fig.4-6 and mainly computation information is listed in Table I Table I suggests that memory consumption is rising linearly with the increase of sub-domain number, and the iterative step goes up at a very slow speed, which demonstrating the favorable numerical scalability of FETI-BI-MLFMA for solving two-dimension extension problems.

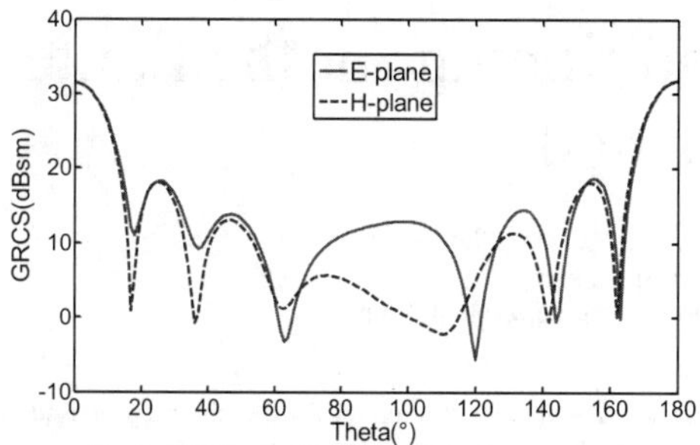

Fig. 4. Bi-static GRCS of the 8×8 patch antenna array

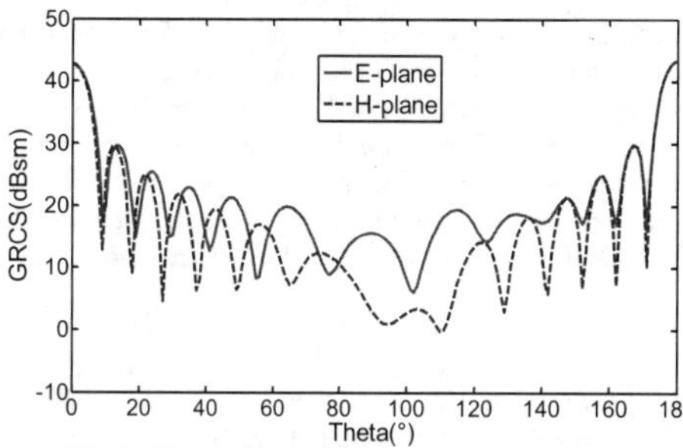

Fig. 5. Bi-static GRCS of the 16×16 patch antenna array

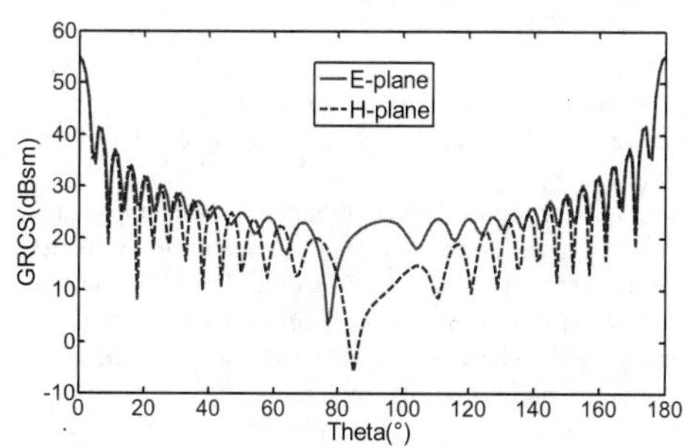

Fig. 6. Bi-static GRCS of the 32×32 patch antenna array

TABLE I
COMPUTATION INFORMATION FOR PATCH ANTENNA ARRAYS

Array size	Iteration number	Memory (GB)	Computing time (min)
8×8	50	0.5	5.2
16×16	87	1.9	23.8
32×32	145	7.6	153.9

Finally, we use the presented method to calculate GRCS by 50×50 patch antenna array on sphere surface 15λ, 20λ, and 30λ away from the center of object. The driving plane wave is at incidence angle $\theta=0^{\circ},\varphi=0^{\circ}$. Fig. 7 shows the computational result.

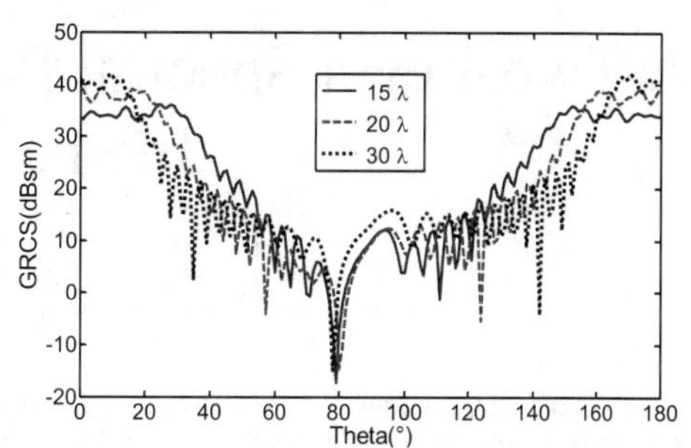

Fig. 7 Bi-static VV-polarized GRCS of the 50×50 patch array antenna in E-plane

ACKNOWLEDGMENT

This work was supported by the National Basic Research Program (973) under Grant 2012CB720702

REFERENCES

[1] B. Despres, P. Joly, and J. E. Roberts, "A domain decomposition method for the harmonic Maxwell equation," in *iterative Methods in linear Algebra. Amsterdam*, The Netherlands: Elsevier, 1992, pp. 475-484

[2] C. Farhat and F. Roux, "A method of finite element tearing and inter connecting and its parallel solution algorithm,"Int. J. Number.Method. Eng., vol. 32, pp. 1205-1227, Oct. 1991.

[3] C. T. Wolfe, U. Navsariwala, and S. D. Gedney, "A parallel finite-element tearing and interconnecting algorithm for solution of the vectorwave equation with PML absorbing medium," *IEEE Trans. AntennasPropagat.*, vol. 48, pp. 278–284, Feb. 2000.

[4] M. N. Vouvakis, Z. Cendes, and J.F.Lee, "A FEM domain decomposition method for photonic and electromagnetic band gap structures," *IEEE Trans. Antennas Propagat.*, vol. 54, no. 2, pp. 721–733, Feb. 2006.

[5] Y. J. Li and J. M. Jin, "A vector dual-primal finite element tearing and interconnecting method for solving 3D large-scale electromagnetic problems," *IEEE Trans. Antennas Propagat.*, vol. 54, pp. 3000-3009, Oct. 2006.

[6] Y. J. Li, J. M. Jin, "A new dual-primal domain decomposition approach for finite element simulation of 3-D large-scale electromagnetic problems," *IEEE Trans. Antennas Propagat.*, vol. 55, pp. 2803-2810, Oct. 2007.

[7] Y. J. Li, J. M. Jin, "Implementation of the Second-Order ABC in the FETI-DPEM Method for 3D EM Problems," *IEEE Trans. Antennas Propagat.*, vol. 56, pp. 5765-5769, Aug. 2008.

[8] X. Q. Sheng, J. M. Song, C. C. Lu, and W. C. Chew, "On the formulation of hybrid finite-element and boundary-integral method for 3D scattering", *IEEE Trans. Antennas Propagat.*, vol. 46, pp.303-311, Mar. 1998.

[9] J. Liu and J. M. Jin, "A highly effective preconditioner for solving the finite element-boundary integral matrix equation for 3-D scattering, *IEEE Trans. Antennas Propagat.*, vol. 50, pp.1212-1221, Sep. 2002.

[10] X. Q. Sheng and E. K. N. Yung, "Implementation and experiments of a hybrid algorithm of the MLFMA-Enhanced FE-BI method for open-region inhomogeneous electromagnetic problems," *IEEE Trans. Antennas Propagat.*, vol. 50, pp.163-167, Feb. 2002.

[11] X. Q. Sheng, W. Song, *Essentials of computational electromagnetic*, 1st ed. Singapore: John Wiley & Sons Singapore Pte. Ltd, 2012

[12] J. M. Taylor,"On the Concept of Near Field Radar Cross Section," *IEEE AP-S Int.Symp.* Dig., 1997, pp. 1172-1175.

A Two-Level Spectral Preconditioning for the Finite Element Method

Zi He, Weiying Ding, Ningye He, and Rushan Chen
Department of Communication Engineering
Nanjing University of Science and Technology, Nanjing, 210094, China

***Abstract*-An efficient auxiliary space preconditioning (ASP) was proposed for the linear matrix equation that was formed by the frequency-domain finite element method. A new two-level spectral preconditioning utilizing auxiliary space preconditioning is presented to solve the linear system. This technique is a combination of ASP and a low-rank update spectral preconditioning, in which the restarted deflated generalized minimal residual GMRES with the newly constructed spectral two-step preconditioning is considered as the iterative method for solving the system. Numerical experiments indicate that the proposed preconditioning is efficient and can significantly reduce the iteration number.**

I. Introduction

The finite element method (FEM) has been applied to the analysis of problems in electromagnetics for more than 35 years. A large number of research papers can be found in the literature [1]-[4]. The application of the finite-element method to electromagnetic problems often yields a sparse, symmetric, and very high-order system of linear algebraic equations. These highly sparse large linear equations can be solved using efficient solution techniques for sparse matrices based on iterative methods.

The Krylov subspace iterative methods like the generalized minimal residual (GMRES) method converges much faster than the other methods. The number of iteration required in the GMRES method can be controlled to some degree by the use of various preconditioning strategies. It is then desirable to precondition the coefficient matrix so that the modified system is well conditioned and can converge to an exact solution in significantly fewer iteration numbers than the original system. Many scholars have done a lot of researches on improving the efficiency of the iterative solution in the past few decades. The incomplete factorizations of the coefficient matrix and its block variants are a widely used class of preconditioners [5]-[7]. The multigrid preconditioning has also widely been used for the FEM. The geometric multigrid (GMG) method is the earliest form of multigrid, originally invented for finite difference methods [8]. The algebraic multigrid (AMG) method [9] was developed to overcome this limitation of GMG. It operates more on the level of the matrix than of the underlying FE mesh, which need not be nested.

In most of the cases, a single preconditioning can improve the iteration convergence speed to a certain extent. We can get more obvious convergence improvements when combining different preconditionings. In this paper, we apply the auxiliary space preconditioning (ASP), which was proposed in [10], as the first-step preconditioning. A spectral preconditioning [11]-[13] was applied in a two-step manner that attempts to further enhance the quality of the one-step preconditioning, resulting in a faster convergence rate.

This paper is organized as follows. Section II gives an introduction to the two-level spectral preconditioning utilizing ASP in detail. Numerical experiments are presented to show the efficiency of the spectral two-step preconditioning in SectionIII. SectionIV gives some conclusions and comments.

II. Hheory

A. The Auxiliary Space Preconditioning

We consider solving a large sparse linear system

$$Ax = b \tag{1}$$

With $A \in C^{n \times n}$, b, $x \in C^n$, which arises from finite-element discretization of the Helmholtz boundary value problems, where A is complex and symmetric and x is a column vector of the unknown values of E on the element edges.

The auxiliary space preconditioner is an approximate inverse of A , used to improve the convergence of an iterative solution to Ax=b using a Krylov method. Let V be the space spanned by the lowest order edge element basis functions on a tetrahedral mesh. There is an associated space N of piecewise linear, scalar functions on the same mesh and it is well known that ∇N is a subspace of [15]. We also need the space of vector "nodal" functions, N^3 . These are vector functions that, unlike edge basis functions, are both normally and tangentially continuous from one element to the next and are fully first-order in each element.

These spaces are linked by the following result [16]: for any $E \in V$ there exists $u \in N^3$ and $\varphi \in N$ such that:

$$E = E_S + \Pi u + \nabla \varphi \tag{2}$$

Where E_S is a "small" component in V. This suggests that it might be possible to solve the problem (1) approximately by solving related problems on the auxiliary spaces N^3 and N.

The operator ∇ becomes the sparse matrix G , which is simply the "node-to-edge" mapping matrix with entries of -1 and 1 per row . To preserve symmetry, we use the transpose, G^T , to map backwards, from V to N.

For N^3 we use the N basis for each Cartesian component of the vector and represent $u \in N^3$ by a column vector in which nodal values of the x , y and z components occupy 3

978-7-5641-4279-7

successive blocks. Then the operator Π becomes a sparse matrix which also has a block form:

$$\Pi = [\Pi_x \quad \Pi_y \quad \Pi_z] \tag{3}$$

Each block has the same dimension and sparsity pattern as G, i.e., two nonzero entries per row. It can be shown that the two nonzero values for row are identical, and, for Π_x, are equal to $(\frac{1}{2}Gx_C)_i$, where x_C is a vector containing the x coordinates of the nodes; similarly for Π_y and Π_z. We use Π^T to map backwards, from V to N^3.

None of these approximations on its own is very effective, but the following combined approach might be better:

$$A^{-1}r \cong R_f(A)DR_b(A)r + \sum_{i=x,y,z} \Pi_i B_i \Pi_i^T r + GB_nG^T r \tag{4}$$

Where $R_f(A) \triangleq (D+L)^{-1}, R_b(A) = (D+U)^{-1}$, $A_n \triangleq G^T AG$, $A_x \triangleq \Pi_x^T A\Pi_x$, $A_y \triangleq \Pi_y^T A\Pi_y$, $A_z \triangleq \Pi_z^T A\Pi_z$, We call the approximate inverses of these four matrices B_n, B_x, B_y and B_z, respectively. D is the diagonal part of A, L and U are the strict lower and upper triangular parts of A.

A W-cycle has been used. Omitting the "Residual update" lines for brevity, this is:

Backward GS: $\Delta x \leftarrow R_b(A)r$

Auxiliary spaces: $\Delta x \leftarrow \sum_{i=x,y,z} \Pi_i B_i \Pi_i^T r + GB_nG^T r$

Backward GS: $\Delta x \leftarrow R_b(A)r$

Auxiliary spaces: $\Delta x \leftarrow \sum_{i=x,y,z} \Pi_i B_i \Pi_i^T r + GB_nG^T r$

Forward GS: $\Delta x \leftarrow R_f(A)r$

Auxiliary spaces: $\Delta x \leftarrow \sum_{i=x,y,z} \Pi_i B_i \Pi_i^T r + GB_nG^T r$

Forward GS: $\Delta x \leftarrow R_f(A)r$

The approximate inverses of A defined by the W-cycle algorithms is the auxiliary space preconditioner considered in this paper.

B. The Spectral Low Rank Preconditioning

Although the ASP described above is very effective as shown in [10], the construction of it is inherently local. When the exact inverse of the original matrix is globally coupled, this lack of global information may have a severe impact on the quality of the preconditioner. We can get more obvious convergence improvements if recovering global information. In this case, some suitable mechanism has to be considered to recover global information.

We firstly let the most of eigenvalues of the system concentrate on the unit 1 by using the ASP, which eliminates the high frequency component of iteration process and accelerates the iteration convergence speed. A spectral preconditioner proposed in [14] can be introduced and used in a two-step way for the above ASP preconditioned system. The purpose here is to recover global information by removing the effect of some smallest eigenvalues in magnitude in the auxiliary space preconditioned matrix, which potentially can slow down the convergence of Krylov solvers.

Suppose $\lambda_1, \lambda_2, \ldots\ldots, \lambda_n$ be the eigenvalues of the ASP preconditioned matrix M_1A from small to large, U be a set of eigenvectors of dimension k associated with the smallest eigenvalues of the ASP preconditioned matrix M_1A.

Define the second spectral preconditioner as:

$$M_2 = I_n + U(1/|\lambda_n| T - I_k)U^H \tag{5}$$

Where $T = U^H(M_1Z)U$, I_n and I_k are unit matrix of dimension N and K respectively, and $\lambda_{k+1}, \lambda_{k+2}, \ldots\ldots, \lambda_n, |\lambda_n|, \ldots\ldots |\lambda_n|$ are the eigenvalues of the coefficient matrix M_2M_1A.

From the above analysis, we can convert the K smallest eigenvalues of the coefficient matrix M_1A 's characteristic spectrum which is based on ASP preconditioner to K arithmetic numbers whose values are $|\lambda_n|$. This process can eliminate negative influences of the K smallest eigenvalues. Combining the second preconditioning with the previously preconditioning in a two-step manner, a new two-step preconditioning is derived and has the form of:

$$M_2M_1Ax = M_2M_1b \tag{6}$$

Supposing that M_1 is a preconditioner of A, M_2 is a preconditioner of M_1A. Therefore, a new two-level spectral preconditioning of multilevel fast multipole method is presented, which is a combination of an ASP and a spectral preconditioner, as follows:

(1) Firstly, construct the auxiliary space preconditioner M_1 using the matrix element of the matrix A, and then solve the K smallest eigenvalues of the linear equations (1) after the preconditioner M_1 by GMRES-DR iterative algorithm.

(2) Secondly, construct the spectrum preconditioner $M_2 = I_n + U(1/|\lambda_n| T - I_k)U^H$ using the information of eigenvectors.

(3) Solve the linear equations (6) by the two-step preconditioner iteration.

III. Numerical Results

Firstly, scattering by a dielectric sphere with radius of 60 mm is considered to show the correctness of the proposed method. There are 108331 unknowns after discretization. The incident plane wave direction is fixed at $\theta^{inc} = 0^\circ, \phi^{inc} = 0^\circ$, the frequency is 1GHz, and the scattering angle is fixed at $\theta_s = 0^\circ - 180^\circ, \phi_s = 90^\circ$. The dielectric constant is $\varepsilon_r = 2\varepsilon_0$. Locally-conformal PML is used in this test.

As shown in Fig.1, the comparison is made for the bistatic RCS of vertical polarization. It can be found that there is an excellent agreement between them and this demonstrates the validation of the proposed algorithm. The convergence history is given in Fig.2.

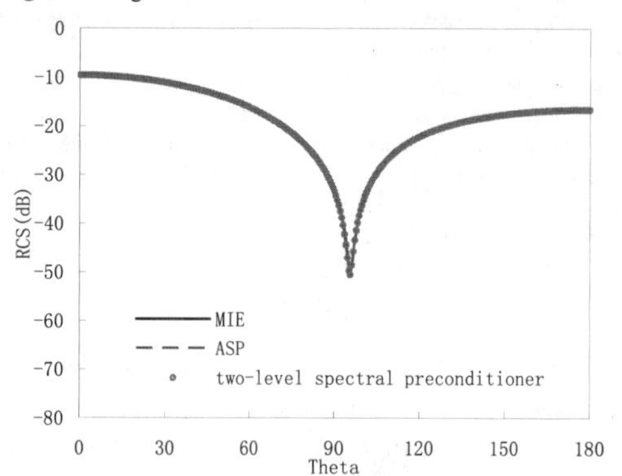

Figure 1. Bistatic RCS of the dielectric sphere

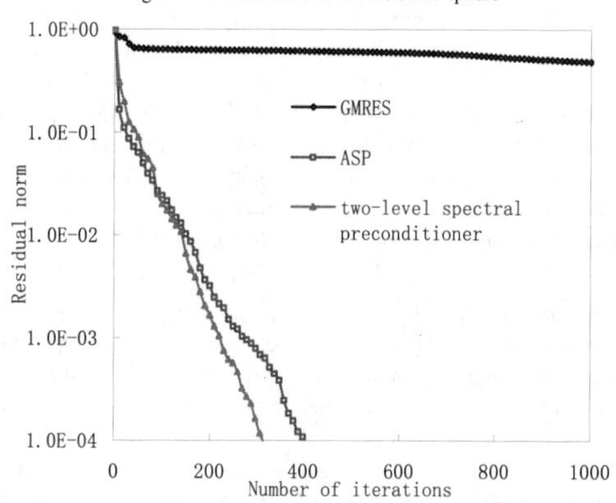

Figure 2. Convergence history of GMRES algorithms for the dielectric sphere

The second example is an analysis of a three-dimensional (3-D) discontinuity of a waveguide partially filled with a dielectric, which is shown in Fig.3.The rectangular waveguide has a width of a = 2 and height of b = 1; the inserted dielectric material slab has dimensions c = 0.888, d = 0.399, and w = 0.8; and the dielectric constant $\varepsilon_r = 6\varepsilon_0$. In this edge-FEM three-dimensional simulation, in order to obtain the reflection coefficient, one block of perfectly matched layer (PML) is placed at the waveguide output to simulate the matched output load. The use of PML in computational domains significantly deteriorates the condition number of the resulting FEM system. A total of 39817 unknown edges are to be solved in a large, sparse matrix equation. The convergence history at 9GHz is given in Fig.4. It can be found that when compared with the ASP preconditioned method, the two-step spectral preconditioned method decreases the number of iterations by a factor of 2.67. Larger improvements can also be found when compared with the GMRES method without preconditioning in terms of iterations.

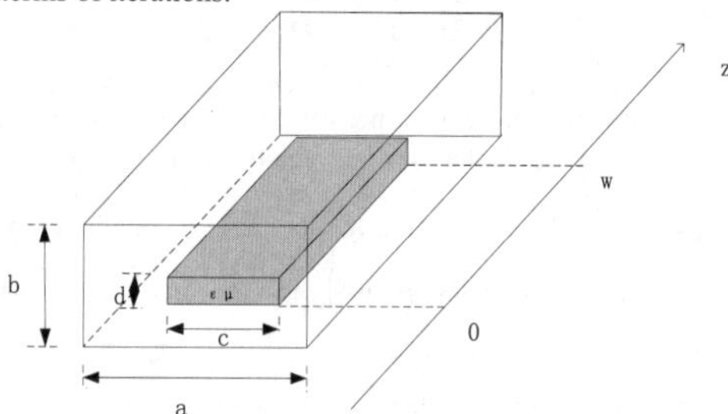

Figure3. Configuration of a dielectric partially filled rectangular waveguide

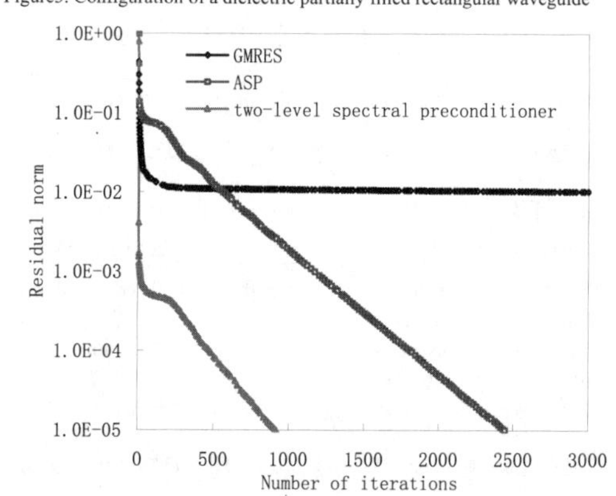

Figure 4. Convergence history of GMRES algorithms for the waveguide

IV. Conclusion

In this paper, a two-level spectral preconditioning utilizing ASP is proposed for FEM. The key of the paper is to combine the spectral preconditioner and the auxiliary space preconditioner in the two-step manner, resulting in faster convergence for GMRES iterations. The right-hand side system is solved by use of the GMRES-DR algorithm and the approximate smallest eigenvector information is obtained for constructing the spectral preconditioner for the system. Numerical experiments are performed and comparisons are made in the numerical results. It can be found that the proposed two-level spectral preconditioner utilizing ASP is more efficient and can significantly reduce the overall simulation time.

Acknowledgment

This work is partially supported by Natural Science Foundation of 61271076, 61171041, 61001009, Jiangsu Natural Science Foundation of BK2012034 and the Major State Basic Research Development Program of China (973 Program: 2009CB320201). The author would like to thank the reviewers for their comments and suggestions.

REFERENCES

[1] J. S. Wang and R. Mittra, “Finite element analysis of MMIC structures and electronic packages using absorbing boundary conditions,” *IEEE Trans. Microwave Theory Tech.*, vol. 42, pp. 441–449, Mar. 1994.

[2] K. Ise, K. Inoue, and M. Koshiba, “Three-dimensional finite-element method with edge elements for electromagnetic waveguide discontinuities,”*IEEE Trans. Microwave Theory Tech.*, vol. 39, pp. 1289–1295, Aug. 1991.

[3] J.-F. Lee and R. Mittra, “A note on the application of edge-element for modeling three-dimensional inhomogeneously-filled cavities,” *IEEE Trans. Microwave Theory Tech.*, vol. 40, no. 9, pp. 1767–1773, 1992.

[4] J.-Y.Wuand R. Lee, “The advantages of triangular and tetrahedral edge elements for electromagnetic modeling with the finite element method,” *IEEE Trans., Microwave Theory Tech.*, vol. 45, pp. 1431–1437, Sep. 1997.

[5] I. Hladik, M. B. Reed, and G. Swoboda, “Robust preconditioners for linear elasticity FEM analyses,” *Int. J. Numer. Meth. Eng.*, vol. 40, pp. 2109–2127, 1997.

[6] M. M. M. Made, “Incomplete factorization-based preconditionings for solving the Helmholtz equation,” *Int. J. Numer. Meth. Eng.*, vol. 50, pp. 1077–1101, 2001.

[7] R. S. Chen, X. W. Ping, Edward K. N. Yung, C. H. Chan, Zaiping Nie,and Jun Hu, “Application of Diagonally Perturbed Incomplete Factorization Preconditioned Conjugate Gradient Algorithms for Edge Finite-Element Analysis of Helmholtz Equations” *IEEE TRANSACTIONS ON ANTENNAS AND PROPAGATION,* VOL. 54, NO. 5, MAY 2006.

[8] J. Gopalakrishnan, J. E. Pasciak, and L. F. Demkowicz, "Analysis of a multigrid algorithm for time harmonic Maxwell equations," *SIAM Journal on Numerical Analysis,* vol. 42, pp. 90-108, 2004.

[9] K. Stuben, "A review of algebraic multigrid," *Journal of Computational and Applied Mathematics,* vol. 128, pp. 281-309, Mar 1 2001.

[10] A. Aghabarati, J. P. Webb, “An Algebraic Multigrid Method for the Finite Element Analysis of Large Scattering Problems”, *IEEE TRANSACTIONS ON ANTENNAS AND PROPAGATION.*

[11] Rui, P.L. and Chen, R.S., “Application of a two-step preconditioning strategy to the finite element analysis for electromagnetic problems,” *Microw. Opt. Technol. Lett.*, vol. 48, no. 8, pp. 1623–1627, 2006.

[12] P.L. Rui, R.S. Chen, Z.H. Fan and D.Z. Ding, “Multi-step spectral preconditioner for fast monostatic radar cross-section calculation,” *ELECTRONICS LETTERS,* Vol. 43 No. 7, 29th March 2007.

[13] Dazhi Z. Ding, Ru-Shan Chen, Z. H. Fan, and P. L. Rui, “A Novel Hierarchical Two-Level Spectral Preconditioning Technique for Electromagnetic Wave Scattering,” *IEEE TRANSACTIONS ON ANTENNAS AND PROPAGATION,* VOL. 56, NO. 4, APRIL 2008.

[14] J. Erhel, K. Burrage, and B. Pohl, “Restarted GMRES preconditioned by deflation,” *Journal of Computational and Applied Mathematics,* vol.69, pp.303-318, 1996.

[15] A. Bossavit and I. Mayergoyz, "Edge-Elements for scattering problems," *IEEE Transactions on Magnetics,* vol. 25, pp. 2816-2821, Jul 1989.

[16] R. Hiptmair and J. C. Xu, "Nodal auxiliary space preconditioning in H(curl) and H(div) spaces," *SIAM Journal on Numerical Analysis,* vol. 45, pp. 2483-2509, 2007.

Preliminary Study of a Ground Penetrating Radar for Subsurface Sounding of Solid Bodies in the Solar System

T. Ito, R. Katayama, and T. Manabe
Graduate School of Engineering
Osaka Prefecture University
1-1 Gakuencho, Naka-ku, Sakai,
Osaka 599-8531 Japan
Email: ss102003@edu.osakafu-u.ac.jp

T. Nishibori and J. Haruyama
Institute of Space and
Astronomical Science (ISAS)
3-1-1 Yoshinodai, Chuo-ku, Sagamihara,
Kanagawa 252-5210 Japan

T. Matsumoto and H. Miyamoto
Graduate School of Frontier Sciences
The University of Tokyo
5-1-5 Kashiwanoha, Kashiwa,
Chiba 277-8561 Japan

***Abstract*—This paper investigates the detectability performance of a breadboard model of a ground penetrating radar (GPR) for subsurface sounding of solid bodies in the Solar System up to a depth of tens of meters. The developed GPR uses a linear FM chirp signal whose frequency is linearly swept from 300 MHz to 900 MHz for 330 μs and can produce a narrower pulse of greater peak amplitude by applying a pulse compression technique. Vivaldi antennas are selected as the antennas of the radar system and designed by a particle swarm optimization (PSO) method to minimize the S_{11} parameter in the operational frequencies. A laboratory experiment for the radar electronics demonstrates that amplitude modulation of a transmitted signal by specific window functions could increase its dynamic range. The S_{11} parameter of the developed antenna is found to be very close to the desired value. Finally, subsurface sounding is simulated by modeling the two-layer subsurface structure. The simulation results reveal that the GPR could observe the subsurface structures up to a depth of 10–20 m from the air and a depth of 15–25 m from the ground.**

I. Introduction

Understanding internal structures at shallow depths is one of the most important keys for unraveling the local geological histories of solid bodies such as the Moon, Mars, and asteroids. A ground penetrating radar (GPR), whose goal is to detect and identify subsurface structures within the ground, is a powerful method for this purpose, particularly when the surface is composed of materials with heterogeneous electromagnetic properties. Regarding Japanese space exploration, Selenological and Engineering Explorer (SELENE) was equipped with Lunar Radar Sounder (LRS) and observed subsurface structures up to a depth of several kilometers with a range resolution of about 75 m [1].

The authors' group has been researching a GPR for future Japanese space missions such as a series of asteroid missions after Hayabusa-2, SELENE-2, and the Mars Exploration with a Lander-Orbiter Synergy (MELOS) mission. The authors' group is developing a GPR that can detect subsurface structures of solid bodies in the Solar System up to a depth of several tens of meters with a resolution of several tens of centimeters, whose depth target has not been targeted in any past space missions.

This paper first explains the principles of subsurface sounding by GPR and then describes the design of a breadboard model of the GPR. In Section IV, the results of verification tests are reported. Finally, the applicability of the GPR to future subsurface missions for solid bodies is discussed.

II. Principles of Subsurface Sounding by GPR

There are two main approaches to subsurface sounding of solid bodies by radar sounders: observation from the air (Fig. 1 left) and that on the ground (Fig. 1 right). Sounding from the air is particularly effective for global observation. Related past projects can be referred to, i.e., Shallow Subsurface Radar (SHARAD) of Mars Reconnaissance Orbiter (MRO) that observed the Martian internal structures up to 1000 m [2] and LRS of SELENE. In contrast, observation on the ground (i.e, by a rover) has never been conducted thus far. The latter approach is effective for local observation. For future missions, the Water Ice Subsurface Deposit Observation on Mars (WISDOM) GPR of the 2018 ExoMars project is planning to search for information about the nature of the subsurface up to a depth of 3 m with a resolution of a few centimeters from a ground-based rover [3].

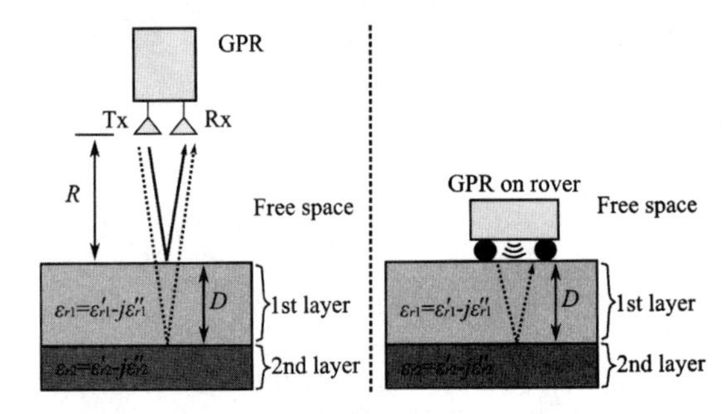

Fig. 1. Two approaches to subsurface sounding of solid bodies by radar sounders. ϵ'_{r1} and ϵ''_{r1} are the real and imaginary parts, respectively, of the complex relative permittivity at the first layer, and ϵ'_{r2} and ϵ''_{r2} are the real and imaginary parts, respectively, of the complex relative permittivity at the second layer.

978-7-5641-4279-7

Depending on the selected approach, a wave received by a GPR is significantly influenced by the dielectric properties comprising the internal structure. For example, the propagation speed of electromagnetic wave at the first layer, c_1, in Fig. 1 is reduced to

$$c_1 = \frac{c_0}{\sqrt{\epsilon'_{r1}}}, \tag{1}$$

where c_0 is the speed of light in vacuum. The transmitted power is attenuated by the subsurface properties as the wave travels. If the first layer is composed of low-loss materials, the power attenuation $A_d(x, f_0)$ is expressed as

$$A_d(x, f_0) = 20\log_{10}\left[\exp\left(-\frac{2\pi f_0 \epsilon''_{r1} x}{c_0\sqrt{\epsilon'_{r1}}}\right)\right] \text{ [dB]}, \tag{2}$$

where x is the traveled distance, and f_0 is the frequency of the transmitted wave. Attenuation is also caused by reflection and scattering at two dielectrically-different subsurfaces. Considering the boundary between the first and second layers to be flat in Fig. 1, for example, the reflection coefficient Γ and permeation coefficient T at this boundary are respectively given by

$$|\Gamma| = \left|\frac{\sqrt{\epsilon'_{r2}} - \sqrt{\epsilon'_{r1}}}{\sqrt{\epsilon'_{r2}} + \sqrt{\epsilon'_{r1}}}\right|, \tag{3}$$

and

$$|T| = 1 - |\Gamma|. \tag{4}$$

If surface roughness exists at the boundary, the wave is scattered and the power is assumed to be attenuated by

$$10\log_{10}\left[\exp\left(-\frac{2\pi f_0 \sigma}{c_0}\right)^2\right] \text{ [dB]}, \tag{5}$$

where σ is the standard deviation of the roughness. The dielectric influences mentioned above should be taken into account for the design of the GPR.

III. Design of the GPR System

The designed GPR system consists of radar electronics and antennas. Table I summarizes the main specifications of the GPR. The GPR system is designed to be able to detect a shallow depth of the solid subsurface (up to tens of meters).

TABLE I. MAIN SPECIFICATIONS OF THE GPR.

Subsystem	Item	Specification
Radar electronics	Mass	Up to 2 kg (TBD)
	Size	30 cm × 30 cm × 10 cm (TBD)
	Power consumption	Up to 30 W during observation
	Pulse processing method	Pulse compression
	Frequency modulation method	Linear FM chirp (+amplitude modulation)
	Transmission power	1 W(average)
	Transmission frequency	300–900 MHz
	Transmission time	0.33 μs
	Transmission repetition time	1 μs
	Observable subsurface depth	Up to tens of meters
	Distance resolution	0.25–1.75 m
Antenna	Antenna type	Vivaldi antenna
	S_{11}	Below -10 dB (bet. 300–900 MHz)

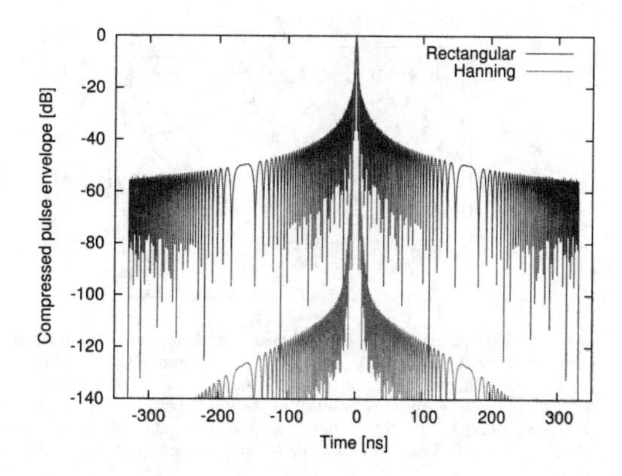

Fig. 2. Two ideal compressed pulses multiplied by rectangular and Hanning windows. While windowing can increase the range side lobe level of a compressed pulse, the pulse is widened, which causes degradation of the distance resolution. In this figure, the pulse widths (main-lobes) of the two compressed signals at -3 dB are 1.7 ns and 5.0 ns for a rectangular window and Hanning window, respectively.

The system is also designed to not require many resources (i.e., mass, space, electrical power, etc.) of the main spacecraft.

The GPR uses a *pulse compression* technique that can produce a narrower pulse of greater peak amplitude by transmitting frequency-modulated pulses and then correlating the received pulses with the modulated transmitted pulse [4]. The GPR uses a linear FM chirp signal to apply this technique with a frequency that is linearly swept from 300 MHz to 900 MHz for 0.33 μs. In addition to pulse compression, the range side lobe level can be reduced by weighting the transmitted or received signals with a window function. Fig. 2 shows two ideal compressed pulses multiplied by two different window functions (rectangular window and Hanning window) as a couple of examples.

Turning to the antenna subsystem, Vivaldi antennas have been selected as both the transmitting and receiving antennas of the GPR system because its sharp directivity and wide bandwidth are suitable for subsurface missions. A particle swarm optimization (PSO) method [5] was applied to the antenna design, and the S_{11} parameter was selected as the optimization parameter. Fig. 3 shows the design parameters of the Vivaldi antenna. As a result of the optimization, the maximum value of S_{11} was -11.13 dB between 300 MHz and 900 MHz, which meets the target value of S_{11} (below -10 dB in the operational frequency range). The detailed antenna design is given in [6].

IV. Experimental Results Using a Breadboard Model of the GPR System

A. The Radar Electronics Experimentation

A breadboard model of a GPR was developed (Fig. 4 left). To investigate the qualities of GPR pulses, a transmission and reception experiment for the radar electronics was conducted by using a delay cable with a length of 45 m and a frequency response of approximately 10 dB/GHz (Fig. 5). The two techniques, *time averaging* for noise reduction and *Wiener-filtering*

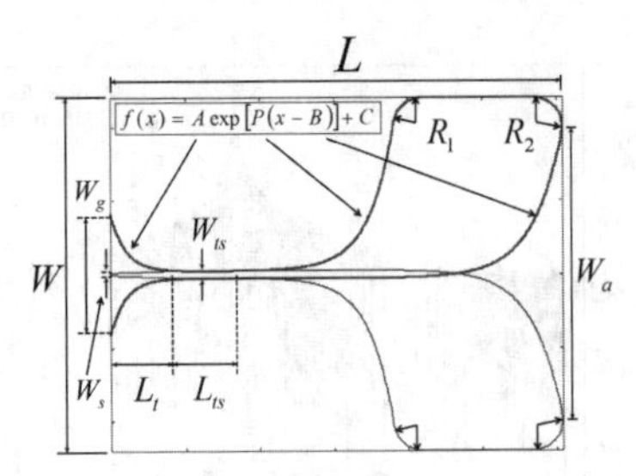

Fig. 3. PSO parameters of the designed Vivaldi antenna. The positional parameters are L = 387.29 mm, L_t = 50.02 mm, L_{ts} = 10.0 mm, P_t = −0.1533 mm^{-1}, P_a = 0.02099 mm^{-1}, P_f = 0.02768 mm^{-1}, and A_f = 0.008168 mm. As a result of the PSO, the following parameters were optimized: W = 352.08 mm, W_a = 293.41 mm, W_s = 3.06 mm, W_{ts} = 4.66 mm, W_g = 119.2 mm, R_1 = 4.66 mm, and R_2 = 29.340 mm.

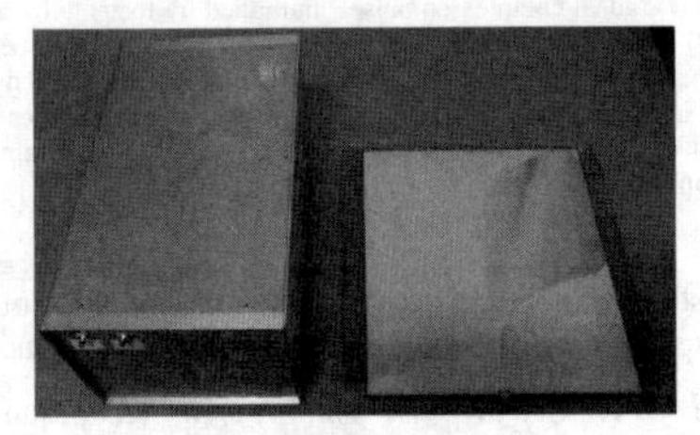

Fig. 4. Developed GPR (left) and Vivaldi antenna (right).

for both noise and distortion reduction, were adopted for signal processing. Rectangular and Hanning window functions were selected to compare the qualities of the compressed pulses.

Fig. 6 shows the two different compressed pulses received after being transmitted through the delay cable. Whereas the range side lobes of the rectangular-windowed compressed pulse (red line in Fig. 6) were very close to those of the ideal ones, the range side lobe levels of the Hanning-windowed pulse were much lower and limited by a noise floor. This result suggests that amplitude modulation of a transmitting signal by specific window functions (Hanning window in this case) can decrease range side lobe levels significantly.

With regard to the signal-to-noise ratio (SNR) of the Hanning-windowed pulse, the measured SNR was 79 dB, which is 10 dB smaller than the ideal one (89 dB). The primary cause of the decrease was a larger noise figure than its target specification, and the secondary one was signal distortion which results in imperfect pulse compression processing. Therefore, the SNR can be further improved by reducing the noise figure at the signal amplifier at the receiver.

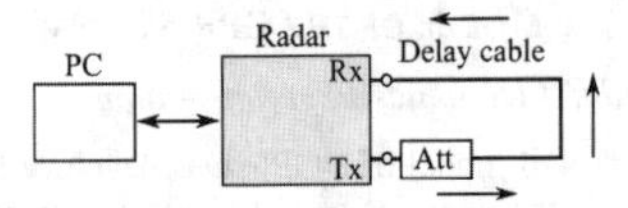

Fig. 5. Experimental configuration using a delay cable. The attenuation between the transport of the radar and the import of the input port of the delay cable was 55 dB.

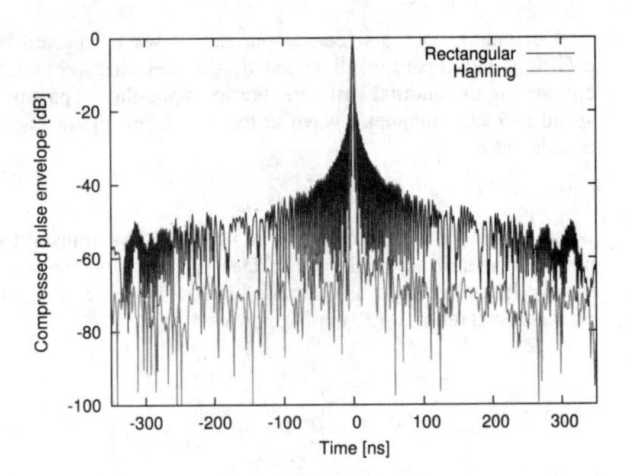

Fig. 6. Two measured compressed pulses multiplied by rectangular and Hanning windows. The signal distortion and noise of both pulses were significantly removed by Wiener-filtering. Comparing the two measured pulses, the range side lobe levels of the Hanning-windowed pulse was about 20 dB lower than those of the rectangular one. In contrast, very good agreement was obtained in the pulse widths between the ideal and measured signals for both rectangular and Hanning-windowed pulses. Time-averaging was executed 50 times, which improved the SNR by 16 dB for the Hanning-windowed pulse.

B. The Antenna Experimentation

A breadboard model of the Vivaldi antenna was manufactured (Fig. 4 right). The S_{11} parameter was measured in an anechoic chamber of the Microwave Energy Transmission Laboratory (METLAB) at the Research Institute for Sustainable Humanosphere (RISH) of Kyoto University to evaluate the performance of the Vivaldi antenna.

Fig. 7 compares the PSO simulation results and the measurements of the S_{11} parameter, and good agreement was obtained between these results. Furthermore, the maximum value of the measurement between 300–900 MHz was -9.46 dB at 590 MHz, which is very close to the target value (below -10 dB).

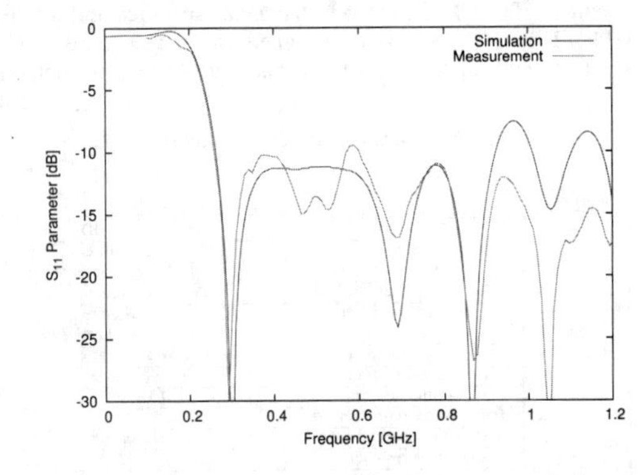

Fig. 7. Comparison of the simulation and measurement results of the S_{11} parameter.

V. Subsurface Sounding Simulation

A. Modeling of Subsurface Sounding

Detectability of subsurface boundaries of solid bodies was investigated by using the performance parameters of the GPR obtained in the previous sections. This simulation considered an observation from a spacecraft at an altitude of 50 m ($R = 50$ m in Fig. 1 left) and that of a rover at an altitude of 1 m. In the simulation, a transmitted signal was assumed to be emitted in one vertical direction to the horizontally flat surface.

The subsurface structure models summarized in Table II were a two-layer model whose shallower layer was a regolith layer with a depth of D and a deeper one was a base rock layer. According to the density model of a lunar regolith, the lunar regolith density approximately ranges from 1.5 $\mathrm{g/cm^3}$ to 2.0 $\mathrm{g/cm^3}$ up to a depth of 50 m [7]. Thus, this simulation considered the maximum and minimum cases. The complex permittivity of the lunar regolith was modeled on the basis of the lunar samples [8]. Similarly, because the dielectric constants of most lunar rock samples range from 4 to 9 [9], the simulation assumed two kinds of lunar base rock ($\epsilon_r' = 9$ for the maximum and $\epsilon_r' = 4$ for the minimum).

Turning to the radar system, a Hanning-windowed signal was selected as a transmission signal, and SNR margin was 10 dB in the simulation. As for the antenna gain, a value of 3.46 dBi was used, which is the antenna gain obtained through the PSO simulation at 600 MHz, on the assumption that the antenna gain is 3.46 dBi equally in the operational frequencies.

B. Simulation Results and Discussion

The detectability of the GPR was analyzed for the four subsurface models (in Table II) and two observation approaches (in Fig. 1). Table III summarizes the maximum thickness of the first layer for detecting the second layer with the GPR.

Table III indicates that the maximum detectable depth from the air is approximately 10–20 m. This result suggests that the developed GPR can be applied to the subsurface sounding missions of small bodies in the Solar System, whose shallow subsurface structure remains relatively unknown. One of the exploration targets could be a small asteroid such as Itokawa, where its regolith layer is estimated to accumulate up to a depth of several meters [10]. Table III also denotes that the maximum detectable depth from the ground is around 15–25 m. A lunar regolith layer whose thickness is 4–5 m in the maria and 10–15 m in the highlands [11] is thus a potential target for future subsurface missions by using the GPR.

TABLE II. Four Subsurface Models. This Simulation Assumed That the Regolith Layer Contained 20 % $TiO_2 + FeO$. Surface Scattering Effects Are Taken into Account ($\sigma = 0.05$ m).

Case	1st layer		2nd layer	
	ϵ_r'	ϵ_r''	ϵ_r'	ϵ_r''
1	2.502	0.0221	4	-
2	2.502	0.0221	9	-
3	3.397	0.0391	4	-
4	3.397	0.0391	9	-

TABLE III. Detectability Analysis Results of the GPR.

Case	Detectable subsurface depth	
	Observation from the air	Observation from the ground
1	19 m	23 m
2	22 m	26 m
3	10 m	14 m
4	14 m	17 m

VI. Conclusion

Experiments with a breadboard model of radar electronics suggested that the range resolution of the radar could increase significantly by changing the weighting amplitude of the transmitted signal. In addition to the radar electronics, a Vivaldi antenna for subsurface space missions was successfully developed, which can be used broadly in the UHF band.

Some of the performance parameters of the radar obtained by experiments were fed back into the simulation for subsurface sounding of solid bodies to acquire a better understanding of the detectability of the radar. The simulation results suggested that the developed GPR could detect a subsurface boundary in solid bodies up to a depth of 10–20 m for air observation and 15–25 m for ground observation. The lunar shallow subsurface as well as small asteroids are potential targets for subsurface missions with the GPR.

For future work, the detectability of the radar will be studied with various subsurface models such as multi-layer structures and interiors with frequency-dependent permittivity.

Acknowledgment

The authors would like to thank the Research Institue for Sustainable Humanosphere, Kyoto University, for letting us use its anechoic chamber. This work is partly supported by JSPS KAKENHI Grant Number 24560527.

References

[1] T. Ono, et al., "Lunar radar sounder observations of subsurface layers under the nearside maria of the moon," *Science*, vol. 323, pp. 909-912, February 2009.

[2] R. Seu, et al., "SHARAD: The MRO 2005 shallow radar," *Planet. Space Sci.*, vol. 52, pp. 157-166, January-March 2004.

[3] V. Ciarletti, et al., "WISDOM GPR Designed for Shallow and High-Resolution Sounding of the Martian Subsurface," *Proc. IEEE*, vol. 99, pp. 824-836, February 2011.

[4] C. E. Cook and M. Bernfeld, *"Radar Signals: An Introduction to Theory and Application," Academic Press*, 1967.

[5] J. Robinson and Y. Rahmat-Samii, "Particle swarm optimization in electromagnetics," *IEEE Trans. Antennas Propagat.*, vol. 52, pp. 397-407, February 2004.

[6] R. Katayama, T. Manabe, and T. Nishibori, "Optimization of Vivaldi antenna for ground penetrating radars to study subsurface structures of solid bodies in the solar system using PSO method," *Proc. IEICE General Conf. 2013*, B-1-112, 2013.

[7] G. R. Olhoeft and D. W. Strangway, "Dielectric properties of the First 100 meters of the Moon," Earth Planet. Sci. Lett., vol. 24, pp. 394-404, January 1975.

[8] G. Heikin, D. Vaniman, B. M. French, and J. Schmitt (Eds.), *Lunar Source Book*, Cambridge University Press, 1991, Figure 9.54.

[9] T. Kobayashi, et al., "Synthetic aperture radar processing of Kaguya Lunar Radar Sounder data for lunar subsurface imaging," *IEEE Trans. Geosci. Remote Sens.*, vol. 50, pp. 2161-2174, June 2012.

[10] H. Miyamoto, et. al., "Regolith on a tiny asteroid: Granular materials partly cover the surface of Itokawa," Proc. of 37th Annual Lunar Planet. Sci. Conf., no. 1686, 2006.

[11] G. Heikin, D. Vaniman, B. M. French, and J. Schumitt (Ed), *Lunar Source Book*,Cambridge University Press, 1991, Section 7.

A 14 GHz Non-Contact Radar System for Long Range Heart Rate Detection

Jee-Hoon Lee[1], Seong-Ook Park[2]
Department of Electrical Engineering, KAIST
291 Daehak-ro, Yuseong-gu, Daejeon 305-701, Republic of Korea
ghoon99@kaist.ac.kr[1], sopark@ee.kaist.ac.kr[2]

Abstract-**This research presents a 14 GHz CW Doppler radar to measure respiration and heart rate. A leakage cancellation technique is used to detect heart rate of human for long range detection. Arctangent demodulation without the dc offset compensation can be applied because of the heterodyne receiver structure and the leakage cancellation technique. HRV analysis of the radar system and the ECG signal which is measured directly are compared. Based on the measurement results, the radar can detect the respiration and heart rate with a small error rate.**

I. INTRODUCTION

Non-contact biosensor system have been used to detect respiration and heart rate since the early 1970s [1]. Doppler radar can detect the heart and respiration rate through clothing as non-invasive method. As interest in health increases, many research institutions have studied about the radar structure, demodulation technique, detection algorithm to achieve accurate performance. Most studies have focused on a simple structure such as direct conversion and a compact size to detect the signal within a few meters. In this paper, we adopt the heterodyne structure and leakage cancellation technique in order to complement the disadvantage of direct conversion and increase the detectable range of radar system.

II. DOPPLER RADAR SYSTEM

A. Radar Structure

Fig. 1 shows the Doppler radar system structure. The 14 GHz transmitted signal reflected by the subject goes to the receiving antenna. The gain of antenna is 16 dBi at 14 GHz. The received signal at the antenna is amplified by LAN and down converted to 11.7 GHz by mixing the signal with 2.3 GHz local oscillator. And it is down converted to 480 MHz intermediate frequency (IF) again. The IF signal is entered into the demodulator. The I and Q channels which are the outputs of demodulator are filtered, amplified, and converted the digital signal at the A/D converter. Finally, it is divided in heart signal and respiration signal and displayed on the computer.

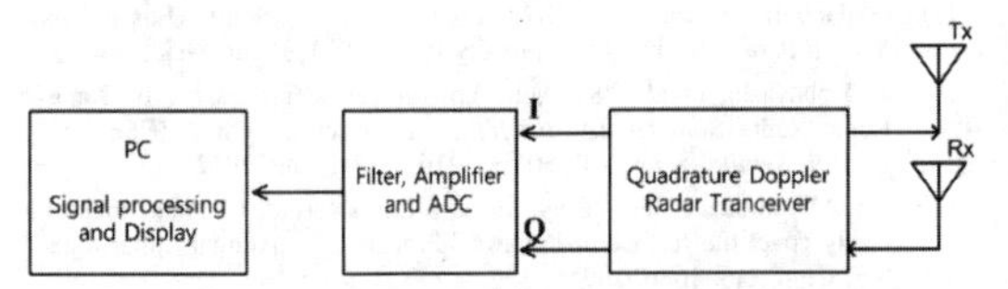

Figure 1. Block diagram of Doppler radar system for detecting heart rate.

B. Arctangent Demodulation

Arctangent demodulation in quadrature receivers can get a high accuracy in demodulation of heart and respiration signals. By applying the arctangent operation to the ratio of I and Q output of radar system, accurate phase demodulation can be achieved regardless of the target's position or motion amplitude [2].

In CW Doppler radar systems the phase displacement $\Delta\varphi$ between the transmitted and received signal is described for range detection [3].

$$\Delta\varphi = \frac{4\pi x(t)}{\lambda} + \varphi_r \tag{1}$$

with x(t) for the time-varying subject range, λ for wavelength, and φ_r for the constant phase shift by the chest movement.

Quadrature channel imbalance and dc offset act as a linear transform on the I and Q components, thus modifying (1) to

$$\begin{aligned}\varphi(t) &= \arctan\left(\frac{B_Q(t)}{B_I(t)}\right) \\ &= \arctan\left(\frac{V_Q + A_e \sin(\theta + \varphi_e + p(t))}{V_I + \cos(\theta + p(t))}\right)\end{aligned} \tag{2}$$

where V_I and V_Q refer to the dc offsets of each channel, and A_e and φ_e are the amplitude error and phase error, respectively [4].

The direct conversion is the simplest architecture, so most of CW radar systems for detecting heart rate have used the direct conversion. But it has the disadvantage which has the dc offset.

Heterodyne architecture can reduce the dc offset, that is V_I and V_Q in (2).

C. HRV Analysis

The heart rate variability (HRV) is the physiological phenomenon of variation in the time interval between heartbeats. It is measured by the variation in the beat-to-beat interval. We compared the value of electrocardiogram (ECG) signal measured by ECG detector with the arctangent demodulation value of I and Q signals measured by radar system. The measured peak points of arctangent and ECG signals can be expressed by mean beat to beat interval, and the standard deviation of normal beat-to-beat interval (SDNN), and the root mean square of successive heartbeat interval differences (RMSSD).

978-7-5641-4279-7

D. Leakage Cancellation

The transmitted power can be more than 100dB higher than the received signal, so if even a small fraction of the transmitted power leaks into the receiver it can saturate the low-noise-amplifier or even damage the sensitive circuitry. This is the disadvantage of CW radar system. For a bi-static radar system, the majority of the leakage comes from the free space coupling between the transmitter antenna and the receiver antenna, where up to 60dB isolation can be achieved [5]. The isolation problem is more dominant when the transmitted power is increased in order to extend the detection range. The arctangent demodulation with dc offset compensation in quadrature Doppler radar has demonstrated in [4]. We can use arctangent demodulation without dc offset compensation because of heterodyne structure which has the low dc offset and leakage cancellation technique which can reduce the dc offset due to the self-mixing.

III. Measurement Results

We obtain the experimental data based on various situations. Fig. 2 shows the measurement results in time domain when the subject is 5m apart from the transmitting and receiving horn antennas. The power at the transmitting antenna is -10dBm, the leakage cancellation technique did not be applied. These signals are obtained simultaneously for the same subject. Data was collected for 20 seconds. The raw data is filtered by the band pass filter with the cut-off frequency of 0.6Hz to 2.5Hz, in order to extract the heart signal information. Fig. 2(c) is the combined arctangent signal of Fig. 2(a) and (b). The magnitude of combined arctangent demodulated output is normalized by the maximum amplitude. Fig. 2(d) shows the ECG signal measured by ECG detector. We can see that heart rate is 24 beats for 20 seconds in both arctangent demodulated signal and ECG. The all peak positions have been successfully detected in Fig. 2(c) and Fig. 2(d).

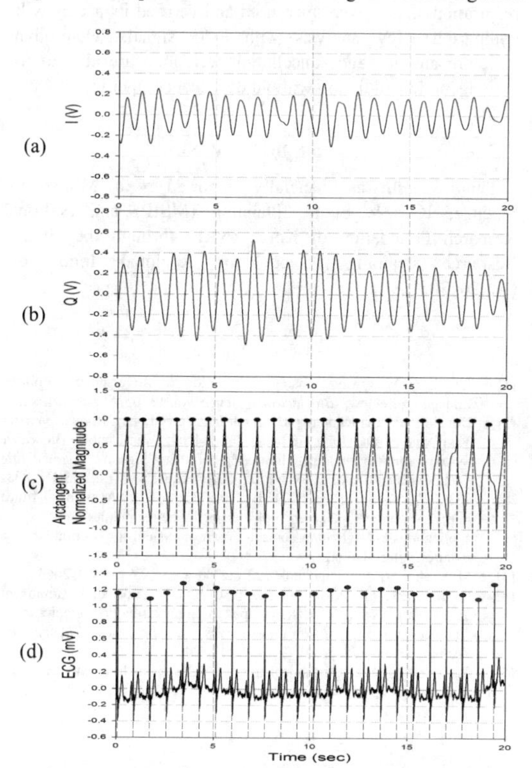

Figure 2. Measured heart signal I (a), Q (b), the combined arctangent demodulated output (c), and ECG signal (d).

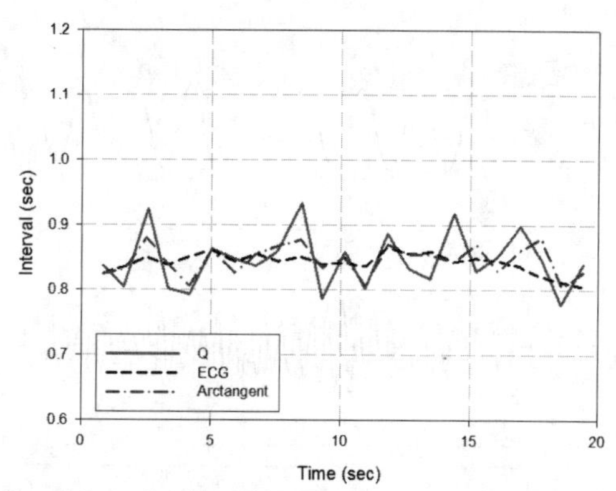

Figure 3. Beat-to-beat interval of Q signal, arctangent signal, and ECG signal.

Fig. 3 shows the beat-to-beat interval of Q, arctangent, and ECG. In order to compare the accuracy improvement of Q with arctangent, HRV analysis is shown in Table I. In case of Q-signal, RMSE is 40.87ms between Q and ECG. But in case of arctangent, RMSE is 21.16ms. The difference between measured by radar and ECG detector is smaller in case of arctangent demodulation. The difference between measured by radar and ECG detector is 19.71ms in case of arctangent demodulation. The error of SDNN and RMSSD is reduced by 16.62ms and 36.11ms, respectively.

TABLE I
HRV Analysis Comparison Q, Arctangent, and ECG

Mean RR (ms)					
Q	ECG	RMSE	Arctan.	RMSE	|Δ|
845.278	841.726	40.87	845.37	21.16	19.71
SDNN (ms)			RMSSD (ms)		
Q	Arctan.	ECG	Q	Arctan.	ECG
43.20	23.58	14.77	72.87	36.76	10.3

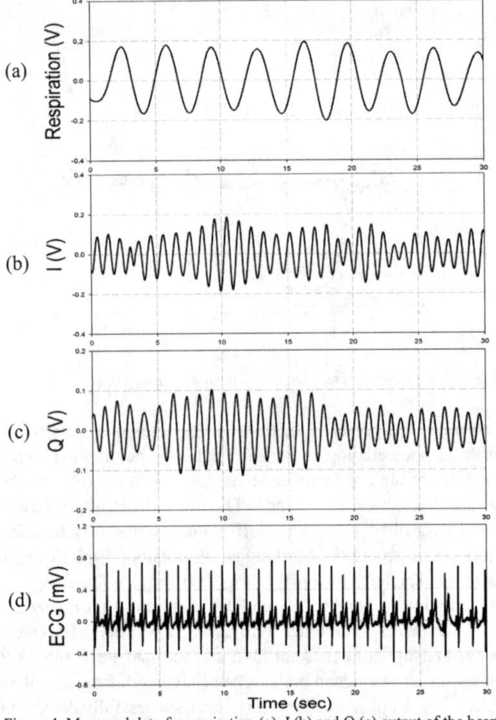

Figure 4. Measured data for respiration (a), I (b) and Q (c) output of the heart signal, and ECG signal (d) at a range of 25m.

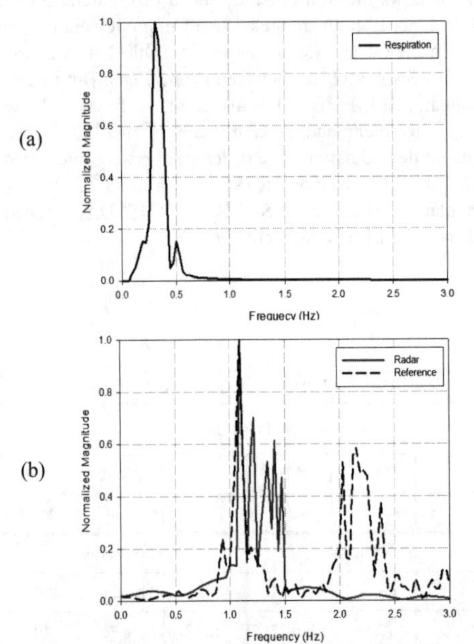

Figure 5. FFT of detected respiration (a) and heart rate (b) at a range of 35m.

To demonstrate the long range detection we increase the output power at the transmitting antenna to 23dBm. When the transmitting power is increased, the isolation between transmitting antenna and receiving antenna is dominant factor of the performance of radar system. By using leakage cancellation technique, we can get 27dB isolation improvement. Fig. 4 shows the measured results during 30 sec for radar signals at the distance of 25m. The raw data from radar have been filtered to respiration signal (a) and heartbeat (b),(c). We can see that heart rate is 33 beats for 30 seconds in both measured signal from radar system and ECG.

The FFT of measured data at the distance of 35m are shown in Fig. 5. The peak position of the respiration spectrum is 0.32Hz, which corresponds to 19.2 breaths per minute. The peak positions of the heart spectrum of radar and ECG are the same as 1.094Hz, which correspond to 65.64 beats per minute.

IV. Conclusion

This paper presents a 14 GHz CW Doppler radar system which can be used for monitoring respiration and heart rate. The Doppler radar system is tested and compared to simultaneous electrocardiogram. An exact heartbeat and respiration signals were measured and verified its accuracy by comparing HRV analysis with ECG signals. Heterodyne structure and leakage cancellation technique are able to use arctangent demodulation without dc offset compensation.

Acknowledgment

This research was financially supported by the Ministry of Science, ICT & Future Planning (MSIP) and National Research Foundation of Korea (NRF) through the Human Resource Training Project for Regional Innovation. (No.2013032190)

References

[1] J. C. Lin, "Microwave Sensing of Physiological Movement and Volume Change - a Review," *Bioelectromagnetics,* vol. 13, pp. 557-565, 1992.

[2] Donald Lie, Ravi Ichapurapu, Suyash Jain, Jerry Lopez, Ronald Banister, Tam Nguyen and John Griswold, "A 2.4GHz Non-Contact Biosensor System for Continuous Monitoring of Vital-Signs," Telemedicine Techniques and Applications, Prof. Georgi Graschew (Ed.), ISBN: 978-953-307-354-5, InTech, DOI: 10.5772/16362. Available from: http://www.intechopen.com/books/telemedicine-techniques-and-applications/a-2-4ghz-non-contact-biosensor-system-for-continuous-monitoring-of-vital-signs

[3] M. Skolnik, "Radar handbook", 2nd de. Norwich, NY:Knovel, 2003

[4] B. K. Park, O. Boric-Lubecke, and V. M. Lubecke, "Arctangent demodulation with DC offset compensation in quadrature Doppler radar receiver systems," *IEEE Transactions on Microwave Theory and Techniques,* vol. 55, pp. 1073-1079, May 2007.

[5] Z. Li and K. Wu, "On the leakage of FMCW radar front-end receiver," in *Global Symposium on Millimeter Waves, 2008*, pp. 127-130.

Interference detection for other systems using MIMO-OFDM signals

Ryochi Kataoka, Kentaro Nishimori, Masaaki Kawahara, Takefumi Hiraguri* and Hideo Makino
Graduate School of Science and Technology, Niigata University, * Nippon Institute of Technology
Ikarashi 2-nocho 8050, Nishi-ku Niigata-shi, 950-2181 Japan
Email : kataoka@gis.ie.niigata-u.ac.jp, nishimori@ie.niigata-u.ac.jp

***Abstract*—This paper proposes an interference detection method in MIMO transmission, which utilizes periodical preamble signals in a frequency domain. In this paper, we assume the collision between the short preamble signal in WLAN system and interfering signal like a microwave oven. In the propose method, second antenna receives the signals while first antenna transmits the preamble signals. Hence, the interference can be detected by observing the subcarriers which are not mapped short preamble signal using the received signal after the FFT processing. Moreover, we utilize the dual polarized antennas to reduce the mutual coupling between the transmitter and receiver. By a computer simulation and measurement with the use of orthogonal polarized antenna, it is shown that the proposed method can successfully detect interference from the other system when the interfering power is greater than the noise power.**

***Index Terms*—MIMO, collision detection, short preamble signals, orthogonal polarized antenna**

I. Introduction

Recently, wireless LAN (WLAN) devices are equipped in many user terminals (UT). Therefore, access points (APs) of WLAN are widely introduced at offices and houses. The interference can be reduced to some extent thanks to the carrier sense [1]. However, the collision occurs between the WLAN devices with IEEE802.11b/g/n standard and products such as microwave oven at 2.4 GHz band. In wired LAN, an access control scheme called carrier sense multiple access/collision detection (CSMA/CD) is introduced [2]. The wired LAN can detect the packet collision by the voltage variation inside the Ethernet cable before packet transmission. Since the wired LAN employs packet contention detection in advance and re-transmission immediately, transmission efficiency is over 90% [3]. On the other hand, the access control method called CSMA/collision avoidance (CSMA/CA) is adopted in WLAN [1]. Unlike wired LAN, it is difficult to detect the packet collision when the signals are transmitted. Hence, the reception characteristic is judged by the reply of acknowledgement (ACK) by a receiving station. Compared to the wired LAN, the transmission efficiency of WLAN is very small and its value is less than 65 %, because the re-transmission cannot be employed when the collision occurs [3]. Hence, the interference detection when the transmission is employed is one of key issues for high efficient communication.

This letter proposes a interference detection method using multiple-input multiple-output (MIMO) transmission [5]. The proposed method utilizes the fact that the second antenna is idle when the first antenna transmit the short preamble signals which are used for timing synchronization with the UT. The signals are mapped in only several subcarriers in short preamble signals of IEEE802.11 based OFDM signals and the signal in a time domain is transformed by IFFT at the transmitter. At the receiver, the correlation calculation is employed and period of the short preamble signals is detected. Conventionally, the signal is discarded after the timing synchronization. On the other hand, in the proposed method, the received signals are transformed in the frequency domain by the FFT processing. The interference can be detected by observing the subcarriers which are not mapped short preamble signal using the received signal after the FFT processing. By a computer simulation, it is shown that the propose method can successfully detect interference from the other system. Moreover, we propose the use of dual polarized antennas at the AP and its coupling characteristic is evaluated.

The remainder of this paper is organized as follows. Section II shows the proposed method using the short preamble in MIMO transmission is explained. Section III shows effectiveness of the proposed method by computer simulation and measurement.

II. Proposed Method

Fig. 1 shows the frame format when applying the proposed method. This format is used in IEEE802.11n based WLAN system. Fig. 2 shows the waveform of short preamble at the transmitter and receiver to show the interference detection using short preamble signals. The system configuration by the proposed method is shown in Fig. 3. In the IEEE802.11a/g/n based WLAN system, short and long preamble signals are used the timing synchronization and the estimation of channel state information (CSI) between the AP and UT, respectively.

In the proposed method, we assume 2×2 MIMO transmission. As can be seen in Figs. 2, the first antenna (Antenna #1) transmits the short preamble signals to the UT. As shown in Fig. 1, we utilize the fact that the short preamble signals are transmitted by only Antenna #1. Hence, there is an opportunity for a second antenna (Antenna #2) to receive the short preamble signals when the transmission by the Antenna #1 to the UT is employed. When the collision arises, the interference arrives at the Antenna #2 when the the Antenna #1 transmits the signals to UT.

978-7-5641-4279-7

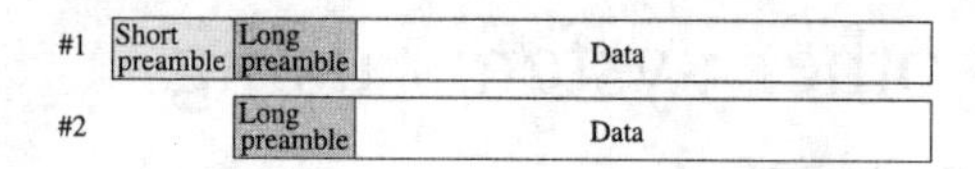

Fig. 1. Flame format of WLAN signals.

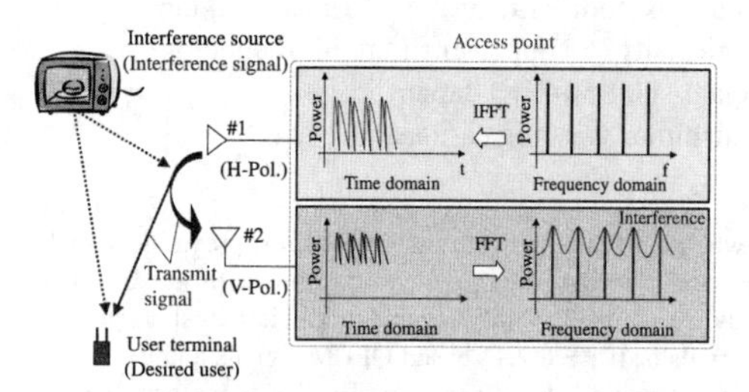

Fig. 2. Waveform of short preamble at transmitter and receiver.

Since the power of desired signal from Antenna #1 to #2 is much larger than that of interference from IS to AP, the isolation is required between Antenna#1 and #2. In this paper, there are following features in the proposed configuration and method.

(A) Isolation between Antennas #1 and #2 by using dual polarized antennas.
(B) Interference detection using received signal after the FFT processing.

As for feature (A), simultaneous transmission (by Antenna #1) and reception (by Antenna #2) using dual polarized antennas is proposed. The dual polarized antennas are generally used for reducing the spatial correlation between two antennas and especially effective in outdoor scenario [5], when the MIMO systems are considered. In the proposed method, the dual polarized antennas are used not only the reduction on the spatial correlation but also the isolation for simultaneous transmission and reception. The basic performance in an actual room is evaluated in Section III.

Regarding the detailed feature of (B), the principle is explained hereafter. Fig. 2 shows the waveform of short preamble at the transmitter and the received signal at the receiver to

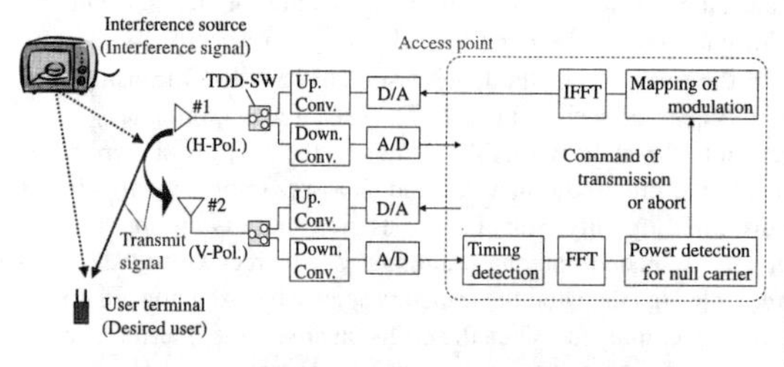

Fig. 3. System configuration by proposed method.

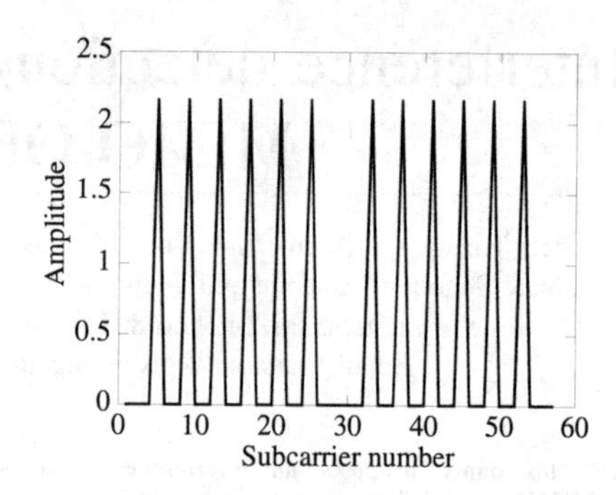

Fig. 4. Amplitude of short preamble signal in the frequency domain.

show interference detection using short preamble signals. As can be seen Fig. 2, the short preamble signal is adopted in IEEE802.11a/n/g based WLAN system. Fig. 4 shows the amplitude of short preamble signal in the frequency domain [1]. As shown in Fig. 4, only twelve signals are mapped. At the transmitter, the signals are mapped for only twelve subcarriers to generate the periodical signal in time domain [1]. The signal is transmitted after the IFFT processing. At the receiver, the timing synchronization between the known transmit and received signals is employed.

At the receiver, the timing synchronization between the known transmit and received signals is employed. The timing synchronization is realized by using the correlation calculation between the received signal, $y(t)$ and $s_p(t)$ which is the short preamble signal in time domain. The correlation value, ρ is denoted as

$$\rho = \frac{\left|\sum_{t=1}^{L} s_p^*(t)y(t)\right|}{\sqrt{\sum_{t=1}^{L}|s_p(t)|^2}\sqrt{\sum_{t=1}^{L}|y(t)|^2}}, \tag{1}$$

where L is the number of symbols that the correlation calculation is employed and $L = 160$ in IEEE802.11n based OFDM signal. The correlation value, ρ is maximized at the initial timing of the short preamble signals.

In the general WLAN system, the short preamble is not used after the timing synchronization. In this paper, we utilize the short preamble signals after the FFT processing. The received signals after the FFT processing contain the desired signal, interference signal and noise. Interference plus thermal noise can be detected by observing the subcarriers which are not mapped short preamble signal using the received signal after the FFT processing. When the interference power is greater than the noise power, the arrival of interference can be judged by checking the power on the subcarriers which are not mapped short preamble signal. The noise power is used as the threshold value. The noise power can be estimated by power on the subcarriers that the signals are not mapped at the transmitter.

Finally, as can be seen Fig. 3, the transmitter stops com-

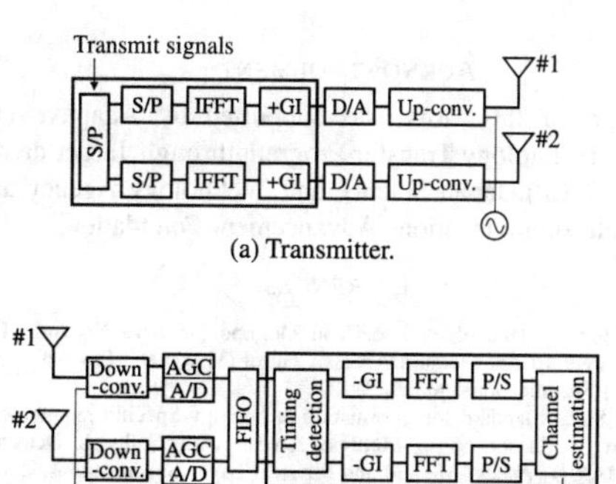

Fig. 5. MIMO-OFDM transceiver.

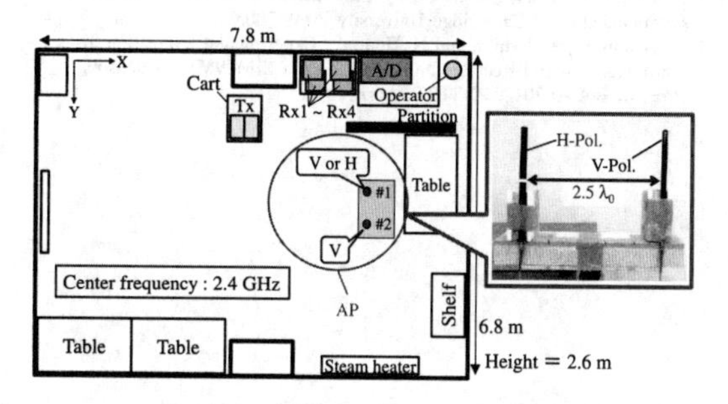

Fig. 6. Measurement environment for mutual coupling effect.

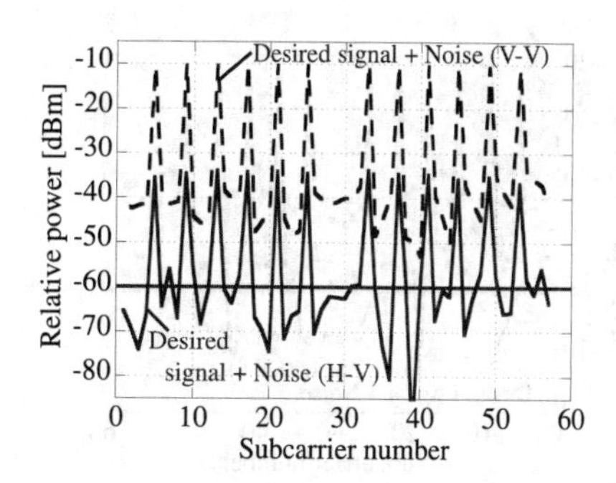

Fig. 7. Received power in frequency domain (Measurement results).

TABLE I
SIMULATION PARAMETERS

Parameter	Value
Desired/Interfering signal	1
SNR	30
SIR average	10-40
The number of rays	100
Propagation path of interference	Rayleigh
number of receive antenna	1
Bandwidth	20MHz
number of FFT points	64
Communication for the number of subcarriers	56
Pilot subcarrier number	4
GI length	16
OFDM symbol length	80
short preamble length	2OFDM symbol
lomg preamble length	2OFDM symbol
pilot subcarrier number	6, 20, 34, 48

municating with UT. Since the proposed method do not need ACK signal unlike CSMA/CA, we confirmed that the transmit efficiency becomes very high and the transmit efficiency by the proposed method is approximately 90% [3].

III. Effectiveness of Proposed Method

Figs. 5 and 6 show MIMO-OFDM transceiver and the measurement environment for mutual coupling effect, respectively. IEEE802.11n based MIMO-OFDM signals can be transmitted and received by the transceiver in Fig. 5(a) and (b) [6]. As shown in Fig. 6, vertical polarized or dual polarized antennas for the transmit and receive antennas are used for the measurement of mutual coupling. The AP is located at the place in Fig. 6. The center frequency is 2.4 GHz. The element spacing is set to be 2.5 λ_0. The transmit power is 0 dBm and the average noise power was -60 dBm.

Fig. 7 shows the measured received power when the short preamble signals are transmitted by actual IEEE802.11n based OFDM signals [6]. The transmit antennas are vertical and horizontal polarized antennas, respectively. The received antenna is vertical polarized antenna. As can be seen in Fig. 7, the power reduction is approximately 10 dB when using vertical polarized antennas due to high mutual coupling. Moreover, the noise power is increased due to the saturation of low noise amplifier at the receiver. On the other hand, 35 dB power reduction can be observed and noise power is not increased when using dual polarized antennas. Hence, it is shown that the use of dual polarized antennas is effective for the proposed interference detection. In order to clarify the basic characteristics of the interference detection by the proposed method, the computer simulation using IEEE802.11n based OFDM signals is employed. The simulation parameters are shown in Table III. The signal to noise power ratio (SNR) is set to be 30 dB and signal to interference power ratio (SIR) is set to be 10 and 40 dB. Since the transmit and receive antenna are closely located with each other, the propagation characteristic of desired signal is assumed be the AWGN channel. The Rayleigh fading is assumed as the propagation channel on the interference by the UT. The other basic parameters are the same with IEEE802.11n standard.

Figs. 8(a) and (b) show the received power in the frequency domain. The solid lines are received powers and the broken lines are the powers of desired signal plus thermal noise. As can be seen Fig. 8(a), it is possible to judge the interference, because the signals can be detected at the null subcarriers. On the other hand, Fig. 8(b) indicates that the interference is hidden in the thermal noise. Therefore, it is shown that the

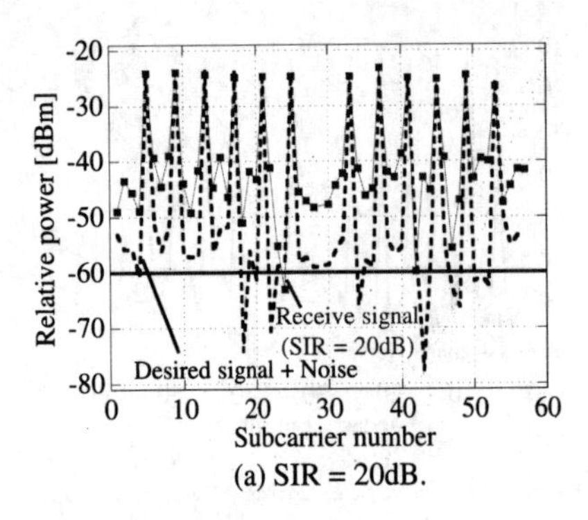

(a) SIR = 20dB.

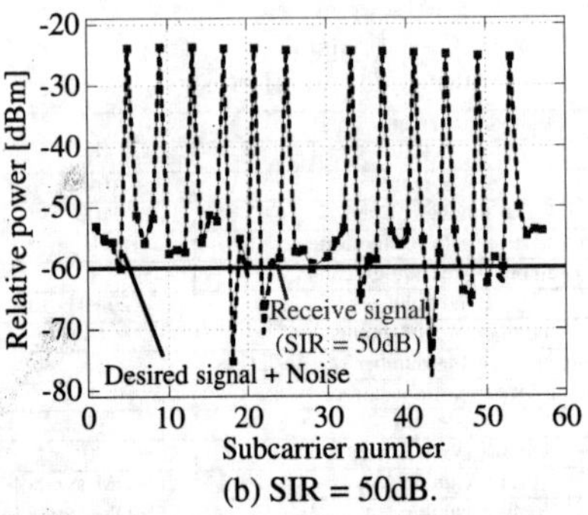

(b) SIR = 50dB.

Fig. 8. interference signal power estimated from the receive signal.

interference can be sufficiently detected when the interference power is greater than the noise power. In the future, we will conduct the measurement regarding the interference detection by using the testbed in [6].

IV. Conclusion

This paper has proposed the interference detection method by utilizing the short preamble signals in the MIMO transmission. The proposed method focuses on IEEE802.11 based short preamble signals which are mapped at only twelve subcarriers in the frequency domain and the simultaneous transmission between transmitter and receiver. By the proposed method, it was shown that the interference from the other system can be detected by observing the null subcarriers which are not mapped the desired signals after the FFT processing. By the measurement using the dual polarized antennas, over 35 dB isolation was obtained. Moreover, it was found that the interference can be detected when the interference power is greater than the noise power via the computer simulation.

Acknowlegement

The part of this work was supported by Adaptive and Seamless Technology Transfer Program through Target-driven R&D (A-STEP), Japan Science and Technology Agency and by the Telecommunications Advancement Foundation.

References

[1] IEEE802.11 Standard for Local and Metropolitan Area Networks Part 11: Wireless LAN Medium Access Control (MAC) and Physical Layer (PHY) Specifications, March 2012.

[2] IEEE 802.3 Standard for Information technology-Specific requirements - Part 3: Carrier Sense Multiple Access with Collision Detection (CSMA/CD) Access Method and Physical Layer Specifications, 2008.

[3] T. Hiraguri, K. Nishimori, T. Ogawa, R. Kataoka, H. Takase, H. Hideo, "Access control scheme for collision detection utilizing MIMO transmission," IEICE ComEX, Vol. 2, No. 4, pp.129-134, April. 2013.

[4] G. J. Foschini and M. J. Gans, "On limits of wireless communications in a fading environment when using multiple antennas," Wireless Personal Commun., vol. 6, pp. 311-335, 1998.

[5] A. Paulaj, R. Nabar, and D. Gore, "Introduction to space-time wireless communications," Cambridge University Press 2003.

[6] K. Nishimori, K. Ushiki, and N. Honma, "Experimental Evaluation Toward Transmit and Receive Diversity Effect in SIMO/MIMO Sensors," Proc. of EuCAP2012, POST2.9, April 2012.

Experimental Study on Statistical Characteristic of MIMO Sensor for Event Detection

Shinji ITABASHI[1], Hiroyoshi YAMADA[2], Kentaro NISHIMORI[3], and Yoshio YAMAGUCHI[4]
[1]Graduate School of Science & Technology, Niigata University
Ikarashi 2-8050, Nishi-ku, Niigata, 950-2181 Japan
[2,3,4]Faculty of Engineering, Niigata University
Ikarashi 2-8050, Nishi-ku, Niigata, 950-2181 Japan

***Abstract*—Recently, researches on radio wave security sensors by using array antenna or MIMO(Multiple-Input Multiple Output) system have been attracting attention. These sensors utilize the property that MIMO channel matrix is sensitive to the propagation environment change in indoor multipath environment. Now the property is used for event detection, such as intrusion, by calculating correlation of MIMO channel matrices. Fundamental study has been performed, however, statistical property, or quantitative detection performance analysis, of the sensor was still remain to be clarified. So, our objective is to derive a probability density function (PDF) of the correlation coefficient and to demonstrate them experimentally. The PDF is a key function to derive detection rate and so on. In the conventional MIMO sensor, correlation coefficient of the channel matrix between with and without an event is used as a measure of events. Therefore, in this report we will derive the PDF of the channel correlation used in the conventional MIMO sensor theoretically and also show the validity by experimental results.**

I. INTRODUCTION

Recently, researches on radio wave security sensors by using array antenna or MIMO (Multiple Input Multiple Output) system have been attracting attention. Various systems and detection algorithms have been proposed and examined, and validity of them is shown by simulations and/or experiments to some extent. However, these considerations were done under some specific system and antenna arrangement in a given propagation environment, hence they may often lack in generality. Actually, the quality assessment in consideration of detection probability analysis in given false alarm rate will be required. We found that the channel coherence defined by the MIMO event detection sensor has the same form as that used in the interferometry for Synthetic Aperture Radar image analysis, then we apply the theory to derive theoretical expression of the PDF. The PDF can be evaluated by histogram of the correlation values obtained by experiments. The PDF is the important function to derive detection probability at given false alarm rate theoretically with given event at the SNR. The validity of the derived theoretical expression for channel coherence in MIMO sensor

Therefore, in this report, we show validity of the PDF of the MIMO sensor experimentally by using histograms of the measured MIMO channels

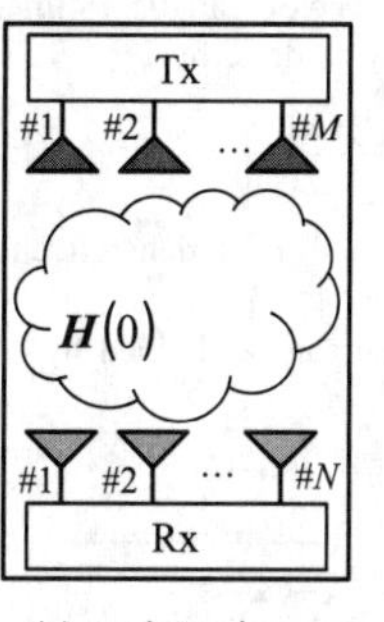

(a) no intrusion

(b) intrusion

Fig. 1. MIMO sensor principle figure

II. INDOOR INVASION DETECTION SYSTEM

In this chapter, we describe signal model for the MIMO sensor and derive a MIMO channel correlation coefficient.

Figure 1 shows an example of MIMO sensor. In this report, we assume that the sensor is operated in an indoor rich-multipath environment, and also assumes that we employ the MIMO sensor with M transmitting and N receiving antennas. Here, we define the channel matrix without and with the event by $\boldsymbol{H}(0)$ and $\boldsymbol{H}(t)$, respectively as shown in Figs. 1(a) and (b). The MIMO sensor assumed in this report catches change of the propagation channel, and detects the event.

A. Signal model for MIMO sensor

In this report, each transmitting antenna transmits independent signals, then the the received data at time t by the m-th $(= 1, 2, \cdots, M)$ transmitting antenna can be given by

$$\boldsymbol{y}_m(t) \;=\; \boldsymbol{h}_m(t)s_m(t)+\boldsymbol{n}_m(t) \tag{1}$$

where $\boldsymbol{h}_m(t)$ is a N dimension propagation channel vector by the m-th transmitting channel, $s_m(t)$ is the transmitting signal (known training symbols), and $\boldsymbol{n}_m(t)$ is the additive Gaussian noise vector having zero mean and power of σ_n^2 in each elements. The MIMO channel matrix can be defined by

$$\boldsymbol{H}(t) \;=\; [\boldsymbol{h}_1(t), \cdots, \boldsymbol{h}_M(t)] \tag{2}$$

The MIMO sensor detects the channel change. When we define the channel matrix without the event as $\boldsymbol{H}(0)$ as shown

978-7-5641-4279-7

in Fig.1(a), the channel variation, $\boldsymbol{H}_d(t)$, can be defined by

$$\boldsymbol{H}_d(t) = \boldsymbol{H}(0) - \boldsymbol{H}(t) \tag{3}$$

In this report, we assume that the MIMO sensor operates at relative high frequency band (e.g. 2.5 GHz), hence change of the propagation channel by intrusion/event intercepts some paths between transmission and reception. Namely, all the intercepted paths in $\boldsymbol{H}_d(t)$ by the intrusion are independent of the remain paths in $\boldsymbol{H}(t)$ not intercepted by the intrusion. As shown in Fig.1, a propagation channel including propagation information is changed by event sensitively. Therefore, the system which detects propagation information from a received signal is required. The channel vector can be estimated by known training symbol(s). It can be done by [3].

$$\hat{\boldsymbol{h}}_m(t) = \frac{E\left[\boldsymbol{y}_m(t)s_m^*(t)\right]}{E\left[s_m(t)s_m^*(t)\right]} \tag{4}$$

where * is complex conjugate, and $E[\cdot]$ denoted ensemble average.

The MIMO channel correlation can be defined by

$$\rho(t) = \frac{\sum_{n=1}^{N}\sum_{m=1}^{M} \hat{h}_{nm}^*(0)\hat{h}_{nm}(t)}{\sqrt{\sum_{n=1}^{N}\sum_{m=1}^{M}|\hat{h}_{nm}(0)|^2}\sqrt{\sum_{n=1}^{N}\sum_{m=1}^{M}|\hat{h}_{nm}(t)|^2}} \tag{5}$$

where $\hat{h}_{nm}(0)$ and $\hat{h}_{nm}(t)$ are the matrix element at the reference and measured MIMO channel matrix of $\boldsymbol{H}(0)$ and $\boldsymbol{H}(t)$, respectively.

III. Statistical Character of a Correlation Coefficient

Theoretical evaluation of (5) is difficult, however the following formula can be easily calculated.

$$\begin{aligned}\bar{\rho} &= \frac{E\left[\sum_{n=1}^{N}\sum_{m=1}^{M} \hat{h}_{nm}^*(0)\hat{h}_{nm}(t)\right]}{\sqrt{E\left[\sum_{n=1}^{N}\sum_{m=1}^{M}|\hat{h}_{nm}(0)|^2\right]}\sqrt{E\left[\sum_{n=1}^{N}\sum_{m=1}^{M}|\hat{h}_{nm}(t)|^2\right]}} \\ &= \frac{P_0 - P_d}{\sqrt{P_0 + P_E}\sqrt{P_0 - P_d + P_E}} \end{aligned} \tag{6}$$

where P_0 and P_d are the channel power of $\boldsymbol{H}(0)$ and $\boldsymbol{H}_d(t)$, respectively, and P_E is the estimated power of the change components. These power components can be given as follows,

$$P_0 = E\left[\|\boldsymbol{H}(0)\|_F^2\right], \tag{7}$$

$$P_d = E\left[\|\boldsymbol{H}_d(t)\|_F^2\right], \tag{8}$$

$$P_E = \frac{M}{N_s}N{\sigma_n}^2, \tag{9}$$

where N_s is snapshot and $||\cdot||_F$ denotes Frobenius norm.

Obviously, the equation (6) is not the correct estimated value for (5). However, this type of correlation is the same form as that in complex coherence in SAR interferometry [2].

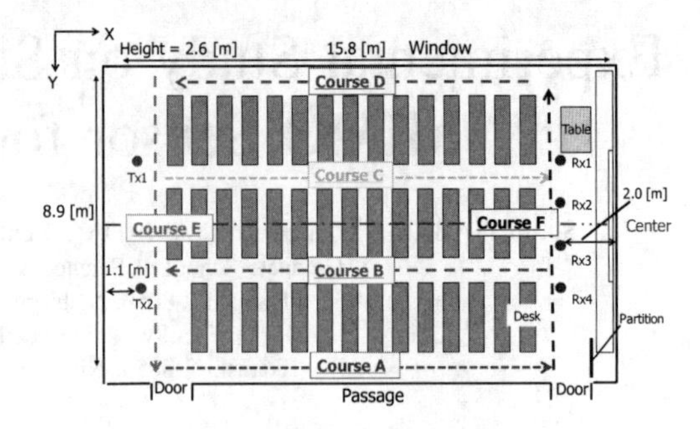

Fig. 2. Measurement environment

TABLE I
Experimental Setup

Tx and Rx antennas ($M \times N$)	2×4
Receiving array shape	Uniform linear array
RF frequency	2.55 [GHz]
Total power	0 [dBm]
Element interval(Rx)	1.2 [m]
Element interval(Tx)	3.6 [m]
Antenna height	1.0 [m]
Training symbol(snaphots)	160 symbols

According to [2], the probability density function (PDF) of (6) can be given by

$$\begin{aligned} f(\rho) = 2(L-1)(1-\bar{\rho}^2)^L\rho(1-\rho^2)^{L-2} \\ \times\ {}_2F_1(L, L; 1; \bar{\rho}^2\rho^2) \end{aligned} \tag{10}$$

where $\Gamma(\cdot)$ and ${}_aF_b$ are the Gamma and generalized hypergeometric function, respectively, and $L = NM$.

IV. Experimental Study

A. Measurement environment

Measurement environment is shown in Fig. 2, and experimental setup is shown in Table I. In this environment, we carried out several experiment with and without a walker. The walking (observation) time at the courses A$\sim$D was 12 seconds each, and the courses E$\sim$F was 8 each.

B. Experimental Results

1) CASE I: No intrusion:

The theoretical PDF is calculated at given SNR and ISR values. However, the received power of experimental data were little bit fluctuated, so we used the data in 1 dB power bandwidth at the target value (target value ± 0.5 dB). The target value is also estimated by making histogram of power distribution for the measured MIMO channel.

The SNR is received power versus the noise power ratio without event (no intrusion), and the ISR is defined by the changed power with the event (intruder) versus received power

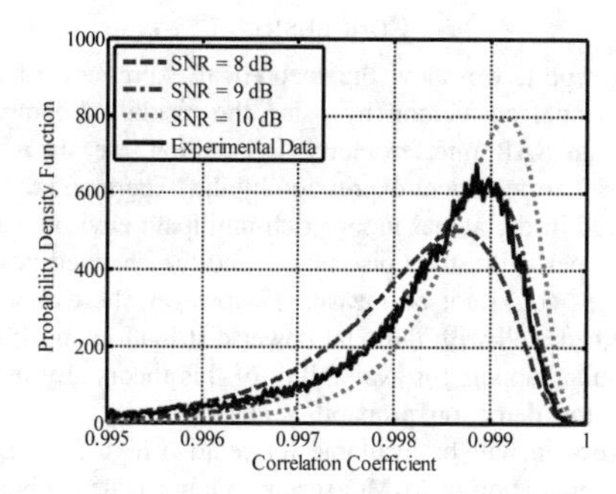

Fig. 3. PDF (w/o intrusion)

without the event. Their definition can be written by

$$\mathrm{SNR} = 10\log_{10}\frac{P_0}{P_N} \quad [\mathrm{dB}] \tag{11}$$

$$\mathrm{ISR} = 10\log_{10}\frac{P_d}{P_0} \quad [\mathrm{dB}] \tag{12}$$

Direct measurement of the noise power for the system used in this experiment was difficult. So, in this experiment, the SNR was experimentally evaluated by the estimated PDF derived by the histogram of MIMO correlation without event. The experimental result and the several theoretical PDFs for SNR=8, 9, and 10 dB, are shown in Fig. 3. From Fig. 3, it turns out that the experimental result is almost coincide with the theoretical PDF of SNR=9 dB. This shows that SNR in the measurement environment in this experiment was 9 dB. Moreover, since the theoretical and the experimental curves agree well, the theoretical PDF in (10) is valid in this indoor MIMO sensor at 2.55 GHz.

2) CASE II: With Intrusion:

Also with intrusion, the same experimental setups as without the intrusion were used. And, in order to check the validity of the theoretical expression for the PDF with intrusion, the histograms are created by the experimental data with each intrusion, and we evaluate the result of each course. Especially, we focus on the results at the course A, B, and F.

In this case with intrusion, direct measurement of P_d is difficult and it may change at some extent with behavior of the walker, then the ISR is also estimated in comparison with theoretical plots of PDF for several different ISRs.

First, the result of the course A is shown in Fig. 4. From this figure, the ISR in this path can be estimated almost -15 dB by the corresponding theoretical curve. The histogram of ISR estimated by the experimental data is also shown in Fig. 5. From Fig. 5, it turns out that peak of ISR in this experiment was -15 dB and it is almost coincide with the estimated value in Fig. 4. In Fig. 4, the experimentally derived PDF looks slightly distorted to the theoretical curve. This is because we evaluate the experimental data including all ISR values in Fig. 5. The number of data is limited in this case, hence we could not evaluate the narrow power-bandwidth ISR data at around -15 dB. However, there results show that the theoretical and experimental distribution is well coincide with each other.

Next, the result of the course B is shown in Fig. 6. From Fig. 6, the estimated ISR in this case becomes -10 dB which was slightly larger change in comparison with that in course A. Likewise in the previous results, the histogram in this experimental data set in Fig. 7 shows the peak at around -10 dB. This coincides the estimated results in Fig. 6.

Finally, we would like to show the results of course F as the differently directed path for course A and B. Result of PDF for this course is shown in Fig. 8.In this case, distortion of the experimental curve is larger than the previous results, and we could hardly estimate the ISR, but the ISR would be in the range between -10 to -15 dB. The cause of distortion can be realized by the histogram shown in Fig. 9. The ISR in this course spreads widely. Obviously, this causes because the course F across the receiving array. The incoming path in each receiving element changes a lot in comparison with the course of along direction.

From these result, we can conclude that the theoretical expression for the PDF is valid with and without the event by the experimental results.

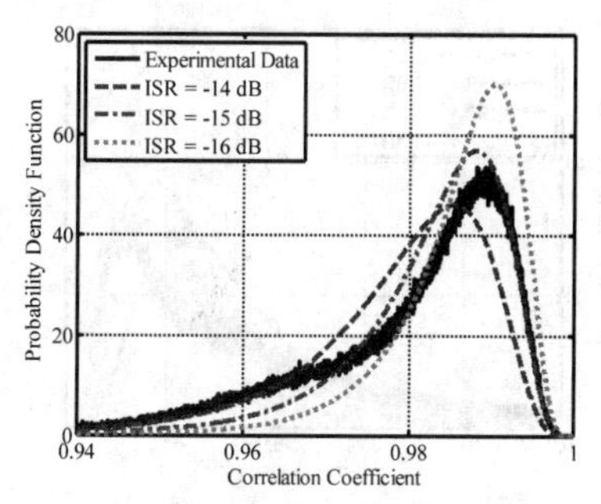

Fig. 4. PDF(w/ intrusion) of Course A

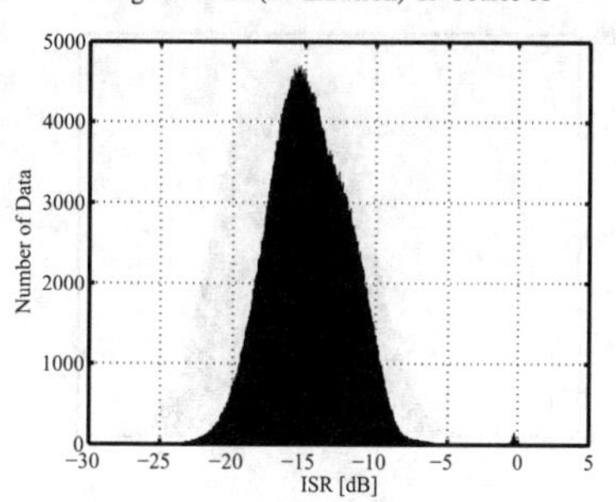

Fig. 5. Histogram of ISR(Course A)

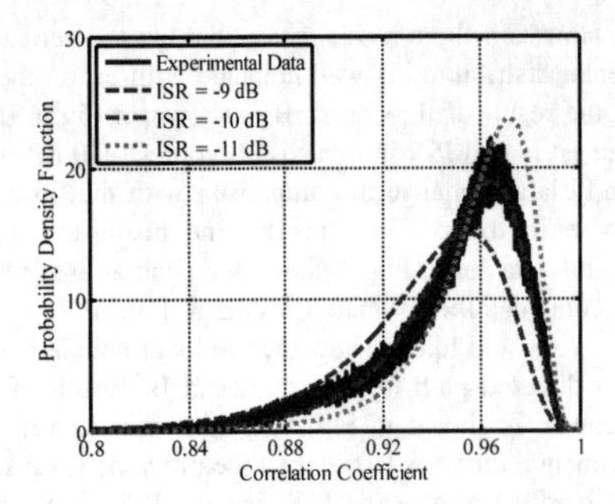

Fig. 6. PDF(w/ intrusion) of Course B

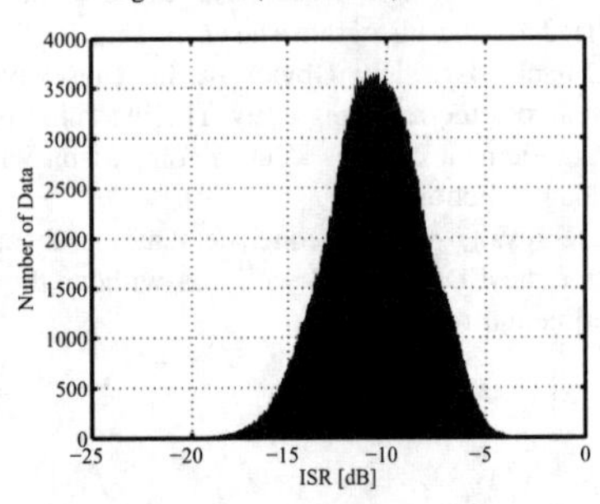

Fig. 7. Histogram of ISR(Course B)

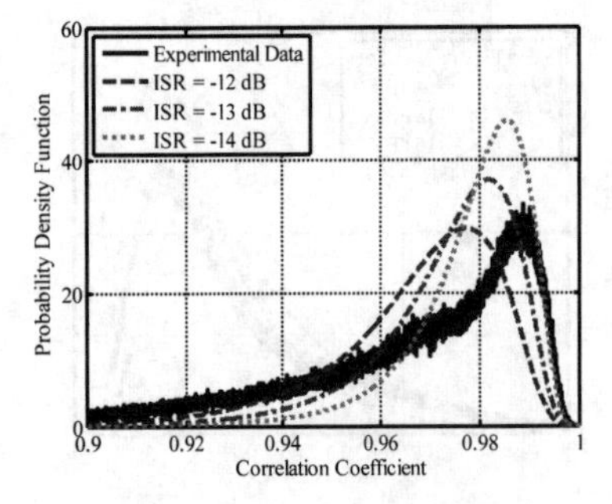

Fig. 8. PDF(w/ intrusion) of Course F

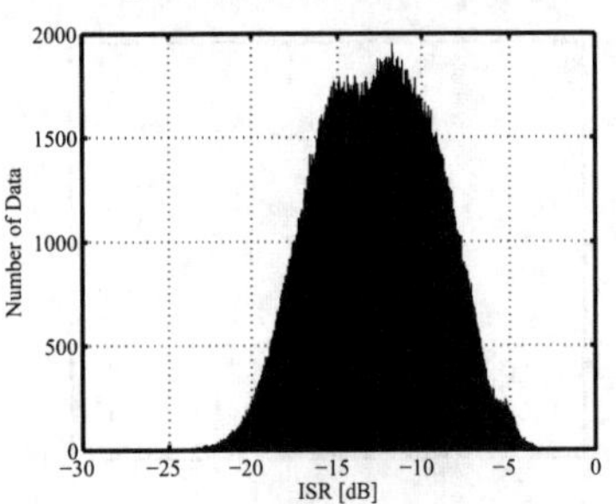

Fig. 9. Histogram of ISR(Course F)

V. CONCLUSION

In this report, we show the theoretical expression of the MIMO channel correlation by using the results of complex coherence in SAR interferometry. In addition, we show the validity of the theoretical expression by the experimental data set acquired in the actual indoor rich-multipath environment.

In the experiment, the noise power and the changed power by the event could not be measured. However, the estimated values agreed well with the total power distribution and ISRs. These results support for availability of this theory. By using this PDF, the detection probability can be easily estimated. This expression will be available for relative high frequency case. For application of MIMO sensor in low frequency below 1 GHz, some modification will be required. This wiil be the one of future works.

ACKNOWLEDGMENT

This work is supported by JSPS KAKENHI, the Grant-in-Aid for Scientific Research (C) (No.23560442).

REFERENCES

[1] S.Ikeda, H.Tsuji, T.Ohtsuki, "Indoor Event Detection with Eigenvector Spanning Signal Subspace for Home or Office Security," IEICE Trans. Commun., vol.E-92-B, no.7, pp.2406-2412, July 2009.

[2] R.Touzi, A.Lopes, J.Bruniquel, P.W.Vachon, "Coherence Estimation for SAR Imagery," IEEE Trans. Geosci. Remote Sensing., vol.37, no.1, pp.135-149, Jan. 1999.

[3] M. Biguesh, A. B. Gershman, "Training-Based MIMO Channel Estimation :A Study of Estimator Treadeoffs and Optimal Training Signals" IEEE Trans. Signal Process., vol.54, no.3, pp.884-893, Mar. 2006.

WP-P

October 23 (WED) PM

Room E

The Design and Simulation of Feederlines of Log-periodic Dipole Antenna in Microwave Band

Yajuan Shi, Dongan Song,Dinge Wen, Hailiang Xiong, Qi Zhang, Mingliang Huang
Science and Technology on Electromagnetic Compatibility Laboratory
China Ship Development and Design Center
Wuhan, China
crystal1988world@126.com

***Abstract*-This paper studied the impact of different distance and angles of feedlines to the log-periodic dipole antenna (LPDA). The study found that in the parallel manner (angle is 0) and different distance between feederlines(d is modified from 2mm to 14mm), VSWR is larger than 3. So the optimization is proposed that the feederlines are inclined by 5° to get wider work band (from 0.47GHz to 3.8GHz) with 8dB gain as well as low VSWR. Thanks to this modification, the LPDA can provide superior performance in the measurement of multiple polarization directions and omni-direction pattern above 2GHz.**

I. INTRODUCTION

Log periodic dipole antenna (namely LPDA) was proposed in 1957, and its characteristics are logarithm cycle changed with frequency. Because the change of the antenna characteristics in a cycle is little, therefore LPDA belongs to the non-frequency-dependent antenna. Because of its simple structure, lower price, good quality, easy to install and so on, the LPDA is the representative of the popular broadband antennas. LPDA is widely used in communications, direction finding, searching and electronic countermeasures within the field of shortwave and FM, as well as microwave-band [1]. LPDA can be used to simultaneously measure the multiple polarization direction and nearly omni-directional pattern, which possess the advantages of the horn antenna and the biconical antenna. The feederlines are important in the design of this antenna. Here the design of feederlines in microwave band is proposed, and the simulation is given by CST [2]. In this paper, we conclude that the LPDA can perform well with good front-to-rear ratio characteristics and high gain, as well as wide work band above 2GHz.

II. THE STRUCTURAL DESIGN OF THE LPDA

LPDA is one of the classic non-repeat forms of frequency varying antenna, its impedance and radiation characteristics are repeated by the logarithm of the frequency. Its main feature is the ultra-wideband and low cross-polarization level. The transmission of electromagnetic energy is from feed point along the antenna structure to push forward. The resonance occurs at the teeth of a quarter-wavelength and then the radiation occurs. The remaining little part of energy will move forward, and then reflect back at the end of the antenna, where the energy is very weak.

When the frequency is changed, the resonant point will move, but the geometric form of the LPDA ensures the characteristics of the antenna will not be affected by the move of the resonant point [3]. The feed way of LPDA is cross power feeding. The feeder is a cross-connect between two adjacent poles, and the purpose is to ensure the poles in the radiation area can obtain appropriate phase relationship, so that the phase of long poles are before the short ones, thereby the result is the end fire pointing to the vertex direction. The feederlines are supported by a parallel dual-pipeline. The coaxial cable outputted by the transmitter is cable through a pipeline, and the skin is connected to the metal tube, as well as the core wire to another metal tube shown in Figure.1. The length of pole arm equals to the half wavelength of the corresponding frequency. The cross-connect of the coaxial cable is to achieve one-way radiation of the short pole directions.

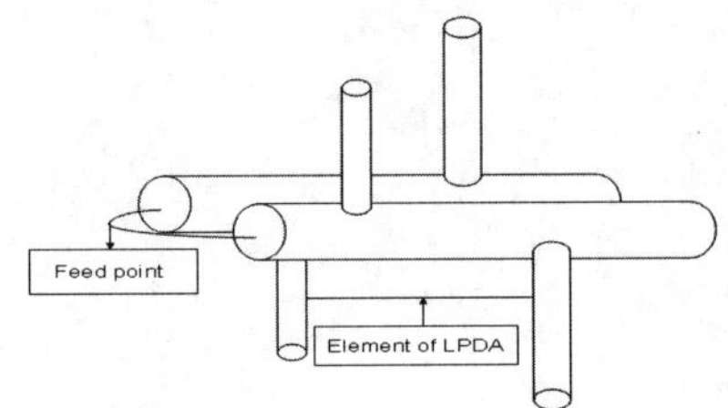

Figure.1 structure of the LPDA

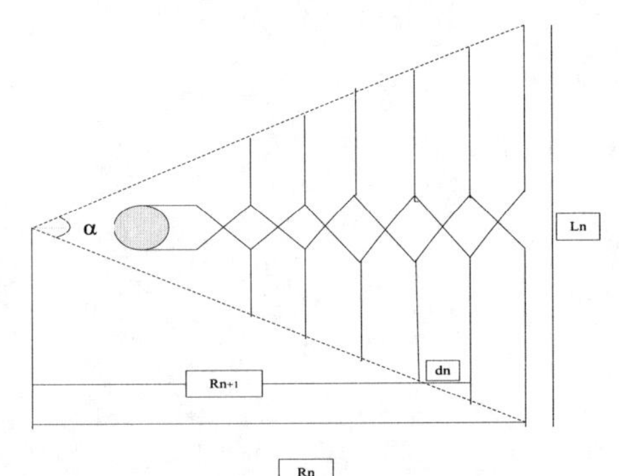

Figure.2 schematic of the LPDA

978-7-5641-4279-7

The feature of pattern, gain and impedance depends on τ and σ.

Scale factor τ is

$$\tau = \frac{R_{n+1}}{R_n} = \frac{L_{n+1}}{Ln} < 1 \qquad (1)$$

The ratio of the distance of successive pole equals to the ratio of the length of the successive pole.

The interval factor of LPDA is:

$$\sigma = \frac{d_n}{2L_n} \qquad (2)$$

The element spacing is:

$$d_n = R_n - R_{n+1} \qquad (3)$$

Because of $R_{n+1} = \tau R_n$, and $R_n = \frac{L_n}{2\tan(\alpha/2)}$,so:

$$d_n = (1-\tau)\frac{L_n}{2\tan(\alpha/2)} \qquad (4)$$

Bring these formulas into the definition of σ, the result is

$$\sigma = \frac{d_n}{2L_n} = \frac{1-\tau}{4\tan(\alpha/2)} \qquad (5)$$

All dimensions into the following proportions:

$$\tau = \frac{R_{n+1}}{R_n} = \frac{L_{n+1}}{L_n} = \frac{d_{n+1}}{d_n} = \frac{\rho_{n+1}}{\rho_n} \qquad (6)$$

In the formula (6), l is the length of the pole, ρ is the radius of the pole, s is the distance between the longest and the shortest poles.

At the end of the feederlines, it can be connected with a resistance equal to its characteristic impedance, as well as can be up to the back of the unit within $\lambda_{\max}/8$ distance. The characteristic impedance of the feeder can affect the input impedance of the antenna. Its relationship is expressed as [4]:

$$Z_c = \frac{Z_0}{\sqrt{1+\frac{Z_0\sqrt{\tau}}{4\sigma Z_a}}} \qquad (7)$$

III. THE DESIGN OF LPDA'S FEEDERLINES

When the distance between feederlines(namely d) is too small, the current amplitude will be inverted, so that in the far-field region it will affect each other. When LPDA works at the high frequency, the beam of the inner core wire will be deflected to the top of the feederline. This assumes that the core wire is self-induced. Thereby the high frequency characteristics of the antenna will be limited. When d increases, it means that the distance of the inner core of the collection line increases, thus the coefficient of self-induction increases too, so does the load impedance Z_l. Therefore, the distance d requires multiple simulations to optimize its design.

In the design, the parameters τ and σ are: 0.928 and 0.0864. The VSWR of LPDA with different d (namely d = 2 mm, 6mm and 14mm respectively) is shown in Figure.3.

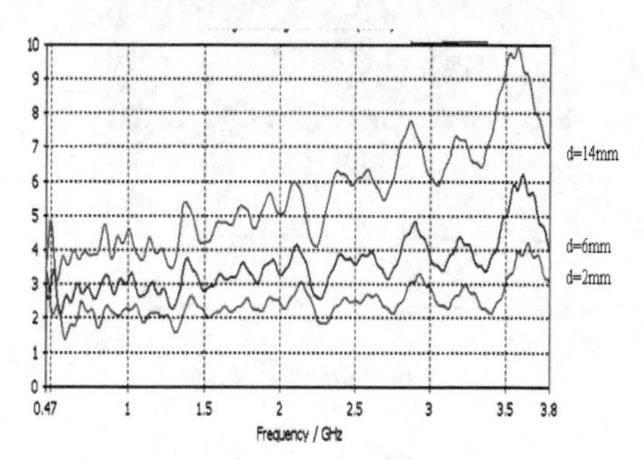

Figure.3. The VSWR of LPDA

In figure.3, even if the distance d = 2 mm, VSWR is also substantially greater than 2 in the whole frequency band. Through the different simulation comparative study, we can conclude that if the feederlines are theoretically parallel, the VSWR is not ideal. Hence a change in the design of the feederlines is put forward. The idea is that the distance between feederlines is linearly varied, the front is closer (less than 2 mm, so here is 1.4mm) ,and the rear is less than 15mm.The reasons to this design are : (1) to achieve the impedance match. A pair of poles on different feederlines is to form a dipole (shown in Figure 4). But the size of C will perform a great influence in the microwave band to the antenna, also C can affect the symmetry of the dipole, as well as the impedance matching, so in the microwave band C should be small enough; (2) to facilitate the matching of the feed point. When the distance between feederlines changed, it will be easier to match with a 50-ohm coaxial cable, the radiated energy at the feed point is reduced, and the pattern will gain better performance; (3) to reduce their own radiation.

Finally, the design of the feederlines is inclined. The value of τ and σ are 0.928 and 0.0864 respectively. The total length of feederlines is 771mm, which front part is 335mm with 8mm radius and the rear part is 436mm with 10mm radius. The LPDA utilizes 33 pairs of poles. The radius of pole is varied from 1.5mm to 6mm. The short circuit is 26.3mm distance to the longest pole. In the modeling process, considering that the current focuses on the model surface, the feeders and pole can be solid metal in the work band(shown in Figure. 4).

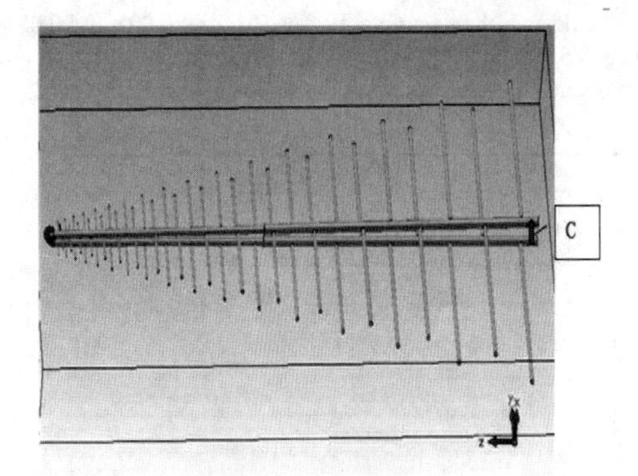

Figure.4 A 3D view of the feederlines

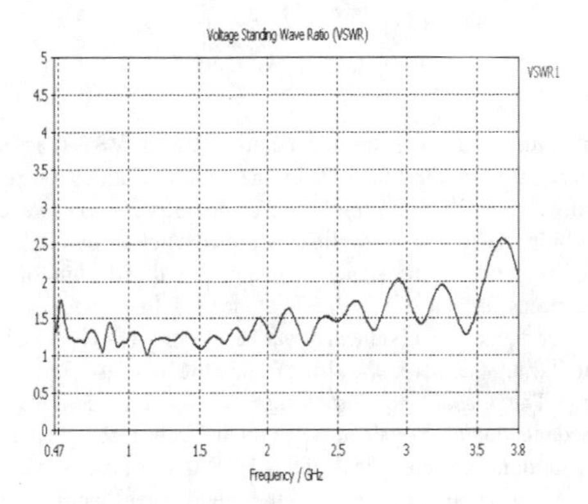

Figure.5. The VSWR of the antenna with inclined feederlines

Figure.5 shows that the antenna with inclined feederlines possesses a wide work band (0.47GHz ~ 3.8GHz) and low VSWR(less than 2).

In Figure.6 the gain of LPDA is larger than 8dB in the full work band.

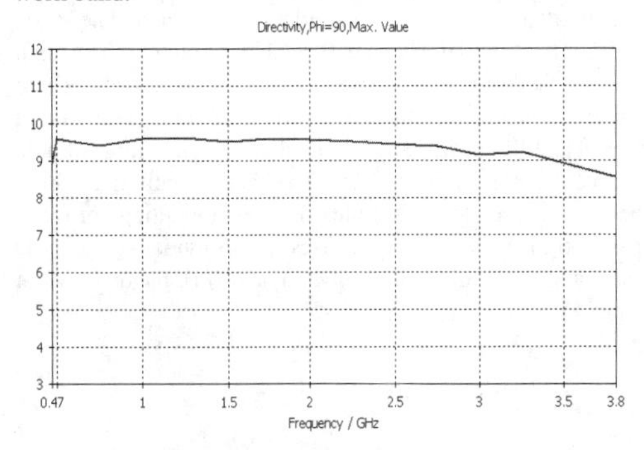

Figure.6. The full band gain of LPDA

The radiation direction of LPDA is toward the short pole. Figure.7-Figure.10 are the patterns of E-plane (red) and H-plane (blue) at 0.47GHz, 0.9GHz,1 .9GHz, 3.8GHz, respectively. The blue line is the 3dB direction.

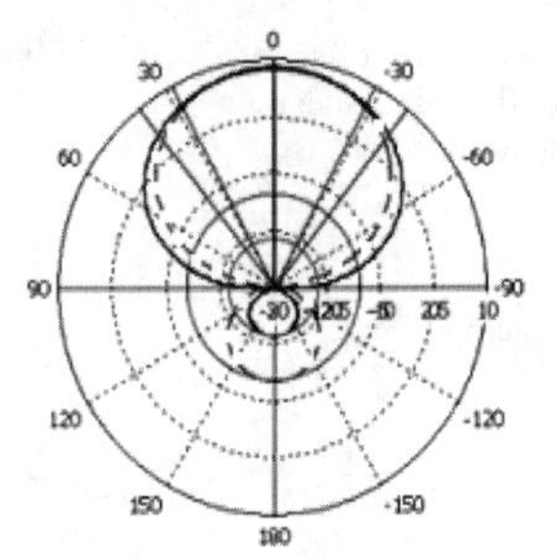

Theta / Degree vs. dB

Figure.7 0.47GHz

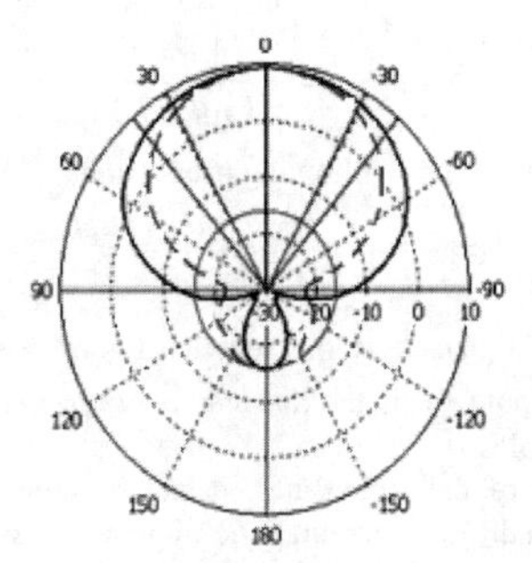

Theta / Degree vs. dB

Figure.8 0.9GHz

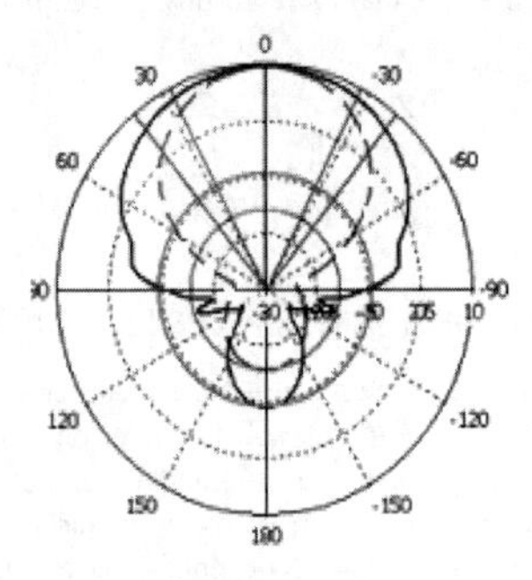

Theta / Degree vs. dB

Figure.9 1.9GHz

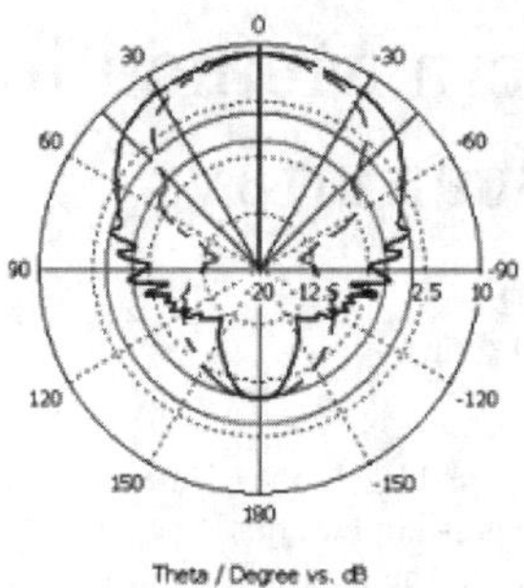

Figure.10 3.8GHz

In Figure.7-figure.10, the 3dB beamwidths of E-plane patterns at different frequencies are about 60 °, which in H-planes are basically less than 80 °. A tail flap is existed in these figures, and the level is -20 ~ -11dB. The ratio of front-and-rear at different frequencies is about 20dB.

IV. MEASURED DATA AND ANALYSIS

A prototype is fabricated as Figure.11, and the VSWR measurement result is shown in Figure.12.

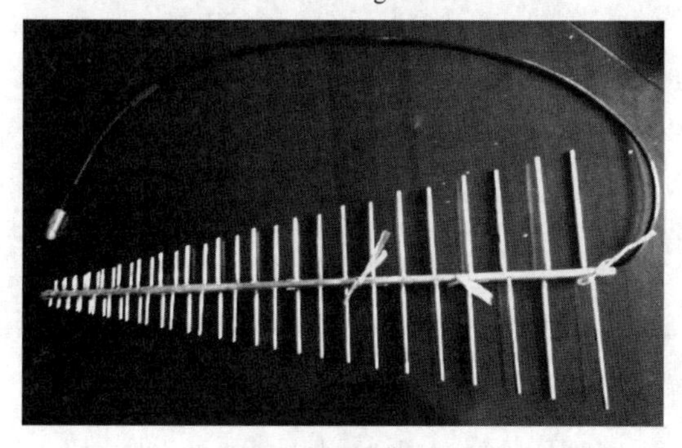

Figure.11 The LPDA model

In Figure.12, the VSWR in the whole work band (0.47GHz-3.8GHz) is less than 2. It can be found the observed performance have a good agreement with the simulated results which confirms the improved structure has excellent performance.

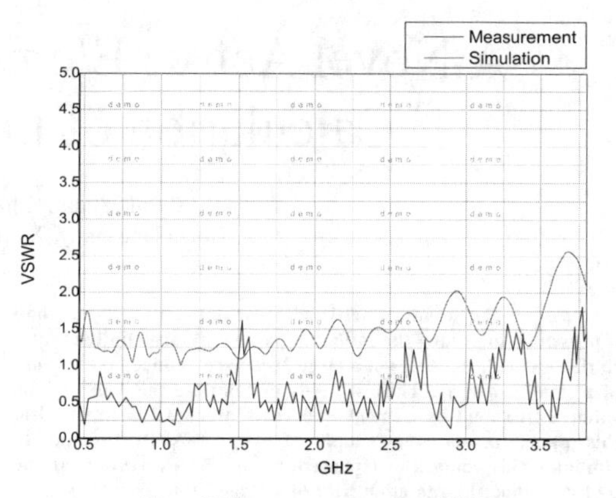

Figure.12 The measurement results of VSWR

V. CONCLSUION

In this paper, different design of feederlines is discussed, we found that parallel manner can't ensure the VSWR and high gain in microwave band by CST simulation. Therefore an improved design of feederlines is proposed. The feederlines are inclined by 5°. The distance between feedlines at front is smaller to ensure the feed and VSWR in high frequency. The distance between the feederlines at the rear is larger to reduce radiation itself. So the high gain and low VSWR, as well as broad work band can be obtained at the same time.

REFERENCES

[1] Xiaoshan Yan, "Structural Design and analysis of the Ultra-shortwave Log-periodic antenna," Electro-Mechanical Engineering, 2003, Vol. 19 No.3: 48-51.

[2] Haixiang Zhang, "Modeling and simulation of Log-periodic dipole antenna radio engineering," 2009, Vol.39 No.10: 42-44.

[3] Warren L.Stutzman, *Antenna Theory and Design*. Posts&Telecom Press, 2006.

[4] Feng Zhang, "Design and simulation of LPD," Journal of Information Engineering University, 2006, Vol.7 No.2: 160-162.

[5] Pual G.Ingerson, "Log-periodic antennas with modulated impedance feeders," IEEE Transactions on antennas and propagation, vol.16 NO.6, Nov. 1968.

A Novel Active Element Pattern Method for Calculation of Large Linear arrays

Shuai Zhang, Shuxi Gong, Huan Hu
Xidian University, Xi'an, Shaanxi 710071 China

Abstract-**The standard active element pattern (AEP) method is presented to predict the radiation pattern of large finite arrays. In this method, the AEPs of a large array are deduced from those of a small subarray. Thus, it needs to compute the AEPs of the subarray element-by-element and costs much CPU time. Also, this paper proposes a simplified AEP (SAEP) method to eliminate this complicated procedure. In the expression of the SAEP method the radiation field of a large finite array is simply related to the total radiation field of two small subarrays. Compared with the standard AEP method, the SAEP method allows us to arrive at a more efficient calculation of the radiation from large arrays. Numerical results show that the SAEP method has the same accuracy as the simulator HFSS while maintaining the simplicity of the pattern multiplication method.**

I. Introduction

Consider a uniform linear array of N identical elements (Fig. 1), ordinary array theory ignores the mutual coupling effects between elements, and expresses the pattern of the array in the well-known pattern multiplication [1] form of an element factors timing an array factor. However, as demand for phased arrays that requires electronic scanning capabilities has increased, needs for accurate analysis is required, such as the method of moments [2], finite element methods [3], etc.. However, even the most efficient numerical method remains infeasible for large arrays. To solve this problem, a powerful and accurate method is presented, depending on the knowledge of the active element pattern (AEP) [4-6]. In this standard AEP method, the AEPs of a large array are deduced from a small array by ignoring the coupling beyond the small array size. Then the large array calculation problem is transformed into that of a small array. However, the standard AEP method needs to compute and store the AEPs of the small array element-by-element.

In this paper, the standard AEP method is modified to eliminate this fussy procedure. In the novel derived expression, the field radiated by a large array is only related to the total radiation field of two small arrays. Compared with the standard AEP method, the proposed one achieves a more efficient calculation of the radiation from large arrays. More importantly, the simplified AEP (SAEP) method significantly extends the range of problems that can be rigorously treated with limited resources. Numerical result shows that the pattern calculated using the SAEP method is in good agreement with the one simulated by the FEM-Based Simulator Ansoft HFSS.

II. Standard Active Element Pattern Method

For the array in Fig. 1, the elements are equally placed along the x-axis with the first one at the origin of the coordinate system. The far field radiated by the array can be obtained by summation of the AEPs of all the elements [4-6]

$$E_N(\theta,\phi)=\sum_{n=1}^{N}E_{N,n}(\theta,\phi) \tag{1}$$

where $E_{N,n}(\theta,\phi)$ is the AEP [4] of the nth element in an N-element array.

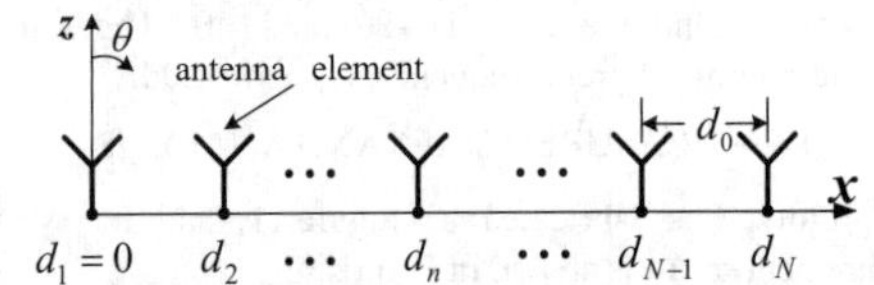

Figure 1. Geometry of a uniform linear array of N identical elements.

As the number of elements in such an array increases, the AEPs of the interior elements (those elements located away from both the ends of the array) become more and more alike [6]. In this case, these AEPs can be assumed to be identical, and the elements of the array can be divided into two *edge* element groups and an *interior* element group (such as the 8-element array shown in Fig. 2). Therefore, the AEPs of the interior elements can be approximated by that of the central element of a small array. Also, the AEP of each edge element of the large array can be approximated by that of this small array, respectively. We shall refer to such a small array as *subarray*. For simplicity, we could illustrate the above procedure for using a 5-element linear array to deduce a linear array of 8 elements as shown in Fig. 2. Then, the radiation pattern of a large array can be calculated by summation of the deduced AEPs of all the elements.

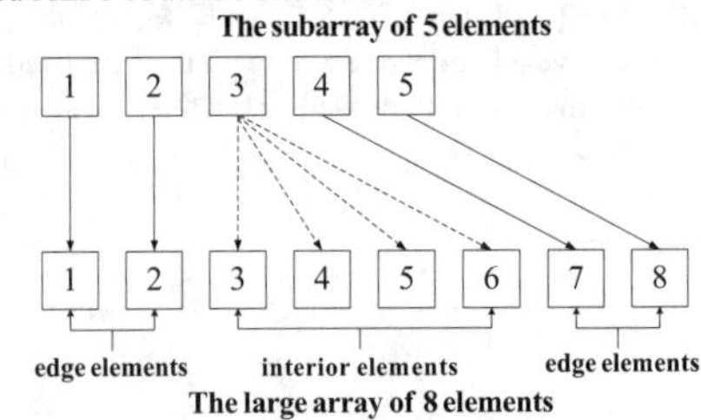

Figure 2. Linear array of 5 elements that is used to deduce an 8-element array

Assume that the number of the subarray elements is M (M is odd), the radiation field of it can be expressed as

978-7-5641-4279-7

$$E_{\mathrm{M}}(\theta,\phi)=\sum_{n=1}^{\mathrm{M}}E_{\mathrm{M},n}(\theta,\phi)=E_l+E_{ce}+E_r$$
$$=\sum_{n=1}^{(\mathrm{M}+1)/2-1}E_{\mathrm{M},n}(\theta,\phi)+E_{\mathrm{M},(\mathrm{M}+1)/2}(\theta,\phi) \quad (2)$$
$$+\sum_{n=(\mathrm{M}+1)/2+1}^{\mathrm{M}}E_{\mathrm{M},n}(\theta,\phi)$$

where E_{ce} is defined as the AEP of the central element of the subarray; E_l and E_r are defined as the summation of the AEPs of the left and right edge elements, respectively.

According to the equivalence method shown in Fig. 2, the far field radiated by a large array of N (N>M) elements can be deduced from the subarray by

$$E_{\mathrm{N}}(\theta,\phi)=E_l+E_{ce}\sum_{n=1}^{\mathrm{N-M+1}}\exp[\mathrm{j}k(n-1)d_0\sin\theta\cos\phi] \quad (3)$$
$$+E_r\cdot\exp[\mathrm{j}k(\mathrm{N-M})d_0\sin\theta\cos\phi]$$

where d_0 is the inter-element spacing; $\mathrm{k}=2\pi/\lambda$ is the free space propagation constant, λ is the wavelength.

III. Simplified AEP Method

Note that the standard AEP method of (3) converts the radiation calculation of a large array into that of a small subarray of M elements. However, it needs to compute and store the AEPs of the subarray element-by-element, thereby leading to much CPU time and fussy operational procedure. These limitations motivate the simplified active element pattern (SAEP) method to the solution of large finite arrays.

A. Derivation of the SAEP Method

Based upon (2) and (3), the difference of the total radiation field between an N-element array and an M-element array is

$$E_{\mathrm{N}}(\theta,\phi)-E_{\mathrm{M}}(\theta,\phi)=\sum_{n=1}^{\mathrm{N}}E_{\mathrm{N},n}(\theta,\phi)-\sum_{n=1}^{\mathrm{M}}E_{\mathrm{M},n}(\theta,\phi)$$
$$=E_{\mathrm{M},(\mathrm{M}+1)/2}(\theta,\phi)\sum_{n=1}^{\mathrm{N-M}}\exp(\mathrm{j2k}nd_0\sin\theta\cos\phi)+ \quad (4)$$
$$\sum_{n=\mathrm{M}-(\mathrm{M}+1)/2+2}^{\mathrm{M}}E_{\mathrm{M},n}(\theta,\phi)\{\exp[\mathrm{j2k}(N-\mathrm{M})d_0\sin\theta\cos\phi]-1\}$$

According to (4), the difference of the radiation field between an (M+1)-element array and an array of M elements is

$$E_{\mathrm{M+1}}(\theta,\phi)-E_{\mathrm{M}}(\theta,\phi)=E_{\mathrm{M},(\mathrm{M}+1)/2}(\theta,\phi)\cdot\exp(\mathrm{j2k}d_0\sin\theta\cos\phi)$$
$$+\sum_{n=\mathrm{M}-(\mathrm{M}+1)/2+2}^{\mathrm{M}}E_{\mathrm{M},n}(\theta,\phi)\cdot[\exp(\mathrm{j2k}d_0\sin\theta\cos\phi)-1] \quad (5)$$

Substituting (5) into (4), we get

$$E_N(\theta,\phi)-E_{\mathrm{M}}(\theta,\phi)=[E_{\mathrm{M+1}}(\theta,\phi)-E_{\mathrm{M}}(\theta,\phi)]\cdot$$
$$\sum_{n=1}^{\mathrm{N-M}}\exp[\mathrm{j2k}(n-1)d_0\sin\theta\cos\phi] \quad (6)$$

Then, the field radiated by a uniform N-element linear array can be written in a simple form as

$$E_{\mathrm{N}}(\theta,\phi)=E_{\mathrm{M}}(\theta,\phi)+[E_{\mathrm{M+1}}(\theta,\phi)-E_{\mathrm{M}}(\theta,\phi)]\cdot$$
$$\frac{1-\exp[\mathrm{j2k}(\mathrm{N-M})d_0\sin\theta\cos\phi]}{1-\exp(\mathrm{j2k}d_0\sin\theta\cos\phi)} \quad (7)$$

Observe that the radiation from the array is simply related to the radiation fields of an M-element and an (M+1)-element small array. Compared with the standard AEP method, the SAEP method eliminates the complicated operation of individual element pattern computation and storage for the subarray. Meanwhile, the calculation results of the two methods are identical since they have the same equivalence method of AEPs as shown in Fig. 2. Thus, the SAEP method allows us to arrive at a fast and accurate calculation of the radiation pattern of a large array when numerical methods and simulation software are infeasible.

B. Verification of the SAEP method

The configuration of the antenna element is shown in Fig. 4. The rectangular patch is printed on the surface of the dielectric substrate with a thickness of 1 mm and a relative permittivity of 2.65. The length L and the width W of the patch are chosen to be 8.7 and 8.4 mm, respectively. The central operating frequency is 10 GHz. The simulator HFSS is employed to compute the radiation field of the subarray and the results of it are taken as reference. In the examples, the pattern calculated using the standard AEP method is not given since it is identical to that of the SAEP method.

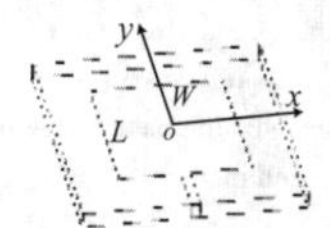

Figure 3. Configuration of the microstrip patch antenna

a. Linear array of 16 elements with $d_0=0.5\lambda$

Given a uniform linear array composed of 16 microstrip patch antennas, we will discuss how many elements are needed in a subarray for satisfied accuracy. The inter-element spacing is $d_0=0.5\lambda$.

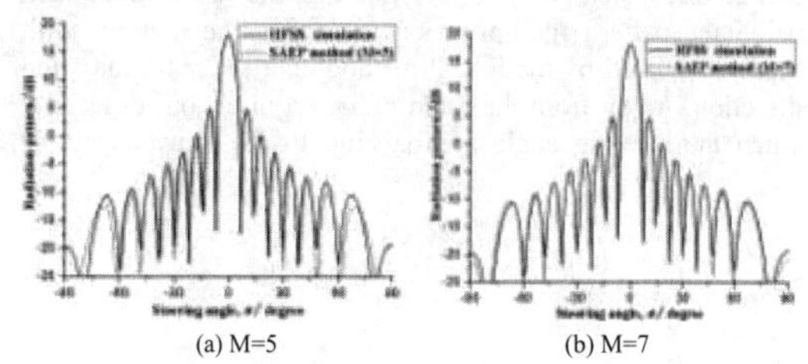

(a) M=5 (b) M=7

Figure 4. Comparison of the radiation pattern simulated by HFSS and calculated by the SAEP method

Fig. 4 (a) shows the comparison of the radiation patterns simulated by HFSS and calculated by the SAEP method while M=5. Observe that the results calculated by the SAEP method are consistent with those simulated by HFSS near the main radiation direction (at boresight of $\theta=0°$). Away from the main radiation direction, the SAEP method fails since the mutual coupling and the array edge effects cause significant variation in AEPs across the array. In order to improve the accuracy of the SAEP method, more elements should be considered as edge elements. Fig. 4 (b) shows the comparison between the radiation pattern calculated by the SAEP method while M=7 and that simulated by HFSS. Observe that the calculated results closely match the simulated ones in the whole region of the θ -angle.

Through the above comparison, we can assume that the SAEP method can obtain satisfactory accuracy while M=7. In this situation, the accuracy of the SAEP method is similar with that of the HFSS simulation software.

b. Large array of 100 elements with $d_0=0.3\lambda$

To demonstrate the superiority of the SAEP method, a linear array consists of 100 microstrip patch antennas (Fig. 3) is considered. The inter-element spacing of the array is chosen as 0.3 λ for stronger mutual coupling effects.

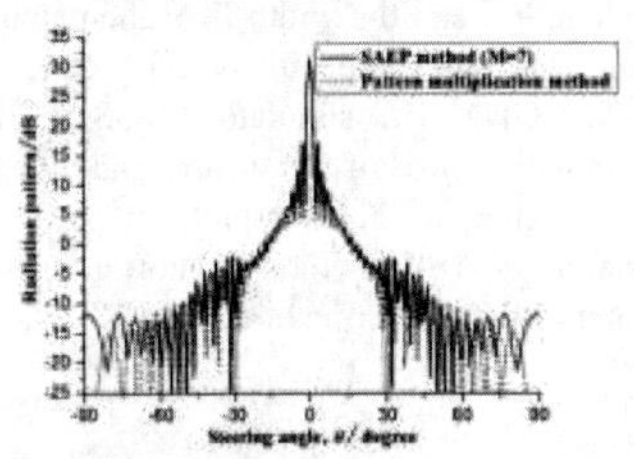

Figure 5. Comparison of the radiation pattern calculated by the SAEP method and by the pattern multiplication method.

For this array, the simulation software and the numerical methods are infeasible with our PC. However, the SAEP method can still fast and accurately calculate the radiation from the array. The radiation pattern of the array calculated by the SAEP method with M=7 is compared to that computed using the pattern multiplication method (in which the radiation from an array is expressed as the product of an element factor and an array factor) in Fig. 5. Note that the results computed using the pattern multiplication method are consistent with those calculated by the SAEP method near the main radiation direction. Away from the main radiation direction, especially when the steering angle approaching to the array edge, the difference between two methods becomes larger and larger since the pattern multiplication method cannot efficiently consider the effects of the mutual coupling between elements and the edge diffraction. The comparisons demonstrate that the SAEP method does lead to a more accurate representation of the radiation from array than the pattern multiplication method.

IV. CONCLUSIONS

This paper proposes a SAEP method for calculating the radiation pattern of large finite arrays. With this method, the radiation field of a large finite array is simply related to the total radiation field of two small subarrays. The examples demonstrate that the SAEP method has the similar accuracy as the HFSS simulation software while maintaining the simplicity of the pattern multiplication method, and its computational cost increases slightly with an increase of the number of array elements. Thus, the SAEP method can fast and accurately solve complex and large array problems using a desktop PC when numerical methods and simulation software are infeasible.

ACKNOWLEDGMENT

This work was supported by the Fundamental Research Funds for the Central Universities of China under Grant K5051302031, and the New Teacher Innovation Foundation of Xidian University under K5051302037.

REFERENCES

[1] C.A. Balanis 2002, *Antenna theory*, 2nd, New York, John Wiley&Sons, Inc. Press.

[2] Ma J., Gong S.-X., and Wang X. (2012), 'Efficient Wide-Band Analysis of Antennas Around a Conducting Platform Using MoM-PO Hybrid Method and Asymptotic Waveform Evaluation Technique ', *IEEE Transactions on Antennas and Propagation*, 60, 6048-6052.

[3] Zhang P.F., Gong S.-X., and Zhao S.F. (2009), 'Fast Hybrid FEM/CRE-UTD Method to Compute the Radiation Pattern of Antennas on Large Carriers', *Progress In Electromag-netics Research*, 89, 75-84.

[4] Plzar D.M. (1994), 'The Active Element Pattern', *IEEE Transactions on Antennas and Propagation*, 42, 1176-1178.

[5] Toh B.Y., Fusco V.F., and Buchanan N.B. (2001), 'Retrodirective Array Tracking Prediction Using Active Element Characterization', *Electronics Letters*, 37, 727-728.

[6] Kelly D.F., and Stutzman W.L. (1993), 'Array Antenna Pattern Modeling Methods that Include Mutual Coupling Effects', *IEEE Transactions on Antennas and Propagation*, 41, 1625-1632.

Design and Research on the Plasma Yagi Antenna

Zhao Hui-chao[1 2] Liu Shao-bin[1] Li Yu-quan[2] Yang Huan[1] Wang Bei-yin[1]
(Nanjing University of Aeronautics and Astronautics, Nanjing 210016, China)[1]
(No. 96625 Troops of PLA, Zhangjiakou 075100,China)[2]

Abstract- **A plasma Yagi antenna was designed based on the traditional theoretical analysis of Yagi antenna, S11 parameters and patterns were simulated by CST software. According to the simulation, we manufacture plasma Yagi antenna principle prototype and S11 parameters are measured by a vector network analyzer ,which is consistent with the simulation results .The results show that the plasma Yagi antenna has good directivity and high gain as the plasma density is high enough. The performance can be compared with the metal antenna.**

I. INTRODUCTION

The plasma antenna is an antenna that use plasma substitute metal as electromagnetic energy transfer medium. Compared to the metal antenna, the plasma antenna has a lot of outstanding advantages, such as stealth, Reconfigurability, low mutual coupling since the plasma has a unique physical properties in electromagnetic wave propagation.

The idea of plasma stealth antenna was firstly proposed by the University of Tennessee in 1996[2], many domestic and foreign research institutions also focus one the development of plasma antennas, but most of them are concentrated in the theoretical investigation [3,4], such as radiation [5] , the input impedance [6,7] , surface wave propagation [8] , the electromagnetic wave dispersion relation [9] . The structure of the plasma antenna which is used to investigate by many researchers is relatively simple [10,11] , or the fluorescent tube is looked directly as a plasma antenna[12].

Based on the traditional Yagi antenna theory, the three-dimensional electromagnetic antenna model is created by the CST (software). Through simulation, the optimal structural parameters are obtained. Then, we manufacture plasma Yagi antenna principle prototype and measure S11 parameters with a vector network analyzer. The measured results are excellent agreement with the simulation results. The results show that the plasma Yagi antenna has good directivity and high gain as the plasma density is high enough.

II. MODEL SIMULATION OF PLASMA YAGI ANTENNA

Plasma is excited by high voltage both ends of discharge tube, since if we use a half- wave dipole as active oscillator, the antenna layout will be very messy complex. Thus, we use folded dipole as active oscillator of plasma Yagi antenna. Discharge pipe at both ends of the excitation electrode are shown in Fig.1.

Figure 1 The excitation electrode of discharge tube

Folded dipole model is plotted in Fig.2 (a), the following part of parcel 5 cm long copper foil for signal coupling. The parameters of plasma Yagi antenna are set according to the metal Yagi antenna, such as size and spacing of active oscillator, director and reflector, reference parameters, which can be adjusted and optimized by the simulation.

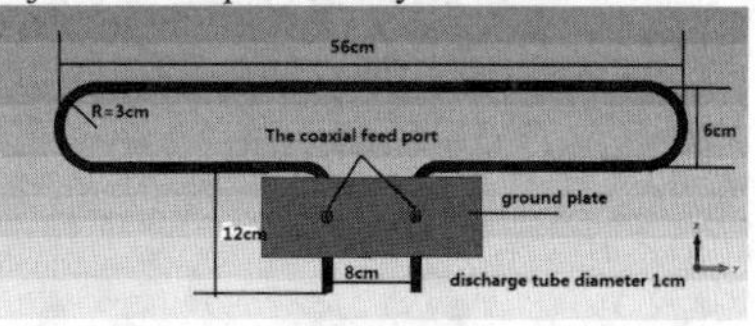

(a) Folded dipole model

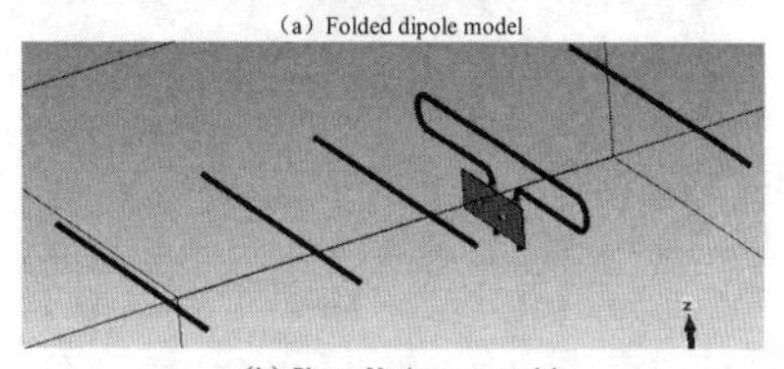

(b) Plasma Yagi antenna model

Figure 2 Folded dipole and Plasma Yagi antenna model

In the CST software, setting the plasma frequency and collision frequency can realize the desired plasma material (Drude model). During the calculation, the implementation of parallel excitation is on the two port model and two ports size is set to 1. The phase is set to be 0 ° and 180 ° and the normalized impedance is set to 50Ω.

During the calculation, the plasma collision frequency is set to be 32MHz. The antenna working frequency is changed with the plasma density as the plasma frequency is covered from 0.5~9GHz. S11 parameters is -42dB as the frequency of electromagnetic wave is 178 MHz, and the plasma frequency is 8GHz. It means that the feed-in signal only 1/10000 is reflected and the signal by feeder terminal unit are fed into the antenna array.

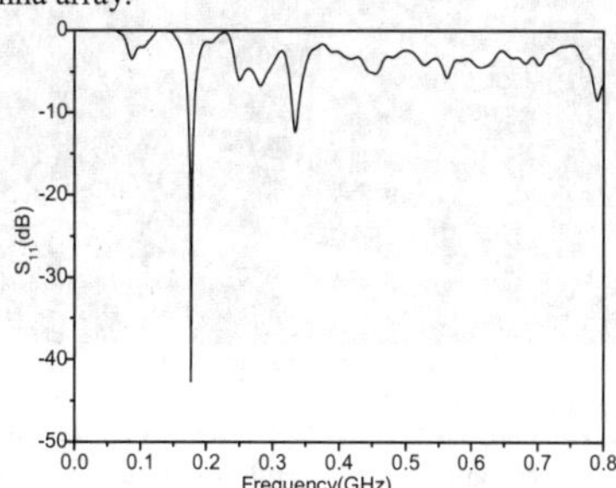

Figure 3 plasma Yagi antenna simulation of S11 parameters

The good performance of plasma Yagi directional radiation can be found in the far field monitoring region as the frequency of electromagnetic wave is 183 MHz. The peak gain of the antenna beam width is 9.4dBi, and 3dB

978-7-5641-4279-7

beamwidth can be found (H surface is 86.7 ° and E surface is 62 °).

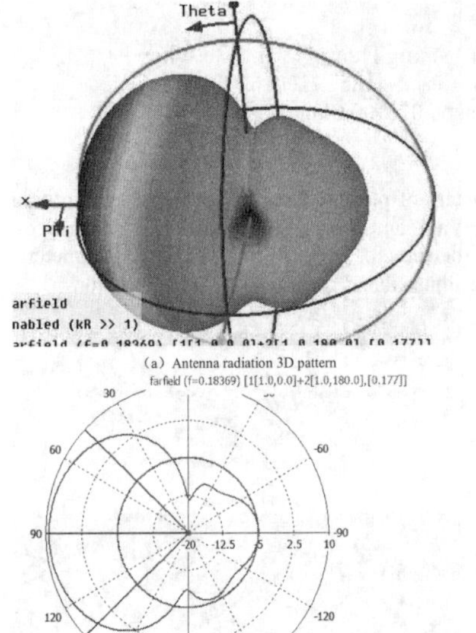

(a) Antenna radiation 3D pattern

(b) Antenna radiation H surface map

Figure 4 Radiation plasma Yagi antenna diagram

III. Fabrication and Measurement of Plasma Yagi Antenna

As shown in Fig.5, we manufactured plasma Yagi antenna folded dipole, reflector, director antenna oscillator unit, and the erection of plasma Yagi antenna measurement platform according to the Yagi antenna theory and CST simulation parameters setting, respectively. The measuring platform can be adjusted freely according to the height of Yagi antenna , each oscillator element spacing and facilitate measurements of plasma Yagi antenna, respectively.

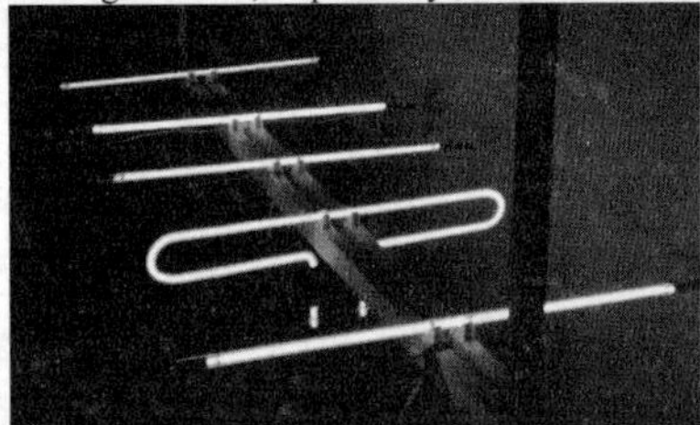

Figure 5 the plasma Yagi antenna

Capacitive coupling is widely used to couple the plasma antenna signal in antenna. One end of the glass pipe wrapped copper ring can be looked as the coaxial feed, and the copper ring and ground is connected by the coaxial line through the SMA head.

The folded dipole requires constant amplitude reversed balanced feed, but the SMA axis is not balanced feed. Thus, the balun is needed to convert the unbalanced transmission line to balance the load. The designed balun is shown in Fig.6. The 50Ω coaxial line and U part of 1/2 electrical length are not in the antenna amplitude reversed feed, but also can improve the performance in accordance with the 4:1 impedance lifting function. The 50Ω coaxial line can match the 200Ω antenna.

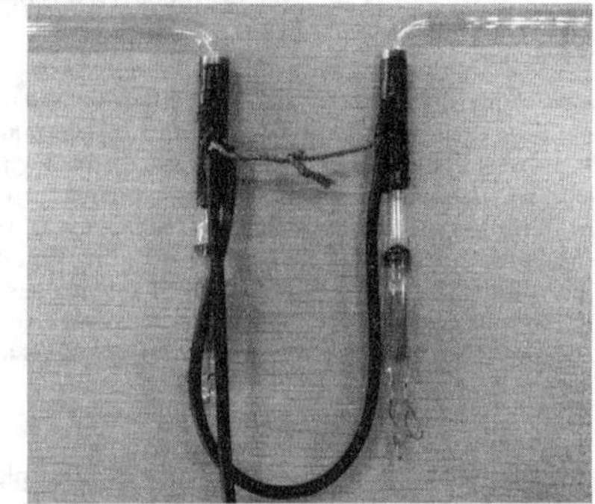

Figure 6 Plasma Yagi antenna folded dipole balun matching graph

Figure 7 Plasma Yagi antenna measurement layout

Both ends of each oscillator with CTP2000K low temperature plasma power are supplied during Plasma Yagi antenna measurement. The matched 50Ω coaxial balun is connected with the folded dipole, and the other end is connected with the AV3620 series of high performance integrated vector network analyzer. Experimental layout is shown in Fig.7.

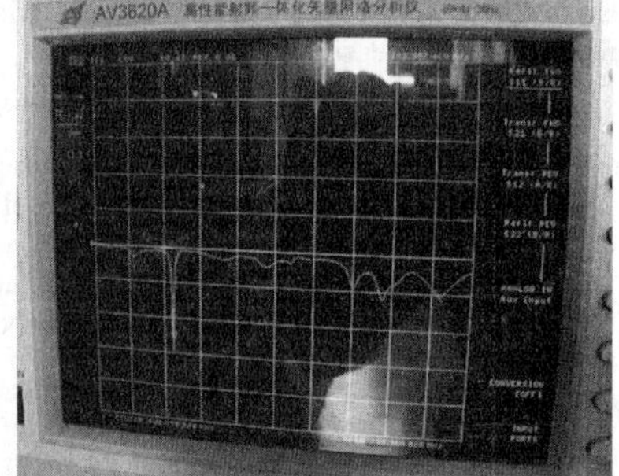

Figure 8 The measurement of S11 parameter of plasma Yagi antenna diagram

As shown in Fig.8, the obvious resonance can be found in the 180.557MHz and min S11 is -30.521dB as plasma Yagi antenna excited and adjusting the input power of each oscillator unit.

IV. PERFORMANCE ANALYSIS OF PLASMA YAGI ANTENNA

Compared CST simulation results with network analyzer measurements, the simulation result is that the S11 is -42dB at 178MHz, and the vector network measurement result is that S11 is -30.521dB at 180.557MHz as shown in Fig.9. The results show that plasma Yagi antenna has an obvious resonance at 180MHz. Therefore, the working frequency of plasma Yagi antenna can be lobtained at 180MHz .

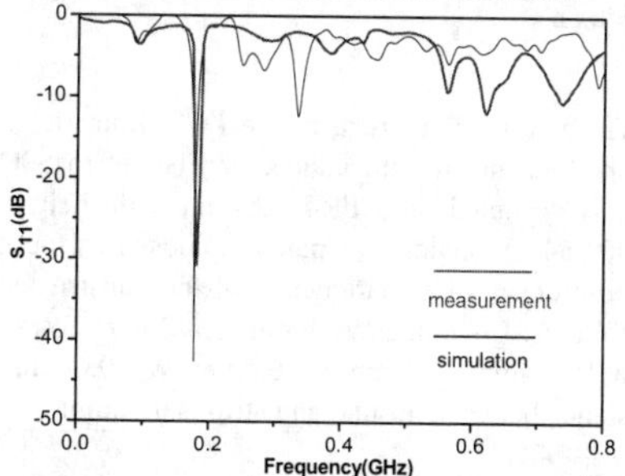

Figure 8 Comparison of simulation and measurement results

Due to the limitation of experimental conditions, the radiation pattern of plasma Yagi antenna is not measured. However, the measured S11 parameter show that the excellent agreement can be found between experimental measurement and simulation results, but the absolute value of measured S11 is smaller compared with simulation. Thus, we can infer that the gain of the plasma antenna is less than 9.4dBi as the numerical simulated by CST.

The generally Yagi antennas have the gain range 5 ~ 20dBi in free space , which compose of 2 to 31 units [13]. Directional High Gain Yagi antenna in market for sale always combine with five units, which is working at the 88 ~ 108MHz with 8dBi [14] . The designed plasma Yagi antenna are five unit and has 9.4dBi gain (measured results is less than this value). It is considerably with metal Yagi antenna gain. The result shows that the plasma antenna can be compared with the performance of the metallic antenna, as plasma density is high enough .

V. CONCLUSION

A plasma Yagi antenna is designed based on the theory of traditional Yagi antenna. The model of plasma Yagi antenna is created in CST, and optimization structure parameters also are obtained by the simulation. We manufacture a plasma Yagi antenna principle prototype and S11 parameters are measured with a vector network analyzer. The measured results are excellent agreement with the simulation results. The measured results also show that the plasma Yagi antenna has good directivity and high gain as the plasma density is high enough, and the performance can be compared with the metal antenna. The plasma antenna has a broader application prospects in the military field because has a lot of outstanding advantages compared to the metal antans, such as stealth, Reconfigurability, low mutual coupling and flexibility and fast control switch properties.

ACKNOWLEDGMENT

The work is supported by Doctoral Fund of Ministry of Education Grant No. 20123218110017, Jiangsu Province Science Foundation Grant No.BK2011727 and Open Topics Foundation of the State Key Laboratory of Millimeter Waves No. K201103.

REFERENCES

[1] Zhao Jian-sen , Zhang Zhi-tao, "Experimental study on impedance and radiation properties of plasma-column antenna," *J. Chinese Journal of Radio Science* , 2012,27 (3) : 372 ~ 373.

[2] Kang, Weng Lock; Rader, Mark; Alexeff, Igor. "Conceptual study of stealth plasma antenna", *IEEE International Conference on Plasma Science Proceedings of the 1996.* Boston, MA, USA: IEEE International Conference on Plasma Science, 1996: 261.

[3] Wu Yang-Cao, "Plasma antennas and stealth technology," *J. Communications and Control* , 2005,29 (2) : 1-4 .

[4] Lee Sheng , " Theoretical analysis of Characteristics of Plasma Antennas droven by Surface Wave," *J. Journal of University of South China(Science and Technology)* , 2007,21 (2) : 37 ~ 40.

[5] Liang Zhi-wei , Zhao Guo-wei , Xu Jie , "Analysis of plasma-column antenna using moment method," *J. Chinese Journal of Radio Science,* 2008,23 (4) : 749 ~ 753 .

[6] Xu Jie , Zhao Guo-wei , Liang Zhi-wei , "Measurement of Column Plasma Radiator Input Impedance," *J. Radio Engineering of China.* 2007,37 (8) : 32 ~ 34.

[7] Xu Jie , Zhao Guo-wei , Liang Zhi-wei , " Measurements and Analysis of Plasma Column Discharge Impedance, " *J. Chinese Journal of Space Science,* 2007,27 (4) : 315 ~ 320 .

[8] Yuan Zhong-cai , Shi Jia-ming , Yu Gui-fang , "Principles and Realization of Surface Wave Driven Plasma Antenna," *J. Electronic Information Warfare Technology,* 2006, 21 (4) : 38 ~ 41.

[9] Yang Lan-lan , Tu Yan, Wang Bao-ping, "Axisymetric Surface-Wave Dispersion in Plasma Antenna," *J. Journal of Vacuum Science and Technology.* 2004,24 (6) : 424 ~ 426 .

[10] Liu Jian, Wang Shiqing, Xu Lingfei. " Experimental study on radiation characteristics of single-source monopole plasma antenna, " *J. High Power Laser and Particle Beams,* 2012,24 (5) : 1137 ~ 1140 .

[11] G. G. Borg, J. H. Harris, N. M. Martin. "Plasma as antennas: Theory, experiment and applications , " *J. Physics of Plasma,* 2000, 7(5): 2198~2202.

[12] V.Kumar, " Study of a fluorescent tube as plasma anternna, " *Progress in electromagnetics research letters,*Vol.24,17-26,2011.

[13] The American Radio League , *The ARRL ANTENNA BOOK.* Beijing : People's Posts and Telecommunications Press , 2010.P308.

A Compact Tri-Band Monopole Antenna for WLAN and WiMAX Applications

Guorong Zou, Linyan Guo, Helin Yang[1]
College of Physical Science and Technology, Central China Normal University, Wuhan 430079, Hubei
People's Republic of China
[1] emyang@mail.ccnu.edu.cn

Abstract- **A compact tri-band planar monopole antenna is proposed in this paper that employs reactive loading and a defected ground plane structure. This proposed antenna operates in two modes. The first resonance exhibits a dipolar mode over the lower WiFi band of 2.40– 2.485 GHz and the upper WiFi band of 5.15 - 5.70 GHz. The second resonance has a monopolar mode over the WiMAX band at 3.30–3.70 GHz. Full-wave analysis shows that the currents of these two modes are orthogonal to each other, resulting in orthogonal radiation patterns in their far field. The defected ground, formed by appropriately cutting a spiral-shaped slot out of one of the CPW (Coplanar Waveguide) ground planes, leads to the second resonance.**

I. INTRODUCTION

The planar monopole is attractive for WLAN antenna design because it has a low profile, broadband, multiband operation. The traditional approach is to use multi-branched strips in order to achieve multiband operation [1], which generally leads to a large volume or requires a large ground plane. Alternatively, the concept of the frequency-reconfigurable multiband antenna [2] has been proposed to develop multiband monopole antennas for WiFi and WiMAX applications [3]. Recently the idea of creating a multiband defected-ground-plane monopole antenna was presented in [4]. And the concept of using slots within printed monopole designs has been investigated by other authors, who have focused on creating a band-reject filtering property in their antennas. For example, in [5] the slots were cut out of the radiating element, while in [6] a single slot was etched out of the ground plane. But these antennas require the use of some vias or many lumped-element components.

In this paper, a compact multiband antenna is proposed. It is easy to fabricate and at a very reasonable cost, thus making it ideal for using in WLAN devices. The resulting defected ground plane creates two orthogonal polarizations and three additional resonances in the input impedance of the antenna, whose locations can be adjusted according to the size and position of the slot. In this work a more detailed explanation of the antenna operation is outlined.

II. ANTENNA DESIGN

The proposed antenna consists of a printed rectangle monopole antenna that has a spiral-shaped slot etched out of its ground plane and bake side with two copper bar loadings, which is shown in Fig. 1. A coplanar waveguide feed is also used. The width of microstrip line (W_2=4mm, L_4=20mm) is determined so that the impedance can be set to 50Ohm. The antenna is designed on a FR4 substrate with height h=1mm, ε=4.0+i0.064. A rectangular patch is chosen as the monopole radiation element. The parameters of the constructed antenna are as follows: L=40mm, W=40mm, L_1=20mm, W_1=15mm, L_2 =15mm, W_2=4mm, g=1mm, L_3=6.5mm, W_3=0.6mm, W_s=7mm. The distance between ground and strip is 0.42mm.

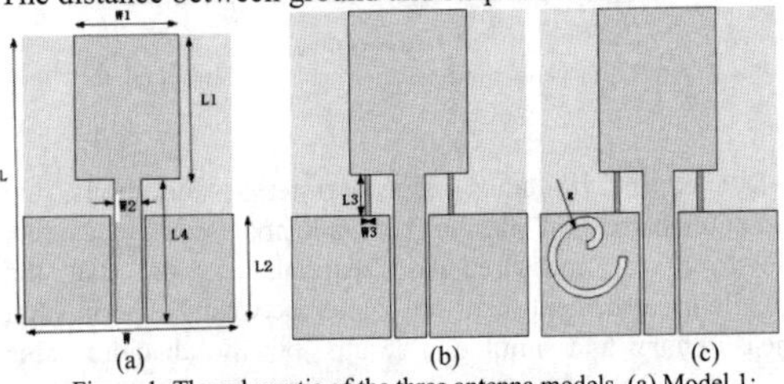

Figure 1. The schematic of the three antenna models. (a) Model 1: conversional monopole antenna. (b) Model 2: dual-band monopole antenna with copper bar loadings. (c) Model 3: tri-band monopole antenna with copper bar loading and a defected ground plane.

III. SIMULATION RESULTS AND DISCUSSIONS

Fig. 2 depicts the simulated S11 parameter based on the standard Finite Difference Time Domain (FDTD) method for three antenna models which have same dimensions. Model 1 is a conversional monopole antenna. Model 2 is a dual-band monopole antenna with two copper bar loadings. And Model 3 is the proposed tri-band monopole antenna with two copper bar loadings and a defected ground plane. As clearly observed from Fig. 2, proposed antenna exhibits a bandwidth of 1100MHz for the lower WiFi band from 1.90 GHz to3.0GHz and a bandwidth from 4.60GHz to 5.70GHz for the higher WiFi band. It also has a bandwidth of 400MHz for the WiMAX band from 3.30GHz to 3.7GHz. Since the operating frequency of the initial design (Model 1) is out of the range of interest for existing WLAN applications, different approaches using copper bar loadings and a defected ground are pursued to create the corresponding second and third resonances at a lower frequency range in order to meet the WLAN specifications.

978-7-5641-4279-7

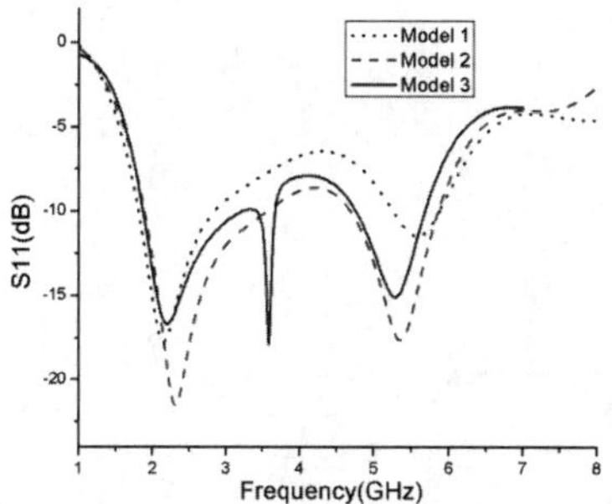

Figure 2. Simulated S11 for Model 1, Model2 and Model 3.

The dual-mode operation mechanism of the Model 3 will be discussed in this section. It can be explained by considering the current distribution on the copper at each of the resonant frequencies, as shown in Fig. 3. These figures were sketched from the surface current distributions. At 5.3GHz, the copper bar loadings was adjusted so that the surface current along the monopole and the bottom thin inductive strip were out of phase. Thus, at this frequency the copper bar loading was used to create a folded monopole, similar to the four-arm folded monopole in [7]. The y-direction currents contribute to radiate, as shown in Fig. 3(a). And at this frequency the currents along the top edges of the two ground planes are out of phase. Therefore, these currents do not contribute to any radiation. At 3.68 GHz, the antenna no longer acts as a folded monopole along the y-axis, but as a dipole oriented along the x-axis. This is a result of the in-phase currents along the top edges of both the ground plane sections (as shown in Fig. 3(b)) which render the ground plane as the main radiating element at this frequency. As can be seen in Fig. 3, there is a strong concentration of the currents along the spiral-shaped slot on the left ground plane. The slot forces the current to wrap around it and thus creates an alternate path for the current on the left ground plane, whose length is approximately at its resonance. It is also noted from Fig. 5 (we will discuss later) that the spiral-shaped slot does not significantly affect the balanced CPW mode, since it is placed far enough away from the CPW.

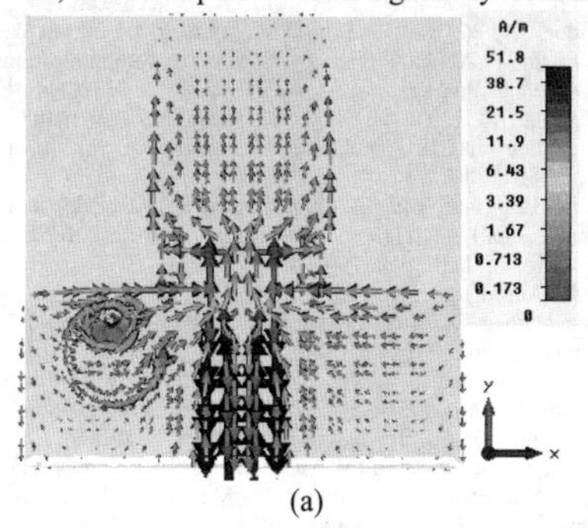

(a)

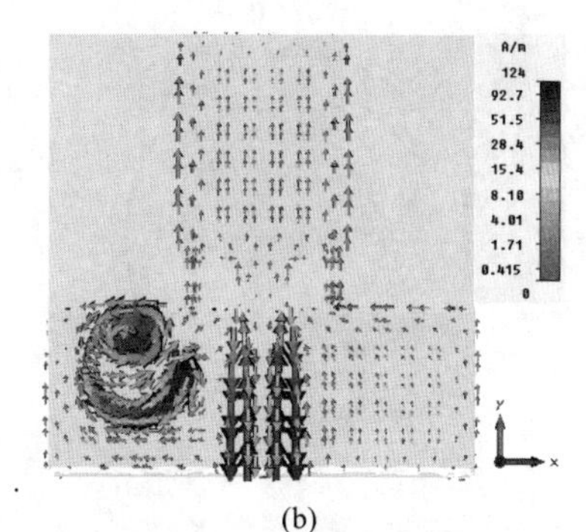

(b)

Figure 3. Simulated surface current distribution on the copper of the Model 3 at the resonant frequencies of 5.3GHz and 3.68 GHz. (a) Folded monopole mode at 5.30 GHz, (b) Dipole mode at 3.68GHz.

The simulated radiation patterns for the proposed tri-band monopole antenna are plotted in Fig. 4 and Fig. 5 for the three principle planes at the frequencies of 5.30GHz and3.68GHz. Fig.4 shows the radiation patterns at 5.30 GHz for the E-plane (xy-plane and yz-plane) and the H-plane (xz-plane). The fact that the antenna exhibits radiation patterns with a horizontal y-directed linear E-field polarization, verifies that the antenna operates in a folded monopole mode around 5.30GHz, due to the y-directed anti-phase currents along the monopole and the thin vertical inductive strip, as shown in Fig. 3(a). The x-directed currents along the thin horizontal inductive strip have a contribution to the cross polarization in the yz-plane. It can also be seen that at this frequency, the slot on the left ground has a minimum contribution to the radiation since the currents are dominated by the y-directed along the monopole. At this frequency, the simulated radiation efficiency is 98.38%. Fig. 4 shows the radiation patterns. At 3.68 GHz, since the spiral-shaped slot which is cut out of the left ground plane results in meandered currents along both the y-direction and the x-direction, which have independent contributions to the radiation. It is observed from Fig. 5 that the antenna exhibits two linear electric fields that are orthogonally polarized in both the y and x directions. In fact, at3.68GHz the width of the ground plane, or equivalently the length of the radiating edge, is approximately equal to $\lambda_g/2$. The simulated radiation efficiency at this frequency is 86.3%.

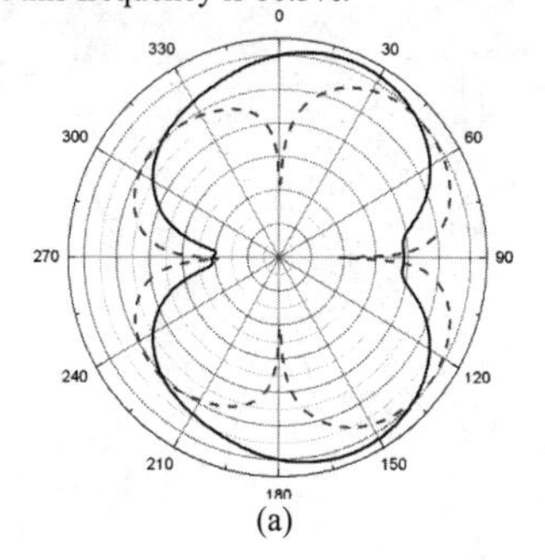

(a)

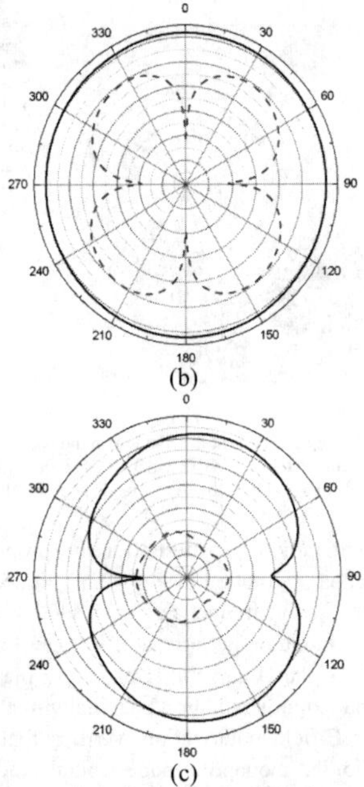

Figure 4. Simulated radiation patterns for the tri-band monopole antenna at 5.30GHz. Solid black line: simulated co-polarization, dash-dot red line: simulated cross-polarization. (a) xy-plane. (b) xz-plane. (c)yz-plane.

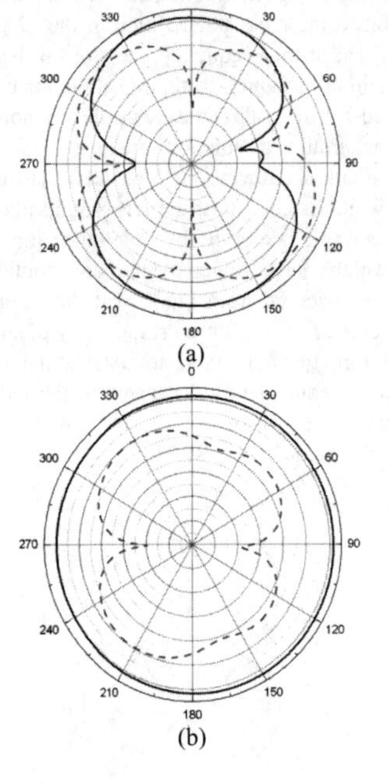

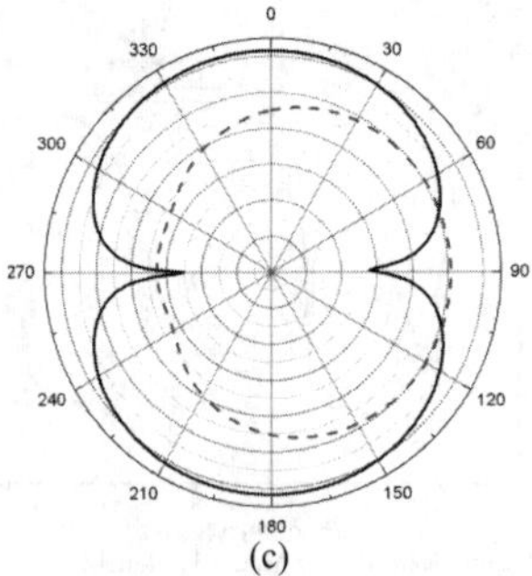

Figure 5.Simulated radiation patterns for the tri-band monopole antenna at3.68GHz. Solid black line: simulated co-polarization, dash-dot red line: simulated cross-polarization. (a) xy-plane. (b) xz-plane. (c)yz-plane.

IV. CONCLUSION

A compact and broadband antenna has been presented, which employs copper bar loadings on a conventional printed monopole design in order to create a dual-mode antenna. It was demonstrated that the addition of the copper bar loadings allows the antenna to be modeled as a short folded monopole at 5.30 GHz while the loading enables the entire top edge of the ground plane to radiate at 3.68 GHz. Thus, the copper bar loaded antenna achieves orthogonal pattern diversity in both the 3.3-3.7GHz WiMAX and 5.15-5.70GHz WiFi bands. Additionally, the antenna maintains a very high efficiency in both the bands of interest. It is therefore well suited for MIMO diversity systems for emerging WLAN applications.

REFERENCES

[1] Y. Ge, K. Esselle, and T. Bird, "Compact triple-arm multiband monopole antenna," inProc. IEEE Int. Workshop on: Antenna Tech-nology Small Antennas and Novel Metamaterials , Mar. 2006, pp.172–175.

[2] S. Yang, C. Zhang, H. K. Pan, A. E. Fathy, and V. K. Nair, "Frequency-reconfigurable antennas for multiradio wireless platforms," IEEE Mi-crow. Mag. , vol. 10, no. 1, pp. 66–83, Feb. 2009.

[3] S. Yang, A. E. Fathy, S. El-Ghazaly, H. K. Pan, and V. K. Nair, "A novel hybrid reconfigurable multi-band antenna for universal wireless receivers," presented at the Electromagnetic Theory Symp., Jul. 2007.

[4] M. A. Antoniades and G. V. Eleftheriades, "A compact monopole antenna with a defected ground plane for multiband applications," in Proc. IEEE Int. Symp. Antenn. Propag., San Diego, CA, Jul. 2008, pp.1–4.

[5] K. Bahadori and Y. Rahmat-Samii, "A miniaturized elliptic-card UWB antenna with WLAN band rejection for wireless communications," IEEE Trans. Antennas Propag. , vol. 55, no. 11, pp. 3326–3332, Nov.2007.

[6] W. S. Chen and Y. H. Yu, "A CPW -fed rhombic antenna with band-reject characteristics for UWB applications," Microw. J., vol. 50, no.11, pp. 102–110, Nov. 2007.

[7] M. A. Antoniades and G. V. Eleftheriades, "A folded-monopole model for electrically small NRI-TL metamaterial antennas," IEEE Antennas Wireless Propag. Lett. , vol. 7, pp. 425–428, Nov. 2008.

[8] Marco A. Antoniades, George V, "A Compact Multiband Monopole Antenna With a Defected Groud Plane", IEEE antennas and wireless propagation letters(2008) 7:652-655.

[9] Macro A. Antioniades, George V. Eleftheriades,"A Folded-Monopole Model for Electrically Small NRI-TL Metamaterial Antennas",IEEE Antennas and wireless propagation letters(2008)7:425-428.

[10] Macro A.Antoniades, George V. Eleftheriades, "A Broadband Dual-Mode Monopole Antenna Using NRI-TL Metamaterial Loading", IEEE Antennas and wireless propagation letters(2009)8:258-261.

Optimization Design of Matching Networks for High-frequency Antenna

Suo Ying[1,2], Li Fazong[1], Li Wei[1,2], Wu Qun[1]
(1.School of Electronics and Information Engineering, Harbin Institute of Technology, Harbin, 150001; 2.Electronic Science and Technology Postdoctoral Station, Harbin Institute of Technology, Harbin, 150001)

Abstract- **Real frequency technique is an efficient method for the optimization design of broadband matching networks. This paper mainly uses the real frequency technique for the design of HF broadband antenna impedance matching network. The network parameters were calculated for two real frequency methods by MATLAB and we compared the two methods of corresponding matching performance. On this basis, some optimization and improvement of the real frequency technique will be done to make real frequency technology more practical and scientific.**

I. INTRODUCTION

With the rapid development of modern wireless communication technology, there are more and more communication system for different purposes. Matching network is used in almost any transceiver system of a communication system. At the same time, the design of matching network has gradually tended to higher frequency and band broadening. Network matching problem now is not only an interesting theoretical problem, but also a widely encountered in practical engineering problems.

Antenna matching theory has experienced a long time. Matching problem was first introduced by Bode in 1945.In 1950, R.M.Fano [7] conducted a general study of the problem. In 1961, Schoeffler proposed the concept of compatibility impedance matching problem, studied the problem from different aspects. H.J.Carlin [4] proposed the real frequency technique in 1979. By using the load data directly, the real frequency method has great advantage in designing impedance matching network. Now the real frequency technique has been made several optimization, and other ways to design matching networks have been developed [8].

II. THE ORIGINAL REAL FREQUENCY METHOD

As figure 1 shows, Vg(s) as the signal source, Rg(s) is the resistance of signal source, Zl=Rl(w)+jXl(w) is input impedance of antenna .Assuming that Zq=Rl(w)+jXl(w) is the driving-point function from the antenna input port to the matching network.

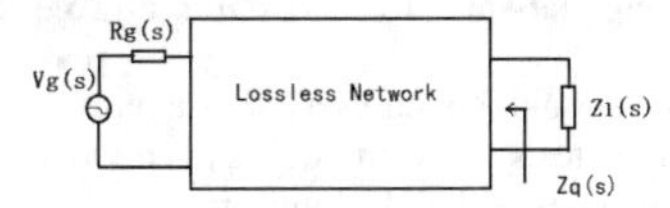

Fig.1. Matching network

Pl is antenna's input power and Pa as the source's average output power.

The TPG is transducer power gain and it can be expressed as follows:

$$T(w)=1-|\rho|^2 \tag{1}$$

ρ is the normalized reflection coefficient, according to the microwave basic knowledge:

$$\rho=\frac{Zq+Zl^*}{Zq+Zl^*} \tag{2}$$

$$Zl^*=\text{Rl(w)-jXl(w)} \tag{3}$$

Thus TPG can be calculated as:

$$T(w)=\frac{4Rq(w)Rl(w)}{|Zq(w)+Zl(w)|^2} \tag{4}$$

TPG can also be shown as:

$$T(w)=\frac{4Rq(w)Rl(w)}{|Rq(w)+Rl(w)|^2+|Xq(w)+Xl(w)|^2} \tag{5}$$

In the above equation, Zl can be measured by mathematical methods, so definite solution about TPG will be got if we know Zq. On the other hand Zq can been assumed to be a minimum reactance function,so the relationship between the real and imaginary parts of Zq, becomes unique. The relation between the real and imaginary parts can be shown in the following integral equations:

$$Rq(w)=-\frac{1}{\pi}\int_{-\infty}^{+\infty}\frac{Xq(\lambda)}{\lambda-w}d\lambda+R(\infty) \tag{6}$$

$$Xq(w)=\frac{1}{\pi}\int_{0}^{+\infty}\frac{dRq(\lambda)}{d\lambda}\ln\left|\frac{\lambda+w}{\lambda-w}\right|d(\lambda) \tag{7}$$

A key step in the real frequency method is the approximation of the resistance function Rq(w) by a number of straight-line segments with break point from 0 to w_n .So we have:

$$Rq(w)=r_0+\sum_{k=1}^{n}a_k(w)r_k \tag{8}$$

Where

$$a_k=\begin{cases}1 & w_k<w\\ \dfrac{w-w_{k-1}}{w_k-w_{k-1}} & w_{k-1}<w<w_k\\ 0 & w<w_{k-1}\end{cases} \tag{9}$$

The r_0 in (8) represent Rq(0)and it is normalized to 1.

With the resistance Rq(w) expressed as in (8), the reactance Xq(w) can be written as a linear combination of the same unknown resistive excursions. We thus have:

978-7-5641-4279-7

$$Xq(w)=\sum_{k=1}^{n} b_k r_k \tag{10}$$

Where

$$b_k=\frac{1}{\pi(w_k-w_{k-1})}\int_{w_{k-1}}^{w_k}\ln\left|\frac{y+w}{y-w}\right|dy \tag{11}$$

Based on the (5)、(8)、(10), the TPG can be expressed as follows :

$$T(w)=\frac{4Rl(w)[r_0+\sum_{k=1}^{n}a_k(w)r_k]}{[Rl(w)+r_0+\sum_{k=1}^{n}a_k(w)r_k]^2+[Xl(w)+\sum_{k=1}^{n}b_k(w)r_k]^2} \tag{12}$$

In (12),T(w) is at most quadratic function of r_k ,and he optimization function can be set to:

$$E=\sum_{k=1}^{M}|1-T(w_k)|^2 \tag{13}$$

Once the r_k is calculated by optimization method, the Rq can be expressed as line form, Xq(w) also can be calculated through Hilbert transform. But in practice, when Rq (w) is expressed as several segments poly line, physically matching network can not be achieved. Rq should be fitted by a rational function. We usually use the following equation fitting Rq(w):

$$\hat{R}=\frac{A_0w^{2k}}{1+B_1w^2+\cdots\cdots+w^{2n}} \tag{14}$$

In (14), if the matching network is a low-pass, then k = 0; if the matching network is a band-pass, then 1 <k <n; if the matching network is a high-pass, then k = n.

By changing the Rq(w) into the graph which can be physically realized, we already know the matching network driving-point function, then Zq(w) can be obtained by GEWERTZ method. Thus, Z (s) has been obtained, and then we will use knowledge of the network, integrated the Zq(w) into its matching network.

III. IMPROVED REAL FREQUENCY METHOD

Improved real frequency method calculates r_k through least square method based on the conjugate matching principle. So it can avoid the initial selection about r_k.

Compared with the real frequency method, the improved method is more simple and available both on the theory and implementation, also avoid the original method of optimization of E function, which is one of the most complex steps in real frequency method.

Being similar to the original method, after a plus an antenna matching network, the transmission power gain is expressed as follow:

$$T(w)=\frac{4Rq(w)Rl(w)}{|Rq(w)+Rl(w)|^2+|Xq(w)+Xl(w)|^2} \tag{15}$$

It is easy to know through the knowledge of antenna that if Zq (w) and Zl(w) conjugate match,antenna get maximum transmission power at this time. This is entirely consistent with the conjugate match of physical meaning .

$$Rq=Rl(w) \tag{16}$$

$$Xq=-Xl(w) \tag{17}$$

Since the introduction of conjugate match, we need' t set function E and initial value of increments of each fold line as the original real frequency method.

Because the matching network design goal has been given by the conjugate matching theory accurately, we can directly use the least square method to calculate Rq (w).In addition, when using least squares method, if we combine the Rq (w) and Xq (w),we could get double expand sampling points, then reach a more accurate solution about r_k.

Based on (8)(10)(16), we can obtain the following matrix:

$$\begin{bmatrix} a \\ b \end{bmatrix}_{2M\times N}\bullet[r]_{N\times 1}=\begin{bmatrix}(R_a-r_0)' \\ -X_a' \end{bmatrix} \tag{18}$$

Thus process of solving the matrix $[r]_{N\times 1}$ is much simpler compared with the original method and in the calculation of the actual programming matrix operations, we can directly call the function obtain the rest. Once we know the solution of the matrix $[r]_{N\times 1}$, we can obtain function Rq(w) by a number of straight-line segments by r_k. After this, the procedures to design the matching network is the same as the original real frequency method.

IV. DESIGN OF MATCHING NETWORK WITH REAL FREQUENCY TECHNIQUE

In the process of designing impedance matching network, we use the sleeve antenna in reference [1], the impedance is shown in Table I:

TABLE I

IMPRDANCE OF ATTENA

f(MHz)	R	X	f(MHz)	R	X
25.0	15.186	-50.945	35.4	107.78	-40.938
25.8	12.864	-28.789	36.2	66.988	-47.633
26.6	13.962	-12.807	37.0	44.611	-35.275
27.4	18.745	-1.006	37.8	38.941	-20.107
28.2	27.146	7.5225	38.6	43.502	-13.500
29.0	34.584	6.9609	39.4	45.232	-18.518
29.8	34.029	3.5625	40.2	34.791	-21.721
30.6	27.544	7.4600	41.0	24.585	-15.210
31.4	25.223	18.500	41.8	18.795	-5.1875
32.2	28.814	33.850	42.6	16.436	6.0557
33.0	43.561	52.582	43.4	16.746	18.769

1 Design matching network by original real frequency method

Firstly the Rq(w) should be represented as a number of line segments. Many tests show, the Rq (w) is divided into five segments may achieve the ideal uniform. Once we get the form of Rq(w), we can calculate r_k using the optimized function E. The results are as follows:

r_1=1.0323; r_2=1.2338; r_3=-0.3771; r_4=0.6973; r_5=-1.5218;
Line segments of Rq (w) is shown in Figure 2:

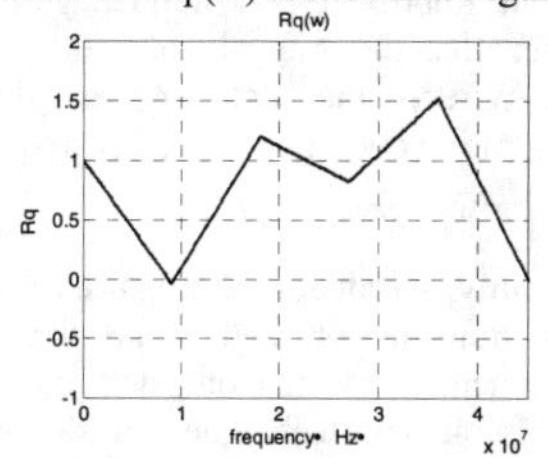

Fig.2. Rq(expressed in the form of fold line)

Now, the Rq(w) has been expressed as the fold line,in this case, Xq(w) can be calculated by using the Hilbert transform. So we can obtain the theoretical transmission power gain of matching antenna .Figure 3 shows:

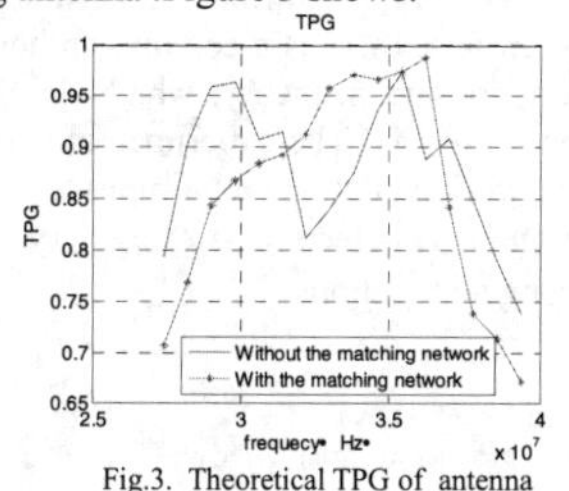

Fig.3. Theoretical TPG of antenna

However, because the Rq is fold line and it can not be implemented in physics, Rq should be fitted by a rational function. Then we obtain Zq(s) through GEWERTZ method.

$$Zq(s)=\frac{0.002889s^2+0.02489s+0.06042}{s^3+8.6133s^2+22.8386s+16.5449}$$

If the Zq(s) is obtained, the specific form of the matching network can be calculated according to the circuit of knowledge. As shown in Figure 4:

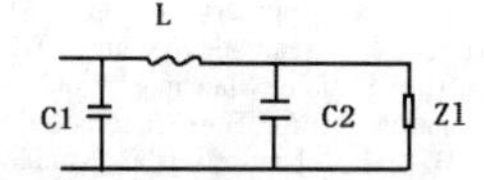

Fig.4. The impedance matching network

Circuit components inside the concrete is as follows:

C1=0.6922uF, L=0.007505uH, C2=6.371uF.

Transmission power gain obtained previously is the theoretical values, now according to the matching network, we calculate the transmission power gain again and get the result shown in Figure 5:

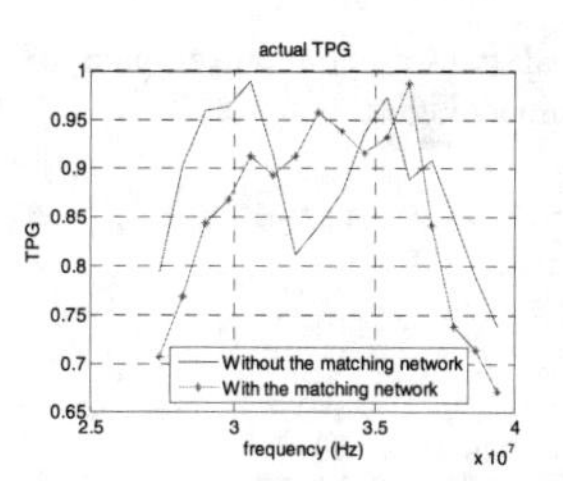

Fig.5.The actual TPG of antenna

There is relation in VSWR and TPG:

$$VSWR=\frac{1+\sqrt{1-TPG}}{1+\sqrt{1-TPG}} \tag{19}$$

Then we can calculate the VSWR of antenna .As shown in figure 6:

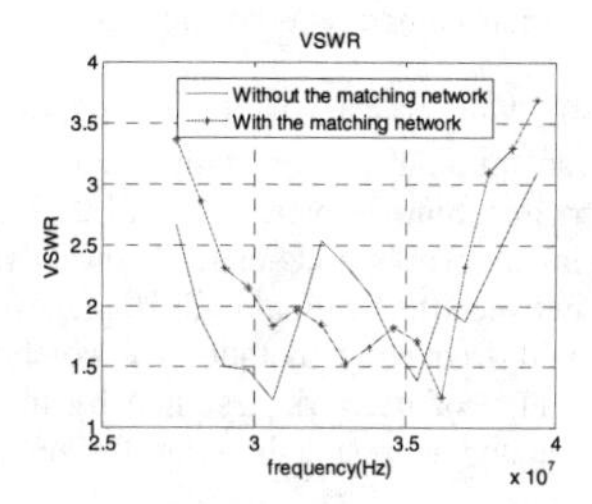

Fig.6. The actual VSWR of antenna

Some conclusions can be obtained based on the figure 3 and figure 4: First of all ,the TPG of the antenna, especially near the center frequency, have significantly increased after adding the matching network, which indicating that the matching network, to a certain extent, improve transmission performance of the antenna and is conducive to antenna transmission power. Secondly, after adding matching network, the actual TPG is basically the same with the theoretical one , and somewhat less than the theoretical value, because the actual line Rq(w) is an ideal fit of the fold line.

2 Design matching network by improved real frequency method

Firstly, r_k can be calculated with the least square method:
r1=-5.0642;r2=8.9973;r3=-3.8411;r4=1.4451;r5=1.4344
Line segments of The Rq (w) is shown in Figure 7:

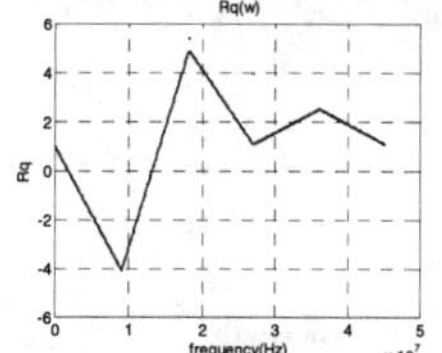

Fig.7. Rq (expressed in the form of fold line)

Then the theoretical transmission power gain of matching antenna can be obtained. Figure 8 shows:

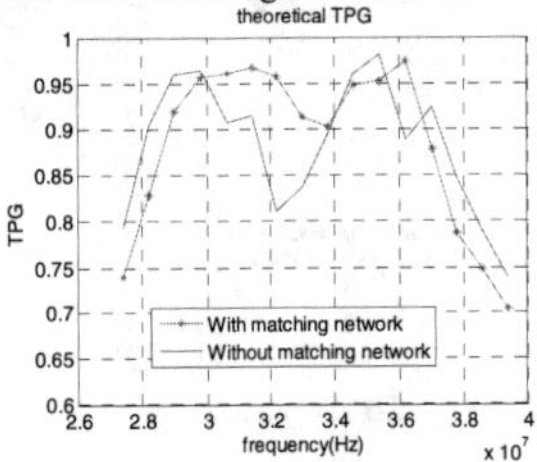

Fig.8. theoretical TPG of antenna

Next we obtain Zq(s) through GEWERTZ method:

$$Zq(s)=\frac{0.002049s^2+0.01808s+0.0460}{s^3+8..824s^2+24.9157s+21.7425}$$

The structure of the matching networks designed by improved real frequency method and what shown in figure 4 are identical. The circuit components inside matching network is as follows:

C1=0.976uF, L=4.17nH, C2=0.1072uF.

The actual transmission power gain can be calculated according to the matching network .To find the differences of the two matching networks and analyze the advantages and disadvantages between the original real frequency method and the improved real frequency method, we combine figure 5 with the actual TPG of network designed by improved real frequency method and get result shown in figure 9:

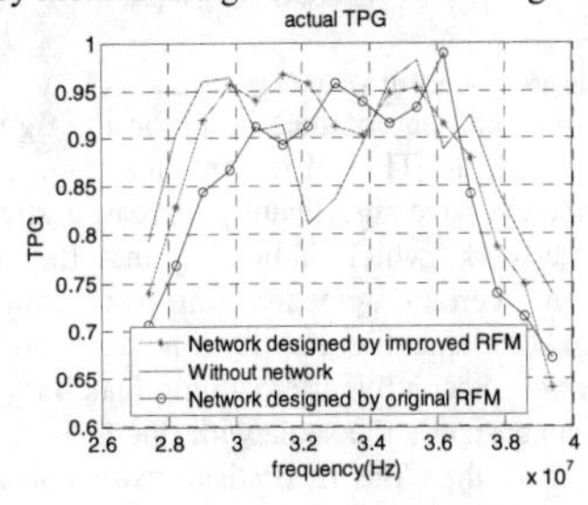

Fig.9. The actual TPG of antenna

Then we can calculate the VSWR of antenna .As shown in figure 10:

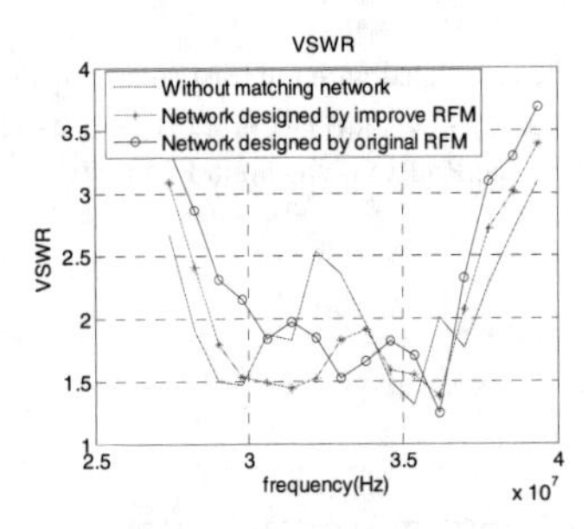

Fig.10. The actual VSWR of antenna

We could know from the figure 10 that the matching network can improve the antenna trans mission power gain in different degree and improved real frequency method has better matching effect than the original real frequency method. The matching can sacrifice the gain of two sides band to improve the transmission power gain of center frequency band.

V. CONCLUSION

This paper mainly introduces the importance of antenna matching network transmission and make the theoretical analysis and experimental research on designing matching network via real frequency technique. In the application research, we design two different impedance matching networks for HF broadband antenna with the original real frequency method and the improved real frequency method. In the past, most researches about real frequency technique stay in the stage of theoretical analysis of transmission power gain. This paper not only calculate both the theoretical TPG and the actual TPG but also analyze the difference between the two matching networks. In fact, there is a certain gap between the actual matching network and in theory, which should caused attention of antenna designers. The experimental results show that the matching network can effectively improve the antenna transmission power gain. This method is using very frequently in modern engineering applications.

ACKNOWLEDGMENT

The authors would like to express their sincere gratitude to funds supported by "the Fundamental Research Funds for the Central Universities" (Grant No. HIT.NSRIF.2014023).

REFERENCES

[1] ChenYi Hong, Sun Yan, "The Real frequency method for the design of broadband antenna impedance matching network," Chinese Journal of Electronics, 1997, pp 19-24.

[2] Ceylan.O, Yagci. H.B, Yarman.S.B, " Wideband matching circuit design for differantial output systems by using real frequency technique," Microwave Symposium (MMS),2010 ,pp. 241 - 244

[3] H.J.Carlin. "A New Approach to Gain Bandwidth Problems," IEEE .Transactions on Circuit and Systems. 1977, pp.170~175 .

[4] H.J.Carlin. "The Double Matching Problem: Analytic and Real Frequency . Solutions," IEEE Tran. Circuits and Sys. 1983,pp.15~28 .

[5] Hongming An B. K. 1. C. Nauwelaers and Antoine R. Vande Capelle. "Broadband microstrip antenna design with the simplified real frequency technique,"IEEE AP. 1994. 2

[6] B. S. Yarman, Metin Sengul, Peter Lindberg, Anders Rydberg. "A Single Matching Design For a Double Band Antenna Via Simplified Real Frequency Technique," APMC 2006: 1325-1328.

[7] R. M. Fano, "Theoretical limitations on the broadband matching of cases for which the objects' dimensions are comparable with the arbitrary impedances," J. Franklin Inst., 1976.

[8] Xiaojun Wang, Zhenyi Niu Yun , "Matching network design for a double band PIFA antenna via simplified real frequency technique," Microwave Technology and Computational Electromagnetics, 2009 , pp 92 - 94 .

The Fast and Wideband MoM Based on GPU and Two-Path AFS Acceleration

Lan Li, Jundong Tan, Zhuo Su and Yunliang Long
Department of Electronics & Communication Engineering
Sun Yat-Sen University, Guangzhou, 510006, P. R. of China
lilan0420@gmail.com

***Abstract*—In this paper, a General Purpose Unit (GPU) accelerated full-wave method of moment (MoM) is combined with a two-path adaptive frequency sampling (AFS) approach to analyze the wideband characteristic of the body-wire structures. An equivalent principle is employed to treat the wire as surface so that the model which is analyzed based on the electric-field integral equation (EFIE) could be purely discretized by triangles, avoiding adopting three different basis functions. Numerical results for a monopole mounted on the square ground plane show the efficiency and accuracy of the proposed methods.**

***Index Terms*-- MoM, body-wire, GPU, two-path AFS**

I. Introduction

In the computational electromagnetics (CEM), the full-wave method of moment has been used extensively because of its accuracy, especially for structures composed of wires and bodies. When solving a MoM problem, one must select suitable basis functions. As to traditional body-wire, it usually employs three different basis functions to represent the wires, bodies and body-wire junctions [1]. It was introduced in [2] that the wire could be treated as surface using an equivalent principle. In this way all elements could be discretized by triangles, so that only one kind of basis function is needed, which will improve the consistency to the numerical implementation. RWG [3] is one of the most widely used basis function to model arbitrary surfaces. However, special attention should be paid to the extra T-junction problem [2] when the triangle belongs to different structures.

Although MoM is an accurate and robust method, it is of very high computation complexity. Parallel computing is an effect way to accelerate MoM computation. Much time could be reduced by using the General Purpose Unit (GPU), which disperses the task for one thread on CPU to thousands of threads on GPU. In fact, many problems in CEM have exploited the powerful computing ability of GPU to get better efficiency, such as FDTD. At the same time, the GPU was engaged to MoM first in [4] using a stream programming language Brook, which poses a programming challenge for researchers. Since the CUDA being proposed by Nvidia at 2006, it has promoted the GPU application in CEM. In [5-6], GPU is adopted to tackle MoM problem with CUDA and high speedup ratios were achieved.

Antenna simulation over a wideband demands a large number of MoM calculation at the sample frequencies. Adaptive frequency sampling (AFS) [7] is a technique to reduce the frequency points when obtaining the broadband performance of antennas. For traditional AFS method, it's necessary to inverse a matrix to get the coefficients of the targeted rational interpolation function. However, when the sampling points increase, it's nontrivial to select linearly independent points to avoid ill condition of the matrix. In [8], the Stoer-Bulirsch AFS (S-B AFS) algorithm has been introduced to avoid the matrix inversion, but it needs to get the EM value at every adapting step and utilizes only one adaptive interpolation model. In this paper, we will extend that method to a two-path AFS approach to get the broadband characteristic of the target model. This improving method could be used to simulate more complex targeted model and faster convergency rate could be achieved.

This paper is organized as follows. Theory of MoM as well as RWG basis functions is presented in Section II. Two methods have been taken to reduce the running time by employing GPU acceleration and AFS method in Section III. At last, the numerical results for a monopole antenna mounted on a conducted square plane are given in Section IV.

II. MOM

The MoM utilized to analyze the body-wire model in this paper is based on the electric-field integral equation (1).

$$\boldsymbol{E}_{tan}^{i} = (j\omega \boldsymbol{A}_S + \nabla \Phi_S)_{tan} \quad r \in S \quad , \qquad (1)$$

Following the MoM procedure, the structure should be first divided into triangles, applied to handle arbitrary shaped objects [3]. As mentioned above, the wire is equivalent to a rectangle surface on condition that the width of the square-W and the radius of the wire-R meet the relationship of R=0.25W. Their height keeps equal. Thus, all elements could be represented by RWG basis function. It is important to notice the T-junction problem in Fig.1caused by body-wire junction. In the programming, a trick to assign different indexes to different structures could remove the T-junctions. Subsequently, the current on the conducted surface may be approximated in terms of f_n as

$$\boldsymbol{J}_s = \sum_{n=1}^{N} I_n \boldsymbol{f}_n \qquad (2)$$

978-7-5641-4279-7

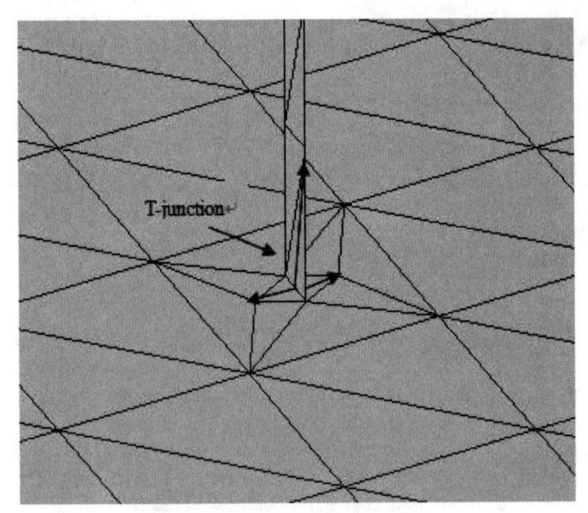

Figure 1. T-junction caused by body-wire junction

f_n is just the RWG basis function which is defined as

$$f_n = \begin{cases} \dfrac{l_n}{2A_n^+}\rho_n^+ & r \ in \ T_n^+ \\ \dfrac{l_n}{2A_n^-}\rho_n^- & r \ in \ T_n^- \\ 0 & otherwise \end{cases} , \tag{3}$$

Then, choosing the f_n as the weight function according to the Galerkin method, we can get (4)

$$\langle E^i, f_m \rangle = j\omega \langle A_s, f_m \rangle + \langle \nabla\Phi_s, f_m \rangle \quad , \tag{4}$$

Substituting the current expansion (2) into (4) yields a matrix form equation in (5). The concrete computing methods for elements in the matrix could refer to [3]. Once obtaining the current on the structure, we can get other performance of radiating/scattering objects.

$$Z \ V = I \tag{5}$$

III. IMPROVING METHODS

A. GPU Acceleration

As mentioned above, GPU is a powerful hardware accelerating tool for CEM problems. Together with CUDA, it's much convenient to parallelize MoM in C++ version. In addition, it was also much more practical and flexible than traditional versions implemented in Fortran and Matlab.

Due to the advantage of the GPU architecture, it's much more suitable to handle with tasks which are more independent and less logical. In addition, the CPU tends to deal with what is more relevant and logical. Therefore, in our programming, the subdivision of model, structure parameters and EM parameters were calculated on the host. At the same time, the device was in charge of matrix filling and matrix equation solving.

Note that the executing time for GPU includes data transfer between GPU and CPU, as well as kernel launching [9]. In order to avoid the bottleneck of data transfer, we could turn to mapped memory, which maps the address on the host to the device address, reducing data transfer latency. Since the accessing time for every kind of memory on the device is different, another problem for GPU implementation is how to select reasonable memory type for every variable. In addition, the register accessing time is the fastest on device, but its amount is limited per block. Therefore, if the block size is too large, the registers for every thread will be insufficient and if the block size is too small, it will lower the level of parallelism. The experiment results show that the 16x8 block size is an appropriate scheme.

In this study, we employed the CULA library to solve matrix equation. It provides two functions – culaDeviceZgesv() and culaZgesv() [10]. The former needs variables allocated on the device and the latter needs variables allocated on the host. Both of them adopt the LU factorization. Thereby the solution has high accuracy.

B. Adaptive Frequency Sampling

When getting the broadband performance of radiating /scattering objects, it always tends to calculating at substantial frequency points, which will consume a lot of time. Due to the AFS strategy, we will obtain the quality of interest by constructing a rational function $R(f)$ in (5), so as to reduce the frequency points computed by EM method. Reference [11] has employed a S-B AFS streme to get $R(f)$, which will avoid inversing a matrix when getting the coefficients of $R(f)$ in traditional AFS technique. In this study, it extends that method to a two-path AFS approach, which can simulate more complex objective function and converge much faster

$$R(f) = \frac{N_0 + N_1 f + \cdots + N_n f^n}{1 + D_1 f + \cdots + D_d f^d}, \tag{6}$$

Unlikely [11], the two-path AFS constructs the rational function $R(f)$ from two directions, that is to say using two efficient recursive tabular algorithms in (7) and (8) in according to [12].

$$R_{j,k}^1 = \frac{(f - f_j)R_{j+1,k-1}^1 + (f_{j+k} - f)R_{j,k-1}^1}{f_{j+k} - f_j} \tag{7}$$

$$R_{j,k}^2 = R_{j+1,k-2}^2 + \frac{f_{j+k} - f_j}{\dfrac{f - f_j}{R_{j+1,k-1}^2 - R_{j+1,k-2}^2} + \dfrac{f_{i+k} - f}{R_{j,k-1}^2 - R_{j+1,k-2}^2}} \tag{8}$$

The procedure for two-path AFS algorithm is illustrated in Fig. 2. It is assumed the simulated frequency band ranges from f_L to f_H. Before the recursive procedure, two initial values calculated by MoM are needed for (7) and (8). The *eps* is the convergence error for the process. After the construction process is terminated, we can use the rational functions (6) to calculate the value at any other frequency points, instead of the time-consuming MoM. Therefore, it will improve the simulation efficiency greatly.

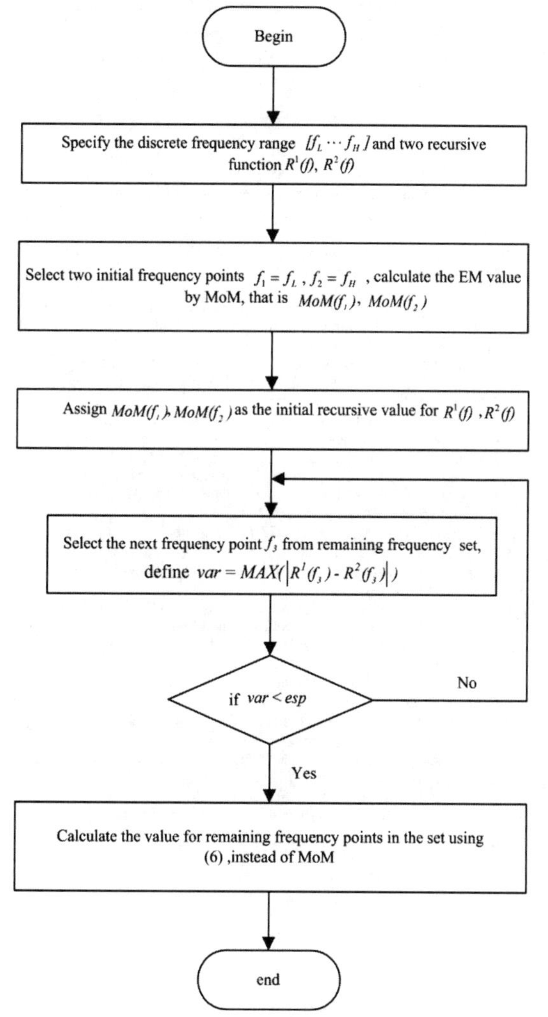

Figure 2 Flowchart of two-path AFS algorithm

IV. RESULT

The experiment for this paper is carried on Dell-Precision T7600 Workstation where the hardware detail of CPU is Xeon E5-2607 with 2.4GHz Frequency and 64G RAM and the GPU type is Tesla C2075. The code is implemented in C++ on Ubuntu 12.04 LTS with double precision to achieve high accuracy.

The benchmark structure considered in this study is a monopole antenna mounted on a square ground as shown in Fig.3. The antenna's width and height is 0.04m and 1m respectively. The ground' edge is 2m.

At first, Fig.4 has displayed the result by the MoM used in this paper for return loss (S11) of the antenna in the frequency

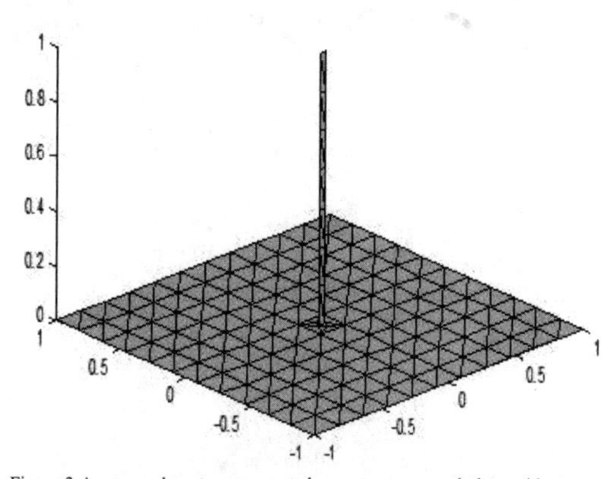

Figure 3 A monopole antenna mounted on a square ground plane with antenna height 1m, width 0.04m and ground edge 2m as the benchmark for this study.

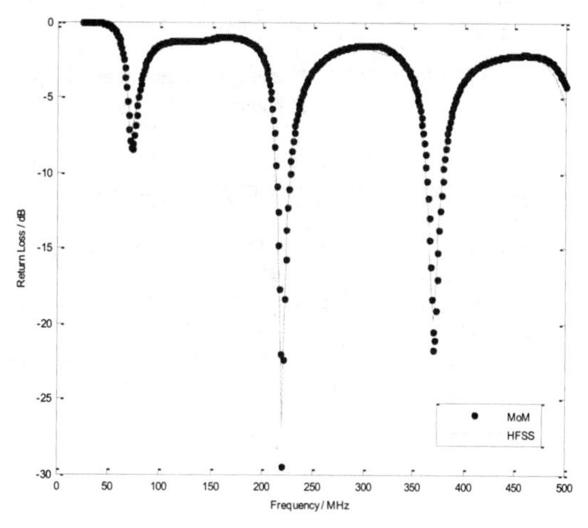

Figure 4 The result of return loss computed by the MoM for 476 frequency points, compared with the result from HFS S.

Table 1 Executing details on CPU versus GPU (1 frequency point)

Element Number	Executing Time / Second		Speedup Ratio
	On CPU	On GPU	
519	12.02	0.56	21.46
687	26.54	0.91	29.16
879	53.51	1.64	32.62
1059	111.81	2.46	45.45
1335	222.53	3.47	64.13
1887	617.78	6.52	94.75

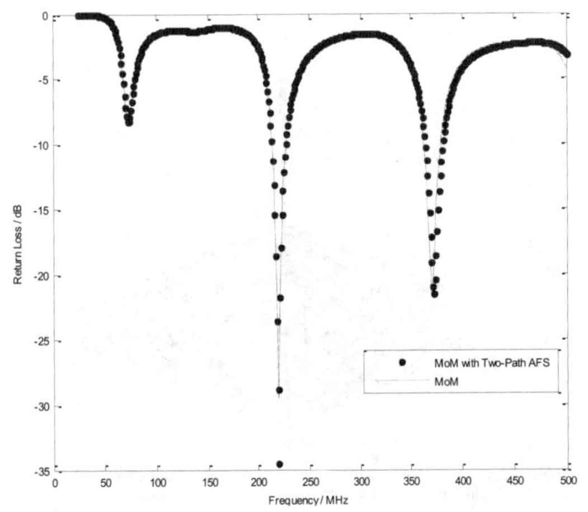

Figure 5 The return loss for the benchmark in the study only 123 frequency points need to be computed by MoM with two-path AFS method getting the same performance effect for 476 frequency points without two-path AFS

range from 25MHz to 500 MHz. As compared, the result from HFSS is also given. Good agreement can be observed from results.

Then, table 1 gives the executing details for the calculation of return loss for one frequency point by MoM. It includes the total running time and speedup ratio on CPU versus GPU with the matrix size increasing. Note that the CUDA timing includes the data transfer between CPU and GPU.

Finally, Fig.5 displays the antenna's S11 performance with two-path AFS method versus without it executed on GPU. We need to compute 476 frequency points for the given frequency range without the Two-Path AFS. However, after taking this technique, only 123 points are needed to get the intact wideband characteristic by MoM. Therefore, it is proved that this method offers a noticeable improvement for the efficiency of calculating antenna's broadband performance.

V. CONCLUSION

This paper has utilized an equivalent principle to model the body-wire structure by purely triangles, avoiding employing three different basis functions. A trick was developed in programming to remove the T-junction caused by body-wire junction. Then two methods were used to improve the executing efficiency by employing GPU acceleration and adaptive frequency sampling method. The experiment results for a monopole mounted against a square ground plan indicate all these methods could achieve good effect.

ACKNOWLEDGMENT

This research was supported by a grant from Natural Science Foundation of China under Contract 61172026. The authors wish to acknowledge the assistance and support of all the fellows and teachers in the Applied Eletromagnetics Lab, Sun Yat-Sen University.

REFERENCE

[1] S. U. Hwu, D. R. Wilton, and S. M. Rao, "Electromagnetic scattering and radiation by arbitrary conducting wire/surface configurations," in IEEE APS Int. Symp. Digest, Syracuse, New York, vol. 2, pp. 890–893, Jun. 1988.

[2] S. Makarov, "MoM antenna simulations with Matlab: RWG basis functions, "IEEE Antennas and Propag.Mag. Vol.43, No 5, pp 100-107, Oct., 2001

[3] S. M Rao and D. R. Wilton, and A. W. Glisson, "Electromagnetic scatterring by surfaces of arbitrary shape" IEEE Trans. Antennas Propag. vol.30, no.3, pp 409-412, May 1982

[4] S. Peng and Z. Nie, "Acceleration of the method of moments calculations by using graphics processing units," *IEEE Trans. Antennas, Propag*, vol. 56, no. 7, pp. 2130–2133, Jul. 2008.

[5] T. Topa, A. Karwowski, and A. Noga, "Using GPU with CUDA to accelerate MoM-based electromagnetic simulation of wire-grid models," IEEE Antennas Wireless Propag. Lett., vol. 10, pp. 342–345, 2011.

[6] T. Topa, A. Noga , and A. Karwowski, "Adapting MoM with RWG basis functions to GPU technology using CUDA", IEEE Antennas Wireless Propag. Lett., vol. 10, pp. 480–483, 2011.

[7] T. Dhaene, J. Ureel, N. Faché, and D. De Zutter, "Adaptive Frequency Sampling Algorithm for Fast and Accurate S-parameter Modeling of General Planar Structures", in IEEE MTT-S Int. Microwave Symp. Dig., 1995, pp. 1427–1430.

[8] A. Karwowski, "Efficient wide-band interpolation of MoM-derived frequency responses using Stoer-Bulirsch algorithm", IEEE International Symposium on Electromagnetic Compatibility, EMC 2009, pp. 249-252, Aug. 2009

[9] "CUDA C Programming Guide," NVIDIA Corporation, Santa Clara, CA,Oct., 2012.

[10] "CULA tools—GPU accelerated LAPACK," EM Photonics, Newark, DE [Online]. Available: http://www.culatools.com/

[11] A. Karwowski, A . noga, T. Topa, "GPU-Accelerated MoM-Based Broadband Simulations Using Stoer-Bulirsch Algorithm", Radioelektronika, IEEE 21st International Conference, pp.1-2, 2011.

[12] J. Stoer and R. Bulirsch, Introduction to Numerical Analysis. Berlin, Germany: Spring-Verlag, 1980.

Circularly Polarized DRA Mounted on or Embedded in Conformal Surface

S.H. Zainud-Deen[1*], N.A. El-Shalaby[2] and K.H. Awadalla[1].
[1]Faculty of Electronic Eng., Menoufia University, Egypt, *anssaber@yahoo.com
[2]Kafer El-Sheikh University, Egypt, noha1511ahm @yahoo.com

***Abstract*- In this paper, the effect of curvature on circular polarization (CP) radiation characteristics of circularly-polarized dielectric resonator antenna (DRA) mounted on or embedded in cylindrical surfaces are studied. The antenna is designed for wireless applications at 9 GHz. In order to eliminate the effect of ground plane curvature on the CP characteristics by varying the DRA element height or embedded the DRA element in the curvature structure. The analysis using the finite element method (FEM) is applied. The results are validated by comparing with that calculated by the finite integration technique (FIT).**

I. INTRODUCTION

In modern communication systems, DRA has been used due to their advantages including its small size, low loss, low cost, light weight, and ease of excitation [1]. In most of DRAs applications, circular polarization antennas are usually preferred. CP antennas are less sensitive to orientation of the transmitter and the receiver [2]. In addition, the CP antenna reduces the propagation loss between the transmitting and the receiving antennas [3-5]. Most of the work on the DRA theory and technology has been concentrated on the performance of the DRA on planar surfaces. However, placing antennas on conformal structures (cylindrical or spherical) plays a considerable role in modern communication systems such as spatial domain multiple access (SDMA), smart antennas, beam-steering array antennas and aerospace applications [6]. The effect of ground plane curvature on the radiation characteristics of a cylindrical DRA mounted on or embedded in a hollow cylindrical ground plane have been investigated in [7, 8]. The radiation characteristics of a hemispherical DRA mounted on or embedded in a spherical ground plane have been investigated in [9-11]. In this paper, the effect of changing the ground plane curvature on the circular polarization properties of the circularly-polarized DRA has been demonstrated. The antenna is mounted on or embedded in a cylindrical ground plane. The DRA elements are designed and analyzed using the finite element method (FEM) [12] and the results are compared with those obtained by the finite integral technique (FIT) [13].

II. NUMERICAL RESULTS

A) DRA MOUNTED ON A CYLINDRICAL GROUND PLANE

A circularly-polarized square DRA with truncated edges mounted on a planer ground plane using a single feeding probe was introduced in [6]. The circular polarization characteristics are produced due to the generation of two orthogonal modes in the antenna structure. The DRA has square cross section with dielectric constant ε_r =9.4, edge length *a* =7.5 mm, and height h=6 mm as shown in Fig.1a. Two triangles are removed from the two corners along one diagonal of the DRA with c= 3.125 mm. A coaxial probe with radius of 0.15 mm, and height h_f =3 mm located at d_f =3 mm is used to excite the antenna. DRA element is mounted on a hollow hemi-cylindrical ground plane with radius R_g= 15 mm, m_g=π R_g, and l_g = 60 mm as shown in Fig.1b. The simulated reflection coefficient against the frequency and the circular polarization radiation pattern components, left-hand E_L and right-hand E_R in x-z plane and y-z plane are calculated at resonance frequency 9.25 GHz for the DRA mounted on hemi-cylindrical ground plane are shown in Fig.2.

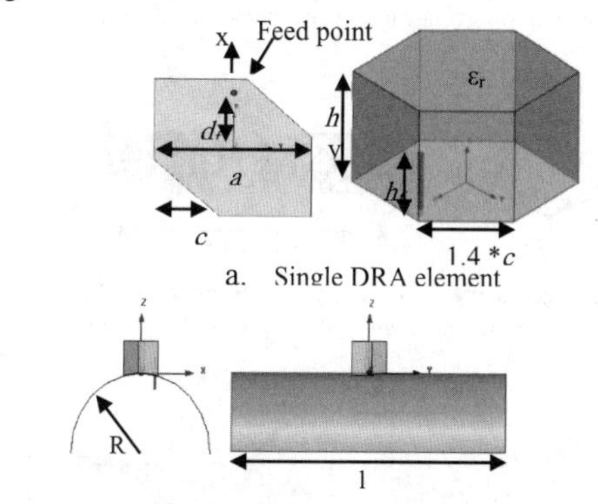

a. Single DRA element

b. Geometry of DRA mounted on metallic hemi-cylindrical ground

Fig.1. The geometry of DRA element mounted on cylindrical ground plane.

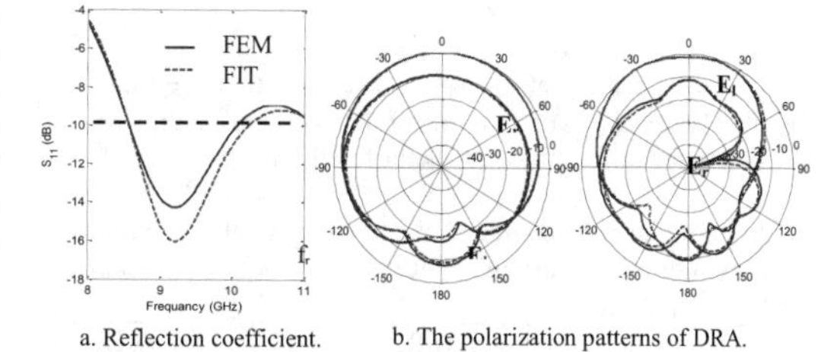

a. Reflection coefficient. b. The polarization patterns of DRA.

Fig.2. Radiation characteristics of DRA mounted on hemi-cylindrical ground plane at f=9.25 GHz.

The impedance matching bandwidth with (S_{11}<-10dB) is 1.5 GHz for the curved ground plane case. The resonance frequency is shifted up for the curved ground plane case to 9.2 GHz due to the change in the size of the DRA to follow the curvature of the ground plane. Also, the change of the feed probe position inside the DRA will launch different modes with higher frequencies in the studied cases. The ratio of E_L/E_R is 9.8 dB for the curved

978-7-5641-4279-7

case. The FEM method (solid line) is used to simulate the radiation characteristics of the DRA structure on different shapes of the ground plane and the results are compared with those determined using the FIT method (dotted line). Good agreement between the results of the two methods is obtained. Figure 3 shows the DRA element mounted on a hemi-cylindrical ground plane with different radii of curvature for a fixed physical area *($l_g \times m_g$)*. The effect of varying the radius of curvature R_g on the reflection coefficient and axial ratio of DRA are demonstrated in Fig.4. It is seen that the axial ratio is reduced with increasing the radius of the hemi-cylindrical ground plane. When R_g=60 mm, the antenna produces the minimum axial ratio at f= 9.25 GHz. The antenna has circular polarization bandwidth (AR<3 dB) over the frequency band from 9.1 GHz to 9.5 GHz within which the antenna matching is still preserved. The impedance matching is increased by increasing the radius of curvature. The resonance frequencies for different values of radius of curvature are decreased by increasing the radius of curvature of the ground plane. A comparison of different cases is presented in Table I.

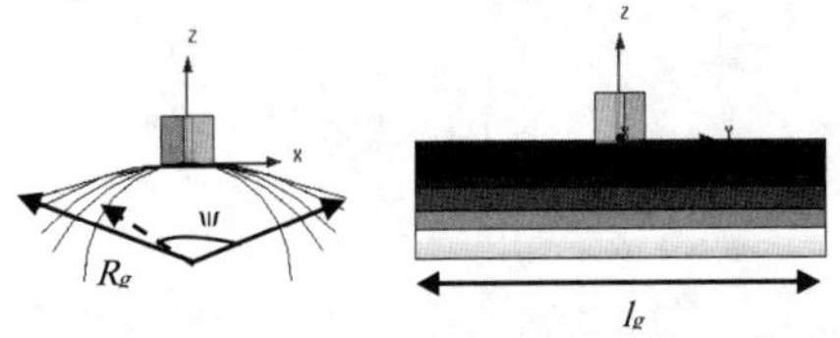

Fig.3. The geometry of DRA element mounted on a hemi-cylindrical ground plane with different radii for ground plane area.

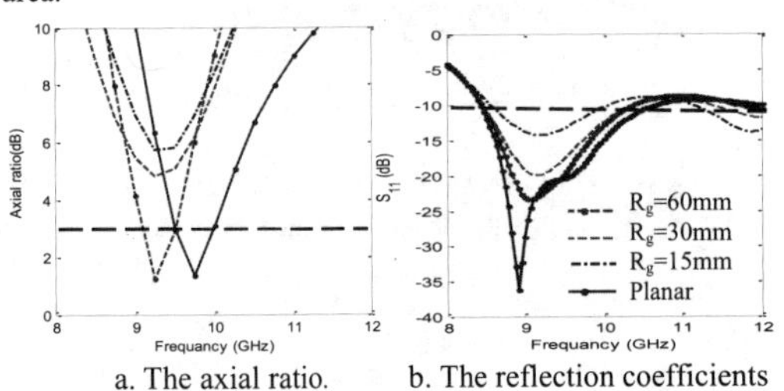

a. The axial ratio. b. The reflection coefficients

Fig.4. The axial ratios and the reflection coefficients versus frequency for different R_g and ψ for fixed l_g and m_g.

Table 1 circular polarization characteristics of DRA mounted on hemi-cylindrical ground plane at different values of R_g and ψ for fixed l_g and m_g.

R_g, ψ	Axial ratio (dB)	CP bandwidth	E_L/E_R
R_g=60 mm *ψ = 45°*	1.23 dB at f=9.25 GHz	(9.1- 9.51) GH	16.4 dB
R_g=30 mm *ψ = 90°*	4.82 dB at f=9.25 GHz	----	9.83 dB
R_g=15 mm *ψ =180°*	5.74 dB at f=9.5 GHz	-----	9.8 dB

To overcome on the degradation of the circular polarization characteristics of the DRA due to the increase in the radius of curvature of the ground plane, the DRA size is modified by varying its height *h* and the coaxial feeding probe length l_p. The effect of radius of the curvature, R_g, after modification (h, l_p) on the reflection coefficient, axial ratio and the polarization patterns at frequency where the axial ratio is at the minimum of each case are demonstrated in Fig.5. The values of the axial ratio, CP bandwidth and E_L/E_R levels at different radii of curvature are listed in Table II. It is seen that the axial ratio is improved with varying the height of DRA and the length of probe according to the ground radius of curvature. The relationship between the height of the DRA element *h* for each radius of the hemi-cylindrical ground curvature R_g to get best circular polarization properties is shown in Fig.6. A mathematical representation of this relation is obtained using a curve fitting technique and can be represented by a straight line relation as

$$h = 0.2R_g + 5.5 \quad (1)$$

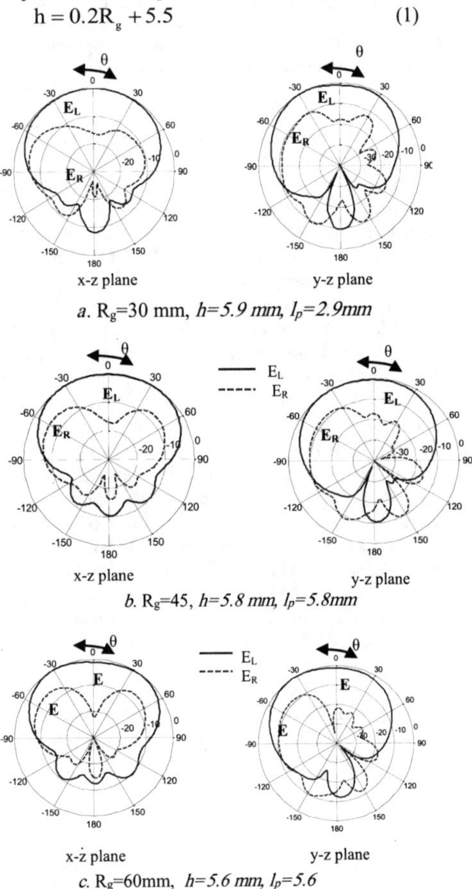

Fig.5. The polarization patterns of DRA for different radii R_g, after modification.

Table II. Comparison of circular polarization characteristics of DRA mounted on hemi-cylindrical ground plane at different values of R_g and ψ for fixed l_g and m_g, after modification *h*, l_p.

R_g, ψ	Axial ratio ((dB))	CP bandwidth	E_L/E_R
R_g=60 mm, ψ =45° h=5.9 mm, l_p=2.9 mm	1.52 dB at f=9.5 GHz	(9.3 -9.75) GHz	21 dB
R_g=45 mm, ψ = 60° h=5.8 mm, l_p=5.8 mm	2.5 dB at f=9.75 GHz	(9.45 -9.85) GHz	17 dB
R_g=15 mm, ψ = 180° h=5.6 mm, l_p=5.6 mm	2.45 dB at f=10.25 GHz	(10.15-10.35) GHz	17 dB

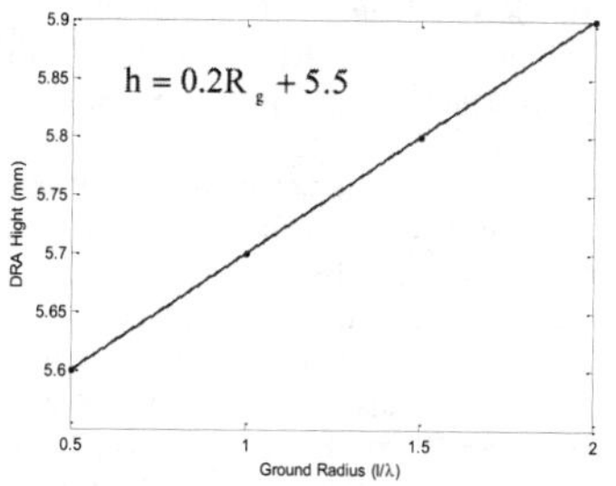

Fig.6. The relation between DRA height and ground plane radius.

B. DRA Element Embedded in a Cylindrical Ground Plane

The previous square DRA element embedded in hemi-cylindrical metallic ground, with a cavity recess height h_c = 7 mm and cavity radius r_c=7 mm, with ground plane radius R_g=30 mm as shown in Fig.7. The simulated reflection coefficient against the frequency and the polarization radiation pattern of the DRA embedded in hemi-cylindrical metallic ground at f= 9.6 GHz are shown in Fig.8. The resonance frequency is shifted up to 9.82 GHz due to the presence of the cavity. The size of DRA does not vary as that happens in case (A). The antenna gives good impedance matching at 9.82 GHz. The impedance matching bandwidth with (S_{11}<-10dB) is 700 MHz. The circular polarization radiation patterns, E_L and E_R, in x-z plane and y-z plane at resonance frequency 9.6 GHz at which the axial ratio has minimum value are shown in Fig.8b. The ratio E_L/E_R=20.13 dB. The FEM method is used to simulate the radiation characteristics of the DRA structure and the results are compared by those calculated by FIT method. Good agreement between the two results methods is obtained.

The effect of radius of curvature, R_g, on the axial ratio, the reflection coefficient, and the circular polarization radiation patterns at the frequency where the axial ratio is at minimum value are demonstrated in Fig.9 and Fig.10. It is seen that the axial ratio is improved by decreasing the radius of the hemi-cylindrical ground plane. When R_g=15 mm the ratio between the circular polarization radiation components at the frequency at which the antenna produces minimum axial ratio value, f=9.82 GHz, is 20.13 dB. The antenna has circular polarization bandwidth (AR<3 dB) over the frequency band from 9.32 GHz to 9.8 GHz within which the antenna matching is still preserved. The impedance matching and resonance frequency are little changed with changing the radius of curvature while the circular polarization patterns nearly the same. Using the embedded cavity, the size of DRA remains constant and not varying by changing R_g. A summary of different cases is presented in Table III.

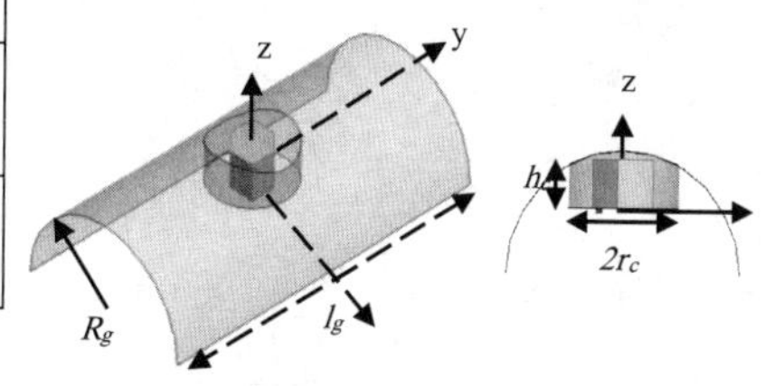

Fig.7. The geometry of DRA element embedded in cylindrical ground plane structure, h_c=7mm, r_c=7mm.

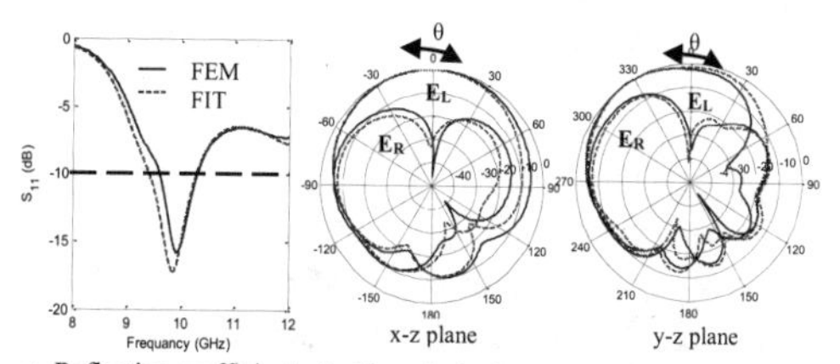

a. Reflection coefficient. b. The polarization radiation patterns at f=9.6 GHz

Fig. 8. The reflection coefficients and the polarization radiation patterns of DRA embedded in cylindrical ground plane, R_g=15 mm, h_c=7 mm, r_c=7 mm.

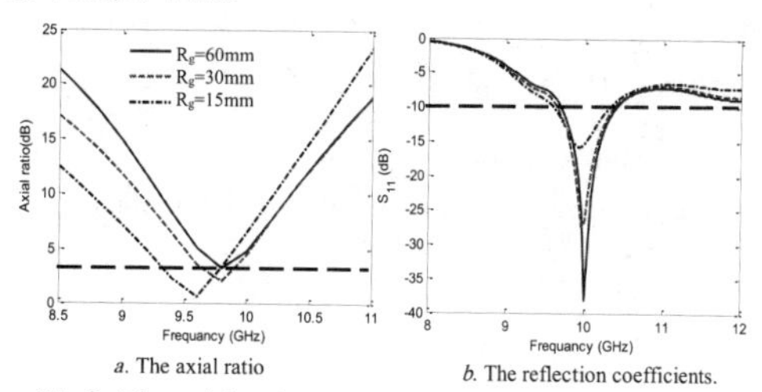

a. The axial ratio b. The reflection coefficients.

Fig.9. The axial ratio and the reflection coefficients versus frequency for different radii.

Table:III. . Comparison of circular polarization characteristics of DRA embedded in hemi-cylindrical ground plane at different values of R_g and ψ for fixed l_g and m_g.

R_g	Axial ratio (dB)	CP bandwidth	E_L/E_R
R_g=60 mm and ψ =45°	3.26 dB at f=9.8 GHz	----	14 dB
R_g=30 mm and ψ =90°	2.02 dB at f=9.8 GHz	(9.68 - 9.88) GHz	18.69 dB
R_g=15 mm and ψ =180°	0.5 dB at f=9.6 GHz	(9.32-9.8) GHz	20.13 dB

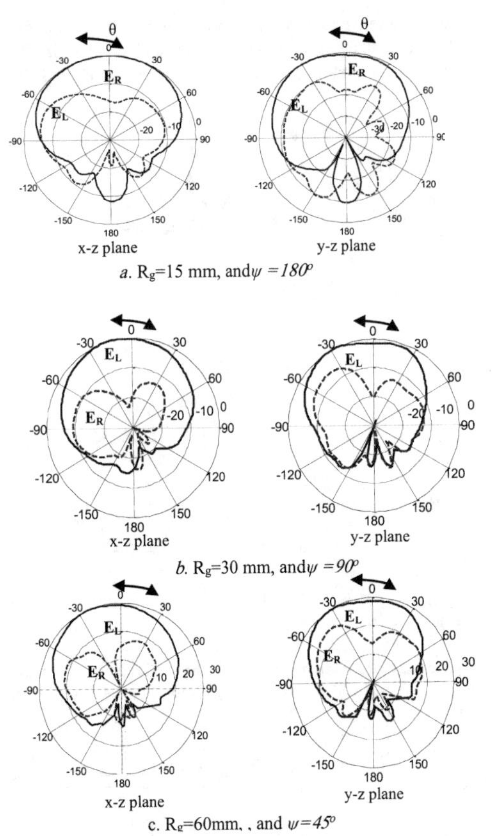

Fig.10. The polarization patterns of DRA for different radii, fixed curvature area after modification.

III. CONCLUSION

In this paper, the effect of curvature on circular polarization radiation characteristics of the circular polarized DRA element mounted on cylindrical surfaces have been demonstrated. The cylindrical curvature of the surface effects on the size of the DRA element, the DRA structure follows the curvature of the ground plane, and the asymmetry becomes more pronounced. The circular polarization is disappearing with decreasing the radius of curvature (increasing the curvature) where E_L/E_R are deteriorating. To reduce this effect by varying the DRA height h and the coaxial feeding probe length l_p to be suitable with the curvature surface. Another solution to overcome the curvature effect on circular polarization characteristics Design embedded structure. This can be achieved by embedding the DRA in a cavity recess in the ground plane in the ground plane. The circular polarization patterns nearly the same with changing the radius of curvature by Using the embedded cavity, the size of DRA remains constant.

IV. REFERENCES

[1] A. Petosa, Dielectric Resonator Antenna Handbook, Artech House, Inc., Norwood, 2007

[2] Kishk, "Performance of planar four element array of single-fed circularly polarized dielectric resonator antenna," Microwave and Optical TechnologyLett., vol. 38, no. 5, pp. 381-384, September 2003., 2008.

[3] G. C. Almpanis, Fumeaux, and R. Vahldieck, "The trapezoidal dielectric resonator antenna," IEEE Transactions on Antennas and Propagation, vol.56, no.9, pp. 2810-2816, September 2008.

[4] B. Li, and K.W. Leung, "On the differentially fed rectangular dielectric resonator antenna," IEEE Trans. Antennas Propagate., vol.56, no.2, pp. 353-359, February 2008.

[5] L. a. L. M. S. Ho Sang, "A study on the enhancement of gain and axial ratio bandwidth of the multilayer CP-DRA.," 5th European Conference, 2010.

[6] S. H. Zainud-Deen, H. A. Malhat, and K. H. Awadalla," Dielectric resonator antenna mounted on a circular cylindrical ground plane," Progress In Electromagnetics Research B, PIER B, vol. 19, pp. 427-444, 2010.

[7] S. H. Zainud-Deen, Noha A. El-Shalaby, and K. H. Awadalla," Radiation characteristics of cylindrical dielectric resonator antenna mounted on superquadric cylindrical Body," Electrical and Electronic Engineering Journal, SAP, vol. 2, no. 3, pp.88-95, 2012.

[8] S. H. Zainud-Deen, Noha A. El-Shalaby, and K. H. Awadalla, ," Hemispherical DRA antennas mounted on or embedded in circulFar cylindrical surface for producing omnidirectional radiation pattern" International Journal of Communication, Network and System Sciences, vol.4, no.9, pp. 601-608, September 2011.

[9] S. H. Zainud-Deen, Noha A. El-Shalaby, and K. H. Awadalla," Hemispherical dielectric resonator antenna mounted on or embedded in spherical ground plane with a superstrate," Electrical and Electronic Engineering Journal, SAP, vol. 1, no. 1, pp.5-11, 2011.

[10] S. H. Zainud-Deen, Noha A. El-Shalaby, and K. H. Awadalla," Hemispherical dielectric resonator antennas over non-planar surfaces for direction finding systems," International Journal of Electromagnetics and Applications SAP, vol. 2, no.6, November 2012.

[11] A. Morris, and A. Rahman, A Practical Guide to Reliable Finite Element Modeling, John Wiley & Sons Ltd, The Atrium, Southern Gate, England, 2008.

[12] Mariana Funieru, Simulation of Electromechanical Actuators using the Finite Integration Technique, Ph. D. Thesis, University of Darmstadt, Bukarest, 2007.

[13] I. Munteanul and T. Weiland, "RF & Microwave Simulation with the Finite Integration Technique from Component to System Design," Scientific Computing in Electrical Engineering Mathematics in Industry, vo. 11, Part III, pp. 247-260, 2007.

A High Isolation MIMO Antenna Used a Fractal EBG Structure

Haiming Wang, Dongya Shen, Teng Guo，Xiupu Zhang
School of Information Science and Engineering
Yunnan University, Kunming, China
shendy@ynu.edu.cn, 441ming@163.com
guoteng208@gmail.com, xzhang@ece.concordia.ca

***Abstract*- To further reduce the coupling of a MIMO antenna containing a mushroom-like EBG structure, we propose a novel MIMO antenna by using a fractal mushroom-like EBG structure to realize a high isolation. The simulated band width of the antenna is 2100 to 2830 MHz (S11<-10 dB), covering the bands of WLAN and WiMax. Compared to the antenna of using mushroom-like EBG structure, the antenna proposed in the paper has higher isolation and smaller dimension.**

***Key words:* - MIMO; Antenna; Fractal; EBG; WLAN; Wimax**

I. INTRODUCTION

The coupling between closely placed antenna elements is an important factor to the antenna performance, especially to the antenna used for the multiple-input-multiple-output (MIMO) wireless communication systems.

It is difficult to get high isolation between closely spaced antenna elements. Many researches have been studied. One solution reported often consisted in moving the antennas with different orientations around the printed circuit board (PCB) [1-3]. The best isolation values are always found when the antennas are spaced by the largest available distance on the PCB.

Other methods for improving the isolation have been reported. In [4], J. OuYang et al presents a slot structure perpendicular to current on the surface of patches to improve the isolation of a pair of closely spaced microstrip antennas on a common ground plane. Besides, a protruding T-shaped stub in the ground plane was used to improve the mutual coupling between antenna elements [5], [6]. Similarly, a T-shaped and dual-inverted-L-shaped ground branch was added to acquire low mutual coupling [7]–[11]. Unfortunately, the high isolation covering the wide band is difficult to realize.

In recent years, the electromagnetic band-gap (EBG) is applied to mitigate mutual coupling between the antennas. Most antennas designed mainly utilize the EBG matrix, offering an efficient means to reduce antenna coupling [12-14]. However, the proposed design occupied a large space between the antennas. In [15], the authors utilize a single lattice mushroom-like EBG between the antennas to reduce the coupling preferably and the size of the antenna.

In this paper, a fractal EBG structure is proposed to improve the Isolation of the MIMO antenna proposed in [15]. Except for the EBG structure, the antenna structure is similar to one in [15], which is based on a G-shaped monopole antenna structure with two back-to-back monopoles. The fractal EBG structure, instead of the lattice mushroom-like EBG is used in the new antenna to increase the isolation of the antenna. Also, several ground branches are introduced to increase the isolation or adjust the resonant frequency. Compared with the antenna in [15], the new MIMO antenna has better performances, higher isolation and the smaller size. The isolation for frequencies across 2100 to 2830 MHz(S11< -10 dB) increased 2dB to 21dB, and the size of the antenna decreases for the smaller EBG used in the antenna. The MIMO antenna designed in this paper can be used to WLAN and WiMax.

II. ANTENNA CONFIGURATION

The geometry of the proposed fractal EBG antenna is demonstrated in Figure 1(a). Figure 1(b) and 1(c) show the front side and back side of the antenna respectively, also labelling all the dimensions in detail. The new antenna is printed on an FR4 substrate with dimensions 99 mm×54 mm × 1.6 mm and relative permittivity of 4.4, which can be considered as the circuit board of a mobile handset. The ground with main sizes 80mm×54mm and two inverted L-shaped branches are printed on the back surface of the substrate. Two monopoles adopting folded techniques are designed to reduce the occupied area. Two symmetric back-to-back G-shaped patches, only 16 mm in height, are printed on the front side of the substrate. For convenience of design, the width of all antenna braches is set to be 2 mm in this paper except for the bottom branch of the G-shaped antenna, which has a width of 1.2 mm. The dual inverted L shaped ground is introduced in this design to obtain wide bandwidth. The detailed function analysis of the ground can be found in [8-11]. The lengths from the feeding point to the end of the metal are smaller than one-quarter wavelength of a conventional straight monopole in free space. This behaviour is largely due to the effect of the microwave substrate supporting the proposed monopole, which leads to decrease resonant length for the proposed monopole. This effect is also helpful for achieving a smaller antenna size for a fixed operating frequency.

Each monopole is directly fed using a 50Ω microstrip line printed on the front surface of the substrate. A Good impedance matching is obtained without adding additional matching circuitry. Between the two monopoles, there is a T-shaped ground plane protruding from the main plane (60×80

978-7-5641-4279-7

mm2). The protruded and main ground planes are both printed on the back surface of the substrate, and the inverted L-shaped ground plane comprises a central vertical strip and a top horizontal strip, both with a constant width of 4 mm. Note that the inverted L-shaped ground plane enhances the isolation between the two ports of the antenna.

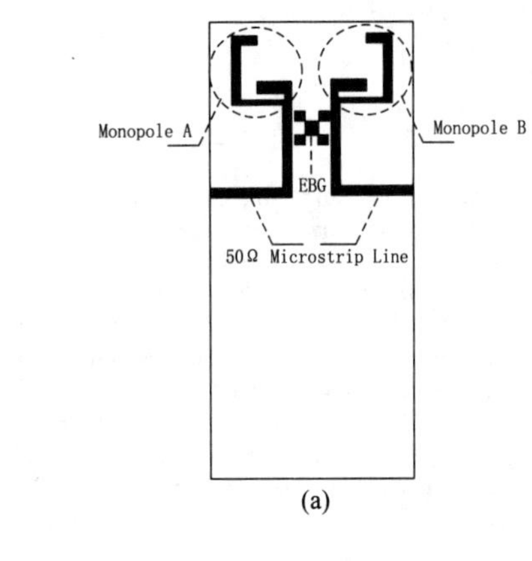

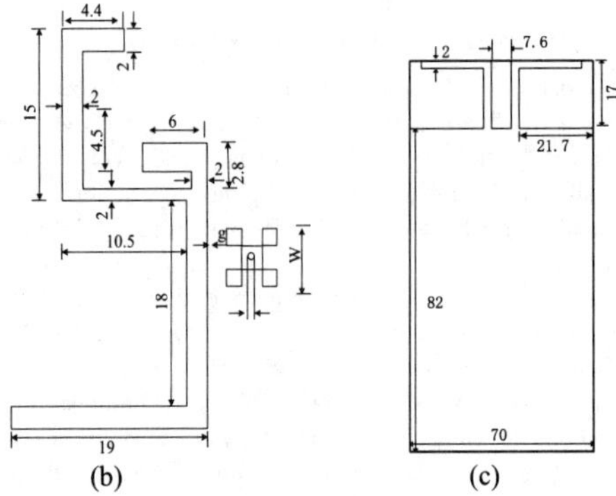

Fig. 1. Geometry of the proposed diversity antenna (a) General view, (b) Dimensions of monopole & EBG structure, and (c) Dimensions of the ground plane.

The EBG, called as fractal EBG, formed by a via-loaded metal patch and a Minkowski fractal dimension, is inserted between two microstrip feed lines of antenna. It can be regarded as an LC resonator with resonant frequency $f_r = 1/2\pi\sqrt{LC}$ .When a single cell is used, a narrow notch at f_r can be engendered.

III. EFFECTS OF THE GEOMETRICAL PARAMETERS OF THE PROPOSED ANTENNA

A. Effects of the Parameter W

The simulated return losses with changing W= 5mm, 5.5mm, 6mm, 6.5mm, 7mm are shown in Figure 2. It is revealed that the parameter W has more effects on the resonant frequency. Within the bandwidth (S11 < -10 dB), there is a notch corresponding to the arch in Figure 2. With the increasing of the parameter W, the value of S11 in the notch decreases, but bandwidth is not significantly changed. When W=6 mm, its return loss close to -22 dB, and the bandwidth keeps the performance all right.

Figure 3 gives the simulated results of the isolation between Port 1 and 2 with changing W= 5mm, 5.5mm, 6mm, 6.5mm, 7mm Within the bandwidth (S11 < -10 dB), with the increasing of the parameter W, the values of S21 is same basically. Considering the S11 and S12 synthetically, the optimal dimension W is chosen as 6mm.

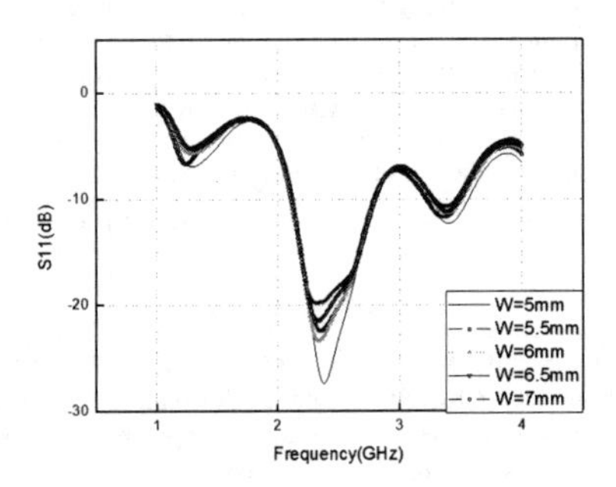

Fig.2. The return loss with different W

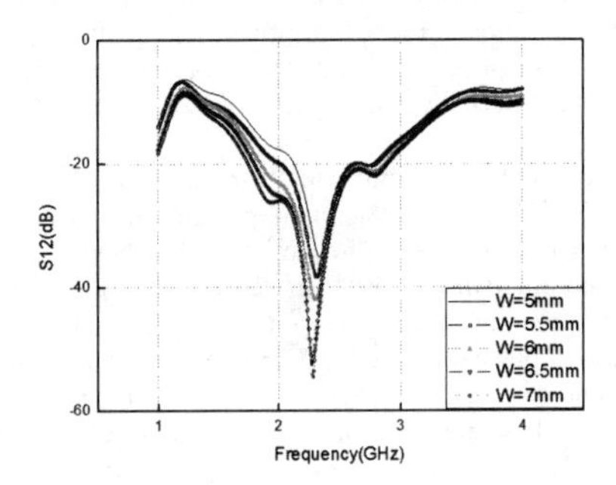

Fig. 3. The isolation between ports with different W

B. Effects of the Parameter g

The simulated return losses and isolation with changing g=0 mm ,0.1 mm ,0.2 mm ,0.3 mm ,0.4 mm are shown in Figure 4 and Figure 5 respectively. It is observed from the two figures, with the increasing of g, the values of the return losses and isolation improved well. Considering the S11, S21 and the size of the antenna synthetically, the optimal dimension g is chosen as 0.3 mm.

IV. RADIATION PATTERNS ANALYSE

The simulated radiation patterns of Port 1 and Port 2 excited at 2470GHz and 2700MHz are given in Figure 6 and Figure 7 respectively. Comparing Figure 6 and Figure 7, one can find that the patterns tend to cover complementary patterns, which provide pattern diversity for the system operation. All these demonstrate that the antenna can overcome the multipath fading problem and enhance the system's performance.

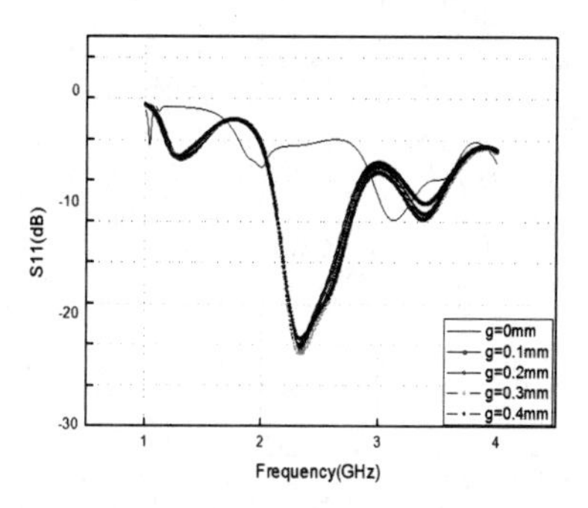

Fig.4. The isolation between ports with different g

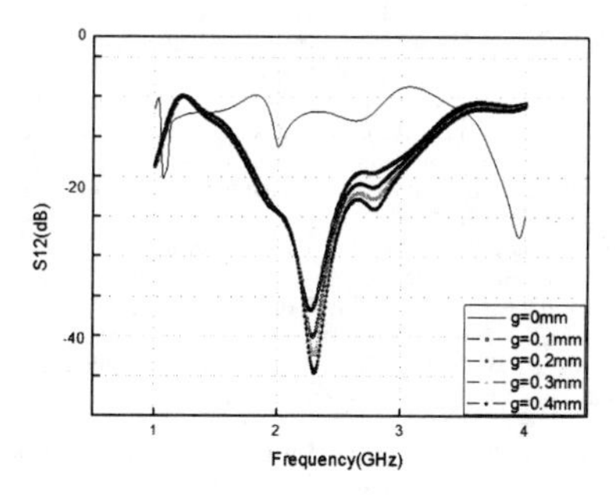

Fig. 5. The isolation between ports with different g

V. THE COMPARISON AND ANALYSIS OF THE ANTENNA PERFORMANCE WITH DIFFERENT EBG STRUCTURE

The comparison is performed between a Minkowski fractal EBG structure antenna (as shown in Figure 1a) and a square (mushroom-like) EBG structure one (as shown in Figure 8 in [15]). The simulated return losses and isolation at the optimal W and g=0.3mm are shown in Figure 9a and 9b, respectively. It can be found that the return losses of both are same at W=6mm of fractal EBG structure antenna and at W=7mm of square EBG antenna, while the isolation with fractal EBG (S12<-21dB) is better than that with the square one(S12<-19dB).

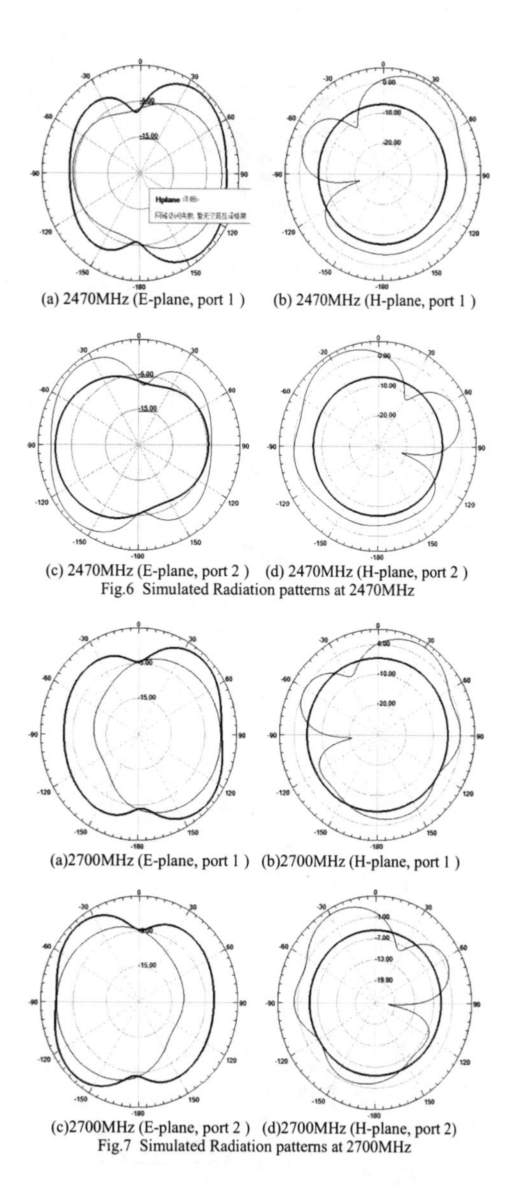

(a) 2470MHz (E-plane, port 1) (b) 2470MHz (H-plane, port 1)

(c) 2470MHz (E-plane, port 2) (d) 2470MHz (H-plane, port 2)

Fig.6 Simulated Radiation patterns at 2470MHz

(a)2700MHz (E-plane, port 1) (b)2700MHz (H-plane, port 1)

(c)2700MHz (E-plane, port 2) (d)2700MHz (H-plane, port 2)

Fig.7 Simulated Radiation patterns at 2700MHz

VI. CONCLUSION

A novel high isolation MIMO antenna has been proposed and studied in this paper. On the basis of simulated results, the optimal values of the parameters are chosen and the radiation mechanisms are studied. The final simulated bandwidth (S11) is 2100 to 2830MHz with high isolation(S12<-21dB) for frequencies across the WLAN and WiMax bands, the size of the antenna has also been decreased effectively. Compared with similar antennas, the proposed antenna can provide higher isolation and smaller dimension in real applications.

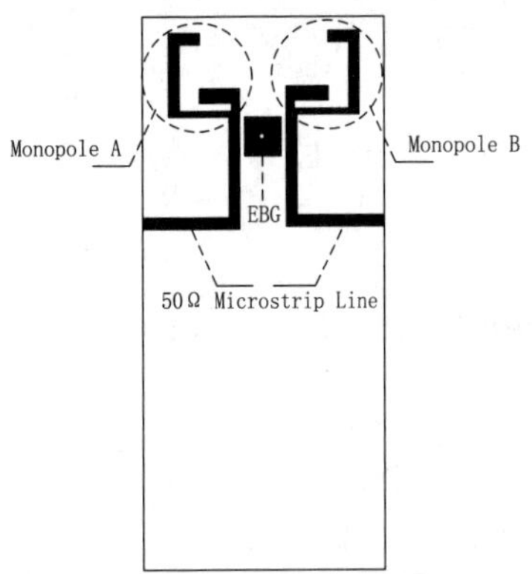

Fig.8. Geometry of square (mushroom-like) EBG structure antenna

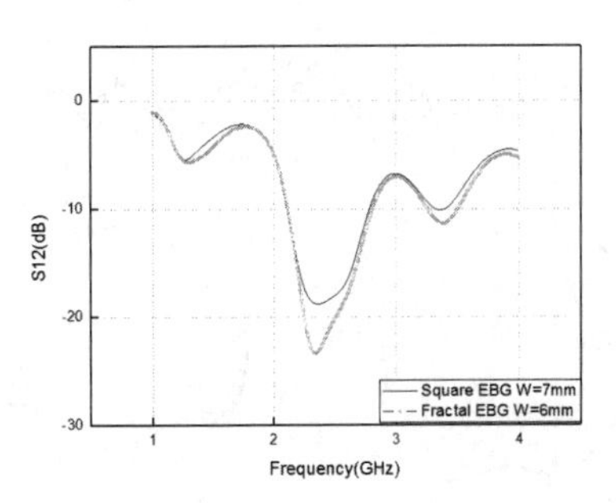

(a) The return loss

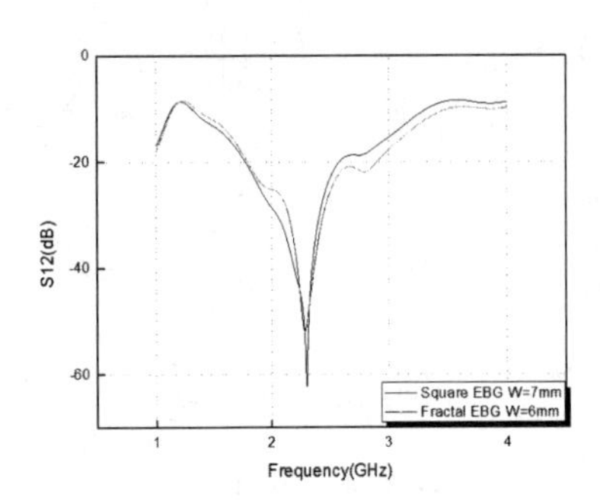

(b) The isolation

Fig. 9 The comparison of the simulated return losses and isolation of both antennas

ACKNOWLEDGMENT

This paper is supported by the Yunnan High-Tech Talents Recruitment Project 2012HA005， National Science and Technology Major Project (NO: 2010ZX03007-001) and Key Project of International Cooperation of Yunnan (2009AC010).

REFERENCES

[1] C. C. Chiau, X. Chen, and C. G. Parini, "A compact four-element diversity-antenna array for PDA terminals in a MIMO system," Microwave and Optical Technology Letters, vol. 44, no. 5, pp. 408 - 412, March 2005.

[2] Jingli Guo, Yanlin Zou, and Chao Liu, "Compact Broadband Crescent Moon-Shape Patch-Pair Antenna," IEEE Antennas and Wireless Propagation Letters, vol. 10, pp. 435 - 437, 2011.

[3] C. Yang, J. Kim, H. Kim, J. Wee, B. Kim, and C. Jung, "Quad-Band Antenna With High Isolation MIMO and Broadband SCS for Broadcasting and Telecommunication Services," IEEE Antennas and Wireless Propagation Letters, vol. 9, pp. 584 - 587, 2010.

[4] J. OuYang, F. Yang, and Z. M. Wang, Reducing Mutual Coupling of Closely Spaced Microstrip MIMO Antennas for WLAN Application[J], IEEE Antennas and Wireless Propagation Letters, vol. 10, pp. 310 - 313, 2011.

[5] G. Chi, B. Li and D. Qi, "Dual - band printed diversity antenna for 2.4/5.2 - GHz WLAN application," Microwave Opt Technol. Lett., vol. 45, no.6, pp.561-563, June 2005.

[6] Teng Guo, Dongya Shen, Shihong Zhu, Xiupu Zhang and Shaojie Li, "A dual-band printed diversity antenna for UMTS and 2.4/5.2-GHz WLAN application," 2011 Global Mobile Congress (GMC), ShangHai,China, vol. 1, pp. 1-8, October 2011.

[7] Yaxing Cai, Zhengwei Du, and Ke Gong, "A novel wideband diversity antenna for mobile handsets," Microwave and Optical Technology Letters, vol. 51, no. 1, pp. 218 - 222, January 2009.

[8] Xuan Wang, Zhengwei Du, and Ke Gong, "A Compact Wideband Planar Diversity Antenna Covering UMTS and 2.4 GHz WLAN Bands," IEEE Antennas and Wireless Propagation Letters, vol. 7, pp. 588 - 5914, 2008.

[9] Zhengyi Li, Xuan Wang, Zhengwei Du, and Ke Gong, "A novel printed dual-monopole array with antenna selection circuit for adaptive MIMO systems," Microwave and Optical Technology Letters, vol. 50, no. 6, pp. 1584 - 1590, June 2008.

[10] Yuan Ding, Zhengwei Du, Ke Gong and Zhenghe Feng, "A Four-Element Antenna System for Mobile Phones," IEEE Antennas and Wireless Propagation Letters, vol. 6, pp. 655 - 658, 2007.

[11] Dongya Shen, Teng Guo, Fuqiang Kuang, Xiupu Zhang, Ke Wu, " A Novel Wideband Printed Diversity Antenna for Mobile Handsets" , Vehicular Technology Conference (VTC Spring), 2012 IEEE 75th

[12] Lei Qiu, Fei Zhao, Ke Xiao, Shun-Lian Chai, and Jun-Jie Mao, "Transmit - Receive Isolation Improvement of Antenna Arrays by Using EBG Structures," IEEE Antennas and Wireless Propagation Letters,

[13] Assimonis, S. D.;Yioultsis, T. V.;Antonopoulos, C. S. "Computational Investigation and Design of Planar EBG Structures for Coupling Reduction in Antenna Applications, " Magnetics, IEEE Transactions on , Vol.48, No.2, 2012

[14] Margaret, D.H. Subasree, M.R., Susithra, S.,Keerthika, S.S.,Manimegalai, B. "Mutual Coupling Reduction in MIMO Antenna System using EBG Structures" , Signal Processing and Communications (SPCOM), 2012 International Conference on

[15] Teng Guo, Dongya Shen, Wenping Ren, Xiupu Zhang, A High Isolation MIMO Antenna for WLAN and WiMax, 2013 IEEE International Symposium on Antennas and Propagation , Orlando, USA, July, 2013 IEEE Press.

[16] Guangtao Wang, Dongya Shen, Xiupu Zhang, An UWB antenna using modified Sierpinski-carpet Fractal Antenna, 2013 IEEE International Symposium on Antennas and Propagation, Orlando, USA, July, 2013 IEEE Press.

Conformal microstrip circularly polarization Antenna array

Xiao-Bo Xuan ,Feng-Wei Yao ,Xiao-qing Tian
Shanghai Key Laboratory of Electromagnetic Effect for Aerospace Vehicles , Shanghai, 200438,China
jojoyao@163.com

Abstract- **In this paper, the conical conformal antenna array is proposed, which employs a feeding network and three double feed square circularly polarization patches. Good agreement between the theoretical and experimental results is obtained .Its measured 3dB axial ratio bandwidth is about 270MHz , covering 1.37GHz-1.64GHz.The measured radiation pattern of antenna is also presented, which is almost onmidirectional.**

I. INTRODUCTION

As the major part of wireless detection system, the design of antenna is essential to the performance of detection system. One of the most important innovations in technology of modern antenna lies in the introduction of the conformal antenna design to many applied areas ,such as radar, data links, mobile cellular base station ,communicative terminals and so on[1].

The conformal antenna usually use the microstrip antenna or slot antenna because of their low profile structure .The microstrip antennas are well known for the characters of lightweight ,low-profile ,low cost advantages and their efficiency is increased by using arrays. Moreover, in the case of using on small vehicles, and considering the aerodynamics character of a flying high-speed aircraft ,microstip conformal arrays are widely used.

There are several papers that contribute to conformal microstrip antenna [2,3,4], but most of them are based on cylinder surface,sometimes a long cone with a small apex angle approximates to a cylinder .Compare to cylinder conformal antennas, the conical conformal ones are much more complex to analyze, design, and produce . On the other hand ,most conformal antenna was studied in the field of narrow circular polarization antenna which adopt single feed in the middle side of patch and two cut corners .

In this paper, a conical conformal circularly polarization array is presented, which is made up of feeding network and 3 square patches with two feed points symmetrically on the two main axis . The simulated and measured radiation patterns at center frequency are both presented. Its structure and experimental results are presented as follow.

II. DESCRIPTION OF THE ARRAY

As shown in Fig.1(a), the circular polarization antenna array has been realized by feeding network of three equal way and three square patches, which are fed by two points symmetrically on the two main axis . To maintain the amplitude and phase relationship between the two points, every patch employ a Wilkinson power divider ,with its output feedlines having a length difference of a quarter-wavelength to produce a 90° phase shift as shown in Fig1(b).A 150Ω chip resistor is added in power divider for achieving good isolation between the two feeds.

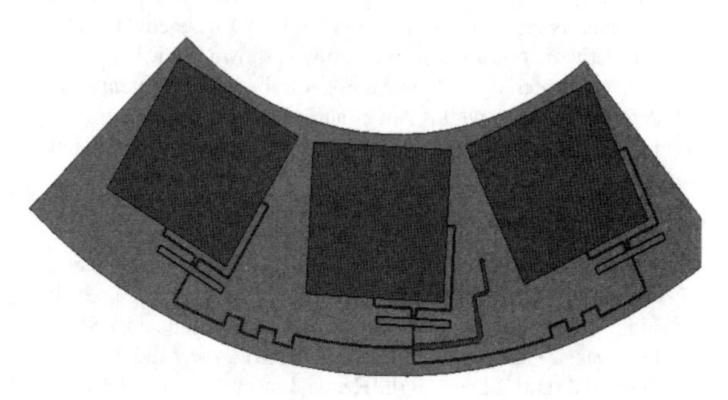

(a) Geometry of planar antenna array

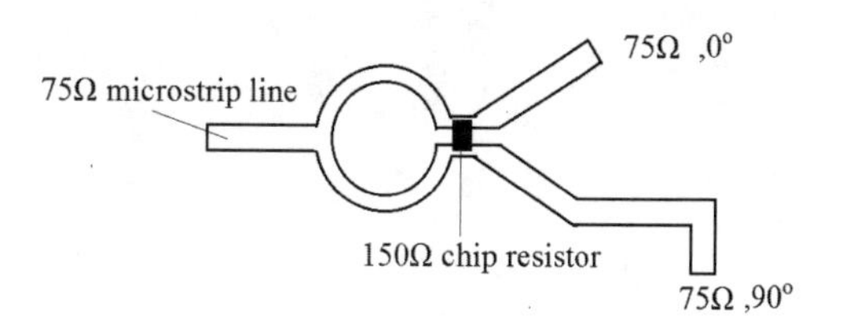

(b)The wilkinson power divider

978-7-5641-4279-7

(c) Geometry of conformal antenna array

Figure 1. Structure of the proposed antenna

Considering the mass of antenna should be reduced and antenna should be curved for conformal of a finite length mental cone, the the thickness and the relative permittivity of substrate is chose to be 0.508mm and 2.33 respectively .The geometry of conformal antenna array is shown in Fig1(c).

After material and thickness of base for antenna is determined,the width of antenna is obtained by formula[6],which is emulated by Ansoft HFSS and modulated to be 61.72mm.

III. SIMULATED AND MEASURED RESULTS

The conical conformal antenna is simulated by using Ansoft HFSS ,the simulated results are shown in Fig.2and Fig.3 ,the central frequency is 1.575GHz,the simulated VSWR less than 2:1 is from 1.33GHz to 2.10GHz, the bandwidth of 3dB axial ratio is about 360MHz,axial ratio is about 1.25dB at $f = 1.575GHz$,and gain of the antenna array is about 0dB at maximum radiation direction, as shown in Fig.4.

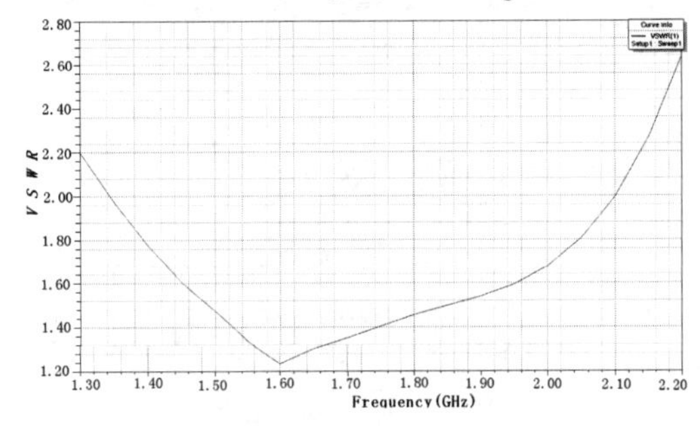

Figure 2 Simulated VSWR vs. frequency

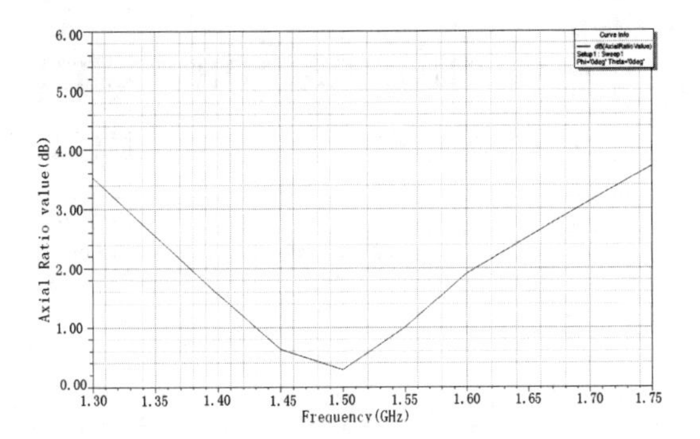

Figure 3 Simulated Axial ratio vs. frequency

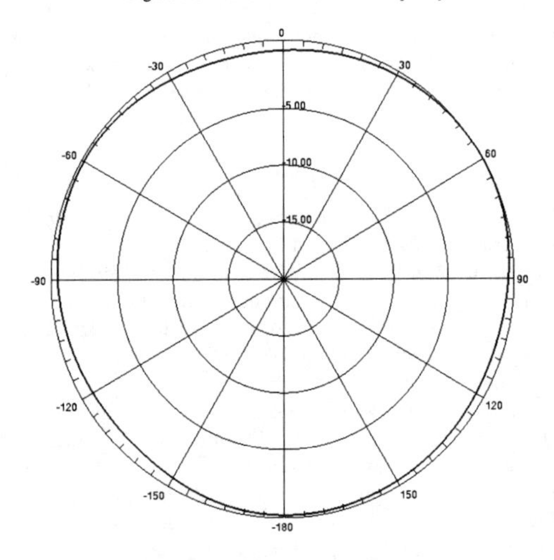

Figure 4 Simulated radiation pattern at f =1.575GHz

A test antenna was fabricated, the VSWR of the conformal microstrip antenna array is measured using the Agilent 8722ES Network Analyzer,the measured VSWR less than 2:1 is from 1.33GHz to 2.05GHz.Some discrepancy between the theory and experiment maybe due mainly to the effect of the SMA connector, which has not been considered in the simulation but inducts a reactance and therefore influences the positions of the resonant points.

The measured axial radio bandwidth became narrow, covering from 1.37 GHz to 1.64GHz.

The radiation patterns were measured in an anechoic chamber.The measured radiation pattern is nearly omnidirectional as shown in Fig.5 and the gain of maximum radiation direction reduce to -0.7dB at $f = 1.575GHz$, and there is about 2dB discrepancy around the cone.

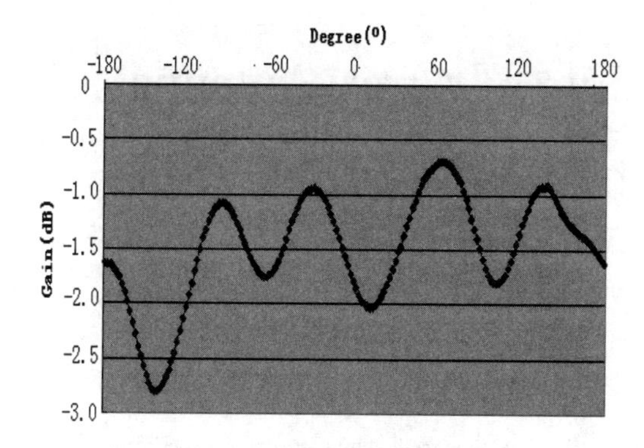

Figure 6 measured radiation pattern at f =1.575GHz

IV. CONCLUSION

This paper proposes a design of dual feed circularly polarized conical conformal antenna array .The performance of conformal antenna is studied and results are verified by experiment. The measured radiation pattern is almost omnidirectional as simulated one. The impedance bandwidth and the measured gain is almost similar with the simulated results. As the antenna structure is simple and its section plane is low, it can be widely used in curving surface and pneumatic structure.

REFERENCES

[1]Jun-wei Niu and Shun-shi Zhong,"Cylindrical conformal bow-tie microstrip antennas small curvature radias",MIcrowave Opt Technol Lett,2003,39(6),pp.511-514

[2]Qun Wu,;Zi-Rui Feng, ,"Design of millimeter-wave conformal microstrip antenna arrays", Microwave Conference, 2008. APMC 2008. Asia-Pacific

[3]Kaifas, T.N.; Sahalos, J.N.,"On the design of cylindrical conformal microstrip antenna arrays", Proceedings of the Second International Symposium of Trans Black Sea Region on Applied Electromagnetism, 2000,pp18

[4] Mao Kanghou; Xue Menglin," A study of conformal microstrip antenna array on a cylinder", Antennas, Propagation and EM Theory, 2000. Proceedings. ISAPE 2000. 5th International Symposium,2000,pp18-21

[5] Liu Zong-quan,Qian Zu-ping,"Design of a circularly polarized conformal microstrip antenna with CPW-fed", 2011 International Conference on Electronics, Communications and Control(ICECC)pp206-209.

[6] Yang-Kai Wang,Hua-Ming Chen,"A circularly polarized square-ring patch antenna for GPS application,"Antenna and propagation Society International Symposium,2008.AP-S 2008,pp1-4.

Detect Phase Shifter Module of Array Antenna Using The Sun

Quan Gang
Troops 63623 of PLA
Jiuquan China

Jiang Wenzi
Troops 63623 of PLA
Jiuquan China

Xu Peng
Troops 63623 of PLA
Jiuquan China

***Abstract*—The Sun is an emission source which has a wide spectrum and high-energy flow, it can be used as a calibration source; The phase shifter module is the minimum detection unit of the phase array antenna, the paper propose two evaluation parameters, which by receiving the energy of the solar radio, then compare the receiving power of each channel, that can get the two evaluation parameters, those two parameters shows the performance of each phase shifter module; Finally, verify the effectiveness of the method by a simulation experiment.**

Keywords-detection of phase shifter module, solar radio, evaluation parameter

I. INTRODUCTION

The antenna of Phased array radar is generally made up of number of phase shifter models, Currently, detect antenna array phase shifter model, depending on the test signal source and the placement of monitoring systems, can be divided into "External Detecting" and "Internal Detecting". When use the "External Detecting" method, the test signal source placed outside of the antenna array, it can be placed to the far field or near field; when use the" Internal Detecting" method, the test signal placed intside of the antenna array, it can be radar transmitters or a special test signal source. The two detection methods need number of hardware, such as network analyzer, recorder of the lobe and anechoic chamber, generally near-field test equipment and accompanying software not equipped at the range and practical application.

The sun is a continuous spectrum radio stars, with a wide frequency spectrum and high radio flux density, is a natural test signal source. With the development of radio astronomy, the sun more and more being used in the calibration and testing of radar and telemetry equipment. Puhakka, perform an experiment that calibrate the radar using the solar radio[1], Helsinki University use the sun on weather radar calibration[2], A.G.Stov use the solar noise calibrate tactical radar[3]. In this paper, the sun is using as a test signal source to detect the phase shifter module, first introduced the radio characteristics of the sun, then two evaluation parameter propose for the phase shifter module, and then describes the detection method, finally, confirm the availability of the method by a simulation experiment.

II. SOLAR RADIO FEATURES

In 1942, scientists have discovered the solar radio burst, the burst radio frequency covered the whole frequency domain, in the microwave band (wavelength 0.3 to 10cm), the solar radio burst has a strong signal power and a slow change [4]. Astronomically solar radiation flux density usually present as Solar Flux, it is defined as the power received in the normal direction of the ground-rays per unit frequency per unit area, and its unit is W / (m2.Hz) [5]:

$$S(f)=\left[S(f_j)/S(f_k)\right]^{\Gamma(f)}S(f_k),\Gamma(f)=\frac{\log(f/f_k)}{\log(f_j/f_k)}$$

$S(f_j)$ and $S(f_k)$ are the flux density at frequencies of f_j and f_k which are measured by the astronomy station.

Somewhere in a moment the sun elevation and azimuth angles can be calculated by the following formula [6]:

$$\sin\varepsilon=\sin\delta\sin I_{at}+\cos\delta\cos I_{at}\cos S\ ,$$

$$\sin\beta=\sin S/\left(\cos S\sin I_{at}-\tan\delta\cos I_{at}\right).$$

Where, δ is the declination of the sun, I_{at} is the local geographical latitude, in north the latitude is positive, in south the latitude is negative, S is the sun corner.

When a receiving system receives the solar radiation, the signal noise ratio can be calculated:

$$SNR=W_S/W_N\ ;\ W_S=A_e\cdot Sv/2\ ;\ W_N=2KT_r/\sqrt{\Delta f\tau}\ .$$

III. SHIFT THE PHASE DETECTOR MODULE EVALUATION

Phase shifter Module is the minimum unit of antenna array that can detected, by comparing the power of the receiving channel, it can determine the module have fault phase shifter or not, but can not obtain the location information of the fault phase shifter, so that if there is a fault phase shifter its coordinates in the module is randomly distributed, in order to evaluate the performance the module, two evaluation parameters are proposed, which are (α_k, β_k).

Let M be the total number of phase shifters of the sub-array, the total number of modules in the sub-array is K, S_K, k=1~K is the number of phase shifter in the module k that work

978-7-5641-4279-7

normally, $S=S_1+S_2+\cdots+S_K$ is the number phase shifter in the sub-array that work normally. Then:

$$\alpha_k = S_k / S, \quad k=1\sim K$$

$$\beta_k = S_k /(M/K) = K S_k / M$$

α_k is the relative parameter of the module, it represents the ratio of the number of normal phase shifters in the module and the total number of phase shifters in the sub-array, it calls as the relative parameter; β_k is the absolute parameter of the module, it shows the ratio of the number of normal phase shifters and the total number of phase shifters in the module, called as absolute parameter.

Where S_k and S can not be obtained directly, while α_k can be calculated by comparing the power that received by the channel. The relative parameter can show the module is good or bad. As long as the absolute parameter obtained, it can determine the number of normal phase shifters in the module, when β_k <90%, this module should be replaced.

IV. Phase Shifter Module Detection Methods

In order to get (α_k, β_k), there are two stages, first stage, to estimate the relative number of normal phase shifter for each sub-array; the second stage, estimate the evaluation parameters of the module.

The steps of the first stage are following:

First, the antenna toward the sun, zero-phase all the antenna array;

Second, at the output port of each receiver channel, estimate the received power, it can use the following formula to calculate the received power:

$$\sigma_n^2 = \sum_{q=0}^{Q-1} \left| x_{nq} \right|^2, n = 1 \sim N$$

Where: n -Receiver channel number; $|x_{nq}|^2$ - in the span of q second, the square of the modulus of the n-th channel output signal sampling (instantaneous solar noise energy); Q is theNumber of samples.

Third, $a_n=\sigma_n^2/\sigma_{sp}^2$ (σ_{sp}^2- power received by standard channel), v_n=SQRT(a_n), when $a_n \leqslant 1$, v_n=1, when $a_n > 1$. Standard channel is that all phase shifter are normal in the sub-array, it can confirm through numerous experiments or that the channel which received the largest power is the standard channel.

The steps of the second stage are following:

First, the antenna toward the sun, zero-phase all the antenna array;

Second, estimate the power of the sub-array received.

Third, phase the module in the sub-array contrarily one by one, that is, a module phase π, the rest phase zero, then estimate the power of the sub-array received, σ_{nk}^2, $k=1\sim K$.

Ignore the Amplitude and phase errors of the shifter, and the SNR of the solar noise power and receiver noise is much larger than 1, the time span of the power receiving is long enough, so that $X_k=S_k/S=\alpha_k$ can meet the linear equations with the restrictions of $X_1+X_2+\cdots+X_K=1$, $AX=a$, that:

$$\begin{pmatrix} 1 & 0 & 0 \\ 0 & \cdot & \cdot \\ 0 & \cdot & 1 \\ 1 & 1 & 1 \end{pmatrix}_{(K+1)\times K} \cdot \begin{pmatrix} X_1 \\ X_2 \\ \cdots \\ X_K \end{pmatrix}_{K\times 1} = \begin{pmatrix} a_1 \\ \cdot \\ a_K \\ a_{K+1} \end{pmatrix}$$

That, a_k=0.5(1-σ_{nk}/σ_n), $k=1\sim K$, $a_{K+1}=1$. Using the Least squares method to solve the above equation can be obtained:

$$X=(A^{\mathrm{T}}A)^{-1}A^{\mathrm{T}}a$$

By calculating:

$$X = \frac{1}{K}\begin{pmatrix} K & -1 & \cdots & -1 & 1 \\ -1 & K & \cdots & \cdots & 1 \\ \cdots & \cdots & \cdots & -1 & \cdots \\ -1 & \cdots & -1 & K & 1 \end{pmatrix} a$$

Then we can get the absolute parameter of any module in any sub-array:

$$\beta_{nk} = 4 v_n X_k$$

V. Numerical Simulation and Analysis

For the purpose of studying the effectiveness of the detection algorithm, a simulation Experiment Conducted, make the following assumptions: phase shifter 2 bits, Δf=300kHz, T_r=200K, G=100dB, T_0=290K, τ=0.1s, S_v=37A.U., A_e=0.3m^2, Q=1000~400, SNR_{out}=14dB.

In order to simulate the actual situation, the following mathematical model approach used to calculate the power of the receiver:

$$U_{nq} = \xi_q(1+\lambda_n)\left[\sum_{v}(1+\varepsilon_{nv})\exp(i\Delta\phi_{nv})\exp(-i\psi_{nv}) + \sum_{\mu}(1+\varepsilon_{n\mu})\exp(i\Delta\phi_{n\mu})\exp(-i\psi_{n\mu})\right] + \zeta_q$$

Where:

$n=1\sim N$, $q=0\sim(Q-1)$;

ξ_q—single form the sun, ξ_q= $SNR_{out}\cdot(N_{in}\cdot G+\Delta N)$, variance σ_ξ^2 = SNR_{out}/M (dB);

ζ_q—receiver noise, $\Delta N = K_{Berzman}\Delta f G$, variance 1dB;

λ_n—normalized relative deviation of the n-th receiver's magnification factor, take ±0.3%;

ε_{nv}, $\Delta\varphi_{nv}$, $\varepsilon_{n\mu}$, $\Delta\varphi_{n\mu}$—modulus value and phase deviations of the Phase shifter transmission coefficient, these deviations are evenly distributed in the range of some value (modulus value approximately ±0.3, phase ±20°).

ψ_{nv}—phase distribution of normal phase shifter, for the detection module, at the first stage ψ_{nv}=0, at the second stage $\psi_{nv}=\pi$; The other three modules, the first and second stages areψ_{nv}=0;

$\psi_{n\mu}$—phase random value of Fault phase shifter, take one of four values 0 ,±2/π, π with probability of 0.25.

Simulations are proposed at the assumption of several different Failure rate: Failure rate, simulation, the results are as follows:

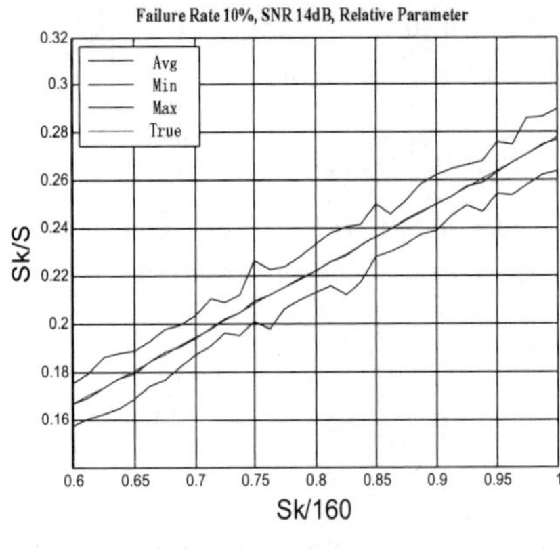

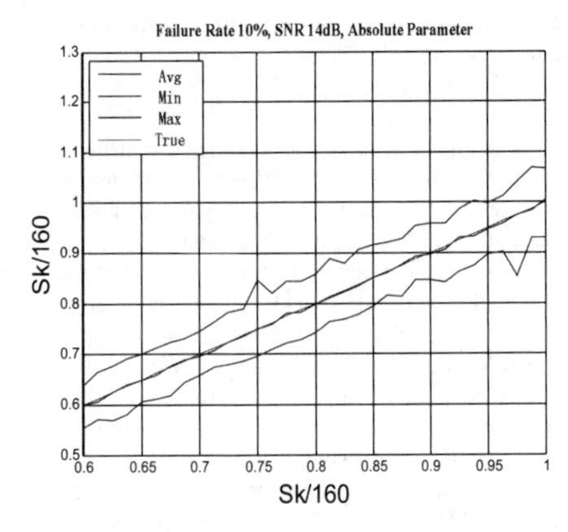

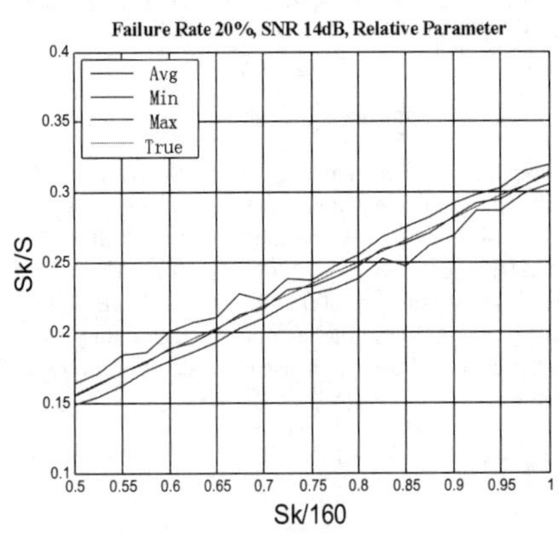

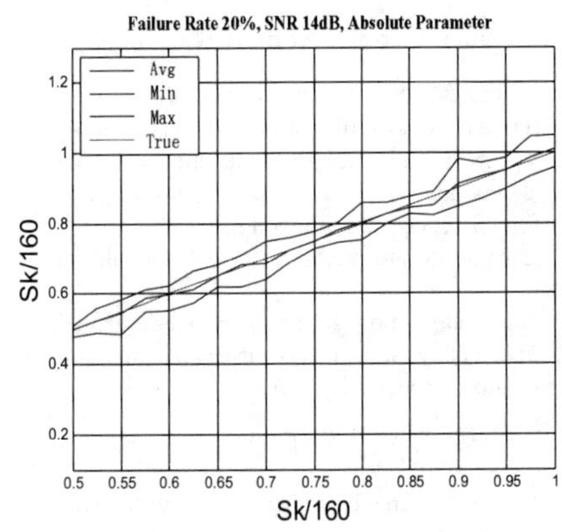

Figure.1 simulation results at failure rate of 10%, 20%

The short dashed line in the figure shows the average of 50 times, the dotted line for the minimum of 50 times, long-dashed line for the maximum of 50 times, solid line is the true value, that the average curve and the true value curve coincide. The variance of the two parameters, as shown in Figure 2.

It can be seen that the root mean square error of the relative parameter is 0.5%, the root mean square error of the absolute parameter is 2.5%, relative parameter has higher accuracy, which is the significance of studying it.

To Improve the accuracy of detection, make more experimental then take the average, for example, take 100 times, ignoring the running time of Program, then the time required is about 1.6~6.4 second.

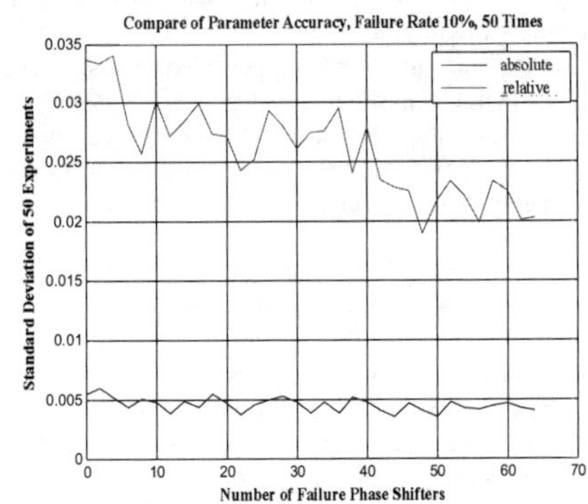

Figure.2 The variance of the two parameters

VI. Conclusion

Detect the phase shifter module use the sun is feasible and accurate, relative parameter and absolute parameter has a certain practical significance for the evaluation of the phase shifter module, the simulation results shows that, the relative parameter is more precise, but it is not a true reflection of the module, the error of the absolute parameter is relatively larger, but after several experiments, the average of the absolute parameter reflect the true state of the fault phase shifter module, therefore, the evaluation of comprehensive the two parameters on the module is more accurate and effective, on the other hand, detect the phase shifter module use the solar noise also fast and easily.

References

[1] P.Puhakka, M.Leskinen, and T.Puhakka Experiments on using the Sun for radar calibration, Proceedings of ERAD (2004): 335–340;

[2] Pekka, V.S.Puhakka, SOLAR CALIBRATIONS WITH DUAL POLARIZATION WEATHER RADAR, University of Helsinki, Finland;

[3] A.G. Stove, Calibration of Tactical Radar Antennas using the Sun - Simulation and Analysis Thales Aerospace Manor Royal Crawley RH10 9PZ;

[4] Wang Shouwan, "Solar radio emission theory", Science Press, Beijing, 1973;

[5] Chen Meixia, the application of the sun in antenna measuring, Gui Lin University of Electronic Technology, 12, 1986;

[6] Ma Jian, high accuracy sun position algorithm and application in solar power generation, Water Resources and Power, Vol 26, No. 2, 2008;

The Numerical Simulation Studies of The Electromagnetic Wave Resistivity Logging-While-Drilling Instrument antennas Testing In The Calibrating Tank

A.Z. Li, J. Zhu, T. T. Zhang, and G.H. Yang
China Petroleum Logging Co. Ltd
Xi'an Shanxi, 710065, China

Abstract : The testing problems for The Electromagnetic Wave Resistivity Logging-While-Drilling Instrument in the calibrating tank are studied. The calibrating tank is a testing and calibrating device in which the water solution resistivity can modulate, wherein the calibrating barrel diameter of 3 meters, 4 meters high. Testing The Electromagnetic Wave Resistivity Logging-While-Drilling Instrument in the device is simulated and computed by the 3-D numerical simulation software MAXWELL, the testing result and the result calculated by MAXWELL are compared and we can find that the error between them is small. The testing charts for The Electromagnetic Wave Resistivity Logging-While-Drilling Instrument in calibrating tank are set up. These charts can guide the instruments testing.

Key words: The calibrating tank, The Electromagnetic Wave Resistivity Logging-While-Drilling Instrument, The MAXWELL simulation software, Testing chart

I. INTRODUCTION

Because the antenna of The Electromagnetic Wave Resistivity Logging-While-Drilling Instrument is unsymmetrical , the three-dimensional numerical simulation calculations is needed. 2010,Gao Jie[1-2] had done a detailed analysis of Numerical simulation in electrical logging, he pointed out that the finite element method applicated better in three-dimensional numerical simulation. The Maxwell Ansoft simulation software based the finite element method, it has convenient and reliable mesh generation, post-processing and post-processor. In this paper, the Maxwell Ansoft simulation software is used for calculating problems. Firstly, the simulation results were compared with the analytical solution, error is small; secondly, the simulation results were compared with the testing results that are observed by The electromagnetic wave resistivity logging-while-drilling instruments, error is small; Last, the testing charts in the calibrating tank are drawed.

II. INTRODUCTION

A. *The calibrating tank*

By putting the electromagnetic wave resistivity logging-while-drilling instruments in the changing conductivity salt test solution in the calibrating tank , we can observe many groups testing results while the conductivity changed like testing in the complex formation. The calibrating tank mainly comprises scale barrel, wooden ladder platform, lifting equipment, solution preparation system. Shown in figure 1. wherein the calibrating barrel diameter of 3 meters, 4 meters high. It filled with saline solution of which the resistivity can be changed by the solution preparation system. The range of solution resistivity is 0.2-2000Ω·m.

When the instruments are testing in the calibrating tank, firstly the instruments antenna testing point must be found, secondly, the relative position between the solution face and the instrument uphole face must be calculated, last, lift the instruments in the calibrating tank by the lifting equipment, note that the instruments antenna must be completely covered by the solution and placed centrally. Connect the cable and computer, open the test software, and test. After the completion of the test, the instrument is hung out, a measurement point of saline solution will be prepared and mixed uniform.

Fig. 1 The calibrating tank

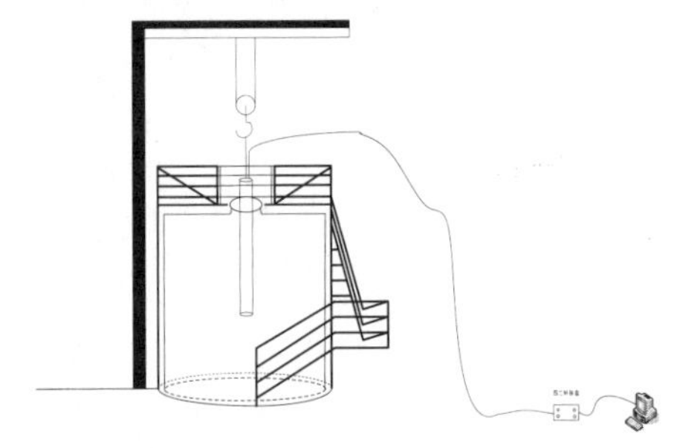

Fig. 2 The Electromagnetic Wave Resistivity Logging-While-Drilling Instrument is working in the calibrating tank.

978-7-5641-4279-7

B. The Electromagnetic Wave Resistivity Logging-While-Drilling Instrument

The Electromagnetic Wave Resistivity Logging-While-Drilling Instrument through the antenna system exciting and receiving high frequency electromagnetic wave. When electromagnetic wave propagats in lossy media, energy will attenuated which causes the phase difference and amplitude attenuation between the two receiving coils[3-4]. The Electromagnetic Wave Resistivity Logging-While-Drilling Instrument uses 2MHz and 400KHz two kinds of frequency, Antenna system uses four transmit（T1、T2*、T3*、T4）and two receive（R1, R2）structure. Shown in figure 3. When testing, T1、T2*、T3*、T4 emit high frequency electromagnetic waves into the formation and the receiving antenna R1、R2, record their phase difference and amplitude ratio.

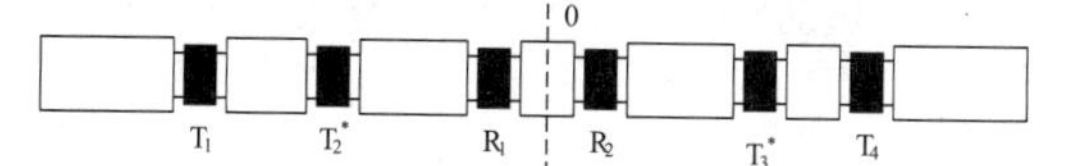

Fig. 3 Antenna system structure diagram of the electromagnetic wave resistivity logging-while-drilling instruments

III. THE ACCURACY VALIDATION OF THE MODEL

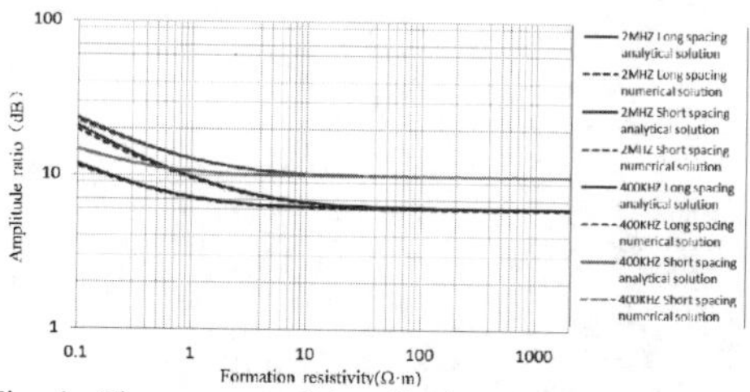

Fig. 4 The contrast of (amplitude ratio) analytical solution and the numerical solution of The Electromagnetic Wave Resistivity Logging-While-Drilling Instrument.

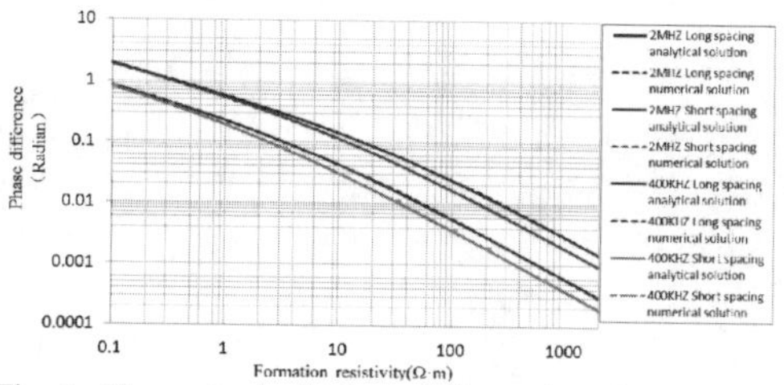

Fig. 5 The contrast of (phase difference) analytical solution and the numerical solution of The Electromagnetic Wave Resistivity Logging-While-Drilling Instrument.

In order to verify the numerical simulation accuracy, assume The Electromagnetic Wave Resistivity Logging-While-Drilling Instrument only has antenna system, we can use the analytical method and the finite element numerical simulation method to calculate the numerical solution and analytic solution to compare. Set the homogeneous formation resistivity 0.1 Ω• m ~2000 Ω•m. The transmitting current of 1 A. Shown in figure 4 and 5, the error between the calculated value and the analytical is small, this verify the correctness of the method and program.

IV COMPARING THE NUMERICAL SIMULATION RESULTS WITH THE ACTUAL MEASURED RESULTS IN THE CALIBRATING TANK

Put The Electromagnetic Wave Resistivity Logging-While-Drilling Instrument in the calibrating tank and test. The resistivity points of solution in the barrel are 0.34、1.28、5.21、13.28、19.417Ω·m. The actual measurement results and numerical simulation results are shown in Table 1. With the increase of salt water resistivity, amplitude ratio resistivity increases, and deviates from the resistivity points of solution; the phase difference resistivity increases also and resistivity deviates from the resistivity points of solution, but phase difference resistivity deviating from the resistivity points of solution is slower than amplitude ratio resistivity.

Table 1 The comparison of actual measurement results and the results of numerical simulation in the tank

The comparison of actual measurement results and the results of numerical simulation in the tank								
Solution resistivity（Ω·m）	2MHZ short spacing amplitude ratio resistivity（Ω·m）		2MHZ short spacing phase difference resistivity（Ω·m）		400KHZ long spacing amplitude ratio resistivity（Ω·m）		400KHZ long spacing phase difference resistivity（Ω·m）	
	Simulation results	Actual testing results	Simulation results	Actual testing results	Simulation results	Actual testing results	Simulation results	Actual testing results
0.34	0.3401	0.35	0.34	0.35	0.3215	0.32	0.3323	0.34
1.28	1.269	1.29	1.2919	1.36	2.3166	2.27	1.1685	1.22
5.21	6.0086	5.84	4.8777	4.91	22.1517	21.5	7.2426	7.42
13.28	28.9576	30.7	13.6942	14.3	73.843	72	22.8988	22.5
19.417	47.6466	49.7	21.3644	22.8277	110.2376	109.0961	35.2427	34.5092

V TESTING CHARTS

The test charts of The Electromagnetic Wave Resistivity Logging-While-Drilling Instrument in scale barrel show in Figure 6 and Figure 7, the horizontal coordinate is saline solution resistivity, the vertical coordinate is amplitude ratio resistivity or phase difference resistivity. As can be seen from Figure 6, in the low resistivity region, water resistivity and amplitude ratio apparent resistivity are close, but in the high resistivity region, solution resistivity deviates from and the amplitude ratio resistivity largely. The face made of the 2MHz short spacing curve and 400KHz short spacing curve is similar to the face which is made of the 2MHz long spacing curve and 400KHz long spacing curve.

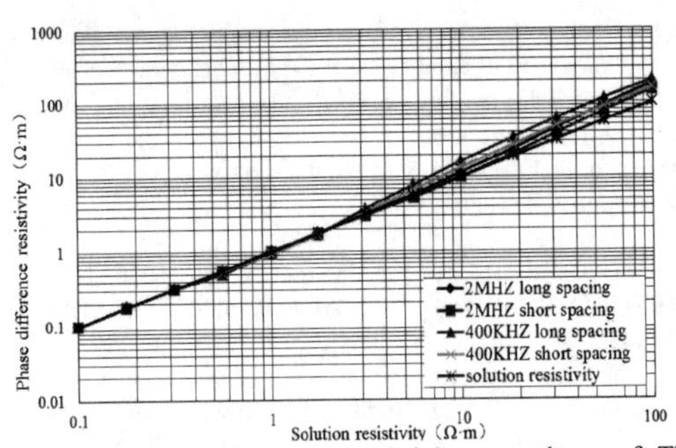

Fig. 6 The amplitude ratio resistivity test charts of The Electromagnetic Wave Resistivity Logging-While-Drilling Instrument in scale barrel.

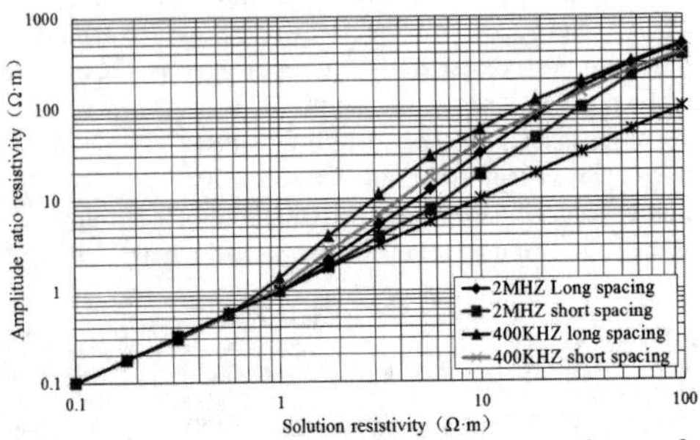

Fig. 7 The phase difference resistivity test charts of The Electromagnetic Wave Resistivity Logging-While-Drilling Instrument in scale barrel.

VI CONCLUSIONS

(1) Putting The Electromagnetic Wave Resistivity Logging-While-Drilling Instrument in the tank for measuring, when in low resistivity, 2MHz and 400KHz amplitude ratio resistivity and phase difference resistivity are approximate with saline solution resistivity, with the salt solution resistivity increasing, amplitude ratio resistivity and phase difference resistivity deviate from the saline solution resistivity, and, the amplitude ratio resistivity deviates faster and greatly.

(2) By test charts of The Electromagnetic Wave Resistivity Logging-While-Drilling Instrument in tank, the arbitrarily saline solution resistivity values can be through interpolation in the chart to find the ideal value measurement. And then compared with the actual measured value, the allowable error within ± 5%, by this way, the instrument stability is confirmed.

REFERENCES

[1] BAKER HUGHES INTEQ “Formation Evaluation Laboratories-Industry Leading Facilities for "Best In Class" MWD Performance”.1998

[2] Gao Jie. ”Analysis of The Current Situation and Development Trend of Numerical Simulation of Electrical Logging”. Well Logging Technology, 2010,34(1):1-5.

[3] Hu Shu. Geophysical Well Logging Instrument [M]. Petroleum Industry Press, 1991,8.

[4] Wang Bin-tao. The Electromagnetic Wave Resistivity Logging-While-Drilling Resistivity Extraction and Simulation Analysis. Science technology and Engineering.2010,5,Vol.10 No.13

A phase-matching method for antenna phase center determination basing upon site insertion loss measurement in OATS

Song Zhenfei* , Xie Ming, Wu Fan, Gao Xiaoxun and Wang Weilong
National Institute of Metrology (NIM), Beijing, 100013, China

***Abstract*-The phase center of directional antennas, such as log-periodic dipole array (LPDA), varies with frequencies and is easily influenced by surroundings. This forms a dominant error for antenna calibration due to an undefined actual distance between the radiation and reception points. This paper concentrates on the phase center determination for broadband wire-element EMC antenna. A novel method named as phase-matching is developed, which uses the measurement data of site insertion loss (SIL) between an antenna under test (AUT) and a reference antenna with a definite phase center in an open area test site (OATS). Precise broadband calculable dipole antennas are utilized in this work. Both the theoretical derivation and validations are presented in this paper. The determination error based upon usages of measurement data is about 60 mm around 250 MHz, and is reduced to 15 mm around 900 MHz.**

I. Introduction

Antenna phase center (PC) is defined as a location from which radiation is considered to emanate. In general, the phase center of a dipole antenna is the center point of radiating elements. A bi-conical antenna has a similar radiation pattern with dipoles, and thus its PC locates at the geometrical center; however, active elements of a log-periodic dipole array (LPDA) antenna shift during a frequency sweep, which causes PCs to change with frequencies. This results in an undefined actual distance between the radiation and reception points, and thus forms a dominant error for LPDA antenna calibration [1-3]. This paper concentrates on the phase center determination for broadband EMC antenna, such as LPDA and bi-log antennas.

Over the decades, a lot of previous work has been focused on phase center determination for antenna calibration purpose. The analytical formula firstly proposed by NPL (National Physical Laboratory, UK)[1] and later appears in a CISPR draft document[3] assumes that phase centers of LPDAs locate exactly at resonant elements and moves linearly with frequencies along its boresight. Such linear interpolation is an approximation within about 50 mm around 200 MHz, and 30 mm around 1 GHz. ETS-Lindgren proposed a CFNSA model [4, 5] for LPDAs; nevertheless, a representation of LPDA pattern with cosine and higher order polynomial functions may not sufficient at high frequencies. In another hand, both the analytical calculation and the CFNSA model are only suitable for commonly-used LPDA antennas. Universal methods which are with no limitation of antenna shape are based on either numerical simulations [6] or phase pattern measurements[7], which are quite time-consuming and not practical for regular antenna calibration.

A novel phase center determination method is developed. Different from the fore-mentioned solutions, the proposed method has no limitation of antenna shape and the phase center variation in multiple dimensions can be considered. Moreover, the site insertion loss (SIL) measurement between an antenna under test (AUT) and a reference antenna can be easily conducted in an Open Area Test Site (OATS).

This paper is organized as follows. A novel phase center determination model including its essential theoretical derivation is presented in Section II. Simulation and measurement validations are discussed in Section III. Section IV ends with conclusions.

II. Phase-Matching Method

A theoretical model for transmission between broadband antennas over an infinite conducting ground plane was proposed by Smith [8] and later became the base of Standard Site Method (SSM) for OATS antenna calibration. The proposed phase center determination method is based on site insertion loss measurements in an OATS. Fig.1 illustrates the measurement model.

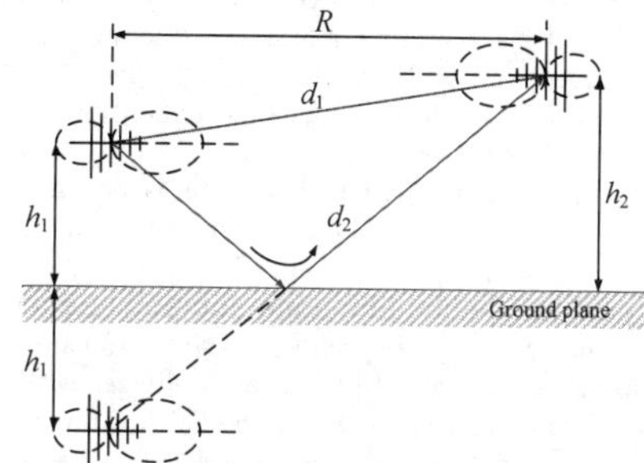

Figure 1. Site Insertion Loss measurement

Basing on the geometrical optics (GO) approximation principle, (1) gives the received electric filed strength by the receiving antenna in condition of horizontal polarization.

$$E_H = \sqrt{30P_T D}\left[\frac{e^{-j\beta d_1}}{d_1} + \frac{|\rho_h| e^{-j\beta d_2} e^{j\phi_h}}{d_2}\right] \quad (\text{mV/m}), \quad (1)$$

where, P_{T} and D are the radiation power and the gain of transmitting antenna, respectively. The complex quantity $|\rho_h|e^{j\phi_h}$ is a reflection coefficient of the ground plane, and

978-7-5641-4279-7

$\beta=2\pi/\lambda$ is the wave number in free space. The antenna height, h_1 and h_2 are defined as heights of antenna phase centers above the ground plane. The quantity R is the separation distance between phase centers of the antennas as they projected on the ground plane. d_1 and d_2 are the propagation distance of direct ray and ground-reflected ray from the transmitting antenna to the receiving antenna, respectively.

SIL is transmission loss between two polarization matched antennas when a direct electrical connection via cables and attenuators between the signal generator output and the measuring receiver input is replaced by transmitting and receiving antennas placed at specified positions on a calibration site[3]. The measurement of SIL is common to all radiated field methods of antenna calibration. The SIL measurement procedures, and measurement uncertainty components, are described in [2]. Equation (2) represents a quantitative relationship between SIL and received electric filed strength, where f_M is the frequencies in mega Hertz; AF_T and AF_R are respectively the free-space antenna factor of the transmitting antenna and the receiving antenna [8].

$$SIL=\frac{79.58\sqrt{30P_T D}AF_T AF_R}{2f_M E_H}. \quad (2)$$

Substituting (1) in (2) gives (3), where the scalar quantity C equals $39.79AF_T AF_R / f_M$.

$$SIL=C\left[\frac{e^{-j\beta d_1}}{d_1}+\frac{|\rho_h|e^{-j\beta d_2}e^{j\phi_h}}{d_2}\right]^{-1}. \quad (3)$$

It is clear in (3) that, two factors contribute phase variations between the transmitting antenna and the receiving antenna when conducting SIL measurement in an OATS. From a phase point of view, the first right hand side term of (1) denotes the phase shifting due to the propagation of direct ray d_1, while the second RHS term implies the phase shifting due to the propagation of ground-reflected ray d_2. The vector superposition of the two terms makes great complexity of the phase of receiving field strength, which also results in difficulties of solving phase center information directly from conventional SIL measurements.

A novel phase center determination model is shown in Fig.2. O_1 and O_2 denote reference points for measuring antenna separation, i.e., pseudo phase centers. The actual phase center P_1 and P_2 can be described by relative coordinate parameters (Δx_1, Δz_1) and (Δx_2,Δz_2), respectively. The coordinate axes of local coordinates O_1-x_1z_1 and O_2-x_2z_2 are collinear with which of the global coordinate O-xz. Since an OATS can represent a half free space, and the measurement model shown in Fig.2 is symmetrical in y direction, the phase center variation in y direction is neglectable.

In the proposed method, heights of two antennas are adjusted in opposite directions, but with same variations, i.e.,

$$h_1'=h_1-\Delta h\,,\ h_2'=h_2+\Delta h\,, \quad (4)$$

where Δh is the height adjustment.

Such specific procedure is developed by intention of the following two aspects.

(1) to keep ground-reflected waves (ray-lengths) constant. The ground-reflected ray can be calculated by (5), with the measurement manner as (4), we can obtain $d_2'=d_2$.

$$d_2'=\sqrt{R^2+(h_1'+h_2')^2}=\sqrt{R^2+(h_1+h_2)^2}=d_2; \quad (5)$$

(2) to keep the reflection coefficient unchanged. The reflect angle γ in Fig.2 keeps unchanged in this height adjustment manner, according to the reflection coefficient calculation (5), this procedure can keep the reflection coefficient unchanged.

$$\rho_h=|\rho_h|e^{j\phi_h}=\frac{\sin\gamma-(K-j60\lambda\sigma-\cos^2\gamma)^{1/2}}{\sin\gamma+(K-j60\lambda\sigma-\cos^2\gamma)^{1/2}}, \quad (6)$$

where K is relative dielectric constant, and σ is the conductivity.

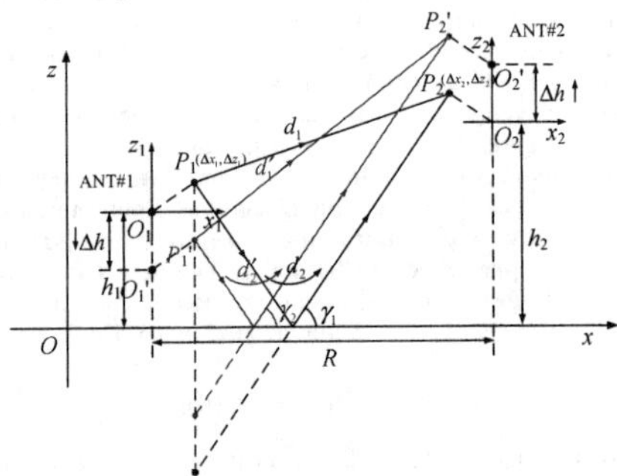

Figure 2. Principle of phase center determination

In this case, equation (7) can be deduced from the theoretical site attenuation model, where C is a constant at a given frequency. With measured SILs of different height configurations, independent equations as (8) can be formed from (7) by matching phases of two sides where the measured SILs are described as $SIL=Ae^{j\varphi}$ and $SIL'=A'e^{j\varphi'}$. The left hand side of (8) can be calculated by the measured SIL magnitude and phase, and in the right hand side , the unknowns d_1 and d_1' are the corresponding direct wave-paths, which are also functions of phase center parameters, Δx_1, Δz_1, Δx_2 and Δz_2. The name of phase matching is given to this method.

$$\frac{1}{SIL'}-\frac{1}{SIL}=C\left[\frac{e^{-j\beta d_1'}}{d_1'}-\frac{e^{-j\beta d_1}}{d_1}\right]. \quad (7)$$

$$\frac{A_2\sin\varphi_1-A_1\sin\varphi_2}{A_1\cos\varphi_2-A_2\cos\varphi_1}=\frac{d_1'\sin\beta d_1-d_1\sin\beta d_1'}{d_1\cos\beta d_1'-d_1'\cos\beta d_1}, \quad (8)$$

where $d_1=\sqrt{(R+\Delta x_1-\Delta x_2)^2+(h_2-h_1+\Delta z_1-\Delta z_2)^2}$

$$d_1'=\sqrt{(R+\Delta x_1-\Delta x_2)^2+(h_2-h_1+\Delta z_1-\Delta z_2+2\Delta h)^2}$$

To simply the calculation, a reference antenna with a definite phase center position is used. High-precision standard broadband calculable dipoles are recommended in this method for their phase centers are exactly located at the centers of radiation elements.

Setting the phase center of a dipole at O_2 in Fig.2, Equation (8) can be simplified by $\Delta x_2=\Delta z_2=0$. SIL measurement in cases of several height configuration results in a certain amount of equations by substituting the measurement data in (8). Theoretically, the left two unknowns of phase center deviation respect to the reference point, i.e., Δx_1 and Δz_1 can be solved by using any two equations as (8). The transcendental equation (8) is difficult to be directly solved by using conventional numerical methods. A specific genetic algorithm is utilized to solve this optimization problem in this work.

With the knowledge of phase center at given frequencies, the antenna factor can be corrected by correction of electric field strength accounting for phase center as (9).

$$\Delta E = 20\log(\frac{R_{phase}}{R}), \qquad (9)$$

where R_{phase} is the separation distance between actual phase centers of the antennas as they projected on the ground plane. For a given frequency, the correction ΔE in dB, is added to the measured field strength.

III. Simulation and Measurement Validation

A. *Validation principle*

Theoretically, phase center of a resonant dipole is located in its geometrical center. In order to validate the proposed phase-matching method, the SIL simulation and measurement of pairs of resonant dipoles are presented in this section. Fig.3 demonstrates the validation principle, where the pseudo phase center P_1' of AUT has deviations of Δx and Δz with respect to the actual phase center P_1, and observation point of reference antenna is exactly its phase center P_2. Since the separation distance R and the height h_1 are measured from AUT's geometrical center, the theoretical value of Δx and Δy are zero.

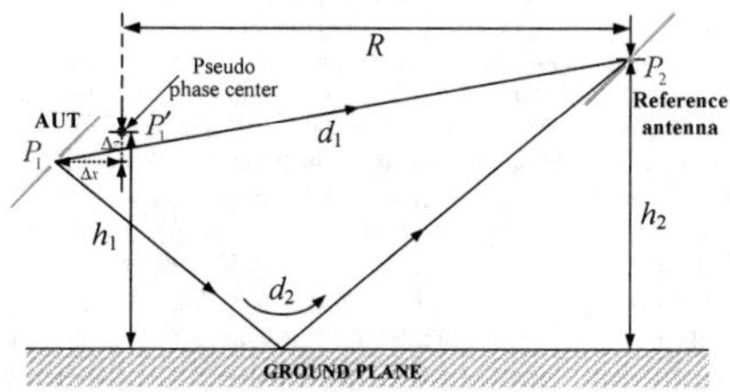

Figure 3. The validation model

B. *Simulation and measurement*

The antenna heights are adjusted keeping h_1+h_2=8m in the validation. The height configuration indicated in Table 1 results in six sets of SIL data including the magnitude and phase information. 15 different equations can be formed by using any combination of two *SIL*s, and finally average values of Δx and Δz can be obtained.

TABLE I

ANTENNA HEIGHT CONFIGURATIONS

	R (m)	h_1 (m)	h_2 (m)
SIL1	5	4	4
SIL2	5	3.8	4.2
SIL3	5	3.6	4.4
SIL4	5	3.4	4.6
SIL5	5	3.2	4.8
SIL6	5	3	5

Simulation experiments using an infinite ground plane and ideal half-wave 250MHz resonant dipoles were performed by Method of Moment (MoM) calculator NEC2. Fig.4 shows the simulated SIL data.

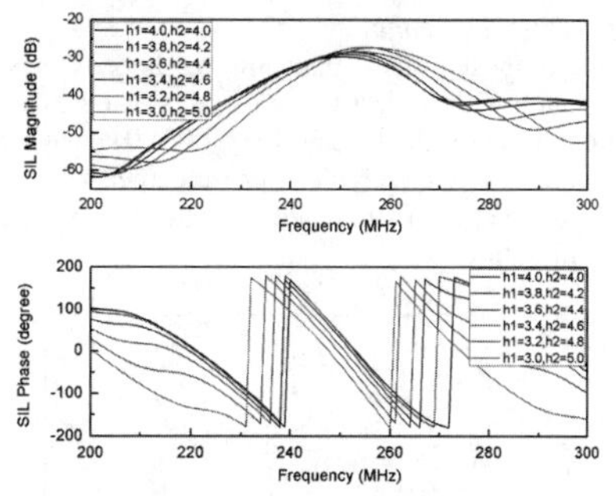

Figure 4. Simulated SIL data of 250MHz resonant dipoles

Corresponding measurement were carried out in a standard OATS of NIM (National Institute of Metrology) in Beijing. The steel ground plane has the dimensions of 60 m by 40 m, and the central (25m by 25m) flatness is within 4 mm.

Fig.5 illustrates the measurement system. Two antenna masts capable of stepping height in small increments (minimum one centimeter) stand over the ground plane. A vector network analyzer (VNA), a motor control unit and other auxiliary device are installed in an underground control room. A serial of NPL broadband calculable dipole antennas (CRD series) which has fixed element lengths are used in the measurement. Fig.6 demonstrates the measurement data of a pair of 250MHz resonant dipoles.

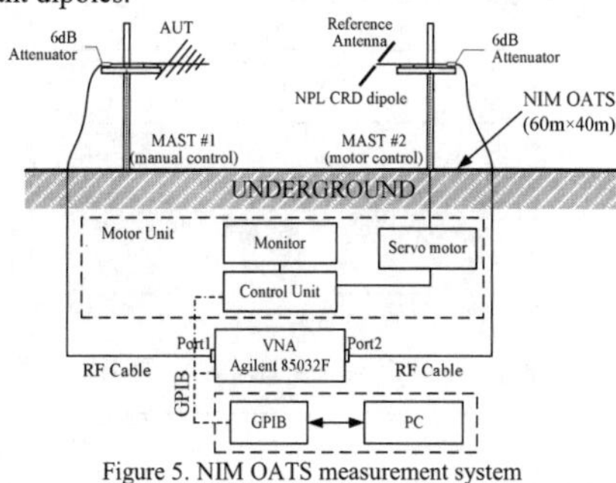

Figure 5. NIM OATS measurement system

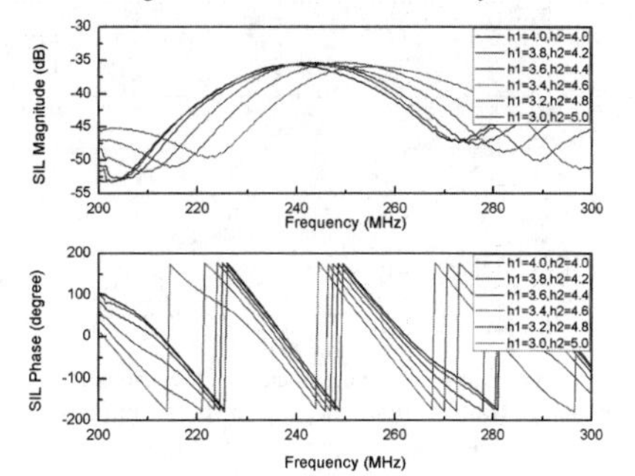

Figure 6. Measured SIL data of 250MHz resonant dipoles

More validations are conducted with 600MHz and 900MHz resonant dipoles.

C. *Phase center determination*

Using the proposed phase matching method, AUT's phase center derivation with respect to the pseudos' can be calculated. Fig.7-9 present the results for 250MHz, 600MHz and 900MHz resonant dipoles, respectively. The phase center determination are based upon the use of simulated *SIL* and measured SIL data for each configration.

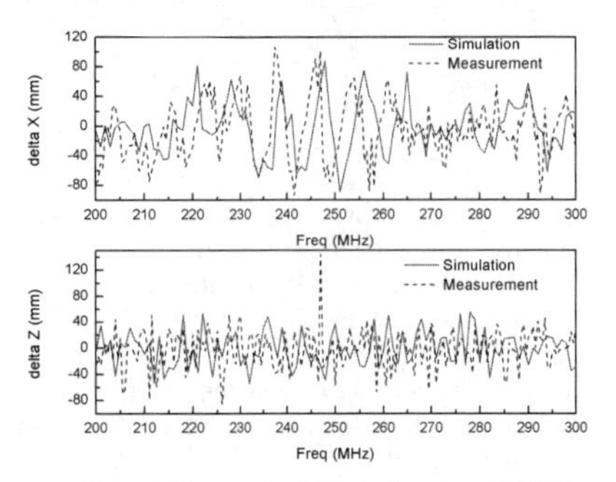

Figure 7. Phase center determination around 250MHz

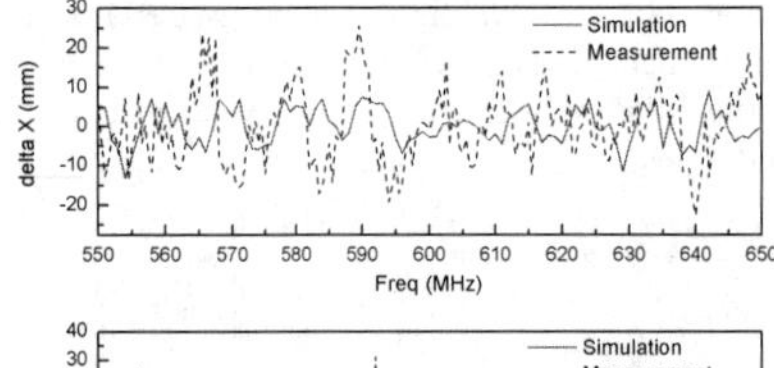

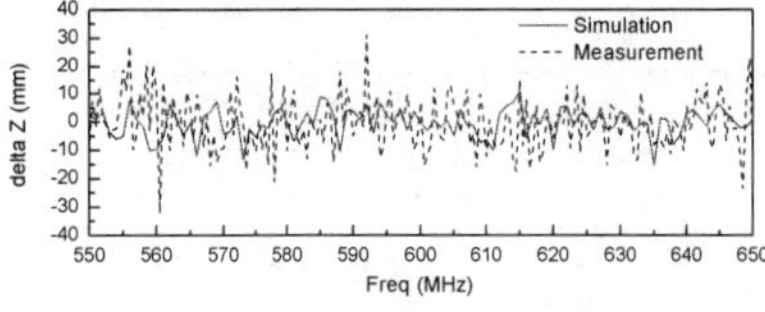

Figure 8. Phase center determination around 600MHz

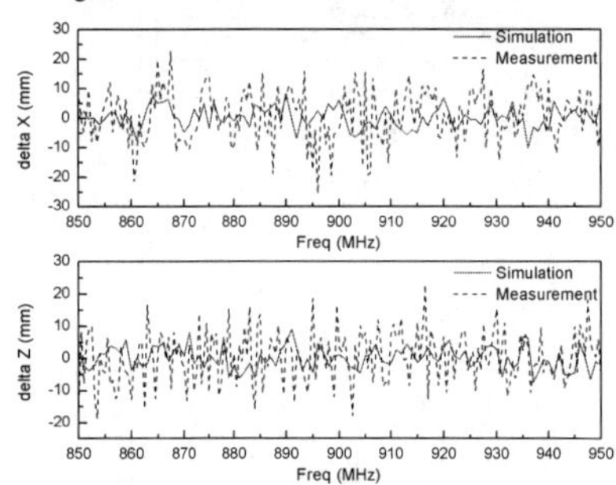

Figure 9. Phase center determination around 900MHz

The determination results based upon the simulated SILs are better than that based upon the measured SILs, due to the measurement uncertainty and system errors. Table 2 gives the detail results. Overall, using the measured SIL data, the determination error is about 60 mm around 250 MHz, and is reduced to 15 mm around 900 MHz.

TABLE II
VALIDATION RESULTS

Frequency (MHz)	By measured SIL		By simulated SIL	
	Δx, mm	Δz, mm	Δx, mm	Δz, mm
200 - 300	± 60	±60	±60	±60
550 - 650	±20	±20	±10	±10
850 - 950	±15	±15	±10	±10

IV. CONCLUSION

This paper presents a novel phase matching method to determine phase centers of broadband wire-element EMC antenna, and it is based upon site insertion loss measurements in OATS. The main advantages respect to conventional methods are as following, this method can determine the phase center derivation from any reference point in two orthogonal dimensions, and it has no limitation of antenna shape. It is based upon actual SIL measurement, so the effect of ground coupling, antenna coupling, and antenna mast can be considered in the phase center determination; meanwhile, the SIL measurement is more easily conducted than other precision phase measurement. As an application of this method, phase center determination of a commonly used LPDA (VUSLP 9111) and the antenna factor correction will be carried out in the future.

REFERENCES

[1] M. J. Alexander, M. J. Salter, D. G. Gentle, D. A. Knight, B. G. Loader, and K. P. Holland, "Calibration and use of antennas, focusing on EMC applications," *A National Measurement Good Practice Guide No. 73*, National Physical Laboratory, Teddington, UK , 2004.

[2] ANSI C63.5-2006, "*American National Standard for Electromagnetic Compatibility–Radiated Emission Measurements in Electromagnetic Interference (EMI) Control–Calibration of Antennas (9 kHz to 40 GHz)*," American National Standards Institute, New York, 2006.

[3] CISPR 16-1-6: 2013, "*Specification for radio distrubance and immunity measuring apparatus and methods - Part 1- 6: Radio disturbance and immunity measuring apparatus - EMC antenna calibration*," (Committee draft for vote, CISPR/A/1027A/CDV), IEC, 2013.

[4] Z. Chen, "Advances in Complex Fit Normalized Site Attenuation using log periodic dipole arrays," *2010 Asia-Pacific Symposium on Electromagnetic Compatibility (APEMC)*, pp. 794-797, 2010.

[5] Z. Chen and M. Foegelle, "Complex fit normalized site attenuation for antennas with complex radiation patterns," *2003 IEEE International Symposium on Electromagnetic Compatibility*, pp. 822- 827, 2003.

[6] V. Rodriguez, "Measurement and Computational Analysis of the Radiation Patterns of EMC Antennas Used in Radiated-Emissions Measurements," *IEEE Antennas and Propagation Magazine*, vol. 53, pp. 103-112, 2011.

[7] S. R. Best and J. M. Tranquilla, "A numerical technique to determine antenna phase center location from computed or measured phase," *The Applied Computational Electromagnetics Society Journal*, vol. 2, 1995.

[8] A. A. Smith, "Standard-Site Method for Determining Antenna Factors," *IEEE Transactions on Electromagnetic Compatibility*, vol. 3, pp. 316 - 322, 1982.

A Novel Method for Antenna Phase Center Calibration

D. Ma, S. G. Yang, Y. Q. Wang, W. R. Huang, L.L.Hou
Southwest Research Institute of Electronic Equipment
NO.496, Cha-dian-zi
Chengdu, 610036 China

***Abstract*-In many antenna application cases, the phase center is not an important parameter, however, in some special cases, antenna phase center should be measured precisely. In this paper, a novel method for antenna phase center calculation is proposed. The calculation examples validate that this method is efficient and useful for actual engineering application.**

I. INTRODUCTION

For an antenna, there are so many parameters to characterize its performances, such as VSWR, gain and axial ratio, etc. In many application cases, the phase center is not an important or critical parameter, however, in some special cases, antenna phase center should be determined precisely and carefully. When an antenna is used as the feed of paraboloid antenna, its phase center should be located in the focus of paraboloid in order to obtain the highest efficiency; If the arrival angle of signal is determined by using interferometer, the phase center of every antenna element need to be determined to measure the length of different base lines; Especially, the phase center of GPS receiver antenna is critical to the location service precision.

In the last years, many literates have studied the antenna phase center calculation or measurement methods. J. Wang and X. Li proposed the method of moving reference point to calculate the phase center of horn antennas[1].This method can find the phase center for all kinds of antenna, but the calculation or measurement efficiency is low. P. A.J.Misaligned proposed an improved method of moving reference point[2]. In this method, the phase center in one section can be calculated by using three pairs of phase and space angle. This method can determine the phase center fast, but the precision is not so high. In order to improve the precision, the advanced method based on generalized least squares is proposed[3,4]. This method can balance the calculation efficiency and precision, so it is becoming the main method for antenna phase center calculation.

In this paper, a novel method based on generalized least squares for antenna phase center calculation is proposed. In this method, the antenna amplitude and phase pattern are both used to calculate the antenna phase center for better precision; In actual engineering application, the antenna has a working frequency bandwidth, however, the antenna phase center is different at different frequency, so an optimization method for finding a best phase center for the whole bandwidth is also proposed in this paper.

II. PRINCIPLE OF PHASE CENTER CALIBRATION

A. Phase Center Calibration Model

The phase center calibration model is shown as Fig.1. The reference point O is the origin of coordinate system. In actual measurement, the reference point O is the rotation center of rotating floor and the $\bar{Z}$ is the transmitting direction of antenna under test. When the phase center calibration is finished, a new reference point O' called as phased center can be determined. The location vector $\bar{D}(\Delta x, \Delta y, \Delta z)$ is the phase center coordinates referred to point O.

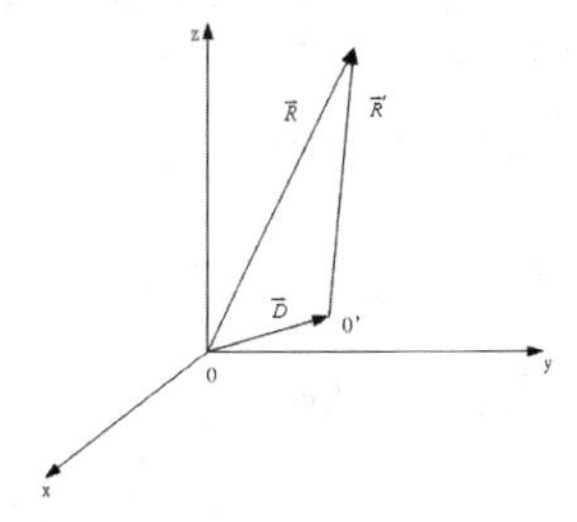

Figure 1. The phase center calibration model.

B. Basic Method of Phase Center Calibration

In the free space, the far field of antenna can be described as following :

$$\bar{E}(\theta,\phi) = \hat{u}E_u(\theta,\phi)e^{j\varphi(\theta,\phi)}e^{-jk\cdot\bar{R}/R} \quad (1)$$

In the equation (1), $\hat{u}$ is the vector of electromagnetic wave polarization. $E(\theta,\phi)$ and $\varphi(\theta,\phi)$ are amplitude and phase pattern respectively. $k = 2\pi/\lambda$ is the wave number. $\bar{R}$ is the distance vector from the reference point O to the observation point. When the reference point is moved to O', antenna far field can be rewritten as equation (2):

$$\bar{E}(\theta,\phi) = \hat{u}E_u(\theta,\phi)e^{-jk\cdot\bar{R}'/R'}e^{j(\varphi(\theta,\phi)-k\bar{R}\cdot\bar{R}')} \quad (2)$$

The phase pattern can be rewritten as

$$\Psi(\theta,\phi) = \varphi(\theta,\phi) - k\bar{R}\cdot\bar{R}' \quad (3)$$

According to the definition of phase center O', the phase pattern $\Psi(\theta,\phi)$ will be a constant. However, the phase pattern can not be a constant in a wide beamwidth for engineering application. There will be an error added on the constant phase pattern, so equation (3) can be expressed as :

$$\Psi(\theta,\phi) = C + \Delta\Psi(\theta,\phi) \quad (4)$$

In order to obtained the constant phase pattern, the $\Delta\Psi(\theta,\phi)$ should be minimal. However, for the most

978-7-5641-4279-7

engineering applications, the far field radiation pattern measured data φ_i is discrete in θ_i direction. To express the macroscopic flat level, the phase center calibration problem can be transferred to search a location to minimize the $\Sigma|\Delta\Psi(\theta,\phi)|^2$.

$$Min(\Sigma[\Delta\Psi(\theta,\phi)]^2) = Min(\Sigma[\Psi(\theta,\phi) - C]^2) = Min(\Sigma[\varphi(\theta,\phi) - k\vec{R}\cdot\vec{R}' - C]^2) \quad (5)$$

The far field radiation pattern is measured in two orthogonal section such as X-O-Z or Y-O-Z for most engineering applications. In order to solve the phase center location, the generalized least squares is proposed. The two- dimension (2-D) expression of can be deduced as:

$$Min(\Sigma[\Delta\Psi(\theta,\phi)]^2) = Min(\sum_{i=1}^{N}[\varphi(\theta_i)-k\cdot\Delta t\cdot\sin(\theta_i)+\Delta z\cdot\cos(\theta_i)-C]^2) \quad (6)$$

where Δt is the landscape orientation component of phase center location.

The derivative as following is operated to obtain the minimum of $\Sigma[\Delta\Psi(\theta,\phi)]^2$.

$$\begin{cases} \dfrac{\partial(\Sigma[\Delta\Psi(\theta,\phi)]^2)}{\partial(\Delta t)}=0 \\ \dfrac{\partial(\Sigma[\Delta\Psi(\theta,\phi)]^2)}{\partial(\Delta z)}=0 \\ \dfrac{\partial(\Sigma[\Delta\Psi(\theta,\phi)]^2)}{\partial(C)}=0 \end{cases} \quad (7)$$

The detailed matrix expression of equation (7) can be shown as (8).

$$\begin{bmatrix} k\sum_{i=1}^{N}\sin^2(\theta_i) & k\sum_{i=1}^{N}\sin(\theta_i)\cos(\theta_i) & \sum_{i=1}^{N}\sin(\theta_i) \\ k\sum_{i=1}^{N}\sin(\theta_i)\cos(\theta_i) & k\sum_{i=1}^{N}\cos^2(\theta_i) & \sum_{i=1}^{N}\cos(\theta_i) \\ k\sum_{i=1}^{N}\sin(\theta_i) & k\sum_{i=1}^{N}\cos(\theta_i) & N \end{bmatrix} \begin{bmatrix}\Delta t \\ \Delta z \\ C\end{bmatrix} = \begin{bmatrix} \sum_{i=1}^{N}\varphi(\theta_i)\sin(\theta_i) \\ \sum_{i=1}^{N}\varphi(\theta_i)\cos(\theta_i) \\ \sum_{i=1}^{N}\varphi(\theta_i) \end{bmatrix} \quad (8)$$

C. Improved Method of Phase Center Calibration

The basic method of phase center calibration is only used the information of phase. In our opinion, the amplitude information can also used to improve the calibration precision of phase center. The amplitude pattern represents the power distribution in the space, so the information of strong power direction is more useful than low ones. According to this opinion, the amplitude pattern $E_u(\theta,\phi)$ can be used as the weight factor. Equation (8) is modified as :

$$\begin{bmatrix} k\sum_{i=1}^{N}E_u(\theta_i)\sin^2(\theta_i) & k\sum_{i=1}^{N}E_u(\theta_i)\sin(\theta_i)\cos(\theta_i) & \sum_{i=1}^{N}E_u(\theta_i)\sin(\theta_i) \\ k\sum_{i=1}^{N}E_u(\theta_i)\sin(\theta_i)\cos(\theta_i) & k\sum_{i=1}^{N}E_u(\theta_i)\cos^2(\theta_i) & \sum_{i=1}^{N}E_u(\theta_i)\cos(\theta_i) \\ k\sum_{i=1}^{N}E_u(\theta_i)\sin(\theta_i) & k\sum_{i=1}^{N}E_u(\theta_i)\cos(\theta_i) & \sum_{i=1}^{N}E_u(\theta_i) \end{bmatrix} \begin{bmatrix}\Delta t \\ \Delta z \\ C\end{bmatrix} = \begin{bmatrix} \sum_{i=1}^{N}E_u(\theta_i)\varphi(\theta_i)\sin(\theta_i) \\ \sum_{i=1}^{N}E_u(\theta_i)\varphi(\theta_i)\cos(\theta_i) \\ \sum_{i=1}^{N}E_u(\theta_i)\varphi(\theta_i) \end{bmatrix} \quad (9)$$

D. Optimized Method of Phase Center Calibration

In practical application, arbitrary electronical equipment has a frequency bandwidth. Antenna phase pattern is different vs. frequency, so the phase center is also changed with frequency. Take the feed of paraboloid antenna for example, the focus of paraboloid is stable at all frequencies, and should overlap with the phase center of feed antenna in order to obtain the highest efficiency. Unfortunately, the feed antenna has a changed phase center with frequency. To obtain the better performance in the whole working bandwidth, an optimized phase center should be found. In this paper, the optimization method called as Hooke-Jeeves method is used to find an equivalent phase center in the whole working bandwidth[5]. The initial value of phase center is the average value of high frequency, low frequency and middle frequency.

III. Phase Center Calibration Example

A classic horn model which can work at L frequency band is shown in Fig.2. To verify the validity of the calibration method proposed in this paper, two examples are shown: one is the comparison between the basic and improved method, the other is the optimized method used to improve the performance of paraboloid antenna.

The phase patterns calibrated by using basic and improved method are shown in Fig.3. It can be found that the improved method considered the factor of amplitude pattern or power distribution in space can obtain better results obviously.

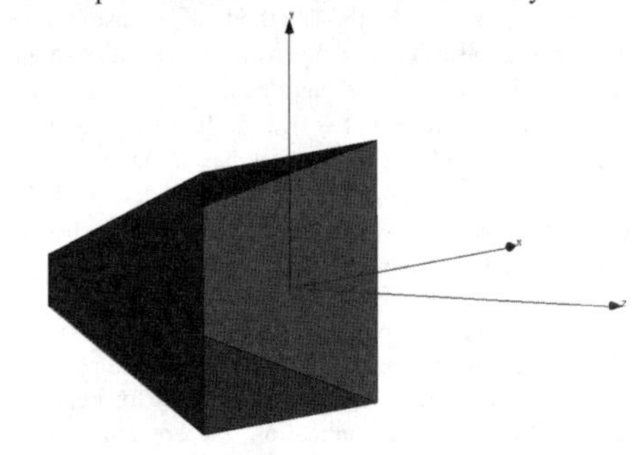

Figure 2. A classic horn model.

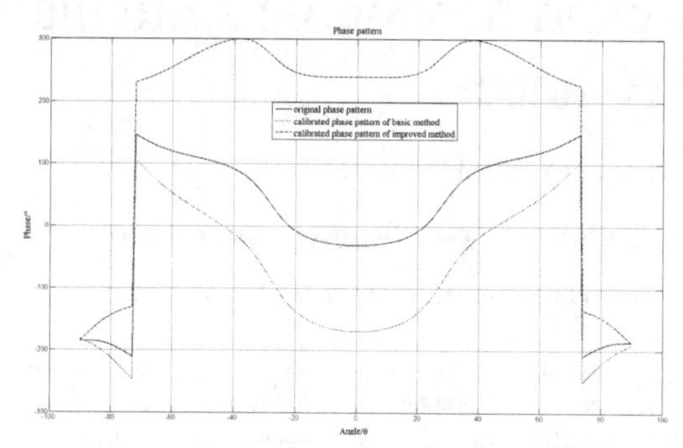

Figure 3. The comparison between the basic and improved method.

An example shown the validity of the optimized method which is used to obtain an equivalent phase center in the whole frequency band will be presented in the meeting in detail. The paraboloid antenna prototype calculated is shown in Fig4. From this example, it can be found that it is important to find an optimized phase center in whole woking frequency band in order to improve the performance of paraboloid antenna, especially, for the wideband applications.

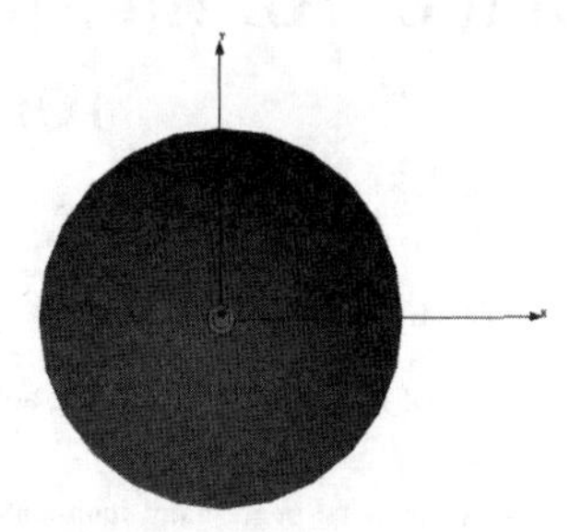

Figure 4. The prototype of paraboloid antenna

IV. CONCLUSIONS

In this paper, a novel method for antenna phase center calibration is proposed. The core of this method is that the position of the phase center should be determined by both antenna amplitude and phase distribution in the space. For the actual engineering applications, all antennas are working at a frequency band and an concept of equivalent phase center is proposed. Two examples have validated the validity of these ideas.

REFERENCES

[1] J. Wang, Y. Xie, and X.Li, "Calculation of phase center for the pyramidal horn with the method of moving reference point," *Journal of University of Electronic Science and Technology of China,* vol.37, No.4, pp. 538-555, Jul. 2008.

[2] Prata A. Jr Misaligned, "Antenna phase-center determination using measured phasepatterns," IPN Progress Report. 42-150,2002.

[3] J. P. Shang, D. M. Fu, and Y. B. Deng, "Research on the accurate measurement method for the antenna phase center," Journal of XIDIAN University, vol.35, No.4, pp. 673-677, Aug. 2008.

[4] X. Chen, G. Fu, and S. X. Gong, etl. "Calculation and analysis of phase center on array antennas," Chinese Journal of Radio Science, vol.25, No.2, pp. 330-335, Aprl. 2010.

[5] Mahdi Pourgholi, V. J. Majd, and M. T. Nabavi, etl., "Synchronous generator's model identification using HookJeeves optimi zation method,"2007 International Conference on Power Engineering, E nergy and Electrical Drives.318-323,APRIL.2007.

Design of An Internal Penta-band Monopole Antenna for Mobile Handset

Naixiao He[#], Yuan Yao[#], Youbo Zhang[#], Junsheng Yu[#], and Xiaodong Chen[*]

[#] School of Electronic Engineering, Beijing University of Posts and Telecommunications, No.10 Xitucheng Road, Beijing, China

[*] School of Electronic Engineering and Computer Science, Queen Mary, University of London, London, UK

henaixiaolele@163.com

***Abstract* — An internal penta-band monopole antenna with compact size and low cost characteristics is proposed, which can cover GSM850, GSM900, DCS1800, PCS1900 and WCDMA2100 bands simultaneously. The antenna comprises a driven monopole and a parasitic shorted patch, which is used to increase the antenna's bandwidth. A prototype is fabricated and measured. Both of the simulated and measured results show the antenna is featured with the mentioned characteristics, which makes the antenna suitable for mobile handset.**

***Index Terms* — Penta-band, Monopole antenna, Handset antenna.**

I. INTRODUCTION

Nowadays, with the increasing development of cellular communication, it is very likely that most electronic devices will include several wireless functionalities. Thus, the mobile handsets with varies type of antennas have been extensively presented. This kind of antennas should be internal, multi-band, low cost, compact size, lightweight and easy fabrication. These desirable features have attracted more and more attentions.

Planar inverted-F antenna (PIFA), which is one of the well-known types of antennas, has many advantages, such as low profile, compact size and multi-band operations[1]. However, this structure suffers from limitation in antenna height to obtain the desired results. As subscribers demand for multi-band operation and the area of the antenna is limited, the traditional type antenna's fabrication is difficult and the production cost becomes high. Compared with PIFA antenna, monopole prevails in ultra thin phones for its very low profile structure, better bandwidth and good efficiency. For such changes, the use of monopole antenna is a good choice.

In the recent published papers [2-5], several internal antennas suitable for mobile phones and smart-phones for multi-band operations were presented. These antennas are preferable for using in mobile systems, including GSM850(824-894MHz), GSM900(880-960 MHz), DCS1800 (1710-1880MHz), PCS1900(1850-1990MHz), WCDMA2100 (1920-2170MHz). However the antennas mentioned above may not be suitable for mobile device for the small size of antenna and its complex electromagnetic environment.

In this paper, we present a compact penta-band monopole antenna with an occupied volume of $25 \times 14 \times 4.5$ mm^3 in the mobile handset which covered the requirement bandwidth of GSM850/900, DCS1800, PCS1900, and WCDMA2100 with sufficient volume. The antenna is composed of a main monopole and a parasitic patch and easily printed on an antenna bracket-a FR4 substrate with low cost . The study is carried out using High Frequency Structure Simulator (HFSS) and experimental results are also given. Details design considerations and the results of the proposed antenna are presented and discussed in the following sections.

II. ANTENNA DESIGN

The geometry and configuration of the proposed penta-band antenna is shown in Fig 1. A 1.5-mm-thick FR4 antenna substrate with a permittivity of 4.4 and loss tangent of 0.02 is used. The substrate dimension is fixed as $48 \times 14 \times 1.5$ mm^3. Taking into considerations the actual mobile phone structure, all the metallic components of the antenna are printed on the side of the substrate with the size of 25×14 mm^2. And the antenna height from the ground is 4.5 mm. Finally, the main PCB of the dimension 112×62 mm^2 is system ground plane.

TABLE I

Dimensions of the proposed antenna for mobile handset

W1	L1	L2	t1	h	L3	h1
14	48	25	3	4.5	20	1.4

Unit :mm

The antenna design mainly comprises of a driven monopole and a parasitic shorted patch. The monopole is a rectangular patch with two pieces of mental branches. The antenna is excited by a coaxial. The shorter branch resonates at the higher band and the longer branch resonates at the lower band. The longer branch always effect both higher and lower frequencies. The parasitic shorted element is electrically connected to the ground through a mental strip and attached to the side of the antenna bracket, which can expand the antenna's bandwidth. Simulator Ansoft HFSS ver. 13 is used to simulate and optimize the antenna design. The finally chosen dimensions of the proposed antenna are illustrated in table I .

978-7-5641-4279-7

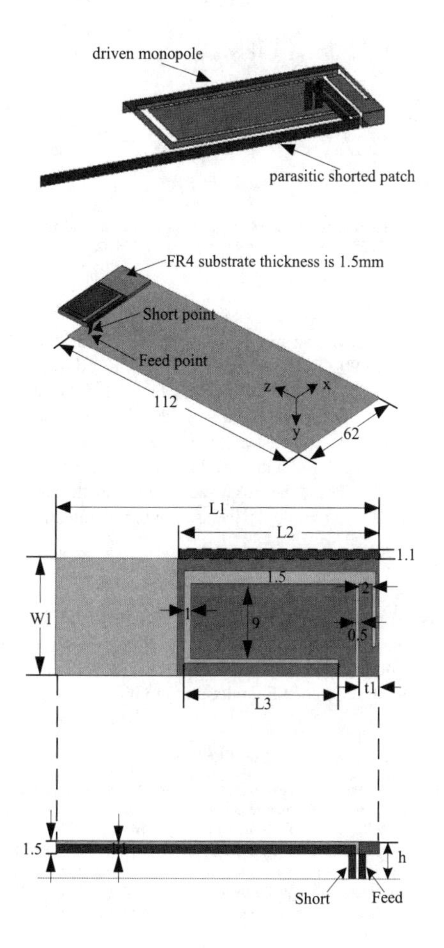

Fig 1. Detailed geometry of the penta-band mobile antenna

III. SIMULATION AND MEASUREMENT RESULTS

The proposed monopole antenna is successfully fabricated and measured. A prototype is fabricated as shown in Fig 2, and the results are presented. The measured results were carried out in anechoic chamber using a vector network analyzer (VNA).

Fig 2. Fabricated penta-band mobile phone antenna

Fig 3 displays the measured and simulated return loss for the proposed antenna. Measured data are seen to substantially agree with the simulated results obtained using Ansoft HFSS. Since the antenna problem of manufacturing precision, the results differ slightly. The measured bandwidths defined by return loss less than -6 dB (widely used for internal mobile handset antennas) are 140MHz (820-960MHz) and 400MHz (1700-2100MHz) respectively. As a results, it can be operated with the GSM850, GSM900, DCS1800, PCS1900 and WCDMA2100 frequency bands.

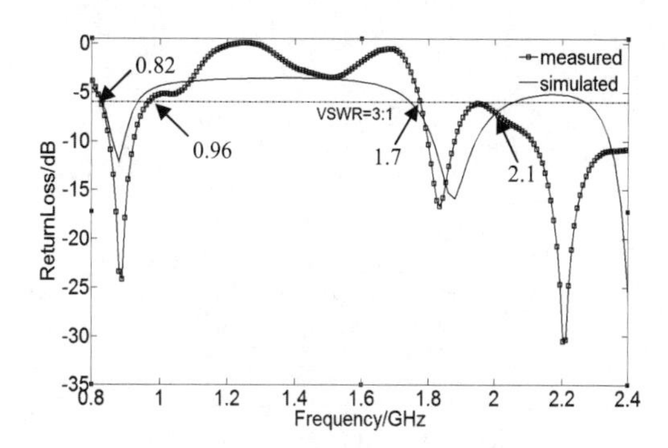

Fig 3. Comparison of simulated and measured reflection coefficient spectra of the proposed penta-band mobile phone antenna

Fig 4 shows the measured radiation patterns of the proposed antenna at f=860, 925 MHz (center frequency of the GSM850 and GSM900 bands). Similar radiation patterns at the two frequencies are observed, both in x-z plane and y-z plane. Similarly, the measured radiation patterns for f=1795, 1920, and 2050 MHz (center frequencies of the DCS, PCS, and WCDMA bands) are presented in Fig 5. For 1920 and 2050 MHz, similar radiation patterns are observed. From these figures, we can see that the antenna has omnidirectional radiation patterns at each operating band, which can meet the demands for mobile devices. In conclusion, the measured radiation patterns for the required bands can acceptable for mobile phone applications.

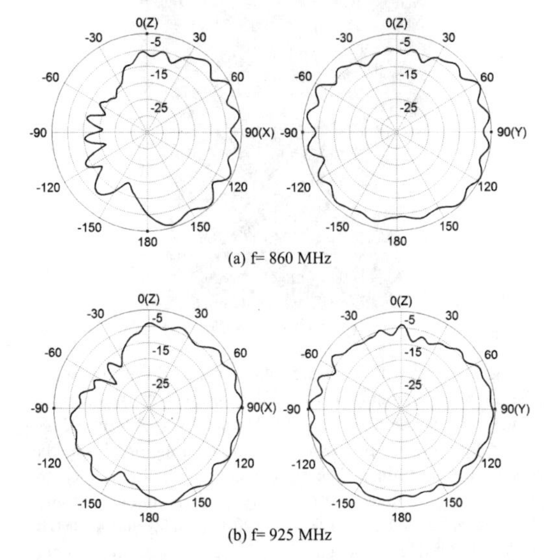

(a) f= 860 MHz

(b) f= 925 MHz

Fig 4 . Measured radiation patterns for the proposed penta-band mobile antenna (a) at 860MHz, (b) at 925MHz

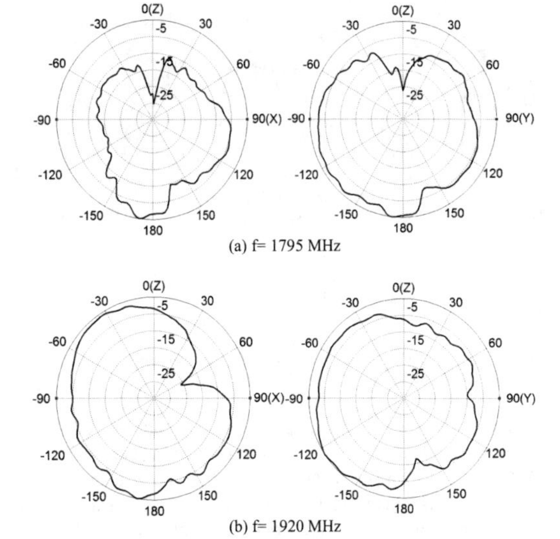

(a) f= 1795 MHz

(b) f= 1920 MHz

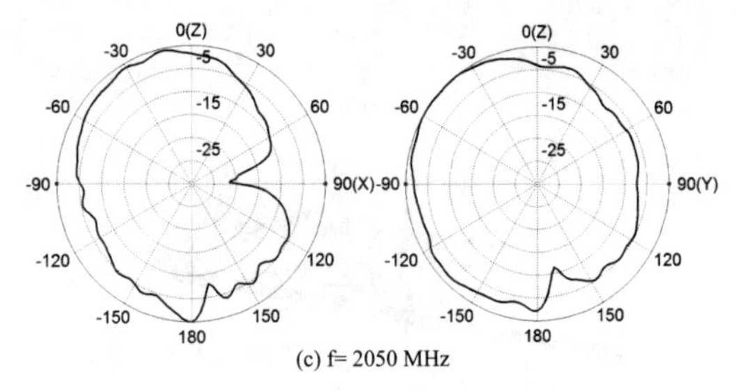

(c) f= 2050 MHz

Fig 5. Measured radiation patterns for the proposed penta-band mobile antenna (a) at 1795MHz, (b) at 1920MHz, (c) at 2050MHz

IV. Conclusion

A compact low cost mobile device antenna for penta-band of GSM850/900/DCS/PCS/WCDMA was designed. A prototype of the antenna design was fabricated successfully. Although simple in structure, both the simulated and measured results demonstrate that the antenna covers the required bands. The antenna also shows good radiation characteristics. Thus, the proposed antenna in this letter can be integrated into mobile device as an internal antenna.

Acknowledgment

This work is supported by the Fundamental Research Funds for the Central Universities, the National Natural Science Foundation of China under Grant No.61201026, and Beijing Natural Science Foundation (4133091).

References

[1] P. Ciais, R. Staraj, G. Kossiavas, C. Luxey, "Design of an internal quad-band antenna for mobile phones," *IEEE Microwave and Wireless Components Letters*, vol. 14, no. 4, pp. 148–150, April 2004.

[2] Kin-Lu Wong, Wei-Yu Chen and Ting-Wei Kang, "On-Board Printed Coupled-Fed Loop Antenna in Close Proximity to the Surrounding Ground Plane for Penta-Band WWAN Mobile Phone," *IEEE Transactions on Antennas and Propagation*, vol. 59, no. 3, pp. 751–757, March 2011.

[3] Chia-Ling Liu, Yi-Fang Lin, Chia-Ming Liang, Shan-Cheng Pan and Hua-Ming Chen, "Miniature Internal Penta-Band Monopole Antenna for Mobile Phones," *IEEE Transactions on Antennas and Propagation,* vol. 58, no. 3, pp. 1008–1011, March 2010.

[4] Chia-Mei Peng, I-Fong Chen, Ching-Chih Hung, Su-Mei Shen, Chia-Te Chien and Chih-Cheng Tseng, "Bandwidth Enhancement of Internal Antenna by Using Reactive Loading for Penta-Band Mobile Handset Application," *IEEE Transactions on Antennas and Propagation,* vol. 59, no. 5, pp. 1728–1733, May 2011.

[5] Ashkan Boldaji, Marek E. Bialkowski and Ahmad Rashidy Razali, "Compact Penta-band Antenna for Portable and Embedded Devices" In *IEEE Proceedings of the Asia-Pacific Microwave Conference(APMC),* December 2011, pp. 1294-1297.

Miniaturized High Gain Slot Antenna with Single-layer Director

Hongjuan Han, Ying Wang*, Xueguan Liu and Huiping Guo
School of Electronic Information Engineering, Soochow University, 215006, Suzhou, China

Abstract- **In this paper, a three-layered slot antenna with compact size and high gain is proposed at 5.4 GHz for WiMAX(5.2~5.8 GHz) applications. Simulated gain of 12dBi and measured gain of 10.48dBi are achieved and the impedance bandwidth is 22.59%. The efficiency is measured of 97.23%. Measured and simulated results are in good agreement. The overall size of the antenna is 100×100 ×21.5 mm^3.**

Index Terms- **slot antenna, high gain, broad bandwidth, high efficiency and WiMAX.**

I. INTRODUCTION

The IEEE 802.16 Working Group has established a standard known as WiMAX (Worldwide Interoperability for Microwave Access). WiMAX has three operating frequency bands, which range from 2.4-2.8 GHz, 3.2-3.8 GHz, and 5.2-5.8 GHz, respectively. Nowadays researches are focusing on miniaturization of the antennas for wireless communications. Antennas for wireless communications are required to exhibit characteristics such as high gain, compact size, high efficiency, low cost, wide bandwidth and so on. Classical Yagi antenna consists of a reflector, a driver and several directors, which has been very popular due to its simplicity as well as its customizable high gain (Gain of a traditional Yagi antenna with three directors can be optimized to 9dBi) [1]. In order to realize a gain of 12dBi, 10 directors are required in [2]. Several microstrip-Yagi or quasi-Yagi antenna structures have been reported in [3]-[6], which have high gain and high front-to-back (F/B) ratio. However, these antennas have narrow bandwidth and large size. A multilayer-stacked microstrip Yagi antenna is presented in [7] and two slot Yagi antennas are proposed in [8]-[9]. In [7], multilayer-stacked microstrip Yagi antennas are proposed. High gain of 11dBi and 10.28dBi can be achieved respectively with adopted four layers of directors for dipole and dual polarization applications. In [8] and [9], the proposed slot Yagi antenna operates at 10 GHz with high gain, narrow bandwidth and large size. Obviously there is an increased demand of antennas with compact size, broad bandwidth and high gain. In this paper, a novel slot Yagi antenna is presented. The proposed antenna is designed for the WiMAX band of 5.2-5.8GHz and detailed discussion is given in the following sections.

II. PRELIMINARY ANTENNA STRUCTURE

The preliminary structure of a slot antenna, named Design 1, is shown in Fig. 1, which consists of a slot, noted as a driver in the following sections and a ground noted as a reflector. In order to achieve good F/B ratio and higher gain for this preliminary structure, the size of the reflector is optimized to 100 ×100mm². Dimensions of this structure are given in Table 1. The antenna is designed on a 1.5mm-thick FR4 substrate with dielectric constant ($\varepsilon r1$) of 4.4. Compared S_{11} between simulation and measured results are shown in Fig.2. The center frequency of the preliminary structure is 5.8 GHz. The measured and simulated bandwidths are 5.32-6.32 GHz and 5.4-6.18 GHz respectively. Fig. 3 shows the simulated radiation patterns of phi=0°and phi=90° at 5.8 GHz. It is noted that the radiation pattern is not acceptable for practical use due to its large side lobes on the direction of theta=±40° as shown in Fig. 3.

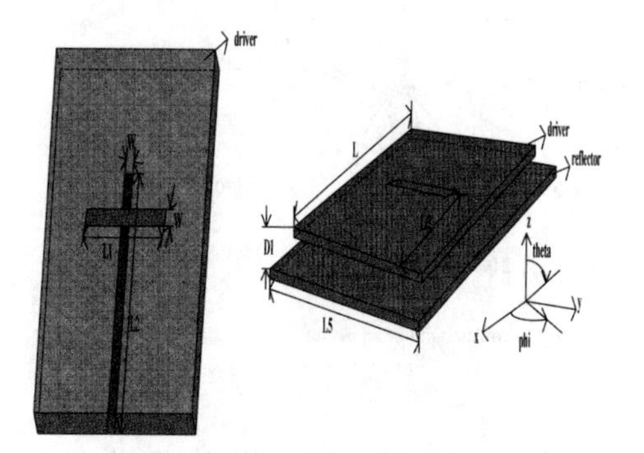

Fig. 1. 3D view of the preliminary structure rectangular

TABLE 1
PARAMETERS OF ANTENNA

Symbol	Value	Unity
H	1.5	mm
L	60	mm
L1	36	mm
L2	38	mm
L3	55	mm
L4	13.75	mm
L5	100	mm
W	3	mm
D1	13.5	mm
D2	6.5	mm
εr1	4.4	/
εr2	2.5	/

978-7-5641-4279-7

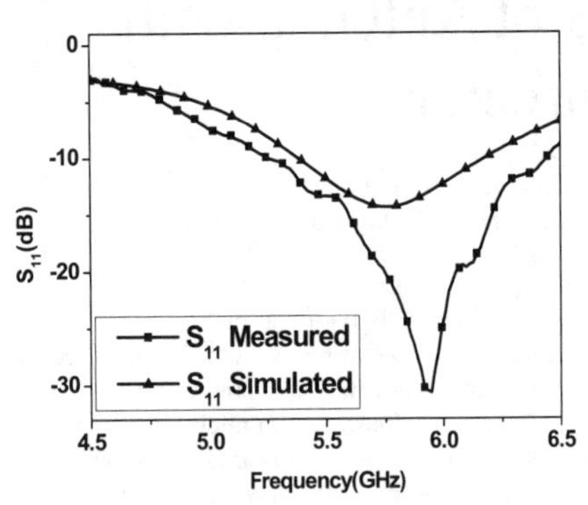

Fig. 2. Measured and simulated S_{11} of basic slot antenna

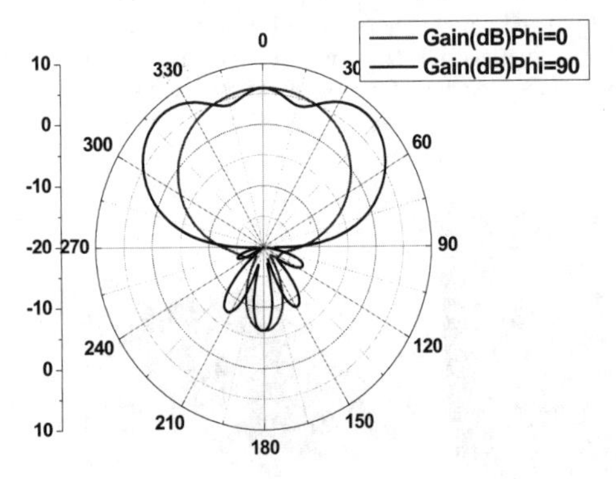

Fig. 3. Simulated radiation pattern of basic slot antenna

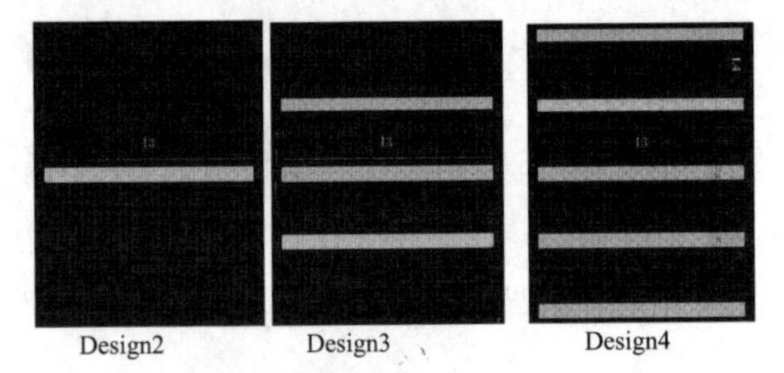

Fig. 4 Structures of the proposed single layer director

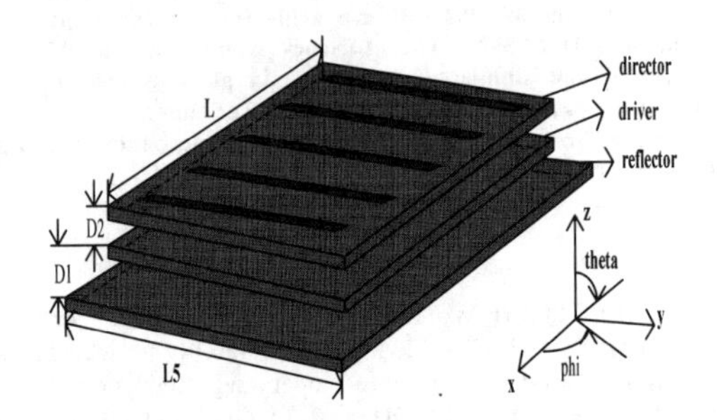

Fig. 5 Structure of the proposed antenna (Design4)

TABLE 2
THE SIMULATED RESULTS OF THE ANTENNA WITH DIFFERENT DIRECTOR

Number of slot in the director	Frequency (GHz)	Bandwidth (GHz)	Gain (dB) Z Axis	Size of the antenna (mm³)
1	4.7	0.06	7.45	100*100*21.5
3	5.4	1.06	11.86	100*100*21.5
5	5.4	1.14	11.98	100*100*21.5

III. Proposed Slot Antenna With Single Layer Director

In order to improve the radiation pattern of the preliminary structure for practical use, a single-layer director with 1, 3 or 5 slots is added on the top of the preliminary structure as shown in Fig. 4 and Fig. 5. As noted in [7], three to four layers of directors are required to achieve high gain. Only one director layer is required to achieve a gain as high as shown in [7]. High performance and compact size are achieved. The optimized distance between two slots is 13.75mm, which is close to λ/4. The size of the director layer is 60(L)×60(L)mm² and the distance from the director to the driver is D_2. The director is designed on the substrate with εr2 of 2.5 and loss tangent of δ =0.001. All the parameters are listed in Table 1, where D_1=13.5mm (0.243 λ), D_2=6.5mm (0.117 λ), L_2=38mm (0.647 λ), L_3=55mm (0.989 λ) and L_4=13.75mm (0.243 λ). The overall height of the proposed antenna is 21.5mm.

IV. Results And Discussions

Performance of the proposed slot antenna with single-layer director is given in Fig. 6- Fig.12. Simulated results of the proposed antenna with one, three and five slots in the director are compared in Table 2. Photograph of the fabricated preliminary structure and the proposed slot antenna with single layer director are given in Fig. 8 and Fig. 9. Comparison between simulated and measured results is shown in Fig. 10, Fig. 11 and Fig.12, which shows good agreement. The proposed design is discussed as following:

a) As shown in Fig. 3 and Fig. 7, director layer with slots helps to reduce the side lobe and improve the radiation pattern. The main lobe focuses on the orthogonal direction of the director layer (theta= 0°).
b) The proposed slot antenna with five slots in the single-layer director has a simulated bandwidth of 20.7% and measured bandwidth of 22.59%, which exhibit wide

band characteristics. However, the bandwidth of the antenna with one slot in the director is only 1.3%.at 4.7GHz as shown in Fig.6. Bandwidth enhancement can be achieved by increased number of slots.

c) As listed in Table 2, gain in the orthogonal direction of the slot antenna could be greatly improved from 5.94dBi (without single-layer director) to 12dBi (with five slots in the director). The measured gain is 10.48dBi at 5.4GHz.

d) As shown in Fig.11, the simulated and measured efficiency of Design4 can be up to 84.5% and 97.2% at 5.4GHz, while the simulated and measured efficiency of design1 are 80.2% and 78.9% at 5.4GHz. Efficiency has been improved with the single-layer director.

e) Fig. 12 presents the simulated and measured *xoy* plane and *yoz* plane radiation patterns of Design4. Good agreement between the measured and simulated radiation patterns has been achieved. The maximum simulated and measured gains are 12dBi and 10.48dBi respectively.

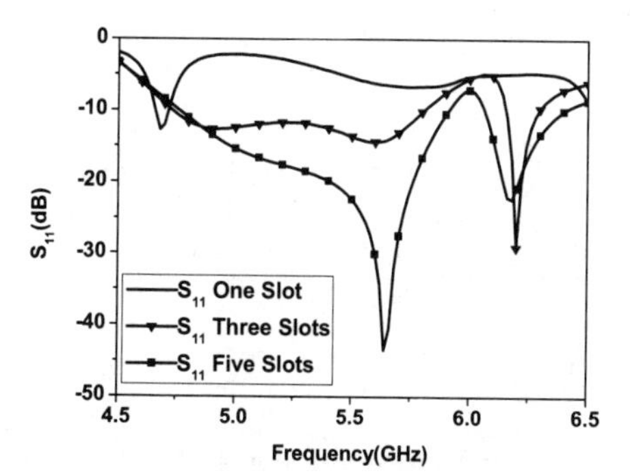

Fig. 6 Comparison of simulated S_{11} for 1, 3 and 5 slots in the Single-layer director

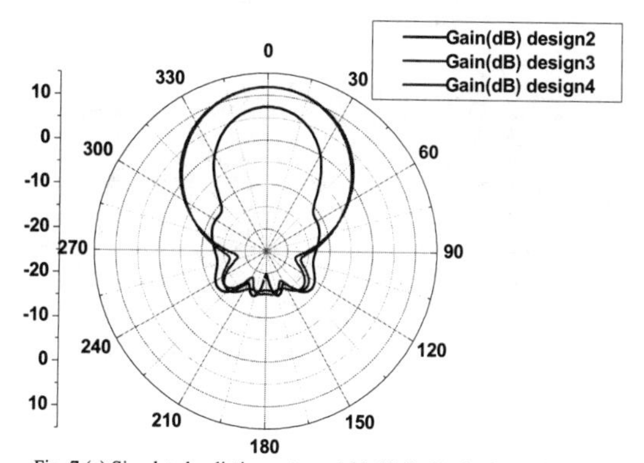

Fig. 7 (a) Simulated radiation patterns (phi=0°) for Design2, Design3 and Design4

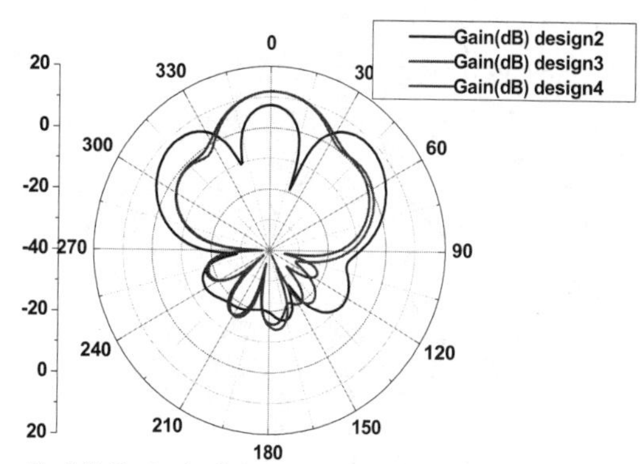

Fig. 7 (b) Simulated radiation patterns (phi=90°) for Design2, Design3 and Design4

Fig. 8. Photograph of the preliminary structure

Fig. 9. Photograph of the proposed slot antenna with single-layer director

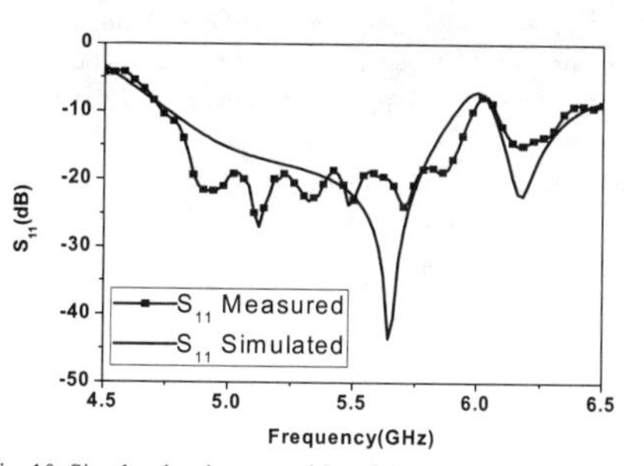

Fig. 10. Simulated and measured S_{11} of the proposed slot antenna with five slots in the director

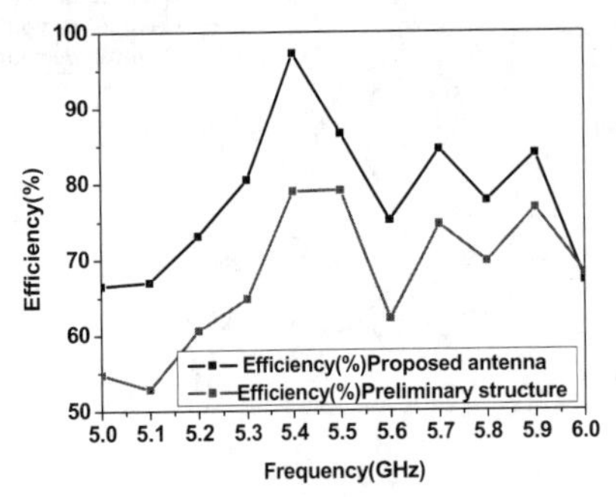

Fig. 11. Measured efficiency for the preliminary slot antenna and the proposed antenna (Design 4)

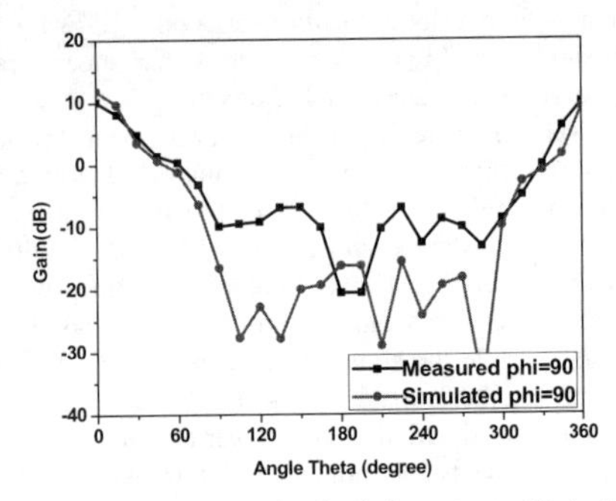

Fig. 12 (b) Measured and simulated radiation patterns of Design4 in *yoz* plane at 5.4GHz

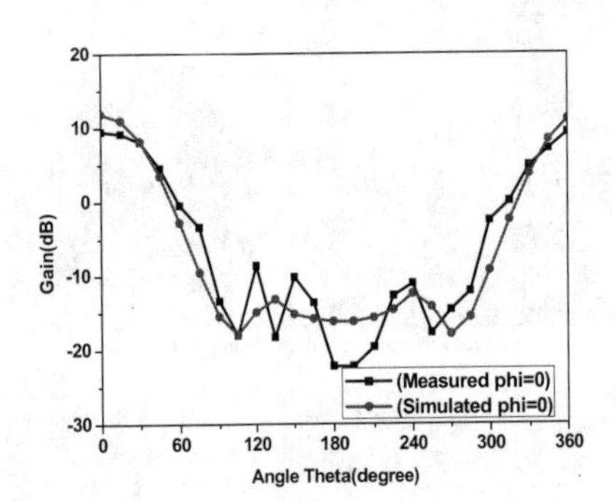

Fig. 12 (a) Measured and simulated radiation patterns of Design4 in *xoz* plane at 5.4GHz

V. CONCLUSION

In this paper, a novel slot antenna with single-layer director is proposed. By adding a director layer with five slots on the top of the basic slot antenna, main lobe of the radiation pattern is shifted to the orthogonal direction theta=0° instead of theta=±40°. Good performance with gain of 12 dBi and efficiency of 97.2% is achieved at 5.4GHz for WiMAX applications. The overall optimum size of the proposed antenna is only $80\times80\times21.5$ mm^3. Bandwidth has also been enhanced up to 22.59% compared to 17.24% for a basic slot antenna. Good agreement between simulated and measured results is achieved.

ACKNOWLEDGMENT

This project is supported by Natural Science Foundation of Jiangsu Province, China under Grant BK2010221. The authors would like to thank their colleagues for their help for this project.

REFERENCES

[1] W. L. Stutzman and G. A. Thiele, *Antenna Theory and Design*, 2nd Ed New York: Wiley, 1998.

[2] R. Carvalho, R. R. Saldanha, B. N. Gomes, A. C. Lisboa and A. X. Martins, "A Multi-Objective Evolutionary Algorithm Based on Decomposition for Optimal Design of Yagi-Uda Antennas," *IEEE Transactions on Magnetics*,vol.48,No.2,February 2012.

[3] Mazen. Alsliety and Daniel. Aloi, "A low profile microstrip Yagi dipole antenna for wireless communications in the 5GHz band," in *proc. Int. Conf. on Electrol/Information Technology,* East Lansing, MI, May 2006, pp.525-528.

[4] Gerald. R. DeJean, Trang. T. Thai, Symeon. Nikolaou, and Manos. M.Tentzeris, "Design and analysis of microstrip Bi-Yagi and Quad-Yagi antenna arrays for WLAN applications," *Antennas Wireless Propag. Lett.* , vol. 6, pp. 244-248, Jun.2007.

[5] H.-D. Lu, L.-M. Si and Y. Liu, "Compact planar microstrip-fed quasi-Yagi antenna," *Electronics Letters*, Vol.48, No. 3, 2nd. February 2012.

[6] William R. Deal, Noriaki Kaneda, James Sor, Yongxi Qian and Tatsuo Itoh, "A New Quasi-Yagi Antenna for Planar Active Antenna Arrays," *IEEE Transactions on Microwave Theory and Techniques*, Vol.48, NO. 6, June 2000.

[7] Oliviver Kramer, Tarek Djerafi and Ke Wu, "Vertically Multilayer-Stacked Yagi Antenna With Single and Dual Polarizations," *IEEE Transactions on Antennas and Propagation*, vol. 58, No.4, April.2010.

[8] A. Y. Simba, M. Yamamoto, T. Nojima and K.Itoh, "Planar-type sectored antenna based on slot Yagi-Uda array," *IEE Proc.-Microw. Antennas Propag,* Vol.152, No.5, October 2005.

[9] Xue-Song Yang, Bing-Zhong Wang, Weixia Wu and Shaoqiu Xiao, "Yagi Patch Antenna With Dual-Band and Pattern Reconfigurable Characteristics," *IEEE Antennas and Wireless Propagation Letters,* Vol.6,2007

A Frequency Reconfigurable Planar Inverted-F Antenna for Wireless Applications

Zheng Xiang[1], Hai-jing Zhou[2], Ju Feng[1], Cheng Liao[1]
1. Institute of Electromagnetics, Southwest Jiaotong University, Chengdu 610031, China;
2. Institute of Applied Physics and Computational Mathematics, Beijing 100088, China;

Abstract-A compact frequency reconfigurable Planar Inverted-F Antenna (PIFA) is presented in this paper. This antenna is suitable for wireless applications. It allows three frequency reconfigurable modes via two switches, which cover five communication bands. The three modes are ISM-LTE, GSM-PCS, and DCS, respectively. The overall area of this antenna is 17×23mm^2. The simulation results of this antenna show good performances in each modes.

I. INTRODUCTION

Nowadays, a growing number of wireless communications equipments appear in people's lives. The widely use in many field makes the miniaturization and multi-band become the mainstream trend of antenna design [1], but it is a huge challenge to design antennas to meet the requirements with a single antenna. As an effective solution, reconfigurable technology is gradually considered by scholars in the whole world. The reconfigurable antenna can work under different electromagnetic environments by changing some parameters of the antenna. At present, reconfigurable technology can be classified into the following three forms. The first form is frequency reconfigurable technology, which can make the antenna work in different frequency. The second form is pattern reconfigurable technology, it means the radiation pattern changes while the frequency band remains the same. The last form is polarization reconfigurable technology [2], which means the polarization could be transformed from right hand (RHCP) to left hand (LHCP) circular. Many methods have been used to realize reconfigurable antenna. PIN switches have been reported in some papers [3] . A reconfigurable method can be used in antenna design by switching different feeding locations [4]. Multi-band antenna can be designed with varactor diodes [5]. Ideal model switches for reconfigurable antennas are in [6].

In this paper, the frequency reconfigurable is considered. Five bands including GSM(0.893 GHz -0.931 GHz)、DCS(1.73 GHz -1.88 GHz)、PCS(1.86 GHz -1.98 GHz)、ISM(2.41 GHz -2.48 GHz) and LTE(2.5 GHz -2.67 GHz) can be generated by using the antenna proposed in the paper. These bands can be shifted by changing the two ideal model switches.

II. ANTENNA DESIGN

A. PIFA Antenna

Planar Inverted-F Antenna (PIFA) is a kind of commonly planar antennas, it is widely used in wireless communication equipment because of its advantages, such as small volume, easy processing and omni-directional radiation [7].

The traditional Inverted-F antenna [7] consists of the radiation patch located above a ground plane, a short circuit and a feeding. For rectangular patch planar Inverted-F antenna, the approximate resonant frequency can be written as

$$f_c = \frac{c}{4\times(a+b)} \tag{1}$$

Where c is the speed of light in a vacuum, a is the length of the rectangular patch, b is the width of rectangular patch. Equation (1) shows that the resonance frequency depends on the radiation piece. The size of the ground plate will affect the performance of low frequency. Usually, the length of the ground plate is 80~120 mm for the band of GSM.

B. Reconfigurable Antenna Structure

Antenna structure is shown in Figure. 1, the antenna applied the meander structure embedded in the patch of the traditional PIFA [8]. The PIFA is shorted to the ground by a 1 mm wide metal strip and fed by a 50 Ω coaxial cable. The ground dimension is $120\times 35\,mm^2$.The distance from the ground to the patch is 10 mm. The switch 1 and switch 2 are two ideal model switches. The dimension of the patch is $17\times 23\,mm^2$. Compared with the size of other applications on the frequency of GSM, the antenna in this paper is smaller. The dimensions of the patch are listed in Table 1.

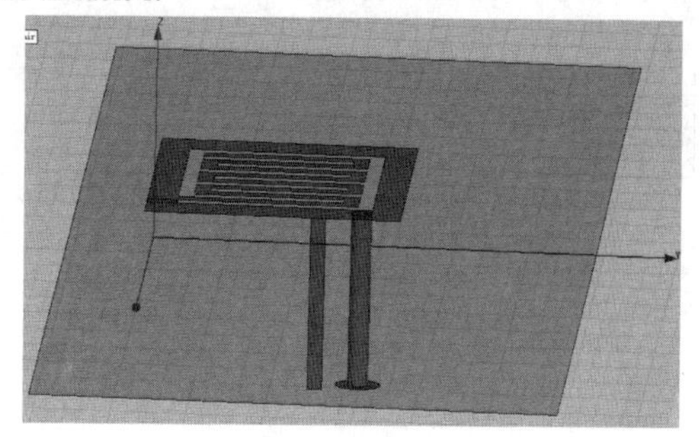

(a) 3D view

* Project supported by the Fundamental Research Funds for The Central Universities (Grant No.SWJTU12ZT08) and the Research Fund of Key Laboratory of HPM Technology (Grant No.2012-LHWJJ.006).

978-7-5641-4279-7

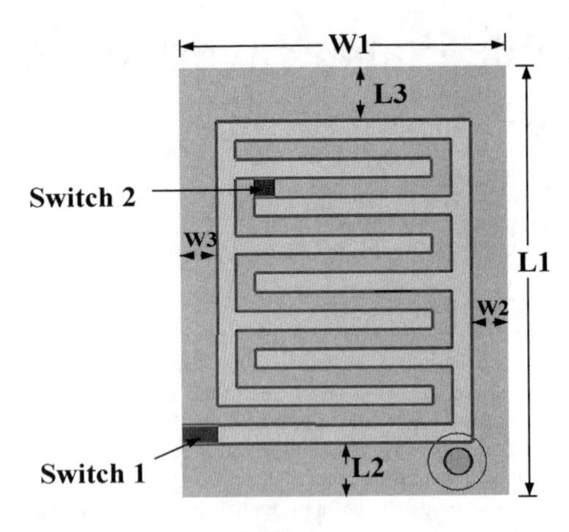

(b) Top view

Figure 1. Structure of the PIFA antenna. (a) 3D view (b) Top view

Table 1. Dimensions of the PIFA antenna.

Parameter name	Parameter value(mm)
L1	23
L2	3
L3	3
W1	17
W2	2
W3	2

III. RESULT AND DISSCUSSION

The three switch conditions corresponding to the frequency are presented in Table 2.The design of slots prompts the antenna to generate five bands. The reflection coefficients (S_{11} <-6 dB) [9] of the PIFA antenna through the switches under the condition of different states are presented in figure 2.

Table 2. Desired Matching for the different Switch Configurations.

Mode	Switch configuration	Frequency band
1	off-on	ISM(2.41-2.48GHz)、LTE(2.5-2.67GHz)
2	off-off	GSM(0.893-0.931GHz)、PCS(1.86-1.98GHz)
3	on-on	DCS(1.73-1.88GHz)

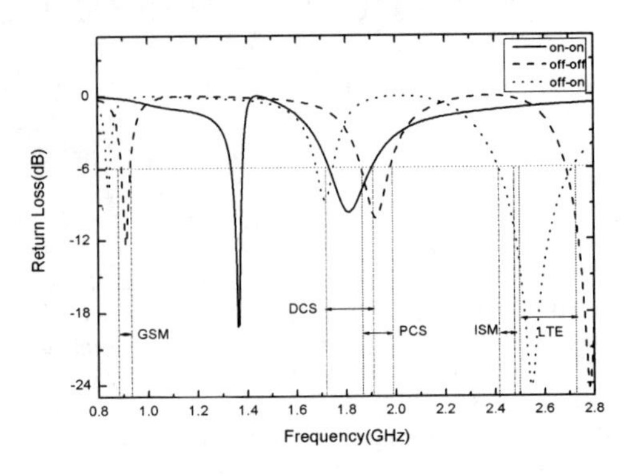

Figure 2. Simulated return loss for three different switch configurations.

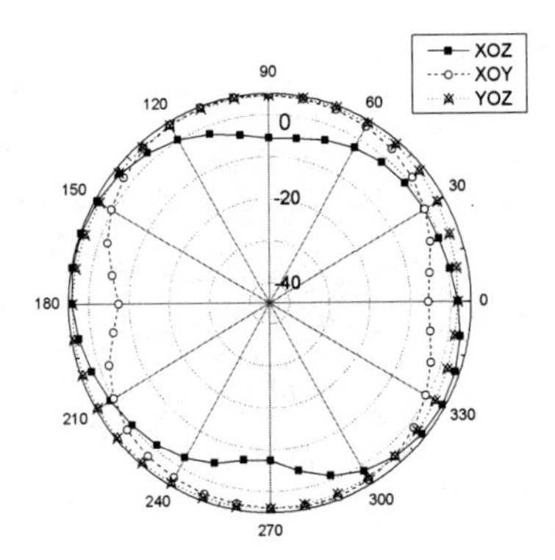

(a)GSM 0.9GHz

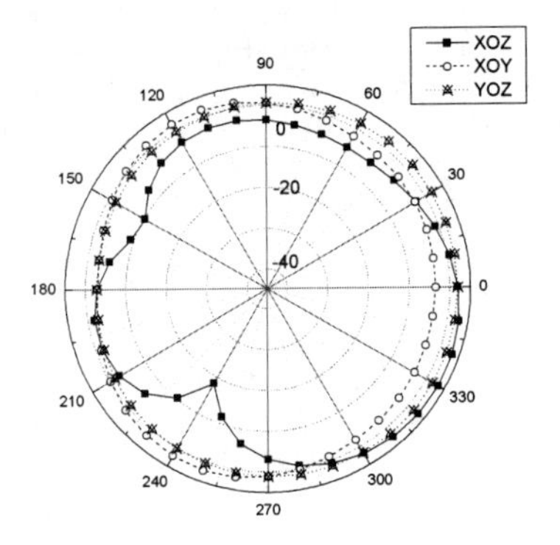

(b) DCS 1.7GHz

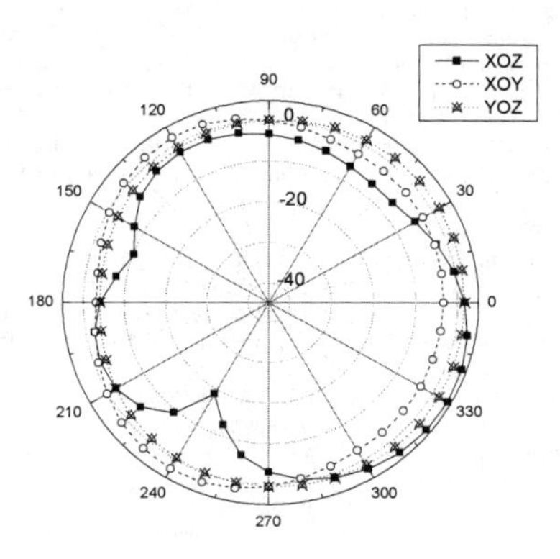

(c) PCS 1.8GHz

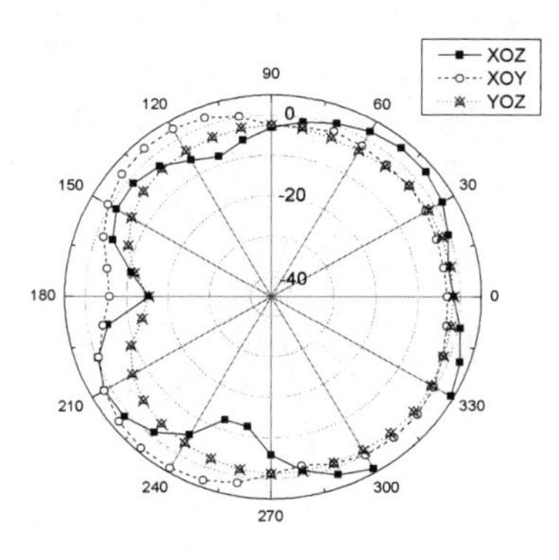

(d) ISM 2.4GHz

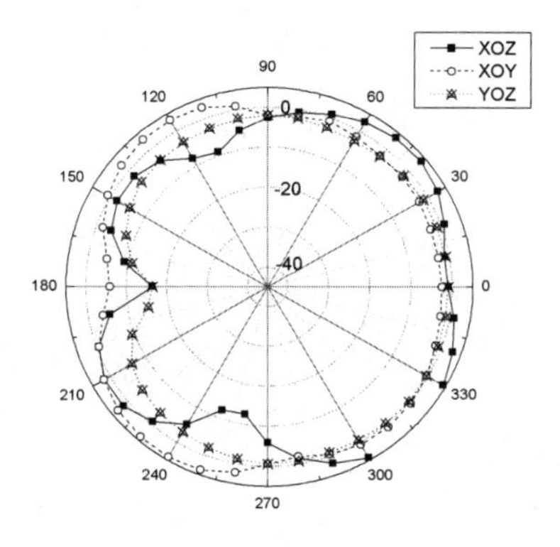

(e) LTE 2.5GHz

Figure 3. Radiation pattern for five bands (a) GSM (b) DCS (c) PCS (d) ISM (e) LTE

The radiation patterns for the five bands are shown in figure. 3 (a)-(e).

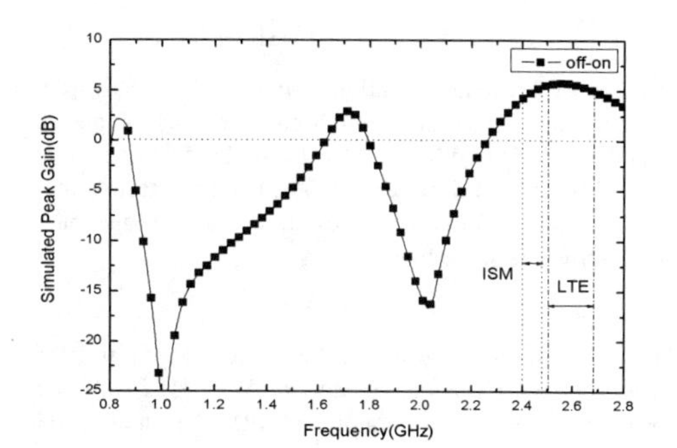

(a) Simulated peak gain for ISM\LTE.

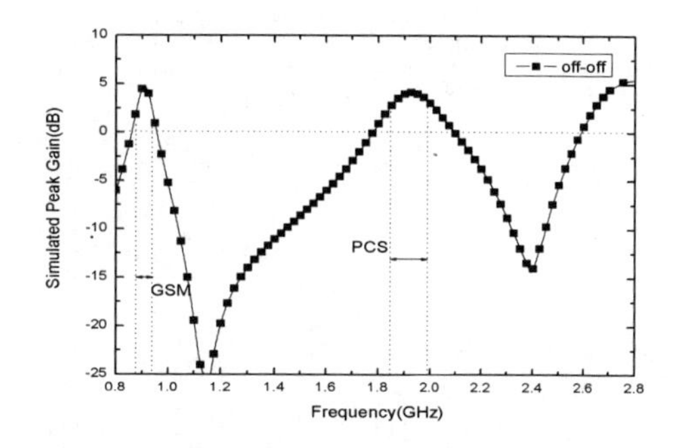

(b) Simulated peak gain for GSM\PCS.

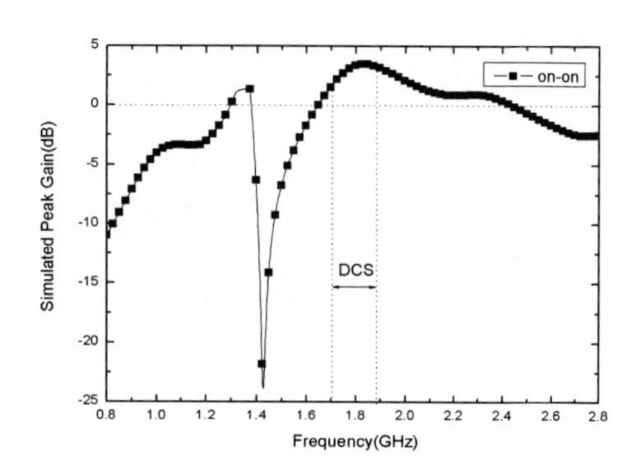

(c) Simulated peak gain for DCS

Figure 4. Peak Gain for three reconfigurations. (a) off-on (b) off-off (c) on-on

The Figure.4(a)-(c) shows the peak gains for the GSM、DCS、PCS、ISM and LTE are 4.76dBi, 3.4dBi, 4.14dBi, 5.1dBi and 5.46dBi.

IV. CONCLUSION

A compact frequency reconfigurable PIFA Antenna with three communication frequency reconfigurable modes is presented. The modes covering GSM、 DCS、PCS、ISM and LTE can be controlled by two ideal switches. The simulation results that the return loss and radiation properties show that the antenna is in good performance in the five bands.

ACKNOWLEDGMENT

This work is supported by the Fundamental Research Funds for The Central Universities (Grant No.SWJTU12ZT08) and the Research Fund of Key Laboratory of HPM Technology (Grant No.2012-LHWJJ.006).

REFERENCES

[1] Pertti Vainikainen, "More than 20 Antenna Elements in Future Mobile Phones, Threat or Opportunity ?", *EUCAP 2009, Berlin*, March 23-27, 2009

[2] Hirazawa K., Haneishi M, "Analysis, design and measurement of small and low profile antennas," *ARTECH HOUSE ONC*, 1992.

[3] Trong Duc Nguyen; Duroc, Y.; Van Yem Vu; Tan Ohu Vuong." Novel Reconfigurable 8 - Shape PIFA Antenna Using PIN Diode" *Advanced Technologies for Communications (ATC), 2011 International Conference on Digital Object Identifier* :10.1109/ATC.2011.6027483 Publication Year: 2011, Page(s): 272-275

[4] Yue Li; Zhijun Zhang; Wenhua Chen; Zhenghe Feng; Iskander,M,F.;An-Ping Zhao "Feeding reconfigurable PIFA for *GSM/PCS/DCS* applications" *Antennas and Propagation Society International Symposium*, 2009.APSURSI,09.IEEE.

[5] AbuTarboush,H.G.; Nilavalan,R.;Peter,T. "PIFA based reconfigurable multiband antenna for wireless applications". *Electromagnetics in Advanced Applications (ICEAA), 2010 International Conference on Digital Object Identifier:10.1109/ICEAA*.2010.5653641 Publication Year: 2010, Page(s); 232-235.

[6] Wang Liping; Liao Cheng; Chang Lei; Gao Qinming. "A Radiation Pattern Reconfigurable Planar Folded Dipole microstrip Antenna Designed for the Same Frequency". *National Microwave Millimeter Wave Conference*, 2011.

[7] LI Xin-yuan, Fu Jia-hui, Zhang Kuang, Hua Jun and Wu Qun, "A Compact Wideband Planar Invered-F Antenna (PIFA) Load with Metamaterial," *2011 Cross Strait Quad-Regional Radio Science and Wireless Technology Conferece*, pp. 549-551,July 2011.

[8] Hattan F. AbuTarboush, R. Nilavalan, H. S. Al-Raweshidy and D. Budimir, "Design of planar inverted-F antennas (PIFA) for multiband wireless applications," *International Conference on Electromagnetics in Advanced Applications*, 2009. ICEAA '09, pp . 78-81, 2009.

[9] Lee,J.H.; Sung,Y. "A recongfigurable PIFA using a pin-diode for LTE/GSM850/GSM900/DCS/PCS/UMTS" *Antennas and Propagation Society International Symposium (APSURSI)*, 2012 IEEE.10.1109/APS.2012.6348487.

Novel Generalized Synthesis Method of Microwave Triplexer by Using Non-Resonating Nodes

Y. L. Zhang, T. Su, Z. P. Li and C. H. Liang
School of Electronic Engineering, Xidian University, Xi'an, 710071, China.
Corresponding author: namarzhang@163.com.

***Abstract*-This paper presents a novel generalized method for synthesizing microwave triplexers composed of TX, MX and RX channel filters. The channel filters are then synthesized by using Non-resonating nodes (NRNs). By introducing NRNs, the topology of the channel filters is simple (inline topology) and the number of the finite transmission zeros can achieve as many as the degree of channel filter. The finite transmission zeros of each channel filter are placed arbitrarily and the topology of triplexer is more flexible. A synthesized example show the validity of the technique presented in this paper.**

I. Introduction

In recent years, a multiplexer is an important component for channel separation in microwave front-end systems for various applications such as communication, radar and other transceiver systems. Multiplexers provide isolation between transmit and receive channels by assigning a different frequency band to each channel and can operate over a wide bandwidth. There have been many efforts to develop various kinds of multiplexers so far, such as duplexers [1-2]. However, The requirements for the antenna combiner networks used in base stations for radio mobile communication have undergone a twofold path: first, selectivity and insertion loss requirement have become more and more stringent; second, complexity (in terms of number of filters and ports) of such devices has increased beyond the typical duplexer structure [3-4]. Because of the interaction of the three filters composing the triplexer, the characteristics of a triplexer are different from the responses of the individual three filters. The complexity of the interaction makes the design of a triplexer complicated.

For the design of microwave triplexers, the traditional approach is to design the three channel filters individually and then to design a distributed network. The distributed network is designed by cut-and-try or optimization method. The convergence could also become problematic due to the large number of "local minima", which characterizes the error function to be minimized. Generally speaking, the distributed network configuration has drawbacks of large volume and time-consuming.

The purpose of this study is to present a general synthesis procedure for triplexer employing RX, MX and TX channel filter with inline topology by introducing the NRNs. The interaction between the channel filters through the specific four-port junction employed in the triplexer is taken into account during the synthesis and the best performances are obtained when the three channels of the triplexer are very close (and even contiguous). The procedure begins with the iterative evaluation of the polynomials associated to the overall triplexer once suitable constraints are imposed on the reflection and transmission parameters of the triplexer. The characteristic polynomials [5-6] of the three channel filters are then evaluated with a polynomial fitting technique, and the synthesis of these filters is realized separately from the triplexer by extracted-pole configuration.

Traditionally, the channel filters are implemented with cross-coupled topology. In this paper, the channel filters are realized by using extracted-pole technique. Because of this, the synthesis of the channel filters is simpler and the topologies of channel filters are inline configuration. Inline configurations in which each finite transmission zero is generated and independently controlled by a dedicated element. A good feature of the extracted-pole configuration is easily tunable of the channel filters (the resonant frequency of the extracted resonators coincides with the imposed finite transmission zeros).

To the best of the authors' knowledge, there is no exact synthesis procedure in the literature for triplexers employing inline configuration filters with NRNs.

This paper is organized as follows. In Section 2, the characteristic polynomials of the triplexer are calculated. It takes the interaction of the channel filters into account, so the characteristic polynomials are accurately obtained. A triplexer example is synthesized to show the validity of the technique in Section 3. A conclusion is drawn in Section 4.

II. Calculation of Triplexer Characteristic Polynomials

A microwave triplexer is generally composed of three bandpass filters with three input ports connected through a four-port junction. The general schematization of the triplexer is shown in Figure 1.

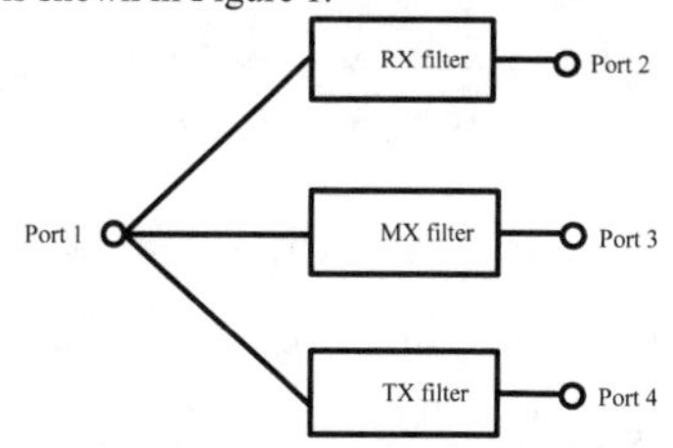

Figure 1 The general schematization of the triplexer

The three low-pass prototype filters (RX, MX and TX) in the normalized frequency domain can be characterized separately from the triplexer through their characteristic polynomials as:

978-7-5641-4279-7

$$S_{11}^{RX}=\frac{F_{RX}(s)}{E_{RX}(s)}\quad S_{21}^{RX}=\frac{P_{RX}(s)}{E_{RX}(s)}=\frac{p_{0RX}P_{RXn}(s)}{E_{RX}(s)}$$
$$S_{11}^{MX}=\frac{F_{MX}(s)}{E_{MX}(s)}\quad S_{21}^{MX}=\frac{P_{MX}(s)}{E_{MX}(s)}=\frac{p_{0MX}P_{MXn}(s)}{E_{MX}(s)} \tag{1}$$
$$S_{11}^{TX}=\frac{F_{TX}(s)}{E_{TX}(s)}\quad S_{21}^{TX}=\frac{P_{TX}(s)}{E_{TX}(s)}=\frac{p_{0TX}P_{TXn}(s)}{E_{TX}(s)}$$

The polynomials $E_{RX}(s)$ and $F_{RX}(s)$ have degree N_{RX} (order of RX channel filter), the polynomials $E_{MX}(s)$ and $F_{MX}(s)$ have degree N_{MX} (order of MX channel filter), the polynomials $E_{TX}(s)$ and $F_{TX}(s)$ have degree (order of TX channel filter). The finite transmission zeros of RX, MX and TX channel filters can completely define the polynomials $P_{RXn}(s)$, $P_{MXn}(s)$, and $P_{TXn}(s)$. The coefficients p_{0RX} , p_{0MX} , and p_{0TX} are determined by the return loss at passband limits [5].

The S-parameters of the triplexers can be defined using four polynomials as follows

$$S_{11}=\frac{N(s)}{D(s)}\qquad S_{21}=\frac{p_{0r}P_r(s)}{D(s)}$$
$$S_{31}=\frac{p_{0m}P_m(s)}{D(s)}\qquad S_{41}=\frac{p_{0t}P_t(s)}{D(s)} \tag{2}$$

The highest degree coefficients of $N(s)$, $D(s)$, $P_r(s)$, $P_m(s)$ and $P_t(s)$ are imposed equal to 1 with p_{0r}, p_{0m} and p_{0t} suitable normalizing coefficients. Note that the roots of $D(s)$ represent the poles of the network, the roots of $N(s)$ are the reflection zeros at the common node of the triplexer (port 1 in Figure 1), and the roots of $P_r(s)$, $P_m(s)$ and $P_t(s)$ are the transmission zeros in the RX, MX and TX path, respectively.

The triplexer polynomials can be derived using the admittances at the input ports of the RX, MX and TX channel filters, taking into account the interaction between the channel filters through junctions. The following expressions can be obtained:

$$\begin{cases} N(s)=S_{TX}S_{RX}S_{MX}-D_{TX}S_{RX}S_{MX}-D_{RX}S_{TX}S_{MX}-D_{MX}S_{TX}S_{RX} \\ D(s)=S_{TX}S_{RX}S_{MX}+D_{TX}S_{RX}S_{MX}+D_{RX}S_{TX}S_{MX}+D_{MX}S_{TX}S_{RX} \\ P_r(s)=P_{RXn}S_{TX}S_{MX},\quad p_{0r}=p_{0RX} \\ P_m(s)=P_{MXn}S_{TX}S_{RX},\quad p_{0m}=p_{0MX} \\ P_t(s)=P_{TXn}S_{RX}S_{MX},\quad p_{0t}=p_{0TX} \end{cases} \tag{3}$$

where

$$\begin{cases} S_{RX}=\dfrac{E_{RX}+F_{RX}}{2},\quad D_{RX}=\dfrac{E_{RX}-F_{RX}}{2} \\ S_{MX}=\dfrac{E_{MX}+F_{MX}}{2},\quad D_{MX}=\dfrac{E_{MX}-F_{MX}}{2} \\ S_{TX}=\dfrac{E_{TX}+F_{TX}}{2},\quad D_{TX}=\dfrac{E_{TX}-F_{TX}}{2} \end{cases} \tag{4}$$

In this paper, once the triplexer polynomials are obtained, the channel filters are synthesized by extracted-pole technique [7]. The example is shown in the next section.

III. SYNTHESIS EXAMPLE

In order to demonstrate the validity of the triplexer synthesis technique presented in this paper, a triplexer is synthesized in the following part.

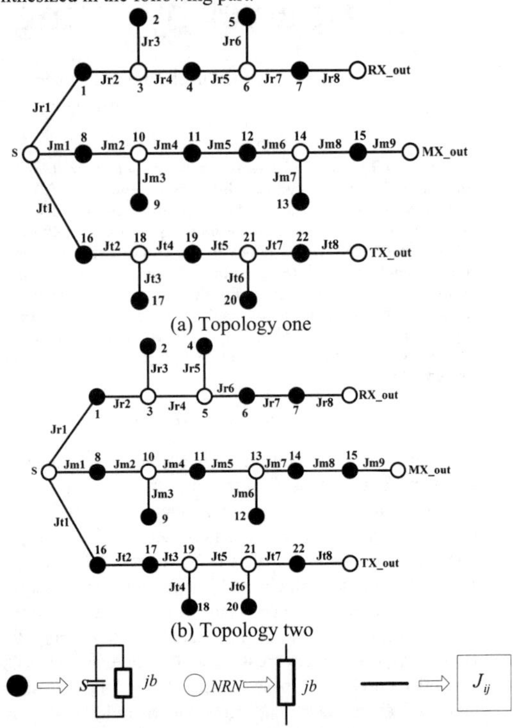

Figure 2 The topology of the triplexer

The topology of the triplexer is shown in Figure 2, By introducing NRNs, the topology of the channel filters is simple (inline topology) and more flexible. The specification of the triplexer is listed in table I.

Table I. The specifications of the triplexer example

	RX channel filter	MX channel filter	TX channel filter
Pass-band	1890MHz-1910MHz	1920MHz-1940MHz	1950MHz-1970MHz
Filter order	5	6	5
Return loss	20 dB	20 dB	20 dB
Finite zeros	1870MHz, 1930MHz	1900MHz, 1960MHz	1940MHz, 1980MHz

The characteristic polynomials of the channel filters are evaluated with technique illustrated in Section II, obtaining the following values (the polynomial coefficients are reported in descending order):

$$F_{RX} = [1 \quad -0.0248+3.8633j \quad -5.8955-0.0446j \quad 0.0127-4.4341j \quad 1.6403-0.0128j \quad 0.0058+0.2383j]$$
$$E_{RX} = [1 \quad 0.5407+3.8633j \quad -5.7496+1.6520j \quad -1.8394-4.1055j \quad 1.4013-0.8823j \quad 0.1536+0.1823j]$$
$$F_{MX} = [1 \quad 0.0515-0.0607j \quad 0.0955-0.0028j \quad 0.0033-0.0040j \quad 0.0022-0.0001j \quad 0 \quad 0]$$
$$E_{MX} = [1 \quad 0.4493-0.0607j \quad 0.1951-0.0226j \quad 0.0447-0.0080j \quad 0.0089-0.0015j \quad 0 \quad 0]$$
$$F_{TX} = [1 \quad -0.0267-3.9016j \quad -6.0181+0.0527j \quad 0.0239+4.5797j \quad 1.7163+0.0067j \quad 0.0046-0.2529j]$$
$$E_{TX} = [1 \quad 0.5233-3.9016j \quad -5.8816-1.6152j \quad -1.8204+4.2687j \quad 1.4872+0.8858j \quad 0.1568-0.1985j]$$

From the characteristic polynomials of the channel filters above evaluated, the normalized low-pass prototypes can be synthesized using, for instance, the technique in [7]. The result is shown as follows:

For figure 2 (a), the result is

$$[Jr1, Jr2, Jr3, Jr4, Jr5, Jr6, Jr7, Jr8] = [0.5318, 1, 3.4252, 0.7573, 1, 4.4467, 1.3714, 0.5072]$$
$$[b1, b2, b3, b4, b5, b6, b7] = [0.922, 1.5133, 17.0456, 0.7488, -0.0105, -28.74, 0.6822]$$
$$[Jm1, Jm2, Jm3, Jm4, Jm5, Jm6, Jm7, Jm8, Jm9] = [0.446, 1, 3.8263, 0.8078, 0.1454, 1, 4.6058, 1.3918, 0.4986]$$
$$[b8, b9, b10, b11, b12, b13, b14, b15] = [0.0365, 0.7454, 21.1157, 0.017, -0.0361, -0.7545, -31.1822, -0.0717]$$
$$[Jt1, Jt2, Jt3, Jt4, Jt5, Jt6, Jt7, Jt8] = [0.5244, 1, 2.1564, 0.7767, 1, 2.6956, 1.3771, 0.4993]$$
$$[b16, b17, b18, b19, b20, b21, b22] = [-0.7829, -0.2596, 11.5653, -0.7557, -1.2443, -18.9269, -0.8553]$$

For figure 2 (b), the result is

$$[Jr1, Jr2, Jr3, Jr4, Jr5, Jr6, Jr7, Jr8] = [0.5318, 1, 3.4252, 1, 0.9782, 0.22, 0.2175, 0.5074]$$
$$[b1, b2, b3, b4, b5, b6, b7] = [0.922, 1.5133, 16.2903, -0.0105, -1.324, 0.7123, 0.7476]$$
$$[Jm1, Jm2, Jm3, Jm4, Jm5, Jm6, Jm7, Jm8, Jm9] = [0.446, 1, 3.8263, 0.8078, 1, 4.8281, 1.0481, 0.2101, 0.4986]$$
$$[b8, b9, b10, b11, b12, b13, b14, b15] = [0.0365, 0.7454, 21.1157, -0.0124, -0.7545, -33.9752, -0.0419, -0.0085]$$
$$[Jt1, Jt2, Jt3, Jt4, Jt5, Jt6, Jt7, Jt8] = [0.5244, 0.1982, 1, 2.7762, 1, 0.4532, 0.2316, 0.4985]$$
$$[b16, b17, b18, b19, b20, b21, b22] = [-0.8694, -0.7138, 17.5572, -0.2596, -0.478, -1.2443, 0.8553]$$

The extraction of a coupling matrix of the triplexer network that contained both resonating nodes and NRNs is fundamentally identical to the synthesis of the standard coupled resonator triplexers [4]. The prototypes must be then de-normalized, imposing the physical configuration of actual resonators and coupling structures. The coefficients are given by the following expressions:

Resonant-Resonant Coupling:

$$k_{ij} = FBW * J_{ij}$$

Resonant-Nonresonant Coupling:

$$k_{ij} = sqrt(FBW) * J_{ij}$$

Nonresonant-Nonresonant Coupling:

$$k_{ij} = J_{ij}$$

The responses of the two circuits in Figure 2 can be calculated from a simple nodal analysis and is shown in Figure 3. We can see that the two responses are plotted simultaneously along with the ideal characteristic polynomials, but they are all indistinguishable. It can be clearly seen that all the specifications of the triplexer are met.

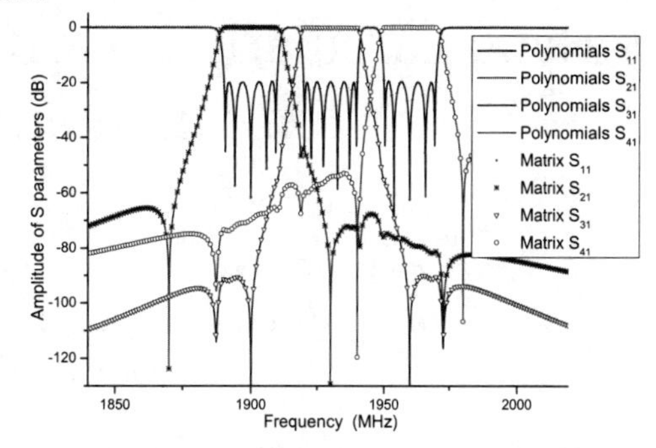

Figure 3 The responses of the two circuits in figure 2

IV. CONCLUSIONS

A novel generalized technique for synthesizing microwave triplexer is presented in this paper. The characteristic polynomials of the triplexer are evaluated and the interactions between the channel filters are taken into account. Then the characteristic polynomials of channel filters are obtained from the triplexer. The extracted-pole technique is used to synthesis the channel filter. The circuit responses (coupling matrix response) agree with the characteristic polynomials well. A synthesized example shows the validity of the new technique.

ACKNOWLEDGMENT

This work was supported by the National High Technology Research and Development Program of China (863 Program) No.2012AA01A308 and the National Natural Science Foundation of China (NSFC) under Project No.61271017.

REFERENCES

[1] Macchiarella, G., Tamiazzo, S. Novel approach to the synthesis of microwave diplexers. IEEE Trans. Microwave Theory Tech., Dec. 2006, vol. MTT-54, no. 12, p. 4281-4290.

[2] K. L. Wu and W. Meng, "A direct synthesis approach for microwave filter with a complex load and its application to direct diplexer design," IEEE Trans. Microwave Theory Tech., vol. 55, no. 5, pp. 1010–1017, May 2007.

[3] G. Macchiarella and S. Tamiazzo, "Design of triplexer combiners for base stations of mobile communications," IEEE MTT-S Int. Microw. Symp. Dig., Anheim, CA, May. 2010. pp. 429-432.

[4] G. Macchiarella and S. Tamiazzo, "Synthesis of star-junction multiplexers," IEEE Trans. Microw. Theory Tech, Vol.58, pp. 3732-3741, Dec. 2010.

[5] Richard J. Cameron, "General coupling matrix synthesis methods for chebyshev filers functions," IEEE Trans. Microwave Theory Tech., Vol. 47, pp.433-442, April. 1999.

[6] Richard J. Cameron, "Advanced coupling matrix synthesis techniques for microwave filter," IEEE Trans. Microwave Theory Tech., Vol. 51, pp.1-10, Jan. 2003.

[7] Smain Amari, Giuseppe Macchiarella, "Synthesis of inline filters with arbitrarily placed attenuation poles by using Non-resonating nodes," IEEE Trans. Microwave Theory Tech., Vol. 53, pp.3075-3081, Oct. 2005.

S-Band Circularly Polarized Crossed Dipole Antenna for Automotive Applications

Jui-Hung Chou*, Ding-Bing Lin**, Che-Hsu Lin***, and Hsueh-Jyh Li****
*Graduate Institute of Communication Engineering, Taiwan University, Taipei, Taiwan
**Department of Electronic Engineering, Taipei University of Technology, Taipei, Taiwan
***Graduate Institute of computer Communication, Taipei University of Technology, Taipei, Taiwan
****Department of Electrical Engineering, Taiwan University, Taipei, Taiwan
d98942001@ntu.edu.tw

***Abstract*- This paper presents a circularly polarized crossed dipole antenna mounted on a 1-m circular ground plane for reception of satellite digital radio signals within the S-band in automotive applications. The main radiator consists of a pair of dipole antennas connected to separate nonresonant lines of unequal length to achieve the current are in phase quadrature and right hand circular polarization (RHCP). A two-ray model utilized to provide a fast estimation method to analyze the radiation pattern characteristic of the proposed antenna. The simulation results show the ground plane influences on the radiation pattern of the antenna at specific elevation angles for satellite broadcasting system.**

Keywords- corssed dipole, automotive, right hand circular polarization (RHCP), satellite broadcasting

I. INTRODUCTION

Nowadays, the satellite broadcasting system for automotive applications are rapidly spreading in America and Europe. One of the important reasons is that the drivers of long-distance commuting already tired of hearing a lot of noise and poor quality of audio and video contents through traditional radio stations. This situation could be improved by using the satellite broadcasting system for broadcasting the high quality video and audio channels. The vehicle manufacturers in manufacturing process of the initial stage to the satellite radio equipment installed on the interior of the vehicle, satellite radio service provider can provide what kind of content has also became the major sticking point to attract customers.

ONDAS Media is the largest supplier of satellite broadcasting services in the Europe. It provides customers with hundreds of channels of digital radio, such as music, news, traffic, sports and data programming all over the Europe [1]. The system mainly operate in the Highly Elliptical Orbit (HEO) based on three satellites which receive the signal within 8-hours each one to achieve the purpose of the signal is not interrupted, different from traditional satellites in geosynchronous orbit (GEO) signal is likely to be blocked. It will provide high-quality service of digital entertainment to 240 million vehicles and 600 million European inhabitants a day in the future. Due to the ONDAS system has many advantages to listen to the radio, this system is subject to the attention of many vehicle manufacturers which have invested in technology of ONDAS satellite radio receiver embedded in the interior of the vehicle. ONDAS system is pursuing frequencies in S-band and L-band, respectively. The S-band comprises two portion of 30 MHz in the 1980 ~ 2010 MHz for the transmit (Tx) band and 2170 ~ 2200 MHz for the receive (Rx) band, allocated by the International Telecommunications Union (ITU) basis to the Mobile Satellite Service (MSS).

Since the receiving antenna design of the ONDAS satellite broadcasting system is not very common in the currently academic literature, the investigation of the Satellite Digital Audio Radio Service (SDARS) is another popular type of Digital Audio Radio Service (DARS) provider operating in the American in this paper, such as the Sirius satellite radio and XM satellite radio. Each satellite broadcasting provider uses different satellite and antenna requirements but their common features of the antenna like omnidirectional radiation pattern characteristics of the antenna at certain range of the elevation angle, especially at low elevation angles, reflections and shadowing effects will significantly change the antenna radiation pattern [2], simple structure, production of low-cost, to allow antennas to be incorporated onto a vehicle roof.

Table I shows the specifications and targets desired of the SDARS and ONDAS antennas for use with mobile satellite communication services, it is observed that the target minimum gain of the antenna is set to +2 dBic for Sirius satellite radio applications at elevation angles between 25^0~90^0 and for XM satellite radio applications at elevation angles between 20^0~60^0, both of them designed with the characteristics of the left hand circular polarization(LHCP), but the demand for gain values at high and low elevation angles of the ONDAS antennas different from the SDARS antennas, the minimum average gain of +4 dBic is required at elevation angle 90^0, +5 dBic is required at elevation angle between 45^0~75^0, and a value of +3 dBic is required at elevation angle 30^0, according to the comparison of the ONDAS antenna and SDARS antenna specifications, the overall required gain of the ONDAS antennas is more stringent than the SDARS antennas. Therefore, the choice of appropriate antenna structure is very important to receive satellite signals for ONDAS satellite broadcasting system. The next section will explore different types of the antennas for satellite broadcasting systems.

978-7-5641-4279-7

TABLE I
SPECIFICATIONS ON THE ANTENNA FOR SDARS AND ONDAS APPLICATIONS

Parameters	SIRIUS	XM	ONDAS
Frequency band (MHz)	2320~2332.5	2332.5~2345	2170~2200
Antenna polarization	LHCP	LHCP	RHCP or LHCP
Antenna gain (dBic)	+2~+4 (EA 90^0-25^0)	+2~+4 (EA 60^0-20^0)	+4 (EA 90^0)
			+5 (EA 75^0-45^0)
			+3 (EA 30^0)

*Note : The antenna is placed on the center of a 1-m diameter ground plane.

Many kinds of antenna have been proposed for satellite mobile communication in pervious literatures. In [3-5], an integrated antenna solution for GPS and SDARS services was proposed, the main radiating elements commonly used patch antennas, stacked in the form of design to reduce the size. The patch antenna is a very promising candidate for satellite broadcasting system due to the structure is easy to fabrication, low profile, low cost, maximum gain in the zenith direction and conformability to vehicle surfaces. Quadrifilar helical antenna also widely used for satellite communication [6-7], it has a good circular polarization characteristic, wide beamwidth and low back-lobe radiation characteristic allows it to be operated with or without a ground plane, especially in vehicles application is a very important property. In [8-9], a circularly polarized beam steering patch array antenna for mobile satellite communications is proposed, the design provides stable beam switching and generates beams that can cover the azimuth angles to achieve average gain and low circular polarization demand, but this way will increase an additional circuit area and cost.

This paper presents a circularly polarized crossed dipole antenna mounted on a 1-m circular ground plane for reception of satellite signals within the S-band (2170 ~ 2200 MHz) in automotive applications. The crossed dipole antenna structure is very suitable for satellite communications owing to it has good circular polarization characteristics, easy to fabrication and a hemispherical quasi-isotropic circularly polarized radiation pattern which can improve antenna gain at low elevation angles, especially mounted onto a vehicle roof will significantly reduce the radiation performance of low elevation angles [10]. The main radiator consists of a pair of dipole antennas connected to separate non-resonant lines of unequal length to achieve the current are in phase quadrature and right hand circular polarization (RHCP). A two-ray model utilized to provide a fast estimation method to analyze the radiation pattern characteristic of the proposed antenna. The simulation results show the ground plane influences on the radiation pattern of the antenna at specific elevation angle for satellite broadcasting system.

II. CIRCULARLY POLARIZED CROSSED DIPOLE ANTENNA STRUCRURE

The geometry of the proposed crossed dipole antenna structure for ONDAS satellite broadcasting system application operation in the S-band (2170 ~ 2200 MHz) is shown in Fig. 1. The proposed antenna consists of a pair of dipole antennas connected to separate nonresonant lines of unequal length of dipole arm A and arm B, fabricated by copper of wifth 0.8 mm and conductivity 5.8×10^7 (S/m), and fed by a 50-Ω coaxial cable with its dimension of 2.2×60 mm^2, to achieve the current are in phase quadrature and RHCP. In addition, because the crossed dipole is center fed directly from the coaxial cable, it will make the outer conductor of the coaxial cable to be a part of the radiating system. Thus, a sleeve balun which apply to proposed structure with its total length of a quarter wavelength of the center frequency 2185 MHz is required. The total length of dipole arm A and arm B is about 69.6 mm and 58 mm, respectively.

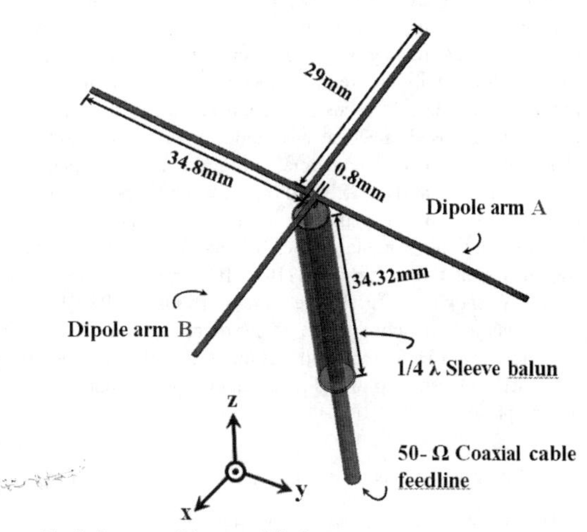

Fig. 1. Geometry of the crossed dipole antenna

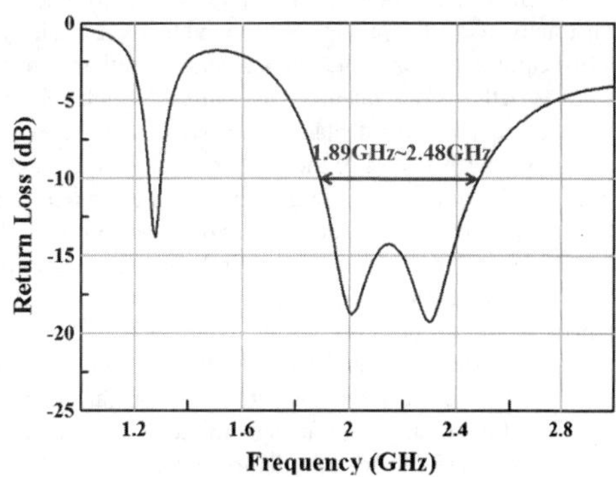

Fig. 2. Simulated return loss for the proposed antenna

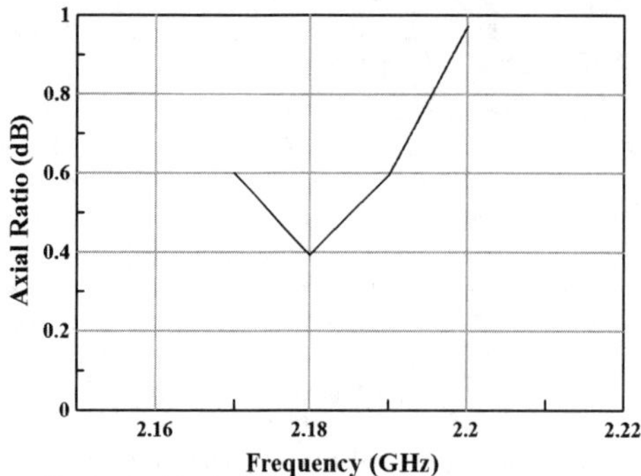

Fig. 3. Simulated axial ratio for the proposed antenna

Fig. 2 shows the simulated return loss for the constructed prototype with dimensions given in Fig. 1. The simulated bandwidth is 1890 ~ 2480 MHz (590 MHz) based on 2:1 VSWR, the impedance bandwidth shows a good impedance matching and could completely operate in S-band for ONDAS satellite broadcasting system. Fig. 3 represents the Axial Ratio (AR) at S-band in the zenith direction (+z direction), it is observed that the AR bands in S-band are lower than 1 dB, thus the proposed antenna has good circular polarization characteristics. The commercial program high frequency structure simulator (HFSS) based on the finite-element method (FEM) is used for analyzing the behavior of proposed model and determining suitable values of parameters [11]. The simulated radiation patterns at 2185 MHz in the x-z and y-z planes are shown in Fig. 4. The pattern are mainly RHCP for +z direction corresponding to the requirement of ONDAS antennas, and LHCP for –z direction, this radiation pattern characteristics can help to design of the antenna mounted on a ground plane in next section.

III. CP Crossed Dipole Antenna Monted on a Ground

The integration of various automobile communication services of antenna is a serious challenge for engineers. One of the most important factors is the mounting environment will significantly affect the properties of the antenna, especially for satellite communications, the average gain of antennas may be defined strictly. Consequently, the antenna mounted on a circular or square ground plane extensively used to examine the radiation performance of the antenna, and the measurement results will be adopted [2]. Fig. 5 shows the crossed dipole antenna mounted on the center of a 1-m circular ground plane as a vehicle roof, where h is the height between the horizontal crossed dipole antenna and the ground plane, due to the presence of the ground, the far field radiation pattern is the resultant of a direct wave and a wave reflected from the ground plane. To acquire the radiation pattern of the far field, it is convenient to simplify and create an equivalent model to analysis of the performance of radiation pattern.

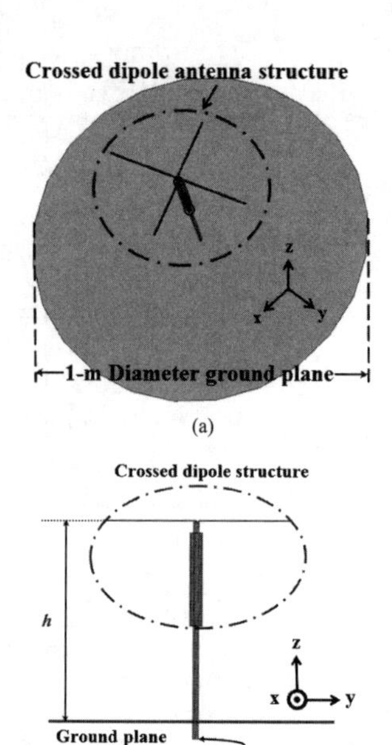

Fig. 5. Geometry of the crossed dipole antenna mounted on the center of a 1-m circular ground plane (a) full view and (b) side view

In Fig. 6, the equivalent two-ray model of antenna mounted above an infinite ground plane is proposed. Herein, it is assumed that the ground plane is perfectly conducting and the tangential electric field should be zero at the surface of ground plane. Therefore, the reflected wave must have a phase reversal of 180° at the reflection point to satisfy the boundary condition. The direct and reflected waves are toward the point G at far filed region. In this model, the ground plane can be replaced by an image of antenna element at the same distance h below the ground plane on the basis of image theory [10]. The current distribution on the antenna and image one are equal in magnitude but opposite phase. Hence the tangential electric field must be zero along the entire ground plane which is the reason that the ground plane can be replaced by image element. To obtain the total field at point G, the overall antenna structure consisting of the antenna and image one should be considered to be in the form of the end-fire array, therefore, the radiation pattern characteristic will be affected by the array factor and pattern factor, as shown in Fig. 7. The gain in field intensity of the antenna can be calculated in following expressions：

$$G_F(\theta) = |2\sin(h_0 \cos\theta)| \quad (1)$$

Fig. 7 and Fig. 8 show the comparison of calculated and simulated radiation pattern by varying the height h to operate at the 2185 MHz, respectively. The simulated results show a good agreement with the calculated results to verify this model is feasible. From the results, if the height h was about 0.25λ or less, the maximum radiation is always in the zenith direction (elevation angle 90^0), as the height increases to 0.3λ, 0.4λ, 0.5λ or 1λ, the maximum radiation is in elevation angle between 0^0-90^0. Fig. 9 show the comparison of the average gain of specific elevation angles by varying the height h between the antenna and the ground plane, if h varies from 0.25λ to 1λ, the average gain of high elevation angle 90^0 will be decreased significantly, furthermore, as the height is nearly 0.5λ or higher, the radiation pattern in certain directions may generate additional null points, this result for ONDAS antenna requirements is not a good situation for reception of satellite signals. If the height selected about 0.25λ or less, the average gain of low elevation angles between 45^0-30^0 cannot meet the specifications of ONDAS antenna. Therefore, the best height of h should be between 0.25λ and 0.5λ, finally, the height is fine tuned to h=0.36λ.

IV. Conclusions

A circularly polarized crossed dipole antenna mounted on a 1-m circular ground plane for reception of satellite digital radio signals within the S-band (2170 ~ 2200 MHz) for ONDAS satellite broadcasting system is proposed. The crossed dipole antenna structure is very suitable for satellite communications owing to it has good circular polarization characteristic and a hemispherical radiation pattern to achieve a stable signal reception. In this paper, a two-ray model utilized to provide a fast estimation method to obtain the desired radiation pattern for satellite broadcasting system applications, the simulated results show a good agreement with the calculated results, and a good performance in the radiation efficiency can be reached, thence, the proposed antenna structure can be a very promising candidate for ONDAS satellite broadcasting system.

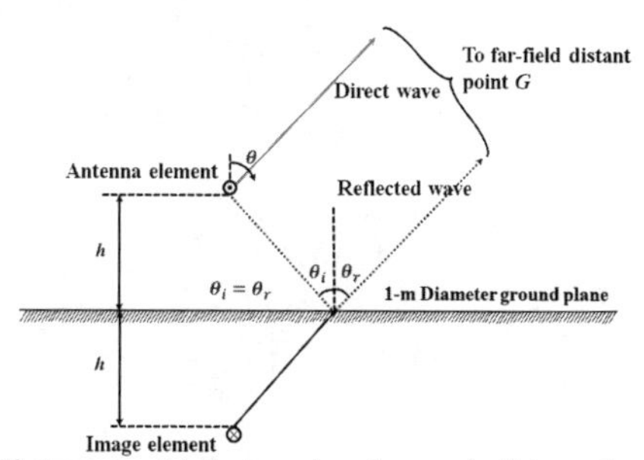

Fig. 6. The two-ray model of antenna above the ground with image element

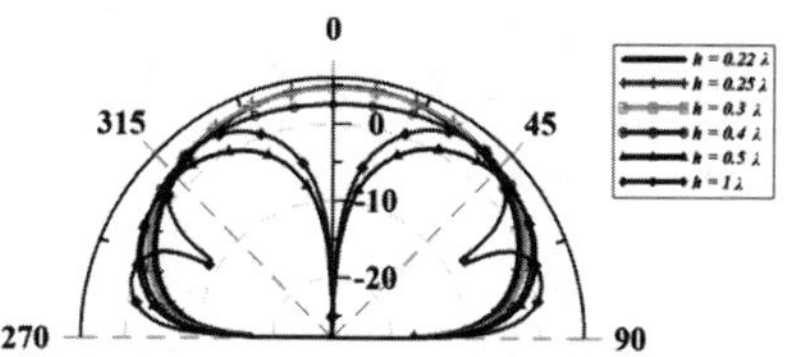

Fig. 7. Calculated radiation pattern for various height of h at vertical-plane

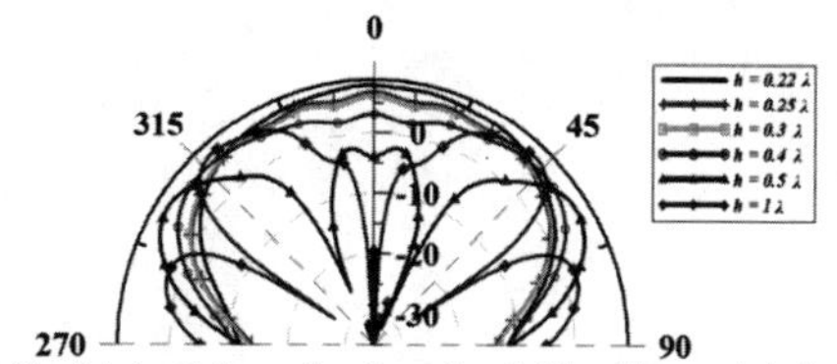

Fig. 8. Simulated radiation pattern for various height of h at vertical-plane

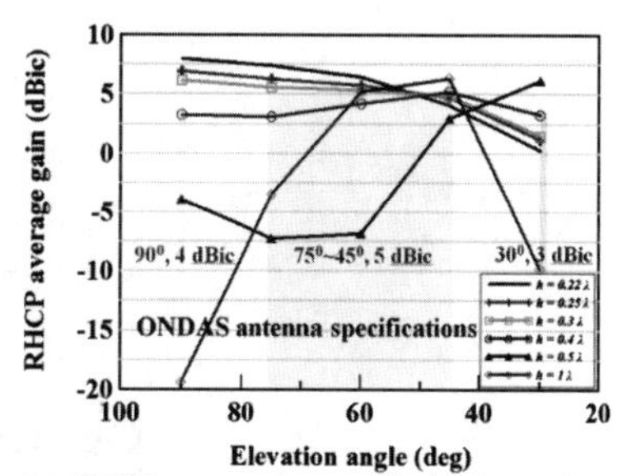

Fig. 9. Simulated RHCP average gain at specific elevation angles

References

[1] ONDAS media. http://www.ondasmedia.com/ceo.htm

[2] R. Kronberger, G.-H. Hassmann, and S. Schulz, "Measurement and analysis of vehicle influences on the radiation pattern of SDARS-antennas," IEEE Antennas and Propagation Society International Symposium, vol. 4, pp. 740–743, 2002.

[3] F. Mariottini, M. Albani, E. Toniolo, D. Amatori, and S. Maci, "Design of a Compact GPS and SDARS Integrated Antenna for Automotive Applications," IEEE Antennas and Wireless Propagation Letters, vol. 9, pp. 405–408, 2010.

[4] K. Geary, J.H. Schaffner, H.-P. Hsu, H.J. Song, J.S. Colburn, and E. Yasan, "Single-Feed Dual-Band Stacked Patch Antenna for Orthogonal Circularly Polarized GPS and SDARS Applications," IEEE 68th Vehicular Technology Conference, pp. 1–5, 2008.

[5] J.M. Ribero, G. Kossiavas, R. Staraj, and E. Fond, "Dual-frequency antenna circularly polarized for GPS-SDARS operation," The Second European Conference on Antennas and Propagation(EuCAP 2007), pp. 1–5, 2007.

[6] S. Shoaib, W.A. Shah, A. Fahim Khan, and M. Amin, "Design and implementation of quadrifilar helix antenna for satellite communication," International Conference on Emerging Technologies(ICET 2010), pp. 230–233, 2010

[7] H. Jing, S. Xin, and Y. Hongchun, "Design of a high gain quadrifilar helix antenna for satellite mobile communication," China-Japan Joint Microwave Conference Proceedings (CJMW 2011) , pp. 1–3, 2011

[8] J.T.S. Sumantyo, K. Ito, and M. Takahashi, "Dual-band circularly polarized equilateral triangular-patch array antenna for mobile satellite communications," IEEE Transactions on Antennas and Propagation , vol. 53, pp. 3477–3485, 2005

[9] M. Konca and S. Uysal, "Circular multi-directional patch antenna array with selectable beams using a novel feed structure and equilateral triangular patches," Mediterranean Microwave Symposium (MMS 2010) , pp. 440–443, 2010

[10] John D. Kraus and Ronald J. Marhefka, Antennas：For All Applications, New York : McGraw-Hill, 2002.

[11] Ansoft HFSS user's Guide-High Frequency Structure Simulator,Ansoft Co., 2003.

TA-1(A)

October 24 (THU) AM

Room A

EurAAP/COST

A General Technique for THz Modeling of Vertically Aligned CNT Arrays

Jiefu Zhang, Yang Hao
School of Electronic Engineering and Computer Science, Queen Mary, University of London
Mile End Road
London E1 4LT, UK

***Abstract*-In this paper a general technique for the simulation of the single- and multi-wall carbon nanotubes is proposed based on a multi-conductor transmission line (MTL) model. The analytical form of the 2-port S-parameter is calculated from the 2n-port transfer matrix of the CNT array, which takes the mutual couplings of CNTs into consideration. The result is compared with that of previously used models. The simulated input impedance and absorption provides information of CNTs for potential THz applications.**

I. INTRODUCTION

Carbon nanotube (CNT) is a seamless hollow cylinder with the wall thickness of only one single carbon atom. Since its discovery, it has attracted an increasing research interest due to its extraordinary electronic and mechanical properties. The vertically aligned CNT arrays can easily be synthesized with chemical vaporation deposition (CVD) and has potential applications as THz antennas, absorbers and VLSI interconnects. To explore its applications requires the study of modeling techniques, which have since been worked on by many researchers. Lüttinger liquid theory was first used as a model of the CNTs' electrical properties in GHz and an RF transmission line (TL) model was derived from the theoretical analysis that predicts the wave velocity and characteristic impedance of a metallic singe wall CNT [1]. Also, from 1-D electron fluid model a similar TL model was derived [2]. To reduce the high DC resistance of an isolated CNT, bundles of parallel CNTs have also been studied and an effective single conductor (ESC) TL model of both a stand-alone multi-wall CNT (MWCNT) and a single-wall CNT (SWCNT) array was derived in [3] and [4]. Finally, a diameter dependent circuit model was proposed in [5].

In the aforementioned ESC models however the magnetic inductance between CNTs are usually neglected, resulting in a simplification of the mutual coupling in the array. There have been studies that suggests when the CNT bundle becomes large, such simplification may not be valid and drew the conclusion that inductances should not be neglected [6]. In this paper we propose a full MTL model analysis and calculate the 2n-port transfer matrix of a CNT array. By incorporating with the terminal conditions the 2-port input impedance, S-parameters and absorption can be solved analytically with all the mutual coupling taken into calculation. Finally, the results of this technique and that of previous studies are compared and the potential application at THz is discussed.

II. MODELING TECHNIQUES FOR CNT ARRAYS

The geometry of the cross section of CNT array above a sufficiently large, perfect conducting ground plane is shown in Fig. 1.

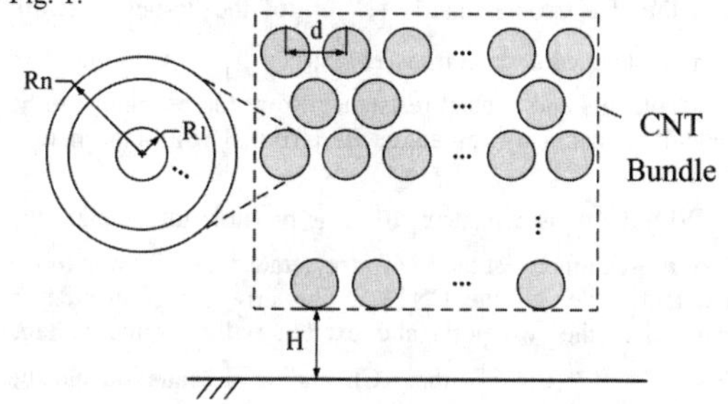

Figure 1. Configuration of MWCNT array above ground

The CNT inside the array can be either single wall or multi wall. In general, a CNT has n concentric shells, with the inner-most shell radius R_1 and the outer-most shell radius R_n. The distance between neighboring shells is set to be 0.34nm, as determined by the Van de Waals gap and the distance between the outer-most shell of the CNT at the bottom of the array and the ground plane is H. If n=1, the CNT becomes a SWCNT, which can be modeled as several 1-D parallel channels connected in parallel. According to different chirality, SWCNTs can be categorized as armchair, zigzag or chiral, whose number of conducting channels N_c depends on the chirality, radius and temperature. Thus, a SWCNT can be either metallic or semiconducting. Armchair SWCNTs are always metallic with $N_c = 2$ and for zigzag and chiral SWCNTs the average number of conducting channels n in an array follows [4].

$$N_c = \begin{cases} k_1 \cdot T \cdot r + k_2 & \text{r>650/T} \\ 2/3 & \text{other} \end{cases}, \tag{1}$$

where $k_1 = 7.74\times10^{-4}$, $k_2 = 0.2$ and T is temperature. In this paper, room temperature (300K) is assumed in all simulations.

The difference from the TL model of CNTs and conventional TL is that apart from the electrostatic inductance

978-7-5641-4279-7

and capacitance, the kinetic inductance and quantum capacitance are also present due to the quantum band structure and 1-D electron movement. The per unit length RLC parameters of each shell can be described as [4]

$$R_i = \frac{h}{4n_i e^2 v_F \tau}, L_{k_i} = \frac{h}{4n_i e^2 v_F}, C_{q_i} = \frac{4n_i e^2}{h v_F}, \quad (2)$$

where τ is the diffusion time defined as the division of the mean free path (MFP) l_{eff} and the Fermi velocity v_F.

The mutual coupling between shells of MWCNT and neighboring SWCNTs in an array can be modeled with the well-defined electrostatic equations. From the MTL analysis [7] the matrix formed telegrapher's equation can be solved and thus the transfer matrix $[\Gamma]_{2n\times 2n}$ of the 2n-port network can be obtained and the transfer matrix $[\Phi]_{2n\times 2n}$ with quantum resistance Rs and contact resistance Rc at the terminal can be calculated using by cascading the terminal resistance matrix and Γ.

With the transfer matrix $[\Phi]$, it is possible to calculate the 2-port S-parameter of the CNT array, under the common mode excitation, i.e. all the CNTs in the array is connected in parallel at the two ports and excited with a same voltage $\vec{V_s} = [V_s \ V_s \ ... \ V_s]^T$, the KCL and KVL equation can be easily written. Then, by further exploiting the common mode excitation conditions, the original KCL and KVL equations that contain 2n unknowns, i.e. the current flowing into and out of each CNT in the array, can be turned into a new set of equations with only 2 unknowns, namely the total current flowing into and out of the entire CNT array, which is sufficient to calculate the S-parameters of the CNT array.

$$\begin{cases} \sum \vec{I}(0) = \dfrac{\sum\sum\left([\Phi_{22}]\cdot[\Phi_{12}]^{-1}\right) + P}{1 + z_0 \sum\sum\left([\Phi_{22}]\cdot[\Phi_{12}]^{-1}\right) + z_0 \cdot P} \cdot V_s \\ \sum \vec{I}(l) = \dfrac{\sum\sum[\Phi_{12}]^{-1}\cdot\left[V_s - z_0 \sum I(0)\right]}{1 + z_0 \sum\sum\left([\Phi_{12}]^{-1}\cdot[\Phi_{11}]\right)}, \end{cases} \quad (3)$$

where z_0 is the reference impedance at the source and load and

$$P = \frac{z_0\left[\sum\sum[\Phi_{21}] - \sum\sum\left([\Phi_{22}]\cdot[\Phi_{12}]^{-1}\cdot[\Phi_{11}]\right)\right]\sum\sum[\Phi_{12}]^{-1}}{1 + z_0 \sum\sum\left([\Phi_{12}]^{-1}\cdot[\Phi_{11}]\right)}.$$

Note that in this process only terminal conditions and matrix transformation techniques are used and no simplification of the mutual coupling is made. Thus the result keeps the full cross-talk information inside the array.

Once the total port currents are solved, the input impedance, S-parameters and absorption for the CNT array can be easily calculated from (3).

III. RESULTS AND DISCUSSION

A. a stand-alone MWCNT

First of all, four stand-alone MWCNTs are simulated using the proposed model and the ESC model. The MWCNTs have a distance of 1um above ground and a length of 10um. The inner: outer radii are 2:4, 5:10, 10:20 and 8:24 respectively. As the estimated input impedance of CNTs is very high at several kilo Ohms, the reference impedance z0 is set to the quantum diffusion resistance of the metallic SWCNT (6.5kΩ) and neglect the contact resistance which appears in reality where the connector touches the CNTs. Thus, a rough matching is achieved to make the S-parameter more visible. The simulation frequency is from 0.1 to 2 THz.

The simulated S11 and S21 results of the above four MWCNTs using MTL (solid) and ESC (dashed) model is shown in Fig. 2 (a) and (b). Clearly, the CNT with larger outer: inner radius ratio tends to give slightly more errors. Nevertheless the two models give almost identical results, indicating that neglecting the mutual coupling in a stand-along MWCNT does not affect much the simulation results. This is expected because the walls in a MWCNT behave like conductors that shield the cross-talk between the inner and outer walls and restrict it only between directly neighboring walls [10]. Thus the influence of the mutual coupling is largely reduced and can be neglected.

The simulated input impedance Z_{in} is shown in Fig. 2 (c). The real and imaginary part both changes periodically across the frequency, showing the Fabry-Perot resonance due to the mismatching at terminals. Within each period, the imaginary part $[\mathrm{Im}]Z_{in}$ cross the zero point twice, where the real part is the

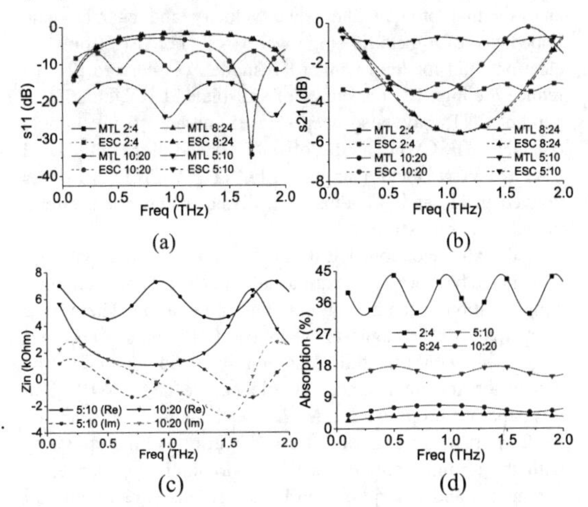

Figure 2. The simulated (a) S11 and (b) S21 results of four MWCNTs with different radius using proposed MTL model and ESC model.

minimum and maximum point, respectively. Also, for these two points, the S11 and S21 in Fig. 2 (a) and (b) reaches either maximum or minimum at the same time, according to the value of the reference impedance z_0 chosen. This is very interesting because while the minimum point of $[\mathrm{Re}]Z_{in}$ can be used as the working frequency for antennas, the maximum point of $[\mathrm{Re}]Z_{in}$ is potentially suitable to be used with THz photomixers, as they generally have a very high and real output impedance.

The absorption in Fig. 2 (d) shows that CNTs have very good absorbing capabilities, due to the large intrinsic impedance of the quantum wire. The smaller the radius of a CNT, the better absorber it is, as the conducting channels are connected parallel, the increase in their number reduces the overall resistance in the transmission. The maximum and minimum of the absorption of each CNT also appears at the two zeros points of the input impedance, where the minimum of $[\mathrm{Re}]Z_{in}$ corresponds to the absorption peak and the minimum to the absorption valley, suggesting that the application with photomixers would have a higher efficiency than with antennas.

B. SWCNT arrays

Next, four SWCNT arrays are simulated using the proposed model and ESC model. The length of the arrays and distances above the ground remains the same as the MWCNT case, and the radius of the SWCNTs inside the array is 10nm. The array sizes are 2×2, 5×5, 10×10 and 20×20, respectively.

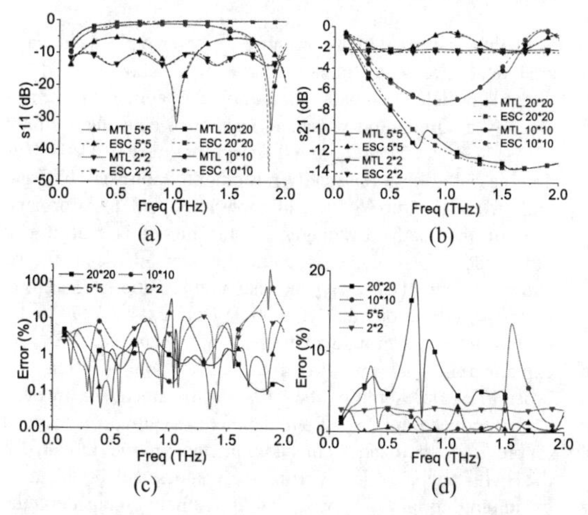

Figure 3. The simulated (a) S11 and (b) S21 results and the relative error of (c) S11and (d) S21 between the proposed MTL model and ESC model for four different SWCNT arrays.

The simulated S-parameters are shown in Fig. 3 (a) and (b). Compared to the results of the stand-alone MWCNT, the SWCNT array shows more errors, due to the non-shielded coupling between every two CNTs inside the array. Several resonant points are present across the frequency, where the difference between two models is the largest. The relative errors are more explicitly shown in Fig. 3 (c) and (d). As seen, the maximum error for S11 occurs at the resonant frequency, where the magnitude of the input impedance gets closest to the reference impedance z_0 while the error for S21 is mainly because of the lacking of consideration of mutual coupling when the wave propagates along the CNT array. The maximum error of S11 increases with the increase of the array size and for a 10*10 array it can reach more than 100%. As the simulated frequency is not high enough for the first resonant point to appear in a 20*20 array, the maximum is not shown, but it can be expected to be even higher than the 10*10 array. Similarly, the error of S21 in the 20*20 array, which is 19.08%, is also the highest in all four arrays simulated. So the conclusion for SWCNT arrays is that the ESC model is not suitable as the large mutual coupling between CNTs is no longer negligible in THz. This conclusion is supported by [6].

IV. CONCLUSION

In this paper a general approach of S-parameter analysis for an arbitrary MWCNT bundle has been proposed based on the MTL modeling techniques. A comparison with previously used ESC model shows that for large CNT arrays the ESC model introduce large errors. The S-parameters analysis shows that there are several resonant points periodically distributed across the frequency band. Due to large input impedance, in terms of interconnection application, the size of the CNT and bundle must be such that the input impedance can achieve a matching to the reference impedance in the desired frequency band. The input impedance simulation provides potential THz application with high output impedance devices. Finally, the absorption analysis shows that for small bundles, one of the potential applications is wideband nano-absorbers that can be more effective than traditional absorbers while still being much more compact in size.

REFERENCES

[1] Burke, P.J., "Luttinger liquid theory as a model of the gigahertz electrical properties of carbon nanotubes," *Nanotechnology, IEEE Transactions on*, vol.1, no.3, pp.129,144, Sep 2002.

[2] A. Maffucci, G. Miano, and F. Villone., "A transmission line model for metallic carbon nanotube interconnects," *Int. J. Circuit Theory Appl.* 36, 1, pp. 31-51, J an 2008.

[3] M. S. Sarto, A. Tamburrano, "Single-Conductor Transmission-Line Model of Multiwall Carbon Nanotubes", IEEE Trans. On Nanotechnology, Vol. 9, No. 1, Jan. 2010.

[4] M.S. Sarto, A. Tamburrano, "Electromagnetic Analysis of Radio-Frequency Signal Propagation Along SWCN Bundles", IEEE Conf. on Nanotechnology, Jun. 2006, Vol. 1, pp. 201-204.

[5] Maffucci, A.; Miano, G.; Villone, F., "A New Circuit Model for Carbon Nanotube Interconnects With Diameter-Dependent Parameters," *Nanotechnology, IEEE Transactions on*, vol.8, no.3, pp.345,354, May 2009.

[6] Nieuwoudt, A.; Massoud, Y., "Understanding the Impact of Inductance in Carbon Nanotube Bundles for VLSI Interconnect Using Scalable Modeling Techniques," *Nanotechnology, IEEE Transactions on*, vol.5, no.6, pp.758,765, Nov. 2006.

[7] C. R. Paul, Analysis of Multiconductor Transmission Lines, John Wiley & Sons, 26 Oct. 2007.

[8] Thomas Ch. Hirschmann, Paulo T. Araujo, Hiroyuki Muramatsu, Xu Zhang, Kornelius Nielsch, Yoong Ahm Kim, and Mildred S. Dresselhaus, "Characterization of Bundled and Individual Triple-Walled Carbon Nanotubes by Resonant Raman Spectroscopy", *ACS Nano* 2013 *7* (3), 2381-2387 .

Efficient numerically-assisted modelling of grounded arrays of printed patches

M. García-Vigueras *, F. Mesa †, R. Rodríguez-Berral †, F. Medina ‡ and J. R. Mosig *.

* Laboratory of Electromagnetics and Acoustics, EPFL, 1015 Lausanne, Switzerland.
† Dept. de Física Aplicada 1, Universidad de Sevilla, 41012 Seville, Spain.
‡ Dept. de Electrónica y Electromagnetismo, Universidad de Sevilla, 41012 Seville, Spain.

***Abstract*— A circuit model is here proposed to characterize grounded arrays of printed patches. This equivalent circuit reproduces the behavior of the structure in a wide frequency range with the only restriction of dealing with single resonant scatterers. Full-wave data is required to build some elements of the equivalent circuit. However, only a reduced number of simulations is needed, in contrast to other available techniques.**

***Index Terms*— Circuit modeling, high impedance surface,frequency-selective surfaces, transmission lines.**

I. INTRODUCTION

Planar periodic arrays of scatterers printed on grounded dielectric substrates are well known for allowing their reflection phase and surface-wave properties to be tuned [1]. Basically, they can be seen as metal-backed printed frequency selective surfaces (FSS) [2]. In principle, with these compound surfaces it is possible to synthesize any desired boundary condition, from perfect electric conductor (PEC) to artificial magnetic conductor (AMC), and due to this reason they are also named as high impedance surfaces (HIS) [1]. Particularly, HIS perform as AMC around their resonance frequency, where they fully reflect incident waves with zero or near-zero phase shift. These composite layers also exhibit electromagnetic band gaps (EBG), which implies that no surface wave propagation is allowed at certain frequencies [3]. These interesting features have triggered the investigation of HIS and their use to suppress surface waves and to enhance the performance of antennas and microwave circuits. As a classical example, the profile of reflector-based antennas can be significantly reduced by employing an AMC ground plane, since it does not reverse the phase of reflected waves [1]. Aperiodic configurations of these surfaces can be found in many other microwave devices such as reflectarrays [4], printed/holographic/modulated leaky-wave antennas [5]–[7] and recently they have been also considered as metasurfaces [8]. In addition, considerable scientific attention has recently been drawn towards periodic planar structures not only as FSS layers [2], but also due to the advent of extraordinary optical transmission [9].

The previous context has motivated extensive research on the circuit modeling of arrays of scatterers printed on grounded dielectrics during the last decades. Analytical expressions have been derived using homogenization procedures (such as in [10]), these approaches are accurate but only in the long wavelength regime. More versatile methods can also be found in the literature, as the one proposed in [11], but, as they are essentially rational fitting procedures, they rely on considerable amounts of full-wave simulations. Recently, a very simple circuit model has been proposed that can predict the response of generic FSS systems [12]. Still, this procedure needs an *a priori* knowledge of the full-wave system response and it is limited to scenarios where high-order modal interaction is not relevant. In order to develop accurate, wideband and versatile models, the physical insight of the structure needs to be captured, which is the goal of the present work. For this purpose, it is of key importance to characterize properly the launch of surface waves in the dielectric slab as well as the interaction between the scattered reactive field and the ground plane.

In this contribution we propose a very simple and quasi-analytical circuit to model arrays of patches printed on grounded dielectric slabs. Some of the authors presented recently a simple and wideband transverse equivalent circuit (TEN) to characterize printed arrays of patches [14]. This TEN is based on a physically meaningful approach [13], and built from sections of transmission lines and transformers. One of the main advantages of this model is that it can be extended in order to account for new elements in the structure without affecting its basic configuration. Therefore, in this contribution, the TEN in [14] is extended in order to account for a ground plane backing the printed FSS. As explained in [14], the aid of full-wave tools is needed in order to extract some of the TEN key parameters. However, a very reduced number of simulations is required, in contrast to previous approaches. In addition, the present model can deal with any electrical size of the patches and polarization of impinging plane waves, and it explicitly considers the dependence with the angle of incidence and the characteristics of the dielectric slab. As the employed approach assumes that the current profile on the patch does not change significantly with frequency, scatterers with more complex geometries can also be considered as long as they are single resonant.

II. DERIVATION OF THE CIRCUIT MODEL

In this section, a simple model is presented to characterize the diffraction at an array of patches printed on a grounded dielectric slab. The geometry of the array unit cell is depicted

978-7-5641-4279-7

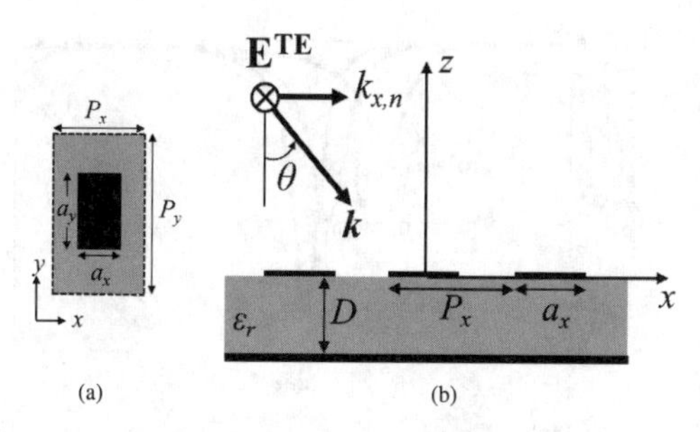

Fig. 1. (a) Patch array unit cell. (b) 2-D sketch of patch array printed on metal-backed dielectric slab. The obliquely incident field is TE-polarized, and impinges in the xz plane.

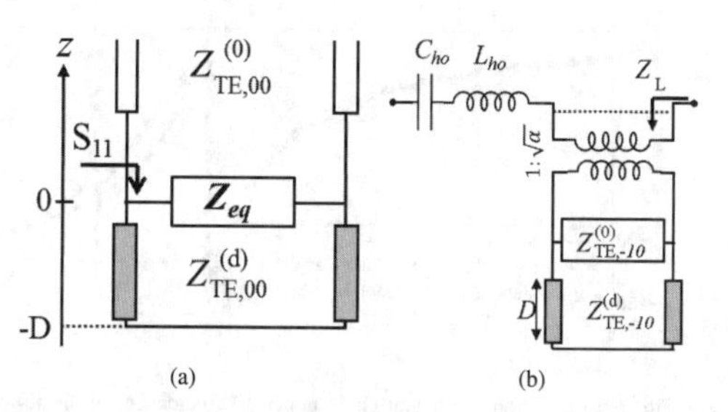

Fig. 2. (a) Tranverse equivalent circuit to model Fig. 1. (b) Equivalent impedance of the discontinuity for Fig. 2(a) (Z_{eq}).

in Fig. 1(a), whereas Fig. 1(b) shows a 2-D scheme of the compound structure illuminated by an oblique TE-polarized plane wave. In order to characterize this scenario we propose the tranverse equivalent circuit (TEN) of Fig. 2(a). This TEN is built following the approach presented by some of the authors in [14]. Thus, the main concepts of the approach will be next briefly summarized due to the lack of space. Basically, the periodicity of the array allows us to express the original scenario as the equivalent problem of a discontinuity in a rectangular waveguide of dimensions $P_x \times P_y$. The incident plane wave is considered as the TE zero-th order Floquet harmonic of the array (TE_{00}), and fundamental mode of the waveguide. Due to the polarization of the impinging wave, the equivalent waveguide is formed by two perfect electric walls parallel to the *xz*-plane, and two Floquet walls parallel to the *yz*-plane. When the impinging wave reaches the discontinuity at $z = 0$ (the metallic patch), it excites all the higher-order harmonics supported by the array. This interaction is modeled by the impedance Z_{eq} detailed in Fig. 2(b). In particular, the excitation of all the high-order TM harmonics is accounted by the the lumped capacitance C_{ho}, whereas L_{ho} accounts for the of all the high-order TE harmonics appart from the one with lowest cutoff frequency (the TE_{-10} [14]). This last harmonic is modeled explicitly by Z_L, which corresponds to the input impedance seen by the TE_{-10} mode at both sides of the discontinuity. The transformer of ratio α accounts for

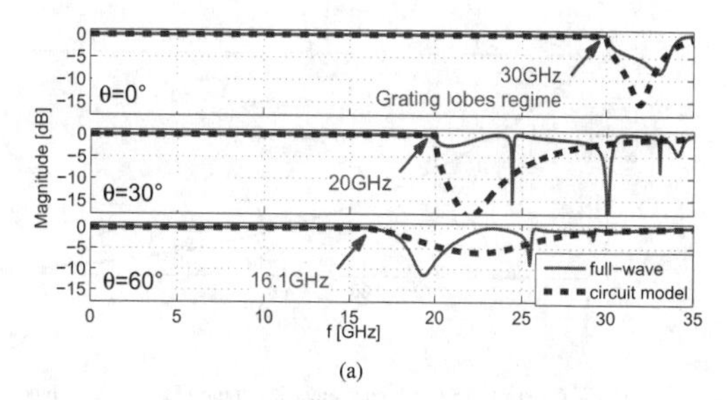

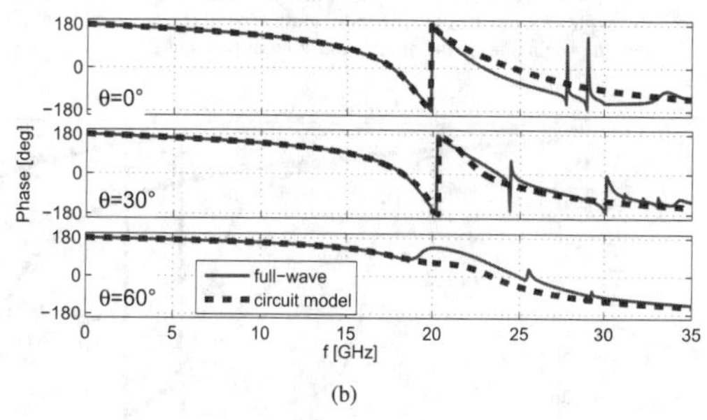

Fig. 3. Reflection coefficient under various angles of incidence for the structure of Fig. 1 with $\varepsilon_r = 4.5$, $d = 1.5\,\text{mm}$, $P_x = P_y = 10\,\text{mm}$, $w_x = 8.75\,\text{mm}$ and $w_y = 0.5\,\text{mm}$.

the degree of excitation of this last harmonic. The value of the the two lumped elements C_{ho} and L_{ho}, together with the ratio α, can be computed from three numerical values of the reflection coefficient provided by a few full-wave simulator.

The main novelty of this contribution with respect to [14] resides in the incorporation of the metallic sheet that is backing the printed array at $z = -D$. This extension has been done straightforward just by short-circuiting the end of the transmission-line sections that account for the dielectric slab. The interaction between the excited harmonics in the array and the ground plane is accuratelly modeled in this simple way thanks to the topology of the circuit model.

Although the proposed approach is very efficient from a numerical point of view, its major strength lies in its capability to provide a convenient frame to study the problem. The simplicity of the equivalent circuits makes it possible to accurately predict the qualitative behavior of the structures, and also to estimate the potential effects of changes in the geometry and characteristics of the repeated element of the periodic surface.

III. RESULTS

The accuracy of the previous simple model is here validated by means of some illustrative examples. The equivalent circuit results are compared with full-wave simulations based on Method of Moments (MoM). Fig. 3 shows the magnitude and phase of the reflection coefficient of a HIS under various

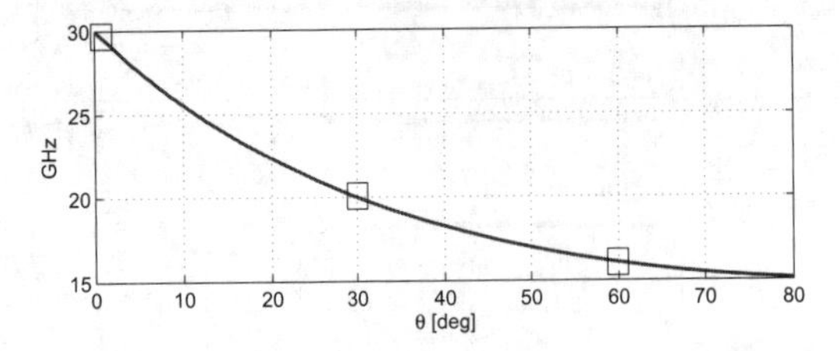

Fig. 4. Onset of the first higher order harmonic (TE_{-10}) as a function of the angle of incidence at a periodic grating with $P_x = P_y = 10$ mm (which sets the beginning of the grating lobes regime in Fig. 3).

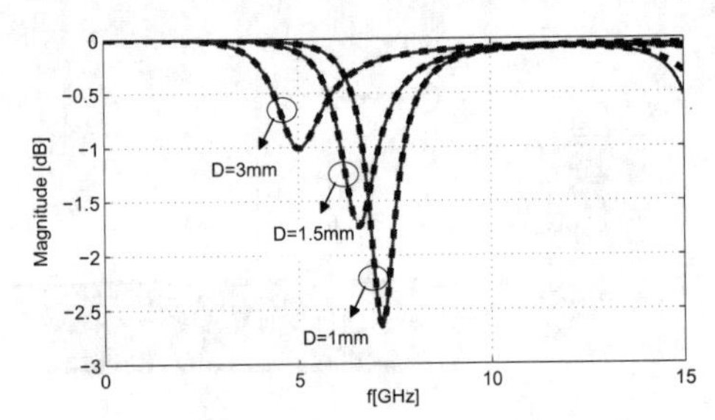

Fig. 6. Reflection coefficient magnitude under normal incidence for the structure of Fig. 1 ($\varepsilon_r = 4.5 - j0.088$, $P_x = P_y = 10$ mm and $w_x = w_y = 8.75$ mm). Solid lines: MoM simulations; dashed lines: circuit model.

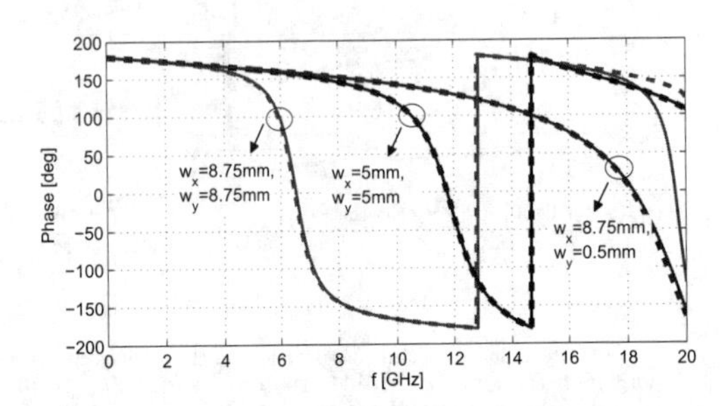

Fig. 5. Reflection coefficient phase under 30° incidence for the structure of Fig. 1 ($\varepsilon_r = 4.5$, $d = 1.5$ mm and $P_x = P_y = 10$ mm). Solid lines: MoM simulations; dashed lines: circuit model.

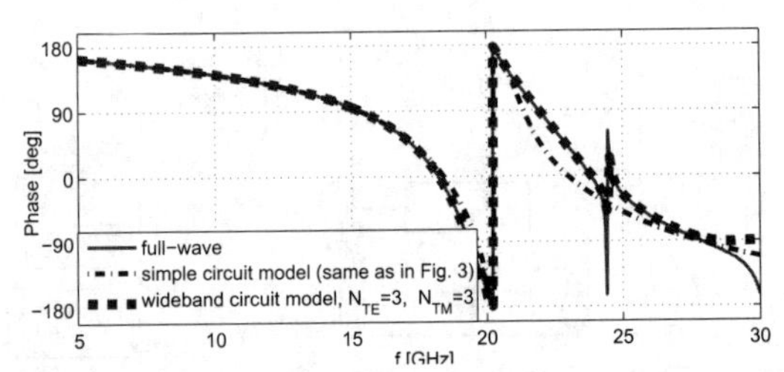

Fig. 7. Reflection coefficient phase under 30° incidence for the structure of Fig. 1 ($\varepsilon_r = 4.5$, $P_x = P_y = 10$ mm, $w_x = 8.75$ mm and $w_y = 0.5$ mm).

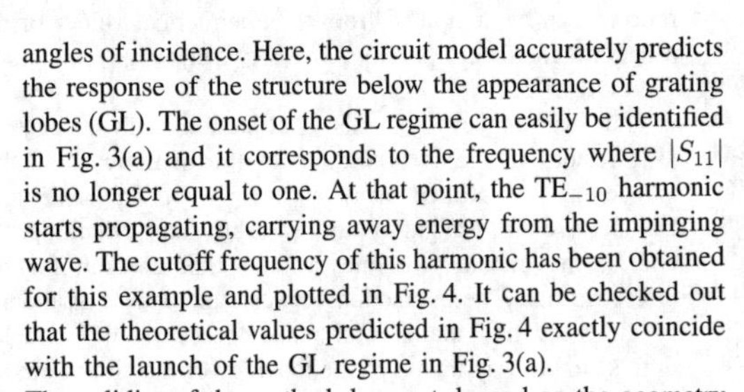

angles of incidence. Here, the circuit model accurately predicts the response of the structure below the appearance of grating lobes (GL). The onset of the GL regime can easily be identified in Fig. 3(a) and it corresponds to the frequency where $|S_{11}|$ is no longer equal to one. At that point, the TE_{-10} harmonic starts propagating, carrying away energy from the impinging wave. The cutoff frequency of this harmonic has been obtained for this example and plotted in Fig. 4. It can be checked out that the theoretical values predicted in Fig. 4 exactly coincide with the launch of the GL regime in Fig. 3(a).

The validity of the method does not depend on the geometry of the patch, and it is proven by the results of Fig. 5. Lossy substrates can be also characterized just by considering their complex dielectric constant in the model. This case is specially interesting for practical purposes, as it allows to analyze the reflection losses of the structure. In addition, the physical insight of the model allows to identify the main causes of this losses. The results plotted in Fig. 6 prove that the circuit model accurately predicts the reflection amplitude reduction below the GL regime due to the losses, even when the substrate is very thin. The thickness of the substrate becomes an important constraint for some models based on effective permittivity approximations [12].

Finally, it should be highlighted that this simple model can be easily extended in order to increase its bandwidth. The previous results were obtained accounting explicitly for the frequency behavior of just one higher order harmonic, the TE_{-10} one. However, accounting for more higher order harmonics, the HIS response can be reproduced at higher frequencies. Note that the individual characterization of each harmonic adds a new unknown to the circuit model (its associated excitation coefficient), and therfore, one more full-wave simulation is needed. As an example, the $\theta = 30^o$ results in Fig. 3(b) are improved in Fig. 7 by considering the frequency behavior of the first three higher order TE and TM harmonics ($N_{TE} = N_{TM} = 3$, following the nomenclature in [14]).

Conclusions

A very simple circuit model is here proposed to characterize the incidence of a TE plane-wave at a patch array printed on a grounded dielectric slab. The guidelines given in [14] are followed in order to build the model. Complex electromagnetic interactions taking place in the structure are accounted for, thus allowing to predict the response of the structure in a wide frequency range. The model can account for lossy dielectrics and patches of different size. TM incidence could be also modeled applying the same following rationale.

So far, the circuit models proposed by the authors have been employed to make plane-wave diffraction analysis of periodic structures, the so-called *study of reflection* [11]. However, these models could also be used to perform *study of dispersion*, obtaining the modal solutions of the structure, which is currently under investigation. Aperiodic structures such as reflectarrays or inhomogeneous metasurfaces could also be synthesized with this technique by making local periodicity assumptions. In addition, due to its capability to seize the underlying physics of the structure, the proposed circuits could also be applied to model devices at higher frequencies. For example, these artificial surfaces could be designed at teraherzs assuming that metals are good conductors in this frequency range.

ACKNOWLEDGMENT

This work was supported by the Spanish Ministerio de Ciencia e Innovación and European Union FEDER funds under Project TEC2010 – 16948.

REFERENCES

[1] D. F. Sievenpiper, L. Zhang, R. F. J. Broas, N. G. Alexopoulos and E. Yablonovich, "High-impedance electromagnetic surfaces with a forbidden frequency band,"*IEEE Trans. Microwave Theory Techniques*, vol. 47, pp. 2059-2074, 1999.

[2] B. Munk, *Frequency Selective Surfaces: Theory and Design*, Edt. John Wiley and Sons, 2000.

[3] S. Fan, P. Villeneuve and J. Joannopoulos, "Large omnidirectional band gaps in metallodielectric photonic crystals,"*Phys. Rev. B.*, vol. 56, pp. 11245-11251, 1996.

[4] D. M. Pozar, S.D. Targonski and H.D. Syrigos, "Design of millimeter wave microstrip reflectarrays,"*IEEE Trans. Antennas Propagat.*, vol. 45, no. 2, pp. 287-296, Feb 1997

[5] A.A. Oliner and Hessel, "Guided waves on sinusoidally-modulated reactance surfaces,"*IRE Trans. Antennas Propagat.*, vol. 7, no. 5, pp. 201-208, December 1959

[6] M. Garcia-Vigueras, P. DeLara-Guarch, J. L. Gomez-Tornero, R. Guzman-Quiros and G. Goussetis, "Efficiently illuminated broadside-directed 1D and 2D tapered Fabry-Perot leaky-wave antennas,"*Proc. 6th European Conference on Antennas Propagat., EuCAP*, pp. 247-251, March 2012

[7] D. Sievenpiper, J. Colburn, B. Fong, J. Ottusch and J. Visher, "Holographic artificial impedance surfaces for conformal antennas,"*IEEE APS/URSI Symposium*, Washington, DC, July 2005.

[8] P.S. Kildal, A.A. Kishk and S. Maci, "Special Issue on Artificial Magnetic Conductors, Soft/Hard Surfaces, and Other Complex Surfaces,"*IEEE Trans. Antennas Propagat.*, vol. 53, no. 1, pp. 2-7, Jan 2005

[9] T. W. Ebbesen, H. J. Lezec, H. F. Ghaemi, T. Thio, and P. A. Wolff, "Extraordinary optical transmission through sub-wavelength hole arrays,"*Nature*, vol. 391, pp. 667–669, Feb. 1998.

[10] O. Luukkonen, C. Simovski, G. Granet, G. Goussetis, D. Lioubtchenko, A. V. Räisänen, and S. A. Tretyakov, "Simple and accurate analytical model of planar grids and high-impedance surfaces comprising metal strips or patches,"*IEEE Trans. Antennas Propagat.*, vol. 56, no. 6, pp. 1624-1632, June 2008.

[11] S. Maci, M. Caiazzo, A. Cucini, and M. Casaletti, "A pole-zero matching method for EBG surfaces composed of a dipole FSS printed on a grounded dielectric slab," *IEEE Trans. Antennas Propag.*, vol. 53, no. 1, pp. 70-81, Jan. 2005.

[12] F. Costa, A. Monarchio and G. Manara "Efficient Analysis of Frequency-Selective Surfaces by a Simple Equivalent-Circuit Model,"*IEEE Trans. Antennas Propagat. Magazine*, vol. 54, no. 4, pp. 35-48, Aug. 2012.

[13] F. Medina, F. Mesa and R. Marqués, "Extraordinary transmission through arrays of electrically small holes from a circuit theory perspective," *IEEE Trans. Microw. Theory Tech.*, vol. 56, no. 12, pp. 3108-3120, Dec. 2008.

[14] M. García-Vigueras, F. Mesa, F. Medina, R. Rodríguez-Berral, J. L. Gómez-Tornero, "Simplified circuit model for metallic arrays of patches sandwiched between dielectric slabs under arbitrary incidence," *IEEE Trans. Antennas Propagat.*, vol. 60, no. 10, pp. 4637-4649, Oct. 2012.

Correlation Between Far-field Patterns on Both Sides of the Head of Two-port Antenna on Mobile Terminal

Ahmed Hussain, Per-Simon Kildal
Antenna Group, Dept. Signals & Systems
Chalmers University of Technology
Gothenburg, Sweden
ahmed.hussain@chalmers.se

Ulf Carlberg, Jan Carlsson
Dept. of Electronics
SP Technical Research Institute of Sweden
Borås, Sweden
ulf.carlberg@sp.se

***Abstract*—We present an electromagnetic simulation of a practical mobile terminal with a two-port MIMO antenna on left and right sides of the head. The computed far-field patterns include the effect of head and hand phantoms. We show that the far-field patterns on the two sides are strongly correlated if they are presented in the coordinate system of the phone, but they are completely uncorrelated in the coordinate system of the environment.**

I. Introduction

Several studies related to different user practices of holding a single-port mobile terminal have been performed during the last decade, see e.g. [1]. However, this has not been extensively done for the latest LTE mobile handsets that have multi-port antennas with MIMO (Multiple-input Multiple-output) technology. It is still unclear how different user practices will affect the performance of these mobile terminals. Therefore, it is important to re-evaluate the impact of different user practices, such as holding handsets on both sides of the head and to check how far-field patterns change from one side of the head to the other. The study of the latter is the purpose of this paper.

Reference [2] proposed new systematic approach to dealing with user statistics via considering an ensemble of arbitrary users in a random Line-of-Sight (LOS) environment. It is argued that the horizontal plane of far-field function of a phone on the left side of the head is similar to the vertical place of far-field function of the same phone on the right side of the head, and vice versa. This makes the phone experience incoming waves in the horizontal plane as if they come from all directions in space, when observed over time from positions on both sides of the head. This means that the environment appears 3-D random, even if it is not. In other words, the purpose of this paper is to show that the far-field functions on the two sides of the head are correlated when presented in the coordinate system of the phone. Both the coordinate system of the environment with vertical v-axis, and the coordinate system of the phone are shown in Fig. 1 (a) for a user holding the phone on the left and right sides. We see that the xyz-coordinate system of phone on both sides of the head is oriented 90 degrees with respect to each other while the uvw-coordinate system of environment is the same on both sides.

Therefore, we will compute the far field functions of a mockup of a multiport terminal on both sides of the head, and we will characterize these to see how they change.

II. EM Simulations Using CST Microwave Studio

EM simulations are performed using CST Microwave Studio. The performance is evaluated in terms of S-parameters and far-field patterns. The multiport mobile terminal model is a practical mockup of a mobile handset with 2-port MIMO antenna. The antennas used in this mobile terminal are Planar Inverted-F Antennas (PIFA) operating in the frequency ranges of 0.7–1.0 GHz and 1.7–3.2 GHz. The PIFA antennas are located along each short side of the mobile terminal. The terminal is located on either side of the head phantom according to the standard cheek position [3].

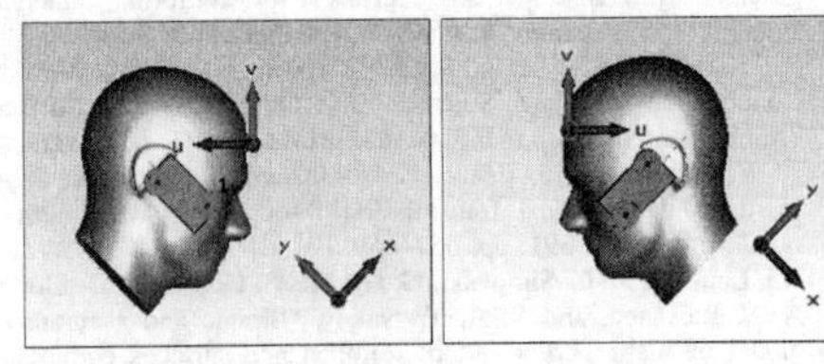

Figure 1 (a). SAM head phantom, hand phantom, and phone mockup in CST on both sides of the head. The environment uvw-coordinate system and phone xyz-coordinate system are shown.

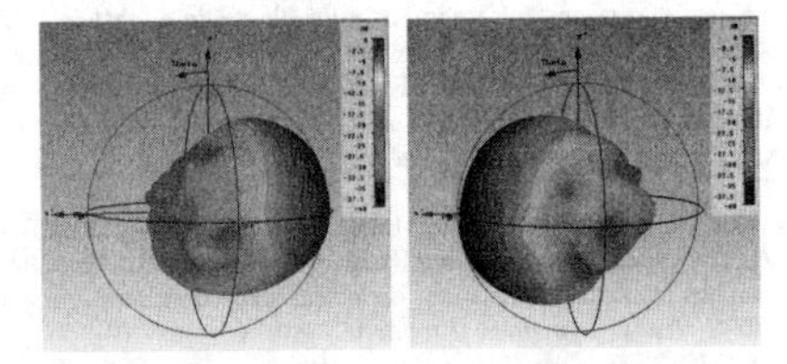
Figure 1 (b). Simulated absolute value of E-field from port 2 on phone mockup on left and right sides of head at 2.5 GHz. Polarization is not visible for such plots.

978-7-5641-4279-7

III. Far-Field Patterns in Different Coordinate Systems

In Fig. 2(a) and 2(b) we compare the simulated far-field patterns for phone on both sides of the head in the coordinate system of the environment. We see that when the phone is on the left side of the head, the horizontal E-field component (i.e. E-phi) in the vertical plane have a similar pattern as the vertical E-field component (i.e. E-theta) in the horizontal plane when the phone is on the right side, and visa versa. In other words, the patterns are diagonally similar in Fig. 2 (a) and Fig. 2 (b), which we want to show.

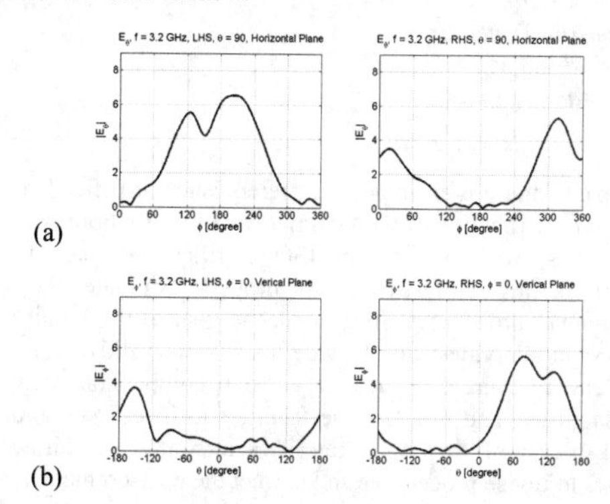

Figure 2 (a). E-theta component in horizontal plane of mobile phone on left and right sides of head phantom, respectively.
Figure 2 (b). E-phi component in vertical plane of mobile phone on left and right sides of head phantom, respectively.

IV. Far-Field Correlations Using VIRM-LAB

We will now study the similarities between the far-field patterns on both sides of the head in more detail, by evaluating different correlation integrals using the MATLAB based tool called ViRM-lab [4]. The formulas to calculate complex and envelope correlation are given in [5].

V. Results & Discussion

The results for complex correlation are presented in Fig. 3. It shows that the far-field patterns on both sides of the head are uncorrelated when they are presented in the coordinate system of the environment; see Fig. 3 (top). In other words, this means that the LOS channel experienced by the mobile terminal on right side of the head will be completely different from the left side of the head. But the same far-field patterns are correlated when they are presented in the coordinate system of the mobile terminal; see Fig.3 (bottom). This also shows that the assumption in the introduction is correct. The results also show that the effect of hand makes the two patterns less similar.

VI. Conclusion

The conclusion is that the shape of the far-field function of the studied mobile terminal remains similar when it is used on either side of the head, if the coordinate system of the presentation is aligned to the mobile terminal. However, the coordinate system is rotated with respect to the environment, so the far-field function in the horizontal plane on one side becomes a far-field function in the vertical plane on the other side, and with orthogonal polarizations. This may not be generally valid, but it is logical and we have proven it for this specific mobile phone mockup. The effect of the hand phantom changes the correlation between the two orthogonal planes on the two sides.

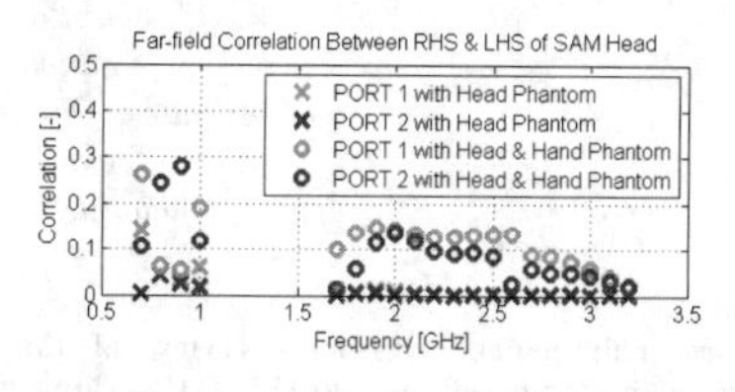

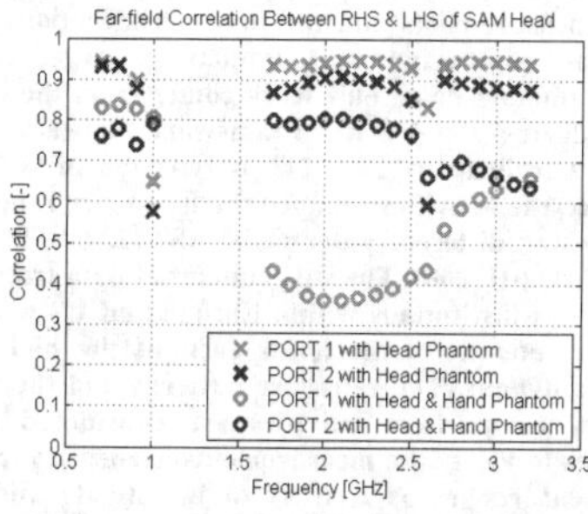

Figure 3. Complex correlation between far-field functions on right and left sides of the head when the coordinate system is aligned to the environment (top) and to the terminal (bottom).

Acknowledgement

The authors acknowledge Sony for providing CST model of a multiport mobile phone mockup. This work is supported in part by Swedish Governmental Agency for Innovation Systems (VINNOVA) within the VINN Excellence Center Chase.

References

[1] M. Pelosi, O. Franek, M.B. Knudsen, G.F. Pedersen, J.B. Andersen, "Antenna Proximity Effects for Talk and Data Modes in Mobile Phones," Antennas and Propagation Magazine, IEEE, vol.52, no.3, pp.15-27, June 2010.

[2] P-S. Kildal, C. Orlenius, J. Carlsson, "OTA Testing in Multipath of Antennas and Wireless Devices With MIMO and OFDM," Proceedings of the IEEE, vol.100, no.7, pp.2145-2157, July 2012.

[3] "IEEE Recommended Practice for Determining the Peak Spatial-Average SAR in the Human Head From Wireless Communications Devices: Measurement Techniques," IEEE Std 1528-2003, 2003.

[4] U. Carlberg, J. Carlsson, A. Hussain, P.-S. Kildal, "Ray based multipath simulation tool for studying convergence and estimating ergodic capacity and diversity gain for antennas with given far-field functions," ICECom, 2010 Conference Proceedings, 20-23 Sept. 2010.

[5] K. Rosengren, P.-S. Kildal, "Radiation efficiency, correlation, diversity gain and capacity of a six-monopole antenna array for a MIMO system: theory, simulation and measurement in reverberation chamber," Microwaves, Antennas and Propagation, IEE Proceedings, vol.152, no.1, pp. 7- 16, Feb. 2005.

Antenna Measurement Intercomparison Campaigns in the framework of the European Association of Antennas and Propagation

L. Scialacqua[1], F. Mioc[1], J. Zhang[2], L. J. Foged[1], M. Sierra-Castañer[3]

[1]*Satimo (Microwave Vision Group)*
Via Castelli Romani 59, I-00040 Pomezia (RM), Italy
[2]*Satimo (Microwave Vision Group)*
17, Avenue de Norvege. 91140 Villebon-sur-Yvette, France
[3]*Technical University of Madrid (UPM)*
Ciudad Universitaria, E-28040 Madrid, Spain

***Abstract*- This paper gives an overview of the ongoing activities in the frame of the EurAPP [1] working group on antennas measurements and the first considerations on useful criteria for comparing and evaluating large amount of measured antenna data. This work comes from the experience acquired during the VI EU Framework network "Antenna Centre of Excellence (ACE) [2]" as reported in [3-9]. During that project, the activities spanned the frequency range from L-Ka band using different antennas (VAST12, SATIMO SH800 and SATIMO SH2000). The vast amount of data from different measurements institutions within Europe and US were used to establish a reference pattern for each of the high accuracy reference antennas. The reference patterns and the data from the facility comparison activities are considered important instruments to verify the measurements accuracies for antenna measurement ranges as well as to investigate and evaluate possible improvements in measurement set-ups and procedures.**

I. INTRODUCTION

Several facility comparison campaigns have been carried out during the last years in the framework of the different European Activities regarding to Antenna Measurements [3-10]. The Antenna Measurement Activity of the Antenna Centre of Excellence [2] of the VI Frame program of the UE, in the period of 2004 to 2007, began to define some reference antennas to be used for these purposes. A high directive reflector antenna, DTU-ESA 12 GHz VAST12 [11], and two dual ridge horns, SATIMO SH800 [12] in L, S and C band (SH800) and SATIMO SH2000 in Ku and Ka bands, were employed. After finishing the works of this network of excellence, the different tasks related to this topic have been continued in the frame of the COST ASSIST (IC0603) [13] and COST-VISTA (IC1102) [14] and now, included in the Antenna Measurement Working group of the EurAAP, where a specific task for Antenna Measurement Intercomparisons is ongoing.

The main lesson of these campaigns is that comparative measurements based on high accuracy reference antennas and involving different antenna measurement systems are important instruments in the evaluation, benchmarking and calibration of the measurement facilities. Regular inter comparisons are also an important instrument for traceability and quality maintenance. These activities promote and document the measurement confidence level among the participants and are an important prerequisite for official or unofficial certification of the facilities. In any case, both type of facilities, with or without ISO certification, need to conduct facility comparisons to properly validate their measurement procedures. In general, the goal of the facility comparison activities: to provide means to validate and document measurement accuracy from comparison with other facilities; and to allow the facilities to investigate and correct in case of "less compliance" or in some cases allow facilities to revise procedures and correct the measurements.

In this paper, we will review the work realized with the three first reference antennas, emphasizing in the achieved conclusions. Then, the ongoing intercomparison campaigns covering different kinds of reference antennas (SATIMO BTS 1940, SATIMO SH800, SATIMO SR-40 and SATIMO LTE MIMO antenna) are presented and the first results explained. Finally, some conclusions and future lines are extracted.

II. PREVIOUS ANTENNA MEASUREMENT CAMPAIGNS IN THE FRAME OF ANTENNA CENTRE OF EXCELLENCE

A. DTU-ESA 12 GHz VAST1 and SATIMO SH800 and SH2000 campaigns.

The VAST-12 comparison campaign took place during 2004 and 2005, with 8 facilities of 6 different institutions providing results (Saab Microwave Systems, France Telecom R&D, RUAG Aerospace Sweden, Technical University of Catalonia, Technical University of Denmark and Technical University of Madrid). The measurements were carried out in compact ranges (CR), far field (FF), spherical near field systems (SNF) and planar near field systems (PNF). DTU leaded this process and two measurements at the beginning and the end of the campaign were performed there. VAST12 is a very high quality antenna with a very directive radiation pattern (Fig.1).

Two wideband dual ridge horns were selected in these cases. In this case, the campaigns were headed by SATIMO. Ridge horns are much smaller and less bulky that the corresponding standard gain horn at comparable frequencies. Carefully designed dual ridge horns have excellent return loss, cross polar and flat gain response (typically 7-15 dBi) in a 1:15 frequency range.

978-7-5641-4279-7

Figure 1. The ESA/DTU VAST12 and the SATIMO SH800 Dual Ridge Horn antennas

The first activity on comparative measurements for SH800 was performed involving different test facilities: DTU-ESA spherical near-field antenna test facility at the Technical University of Denmark (DTU), in both of the SATIMO multi probe spherical near field systems (SG-64) in Atlanta (USA) and Paris (France), in the spherical near field system of Technical University of Madrid (Spain) and the combined farfield/spherical near field test range of Saab Ericsson Space (Sweden) and the far field ranges of IMST (Germany) and National Centre for Scientific Research (Greece). The data collection and processing was conducted by SATIMO in cooperation with the other participants and documented in an ACE report [2]. The traditional comparison of data involved the comparison of boresight gain and directivity values for different frequencies; however, the measurement differences and their sources are often better understood by direct inspection and comparison of the patterns. Since the direct comparison of large amount of measured pattern data is unfeasible by inspection of pattern differences alone, a statistical approach was implemented that allows the comparison of data in a simple form.

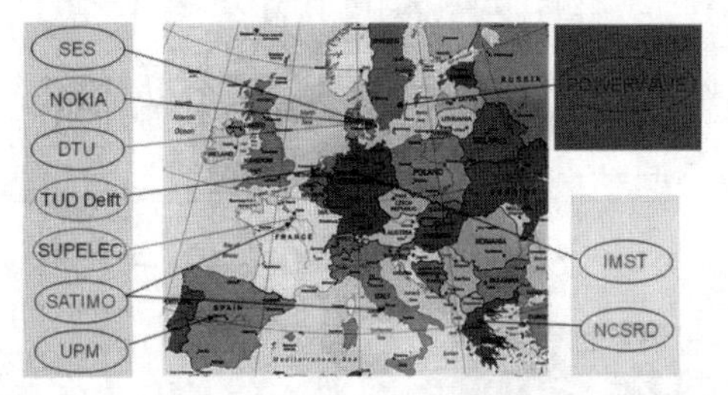

Figure 2. Participants for the SATIMO SH800 Intercomparison Campaign.

The next antenna, the SH2000 (2-32GHz) is dual ridge horn that combines a stable gain performance and low VSWR with wide band frequency operation. The horn is single linearly polarized with high cross-polar discrimination and is often used as a reference antenna for gain calibration of antenna measurement systems or as a wideband probe in classical far field test ranges. The horn is specifically designed to maintain a well-defined smooth radiation pattern in the direction of the boresight axis throughout the operational bandwidth. The horn is equipped with a high precision female 3.5 mm connector intermateable with SMA and K connectors. For the SH2000, 8 frequencies were selected in Ku and Ka bands and 11 different test facilities were involved.

B. Reference values.

This reference value and its uncertainty were obtained using the measured values and their uncertainties (1). In the case of single point values, the expressions are:

$$X_{typ} = \frac{\sum_{i=1}^{N}\left(\frac{x_i}{u_i^2}\right)}{\sum_{i=1}^{N}\left(\frac{1}{u_i^2}\right)} \qquad u_{typ} = 1 + \sqrt{\frac{1}{\sum_{i=1}^{N}\frac{1}{(1-u_i)^2}}} \tag{1}$$

The uncertainty associated with the weighted mean is "improved" if the measurements are truly independent. The previous formulas give RSS values corresponding to the 1σ value, with 69% confidence level assuming a normal distribution. From the reference pattern, the standard deviation of the differences for each measurement and in each direction was calculated. This value expresses the effective variation over the 45° forward cone giving an indication of the measurement error level in a single value. The procedure is expressed in (2), where directivity data for each angular position in linear scale is normalized respect the boresight value and the reference value:

$$f(\theta) = \left(\frac{Dir_{co,xp} - Dir_{ref\_co,xp}}{Dir_{ref\_co,xp}}\right)\cdot\left(\frac{Dir_{co,xp}}{Dir_{co,boresight}}\right) \tag{2}$$

The resulting number express the equivalent signal-to-noise level in which all deviations with respect to the reference pattern has been converted into an equivalent "noise". The calculated co-polar standard variation for each facility with respect to the weighted mean reference pattern is shown in Fig. 6. The standard deviation σ is very useful to quantify the range in which measurements errors are distributed. It expresses the 68.3% confidence that the measurements errors are within this level (the 99.7% confidence level is 3σ). The standard deviation expresses only the variation, but it does not consider a general shift. This also means that this value "cleans" the comparison from differences caused by pattern difference in the antenna back-lobe (usually due to differences in the measurement set-up). The impact of this is often very small in high gain measurements, but can be a significant contribution for medium and low gain antennas, as in this case.

III. Ongoing Intercomparison Campaigns

Four antennas have been selected for the new campaigns in the frame of the European Association on Antennas and Propagation. These antennas cover a broad frequency band, different applications and different radiation patterns.

A. Intercomparison campaign with SATIMO BTS1940 BTS antenna.

In 2009, a new intercomparison campaign using the BTS1940 (Fig. 3) array began. This antenna is a linear array reference antenna with dual slant +45°/-45° or H/V polarized working in GSM1800 and UMTS bands (1710 to 2170 MHz). The array is specifically designed to achieve excellent crosspolar discrimination and to maintain a well defined radiation pattern in the direction of the boresight axis throughout the operational bandwidth. The BTS1940 antennas are equipped with high precision female N type connectors for superior repeatability and durability. The nominal impedance is 50 ohm with return loss values better than -15 dB. The SATIMO linear array antenna BTS1940 has been measured in the reference coordinate system shown on the antenna in Fig. 7. The BTS1940 measurement stage is finished, and now SATIMO is processing the results. The first results (Fig. 3 and Table I) show good agreement in the comparison of radiation pattern comparisons at 1920 MHz and directivity for the first three facilities for five facilities.

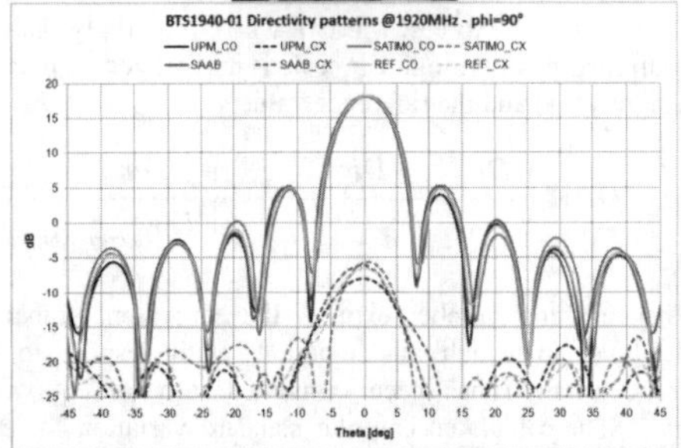

Figure 3: SATIMO BTS1940 in test configuration BTS 1940 and vertical pattern at SATIMO, UPM and SAAB compared with weighted reference value.

TABLE I

BORESIGHT DIRECTIVITY OF THE PARTICIPANT FACILITIES FOR BTS1940

	Boresight Directivity [dBi]				
	UPV	UPM	SAAB	Huawei	SATIMO
Freq [MHz]	SNF	SNF2	CR	SNF3	SNF4
1710	17,40	17,28	17,35	17,38	17,34
1795	17,77	17,62	17,73	17,68	17,63
1880	18,14	18,02	18,05	18,01	18,02
1920	18,33	18,17	18,22	18,17	18,15
2170	18,81	18,76	---	18,93	18,87
2200	18,79	18,77	---	---	18,85

B. Intercomparison campaign with SH800 dual ridge antenna.

The second antenna is again the SH800. In this case, the antenna has been modified in order to have a more stable setup. In particular, an absorber plate has been added behind the antenna to eliminate the sensibility to measurement setup and a cable has been added in order to have the same interface. The measured parameters are peak gain (IEEE definition) at discrete frequencies, directivity and gain patterns (Ludwig III co-polar and cross-polar) for 4 cuts (0°, 45°, 90° & 135°) and return losses. The participants must give the data in an appropriate format; give the description of measurement facility, the mechanical and electrical set-up description, the measurement procedure, the mechanical/electrical alignment, AUT alignment, and the uncertainty budget for the measurements. Fig. 4 shows the measurement results in the Stargate SG64 at SATIMO Paris facilities.

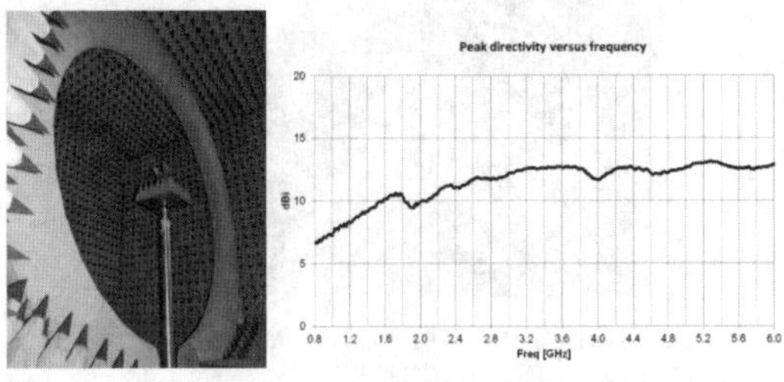

Figure 4. SATIMO SH800 antenna on SATIMO-Paris SG64 System and peak directivity measurement results.

C. Intercomparison campaign with SATIMO SR-40 antenna

In order to cover also higher frequencies and higher directivity antennas, the SATIMO SR-40 reference antenna has been selected. The frequencies to be measured in this case are 10.7-14.5GHz, 18-20GHz, and 28-31GHz, with a step of 100 MHz and 38GHz. The same antenna parameters and the same information than in the previous campaign is required for this measurement. Also in this case, the acquisitions are not completed, but preliminary results can be shown at this moment. Fig. 5 shows the pattern comparison (horizontal plane) at 10.7 GHz among 3 facilities and the reference value. Fig. 6 shows the averaged difference between each facility result and the reference value.

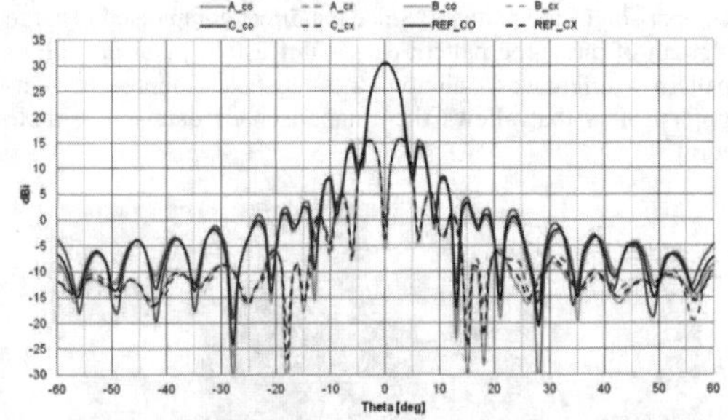

Figure 5. SATIMO SR40 first measurement results for pattern at 10.7 GHz.

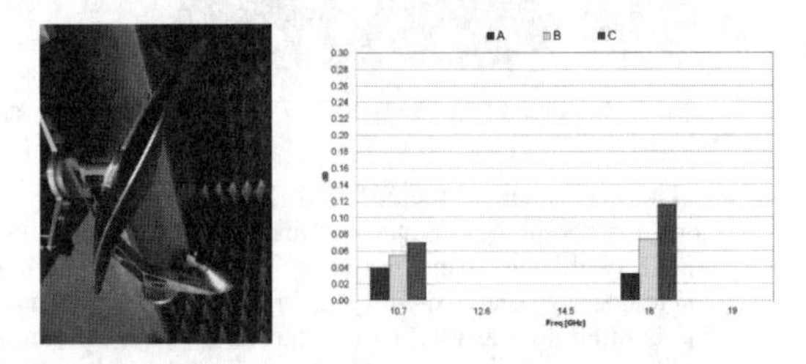

Figure 6. SATIMO SR40 antenna and statistical comparison of the co-polar radiation pattern at 10.7 and 18 GHz at 3 different facilities.

D. Intercomparison campaign for LTE MIMO Antenna.

Finally, in order to cover also small antenna measurements, a new campaign has been defined, and measurements will begin during this year 2013. In this case the CTIA 2x2 MIMO reference antennas (Fig. 7) have been proposed as reference antennas, during the WG5 progress meeting at

EuCAP2013 conference in Göteborg, Sweden, to be measured by the institutions of the Small Antenna Measurements Group. In particular, different designs (good performance, nominal and bad) have been done in order to investigate the effect of MIMO. The frequency bands are: LTE Band 2 (1930-1990MHz), LTE Band 7 (2620-2690 MHz) and LTE Band 13 (746-756MHz).

Figure 7. SATIMO CTIA 2x2 MIMO reference antenna in the Stargate SG64 System (SATIMO-Paris).

IV. CONCLUSIONS

The previous campaigns gave us a set of important conclusions to be considered during the antenna measurement intercomparison processes, summarized in:

- A very precise definition of the setup is necessary in order to assure that all participants are performing the measurements in the same way.
- The organization of the campaigns requires the exact definition of the objectives to be pursued, in order to prepare in advance the test plan, test procedure and the procedure for giving the results. Also, the deadlines for each laboratory must be assured in order not to extend the campaign for years.
- A unified procedure for getting the uncertainty for a specific measurement has to be agreed.
- In any case, this exercise is very useful for improving the quality of the facility (detecting errors) and the measurement capability of the participants.
- These campaigns have become very useful for the facilities that have the ISO17025 accreditation or are in the process. Also, regular inter comparisons are also an important instrument for traceability and quality maintenance.
- The reference value has to be obtained considering the uncertainty of each measurement and facility.

EurAAP Working Group about Antenna Measurements invites other laboratories to participate in the future campaigns.

ACKNOWLEDGMENT

This work has been carried under the Working Group of Antenna Measurements or EurAAP.

REFERENCES

[1] EurAAP, "European Association on Antennas and Propagation", www.euraap.org/

[2] Antenna Centre of Excellence, www.antennasvce.org

[3] L. J. Foged, Ph. Garreau, O. Breinbjerg, S. Pivnenko, M. Castañer, J. Zackrisson "Facility comparison and evaluation using dual ridge horn", AMTA 2005, October 30-November 4, Newport RI.

[4] L. J. Foged, P. O Iversen, L. Duchesne, O. Breinbjerg, S. Pivnenko "Comparative Measurement of standard gain horns" 28th ESA antenna workshop on Space Antenna Systems and Technologies, 31 May - 3 June 2005, Noordwijk, The Netherlands.

[5] Foged L.J.; Breinbjerg, O.; Pivnenko, S.; Di Massa, G.; Sabatier, C.; "Antenna measurement facility comparison within the European Antenna Centre of Excellence" European Microwave Conference, 4-6 Oct. 2005.

[6] A. Giacomni, B. Bencivenga, L.Duchesne, L. J. Foged "Determination of high accuracy performance data for dipole reference antennas", European AMTA conference, Munich 2006.

[7] Foged, L. J.; Bencivenga, B.; Durand, L.; Breinbjerg, O.; Pivnenko, S.; Sabatier, C.; Ericsson, H.; Svensson, B.; Alexandridis, A.; Burgos, S.; Sierra-Castaner, M.; Zackrisson, J.; Boettcher, M.; "Error calculation techniques and their application to the Antenna Measurement Facility Comparison within the European Antenna Centre of Excellence", The second European conference on Antennas and propagation, EuCap2007, 11-16 Nov. 2007

[8] L. J. Foged, B. Bencivenga, O. Breinbjerg, S. Pivnenko, M. Sierra-Castañer, "Measurement facility comparisons within the European antenna centre of excellence", ISAP2008, International symposium on Antennas and propagation. 27-30 October 2008. Taipei, Taiwan.

[9] L. J. Foged, B. Bencivenga, L. Scialacqua, S. Pivnenko, O. Breinbjerg, M. Sierra-Castañer, P. C. Almena, E. Séguenot, C. Sabatier, M. Böttcher, E. Arnaud, T. Monediere, H. Garcia, D. Allenic, G. Hampton, A. Daya, " Facility comparison and evaluation using dual ridge horns", 3rd European conference on Antennas and propagation, EuCap2009, Berlin, March 23 - 27 2009.

[10] L. J. Foged, M. Sierra Castañer, L. Scialacqua "Facility Comparison Campaigns within EurAAP", 5th European conference on Antennas and propagation, EuCAP2011, Rome, April 2011.

[11] J.E. Hansen, "Definition, Design, Manufacture, Test and Use of a 12 GHz Validation Standard Antenna, Executive Summary". Technical University of Denmark, Lyngby, Denmark, 1997, Tech. Report R672.

[12] L. J. Foged, A. Giacomini, L. Duchesne, E. Leroux, L. Sassi,, J. Mollet, "Advanced modelling and measurements of wideband horn antennas", ANTEM Saint-Malo, June 2005

[13] COST ASSIST IC0603,"Antenna Systems and Sensors for Information Society Technologies", www.cost-ic0603.org

[14] COST VISTA IC1102, "Action on Versatile, Integrated and Signal-aware Technologies for Antennas, www.cost-vista.eu

[15] S. Pivnenko, J.E. Pallesen, O. Breinbjerg, M. Sierra-Castañer, P. Caballero, C. Martínez, J.L. Besada, J. Romeu, S. Blanch, J.M. González-Arbesu, C. Sabatier, A. Calderone, G. Portier, H. Eriksson, J. Zackrisson. "Comparison of Antenna Measurement Facilities With the DTU-ESA 12 GHz Validation Standard Antenna Within the EU Antenna Centre of Excellence". IEEE TAP. July 2009. Vol. 57. N. 7. Pp. 1863- 1878.

[16] S. Pivnenko, O. Breinbjerg, S. Burgos, M. Sierra-Castañer, H. Eriksson, "Dedicated Measurement Campaign for Definition of Accurate Reference Pattern of the VAST12 Antenna". 2008 AMTA Symposium Proceedings.

Capacitively-Loaded THz Dipole Antenna Designs with High Directivity and High Aperture Efficiency

Ning Zhu, and Richard W. Ziolkowski
Department of Electrical and Computer Engineering, University of Arizona
1230 E. Speedway Blvd.
Tucson, AZ 85721 USA

***Abstract*-Several linearly polarized photoconductive terahertz antenna designs and their performance characteristics are reported. A single capacitively-loaded dipole antenna is introduced first; it has a 12.8 dB directivity, a 11.6 dB realized gain, and a 82% radiation efficiency. By connecting two single capacitively-loaded dipole antennas to form a linear array, higher directivity (14.8 dB), lower sidelobe level (-18 dB) and larger front-to-back-ratio (14 dB) values are achieved. Moreover, by incorporating a meta-film structure as a superstrate of the antenna array, some additional interesting behaviors are achieved.**

IV. INTRODUCTION

Terahertz (THz) is the set of frequencies between the microwave and optical frequencies. Because THz imaging systems are able to provide images with higher resolution than microwave imaging systems and are non-ionizing so they also offer less potential harm to the human body than X-ray imaging systems, they have drawn greater attention recently, for instance, in quality control, non-destructive evaluation, medical, and security applications. As one of the most important components in a THz imaging system, the antenna is playing both impedance matching and power radiating roles. Some representative THz antenna examples can be found in [1-4]. While both photoconductive and photomixer style THz sources are being developed, we emphasize the former in all of the designs reported herein.

As the semiconductor across the photoconductive switch region (gap) is exposed to a laser beam, electron-hole pairs will be generated in it. If there is an external electric field across this photoconductive switch gap, usually generated between a DC voltage and ground, a current will be formed across it. If the laser signal has a short enough time period, i.e., roughly 100 femto-seconds, a THz signal will be generated by the resulting photoconductive current.

A single capacitively-loaded THz dipole antenna is initially designed to achieve the high directivity and high radiation efficiency properties desired for eventual application in a THz spectral imaging system. The capacitively-loaded structure also incorporates the necessary DC bias and ground lines. By connecting the DC bias and ground lines of two single capacitively-loaded THz dipole antennas together, an antenna array is generated. The impedance matching and performance characteristics of the array are fine tuned to achieve even more appealing radiation properties. In addition, the impact of the introduction of a meta-film structure-based superstrate into this array also will be analyzed. Since the aperture efficiencies for the original capacitively-loaded THz dipole antenna and the corresponding array are already over 100%, there is not much room to increase the directivity. Nevertheless, the presence of the superstrate provides additional design degrees of freedom which allow the increase of the front-to-back ratio value of the array at a trade-off cost of slightly increasing the side lobe levels of its radiation patterns. All the antennas in this paper were simulated with ANSYS/ANSOFT's high frequency structure simulator (HFSS).

V. SINGLE CAPACITIVELY-LOADED DIPOLE ANTENNA DESIGN

The single capacitively-loaded dipole antenna and its dimensions are shown in Fig. 1. The radiating element and ground plane are taken to be 0.35-μm-thick Ti-Au ($\sigma = 1.6 \times 10^7$ S/m). Both the substrate and the superstrate were selected to be GaAs ($\varepsilon_r = 12.9$, $\mu_r = 1.0$, loss tangent = 0.006). The substrate and superstrate thicknesses are 60 μm and 90 μm, respectively. The photoconductive gap region is located in the center of the figure. Because the femto-second laser beam must expose this gap region to generate the necessary electron-hole pairs, a truncated cone (frustrum) structure was cut through the ground plane and partially through the GaAs substrate to the expose the gap region. It has a 15 μm upper radius, a 20 μm lower radius, and 58 μm height. In order to assemble several single antennas into a THz antenna array, the DC bias and ground lines are designed to connect the edges of each single antenna. The four patch-like structures on the DC bias and ground lines represent the capacitive loadings of the dipole arms; they are introduced to achieve impedance matching. It is assumed that the power level of the incident laser beam will generate an effective 50 Ω source impedance. The design can be readjusted for other power levels and, hence, source impedances.

The simulation results show that the peak directivity of this single capacitively-loaded dipole antenna is 12.8 dB and the realized gain is 11.6 dB at 1.025 THz, a single-mode resonance with reasonable impedance matching. The radiation efficiency is 82% and the aperture efficiency is 144% [5]. Moreover, the 10dB bandwidth of this antenna is 0.071 THz at 1.025 THz, giving a fractional 10dB bandwidth: FBW_{10dB} =

978-7-5641-4279-7

7.1%. The corresponding far-field radiation pattern is shown in Fig. 2. One observes that the single THz antenna has a -7.7 dB sidelobe level and a 10 dB front-to-back ratio.

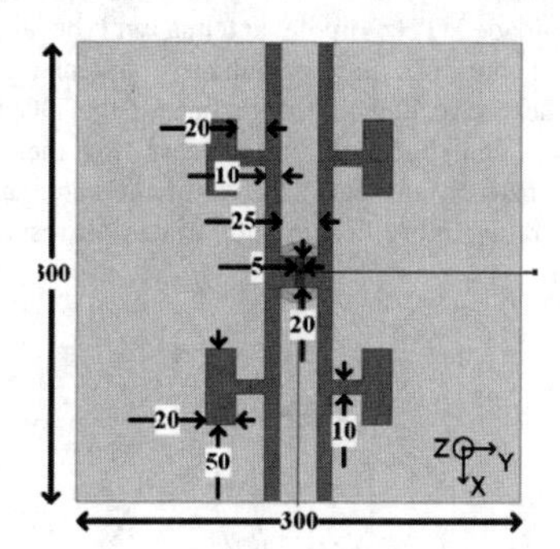

Figure 1. Single capacitively-loaded dipole antenna design. All dimensions are in μm.

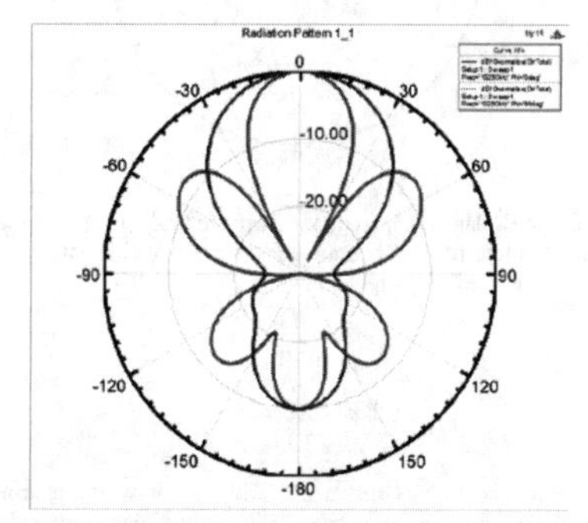

Figure 2. The E-plane (XZ) (red) and H-plane (YZ) (magenta) radiation patterns of the single capacitively-loaded THz dipole antenna at 1.025 THz.

VI. Capacitively-Loaded Dipole Antenna Array Design

By connecting the DC bias and ground lines of two single capacitively-loaded THz dipole antennas, one can generate a two element antenna array. This antenna array was then fine-tuned to achieve yet better antenna performance characteristics. The structure and dimension of the fine-tuned antenna array are shown in Fig. 3.

The simulation results predict that the peak directivity and realized gain are both increased by approximately 2 dB. A 77% radiation efficiency and a 120% aperture efficiency have been achieved. This aperture efficiency is slightly lower than the value for the single element system due to the increase in losses associated with the additional materials being present. Furthermore, the 10 dB bandwidth of the antenna array is now about 0.04 THz, which yields FBW_{10dB} = 4.0%. Fig. 4 shows the simulated radiation patterns of the antenna at its resonance frequency, 1.0 THz. These patterns have a -18 dB sidelobe level and 14 dB front-to-back ratio.

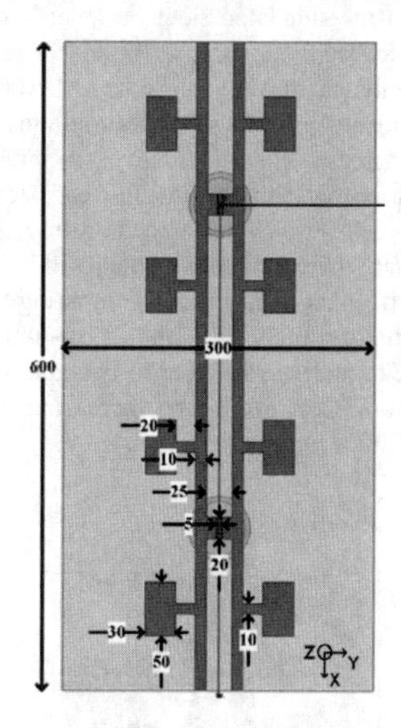

Figure 3. The two element capacitively-loaded THz dipole antenna array. All dimensions are in μm.

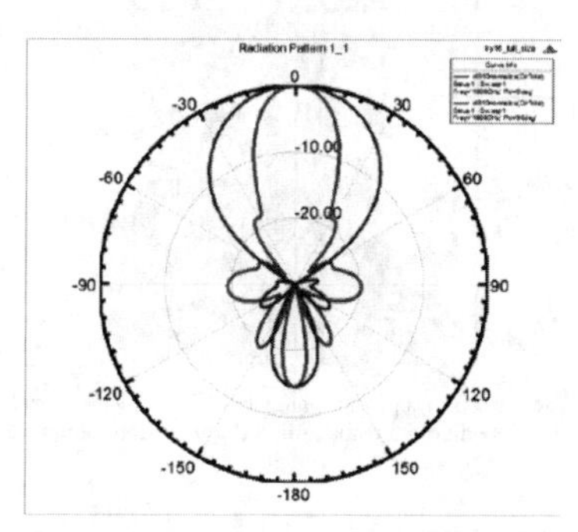

Figure 4. The E-plane (XZ) (red) and H-plane (YZ) (magenta) radiation patterns of the capacitively-loaded THz dipole antenna array at 1.0 THz.

VII. Superstrate-Augmented Two-Element Capacitively-Loaded Dipole Antenna Array Design

A meta-film structure was selected as a superstrate for the two element capacitively-loaded THz dipole antenna array. It is based on the "I" elements used for an earlier epsilon-negative (ENG) metamaterial design [6]. Fig. 5 shows one example of this meta-film superstrate-augmented THz antenna array.

Some interesting behaviors were observed with the presence of the superstrate. The HFSS-predicted peak directivity and realized gain are 15.2 dB and 14.0 dB, respectively. Furthermore, a 77% radiation efficiency and 132% aperture efficiency have been achieved in this case. In addition, the

FBW_{10dB} = 4.0% at its resonance frequency 1.0 THz. The radiation pattern of this antenna at 1.0 THz is shown in Fig. 6. It illustrates that the simulated sidelobe level is now -15 dB, while the front-to-back ratio is now 29 dB. Since the original two element antenna array achieved over 100% aperture efficiency, this result confirms the expectation that it would be difficult to increase the directivity simply by using metamaterial-based superstrate. Moreover, because of this large aperture efficiency, the antenna array exhibits the expected relatively narrow directivity bandwidth. Nevertheless, this is a trade-off. In particular, when comparing Fig. 6 to Fig. 4, one immediately finds that the meta-film superstrate structure is able to increase the front-to-back ratio value of the array at a trade-off cost of slightly increasing the side lobe levels of its radiation patterns.

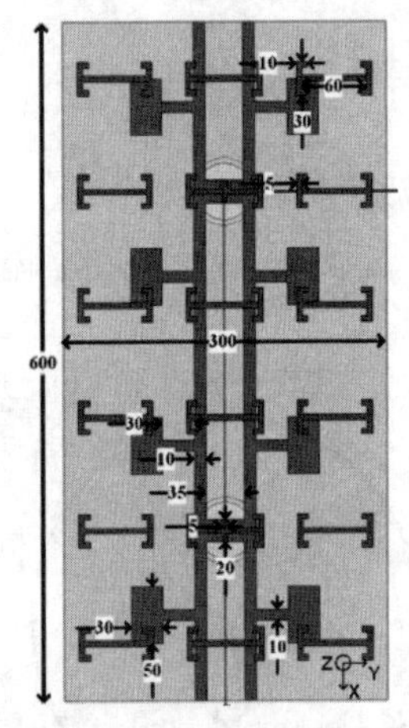

Figure 5. The meta-film superstrate augmented, two-element capacitively-loaded THz dipole antenna array. All dimensions are in μm.

V. CONCLUSIONS

In this paper, we designed several capacitively-loaded THz dipole-based antennas. All of these antennas were shown to achieve high directivity, resulting in over 100% aperture efficiency at resonance frequencies near 1.0 THz. It was demonstrated that a two-element array based on this single dipole element achieved about a 2 dB increase in directivity and gain. By augmenting the two-element array with a meta-film superstrate structure, the resulting antenna system was shown to have a further increase in its front-to-back ratio value of the array at a trade-off cost of slightly increasing the side lobe levels of its radiation patterns.

These simulation results have confirmed that the single capacitively-loaded THz dipole antenna will be an eligible candidate for our THz photoconductive antenna array. In particular, they have demonstrated that all the DC bias and ground lines can be easily connected together without impacting negatively the performance of the antenna system. We hope to report more design and simulation results in our presentation.

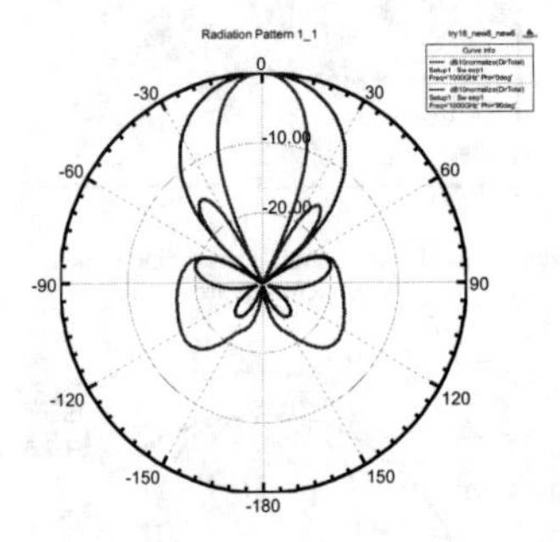

Figure 6. The E-plane (XZ) (red) and H-plane (YZ) (magenta) radiation patterns of the meta-film superstrate-augmented two element capacitively-loaded THz dipole antenna array at 1.0 THz.

REFERENCES

[1] Z. Popovic, and E. N. Grossman, "THz metrology and instrmentation," *IEEE Transaction on Terahertz Science and Technology*, vol. 1, No. 1, pp. 133–144, September 2011.

[2] K. Han, T. K. Nguyen, I. Park, and H. Han, "Terahertz Yagi-Uda antenna for high input resistance," *J. Infrared Milli Terahz Waves*, vol. 18, pp. 441–454, 2010.

[3] R. Singh, C. Rockstuhl, C. Menzel, T. P. Meyrath, M. He, H. Giessen, F. Lederer, and W. Zhang, "Spiral-type terahertz antennas and the manifestation of the Mushiake principle," *Optics Express*, vol. 17, pp. 9971–9980, May 2009.

[4] N. Zhu, and R. W. Ziolkowski, "Progress toward THz antenna designs with high directivity and high efficiency," accepted by *IEEE Antennas Propagat. Society Int. Symp.*, Orlando, FL, 2013.

[5] C. A. Balanis, *Antenna Theory*, 3rd Ed. New York: Wiley, 2005.

[6] R. W. Ziolkowski, "Design, fabrication, and testing of double negative metamaterials," *IEEE Trans. Antennas Propag.*, vol. 51, no. 7, pp. 1516-1529, July 2003.

TA-2(A)

October 24 (THU) AM

Room A

Wireless Power Transmis.

Theoretical Analysis, Design and Optimization of Printed Coils for Wireless Power Transmission

Jia-Qi Liu[1], Yi-Yao Hu[1], Yin Li[1], Joshua Le-Wei Li[1,2]

[1]Institute of Electromagnetics and School of Electronic Engineering
University of Electronic Science and Technology of China, Chengdu, China 611731
[2]Advanced Engineering Platform and School of Electronic Engineering
Monash University, Sunway/Clayton, Selangor 46150/Victoria 3800, Malaysia / Australia

***Abstract*-Printed coils design and optimization are crucial and important for wireless power transmission (WPT). Currently the methods to design printed coils are left with a number of choices that each has advantages and disadvantages in different situations. In this paper, both electromagnetic theory analysis and equivalent circuit modeling have been used together in theoretical analysis, design and optimization of the printed coils system. With these design procedure, we can obtain the optimal parameters like coupling coefficient *K*, the self and mutual Inductance and others in printed coils through electromagnetic theory analysis. And the received power and power transmission efficiency of printed coils system can be achieved by the formula derivation and Matlab simulation more precisely than any previous method for the printed coils system in magnetically resonant coupling WPT system consisting of coils and reactive elements.**

I. INTRODUCTION

Nowadays, the WPT applications are not yet common for the miniature and expensive problem, but statistically indisputable reality is that WPT technology will meet more and more broad market demands and applied prospect because its portability, durative and water tightness advantages. For Magnetically resonant coupling is demonstrated for a steady and feasible approach for WPT [1]-[3], the research on printed coils in Magnetically resonant coupling WPT system is being more and more important. In previous research, when deciding to make printed coils directly, or analysis printed coils just by theory analysis or equivalent circuit modeling, it is imperative to understand the limitations this can place on the overall WPT system because each method has its limitations and the printed coils system consisting of coils and reactive elements.

In this paper, electromagnetic theory analysis and equivalent circuit modeling are presented. By electromagnetic theory analysis, the self and mutual Inductance of coils can be determined though formula derivation in given the number of turns, the width of the metal trace, the turn spacing in coil and the distance between two coils. Furthermore, we can obtain the coupling coefficient *K* and resonant frequency of printed coils through the self and mutual Inductance, which indicates the degree of coupling between the coil pairs.

In equivalent circuit modeling, we can design and optimize whole coils system because the printed coils system consisting of coils and reactive elements what used in magnetically resonant coupling WPT system. And printed coils are mainly presented inductance. The resonant frequency, coupling coefficient *K*, the value of Q, the series resistance and other parameters in equivalent circuit can be obtained through formula derivation and Matlab simulation [4]. Finally, we can determine the optimal working mode of whole printed coils system exactly and embed in WPT systems because these two design approaches complement each other.

II. ELECTROMAGNETIC THEORY AND ANALYSIS

A. Self and Mutual Inductance

The conducting loop of wire pass a steady direct current (DC) I, which will produce a magnetic flux density B that circulates about the wire as shown in Figure 1. Besides, Vector magnetic potential A is defined by:

$$B = \nabla \times A \tag{1}$$

So the inductance of the current loop is defined as:

$$L = \frac{\psi}{I} \tag{2}$$

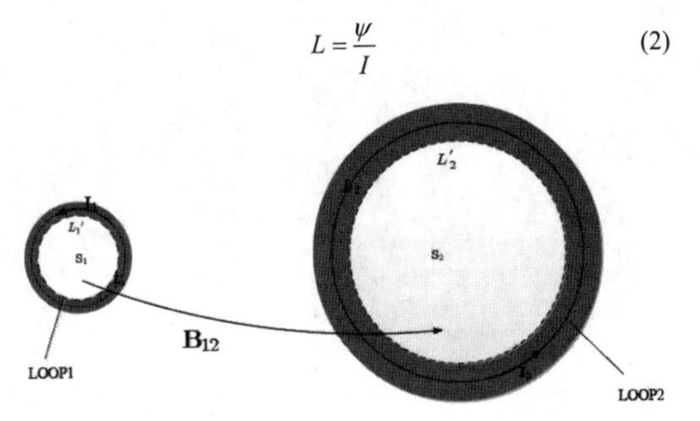

Figure 1. Mutual inductance between two circular loops

When the loop1 carrying a current I_1, the magnetic flux density through loop2 is B_{12}, which is caused by the current of Loop1 as shown in Figure 1. The mutual inductance between the two loops is:

$$M_{12} = \frac{\psi_{12}}{I_1} \tag{3}$$

978-7-5641-4279-7

The vector magnetic potential around the perimeter of that loop1:

$$\psi_{12} = \oint A_{12} \cdot dL_2' \tag{4}$$

Where c_2' is the inner edge of that wire that bounds the surface of the loop, c_1 is the center of the loop. The self inductance of a loop can be determined by letting the two loops be coincident.

$$M_{12} = L_1 = M_{11} = \frac{\mu_0}{4\pi} \oint_{c_2'} \oint_{c_1} \frac{dL_1 \cdot dL_2'}{R_{12}} \tag{5}$$

B. Theoretical Analysis

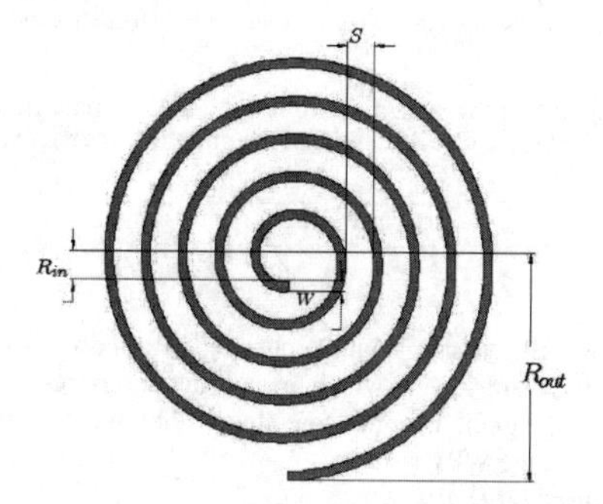

Figure 2. Geometrical parameters of a circular spiral coil

The geometry parameters of the circular spiral inductor are the number of turns *n*, the width of the metal trace W, the turn spacing S, the inner diameter R_{in} and the outer diameter R_{out} as shown in Figure 2.

The number of turns in a spiral coil can then be calculated from:

$$N = \frac{R_{out} - R_{in}}{W + s} + 1 \tag{6}$$

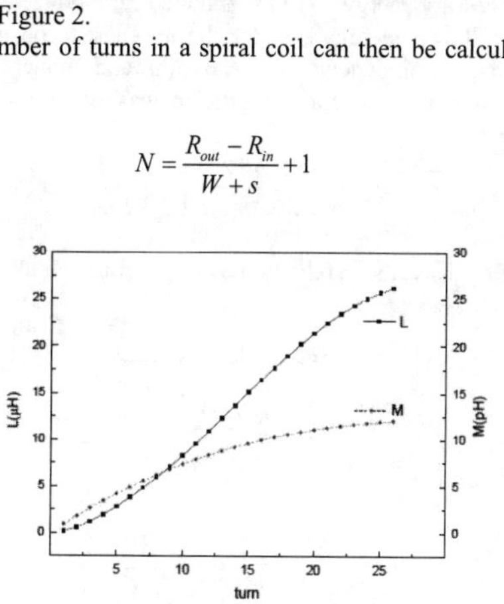

Figure 3. Mutual inductance and self inductance plot for the same R_{out}

Figure 3 shows the self inductance is increased with turn. The parameters of the circular spiral Loop1 is R_{out} =27.28 mm; turn = 1, Loop2 is R_{out} = 27.28 mm, turn from 1 to 26. From the curves, the self inductance is enhanced fast than the mutual inductance.

The coupling coefficient *K* between two coils with the self inductance L_1 and L_2 is defined as:

$$K = \frac{M}{\sqrt{L_1 \cdot L_2}} \tag{7}$$

Where M is mutual inductance of the two coils. So we can achieve maximum received power because the obtained coupling coefficient *K* indicates the degree of coupling between the coil pairs.

In case of spiral, however, the results in Figure 4 showed that as was increased, coupling coefficient *K* on increasing until turn = 9.

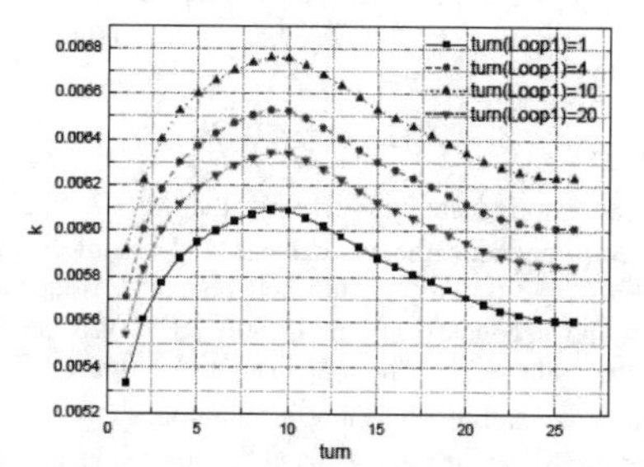

Figure 4. The coupling coefficient k between two spiral

III. EQUIVALENT CIRCUIT AND MODELING

A. Equivalent Circuit

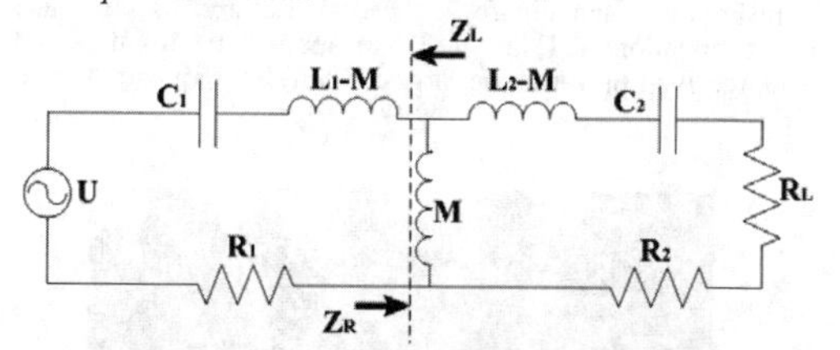

Figure 5. T-type equivalent circuit of WPT system

The printed coils system used in magnetically resonant coupling WPT system consist of coils and reactive elements. So it is incomplete and inaccurate analysis if we design printed coils only by electromagnetic theory analysis. There are four basic types of magnetically resonant coupling circuit designs including SS, PP, PS, SP [5]. We prefer to use SS configuration where the coils and capacitors are connected in series. In order to study this circuit model more conveniently, T-type equivalent circuit is used in Figure 5. Here, L_1 and L_2 represent the inductances of coils. M is mutual inductance. C_1

and C_2 are capacitors what connect to coils in series. R_1 and R_2 are the primary and secondary circuits' losses.

B. Received Power

Some researches of WPT systems focused on the received power [6]. We assume that the internal losses R_1 and R_2 are small enough so that we can ignore their contribution in the equivalent circuit shown in Figure 5.

If we want to maximize the power received by load R_L, Z_R and Z_L must satisfy:

$$Z_R = Z_L^* \tag{8}$$

Where

$$Z_R = \frac{(j\omega(L_2 - M) + \frac{1}{j\omega C_2} + R_L) * j\omega M}{(j\omega(L_2 - M) + \frac{1}{j\omega C_2} + R_L) + j\omega M} \tag{9}$$

$$Z_L = j\omega(L_1 - M) + \frac{1}{j\omega C_2} + R_S \tag{10}$$

We can get inductances of coils that $L_1 = L_2 = 27.6\,\mu H$ from electromagnetic theory analysis and assume capacitors that $C_1 = C_2 = 3.671\,nF$, the self-resonate frequency of primary and secondary circuit is 500 KHz. We can obtain Figure 6 when $R_S = R_L = 3\,ohm$ and Figure 7 when $R_S = R_L = 6\,ohm$ through Matlab simulation.

Equation (8) can be rewritten and becomes (11) and (12).

$$real(Z_L) - real(Z_R) = 0 \tag{11}$$

$$imag(Z_L) + imag(Z_R) = 0 \tag{12}$$

In Figure 6 and Figure 7, the dark red area is the place where Equations (11) and (12) are applied. In this dark red area, received power is the largest when the frequency is 500 KHz.

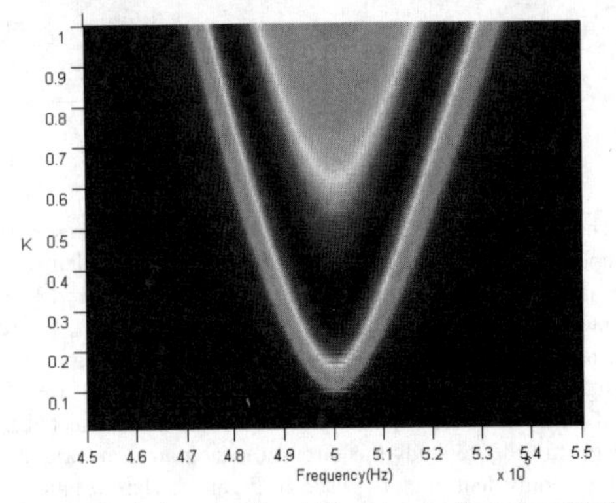

Figure 6. The absolute value of received power versus frequency and K when $R_S = R_L = 3\,ohm$

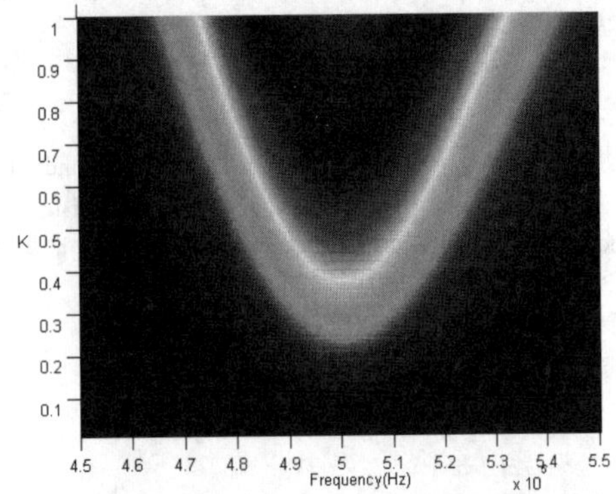

Figure 7. The absolute value of received power versus frequency and K when $R_S = R_L = 6\,ohm$

Considering the system is working at the resonate frequency of 500 KHz, we can achieve that the coupling coefficient K as:

$$K = \sqrt{\frac{Rs + R_1}{\omega L_1}}\sqrt{\frac{R_L + R_2}{\omega L_2}} = \frac{1}{\sqrt{Q_1 Q_2}}. \tag{13}$$

Above all, we can achieve more received power when the Q_1 and Q_2 are lower, which means the series resistance in equivalent circuit become low the further the transmission distance in the WPT system.

C. Power Efficiency

The system may not achieve maximum power transmission efficiency when it leads maximum transmission power. In the current study, most research groups focus on improving the power transmission efficiency, which can eliminate unnecessary energy consume and the heat produced by circuit.

As shown in Figure 5, for the no energy consumption in reactance components like capacitor and inductor, we can achieve power transmission efficiency as follows:

$$\eta = \frac{real(Z_R)}{real(Z_R) + real(Z_L)} \frac{R_L}{R_2 + R_L}. \tag{14}$$

Obviously, the maximum power transmission efficiency is:

$$\eta = \frac{1}{1 + (\frac{1}{K^2 Q_1 Q_2})^2} \frac{R_L}{R_2 + R_L}; \tag{15}$$

When

$$\omega = \frac{1}{\sqrt{L_1 C_1}} = \frac{1}{\sqrt{L_2 C_2}} \tag{16}$$

Where

$$K = \frac{M}{\sqrt{L_1 L_2}}, Q_1 = \frac{\omega L_1}{R_s + R_1}, Q_2 = \frac{\omega L_2}{R_L + R_2}. \quad (17)$$

Similarly, we can get inductances of coils that $L_1 = L_2 = 27.6\,\mu H$ from electromagnetic theory analysis and assume capacitors that $C_1 = C_2 = 3.671\,nF$, the self-resonate frequency of primary and secondary circuit is 500 KHz. We can obtain Figure 8 when $R_1 = R_2 = 0\,ohm$ and Figure 9 when $R_1 = R_2 = 3\,ohm$ through Matlab simulation.

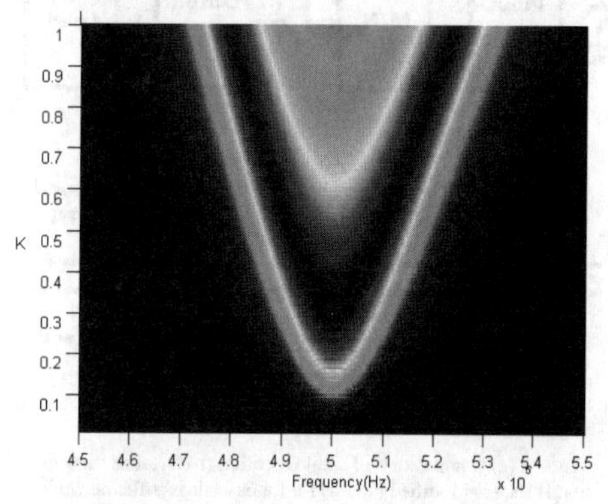

Figure 8. The absolute value of power transmission efficiency versus frequency and K when $R_1 = R_2 = 0\,ohm$

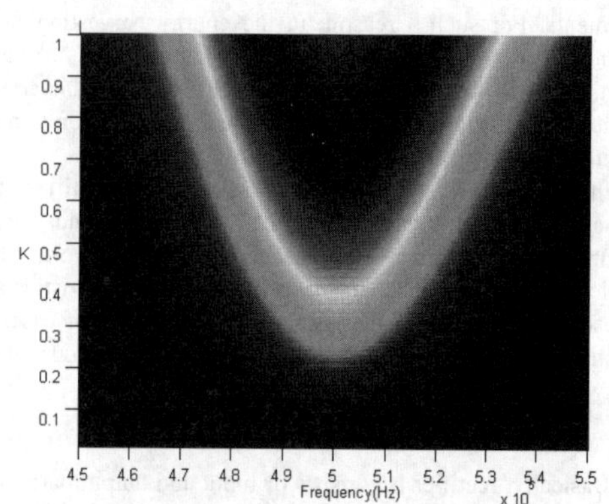

Figure 9. The absolute value of power transmission efficiency versus frequency and K when $R_1 = R_2 = 3\,ohm$

In Figure 8, the maximum power transmission efficiency approaches one hundred percent. Besides, the maximum power transmission efficiency approaches thirty-five percent as shown in Figure 9. The dark red area is the place where we can achieve maximum power transmission efficiency in these figures.

Finally, we can achieve maximum power transmission efficiency when the K tends to 1 and resonant frequency is 500 KHz.

IV. Conclusion

In this paper, we have designed and optimized whole printed coil system base on electromagnetic theory analysis and equivalent circuit modeling. The coupling coefficient K of printed coils what indicate the degree of coupling between the coil pairs, the self and mutual Inductance of printed coils are analyzed and optimized by electromagnetic theory analysis. Furthermore, in the process of equivalent circuit modeling, we have taken detailed method that how to achieve received power and power transmission efficiency in printed coils system through the formula derivation and Matlab simulation. In the next stage, how to design and optimize the multi-coils [7] of PCB simultaneously is our major direction.

Acknowledgment

The authors are grateful to the partial financial supports by (a) Project No. 61171046 from National Science Foundation of China, and also (b) the "*Program for Changjiang Scholars and Innovation Team in University* (PCSIRT)'" in terms of a Project "*Computational Electromagnetics and Its Applications to Microwaves*" from Ministry of Education, China.

References

[1] A. Kurs, A. Karalis, R. Moffatt, J. D. Joannopoulos, P. Fisher, and M. Soljacic, "Wireless power transfer via strongly coupled magnetic resonances," *Science,* vol. 317, pp. 83-86, Jul. 2007.

[2] Z. N. Low, R. Chinga, R. Tseng, and J. Lin, "Design and test of a high-power high-efficiency loosely coupled planar wireless power transfer system," *IEEE Trans. Ind. Electron.* , vol. 56, pp. 1801–1812, May 2009.

[3] J.-Q. Liu; L. Wang; Y.-Q. Pu; Li, J.L.; K. Kang, "A magnetically resonant coupling system for wireless power transmission," *Antennas, Propagation & EM Theory (ISAPE), 2012 10th International Symposium one,* pp. 1205–1209, 22- 26 Oct. 2012.

[4] MATLAB Web Site: http://www.mathworks.com/

[5] L. Chen, S. Liu, Y. Zhou and T. Cui, "An optimizable circuit structure for high-efficiency wireless power transfer," *IEEE Trans. Ind. Electron.* , vol.60, no.1, pp. 339–349, Jan. 2013.

[6] T. Sekitani, M. Takamiya, Y. Noguchi, S. Nakano, Y. Kato, T. Sakurai and T. Someya, "A large-area wireless power-transmission sheet using printed organic transistors and plastic MEMS switches," *Nature materials,* vol. Nature Materials 6, pp. 413 - 417, Apr. 2007.

[7] B. Cannon, J. Hoburg, D. Stancil, and S. Goldstein, "Magnetic resonant coupling as a potential means for wireless power transfer to multiple small receivers," *IEEE Trans. Power Electron.*, vol. 24, pp. 1819-1825, Jul. 2009.

Rectifier Conversion Efficiency Increase in Low Power Using Cascade Connection at X-band

JoonWoo Park[1], YoungSub Kim[1], YoungJoong Yoon[1], Jinwoo Shin[2] and Joonho So[2]
[1]The School of Electrical and Electronic Eng., Yonsei University, Seoul, Republic of Korea
[2]The Agency of Defense Development, Daejeon, Republic of Korea
Hypercube6174@gmail.com, mashell@naver.com, yjyoon@yonsei.ac.kr, sjinu@add.re.kr, joonhoso@add.re.kr

***Abstract*-**This paper presents an efficiency increase method using the cascade connected rectifier network for low power operation in X-band. The presented cascade rectifier network is better for the large number of rectifier array, especially low power applications and has a simpler network than series connection network while producing a higher voltage than parallel connection network. The result shows 12% and 38% more power gained compared to series and parallel connection at 0 dBm and -3 dBm of input power respectively. Simulation conducted using Avago Technology HSMS-8101 schottky barrier diode and Taconic TLC-30-0200 substrate.

I. INTRODUCTION

Recently, wireless power transmission (WPT) and energy harvesting technology is being spotlighted as energy transmission and saving. [1]-[5] The key point of WPT application is composition of transmitting and receiving antenna array and RF-to-DC conversion efficiency. Especially, RF-to-DC conversion efficiency is the key factor that determines overall system efficiency. When low power is converted to DC, if received power could not overcome forward voltage drop of the diode, then RF-to-DC conversion efficiency becomes extremely low. Moreover, when low RF power is converted to DC, it is hard to gain high DC output voltage. This makes forward voltage drop become significant factor for conversion efficiency. To overcome this situation, there exists many kind of series connection to get high DC voltage. [4], [6] But in case of large number of rectifiers are connected, it is hard to realize series connection since its ground must be separated from each other or rectifying network should exists same plane of antenna to gives bias voltage to next antenna.

In transmitting and receiving array system, since received power pattern at receiving antenna is usually not uniform, elements which are positioned near edge of the antenna array receive relatively less power and it is inevitable in wireless power transmission system. Moreover, in high power application, matching point of diode is varies with input power. And to get a high power, rectifier is designed and matched for maximum power that usually received at the center of receiving array. In 2D array, since the bigger the array size, the number of element which are positioned near edge becomes larger, increase the efficiency of these elements is helpful for overall system efficiency. Presented cascade connection could increase conversion efficiency at low voltage that forward voltage drop of the diode is nonlinear and therefore helps increase the conversion efficiency of these elements. For such a reason, in the energy harvesting field which usually converts low power to DC power, this cascade rectifier network could increases RF-to-DC conversion efficiency of elements which are positioned not only near center but also positioned near edge.

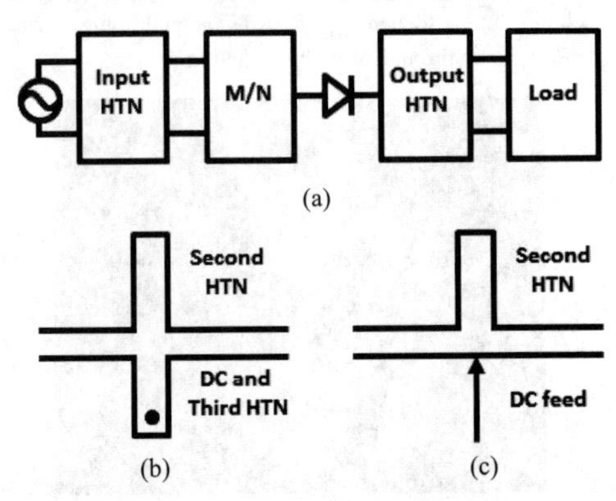

Figure 1. (a) Composition of single rectifier, (b) Detailed view of input HTN, (c) Modified input HTN for cascade rectifier network.

In this paper, characteristics of single rectifier for cascading network, basic theory of parallel and series connection, properties of cascade rectifier network, and efficiency increase mechanism using diode characteristics are presented. Also, this paper deals with advantage of cascade rectifier network beside the parallel and series connection network.

II. RECTIFIER DESIGN

Basically, rectifier is consists of input and output harmonic termination network (HTN), matching network, and load as shown in Fig. 1 (a). HTN increases conversion efficiency by eliminating harmonics. In order to implement cascade network, DC biasing point is needed between RF feeding (or antenna) and rectifying diode. But basically, at input network, there is grounded stub as shown in Fig. 1 (b) that eliminates DC harmonic which is generated from nonlinear characteristics of diode, and this stub must be removed for DC bias. Thus, from second rectifier of cascade network, rectifier is redesigned by leaving the stub that eliminates

978-7-5641-4279-7

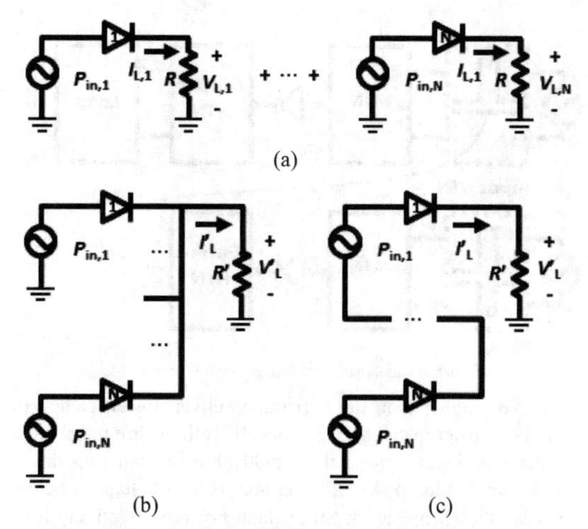

Figure 2. (a) Rectifier array which gets different input power each other and matched to same load resistance R, (b) parallel connection, (c) series connection.

second harmonic, and removing the stub that eliminates DC and third harmonic as shown in Fig. 1(c). The first rectifier of cascade network is same as basic single rectifier. In case of third harmonic is dominant over second harmonic, better efficiency can be achieved by leaving the stub that eliminates third harmonic only, and removing the stub that eliminates DC and second harmonic.

In this paper, every simulation is conducted by basic halfwave rectifier which is consists of single diode and single capacitor. And halfwave rectifier shows better conversion efficiency relatively low input power. [6]

The definition of the RF-to-DC conversion efficiency is [1]:

$$\eta_{\text{RF-to-DC}} = \frac{P_{\text{DC}}}{P_{\text{in}}} = \frac{(V_{\text{DC}})^2 / R_{\text{LOAD}}}{P_{\text{in}}} \tag{1}$$

Where P_{in} is input power to the rectifier, R_{LOAD} is the load resistance of the rectifier, and V_{DC} is the DC output voltage.

Since input power is not small signal, matching point varies with input power and very sensitive to circuit size and organization especially for high frequency applications. So, insensitive structure is required for stable fabrication and measurement. Radial stubs are recommended. In this paper, rectifier is matched to 0 dBm of input to shows increase of conversion efficiency better.

III. Parallel and Series Connection

The basic theory of parallel and series connection of rectifier can be explained by assuming output power out diode as current source. Suppose each rectifier is identical and matched to load resistance R for the maximum power transfer. Each rectifier gets different input power P_{in} as shown in Fig. 2 (a). Then the load resistance and the load voltage for the maximum power transfer for each connection method could be calculated as shown in below.

A. Parallel connection

For parallel connection, as shown in Fig. 2 (b), when the sum of both current and power from each rectifier are transferred to load resistance, the maximum power transfer occurs. Load resistance and load voltage of parallel connection network can be calculated by following equations.

$$R' = \frac{\sum_i V_{L,i}^2}{(\sum_i V_{L,i})^2} R \tag{2}$$

$$V_L' = \frac{\sum_i V_{L,i}^2}{\sum_i V_{L,i}} \tag{3}$$

Note that, as more rectifiers are connected in parallel, load resistance becomes smaller and it causes more current flows at load resistance. Thus, for a large number of arrays, overcurrent may flows across the load resistance.

B. Series connection

For series connection, as shown in Fig. 2 (c), when the sum of both voltage and power from each rectifier are transferred to load resistance, the maximum power transfer occurs. Load resistance and load voltage of series connection network can be calculated by following equations.

$$R' = \frac{(\sum_i V_{L,i})^2}{\sum_i V_{L,i}^2} R \tag{4}$$

$$V_L' = \sum_i V_{L,i} \tag{5}$$

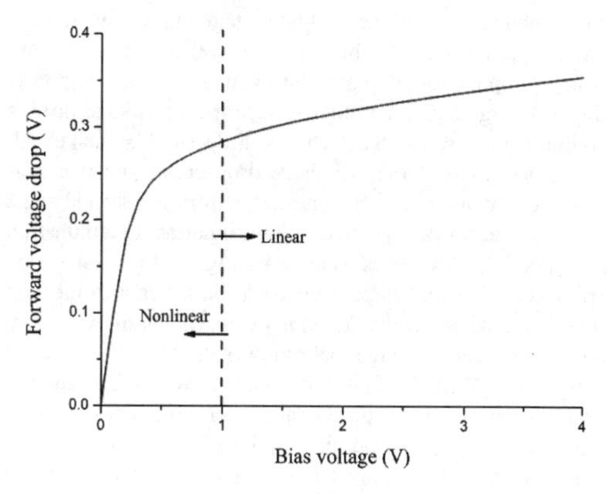

Figure 3. Forward voltage drop versus bias voltage.

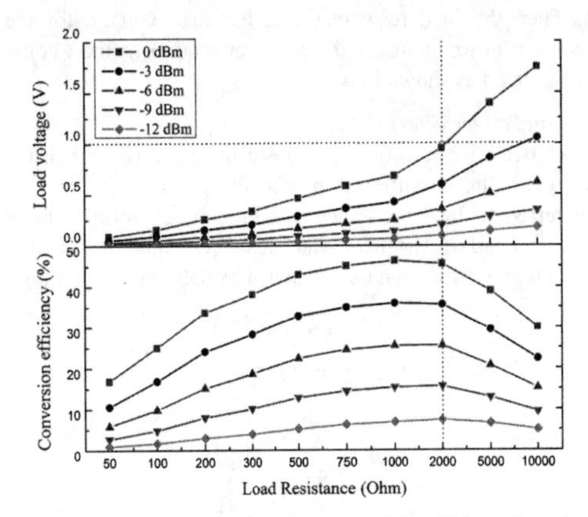

Figure 4. Load voltage and maximum conversion efficiency for each load resistance for each input power.

Note that the more rectifiers are connected in series, the higher the load resistance is.

When consider the maximum conversion efficiency, the overall output voltage of parallel connection cannot be higher than the maximum output voltage of rectifier in parallel connection network. For the series connection network, the output voltage could be higher, but each ground of rectifier must be separated from each other. If connection efficiency is considered, the overall conversion efficiency of network could be lower. Because for the parallel and series connection, power across the load resistance is calculated by linear summation of each rectifier, conversion efficiency lies between minimum and maximum efficiency of rectifier in the connection network.

IV. CASCADE CONNECTION

The key point of increase of conversion efficiency in cascade network is nonlinearity of diode. If forward voltage drop lies in linear region, it is hard to expect increase of conversion efficiency since output voltage of previous stage is dropped again when it biased to diode, and this forward voltage drop is proportional to bias voltage in linear region as shown in Fig. 3. On the other hand, forward voltage drop is nonlinear for low DC bias. That is, little DC bias makes RF signal overcomes forward voltage drop easier and it causes increase of conversion efficiency. But from the second stage of cascade network, since DC and third harmonic elimination stub does not exists as shown in Fig. 1 (c), conversion efficiency of second stage itself could be lower than the first stage of cascade network. Moreover, DC harmonic from diode of the second stage makes lower the DC bias which is output voltage of the first stage. But power of harmonic is proportional to input power and then efficiency loss is relatively low at low input power, therefore cascade network will increase the conversion efficiency. Thus, unlike conversion efficiency is bounded in parallel and series connection, overall conversion efficiency of cascade network

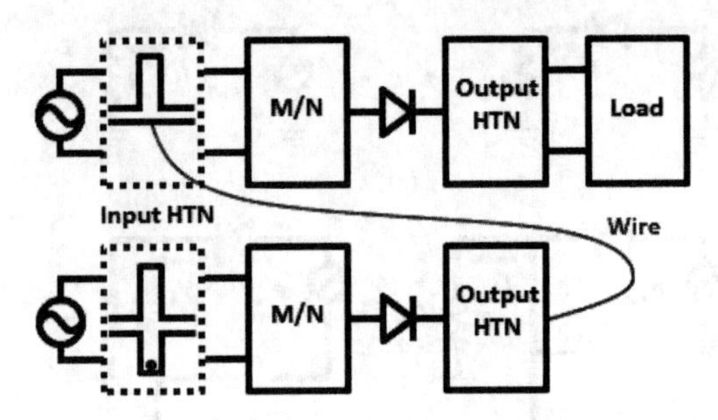

Figure 5. Composition of 2-stage cascade rectifier.

could be higher than the maximum conversion efficiency of single rectifier in cascade network. If rectifiers are matched to 0 dBm of input power, then 2000 ohm of load impedance makes maximum power transfer and its load voltage is below 1 V as shown in Fig. 4. Since input power is relatively low, unlike parallel and series connection, variation of optimum load resistance of cascade connection is small. Cascade network can be composed by removing load resistance of the first stage and DC HTN of the second stage, and biasing DC output voltage of the first stage to the second stage as shown in Fig. 5. Three and more cascade network could be composed by biasing DC output voltage of previous stage to next stage. Wire is used for DC feeding method since in X-band, it is hard to use chip inductor as RF choke. The equivalent circuit model for long straight wire over the ground is series connected resistor and inductor. And its value for equivalent circuit model can be calculated by following formula. [8]

$$R = \rho \frac{l}{A} \tag{6}$$

$$L = \frac{\mu}{2\pi} \ln\left(\frac{4h}{d}\right) \tag{7}$$

Where, ρ is resistivity, A is cross section area of wire, d is diameter of wire and h is distance between ground and the center of wire. Assume that $h > 1.5d$.

Since distance between rectifiers is usually short, resistance of wire is negligible. And since wire over the ground prevents RF guiding, the wire could be good RF choke in high frequency operations. Therefore, only DC power exists in wire. Although wire is open circuit in the view point of RF, when it is biased to rectifier, there is RF power at DC bias point, and it may influence matching network and harmonic termination. Conversion efficiency for each input power and each number of cascade stage is shown in Table I. Load resistance of every cascade network in Table I is fixed to 2000 Ohm. Note that since input power is low, conversion efficiency is relatively low and thus load voltage is also low. But unlike conversion efficiency of parallel and series

TABLE I
SUMMARY OF CASCADE RECTIFIER

Input power	Output voltage and Conversion efficiency			
	Single	2 stage	3 stage	4 stage
0 dBm	0.79 V / 31%	1.18 V / 35%	1.42 V / 34%	1.58 V / 31%
-3 dBm	0.41 V / 16%	0.68 V / 23%	0.83 V / 23%	0.93 V / 22%
-6 dBm	0.17 V / 6%	0.33 V / 11%	0.42 V / 12%	0.48 V / 11%
-9 dBm	0.05 V / 0.9%	0.12 V / 2.7%	0.16 V / 3.2%	0.18 V / 3.2%

connection is remains same or degraded as more rectifiers are connected, cascade connection could increases conversion efficiency as connected more. For example, output power of 2 stage parallel or series network for input power of -3 dBm is 168 μW, but in cascade connection, 231 μW of output power could be achieved and its 38% more than parallel and series network.

As input power is lower, the increase rate of conversion efficiency is higher. And it is due to nonlinear characteristic of diode as explained previously. Note that, increase the number of stage is not always causes increase of conversion efficiency. Because DC harmonic that caused by each stage is lowers the output voltage or DC bias. For example, DC harmonic caused by the third stage could lower the output voltage of the second stage, and this makes degradation of output voltage of the first stage. Therefore, as more rectifiers are connected in cascade, the output voltage of the first stage keeps degrading. After all, the first stage could not act as rectifier at all and then overall conversion efficiency of cascade network is degraded rapidly. Since DC harmonic is proportional to input power, increase of the number of cascade stage is not necessary as input power of DC bias is getting higher.

V. CONCLUSION

In this paper, increase conversion efficiency method using cascade connection is presented. The maximum increase of conversion efficiency is 12%(2 stage), 38%(2 stage), 103%(3 stage), 340%(3 stage) for 0 dBm, -3 dBm, -6 dBm, -9 dBm of input power, respectively, while conversion efficiency of parallel and series connection does not increase. At cascade network, increase of conversion efficiency is high, but since input power is low, the output power itself is not that high. Therefore gathering these power helps increase of overall system efficiency especially energy harvesting or WPT applications.

REFERENCES

[1] G. Monti, L. Tarricone, and M. Spartano, "X-Band Planar Rectenna," *IEEE Antennas Wireless Propag. Lett.*, vol. 10, pp. 1116-1119, 2011.
[2] W.C. Brown, "Rectenna technology program: Ultra light 2.45 GHz rectenna and 20 GHz rectenna," Raytheon Company, 1987.
[3] S. S. Mohammed, K. Ramasamy, T. Shanmuganantham, "Wireless Power Transmission – A Next Generation Power Transmission System," *2010 International Journal of Computer Applications.*, vol. 1, no. 13, pp. 100-103, 2010.
[4] C. Walsh, S. Rondineau, M. Jankovic, G. Zha, Z. Popovic, "A conformal 10 GHz Rectenna for Wireless Powering of Piezoelectric Sensor Electronics," *2005 IEEE MTT-S Int. Microwave Symp. Dig.*, pp. 143-146, June 2005.
[5] Gianfranco Andia Vera, "Efficient Rectenna Design for Ambient Mocriwave Energy Recycling," July 2009.
[6] Tan, Lee Meng Mark, "Efficient Rectenna Design for Wireless Power Transmission for MAV Applications," Naval Postgraduate School, December 2005.
[7] Alirio B., Ana C., Nuno B. C., and Apostolos G., "Optimum Behavior," *Microwave Magazine, IEEE,* March/April 2013
[8] Henry W. Ott. "Noise Reduction Techniques in Electronic Systems," 2nd ed. Wiley-Interscience publication, 1988.

Interference Reduction Method Using a Directional Coupler in a Duplex Wireless Power Transmission System

K. Nishimoto[1], K. Hitomi[2], T. Oshima[1], T. Fukasawa[1], H. Miyashita[1], Y. Takahashi[2], and Y. Akuzawa[2]
[1]Mitsubishi Electric Corporation
Kamakura, Kanagawa, Japan
[2]Mitsubishi Electric Engineering Co., Ltd.
Kamakura, Kanagawa, Japan

***Abstract*-We propose a method for reducing interference in a duplex wireless power transmission system that uses electromagnetic resonant coupling. Duplex wireless power transmission acts as a directional coupler if it has no loss and is perfectly matched. The proposed method uses this characteristic and the interference is decreased by connecting a directional coupler to the input (or output) ports of a duplex wireless power transmission system. We validate our method by performing simulations for two configurations of duplex wireless power transmission: two sets of oppositely arranged antenna pairs aligned horizontally, and two sets of coaxially arranged antenna pairs aligned vertically. In addition, we show that the location of the isolation port changes depending on the transmission distance in the former configuration.**

I. INTRODUCTION

Recently, wireless power transmission systems that use electromagnetic resonant coupling have been widely researched [1]–[8]. In these systems, the transmission distance becomes shorter, but the transmission efficiency becomes higher than in wireless power transmission systems that use microwaves [5]. In this paper, a wireless power transmission system that uses electromagnetic resonant coupling is applied to multiplex power transmission. In a multiplex wireless power transmission system, multiple pairs of transmitting and receiving antennas transmit energy independently of each other. This can provide advantages such that the realization of a redundant system. To scale down a multiplex wireless power transmission system, multiple transmitting and receiving antennas are placed closely. Therefore, each antenna couples strongly to two or more antennas, and the interference between multiple transmissions within the system increases.

We propose a method for reducing the amount of interference by connecting a directional coupler to the input (or output) ports of a duplex wireless power transmission system. First, we describe the characteristics of duplex wireless power transmission. Next, the interference reduction method using a directional coupler is detailed. Moreover, we validate our method by performing simulations for the two configurations of the duplex wireless power transmission: two sets of oppositely arranged antenna pairs aligned horizontally, and two sets of coaxially arranged antenna pairs aligned vertically.

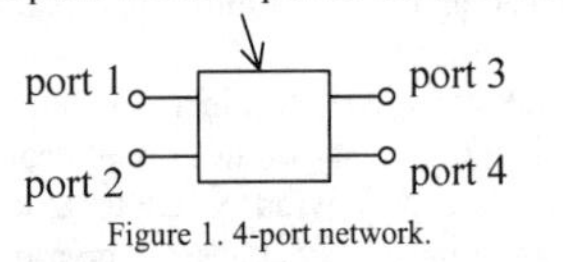

Figure 1. 4-port network.

II. CHARACTERISTICS OF DUPLEX WIRELESS POWER TRANSMISSION

A duplex power transmission system is composed of two transmitting antennas and two receiving antennas, and it can be expressed as a 4-port network, as shown in Fig. 1. If a 4-port network has no loss and is perfectly matched, it becomes a directional coupler from the condition that the S-matrix is unitary and $S_{11} = S_{22} = S_{33} = S_{44} = 0$ [9]. Therefore, we consider that the duplex wireless power transmission that has no loss and is perfectly matched acts as a directional coupler.

We now show the relationship between transmission phases. When $S_{21} = S_{43} = 0$ in a directional coupler, the S-matrix is:

$$[S]=\begin{pmatrix} 0 & 0 & S_{31} & S_{41} \\ 0 & 0 & S_{32} & S_{42} \\ S_{31} & S_{32} & 0 & 0 \\ S_{41} & S_{42} & 0 & 0 \end{pmatrix}. \tag{1}$$

As $[S]$ is unitary, we have:

$$|S_{31}| = |S_{42}|,\ |S_{41}| = |S_{32}|, \text{ and} \tag{2}$$

$$S_{31}S_{32}^{*} + S_{41}S_{42}^{*} = 0. \tag{3}$$

At this point, we move the port positions by connecting transmission lines to the ports. First, the positions of ports 1, 3 and 4 are chosen in such a way that S_{31} is pure real and positive and that the phase of S_{41} is θ. Then, the position of port 2 is chosen in such a way that S_{42} is pure real and positive. Therefore, from (2),

$$S_{31} = a,\quad S_{41} = be^{j\theta},\quad S_{42} = a,\quad S_{32} = be^{j\phi}, \tag{4}$$

where a and b are pure real numbers, and ϕ is the phase of S_{32}. Substituting (4) into (3), we have:

$$e^{-j\phi} = e^{j(\theta-\pi)}. \tag{5}$$

Then,

$$\theta + \phi = \pi. \tag{6}$$

978-7-5641-4279-7

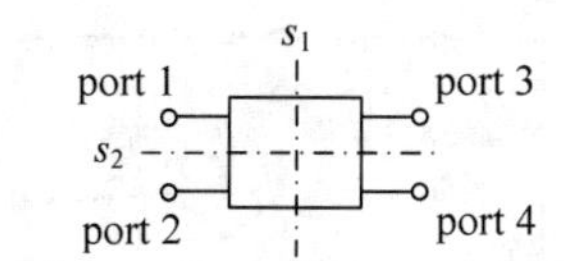

Figure 2. 4-port network with two symmetry planes.

For example, $\phi = -\pi/2$ when $\theta = -\pi/2$, and $\phi = \pi/2$ when $\theta = \pi/2$. Therefore, the differences between the transmission phases from one port to the other two ports (except for the isolation port) can be set to $\pm \pi/2$ by moving the port positions in an arbitrary directional coupler.

As a special case, we consider the 4-port network that has two symmetry planes, as shown in Fig. 2. The 4-port network is symmetric with respect to the planes s_1 and s_2. From this symmetry, the S-matrix is:

$$[S]=\begin{pmatrix} 0 & 0 & v & w \\ 0 & 0 & w & v \\ v & w & 0 & 0 \\ w & v & 0 & 0 \end{pmatrix}, \tag{7}$$

where $v=|v|e^{j\varsigma}$ and $w=|w|e^{j\xi}$. As $[S]$ is unitary, we have:

$$\varsigma - \xi = \pm\pi/2. \tag{8}$$

Therefore, the duplex wireless power transmission with two symmetry planes acts as a directional coupler in which the differences between the transmission phases from one port to the other two ports are $\pm \pi/2$.

III. Interference Reduction Method Using a Directional Coupler

Fig. 3 shows the configuration of the proposed interference reduction method using a directional coupler. Let Z_0 be the impedance of each port. Because the duplex wireless power transmission that has no loss and is perfectly matched becomes a directional coupler, we assume that it is a directional coupler with a coupling coefficient k in which $S_{21} = S_{43} = 0$. The input (or output) ports of the duplex wireless power transmission system are connected to a directional coupler by two transmission lines whose characteristic impedances are Z_0. As detailed in Sec. II, the differences between the transmission phases from one port to the other two ports can be set to $\pm \pi/2$ by moving the port positions in an arbitrary directional coupler. Then, the lengths of the transmission lines l_1 and l_2 are determined such that $\alpha = \arg(S_{51})-\arg(S_{61})$ becomes $\pm \pi/2$. In the additional directional coupler, the coupling coefficient is k', $S_{65} = S_{87} = 0$, and $\beta = \arg(S_{75})-\arg(S_{85}) = \pm\pi/2$. There are four combinations of α and β: $(\alpha, \beta) = (\pi/2, \pi/2)$, $(\pi/2, -\pi/2)$, $(-\pi/2, \pi/2)$, and $(-\pi/2, -\pi/2)$. Let a_1 be the input wave into port 1, while b_1 and b_2 are the output waves from ports 7 and 8 respectively. When $(\alpha, \beta) = (\pi/2, \pi/2)$,

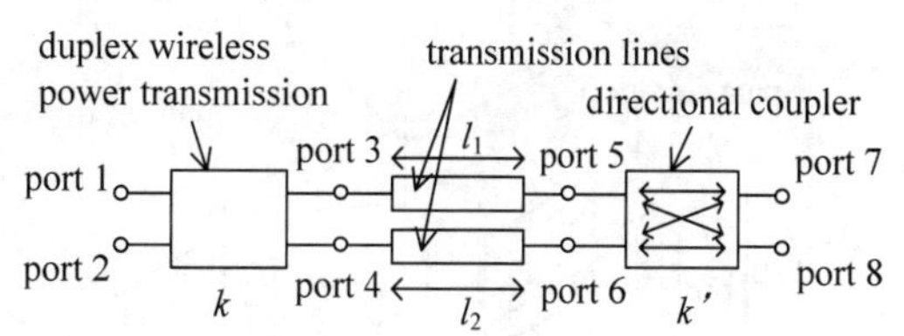

Figure 3. Configuration of the proposed method.

$$\begin{aligned} b_7 &= \left(-\sqrt{1-k^2}\sqrt{1-k'^2} + kk'\right)a_1 \\ b_8 &= j\left(k'\sqrt{1-k^2} + k\sqrt{1-k'^2}\right)a_1 \end{aligned}. \tag{9}$$

As k is pure real and positive, b_7 can be 0, and then we have:

$$k' = \sqrt{1-k^2}. \tag{10}$$

When $(\alpha, \beta) = (\pi/2, -\pi/2)$,

$$\begin{aligned} b_7 &= j\left(\sqrt{1-k^2}\sqrt{1-k'^2} + kk'\right)a_1 \\ b_8 &= \left(-k'\sqrt{1-k^2} + k\sqrt{1-k'^2}\right)a_1 \end{aligned}. \tag{11}$$

As k is pure real and positive, b_8 can be 0, and then we have:

$$k' = k. \tag{12}$$

Similarly, from the condition that S_{71} or S_{81} becomes 0, (12) is derived when $(\alpha, \beta) = (-\pi/2, \pi/2)$, and (10) is derived when $(\alpha, \beta) = (-\pi/2, -\pi/2)$. In summary, we obtain:

$$k' = \begin{cases} k & \alpha - \beta = \pm\pi \\ \sqrt{1-k^2} & \alpha - \beta = 0 \end{cases}. \tag{13}$$

By deciding k' and $\alpha-\beta$ according to (13), the interference between the two transmissions can be reduced. In this method, we have to use a low-loss directional coupler in order not to decrease the transmission efficiency.

IV. Simulated Results

A. *Two sets of oppositely arranged antenna pairs aligned horizontally*

The proposed method is studied by simulations. The simulated frequency is 6.78 MHz, because this frequency is used by A4WP (Alliance for Wireless Power) and CEA (Consumer Electronics Association) 2042.2. Fig. 4 illustrates the simulated model that horizontally aligns two sets of oppositely arranged antenna pairs. We place antennas 1 and 3 opposite to each other, and place antennas 2 and 4 opposite to each other. We separate antennas 1 and 2 by a distance of 20 mm. Antennas 1 and 2 are designated as transmitting antennas, and antennas 3 and 4 are designated as receiving antennas. Each antenna consists of a feed loop and a parasitic loop with a capacitance C, and is assumed to be a perfect conductor. The directions of ports 1 and 3 are opposite to the directions of ports 2 and 4. This model is symmetric with respect to the *yz* and *zx* planes. While changing the distance between the transmitting and receiving antennas d, the simulations are performed using the FEM (Finite Element Method). At each d, we adjust the distance between the feed and parasitic loops d_r and C so that the amount of reflection decreases.

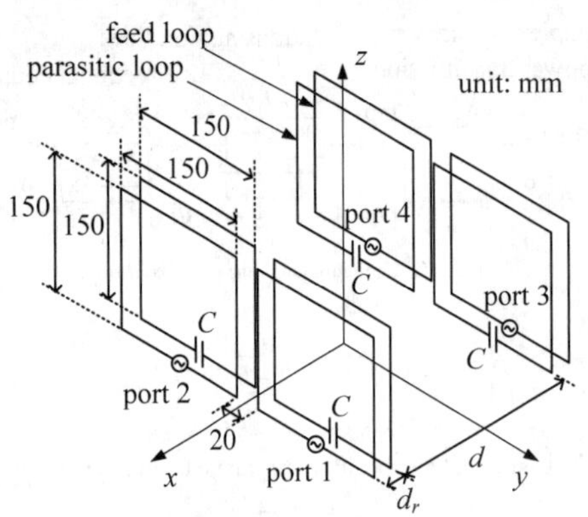

Figure 4. Simulated model that horizontally aligns two sets of oppositely arranged antenna pairs.

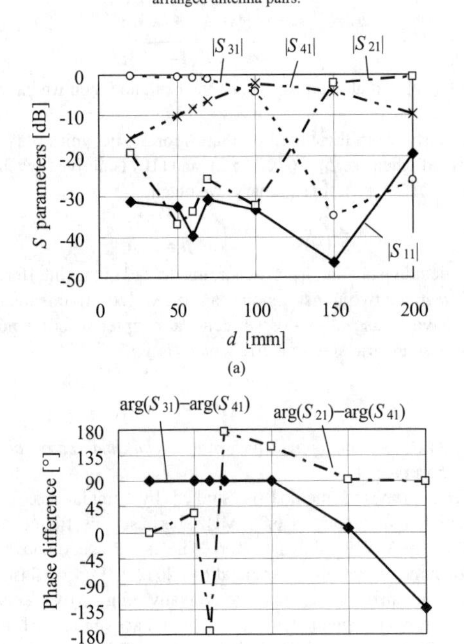

Figure 5. Calculated S-parameters for the system without a directional coupler: (a) amplitude and (b) phase.

Fig. 5 shows the calculated S-parameters for the system without a directional coupler. As the model is symmetric, only the case in which port 1 is an input port is shown. In Fig. 5, $|S_{11}|$ is less than –19.5 dB. When $d \leq$ 125 mm, $|S_{21}|$ is less than $|S_{31}|$ and $|S_{41}|$, and is also less than –19.3 dB. On the other hand, when $d \geq$ 125 mm, $|S_{31}|$ is less than $|S_{21}|$ and $|S_{41}|$, and is also less than –26.0 dB. Therefore, port 2 (the adjacent antenna) becomes an isolation port at $d \leq$ 125 mm, and port3 (the opposite antenna) becomes an isolation port at $d \geq$ 125 mm. In effect, the location of the isolation port changes depending on the transmitting distance. In Fig. 5(b), arg(S_{31})–arg(S_{41}) is about 90° at $d \leq$ 125 mm, and arg(S_{21})–arg(S_{41}) is about 90° at $d \geq$ 125 mm. Therefore, the differences between the transmission phases from one port to the other two ports (except for the isolation port) are shown to be approximately 90°. This is because the model in Fig. 4 has two symmetry planes.

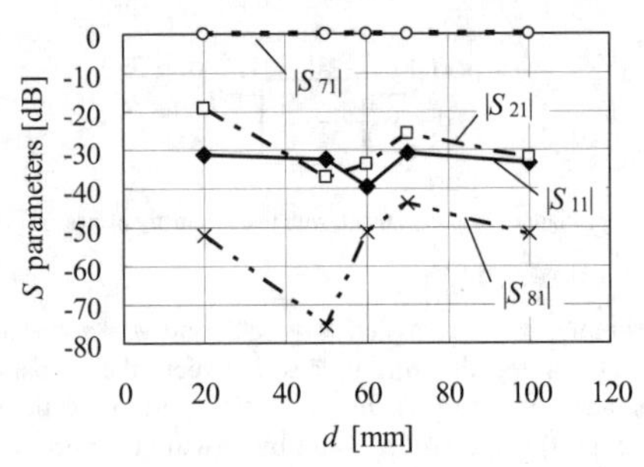

Figure 6. Calculated S-parameters for the system with a directional coupler.

When port 2 is an isolation port, $|S_{31}|$ and $|S_{41}|$ are greater than −10.0 dB for 50mm $\leq d \leq$ 100mm. Then, a directional coupler is added when $d \leq$ 125 mm, as shown in Fig. 3. Because arg(S_{31})–arg(S_{41}) $\approx \pi/2$, the choice $l_1 = l_2$ gives α = arg(S_{51})–arg(S_{61}) $\approx \pi/2$. According to (13), we choose $k' = k = |S_{41}|$, and β = arg(S_{75})–arg(S_{85}) = $-\pi/2$. Fig. 6 shows calculated S-parameters for the system with a directional coupler. We find that $|S_{71}| \geq -0.05$dB, and $|S_{81}| \leq -40$dB for 20mm $\leq d \leq$ 100mm. In addition, $|S_{82}| \geq -0.05$dB, and $|S_{72}| \leq -40$dB. From these results, we confirm that the interference between the two transmissions is reduced from above –10 dB to below –40dB by adding a directional coupler.

B. Two sets of coaxially arranged antenna pairs aligned vertically

Fig. 7 illustrates a simulated model that vertically aligns two sets of coaxially arranged antenna pairs. We stack antennas 1 and 3 coaxially, and stack antennas 2 and 4 coaxially. We arrange antennas 1 and 2 at an interval of d on the center axis. Antennas 1 and 2 are transmitting antennas, and antennas 3 and 4 are receiving antennas. The directions of ports 1–4 are the same. This model is symmetric with respect to the yz and zx planes. While changing d, the simulation is performed by using the FEM. At each d, we adjust d_{rs}, d_{rl}, C_s, and C_l so that the amount of reflection decreases.

Fig. 8 shows the calculated S-parameters for the system without a directional coupler. In Fig. 8, $|S_{11}|$ is less than –25.0 dB. In addition, we find that $|S_{21}|$ is less than $|S_{31}|$ and $|S_{41}|$, and is also less than –14.5 dB. Therefore, port 2 is an isolation port. In Fig. 8(b), arg(S_{31}) –arg(S_{41}) is about 90°, because the model in Fig. 7 has two symmetry planes.

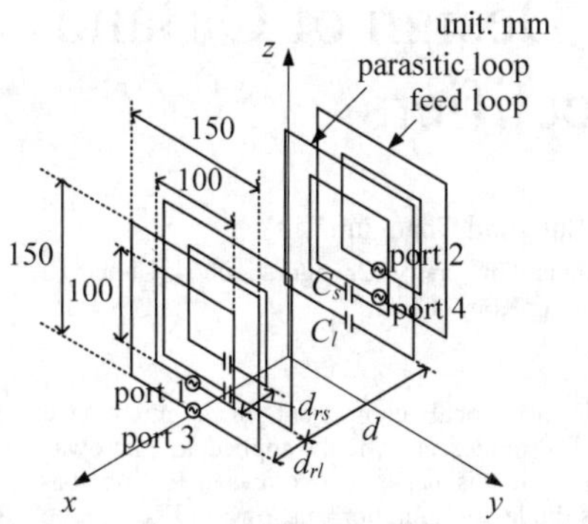

Figure 7. Simulated model that vertically aligns two sets of coaxially arranged antenna pairs.

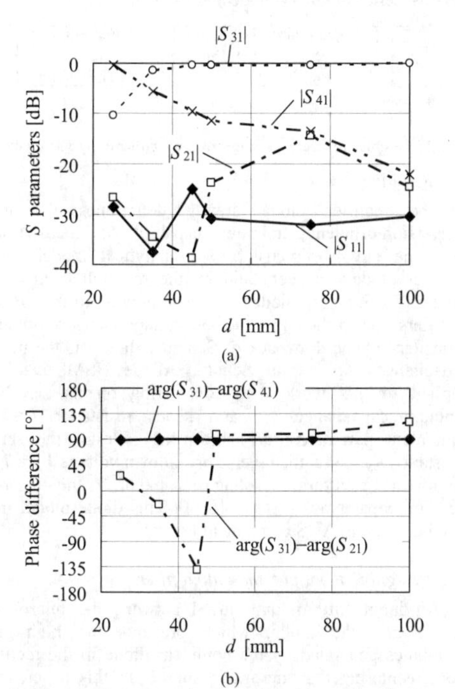

Figure 8. Calculated S-parameters for the system without a directional coupler: (a) amplitude and (b) phase.

For 25mm $\leq d \leq$ 50mm, $|S_{31}|$ and $|S_{41}|$ are greater than –11.4 dB. Then, a directional coupler is added, as shown in Fig. 3. Because $\arg(S_{31}) - \arg(S_{41}) \approx \pi/2$, the choice $l_1 = l_2$ gives $\alpha \approx \pi/2$. According to (13), we choose $k' = k = |S_{41}|$, and $\beta = -\pi/2$. Fig. 9 shows the calculated S-parameters for the system with a directional coupler. We find that $|S_{71}| \geq -0.02$dB, and $|S_{81}| \leq -40$dB for 25mm $\leq d \leq$ 50mm. In addition, $|S_{82}| \geq -0.02$dB, and $|S_{72}| \leq -40$dB. From these results, we confirm that the interference between the two transmissions is reduced

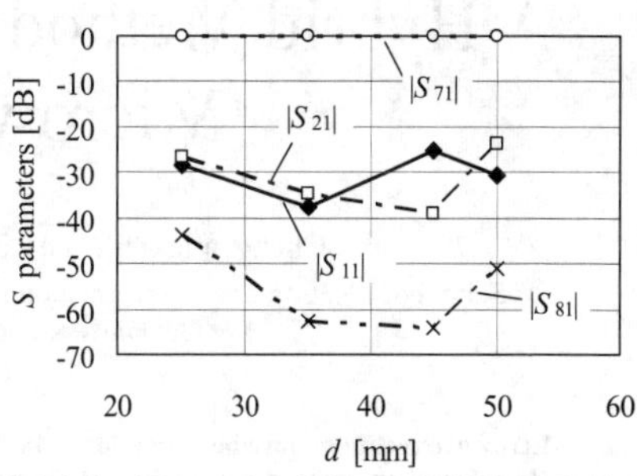

Figure 9. Calculated S-parameters for the system with a directional coupler.

from above –11.4 dB to below –40dB by adding a directional coupler.

V. Conclusions

We have proposed a method for reducing interference in a duplex wireless power transmission system by adding a directional coupler. Furthermore, we have validated our method by performing simulations for two configurations of duplex wireless power transmission: two sets of oppositely arranged antenna pairs aligned horizontally, and two sets of coaxially arranged antenna pairs aligned vertically. Moreover, we have shown that the location of the isolation port changes depending on the transmission distance in the former configuration.

References

[1] A. Kurs, A. Karalis, R. Moffatt, J. D. Joannopoulos, P. Fisher, and M. Soljacic, "Wireless power transfer via strongly coupled magnetic resonators," *Science Express*, vol. 317, no. 5834, pp. 83-86, Jul. 2007.

[2] A. Karalis, J.D. Joannopoulos, and M. Soljačić, "Efficient wireless non-radiative mid-range energy transfer," *Ann. Phys.*, vol. 323, no. 1, pp. 34-48, Jul. 2008.

[3] H. Hirayama, T. Ozawa, Y. Hiraiwa, N. Kikuma, and K. Sakakibara, "A consideration of electro-magnetic-resonant coupling mode in wireless power transmission," *IEICE Electronics Express*, vol. 6, no. 19, pp. 1421-1425, Oct. 2009.

[4] Q. Yuan, Q. Chen, and K. Sawaya, "Maximum transmitting efficiency of wireless power transfer system with resonant/non-resonant transmitting/receiving elements," *2010 IEEE AP-S*, Jul. 2010.

[5] H. Shoki, "Issues and initiatives for practical use of wireless power transmission technologies in Japan," *IEEE IMWS-IWPT 2011*, pp. 87-90, May 2011.

[6] N. Inagaki and S. Hori, "Classification and characterization of wireless power transfer systems of resonance method based on equivalent circuit derived from even- and odd mode reactance functions," *IEEE IMWS-IWPT 2011*, pp. 115-118, May 2011.

[7] T. Imura and Y. Hori, "Maximizing air gap and efficiency of magnetic resonant coupling for wireless power transfer using equivalent circuit and Neumann formula," *IEEE Trans. Ind. Electron.*, vol. 58, no. 10, pp. 4746-4752, Oct. 2011.

[8] I. Awai and T. Ishizaki, "Superiority of BPF theory for design of coupled resonator WPT systems," *APMC 2011*, pp. 1889-1892, Dec. 2011.

[9] C.G. Montgomery, R.H. Dicke, and E.M. Purcell, *Principles of Microwave Circuits*. London, UK: The Institution of Engineering and Technology, 2007.

A Hybrid Method on the Design of C Band Microwave Rectifiers

Chengyang Yu[1], Biao Zhang[1], Sheng Sun[2] and Changjun Liu[1]
[1]School of Electronics and Information Engineering, Sichuan University, Chengdu 610064, China
[2]The University of Hong Kong, Hong Kong, China
e-mail: cjliu@ieee.org

***Abstract*—Microwave rectifiers have been developed in various forms since the microwave power transmission (MPT) began to attract researchers' attention. A hybrid simulation method is implemented by the combination of IE3D and ADS simulation to realize a fast and accurate rectifier design in this paper. A 5.8 GHz microstrip rectifier based on HSMS 286 Schottky diode is realized and fabricated based on the proposed method for demonstration. Microstrip structures are light and easy to be integrated into rectennas in a MPT system. The whole circuit is compact with a dimension of 55 mm by 18 mm. The measured MW-to-DC conversion efficiency is 68%, which is obtained at an input microwave power of 16 dBm. The simulated and measured results agree well, which proves the validity of the proposed design method.**

Keywords—microwave rectifier; hybrid simulation; microwave power transmission

I. INTRODUCTION

In wireless power transmission (WPT) systems, power is transmitted wirelessly from one place to another where it is then converted to DC output power [1]. In general, microwave (MW) or laser is used as the energy carrier to realize the system. A laser based power transmission system is very compact due to its natural narrow beams. It has extremely crucial environmental conditions. The power transmission efficiency suffers from rain, fog, dusts and so on. On the other hand, microwave power transmission may keep its transmission efficiency in various environments, and becomes a research hotspot [2]. The representative progress can date back to the work done by Brown W.C. in 1963 [3], which realized a rectifying antenna with conversion efficiency of almost 50% using Schottky diodes.

In order to operate the cost of a MPT system effectively, rectifiers with a high MW-DC conversion efficiency have been intensively researched and developed around the world [4]-[7]. Accurate and fast designs on microwave rectifiers are very important. A hybrid simulation method is implemented by the combination of Mentor Graphics IE3D and Agilent Advanced Design System (ADS) to realize the goal in this paper. The hybrid design methods exhibits both the advantages of the IE3D in microstrip circuit design and ADS in active nonlinear components. A rectifier at 5.8 GHz is designed with the proposed method, and it is fabricated and measured to demonstrate its validity.

II. ANALYSIS AND DESIGN

This work was supported in part by the NFSC 61271074, 863 Program2012AA120605, and National Key Basic Research Program of China 2013CB328902

978-7-5641-4279-7

As a critical component in a microwave rectifier, Schottky diodes are widely applied to microwave rectifier designs. In this paper, a microwave rectifier based on the series diode configuration is shown in Fig. 1. It consists of a few key components: input and output filter, matching circuits, series Schottky diode and DC load.

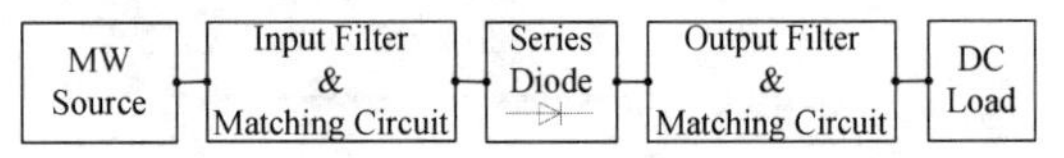

Fig. 1 The structure diagram of a microwave rectifier using a series diode

A. Rectifing diode

The rectifier diode greatly determines the total conversion efficiency and power capability of a rectifier, and even the circuit dimension. An analytical model for a rectifier diode has been built by the research team of K. Chang [8]. Schottky diodes are most suitable to microwave rectifiers due to their low turn-on voltage and low junction capacitance. The drawback of Schottky diodes is the power capacitance. An Avago Schottky diode HSMS-286C is applied to this work with considering on its zero-bias junction capacitance C_{J0} and series resistance R_S. Its equivalent parameters are shown as follows: the series resistance $R_S = 6\ \Omega$, the reverse breakdown voltage $V_B = 7$ V, the forward bias turn-on voltage $V_{BI} = 0.65$ V, the zero-bias junction capacitance $C_{J0} = 0.18$ pF. This diode model may also be found in ADS component library.

B. Designing the input and output filter

An input filter is introduced between the microwave source and the diode, which prevents the high-order harmonics generated by the nonlinear diode in the rectifier from reentering the microwave source. In this paper, split ring resonators based on defected ground structure (DGS) are utilized as the input filter to pass the power at 5.8 GHz and block its harmonics. Two Murata high frequency chip capacitors (GQM2195C2A3R3CB01) are applied to output DC filter, which avoid microwave power dissipation on the DC load. The input and output filter work cooperatively that the harmonics are reflected between the input filter, i.e. the split ring resonators, and the output DC filter, i.e. the two capacitors, until they are eventually converted to DC by the diode as shown in the harmonic recycling theory.

C. Design impedance matching circuits

The input impedance of a rectifier diode is calculated to obtain its imaginary part based on the closed form equations in [9]. The imaginary impedance is mainly generated from the junction capacitance C_J. This part can be canceled out by adjusting dimensions of the microstrip between chip capacitors and rectifier diode. Then, a quarter-wavelength impedance transformer is designed to match the real part of the diode impedance into the input filter. Furthermore, a short-ended shunt stub is placed after the input filter, which protects the MW source from the DC voltage of the diode and fine-tunes the matching circuit.

III. HYBRIDE SIMULATION DESIGN

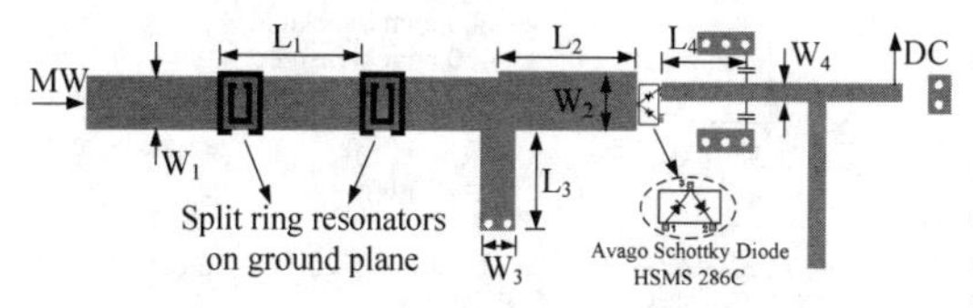

Fig. 2 Configuration of the proposed 5.8 GHz rectifier

Fig. 2 shows the configuration of the proposed 5.8 GHz rectifier. An HSMS 286C diode package contains a pair of Schottky diodes in series. We have used only one of them. This circuit is printed on a F4B-2 substrate with a dielectric constant of 2.65 and a thickness of 1 mm. Its metal layer is 17 □m copper. The new hybrid simulation method mainly includes two aspects: microstrip circuit designs with Mentor Graphics IE3D, and co-simulation with Agilent ADS invoking the data obtained from IE3D. The main processes are showed as follows.

The IE3D, which has an excellent accuracy for frequency domain analysis on microstrip circuits, are used to accurately simulate and optimize the microstrip sections of the rectifier. In other words, the IE3D is applied to passive parts of the rectifier. The diode and capacitors are replaced by ports in IE3D. The simulated S-parameters in the text format (.SP) of the multiport passive circuit are obtained. As shown in in Fig. 3, all the ports including the lumped components, diode, and feeding points are defined as Localized for MMIC ports. The main optimized dimensions are shown in Table. I.

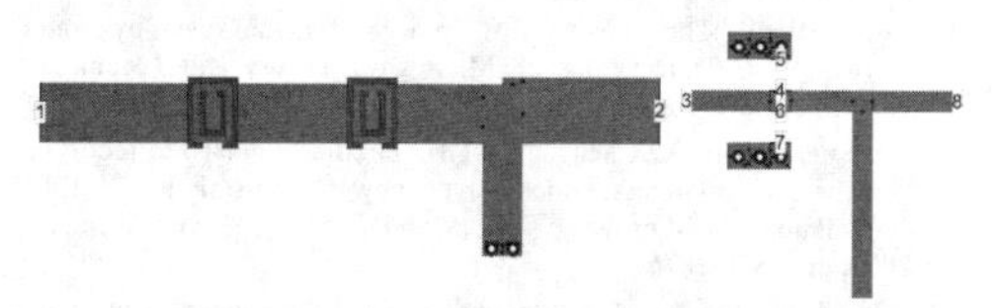

Fig. 3 Simulation view in Mentor Graphics IE3D

TABLE I. SUMMARY OF THE RECTIFIER DIMENSIONS

Item	L_1	L_2	L_3	L_4	W_1	W_2	W_3	W_4
(mm)	7.5	7.4	5.1	4.4	2.7	3	1.9	0.9

Then, the S-parameters of the microstrip circuits of the rectifier are treated as a multiport component. It is used in Agilent ADS along with the rectifier diode model of HSMS 286, lumped chip capacitors GQM2195C2A3R3CB01, and DC load to simulate the entire circuit. The detailed simulation configuration is shown in Fig. 4. In ADS, the harmonic balance simulation is a frequency-domain analysis technique for simulating nonlinear circuits and systems. It is well-suited for simulating analog RF and microwave circuits such as rectifiers.

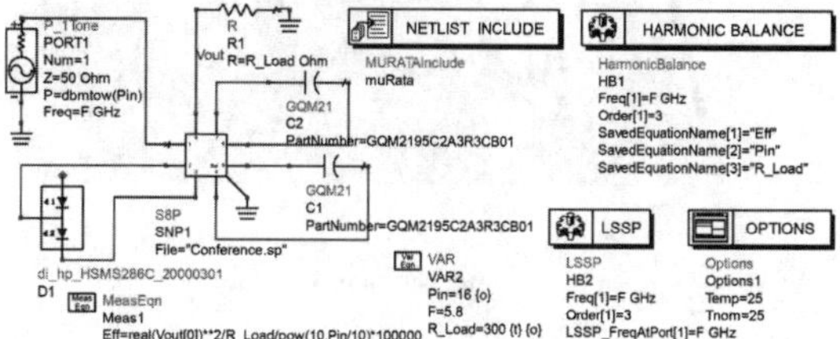

Fig. 4 Simulation view in Agilent ADS

We optimize the MW-to-DC efficiency in ADS by means of adjusting the source power, load impedance, and the capacitors in ADS, and regulating the width and length of microstrip lines in IE3D. The optimal MW-to-DC efficiency is achieved after a few of iteration operations.

IV. MEASUREMENTS AND RESULTS

On the basis of simulation and optimization, a 5.8 GHz microwave rectifier is fabricated, as shown in Fig. 5(a), and measured. Fig. 5(b) and (c) show the top and bottom sides of the rectifier. An Agilent E8267C vector signal generator, which has a maximum output power of 20 dBm, is applied as the microwave source. An Agilent 34970A data acquisition is placed with a standard resistor box to monitor and display the output DC voltage at the DC load. The measured conversion efficiency is defined as

$$\eta = \frac{(V_{DC})^2}{R_L} \times \frac{1}{P_{MW}} \times 100\% \quad .!$$

where V_{DC} is the measured voltage across the DC load R_L and P_{MW} is the output microwave power generated by the Agilent E8267C generator.

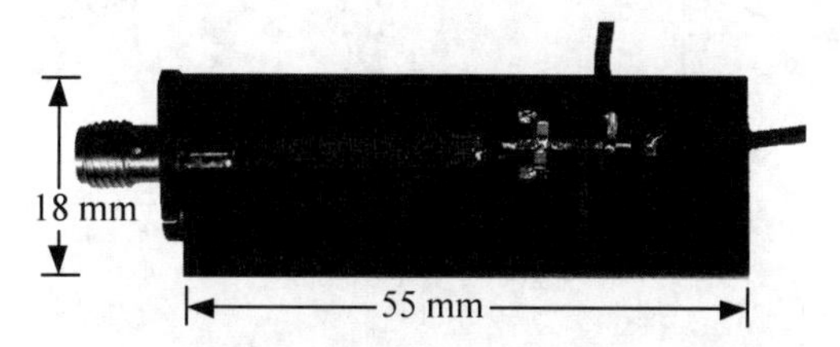

(a) Fabricated rectifier

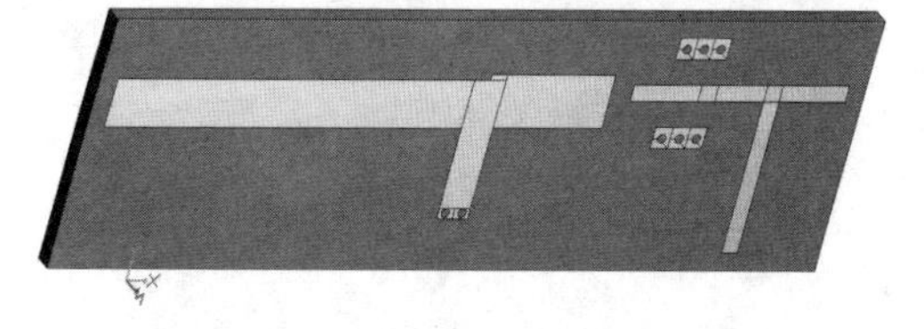

(b) Top view

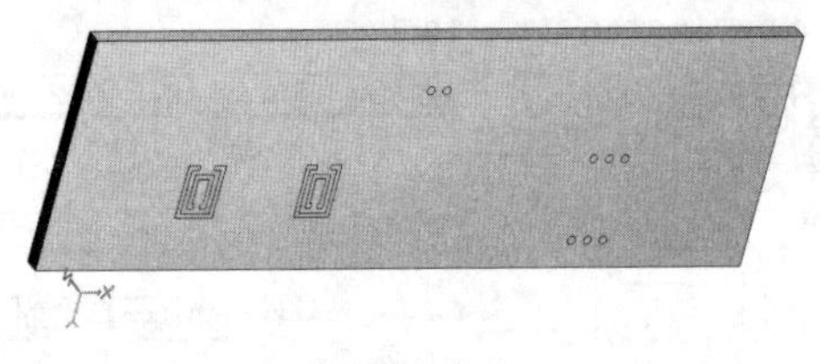

(c) Bottom view

Fig. 5 Photograph of the fabricated microwave rectifier

A comparison between the measured and simulated output voltages is shown in Fig. 6, where the simulated results show a very agreement to the measured results, when the microwave power is less than 17 dBm. The greater difference at a higher input power is mainly due to the breakdown voltage of a real rectifier diode. The nonlinearity is much higher when the output DC voltage is close its breakdown voltage. Furthermore, Fig. 7 shows a comparison between the measured and simulated MW-DC conversion efficiency. The difference in output voltage is apparently amplified in the efficiency comparison since there is a quadratic relationship between them. However, experimental results are still good and both the measured and simulated data show an equal optimal load resistor. The measured highest conversion efficiency is 68% an input power of 16 dBm, where the optimal load resistor is 300 ohm.

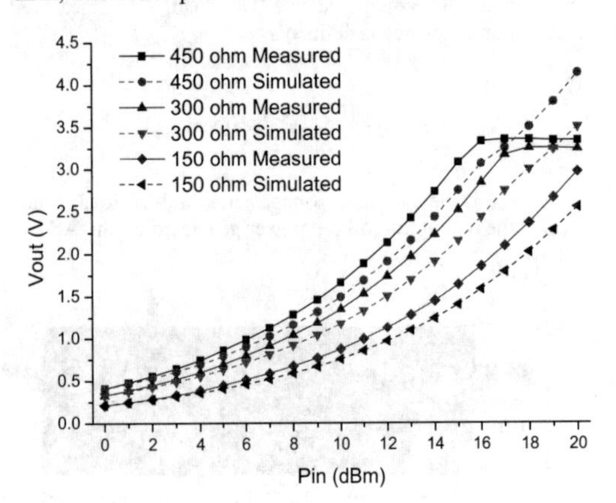

Fig. 6 Comparison of measured and simulated output voltage

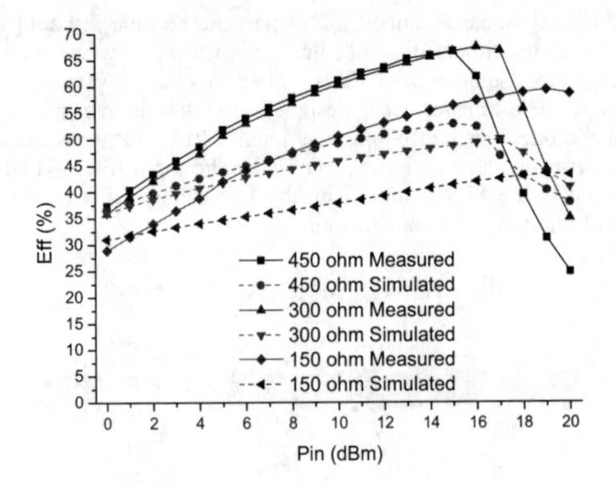

Fig. 7 Comparison of measured and simulated MW-DC conversion efficiency

V. CONCLUSIONS

In the design of microwave rectifiers, this paper proposed a hybrid design method using Mentor Graphics IE3D electromagnetic simulation software and Agilent ADS circuit simulation software in order to achieve a fast and accurate simulation. A 5.8 GHz microwave rectifier was then designed with the proposed method, and fabricated and measured. Experimental measurements showed that the rectifier achieved a maximum MW-DC conversion efficiency of 68% when the input power was 16 dBm and the optimal load resistor was 300 ohm. The measurements showed a good match to the simulated results at low power level, which validate the proposed design method.

REFERENCES

[1] Tesla, N., "The transmission of electric energy without wires," The thirteenth Anniversary Number of the Electrical World and Engineer, March 5, 1904.

[2] Matsumoto, H., "Microwave power transmission from space and related nonlinear plasma effects," The Radio Science Bulletin, no.273, 1995, pp. 11-35.

[3] Brown W. C., "The history of power transmission by radio waves," IEEE Transactions on Microwave Theory and Techniques, vol. 32, no. 9, 1984, pp. 1230-1242.

[4] Strassner, B. and K. Chang, "5.8-GHz circularly polarized rectifying antenna for wireless microwave power transmission," IEEE Transactions on Microwave Theory and Techniques, vol. 50, no. 8, 2002, pp. 1870-1876.

[5] Saka, T., Y. Fujino, M. Fujita, and N. Kaya, "An experiment of a C band rectenna," Proceedings of the Conference on Space Power Systems, Energy and Space for Humanity, 1997, pp. 251-253.

[6] Zibitou, J., M. Latrach, and S. Toutain, "Wide band power rectenna with high sensitivity detection," Proceedings of the 4th International Conference on Solar Power from Space (SPS'04), 2004, pp. 251-255.

[7] C. Yu, C. Liu, B. Zhang, X. Chen, and K. Huang, "An intermodulation recycling rectifier for microwave power transimisson at 2.45 GHz," Progress in Electromagnetics Research, vol. 119, 2011, pp. 435-447.

[8] Jonathan Hansen and K. Chang, "Diode modeling for rectenna design," Antennas and Propagation (APSURSI), July2011, pp. 1077-1080.

[9] J. O. McSpadden, L. Fan, and K. Chang, "Design and experiments of a high-conversion-efficiency 5.8-GHz rectenna," IEEE Trans. on Microwave Theory and Techniques, vol. 46, no. 12, pp. 2053-2059.

Analysis of Near-Field Power Transfer of Multi-Antenna Using Multiport Scattering Parameters

Mingda Wu[1], Qiang Chen[1], Qiaowei Yuan[2]
Department of Communications Engineering, School of Engineering, Tohoku University[1]
Sendai National College of Technology[2]

***Abstract*-Power transfer efficiency (PTE) of wireless power transfer (WPT) using near-field coupling of multi-antenna is numerically analyzed by using the scattering parameters after modeling the whole multi-antenna WPT system as a multi-port network. Because the multi-port scattering parameters can easily be measured by using a vector network analyzer and calculated by a full-wave numerical analysis, it is convenient to investigate the relationship between the PTE and the geometry of the multi-antenna WPT system, such as the relative position of the transmitting antenna and the receiving antennas, the geometry of antennas and copper loss in the antennas.**

I. Introduction

Wireless power transfer (WPT) attracts great attention because of its potential application to charge laptop computers, mobile phones, portable audio players and other electronic devices without cords [1]-[6]. Researches even showed the application prospect of charging motor vehicles wirelessly. It was also experimentally demonstrated that very efficient power transmission can be achieved by using the strongly-coupled resonance method [1]. It was shown that the strongly-coupled resonance method can transmit energy for a longer distance than the preciously used near-field induction method [2]. Again, the strongly-coupled resonance method was shown to be more efficient than the far-field radiation method [3], where most energy is wasted due to the transmission loss [4]-[5]. Because of the existence of the electromagnetic hazard on human body and the multi-user situation, a more efficient multi-user WPT system is urgently concerned.

The power transfer efficiency (PTE) is one of the most important parameters to evaluate the performance of a WPT system. In order to develop a WPT system with a high PTE, it is required to have an efficient method to calculate the PTE by analyzing electromagnetically the WPT system. If the transmitting antenna and receiving antenna are described as a two-port network circuit, the power transfer between the transmitting antenna and receiving antenna in the WPT system can be indicated by using the scattering parameters of the network circuit and further the PTE can be calculated by using the scattering parameters. Because the scattering parameters of a WPT system can be measured by a vector network analyzer and calculated by a full-wave numerical analysis, the scattering parameters are a very efficient tool in analyzing and designing the antennas and RF modules for a WPT system.

A fundamental study focused on the PTE of a WPT system composed of dipole and loop antennas as the transmitting and receiving antennas was carried out by the present authors, where a two-port scattering parameters calculated by the method of moments (MoM) were used to analyze the system and it was found the largest PTE was obtained when the near-field coupled antennas of both transmitting and receiving sides were conjugate-matched with the impedance of the transmitting and receiving circuits, respectively [6]. The optimum load for maximum transfer efficiency of a practical WPT system was derived when the WPT system was equivalent to a 2-port lossy network [7] also by the present authors.

This paper shows the result of the continued study published in [8], where multi-port scattering parameters are applied to the analysis of multi-antenna in a WPT system corresponding to multi-user situations. The expression of PTE is defined in term of the multi-port scattering parameters. Some numerical simulations are shown to demonstrate it is easy to evaluate the PTE for various models of the antenna geometries, locations of transmitting and receiving antennas in a multi-user WPT system by using the multi-port scattering parameters.

II. Analysis of Multi-port Network For Multi-user WPT system

A generalized 4-port network is shown in Fig. 1, which represent a 4-antenna WPT system with one transmitting antenna and three receiving antennas. 4 ports are named A, B, C and D, respectively. By using MoM, 4×4 scattering parameters of this 4-port network can be got. The reflection coefficients at every port are labeled in the Fig. 1, and the reference impedance is Z_0=50 Ω. P_A is the power available from the source, P_B, P_C and P_D is the power delivered to each load. The a_1, a_2, a_3 and a_4 in the Fig. 1 are the incident waves of port A, B, C and D, respectively. Besides, b_1, b_2, b_3 and b_4 are the reflected waves of each load. Therefore, if

978-7-5641-4279-7

these 4 ports are impedance matched simultaneously, four equations can be obtained:

$$\Gamma_A=\Gamma_1^*,\ \Gamma_B=\Gamma_2^*,\ \Gamma_C=\Gamma_3^*,\ \Gamma_D=\Gamma_4^* \tag{1}$$

Fig. 2 is the signal flow graph of this 4-port network. From the Mason's gain formula and the definition of reflection coefficient, follow equations can be obtained:

$$\Gamma_1=\frac{b_1}{a_1}=f_1(\Gamma_A,\Gamma_B,\Gamma_C,\Gamma_D)=\Gamma_A^* \tag{2}$$

$$\Gamma_2=\frac{b_2}{a_2}=f_2(\Gamma_A,\Gamma_B,\Gamma_C,\Gamma_D)=\Gamma_B^*$$

$$\Gamma_3=\frac{b_3}{a_3}=f_3(\Gamma_A,\Gamma_B,\Gamma_C,\Gamma_D)=\Gamma_C^*$$

$$\Gamma_4=\frac{b_4}{a_4}=f_4(\Gamma_A,\Gamma_B,\Gamma_C,\Gamma_D)=\Gamma_D^*$$

In equation (2), Γ_1, Γ_2, Γ_3 and Γ_4 are expressed by functions of Γ_A, Γ_B, Γ_C and Γ_D. Solving equation (2), Γ_A, Γ_B, Γ_C and Γ_D can be figured out. From the relationship between the reflection coefficient and the impedance load below,

$$Z_L=\frac{1+\Gamma_L}{1-\Gamma_L} \tag{3}$$

the optimal load impedance Z_A, Z_B, Z_C and Z_D that make the 4-port network impedance matched can be figured out.

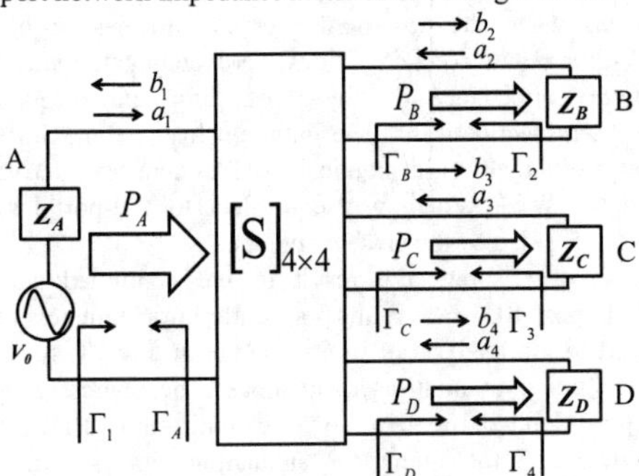

Fig. 1. 4-port network

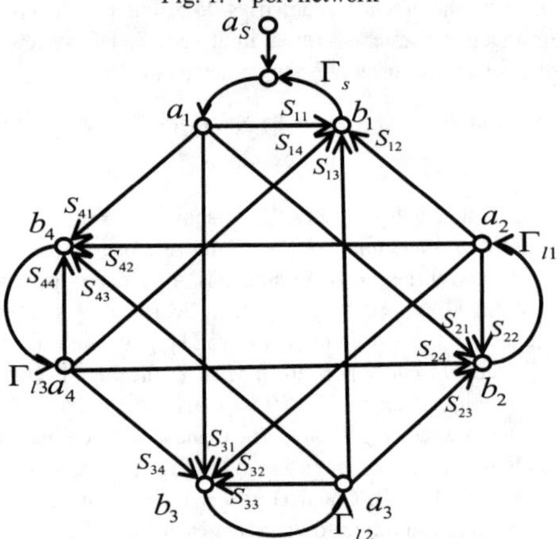

Fig. 2. Signal flow graph of 4-port network

Power flow at each port is expressed as:

$$P_A=(1-|\Gamma_A|^2)\frac{|a_s|^2}{|1-\Gamma_A\Gamma_A^*|^2}=\frac{|a_s|^2}{1-|\Gamma_A|^2} \tag{4}$$

$$P_B=|b_2|^2-|a_2|^2=(1-|\Gamma_B|^2)|b_2|^2$$

$$P_C=|b_3|^2-|a_3|^2=(1-|\Gamma_C|^2)|b_3|^2$$

$$P_D=|b_4|^2-|a_4|^2=(1-|\Gamma_D|^2)|b_4|^2$$

So the individual PTE to each load can be expressed as

$$\eta_B=\frac{P_B}{P_A}=\frac{(1-|\Gamma_B|^2)|b_2|^2}{\frac{|a_s|^2}{1-|\Gamma_A|^2}}=(1-|\Gamma_A|^2)(1-|\Gamma_B|^2)\left|\frac{b_2}{a_s}\right|^2 \tag{5}$$

$$\eta_C=\frac{P_C}{P_A}=\frac{(1-|\Gamma_C|^2)|b_3|^2}{\frac{|a_s|^2}{1-|\Gamma_A|^2}}=(1-|\Gamma_A|^2)(1-|\Gamma_C|^2)\left|\frac{b_3}{a_s}\right|^2$$

$$\eta_D=\frac{P_D}{P_A}=\frac{(1-|\Gamma_D|^2)|b_4|^2}{\frac{|a_s|^2}{1-|\Gamma_A|^2}}=(1-|\Gamma_A|^2)(1-|\Gamma_D|^2)\left|\frac{b_4}{a_s}\right|^2$$

$$\eta_t=\eta_B+\eta_C+\eta_D$$

In the equation (5), η_B, η_C, η_D are the individual PTE to the load Z_B, Z_C, Z_D, respectively. And η_t is the total PTE of this 4-port system. Using Mason's gain formula again, the ratios at the right part of the equation (5) can be expressed by functions of Γ_A, Γ_B, Γ_C and Γ_D:

$$\frac{b_2}{a_s}=f_5(\Gamma_A,\Gamma_B,\Gamma_C,\Gamma_D) \tag{6}$$

$$\frac{b_3}{a_s}=f_6(\Gamma_A,\Gamma_B,\Gamma_C,\Gamma_D)$$

$$\frac{b_4}{a_s}=f_7(\Gamma_A,\Gamma_B,\Gamma_C,\Gamma_D)$$

where, Γ_A, Γ_B, Γ_C, Γ_D can be calculated from equation (2). That is the whole process in calculating the PTE and optimal loads of an arbitrary 4-port network when knowing its scattering parameters.

III. Power Transfer Efficiency of Multi-Antenna WPT

A WPT system composed of one transmitting antenna and three receiving antennas is shown in Fig. 3 as an analysis model for following numerical simulation. The transmitting and receiving antennas have the same antenna length l and conductivity σ. Transmitting and receiving antennas are all placed along the z-axis. Three receiving antennas are placed at a distance d from the transmitting antenna with an angle θ = 120°. The radius of the dipole antenna is 1 mm.

For comparison, a 2-antenna WPT model is built, show in Fig. 4. The antenna elements of this 2-antenna WPT system have same geometry with the antenna elements in Fig. 3. The distance between two dipole antennas is d. The calculation method of PTE for 2-antenna model under impedance matching is introduced in the previous research by the present authors [6].

Fig. 5 is the frequency characteristic of PTE of this two model, when $l = 30$ cm, $d = 50$ cm, $\sigma = \infty$.

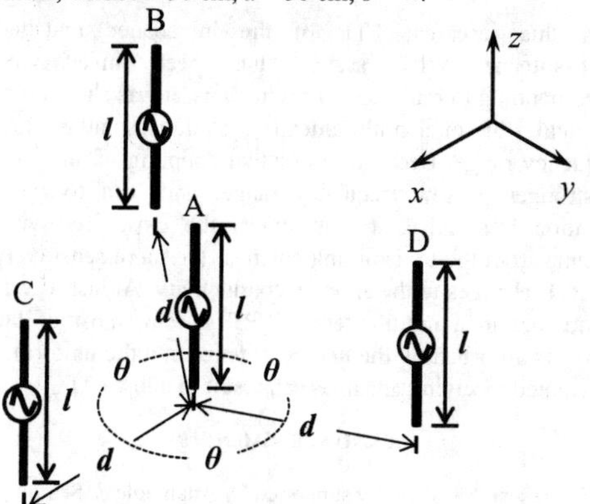

Fig. 3. Analysis model: WPT with one transmitting and three receiving dipole antennas

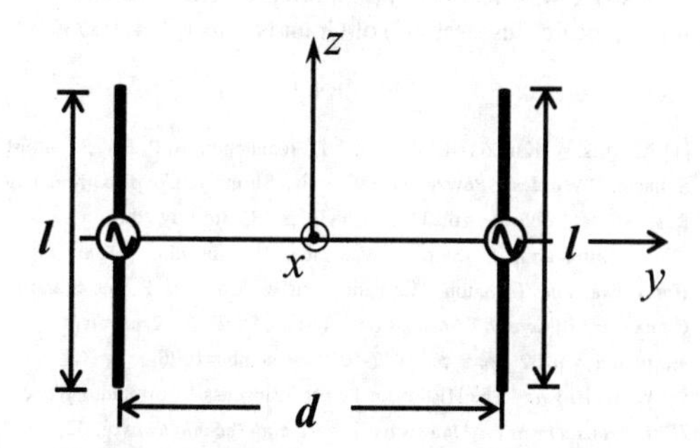

Fig. 4. 2-port WPT system for comparison

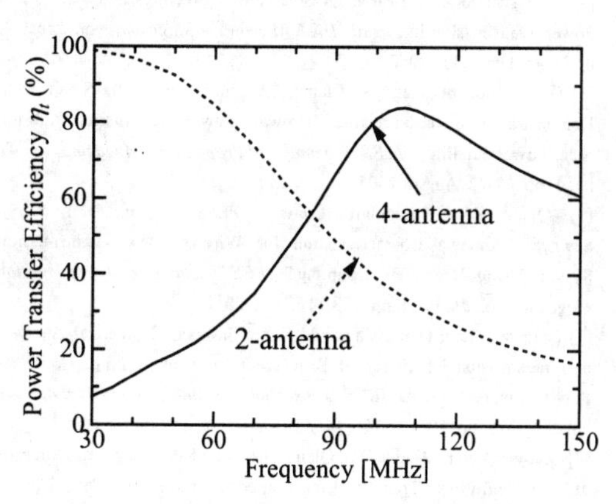

Fig. 5. Frequency characteristics of PTE of 4-port and 2-port

From Fig. 5 it is found that the total PTE η_t of 4-antenna model is increased till 107 MHz, and then is decreased when the frequency becomes higher than 107 MHz. The PTE of 2-antenna model decreases all along from 30 MHz to 150 MHz. Comparing the solid line and dot line in Fig. 5, it is found 2-antenna WPT system performs better than the 4-antenna WPT system in low frequency range, 30 MHz to 85 MHz in this case. However in the high frequency range, 85 MHz to 150 MHz in this case, 4-antenna model takes its advantage to 2-antenna model in terms of the PTE.

This result might because when frequency gets lower, the wave length becomes longer. Therefore, antennas of this 4-antenna model are seen much closer to each other compared with wavelength. The combination of these 4 dipole antennas will radiate more like a single and large antenna. At last, large radiation loss cause the low PTE in low frequency range.

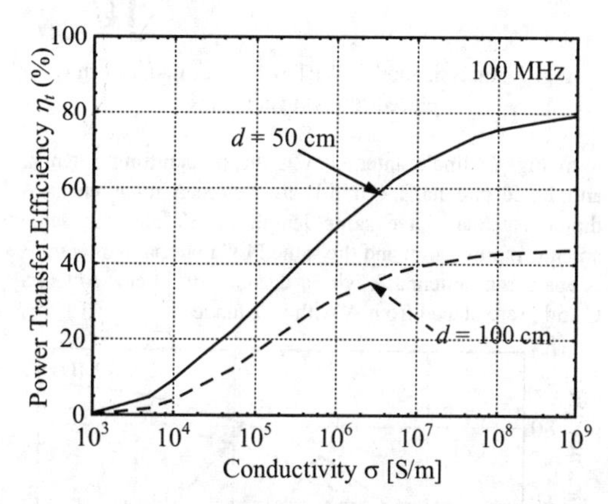

Fig. 6. PTE of 4-antenna model with changing antenna conductivity σ

Another simulation analysis was conducted when the antenna conductivity σ of the 4-antenna model is changed where $l = 30$ cm, $d = 50$ or 100 cm and at frequency 100 MHz. The result is shown in Fig. 6.

From Fig. 6, it is found that the total PTE η_t gets larger with the increasing of conductivity σ. However, comparing the solid line and the dash line, it's found that in the high conductivity range (σ from 10^8 to 10^9 in this case), the solid line changes larger but the dash changes little. That means the PTE of the multi-antenna system with long distance between the transmitting antenna and the receiving antenna is not sensitive to the change of the antenna conductivity σ.

In the above numerical simulation, Discussions of 4-antenna WPT system with 3 uniformly spaced receiving antennas have been covered. We are interested in the situation where the antennas are not uniformly placed, for example, the model of 3 receiving antennas which have different distance towards the transmitting antenna. In order to simplify the problem, we build a model shown in Fig. 7.

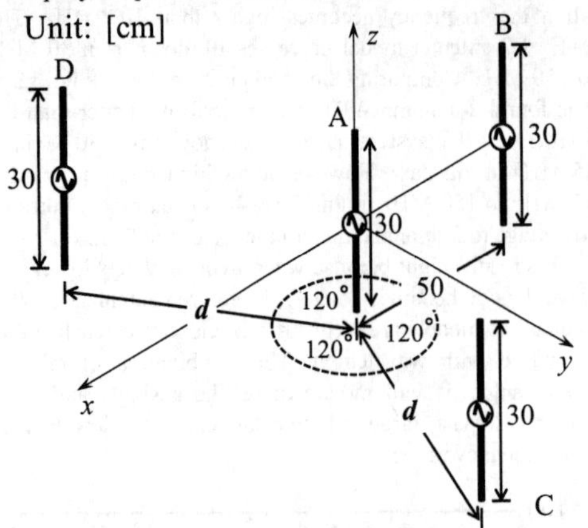

Fig. 7. Analysis model: WPT with three nonuniformly placed receiving antennas

In Fig. 7, dipole antenna A is the transmitting antenna, and dipole antenna B, C and D are the receiving antennas. 4 dipole antennas have same length of 30 cm, the same antenna radius 1 mm and the same PEC material. Antenna B is apart from antenna A with a distance of 50 cm. Antenna C and D are placed from A with a distance *d*.

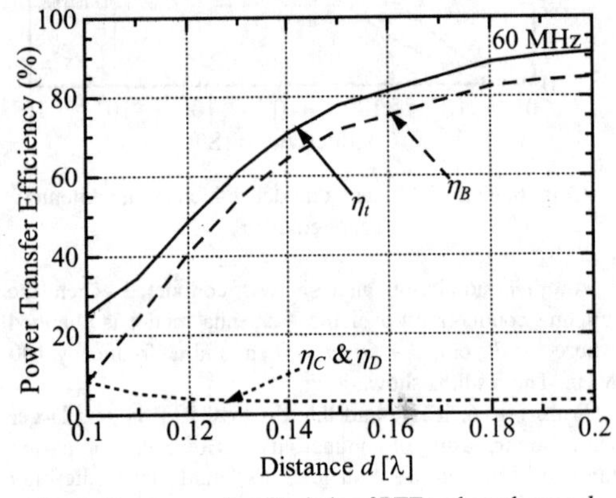

Fig. 8. Frequency characteristic of PTEs when change *d*

When the frequency is 60 MHz, the distance *d* is changed from 50 cm (0.1λ) to 100 cm (0.2λ), in order to find out the relationship between PTE and distance between transmitting and receiving antennas. The result is shown in Fig. 8.

From Fig. 8, it is found when *d* changes from 0.1λ to 0.2λ, the PTE of antenna B gets larger and the PTEs of antenna C and antenna D get smaller. The total PTE of this 4-antenna system, which is the sum of η_B, η_C and η_D, also become larger. This means a 4-antenna WPT system with uniformly placed receiving antennas is not suitable to obtain the maximum PET.

IV. Conclusions

In this research, PTE of the impedance matched multi-antenna WPT system has been investigated corresponding to multi-antenna situations. It was shown that the total PTE of a multi-antenna system is small in low frequency range. Because the mutual coupling of antennas is stronger in low frequency range, and lead to more radiation loss. Also, it was shown the closer receiving antenna from the transmitting antenna, the more sensitively the PTE changes to the antenna conductivity. At last, it was found that in a multi-antenna WPT system, most of the power is absorbed by the nearest antenna and the uniformly distributed receiving antennas can result in a low PTE.

Acknowledgment

This research was partly supported by Adaptable & Seamless Technology Transfer Program through Target-driven R&D (A-STEP) of The Japan Science and Technology Agency (JST) This work was also partly supported by JSPS Grant-in-Aid for Scientific Research (C) of Grant Number 25420353.

References

[1] A. Kurs, A. Karakis, R. Moffatt, J. D. Joannopoulos, P. Fisher, and M. Soljacic, "Wireless Power Transfer via Strongly Coupled Magnetic Resonances," *Science*, vol. 317, no. 5834, pp. 83-86, July 2007.

[2] J. Murakami, F. Sato, T.Watanabe, H. Matsuki, S.Kikuchi, K. Harakaiwa, and T. Satoh, "Consideration on Cordless Power Station - Contactless Power Transmission System," *IEEE Transactions on Magnetics*, vol. 32, no. 5, pp. 5017-5019, September 1996.

[3] W. C. Brown, "The History of Power Transmission by Radio Waves," *IEEE Transactions on Microwave Theory and Techniques*, vol. 32, no. 9, pp. 1230- 1242, September 1984.

[4] H. Matsumoto, "Research on Solar Power Satellites and Microwave Power Transmission in Japan," *IEEE Microwave Magazine*, vol. 3, no. 4, pp.36-45, December 2002.

[5] C. T. Rodenbeck and K. Chang, "A Limitation on the Small-Scale Demonstration of Retrodirective Microwave Power Transmission from the Solar Power Satellite, " *IEEE Antennas and Propagation Magazine*, vol. 47, no. 4, pp. 67-72, August 2005.

[6] Qiang Chen, Kazuhiro Ozawa, Qiaowei Yuan, and Kunio Sawaya, "Antenna Characterization for Wireless Power-Transmission System Using Near-Field Coupling," *IEEE Antennas and Propagation Magazine*, vol. 54, no. 4, pp. 108-116, Aug. 2012.

[7] Qiaowei Yuan, Qiang Chen and Kunio Sawaya, "Numerical Analysis on Transmission Efficiency of Evanescent Resonant Coupling Wireless Power Transfer System," *IEEE Transactions on Antennas and Propagation*, vol. 58, no. 5, pp. 1751 - 1758, May 2010.

[8] Qiaowei Yuan, Mingda Wu, Qiang Chen, and Kunio Sawaya, "Analysis of Near-Field Power Transfer Using Scattering Parameters," *Proc. The 7th European Conference on Antennas and Propagation (EuCAP2013)*, pp.2965-2967, 2013.

TA-1(B)

October 24 (THU) AM

Room B

Computational EM

Fast Broad-band Angular Response Sweep Using FEM in Conjunction with Compressed Sensing Technique

Lu Huang, Bi-yi Wu, Xin-qing Sheng
[1,2,3] School of Information Science and Technology, Beijing Institute of Technology
Beijing, 100081, China

Abstract- **The Compressed Sensing (CS) technique is applied to the finite element method (FEM) for fast radar cross section (RCS) calculation over a wide incident angular band. Rather than point-by-point solving multiple right-hand vectors of each incident angle, the right-hand sides are first compressed and then recovered to find each incident angle response in CS procedure. The numerical example shows the proposed approach leads a magnitude decrease of computation time without losing accuracy.**

I. INTRODUCTION

The finite-element method (FEM) is a full-wave numerical method that discretizes the variational formulation of a functional[1,2],has been widely used in electromagnetic analysis. For many practical applications such as radar imaging, our interests often a wide range of angles. In order to alleviate the computational burden and improve efficiency in this circumstance, some fast methods like interpolation [3] and extrapolation [4, 5] have been put forward to accelerate wide incident angular band calculation rather than point-by-point sweep. In this work, our renewed interest is to employ the novel compressed sensing technique for this scenario.

Compressed sensing was first developed in [6,7] by Candès et al in 2006,and applied in many fields recently, like CS radar[8], wireless sensor networks[9], magnetic resonance imaging[10]. It points out that if a sparse or compressible high-dimensional signal could be projected onto a low-dimensional space, with the sparse signal's priori conditions, the original signal could be recovered by a linear or non-linear reconstruction model.

Indeed, from the view of signal and system, the incident wave from different angler imposed on targets can be seen as a series of input signals, and the scatter fields can be viewed as the system response. So we can first compress these signals, subsequently solve the response of compressed signals and then recover to find the original response. If the computational burden of compression and recovery is minor, the angular response sweep calculation can be accelerated by using compressed sensing technique.

The rest of this paper is organized as follow. The total field formulations for scattering calculation using FEM are given in section II. A brief introduction of CS theory and some compression and recovery algorithms are presented in Section III. In section IV, the validity and efficiency of proposed method are demonstrated by numerical example. Some concluding remarks and observations are given in last section.

II. FEM TOTAL FIELD FORMULATION FOR SCATTERING

For scattering applications, the FEM can be formulated in terms of either the total field or the scatter field. In the scatter field formulation, the incident field is localized to the scatters throughout the computation space and generates a dense right-hand side vector. While in the case of total field formulation, a function is typically defined that collocates an impressed source with an analytic (typically either first- or second-order) absorbing boundary condition on the boundary surface, thus the right-hand side vector is sparse owing to only minor unknowns on the boundary surface. For the sake of alleviating the computational burden of compressing and recovering in CS procedure, we choose the total field formulation.

The typical geometry of interest of FEM is shown in Fig.1, the surface S_{ab} defines the termination of the overall finite element region, and on this surface, the boundary value problem satisfies the Sommerfeld condition[2] given by (2.1)

$$\hat{\mathbf{r}}\times(\nabla\times\mathbf{E}^{sc}) = -jk_0\hat{\mathbf{r}}\times(\hat{\mathbf{r}}\times\mathbf{E}^{sc}) \quad , \ r\in S_{ab} \quad (2.1)$$

where $\mathbf{E}^{sc}$ is the scattered electric field. According to $\mathbf{E}^{sc} = \mathbf{E} - \mathbf{E}^{inc}$, the formula (2.1) can be written as

$$\begin{aligned}&\hat{\mathbf{r}}\times(\nabla\times\mathbf{E}) + jk_0\hat{\mathbf{r}}\times(\hat{\mathbf{r}}\times\mathbf{E}) \\ &= \hat{\mathbf{r}}\times(\nabla\times\mathbf{E}^{inc}) + jk_0\hat{\mathbf{r}}\times(\hat{\mathbf{r}}\times\mathbf{E}^{inc})\end{aligned} \quad , \ r\in S_{ab} \quad (2.2)$$

The total-field inside the FEM region satisfies the second-order wave equation derived from Maxwell equations as

$$\nabla\times\frac{1}{\mu_r}\nabla\times\mathbf{E} - k_0^2\varepsilon_r\mathbf{E} = 0 \quad , \ r\in V_t \quad (2.3)$$

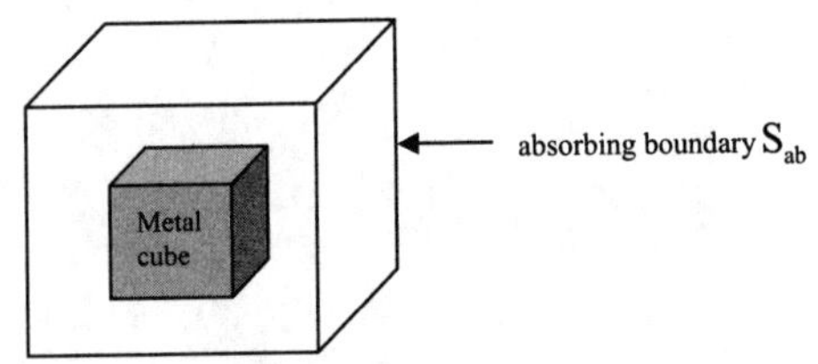

Figure 1. FEM domain

978-7-5641-4279-7

In a finite-element-based solution, the weak form of (2.3) can be obtained by Ritz procedure with edge-element base functions

$$\int_{V_t}\left\{\frac{1}{\mu_r}(\nabla\times\mathbf{N})\cdot(\nabla\times\mathbf{E})-k_0^2\varepsilon_r\mathbf{N}\cdot\mathbf{E}\right\}dV_t$$

$$-\int_{S_{ab}}\frac{jk_0}{\mu_r}\left[(\hat{\mathbf{n}}_{S_{ab}}\times\mathbf{N})\cdot(\hat{\mathbf{n}}_{S_{ab}}\times\mathbf{E})\right]dS_{ab}$$

$$=\frac{-1}{\mu_r}\int_{S_{ab}}\mathbf{N}\cdot\left[\hat{\mathbf{n}}_{S_{ab}}\times(\nabla\times\mathbf{E}^{inc})+jk_0\hat{\mathbf{n}}_{S_{ab}}\times(\hat{\mathbf{n}}_{S_{ab}}\times\mathbf{E}^{inc})\right]dS_{ab}$$

III. Compressed Sensing Theory

The core idea of CS theory contains three aspects, sparse representation, measurement matrix and reconstruction algorithms. Firstly, if the original signal $X\in\mathbb{R}^N$ is sparse or compressible based on the orthogonal basis Ψ like the Fourier or wavelet coefficients of smooth signal etc. [11], we need to work out transform coefficient Θ,

$$\Theta=\Psi^T X. \tag{3.1}$$

Secondly, to ensure its convergence, a measurement matrix Φ should be given with dimension of $M\times N$ based on irrelevance of Ψ, which makes matrix $\mathbf{A}^{CS}$

$$A^{CS}=\Phi\Psi^T \tag{3.2}$$

to satisfy Restricted Isomerty Property (RIP)[12].Common choices of measurement matrix Φ include Gaussian random matrix[13], Bernoulli matrix[13] etc.. Gaussian random matrix has a series of advantages that the entries independently subject to a distribution, and it is irrelevant to most sparse signals, the number of measurements for accurate recovery is smaller. Then the measurement set from the function

$$Y=\Phi\Theta=\Phi\Psi^T X \tag{3.3}$$

can be obtained.

At last, l_0-regularization or l_1-regularization methods are used to acquire the approximation or the precise vector $\overline{X}$ of the original signal X, which is the sparsest vector based on Ψ, that is

$$\min\left\|\Psi^T X\right\|_{0or1} \quad \text{s.t.} \quad A^{CS}X=\Phi\Psi^T X=Y$$

Two major approaches of sparse recovery are greedy algorithm and convex programming. The greedy algorithm, like orthogonal matching pursuit (OMP)[14], regularized orthogonal matching pursuit(ROMP)[15], computes the support of X iteratively, finding one or more new elements and subtracting their contribution from the measurement vector Y at each step. Greedy methods are usually fast and easy to implement. Whereas, the convex programming, such as iterative threshold method [16], is to solve a convex program whose minimizer is known to approximate the target signal. OMP is mainly to pick a coordinate of $\Phi^* Y$ of the biggest magnitude. While ROMP, a variant of OMP, combines the speed and ease of implementation of the greedy methods with the strong guarantees of the convex programming methods. The flow chart as shown in Fig.2, illustrates the CS procedure.

IV. Numerical Example

To demonstrate the validity and efficiency of the proposed method, we calculate the monostatic RCS of a metal cube with 0.55 wavelength. The incident wave is vertical polarized, and angle θ ranges from 0 to 360.The measurement matrix we choose is a sparse column matrix based on Gaussian random matrix, and the compressed ratio is 0.2. OMP and ROMP are used as the reconstruction algorithms. The results are presented in Fig.3 and Fig.4 compared with point-by-point calculation. It can be seen that both two reconstruction algorithms have a good accuracy.

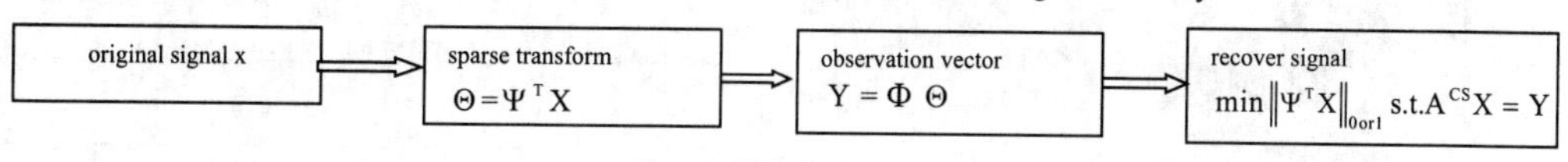

Figure 2. CS theory frame

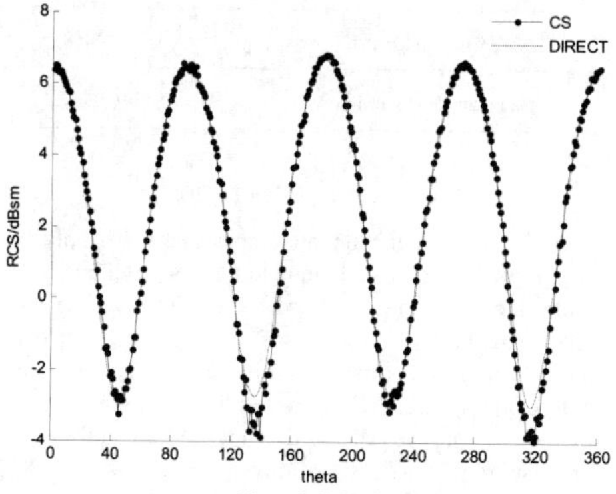

Figure 3. RCS via OMP

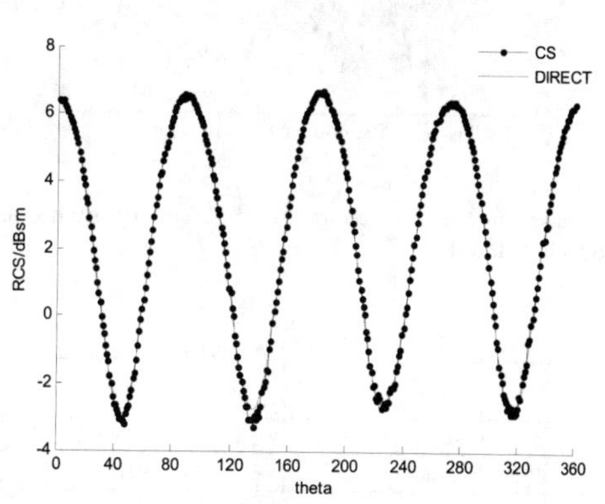

Figure 4. RCS via ROMP

The total CPU time of these two reconstruction algorithms and direct angular sweep method are given in Table I.

TABLE I
CPU TIME FOR DIFFERENT METHODS

Method	Time(s)	Speedup
DIRECT	1339.430	1
CS-OMP	468.826	2.86
CS-ROMP	318.486	4.21

Furthermore, the CPU time distribution of each part using CS_OMP and CS_ROMP are also given in Fig.5 and Fig. 6 respectively.

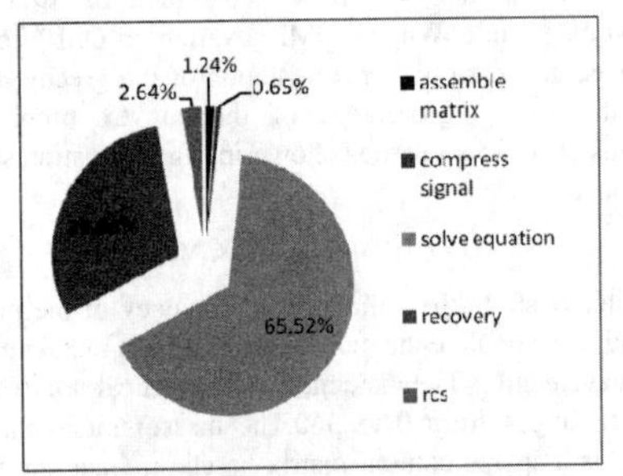

Figure 5. Time distribution chart by OMP

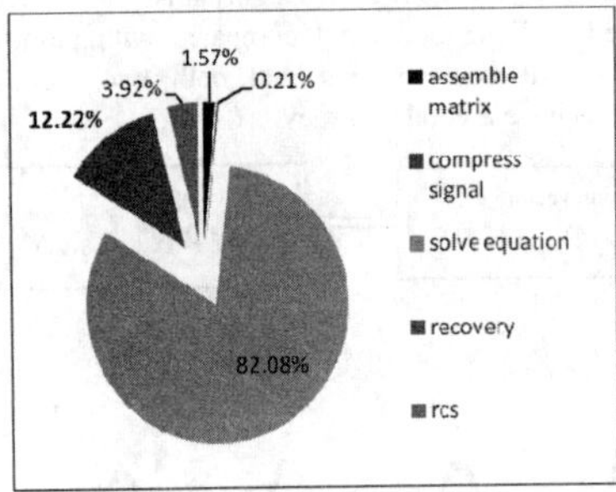

Figure 6. Time distribution chart by ROMP

In addition, relative errors for different algorithms are shown in Table II.

TABLE II
ERROR FOR DIFFERENT ALGORITHMS

Method	Error
CS-OMP	0.0746
CS-ROMP	0.0328

It can be concluded that ROMP shows good characteristics in recovering compared to OMP. In the time distribution chart, the CPU time for solving equation occupies a large proportion, while compression and recovery just take up much smaller part. So there is a foreseeable larger speedup when solving equation becomes a dominant part in whole FEM.

Besides, we also investigated the performance of different measurement matrices based on the same reconstruction algorithm-ROMP. In Fig.7, the recovered RCS by Band, Column, Gauss measurement matrices are presented. The relative error compared to direct method in Table III shows the sparse column matrix is more accurate than the other two.

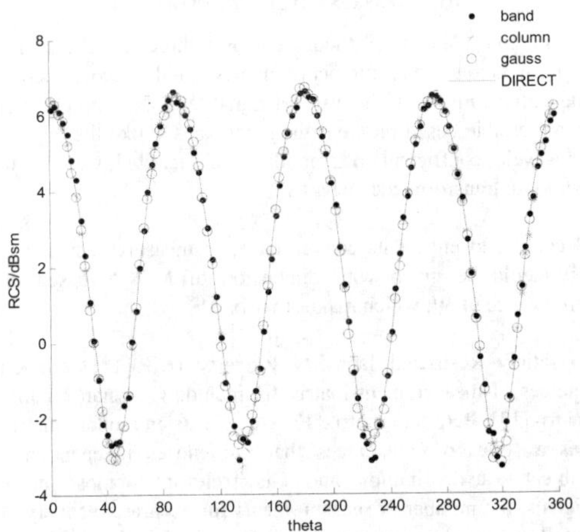

Figure 7. RCS for different measurement matrices by ROMP

TABLE III
ERROR FOR DIFFERENT MEASUREMENT MATRIX

Method	Error
Sparse Column Matrix	0.0328
Sparse Band Matrix	0.0458
Gaussian Random Matrix	0.0338

V. CONCLUSION

In this work, our innovative approach is to apply the novel compressed sensing technique to FEM to calculate RCS over a wide incident angular band. According to the numerical results, if the CPU time cost in solving equation occupies a large part of the total, this new approach could lead to a magnitude speedup. From a certain viewpoint, the proposed method improves the efficiency significantly under the premise of guaranteeing validity.

REFERENCES

[1] Xin-Qing Sheng, Wei Song, *Essentials of Computational Electromagnetics*, 1st ed., Singapore: Wiley, 2012.

[2] J.M.Jin, *The Finite Element Method in Electromagnetics*, 2nd ed., New York: Wiley, 2002.

[3] Ma J.F., Mittra R., Prakash V.V.S.,"Efficient RCS computation for incident angle sweep with the use of the FMM and MNM techniques", *Microwave and Optical Technology Letters*, v 34, n 4, p 273-276, August 20, 2002.

[4] Lu C.C.,"A simple extrapolation method based on current for rapid frequency and angle sweep in far-field calculation of an integral equation algorithm," *IEEE/ACES International Conf. on Wireless Communications and Applied Computational Electromagnetics*, p 333-6, 2005.

[5] Peng Zhen, Sheng XinQing,"Application of asymptotic waveform approximation technique to hybrid FE/BI method for 3D scattering," *Sci. China Ser. F-Inf. Sci*,vol.50,no1,124-134,2007.

[6] Candès, E.J., Romberg JK, Tao T.," Stable signal recovery from incomplete and inaccurate measurements," *Communications on pure and applied Mathematics*, 2006, 2107-1223.

[7] D L.Donoho,"Compressed sensing," *IEEE Trans. on Information Theory*,2006, 52(4):1289-1306.

[8] S Bhattacharya, T Blumensath, B Mulgrew,"Fast encoding of synthetic aperture radar raw data using compressed sensing," *IEEE Workshop on Statistical signal Processing, Madison,Wisconsin*,2007.448-452.

[9] W Bajwa, J Haupt, A Sayeed, "Compressive wireless sensing," *Proceedings of the fifth International Conference on Information Processing in Sensor Networks, IPSN'06, New York: Association for Computing Machinery*, 134-142, 2006.

[10] Haldar, J.P.,Hernando D. ,Zhi-Pei Liang, "Compressed-Sensing MRI with Random Encoding," *IEEE Transactions on Medical Imaging*, 2011.

[11] E J Candès, Tao, "Near optimal signal recovery from random projections Universal: encoding strategies," *IEEE Trans. on Information Theory*,2006, 52(12):5406-5425.

[12] Candès E J, "The Restricted Isometry Property and its implications for compressed sensing," *Academie des sciences*, 2006, 346(I):598-592.

[13] Emmanuel Candès, Justin Romberg, Terrence Tao,"Robust uncertainty principles: exact signal reconstruction from highly incomplete frequency information," *IEEE Trans. on Information Theory*,2006, 52(12):489-509.

[14] J A Tropp,A C Gilbert, "Signal recovery from random measurements via orthogonal matching pursuit," April, 2005, www-personal.umich.edu/_jtropp/papers/TG05-Signal-Recovery.pdf.

[15] Deanna Needell, Roman Vershynin, "Signal Recovery from Incomplete and Inaccurate Measurements via Regularized Orthogonal Matching Pursuit," *IEEE Journal of selected topics in signal processing*, vol. 4, no. 2, April 2010.

[16] I Daubechies, M Defrise, C De Mol," An iterative thresholding algorithm for linear inverse problems with a sparsity constraint," *Comm. Pure. Appl. Math.*, 2004, 57(11):1413-1457.

Finite Macro-Element Method for Two-Dimensional Eigen-Value Problems

H. Zhao[1] and Z. Shen[2]
[1]A*Star Institute of High Performance Computing, Singapore
[2]School of Electrical and Electronic Engineering, Nanyang Technological University, Nanyang Avenue, Singapore 639798
ezxshen@ntu.edu.sg

Abstract — This paper presents a finite macro-element method based on a new macro basis function constructed from the modal solutions of Helmholtz equation. Different from existing basis functions, the proposed macro basis function satisfies Helmholtz equation automatically, and it allows the use of coarse elements. As an example, the proposed method is applied to calculate the cutoff wavenumbers of waveguides, and its advantages are demonstrated.

I. INTRODUCTION

In the finite element method (FEM), the maximum element size is normally very small compared to the operating wavelength [1]. In some cases, the required maximum element size can be as small as one hundredth of the wavelength [2]. This makes the number of unknowns very large for high frequency problems and therefore the conventional finite element method can only be used for modeling electromagnetic problems of relatively small electrical size. It was shown that larger element size can be used if the order of the basis function is increased under the same accuracy [3]. However, the maximum element size allowed is still quite small in term of wavelength and the matrix size is very big as the frequency increases.

The reason behind the constraint on the element size in the finite element method might be due to the basis function used, which doesn't satisfy the Maxwell's equations. In FEM, it is assumed that the spatial variation of electromagnetic fields can be accurately represented by the polynomial basis function within the element. If the basis function doesn't satisfy the Maxwell's equations, the element size is required to be very small to minimize the errors.

In order to overcome the constraint on the element size, a new macro basis function is proposed in this work. The new basis function satisfies the Helmholtz equation automatically because it is based on the modal solutions of the Helmholtz equation, and the element size is no longer required to be a small fraction of the wavelength. The finite macro-element method is then developed based on the new macro basis function. As an example, we apply the finite macro-element method to calculate the cutoff wavelengths of canned and ridged waveguides. Through this example, the advantages of the proposed method are demonstrated.

II. THE MACRO BASIS FUNCTION

Consider the two-dimensional transverse electric case of a closed waveguide problem. The Helmholtz equations governing the normal magnetic field can be written as

$$\frac{\partial^2 H_z}{\partial x^2}+\frac{\partial^2 H_z}{\partial y^2}+k_c^2 H_z=0 \qquad (1)$$

where k_c is the cutoff wavenumber. The modal solution to (1) can be obtained for a rectangular element shown in Fig. 1, where D_x and D_y are dimensions of the rectangular element in the x- and y-directions, respectively.

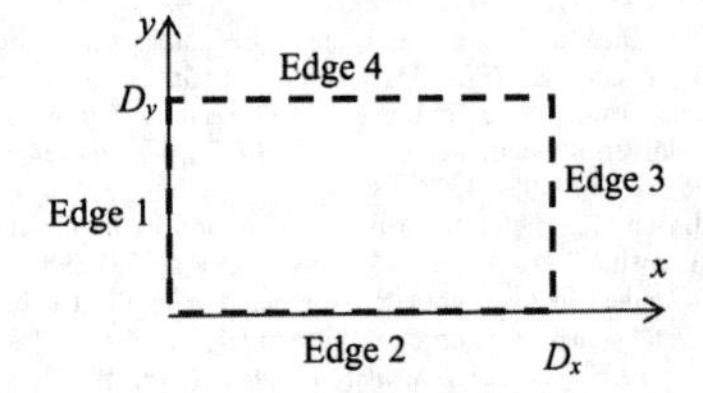

Fig. 1. A rectangular element.

Based on the modal solution to (1), the electric field inside the rectangular element can be expanded as

$$\vec{E}(x,y)=\sum_{i=1}^{4}\sum_{n=0}^{N_k}\vec{B}_{i,n}(x,y)E_{i,n}\,, \qquad (2)$$

where

$$\vec{B}_{1,n}(x,y)=\frac{n\pi}{D_y}S_{xn}\cos[k_{xn}(D_x-x)]\sin\frac{n\pi y}{D_y}\hat{x}+ k_{xn}S_{xn}\sin[k_{xn}(D_x-x)]\cos\frac{n\pi y}{D_y}\hat{y}, \qquad (3a)$$

$$\vec{B}_{2,n}(x,y)=k_{yn}S_{yn}\sin[k_{yn}(D_y-y)]\cos\frac{n\pi x}{D_x}\hat{x}+ \frac{n\pi}{D_x}S_{yn}\cos[k_{yn}(D_y-y)]\sin\frac{n\pi x}{D_x}\hat{y}, \qquad (3b)$$

$$\vec{B}_{3,n}(x,y)=-\frac{n\pi}{D_y}S_{xn}\cos[k_{xn}x]\sin\frac{n\pi y}{D_y}\hat{x}+ k_{xn}S_{xn}\sin[k_{xn}x]\cos\frac{n\pi y}{D_y}\hat{y}, \qquad (3c)$$

978-7-5641-4279-7

$$\vec{B}_{4,n}(x,y)=k_{yn}S_{yn}\sin[k_{yn}y]\cos\frac{n\pi x}{D_x}\hat{x}-\frac{n\pi}{D_x}S_{yn}\cos[k_{yn}y]\sin\frac{n\pi x}{D_x}, \quad (3d)$$

$$k_{xn}=\sqrt{k_c^2-(n\pi/D_y)^2}\quad ,k_{yn}=\sqrt{k_c^2-(n\pi/D_x)^2},$$

$$S_{\xi n}=\begin{cases}1, \text{if } k_{\xi n}=s\pi/D_\xi\,(s \text{ is an integer})\\ \csc(k_{\xi n}D_\xi)/k_{\xi n}, \text{otherwise}\end{cases}.$$

$S_{\xi n}$ is the normalization factor to ensure the continuity of tangential fields, where ξ can be x or y. $E_{i,n}$ is the n-th modal expansion coefficient for the tangential electric field along the i-th edge. It can be shown that basis functions in (3) have the following three properties. 1) It can be shown that $\nabla\cdot\vec{B}_i^n(x,y)=0$. Namely, basis functions in (3) are divergence free; 2) The tangential electric field along the i-th edge of the rectangular element is only determined by $E_{i,n}$, which ensures the continuity of the tangential electric field; 3) The normal magnetic field derived from (2) satisfies the Helmholtz equation. The third property is important because it allows the element to be arbitrarily large. Equation (3) defines the macro basis function for one element. For multiple elements, the macro basis function may be defined on an edge. Fig. 2 illustrates two cases of edge directions. Depending on the direction of the m-th edge, the macro basis function can be defined as

$$\vec{B}'_{m,n}(x,y)=\begin{cases}\vec{B}^{m-}_{4,n}(x_m^-,y_m^-), \text{if } \hat{t}_m\times\hat{x}=0 \text{ and } (x,y)\in\Omega_{m-}\\ \vec{B}^{m-}_{2,n}(x_m^+,y_m^+), \text{if } \hat{t}_m\times\hat{x}=0 \text{ and } (x,y)\in\Omega_{m+}\\ \vec{B}^{m+}_{3,n}(x_m^-,y_m^-), \text{if } \hat{t}_m\times\hat{y}=0 \text{ and } (x,y)\in\Omega_{m-}\\ \vec{B}^{m+}_{1,n}(x_m^+,y_m^+), \text{if } \hat{t}_m\times\hat{y}=0 \text{ and } (x,y)\in\Omega_{m+}\end{cases}, \quad (4)$$

where $(x_m^\pm,y_m^\pm)$ is the local coordinate originated at the left and lower corners of $\Omega_{m\pm}$. $\hat{t}_m$ is the tangential vector along the m-th edge. The superscript $m\pm$ of $\vec{B}$ means the geometry parameters of $\Omega_{m\pm}$ is substituted in (3).

III. THE FINITE MACRO-ELEMENT METHOD

Using the macro basis function in (4), the electric field is expanded as

$$\vec{E}(x,y)=\sum_{m=0}^{M}\sum_{n=0}^{N_h}\vec{B}'_{m,n}(x,y)w_{m,n}, \quad (5)$$

where $w_{m,n}=0$ if the m-th edge falls on a boundary of perfect electric conductor. Substituting the field expansion into the following equation

$$\nabla_t\times\nabla_t\times\vec{E}-k_c^2\vec{E}=0, \quad (6)$$

and applying the variational principle, the following matrix equation can be derived

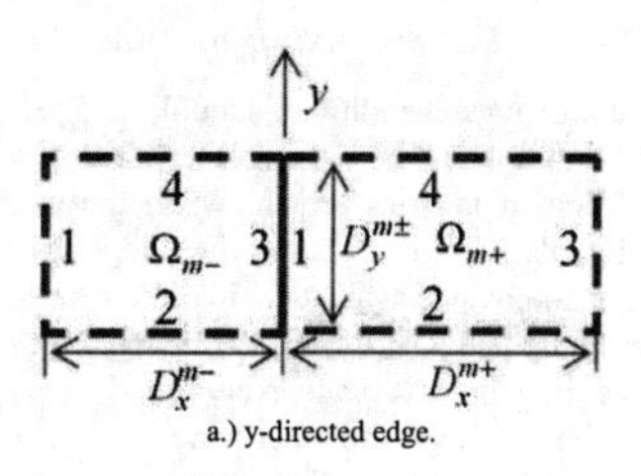

a.) y-directed edge.

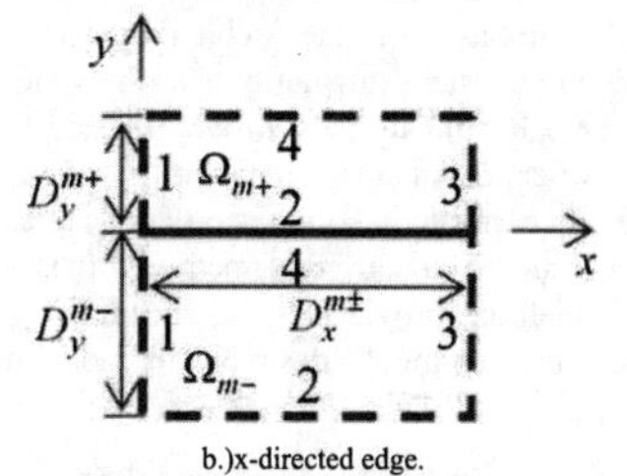

b.)x-directed edge.

Fig. 2 Two cases for the direction of the m-th edge.

$$\mathbf{A}\mathbf{x}=\mathbf{0}, \quad (7)$$

where

$$A_{i,j}=\iint_{\Omega_p}\nabla_t\times\vec{B}'_{p,q}(x,y)\cdot\nabla_t\times\vec{B}'_{m,n}(x,y)d\Omega-k_c^2\iint_{\Omega_p}\vec{B}'_{p,q}(x,y)\cdot\vec{B}'_{m,n}(x,y)d\Omega, \quad (8)$$

$$i=p(N_h+1)+q, j=m(N_h+1)+n. \quad (9)$$

The integrands in the first term of (8) are inner product of the curls of macro basis functions. According to the definition of the macro basis function in (4), the curls of $\vec{B}'_{m,n}$ can be easily obtained from the curls of the basis functions defined in (3), which are

$$\nabla_t\times\vec{B}_{1,n}(x,y)=-k_c^2S_{xn}\cos[k_{xn}(D_x-x)]\cos\frac{n\pi y}{D_y}\hat{z}, \quad (10a)$$

$$\nabla_t\times\vec{B}_{2,n}(x,y)=k_c^2S_{yn}\cos[k_{yn}(D_y-y)]\cos\frac{n\pi x}{D_x}\hat{z}, \quad (10b)$$

$$\nabla_t\times\vec{B}_{3,n}(x,y)=k_c^2S_{xn}\cos[k_{xn}x]\cos\frac{n\pi y}{D_y}\hat{z}, \quad (10c)$$

$$\nabla_t\times\vec{B}_{4,n}(x,y)=-k_c^2S_{yn}\cos[k_{yn}y]\cos\frac{n\pi x}{D_x}\hat{z}. \quad (10d)$$

Depending on the local numbers of edges p and q, the integrands in (8) may take different forms. For all cases, the integrands can be simplified to product and sum of trigonometry functions and the integral can be computed analytically. Since there are many combinations of the four basis functions in (3), the expression for the elements of matrix $\mathbf{A}$ is very lengthy and it is omitted here. Once matrix $\mathbf{A}$ is available, the cutoff wavenumber k_c can be found by enforcing the determinant of matrix $\mathbf{A}$ to zero.

IV. Simulation Results

The cutoff wavelengths of double ridged and vaned waveguides are computed using the finite macro-element method. Their geometries are shown in the inset of Fig. 3, where the dashed lines are the edges of elements. Convergence behavior against N_h is plotted in Fig. 3, where errors are computed against results obtained with N_h=32. For the double ridged waveguide, very good accuracy can be obtained with N_h=1. On the other hand, N_h=7 should be used for the vaned waveguide. The convergence for the vaned waveguide is not as fast as the double ridged waveguide. This may be due to the field singularity near the vane edge, which should be dealt with using a different basis function. The cutoff wavelength of the dominant mode in the vaned waveguide is computed as the vane depth varies, and Fig. 4 compares results from different methods. It is seen that the proposed method provides good accuracy with coarse elements, which is highly desirable in modeling large and uniform structures.

V. Concluding Remarks

This paper has proposed a finite macro-element method for electromagnetic boundary-value problems. The modal solution to the Helmholtz equation has been used as the macro-basis function in the proposed method. Unlike conventional basis functions, the macro-basis function proposed in this paper automatically satisfies the Helmholtz equation, which allows a large element size. Simulation results have been presented to illustrate the advantages of the finite macro-element method. It has been observed that the finite macro-element method predicts the eigen-values of various waveguides accurately using very coarse elements. One limitation of the proposed method is the number of modal functions increases when field singularities exist. Future work includes combining the finite macro-element method with the conventional FEM to model structures with both regular and irregular geometries (e.g. a reverberation chamber).

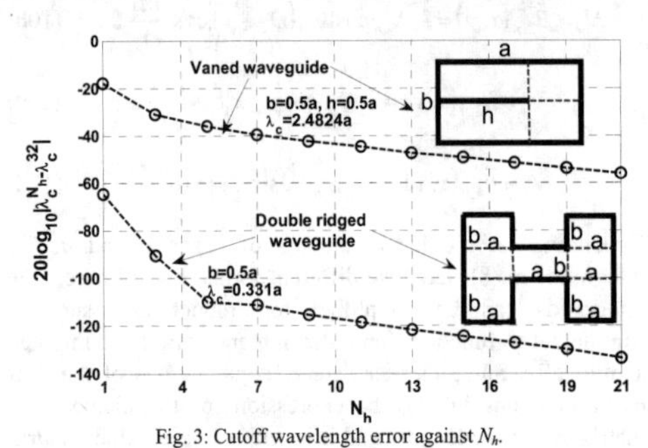

Fig. 3: Cutoff wavelength error against N_h.

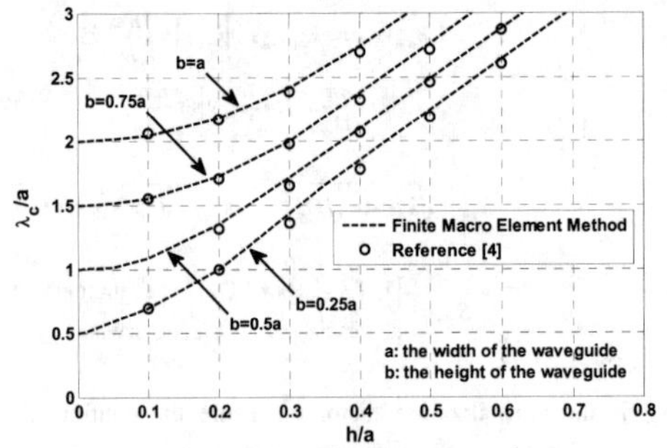

Fig. 4: Cutoff wavelength of vaned waveguide.

References

[1] J. Jin, *The Finite Element Method in Electromagentics*, NY: John Wiley & Sons, 1993.

[2] J. Liu and J. M. Jin, "A special higher order finite-element method for scattering by deep cavities," *IEEE Trans. Antennas Propagat.*, vol. 48, no. 5, pp. 694–703, May 2000.

[3] J. M. Jin, J. Liu, Z. Lou, and C. S. T. Liang, "A fully high-order finite-element simulation of scattering by deep cavities," *IEEE Trans. Antennas Propagat.*, vol. 51, no. 9, pp. 2420–2429, Sep. 2003.

[4] P. Silvester, "A general high-order finite-element waveguide analysis program," *IEEE Trans. Microwave Theory Tech.*, vol.17, pp. 204-210, Apr. 1969.

Accelerated Plasma Simulations using the FDTD Method and the CUDA Architecture

Wei Meng[#1], Yufa Sun[#2]
[#]Key Lab of Intelligent Computing & Signal Processing, Ministry of Education, Anhui University
Hefei, P.R.China
[1]mengwee@sohu.com
[2]yfsun@ahu.edu.cn

***Abstract*-This letter presents the graphic processor unit (GPU) implementation of the finite-difference time-domain (FDTD) method for the solution of the two-dimensional electromagnetic fields inside dispersive media. An improved Z-transform-based finite-difference time-domain (ZTFDTD) method was presented for simulating the interaction of electromagnetic wave with unmagnetized plasma. By using the newly introduced Compute Unified Device Architecture (CUDA) technology, we illustrate the efficacy of GPU in accelerating the FDTD computations by achieving significant speedups with great ease and at no extra hardware cost. The effect of the GPU-CPU memory transfers on the speedup will be also studied.**

***Keywords*-GPU;FDTD;Z-transform;CUDA**

I. INTRODUCTION

Modelling the electromagnetic radiation can be done using a variety of approaches, however, the finite-difference time-domain (FDTD) approach is perhaps the most common [1]. The popularity of this method continues to rise due to its simplicity and robustness in solving complex electromagnetic problems [2]. Z-Transform-based finite-difference time-domain (ZTFDTD) method has universal applicability for various dispersive medium [3]. Although the method has existed for over two decades, enhancements to improve the computational time are continuously being published [4], of special interest are parallel implementations.

Present-day computers with powerful graphics processing unit (GPU) show considerable promise of increased performance for the electromagnetic model. Order of magnitude increases in computing speed have been reported for wave equation solutions, for example [5]. The increased performance can potentially be realized for a modest price without the complications of parallel processing on multi-core machines. Problem-independent parallelization is can be built into the basic model. GPU promise a relatively low cost boost which leverages the vast development driven by the game industry. Potential benefits dictate investigation of efficacy in electromagnetic plasma calculation models.

This article aims to provide empirical results into the efficacy of using a rather recent and somewhat different approach in adapting the ZTFDTD method for parallel computing. The approach makes use of the CUDA architecture. CUDA is the computing engine in NVIDIA GPU. GPU have a parallel "many-core" architecture, each core is capable of running thousands of threads simultaneously. NVIDIA CUDA GPU provide a cost-effective alternative to traditional supercomputers and cluster computers. To place this work in the context of various dispersive medium applications, a ZTFDTD program is developed to operate on the NVIDIA CUDA architecture. This program is subsequently used to model a variety of electromagnetic plasma calculation models. Experiments are conducted to quantify the speedup obtained.

The devices supporting CUDA include NVIDIA G80 series or above. In this paper, A GTX650Ti video card with a GK106 core is used to execute ZTFDTD program with CUDA. GTX650Ti GPU contains 4 multiprocessors (SMX), each of which composed by 192 streaming processors and other devices, such as 48K shared memory, instruction cache. And on the video card is 1GB memory can be used as global memory or texture in CUDA. Kepler GK106 supports the new CUDA Compute Capability 3.1 [6].

II. OVERVIEW OF THE CUDA ARCHITECTURE

A. *CUDA Programming Model*

CUDA is the hardware and software architecture introduced by NVIDIA in November 2006 [7] to provide developers with access to the parallel computational elements of NVIDIA GPU. The CUDA architecture enables NVIDIA GPU to execute programs written in various high-level languages such as C, Fortran, OpenCL and DirectCompute. Compared to the CPU, the GPU devotes more transistors to data processing rather than data caching and flow control. This allows GPU to specialize in mathematical-intensive, highly parallel operations compared to the CPU which serves as a multi-purpose microprocessor.

CUDA has a single-instruction multiple-thread (SIMT) execution model where multiple independent threads execute concurrently using a single instruction [8]. CUDA GPU has a hierarchy of grids, threads and blocks as shown in Fig. 1. A thread can be considered the execution of a kernel with a given thread index. Each thread uses its unique index to access elements of the dataset passed to the kernel. Thus the collection of all the threads cooperatively processes the entire dataset concurrently. A block is a group of threads while a grid is a group of blocks. Each thread has its own private memory. Shared memory is available per-block and global

978-7-5641-4279-7

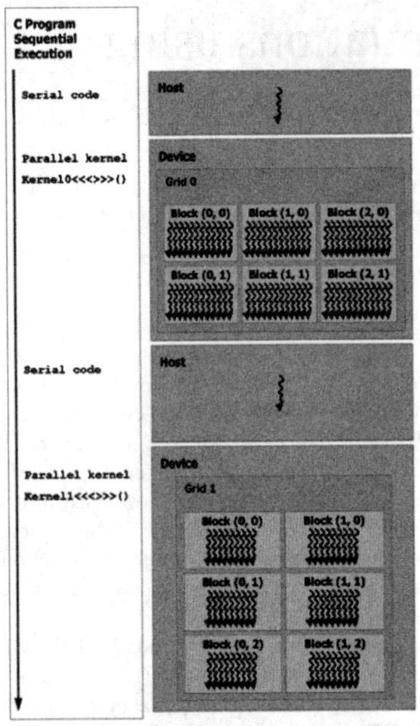

Figure 1. CUDA heterogeneous programming model.

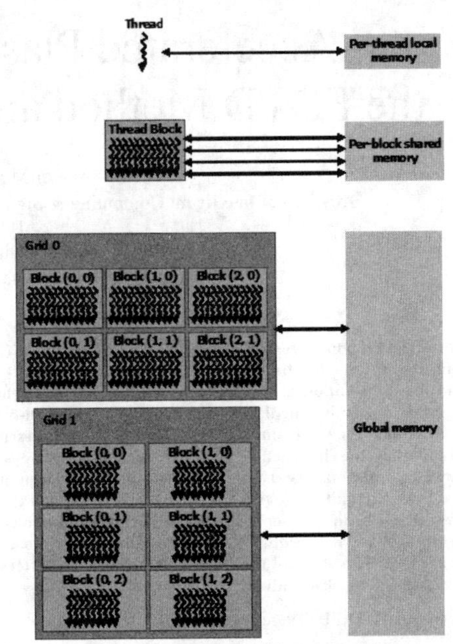

Figure 2. CUDA memory hierarchy model

memory is accessible by all threads. This multi-threaded architecture model puts focus on data calculations rather than data caching. Thus, it can sometimes be faster to recalculate rather than cache on a GPU. A CUDA program is called a kernel which is invoked by a CPU program.

B. *CUDA Memory Model*

The CUDA programming model assumes that CUDA threads execute on a physically separate device (GPU). The device is a co-processor to the host (CPU) which runs the program. CUDA also assumes that the host and device both have separate memory spaces: host memory and device memory, respectively. Because host and device both have their own separate memory spaces, there is potentially a lot of memory allocation, reallocation and data transfer between host and device as shown in Fig. 2. Thus, memory management is a key issue in general purpose GPU computing. Inefficient use of memory can significantly increase the computation time and mask the speedup obtained by the data calculations. While global memory is located off-chip and has the longest latency, there are techniques available that can reduce the amount of GPU clock cycles required to access large amounts of memory at one time. This can be done through memory coalescing. Memory coalescing refers to the alignment of threads and memory. For example, if memory access is coalesced, it takes only one memory request to read 64-bytes of data. On the other hand, if it is not coalesced, it could take up to 16 memory requests depending on the GPU's compute capability.

Choice of the grid and block configurations is crucial in obtaining purely coalesced memory access.

III. CUDA Implementation

A. *ZTFDTD Formulation*

In the following, we present the analysis of TM-polarized plane wave penetration through two-dimensional dispersive media. As such, we follow the procedure outlined in [9] for ZTFDTD for the unmagnetized plasma along with the UPML boundary condition, which is not only efficient in minimizing the memory requirements, but also the most accurate form of absorbing material. This, in turn, yields the following form of the FDTD update equation for the *x*-component of the electric field:

$$E(k)^n = \frac{(\Delta t/\varepsilon_0)\cdot\left[\nabla\times H - e^{-\nu_c\cdot\Delta t}\cdot J(k)^{n-1}\right] + E(k)^{n-1}}{1+\omega_p^2\Delta t^2} \tag{1}$$

$$J(k)^n = e^{-\nu_c\cdot\Delta t}J(k)^{n-1} + \omega_p^2\varepsilon_0\Delta t\cdot E(k)^n \tag{2}$$

where , Δt is the timestep, ε_0 is the permittivity of vacuum, ν_c is the electron collision, ω_p is the plasma angular frequency, $J(k)^n$ is the auxiliary array for $E(k)^n$.

B. *CUDA ZTFDTD Program*

In this paper, we present ZTFDTD Parallel implementation on GPU. According to the basic FDTD algorithm on CPU, we present its GPU implementation shown in Fig. 3.

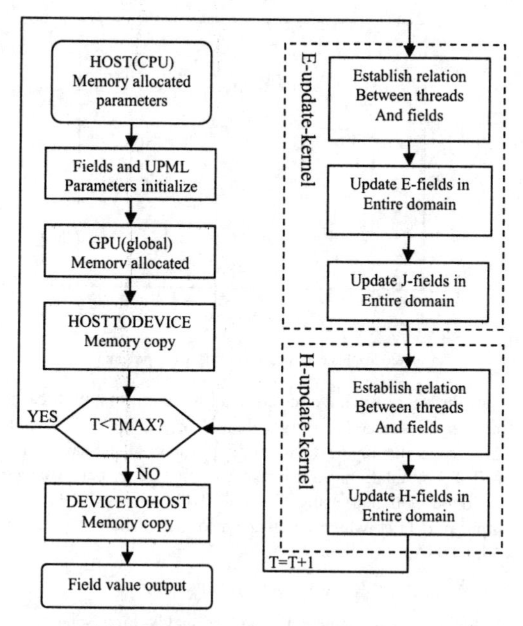

Figure 3. Flowchart of ZTFDTD algorithm on GPU.

Only procedures in the dot line frame are operated on GPU and when a kernel called a grid of threads is established. Because there is not any synchronization function for threads of whole grid while we have to use the value of former step, two kernels are launched sequentially to give a barrier for all threads. The number of threads is decided by the number of Yee cells (field size) because we make one thread to process one cell's spatial calculation.

IV. ZTFDTD Simulation Of CUDA

The ZTFDTD experiments based on CUDA are taken to research the parallel performance to solve complex computational problems of simulating the interaction of electromagnetic wave with unmagnetized plasma.

A. *Specifications of the Test Platform*

As mentioned before, one Geforce GTX650Ti video card, contained in a PC with Intel Core i3-2120(3.3GHz) CPU and 4GB RAM, is used for our simulation based-on CUDA. And for comparison, the same simulation with standard C code is operated on a computer with Intel Quad-Core i7-3820(3.6GHz) CPU computer and 64GB RAM (see Table I). Performance results and comparison are obtained by calculating time for specific number of time steps (iterations). And runtime of CPU program is only ZTFDTD iteration time, not including the memory allocating and initialing time. For CUDA, it also includes allocating time for global memory, time for copying memory form CPU RAM to GPU RAM and back, but these procedures execute only once. Then speedup is defined as the result GPU runtime divided by CPU runtime. The CUDA compilation is Microsoft Visual studio 2010 with CUDA v5.0 (64bit).

TABLE I
HARDWARE SPECIFICATIONS

Compute Device	i7-3820	GTX650Ti
Frequency(MHz)	3600	925
core /stream processor	4	768
floating point calculations per second(GFlops)	57.6	2131.2

B. *Algorithm to Achieve*

Taking the two-dimensional TM wave as an example, we simulate a plane wave impinging on a unmagnetized plasma cylinder. Incident wave is a simple Gaussian pulse source generated in the middle of the problem space and 8-point UPML as the absorbing boundaries shown in Fig. 4.

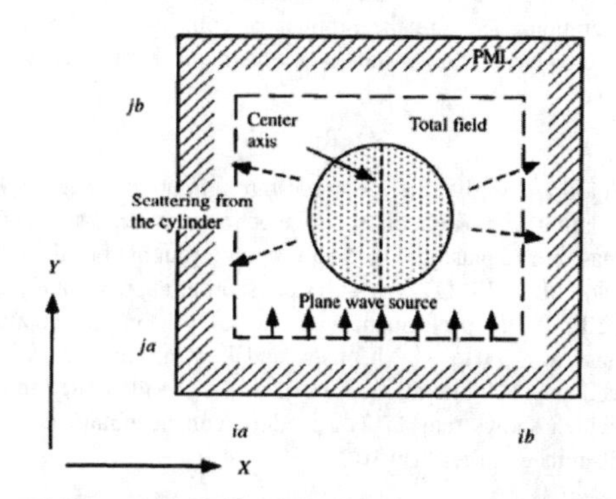

Figure 4. Diagram of the simulation of a plane wave striking a unmagnetized plasma cylinder [10].

To simulate a plane wave interacting with a unmagnetized plasma cylinder using parallel algorithm, we use each thread to calculate an electric field in Yee cell according to the parallel property of ZTFDTD. CUDA implement programs with the unit of warp. A half-warp is either the first or second half of a warp, which is an important concept for memory accesses because in order to hide the latency effectively, half of a warp should be performed at least at a time. A warp contained 32 threads. So, we make each Block has 32 * 16 Threads. When running, each thread in the block will have 512 threads to execute at the same kernel.

C. *Texture memory Optimization*

Optimizing the CUDA algorithm most often involves optimizing data accesses, which includes the use of the various CUDA memory spaces. Texture memory provides great capabilities including the ability to cache global memory .In order to speedup the parallel algorithm, a texture optimization algorithm was presented [11]. Texture cache is on-chip memory, which has good processing speed. Texture memory is designed originally for dealing with graphics which have a large number of spatial localities when we access memory. According to the characteristics of texture memory that is read-only memory, we put the parameters of UPML boundary conditions and dielectric cylinder into the texture memory and realize optimization.

D. Performance Calculation

Throughout this article, speedups and throughputs are used to quantify. Speedup is defined as the improvement in simulation time with respect to the ZTFDTD program being computed using only on the CPU. Thus:

$$Speedup = \frac{CUP\ Execution\ Time}{GUP\ Execution\ Time} \tag{3}$$

Throughput is defined as number of finite difference points that are updated per second:

$$Throughput(MCell/s) = \frac{I \cdot J \cdot T}{10^6 \tau} \tag{4}$$

Where τ is the execution time in seconds and T the number of iterations. I, J are the number of cells in two directions respectively. In this instance we have used mega-cells per second.

V. RESULTS

Fig. 5 and 6 give the measured throughput and speedup for simulating the interaction of electromagnetic wave with unmagnetized plasma. Fig. 5 shows the throughput of the CPU version of the FDTD code stays constant at approximately 2.3 MCells/s. The performance of the CUDA implementation varies from 33 to 51 MCells/s, and kept around 50 at last. Speedup of 5000 iterations for every case are illustrated in Fig. 6, which shows that 17-21 speedups can be obtained when cells number larger than 10^6.

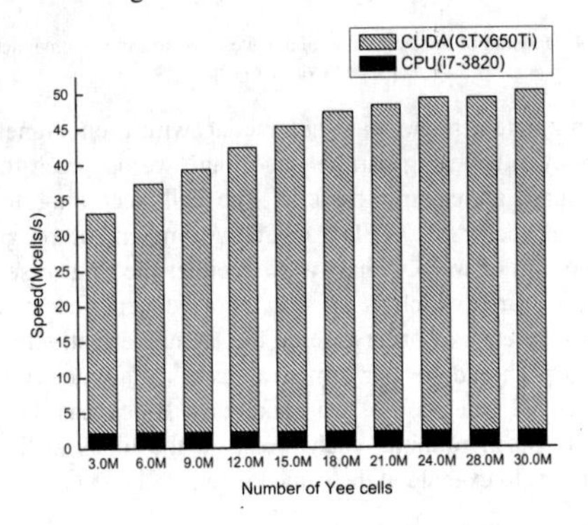

Figure 5. Throughput of CPU/CUDA ZTFDTD program.

VI. CONCLUSION

Using GTX650Ti video card, CUDA based code has shown good performance compared with C code operated on i7-3820 CPU in 2D FDTD simulations. The speedups are archived 14

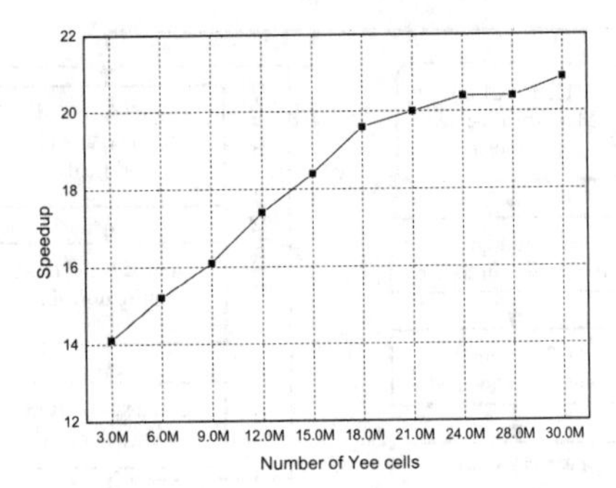

Figure 6. Obtained speedup of the CUDA program.

in all the cases we have tested. And for more common cases (number of Yee cells is larger than 10^7), it is up to 21 at last. The card is displaying for operating system at the same time of FDTD simulation. So it is expected that better performance would be obtained using Tesla C2070 GPU without video output port [12], which is designed for scientific computation with CUDA.

ACKNOWLEDGMENT

This work was financed by the National Natural Science Foundation of China (Grant No.61172020), the 211 Project of Anhui University (Grant No.ZYGG201202) and the academic innovation Foundation of Anhui University (Grant No.01001770-10117700481).

REFERENCES

[1] Ge Debiao and Yan Yubo, *Finite-Difference Time-Domain Method for Electromagnetic Waves*, 3rd ed., Xi'an: Xidian University Press, 2011.

[2] A. Taflove and S. C. Hagness, *Computational Electrodynamics: The Finite-Difference Time-Domain Method*, 3rd ed. Norwood, MA: Artech House, 2005.

[3] Sullivan D M, "Frequency-dependent FDTD methods using Z transforms," IEEE Trans. Antennas Propagat, pp.1223-1230, 1992.

[4] Liu Shaobin, Liu Song and Hong Wei, *The Frequency-Dependent Finite-Difference Time-Domain Method*, Beijing: Science Press, 2010.

[5] P. Micikevicius, "3D Finite Difference Computation on GPUs using CUDA", ACM International Conference report, 2009.

[6] NVIDIA Corporation Technical staff, *Kepler GK110 Architecture Whitepaper*, NVIDIA Corporation, 2012.

[7] D. B. Davidson, "Development of a CUDA Implementation of the 3D FDTD Method," IEEE Antennas and Propagation Magazine, Vol. 54, No. 5, Oct. 2012.

[8] NVIDIA Corporation Technical staff, *NVIDIA CUDA C Programming Guide 5.0*, NVIDIA Corporation, 2012.

[9] Liu Jianxiao, Yin Zhihui "Improved Z-FDTD Method Analysis of 3D Target with Non-Magnetized Plasma," Journal of Microwaves ,Vol.28, No.2 Apr.2012.

[10] Sullivan D M. *Electromagnetic Simulation Using the FDTD Method.* New York: IEEE Press, 2000, pp.49-62.

[11] J. Sanders and E. Kandrot, *CUDA By Example an Introduction to General-Purpose GPU Programming*. American: Pearson Education, 2010, pp.84-101.

[12] Tomoaki Nagaoka and Soichi Watanabe, "Accelerating three-dimensional FDTD calculations on GPU clusters for electromagnetic field simulation," EMBC, 2012, pp.5691 – 5694.

A Near-Surface Interpolation Scheme Based on Radial Basis Function

Can Lin PAN, Ming ZHANG and Ya Ming Bo
College of Electronic Science & Engineering,
Nanjing University of Posts and Telecommunications, Nanjing 210003, China
ymbo@njupt.edu.cn

Abstract- **The interpolation points adopted in the radial basis function (RBF) can be scattered. Based on the above fact, a near-surface interpolation scheme is proposed to combine RBFs for the scattering problems modeled with surface integral equations. The interpolation efficiencies of different RBFs with the proposed and Tartan grid schemes are compared to approximate the interactions between well-separated groups. It can be seen from the numerical results that the number of interpolation points is reduced significantly for all four RBFs, and the accuracy of the Gaussian RBF is better than the other three RBFs for different sizes of groups. The proposed scheme with GA RBF can be employed to build a fast solver.**

I. INTRODUCTION

The method of moments (MoM) using RWG basis functions [1-2] is widely used to solve 3-D electromagnetic scattering problems. Surface discretization is required in the MoM as opposed to volumetric discretization in the finite element method (FEM) and the finite difference time domain (FDTD) method. However, it is well known that the matrix given by the moment method is dense, which results in that high time cost is needed for the matrix-vector multiplications in an iterative solver. Hence, various fast algorithms [2] have been developed to accelerate the computing proceduce. The key step in these fast algorithms is to find an effective approximation of the interaction between two well-separated groups.

H-matrix [3] and H^2-matrix [4] techniques proposed recently can be regarded as general mathematical frameworks to combine various matrix approximations to accelerate an iterative solver. Lagrange polynomials interpolation can be employed to build a degenerate approximation of the interaction between two well-separated groups [5]. Similar works can be found in the multilevel Green's function interpolation method (MLGFIM) [6-9], a kernel independent approach, which has been successfully used to solve low-frequency [6] and full-wave electromagnetic problems [7]. Radial basis functions (RBFs) have been introduced to instead of the Lagrange polynomials to improve the computational performances of MLGFIM [7]. Furthermore, two different staggered interpolation point schemes [7-8] have been designed to reduce the number of interpolation points.

However, for the scattering problems modeled with surface integral equations, the unknowns are only distributed on the surface. The improvement of the computational efficiency is limited with the existing interpolation schemes, which are similar to the uniformly-spaced rectangular grids in adaptive integral method (AIM) [10]. To further reduce the number of interpolation points, a near-surface interpolation scheme is developed in this paper. The interpolation efficiencies of different RBFs with the proposed and Tartan grid schemes are compared. The shape parameter which is sensitive to the cube length and the number of interpolation points is also discussed.

II. THEORY

A. Radial Basis Function Interpolation

With the framework of H-matrix, a 3-D arbitrarily shaped object is recursively divided into smaller groups of different sizes. The well-separated groups are obtained according to their sizes and the distance of the two groups. Then the interaction between these groups can be approximated using interpolation technique

$$\tilde{G}\left(\boldsymbol{r}_i,\boldsymbol{r}_j'\right)=\sum_{p=1}^{K}\sum_{q=1}^{K}w_p\left(\boldsymbol{r}_i\right)w_q\left(\boldsymbol{r}_j'\right)g\left(\boldsymbol{r}_{i,p},\boldsymbol{r}_{j,q}'\right) \quad (1)$$

where $g(\boldsymbol{r},\boldsymbol{r}')=e^{-jkR}/4\pi R$ is the free space Green's function, $w_p(\boldsymbol{r}_i)$ and $w_q(\boldsymbol{r}_j')$ are the pth and qth interpolation functions in the field group i and source group j, $\boldsymbol{r}_{i,p}$ and $\boldsymbol{r}_{j,q}'$ are the pth and qth interpolation points, respectively. Additionally, K is the number of interpolation points. Once the degenerate approximation is obtained, it can be integrated into H-matrix or H^2-matrix frameworks to establish a fast solver.

A function $f(\boldsymbol{r})$ interpolated with RBFs can be expressed as follows:

$$f(\boldsymbol{r})=\sum_{i=1}^{K}\beta_i\varphi_i(\boldsymbol{r}) \quad (2)$$

where $\{\varphi_i\}_{i=1}^{K}$ is a set of radial basis functions and $\{\beta_i\}_{i=1}^{K}$ are the corresponding coefficients. Several infinitely smooth RBFs are listed in Table I, c is the shape-controlling parameter of the functions.

Although it has been verified that RBFs interpolation can provide better interpolation accuracy than Lagrange polynomials interpolation, RBFs do not satisfy the Kronecker delta condition, i.e.,

$$\Phi_j(\boldsymbol{r}_i)=\begin{cases}1, & i=j\\ 0, & i\neq j\end{cases} \quad (3)$$

978-7-5641-4279-7

TABLE I
INFINITE SMOOTH RBFs

Type of RBFs	$\varphi_i(\boldsymbol{r})$
Gaussian (GA)	$e^{-c\lvert\boldsymbol{r}-\boldsymbol{r}_i\rvert^2}$
Inverse quadratic (IQ)	$1/\left(\lvert\boldsymbol{r}-\boldsymbol{r}_i\rvert^2+c^2\right)$
Inverse multiquadric (IMQ)	$1/\sqrt{\lvert\boldsymbol{r}-\boldsymbol{r}_i\rvert^2+c^2}$
Multiquadric (MQ)	$\sqrt{\lvert\boldsymbol{r}-\boldsymbol{r}_i\rvert^2+c^2}$

which is important for the construction of high quality interpolation, and makes it easier to get the coefficients in (2). In order to obtain the orthogonal RBFs, a normalized orthonormalization can be utilized to get a set of auxiliary functions $\{\psi_i\}_{i=1}^{K}$, which is a linear combination of the RBFs and satisfies Eq. (3).

$$\psi_i(\boldsymbol{r})=\sum_{j=1}^{K}\gamma_{i,j}\varphi_j(\boldsymbol{r}) \tag{4}$$

Substituting (4)into (3), a matrix equation can be obtained,

$$\overline{\overline{\gamma}}\cdot\overline{\overline{\varphi}}=\begin{bmatrix}\gamma_{1,1} & \cdots & \gamma_{1,K}\\ \vdots & \ddots & \vdots\\ \gamma_{K,1} & \cdots & \gamma_{K,K}\end{bmatrix}\begin{bmatrix}\varphi_1(\boldsymbol{r}_1) & \cdots & \varphi_1(\boldsymbol{r}_K)\\ \vdots & \ddots & \vdots\\ \varphi_K(\boldsymbol{r}_1) & \cdots & \varphi_K(\boldsymbol{r}_K)\end{bmatrix}=\overline{\overline{I}} \tag{5}$$

where $\overline{\overline{I}}$ is a *K-by-K* unitary matrix. The singular valued decomposition (SVD) algorithm [11] is adopted to get the coefficients $\gamma_{i,j}$ because the matrix $\overline{\overline{\varphi}}$ consisted of RBFs is ill conditioned for a large matrix size *K*. Then the normalized orthogonal auxiliary functions $\psi_i(\boldsymbol{r})$ are obtained and employed to interpolate the function $f(\boldsymbol{r})$.

$$f(\boldsymbol{r})=\sum_{i=1}^{K}f(\boldsymbol{r}_i)\psi_i(\boldsymbol{r}) \tag{6}$$

In this way, the functions $w_p(\boldsymbol{r}_i)$ and $w_q(\boldsymbol{r}'_j)$ in (1) can be expressed in the interpolation form [7-8].

B. Near-Surface Interpolation Scheme

From the interpolation theory, it is known that the function value of a point is a linear combination of interpolation basis functions, and such a value can mainly be determined by contributions of the nearby interpolation points.

Fig. 1(a) shows 2-dimensional Tartan grid. The number of the interpolation points marked with small dots within Tartan grid is determined by the size of the group region. The two different staggered Tartan grids proposed in [7-8] involve all the interpolation points in the region. However, for solving surface integral equations, only part of the points is near the surface, which can be seen from Fig. 1(b). The surface is represented by the curve. Most of the interpolation points within Tartan grid are far away from the surface. It is found that good interpolation accuracy can still be achieved without the interpolation points far away from the surface. Hence, a near-surface interpolation scheme is proposed, which is illustrated with bigger black dots in Fig. 1(b). The required interpolation points for interpolation can be limited with the condition

$$\lvert\boldsymbol{r}_i-\boldsymbol{r}_s\rvert\le\alpha d \tag{7}$$

where $\boldsymbol{r}_i$ and $\boldsymbol{r}_s$ are the points from Tartan grid and the surface, d is the distance of two adjacent interpolation points, α is a factor used to control the bounds of interpolation points. With the increase of the factor α, the proposed scheme behaves more similar to Tartan grid. A small α may give low interpolation accuracy. It is recommended that the factor α is set to 1. 02.

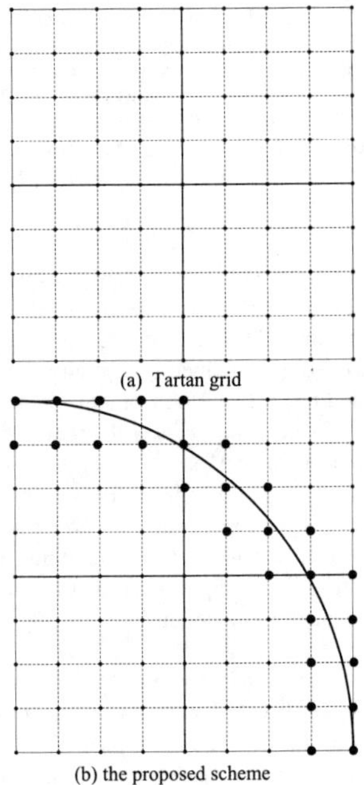

(a) Tartan grid

(b) the proposed scheme

Figure 1. 2-dimensional interpolation point schemes

III. Numerical Results And Discussion

The accuracy and efficiency of the near-surface interpolation scheme are verified with the test examples in this section. The results of different RBFs with Tartan grid are given as a reference, and then the interpolation relative errors of different RBFs with the proposed scheme are calculated.

In the first example, the interpolation efficiencies of different RBFs with Tartan grid are compared for groups of different electrical sizes. The results are listed in Table II. The interpolation error threshold is set to 0.02. The interpolation relative error is calculated with the following formula,

$$\varepsilon \approx \frac{\left\| G(r,r') - \tilde{G}(r,r') \right\|_F}{\left\| G(r,r') \right\|_F} \tag{8}$$

where $\tilde{G}(r,r')$ is the interpolated approximation of $G(r,r')$, $\|\bullet\|_F$ denotes the Frobenius norm.

From Table II, it is found that the interpolation with the Gaussian (GA) RBF obtains better interpolation accuracy than the other three kinds of RBFs, which is consistent with the results in [7-8].

The efficiency comparisons of different RBFs with the near-surface interpolation scheme are listed in Table III. In this example, spheres with diameters of different wavelengths are divided uniformly into 8 groups of different sizes. It can also be found that the interpolation with the GA RBF obtains the best performance, and less interpolation point numbers are needed for all four RBFs to satisfy the error threshold.

It can be seen from Table II and III that the numbers of interpolation points are reduced drastically for all RBFs with the proposed scheme, and the increase of the interpolation point number of the proposed scheme is much slower than the Tartan grid with the increase of group sizes, which can be seen explicitly in Fig. 2. When the group lengths are larger than 2.5λ, it is more difficult for Tartan grid to obtain the pseudo inverse using the singular valued decomposition (SVD) algorithm than the proposed scheme. Hence, the near-surface interpolation scheme with the GA RBF is more suitable for interpolation of large groups.

Fig. 3 shows the interpolation error of the GA RBF versus the shape parameter c of groups with length of 2 wavelengths. A sphere with a diameter of 4 wavelengths is demonstrated,

TABLE II

EFFICIENCY COMPARISONS OF DIFFERENT RBFS

Type of basis function	Length of groups (λ)	Number of interpolation points	Optimized shape parameter	Interpolation relative error
GA RBF	0.5	64	2.80	0.00426
	1.0	125	2.40	0.0140
	1.5	343	1.50	0.00434
	2.0	512	1.40	0.00827
	2.5	729	1.30	0.0132
IQ RBF	0.5	64	1.00	0.00618
	1.0	216	1.60	0.00362
	1.5	343	1.70	0.00841
	2.0	512	1.90	0.0154
	2.5	1000	2.60	0.00837
IMQ RBF	0.5	64	0.90	0.00643
	1.0	216	1.50	0.00379
	1.5	343	1.70	0.00862
	2.0	512	1.80	0.0160
	2.5	1000	2.50	0.00870
MQ RBF	0.5	64	0.70	0.00725
	1.0	216	1.30	0.00422
	1.5	343	1.50	0.00961
	2.0	512	1.60	0.0175
	2.5	1000	2.20	0.00969

TABLE III

EFFICIENCY COMPARISONS OF DIFFERENT RBFS WITH NEAR-SURFACE INTERPOLATION SCHEME

Type of basis function	Length of groups (λ)	Number of interpolation points	Optimized shape parameter	Interpolation relative error
GA RBF	0.5	43	2.90	0.00468
	1.0	73	2.30	0.0116
	1.5	145	1.40	0.0131
	2.0	251	1.20	0.0188
	2.5	371	1.20	0.0167
	3.0	515	1.10	0.0144
	4.0	885	1.00	0.0200
IQ RBF	0.5	43	1.20	0.0119
	1.0	106	1.60	0.0124
	1.5	184	2.30	0.0133
	2.0	304	2.50	0.0190
	2.5	427	2.40	0.0186
	3.0	609	2.30	0.0179
	4.0	992	2.20	0.0197
IMQ RBF	0.5	43	1.00	0.0127
	1.0	106	1.50	0.0130
	1.5	184	2.30	0.0134
	2.0	304	2.30	0.0200
	2.5	427	2.10	0.0176
	3.0	609	2.00	0.0178
	4.0	992	2.00	0.0203
MQ RBF	0.5	43	0.90	0.0146
	1.0	106	1.40	0.0145
	1.5	184	2.10	0.0149
	2.0	304	2.10	0.0221
	2.5	427	1.90	0.0185
	3.0	609	1.80	0.0191
	4.0	992	1.60	0.0220

and similar results can be observed for other electrical sizes. The interpolation point numbers of the proposed scheme are about 251, which vary slightly for different groups. Although the surface distributions in 8 groups are different, the same shape parameter can be adopted to satisfy the error threshold, which is very important for the proposed scheme to be employed to build a fast solver.

IV. CONCLUSION

In this paper, a near-surface interpolation scheme is presented based on the radial basis functions, which can reduce the interpolation points significantly. The interpolation efficiencies of the proposed and Tartan grid schemes are compared with different RBFs. Numerical results show that the performance of the Gaussian RBF is better than the other three RBFs when they are used to approximate the free space Green's function. The presented interpolation technique can be employed to construct a degenerate kernel for electromagnetic surface integral equations, for establishment of a fast solving algorithm.

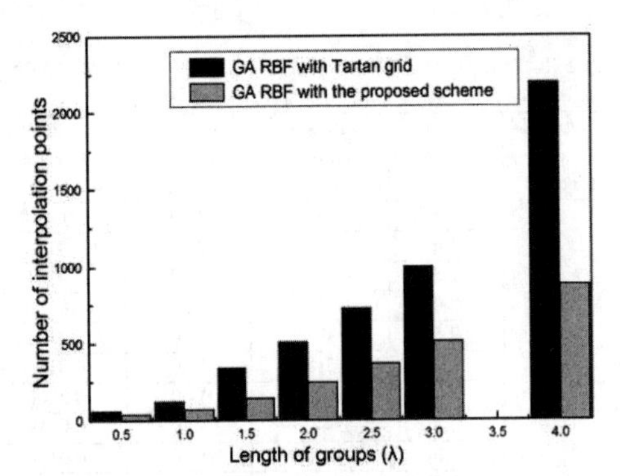

Figure 2. Variation of the number of interpolation points with groups of different wavelengths

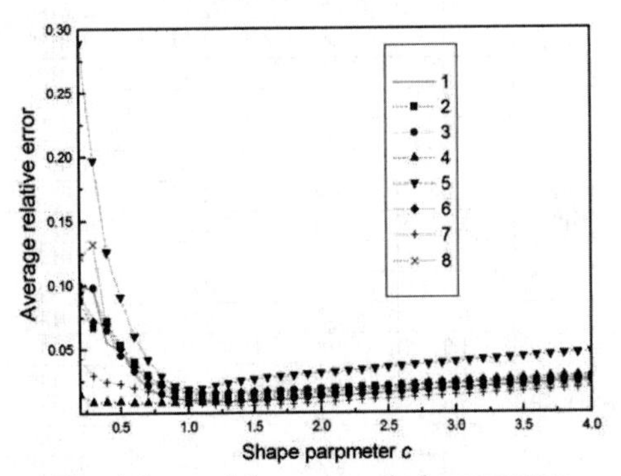

Figure 3. Average relative error versus the shape parameter c

ACKNOWLEDGMENT

This work is supported by the National Natural Science Foundation of China under grant No. 61071021.

REFERENCES

[1]. S. M. Rao, D. R. Wilton, A. W. Glisson, "Electromagnetic scattering by surfaces of arbitrary shape," *IEEE Trans on Antennas Propag.*, vol. 30 pp. 409~418 March 1982.

[2]. W. C. Chew, J. M. Jin, E. Michielssen, and J. M. Song, *Fast and efficient algorithms in computational electromagnetics*. New York: Artech House, 2001.

[3]. W. Hackbusch, "A sparse matrix arithmetic based on H-matrices, Part I: Introduction to H-matrices," *Computing*, vol. 62, pp. 89-108, Feb. 1999.

[4]. S. Borm, "H^2-matrices—multilevel methods for the approximation of integral operators," *Comput. Visual. Sci.*, Vol. 7, pp. 173~181, 2004.

[5]. W. Chai, D. Jiao, "An H^2-Matrix-Based Integral-Equation Solver of Reduced Complexity and Controlled Accuracy for Solving Electrodynamic Problems," *IEEE Trans on Antennas Propag.*, vol. 57, pp. 3147-3159, Oct. 2009.

[6]. H. G. Wang, C. H. Chan, and L. Tsang, "A new multilevel Green's function interpolation method for large-scale low-frequency EM simulations," *IEEE Trans. Comput.-Aided Des. Integr. Circuits Syst.*, vol. 24, pp. 1427–1443 Sept. 2005.

[7]. H. G. Wang, and C. H. Chan, "The implementation of multilevel Green's function interpolation method for full-wave electromagnetic problems," *IEEE Trans. Antennas Propag.*, vol. 55, pp. 1348-1358, May 2007.

[8]. Y. Shi, and C. H. Chan, "Comparison of interpolating functions and interpolating points in full-wave multilevel Green's function interpolation method," *IEEE Trans. Antennas Propag.*, vol. 58, pp. 2691–2699, May 2010.

[9]. Y. Shi, and C. H. Chan, "Improved 3D full-wave multilevel Green's function interpolation method," *Electron Lett.*, vol. 47, pp. 174–175 Feb. 2011.

[10]. E. Bleszynski, M. Bleszynski, and T. Jaroszewicz, "AIM: Adaptive integral method for solving large - scale electromagnetic scattering and radiation problems," *Radio Science*, vol. 31, pp. 1225-1251, May 1996.

[11]. W. H. Press, S. A. Teukolsky, W. T.Vetterling, et al, *Numerical recipes 3rd edition: The art of scientific computing*. Cambridge university press, 2007.

1D Modified Unsplitted PML ABCs for truncating Anisotropic Medium

Zhichao Cai[1)], Shuibo Wang[1)], Haochuan Deng[2)], Lixia Yang[1)], Xiao Wei[2)], Hongcheng Yin[2)]
(1. Department of Communication Engineering, Jiangsu University, Zhenjiang 212013, China.
E-mail:zhichaocai@yeah.net ;
2. Science and Technology on Electromagnetic Scattering Laboratory, P.O.Box142-207, Beijing 100854, China.)

***Abstract*-Based on the FDTD method in anisotropic medium, the implementation of the modified NPML absorbing boundary conditions for truncating anisotropic medium is presented. By using the partial derivatives of space variables stretched-scheme in the coordinate system, the programming complexity is reduced greatly. According to one dimensional numerical simulation analysis, the modified NPML absorbing boundary condition is validated.**

I INTRODUCTION

In order to simulate the open-domain electromagnetic scattering questions, the FD-TD methods introduce absorbing boundary conditions (ABCs) which is used to truncate an infinite problem space to a finite computation domain for us to simulate the electromagnetic scattering question. In 1981, Mur put forwarded the Mur ABCs with the FDTD discrete form in the computational domain truncation at the boundary [1]. This is an effective FDTD absorbing boundary condition and widely available. In 1994, 1996 Berenger proposed extended Maxwell's equations to splitted field form and constitutes a perfectly matched layer (PML), which is a highly effective absorbing boundary condition has been adapted in a variety of ways [2,3]. The Uniaxil Perfectly Matched Layer (UPML) were presented [4~7], which has been used successfully in the FD-TD computation for open-region electromagnetic problems.

Cummer has introduced a new kind of PML formulation named as Nearly Perfectly Matched Layer (NPML) [8], and was obtained by inserting the stretching factor of the PML within the derivatives on space of the curls. Hu and Cummer observed that the reflection from a NPML is as low as that from a normal PML [9]. The first implementation of the NPML for truncating nonlinear dispersive FDTD grids is presented by O. Ramadan [10]. Yang etc studied a novel non-splitted field perfectly matched layer ABC to truncate anisotropic magnetized plasma in FDTD computation [11]. The advantage of NPML is a simplicity ABC, and has the important application value in the truncated anisotropic medium.

Based on the FDTD method in anisotropic medium, a modified NPML ABC, this paper introduced a modified NPML ABC for truncating anisotropic medium. The programming complexity is reduced greatly by using the partial derivatives of special variables stretched-scheme in the coordinate system. The validity of this modified NPML ABC is proved through theoretical analysis and one-dimensional numerical simulation analysis.

II FDTD FORMULATION OF MODIFIED NPML ABSORBING BOUNDARY CONDITION FOR TRUNCATING ANISOTROPIC MEDIUM

For a homogeneous anisotropic medium, Maxwell's equations in time domain are given as follows:

$$\nabla \times \boldsymbol{H} = \varepsilon \frac{\partial \boldsymbol{E}}{\partial t} + \boldsymbol{\sigma E} \tag{1}$$

$$\nabla \times \boldsymbol{E} = -\boldsymbol{\mu} \frac{\partial \boldsymbol{H}}{\partial t} - \boldsymbol{\sigma_m H} \tag{2}$$

where $\boldsymbol{E}$ is the electric field, $\boldsymbol{H}$ is the magnetic field, the permittivity tensor is given by $\boldsymbol{\varepsilon}=[\varepsilon_{ij}]$ and the permeability tensor is given by $\boldsymbol{\mu}=[\mu_{ij}]$, $i,j=1,2,3$. The conductivity tensor is given by $\boldsymbol{\sigma}=[\sigma_{ij}]$ and the magneto conductivity tensor is given by $\boldsymbol{\sigma_m}=[\sigma_{mij}]$, $i,j=1,2,3$.

Suppose the 1D TEM electromagnetic waves with fields vary in the -z direction of anisotropic medium, i.e. $\partial/\partial x=0$, $\partial/\partial y=0$. In Cartesian coordinates, Eq.(1) and Eq.(2) can be written as follows:

$$-\frac{\partial H_y}{\partial z} = \varepsilon_{11}\frac{\partial E_x}{\partial t} + \varepsilon_{12}\frac{\partial E_y}{\partial t} + \sigma_{11}E_x + \sigma_{12}E_y \tag{3}$$

$$\frac{\partial H_x}{\partial z} = \varepsilon_{21}\frac{\partial E_x}{\partial t} + \varepsilon_{22}\frac{\partial E_y}{\partial t} + \sigma_{21}E_x + \sigma_{22}E_y \tag{4}$$

$$-\frac{\partial E_y}{\partial z} = -\mu_{11}\frac{\partial H_x}{\partial t} - \mu_{12}\frac{\partial H_y}{\partial t} - \sigma_{m11}H_x - \sigma_{m12}H_y \tag{5}$$

$$\frac{\partial E_x}{\partial z} = -\mu_{21}\frac{\partial H_x}{\partial t} - \mu_{22}\frac{\partial H_y}{\partial t} - \sigma_{m21}H_x - \sigma_{m22}H_y \tag{6}$$

Based on the stretching coordinate transform, replacing ∂z by $\partial \tilde{z} = (1+\sigma_z/j\omega)\partial z$, then Eqs. (3)~(6) become

$$-\frac{\partial H_y}{\partial \tilde{z}} = \varepsilon_{11}\frac{\partial E_x}{\partial t} + \varepsilon_{12}\frac{\partial E_y}{\partial t} + \sigma_{11}E_x + \sigma_{12}E_y \tag{7}$$

$$\frac{\partial H_x}{\partial \tilde{z}} = \varepsilon_{21}\frac{\partial E_x}{\partial t} + \varepsilon_{22}\frac{\partial E_y}{\partial t} + \sigma_{21}E_x + \sigma_{22}E_y \tag{8}$$

978-7-5641-4279-7

$$-\frac{\partial E_y}{\partial z}=-\mu_{11}\frac{\partial H_x}{\partial t}-\mu_{12}\frac{\partial H_y}{\partial t}-\sigma_{m11}H_x-\sigma_{m12}H_y \quad (9)$$

$$\frac{\partial E_x}{\partial z}=-\mu_{21}\frac{\partial H_x}{\partial t}-\mu_{22}\frac{\partial H_y}{\partial t}-\sigma_{m21}H_x-\sigma_{m22}H_y \quad (10)$$

According to Ref. [9], we can obtain

$$\frac{\partial H_y}{\partial z}=\frac{\partial H_y}{(1+\sigma_z/j\omega)\partial z}=\frac{\partial\left(H_y/(1+\sigma_z/j\omega)\right)}{\partial z}$$

Let $\tilde{H}_{yz}=H_y/(1+\sigma_z/j\omega)$, Eq.(7) becomes

$$-\frac{\partial \tilde{H}_{yz}}{\partial z}=\varepsilon_{11}\frac{\partial E_x}{\partial t}+\varepsilon_{12}\frac{\partial E_y}{\partial t}+\sigma_{11}E_x+\sigma_{12}E_y \quad (11)$$

Similarly, applying this coordinate transformation and introducing new variables $\tilde{E}_{xz}$, $\tilde{H}_{yz}$ and $\tilde{H}_{xz}$ to Eqs.(8) ~(10), then we have

$$\frac{\partial \tilde{H}_{xz}}{\partial z}=\varepsilon_{21}\frac{\partial E_x}{\partial t}+\varepsilon_{22}\frac{\partial E_y}{\partial t}+\sigma_{21}E_x+\sigma_{22}E_y \quad (12)$$

$$-\frac{\partial \tilde{E}_{yz}}{\partial z}=-\mu_{11}\frac{\partial H_x}{\partial t}-\mu_{12}\frac{\partial H_y}{\partial t}-\sigma_{m11}H_x-\sigma_{m12}H_y \quad (13)$$

$$\frac{\partial \tilde{E}_{xz}}{\partial z}=-\mu_{21}\frac{\partial H_x}{\partial t}-\mu_{22}\frac{\partial H_y}{\partial t}-\sigma_{m21}H_x-\sigma_{m22}H_y \quad (14)$$

According to Eqs. (11)-(14), we obtain FDTD iterative formula of the electric and magnetic field yield as follows:

$$E_x^{n+1}(k)=\frac{1}{\delta}\left\{\begin{aligned}&\left(\frac{\varepsilon_{22}}{\Delta t}+\frac{\sigma_{22}}{2}\right)\left[\frac{\tilde{H}_{yz}^{n+0.5}(k-\frac{1}{2})-\tilde{H}_{yz}^{n+0.5}(k+\frac{1}{2})}{\Delta z}+\left(\frac{\varepsilon_{11}}{\Delta t}-\frac{\sigma_{11}}{2}\right)E_x^n(k)+\left(\frac{\varepsilon_{12}}{\Delta t}-\frac{\sigma_{12}}{2}\right)E_y^n(k)\right]\\&+\left(\frac{\varepsilon_{12}}{\Delta t}+\frac{\sigma_{12}}{2}\right)\left[\frac{\tilde{H}_{xz}^{n+0.5}(k-\frac{1}{2})-\tilde{H}_{xz}^{n+0.5}(k+\frac{1}{2})}{\Delta z}+\left(\frac{\sigma_{21}}{2}-\frac{\varepsilon_{21}}{\Delta t}\right)E_x^n(k)+\left(\frac{\sigma_{22}}{2}-\frac{\varepsilon_{22}}{\Delta t}\right)E_y^n(k)\right]\end{aligned}\right\} \quad (15)$$

where $\delta=\left(\frac{\varepsilon_{11}}{\Delta t}+\frac{\sigma_{11}}{2}\right)\left(\frac{\varepsilon_{22}}{\Delta t}+\frac{\sigma_{22}}{2}\right)-\left(\frac{\varepsilon_{12}}{\Delta t}+\frac{\sigma_{12}}{2}\right)\left(\frac{\varepsilon_{21}}{\Delta t}+\frac{\sigma_{21}}{2}\right)$.

Similarly, we consider the same derivation with $E_y^{n+1}(k)$, $H_x^{n+0.5}(k+0.5)$, and $H_y^{n+0.5}(k+0.5)$.

Now, we consider the new variable $\tilde{H}_{xz}$,

$$\tilde{H}_{xz}=\frac{H_x}{s_z}=\frac{H_x}{1+\frac{\sigma_z}{j\omega\varepsilon_0}} \quad (16)$$

Then Eq.(16) is rewritten as follows:

$$j\omega\varepsilon_0\tilde{H}_{xz}+\sigma_z\tilde{H}_{xz}=j\omega\varepsilon_0 H_x \quad (17)$$

By transformation, Eq.(17) becomes

$$\frac{\partial \tilde{H}_{xz}}{\partial t}+\sigma_z\tilde{H}_{xz}=\frac{\partial H_x}{\partial t} \quad (18)$$

Using the center differential scheme, we can discrete Eq.(18) as follows:

$$\tilde{H}_{xz}\Big|_{k+\frac{1}{2}}^{n+0.5}=\frac{1-\frac{\sigma_z\Delta t}{2}}{1+\frac{\sigma_z\Delta t}{2}}\tilde{H}_{xz}\Big|_{k+\frac{1}{2}}^{n-0.5}+\frac{1}{1+\frac{\sigma_z\Delta t}{2}}\left(H_x\Big|_{k+\frac{1}{2}}^{n+0.5}-H_x\Big|_{k+\frac{1}{2}}^{n-0.5}\right) \quad (19)$$

Similarly, we have

$$\tilde{E}_{xz}\Big|_k^{n+1}=\frac{1-\frac{\sigma_z\Delta t}{2}}{1+\frac{\sigma_z\Delta t}{2}}\tilde{E}_{xz}\Big|_k^n+\frac{1}{1+\frac{\sigma_z\Delta t}{2}}\left(E_x\Big|_k^{n+1}-E_x\Big|_k^n\right), \quad (20)$$

$$\tilde{E}_{yz}\Big|_k^{n+1}=\frac{1-\frac{\sigma_z\Delta t}{2}}{1+\frac{\sigma_z\Delta t}{2}}\tilde{E}_{yz}\Big|_k^n+\frac{1}{1+\frac{\sigma_z\Delta t}{2}}\left(E_y\Big|_k^{n+1}-E_y\Big|_k^n\right), \quad (21)$$

$$\tilde{H}_{yz}\Big|_{k+\frac{1}{2}}^{n+0.5}=\frac{1-\frac{\sigma_z\Delta t}{2}}{1+\frac{\sigma_z\Delta t}{2}}\tilde{H}_{yz}\Big|_{k+\frac{1}{2}}^{n-0.5}+\frac{1}{1+\frac{\sigma_z\Delta t}{2}}\left(H_y\Big|_{k+\frac{1}{2}}^{n+0.5}-H_y\Big|_{k+\frac{1}{2}}^{n-0.5}\right). \quad (22)$$

III VALIDATION AND NUMERICAL RESULTS

Suppose the TEM wave vertically incident on a half-space anisotropic medium, the reflection of propagating wave has been computed. In the simulations, the incident wave is a Gaussian pulse plane wave whose frequency spectrum from 0 GHz to 1 GHz. The space cells size Δz=75μm and the time step $\Delta t=\Delta z/2c$ is 0.125ps, where c is the velocity of electromagnetic wave. The modified NPML ABCs are applied to both ends of computation domain. The thickness of the NPML ABC is $6\Delta z$.

In the first simulation, we validate the reflection coefficients for uniaxial anisotropic media. In FDTD computation, ε_{11}=4, ε_{22}=4, ε_{33}=16, the reflection coefficients of uniaxially anisotropic media are computed by using the modified NPML ABCs. They are compared with those of the analytical solution [12], shown in Figure.1. Numerical results are presented to verify the effectiveness of the proposed method, and demonstrate that the modified NPML ABCs method is in good agreement with the analytical solution.

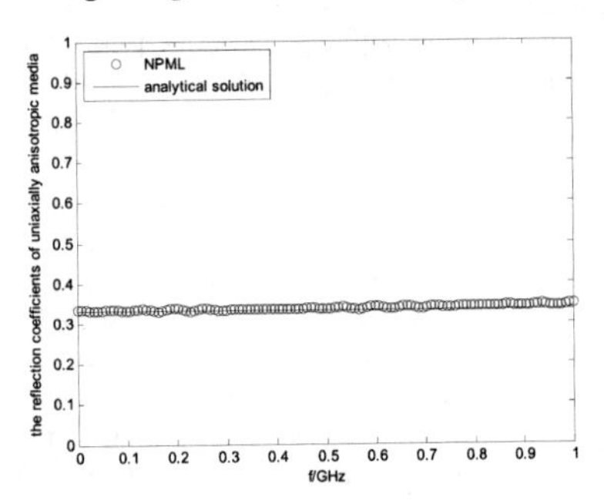

Figure 1 The reflection coefficients of uniaxially anisotropic media

In the second simulation, we let the parameters ε_{11}=9, ε_{22}=4, ε_{33}=16. By Numerical simulation, we obtain the Fig.2. The reflection coefficients of biaxial anisotropic media computed by using the modified NPML ABCs are compared with those of the analytical solution [13]. The results also show the modified NPML ABCs method is good agreement with the analytical solution, and the validity of the modified NPML ABC is proved, too. The results in Figure 2 also show that the modified NPML ABCs perform very well for the biaxial anisotropic media.

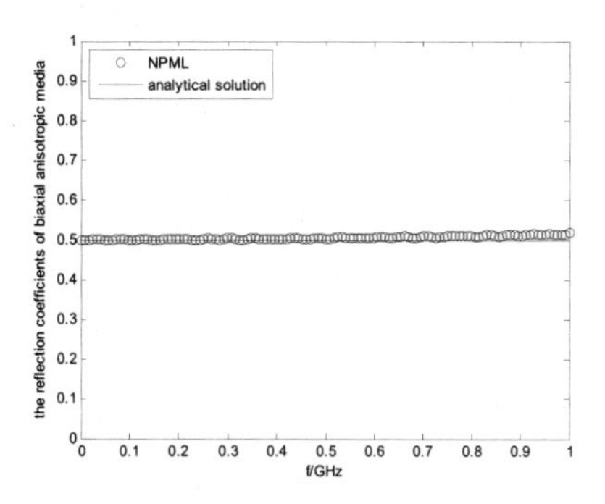

Figure2 The reflection coefficients of biaxial anisotropic media

IV CONCLUSION

In this paper, we have shown that based on the FDTD method for electromagnetic scattering by anisotropic medium and NPML, the modified NPML absorbing boundary conditions are presented, which can be used to truncate the anisotropic medium. Two numerical examples were given to illustrate that the proposed formulations can achieve a relatively simple programming complexity and the numerical examples also show that the modified NPML ABCs provide good absorbing performance. Therefore, the modified NPML ABCs have a large advantage in truncating anisotropic dispersion medium.

ACKNOWLEDGMENT

The authors would like to acknowledge the financial support from the National Natural Science Foundation of China (Grand No. 61072002), the Ph.D. Programs Foundation of Ministry of Education of China (Grand No. 20093227120018), the eighth "Elitist of Liu Da Summit" project of Jiangsu Province in 2011(Grand No.2011-DZXX-031), the Postdoctoral Science Foundation in Jiangsu (Grand No.1201001A), the Undergraduate Research Foundation of Jiangsu University (Grand Nos. 11A130 and 12A161).

REFERENCES

[1] G. Mur, "Absorbing boundary conditions for the finite-difference approximation of the time-domain electromagnetic-field equations," *IEEE Trans. Electromagn. Compat.*, vol. 25, no. 3, pp. 389-392, 1981.

[2] J. P. Bérenger, "A perfectly matched layer for the absorption of electromagnetic waves," *J. Comput. Phys.*, vol. 114, pp. 185-200, 1994.

[3] J. P. Bérenger, "Perfectly matched layer for the FDTD solutions of wavestructure interaction problems," *IEEE Trans. Antennas Propagat.*, vol. 44, pp. 110-117, 1996.

[4] D. B. Ge, Y. B. Yan, "*Finite-Difference Time-Domain Method for Electromagnetic Waves*," Xi'an：Xidian University Press, 2011.

[5] Z. S. Sacks, D. M. Kingsland, R. Lee, and J. F. Lee, "A perfectly matched anisotropic absorber for use as an absorbing boundary condition," *IEEE Trans. Antennas Propagat.*, vol. 43, pp. 1460-1463, 1995.

[6] D. C. Katz, E. T. Thiele, and A. Taflove, "Validation and extension to three dimensions of the Bérenger absorbing boundary condition for FDTD meshes," *IEEE Microwave Guided Wave Lett.*, vol. 4, pp. 268-270, 1994.

[7] S. D. Gedney, "An anisotropic PML absorbing media for the FDTD simulation of fields in lossy and dispersive media," *Electromagn.*, vol. 16, pp. 399-415, 1996.

[8] S. A. Cummer, "A simple, nearly perfectly matched layer for general electromagnetic media," *IEEE Microwave Wireless Components Lett.*, vol. 13, pp. 128-130, 2003.

[9] W. Y. Hu, S. A. Cummer, "The nearly perfectly matched layer is a perfectly matched layer," *IEEE ANTENNAS AND WIRELESS PROPAGATION Lett.*, vol. 3, pp. 137-140, 2004.

[10] O. Ramadan, "On the accuracy of the nearly PML for nonlinear FDTD domains," *IEEE Microwave Wireless Components Lett.*, vol. 16, pp. 101-103, 2006

[11] L. X. Yang, Q. Liang, P. P. Yu and G. Wang, "A novel 3D non-splitted field perfectly matched layer absorbing boundary condition in FDTD computation," *Chin. J. Radio Sci.*, vol. 26, no. 1, pp. 67-72, 2011.

[12] H. C. Chen, Theory of Electromagnetic Waves: A Coordinate Free Approach. New York: McGraw-Hill, 1983, pp. 280.

[13] Y. Z. Jia, "The Electromagnetic-Wave Propagation in Anisotropic medium," (MD) China University of Petroleum. 2008.

TA-2(B)

October 24 (THU) AM

Room B

EM Scattering

Enhancement of Near Fields Scattered by Metal-Coated Dielectric Nanocylinders

Pei-Wen Meng, Kiyotoshi Yasumoto, and Yun-Fei Liu
College of Information Science and Technology, Nanjing Forestry University
Nanjing, 210037, China
E-mail: dorismpw@163.com, kyasumoto@kyudai.jp, lyf@njfu.edu.cn

***Abstract*-The scattering of TE polarized plane wave by electrically small metal-coated dielectric cylinders is investigated with a particular emphasis on the enhancement of the near fields. If the wavelength of illumination is properly chosen, two unique near field distributions can be formed along the cylinders. It is shown that the enhanced near fields are localized around two interfaces of the coating metal layer and closely related to the surface plasmon resonances.**

I. Introduction

With the rapid development of nanoscience and nanotechnology, the interaction of light with nanoscaled objects remains as an important topic in recent years because of their promising applications to the optical sensors, imaging, and integrated devices. Enhanced backscattering by multiple nanocylinders under plasmon resonance was theoretically studied in [1]. The visible-light absorption by Si nanowires was discussed in [2]. This explains the optical ignition phenomena. The anomalous light scattering and the peculiarities of the energy flux around a thin wire with surface plasmon resonance were considered in [3]. It presents light scattering by small wires with weak dissipation near plasmon resonant frequencies. The effects of anomalous scattering are too complicated to see in the far field for the majority of metals such as gold, silver and platinum, whereas they can be quite pronounced in the near field region.

In this paper, the scattering of TE polarized plane wave by metal-coated dielectric nanocylinders is investigated with a particular emphasis on the enhancement of the near fields. If the wavelength of illumination is properly chosen, two unique near field distributions can be excited along the cylinders. Numerical results demonstrate that the enhanced near fields are localized around two interfaces of the coating metal layer and closely related to the surface plasmon resonances.

II. Formulation of The Problem

The cross section of coaxial cylindrical structures to be considered here is shown in Fig. 1. The coaxial cylinder with outer radius r_1 consists of a circular dielectric core with radius r_2 and a metal coating layer of thickness $r_1 - r_2$. The cylinder is infinitely long in the z direction and placed in free space. The material constants of the outer free space, coating metal, and dielectric core are denoted by (ε_0, μ_0), (ε_M, μ_0), and (ε, μ_0), respectively. Fig. 1 shows the configurations of (a) a single coaxial cylinder and (b) two identical coaxial cylinders

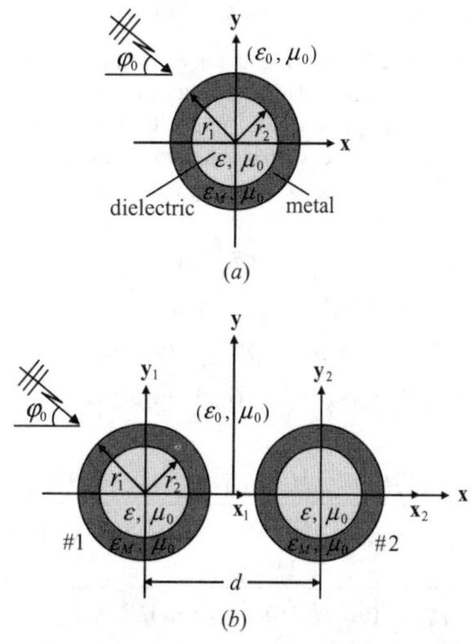

Fig. 1. Cross section of coaxial circular cylinders illuminated by a TE plane wave which is incident normally to the cylinder axis; (a) single cylinder system and (b) two cylinder system.

separated by a distance d along the x axis. The cylindrical structures are illuminated by a plane wave of unit amplitude which propagates normally to the cylinder axis. The angle of incidence of the plane wave is φ_0 with respect to the x axis. The scattering problem is two-dimensional and hence the electric and magnetic fields are decomposed into TE-wave and TM-wave. Since we are interested in the scattering problem related to the plasmon resonances, we focus our investigation on the scattering of TE wave with (H_z, E_x, E_y) component.

Let us consider first the scattering by a single coaxial cylinder shown in Fig. 1(a). The reflection and transmission of the standing and outgoing cylindrical waves at two cylindrical interfaces ($\rho = r_1$ and $\rho = r_2$) are solved separately. This leads to the reflection and transmission matrices for cylindrical harmonic waves at each of the interfaces, which are concatenated to obtain the generalized reflection and transmission matrices over two interfaces. If we denote by the column vectors the set of cylindrical basis function and the set of amplitude coefficients, the solutions to the H_z field in three regions of Fig. 1(a) are obtained, respectively, as follows:

978-7-5641-4279-7

$$H_z = \mathbf{\Phi}_0^T \cdot \boldsymbol{p} + \mathbf{\Psi}_0^T \cdot \mathbf{T} \cdot \boldsymbol{p} \quad \text{for } \rho > r_1 \tag{1}$$
$$H_z = \mathbf{\Phi}_M^T \cdot \mathbf{B} \cdot \boldsymbol{p} + \mathbf{\Psi}_M^T \cdot \mathbf{C} \cdot \boldsymbol{p} \quad \text{for } r_2 < \rho < r_1 \tag{2}$$
$$H_z = \mathbf{\Phi}^T \cdot \mathbf{D} \cdot \boldsymbol{p} \quad \text{for } 0 < \rho < r_2 \tag{3}$$

with

$$\left.\begin{array}{l} \mathbf{\Phi}_0 = [J_m(k_0\rho)e^{im\varphi}],\ \mathbf{\Psi}_0 = [H_m^{(1)}(k_0\rho)e^{im\varphi}] \\ \mathbf{\Phi}_M = [J_m(k_M\rho)e^{im\varphi}],\ \mathbf{\Psi}_M = [H_m^{(1)}(k_M\rho)e^{im\varphi}] \\ \mathbf{\Phi} = [J_m(k\rho)e^{im\varphi}] \end{array}\right\} \tag{4}$$

$$\boldsymbol{p} = [p_m],\ p_m = i^m e^{-im\varphi_0} \tag{5}$$
$$k_0 = \omega\sqrt{\varepsilon_0\mu_0},\ k_M = \omega\sqrt{\varepsilon_M\mu_0},\ k = \omega\sqrt{\varepsilon\mu_0} \tag{6}$$

where J_m $(m = 0, \pm1, \pm2, \cdots)$ is the m-th order Bessel function, $H_m^{(1)}$ is the m-th order Hankel function of the first kind, p_m denotes the amplitude coefficient of the incident plane wave expressed by the cylindrical harmonic expansion. In Eqs.(1)-(3), **T**, **B**, **C**, and **D** are diagonal matrices which are defined as follows:

$$\mathbf{T} = [T_m\delta_{mn}],\ T_m = R_{12,m} + \eta_M{}^2 F_{12,m}{}^2 (1 - R_{21,m}R_{23,m})^{-1} R_{23,m} \tag{7}$$
$$\mathbf{B} = [B_m\delta_{mn}],\ B_m = (1 - R_{21,m}R_{23,m})^{-1} F_{21,m} \tag{8}$$
$$\mathbf{C} = [C_m\delta_{mn}],\ C_m = (1 - R_{21,m}R_{23,m})^{-1} R_{23,m}F_{21,m} \tag{9}$$
$$\mathbf{D} = [D_m\delta_{mn}],\ D_m = (1 - R_{21,m}R_{23,m})^{-1} F_{32,m}F_{21,m} \tag{10}$$

where

$$R_{21,m} = -\frac{\eta_M H_m^{(1)}(u_M) H_m^{(1)\prime}(u_0) - H_m^{(1)\prime}(u_M) H_m^{(1)}(u_0)}{\eta_M J_m(u_M) H_m^{(1)\prime}(u_0) - J_m'(u_M) H_m^{(1)}(u_0)} \tag{11}$$
$$F_{12,m} = i\frac{2/(\pi\eta_M)}{[\eta_M J_m(u_M) H_m^{(1)\prime}(u_0) - J_m'(u_M) H_m^{(1)}(u_0)]} \tag{12}$$
$$R_{12,m} = -\frac{\eta_M J_m(u_M) J_m'(u_0) - J_m'(u_M) J_m(u_0)}{\eta_M J_m(u_M) H_m^{(1)\prime}(u_0) - J_m'(u_M) H_m^{(1)}(u_0)} \tag{13}$$
$$F_{21,m} = \eta_M{}^2 F_{12,m} \tag{14}$$
$$R_{23,m} = -\frac{\xi_M J_m(w_M) J_m'(w) - J_m'(w_M) J_m(w)}{\xi_M H_m^{(1)}(w_M) J_m'(w) - H_m^{(1)\prime}(w_M) J_m(w)} \tag{15}$$
$$F_{32,m} = -i\frac{2/(\pi\xi_M)}{\xi_M H_m^{(1)}(w_M) J_m'(w) - H_m^{(1)\prime}(w_M) J_m(w)} \tag{16}$$
$$u_M = k_M r_1,\ u_0 = k_0 r_1,\ w_M = k_M r_2,\ w = k r_2 \tag{17}$$
$$\eta_M = \sqrt{\varepsilon_M/\varepsilon_0},\ \xi_M = \sqrt{\varepsilon_M/\varepsilon}\,. \tag{18}$$

For the two-cylinder system shown in Fig. 1(b), we have to take into account the multiple interactions of the fields scattered from individual cylinders. The interactions can be easily calculated by using the T-matrix **T** of the single coaxial cylinder which are defined by Eqs.(1) and (7). Employing two local coordinate systems (ρ_1, φ_1) and (ρ_2, φ_2) whose origins are located at each center of two coaxial cylinders, after straightforward manipulations, the H_z fields in five different regions of the two-cylinder system are derived as follows:

$$H_z = \mathbf{\Phi}_0^T \cdot \boldsymbol{p} + \mathbf{\Psi}_{0,1}^T \cdot \boldsymbol{a}_1 + \mathbf{\Psi}_{0,2}^T \cdot \boldsymbol{a}_2 \quad \text{for } r_1 < \rho_1, \rho_2 \tag{19}$$
$$H_z = \mathbf{\Phi}_{M,j}^T \cdot \mathbf{B} \cdot \boldsymbol{q}_j + \mathbf{\Psi}_{M,j}^T \cdot \mathbf{C} \cdot \boldsymbol{q}_j\ (j=1,2) \quad \text{for } r_2 < \rho_j < r_1 \tag{20}$$
$$H_z = \mathbf{\Phi}_j^T \cdot \mathbf{D} \cdot \boldsymbol{q}_j\ (j=1,2) \quad \text{for } 0 < \rho_j < r_2 \tag{21}$$

with

$$\left.\begin{array}{l} \mathbf{\Psi}_{0,j} = [H_m^{(1)}(k_0\rho_j)e^{im\varphi_j}] \\ \mathbf{\Phi}_{M,j} = [J_m(k_M\rho_j)e^{im\varphi_j}],\ \mathbf{\Psi}_{M,j} = [H_m^{(1)}(k_M\rho_j)e^{im\varphi_j}] \\ \mathbf{\Phi}_j = [J_m(k\rho_j)e^{im\varphi_j}] \quad (j=1,2) \end{array}\right\} \tag{22}$$

$$\boldsymbol{a}_1 = (\mathbf{I} - \mathbf{T}\boldsymbol{\alpha}_{12}\mathbf{T}\boldsymbol{\alpha}_{21})^{-1}(e^{-ik_0 d\cos\varphi_0/2}\mathbf{I} + e^{ik_0 d\cos\varphi_0/2}\mathbf{T}\boldsymbol{\alpha}_{12})\mathbf{T}\cdot\boldsymbol{p} \tag{23}$$
$$\boldsymbol{a}_2 = (\mathbf{I} - \mathbf{T}\boldsymbol{\alpha}_{12}\mathbf{T}\boldsymbol{\alpha}_{21})^{-1}(e^{ik_0 d\cos\varphi_0/2}\mathbf{I} + e^{-ik_0 d\cos\varphi_0/2}\mathbf{T}\boldsymbol{\alpha}_{12})\mathbf{T}\cdot\boldsymbol{p} \tag{24}$$
$$\boldsymbol{\alpha}_{12} = [\alpha_{12,mn}],\ \alpha_{12,mn} = H_{m-n}^{(1)}(k_0 d/2) \tag{25}$$
$$\boldsymbol{\alpha}_{21} = [\alpha_{21,mn}],\ \alpha_{21,mn} = (-1)^{m-n} H_{m-n}^{(1)}(k_0 d/2) \tag{26}$$
$$\boldsymbol{q}_1 = e^{-ik_0 d\cos\varphi_0/2}\boldsymbol{p} + \boldsymbol{\alpha}_{12}\cdot\boldsymbol{a}_2 \tag{27}$$
$$\boldsymbol{q}_2 = e^{ik_0 d\cos\varphi_0/2}\boldsymbol{p} + \boldsymbol{\alpha}_{21}\cdot\boldsymbol{a}_1 \tag{28}$$

where $\boldsymbol{\alpha}_{12}$ ($\boldsymbol{\alpha}_{21}$) are the translation matrix which transform $\mathbf{\Psi}_{0,2}$ ($\mathbf{\Psi}_{0,1}$) to $\mathbf{\Phi}_{0,1}$ ($\mathbf{\Phi}_{0,2}$) according to the Graf's addition theorem.

III. Numerical Results and Discussions

Equations (1)-(3) and (19)-(21) are used to calculate the near field of H_z for several configurations of metal-coated coaxial cylindrical geometry. It is known that the permittivity of a metal in optical region takes a complex value with a negative real part and a small imaginary part. The value of the real part of the permittivity strongly depends on the wavelength. The proper evaluation of $\varepsilon_M(\lambda)$ for metal is crucial in the analysis. We employ here the Drude-Lorentz model [4] which expresses $\varepsilon_M(\lambda)$ in the following form:

$$\frac{\varepsilon_M(\omega)}{\varepsilon_0} = \varepsilon_\infty - \frac{\omega_{p,D}^2}{\omega(\omega + i\nu_D)} - \Delta_L \frac{\omega_{p,L}^2}{\omega^2 - \omega_{p,L}^2 + i\nu_L\omega} \tag{29}$$

where ε_∞ is the relative permittivity in $\omega \to \infty$, ω_p is the plasma frequency, ν is the collision frequency, Δ_L is the weighting factor for Lorentz mode, and the subscripts D and L are referred to Drude model and Lorentz model, respectively. In the numerical examples, we assumed Ag for the metal and calculated $\varepsilon_M(\lambda)$ using Eq.(29) and the parameters given in TABLE I.

The near field distributions of $|H_z|$ calculated for three different configurations of single cylinder are compared in Fig. 2. The angle of incidence of TE plane wave is $\varphi_0 = 0^\circ$ and the wavelength is $600nm$. The cylinder is (a) a pure dielectric with $r_1 = 40nm$ and $\varepsilon/\varepsilon_0 = 6.5$, (b) a pure metal (Ag) with $r_1 = 40nm$ and $\varepsilon_M/\varepsilon_0 = -13.862 + i0.968$, and (c) a metal

TABLE I

Typical parameters for Ag [4]. All frequencies are given in *THz*.

ε_∞	$\omega_{p,D}/2\pi$	$\omega_{p,L}/2\pi$	$\nu_D/2\pi$	$\nu_L/2\pi$	Δ_L
3.91	13,420	6,870	84	12,340	0.76

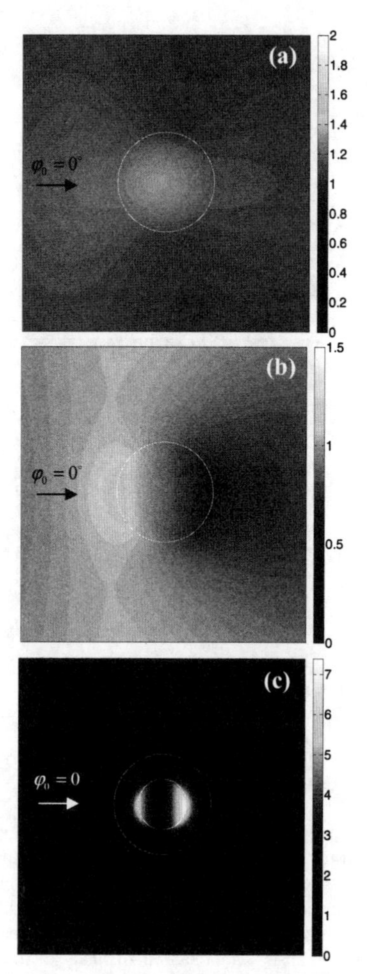

Fig. 2. Comparison of near field patterns of $|H_z|$ for three different configurations of single nanocylinder; (a) a pure dielectric cylinder with $r_1 = 40nm$ and $\varepsilon/\varepsilon_0 = 6.5$, (b) a pure metal (Ag) cylinder with $r_1 = 40nm$ and $\varepsilon_M/\varepsilon_0 = -13.862+i0.968$, and (c) a metal (Ag)-coated dielectric cylinder with $r_1 = 40nm$, $r_2 = 20nm$, $\varepsilon/\varepsilon_0 = 6.5$, and $\varepsilon_M/\varepsilon_0 = -13.862+i0.968$. The wavelength is $600nm$.

(Ag)-coated dielectric with $r_1 = 40nm$, $r_2 = 20nm$, $\varepsilon/\varepsilon_0 = 6.5$, and $\varepsilon_M/\varepsilon_0 = -13.862+i0.968$. The circles depicted by white lines indicate the boundary surfaces of the single and coaxial cylinders. The near field distributions of Figs. 2(a) and 2(b) are conventional, in which we can hardly observe any enhancement of a localized field. In contrast, the near field of the metal-coated coaxial cylinder shows a unique and interesting feature as shown in Fig. 2(c). The incident TE wave penetrates through the metal layer of thickness $r_1 - r_2 = 20nm$ and excites a strong field inside the coaxial cylinder. The excited field is localized in two sides of the cylinder

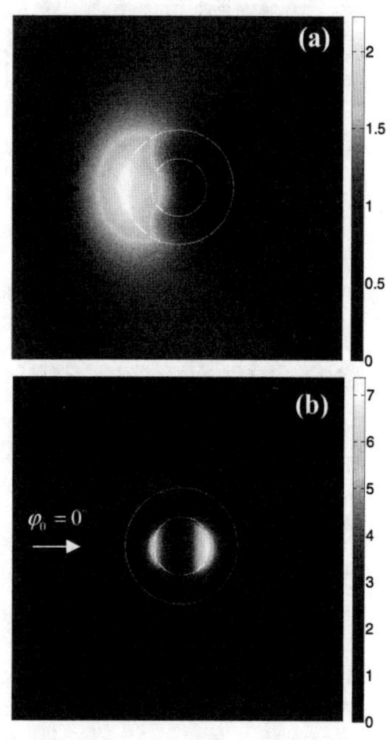

Fig. 3. Near field patterns of $|H_z|$ for the metal (Ag)-coated dielectric nanocylinder with $r_1 = 40nm$, $r_2 = 20nm$, and $\varepsilon/\varepsilon_0 = 6.5$. Two different wavelengths are considered; (a) $\lambda = 314nm$, $\varepsilon_M/\varepsilon_0 = -1.015 + i0.544$ and (b) $\lambda = 600nm$, $\varepsilon_M/\varepsilon_0 = -13.862 + i0.968$.

parallel to the propagation direction of the incident wave and along the interface between the dielectric core and the coating metal (Ag) layer. From the field profile of Fig. 2(c) it follows that the surface plasmon resonance at the inner interface of the metal layer occurs when the wavelength is $\lambda = 600nm$.

In Fig. 3, the near field distributions of $|H_z|$ for the Ag-coated dielectric nanocylinder with $r_1 = 40nm$, $r_2 = 20nm$, and $\varepsilon/\varepsilon_0 = 6.5$ are shown for two different wavelengths. The wavelength and the corresponding $\varepsilon_M(\lambda)$ of Ag are (a) $\lambda = 314nm$, $\varepsilon_M/\varepsilon_0 = -1.015+i0.544$ and (b) $\lambda = 600nm$, $\varepsilon_M/\varepsilon_0 = -13.862+i0.968$. For comparison we have reproduced Fig. 2(c) as Fig. 3(b). We can see that the near field pattern in Fig. 3(a) is quite different from that of Fig. 3(b). A strong field is excited only in the illuminated side of the cylinder and along the interface between the metal layer and the outer free space. From the field profile it follows that the surface plasmon supported by the interface between the metal and free space resonates to the incident TE plane wave of $\lambda = 314nm$.

In order to discuss the localized nature of the enhanced field shown in Fig. 2(c) and Fig. 3, we have analyzed the scattered field by a two-cylinder system. Two-identical Ag-coated dielectric nanocylinders with $r_1 = 40nm$, $r_2 = 20nm$, and $\varepsilon/\varepsilon_0 = 6.5$ is placed in parallel to each other. The distance between the centers of two cylinders is $d = 100nm$. Figure 4 shows the near field distributions of $|H_z|$ for two different wave-lengths.

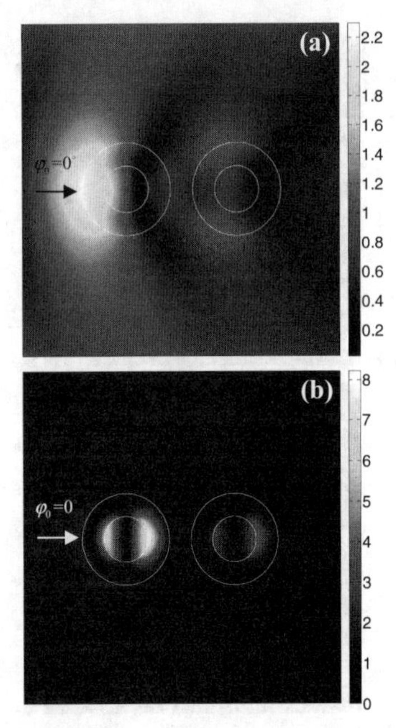

Fig. 4. Near field patterns of $|H_z|$ for two-identical metal (Ag)-coated dielectric nanocylinders with $r_1=40nm, r_2=20nm$, and $\varepsilon/\varepsilon_0=6.5$. The distance between the centers of two cylinders is $100nm$ and the incident angle of plane wave is $\varphi_0=0^\circ$. Two different wavelengths are considered; (a) $\lambda=314nm$, $\varepsilon_M/\varepsilon_0=-1.015+i0.544$ and (b) $\lambda=600nm$, $\varepsilon_M/\varepsilon_0=-13.862+i0.968$.

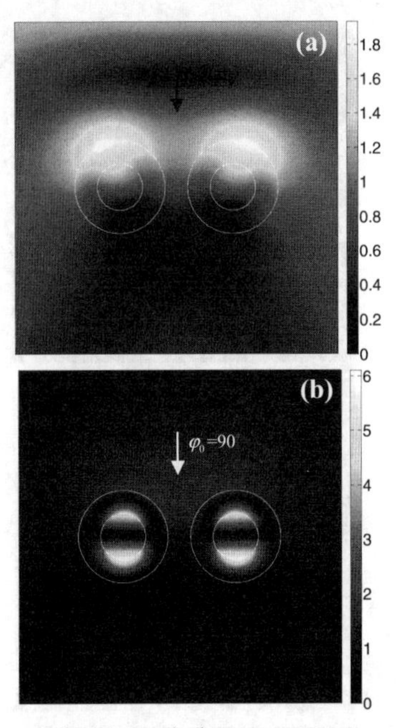

Fig. 5. Near field patterns of $|H_z|$ for two-identical metal (Ag)-coated dielectric nanocylinders. The incident angle of plane wave is $\varphi_0=90^\circ$. The others are the same as those in Fig.4.

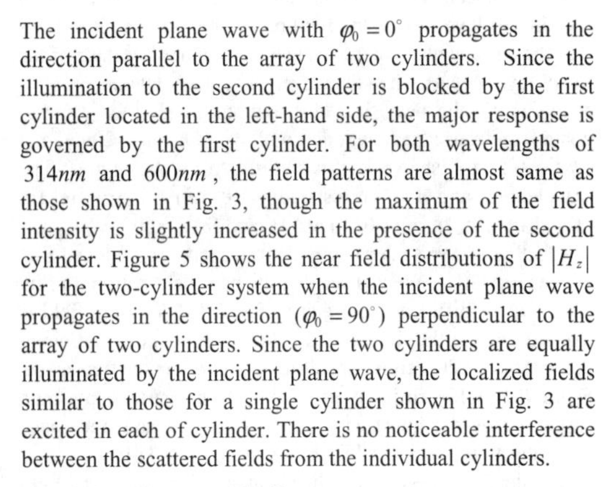

The incident plane wave with $\varphi_0=0^\circ$ propagates in the direction parallel to the array of two cylinders. Since the illumination to the second cylinder is blocked by the first cylinder located in the left-hand side, the major response is governed by the first cylinder. For both wavelengths of $314nm$ and $600nm$, the field patterns are almost same as those shown in Fig. 3, though the maximum of the field intensity is slightly increased in the presence of the second cylinder. Figure 5 shows the near field distributions of $|H_z|$ for the two-cylinder system when the incident plane wave propagates in the direction ($\varphi_0=90^\circ$) perpendicular to the array of two cylinders. Since the two cylinders are equally illuminated by the incident plane wave, the localized fields similar to those for a single cylinder shown in Fig. 3 are excited in each of cylinder. There is no noticeable interference between the scattered fields from the individual cylinders.

IV. CONCLUSION

The scattering of TE plane wave by metal-coated dielectric nanocylinders has been analyzed by using the cylindrical wave expansion and the T-matrix of a circular cylinder. It was shown that two unique near field distributions localized along the nanocylinders are excited when the wavelength of illumination resonates to surface plasmons on either side of the metal layer. For full understanding of light interaction with nanocylinders, it is also required to analyze the scattering cross section, the absorption cross section and their wavelength dependency. This study is under consideration.

REFRENCES

[1] H. Y. She, L. W. Li, S. J. Chua, W. B. Ewe, O. J. F. Martin, and J. R. Mosig, "Enhanced backscattering by multiple nanocylinders illuminated by TE plane wave," *J. Appl. Phys.*, vol. 104, 064310, 2008.

[2] G. Ding, C. T. Chan, Z. Q. Zhang, and P. Sheng, "Resonance-enhanced optical annealing of silicon nanowires," *Phys. Rev.B*, vol. 71, 205302, 2005.

[3] B. S. Luk'yanchuk and V. Ternovsky, "Light scattering by a thin wire with a surface plasmon resonance: Bifurcations of the Poynting vector field," *Phys. Rev. B*, vol. 73, 235432, 2006.

[4] T. Laroche, and C. Girard, "Near-field optical properties of single plasmonic nanowires," *Appl. Phys. Lett.*, vol. 89, 233119, 2006.

Diffraction Components given by MER Line Integrals of Physical Optics across the Singularity on Reflection Shadow Boundary

Pengfei Lu, Makoto Ando
Department of Electrical and Electronic Engineering, Tokyo Institute of Technology
2-12-1-S3-19 O-okayama
Meguro, Tokyo, 152-8552 Japan

***Abstract*-It is investigated about the diffracted field of the accuracy by Modified Edge Representation in surface-to-line integral reduction of Physical Optics. This components is compared between two results, MER line and PO surface integrals. The performance of stability from MER on RSB is confirmed, and this characteristic is shown in this paper.**

I. Introduction

The field generated from the scatterer is the main topic in the electromagnetics. Physical Optics (PO) one of the widely used high frequency asymptotic techniques in scattered field analysis, it is given by the radiation integrals over the scatterer. The surface-to-line integral reduction is important for reducing the computation and extracting the diffraction components. The Modified Edge Representation (MER) [1] is the concept to be used in the surface-to-line integral reduction [2], [3] for computing the scattered field derived from PO surface radiation [4] integrals. If Stationary Phase Point (SPP) exists on the scatterer, PO surface integrals will be reduced into two MER line integrals; one is diffraction components given by the integration along the periphery of the scatterer, another is the infinitesimally small indentation around inner SPP producing the reflected field [5]. The reflection components from SPP for various curvature of scatterer equals to Scattering Geometrical Optics (SGO) by using correction term has been shown in [6]. In this paper, the stability of diffraction at RSB calculated by MER has been stated.

II. Modified Edge Representation and Physical Optics

A. Modified Edge Representation Line Integration

MER line integration is an alternative methodology for the PO radiation pattern calculation of surfaces as shown in Fig. 1.

The wave diffracted from periphery is defined by the MER line integration as,

$$E_{MER}^{per} = j\frac{1}{4\pi}\oint_{\Gamma}\hat{r}_o\times\left[\hat{r}_o\times\eta J_{MER}+M_{MER}\right]\frac{e^{-jkr_o}}{r_o}dl \tag{1}$$

the MER unit vector $\hat{\tau}$ is defined as,

$$(\hat{r}_i+\hat{r}_o)\cdot\hat{\tau}=0;\hat{n}\cdot\hat{\tau}=0 \tag{2}$$

equivalent electric and magnetic line currents along the actual edge $\hat{t}$ are defined using the modified edge vector $\hat{\tau}$ as,

$$J_{MER}=\frac{\{\hat{r}_o\times(\hat{r}_o\times J_{PO})\}\cdot\hat{\tau}}{j\left(1-(\hat{r}_o\cdot\hat{\tau})^2\right)(\hat{r}_i+\hat{r}_o)\cdot(\hat{n}\times\hat{\tau})}\hat{t}$$

$$M_{MER}=\frac{\eta(\hat{r}_o\times-J_{PO})\cdot\hat{\tau}}{j\left(1-(\hat{r}_o\cdot\hat{\tau})^2\right)(\hat{r}_i+\hat{r}_o)\cdot(\hat{n}\times\hat{\tau})}\hat{t} \tag{3}$$

$$J_{PO}=2\hat{n}\times H^i$$

It was also confirmed that PO surface integration can be well approximated to MER line integration if only the radiation term is used in the definition of H^i in (3).

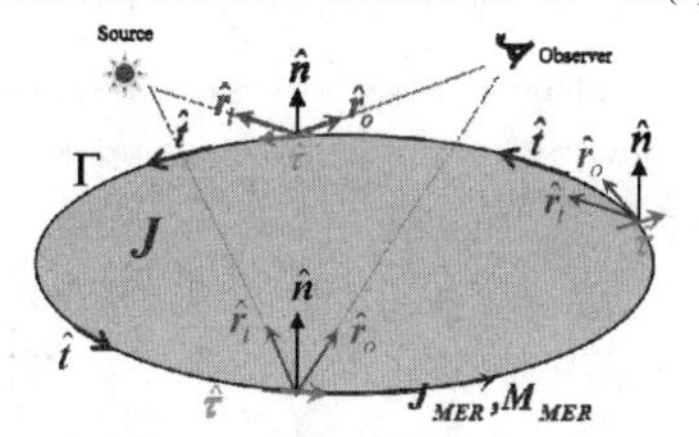

Figure 1 Parameters used in MER line integration.

B. Physical Optics

In PO, the scattering wave E_{PO}^S is expressed as the surface radiation integral of the currents on the scatterer as,

$$E_{PO}^S=j\frac{k\eta}{4\pi}\int_S\hat{r}_o\times\hat{r}_o\times J_{PO}\frac{e^{-jkr_o}}{r_o}dS$$
$$J_{PO}=2\hat{n}\times H^i \tag{4}$$

where J_{PO} is PO surface current.

III. Agreement between MER and PO in Diffraction and its Dependence upon Integral Path of Periphery

A. Agreement between MER and PO in diffracted field

The diffraction components calculated by E_{MER}^{per} and $E_{PO}^S-E_{SGO}$ are compared in Fig. 3 from the convex scatterer model as shown in Fig. 2, and the difference appears if the observer located at RSB.

978-7-5641-4279-7

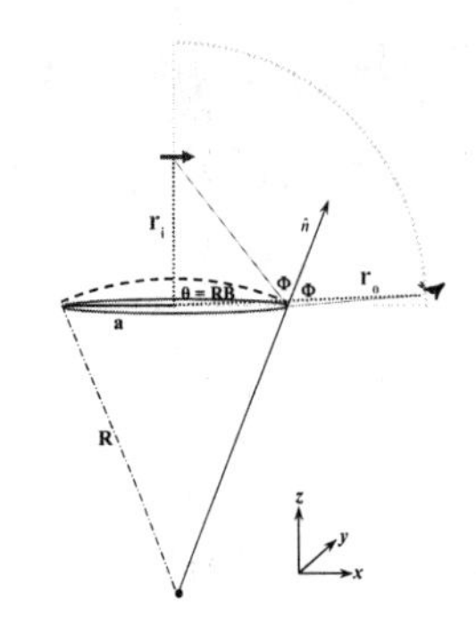

Figure 2 Convex scatterer from partial sphere.

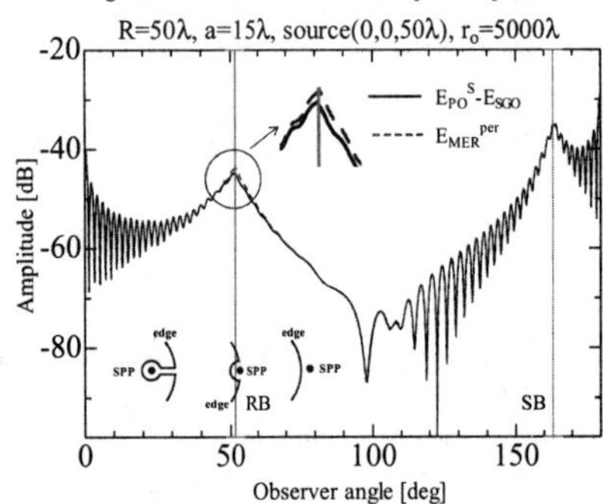

Figure 3 Diffraction field generated from convex scatterer.

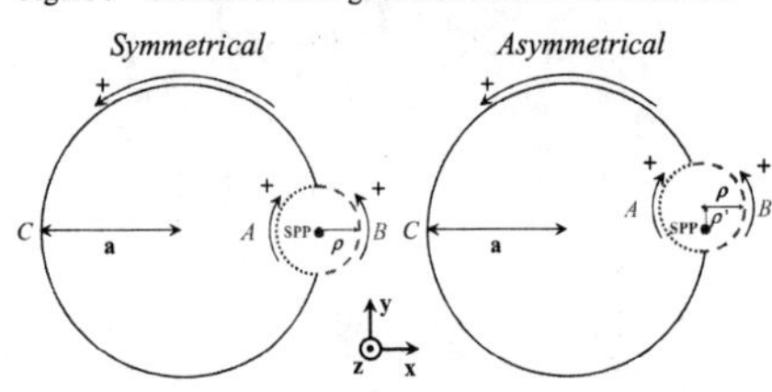

Figure 4 Top view from Fig.2 (a), (b).

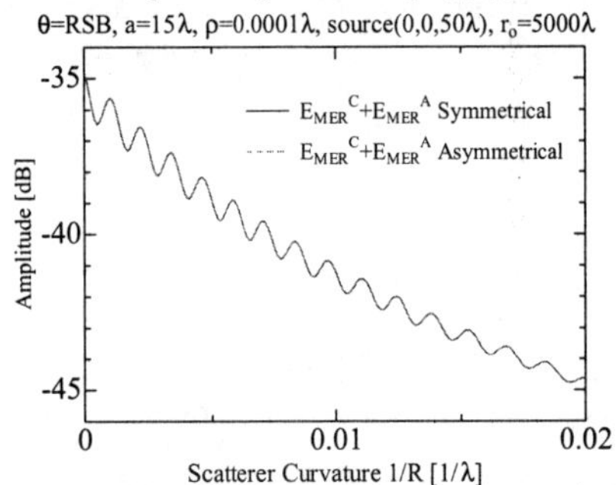

Figure 5 Diffracted field generated from convex scattered at RB.

B. Comparison between MER and PO in scattered field at RSB along different paths

The reflection point (RP) is the singularity for MER, the integral path along the periphery should bypass the RP if it is located on the edge. The amplitudes of diffracted field observed on RSB given by PO and MER along A respectively for various surface curvature are drawn in Fig. 5 from the top view of the scatterer shown in Fig 4 (a) and (b). The fields given by different integral paths are compared, both symmetrical and asymmetrical to SPP. The errors of MER should be compensated in the future. E^{A}_{MER} is almost identical with SGO/2 as shown in Fig. 6, which is much bigger than E^{C}_{MER} and is dominant in the diffraction. It is confirmed that the field given by symmetrical and excluding the SPP integral path calculated by MER contributes more agreeable field to PO. With the increasing of the surface curvature, the error between MER and PO is arising.

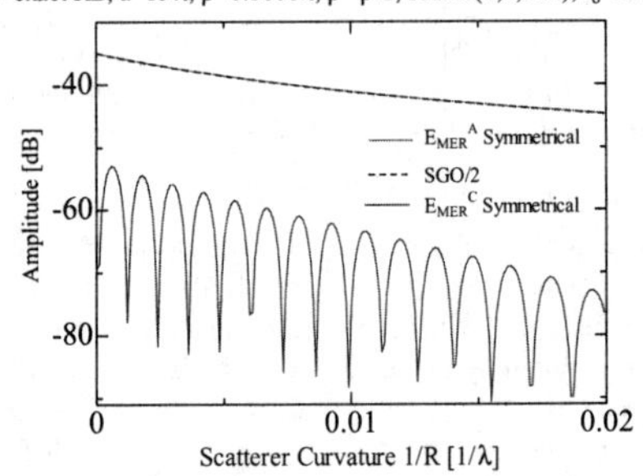

Figure 6 Various components generated from convex scattered at RB.

IV. Conclusion

The diffraction components calculated by PO and MER have been compared. It is investigated that the integral path along the periphery which gives the fields by MER should bypass the RP if it is located on the edge.

Acknowledgment

This work was conducted in part as the Research and Development for Expansion of Radio Wave Resources, the Ministry of Internal Affairs and Communications.

References

[1] T. Gokan, M. Ando and T. Kinoshita, "A new definition of equivalent edge currents in a diffraction analysis," IEICE Technical Report, AP89-64, Dec. 1989.

[2] Michaeli A., "Equivalent edge currents for arbitrary aspects of observation", IEEE Trans. Antennas Propag., 32, 252-258, 1984.

[3] Johansen P. M. and O. Breinbjerg, "An exact line integral representation of the physical optics scattered field: The case of a perfectly conducting polyhedral structure illuminated by electric Herzian dipoles", IEEE Trans. Antennas Propag., 43(7), 689-696, 1995.

[4] Silver S., *Microwave Antenna Theory and Design*, McGraw- Hill, New York, 1949.

[5] T. Murasaki and M. Ando, "Equivalent edge currents by the modified edge representation: physical optics components," IEICE Trans. Electron, vol.E75-C, No.5, pp.617-626, May, 1992.

[6] Pengfei Lu, Makoto Ando, "Difference of Scattering Geometrical Optics Components and Line Integrals of Currents in Modified Edge Representation", AGU Journals, Radio Sci., doi:10.1029/2011RS004899, 2012.

Near-Field Scattering Characters of the Ship

Chonghua Fang#, Xuemei Huang#, Qiong Huang#, Hui Tan# and Jing Xiao*
#Science and Technology on Electromagnetic Compatibility Laboratory, China Ship Development and Design Centre
Wuhan, China
[1]aiweng@yeah.net
[2]aiweng@yeah.net
[3]27634073@qq.com
[4]tanhuilucky@126.com
*Business management department, Hubei science and technology vocational college
Wuhan, China
[5]little-moon@163.com

Abstract— **The near-field scattering of ships is a very useful characteristic parameter for theory analysis and measurement. With the MLFMM, the near-field scattering of a typical simplified ship have been simulated to analyze the near-field scattering characters of the ship in free space in different detecting distances and incident angles. And the results show that with reduction of range, the difference between results of near-field and far-field scattering become very prominent. In addition, the required distance between antennas and ships may be far less than traditional far-field distance by comparing mean of scattering results. Accordingly to the phenomena of dominant lobe division of near-field scattering patterns, we developed a new expression to estimate the width of dominant lobe division. Finally, the viability of this expression is also given.**

Keywords**—MLFMM, near-field scattering, ship**

I. INTRODUCTION

The electromagnetic wave scattering is of great significance to the research of modern targets, such as ships. The research on this topic was mostly centered on far-field analysis: assume an incident plane wave, compute its scattered field due to the scatter, and evaluate the radar cross section (RCS) of the scatterer.

However, in practical applications, there are many situations that the distance between the radar and the scatterer is not large enough to treat the field arriving the scatterer as a plane wave. In these conditions the far-field analysis is not valid while the near-field analysis is necessary. Therefore, in some kind of scenario, it can be more appropriate to deal with the near field scattering characteristics of the targets. For example, most of the times, ships are observed in near field condition.

The few near-field scattering studies that the authors know are the NcPTD code and the Cpatch code developed in DEMACO [1]. Then, Shyh-Kang Jeng has investigated the near-field scattering by Physical Theory of Diffraction and Shooting and Bouncing Rays [2].

However, these studies mainly based on the high frequency methods instead of numerical method, such as MLFMM. It is well-known that the prediction precision based MLFMM is more than that based high frequency methods. Obviously, higher prediction precision can ensure that characteristic analysis is more effective. Also, to our eyes, the near-field scattering of ship by MLFMM have not be found in academic journals.

This paper presents some results of near-field scattering of the ship based MLFMM. In addition, the variation characteristic of near-field scattering versus detecting distance and different elevation angles has been investigated. Moreover, the comparison of means of near-field scattering of the ship in different distances is obtained. For the sake of analysis, some predicted results and discussions about typical construction of the ship will also be given.

II. STATEMENT OF PROBLEM

Let's consider the near-field scattering problem of a typical simplified ship in free space in Fig. 1. To calculate the EM fields scattered by the ship we have used the MLFMM [3]. Then we subdivide the ship surface in N sub-surfaces (triangular meshes), in such a way that all the elementary surfaces are in the far field of the detecting antenna and target. So, the electric and magnetic fields scattered by a perfectly conducting target are given by electric-field integral equation (EFIE).

$$\hat{t}\bullet\int_S\left[\bar{J}(r')+\frac{1}{k^2}\nabla\bullet\bar{J}(r')\nabla\right]\frac{e^{jkR}}{R}d's=\frac{4\pi j}{k\eta}t\bullet\vec{E}^i(r') \quad (1)$$

where $\bar{J}$ denotes the surface currents, $\hat{t}$ is the unit vector tangent to the surface of the PEC, $\vec{E}^i$ is the incident electric field, η represents the free space impedance and k refers to the wavenumber.

Figure. 1 A typical simplified ship (length 34m, width 7m, height 8m)

Then, the traditional far scattering (RCS) is followed.

978-7-5641-4279-7

$$\sigma_{FAR} = \lim_{r\to\infty} 4\pi r^2 \frac{\left|\vec{E}^s\right|^2}{\left|\vec{E}^i\right|^2} = \lim_{r\to\infty} 4\pi r^2 \frac{\left|\vec{H}^s\right|^2}{\left|\vec{H}^i\right|^2} \qquad (2)$$

It is well-known that the classical definition of RCS is a far field parameter. It supposes that the wave front is plane on the entire ship surface. In near field we should deal with limitations due to the directivity of the antenna diagram and to the spherical shape of the wave front. In this study we only focus on the effect of the wave front shape. We suppose that the antenna radiation is isotropic, and then the exact expression of scattering in near field conditions is this expression.

$$\sigma_{NEAR} = 4\pi r^2 \frac{\left|\vec{E}^s\right|^2}{\left|\vec{E}^i\right|^2} = 4\pi r^2 \frac{\left|\vec{H}^s\right|^2}{\left|\vec{H}^i\right|^2} \qquad (3)$$

Fig. 2 shows the monostatic scattering of this ship, versus azimuth angle, elevation angle, in near field at 150, 200, 500, 2000 meters (namely approximated far field range) respectively. For the sake of highlight, the scale in this picture is not well-proportioned.

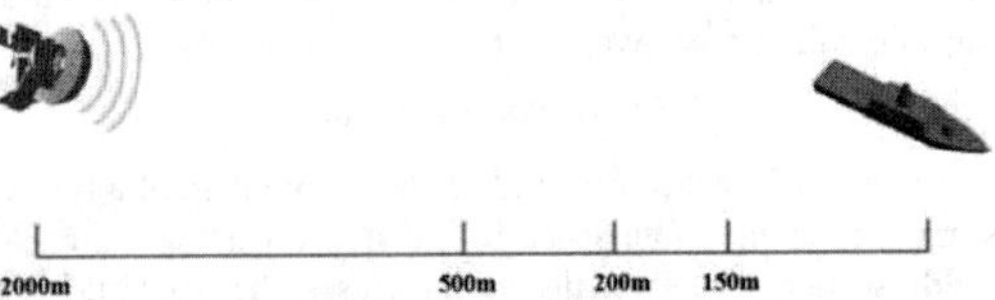

Figure. 2 The schematic drawing of detecting the ship with different distances.

III. Results and Analysis

A. Curves of near-field and far-field scattering

According to the geometrical configurations of Fig. 2, Fig. 3-6 plot the near-field scattering of the typical simplified ship in free space, at 300MHz and perpendicular polarization, at different elevation angles (90° and 87°) and azimuth angles (0°-180°) for several ranges R from near field to far field. Being confined to computer resource, here we only consider one frequency point.

From these pictures, we observe that the near field effects are mainly materialized by the shift and the spreading of some echoes. For example, the echo generated by the shipboard leading edge appears at 90° in far field; whereas in near field, this echo appears shifted and spread in a certain angular domain. These results show that with reduction of range, the difference between results of near-field and far-field scattering become very prominent. In addition, when the elevation angle is changed, the trend of the difference is still pronounced. Moreover, it is noteworthy that peaks of curves of elevation angle 87° are more than those of elevation angle 90°. Obviously, this comparison result is attributed to the fact that at elevation angle 87°, the incident rays are close to the normal direction of superstructure and cannon.

B. Comparison of mean of scattering results

For the sake of comparison of mean of scattering results, the statistical method of the typical means of scattering results of the ship is followed.

- Get ride of the results of four angle domains: $\pm 4.5^\circ$ of elevation angles of 0°, 90°, 180° and 270°.
- Transform the residual results from log space to linear space.
- Take the arithmetic mean of linear results.

Fig. 7-8 show the typical means of scattering results with different distances of the typical simplified ship. From these pictures, we can see that at the condition of error margin with 1dB, even rather nearer distance, such as 500m and the quarter of far-field distance, can satisfied with the traditional far-field requirement of measurement. So, it means that as for complex targets, such as ships, the required distance for RCS measurement between antennas and them may be far less than traditional far-field distance. Of course, it appears that the differences of typical means of scattering results grow as the distance shrinks further.

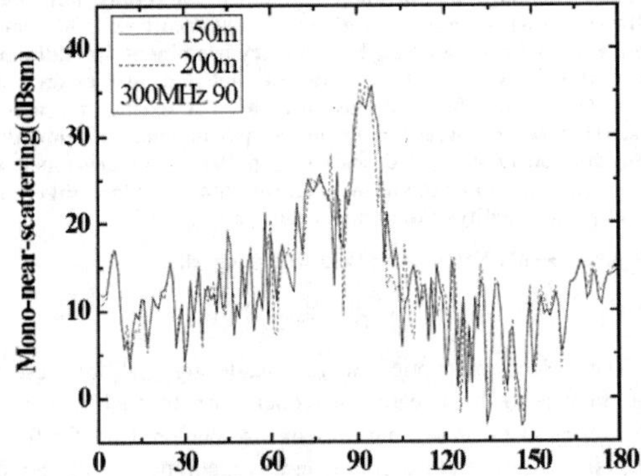

Figure. 3 Near-field scattering results of the typical simplified ship (150-200m, 300MHz, 90°)

C. Analysis for dominant lobe division.

As we known, far-field scattering curves of the flat always exhibit the acute character at the normal direction. Moreover, as for the near-field scattering of the flat, the curve character may be changed into the configuration of dominant lobe division [4], so does the shipboard. As illustrated in Fig 3-6, we can see that the dominant lobe division occurs in different distances around the shipboard, namely it is a frequent phenomena in near-field scattering. In addition, shorter distance the more remarkable it exhibits. The picture of detail with enlarged scale of these phenomena is shown in Fig. 9, while the double-arrow lines denote the width of dominant lobe division.

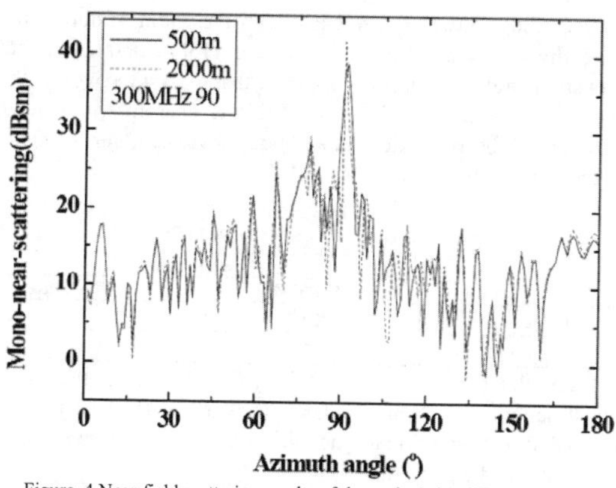

Figure. 4 Near-field scattering results of the typical simplified ship (500-2000m, 300MHz, 90°)

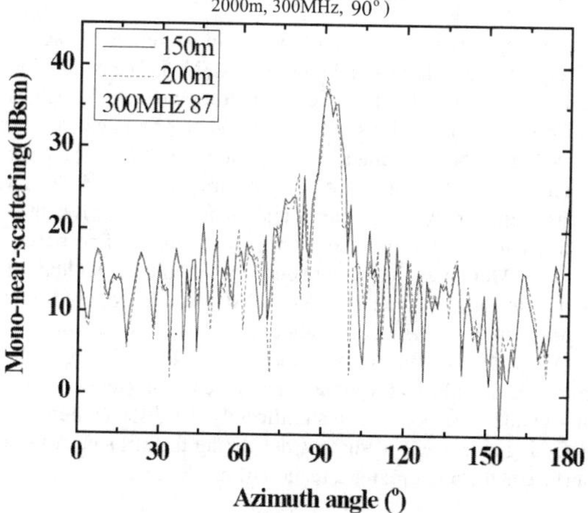

Figure. 5 Near-field scattering results of the typical simplified ship (150-200m, 300MHz, 87°)

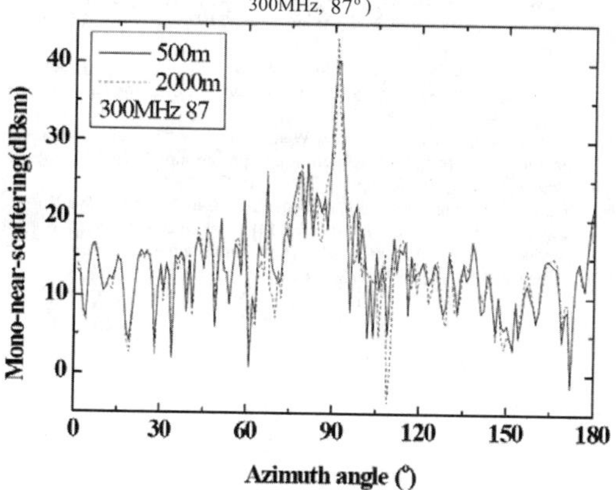

Figure. 6 Near-field scattering results of the typical simplified ship (500-2000m, 300MHz, 87°)

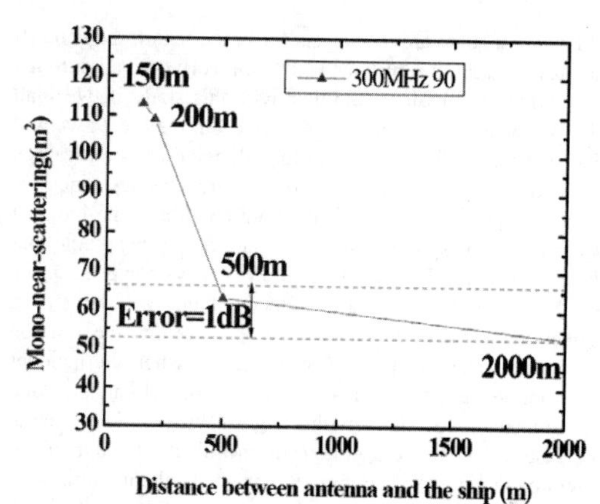

Figure. 7 The typical means with different distance of the typical simplified ship. (300MHz, 90°)

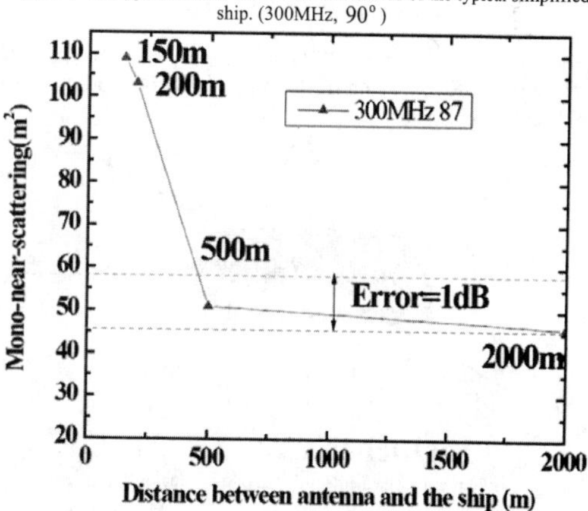

Figure. 8 The typical means with different distance of the typical simplified ship. (300MHz, 87°)

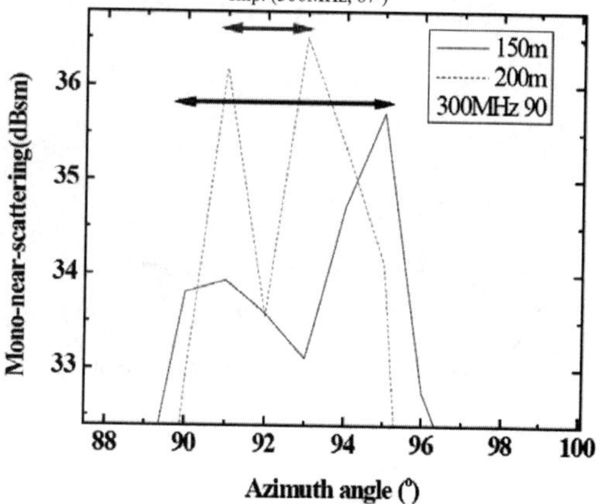

Figure. 9 The phenomena of dominant lobe division of near-field scattering patterns (detail with enlarged scale).

To give a quantitative analysis for it, we regard the conduct flat as the research subject and use the theory derivation to get the analysis expression. And then with this analysis, we shall give a coarse explanation for the ship. From the Fig. 9, we will find that the width of dominant lobe division can be regarded as the function of distance, size of flat, etc. Besides, since the distances do not satisfy with far field condition, the incident wave should be spherical wave. In our study, we assume that the size of flat is D in the direction of analysis and distances are R. As for the normal incidence, there will be strong reflection, just like far-field scattering. As for the oblique incidence, since the spherical wave effect, when the incident ray inclines to the normal direction, the dominant lobe division will exist in certain angle. With incident angle increase, while the incident ray is perpendicular to edge of flat, it will be in the end. Therefore, it is the critical condition now. If we assume incline angle to the normal indirection to be θ, we can obtain the new formula as followed:

$$\theta = \arcsin(D/2R) \tag{4}$$

Here θ is also the width of dominant lobe division of near-field scattering curve.

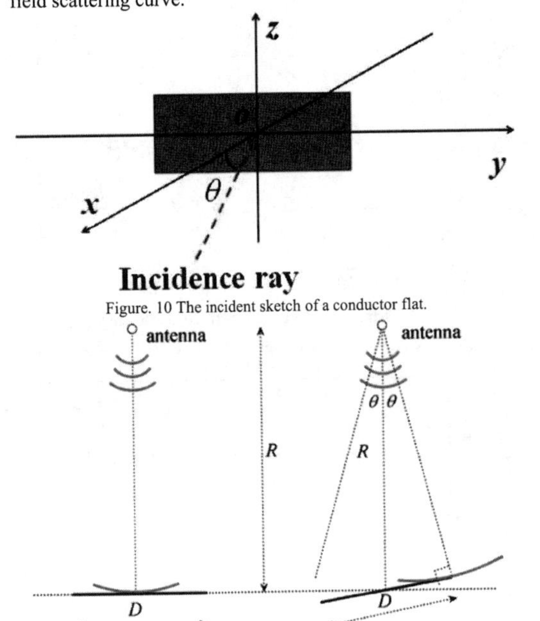

Figure. 10 The incident sketch of a conductor flat.

Figure. 11 The incident sketch of analysis of dominant lobe division.

According to the (4), we can obtain the width of dominant lobe division with different distances (150m and 200m at elevation angle 90°). It is very clear that they agree well with the results from MLFMM from Table I. Therefore, our new formula has been verified. In addition, obviously, our formula is much faster than that of MLFMM.

TABLE I
THE DOMINANT LOBE DIVISION WIDTH WITH DISTANCES

Case	Comparison	
	MLFMM	The new formula
150m, 90°	1.25°	1.3°
200m, 90°	2.5°	2.6°

Using this theory analysis, we may find that due to the facet construction of the ships, when incident ray is close to the normal direction of that, the phenomena of dominant lobe division may be exhibited.

IV. CONCLUSIONS

We proposed a simulation of near-field scattering characters of a typical simplified ship based on MLFMM. The simulation results shown that the near-field scattering characters, such as the dominant lobe division, become very prominent as the detecting distances reduce from 2000m to 150m. And we observed that for the ships, the required distance for RCS measurement between antennas and them may be far less than traditional far-field distance by comparing mean of scattering results. Moreover, we compared the predicted widths of dominant lobe division with MLFMM and the proposed new expression, and verified the validation the new expression.

Our study show the need to study near-field scattering of the ship that would exhibit more complicated characters in space and would possibly reduce significantly the distance between detecting antennas and ships. Considering the influence of sea surface is the other major direction of future research.

REFERENCES

[1] S. W. Lee, H. T. G. Wang, and G. Labarre, "Near-field RCS computation," *Appendix in the Manual for NcPTD-1.2*, S. W. Lee, writer. Champaign, IL: DEMACO, 1991.

[2] Shyh-Kang Jeng, "Near-Field Scattering by Physical Theory of Diffraction and Shooting and Bouncing Rays," *IEEE Antennas Propagat. Mag*, vol. 46, no. 4, April 1998.

[3] R. Coifman, V. Rokhlin, and S. Wandzura, "The fast multipole method for the wave equation: A pedestrian prescription," *IEEE Antennas Propagat. Mag.*, vol. 35, pp. 7–12, June 1993.

[4] ChongHua Fang, Qi Zhang, QiFeng Liu, XiaoNan Zhao, WeiBo Yu, "Prediction Formula for Flat-roofed Effect in Near-Field Scattering of Conductor Flat," in *2011 4th IEEE International Symposium on Microwave, Antenna, Propagation* .

Hybrid SPM to investigate scattered field from rough surface under tapered wave incidence

Qing Wang[1], Xiao-Bang Xu[2], Zhenya Lei[1], Yongjun Xie[3]

[1]National Laboratory of Science and Technology on Antennas and Microwaves, Xidian University, Xi'an 710071, P.R. China, wangqing@mail.xidian.edu.cn, zylei@xidian.edu.cn

[2]Holcombe Department of Electrical and Computer Engineering, Clemson University, Clemson 29634, USA, ecexu@exchange.clemson.edu

[3]School of Electronic and Information Engineering, Beihang University, Beijing 100191, P.R. China, yjxie@buaa.edu.cn

***Abstract*—We propose a novel time-efficient analytical model to investigate the near-zone scattered field from a conducting random rough surface under tapered wave incidence. The proposed method is based on the small perturbation method (SPM), and combined with the principle of stationary phase (POSP), the Lagrange polynomial and the Monte Carlo simulation. By comparing with the method of moment (MoM), the running time of proposed method is about 200 times less than that of MoM under the same accuracy, indicating that the proposed method is suitable for applications where computational time is urgent.**

I. INTRODUCTION

The analysis of electromagnetic wave scattered from random rough surfaces is of great interest in many applications including radio communications and remote sensing. Small perturbation method (SPM) [1] is one of the most commonly used analytical methods. It is valid to a surface that is slightly rough ($kh \ll 1$) and whose surface slope is smaller than unity, where k is the free-space wavenumber, and h is the root-mean-squared (RMS) height of the random rough surface. In numerical simulations, rough surface is truncated at edges where $x = \pm L/2$, which means that surface current is forced to be zero for $|x| > L/2$ [2]. To prevent current discontinuity at edges, a tapered incident wave was introduced [3, 4] in spatial or spectral domain.

Different from previous studies, which was focused on far-zone field, in this paper we present a time-efficient analysis of near-zone scattered field from a random rough surface. Two coefficients are defined, namely the coherent scattered taper function T_{coh} and incoherent one T_{incoh} . T_{coh} is modified for coherent wave by Lagrange polynomial, and for the incoherent wave, and the POSP is used to deal with an infinite complex integral. Finally a Monte Carlo simulation is used to determine the scattered field statistics as a function of different rough surface realizations.

II. DESCRIPTION OF THE METHOD

The geometry of the problem is shown in Figure 1. A TE tapered wave, $\vec{\psi}_{inc}(\vec{r})$ [3, 5] is incident with time dependence $\exp(-j\omega t)$, impinging upon a 1-D conducting Gaussian-distributed rough surface, with a random height profile $z = f(x)$, correlation length l , surface length L, RMS height h, and the spectral density $W(k_x) = h^2 l/(2\sqrt{\pi}) \cdot \exp(-k_x^2 l^2/4)$. The expression for $\vec{\psi}_{inc}(\vec{r})$ is [4]

$$\begin{aligned}\vec{\psi}_{inc}(x,z) &= T(x,z)\exp\left(-j\vec{k}_i \bullet \vec{r}\right)\hat{y} \\ &\equiv T(x,z)\exp\left[-j\left(k_{ix}x - k_{iz}z\right)\right]\hat{y},\end{aligned} \tag{1}$$

where $k_{ix} = k\sin\theta_i, k_{iz} = k\cos\theta_i$, θ_i is the incident angle, and $T(x,z)$ is the incident taper function

$$T(x,z) = \exp\left[i\left(k_{ix}x - k_{iz}z\right)w(\vec{r}) - \left(x + z\tan\theta_i\right)^2/g^2\right] \tag{2}$$

where g is the tapering parameter, and $w(\vec{r}) = \left\{2\left[\left(x + z\tan\theta_i\right)^2 - 1\right]/g^2 - 1\right\}/\left(kg\cos\theta_i\right)^2$.

The height of the random rough surface is used as a small parameter. For the scattered electric field, we write it as a perturbation series [1], $\psi_s = \psi_s^{(0)} + \psi_s^{(1)} + \psi_s^{(2)} + \cdots$. Since the Dirichlet boundary condition requires that on

978-7-5641-4279-7

the PEC surface $z=f(x)$, $\psi_{inc}+\psi_s=0$, and there is a taper function in the incident wave expression in (1), therefore a scattered taper function should be considered

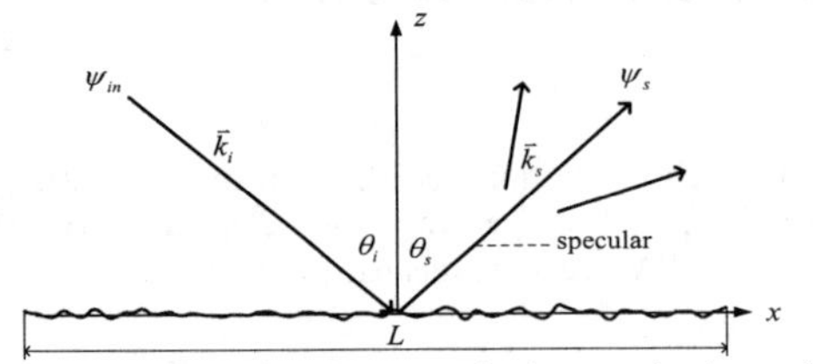

Figure 1 Geometry for wave scattering by a randomly rough.

into the expression of scattered wave. In the following, two different scattered taper functions are introduced

$$\begin{aligned}\psi_s(\vec{r})&=\psi_{coherent}(\vec{r})+\psi_{incoherent}(\vec{r})\\&=T_{coherent}\psi'_{coherent}(\vec{r})+T_{incoherent}\psi'_{incoherent}(\vec{r})\end{aligned} \quad (3)$$

where $\psi'_{coherent}(\vec{r})$ and $\psi'_{incoherent}(\vec{r})$ are the coherent and incoherent wave under plane wave incidence using the traditional SPM. $T_{coherent}$ and $T_{incoherent}$ are defined as the coherent and incoherent taper function, respectively. Their expressions are $T_{coherent}=\Delta\bullet T$ and $T_{incoherent}=T$, where Δ is a polynomial that only related to the incident angles. From [1], the coherent and incoherent wave for the tradition SPM are

$$\psi'_{coherent}(\vec{r})=-e^{ik_{ix}x+ik_{iz}z}\left(1-2k_{iz}\int_{-k}^{k}dk_x k_z W(k_{ix}-k_x)\right) \quad (4)$$

and

$$\psi'_{incoherent}(\vec{r})=\int_{-\infty}^{\infty}dk_x e^{ik_x x+ik_z z}2ik_{iz}F(k_x-k_{ix}) \quad (5)$$

where $k_z=\sqrt{k^2-k_x}$, and $F(k_x)$ is the Fourier transform of $f(x)$. With the two scattered taper functions, the scattered field under tapered wave incidence satisfies the wave equation to order of $O\left[1/(k^2g^2)\right]$.

Since (5) involves complex integral, and the wavenumber k is chosen to be large enough in our case, the POSP is used[7]. The points of stationary phase in (5) are $k_{x0}=\pm xk/\sqrt{f^2+x^2}$. The integral (5) can be approximately evaluated by considering the sum of the contributions from each of these points, which is $I(k)$:

$$\begin{cases}I_1(k;x,z)\\ \sim 2ik_{iz}F(k_{x0}-k_{ix})\cdot\exp\left[i(k_{x0}x+k_{z0}z-\pi/4)\right]\\ \cdot\sqrt{2\pi/\left|k^2/(k^2-k_{x0}^2)^{\frac{3}{2}}\right|}\\ I_2(k;x,z)\\ \sim 2ik_{iz}F(-k_{x0}-k_{ix})\cdot\exp\left[i(-k_{x0}x+k_{z0}z-\pi/4)\right]\\ \cdot\sqrt{2\pi/\left|k^2/(k^2-k_{x0}^2)^{\frac{3}{2}}\right|}\end{cases} \quad (6)$$

$$I(k)=I_1(k)+I_2(k). \quad (7)$$

TABLE I. THE RELATIONSHIP BETWEEN Δ AND θ_i

θ_i	10°	20°	30°	40°	50°	60°	70°	80°
Δ	1.18	1.17	1.14	1.109	1.105	1.04	1.02	0.991

At first, when we simply set the coherent wave tapered function as T, the results of the scattered field from different incident angles always have the same shape but with differ magnitudes compared with that of MoM (see Figure 2). It turns out that a relationship between the incident angles θ_i and the coherent scattered field amplitude Δ leads to this difference (see Table I). Then, the additional coefficients for the coherent wave were taken into account. The Lagrange polynomial [8] $\Delta(\theta_i)=\sum_{n=0}^{7}a_n\theta_i^n$ is used to fit this relationship at knots $\theta_{i(i=0\sim7)}$ = [10° , 20° , 30° , 40° , 50° , 60° , 70° , 80°], yielding coefficients $a_{i(i=0\sim7)}$ = [0.8590, 0.7703, -0.6920, 0.3105, -0.0782, 0.0111, -8.25e-4, 2.5e-05]/10,. Later we test its reliability in a large group of different incident angles, the results show that this polynomial approximation yields error less than 0.8% compared with MoM data.

To obtain the statistical average of our method, Monte Carlo simulation is used [6]. In each simulation trial, the scattered field is a function with respect to each sampled rough surface. The statistical average of the field is determined by

$$\psi^m_{average}=\left(\psi^m+\psi^{m-1}_{average}\bullet(m-1)\right)/m, \quad (1\le m\le M) \quad (8)$$

where m is the index Monte Carlo trial, and M is the total number of trials. From (8), the convergence rate for

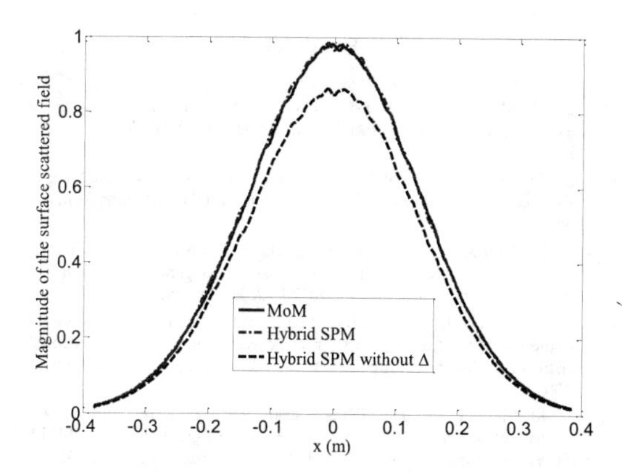

Figure 2 Magnitude of the surface scattered field with the incident wave angle of 30°.

the scattered field can be estimated by

$$convergence(m) = \left(\psi^{m+1}_{average} - \psi^{m}_{average}\right) / \psi^{m}_{average} \quad (9)$$

III. RESULTS

The programs were written in Matlab, and ran on a PC with 3.20GHz Intel Core(TM) i3 and 1.74GB memory. The wavelength is $\lambda = 0.03m$ in our case. The parameters of the random rough surface are selected as: $L = 25.6\lambda, h = 0.05\lambda, l = 0.35\lambda$. The tapering parameter is set to be $g = L/4$. The total number of points is $N = 528$, and N denote the number of times of Monte Carlo trials.

We compare the surface electric fields using the hybrid SPM with that obtained from the MoM. Such comparison results are given in Figure 2 and Figure 3 with incident angles $\theta_i = 30^\circ, 40^\circ, 60^\circ$. It is shown that the two curves obtained using hybrid SPM and MoM overlap each other well. The comparisons for computation time and accuracy are given in Table 2, where accuracy is calculated as

Accuracy=

$$\max\left[\left\| E_{sc(MoM)} \right| - \left| E_{sc(Hybrid\ SPM)} \right\| / \left| E_{sc(MoM)} \right| \cdot 100\right] \quad (10)$$

It is shown that our proposed method can accelerate the computation with a factor of 200 over MoM. This is because the adoption of the POSP into our procedure makes the computation more efficient.

TABLE II. COMPARISONS OF COMPUTATION TIME (IN SECOND) AND ACCURACY BETWEEN MOM AND HYBRID SPM

θ_i	10°	20°	30°	40°
MoM (*s*)	29.8632	30.0314	30.1508	30.1236
Hybrid SPM (*s*)	0.1462	0.1416	0.1491	0.1651
Accuracy(%)	2.3e-03	0.42	0.39	0.9
θ_i	50°	60°	70°	80°
MoM (*s*)	30.0075	30.1129	30.0138	30.2286
Hybrid SPM (*s*)	0.15670	0.1435	0.1389	0.1557
Accuracy(%)	0.29	1.7e-03	3.9e-02	0.52

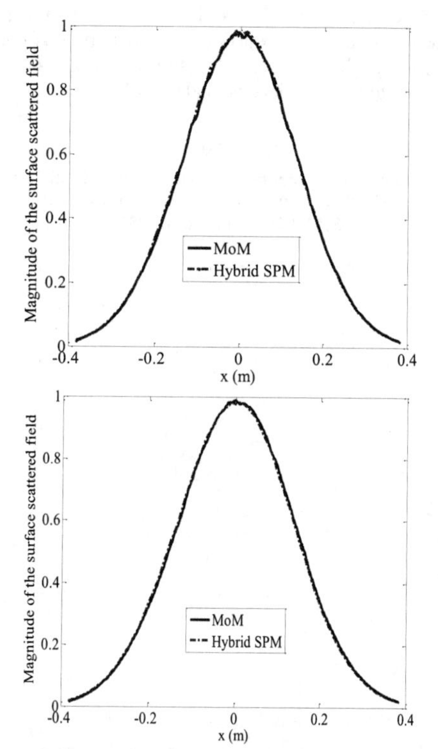

Figure 3 Scattered field with different incident angles (left figure $\theta_i = 40^\circ$, right one $\theta_i = 60^\circ$)

To check the convergence of the Monte Carlo simulation, 50 Gaussian random rough surfaces were generated, and the convergence rate is calculated using (9), and simulation results show that the computations converge after about 25 Monte Carlo realizations for both methods.

IV. CONCLUSION

In this paper, a hybrid model using SPM combined with POSP, Lagrange polynomial and Monte Carlo simulation is proposed for investigating the near field scattered by a conducting random rough surface under tapered wave incidence. The method is validated by comparing with MoM data in terms of computation time and accuracy, and the convergence rate has also been studied. It is shown that our proposed method can accelerate the computation with a factor of 200 over MoM. Therefore it is expected that the proposed method might be useful for fast and accurate computing for applications such as subsurface investigations involving a random rough surface. The future work could be extension to the two-dimensional model.

ACKNOWLEDGEMENT

This work was supported in part by the US NSF (Award number: 0821918), fundamental research funds for the central universities No. 5051302036 and No. K5051302029.

REFERENCES

[1] L. TSANG, J. A. KONG, and K. H. DING, Scattering of electromagnetic waves, theories and application: John Wiley & Sons Inc., 2000.

[2] L. TSANG, J. A. KONG, K. H. DING, and C. O. AO, Scattering of Electromagnetic Waves, Vol. II, Numerical Simulations: John Wiley & Sons Inc., 2001.

[3] E. I. THORSORS, "The validity of the Kirchhoff approximation for rough surface scattering using a Gaussian roughness spectrum," J. Acoust. Soc. Am, vol. 83, no. 1, pp. 78-92, January 1988.

[4] H. YE, Y. Q. JIN, "Parameterization of the tapered incident wave for numerical simulation of electromagnetic scattering from rough surface," IEEE Trans Antennas Propagation, vol. 53, no. 3, pp. 1234-1237, March 2005.

[5] L. KUANG, Y. Q. JIN, "Bistatic scattering from a three-dimensional object over a random rough surface using the FDTD algorithm," IEEE Trans. Antennas Propagation, vol. 55, no. 8, pp. 2302-2312, August 2007.

[6] R. L. WAGNER, J. M. SONG, W. C. CHEW, "Monte Carlo simulation of electromagnetic scattering from two dimensional random rough surfaces," IEEE Trans. Antennas Propagation, vol. 45, no. 2, pp. 235-245, February 1997.

[7] G. CUMMING, F. H. Wong, Digital Pocessing Of Synthetic Aperture Radar Data: Algorithms and Implementation: Artech House, 2005.

[8] YANG, W.Y. and etc.(why and etc?): 'Applied Numerical Methods Using Matlab', (John Wiley & Sons Inc., 2005).

Study on the Optical Properties of Nanowires Using FDTD Method

Xiang Huang[a], Liang Yu[a], Jin-Yang Chu[a], Zhi-Xiang Huang[a*], Xian-Liang Wu[a,b*]
[a]Key Laboratory of Intelligent Computing and Signal Processing, Anhui University, Hefei 230039,China
[b]School of Electronic and Information Engineering, Hefei Normal University, Hefei 230061,China
*Email: zxhuang@ahu.edu.cn, xlwu@ahu.edu.cn

***Abstract*- Finite-difference time-domain studies of a variety of cylinder arrays with nanometer-scale diameters (nanowires) interacting with light are presented. Scattering and absorption cross sections for metallic nanoscale objects can be obtained from such calculation. The method is verified by comparing the analytical results for cylindrical nanowires. Calculate the optical properties of Ag nanowires and study the explicit time-domain behavior of more cylinders.**

***Keywords*- Finite-Difference Time-Domain Method, Absorption Cross Section, Scattering Cross Section, Extinction Cross Section**

I. INTRODUCTION

The finite-difference time-domain (FDTD) method, since its introduction by Yee[1], has been widely used to obtain numerical solutions of Maxwell's equations for a broad range of problems. The applications of FDTD in electrodynamics include antenna and radar design, electronic and photonic circuit design, microwave tomography, cellular and wireless network simulation, mobile phone safety studies, and many more. The method is not limited to electrodynamics and can be used to solve other spatiotemporal partial differential equations such as those occurring in acoustics. The explicit nature of FDTD formulation, its simplicity, accuracy and robustness, together with a well established theoretical framework have contributed to a seemingly unending popularity of the method.

Nanowires, unlike other low-dimensional systems, have two quantum-confined directions but one unconfined direction available for electrical conduction. At the same time, owing to their unique density of electronic states, in the limit of small diameters, nanowires are expected to exhibit significant different optical, electrical, and magnetic properties. Here we show how finite-difference time-domain FDTD calculations can be used to study light interacting with arrays of cylinders with nanoscale diameters, and we carry out a variety of studies of various configurations. The cylinders can be viewed as metallic nanowires and exhibit optical behavior similar to metal nanoparticles (MNP's). In particular, surface plasmon polarizations (SPP's), resonance interactions of light with electronic charge density near the metal surface, [2] can play an important role.

Section II outlines our theoretical and computational methods. Section III presents cross-section results for Ag cylinders. Section IV concludes the paper.

II. THEORETICAL AND NUMERICAL CALCULATION MODEL

A. *FDTD Algorithm of Metal Nanowires*

The interaction of light with matter in the classical continuum limit is described by Maxwell's equations. Frequency-domain solutions to Maxwell's equations for light interacting with materials are constructed by allowing spatial (x,y,z) and possibly light frequency variation in the dielectric constant ε. In a region of space (x,y,z) occupied by a metal, ε can be complex valued and frequency dependent: $\varepsilon=\varepsilon_0\varepsilon_r(\omega)$,where ε_0 is the permittivity of free space. In classical electrodynamics, materials are described through a dielectric function ε that relates the electric displacement field **D** to the electric field **E** at a given frequency of light ω

$$\mathbf{D}(\omega)=\varepsilon_0\varepsilon_r(\omega)\mathbf{E}(\omega) \tag{1}$$

The dielectric function of a metal like Ag is well-described in the classical continuum limit by three separate components,

$$\varepsilon_r(\omega)=\varepsilon_\infty+\varepsilon_{\mathrm{int}\,er}(\omega)+\varepsilon_{\mathrm{int}\,ra}(\omega) \tag{2}$$

with, ε_∞, a contribution from *d*-band to *sp*-band (conduction band) inter band electron transitions, $\varepsilon_{\mathrm{int}\,er}(\omega)$, and a contribution due to *sp*-band electron excitations, $\varepsilon_{\mathrm{int}\,ra}(\omega)$.

$\varepsilon_{\mathrm{int}\,er}(\omega)$ can be physically described using a multipole Lorentz oscillator model[3].

$$\varepsilon_{\mathrm{int}\,er}(\omega)=\sum_j\frac{\Delta\varepsilon_{Lj}\omega_{Lj}^2}{\omega_{Lj}^2-\omega\left(\omega+i2\delta_{Lj}\right)} \tag{3}$$

where j is an index labeling the individual *d*-band to *sp*-band electron transitions occurring at ω_{Lj}. In order to fit the two interband transitions in Ag at designed band of frequencies[4],we choose $j=2$.

$\varepsilon_{\mathrm{int}\,ra}(\omega)$ is responsible for the plasmonic optical response of metals. It can be described by the hydrodynamic Drude model[5].

$$\varepsilon_{\mathrm{int}\,ra}(\omega)=-\frac{\omega_D^2}{\omega\left(\omega+i\gamma\right)} \tag{4}$$

where ω_D is the plasma frequency, γ is the collision frequency.

Insert Eqs. (1) and (2) into the Ampere law for a time-harmonic field, lead to

$$-i\omega\varepsilon_0\varepsilon_\infty\mathbf{E}(\omega)+\sum_j\mathbf{J}_{Lj}(\omega)+\mathbf{J}_{HD}(\omega)=i\boldsymbol{k}\times\mathbf{H}(\omega) \tag{5}$$

with

978-7-5641-4279-7

$$\mathbf{J}_{Lj}(\omega) = -i\omega\varepsilon_0 \frac{\Delta\varepsilon_{Lj}\omega_{Lj}^2}{\omega_{Lj}^2 - \omega(\omega + i2\delta_{Lj})}\mathbf{E}(\omega) \tag{6}$$

$$\mathbf{J}_{HD}(\omega) = i\omega\varepsilon_0 \frac{\omega_D^2}{\omega(\omega + i\gamma)}\mathbf{E}(\omega) \tag{7}$$

Conveniently exploiting the differentiation theorem for the Fourier transform, we perform an inverse Fourier transformation of each term of (6) and (7). Thus, we can get

$$\frac{\partial^2}{\partial t^2}\mathbf{J}_{Lj}(t) + 2\delta_{Lj}\frac{\partial}{\partial t}\mathbf{J}_{Lj}(t) + \omega_{Lj}^2\mathbf{J}_{Lj}(t) = \varepsilon_0\Delta\varepsilon_{Lj}\omega_{Lj}^2\frac{\partial}{\partial t}\mathbf{E}(t) \tag{8}$$

$$\frac{\partial^2}{\partial t^2}\mathbf{J}_{\mathrm{HD}}(t) + \gamma\frac{\partial}{\partial t}\mathbf{J}_{\mathrm{HD}}(t) = \varepsilon_0\omega_D^2\frac{\partial}{\partial t}\mathbf{E}(t) \tag{9}$$

Eqs. (8) and (9) can be solved self-consistent with Maxwell's equations. The inverse Fourier-transformed form of Eq.(5) is

$$\varepsilon_0\varepsilon_\infty\frac{\partial}{\partial t}\mathbf{E}(t) + \sum_j \mathbf{J}_{Lj}(t) + \mathbf{J}_{HD}(t) = \nabla\times\mathbf{H}(t) \tag{10}$$

Specializing to the transverse electric (*TE*) case with cylindrical symmetry, we drop the *z* coordinate in all equations and deal with just three of the six field components, E_x,E_y, and H_z.

B. *Cross-Section Formulas*

The optical properties of most of the nanostructures are determined from intensity profiles of the electromagnetic fields, or from their optical cross sections[6]. A cross section is the effective area that governs the probability of some scattering or absorption event.

We use a simple, direct procedure, which is both feasible and sufficiently accurate for our purposes, to estimate cross sections[7]. From the above description, we can get real-valued time-dependent electric-field and magnetic-field vectors $\mathbf{E}(t) = \mathbf{E}(x,y,z,t)$ and $\mathbf{H}(t) = \mathbf{H}(x,y,z,t)$, consistent with an appropriate initial pulse or a source, and scattering off one or more particles, the complex, frequency-resolved total fields are

$$\mathbf{E}(\omega) = \int_0^\infty dt\exp(i\omega t)\mathbf{E}(t) \tag{11}$$

$$\mathbf{H}(\omega) = \int_0^\infty dt\exp(i\omega t)\mathbf{H}(t) \tag{12}$$

A comparable calculation, but we must know the incident time-dependent fields $\mathbf{E}_{inc}(t)$ and $\mathbf{H}_{inc}(t)$, with the frequency-resolved incident fields

$$\mathbf{E}_{inc}(\omega) = \int_0^\infty dt\exp(i\omega t)\mathbf{E}_{inc}(t) \tag{13}$$

$$\mathbf{H}_{inc}(\omega) = \int_0^\infty dt\exp(i\omega t)\mathbf{H}_{inc}(t) \tag{14}$$

The scattered fields are then given by

$$\mathbf{E}_{sca}(\omega) = \mathbf{E}(\omega) - \mathbf{E}_{inc}(\omega) \tag{15}$$

$$\mathbf{H}_{sca}(\omega) = \mathbf{H}(\omega) - \mathbf{H}_{inc}(\omega) \tag{16}$$

Cross sections means per unit length of the cylinder axis, which therefore have units of length as opposed to length, are calculated. The corresponding scattering cross section is

$$\sigma_{sca}(\omega) = \frac{P_{sca}(\omega)}{I_{inc}(\omega)} \tag{17}$$

where $P_{sca}(\omega)$ is the absorbed power per unit area, and $I_{inc}(\omega)$ is the magnitude of the incident power. In terms of the (time-averaged) Poynting vector associated with the scattered fields, $\mathbf{S}_{sca}$, and employing cylindrical coordinates (r,φ,z),

$$P_{sca}(\omega) = r_\infty\int_0^{2\pi} d\varphi\mathbf{S}_{sca}(\omega)\cdot\mathbf{r}\,|_{r=r_\infty} \tag{18}$$

Where the path of integration is along a circle of large radius r_∞ surrounding the cylinder for any value of *z*, and

$$\mathbf{S}_{sca}(\omega) = \frac{1}{2}\mathrm{Re}\left[\mathbf{E}_{sca}(\omega)\times\mathbf{H}_{sca}^*(\omega)\right] \tag{19}$$

The absorption cross section is given by

$$\sigma_{abs}(\omega) = \frac{P_{abs}(\omega)}{I_{inc}(\omega)} \tag{20}$$

with

$$P_{abs}(\omega) = r_\infty\int_0^{2\pi} d\varphi\mathbf{S}_{abs}(\omega)\cdot\mathbf{r}\,|_{r=r_\infty} \tag{21}$$

$$\mathbf{S}_{abs}(\omega) = \frac{1}{2}\mathrm{Re}\left[\mathbf{E}(\omega)\times\mathbf{H}(\omega)^*\right] \tag{22}$$

The extinction cross section is a sum of the two,

$$\sigma_{ext}(\omega) = \sigma_{sca}(\omega) + \sigma_{abs}(\omega) \tag{23}$$

A useful check of the numerical calculations is to determine the extinction cross section directly via the formula

$$\sigma_{ext}(\omega) = \frac{P_{ext}(\omega)}{I_{inc}(\omega)} \tag{24}$$

where

$$P_{ext}(\omega) = r_\infty\int_0^{2\pi} d\varphi\mathbf{S}_{ext}(\omega)\cdot\mathbf{r}\,|_{r=r_\infty} \tag{25}$$

$$\mathbf{S}_{ext}(\omega) = \frac{1}{2}\mathrm{Re}\left[\mathbf{E}_{inc}(\omega)\times\mathbf{H}_{sca}(\omega)^* + \mathbf{E}_{sca}(\omega)\times\mathbf{H}_{inc}(\omega)^*\right] \tag{26}$$

C. *Numerical Details*

Infinite metallic nanowires can be considered as a two-dimensional model. Two-dimensional nature of the problem, allow we use large, dense grid, without too much calculation burden. To assure good convergence, we have actually used $\Delta x = \Delta y = \Delta = 0.1nm$ in all the results presented here. We generally consider grids in *x* and *y* ranging from -1000 to 1000 nm, with the silver cylinder structures centered about the origin. But for a single and two nanowires, to shorten the calculation time, we can choose 200 and 400 grid points in each direction. In order to ensure a stable simulation, time steps τ must under the Courant stability limit $\tau_C = \Delta/(3^{1/2}c)$; however for good convergence we chose to use steps $\tau = \tau_C/2$. 150×10^{-15} *s* is a good length of time to get accurate Fourier transformed fields.

In order to obtain accurate Fourier-transformed fields necessary for such calculations, incident Gaussian damped sinusoidal pulses were introduced into the computational domains using the total-field scattered-field technique[8].

We do calculations by treating the metal as silver with a Drude plus two-pole Lorentz form for its dielectric constant[9]. We find, $\varepsilon_\infty = 2.3646$, ω_D =8.7377eV, γ = 0.07489eV , $\Delta\varepsilon_{L1}$ = 0.3150, ω_{L1} = 4.3802eV, δ_{L1} = 0.28eV, $\Delta\varepsilon_{L2}$ = 0.8680, ω_{L2} = 5.183eV, δ_{L2} = 0.5482eV provide a good description of empirical dielectric constant data for silver over the λ =250nm – 1000nm range of interest.

III. NUMERICAL CALCULATION MODEL

A. *Isolated Ag Cylinder*

Before studying on the properties of the cylinder surface-plasmon resonance, we should know the accuracy of our FDTD calculations. So we first calculate *TE* scattering off a single Ag cylinder. Then compare with the analytical solution for this case.

Fig.1 displays as curves the analytical cross sections for TE scattering off an Ag cylinder with radius *a*=25 nm. The symbols in the figure, are now the FDTD cross sections estimated from a single propagation as discussed in Sec. II, using the same Drude and plus two-pole Lorentz dielectric constant model. The analytical and FDTD cross sections agree to within 10% or better. Because the FDTD method has been previously used for the calculation of optical scattering cross sections, we believe this is demonstration that the FDTD method can reproduce metal nanowires scattering cross sections with reasonable accuracy.

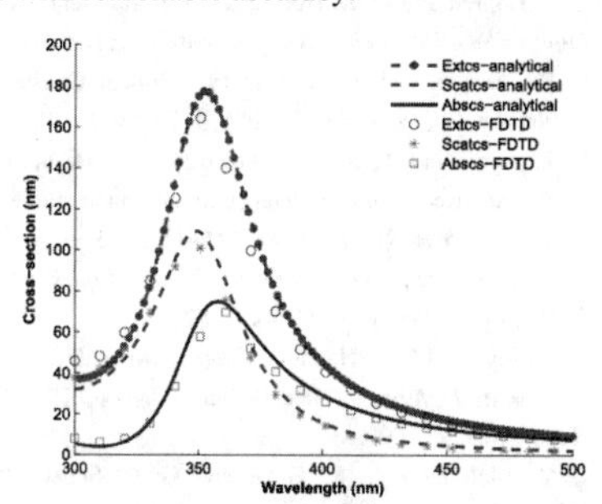

Figure 1. Comparison of analytical (smooth curves) and FDTD-based (symbols) cross sections for a single Ag cylinder of radius a =25 nm. Extcs is short for extinction cross section. Scatcs is short for scattering cross section. Abscs is short for absorption cross section.

The large peaks in Fig.1 are due to the surface Plasmon polarizations(SPP)resonance. FDTD algorithm can adequately reproduce the trends in the cross sections. So we can study on optical properties of nanowires using FDTD.

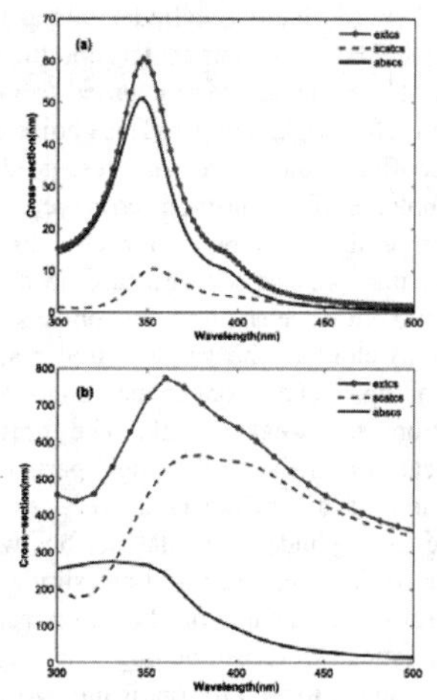

Figure 2. Comparison cross sections for a single Ag cylinder, (a) radius a = 15 nm and (b) radius a = 100 nm.

We study on a single Ag cylinder with different radius using FDTD method. We can see from Fig.2 that, as the cylinder radius *a* increases, the resonance peak redshifts and broadens, and scattering becomes more dominant than absorption. As radius a = 15 nm, the resonance occurs at $\lambda \approx$ 350 nm and near this wavelength $\sigma_{sca} / \sigma_{abs} \approx 0.2$, absorption is more intense than scattering. When *a*=100nm the resonance has redshifted to $\lambda \approx$ 360 nm with scattering, becoming more pronounced than absorption, $\sigma_{sca} / \sigma_{abs} \approx 2.3$. σ_{sca} increases from 11 to 554 nm between a = 15 and 100 nm, while σ_{abs} increases from 50 to 274 nm over the same *a* range.

B. *Linear Arrays of Ag Cylinders*

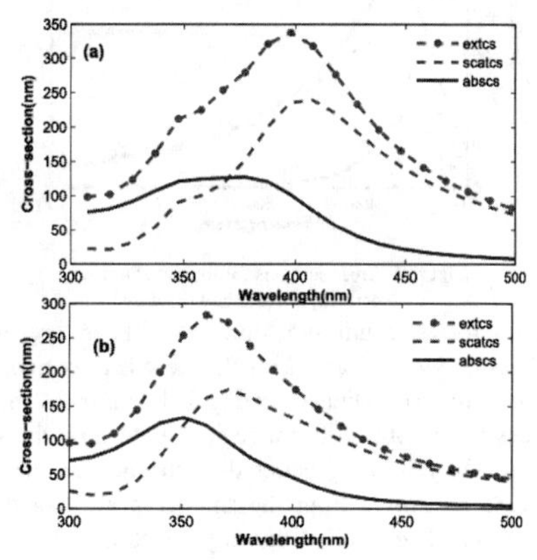

Figure 3. Cross sections for two Ag cylinders (*a*=25nm). (a) *d*/*a*=2.2 and (b) *d*/*a*=3

Fig.3(a) presents cross sections for two cylinders. We choose 400 grid points in each direction and centered at the middle of the grid, each with radius a = 25 nm. There is just a

5-nm space between the two cylinders along y, and the ratio of the distance d between their centers and the cylinder radius is $d/a = 2.2$. The cylinders were exposed to y-polarized light moving from left to right along x. This choice of polarization, given the configuration of the particles, is ideally suited to exciting coupled surface-plasmon resonances consistent with induced (and oscillating) dipoles in each cylinder along the y axis. In particular, we find two structures in the cross sections in the 300–500 nm wavelength region, one a shoulder or weak ($\lambda = 350$ nm) close to the single-cylinder surface-plasmon resonance of Fig.1, and stronger maximum redshifted ($\lambda = 400$ nm) from the weaker peak. The presence of extra resonance features relative to the single-particle case is due to the interaction of the cylinders at very short separations. Comparable two cylinders calculations but with $d/a = 3$ in Fig.3(b), shows just one peak and are similar to the single-cylinder results. The nature of the resonance structures in Fig.3 is interesting. Whereas one might naively think the 350 nm structure, owing to its position, is the two-cylinder analog of the single-cylinder dipolar resonance, plots(not shown) of the charge density indicate that it is of mixed dipolar and quadrupolar character, whereas the larger peak to the red of the shoulder is a more of a pure dipolar excitation.

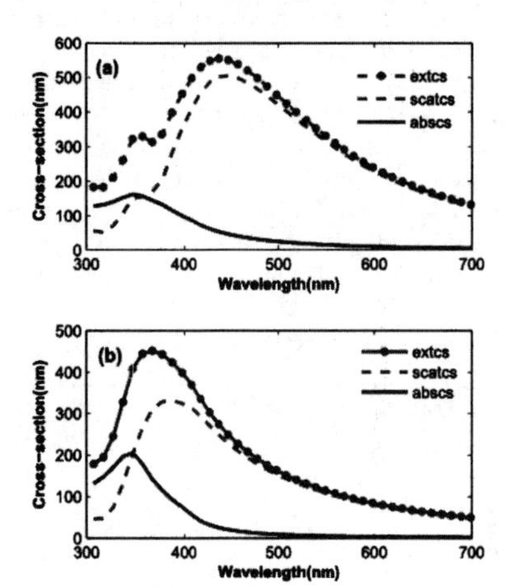

Figure 4. Cross sections for four a=25nm Ag cylinders. (a) d/a=2.2 and (b) d/a=3

Results for a=25 nm and four Ag cylinders are given in Fig.4 for both $d/a = 2.2$ and $d/a = 3$. There is a continuation of the pattern for two cylinders in Fig.3. For d/a=2.2 there are two peaks in the 300–500 nm region, with now the smaller peak near 350 nm being better defined and the larger peak being more redshifted from the smaller peak than the two-cylinder result. The width of the larger peak is also larger than the two-cylinder result. As with the two-cylinder result, the $d/a = 3$ cross sections are simpler and more like the single results, although both the magnitude and widths of the corresponding structures are larger than the single results.

IV. Conclusion

We presented an FDTD approach to studying the interaction of light with nanoscale radius metallic cylinders or nanowires. We obtained reasonably accurate cross sections for single- and multiple-cylinder arrays, confirming the reliability of our approach. There are many directions for future work. It is likely that the intensity of light developing and propagating between the nanowires can be significantly enhanced by varying the distance between the nanowires, cylinder radius, and of course the wavelength and propagation direction of the incident light.

Acknowledgment

This work was supported by the National Natural Science Foundation of China under Grant (Nos. 60931002, 61101064, 51277001, 61201122), DFMEC (No. 20123401110009) and NCET (NCET-12-0596) of China, Distinguished Natural Science Foundation (No.1108085J01), and Universities Natural Science Foundation of Anhui Province (No. KJ2011A002), Graduate Academic Innovate Research Foundation (No.60931002, 61101064), and the 211 Project of Anhui University.

References

[1] K. S. Yee, "Numerical solution of initial boundary value problems involving Maxwell's equations in isotropic media", *IEEE Transactions on Antennas and Propagation*, vol.14, 302–307, 1966.

[2] U. Kreibig and M. Vollmer, *Optical Properties of Metal Clusters*, Springer:Berlin, 1995.

[3] C. F. Bohren and D. R. Huffman, *Absorption and Scattering of Light by Small Particles*, Wiley & Sons, Inc.: New York, 1983.

[4] P. B. Johnson and R. W. Christy, "Optical Constants of the Noble Metals", *Phys. Rev. B*, vol.6, 4370, 1972.

[5] S. K. Gray and T. Kupka, "Propagation of light in metallic nanowire arrays: Finite-difference time-domain studies of silver cylinders", *Phys. Rev. B*, vol.68, 045415, 2003.

[6] A. D. Boardman, *Electromagnetic Surface Modes*, edited by A. D. Boardman ,Wiley: NewYork, 1982.

[7] A. Taflove and S. C. Hagness, *Computational Electrodynamics: The Finite-Difference Time-Domain Method*, 2nd ed., Artech House, Boston, 2000.

[8] J. M. McMahon, S. K. Gray, and G. C. Schatz, "Nonlocal Optical Response of Metal Nanostructures with Arbitrary Shape", *Phys. Rev. Lett*, vol.103, 097403,2009.

[9] T. W. Lee and S. Gray, "Subwavelength light bending by metal slit structures," *Opt. Express*, vol.13, 9652-9659,2005.

TA-1(C)

October 24 (THU) AM

Room C

WLAN Antennas

A Compact Microstrip-Line-Fed Printed Parabolic Slot Antenna for WLAN Applications

Wanwisa Thaiwirot and Norakamon Wongsin
Faculty of Engineering, King Mongkut's University of Technology North Bangkok
1518 Pracharat 1 Road, Wongsawang, Bangsue, Bangkok, 10800 Thailand
E-mail: wanwisat@kmutnb.ac.th

***Abstract*-This paper presents a compact microstrip-line-fed printed parabolic slot antenna for wireless local area network (WLAN) applications in IEEE 802.11/b/g/a. The proposed antenna consists of a microstrip-fed-line and parabolic slot with a rectangular slot. By introducing rectangular slot with parabolic slot and a loading strip protruded into the rectangular slot, the proposed antenna can achieve two operating bandwidths. The measured impedance bandwidth of proposed antenna covers the operating bands of 2.4-2.84 GHz and 4.13-5.83 GHz, which covers the required bandwidth of WLAN applications. In measured results, good radiation characteristics are obtained.**

I. INTRODUCTION

Recently, wireless communications have developed rapidly, which leads to great demand for developments in novel wireless products. Antennas are important and indispensable elements of any wireless communication systems. In recent years, printed slot antennas are getting more popular because they usually have a wide impedance bandwidth and can be applied to various wireless communication systems such as WiMAX (Worldwide Interoperability for Microwave Access) and WLAN (Wireless Local Area Network). Furthermore, they also have the advantages of low profile, lightweight, east of fabrication and integration with other devices or RF circuitries. A conventional narrow slot antenna has limited bandwidth, whereas wide slot antennas exhibit wider bandwidth. Many previous studies of the different printed wide slot antennas fed by a microstrip line and coplanar waveguide have been reported [1-2]. By using different tuning techniques or employing different slot shapes such as rectangular, circular, arc-shape, annular-ring, U-shape etc., different slot antennas achieved wideband or ultra-wideband performance [3-4]. In [5], a polygonal slot antenna fed by microstrip line has been proposed. By employing a rectangular slot with a triangular slot on the bottom, the antenna can obtain 104% impedance bandwidth (1.85-5.83 GHz). However, the antenna does not possess a compact profile having a dimension of $100\times80\,\text{mm}^2$. A printed wide-slot antenna fed by a microstrip line has been introduced [6]. This antenna consists of an arc-shape slot and a square-patch feed and can achieve 120% impedance bandwidth (1.82-7.23 GHz). The bandwidth is good enough but the size is very large ($110\times110\,\text{mm}^2$). However, these printed slot antennas mentioned above have larger dimension and may not be suitable for applications that require miniaturized antennas. Moreover, a WLAN antenna should be capable of operating at 2.400-2.484 GHz (specified by IEEE 802.11b/g) and 5.150-5.350/5.725-5.825 GHz (specified by IEEE 802.11a).

In this paper, a microstrip-line-fed printed parabolic slot antenna that achieves a compact size is proposed. By using parabolic shaped slot with rectangular slot and T-shaped feed, the antenna can obtain dual band operation suitable for WLAN applications. Details of the antenna design and both simulated and measured results are presented and discussed.

II. ANTENNA DESIGN

The geometry of the proposed antenna is shown in Fig.1. The antenna is printed on the inexpensive FR4 substrate of thickness h = 1.6 mm, with relative permittivity $\varepsilon_r = 4.4$. The slot and the feeding line are printed on different sides of FR4 substrate. The microstrip fed line has a width W_f = 3.2 mm and a length L_f = 14 mm with a tuning stub width W = 2 mm and L = 8 mm. The slot in the ground plane consists of two sections, the rectangular section with dimensions of $6\times30\,\text{mm}^2$ ($W_s\times L_s$) and the parabolic section with F = 6 mm. Parameter F is used to define the parabolic profile of the slot antenna. By combining rectangular slot and parabolic slot with optimal length F, the proposed antenna can provide the two operating bands. For achieving efficient excitation and good impedance matching, the loading strip is protruded into the rectangular slot, the width and length of the protruded loading strip are denoted as W_2 and L_2, respectively as shown in Fig.1(b). The overall size of the proposed antenna is $30\times35\,\text{mm}^2$ ($W_g\times L_g$).

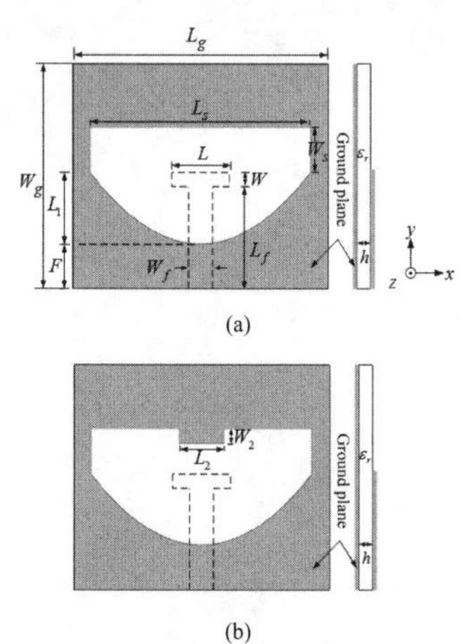

Figure 1. Geometry of the proposed antenna: (a) without a loading strip, (b) with a loading strip.

978-7-5641-4279-7

The equation of parabolic shape slot antenna is expressed in (1) with $-L_s/2 \le x \le L_s/2$.

$$y = \frac{x^2}{4F} + F \tag{1}$$

where F is parameter to define the width of parabolic slot (L_1) and position of the slot.

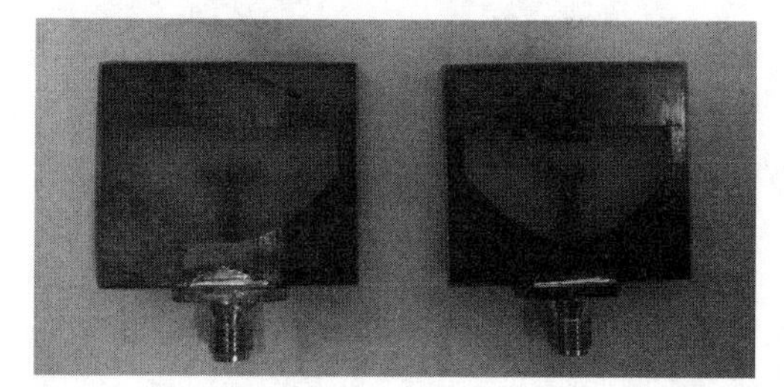

Figure 2. Photograph of fabricated antennas with and without loading strip.

III. EXPERIMENTAL RESULTS AND DISCUSSION

Fig.2 shows the prototypes of parabolic shape slot antennas with and without loading strip. The return loss of proposed antenna was simulated by CST microwave studio and measured by vector network analyzer. In this section, some sensitive parameters are studied numerically in order to demonstrate parameters that affect the performance of printed parabolic slot antenna. In the simulation only one parameter is varied each time, whereas the others are kept constant. The case of the proposed antenna without loading strip, the microstrip-line length L_f is an important parameter. By tuning the length of microstrip-line, the lower resonant frequency is shifted and impedance matching of the upper frequency is affected as shown in Fig.3. It is found that the length of L_f = 14 mm can obtain a good impedance bandwidth and cover two operating bands. The length L_s and width W_s of rectangular slot have been also investigated. Fig. 4 shows the return loss of varying the rectangular slot length as L_s = 26, 28, 30 mm with a fixed W_s = 6 mm. It can be observed that with increasing length of L_s, which the surface current path can be extended, the resonant frequency shifts down and impedance bandwidth of the upper band is improved. Fig. 5 shows the return loss for different widths of W_s, where the length of rectangular slot L_s = 30 mm is chosen. It is seen that with decreasing W_s, the impedance matching of the lower band is improved, while the upper resonant frequency shifts down. The parameter F, which is parameter to define the parabolic shape profile, also influences impedance matching of the proposed antenna. Fig. 6 shows the return loss of different lengths of F. When the parameter F = 5 mm, the 10-dB return loss from 2.38 to 5.41 GHz covers 2.4/5.2 GHz, however, it cannot covers at 5.8 GHz band for WLAN operation. When the parameter F = 9 mm, the resonance mode shifts to higher frequency bands and impedance bandwidth is not enough for WLAN application. It is seen that the length of F = 6 mm obtains dual band operation which covers 2.40-2.87 GHz for lower band and covers 3.45-5.87 GHz for upper band.

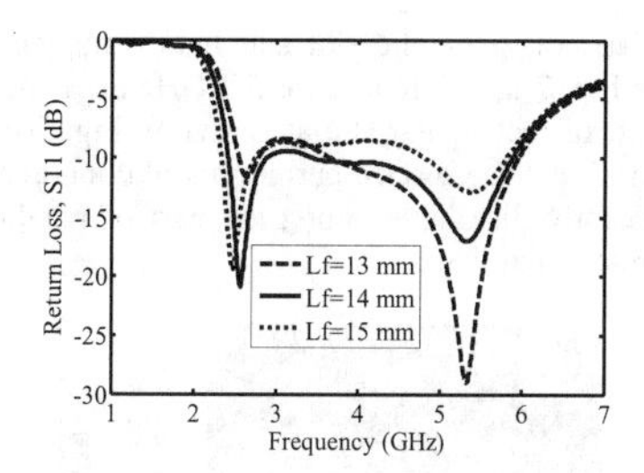

Figure 3. Simulated return loss for various lengths L_f of proposed antenna without loading strip.

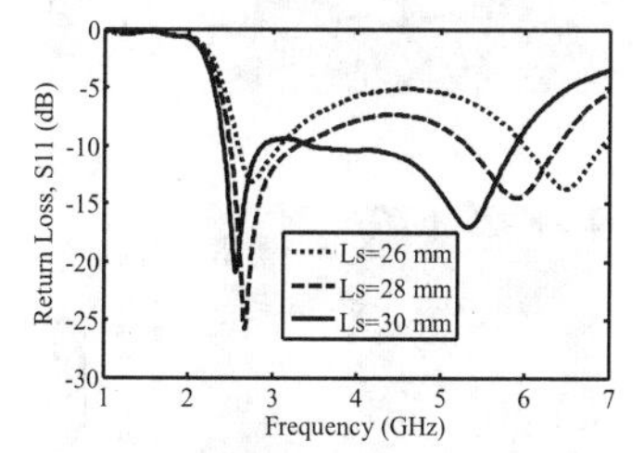

Figure 4. Simulated return loss for various lengths L_s of proposed antenna without loading strip.

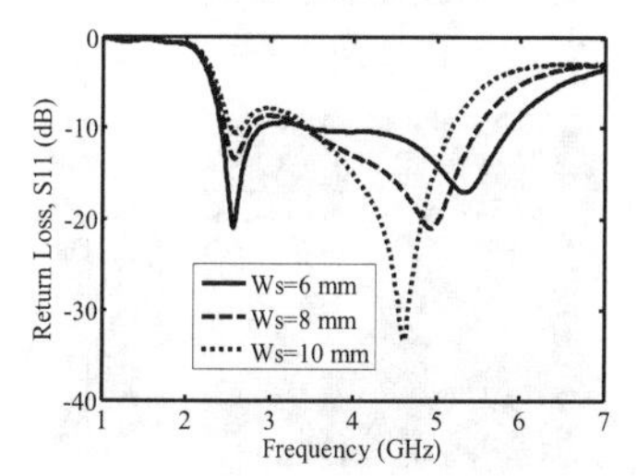

Figure 5. Simulated return loss for various lengths W_s of proposed antenna without loading strip.

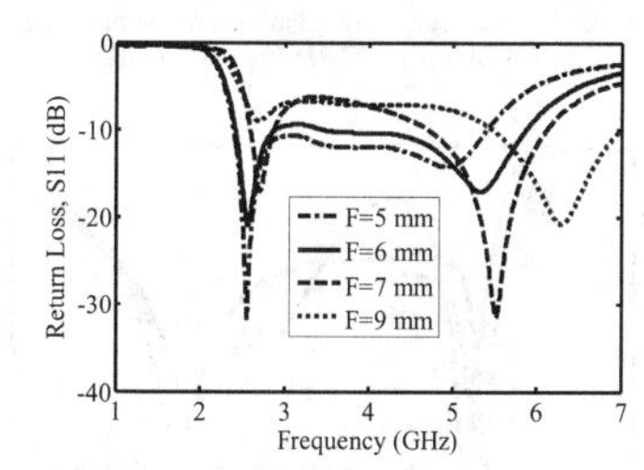

Figure 6. Simulated return loss for various lengths F of proposed antenna without loading strip.

For achieving efficient excitation and good impedance matching, the loading strip is protruded into the rectangular slot. It is found that the length of W_2 has a significant effect on the impedance bandwidth, while L_2 has effect on impedance matching. The optimal length W_2 and L_2 of the loading strip for dual operating bands are found to be 2 mm and 6 mm, respectively. Fig.7 shows the surface current distributions of the proposed antenna at frequencies of 2.4 GHz, 5.2 GHz and 5.8 GHz. The surface current distributions for 2.4 GHz are concentrated on the part of rectangular slot, parabolic slot and microstrip fed line as shown in Fig. 7(a), while that for 5.2 GHz are concentrated on the part of rectangular slot as shown in Fig. 7(b). It can be observed that the surface current distribution at 5.8 GHz is significantly distributed along the part of parabolic slot and microstrip fed line.

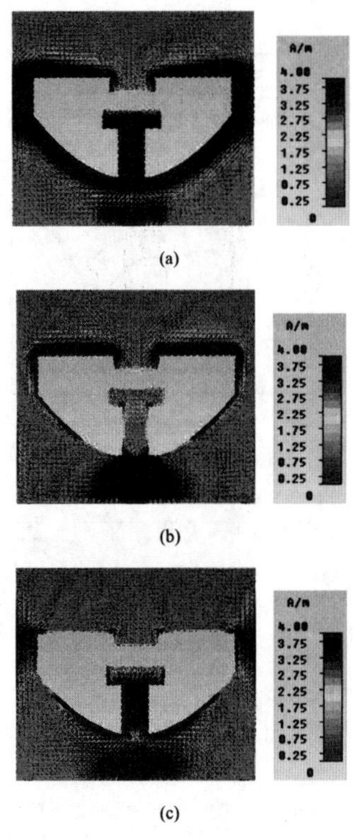

Figure 7. Simulated surface current distributions of the proposed antenna at (a) 2.4 GHz (b) 5.2 GHz and (c) 5.8 GHz.

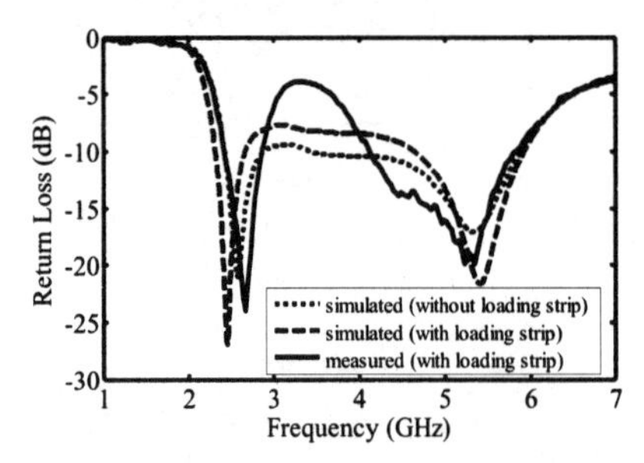

Figure 8. Simulated and measured results of return losses of proposed antenna: W_g =30 mm, L_g = 35 mm, W_s = 6 mm, L_s = 30 mm, F = 6 mm, L_1 = 15.4 mm, W_f = 3.2 mm, L_f = 14 mm, W = 2 mm, L = 8 mm, W_2 = 2 mm, L_2 = 6 mm.

Fig.8 shows the measured and simulated results of return losses of the proposed antenna. It can be seen that the measured return loss of the proposed antenna shifts to the right for the lower operating band and shifts to the left for the upper operating band. The disagreement between simulation and measurement is mainly due to the fabrication tolerance. However, the measured result of proposed antenna still covers the two operating bands of 2.40-2.84 GHz and 4.13-5.83 GHz for 2.4/5.2/5.8 GHz WLAN application.

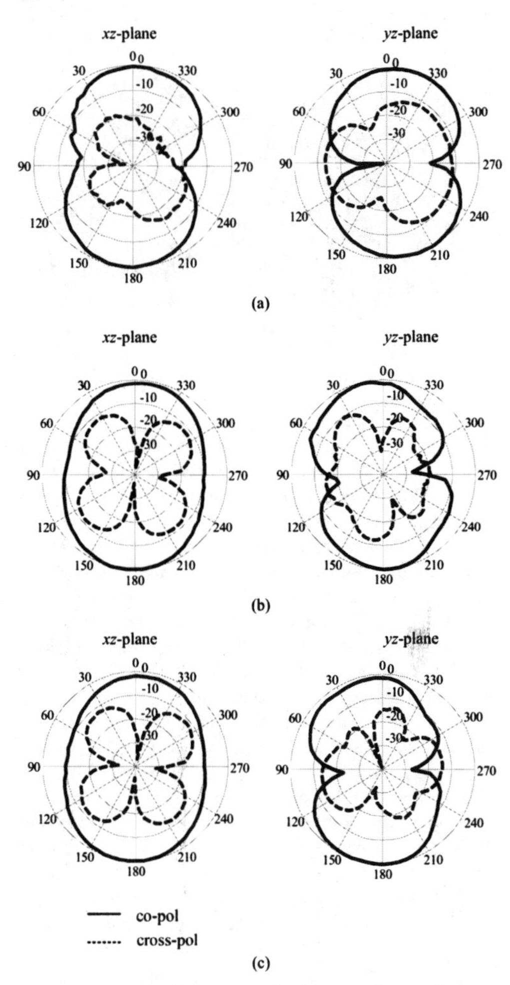

Figure 9. Measured radiation patterns for the proposed antenna in the *xz*-plane and *yz*-plane at (a) 2.4 GHz (b) 5.2 GHz and (c) 5.8 GHz.

The radiation characteristics of the proposed antenna with loading strip at operating frequencies within the impedance bandwidth are also investigated. Fig. 9 shows the measured radiation patterns in the *xz*-plane and *yz*-plane with both co- and cross polarizations at 2.4 GHz, 5.2 GHz and 5.8 GHz. The measured results show that the radiation patterns of the antenna are still similarly to bidirectional radiation pattern at two operating bands. The measured antenna gains are about 2.48 dBi, 3.15 dBi and 3.23 dBi at 2.4 GHz, 5.2 GHz and 5.8 GHz, respectively.

IV. Conclusion

A compact microstrip-line-fed parabolic slot antenna for dual band operations has been presented and implemented. By introducing rectangular slot with parabolic slot and a loading strip protruded into the rectangular slot, the proposed antenna can obtain two separate impedance bandwidths for 2.4 and 5.2/5.8 GHz bands. In the experimental results, bidirectional radiation pattern and sufficient antenna gain of operating frequencies across the two bands can also obtained. In addition, the proposed antenna also has simple structure, compact size and good radiation performances suitable for WLAN applications.

Acknowledgment

I would like to acknowledge the financial support from faculty of engineering, king mongkut's university of technology north bangkok.

References

[1] W.S. Chen and F. M. Hsieh, "A broadband design for a printed isosceles triangular slot antenna for wireless communications," Microwave Journal, vol.48, pp.98-112, 2005.

[2] S.W., Qu, C. Ruan and B. Z. Wang, "Bandwidth enhancement of wide-slot antenna fed by CPW and microstrip line," IEEE Antenna and Wireless Propagation Letters, vol. 5. pp. 15-17, 2006.

[3] M.A. Saed, "Broadband CPW-fed planar slot antennas with various tuning stubs," Progress in electromagnetics research, vol. 66, pp. 199-212, 2006.

[4] M. Gopikrishna, D.D. Krishna and C.K. Anandan, P. Mohanan and K.P. Vasudevan, "Design of a compact semi-elliptic monopole slot antenna for UWB systems,". IEEE Transactions on Antennas and Propagation, vol. 57. pp.1834-1837, 2009.

[5] W.S. Chen and B.S. Kao, "A Triple-band polygonal slot antenna for WiMAX applications," Microwave Journal, Vol. 50, pp. 134-143, 2007.

[6] Y.F. Liu, K.L. Lan, Q. Xue and C. H. Chan, "Experimental studies of printed wide-slot antenna for wide-band applications," IEEE Antenna and Wireless Propagation Letters, vol. 3. pp. 273-275, 2004.

Dual-band Circularly Polarized Monopole Antenna for WLAN Applications

Hao-Shiang Huang and Jui-Han Lu
Department of Electronic Communication Engineering
Kaohsiung Marine University
Kaohsiung, Taiwan 811

***Abstract*- A novel planar dual-band monopole antenna with circular polarization (CP) operation is proposed. By appropriately introducing dual strip-sleeves shorted at the ground plane, the proposed dual-band CP design can easily be achieved with the impedance bandwidth (RL □10 dB) of about 266 / 980 MHz and the 3 dB axial-ratio (AR) bandwidth of about 103 / 700 MHz for 2.4 / 5.2 GHz wireless local area network (WLAN) applications. The measured peak gain and radiation efficiency are about 4.1 / 3.3 dBic and 94 / 84 % across the operating bands, respectively, with nearly bidirectional patterns in the XZ- and YZ-planes.**

I. INTRODUCTION

Owing to tremendous growth in wireless communication technology, especially for the IEEE 802.11a/b/g WLAN standards in the 2.4 GHz (2400–2484 MHz), 5.2 GHz (5150–5350 MHz) and 5.8 GHz (5725–5825 MHz) bands, the dual-band printed monopole antenna (MA) has attracted high attention because it has the merit of low profile and can provide the feature of multi-band operation. Meanwhile, circularly polarized (CP) antennas can reduce the loss caused by the multi-path effects between the reader and the tag antenna. Several CP monopole antennas have been investigated, such as the annular ring monopole antenna [1, 2], a circular monopole antenna [3], the rectangular patch monopole antenna with a L-shaped slit inset into the ground plane [4, 5], an asymmetrical dipole antenna [6] and an asymmetric-fed rectangular patch monopole antenna with the stub shorted at the ground plane [7]. However, the above mentioned CP antennas are focused on the single-band operation and they have disadvantage of bulky volume [1] or complex structure [2, 3]. Moreover, there is an increasing demand for antennas having more compact size to be suitably embedded in the practical portable devices for multi-input / multi-output (MIMO) system. Therefore, in this article, we propose a novel planar dual-band monopole antenna with circular polarization operation for WLAN communication. By introducing dual asymmetrical strip-sleeves shorted at the ground plane [8] to replace the designs by using the inset slit [2, 4-5], the proposed dual-band CP design can provide the impedance bandwidth (RL □10 dB) of about 266 / 980 MHz and the 3 dB axial-ratio (AR) bandwidth of about 103 / 700 MHz for 2.4 / 5.2 GHz wireless local area network (WLAN) applications, respectively. Good symmetry of bidirectional CP radiation has been observed. Details of the proposed antenna design and experiment results are presented and discussed.

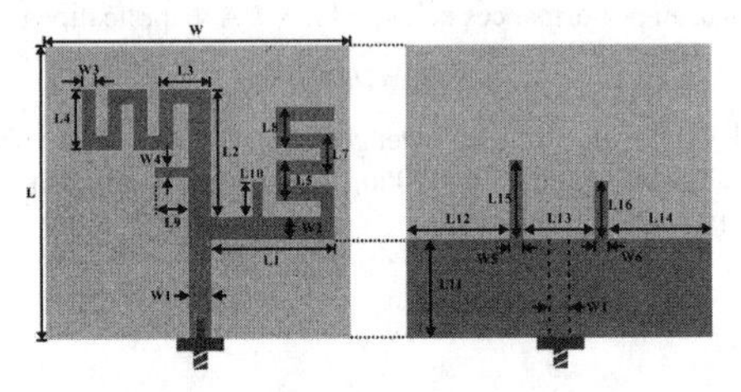

Figure 1. Geometry of the proposed dual-band CP monopole antenna with dual asymmetrical strip-sleeves for WLAN application.

II. ANTENNA DESIGN

Figure 1 illustrates the geometrical configuration of the proposed planar dual-band CP antenna for WLAN application. A 50Ω microstrip line is etched as the feeding structure on the inexpensive FR-4 substrate with the overall volume of $40 \times 40 \times 0.8$ mm^3, dielectric constant $\varepsilon_r = 4.4$ and loss tangent tan δ = 0.0245. The proposed antenna consists of a pair of orthogonal F-shaped meander monopoles with dual asymetrical strip-sleeves shorted at the ground plane. For compact operation, the longer arm of the F-shaped monopole strip is meandered to obtain less dimension, which is used to generate the fundamental resonant mode at approximately 2450 MHz. The longer strip-sleeve with the dimension of L15 x W5 is introduced to disturb the surface current on the ground plane, which is different from the CP design using the L-shaped slit inset into the ground plane [2, 4-5]. Meanwhile, the right-hand circular polarization (RHCP) can be obtained by exciting the two orthogonal linearly polarized modes with a 90° phase offset, which is due to this proposed strip-sleeve perpendicular to the x-directional F-shaped monopole strip. Similarly, The shorter arm of the F-shaped strip is used to generate the fundamental resonant mode at approximately 5200 MHz. The shorter strip-sleeve with the dimension of L16 x W6 is used to excite the left-hand circular polarized (LHCP) wave in the $+z$ direction. Moreover, the antenna is optimized based on the above guidelines and by using Ansoft HFSS, a commercially available software package based on the finite element method [9]. Return loss is measured with an Agilent N5230A vector network analyzer.

978-7-5641-4279-7

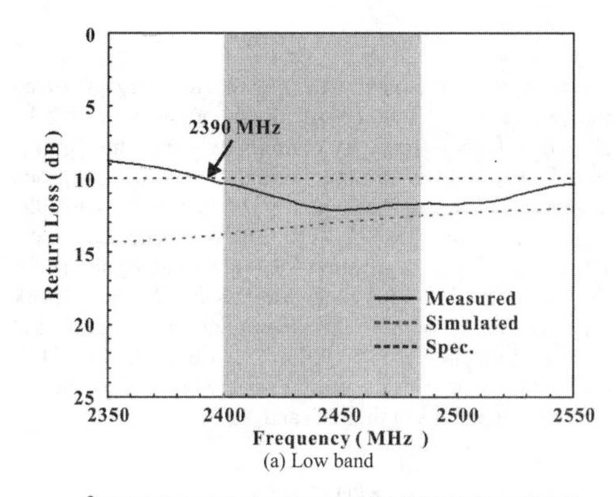

(a) Low band

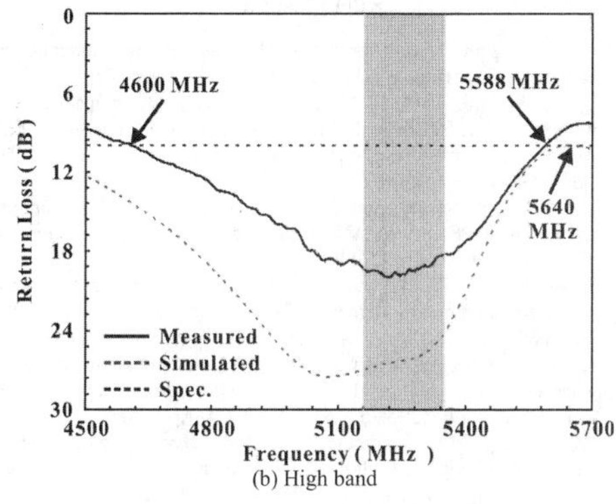

(b) High band

Figure 2. Simulated and measured return loss against frequency for the proposed compact dual-band CP monopole antenna.

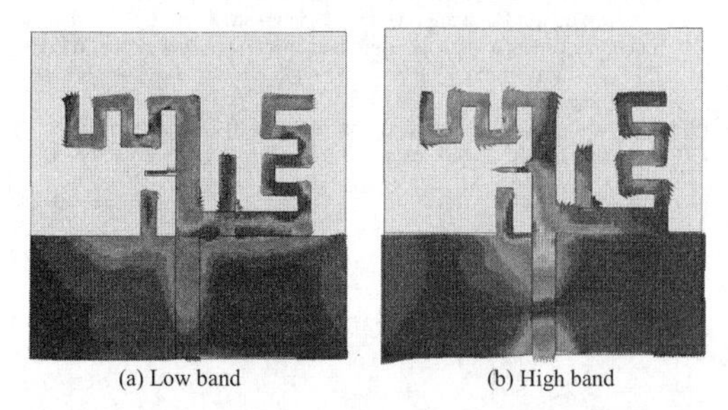
(a) Low band (b) High band

Figure 3. Simulated and measured axial ratio (AR) against frequency for the proposed compact dual-band CP monopole antenna.

III. Results And Discussion

Figure 2 summarizes the simulation and experimental results for return loss in the proposed dual-band CP monopole antenna. The lower band reveals a measured 2:1 VSWR (10-dB return loss) bandwidth of 266 MHz (2400–2660 MHz), whereas the upper band has a bandwidth of 980 MHz (4615–5595 MHz). Dual bands can comply with the bandwidth requirements of the desired dual-band WLAN (2.4/5.2 GHz) application. Figure 3 shows the simulated surface current distributions on the proposed CP antenna at dual operating bands. Figure 4 shows the related simulated and experimental results of the axial ratio (in the boresight direction) for the proposed dual-band CP antenna of Figure 1. From the related results, the measured operating bandwidth (3-dB AR) can reach about 103 MHz (2382–2485 MHz) and 700 MHz (4810–5510 MHz) at 2.4 / 5.2 GHz bands, respectively, and agrees well with the HFSS simulated results. Figure 5 presents the measured antenna gain and efficiency (mismatching loss included, [10]) for the proposed compact printed antenna. This figure also shows the simulation results for comparison. For frequencies over the 2.4 GHz bands, the measured antenna gain is approximately 3.5 ~ 4.1 dBi. Meanwhile, that for the 5.2 GHz band ranges from approximately 2.5 to 3.3 dBi. The measured antenna efficiency is around 91 ~ 94 % over the 2.4 GHz bands, while that over the 5.2 GHz bands is approximately 60 ~ 84 %. The CP radiation patterns measured at 2400/5200 MHz are plotted in Figure 6, and good symmetry of bidirectional radiation has been observed. Results show the coherent agreement between the measured and simulated results. Since a CP monopole antenna radiates a bidirectional wave, the radiation patterns on both sides of the proposed CP monopole antenna are almost the same, in which a contrary circular polarization is produced; the front-side radiates LHCP while the back-side radiates RHCP. By verification, this antenna structure has successfully achieved a cross polarization discrimination of 20 dB on a wide azimuth range.

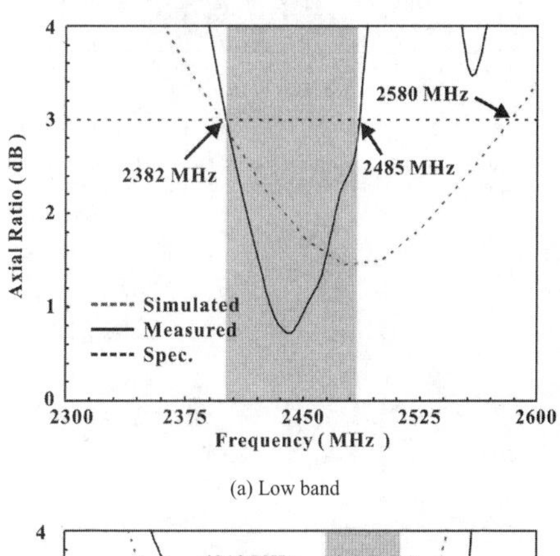

(a) Low band

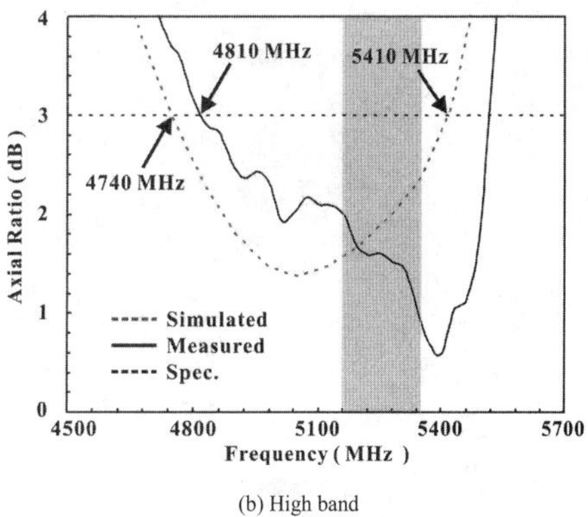

(b) High band

Figure 4. Simulated and measured axial ratio (AR) against frequency for the proposed compact dual-band CP monopole antenna.

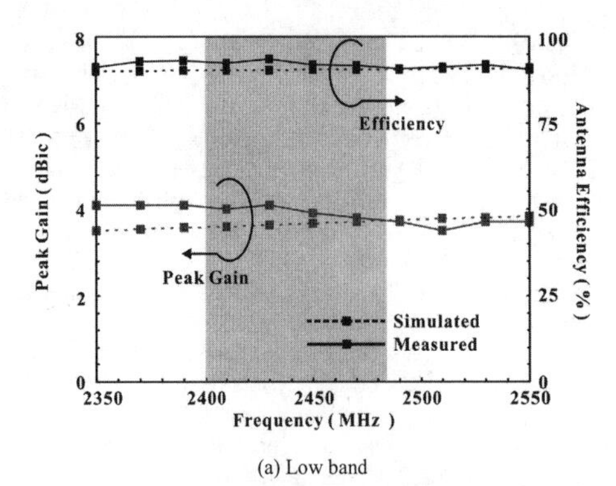

(a) Low band

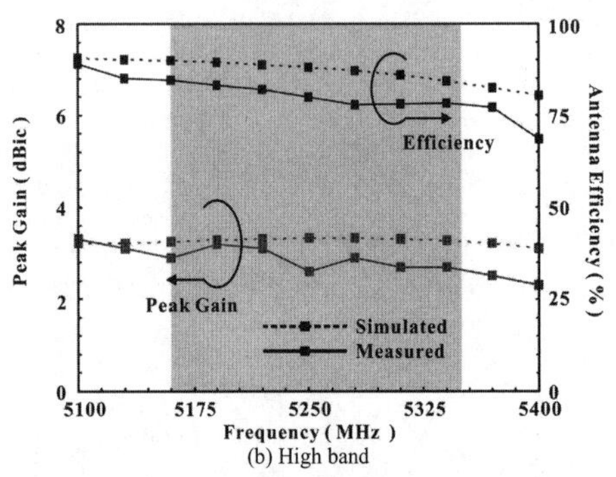

(b) High band

Figure 5. Measured and simulated antenna gain and efficiency for the proposed compact printed antenna studied in Figure 2.

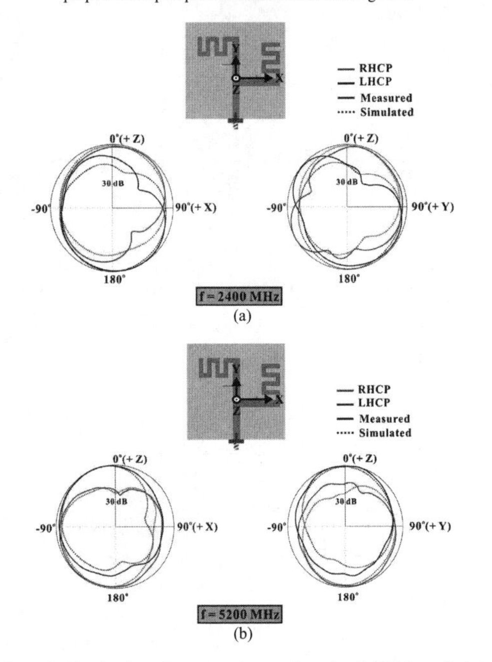

Figure 6. Simulated and measured two-dimensional (2-D) radiation patterns for the proposed compact dual-band CP monopole antenna.

IV. CONCLUSIONS

A novel planar compact dual-band circularly polarized monopole antenna is proposed for the application of 2.4 / 5.2 GHz WLAN system. By appropriately introducing dual shorted strip-sleeves on the ground plane, the proposed dual-band CP design can easily be achieved with the impedance bandwidth (RL □10 dB) of about 266 / 980 MHz and the 3 dB axial-ratio (AR) bandwidth of about 103 / 700 MHz for 2.4 / 5.2 GHz wireless local area network (WLAN) applications. The measured peak gain and radiation efficiency are about 4.1 / 3.3 dBic and 94 / 84 % across the operating band, respectively, with nearly bidirectional pattern in the XZ- and YZ-plane.

REFERENCES

[1] S. P. Phang and N.W. Chen, ''Annular ring monopole antenna with switchable polarization'', *2012 IEEE AP-S Int. Symp. Dig.*.

[2] G. Li, H. Zhai, T. Li, L. Li and C. H Liang, ''A compact antenna with broad bandwidth and quad-sense circular polarization'', *IEEE Antennas and Wireless Propagation Letters*, vol. 11, pp. 791-794, July 2012.

[3] S.A. Rezaeieh, ''Dual band dual sense circularly polarised monopole antenna for GPS and WLAN applications'', *Electronics Letters*, vol. 47, pp. 1212-1214, Oct. 2011.

[4] J. W. Wu, C. F. Jou and C. J. Wang, ''A dual-band circularly polarized monopole antenna for WLAN application'', *2009 International Workshop on Antenna Technology*, pp. 346-349.

[5] C. F. Jou, J. W. Wu and C. J. Wang, ''Novel broadband monopole antennas with dual-band circular polarization'', *IEEE Trans. Antennas Propagat.*, vol. 57, pp. 1027-1034, April 2009.

[6] X. L. Bao, M. J. Ammann and P. McEvoy, ''Microstrip-fed wideband circularly polarized printed antenna'', *IEEE Trans. Antennas Propagat.*, vol. 58, pp. 3150-3156, Oct. 2010.

[7] J. W. Wu, J. Y. Ke, C. F. Jou and C. J. Wang, ''Microstrip-fed broadband circularly polarised monopole antenna'', *IET Microwaves, Antennas & Propagation,* vol. 4, pp. 518-525, April 2010.

[8] J. W. Wu, Y. D. Wang, H. M. Hsiao and J. H. Lu, "T-shaped monopole antenna with shorted L-shaped strip-sleeves for WLAN 2.4/5.8 GHz operation," *Microwave Opt. Technol. Lett.*, vol. 46, pp. 65-69, July 2005.

[9] http://www.ansoft.com/products/hf/hfss, Ansoft Corporation HFSS.

[10] Y. Huang and K. Boyle, *Antennas from Theory to Practice*, pp. 124, John Wiley & Sons Ltd, 2008.

Tapered Slot Antenna with Squared Cosine Profile for WLAN Applications

Yosita Chareonsiri[1], Wanwisa Thaiwirot[2] and Prayoot Akkaraekthalin[3]
Faculty of Engineering, King Mongkut's University of Technology North Bangkok
1518 Pibulsongkram Road, Bangsue, Bangkok, 10800 Thailand
E-mail: [1]pockly_white@hotmail.com, [2]wanwisat@kmutnb and [3]prayoot@kmutnb.ac.th

***Abstract*-This paper presents a tapered slot antenna with squared profile for wireless local area network (WLAN) applications. The proposed antenna is composed of a feed line, a microstrip-slotline transition and a radiating slot by using squared profile. The measured results for the optimized design show the bandwidths for the $S_{11} \leq -10$ dB are about 830 MHz (from 1.97 to 2.80 GHz) and 2850 MHz (from 3.97 to 6.82 GHz), which can cover the 2.4/5.2/5.8 GHz WLAN operating bands and 2.5/5.5 GHz for worldwide interoperability for microwave access (WiMAX) bands. The performances of the proposed antenna are demonstrated along with the measured and simulated results.**

I. INTRODUCTION

As wireless communications systems, especially, the wireless local area network (WLAN) and worldwide interoperability for microwave access (WiMAX) communications have developed and expanded rapidly, the many types of antennas to be with wideband or multiband performances to satisfy WiMAX (2.5-2.69 GHz, 3.4-3.69 GHz and 5.25-5.85 GHz) and WLAN (2.4-2.484 GHz, 5.15-5.35 GHz and 5.725-5.825 GHz) have been proposed. It was found that the planar antennas still play a key role in wireless communications systems. The tapered slot antennas (TSAs) are typical examples of planar antennas. The designs of several different TSA have been reported such as linearly tapered slot antenna (LTSA), exponentially tapered slot antenna (ETSA or Vivaldi antenna), constant width slot antenna (CWSA) and broken linearly tapered slot antenna (BLTSA) [1]. TSAs offer a wide operating bandwidth depending on its tapered radiator profile. Compared to other wideband or multiband antennas, tapered slot antennas have nearly symmetrical radiation patterns and moderate gain. They also have the advantage of easy fabrication and geometric simplicity. Because of their potential modern wideband or multiband applications, TSAs have been increasingly studied in recent years [2]-[3]. The performance of the antennas can be improved by change the radius of exponential taper curvature in the case of Vivaldi antenna [4] or design of new tapered profile [5].

In this paper, the new tapered profile of TSA, which can operates in WLAN applications, is proposed. By using squared cosine profile, the antenna can provides appropriate characteristics and can candidates with exponential profile of TSA. The performance of the antenna is verified by experimental data which is obtained from fabrication and measurement. The CST microwave studio software is employed to perform the design and optimization processes. Parametric studies and radiation characteristics are proposed. The measured results of the fabricated prototype are compared with the simulated ones, which show that the antenna can cover 2.4/5.2/5.8 WLAN bands.

II. ANTENNA CONFIGURATION

Fig.1. shows the geometry of the proposed TSA, which is printed on the inexpensive FR4 substrate with the thickness of h = 1.6 mm and the relative permittivity $\varepsilon_r = 4.4$. The basic antenna structure consists of a feed line, a microstrip-slotline transition and a radiating slot. The total size of the antenna is about $90 \times 63 \times 1.6\,\mathrm{mm}^3$ ($L_a \times W_a \times h$). As shown in Fig.1, the microstrip-slotline transition is connected to the feed line and the radiating slot, respectively. The microstrip-slotline transition can offer a frequency independent transition characteristic of electromagnetic fields from microstrip feed line to slotline, and vice versa [6]. The squared cosine profile of the radiating slot is defined by equation,

$$x = -C_1 \cos^2\left(\frac{\pi y}{W}\right) + C_2 \tag{1}$$

where C_1 is a scale factor and C_2 is offset factor of tapered profile, which can be calculated by given the start point (x_1, y_1) and end point (x_2, y_2) of tapered profile. The taper length, L is $x_2 - x_1$ and the aperture width W is $2(y_2 - y_1) + W_s$, where W_s is the slotline width.

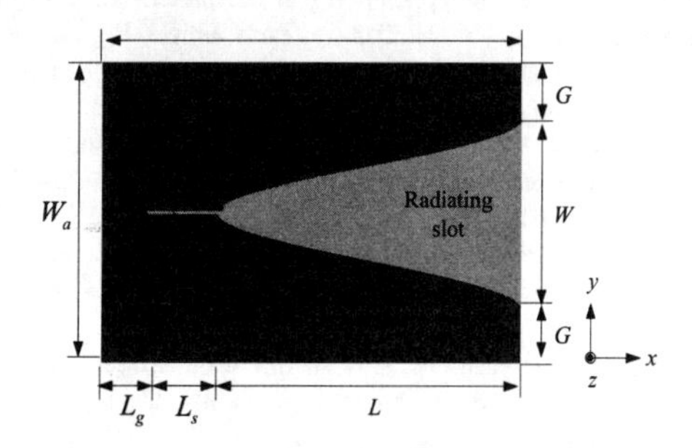

Figure 1. Geometry of the proposed antenna.

978-7-5641-4279-7

The squared cosine profile can be expressed by varying $Ws/2 \leq y \leq W/2$. The width of the antenna W_a is set to close the half free space wavelength at 2.4 GHz, which is lowest frequency for antenna design. To investigate the performance of the proposed antenna, the commercially simulation software CST was used for numerical analysis and to obtain the optimal antenna geometric parameters as summarized in Table I.

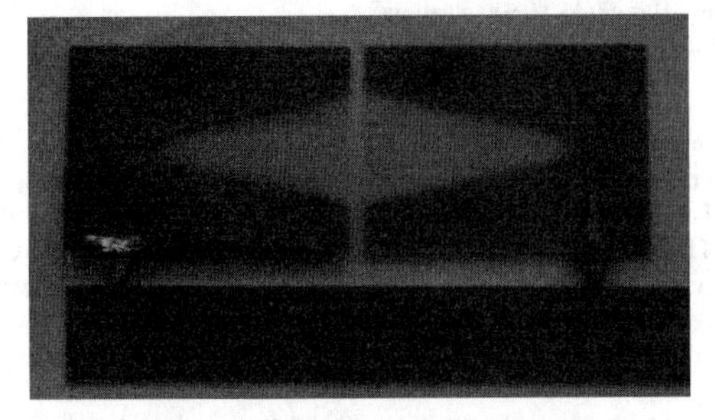

Figure 2. Photograph of fabricated antenna.

TABLE I. OPTIMAL PARAMETERS OF THE TSA.

Parameters	Values (mm)	Parameters	Values (mm)
W_s	1	L_g	10
W_a	63	L_s	15
L_a	90	W_f	3
W	40	L_f	40
L	65	G	11.5

III. RESULTS AND DISCUSSION

Based on the design dimensions as shown in Table I, an experimental prototype of the proposed antenna was fabricated as illustrated in Fig. 2. By using CST software, the surface current distributions on the TSA with squared cosine profile at 2.4 GHz, 5.2 GHz, 5.5 GHz and 5.8 GHz are shown in Fig.3. As shown in Fig. 3(a), the surface current at 2.4 GHz is significantly distributed along the end of slotline and narrow part of squared cosine profile. The current distribution as shown in Fig. 3(b)-(c), it is observed that the current distribution is relatively constant at 5.2 GHz, 5.5 GHz and 5.8 GHz. It may be concluded that the pattern at these three frequencies will be similar to each other. Fig. 4 shows measured and simulated return loss of proposed antenna. The measured impedance bandwidths at $S_{11} \leq -10$ dB are about 830 MHz (1.97-2.80 GHz) and 2850 MHz (3.97-6.82 GHz), which can cover the WLAN bands in the 2.4 GHz (2.4-2.484 GHz), 5.2 GHz (5.15-5.35 GHz) and 5.8 GHz (5.725-5.825 GHz). Moreover, the bandwidths of the proposed antenna can also cover the WiMAX bands in the 2.5 GHz (2.5-2.69 GHz) and 5.5 GHz (5.25-5.85 GHz). As shown in this figure, the agreement between the simulated and measured results is achieved. The disagreement is mainly caused by the fabrication error.

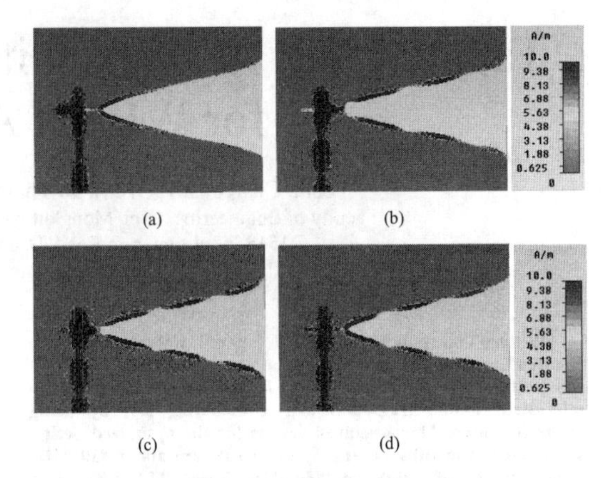

Figure 3. Simulated current distributions of the proposed antenna at (a) 2.4 GHz, (b) 5.2 GHz, (c) 5.5 GHz, (d) 5.8 GHz.

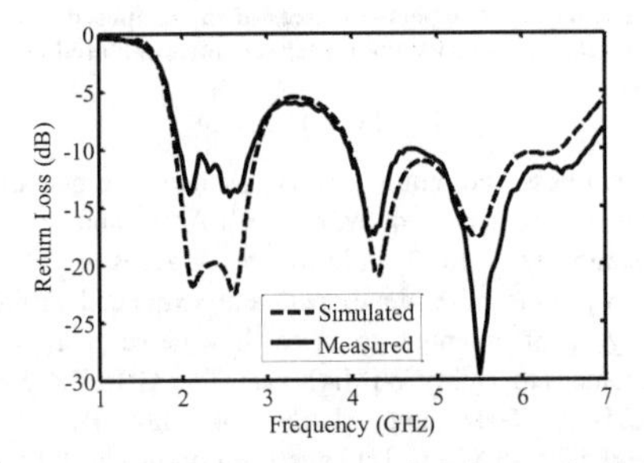

Figure 4. Measured and simulated return loss of proposed antenna.

Further, parametric studies are performed to understand the effects of various parameters and to optimize the final design. Fig. 5 shows the simulated return loss with different aperture widths, W = 36, 40, and 44 mm. As the aperture width is increased, it is seen that the impedance matching characteristics at lower resonant frequency is improved while the upper resonant frequency shows a worse matching with corresponding resonant frequency remaining unchanged. Next, the effect of the taper lengths L on the return loss is investigated by varying the taper lengths L = 55, 60, and 65 mm as illustrated in Fig. 6. As taper length L is increased, both lower and upper resonant frequencies are moved towards the lower frequency. This parameter controls the operating frequency bands. An important feature of the proposed antenna is the influence of the impedance matching caused from the coupling effect between microstrip-slotline transition and microstrip feed line. For this reason, the effects of feed line length L_f = 36, 38, 40, and 42 mm on the return loss of the proposed antenna are also studied and illustrated in Fig. 7. Large change due to the variation in L_f is observed. The impedance bandwidth changes significantly with varying the parameter L_f. This is due to the sensitive of the impedance matching to the parameter L_f.

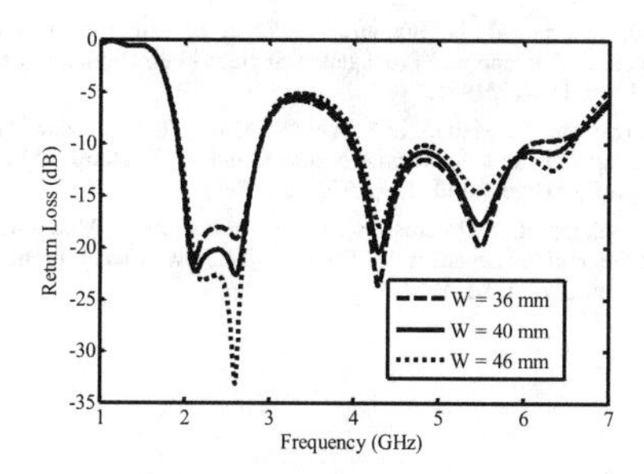

Figure 5. Simulated return loss for various aperture widths W of proposed antenna.

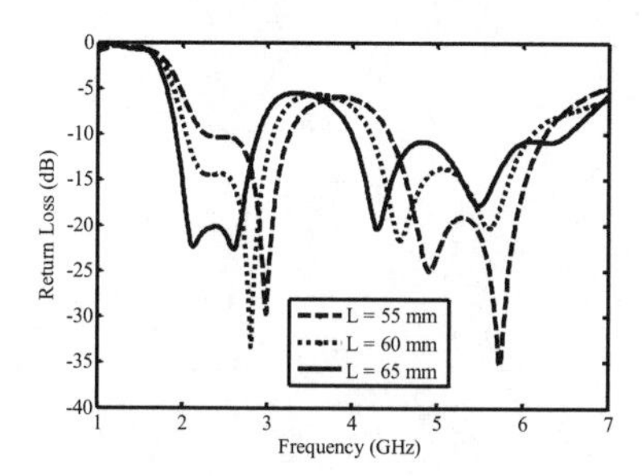

Figure 6. Simulated return loss for various taper lengths L of proposed antenna.

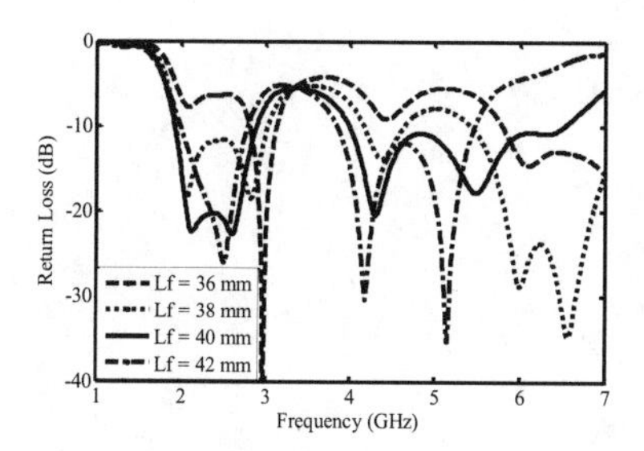

Figure7. Simulated return loss for various feed line lengths L_f of proposed antenna.

The measured and simulated radiation patterns of the proposed antenna in *xy*-plane and *xz*-plane at frequencies 2.4, 5.2, and 5.8 GHz are plotted in Fig. 8. As shown in the figure, the proposed antenna has endfire characteristic with the main lobe in the axial direction of the tapered slot (*x*-direction). It is observed that the radiation patterns of the proposed antenna are in symmetry. The peaks of measured antenna gains are about 3 dBi at 2.4 GHz and about 10 dBi at 5.2/5.8 GHz for WLAN applications.

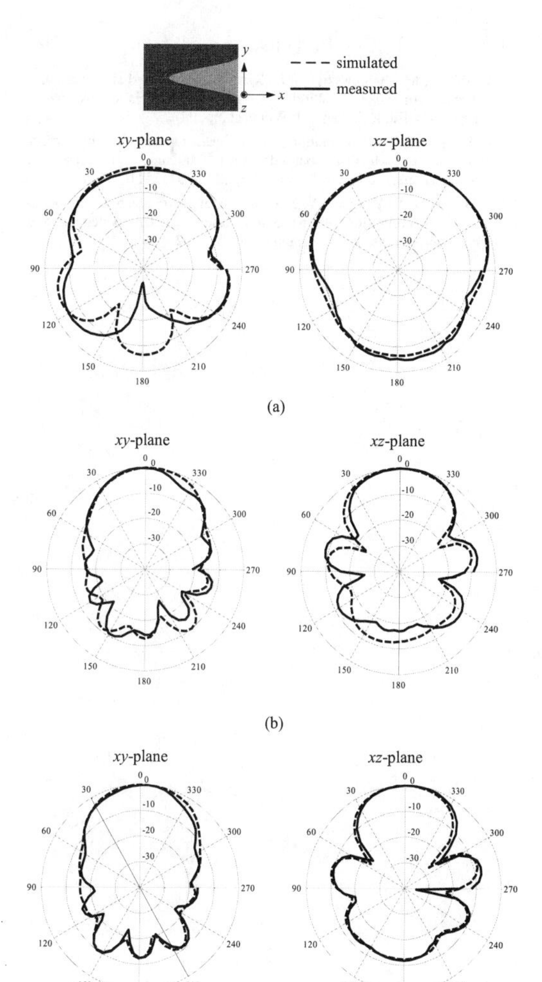

Figure 8. Measured radiation patterns for the proposed antenna at (a) 2.4 GHz (b) 5.2 GHz and (c) 5.8 GHz.

IV. CONCLUSION

In this paper, the tapered slot antenna with squared cosine profile has been presented. The simulated and measured results in terms of return loss, radiation pattern and gain have been compared and analyzed, which show this antenna can support the 2.4/5.2/5.8 GHz WLAN bands and 2.5/5.5 GHz WiMAX bands. Moreover, the proposed antenna with squared profile has moderately high directivity and symmetrical radiation pattern. Therefore, it might be suitable for the WLAN applications.

REFERENCES

[1] A. Abbosh, M. Bialkowski, and H. Kan, "Planar tapered slot antennas," in Printed Antennas for Wireless Communications Handbook (by R. Waterhouse), Ed. Hoboken, NJ: Wiley, ch.6, 2007.

[2] A. Kedar and K. S. Beenamole, "Wide beam tapered slot antenna for wide angle scanning phased array antenna," Progress In Electromagnetics Research B, vol. 27, pp. 235-251, 2011.

[3] D. H. Lee, H. Y. Yang and Y. K. Cho, "Tapered slot antenna with band-notched function for ultrawideband radios," IEEE Antennas and Wireless prop. Letters, vol. 11, pp. 682-685, 2012

[4] R. Q. Lee and R. N. Simons, "Effect of curvature on tapered slot antennas," Antenna and Propagation Society International Symposium, vol. 1, pp. 188-19, 1996.

[5] S. Sugawara, Y. Maita, K. Adachi, K. Mori and K. Mizuno "A mm-wave tapered slot antenna with improved radiation pattern," Microwave Symposium Digest, Vol. 2, pp. 959-962, 1997.

[6] B. Schuppert, "Microstrip/slotline transition: Modeling and experimental investigation," IEEE Trans. Microw. Theory Tech., vol.36, no. 8, pp. 1272-1282, 1987.

A Triple Band Arc-Shaped Slot Patch Antenna for UAV GPS/Wi-Fi Applications

Jianling Chen[1], Kin-Fai Tong[2], Junhong Wang[1]
[1] Department of Electronics and Information Engineer
Beijing Jiaotong University
Beijing, China
j.chen@ee.ucl.ac.uk
[2]Department of Electronic and Electrical Engineer
University College London
London, UK

***Abstract* - In this paper, a triple band capacitive-fed circular patch antenna with arc-shaped slots is proposed for 1.575 GHz GPS and Wi-Fi 2.4/5.2 GHz communications on unmanned aerial vehicle (UAV) appliactions. In order to enhance the impedance bandwidth of the antenna, a double-layered geometry is applied in this design with a circular feeding disk placed between two layers. The antenna covers 2380 - 2508 MHz and 5100 - 6030 MHz for full support of the Wi-Fi communication between UAV and ground base station. The foam-Duroid stacked geometry can further enhance the bandwidths for both GPS and Wi-Fi bands when compared to purely Duroid form. The simulation and measurement results are reported in this paper.**

I. INTRODUCTION

Modern Unmanned Aerial Vehicles (UAV) usually carries a large number of surveillance sensors and communication modules for information gathering and surveillance purposes [1]. It requires extensive communications and data transferring between the sensors, such as GPS receiver and accelerometer, and the Ground Base Station (GBS) [2]. Therefore, antenna systems integrated on the UAV device are critical to establish reliable radio communication links. For this application, patch antenna becomes a good candidate due to its simple geometry, low profile and inexpensive cost. In [3], a directional triple band planar antenna is proposed for WLAN/WiMax access point applications. The triple-band planar antenna consists of a top-loaded dipole for the 2.4 GHz band, two shorter pairs of dipoles are for the 3.5 GHz and 5 GHz bands. In [4], the antenna operates at 1.6 GHz and 2.05-2.29 GHz. Modifying the subtending angle of the arc-shaped slot allows tuning of the frequency bands and the band spacing ratio. In [5], a triple band H-shaped slot antenna fed by microstrip coupling is proposed, exciting a monopole mode, a slot mode, and their higher-order modes, to cover GPS (1.575 GHz) and Wi-Fi (2.4-2.485 GHz and 5.15-5.85 GHz), respectively. In [6-8], different antennas have been report to realize tri-band (WLAN/WiMax). They were developed for user terminals, such as mobile handsets or laptop computers with omnidirectional and low radiation gain. However, the antenna for UAV application needs wide bandwidth and high gain. In this paper, a tri-band patch antenna design with a capacitive feeding disk and double-layer stacked substrates is presented. It can be mounted on the side of UAV. With using the capacitive feeding disc and stacked substrate, the impedance bandwidths at both frequencies are much improved. It has wider bandwidth than the antenna in [3] and higher gain than the antenna in [5] to fully support the UAV GPS and Wi-Fi communications.

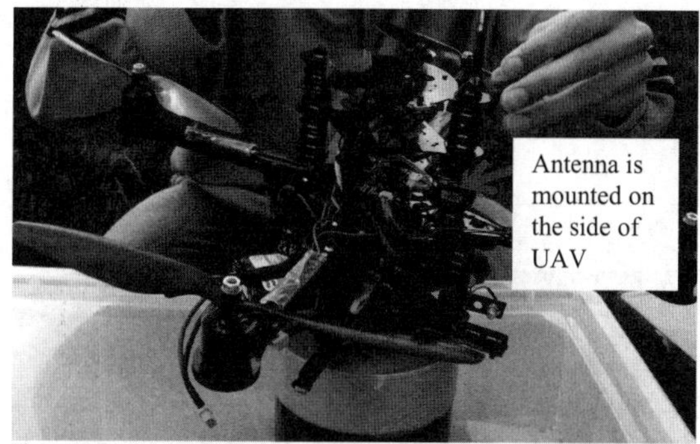

Figure 1. Quadrocopter with three antenna communication systems used in the project

II. ANTENNA CONFIGURATION

Figure 1 illustrates the UAV with the proposed patch antenna will be mounted on. Three separate communication systems are applied on the UAV – GPS, Wi-Fi 2.4/5.2GHz and a remote control unit operates at 2.4GHz. Therefore, the antenna needs supporting a triple band operation. Several antenna types have been attempted, including spiral antenna, monopole antenna and patch antenna. Amongst these antenna types, patch antenna is least susceptible to the rest of UAV device because of its full ground plane. The ground plane shields most RF interferences and demonstrates stable and efficient communication links. Figure 2 shows the configuration of the patch antenna. It is composed of a circular patch, a circular feeding disk stacked between two substrate layers and a square ground plane of 50×50 mm. The bottom substrate is a 7mm-thick polystyrene foam. The copper feeding disc with impedance matching of each resonance can be adjust-

978-7-5641-4279-7

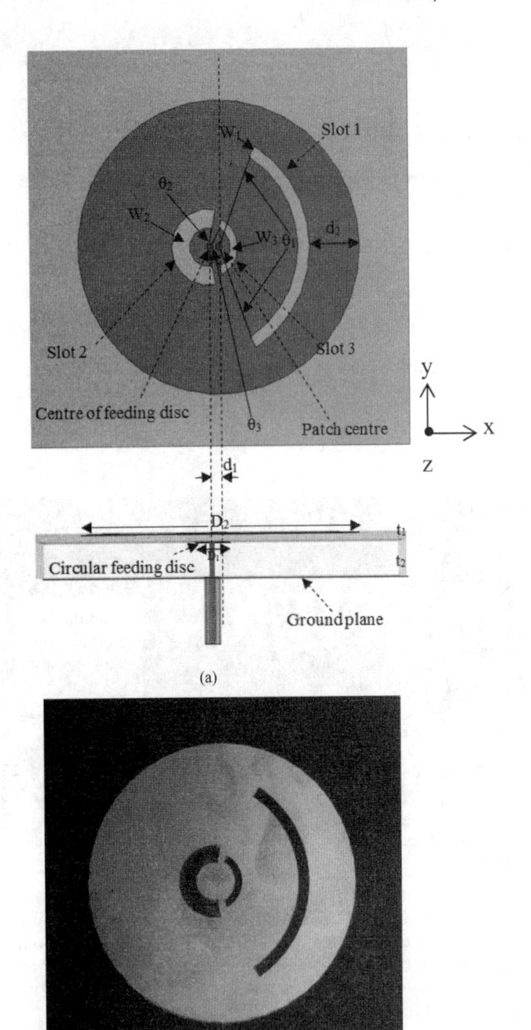

Figure 2. Geometry and the fabricated prototype of the proposed antenna

ed. The diameter D_1 = 10 mm is placed on the top of the foam. The top substrate layer stacked on the polystyrene foam is a single sided Rogers Duroid 5880 laminate with the thickness of 1.575 mm. The relative permittivity (ε_r) of the laminate is 2.2. Dimensions of both substrate layers are 50x50mm, with same sized ground plane on the back of the form. The circular feeding disk between these two substrate layers is connected to the center conductor of a coaxial cable, which is located at the distance of d_1 = 2.4mm with respect to the center of the circular patch, whose diameter is D_2 = 74.4 mm.

Figure 2 (b) shows a photo of the fabricated antenna prototype. In order to excite three resonances for the antenna, three arc-shaped slots are embedded on the circular patch, named Slot 1, 2, 3 in Figure2. By controlling the slot width, the

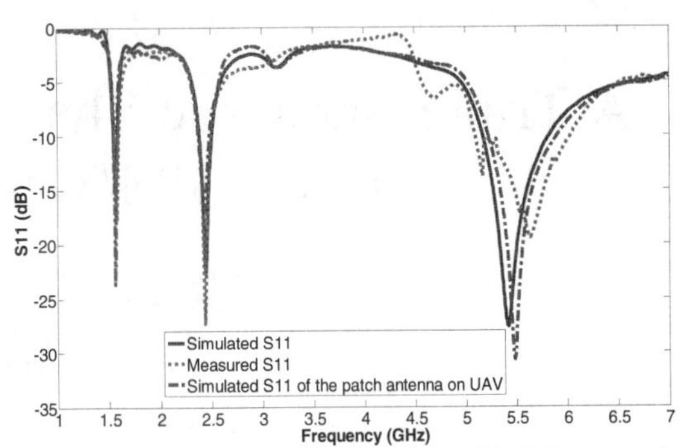

Figure 3. Measured S11 in free space against simulated S11 in free space and on UAV

optimized slot widths in this design are 3mm, 4.7mm, and 2mm, respectively. Slot 1 is located at the centre of the circular patch and subtended by an angle θ_1 = 127.5° with distance d_2 = 13.3 mm to the edge of the patch, while Slot 2 and Slot 3 are centered with respect to the feeding disc centre and subtended by angle θ_2 = 205° and θ_3 = 122°. The radius (from the inner edge to the centre of the feeding disc) of Slot 2 and Slot3 are 5.65 mm and 4.83 mm, respectively.

III. SIMULATION AND MEASUREMENT RESULT

A. *Simulated and Measured S11*

With the increase in bandwidth by applying the capacitive feeding disk, the antenna can cover 2380 to 2508 MHz and 5100 to 6030 MHz for fully support of the Wi-Fi communication between UAV and GBS. The foam-Duroid stacked geometry further enhances the bandwidths compared to that with only single-layered Duroid substrate.

Figure 3 shows the simulated and measured S11 of the proposed patch antenna. As presented, the antenna operates at 1.575 GHz with 45 MHz bandwidth, 2.45 GHz with 128 MHz bandwidth and 5.2 GHz with 930 MHz bandwidth, which is sufficient for both GPS and dual-band Wi-Fi operations. The simulation and measurement results show good agreement. At 5.2 GHz, the measured bandwidth is better than the simulated. In addition, the effect on the antenna performance from UAV shown in Figure 1 is considered in the investigation. Figure 3 shows the comparison of S11 for the antenna in free space and mounted on the UAV. It can be seen that when the antenna is fixed on UAV, at 5.2 GHz band the S11 is slightly offset to higher frequency. There is no noticeable frequency shift at the other bands due to the full sized antenna ground plane. Therefore, it effectively solves the problem when using monopole antenna for Wi-Fi communication.

B. *Impedance matching analysis*

Figure 4 shows real part in Z-parameters of the antenna with slot 3 and antenna without slot 3. At 5.2 GHz band, the real part is higher than that when the slot 3 is removed. Figure 4 also shows imaginary part in Z-parameters of the antenna with slot 3 and antenna without slot 3. It can be seen the inductance of the antenna with slot 3 is higher at 5.1 GHz than that of the antenna without slot 3. At 5.5 GHz, the inductance of the antenna with slot 3 is lower than that of the antenna without slot 3. So the slot 3 gives wider impedance match for the antenna.

C. *Surface current distributions*

Figure 5 shows current distributions on the patch at 1.575, 2.4 and 5.2 GHz. For the case of surface current shown in Figure 5 (a), the resonant mode is perturbed TM_{11} mode, it can be observed that the excited current distribution is very similar to that of the TM_{11} mode of the case without the slot. With the presence of the slot, the fundamental mode TM_{11} of the circular mode has been slightly perturbed because the slot is located close to the patch boundary, where the excited patch surface current for the TM_{11} mode had a minimum value. In Figure 5 (b), the second resonant mode excited in the present design is the perturbed TM_{01} mode. In Figure 5 (c), it can be seen the surface current is mostly concentered around slot 2 and 3. In this mode, the slot is radiating rather than the patch radiating.

D. *Simulated radiation patterns*

Simulated radiation patterns of antenna standalone and on the UAV in two principle planes (E-plane and H-plane) are plotted in Figure 6 and Figure 7 at 1.575, 2.45 and 5.5 GHz. It can be seen in Figure 6 that the cross-polarization across E-plane is very low. It has symmetrical co- polarization pattern and cross-polarization pattern across H-plane. In Figure 6 (c) the main lobe of co-polarization across E-plane is offset 20° due to the affect ion of Slot 2 and Slot 3.

In Figure 7, cross-polarization across E-plane is much higher than that in Figure6. As the antenna is mounted on the UAV, the radiation pattern is affected by the presence of the UAV. Same reason causes the unsymmetrical co and cross-polarization across H-plane. The main lobe and 3 dB beam width is similar with Figure 6 (c).

IV. CONCLUSION

The performance of a circular patch antenna with a capacitive feeding disk and arc-shaped slots is reported. Adding the slots and modifying their dimensions provide a tri-frequency band antenna operating at GPS 1.575 GHz and Wi-Fi 2.4/5.2 GHz. The antenna can be mounted on the side of an UAV to support its communication systems. In order to minimize interferences from the rest of the UAV structure, patch antenna becomes a good candidate to offer an independent working environment due to its full-sized ground plane. The bandwidth of the proposed antenna, benefiting from the capacitive feeding and the stacked substrates, has achieved 45 MHz at GPS band, 128 MHz at 2.4 GHz band and 930 MHz at 5.2 GHz, giving the antenna gain of 7.23 dB at 1.575 GHz, 8.34 dB at 2.45 GHz and 9.28 dB at 5.5 GHz, respectively. The radiation pattern will be measured and presented in presentation.

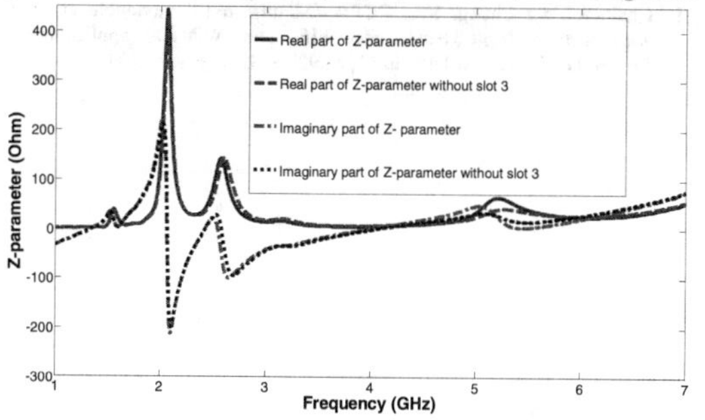

Figure 4. Comparied Z-parameters between the antenna and antenna without slot 3

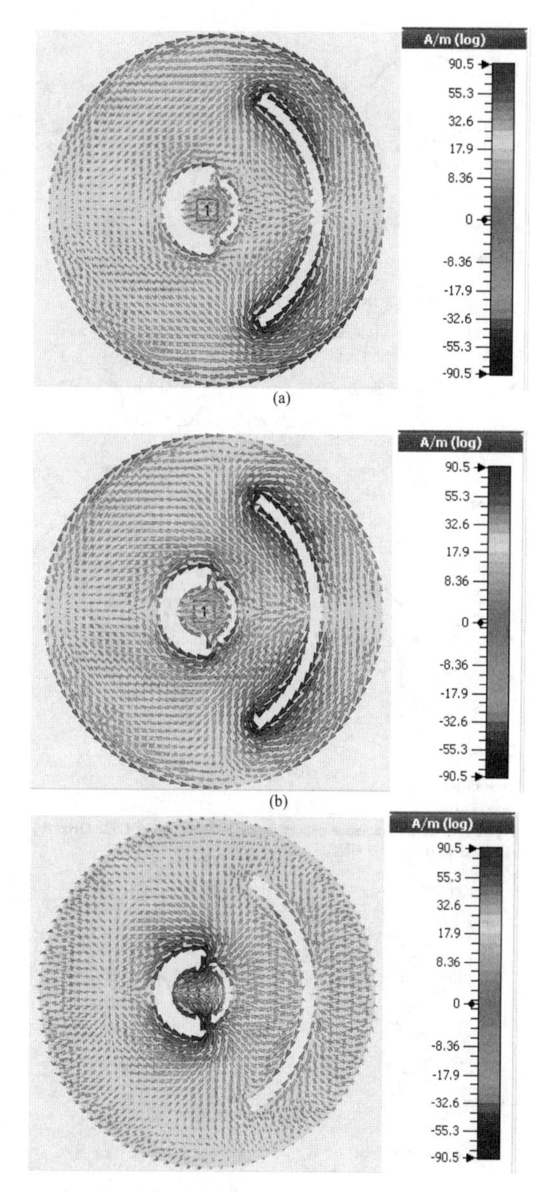

Figure 5. Simulated surface current distribution of frequency (a) f_1 = 1.575 GHz (b) f_2 = 2.4 GHz and (c) f_3 = 5.5 GHz

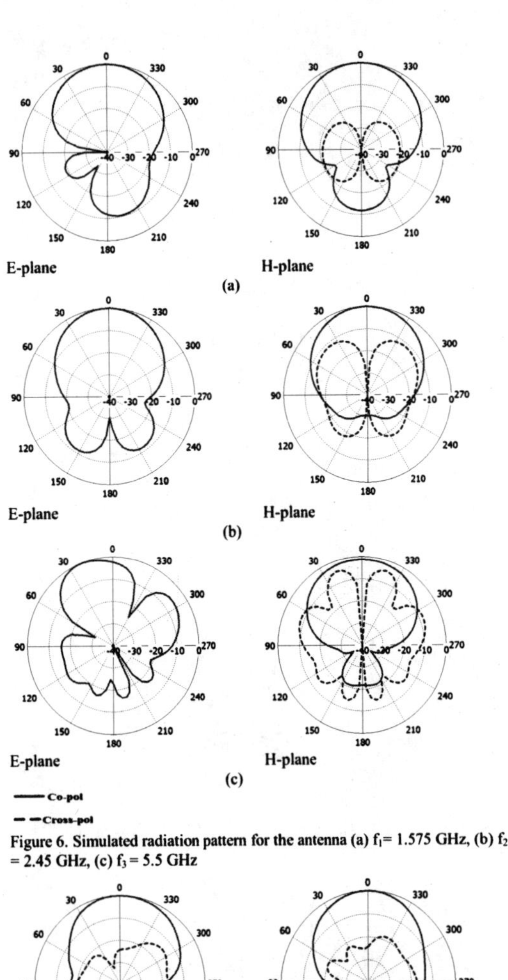

Figure 6. Simulated radiation pattern for the antenna (a) f_1= 1.575 GHz, (b) f_2 = 2.45 GHz, (c) f_3 = 5.5 GHz

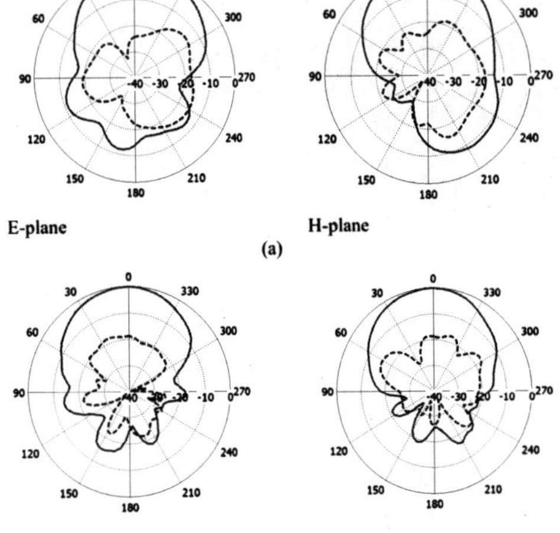

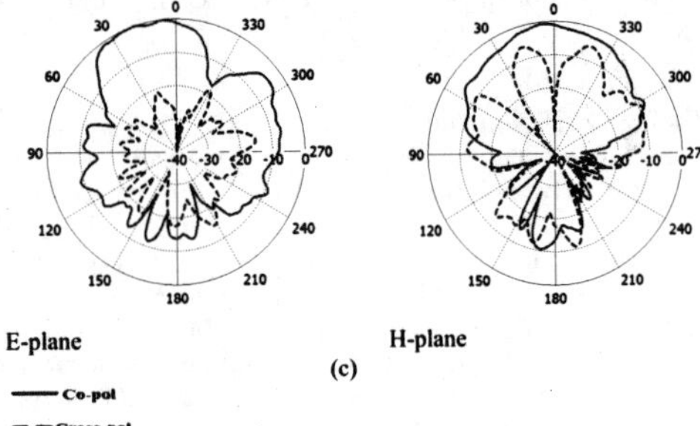

Co-pol

Cross-pol

Figure 7. Simulated radiation pattern for the antenna on UAV (a) f_1= 1.575 GHz, (b) f_2 = 2.45 GHz, (c) f_3 = 5.5 GHz

V. ACKNOWLEDGEMENT

This work is supported in part by the National Program on Key Basic Research Project under Grant No.2013CB328903, and in part by the National Natural Science Foundation of China under Grant No.61271048.

REFERENCES

[1] Williams, W.; Burton, C., "Lightweight Agile Beam Antennas for UAVS," *Military Communications Conference, 2006.* MILCOM 2006. IEEE , vol., no., pp.1,5, 23-25 Oct. 2006.

[2] Chen-Mou Cheng; Hsiao, Pai-Hsiang; Kung, H. T.; Vlah, D., "Performance Measurement of 802.11a Wireless Links from UAV to Ground Nodes with Various Antenna Orientations," *Computer Communications and Networks*, 2006. ICCCN 2006. Proceedings.15th International Conference on , vol., no., pp.303,308, 9-11 Oct. 2006

[3] Li, R.L.; Quan, X.L.; Cui, Y.H.; Tentzeris, M.M., "Directional triple-band planar antenna for WLAN/WiMax access points," *Electronics Letters* , vol.48, no.6. pp.305,306, March 15 2012.

[4] Hsieh, G.-B.; Wong, K.-L, "Inset-microstrip-line-fed dual-frequency circular microstrip antenna and its application to a two-element dual-frequency microstrip array," *Microwaves, Antennas and Propagation,* IEE Proceedings , vol.146, no.5, pp.359,361, Oct

[5] Chang, T.; Kiang, J., "Compact Multi-band H-Shaped Slot Antenna," *Antennas and Propagation, IEEE Transactions on* , vol.PP, no.99, pp.1,4, 29-31 Oct. 2010

[6] Yen-Chi Shen; Yu-Shin Wang; Shyh-Jong Chung, "A printed triple-band antenna for WiFi and WiMAX applications," *Microwave Conference, 2006. APMC 2006. Asia-Pacific* , vol., no., pp.1715,1717, 12-15 Dec. 2006

[7] Pingan Liu; Yanlin Zou; Baorong Xie; Xianglong Liu; Baohua Sun, "Compact CPW-Fed Tri-Band Printed Antenna With Meandering Split-Ring Slot for WLAN/WiMAX Applications," *Antennas and Wireless Propagation Letters, IEEE* , vol.11, no., pp.1242,1244, 2012

[8] Chaimool, S.; Chung, K.L., "CPW-fed mirrored-L monopole antenna with distinct triple bands for WiFi and WiMAX applications," *Electronics Letters* , vol.45, no.18, pp.928,929, August 27 2009

Dual-Band Printed L-Slot Antenna for 2.4/5 GHz WLAN Operation in the Laptop Computer

Saran Prasong, Rassamitut Pansomboon, and Chuwong Phongcharoenpanich
Faculty of Engineering, King Mongkut's Institute of Technology Ladkrabang, Bangkok 10520, Thailand
Email: pchuwong@gmail.com

***Abstract*— This paper presents a dual-band L-shaped slot antenna for laptop computer operated in WLAN system of 2.4/5.2/5.8 GHz. The proposed antenna is formed by L-shaped slot and fed by strip line structure. The antenna size is relatively compact with the dimension of 15 mm × 60 mm × 0.8 mm. The antenna is compatible to embed at the top of the display panel for laptop computer. The antenna can be operated from 2.3 GHz to 2.6 GHz and from 5.0 GHz to 6.0 GHz that can cover WLAN system of 2.4 GHz, 5.2 GHz and 5.8 GHz.**

I. INTRODUCTION

Presently, the wireless local area network (WLAN) technology is very important for laptop computer because WLAN system is very easy to connect for internet access. The internal laptop antenna is essential part of wireless communication system to connect with other devices and internet [1-2]. The internal laptop antenna is required to possess compact size and nearly omnidirectional pattern. The printed slot antenna is the good candidate for internal laptop antenna because the antenna has thin structure and low profile suitable for the design and the fabrication. The antenna is embedded at the top edge of the display panel of laptop computer.

In this paper, the dual-band printed L-shaped slot antenna is proposed. The antenna is formed by L-shaped slot with strip line [3-4]. The size of antenna is relatively compact with width of 15 mm, length of 60 mm and thickness of 0.8 mm for dual-band WLAN 2.4/5 GHz operated in laptop computer. The antenna is printed on both sides of the FR-4 substrates. The ground plane of laptop antenna is integrated with the antenna for simulation. The antenna structure is described, and the simulated and measured results of the antenna are presented and discussed in the next section.

II. ANTENNA STRUCTURE

The geometry of dual band printed L-shaped slot antenna is shown in Figure 1. The antenna is printed on both sides of FR-4 substrate with relative permittivity of 4.3.

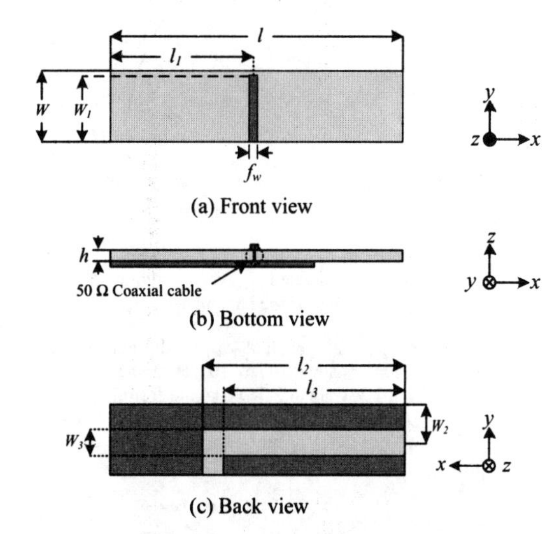

Figure 1. Geometry of the proposed antenna.

The parametric description and physical size of the proposed antenna is shown in Table I. The antenna size is small with 15 mm×60 mm×0.8 mm.

TABLE I
The associated dimension of the proposed antenna.

Parameter	Physical Size (mm)
f_w	1.65
h	0.80
l	60.00
l_1	30.80
l_2	47.00
l_3	44.00
w	15.00
w_1	13.75
w_2	7.50
w_3	8.60

978-7-5641-4279-7

The antenna structure consists of L-shaped slot fed by strip line via 50 Ω coaxial cable. In this study, the antenna is to be mounted at the center of the top edge of the ground plane with the length of 26 mm and width of 20 mm. The ground plane is modeled using 0.2 mm-thick copper plate and treated as the system ground plane or supporting metal frame for display panel.

III. SIMULATED RESULTS

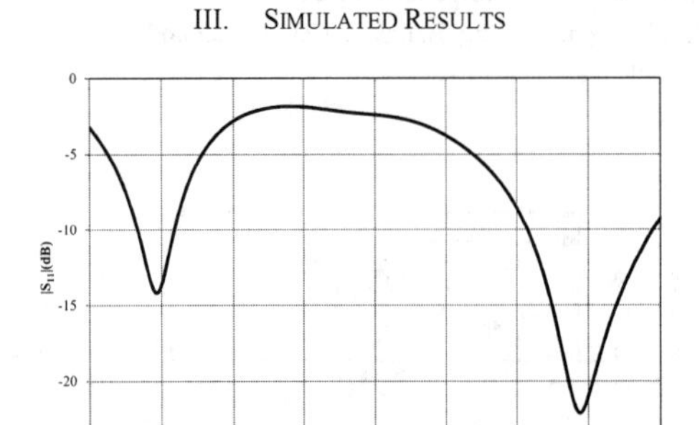

Figure 2. The simulated $|S_{11}|$ versus frequency of the proposed antenna.

Figure 2 shows the simulated $|S_{11}|$ of the antenna. The $|S_{11}|$ of the antenna is lower than -10 dB at the dual-band frequencies from 2.33GHz to 2.55 GHz and from 5.1 GHz to higher than 6 GHz. From the $|S_{11}|$ result, the antenna can be operated to cover WLAN system of 2.4 GHz, 5.2 GHz and 5.8 GHz.

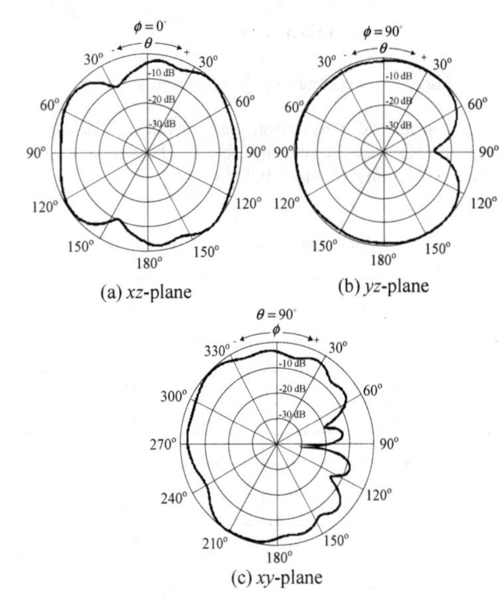

(a) At the frequency of 2.45 GHz

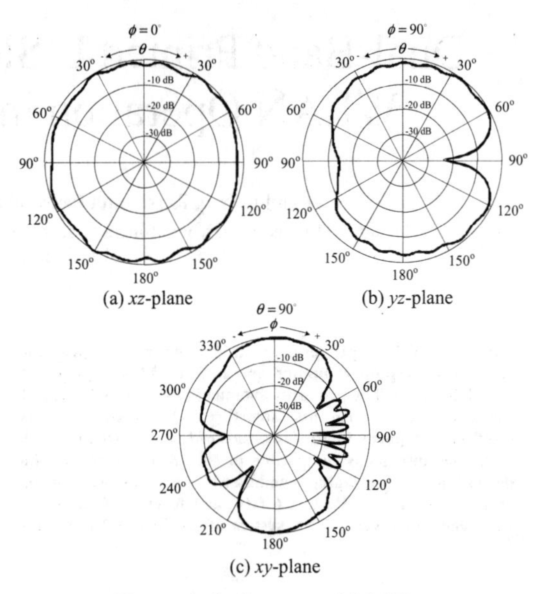

(b) At the frequency of 5.2 GHz

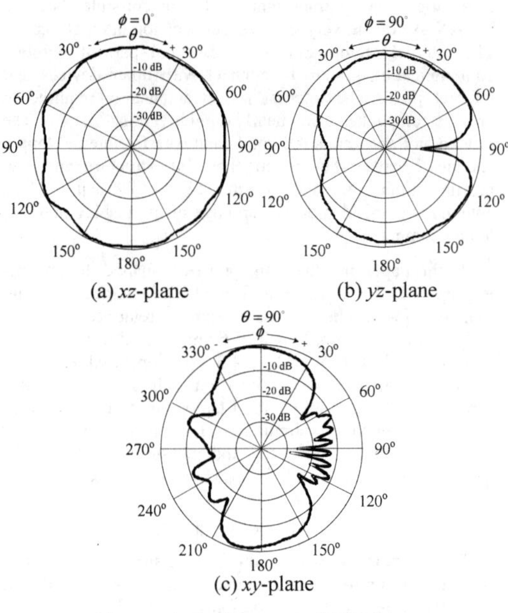

(c) At the frequency of 5.8 GHz

Figure 3. The simulated radiation pattern of proposed antenna.

Figure 3 shows the simulated radiation pattern of the antenna. The radiation pattern of the proposed antenna is almost omnidirectional beam.

TABLE II

The simulated gain of the proposed antenna.

Frequency(GHz)	Gain (dBi)
2.45	3.23
5.20	4.05
5.80	4.41

Table II shows the simulated gain of the antenna. The simulated maximum gain at 5.8 GHz is 4.41 dBi.

IV. Results And Discussion

The photograph of the prototype antenna is depicted in Figure 4. The comparison between simulation and measurement results are shown. The simulated results show the $|S_{11}|$ of the proposed antenna with the structure given in Figure 1. The simulated and measured results of $|S_{11}|$ are shown in Figure.5, where the acceptable reflection is considered at $|S_{11}|$ < -10 dB. The covered bandwidth is ranging from 2.31 GHz to 2.54 GHz and from 5.12 GHz to 5.94 GHz.

Figure 4 The photograph of the prototype antenna.

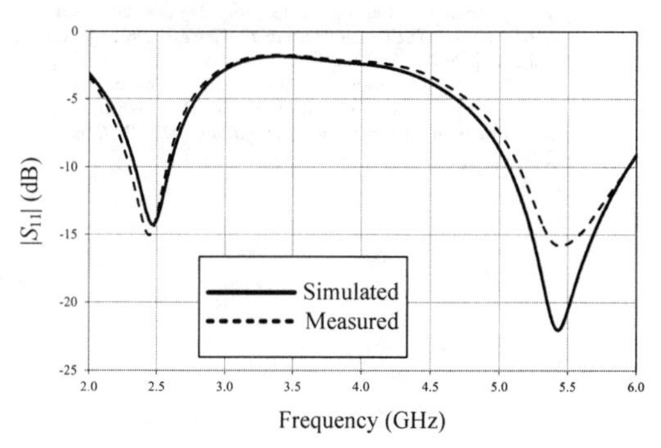

Figure 5 The compared $|S_{11}|$ from the simulation and measurement.

Figure. 6 shows the radiation pattern at 2.45 GHz, 5.2 GHz, 5.5GHz and 5.8 GHz which are frequencies of WLAN system. The radiation pattern from the simulation and measurement is almost omnidirectional pattern.

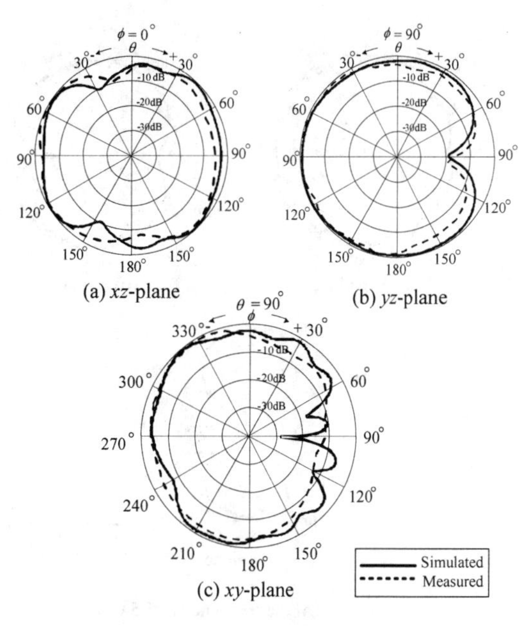

(a) *xz*-plane

(b) *yz*-plane

(c) *xy*-plane

(a) At the frequency of 2.45 GHz

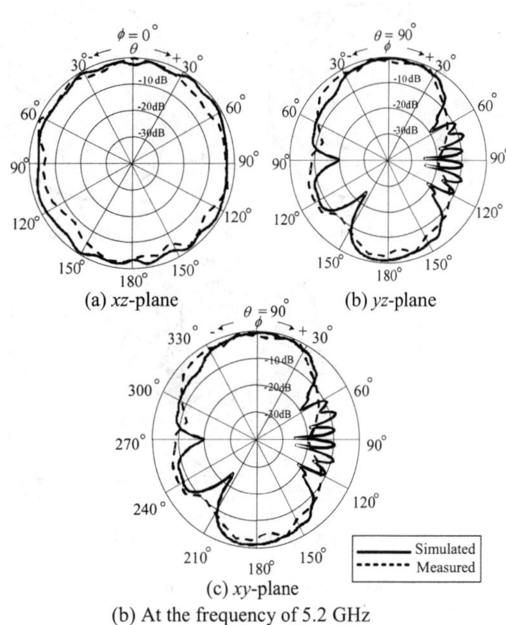

(a) *xz*-plane

(b) *yz*-plane

(c) *xy*-plane

(b) At the frequency of 5.2 GHz

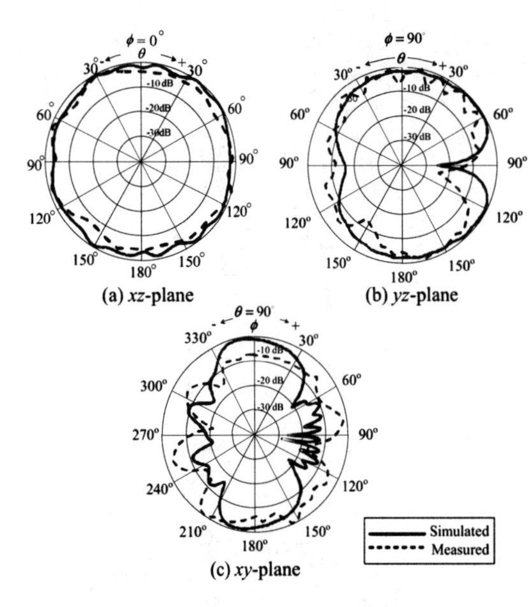

(c) At the frequency of 5.5 GHz

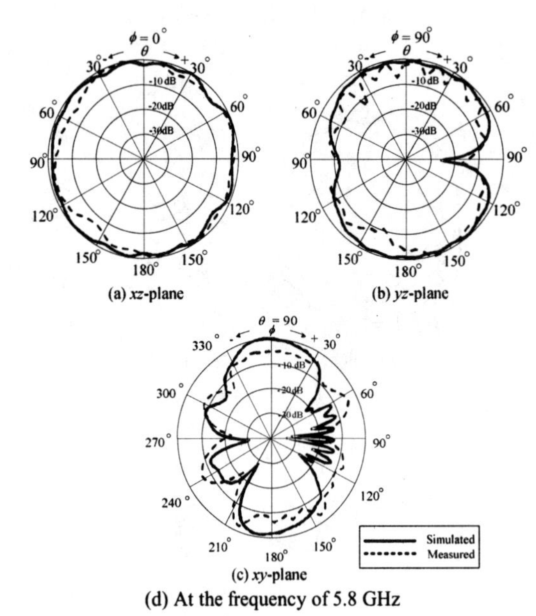

(d) At the frequency of 5.8 GHz

Figure 6 The radiation pattern from the comparison between the simulation and the measurement.

TABLE III

The measured gain of the proposed antenna.

Frequency (GHz)	Gain (dBi)
2.45	2.13
5.20	1.67
5.50	1.73
5.80	1.24

Table III shows the measured gain of the antenna. The measured gain at 2.45 GHz, 5.2 GHz and 5.8 GHz are 2.13 dBi, 1.67 dBi and 1.24 dBi, respectively.

V. Conclusion

A dual-band printed L-shaped slot antenna for laptop computer has been proposed. The proposed antenna is formed by an L-shaped slot and fed by strip line. The antenna is printed on both sides of FR-4 substrate. The proposed antenna is mounted on the ground plane of laptop computer for simulation. The measured $|S_{11}|$ is lower than -10 dB. The covered bandwidth is ranging from 2.31 GHz to 2.54 GHz and from 5.12 GHz to 5.94 GHz. The radiation patterns are almost omni-directional beam in *E*-plane. The measured maximum gain at 2.45 GHz is 2.13 dBi. Therefore, the proposed antenna can be operated for dual-band to cover WLAN 2.4/5.2/5.8 GHz. Accordingly, the antenna is suitable to mount on the ground plane of laptop computer.

References

[1] K. L. Wong and L. C. Lee, "Multiband printed monopole slot antenna for WWAN Operation in the laptop computer," *IEEE Transactions on Antennas and Propagation.*, vol. 57, pp.324-330, Feb. 2009.

[2] W.J.Lin and K.L. Wong,"Printed WWAN/LTE slot antenna for tablet computer application," in *Proc. Asia Pacific Microwave Conference (APMC)*, pp.821-824, 2011.

[3] H.-W. Liu, S.-Y. Lin, and C.-F. Yang "Compact inverted-F antenna with meander shorting strip for laptop computer WLAN applications," *IEEE Antennas and Wireless Propagation Letters*, vol.10, pp. 540-543, 2011.

[4] P. Mousavi, B. Miners, and O. Basir, "L-shaped monopole slot antenna for circular polarization," in *Proc. IEEE International Symposium on Antennas and Propagation (APSURSI)*, pp. 796 - 797, 2011.

TA-2(C)

October 24 (THU) AM

Room C

Patch Antennas

Analysis of L-Probe Fed-Patch Microstrip Antennas in a Multilayered Spherical Media

TaoYu, Chengyou Yin

National Key Laboratory of Pulsed Power Laser Technology of Electronic Engineering Institute,

Address: 460 Huangshan Rd. Hefei, Anhui, China, 230037

***Abstract*-In this paper, the L-probe fed-patch microstrip antennas in a multilayered spherical media are analyzed based on the method of moments (MoM). Firstly, the structures of microstrip antennas in three-layered spherical media fed by L-probe are established as similar as the plane case. Then, the method of moments is employed to solve the mixed potential integral equation (MPIE). Numerical results for the input impedance, radiation patterns are showed good agreement with ones come from commercial software FEKO. Finally, the effects of all kind of geometric parameters on the characteristics of antenna are investigated for the next design purpose.**

I. INTRODUCTION

As we know, the microstrip antenna is a kind of resonance antenna and the narrow bandwidth is the major weakness. Much effort has been done to widen the bandwidth of microstrip in plane case in past decades, including impedance matching technology [1] and using thick substrates [2]. The feeding approach employing L-probe is an effective method to ameliorate bandwidth, which is especially convenient for the coplanar case. The microstrip antenna with an L-probe in planar case has been studied in [3]-[5]. A modified L-probe fed microstrip antenna has been presented in [6].

Compared with in planar case, only a few literatures relate to the microstrip antenna with L-probe in coplanar case [7]. The conformal microstrip antennas have a major advantage that they have a wider angular radiation pattern than the planer counterparts. It is worthwhile to research the characteristics of the antennas with L-probe in conformal case.

In this paper, the L-probe patch antenna in a multilayered spherical media is studied by employing the method of moments (MoM). The structure of antenna is obtained with the aid of commercial software FEKO, which is divided into a mesh of triangular patches. The surface current density can be calculated by solving the mixed potential integral equation (MPIE) based on the Rao-Wilton-Glisson (RWG) basis function. Characteristics including input impedance, radiation pattern are presented, which agree well with those from FEKO. Finally, extensive numerical results of antenna parametric study are showed in order to provide some design information.

II. THEORY

A. *Structure of Antenna*

The structure of antenna has three layers as shown in Fig.1. The innermost layer is a perfectly conducting spherical core, which can be modeled assuming the permeability tends to zero and permittivity tends to infinity [8]. The outermost layer assumed to be free space. The rectangular patch locates at interface of medium layer and free space and the conformal L-probe positions at the medium layer. Detailed parameters are shown in Fig.1. In this paper, nonmagnetic materials have been considered, that is $\mu_r = 1$.

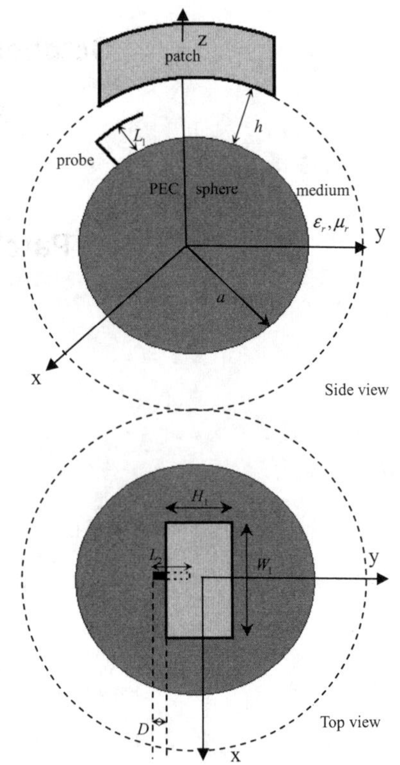

Figure 1. Structure of antenna with L-probe

B. *Mixed potential integral equation*

In this paper, MoM is employed to solve the problem because of its accuracy and efficient. In order to make analysis process simplification, the L-probe can be treated as an L-trip [9], which is connected to the PEC spherical core. A delta gap voltage source has been placed at the base of the feeding trip. The tangential electric field can be achieved employing the mixed potential integral equation:

978-7-5641-4279-7

$$\boldsymbol{E}^{i}\big|_{t}=-j\omega\mu_{f}\iint_{s'}\bar{\boldsymbol{G}}_{A}\bullet\boldsymbol{J}_{s}(r')ds'-\nabla\iint_{s'}\nabla'G_{\varphi}\bullet\boldsymbol{J}_{s}(r')ds'\big|_{t} \quad (1)$$

where μ_f represents the permeability of the layer of field point. s' is the surface area of the antenna conducting body except the ground sphere as shown in Fig.2, and $\boldsymbol{J}_s(r')$ is the current density of the surface area. The contribution of the ground sphere has been incorporated in the Green's functions including magnetic vector potential Green's function $\bar{\boldsymbol{G}}_A$ and electric scalar potential Green's function G_φ.

Solving the dyadic Green function in spherically multi-layered media is indispensable in order to make the analysis results accurate. A general expression of the dyadic Green's function for the electromagnetic fields in an arbitrary multi-layered medium has been achieved in [10]. Further development of Green's function has been proposed in [8], where the electric field Green's function was transformed to the mixed potential Green's function. Take G_φ component for example:

$$G_{\varphi}=-\frac{\omega\mu_{f}}{4\pi k_{f}}\sum_{n=1}^{\infty}\frac{2n+1}{n(n+1)}\times$$
$$\begin{pmatrix}\partial h_{n}^{(2)}(k_{s}r)\left[A_{n}^{TM}\partial h_{n}^{(2)}(k_{f}r')+B_{n}^{TM}\partial j_{n}(k_{f}r')\right]\\+\partial j_{n}(k_{s}r)\left[C_{n}^{TM}\partial h_{n}^{(2)}(k_{f}r')+D_{n}^{TM}\partial j_{n}(k_{f}r')\right]\end{pmatrix}\bullet P_{n}(\cos\gamma) \quad (2)$$

meaning of the denotations can be found in [8]. From the expressions, we can see that the above expression is an infinite summation of spherical harmonics. In order to accelerate the convergence of the infinite series, asymptotic extraction approach has been adopted in [11]-[12].

The surface area of the antenna conducting body has been divided into a mesh of triangular patches as shown in Fig.2, and the RWG triangular function is the most appropriate basis function [13]. Then, we expand the current $\boldsymbol{J}_s(r')$ via a sum of weighted basis functions:

$$\boldsymbol{J}_{s}(r')=\sum_{n=1}^{N}a_{n}\boldsymbol{f}_{n}(r') \quad (3)$$

and the method of Galerkin has been adopted in this paper. Every element of the impedance matrix can be obtained by the expression below:

$$Z_{mn}=\iint_{s}\iint_{s'}\begin{bmatrix}j\omega\mu_{f}\boldsymbol{f}_{m}(r)\bullet\bar{\boldsymbol{G}}_{A}\bullet\boldsymbol{f}_{n}(r')+\\(\nabla\bullet\boldsymbol{f}_{m}(r))G_{\varphi}(\nabla'\bullet\boldsymbol{f}_{n}(r'))\end{bmatrix}dsds' \quad (4)$$

where m and n refer to the test and expansion function.

The electric field can be expressed as:

$$\boldsymbol{E}_{n}=\iint_{s'}j\omega\mu_{f}\bar{\boldsymbol{G}}_{A}\bullet\boldsymbol{f}_{n}(r')+\nabla(\nabla'G_{\varphi}\bullet\boldsymbol{f}_{n}(r'))ds' \quad (5)$$

A problem that needs special attention is the connection of the feeding trip to the perfectly conducting sphere. The delta gap voltage source has been placed at the connection, and more detailed analysis can be found in [14].

C. Numerical results

An L-probe fed-patch microstrip antenna in multilayered

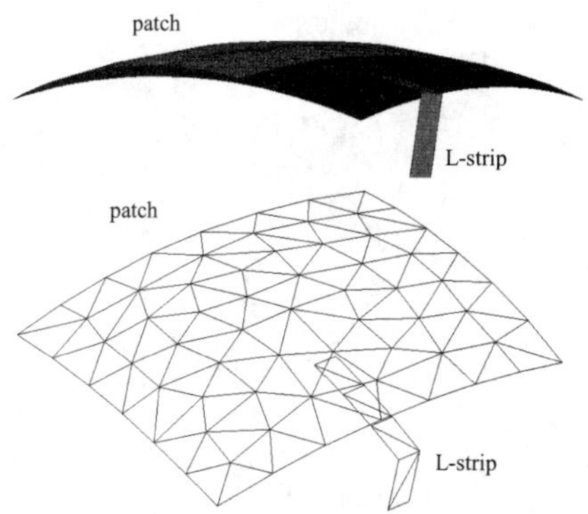

Figure 2. Surface area of antenna and mesh of the antenna

media has been calculated, which has the following parameters: $W_1=30mm$, $H_1=25mm$, $a=50mm$, $L_1=4.95mm$, $L_2=9.5mm$ and $D=2mm$. The spherical substrate has a relative permittivity of $\varepsilon_r=1.01$ and a thickness of $h=6.6mm$. The L-probe has been modeled as a strip of 2 mm width. The numerical results as shown in Fig. 3-6 are verified by comparison with the results from commercial software FEKO.

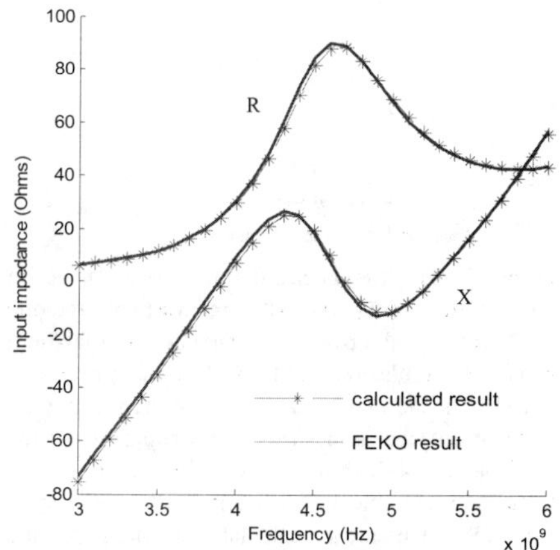

Figure 3. Input impedance versus frequency

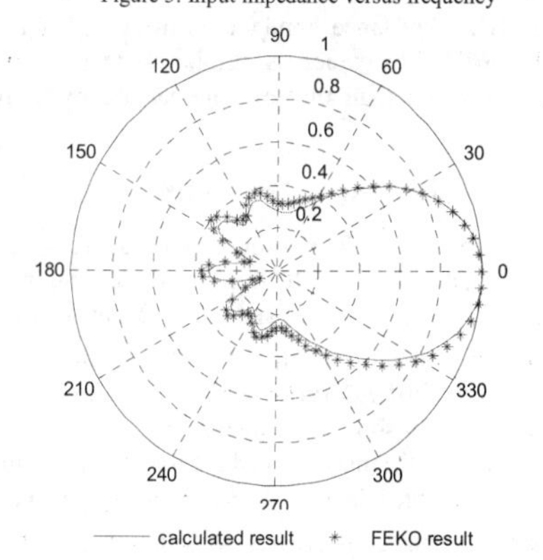

Figure 4. Far-field radiation pattern at 4.5 GHz ($\varphi=90^\circ$)

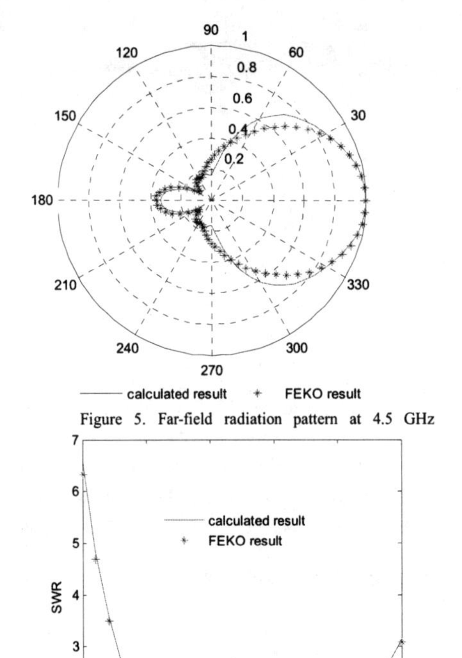

Figure 5. Far-field radiation pattern at 4.5 GHz

Figure 6. SWR of the antenna

From above figures, we can see that the calculated result of input impedance show good well agreement with the one from FEKO, and a small error is observed between far-field radiation pattern calculated and FEKO. The main reason is that the meshes of the two antennas are different and only surface area of antenna is meshed in this paper, while the whole antenna is meshed in FEKO. The error can be ignored that is acceptable in engineering designs. Fig. 6 shows the SWR of the antenna calculated by this paper and FEKO. From the figure, we can see the antenna attains an about 1.7 GHz impedance bandwidth (from 3.95 GHz to 5.65 GHz, $SWR \leq 2$), which is similar to that in planar case [2], and wider than that of antenna fed by straight probe.

III. PARAMETRIC STUDY

In this section, we will investigate the variation of the antenna characteristics with the antenna parameters including W_1, H_1, L_1, L_2 and D. Because of the limited paper space, only the input impedances are shown, and the far-field radiation pattern and SWR of the antennas are not shown in this paper.

The variation of the input impedance with the length of patch W_1 is illustrated in Fig. 7. From the figure, we can see that input resistance decreases with increasing W_1 and the input reactance decreases before the resonant frequency and increases after the resonant frequency with increasing W_1.

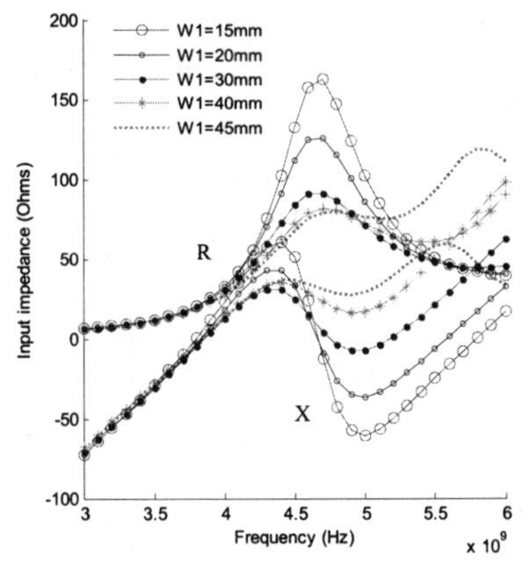

Figure 7. Input impedance versus frequency for different W_1

($H_1 = 25mm$, $L_1 = 4.95mm$, $L_2 = 10mm$, $h = 6.6mm$ and $D = 2mm$)

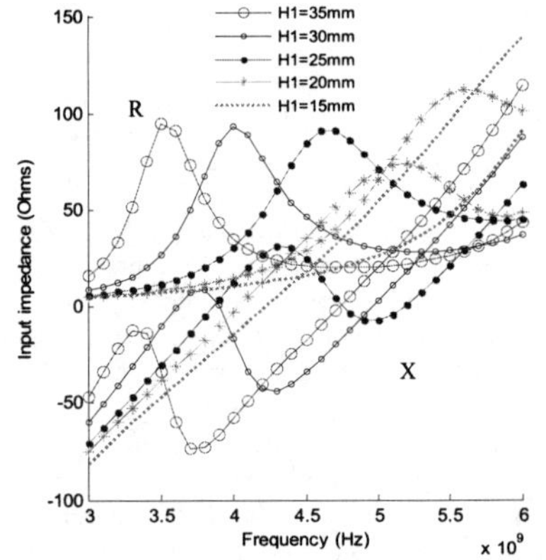

Figure 8. Input impedance versus frequency for different H_1

($W_1 = 30mm$, $L_1 = 4.95mm$, $L_2 = 10mm$, $h = 6.6mm$ and $D = 2mm$)

The variation of the input impedance with the patch width H_1 is shown in Fig. 8. It can be observed that H_1 determined the resonant frequency of the antenna. The resonant frequency becomes high with decreasing H_1.

Fig. 9 and Fig. 10 show the variation of the input impedance with the vertical length L_1 and the horizontal length L_2 of the L-strip, respectively. We can see from the figures that both the input resistance and reactance increase with L_1, L_2. At the resonant frequency, the input resistance increases clearly with increasing L_1, and the reactance increases clearly with increasing L_2.

The effect of changing the position D of the L-strip from the patch is studied as shown in Fig. 11. $D > 0$ denotes that the L-strip is partly covered by the patch, and $D < 0$ denotes the strip is covered by the patch in whole. We can see that both the resistance and reactance change slightly when $D \geq 0$, and decrease clearly when $D < 0$.

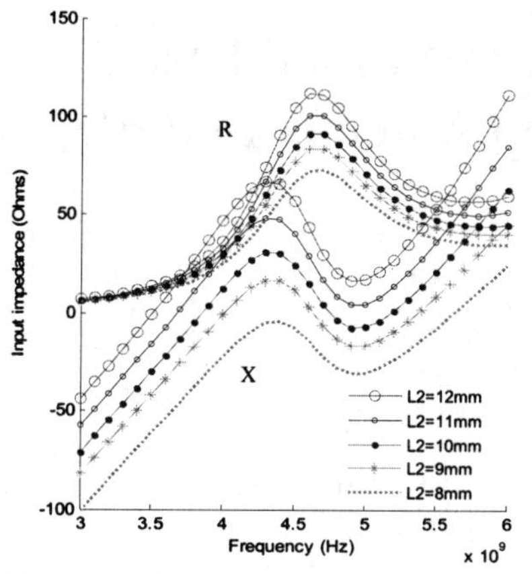

Figure 10. Input impedance versus frequency for different L_2

($W_1 = 30mm$, $H_1 = 25mm$, $L_1 = 4.95mm$, $h = 6.6mm$ and $D = 2mm$)

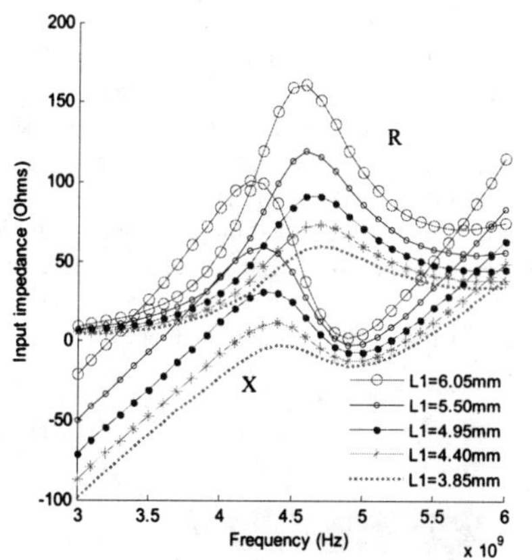

Figure 9. Input impedance versus frequency for different L_1

($W_1 = 30mm$, $H_1 = 25mm$, $L_2 = 10mm$, $h = 6.6mm$ and $D = 2mm$)

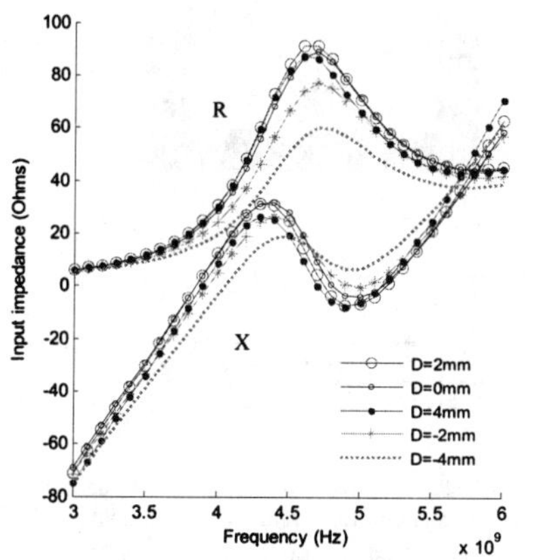

Figure 11. Input impedance versus frequency for different D

($W_1 = 30mm$, $H_1 = 25mm$, $L_1 = 4.95mm$, $L_2 = 10mm$ and $h = 6.6mm$)

IV. CONCLUSION

In this paper, the L-probe patch antenna in a multilayered spherical media is studied by employing the method of moments. The RWG function is regarded as bases function and test function for solving the mixed potential integral equation. Numerical results for input impedance, far-field radiation pattern are presented and verified by comparison with the results from commercial software FEKO. The effects of various parameters on the characteristics of the L-strip patch microstrip antenna have been investigated. From results obtained, we can see that the input impedance is sensitive to W_1, L_1 and L_2. The resonant frequency is determined by the parameter H_1. The effects of various parameters here are similar to that in planar case.

REFERENCES

[1] H. F. Pues and A. R. Van de Capelle, "An impedance matching technique for increasing the bandwidth of microstrip antennas," *IEEE Trans. Antennas Propagat.*, vol. 37, pp.1345-1354, Nov. 1989.

[2] E. Chang, S. A. Long, and W. F. Richards, "Experimental investigation of electrically thick rectangular microstrip antennas," IEEE Trans. Antennas Propagat., vol. AP-43, pp. 767-772, 1986.

[3] Y. X. Guo, C. L. Mark and K. M. Luk, "Analysis and design of L-probe proximity fed-patch antennas," *IEEE Trans. Antennas Propagat.*, vol. 49, pp. 145-149, Feb. 2001.

[4] C. L. Mark, K. M. Luk and K. F. Lee, "Experimental Study of a microstrip patch antenna with an L-shaped probe," *IEEE Trans. Antennas Propagat.*, vol. 48, pp. 777-783, May 2000.

[5] J. Park, H. Na and S. Baik, "Design of a modified L-probe fed microstrip patch antenna," *IEEE Antennas Wire. Propagat. Let.*, vol. 3, pp. 117-119, 2004.

[6] H. Wong, K. L. Lau and K. M. Luk, "Design of dual-polarized L-probe patch antenna arrays with high isolation," *IEEE Trans. Antennas Propagat.*, vol. 52, pp. 45-52, Jan. 2004.

[7] X. Y. Cao, P. Li and K. M. Luk, "Efficient analysis of L-probe coupled patch antenna arrays mounted on a finite conducting cylinder," *Micro. Opt. Techno. Let.*, vol. 41, pp. 403-407, 2004.

[8] S. K. Khamas, "Electromagnetic radiation by antennas of arbitrary shape in a layered spherical media," *IEEE Trans. Antennas Propagat.*, vol. 57, pp. 3827-3834, Dec. 2009.

[9] C. Butler, "The equivalent radius of a narrow conducting trip," *IEEE Trans. Antennas Propagat.*, vol. AP-30, pp. 755-758, Jul. 1982.

[10] L. W. Li, P. S. Kooi, M. S. Leong and T. S. Yeo, " Electromagnetic dyadic Green's function in spherically multilayered media," *IEEE Trans. Micro. Theory and Techniques,* vol. 42, pp. 2302-2310, Dec. 1994.

[11] S. K. Khamas, "Asymptotic extraction approach for antennas in a multilayered spherical media," *IEEE Trans. Antennas Propagat.*, vol. 58, pp. 1003-1008, Mar. 2010.

[12] S. K. Khamas, "A Generalized asymptotic extraction solution for antennas in multilayered spherical media," *IEEE Trans. Antennas Propagat.*, vol. 58, pp. 3743-3747, Nov. 2010.

[13] S. M. Rao, D. R. Wilton and A. W. Glisson, "Electromagnetic scattering by surfaces of arbitrary shape," *IEEE Trans. Antennas Propagat.*, vol. AP-30, pp. 409-418, May 1982.

[14] T. Yu, C. Y. Yin and H. Y. Liu, "Full-wave analysis for antennas in the presence of the metal sphere based on RWG-MoM," *Journal of Microwaves (Chinese)*, vol. 28, pp. 23-28, Feb. 2012.

Research on Circularly Polarized Small Disk Coupled Square Ring Microstrip Antenna for GPS Application

PENG Cheng, YU Tongbin, LI Hongbin, CAO Wenquan

***Abstract*—A circularly polarized (CP) small Disk coupled square ring microstrip antenna is presented in this paper. The CP radiation of the presented antenna is achieved by using two disks under the square ring fed by two probes with 90-degree phase shift. The resonant frequency of the antenna is at 1.575GHz, Which is lowered by about 23.4% as compared to that of the normal patch antenna. The axial ratio (AR) of the antenna is lower than 3dB and the maximum gain is more than 3dB at the resonant frequency, and the isolation between the two disks is lower than -18dB during the operating band. The measured results of the fabricated antenna agree well with the simulated results.**

***Index Terms*—Circularly polarized, small Disk coupled, Square ring, microstrip antenna**

I. Introduction

MICROSTRIP antenna has been widely used in the cellular communication systems. As the integration of the communication equipment develops, the technique of miniaturized micro-strip antenna has been paid more attention.

The square-ring microstrip antenna is usually smaller than the normal patch antenna (because the current path has been lengthened) [1], which has been paid more and more attention these years. But, as the size of the antenna reduces, the input impedance grows fast. So it is necessary that the inside diameter W1 of the square ring is less than 40% of the outside diameter W2 for the probe-feed method to achieve 50Ω input impedance match[2], and this becomes a problem for the miniaturized micro-strip antenna. Reference [3] added a Cross strip which is fed by probes to achieve 50Ω input impedance match, and the center frequency was lowered by about 22% with the same antenna size. Reference [4] used stacking technique to get input impedance match and to enhance the bandwidth, but it would enlarge the size of the antenna. Reference [5] used the circumferential variation of current on the ring to achieve 50Ω input impedance match, but it did not give antennas for circular polarization. Reference [6] put a rectangular disk under the square-ring patch for the coupling-feed, but the size of the antenna is too big and circular polarization is hard to achieve..

This paper uses dual feed method to design a circularly polarized microstrip antenna [7], and we found that the disk coupled method can easily achieve 50Ω input impedance and reduce the size of the square ring further more.

978-7-5641-4279-7

II. Antenna Design

The circularly polarized small disk coupled microstrip antenna uses double layer structure as showed in Fig 1. The size of the substrate and the ground plane is $Lg \times Lg$, the height of the bottom substrate is $h1$, the height of the top layer is $h2$, and both of the two layers have the relative permittivity ξ. The square ring is on the surface of the top layer, with the inner side length W1 and the outer side length $W2$. In order to make the antenna radiate circularly polarized electromagnetic wave, we use the dual feed method [8]. The square ring is coupled with two circular disks fed by probe, and each disk has a radius r. The distance between the center of the disk and the center of the square ring side nearby is d, which is one of the key parameter for the 50Ω input impedance match.

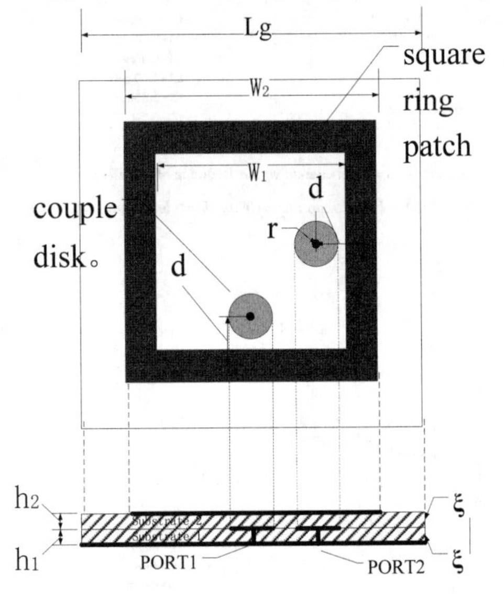

Fig. 1. The structure of the small disk coupled microstrip antenna.

The antenna has been built and simulated using ANSOFT HFSS software. By adjusting the parameter d, we could easily achieve the 50Ω input impedance when W_1、W_2 and r (the radius of the circular disks) is assumed. In order to make the antenna operating at GPS band(the center frequency is at 1.575GHz), the antenna size is optimized as Lg=50mm,

W_1=24mm, W_2=35mm, h_1=2mm, h_2=1mm, ξ=3.5, r=3.9. The input impedance of the antenna changes with d, as we can see from Table I, and we can find the best position of the two disks to achieve 50Ω input impedance.

TABLE I

d/mm	Resonant frequency/GHz	-10dB bandwidth/MHz	Minimum S11/dB
1	1.58	0	-2.01
2	1.58	0	-2.23
3	1.58	0	-3.76
4	1.58	0	-4.77
5	1.58	0	-8.64
6	1.58	18	-13.51
7	1.59	0	-4.60
8	1.60	0	-1.53

In Table I we find that we can get good impedance match when d is close to 6mm. When we optimized d as 5.85mm, we get the minimum S_{11} of -21.59dB and the -10dB impedance bandwidth of 20MHz.

The disk coupled method is a good solution to achieve 50Ω input impedance. But when we use dual feed method to get circular polarization antenna, the mutual coupling between the two capacitive disks becomes a problem [8]. The isolation between the two probes changes with the radius of the disks, and it is hard to get good performance of the antenna. Therefore, r is another key parameter that can both affect the isolation between the two probes and the input impedance match.

TABLE II

r/mm	Maximum S_{12}/dB	-10dB bandwidth/MHz	Minimum S_{11}/dB
0.9	-40.01	0	-0.36
1.9	-29.35	0	-1.22
2.9	-23.71	0	-5.89
3.9	-18.64	20	-21.59
4.9	-10.01	8	-10.58
5.9	-20.01	0	-5.28

From Table II we can find that when the position of the feeding probes are assumed, S_{12} changes with the size of the disk. When r increases, the absolute value of S_{12} reduces. The reason is that the mutual coupling becomes more intensive when the disks are close to each other. When the radius of the two disks is bigger than 4.9mm, the absolute value of S_{12} increases again. When r grows more than 6mm, the two disks will connect to each other. The size of the disk can also affect the 50Ω input impedance match [9]. When the radius of the two disks is set as r=3.9mm the best 50Ω input impedance match is achieved, while the maximum S_{12} is -18.64dB. Since this is enough for the dual-feeding network, so we choose r as 3.9mm

III. SIMULATED AND MEASURED RESULTS

After the parameters of the antenna are chosen according to simulated results, the antenna is fabricated. The fabricated antenna and its feeding network can be seen clearly in Fig 2(a) and Fig 2(b). The antenna use two disks fed by one willkins power divider under the ground plane.

Fig 2(a) Th e patch

Fig 2(b)The feeding network

Fig. 2. The small disk coupled square ring microstrip antenna and its feeding network. There are two disks fed under the square ring. It is multi-layer structure

Fig.3 shows the simulated and measured plots of the gain and the axial ratio of the fabricated antenna. The measured results agree well with the simulated results. From Fig3(a) we can see that the 3dB-beam width has reached 120 degree while the maximum gain is about 3dB. The maximum gain of the fabricated antenna is lower than that of the normal patch antenna, this may be caused mainly by beam width enlarging and size reduction. We can also see that the measured result is a bit lower than the simulated result, it may be caused by the imprecise fabrication. Fig3(b) shows that the AR is lower than 3 dB when theta is from -60 degree to +60 degree.

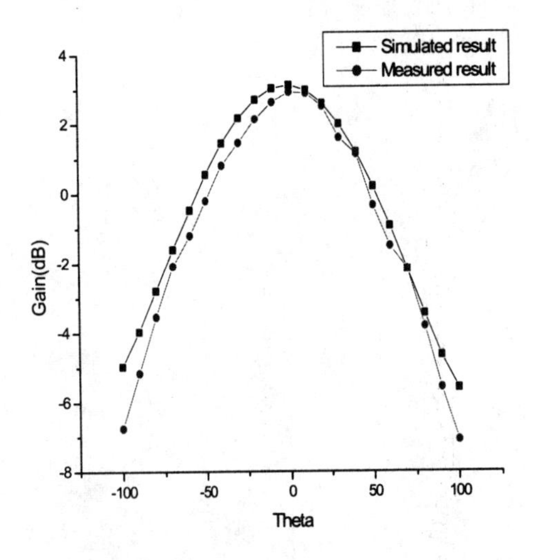

Fig 3(a) Simulated and measured gain at resonant frequency

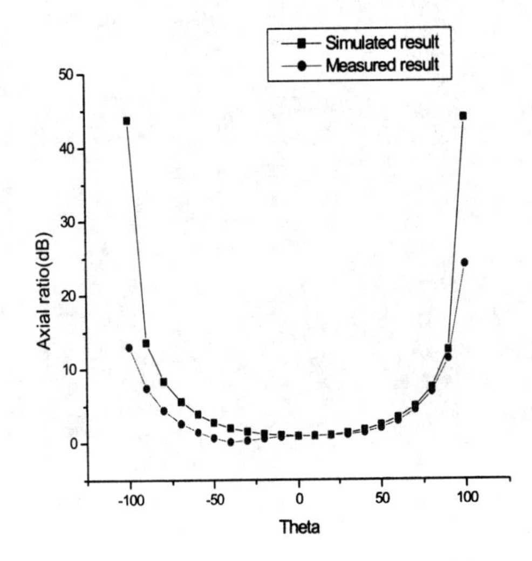

Fig 3(b) Simulated and measured AR at resonant frequency

Fig 3 Simulated and measured plots of the gain and AR at resonant frequency

A comparison is made between the small disk coupled square ring and the normal rectangular patch antenna by the same size, Fig 4 shows the different reflection coefficient from input port(S_{11}):

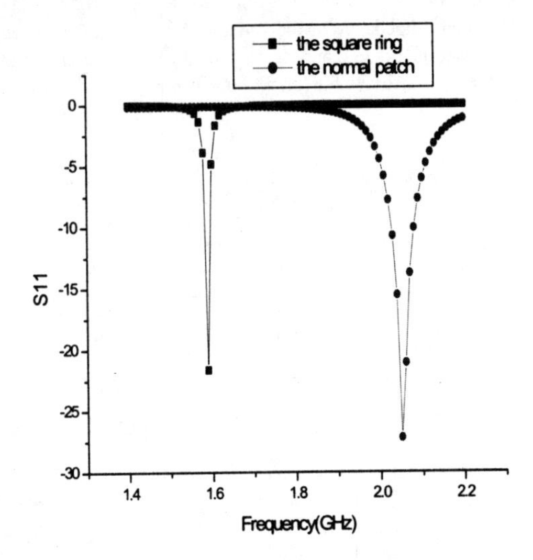

Fig 4 The reflection coefficient of the two antennas

From Fig 4 we can see that the center frequency of the small disk coupled square ring (1.575GHz) is lowered by about 23.4% as compared to that of the normal patch (2.05GHz). This means that a good antenna size reduction can be obtained. Though the -10dB bandwidth becomes narrower (from 50MHz to 20MHz), it is still enough for the GPS Applications (only need 5MHz). Working at the GPS band (the center frequency is at 1.575GHz) [10], the small disk coupled square ring microstrip antenna has a 27% size reduction compared with the normal patch antenna.

IV. CONCLUSION

A successful design of circularly polarized small disk coupled square ring microstrip antenna is presented in this paper. Compared with the normal patch antenna, the antenna gets a 27% size reduction and wider beam width. The input impedance match in the square ring microstrip antenna is achieved by using two disks fed by two probes. The mutual coupling between the two disks is analyzed in detail. The proposed antenna has been fabricated and tested, the results agree well with the simulated results.

REFERENCES

[1] S. ROW, "Design of square-ring microstrip antenna for circular polarization" *Electron. Lett.*,vol.40,pp.93-95,Feb.2004.
[2] P.Moosavi and L.Shafai. "Characteristics and design of microstrip square ring antennas" , in Proc. 5th Iranian Conf. Electrical Engineering-ICEE97, Tehran, Iran, May 7-9, 1997
[3] Wen-Shyang Chen, Chun-Kun Wu and Kin-Lu Wong, "Square-Ring Microstrip Antenna with a cross strip for compact circular polarization operation", *IEEE Trans, Antenna propgat*, vol, 47,pp, 1566-1567, Oct, 1999.
[4] Pedram Moosavi Bafrooei,and Lotfollah Shafai, "Characteristics of Single- and Double-Layer Microstrip Square-Ring Antennas," *IEEE Trans, Antenna Propagat*, vol,47, pp, 1633-1639, Oct, 1999.
[5] R.Garg and V.S.Reddy, "Edge feeding of microstrip ring antennas," *IEEE Trans. Antennas Propagat.* vol,51, pp, 1941-1946, Oct, 1999
[6] Amit A. Deshmukh ,Ray, K. P. Ray, and Chine, P. N, "Broadband reactively coupled ring microstrip antenna," Applied Electromagnetic Conference (AEMC),pp,1-4,Dec,2009.
[7] Agarwal, Kshitiz: Rao, G.purnachandra; Kartikeyan, M.V; Thumm, M,K, "Wideband dual feed electromagnetically coupled circularly polarized microstrip patch antenna," *Infrared Millimeter wave and 14th International conference on teraherz electronics 2006.* pp:438-438, 12-22 sept. 2006
[8] C. J. Kaufman,LI jian-feng, SUN Bao-hua, ZHANG jun, LIU Qi-zhong, " Dual frequency and dual circular polarization antenna with high isolation for satellite communication," *ACTA ELECTRONICA SINICA*, Dec, 2009: Vol. 37 No. 12.
[9] Wenquan Cao, Bangning Zhang, Tongbin Yu, Hongbin Li "A single-feed broadband circular polarized rectangular microstrip antenna with chip-resistor loading " *IEEE Trans, Electron Letterss.*, vol. 53, no. 3, pp. 1227–1229, March. 2010.
[10] Shun-Yun Lin; Kuang-Chih Huang "A compact microstrip antenna for GPS and DCS application," *IEEE Trans, Antenna propgat,.*, vol. 53, no. 3, pp. 1227–1229, March. 2005.

A 35GHz Stacked Patch Antenna with Dual-Polarized Operations

Xue-Xia Yang, Guan-Nan Tan, and Ye-Qing Wang
School of Communications and Information Engineering
Shanghai University
Shanghai 200072, China

***Abstract*—A slot-coupled stacked patch antenna operating at 35GHz with dual-polarized operations is presented in this paper. To achieve wide bandwidth and high gain at 35GHz, four parasitic patches are located above the active element. The measured results show that the frequency bandwidth of S_{11} less than -10dB covers 30-40GHz and the isolation between two polarization ports is higher than 20dB at the center frequency of 35GHz. The peak gain at is 9dBi. Both theoretical and experimental results of reflection coefficient, isolation, and radiation patterns are presented and discussed.**

***Index Terms*— stacked patch antennas, dual-polarization, millimeter wave**

I. Introduction

Since 1960s, space solar power transmission (SPT) and microwave wireless power transmission (WPT) have become an interesting topic for an energy transmission [1]. It involves the conversion of DC power into microwave power at the transmitting end and forming the microwave power into electronically steerable microwave beams. The microwave energy is launched into space through a transmitting antenna. At the receiving end, the receiving antenna captures it back and the energy is then converted back into DC power [1, 2]. So, high gain receiving antenna is very important for the WPT system. Recently, microwave power transmission at millimeter wave is noticeable in biomedicine, fractionated reconfigurable satellites.

The microstrip antennas are popularly applied in communication and radar systems because of many good characteristics, such as low profile, ease to be integrated, and low cost [3]. By selecting the suitable dielectric constant and thickness of the substrate, microstrip antennas can be designed to operate on millimeter-wave [4-6]. In [4], a 2×2 microstrip patch antenna array operating at 60GHz is designed. However, the bandwidth is too narrow. In [5], a wideband 60GHz microstrip grid-array antenna is proposed based on the LTCC technology, which is relatively expensive now.

A microstrip antenna operates on resonant mode so its operation bandwidth is narrow. Some improvements have been suggested to expand the impedance bandwidth. In reference [7], a parasitic center patch is used to enhance the S11≤-10dB bandwidth of a printed wide-slot antenna to 80% (2.23-5.35GHz). In reference [8], a wide band dual-beam microstrip antenna is proposed by using the U-slot technique. The antenna operating frequency range is 5.18-5.8GHz with VSWR less than 2, which corresponds to 11.8% impedance bandwidth at 5.5GHz.

Usually, the WPT systems require communication function [9, 10]. In [9], a two-port with dual-polarization printed microstrip rectenna is presented. One port is used for power receiving at 5.78GHz, and the other is used for data communication at 6.1GHz. In [10], a triple-band antenna is suggested for biotelemetry with data telemetry (402MHz), wireless powering transmission (433MHz), and wake-up controller (2.45GHz).

In this paper, a 35GHz slot-coupled stacked antenna operating on two orthogonal linear polarization states is proposed. The simulation and measurement show that the antenna has a wide bandwidth, high gain and good isolation between two polarizations. This dual-polarized patch antenna can be used as the receiving antenna in a wireless microwave power transmission system with communication function.

II. Description of Antenna

The geometry of the proposed antenna is shown in Figure 1. There are three dielectric substrate layers, which are labeled as layer a, b and c. The three substrate layers are all Rogers RT/duroid 5880 with the relative permittivity of 2.2 and the dielectric loss tangent $\tan\delta$ of 0.0009. Three layers have different thickness of h_a, h_b and h_c. The active patch is placed on the top of layer b. Four parasitic patches on the top of layer c are placed over the four corners of the active patch. The antenna has two feed ports of port 1 and port 2. The characteristic impedance of the two feed lines on the bottom of layer is 50 Ω. Two perpendicular slots on the ground plane are used to couple two orthogonal linear polarization waves. The bottoms of two feed lines are bended to realize impedance match.

Not only the bandwidth but also the gain of this patch antenna is improved by two methods. The first one is the stacked multi-layer patch structure. The second one is the four parasitic patches. The perpendicular coupling slots leading to a high isolation between two linear polarization ports and low cross-polarization.

The antenna performances is determined by many geometrical parameters, such as the sizes and position of two slots, the side lengths of the active and parasitic patches, and the distance between parasitic patches and feed lines.

This work was supported by the National Natural Science Foundation of China (No. 61271062), and STCSM (08DZ2231100).
978-7-5641-4279-7

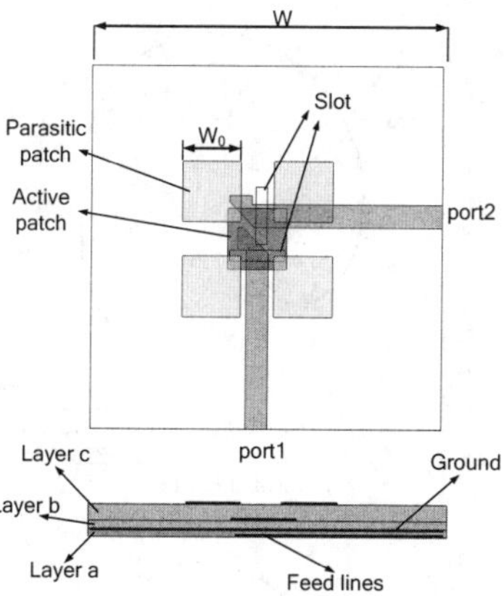

Figure 1 The structure of the proposed antenna

III. Simulation and Measurement

The optimized dimension of the antenna is shown in table 1. The slots dimensions are 0.4mm×2mm. The size of the parasitic patch is the same as the active patch. The width of the feed line is 0.8mm. The proposed antenna is optimized and simulated using the 3D-fullwave simulator of Ansoft HFSS 12.

Table 1 Dimension of the proposed antenna

w	w_0	h_a	h_b	h_c
26	2.1	0.254	0.254	0.508

The Fabricated antenna is presented in Fig. 2. The ground is larger than layer b to measure the antenna conveniently. The microstrip feed line of the antenna was connected to an end launch connector 149-02A-5 of Southwest Microwave. The reflection coefficient was measured by the vector network analyzer of Agilent PNA-N5227A.

The simulated and measured reflection coefficient S_{11} and S_{22} for two feed ports are shown in Fig. 3. The bandwidths of S_{11} and S_{22} less than -10dB cover from 30GHz to 40GHz, which is about 28%. Due to a slight asymmetry of the two feed lines and slots, S_{11} and S_{22} curves are not exactly the same. The simulated and measured isolation between two polarization ports are presented in Fig. 4. Due to limitation of test fixture, the measured S_{21} has many ripples. But the trend is relatively the same. At 35GHz, the isolation is about 22dB. Throughout the frequency range from 30GHz to 40GHz, the isolations are all higher than 20dB. The high isolation of this dual-polarized antenna is very useful for many practical applications.

(a) Front view

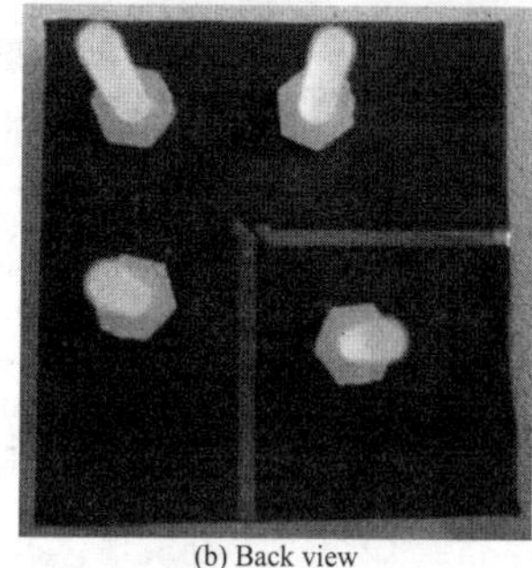
(b) Back view

Fig. 2 Fabricated antenna

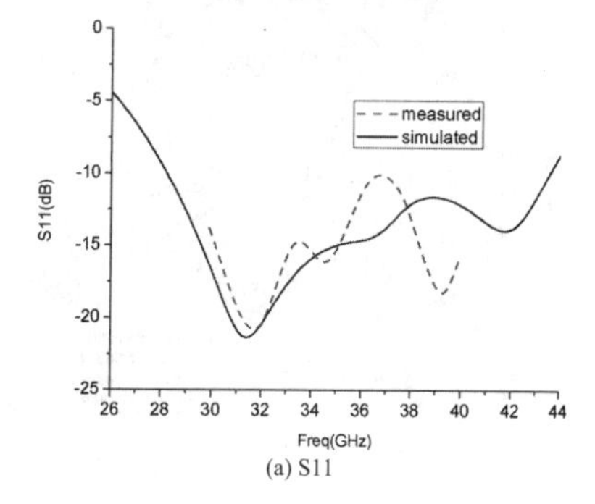

(a) S11

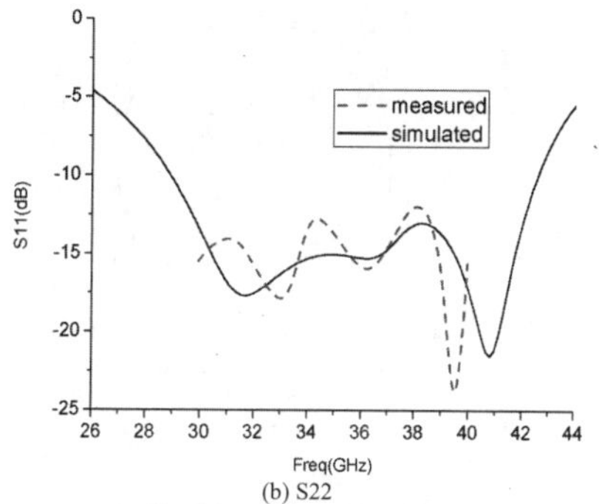

(b) S22

Fig. 3 S_{11} and S_{22} vs. frequency

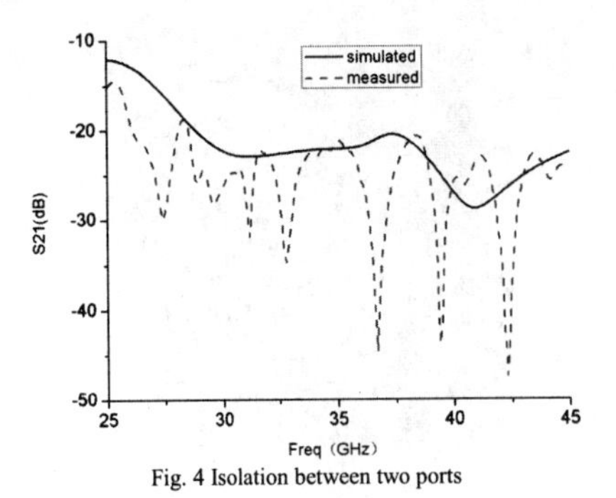

Fig. 4 Isolation between two ports

The radiation characteristics were measured in the anechoic chamber. The simulated and measured E plane and H plane radiation and cross-polarization patterns at the frequency of 35GHz of port1 and port2 are presented in Fig. 5, Fig. 6, respectively. It can be seen that the simulated results almost coincide with the measured ones . The cross-polarization levels are lower about -20dB than the main polarization. The simulated and measured antenna gain between 30GHz and 40GHz of port1 and port2 are shown in Fig. 7. The measured gains of the two ports are all about 9dBi at 35GHz.

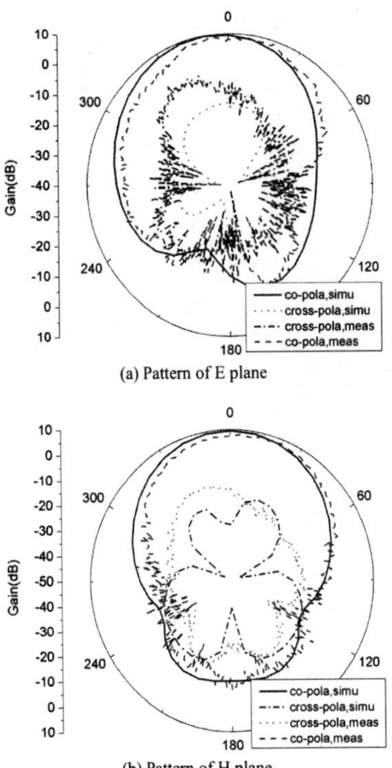

(a) Pattern of E plane

(b) Pattern of H plane

Fig. 5 Antenna pattern of port1 at 35GHz

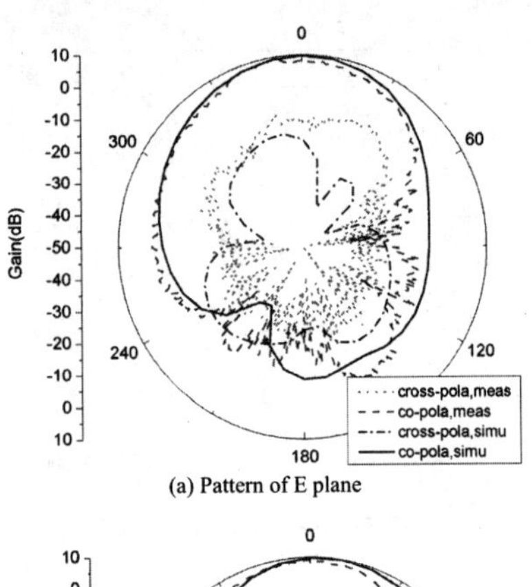

(a) Pattern of E plane

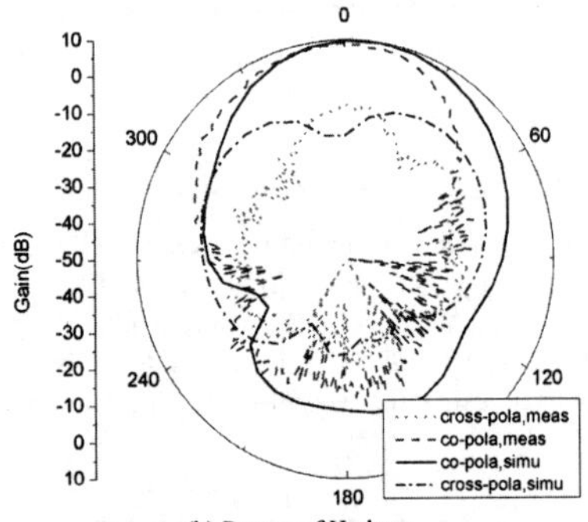

(b) Pattern of H plane

Fig. 6 radiation patterns of port2 at 35GHz

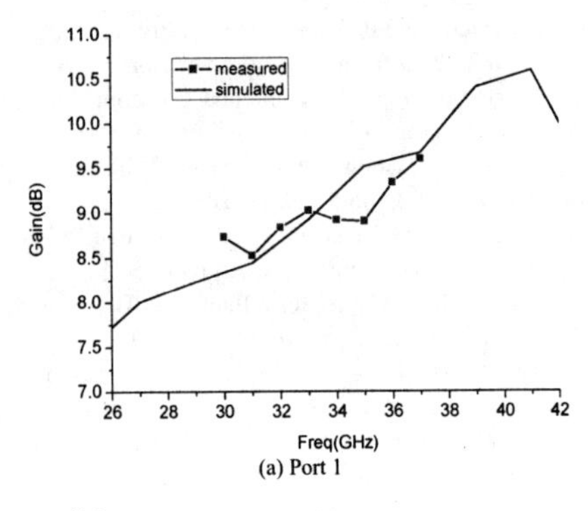

(a) Port 1

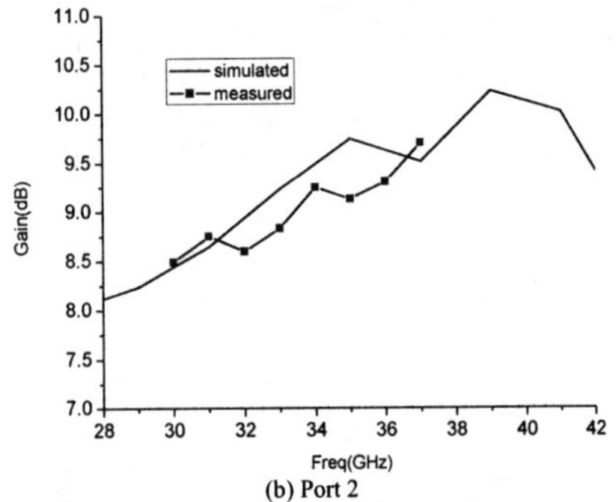

(b) Port 2

Fig. 7 Measured and simulated gains vs. frequency

IV. CONCLUSIONS

This paper proposed a 35GHz slot-coupled stacked patch antenna operating on dual orthogonal linear polarization states for wireless microwave power transmission. One linear polarized wave excited by port 1 is used to receiving microwave power, while another one activated by port 2 could be used to communicate. This antenna has good performances of wide-bandwidth, high isolation, low cross-polarization level and high gain. The measured reflection coefficient exhibits an impedance bandwidth over 30~40GHz and the isolation between two polarization ports is better than 20dB over the bandwidth. The cross-polarization levels in both E and H planes are lower than -22dB. This dual-polarization antenna is easy to be expanded to a large-scale array.

REFERENCES

[1] B. Strassner and K. Chang, "Microwave Power Transmission: Historical Milestones and System Components," Proceedings of the IEEE, vol. 101, no. 6, pp. 1379-1396, 2013.

[2] R. M. Dickinson, "Power in the sky: Requirements for microwave wireless power beamers for powering high-altitude platforms," Microwave Magazine, IEEE, vol. 14, no. 2, pp. 36-47, 2013.

[3] Kin-Lu Wong, Compact and broadband microstrip antennas, New York, John Wiley & Sons Inc., 2002.

[4] B. Biglarbegian, M. Fakharzadeh, D. Busuioc, M. R. Nezhad-Ahmadi, and S. Safavi-Naeini, "Optimized Microstrip Antenna Arrays for Emerging Millimeter-Wave Wireless Applications," IEEE Transactions on Antennas and Propagation, vol. 59, no. 5, pp. 1742-1747, 2011.

[5] S. Zhang and Y. P. Zhang, "Analysis and synthesis of millimeter-wave microstrip grid-array antennas," IEEE Antennas and Propagation Magazine, vol. 53, no. 6, pp. 42-55, 2011.

[6] A. Bondarik, J. Dong-Suk, K. Joung-Myoun, and H. Y. Je, "60 GHz system-on-package antenna array with parasitic microstrip antenna single element," Asia-Pacific Microwave Conference 2008, APMC 2008, pp. 1-4, Dec. 2008.

[7] Y. Sung, "Bandwidth Enhancement of a Microstrip Line-Fed Printed Wide-Slot Antenna With a Parasitic Center Patch," IEEE Transactions on Antennas and Propagation, vol. 60, no. 4, pp. 1712-1716, 2012.

[8] A. Khidre, L. Kai-Fong, A. Z. Elsherbeni, and Y. Fan, "Wide Band Dual-Beam U-Slot Microstrip Antenna," IEEE Transactions on Antennas and Propagation, vol. 61, no. 3, pp. 1415-1418, 2013.

[9] Xue-Xia Yang, Chao Jiang, Atef Z. Elsherbeni, Fan Yang, and Ye-Qing Wang, "A Novel Compact Printed Rectenna for Data Communication Systems", IEEE Transactions on Antennas and Propagation, Vol. 61, no. 5, pp. 2532-2539, 2013.

[10] H. Fu-Jhuan, L. Chien-Ming, C. Chia-Lin, C. Liang-Kai, Y. Tzong-Chee, and L. Ching-Hsing, "Rectenna Application of Miniaturized Implantable Antenna Design for Triple-Band Biotelemetry Communication," IEEE Transactions on Antennas and Propagation, vol. 59, no. 7, pp. 2646-2653, 2011.

Design of a Circularly Polarized Elliptical Patch Antenna using Artificial Neural Networks and Adaptive Neuro-Fuzzy Inference System

Aarti Gehani, Jignesh Ghadiya, and Dhaval Pujara
Deptt. of Electronics & Communication Engg.
Institute of Technology, Nirma University,
Ahmadabad-382481, Gujarat, India
E-mail: aarti.gehani@nirmauni.ac.in, dhaval.pujara@nirmauni.ac.in

Abstract—**In the present paper, a circularly polarized elliptical patch antenna is designed using artificial neural networks (ANN) and adaptive neuro-fuzzy inference system (ANFIS). The resonant frequency and gain are taken as the input while the radius of semi-major axis and heights of the substrate are taken as the outputs. From the numerical results, it is observed that the errors in the output parameters are less in case of an ANFIS model as compared to that of an ANN model. With ANFIS model, a parametric study with different types and numbers of membership functions is also presented.**

I. Introduction

The recent development in the wireless communication has given rise to the need of low weight, small size, cheap and easily installable antennas. Microstrip patch antennas fulfill all such requirements. They are low profile antennas and are suitable for planar as well as non-planar surfaces. Depending upon the resonant frequency and the excited mode, microstrip antennas are capable of providing linear as well as circular polarization. Circular polarization is preferred in many practical applications, such as radar, communication, navigation system, etc. Circular polarization may be obtained by using multiple feeds [1] or by altering the shape of microstrip patch antenna [2]. It is reported in the open literature [3] that, an elliptical patch antenna provides circular polarization with only a single feed. However, elliptical geometry is the least analyzed geometry, due to the involvement of complex Mathieu functions in the mathematical analysis. The conventional techniques for designing and analyzing elliptical microstrip antennas are difficult to implement and are time consuming. In order to avoid the complexity of the conventional techniques, two soft computing techniques viz. artificial neural networks [4] and adaptive neuro-fuzzy inference system [5] are proposed in the present paper.

In [6], the synthesis problem of an elliptical microstrip patch antenna is discussed, considering single material with resonant frequency as the output. The present work aims at the comparison of the results obtained using ANN and ANFIS for circularly polarized elliptical patch antenna. Moreover, for ANFIS, a study of effect of number of membership functions and type of membership functions is also carried out. Section II discusses the development of the CAD model and Section III presents the results and the analysis.

II. Elliptical Patch Antenna

Figure 1 shows the geometry of the elliptical patch antenna under consideration, where a is the semi-major axis, b is the semi-minor axis and a_{eff} is the effective semi-major axis. In order to obtain circular polarization, the feed was located at 45° with respect to the major axis. By changing the feed location, the relative amplitudes of x directed and y directed electric fields is changed. When the feed is moved towards the major axis, it enhances the excitation of current flowing parallel to major axis and thus results in greater amplitude of E_x [7]. The elliptical antenna is inherently elliptical polarized antenna but when the antenna is coupled through a feed located at 45° azimuthal angle then circular polarization is obtained.

III. Development of the CAD Model

In the present paper, for designing of elliptical patch antenna, two CAD models have been developed. The first model is based on ANN, while the second model is based on ANFIS concept. The results obtained using both the models are then compared.

As shown in Figure 2, the resonant frequency (f_r) and the gain (g) are taken as the input, while the radius of major-axis (a) and height of the substrate (h) are considered as the outputs.

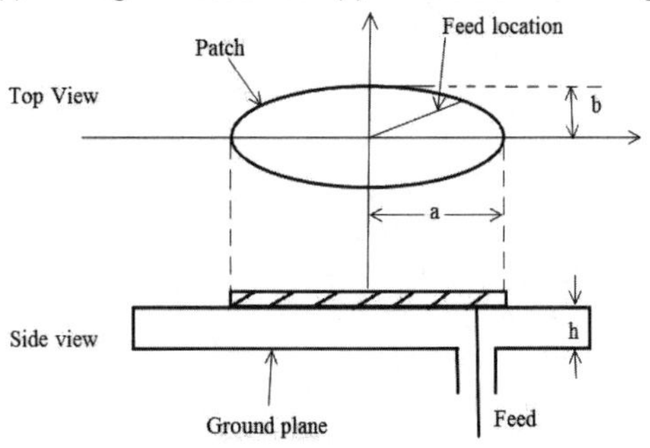

Figure 1. Elliptical patch antenna geometry under consideration

978-7-5641-4279-7

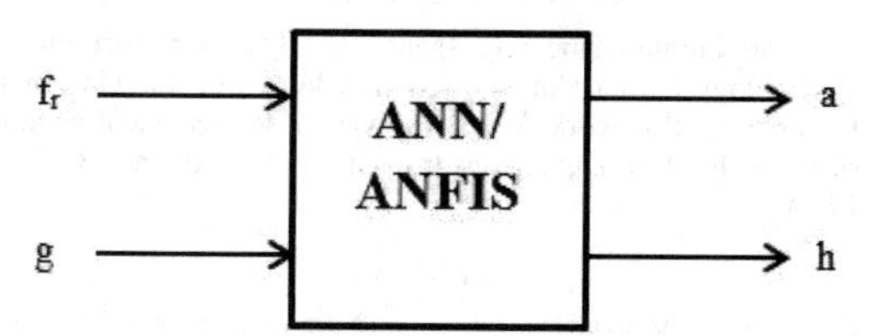

Figure 2. CAD Model for design of Elliptical Patch Antenna

ANNs are non-linear mapping structures, which are principally based onthe functions of the human brain. It consists of simple computational units called neurons, which are highly interconnected.The connections between the neurons are called as links. Every link has a weight parameter associated with it. Each neuron receives stimulus from the neighboring neuron connected to it, process the information and produces an output. The structure of ANN consists of an input layer, an output layer and one or many hidden layer/s. The number of hidden layers and number of neurons therein does not follow any thumb rule. However, they should be chosen judiciously so that the network is neither under trained nor over-trained. In the present paper, radial basis function type artificial neural network (RBF-ANN) is used to obtain the geometrical parameters of elliptical patch antenna using spread constant = 1. Different values of spread constants were verified but the most appropriate was spread constant = 1.

ANFIS is a very powerful soft-computing technique which combines the best features of Fuzzy Inference Systems (FISs) and ANN. By using a hybrid learning procedure, the ANFIS can construct an input-output mapping based on both human knowledge (in the form of fuzzy if-then rules) and stipulatedinput-output data pairs. Also ANFIS combines the gradient descent method and the least squares estimate (LSE) to identify its parameters. LSE is used in the forward pass while gradient descent is used in the backward pass.ANFIS structure consists of five different layers. The first layer is the input data which is fuzzified and the neuron values are represented by parameterized membership function. The second layer deals with calculation of activation of fuzzy rules via differentiable T-norms. The third layer realizes a normalization operation over the matching values of the rule. The fourth layer multiplies the normalized activation level and the output of the respective rule while the last and the fifth layer produce the output by an algebraic sum. Unlike ANN, ANFIS does not require selection of number of layers a priory. However, the only structural limitation is that, it should be feed-forward type. The present ANFIS model employed '*gbell*' type membership function. Also, five membership functions have been considered after rigorous simulations.

III. Results and Discussion

For the purpose of training and testing, data set was generated by simulating elliptical patch antenna using full wave solver (FWS). For obtaining circular polarization, eccentricity 'e' was kept constant as suggested in [7]. Out of the total generated data set, 70% data was utilized in training the networks, while remaining 30% data was used in testing the models.

Sample results for two different materials, Rogers3003 and FR4 for both ANN and ANFIS models are presented in Table I and Table II, respectively. In case of ANFIS model, the study of different membership functions (*gbell*, *gauss*, *pimf*) was also carried out and the results are summarized in Table III. Moreover, for ANFIS model, the effect of number of membership function on the result is summarized in Table IV.

The formula used for calculating the percentage error is:

$$\%\ \text{Error} = \frac{\text{Approximate value} - \text{Actual value}}{\text{Actual value}} \times 100 \quad (1)$$

where, the value obtained from FWS is considered as actual value and that obtained from the CAD model (ANN and ANFIS) is considered as approximate value.

Table I
Results of RBF-ANN and ANFIS Model for FR4 Material

Input		Output (FWS)		% Error (RBF-ANN)		% Error (ANFIS)	
f_r (GHz)	g (dBi)	a (mm)	h (mm)	a	h	a	h
1.14	0.04	36.5	0.762	0.05	0.65	0	0
1.19	0.03	35.0	0.762	0.20	0.52	0	0
1.69	1.33	24.5	1.524	0.36	0.91	0	0
1.22	2.44	34.0	1.524	0.64	0	0	0

Table II
Results of RBF-ANN and ANFIS Model for Rogers3003 Material

Input		Output (FWS)		% Error (RBF-ANN)		% Error (ANFIS)	
f_r (GHz)	g (dBi)	a (mm)	h (mm)	a	h	a	h
1.55	4.37	33	0.508	1.3	0.78	0	0
1.25	5.54	40	0.762	0.52	0.26	0.32	0
1.18	5.65	43	0.762	0.72	0.39	0.18	0.13
1.08	6.83	46	1.524	0.04	1.15	0.04	0.06

Table III
Study of Membership Functions in ANFIS Model for Input-Output Pairs of Table II

% Error for '*gbell*'		% Error for '*gauss*'		% Error for '*pimf*'	
a	h	a	h	a	h
0	0	0	0	0	0
0.32	0	0.2	0	0.22	0
0.18	0.13	0.09	0	0.09	0.13
0.04	0.06	0	0.06	0.02	0.06

TABLE IV
EFFECT OF CHANGE IN NUMBER OF MEMBERSHIP FUNCTIONS IN ANFIS MODEL FOR FR4 MATERIAL

Input		Output (FWS)		% Error for 3 membership functions		% Error for 5 membership functions	
f_r (GHz)	g (dBi)	a (mm)	h (mm)	a	h	a	h
1.14	0.04	36.5	0.76	0.11	0.45	0	0
1.19	0.03	35.0	0.76	0.03	0.35	0	0
1.69	1.33	24.5	1.52	0.03	0.18	0	0
1.22	2.44	34.0	1.52	0.03	0.07	0	0

IV. CONCLUSION

Comparison of results generated using RBF-ANN and ANFIS shows that the percentage error obtained in ANFIS model is less as compared to that obtained by RBF-ANN model. For the present design, '*gauss*' type of membership function gives relatively better results. Also, in case of ANFIS, as the number of membership function increases, the error reduces. The ANFIS model requires less time and is accurate in prediction. Thus, for antenna problems, ANFIS technique is simple, easy to apply, very useful and efficient once trained.

ACKNOWLEDGEMENT

The authors sincerely thank, the Director, Institute of Technology Nirma University, Ahmedabad for allowing them to carry out this work. We acknowledge the technical support provided by Prof. Dipak adhyaru, Mr. Anuj Modi and Mr. Jigar Mehta.

REFERENCES

[1] G. Stanford, "Conformal microstrip phased array for aircraft tests with ATS-6," IEEE Transactions on Antennas and Propagation, vol 6, no.5, pp. 642-646, September 1978
[2] W. Richards and Y. Lo, "Design and theory of circularly polarized microstrip antennas," Antennas and Propagation Society International Symposium, vol. 17, pp. 117-120, June 1979
[3] I. P. Yu, "Low profile circularly polarized antenna," NASA Rep. N78-15332,1978
[4] Simon Haykin, "Neural Networks A Comprehensive Foundation," 2nd Ed., 1999
[5] Jang, J.-S. R., "ANFIS: Adaptive-Network-based Fuzzy Inference Systems," IEEE Transactions on Systems, Man, and Cybernetics, vol. 23, no. 3, pp. 665-685, May 1993.
[6] A. Agrawal, D. Vakula and N. V. S. N. Sarma, "Design of Elliptical Microstrip Patch Antenna Using ANN," PIERS Proceedings, pp. 264-268, Sept. 2011
[7] L. Shen, "The elliptical microstrip antenna with circular polarization," IEEE Transactions on Antennas and Propagation, vol 29, no.1, pp. 90-94, January 1981.

Circularly Polarized Microstrip Antenna Based on Waveguided Magneto-Dielectrics

Xin Mi Yang, Hui Ping Guo, Xue Guan Liu
School of Electronics and Information Engineering, Soochow University
Suzhou Jiangsu 215021, China

Abstract- **In this paper, we make an investigation on the waveguided magneto-dielectric materials (WG-MDM) based on the embedded meanderline structure. As verified by retrieval results of effective medium parameters, the WG-MDM exhibits nearly the same magneto-dielectric property with respect to the orthogonally applied TEM wave excitations. The WG-MDM is utilized as the artificial substrate of square microstrip patch antenna and the patch is perturbed at two corners to realize circularly polarized radiation. Simulation results show that the impedance and axial ratio bandwidths of the resulting novel circularly polarized patch antenna have increased by 78% and 71%, compared with those of pure-dielectric-substrate-based patch antenna with the same size.**

***Index Terms*—Circular polarization, microstrip atenna, waveguided metamaterial, magneto-dielectrics.**

I. INTRODUCTION

Bandwidth improvement of microstrip antennas is attractive in the antenna community. Recently, increasing the bandwidth of patch antenna by loading artificial magneto-dielectric substrate has aroused interest among researchers [1]-[12]. Generally, replacing conventional dielectric substrate by magneto-dielectric substrate will decrease the electromagnetic energy stored under the patch at resonance, and accordingly leads to lower quality factor and broader bandwidth of patch antenna [1],[2]. Compared with many other broadbanding techniques, the magneto-dielectric substrate technique has theoretically no influence on the radiation chracteristics of microstrip patch and hence may be preferred in many applications. However, researches about this technique have been limited to broadbanding of linearly polarized microstrip antennas up to now. It is due to the fact that common artificial magneto-dielectric materials are not suitable for creating two orthogonal degenerate TM modes required by circularly polarized radiation, since those materials are composed of anisotropic inclusions such as split ring resonators [13], [14], [5],[6] and spiral resonators [13], [15], [4].

This paper introduces a waveguided metamaterial which exhibits nearly the same magneto-dielectric property with respect to the orthogonally applied TEM wave excitations in microstrip plane. This waveguided magneto-dielectric material (WG-MDM) is used to fill in the volume between a perturbed square patch and a ground plane and the resulting novel circularly polarized patch antenna proves to have both improved impedance bandwidth and axial ratio bandwidth. To our knowledge, it is the first report on broadbanding of circularly polarized patch antenna by artificial magneto-dielectric loading.

The proposed patch antenna loaded with the WG-MDM comprises three layers: a normal dielectric layer 1.43mm thick and two metallic layers attached to the two sides of the dielectric layer. The permittivity and loss tangent of the dielectric layer are 2.65 and 0.001, respectively. Fig. 1(a) and (b) illustrate the metal arrangement on the top and the bottom side of the dielectric layer. The circular polarization characteristic is realized by perturbing the square patch at two diagonal corners (Fig. 1(a)). The effective magneto-dielectric material loading is implemented by etching periodic array composed of electrically small planar unit cells in the ground plane right under the patch (Fig. 1(b)). The patch is singly fed by microstrip and a matching line is inserted between the patch and the 50Ω feeding line to achieve good matching (Fig. 1(a)). The whole structure is compact and simple and could be fabricated on copper clad laminate with normal PCB process.

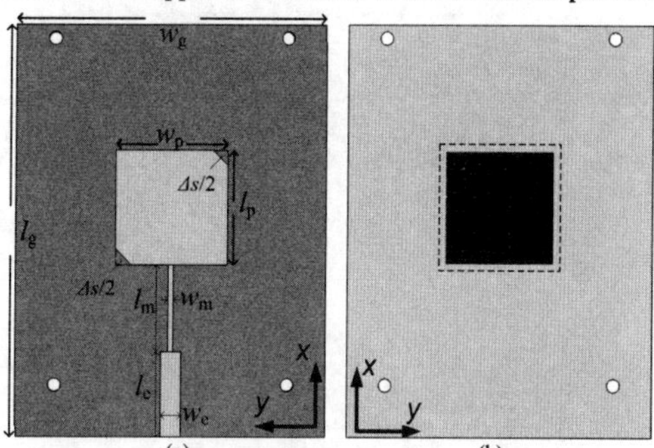

Figure 1. Circularly polarized patch antenna loaded with waveguided magneto-dielectric material.(a) Top view of perturbed square patch and feeding network. (b) Bottom view of ground plane with periodic etching. The region of etching is indicated by a dashed square.

II. WAVEGUIDED MAGNETO-DIELECTRIC MATERIAL

The planar unit cell referred in the last section is called embedded meander line (EML). It characterizes a meander line embedded in a square area defect in the ground, as shown by Fig.2. Dimensions of the EML in this paper are listed in Table I. The periodicity of the EML array is 4.2 mm along both x and y directions. As stated in [12], when the size of EML element is much smaller than the wavelength, the volume occupied by the EML array right under the patch can

978-7-5641-4279-7

be taken as an effective magneto-dielectric medium. This effective medium belongs to a special category of metamaterial: waveguided metamaterial, since it resides in a planar waveguide environment constituted by the ground plane and the upper patch [12],[16],[17].

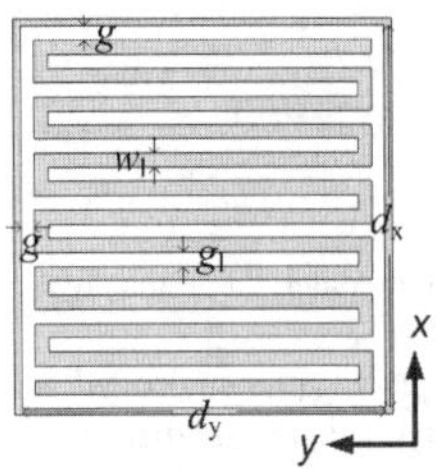

Figure 2. Depiction of the unit cell of embedded meander line structure.

TABLE I
DIMENSIONS OF EML UNIT CELL

d_x	d_y	g	w_l	g_l
4.05mm	4.05mm	0.15mm	0.15mm	0.15mm

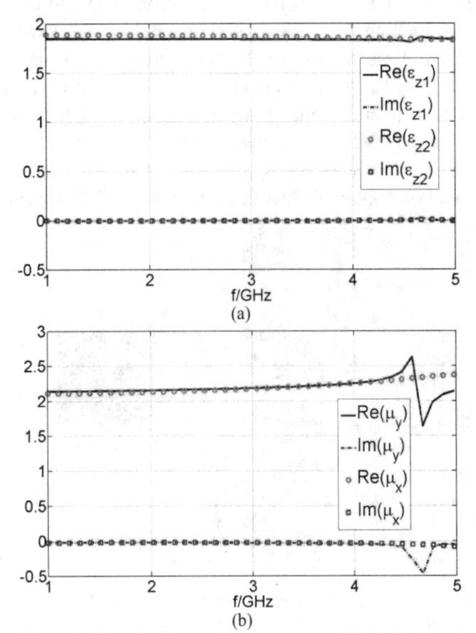

Figure 3. The effective medium parameters of the EML-based waveguided metamaterial, where ε_{z1} and μ_y are retrieved when the incident wave is from x direction and ε_{z2} and μ_x are retrieved when the incident wave is from y direction.

Though the EML is an asymmetric structure, the EML-based waveguided magneto-dielectric material (WG-MDM) exhibits nearly isotropic magneto-dielectric property within a broad band with respect to the plane waves travelling along two orthogonal directions (i.e. x and y directions) in the microstrip plane. As verification, the effective medium parameters of the WG-MDM with those two excitation manners are retieved seperately and compared in Fig. 3. It is apparent that the two sets of effective medium parameters are very similar to each other in the band below 4 GHz. Here the effective medium parameters are retrieved using the technique introduced in [17],[18].

Fig.3 also shows that the self resonance of the WG-MDM is very weak and the effective medium parameters are almost non-dispersive in the band below 4 GHz. Besides, though the defect in the ground plane causes energy leakage, the effective medium loss (seen from the imaginary part of the retrieved medium parameters) of the WG-MDM is resonably low in this band. These two cases are advantageous to improving the bandwidth and efficiency of the WG-MDM-based antenna.

III. DESIGN OF CIRCULARLY POLARIZED PATCH ANTENNA

The design of the proposed circularly polarized patch antenna shown in Fig.1 contains three main stages which could be carried out with the aid of numerical simulation tool.

1). Determine the scale of the EML array and the size of the square patch according to the working frequency of the antenna.

2). Find the unloaded quality factor Q_0 of the square patch before perturbation and determine the amount of perturbation Δs initially from the following expression [19]:

$$2\Delta s/S=1/Q_0. \quad (1)$$

where S is the area of the unperturbed patch. Then slightly adjust Δs to get a satisfactory circularly polarization characteristic along broadside direction (i.e. z direcion) around the working frequency.

3). Determine the input impedance of the perturbed square patch and the dimensions of the matching line according to this input impedance.

In our present design, the scale of the EML array is chosen as 5×5 and the size of the square patch right over the array is set as 21.5mm × 21.5mm. Such configuration makes the antenna operate at about 3.61GHz. Besides, the unloaded quality factor Q_0 of the unperturbed patch is 26.9 and the optimized value of Δs is 6.76mm^2. Accordingly, the input impedance of the perturbed patch is found to be (203.92 + j63.96)Ω at 3.61GHz and the width and length of the matching line are determined as 0.88mm and 16.3mm subsequently. Note that the working frequency is located in the useful band of the WG-MDM and the effective medium parameters at this working frequency are ε_{z1}=1.84-j0.003, ε_{z2}=1.86-j0.005 and $\mu_y\approx\mu_x$=2.22-j0.03. All the antenna dimensions involved in the three stages are listed in Table II.

TABLE II
DIMENSIONS OF CIRCULARLY POLARIZED PATCH ANTENNA

w_p, l_p	Δs	w_m	l_m	w_c	l_c
21.5mm	6.76mm^2	0.88mm	16.3mm	3.85mm	16mm

It should be noted that since loading the EML array in the ground would inevitably causes energy leakage and increase of back radiation, an additional metal shield plate is added beneath and parallel to the antenna ground, as shown in Fig. 4. The distance between the shield and the antenna ground is

5mm. The overall size of the antenna loaded with the WG-MDM is 60 mm×77.3 mm.

Figure 4. The EML loaded circulaly polarized patch antenna with an additional shield metal plate.

For comparison, a circularly polarized control antenna is also designed. The geometry of the control antenna is almost the same as that of the antenna with the WG-MDM except that the volume right under the patch of the control antenna is filled with an independent piece of pure dielectrics instead of artificial magneto-dielectrics, as shown in Fig.5. The permittivity of the dielectric is tuned to be 3.6 (with the loss tangent of 0.001) so that the control antenna has approximately the same operating frequency, the same patch size, and the same overall size as the antenna with the WG-MDM does.

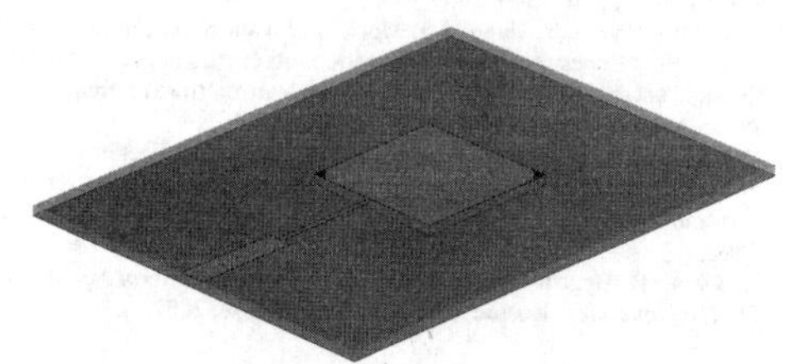

Figure 5. Geometry of the control antenna.

IV. Simulation Results and Discussion

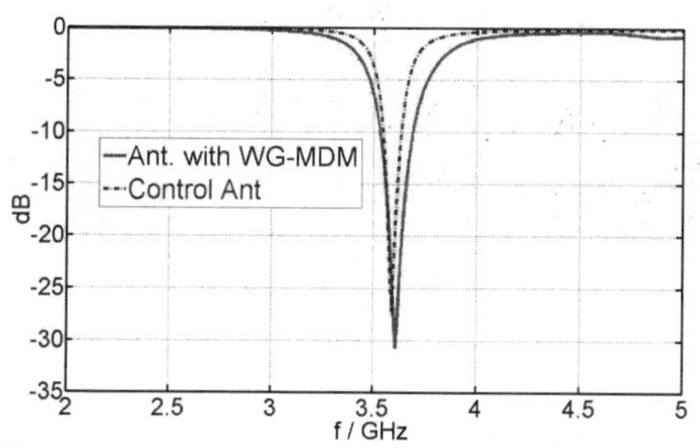

Figure 6. The simulated reflection coefficients of the antennas with and without the WG-MDM.

The simulated reflection coefficients and the axial ratios of the circularly polarized antenna with and without the WG-MDM are shown in Fig.6 and Fig.7, respectively. From Fig.6, the relative -10dB impedance bandwidths of the proposed antenna and the control antenna are 4.43% and 2.49%, respectively. Hence the improvement factor of impedance bandwidth of the proposed antenna over the control antenna is about 1.78. From Fig.7, it is observed that the relative -3dB axial ratio bandwidth of the antenna with the WG-MDM is 1.5%, which is 1.71 times that of the control antenna: 0.88%.

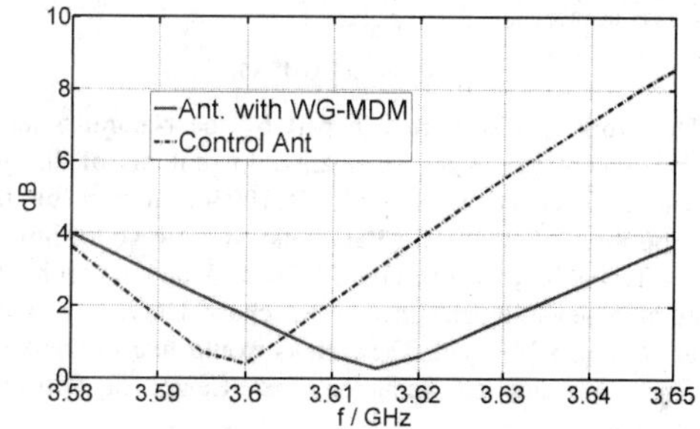

Figure 7. The simulated axial ratios of the antennas with and without the WG-MDM.

Fig.8 demonstrates the simualted radiation patterns of the antenna with the WG-MDM at 3.615GHz and the simualted patterns of the control antenna at 3.6GHz, in both *xoz* plane and *yoz* plane. Apparently, the radiation characteristics of the antenna with WG-MDM could compare favorably with the control antenna.

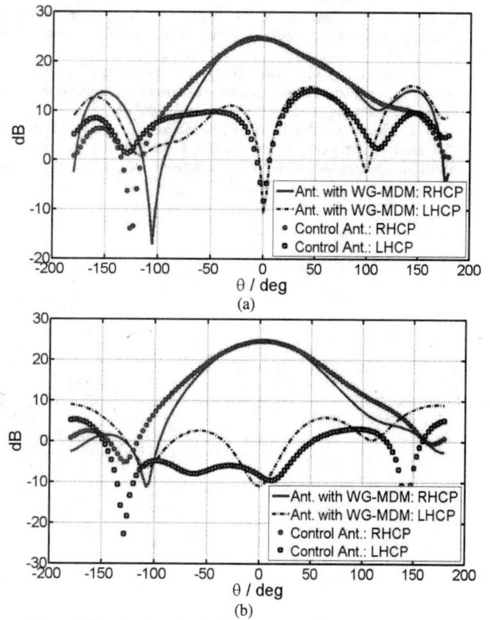

Figure 8. The simulated radiation patterns of the antennas with and without the WG-MDM in (a) *xoz* plane (b) *yoz* plane.

V. Conclusion

We have successfully improved both the impedance and axial ratio bandwidth of circulaly polarized patch antenna by loading an isotropic waveguided magneto-dielectric material based on the EML structure. The proposed broadbanding technique of WG-MDM loading is promising since it influences little on the radiation characteristic and the resulting novel circulaly polarized patch antenna is compact and easy to fabricate.

ACKNOWLEDGMENT

This work was supported in part by the Natural Science Foundation of the Higher Education Institutions of Jiangsu Province under Grant No. 12KJB510030, in part by the Suzhou Key Laboratory for Radio and Microwave/Millimeter Wave Technology under Grant SZS201110, and in part by the Natural Science Foundation of Soochow University under Grant No. SDY2011A12. The authors would like to thank the staffs of the Jointed Radiation Test Center of Soochow University.

REFERENCES

[1] R. C. Hansen and M. Burke, "Antenna with magneto-dielectrics," *Microw. Opt. Technol. Lett.*, vol. 26, no. 2, pp. 75–78, 2000.

[2] P. Ikonen and S. A. Tretyakov, "On the advantages of magnetic materials in microstrip antenna miniaturization," *Microw. Opt. Technol. Lett.*, vol. 50, no. 12, pp. 3131–3134, 2008.

[3] H. Mosallaei and K. Sarabandi, "Magneto-dielectrics in electromagnetics: Concept and applications," *IEEE Trans. Antennas Propag.*, vol.52, pp. 1558–1567, 2004.

[4] K. Buell, H. Mosallaei, and K. Sarabandi, "A substrate for small patch antennas providing tunable miniaturization factors," *IEEE Trans. Microw. Theory Tech.*, vol. 54, pp. 135–146, 2006.

[5] H. Mosallaei and K. Sarabandi, "Design and modeling of patch antenna printed on magneto-dielectric embedded-circuit metasubstrate," *IEEE Trans. Antennas Propag.*, vol. 55, pp. 45–52, 2007.

[6] P. M. T. Ikonen, S. I. Maslovski, C. R. Simovski, and S. A. Tretyakov, "On artificial magnetodielectric loading for improving the impedance bandwidth properties of microstrip antennas," *IEEE Trans. Antennas Propag.*, vol. 54, pp. 1654–1662, 2006.

[7] P. M. T. Ikonen, K. N. Rozanov, A. V. Osipov, P. Alitalo, and S. A. Tretyakov, "Magnetodielectric substrates in antenna miniaturization: Potential and limitations," *IEEE Trans. Antennas Propag.*, vol. 54, pp.3391–3399, 2006.

[8] L. Kempel, B. Shanker, J. Xiao, and S. W. Schneider, "Radiation by a magneto-dielectric loaded patch antenna," in *Proc. Antennas Propag. Conf.*, Loughborough, U.K., Nov. 2009, pp. 745–748.

[9] A. Foroozesh and L. Shafai, "Size reduction of a microstrip antenna with dielectric superstrate using meta-materials: Artificial magnetic conductors versus magneto-dielectrics," in *Proc. Antennas Propag. Soc. Int. Symp.*, Jul. 2006, pp. 11–14.

[10] S. M. Han, K. S. Min, and T. G. Kim, "Study on miniaturization and broadband of patch antenna using magneto-dielectric substrate," in *Proc. Asia–Pacific Microw. Conf.*, Dec. 2008, pp. 1–4.

[11] A. Louzir, P. Minard, and J. F. Pintos, "Parametric study on the use of magneto-dielectric materials for antenna miniaturization," in *Proc. Antennas Propag. Soc. Int. Symp.*, Jul. 11–17, 2010, pp. 1–4.

[12] X. M. Yang, Q. H. Sun, Y. Jing, Q. Cheng, X. Y. Zhou, H. W. Kong, and T. J. Cui, "Increasing the bandwidth of microstrip patch antenna by loading compact artificial magneto-dielectrics," *IEEE Trans. Antennas Propag.*, vol. 59, no. 2, pp. 373–378, Feb. 2011.

[13] J. B. Pendry, A. J. Holden, D. J. Robbins, W. J. Stewart, "Magnetism from conductors and enhanced nonlinear phenomena," *IEEE Trans. Microwave Theory Tech.*, 47(11): 2075-2084, 1999.

[14] P. Gay-Balmaz, O. J. F. Martin, "Electromagnetic resonances in individual and coupled split-ring resonators," *J. Appl. Phys.*, 92: 2929, 2002.

[15] J. D. Baena, R. Marqu´es, F. Medina, "Artificial magnetic metamaterial design by using spiral resonators," *Phys. Rev. B*, 69(1): 014402, 2004.

[16] R. Liu, X. M. Yang, J. G. Gollub, J. J. Mock, T. J. Cui, and D. R. Smith, "Gradient index circuit by waveguided metamaterials," *Appl. Phys. Lett.*, vol. 94, no. 7, p. 073506, 2009.

[17] R. Liu, Q. Cheng, T. Hand, J. J. Mock, T. J. Cui, S. A. Cummer, and D. R. Smith, "Experimental demonstration of electromagnetic tunneling through an epsilon-near-zero metamaterial at microwave frequencies," *Phys. Rev. Lett.*, vol. 100, no. 2, p. 023903, 2008.

[18] D. R. Smith, S. Schultz, P. Markos, C. M. Soukoulis, "Determination of effective permittivity and permeability of metamaterials from reflection and transmission coefficients," *Phys. Rev. B*, vol. 65, no. 19, p. 195104, 2002.

[19] R. Garg, P. Bhartia, I. Bahl, and A. Ittipiboon, *Microstrip Antenna Design Handbook.* Boston, London: ArtechHouse, 2001.

TA-1(D)

October 24 (THU) AM

Room D

Measurements

Fast Measurement Technique Using Multicarrier Signal for Transmit Array Antenna Calibration

Kazunari Kihira, Toru Takahashi, and Hiroaki Miyashita
Information Technology R&D Center, Mitsubishi Electric Corporation
5-1-1 Ofuna, Kamakura, Japan

Abstract- **Calibration is a key technology to realize the desired radiation pattern in phased array antennas. Various techniques have been proposed for array antenna calibrations. The main objective is to obtain the complex electric fields (amplitude and phase) of the individual antenna elements and to compensate for the element-to-element variations. A new technique is required for a fast and efficient measurement operation since the measurement time increases in proportion to the number of antenna elements and/or measurement frequencies. In this paper, we propose a novel measurement method using multicarrier signal for phased array antenna calibration. The proposed method can simultaneously measure the complex electric fields of all antenna elements, therefore fast calibration is possible. The experimental result in an anechoic chamber is also demonstrated.**

I. INTRODUCTION

Phased array antennas based on Digital BeamForming (DBF) have been investigated for a variety of wireless systems [1]-[3]. This antenna system, in which the beamforming is performed directly on a digital level, allows the most flexible and powerful control of the radiation pattern of the antenna. Therefore, each radiator (antenna element) must be equipped with its own receiver or transmitter. For proper operation, all channels (antenna plus microwave circuit) of the DBF antenna should exhibit defined amplitudes and phases. Calibration is a key technology to realize the desired radiation pattern. Many measurement techniques have been proposed for phased array calibrations. The main objective is to obtain the complex electric fields (amplitude and phase) of the individual antenna elements and to compensate for the element-to-element variations. Unlike a single-element measurement, where the electric fields of the individual elements are determined with only a single element illuminated, the measurement techniques described in [4]–[8] measure the in-situ electric fields with the entire array radiating and include some error terms such as T/R module variations, feed-circuit variations, mutual coupling, and diffraction due to antenna structures etc.

For phased array calibrations, an important issue is to overcome the large measurement times which increases in proportion to the number of antenna elements and/or measurement frequencies. Thus, it is required for a fast and efficient measurement operation. In this paper, a novel calibration method using multicarrier signal is proposed. The proposed method can simultaneously measure the complex electric fields of all antenna elements, therefore fast calibration is possible.

In the next section, the principle of the proposed measurement method using multicarrier signals is presented. Section III shows the experimental results in an anechoic chamber and section IV describes the conclusions.

II. PROPOSED METHOD

A. *System Configuration*

Figure 1 shows the system configuration of the proposed method. Each antenna element is equipped with high power amplifier (HPA), and up-converter (U/C) and digital to analog converter (D/A). In digital signal processing part, it has a function of inverse discrete fourier transform (IDFT), and subcarrier arrangement and multicarrier signal generation for calibration. In receiver, for the measurement of all channels of transmit array antenna, Rx antenna is equipped with low noise amplifier (LNA), and down-converter (D/C) and analog to digital converter (A/D). It has a function of the demultiplexing of subcarriers and calculation of the relative amplitude and phase between antenna elements.

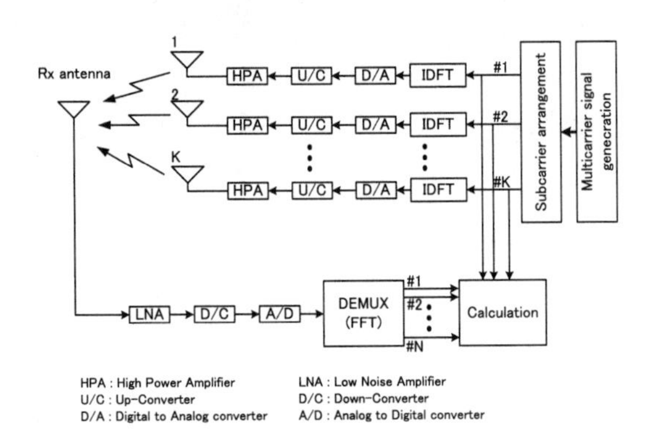

Figure 1. System configuration of the proposed method

978-7-5641-4279-7

B. Procedure in Transmitter

First, a multicarrier signal (called as reference signal) for calibration is generated in transmitter as shown in Figure 2. N signals $c(n)$ (n=1, 2,...,N=MK) are arranged in a measurement frequency band. K is the number of antenna elements and M is that of measurement frequencies, respectively. The reference signal is designed as an orthogonal frequency division multiplexing (OFDM) signal, that is, each subcarrier is orthogonal to the others.

Next, Subcarrier signals $c(n)$ are assigned to each antenna element by every K as shown in Figure 3. This arrangement can be uniformly assigned subcarriers in the measurement frequency band, and does not produce interference to the reference signal between antenna elements. After that, the reference signal of each antenna element is transformed to the time-domain signal by the inverse discrete fourier transform (IDFT) as follows.

$$s_k(t)=\frac{1}{N}\sum_{n=0}^{N-1}c_k(n)e^{j\frac{2\pi}{N}nt} \tag{1}$$

where

$$c_k(n)=\begin{cases}c(n) & n=k,k+K,\cdots,k+(M-1)K\\ 0 & n\neq k,k+K,\cdots,k+(M-1)K\end{cases} \tag{2}$$

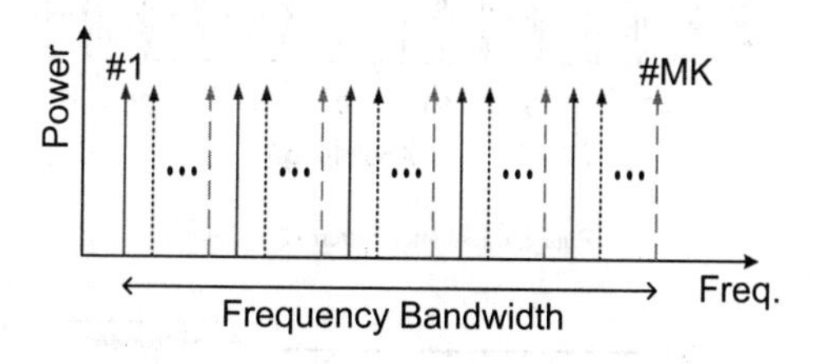

Figure 2. Subcarrier arrangement of the reference signal

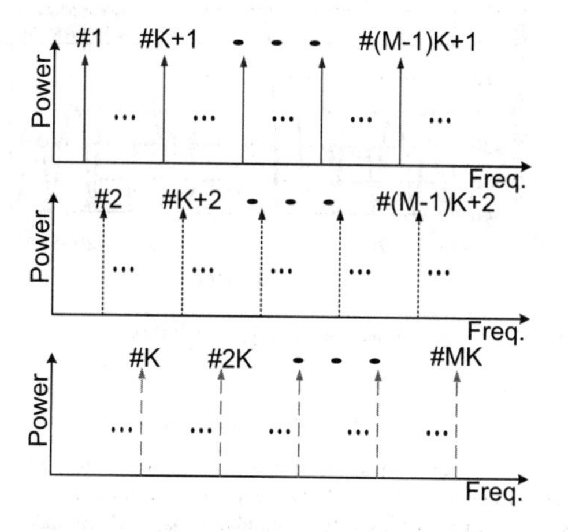

Figure 3. Subcarrier arrangement for each antenna element

C. Procedure in Receiver

The received microwave signal of RX antenna is converted into a complex digital signal on baseband. The signal is transformed to frequency-domain signal by DFT. Since the reference signal of each antenna element is orthogonal to the others, it is easy to separate those signals. Therefore, the relative complex amplitudes are obtained by,

$$\beta_{k,m}=\frac{x(k+(m-1)K)}{c(k+(m-1)K)} \tag{3}$$

where $x(n)$ is the received complex signal after DFT.

III. Experimental Results

A. Measurement Condition

Table 1 shows the measurement conditions in an anechoic chamber. The proposed measurement method is evaluated with some radiation patterns after calibration. To compare with the proposed method, the results of the single-element and single-frequency measurement method (conventional method) are also shown. The transmit power are the identical values for each antenna element in both methods. The proposed method can reduce measurement time to 1/48 compared with the conventional one.

TABLE I
Measurement conditions

Array configuration	16 element linear array
Element spacing	0.76 [wavelength]
Center frequency	2.185 [GHz]
Data size for calibration	2^18 [sample]
Subcarrier interval	2.5 [kHz]
Number of subcarrier each element	3
Frequency bandwidth	10 [MHz]

B. Spectrum of Received Reference Signal

Figure 4 shows the spectrum of the received signal for measuring by the proposed method. It turns out that subcarriers are in three points (0MHz and ±5MHz). Furthermore, Figure 5 is a figure which expanded near 0 MHz of Figure 4. It turns out that the subcarriers of each antenna element are detectable at intervals of 2.5 kHz.

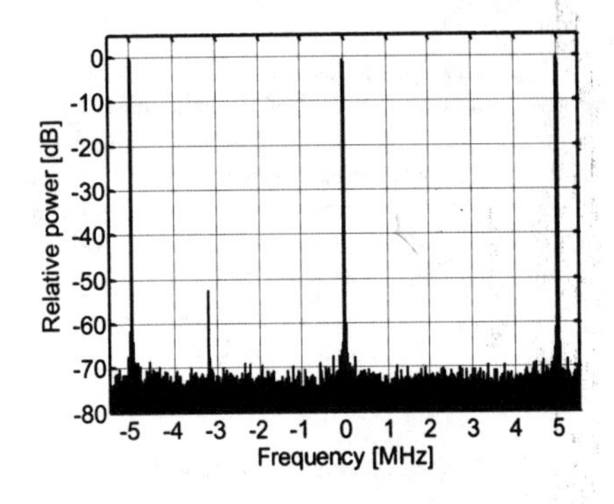

Figure 4. Spectrum of the received signal

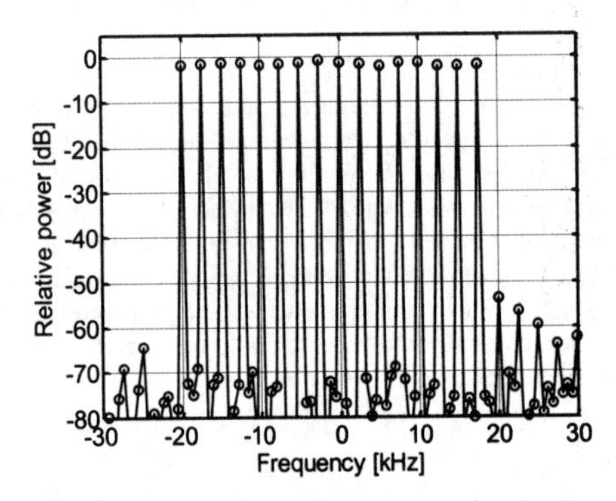

Figure 5. Spectrum of the received signal (expanded near 0MHz)

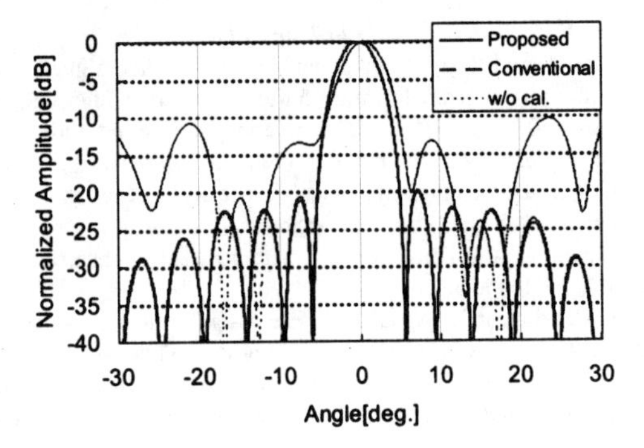

Figure 6. Radiation pattern (2.185GHz)

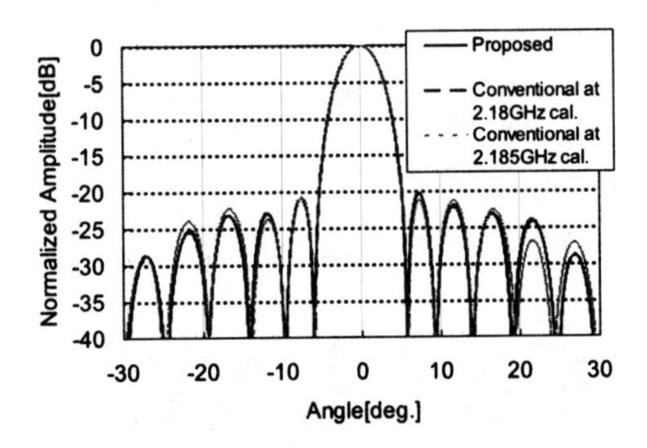

Figure 7. Radiation pattern (2.18GHz)

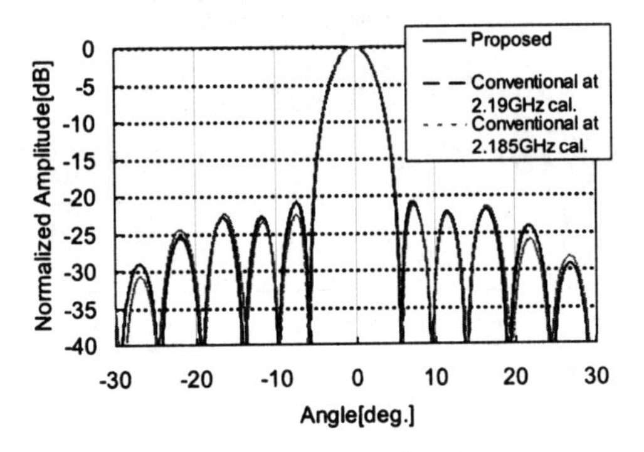

Figure 8. Radiation pattern (2.19GHz)

C. Radiation Pattern

Figure 6 shows the measured radiation pattern at center frequency (2.185GHz). The nominal excitation amplitudes and phases of each antenna element were designed to form a main lobe to 0 degree and to achieve a sidelobe level of -20dB. The pattern before calibration (w/o cal.) causes the beam pointing error and the sidelobe level is very high. On the other hand, the pattern using the proposed method has good performance corresponding to the nominal design. The result of the proposed method and that of the single-element method are in good agreements. Therefore, the proposed measurement method is experimentally validated.

Figure 7 shows the radiation patterns at the lower frequency. The result of the proposed method and that of the single-element method at 2.18GHz measurement also are in good agreements. In conventional method, when using the measurement values at the center frequency (2.185GHz), the sidelobe characteristic has deteriorated. Thus, the proposed method that can simultaneously measure the complex electric fields of multiple frequencies is very effective. Figure 8 also shows the radiation patterns at the higher frequency.

IV. Conclusions

A fast measurement technique for phased array antenna calibration was presented. The approach exploits multicarrier signal to execute all elements simultaneously, which makes the method N times faster than single-element measurement, where N is the number of the products of antenna elements and frequency points. Moreover, measured results demonstrated the effectiveness of the proposed method.

REFERENCES

[1] H.Steyskal, "Digital Beamforming Antennas - An Introduction," Microwave Journal, vol.30, no.1, pp.107–124, Jan. 1987.

[2] J. Litva and K. Y. Lo, Digital Beamforming in Wireless Communications, Artech House, 1996.

[3] A.Garrod, "Digital Modules for Phased Array Radar," IEEE InternationalRadar Conference, RADAR-95, 1995.

[4] S. Mano and T. Katagi, "A method for measuring amplitude and phase of each radiating element of a phased array antenna," Trans. IECE, vol. J65-B, no. 5, pp. 555–560, May 1982.

[5] K.M.Lee, R.S.Chu, and S.C.Liu, "A built-in performance monitoring/ fault isolation and correction (PM/FIC) system for active phased-array antennas," IEEE Trans. Antennas Propag., vol.41, pp.1530–1540, Nov. 1993.

[6] G.A.Hampson and A.B.Smolders, "A Fast and Accurate Scheme for Calibration of Active Phased-Array Antennas," IEEE AP-S Int. Symp. Digest, pp.1040–1043, 1999.

[7] S.D.Silverstein, "Application of Orthogonal Codes to the Calibration of Active Phased Array Antennas for Communication Satellites," EEE Trans. SP., vol.45, no.1, pp.206–218, Jan. 1997.

[8] T.Takahashi, Y.Konishi, S.Makino, H.Ohmine and H.Nakaguro, "Fast Measurement Technique for Phased Array Calibration," IEEE Trans. Antennas Propag., vol.56, no.7, pp.1888–1899, July 2008.

Evaluation of RCS Measurement Environment in Compact Anechoic Chamber

Naobumi Michishita, Tadashi Chisaka, and Yoshihide Yamada
National Defense Academy
1-10-20 Hashirimizu
Yokosuka, 239-8686 JAPAN

***Abstract*- To enhance the dynamic range of monostatic RCS measurement in compact anechoic chamber, this paper presents the quantitative evaluation of simple CW measurement method by using transmitting and receiving antennas. The mutual coupling is reduced by several approaches such as the separation between antennas, suppressing the sidelobe level, and inserting the shielded conductor. The measurable maximum and minimum RCS values are clarified. The effect of the arrangement of wave absorber in target side is also shown.**

I. INTRODUCTION

Recently, the monostatic RCS of a patch antenna with different terminal loads has been investigated [1]-[3]. We have verified a RCS reduction of 15 dB by applying high resistive load of 500Ω through both simulation and measurement [4]. In the monostatic RCS measurement of the patch antenna with resistive load, the minimum RCS value becomes about −30 dBsm. To measure the low receiving level, the arrangement of transmitting and receiving antennas and the appropriate employment of wave absorbers are important. In this paper, the effect of RCS measurement environment in compact anechoic chamber is estimated quantitatively by electromagnetic simulation. Typical RCS measurement techniques are the continuous wave (CW) method by using simple transmitting and receiving antennas, and the pulsed measurement method by using time domain function in the network analyzer. To evaluate the simplified RCS measurement method, the CW method is employed in this paper.

II. MAJOR FACTOR OF MUTUAL COUPLING BETWEEN TRANSMITTING AND RECEIVING HORNS

The major factor of mutual coupling between transmitting and receiving horn antennas is investigated. Figure 1 shows the measurement environment for monostatic RCS of a patch antenna with resistive load in compact anechoic chamber. The mutual coupling derived from Friss formula is given by

$$\frac{P_r}{P_t} = G_t(\theta = 90°)G_r(\theta = 90°)L(d) \qquad (1)$$

where P_t and P_r are transmitting and receiving power, G_t and G_r are transmitting and receiving antenna gain, d is the distance between antennas, and $L(d) = (\lambda / 4\pi d)^2$ is free space path loss. Figure 2(a) shows the configuration of the transmitting and receiving horns. In this paper, horizontal polarization is employed. Figure 2(b) shows the radiation pattern of the horn antenna at 3 GHz. The simulated results are obtained by moment method. The antenna gain is 15 dBi and the sidelobe level at θ = 90 deg. is −17.5 dBi. Table I shows the mutual coupling between horns when d is varied. Since transmitting and receiving horns are identical, G_t + G_r becomes −35 dB. Table I also shows S_{21} obtained by electromagnetic simulation. The mutual coupling calculated by Eq. (1) and simulated S_{21} are almost identical. Figure 3 shows S_{21} characteristics when d is varied. S_{21} of −100 dB can be achieved with d = 14 m. Therefore, The reduction of G_t + G_r or $L(d)$ are required for mutual coupling suppression.

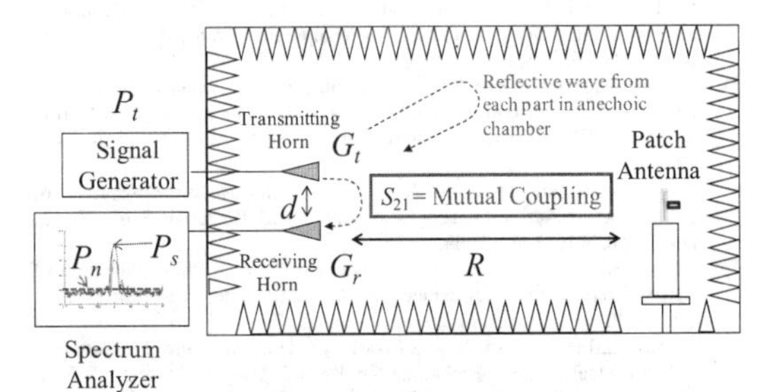

Figure 1. Measurement environment for monostatic RCS of patch antenna with resistive load in anechoic chamber.

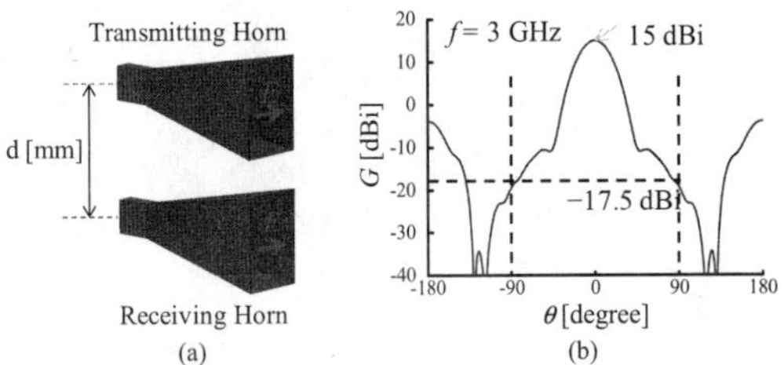

Figure 2. (a) Configuration of transmitting and receiving horns. (b) Radiation pattern of horn antenna

TABLE I
MUTUAL COUPLING BETWEEN TRANSMITTING AND RECEIVING HORNS AND SIMULATED S21

d	$G_t + G_r$	$L(d)$	P_r/P_t	S_{21}
300 mm		−31.5 dB	−66.5 dB	−67.0 dB
600 mm	−35 dB	−37.5 dB	−72.5 dB	−71.8 dB
900 mm		−41.0 dB	−76.0 dB	−76.1 dB

978-7-5641-4279-7

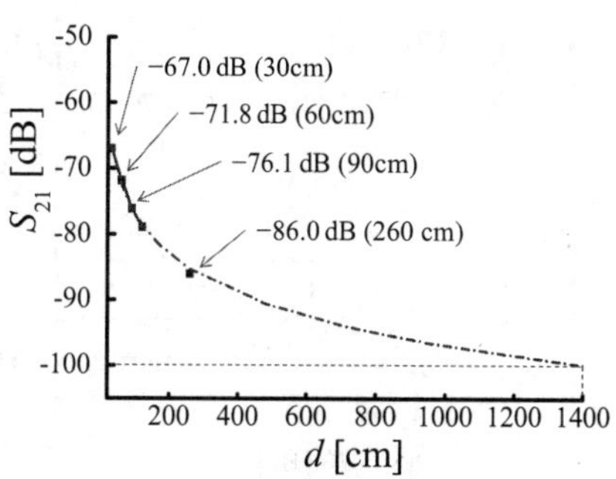

Figure 3. S_{21} characteristics when distance d between horns is varied.

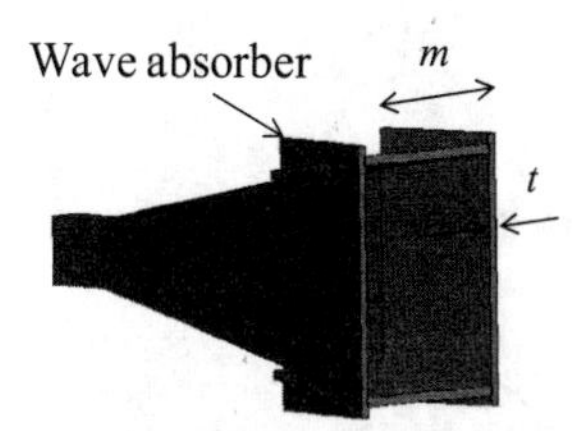

Figure 4. Horn antenna with wave absorber for reduction of sidelobe level.

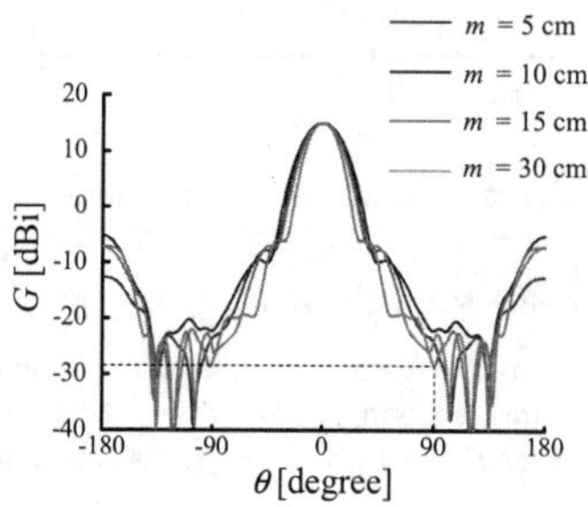

Figure 5. Radiation pattern of horn antenna with $t = 1$ cm when m is varied.

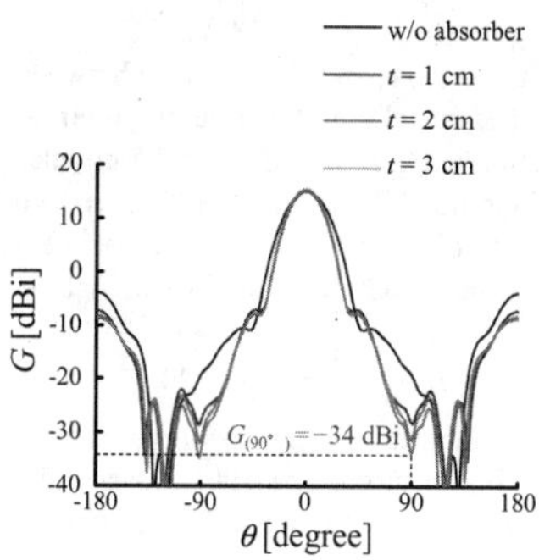

Figure 6. Radiation pattern of horn antenna with m = 15 cm when t is varied.

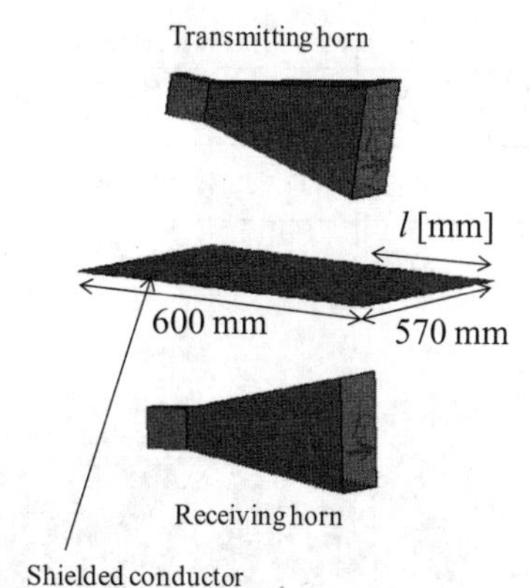

Figure 7. Configuration of shielded conductor inserted between horns.

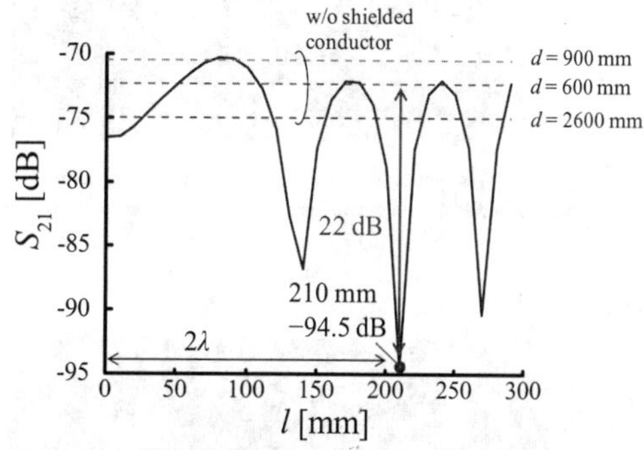

Figure 8. S_{21} characteristics when l is varied.

III. Mutual Coupling Reduction Techniques

First, the arrangement of wave absorber is examined for reduction of sidelobe level to reduce $G_t + G_r$. Figure 4 shows the configuration of horn antenna with wave absorber. The wave absorber is arranged at the edge of the horn antenna. Figure 5 shows the radiation pattern with $t = 1$ cm when m is varied. The sidelobe level at $\theta = 90$ deg. can be suppressed at $m = 15$ cm. Figure 6 shows the radiation pattern with $m = 15$ cm when t is varied. The sidelobe level at $\theta = 90$ deg. becomes −34 dBi at $t = 3$ cm.

Next, the arrangement of shielded conductor is examined for reduction of path loss to reduce $L(d)$. Figure 7 shows the configuration of shielded conductor inserted between horns. The position of the shielded conductor is middle of two antennas. Figure 8 shows S_{21} characteristics when l is varied. The mutual coupling reduction of 22 dB is achieved with $l = 2\lambda$ due to standing wave occurred by diffraction at shielded conductor.

TABLE II
MUTUAL COUPLING REDUCTION EFFECTS

Method	$G_t + G_r$	$L(d)$	P_r/P_t	S_{21}
Horn only	−35 dB	−37.5 dB	−72.5 dB	−72.3 dB
Sidelobe reduction	−68 dB	−37.5 dB	−105.5 dB	−78.0 dB
Shielded conductor	−35 dB	−59.5 dB	−94.5 dB	−94.5 dB

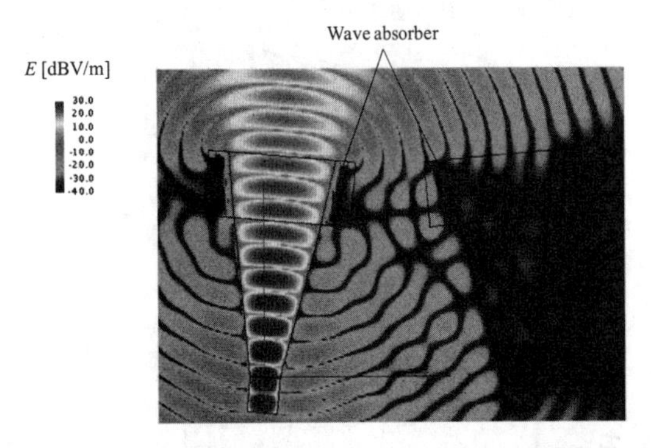

Figure 9. Near field distribution of two horn antennas with wave absorber.

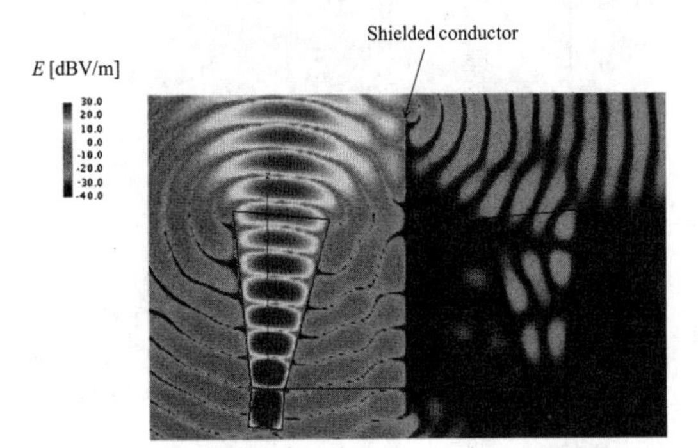

Figure 10. Near field distribution of two horn antennas with shielded conductor.

Table II shows the mutual coupling reduction effect. In the case of the sidelobe reduction method, the mutual coupling calculated from Eq. (1) is −105.5 dB, which is not identical to the simulated S_{21} of −78.0 dB. In the case of the shielded conductor method, both P_r/P_t and S_{21} agree well.

Figures 9 show the near field distribution of two horn antennas with wave absorber. The discrepancy of P_r/P_t and S_{21} seems to be due to approaching each aperture planes by adding wave absorber. Figures 10 show the near field distribution of two horn antennas with shielded conductor. Although strong electric field intensity is confirmed inside horn antenna, the feed position becomes standing wave node.

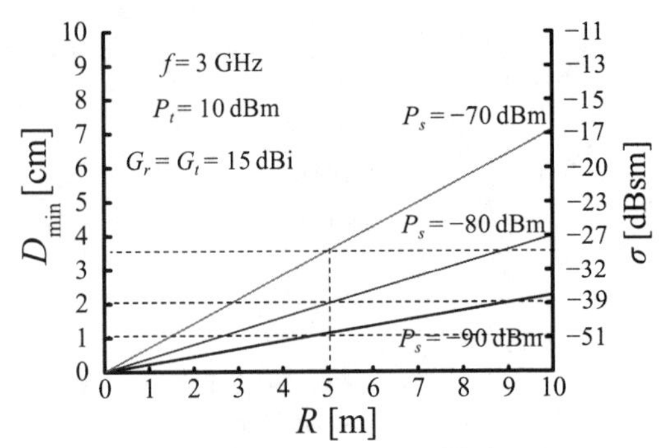

Figure 11. Measurable minimum RCS values.

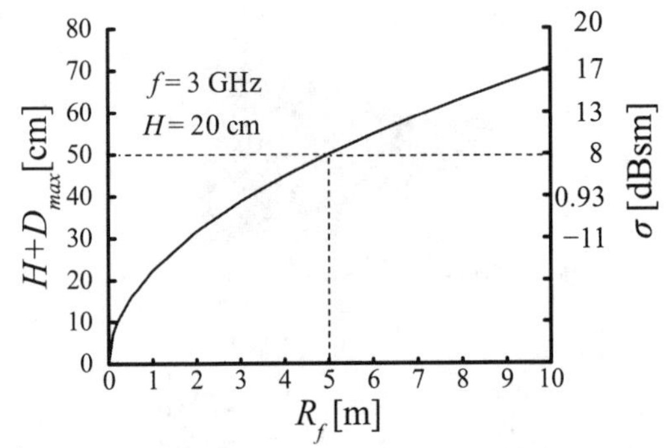

Figure 12. Measurable maximum RCS values.

IV. MEASURABLE MAXIMUM AND MINIMUM RCS VALUES

The minimum S_{21} of −94.5 dB estimated in previous section corresponds to the receiving power P_r = −84.5 dBm when transmitting power P_t = 10 dBm. The radar equation is given by

$$P_r = P_t G_t G_r \left(\frac{D}{4R} \right)^4 \tag{2}$$

where D is the target size, and R is distance between horns and target. Figure 11 shows the measurable minimum RCS values. σ is the theoretical RCS value of D of the circular disk. The measurable minimum RCS value is −40 dBsm with R = 5 m and Pr = −84.5 dBm at 3 GHz. Figure 12 shows the measurable maximum RCS values. The far field condition is given by

$$H + D_{max} = \sqrt{R_f / 20} \tag{3}$$

where H is the aperture size of horn, and D_{max} is the measurable maximum target size. The measurable maximum RCS value and D_{max} are 8 dBsm and 30 cm with R_f = 5 m and H = 20 cm.

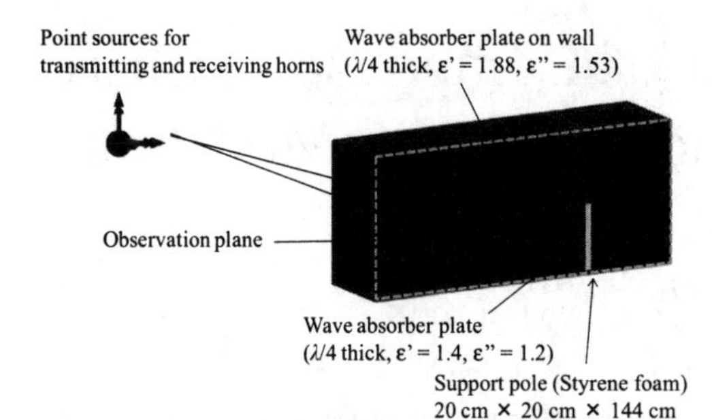

Figure 13. Configuration of anechoic chamber.

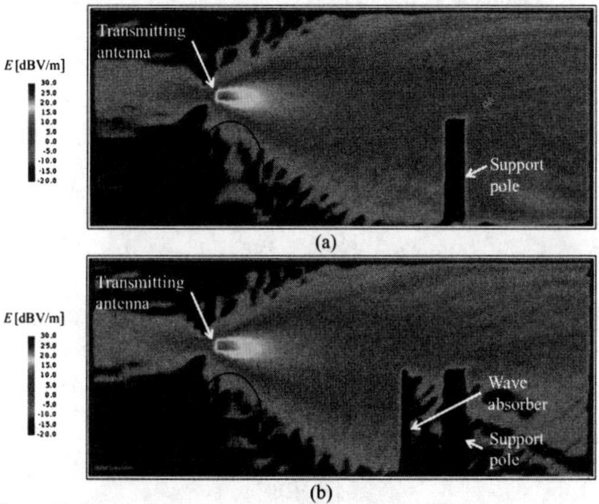

Figure 14. Electric field distributions at observation plane (a) with support pole and (b) wave absorber plate placed in front of support pole.

TABLE III
MUTUAL COUPLING REDUCTION EFFECTS IN SEVERAL ENVIRONMENT CONDITIONS

Conditions	P_r/P_t
Free space	−73.6 dB
In anechoic chamber	−71.6 dB
Existing support pole	−59.6 dB
Put wave absorber	−64.4 dB

V. GO SIMULATION IN ANECHOIC CHAMBER

Figure 13 shows the configuration of anechoic chamber. The effects of the measurement environment are estimated thorough GO simulation. The length, width, and height of the compact anechoic chamber are 7 m × 3.705 m × 2.915 m. In GO simulation, the far field radiation patterns of the horn antenna are employed as point sources at the transmitting and receiving horns. And, the number of reflection/transmission interactions for each ray is 3. Typical pyramidal wave absorber put on the wall in compact anechoic chamber is modeled as the single layer of the dielectric material with λ/4 thick. The reflection reduction of this dielectric sheet is 59 dB in oblique incidence of 45°. There are support pole composed of styrene foam and the wave absorber for normal incidence in the target side.

Figure 14 shows the electric field distributions at longitudinal plane at center of the anechoic chamber with the support pole and the wave absorber plate placed in front of the support pole. The electric field intensities at receiving area are affected by the support pole and wave absorber. Table III shows the mutual coupling reduction effect in several environment conditions. The existing of the support pole increases the mutual coupling of 12.0 dB. The reflection from the support pole can be reduced by placing the wave absorber in front of the support pole.

VI. CONCLUSION

This paper presents the evaluation of the RCS measurement environment in compact anechoic chamber. The electromagnetic simulation results help us to achieve the mutual coupling reduction of the transmitting and receiving horns. Further improvement of the measurable minimum RCS values is also important to measure the lower RCS values in future.

REFERENCES

[1] Y. Liu, S.-X. Gong, and D.-M. Fu, "A novel model for analyzing the RCS of microstrip antenna," *IEEE AP-S Int. Symp.*, Columbus, OH, vol.4, pp.835-838, June 2003.

[2] D.M. Pozar, "Radiation and scattering from a microstrip patch on a uniaxial substrate," *IEEE Trans. Antennas Propagat.*, vol.35, no.6, pp.613-621, June 1987.

[3] Y. Inasawa, Y. Nishioka, N. Yoneda, and Y. Konishi, "Radar Cross Section Analysis of a Patch Antennas Considering Terminal Conditions," *IEICE Communications Society Conference*, C-1-4, Sep. 2011 (in Japanese).

[4] T. Chisaka, N. Michishita, and Y. Yamada, "Reduction of RCS Values of a Patch Antenna by Resistive Loading," *IEEE AP-S Int'l Symp.*, Orlando, FL, July 2013.

Stable Parameter Estimation of Compound Wishart Distribution for Polarimetric SAR Data Modeling

Yi Cui, Hiroyoshi Yamada, and Yoshio Yamaguchi
Faculty of Engineering, Niigata University
8050 Ikarashi, 2-no-cho, Nishi-ku, Niigata-shi, 950-2181 JAPAN

***Abstract*—This paper investigates the parameter estimation of compound Wishart distribution for statistical modeling of polarimetric synthetic aperture radar data. We show that the recently proposed method of matrix log-cumulant may lead to non-invertible equations and cause unstable performances. In order to overcome such difficulty, we proposed a Bayesian-based method to re-estimate the log-cumulants. Simulation experiment demonstrates that the proposed algorithm provides both improved and stabler results. Finally, we present an application of the new method for texture analysis of the Germany F-SAR polarimetric data.**

I. INTRODUCTION

Polarimetric synthetic aperture radar (POLSAR) is an important microwave remote sensing system. It is able to provide a high-resolution map of the ground terrains and at the same time polarimetric scattering matrices. However, since POLSAR is a coherent system, the acquired scattering matrix is subject to a noise-like phenomenon, often called speckle in radar community [1]. Although strictly speaking, speckle is not really noise but a useful signal, it becomes a nuisance in the context of single-dataset acquisition and should be removed for quantitative applications.

POLSAR data are characterized by both polarimetric and spatial information. The former is affected by speckle and the latter is related to the spatial pattern of the scene. In order to statistically model their joint behaviors, a product distribution has been popularly used [2]. It says that the measured observable is the multiplication of one Wishart-distributed random matrix and one positive random variable, which independently account for the speckle structure and texture variation (see Section II-B). Under this scheme, several compound distributions have been derived. Particularly, the K-Wishart distribution has been proved useful in a number of applications such as image classification or segmentation [3].

Clearly, the accuracy of applying the compound Wishart distribution depends on the accuracy of parameter estimation. Recently, based on the matrix-variate Mellin transformation, Anfinsen et al. [2] proposed a new parameter estimation method using matrix log-cumulants. It appears to be promising because of the incorporation of full polarimetric information. However, as we will point out, this method may sometimes suffer from non-invertible problem and so unstable performances. This paper proposes a solution to overcome such difficulty. It is organized as follows. In Section II, a brief introduction of POLSAR data and the associated compound Wishart distribution is provided. In Section III, we describe the parameter estimation using the method of matrix log-cumulant and propose the improved algorithm. In Section IV, experiments results with simulated and real POLSAR data are given. Section V is the conclusion.

II. POLSAR DATA AND COMPOUND WISHART DISTRIBUTION

A. POLSAR Data

In the horizontal/vertical (H/V) polarization basis, the POLSAR system measures, for each resolution cell, the scattering matrix:

$$\mathbf{S} = \begin{bmatrix} S_{\mathrm{HH}} & S_{\mathrm{VH}} \\ S_{\mathrm{HV}} & S_{\mathrm{VV}} \end{bmatrix}, \tag{1}$$

where S_{AB} represents the scattering coefficient in the A-receive/B-transmit channel. Under the backscattering reciprocal condition, we have $S_{\mathrm{HV}} = S_{\mathrm{VH}}$ such that (1) can be equally written by the vector:

$$\mathbf{k} = \begin{bmatrix} S_{\mathrm{HH}} \\ \sqrt{2}S_{\mathrm{HV}} \\ S_{\mathrm{VV}} \end{bmatrix}, \tag{2}$$

Both $\mathbf{S}$ and $\mathbf{k}$ represent the single-look data format of the POLSAR measurement. The multi-look covariance matrix is defined as:

$$\mathbf{Z} = \frac{1}{L}\sum_{i=1}^{L} \mathbf{k}_i \mathbf{k}_i^{\mathrm{H}} \tag{3}$$

where $\mathbf{k}_i$ is the i-th independent look and $\mathbf{k}_i^{\mathrm{H}}$ denotes its conjugate transpose. According to (3), $\mathbf{Z}$ is a 3×3 Hermitian and positive-definite matrix. In this paper, we focus on statistical modeling of such kind of data.

B. Product Model

The product model for the multi-look covariance matrix is given by [2]:

$$\mathbf{Z} = \gamma \mathbf{W}, \tag{4}$$

where $\mathbf{W}$ is a Hermitian and positive-definite random matrix and γ is a positive random variable that is independent of $\mathbf{W}$. $\mathbf{W}$ contains the speckle and polarimetric information and has a probability density function (pdf) of the Wishart distribution [4]:

$$p_{\mathbf{W}}(\mathbf{W}) = \frac{L^{Ld}}{\Gamma_d(L)} \frac{|\mathbf{W}|^{L-d}}{|\mathbf{\Sigma}|^d} \exp\left[-\mathrm{Tr}\left(L\mathbf{\Sigma}^{-1}\mathbf{W}\right)\right], \tag{5}$$

978-7-5641-4279-7

where d is the dimension of $\mathbf{W}$ which equals 3 as in this paper; $\Gamma_d(L) = \pi^{d(d-1)/2} \prod_{i=0}^{d-1} \Gamma(L-i)$ is the multivariate generalization of the gamma function; $\mathbf{\Sigma} = \mathbb{E}(\mathbf{W})$ is the mean covariance matrix. On the other hand, γ is a unit-mean positive random variable which characterizes textures of the scene. Its pdf can be set flexibly but the gamma distribution [5] as given in (6) has proved to be both useful and tractable.

$$p_\gamma(\gamma) = \frac{\nu^\nu}{\Gamma(\nu)} \gamma^{\nu-1} \exp(-\nu\gamma). \tag{6}$$

As a result of (4)–(6), the pdf of the observed covariance matrix $\mathbf{Z}$ becomes [4]:

$$\begin{aligned} p_{\mathbf{Z}}(\mathbf{Z}) =& \frac{2(L\nu)^{\frac{\nu+Ld}{2}}}{\Gamma_d(L)\Gamma(\nu)} \frac{|\mathbf{Z}|^{L-d}}{|\mathbf{\Sigma}|^L} \left[\mathrm{Tr}\left(\mathbf{\Sigma}^{-1}\mathbf{Z}\right)\right]^{\frac{\nu-Ld}{2}} \\ &\times K_{\nu-Ld}\left[2\sqrt{L\nu\mathrm{Tr}\left(\mathbf{\Sigma}^{-1}\mathbf{Z}\right)}\right], \end{aligned} \tag{7}$$

where $K_\nu(\cdot)$ represents the modified Bessel function of the second kind. The above distribution is often called the compound K-Wishart distribution due to its generalization of the K-distribution in the single-variable case [2].

III. Parameter Estimation of K-Wishart Distribution

According to (7), it can be seen that the compound K-Wishart distribution contains three parameters: L, $\mathbf{\Sigma}$, and ν. In SAR statistics, L is often called the equivalent number of looks (ENL) indicating the speckle level of the multi-look data. It can be convenient estimated from a homogeneous (non-textured) area [6] and maintains the same for a given system. Hence throughout this paper, L is assumed to be a known constant. Then the remaining task is to estimate $\mathbf{\Sigma}$ and ν.

A. Estimation of Mean Covariance Matrix

Suppose $\mathbf{Z}_i(i = 1, 2, , n)$ are identically and independently distributed (i.i.d.) K-Wishart samples. According to (4), it is easy to obtain:

$$\mathbb{E}(\mathbf{Z}) = \mathbb{E}(\gamma \cdot \mathbf{W}) = \mathbb{E}(\gamma) \cdot \mathbb{E}(\mathbf{W}) = \mathbf{\Sigma}. \tag{8}$$

where the fact that γ is a unit-mean variable independent of $\mathbf{W}$ is used. From (8) it is clear that $\mathbf{\Sigma}$ can be straightforwardly estimated by:

$$\hat{\mathbf{\Sigma}} = \frac{1}{n} \sum_{i=1}^{n} \mathbf{Z}_i. \tag{9}$$

B. Estimation of Shape Parameter

In [2], Anfinsen et al. proposed that by the matrix-variate Mellin transform, the shape parameter ν of the K-Wishart distribution can be related to the second-order matrix log-cumulant as follows:

$$\psi^{(1)}(\nu) = \frac{\hat{\kappa}_2\{\mathbf{Z}\} - \sum_{i=0}^{d-1} \psi^{(1)}(L-i)}{d^2}, \tag{10}$$

where $\psi^{(1)}(\cdot)$ is the trigamma function and the second-order log-cumulant is estimated by:

$$\hat{\kappa}_2\{\mathbf{Z}\} = \frac{1}{n} \sum_{i=1}^{n} (\ln |\mathbf{Z}_i|)^2 - \left(\frac{1}{n} \sum_{i=1}^{n} \ln |\mathbf{Z}_i|\right)^2. \tag{11}$$

Inversion of ν from (10) can be numerically accomplished by, e.g., the Newton method. However, since $\psi^{(1)}(\nu) > 0$ for all $\nu > 0$ [7], the solution to (10) exists if and only if

$$\hat{\eta} = \hat{\kappa}_2\{\mathbf{Z}\} - \sum_{i=0}^{d-1} \psi^{(1)}(L-i) > 0. \tag{12}$$

Unfortunately, this condition cannot be always satisfied due to the estimation uncertainty in (11). The situation becomes especially worse with small samples. For example, simulation by the true parameters of $L = 3$, $\nu = 10$, and

$$\mathbf{\Sigma} = \begin{bmatrix} 11.9 & -2.5 + 1.0i & -0.8 - 1.0i \\ -2.5 - 1.0i & 3.4 & 0.2 + 0.3i \\ -0.8 + 1.0i & 0.2 - 0.3i & 1.3 \end{bmatrix} \tag{13}$$

indicates that with 49 independent samples, around 15% of the cases lead to the invertible equation of (10). This phenonmenon necessitates to modify $\hat{\eta}$ in order to ensure its non-negativeness. The simplest method is to threshold it to zero whenever it becomes negative. It means forcing $\nu = \infty$ while the K-Wishart distribution degenerates to the standard Wishart distribution. However, such treatment may lead to very unstable results, as will be seen in Section IV-A.

In order to overcome the aforementioned problem, we propose to re-estimate the posterior mean of $\hat{\eta}$ with Bayesian method. Specifically, the posterior pdf of $\hat{\eta}$ is:

$$p(\eta_m|\hat{\eta}) = \frac{p(\hat{\eta}|\eta_m)p(\eta_m)}{p(\hat{\eta})}, \tag{14}$$

where η_m stands for the mean of $\hat{\eta}$. Clearly, in order to impose the non-negative constraint, a reasonable choice for $p(\eta_m)$ is the uniform distribution:

$$p(\eta_m) = \frac{1}{b}, 0 \le \eta_m \le b, \tag{15}$$

where b is an arbitrarily large positive number. Next we turn the attention to the conditional pdf $p(\hat{\eta}|\eta_m)$. Considering (11) and (12), we may approximate $p(\hat{\eta}|\eta_m)$ with a Gaussian distribution, an assumption justified by the central limit theorem:

$$p(\hat{\eta}|\eta_m) = \frac{1}{\sqrt{2\pi\sigma^2}} \exp\left[\frac{(\hat{\eta} - \eta_m)^2}{2\sigma^2}\right]. \tag{16}$$

Consequently, in order to determine the functional form of $p(\hat{\eta}|\eta_m)$ it is necessary to obtain σ^2, i.e., the variance of $\hat{\eta}$ (or equivalently, the variance of $\hat{\kappa}_2\{\mathbf{Z}\}$). This can be accomplished by a moment approach [8]. Specifically, if we let $z_i = \mathbf{Z}_i(i = 1, 2, ..., n)$, the variance of $\hat{\kappa}_2\{\mathbf{Z}\}$ can be estimated by [8]:

$$\hat{\sigma}^2 = \left(\frac{1}{n} - \frac{2}{n^2}\right) \hat{\xi}_z^4 + \left(\frac{4}{n^2} - \frac{1}{n}\right) \hat{\sigma}_z^4. \tag{17}$$

where $\hat{\sigma}_z^2$ and $\hat{\xi}_z^4$ represent the estimation of the variance and kurtosis of z_i which are respectively given by:

$$\hat{\sigma}_z^2 = \frac{1}{n} \sum_{i=1}^{n} \left(z_i - \frac{1}{n} \sum_{j=0}^{n} z_j\right)^2, \tag{18}$$

$$\hat{\xi}_z^4 = \frac{1}{n}\sum_{i=1}^{n}\left(z_i - \frac{1}{n}\sum_{j=0}^{n} z_j\right)^4. \quad (19)$$

Finally, we calculate the posterior mean:

$$\bar{\eta}_m = \int \eta_m p(\eta_m|\hat{\eta})d\eta_m. \quad (20)$$

Since we do not have any prior knowledge of b other than $b > 0$, we allow $b \to +\infty$ in (20) and perform the integration to obtain:

$$\bar{\eta}_m = \hat{\eta} + \frac{\exp\left(-\frac{\hat{\eta}^2}{2\hat{\sigma}^2}\right)}{\sqrt{2\pi}\Phi\left(\frac{\hat{\eta}}{\hat{\sigma}}\right)} \cdot \hat{\sigma}. \quad (21)$$

where $\Phi(\cdot)$ is the cumulative function of the standard normal distribution. It can be verified that $\eta_m > 0$ for any $\hat{\eta} \in \mathbb{R}$ and $\hat{\sigma} \in \mathbb{R}^+$. Therefore, instead of solving (10) we solve (22) as below for the shape parameter.

$$\psi^{(1)}(\nu) = \frac{\eta_m}{d^2}. \quad (22)$$

To sum, the parameter estimation algorithm for the K-Wishart distribution is re-stated in Algorithm 1.

Algorithm 1 Parameter Estimation of K-Wishart Distribution

1: **input**: $\mathbf{Z}_1, \mathbf{Z}_2, ..., \mathbf{Z}_n$
2: $\hat{\mathbf{\Sigma}} = \frac{1}{n}\sum_{i=1}^{n} \mathbf{Z}_i$
3: $\hat{\kappa}_2\{\mathbf{Z}\} = \frac{1}{n}\sum_{i=1}^{n} (\ln|\mathbf{Z}_i|)^2 - \left(\frac{1}{n}\sum_{i=1}^{n} \ln|\mathbf{Z}_i|\right)^2$
4: $\hat{\eta} = \hat{\kappa}_2\{\mathbf{Z}\} - \sum_{i=0}^{d-1} \psi^{(1)}(L-i)$
5: $\hat{\sigma}_z^2 = \frac{1}{n}\sum_{i=1}^{n}\left(\ln|\mathbf{Z}_i| - \frac{1}{n}\sum_{j=1}^{n}\ln|\mathbf{Z}_j|\right)^2$
6: $\hat{\xi}_z^4 = \frac{1}{n}\sum_{i=1}^{n}\left(\ln|\mathbf{Z}_i| - \frac{1}{n}\sum_{j=1}^{n}\ln|\mathbf{Z}_j|\right)^4$
7: $\hat{\sigma}^2 = \left(\frac{1}{n} - \frac{2}{n^2}\right)\hat{\xi}_z^4 + \left(\frac{4}{n^2} - \frac{1}{n}\right)\hat{\sigma}_z^4$
8: $\bar{\eta}_m = \hat{\eta} + \left\{\exp\left(-\frac{\hat{\eta}^2}{2\hat{\sigma}^2}\right) / \left[\sqrt{2\pi}\Phi\left(\frac{\hat{\eta}}{\hat{\sigma}}\right)\right]\right\} \cdot \hat{\sigma}$
9: $\psi^{(1)}(\nu) = \bar{\eta}_m/d^2 \Longrightarrow \hat{\nu}$
10: **output**: $\hat{\mathbf{\Sigma}}$, $\hat{\nu}$

IV. Experimental Results

A. Simulation Evaluation

In this subsection, simulated K-Wishart samples are generated to evaluate the proposed parameter estimation algorithm. The sample size is fixed to 49 (corresponding to a 7×7 moving window if adaptive estimation is required in POLSAR imagery). We choose $L = 3$ and $\mathbf{\Sigma}$ given in (13) as true parameters. Using both the original and proposed methods, the bias and standard deviation of $\hat{\nu}$ are shown in Fig. 1 and Fig. 2 when its true values varies from 5 to 50 . It should be noted that by the orignal method, we mean estimation of the shape parameter directly from (10) and those cases when $\hat{\eta} < 0$ are discarded. From Fig. 1 and Fig. 2, large performance fluctuations can be observed for the orignal method whereas a significantly improved bias and standard deviation with much stabler behavior are seen for the proposed method.

The reason behind the performance differences in Fig. 1 and Fig. 2 can be explained by (21). Especially, the second term on the right hand side of (21) may be considered as a variance stablizer. It reduces the uncertainty of $\hat{\eta}$ due the introduction of the prior distribution. Such reduction propagates through (22) and finally leads to a stabler solution of $\hat{\nu}$.

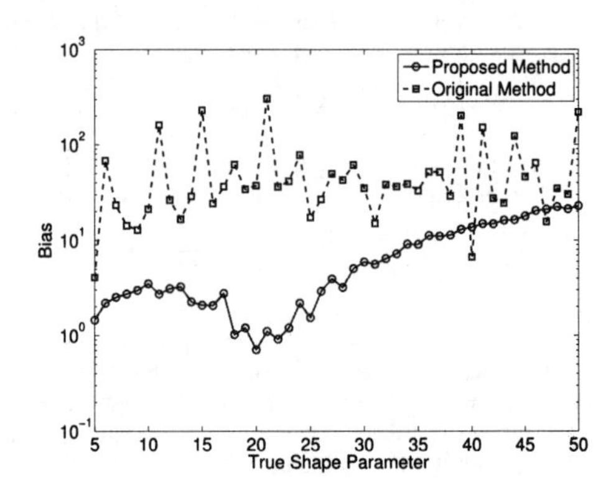

Fig. 1. Bias of the estimated shape parameter.

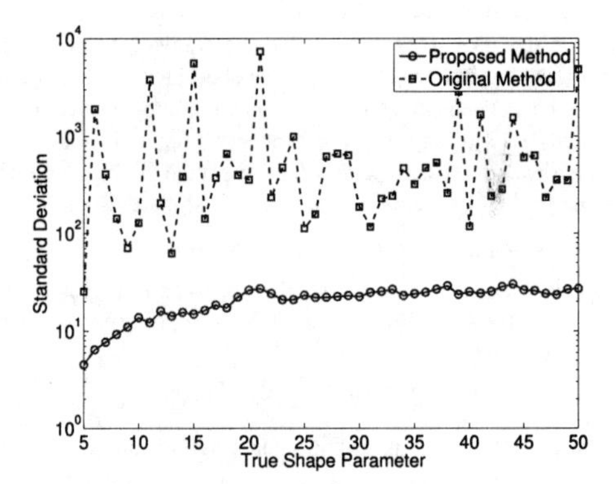

Fig. 2. Standard deviation of the estimated shape parameter.

B. Application with Real POLSAR Data

We provide one application of the proposed parameter estimation algorithm for texture analysis of real POLSAR images. The data used here is acquired by the Germany F-SAR system at the S-band. Fig. 3 displays the image color-coded in Pauli-basis. Multi-look preprocessing has been performed by combining 3×3 pixels to formulate the covariance matrix as given by (3). Therefore, the nominal number of looks is nine but it should be kept in mind that the ENL can be smaller than this value due to the oversampling of the data. Nonetheless, we will for the moment assume that $L = 9$. Later we will see how to obtain a more accurate of the ENL estimation by using the texture analysis result.

We assume that the multi-look covariance data follows the K-Wishart distribution. The shape parameter of the K-Wishart distribution ν is then adaptively estimated by using a 7×7 moving window. Fig. 4 shows the map of the estimated shape parameters using Algorithm 1 for each position of the image. Comparing Fig. 3 and Fig. 4, it is easy to see that high values are identified in homogeneous areas which indicates negligible texture effects whereas in urban and forested areas, low values are found, revealing the rich texture information therein.

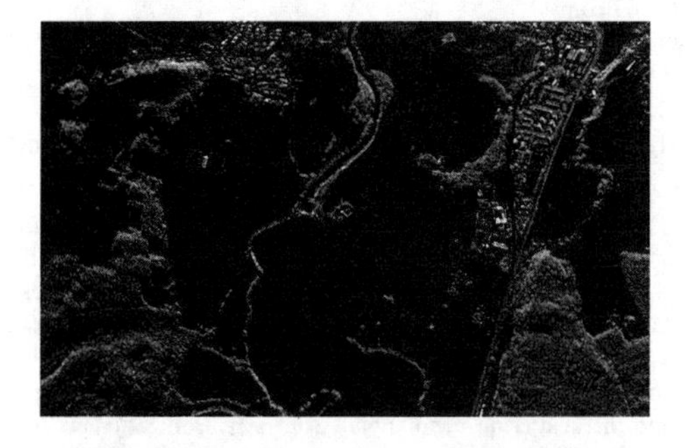

Fig. 3. Pauli-basis color display of F-SAR data.

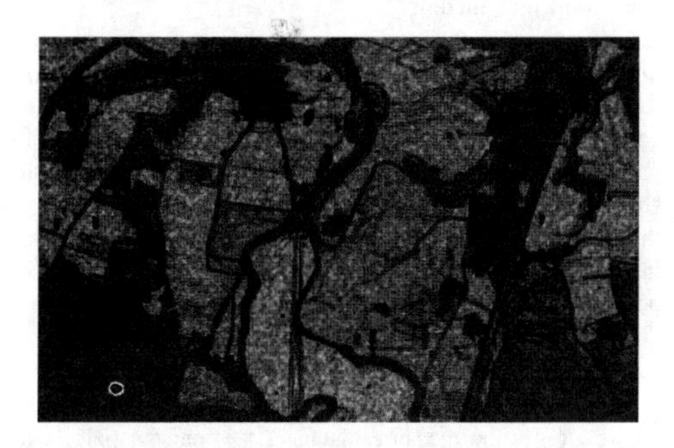

Fig. 4. Map of shape parameter estimation (in log-scale).

By thresholding the shape parameter homogeneous areas can be identified. For example, those areas where $\nu > 5L$ can be considered as homogeneous and we may use the pixels therein to refine the ENL estimation results obtained by [9]. Fig. 5 shows the histogram of the locally estimated ENL within the areas satisfying $\nu > 5L$. We see that the mode of such histogram approximately corresponds to 6.1. It is this value that is the true ENL of the synthesized multi-look data.

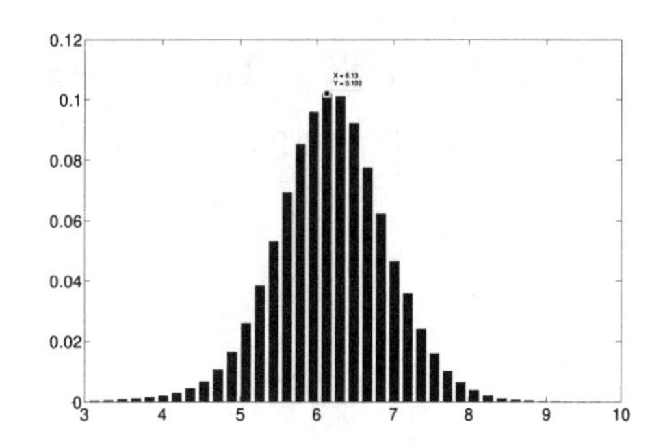

Fig. 5. Histogram of local ENL estimates in homogeneous areas.

V. Conclusion

In this paper, we have proposed a modified method of matrix log-cumulant for parameter estimation of the K-Wishart distribution that is significant in statistical POLSAR data modeling. The new approach applies Bayesian estimation of the second-order log-cumulant which ensures an always invertible equation for the shape parameter. Importantly, using simulated K-Wishart samples we have shown that the new method presents improved and stabler performance than the original one. Finally, we have demonstrated a promising application of the proposed algorithm for texture analysis of the real POLSAR imagery. It is found that with the extra texture information, a more accurate ENL estimation can be obtianed.

References

[1] F.T. Ulaby, R.K. Moore, and A. K. Fung, *Microwave Remote Sensing: Active and Passive, vol. II, Radar Remote Sensing and Surface Scattering and Emission Theory*, Inc., Dedham, Massachusetts, 1986

[2] S.N. Anfinsen and T. Eltoft, "Application of the matrix-variate Mellin transform to analysis of polarimetric radar images," *IEEE Trans. Geosci. Remote Sens.*, vol. 49, pp. 2281-2295, 2011.

[3] A.P. Doulgeris, S.N. Anfinsen, and T. Eltoft, "Automated non-Gaussian clustering of polarimetric synthetic aperture radar images," *IEEE Trans. Geosci. Remote Sens.*, vol. 49, pp. 3665-3676, 2011

[4] J.S. Lee and E. Pottier, *Polarimetric Radar Imaging: From Basics to Applications*, CRC Press, 2009.

[5] C. Oliver and S. Quegan, *Understanding synthetic aperture radar images*. SciTech Publishing, 2004.

[6] Y. Cui, G. Zhou, J. Yang, and Y. Yamaguchi, "Unsupervised estimation of the equivalent number of looks in SAR images," *IEEE Geosci. Remote Sens. Letters*, vol. 8, pp. 710-714, 2011.

[7] F.W.J. Olver, D.W. Lozier, R.F. Boisvert, and C.W. Clark, *NIST Handbook of Mathematical Functions*, Cambridge University Press, 2010.

[8] D.A. Abraham and A.P. Lyons, "Reliable methods for estimating the K-distribution shape parameter," *IEEE J. Oceanic Eng.*, vol. 35, pp. 288-301, 2010.

[9] S.N. Anfinsen, A.P. Doulgeris and T. Eltoft, "Estimation of the equivalent number of looks in polarimetric synthetic aperture radar imagery," *IEEE Trans. Geosci. Remote Sens.*, vol. 47, 3795-3809, 2009.

Narrow Pulse Transient Scattering Measurements and Elimination of Multi-path Interference

Zichang Liang, Wei Gao, Jinpeng Fang
Science and Technique on Electromagnetic Scattering Laboratory
Yangpu, Shanghai 200438, China
machlia@163.com
generalgaowei@163.com
davik.king@gmail.com

Abstract- **Via sub-ns pulse source and sampling oscilloscope, synchronous acquisition of target scattering echo pulse is achieved by adopting the dual-receiving antenna. The target scattering stretch resulted from groud multi-path effect is avoided by adjusting the measurement height. Ultra-wideband (UWB) radar cross section (RCS) distribution of the target is obtained. Compared to the calibration targets' theoretical RCS, the measured results' average error is less than 1 dB, and the time-domain inverse synthetic aperture radar (ISAR) imaging demonstrate the effectiveness of our measurement system and the data processing method.**

I. INTRODUCTION

Ultra-wideband (UWB) narrow pulse transient scattering measurement system mainly consists of UWB pulse emission source, high-bandwidth digital oscilloscope and UWB antenna [1-5]. With the development of UWB pulse power technique, UWB antenna technique and data collection technique, UWB narrow pulse transient scattering measurement technique has gradually matured.

Due to the fine time and spatial resolution of narrow pulse transient scattering measurement, transient scattering echo measurement of complex targets can distinguish the different parts of the scattering echo. For single transient scattering measurement echo contains abundant information of the target from low frequency to high frequency, scattering cross section, impulse response, scattering center, and pole distribution of the target can be extracted. Compared to conventional frequency RCS amplitude and phase measurement technique, UWB scattering time domain measurement technique has a finer resolution and need no requirement of anechoic chamber.

In this paper, via sub-ns pulse source and sampling oscilloscope, synchronous acquisition of target scattering echo pulse is achieved by adopting the dual-receiving antenna. The target scattering stretch resulted from ground multi-path effect is avoided by adjusting the measurement height. Ultra-wideband (UWB) radar cross section (RCS) distribution of the target is obtained. Compared to the calibration targets' theoretical RCS, the measured results' average error is less than 1 dB, and the time-domain inverse synthetic aperture radar (ISAR) imaging demonstrate the effectiveness of our measurement system and the data processing method.

II. TIME DOMAIN SCATTERING MEASUREMENT SYSTEM AND METHOD

Because real-time oscilloscope acquisition data accuracy and bandwidth is relatively low, the target scattering characteristics of the transient measurement commonly used sampling oscilloscope. So sampling oscilloscope is commonly used in transient measurement of target scattering characteristics. The latter requires acquisition trigger signal and the scattering echo signal strict synchronization. Usually people use pulse source to directly provide synchronous trigger signals, but there are still some problems such as trigger timebase jitter and transmit power deterioration in the synchronization method.

UWB narrow pulse transient scattering measurement system introduced in this paper mainly consists of sub-ns pulse source, sampling digital oscilloscope, UWB antenna and other components, as depicted in Fig. 1. To eliminate the effect of groud multipath reflection, the entire measurement system should be appropriately elevated. Generally speaking, to distinguish the echo of a target with a length L and groud multipath, the height from the test target to the ground should be satisfied as below,

$$H > \sqrt{\frac{R_x(2L + c\tau)}{2}} \tag{1}$$

where, R_x is the measure distance, τ is narrow pulse width and c is the light speed. In this measurement, the system height is set as 5m, so the target to be measured can be as long as 1m.

978-7-5641-4279-7

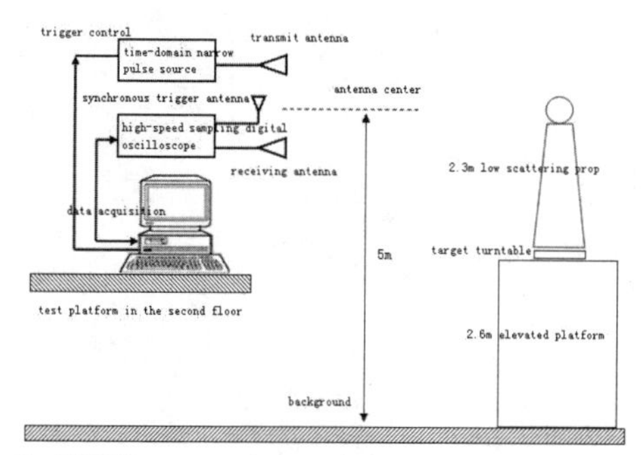

Fig.1 UWB narrow pulse transient scattering measurement system

Meanwhile, to improve the measurement stability and system sensitivity, three antennas is adopted. Besides transmit antenna and receiving antenna, a third antenna is used to provide the oscilloscope trigger signal. Compared to use splitter signal trigger of receiving antenna, the proposed method not only ensure the synchronization of the trigger signal, but also reduce splitter attenuation of the effective signal.

During the process of target scattering measurement, scattering echo time-domain waveform of pulse source signal, target and background are tested, respectively. Background echo interference is offset via the pulse alignment method. Via repeated measure average (32 times) method, the measured data accuracy can be improved.

Relative calibration is adopted in RCS calibration of narrow pulse time-domain scattering measurement. It can be expressed as below,

$$\sigma_{ta}(\theta, f) = \frac{\left|\mathrm{FFT}(E_{ta}(\theta,t))\right|^2}{\left|\mathrm{FFT}(E_{ca}(\theta,t))\right|^2} \times \sigma_c(\theta, f) \qquad (2)$$

where, E_{ta} is target measured time-domain response versus azimuth and relative time, E_{ca} is measured time-domain response of calibration target and σ_c is theoretical RCS of calibration target.

The target with RCS changing relatively slowly in the whole measured bandwidth is properly chosen as calibration target. Here, the metallic plate with a length of 20cm is chosen as calibration target. Fig. 2 illustrates RCS distribution as a function of frequency (0.5GHz-4GHz) calculated by the method of moment (MoM), which changes relatively slowly.

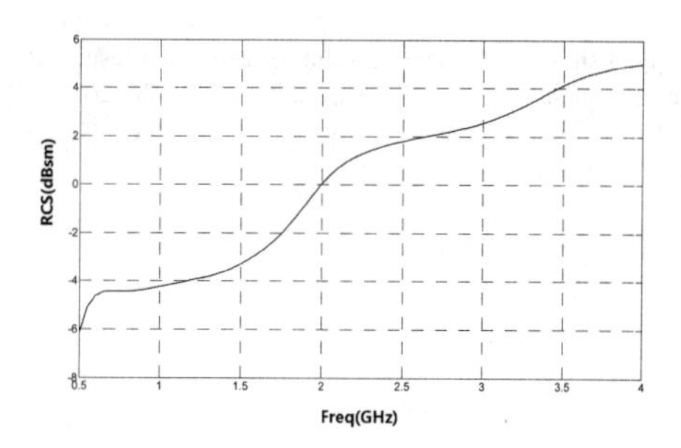

Fig.2 RCS of a 20 cm PEC plate as a function of frequency

III. Time Domain Scattering Measure Result and Comparison

Time-domain echo of a 20 cm PEC plate is measured, as depicted in Fig. 3(a). Via the RCS distribution in Fig. 2, incident waveform of measurement system is obtained by Fast Fourier Transform (FFT) and inverse FFT (IFFT), which agree well with the direct measurement result, as depicted in Fig. 3(b).

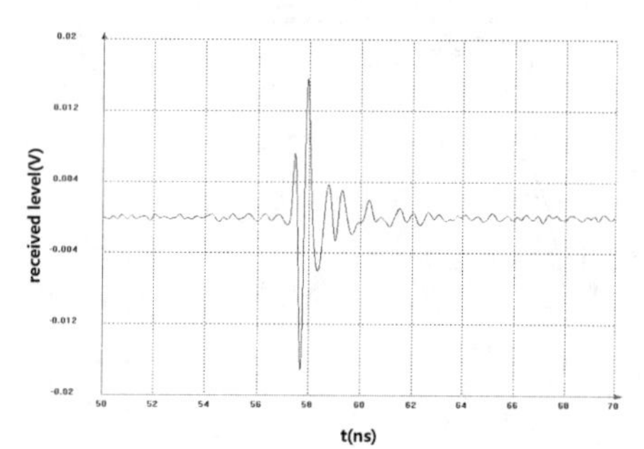

(a) time-domain echo of a 20 cm PEC plate

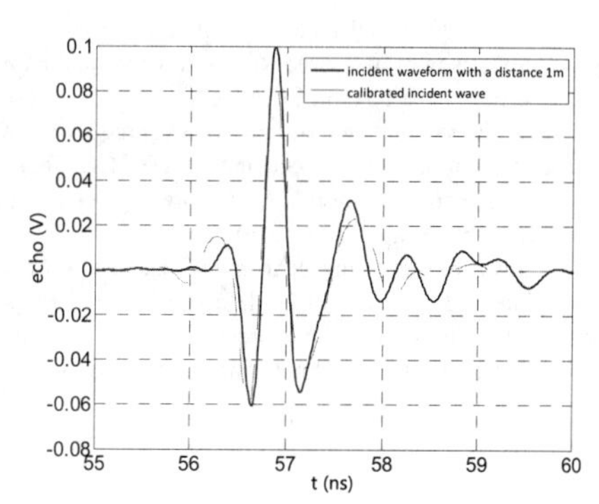

(b) incident waveform of time-domain scattering measurement

Fig. 3 time-domain echo of a 20 cm PEC plate

Fig. 4 shows the scattering echo measurement result of a 10 cm PEC plate. Its UWB RCS is calibrated by the 20 cm PEC plate calibration, and compared with the theoretical computational result.

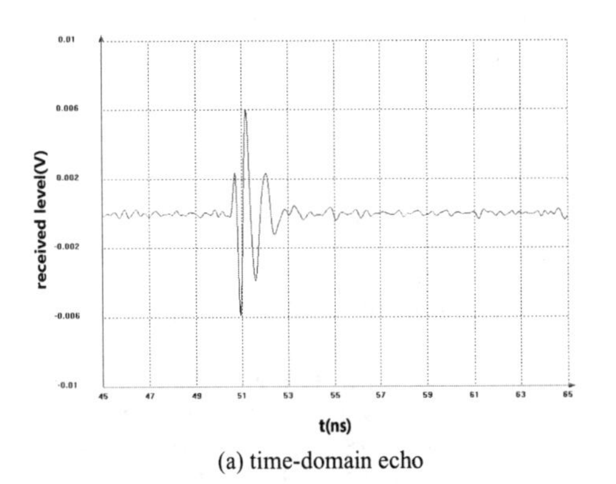

(a) time-domain echo

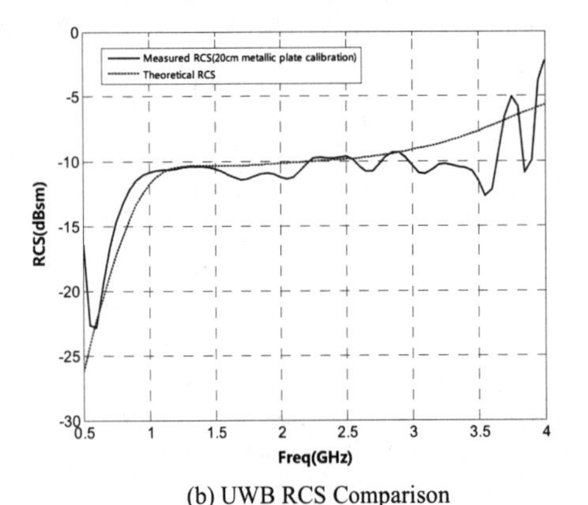

(b) UWB RCS Comparison

Fig. 4 Comparison of measured RCS and theoretical RCS of a 10 cm PEC plate

Fig. 4(b) shows that measured RCS agree well with theoretical RCS in the center frequency 1GHz-3GHz, and the average error is approximate 0.5dB. The average error in the remained frequency is approximate 1dB.

Fig. 5 shows the scattering echo measured results of a 10 cm PEC plate and a 20 cm PEC plate. Their down range is 0.44m (X direction) and their cross range is 0.15m (Y direction). The two plates can be distinguished obviously.

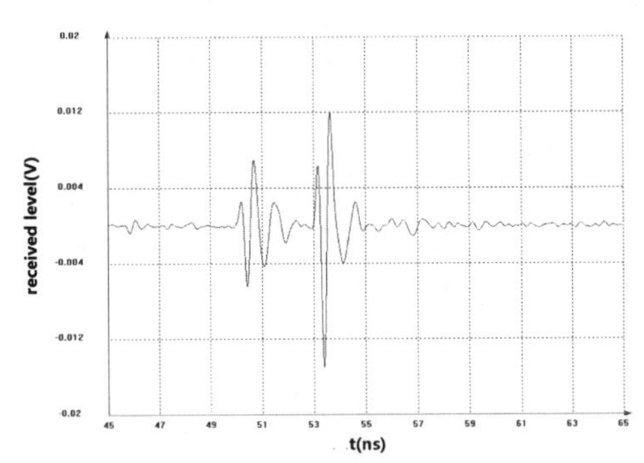

Fig. 5 scattering echo measured results of a 10 cm PEC plate and a 20 cm PEC plate

Fig. 6 is 2-D ISAR imaging[6,7] of the two composite plates using the time-domain scattering echo measured data, where the maximum rotating angle is 20°, the interval angle is 2°. There are two obvious scattering brightness, and their down range is 0.15m and their cross range is 0.45m. They agree well with the actual relative distance of the two composite plates.

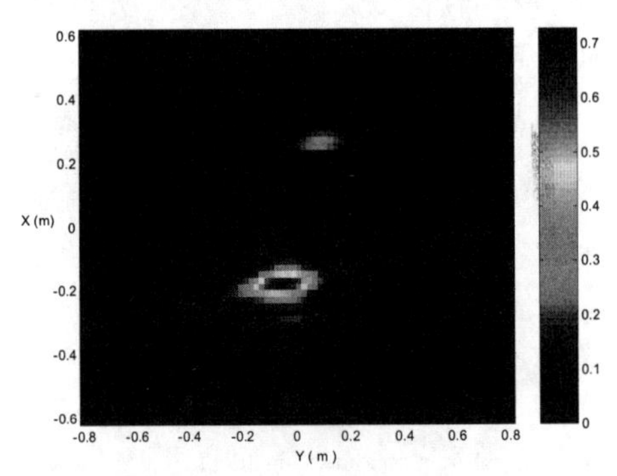

Fig. 6 ISAR of two plates

IV. CONCLUSION

Ground multipath effect is analyzed and improved by the sub-ns narrow pulse transient scattering measurement system. Synchronous acquisition of target scattering echo pulse with no gain attenuation is realized by adopting the dual-receiving antenna. Time-domain scattering echoes of metallic plate and their composite ones are measured and analyzed. The processed RCS error is less than 1dB compared to theoretical computational RCS, and ISAR of the measured target is obtained. The results demonstrate the effectiveness of our measurement system and the data processing method.

REFERENCES

[1] R. J. Fontana, E. Richley and J. Barney. Commercialization of an Ultra Wideband Precision Asset Location System[C]. IEEE Conference on Ultra Wideband Systems and Technologies, 2003, Reston.

[2] R. Fontana, A. Ameti, E. Richley, L. Beard and D. Guy. Recent Advances in Ultra Wideband Communications Systems[C]. Proceedings IEEE Conference on Ultra Wideband Systems and Technologies, 2002.

[3] GUAN Xin-pu, WANG Shao-gang, WANG Dang-wei, SU Yi and MAO Jun-jie. Measurement in Time Domain for Ultra-wideband Electromagnetic Scattering Signals. Journal of Microwaves, Vol. 24, No. 1, 2008.

[4] Rothwell E J, Chen K M, Nyquist D P, et al. Measurement and processing of scattered ultra wideband/shortpulse signals. Proceedings of SPIE the International Society for Optical Engineering, 1996.

[5] M. A. Morgan. Ultra-wideband impulse scatting measurements. IEEE Trans. Antenna Propagate, 1994.

[6] R. J. Sullivan. Radar Foundation for Imaging and Advanced Concepts. SciTech Publishing, 2004.

[7] M. Soumekh. Synthetic Aperture Radar Signal Processing with MATLAB Algorithms. Wiley-Interscience, 1999.

A Composite Electromagnetic Absorber for Anechoic Chambers

Weijia Duan(1), Han Chen(1), Mingming Sun(1), Yi Ding(1), Xiaohan Sun(1)*, Chun Cai(2) and Xueming Sun(2)
(1) National Research Center for Optical Sensing/Communications Integrated Networking, Southeast University, Sipailou 2, Nanjing 210096, China
(2) Shenzhen Academy of Metrology & Quality Inspection, The Quality Inspection Building, Middle of Longzu Avenue, Shenzhen 518055, China
*Corresponding Author: xhsun@seu.edu.cn

Abstract- **A composite electromagnetic absorber for anechoic chambers is proposed with the structure consisting of a pyramidal foam and a lossy frequency selective surface (FSS) structure. Simulation results show that the period of absorber array and the pyramid base impact directly on absorption nulls, and the effects of pattern size, surface impedance and substrate thickness of FSS are more complicated due to interactions between the pyramidal foam and FSS structure. Two examples are discussed and the results show that improved absorption below 3GHz can be achieved by properly selecting parameters.**

I. INTRODUCTION

Pyramidal foams are the most popular radar absorbing material used in electromagnetic anechoic chambers. They are not only cost-effective and lightweight, but also capable of providing good broadband microwave absorbing performance at wide incident angles[1]. Since the foams absorb wave energy by multiple reflections between the slopes of adjacent pyramids, the pyramids should be relatively large compared to wavelengths[2]. Their performance on long wavelengths will degrade if the thickness of pyramids is limited. The lossy frequency selective surface (FSS), on the other hand, has been used in designs of thin absorbers due to its ease of manufacturing and compact size of resonance cell[3]. Typically it is formed by periodic arrays of resistive elements patterned on a dielectric sheet. Although in theory its bandwidth position can be set at any frequency range, most researchers focus their work on bandwidth and absorption improvement at X band and Ku band for the applications in the field of radar[4, 5]. So far few reports have been found for absorption below 3GHz or applications in anechoic chambers.

In this paper, a thin composite foam-FSS absorber structure is proposed. The effects of absorber parameters including periods, thickness of base and substrate, pattern shape and size, are simulated and the results are discussed. It is shown that by carefully choosing parameters, the composite absorber can achieve better absorption performance than an ordinary foam or FSS absorber alone. Two examples of configurations with improved absorption at different frequency ranges below 3GHz are also presented. The design is suitable for anechoic chamber or environment of narrow band measurement.

This work is supported by the Special Fund for Quality Supervision, Inspection and Quarantine Research in the Public Interest (No.201110044).

978-7-5641-4279-7

II. STRUCTURE AND PARAMETERS

The composite absorber comprises two parts, as is illustrated in Fig. 1(a). The top part is a straight square pyramidal absorber with a taper height of L and a base thickness of D. The bottom part is an FSS structure composed of a resistive sheet with the shape of a square ring on a grounded square substrate of the thickness d. The square ring has an outer side length of a and inner side length of b, as is shown in Fig. 1(b). The surface impedance of the square ring is Sr. The square substrate has a side length of A as same as the side length of the pyramidal base.

III. SIMULATIONS AND DISCUSSIONS

To study the effects of absorber parameters on its reflectivity performance, simulations using finite element method are performed. We choose to set the taper length to a fixed value L=100mm. A commercial absorber material from Eccosorb is selected for both the pyramidal foam and the FSS substrate. Its relative permittivity and permeability around 3GHz is $\varepsilon_r = 6-i$ and $\mu_r = 1.7+1.2i$ respectively.

In this section, the effects of parameters on the absorber's performance below 5GHz will be simulated and discussed. Results are shown from Fig. 2 to Fig. 7. Two examples of configuration are illustrated in Fig. 8 and Fig. 9.

Fig. 2 indicates the significant effects caused by side length

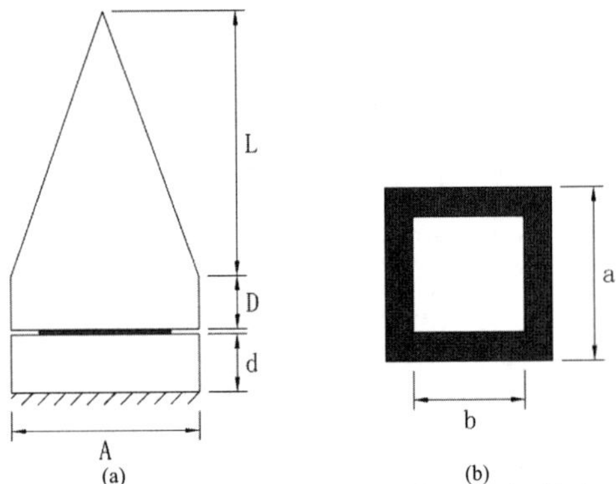

Fig.1 Schematic diagram of the proposed absorber. (a) is the side view of the absorber. (b) is the top view of the FSS pattern.

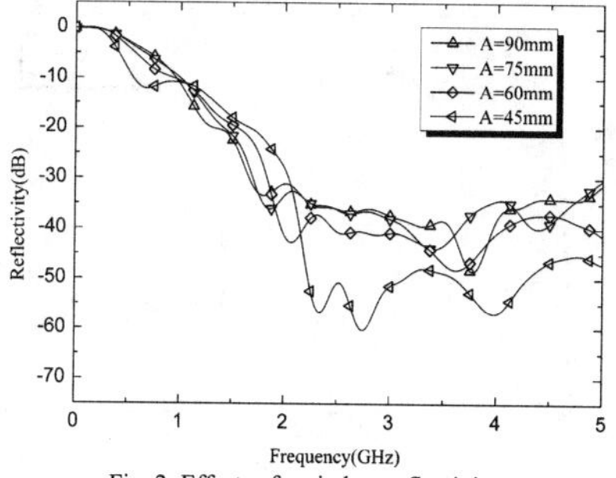

Fig. 2. Effects of period on reflectivity.

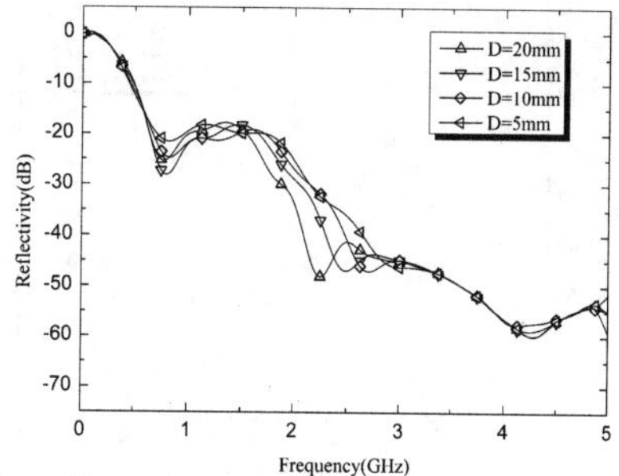

Fig. 3. Effects of base thickness on reflectivity.

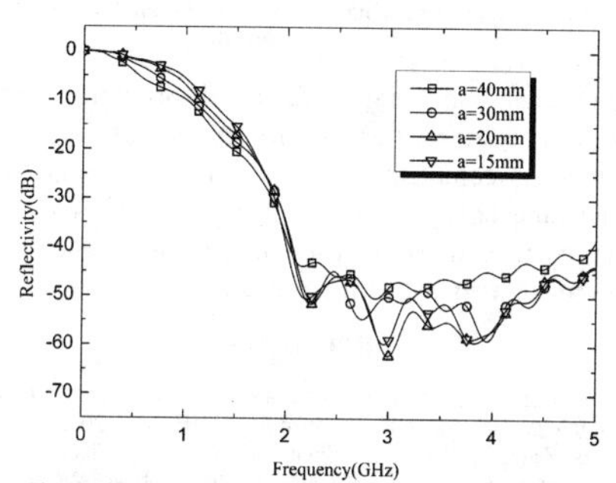

Fig. 4. Effects of side length of outer ring on reflectivity.

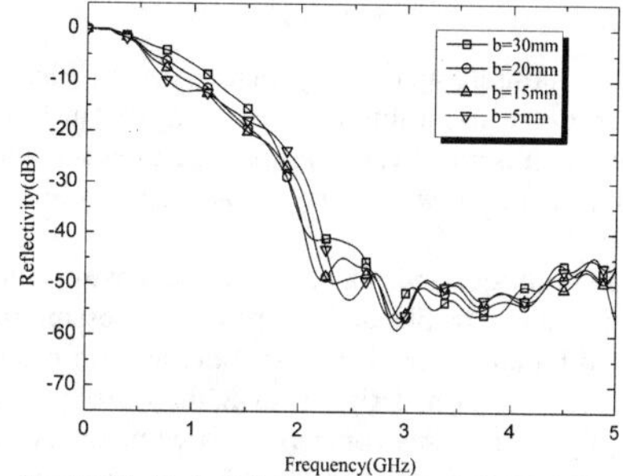

Fig. 5. Effects of side length of inner ring on reflectivity.

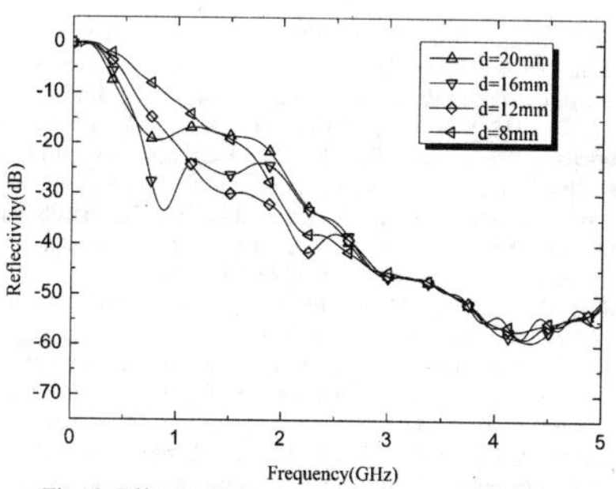

Fig. 6. Effects of substrate thickness on reflectivity.

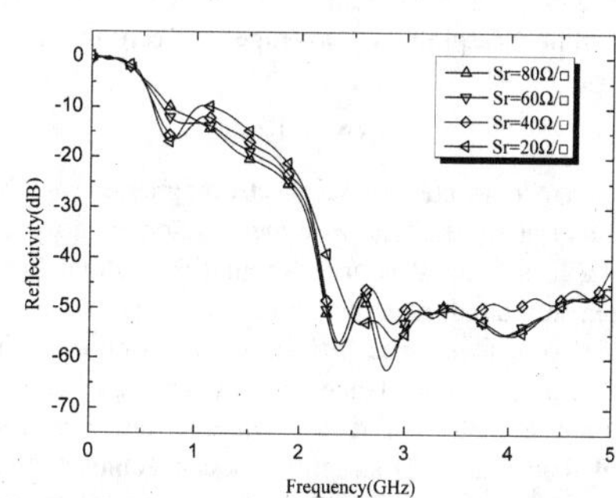

Fig. 7. Effects of surface impedance of FSS on reflectivity.

A. For the pyramidal part, as the side length increases, the taper angle also increases. This will slightly degrade the performance because incident waves reflect less times between adjacent taper slopes, resulting in less absorption. Moreover, the side length *A* is also the period of the lattice and it affects the bandgap of periodic structures[6]. It can be seen from Fig. 5 that an apparent gap appears between 2GHz and 3GHz when the period *A* is set to 45mm.

Fig. 3 shows the effects of thickness of pyramid base. While a thicker base can improve absorption due to longer lossy path the waves travel, it also shifts the position of absorbing null towards longer wavelength. That is probably because the null is formed by cancelling of transmitted and reflected waves between surfaces[7].

Fig. 4 and Fig. 5 show the influence of square ring on reflectivity. It appears that a larger ring leads to slightly less reflection. This might be explained as that large area of resistive patch could cause more wave loss under some circumstances. However, it is quite difficult to find more significant changes, because the size and shape of the square ring affects the FSS structure in many respects, such as the capacitance between adjacent rings, the inductance of the ring, the surface impedance of the FSS[5]. In fact, the design and analysis of FSS structure alone is very complicated. With the pyramidal foam combined to it, the whole absorber is added with more complication. In Fig. 6 we can see how the thickness of substrate affects the null. It should be noted that the shift of null position is not as regular as in Fig. 3. This might be explained as the result of multi-layer interactions

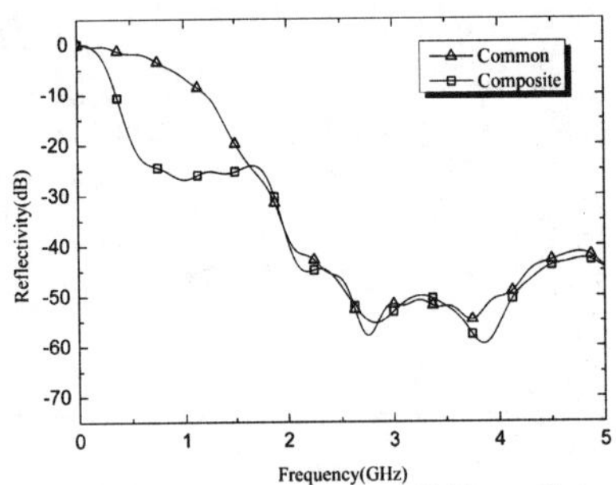

Fig. 8. Comparison between a common pyramidal foam and a composite absorber. (L=100mm, A=46mm, a=38mm, b=23mm, D=5mm, d=19mm, Sr=60 Ω/□)

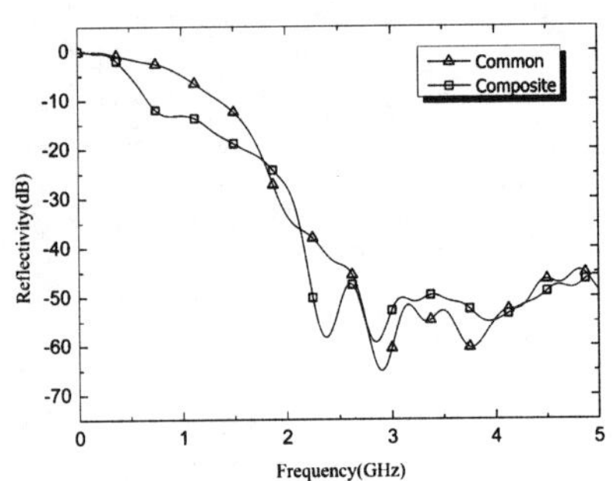

Fig. 9. Comparison between a common pyramidal foam and a composite absorber. (L=100mm, A=44mm, a=34mm, b=8mm, D=1mm, d=9mm, Sr=60 Ω/□)

between the substrate and the pyramid. Also in Fig. 7, the surface impedance has a different effect with what has been reported in [8]. It is not obvious from the figure that maximum bandwidth is achieved when surface impedance is around 50 Ω/□.

Although the design and analysis of the proposed composite absorber is quite complicated, it provides possibilities to absorbers with improve performance. Here are two examples. In Fig. 8 we present a design with 25dB reflectivity from 500MHz to 1.5GHz. As a comparison, a common pyramidal absorber of the same material and total height is also presented. In the design of Fig. 9 we manage to introduce a null around 2.4GHz while keeping an average reflectivity at other frequencies.

IV. Conclusion

In this paper a composite electromagnetic absorber is proposed and analyzed. The absorber comprises a pyramidal foam and a lossy FSS structure. Simulation results show that while the period of absorber array and the base of the pyramidal foam have direct effects on absorption nulls, the influence of surface impedance and substrate thickness could be far more complicated due to interactions between the pyramidal foam and FSS structure. It can achieve improved absorption performance at low frequencies by proper selection of absorber parameters. Two examples of configurations are also presented, showing different absorption improvement below 3GHz compared with common foam of the same height. The proposed composite absorber might find applications in compact anechoic rooms or narrow band measurement. Further work can be focused on simplification of design procedure and manufacturing process.

References

[1] L. H. Hemming, *Electromagnetic Anechoic Chambers: A Fundamental Design and Specification Guide*: Wiley-IEEE Press, 2002.

[2] Y. B. Feng, T. Qiu, C. Y. Shen, and X. Y. Li, "Electromagnetic and Absorption Properties of Carbonyl Iron/Rubber Radar Absorbing Materials," *IEEE Transactions on Magnetics,* vol. 42, pp. 363-368, Mar 2006.

[3] D. J. Kern and D. H. Werner, "A Genetic Algorithm Approach to The Design of Ultra-Thin Electromagnetic Bandgap Absorbers," *Microwave and Optical Technology Letters,* vol. 38, pp. 61-64, Jul 2003.

[4] H.-T. Liu, H.-F. Cheng, Z.-Y. Chu, and D.-Y. Zhang, "Absorbing Properties of Frequency Selective Surface Absorbers with Cross-Shaped Resistive Patches," *Materials & Design,* vol. 28, pp. 2166-2171, 2007.

[5] F. Costa, S. Genovesi, and A. Monorchio, "On The Bandwidth of High-Impedance Frequency Selective Surfaces," *IEEE Antennas and Wireless Propagation Letters,* vol. 8, pp. 1341-1344, 2009.

[6] S. R. A. Dods, "Bragg Reflection Waveguide," *J. Opt. Soc. Am. A,* vol. 6, pp. 1465-1476, Sept 1989.

[7] E. F. Kuester and C. L. Holloway, "A Low-Frequency Model For Wedge or Pyramid Absorber Array. 1. Theory," *IEEE Transactions on Electromagnetic Compatibility,* vol. 36, pp. 300-306, Nov 1994.

[8] F. Costa, A. Monorchio, and G. Manara, "Analysis and Design of Ultra Thin Electromagnetic Absorbers Comprising Resistively Loaded High Impedance Surfaces," *IEEE Transactions on Antennas and Propagation,* vol. 58, pp. 1551-1558, May 2010.

TA-2(D)

October 24 (THU) AM

Room D

Radio Propagation

A 3-D FDTD Scheme for the Computation of HPM Propagation in Atmosphere

Ke Xiao, Shunlian Chai, Haisheng Zhang, Huiying Qi, and Ying Liu
College of Electronic Science and Engineering, National University of Defense Technology,
Changsha, Hunan 410073 China

***Abstract-* Focus on the simulation of the dynamics of high power microwave (HPM) ionization of air at atmosphere, we analyze the physical mechanism in the formation of plasma. The magnetohydrodynamic formulations, formed by Maxwell equations coupled with simple fluid description, are applied to describe the ionization procedure. Then, we present a three dimensional (3-D) finite-difference time domain (FDTD) scheme to deal with such problems, of which the computational procedure is discussed in detail. At last, numerical experiments are provided, and some parameters about plasma are analyzed. The theory also predicts the pulse breakdown threshold accurately.**

I. INTRODUCTION

Air ionization using high power microwave (HPM) source in a range of pressures and frequencies has been extensively investigated. Most of work was motivated by the need to avoid breakdown in RF transmission systems in waveguides and near antennas, some of the applications involve earth-space communications and creation of artificial ionized layers in the atmosphere, as well as antennas for high-power microwave transmission [1][2].

Early studies [3][4] of air ionization focused on the determination of the breakdown field as a function of several parameters such as pressure, frequency, and pulse duration, the approaches generally use the effective field concept, where it was calculated assuming a simple relationship between momentum transfer collision frequency and pressure, besides, all calculations were made assuming that diffusion is negligible. However, it is difficult for us to use the effective field model to investigate the ionization procedure, since the complex coupling between the wave and the plasma can not be described analytically and numerically.

In order to investigate the mechanism of ionization under HPM, we should describe the procedure using exact physical and computational model based on kinetic theory [5], one is the microcosmic model describing the motion of particles, which is simple but the interaction between particles can not be considered. Another is magnetohydrodynamic equations (MHD) based on macroscopic formulations of plasma. Based on the MHD theory, Yee et al. [6] provided a one-dimensional method based on fluid equations and Maxwell equations to investigate the dynamic behavior of intense electromagnetic pulse propagating through atmosphere. In [7], a simple 1D model consists of a 1D plasma fluid model and a 1D wave equation was developed to investigate the experimental observations. The pattern description, however, needs at least a two-dimensional (2D) approach, such studies includes works by J. P. Boeuf et al. [8][9], where coupling Maxwell equations with plasma fluid equations were applied to describe the formation of plasma under conditions similar to the experiments[10]. Recently, differential solver such as finite-difference time domain (FDTD) method has been proposed to analyze the hybrid equations [9][11]. Briefly, there are always two ways to consider the model of plasma in FDTD, one way is to treat the generated plasma in atmosphere as air filled with induced plasma currents [9][11], another way is the usage of equivalent dielectric medium theory[5].

In this paper, we use the coupling Maxwell equations with plasma fluid equations to describe the procedure of HPM propagating in atmosphere, a continuity equation is also applied to describe the electron density. Then, a 3-D FDTD method is provided to solve the hybrid equations, afterwards, power thresholds for breakdown of incident HPM pulses, as well as plasma parameters (e.g. electron density, plasma frequency) are to be analyzed, and by comparing to the traditionally calculated results, the FDTD method is demonstrated to be more accurate.

II. PHYSICAL AND COMPUTATIONAL MODEL

We know that, under the condition of HPM propagating in atmosphere, the quiver energy of free electrons increases, as well as the collisions between charged particles and neutral molecules, where some kinetic energy is transformed to thermal energy since inelastic collisions exists. In the process, when the quiver energy of bound electron is higher than the energy required for ionizing, the bound electrons in neutral molecules may acquire sufficient energy to escape and become free ones, which we called ionization.

A. Magnetohydrodynamic Formulations

To analyze the process, Maxwell equations are coupled with the Boltzmann equation and air plasma transport equations through the electron current density

$$\nabla\times\vec{E}=-\mu_0\frac{\partial\vec{H}}{\partial t} \tag{1}$$

$$\nabla\times\vec{H}=\varepsilon_0\frac{\partial\vec{E}}{\partial t}+\vec{J}_e \tag{2}$$

$$\vec{J}_e=-eN_e\vec{v}_e \tag{3}$$

where, $\vec{E}$ and $\vec{H}$ represent the electric and magnetic field of HPM, $\vec{J}_e$ is the plasma current density induced by the incident waves, N_e is the electron density, e denotes the electron charge, $\vec{v}_e$ is the electron mean velocity, μ_0 and ε_0 express the magnetic permeability and electric permittivity

978-7-5641-4279-7

of vacuum, respectively, where the ion contribution to the current density is neglected.

In order to obtain the plasma current density in (3), N_e and $\vec{v}_e$ should be calculated. From the simplified electron momentum transfer equation as shown in (4), an approximation to the mean electron velocity can be derived.

$$m_e \frac{d\vec{v}_e}{dt} = -e\vec{E} - m_e \vec{v}_e v_m \tag{4}$$

where, m_e is electron mass, v_m represents the momentum transfer collision frequency. The time evolution of the free electrons density N_e in atmosphere can be represented as follows associated with ionization, diffusion, attachment and recombination [9]

$$\frac{dN_e}{dt} = v_i N_e - v_d N_e - v_a N_e - v_r N_e^2 + v_{r0} N_{e0}^2 \approx v_{net} N_e \tag{5}$$

where, v_{net} denotes the net ionization frequency, and we use the approach in [12] to express v_{net} as follows:

$$v_{net} = 9.0\times10^{-9} N_e \frac{\tilde{\varepsilon}}{1+v_m^2/\omega^2} \exp\left[-\sqrt{\frac{2\text{eV}}{\tilde{\varepsilon}/\left(1+v_m^2/\omega^2\right)}}\right] \tag{6}$$

where, v_m is estimated by $3\times10^{-7}N_e$ [12], and $\tilde{\varepsilon}$ represents the quiver energy of an electron:

$$\tilde{\varepsilon} = \frac{1}{2} m_e \frac{e^2 E_0^2}{m_e^2 \omega^2} \tag{7}$$

where E_0 is the peak amplitude of an oscillating electric field.

B. *FDTD Solver for Hybrid Equations*

To solve the magnetohydrodynamic equations numerically, it is needed to transform them to an equivalent set of equations on discrete time, electromagnetic fields and currents (relate to average velocity of electron and its density). One of the powerful method is the FDTD method, which is an explicit time-domain method using centered finite differences on the Yee space lattice [13]. As it is known that, in the total-field FDTD formulation, the incident wave is propagated through the grid, while in a scattered-field FDTD codes used here, the incident wave can be generated accurately via an exact analytical function at each location. The electric and magnetic field update equations in (1) and (2) are obtained by decomposing the field components into incident and scattered terms as

$$\boldsymbol{E}^t = \boldsymbol{E}^i + \boldsymbol{E}^s \tag{8a}$$

$$\boldsymbol{H}^t = \boldsymbol{H}^i + \boldsymbol{H}^s \tag{8b}$$

where subscripts i, s, and t stand for incident, scattered, and total fields, respectively. Then, the magnetohydrodynamic formulations can be discretized using leapfrog approximations. Without loss of generality, we consider the calculation of field component in x-direction, the difference scheme for Maxwell equations can be expressed as

$$H_x^s\Big|_{i,j+\frac{1}{2},k+\frac{1}{2}}^{n+1/2} = H_x^s\Big|_{i,j+\frac{1}{2},k+\frac{1}{2}}^{n-1/2} - \frac{\Delta t}{\mu_0}\left[\frac{E_z^s\Big|_{i,j+1,k+\frac{1}{2}}^{n} - E_z^s\Big|_{i,j,k+\frac{1}{2}}^{n}}{\Delta y} - \frac{E_y^s\Big|_{i,j+\frac{1}{2},k+1}^{n} - E_y^s\Big|_{i,j+\frac{1}{2},k}^{n}}{\Delta z}\right] \tag{9}$$

$$E_x^s\Big|_{i+\frac{1}{2},j,k}^{n+1} = E_x^s\Big|_{i+\frac{1}{2},j,k}^{n} + \frac{\Delta t}{\varepsilon_0}\left[\frac{H_z^s\Big|_{i+\frac{1}{2},j+\frac{1}{2},k}^{n+1/2} - H_z^s\Big|_{i+\frac{1}{2},j-\frac{1}{2},k}^{n+1/2}}{\Delta y} - \frac{H_y^s\Big|_{i+\frac{1}{2},j,k+\frac{1}{2}}^{n+1/2} - H_y^s\Big|_{i+\frac{1}{2},j,k-\frac{1}{2}}^{n+1/2}}{\Delta z} + eN_e\Big|_{i+\frac{1}{2},j,k}^{n} v_{ex}\Big|_{i+\frac{1}{2},j,k}^{n+1/2}\right] \tag{10}$$

Approximation for electron momentum equation (4) can be written as

$$v_{ex}\Big|_{i+\frac{1}{2},j,k}^{n+1/2} = -\frac{2e\Delta t}{m(2+v_m\Delta t)} E_x^t\Big|_{i+\frac{1}{2},j,k}^{n} + e^{-v_m\Delta t} v_{ex}\Big|_{i+\frac{1}{2},j,k}^{n-1/2} \tag{11}$$

The continuity equation (5) is solved using a simple analytical scheme as follows

$$N_e\Big|_{i,j,k}^{n+1} = N_e\Big|_{i,j,k}^{n} e^{v_{net}\Delta t} \tag{12}$$

We use the same grid spacing, noted as Δ in the x-, y- and z-directions, and it is chosen as $\Delta \le \lambda/12$. Additionally, by considering the stability for hybrid equations, the time step Δt of the FDTD scheme should satisfy

$$\Delta t \le \min\left\{\frac{\Delta}{v_C\sqrt{3}}, \frac{1}{10v_m}, \frac{1}{v_C}\left(\frac{1}{\Delta^2} + \frac{e^2\mu_0 N_e}{4m_e}\right)^{-1/2}, \frac{1}{10v_{net}}\right\} \tag{13}$$

where v_c is the velocity of light in vacuum.

Convolutional PML (CPML) is applied here as absorbing boundary condition (ABC) [14], which is based on a recursive-convolution technique.

III. Results and Discussion

In this section, we show the 3-D simulation results for ionization performance in atmosphere around altitude of 60km, with 300MHz incident HPM transmit vertically along the z-direction, the magnitude the sinusoidal wave is E_0, the pulse width is 1μs. The simulation domain is $1\lambda\times1\lambda\times30\lambda$ ($z\times y\times z$), the grid size is set to $\Delta = \lambda/20$, and the time step $\Delta t = 0.9\Delta/v_c/\sqrt{3}$, 10-cell CPML are used as ABCs, and 8 cells are defined between the ABCs and plasma domain.

Calculation with four different E_0 (5×10^4, 8×10^4, 9×10^4, 10^5 V/m) have been performed, the electron density is compared with the breakdown threshold ($N_e = 10^8 N_{e0}$), from the results, the threshold electric field for breakdown is 9×10^4 V/m $< E_0 < 10^5$ V/m.

Fig. 1 shows the electron density when $E_0=10^5$ V/m after one pulse, the peak electron density has reached over 10^{14} cm^{-3}, which exceeds the breakdown threshold ($10^8 N_{e0} \approx 10^{9.9}$ cm^{-3}). Fig. 2 shows the calculated plasma frequency, and the peak point reached about 15MHz. Fig. 3 depicts the power density of total field, where in the ionized region, for the effects of absorbing and reflection, the power density of waves has been disturbed obviously, it costs about 12.28 minutes for the calculation on a PC workstation.

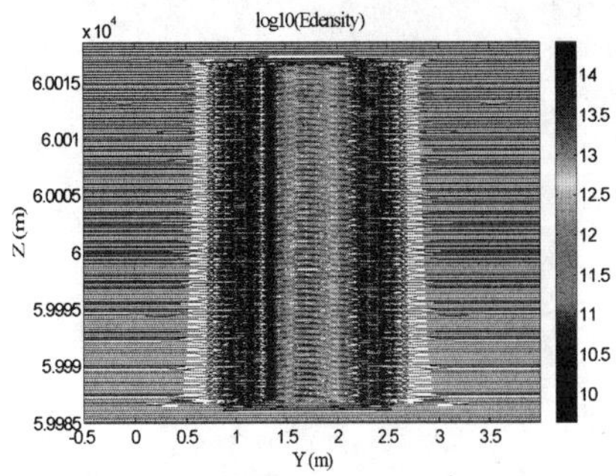

Figure 1. Evolution of electron density (cm^{-3}) in plane x=0.

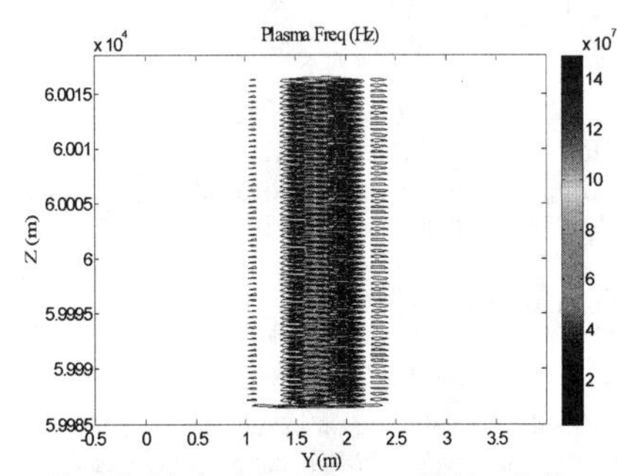

Figure 2. Calculated plasma frequency (Hz) in x=0 plane.

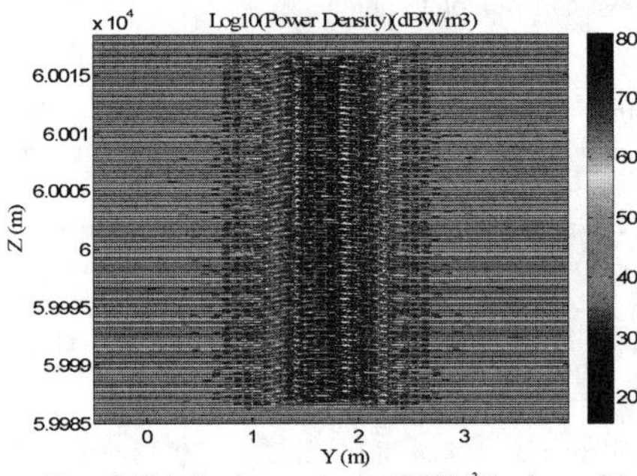

Figure 3. Calculated power density (dBW/m^3) in plane x=0.

It can be concluded from the results that, the threshold of power density for breakdown is 67.3 dBW/m^3 $<P_d<$68.2 dBW/m^3, while, by using the traditional ionization theory based on effective field approach and equivalent DC experiments [3][4], the air-breakdown-threshold is about 64.8 dBW/m^3 which is less than the 3-D FDTD results. However, there are many limitations to the effective field model, since diffusion is negligible and the dense coupling between wave and plasma can not be considered numerically, besides, the results in chamber were obtained under certain conditions, e.g. restricting the range of v_m/ω and E_e/P. So the results obtained by FDTD method are more accurate.

IV. CONCLUSION

In this paper, we use the magnetohydrodynamic formulations as the physical model to consider the ionization, in which, coupling Maxwell equations with plasma fluid equations are applied to describe the procedure. Then, in order to describe the formation of 3-D plasma patterns during HPM propagation in atmosphere, we develop a 3-D FDTD solver to simulate the time-vary procedure. An example is provided to demonstrate the efficiency of the method, where, a simulation domain with size $1\lambda\times1\lambda\times30\lambda$ near the attitude of 60km is defined, a sinusoidal plane wave is assumed to propagate along z-direction, CPML is set as the ABCs. By changing the input power density and observing the calculated parameters involving the electron density and plasma frequency, we indicate a more reasonable breakdown threshold, compared to the traditional results.

REFERENCES

[1] J. Benford, J. A. Swegle and E. Schamiloglu, *High Power Microwave*, 3rd ed., CRC Press, Taylor & Francis Group, 2007.

[2] M. Löfgren, D. Anderson, et. al, "Breakdown-induced distortion of high-power microwave pulses in air," *Phys. Fluids B*, vol. 3, pp. 3528-3531, 1991.

[3] P. Felsenthal, "Nanosecond pulse microwave breakdown in air," *Journal of Applied Physics*, vol. 37, pp. 4557-4560, 1966.

[4] G. N. Hays, L. C. Pitchford, and J. B. Gerardo, "Ionization rate coefficients and induction times in Nitrogen at high values of E/N," *Phys. Rev. A*, vol. 36, pp. 2031-2040, 1987.

[5] J. A. Bittencourt. *Fundamentals of Plasma Physics*. New York: Springer, 2004.

[6] J. H. Yee, R. A. Alvarez, D. J. Mayhall, et al. "Theory of intense electromagnetic pulse propagation through the atmosphere," *Physics of Fluids*, vol. 29, pp. 1238-1244, 1986.

[7] S. K. Nam and J. P. Verboncoeur, "Theory of filamentary plasma array formation in microwave breakdown at near-atmospheric pressure," *Phys. Rev. Lett.*, vol. 103, 055004, July 2009.

[8] J. P. Boeuf, B. Chaudhury and G. Q. Zhu, "Theory and modeling of self-organization and propagation of filamentary plasma arrays in microwave breakdown at atmospheric pressure," *Phys. Rev. Lett.*, vol. 104, 015002, 2010.

[9] B. Chaudhury, J. P. Boeuf, "Computational studies of filamentary pattern formation in high power microwave breakdown generated air plasma," *IEEE Trans Plasma Sci.* vol. 38, pp. 2281-2288, 2010.

[10] Y. Hidaka, E. M. Choi, I. Mastovsky, et al., "Observation of large arrays of plasma filaments in air breakdown by 1.5-MW 110-GHz gyrotron pulses," *Phys. Rev. Lett.*, vol. 100, 035003, Jan 2008.

[11] T. Tang, C. Liao, "Analysis of reflection properties of high power microwave propagation in mixture-atmosphere," *J. Electromagnetic Analysis & Application*, vol. 2, pp. 543-548, 2010.

[12] K. Tsang, "RF ionization of the lower ionosphere," *Radio Sci.* vol. 26, pp. 1345-1360, May 1991.

[13] K. S. Yee, "Numerical solution of initial boundary value problems involving Maxwell's equations in isotropic media," *IEEE Trans. Antennas Propagat.*, vol. 14, no. 3, pp. 302-207, 1966.

[14] J. A. Roden, and S. D. Gedney, "Convolutional PML (CPML): An efficient FDTD implementation of the CFS-PML for arbitrary media," *Microwave Optical Tech. Lett.*, vol. 27, pp. 334-339, 2002.

Analysis of Schumann Resonances based on the International Reference Ionosphere

Yi Wang, Xiao Yuan, Qunsheng Cao
College of Electronic and Information Engineering
Nanjing University of Aeronautics and Astronautics (NUAA), Nanjing 210016, China
Email: jflsjfls@nuaa.edu.cn

***Abstract*—Schumann Resonances (SRs) are simulated and analyzed in this work using the geodesic finite-difference time-domain (GFDTD) method coupled with the international reference ionosphere (IRI). Through simulations, global lightning activities are considered as excitations and real environmental parameters of the Earth-ionosphere system are introduced. Simulation results obtained with analytic ionosphere models are also included in order to compare with our simulation results.**

I. Introduction

Schumann Resonances (SRs) are the resonant electromagnetic (EM) waves in the Earth-ionosphere system, which are mainly caused by global lightning discharges. Because SRs are easily effected by local environment and global properties (such as lightning distribution), they are considered to be effective indicators of global phenomena such as EM distribution, thunderstorm activities and temperature variations [1], [2], [3], [4]. However the lack of efficient simulation tools and experimental data before 2000s slows the application study of SRs. In the recent years, with the development of modern computation/experiment techniques the study of SRs are again valued.

The study of SRs can trace back to 1950s [5], since then researchers had studied this phenomena using mainly the analytic method [6], [7]. It is not until recent years numerical methods have been applied to such study. With the help of modern measurement techniques, the study of SRs applications are becoming possible, such as global lightning triangulations and global climate variation detections.

Although numeric EM simulation techniques such as the finite-difference time domain (FDTD) method can provide enough accuracy for most EM application studies [8], the uncertainty of parameters of the whole Earth-ionosphere system makes SRs simulations difficult. Nowadays, the accurate simulation of global parameters, especially global ionosphere conductivity distributions play a key role in most present SRs application studies, as they vary with many parameters such as time, location and temperature [9]. Many recent SRs simulations relies on the analytic ionosphere models such as the "Knee" model or "two-exponential" model [10], [1], however such models can only reveal approximate average ionosphere conductivity distributions, thus cannot provide enough accuracy for further SRs application studies. The development of the international reference ionosphere (IRI) [9] provides a good tool for such problem.

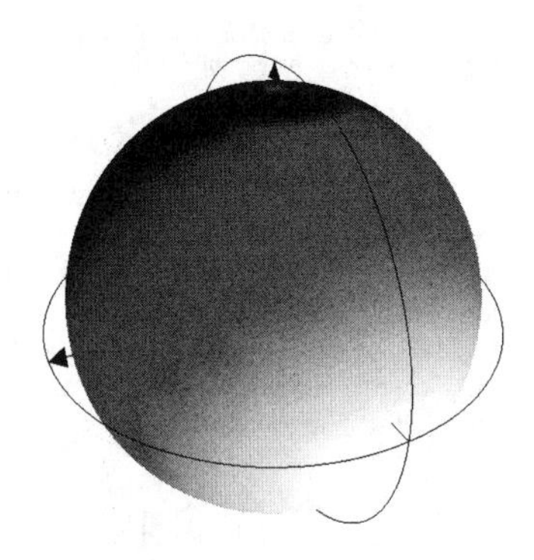

Fig. 1. Ionosphere conductivity distribution at height 80km. Generated at UT=0.0 from the international reference ionosphere (IRI).

In this work we apply the geodesic FDTD (GFDTD) method to simulate extremely low frequency (ELF) EM wave propagation in the Earth-ionosphere system, to study the SRs based on the ionospehre conductivity data from IRI, because the geodesic FDTD model has many advantages over other time domain EM solutions such as the optimized stability property of the FDTD algorithm [11]. Global parameters such as lightning discharges, lossy Earth's crust and geodesic information are also included in the simulations. We have also compared the results simulated with IRI and with different analytical ionosphere models.

II. Method

A. Ionosphere Conductivity Model

The data of IRI are generated from the worldwide network of ionosondes, which can provide an empirical standard model of the ionosphere. Because conductivity profiles are not directly provided by IRI, calculations have to be made to obtain these data [1]. Fig. 1 presents typical global conductivity distributions generated from IRI. The observation height is 80km and at UT=0.0 on Jan. 1st, 2010. From this figure the diurnal variations and polar anomalies of the ionosphere are clearly presented.

The comparison of ionosphere conductivity generated from IRI and from analytic "Knee" ionosphere models are presently

978-7-5641-4279-7

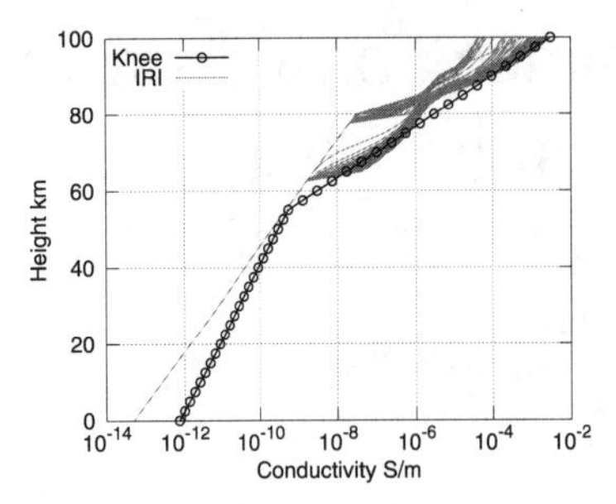

Fig. 2. Conductivity profiles of different ionosphere models. Knee: the "Knee" model. IRI: global conductivity profile variations obtained on Jan. 1st, 2010.

in Fig. 2, in which the vertical conductivity profiles are generated. In this figure, global conductivity profile variations are presented to compare with the "Knee" conductivity model, which was proposed in [10] and widely adopted in many ionosphere studies. Through comparison the advantage of using IRI data is clearly presented as the inhomogeneity of global conductivity distributions can be fully considered.

B. Excitations

Global lightning activities are believed to be the common reason of SRs as they are continuous and possess tremendous power. In this work artificial lightning currents from global lightning centers (Southeast Asia, Africa, and South America [1]) are modeled to be excitations of SRs. The form of the lightning currents is described in [1].

C. Simulation Model and Method

The GFDTD method is applied to study SRs in this work. In this method the Earth-ionosphere system model is constructed by the alternating planes of transverse-magnetic (TM) and transverse-electric (TE) field components, which are composed of triangular cells and hexagonal cells (including 12 pentagonal cells), respectively. Then integral form of the Maxwell's equation is introduced to the model to complete the FDTD process. In this work, the horizontal direction (tangential direction of the Earths surface) resolution is about 250 km and the radial (vertical) direction resolution is 5km. This mesh can provide enough accuracy for the frequency band 0.0150 Hz, to calculate the SRs.

The calculation region of this work extends to a depth of 50 km into the lithosphere (to account for the lossy ground) and to an altitude of 100 km above sea level (the top of the ionosphere D region). To accurately describe the EM environments of the lossy Earth-ionosphere system, topographic data from the NOAA-NGDC [12] are used. In the crust, the permittivity and conductivity are set according to reference models [13], [14].

III. Results and Discussion

The waveforms presented in Fig. 1 are excited using global lightning currents and obtained at $UT = 0$, 32°N,118°E

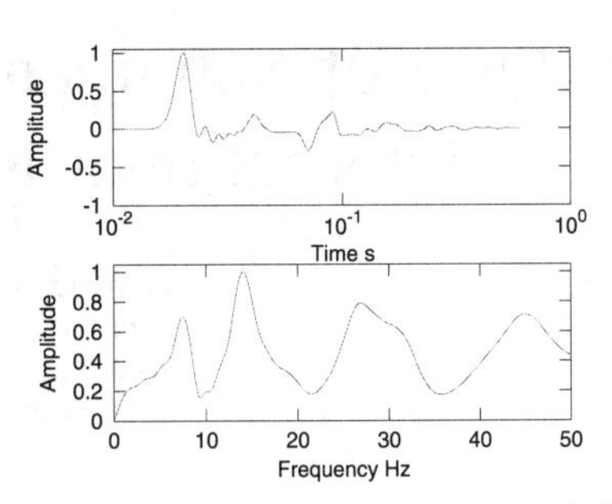

Fig. 3. Time domain and frequency domain waveforms obtained at $UT = 0$, 32°N,118°E (Nanjing).

TABLE I
Comparison of resonant frequencies and qualify factors obtained using different ionosphere conductivity models.

	Analytic	IRI		Knee	
	f_n Hz	f_n Hz	Q	f_n Hz	Q
1	7.7	7.81	5.45	7.77	3.80
2	14.0	13.97	6.09	14.16	4.59
3	20.2	20.13	6.52	20.50	5.55
4	26.5	26.24	6.54	26.77	6.46
5	32.8	32.80	7.78	32.99	7.23

(Nanjing). The result is simulated using the conductivities from IRI. From this figure, the attenuation of waveforms with time are clearly shown (up sub figure). Because the Earth-ionosphere system is a lossy cavity, the SRs can not be accurately obtained using just the Fourier transform (as shown in Fig. 1, down sub figure). In this work, Prony's method is applied to obtain the resonant frequencies and quality factor.

Table 1 summarizes the resonant frequencies and qualify factors obtained using different ionosphere conductivity models, compared against analytical results [10]. From this table we can see the frequencies obtained with different models are very similar (the errors are within 1%). Although the differences are slight, they contain information about the local environment and ionosphere properties for further study. It is noted that the average experimental value of the first SR is about 7.8Hz, which means the FDTD results are more accurate than analytic one. The comparison of qualify factors shows differences at the first two resonant frequencies, in which the qualify factor obtained using IRI data is higher than using "Knee" model. Because the simulation results using IRI consider more details of the ionosphere conductivity, we believe these value are more accurate in describing local environment.

A study of global distribution of SRs are presented in Fig. 4. In this simulation, EM waves are observed along the equator at 0°N0°E, 0°N10°E, ..., 0°N350°E (0°N10°W), respectively. Frequency domain waveforms are obtained in 0.01-50Hz, $\triangle f = 0.06$Hz. Similarly, conductivity data from IRI are used to study SR variations. In this figure, several

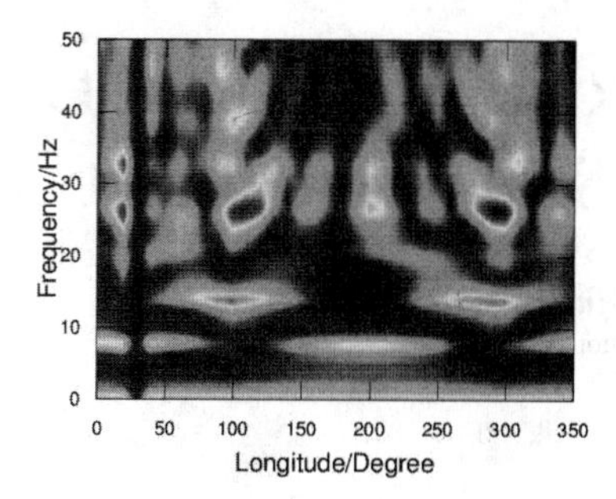

Fig. 4. Global distribution of SRs, $\triangle f = 0.06$Hz.

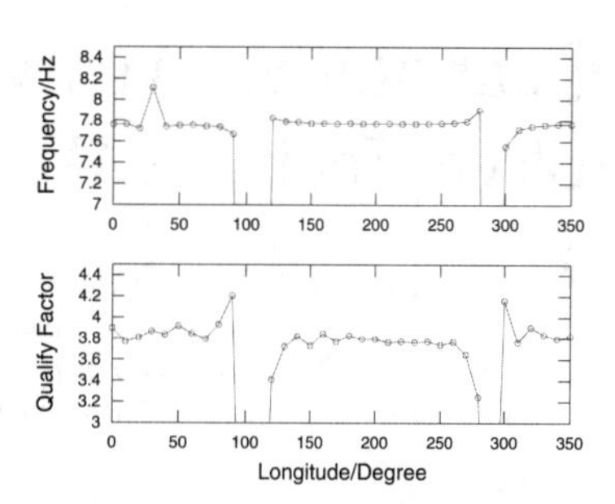

Fig. 5. Global distribution of SRs, $\triangle f = 0.06$Hz.

phenomena are worth noticing: 1, the resonant phenomena are clearly shown with asymmetry, which is obviously caused by the inhomogenous ionosphere conductivities (unlike analytic models such as the "Knee" model); 2, the regular distribution of resonant frequencies (especially the first two ones) is because of the property of ELF waves and the source locations (see previous introduction) and the discontinuity appears at 30°E is also because of the lightning source; 3, the resonances frequencies above 20Hz are not very obvious and vary a lot at specific regions as a result of the mixture of different resonant modes.

Finally, a quantitatively description of the SR variation along the equator are presented in Fig. 5, in which only the first resonant frequency is considered. This figure clearly shows that the first SR and its qualify factor slightly varies with longitude, except for the lightning source region where the first SR can not be obtained. The results of Fig. 4 and Fig. 5 can be good references for further study of SRs applications.

IV. Conclusion

In this work we have applied the GFDTD method to simulate SRs of the Earth-ionosphere system. Conductivity values from the IRI are introduced to study the variation of SR frequencies and qualify factors. Analytic model of "Knee" profile and analytic SR values are also presented to compare with our simulation results. Through simulations, SR frequency variations and global SR distributions are obtained for further study of the Earth's environment and the ionosphere.

The introduction of IRI into the GFDTD method provides a more rigorous model of the Earth-ionosphere system, thus improves the accuracy of EM wave propagation study. In ongoing works the applications of SRs as global indicators are to be studied using the GFDTD coupled with IRI.

Acknowledgment

This work was partly supported by the Natural Science Foundation of China (BK2009368), the Jiangsu Planned Projects for Postdoctoral Research Funds (1201006C) and the Jiangsu Province meteorological detection and information processing Key Laboratory (KDXS1202).

References

[1] H. Yang, "Three Dimensional Finite Difference Time Domain Modeling of Schumann Resonances on Earth and Other Planets of the Solar System," Ph.D. dissertation, The Pennsylvania State University, 2007.

[2] C. Price, O. Pechony, and E. Greenberg, "Schumann Resonances in Lightning research," *Geophysics*, vol. 1, pp. 1 –15, 2007.

[3] K. Schlegel and M. Füllekrug, "50 Years of Schumann Resonance," *originally published in Physik in unserer Zeit. Translation: Catarina Geoghan, 2007*, vol. 33, no. 6, pp. 256–26, 2002.

[4] Y. Wang and Q. Cao, "Analysis of seismic electromagnetic phenomena using the FDTD method," *IEEE Transactions on Antennas and Propagation*, vol. 59, no. 11, pp. 4171–4180, 2011.

[5] S. W.O., "On the radiation free self-oscillations of a conducting sphere, which is surrounded by an air layer and an ionospheric shell." *Z Naturforschung (in German)*, vol. 7a, pp. 149–154, 1952.

[6] D. D. Sentman, "Approximate Schumann resonance parameters for a two-scale-height ionosphere," *Journal of Atmospheric and Terrestrial Physics*, vol. 52, no. 1, pp. 35–46, 1990.

[7] ——, "Schumann resonance effects of electrical conductivity perturbations in an exponential atmospheric / ionospheric profile," *Journal of Atmospheric and Terrestrlal Physics*, vol. 45, no. 1, pp. 55–65, 1983.

[8] A. Taflove and S. C. Hagness, *Computational electrodynamics: the finite-difference time-domain method.* Artech House, 2005.

[9] D. Bilitza and B. Reinisch, "International Reference Ionosphere 2007: Improvements and new parameters," *Advances in Space Research*, vol. 42, no. 4, pp. 599–609, Aug. 2008. [Online]. Available: http://linkinghub.elsevier.com/retrieve/pii/S0273117708000288

[10] V. Mushtak, "ELF propagation parameters for uniform models of the Earthionosphere waveguide," *Journal of Atmospheric and Solar-Terrestrial Physics*, vol. 64, no. 18, pp. 1989–2001, Dec. 2002.

[11] J. J. Simpson, "Current and Future Applications of 3-D Global Earth-Ionosphere Models Based on the Full-Vector Maxwells Equations FDTD Method," *Surveys in Geophysics*, vol. 30, no. 2, pp. 105–130, Mar. 2009.

[12] "GEODAS Grid Translator - Design-a-Grid," 2011. [Online]. Available: http://www.ngdc.noaa.gov/mgg/gdas/gd_designagrid.html

[13] A. Martinez, A. P. Byrnes, K. G. Survey, and C. Avenue, "Modeling Dielectric-constant values of Geologic Materials: An Aid to Ground-Penetrating Radar Data Collection and Interpretation," *Current Research in Earth Sciences*, vol. Bulletin 2, no. Part 1, 2001.

[14] J. F. Hermance, "Electrical conductivity models of the crust and mantle," *Global Earth Physics: A Handbook of Physical*, pp. 190–205, 1995.

Quantitative Analysis of Rainfall Variability in Tokyo Tech MMW Small-Scale Model Network

Hung V. Le* Takuichi Hirano Jiro Hirokawa Makoto Ando
Tokyo Institute of Technology
2-12-1-S3-19, O-okayama, Meguro-ku, Tokyo, 152-8552, JAPAN
Email: *vuhung@antenna.ee.titech.ac.jp

***Abstract*—Millimeter-wave band is largely affected by attenuation due to rain. While calculating link budget for wireless systems using this frequency band, behavior of rain, attenuation due to rain and amount of degradation must be accurately understood. Evaluation of localized behaviors of rain and its effect on wave propagation in Tokyo Tech millimeter-wave model mesh network is presented in this paper. A quantitative analysis of rainfall variability using variogram in Tokyo Tech millimeter-wave (MMW) small-scale model network is reported. The unique effects due to highly localized behaviors of the heavy rain have become clear.**

I. Introduction

To cope with the recent demands of broadband communications, millimeter-wave has been receiving substantial attention because of its high-speed data transmission capability, wide bandwidth and generation of new frequency resource. Recent reports on the global warming effects on weather suggests the guerilla types of rain which is extremely localized on the order of less than 1 km [1]. For such small cell sizes, millimeter-wave wireless communication systems are more advantageous. Though, one of the biggest disadvantage is large attenuation due to rain. To overcome this, we could have power margin so that attenuation due to very strong rain is focused. An important aspect of intensive rain is localized behavior in space and time. Of special note is the correlation coefficient, regarded as an essential parameter for diversity [2]. Detailed statistical analysis and discussions on the effects of localized rain on microwave and millimeter-wave links are found in [3]. Detailed analysis and discussions on the spatial correlation characteristics of rainfall within several kilometers are found in [4]. However, these studies are only focus on the spatial correlation of rainfall and it didn't point out the relation between correlation and rain attenuation. This paper discusses the propagation characteristics measured in Tokyo Tech millimeter-wave model network taking localized rain effects into account and evaluates the localized behaviors of rain, its effects on the link attenuation in the small network. A quantitative analysis of rainfall variability using variogram in Tokyo Tech millimeter-wave small-scale model network is presented. Experimental data of 3 years is used to support this analysis.

II. Experiment Setup

Tokyo Tech millimeter-wave model network consists of 9 fixed wireless access (FWA) links, 6 base stations as shown in Fig. 1. The FWA lines are connected with each other using network switches at 6 FWA base stations on the rooftops of 6 buildings. The shortest link is 77 m and the longest one is 1020 m. Some basic research of millimeter-wave propagation characteristics using this network were reported in [5], [6]. Fig. 2 shows photographs of wireless terminals for 25 GHz and 38 GHz, respectively. High gain antennas are used for the FWA terminals which have specifications listed in Table I.

Rain rate, Rx Level, BER are recorded every 5 seconds. Rainfall intensity is measured by tipping-bucket rain gauges installed at all base stations with 0.2 mm resolution as the average of 1-minute time intervals. In this research, 3 year 3 months' data (4/2009-5/2012) for 25 GHz and 2 year 3 months' data (3/2010-5/2012) for 38 GHz have been used for analysis.

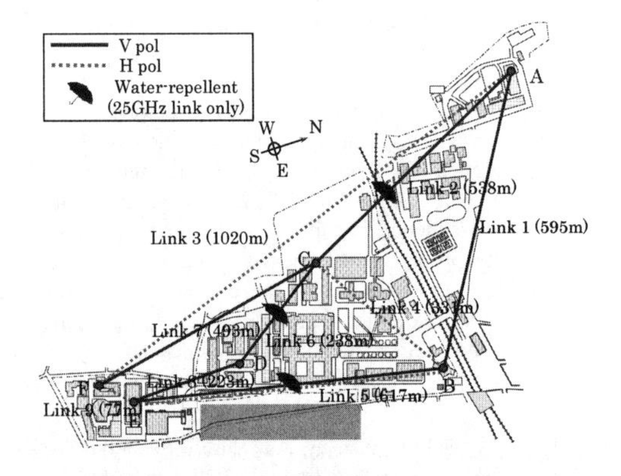

Fig. 1: Tokyo Tech MM-wave model network.

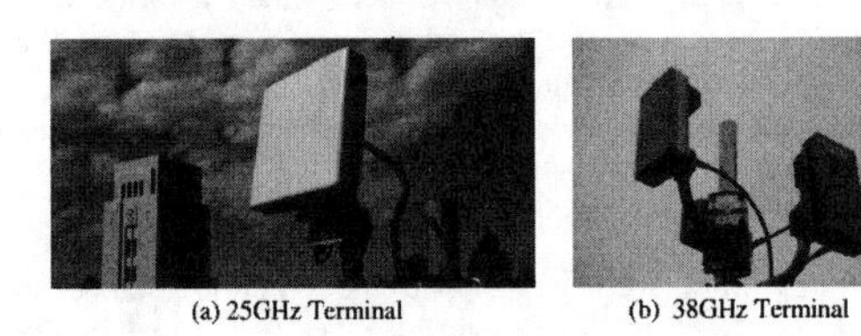

(a) 25GHz Terminal (b) 38GHz Terminal

Fig. 2: Photographs of the wireless terminals.

978-7-5641-4279-7

TABLE I: Wireless terminal specification

RF	25GHz	38GHz
Bandwidth	20MHz	200MHz
Duplex Scheme	TDD	TDD
Modulation Scheme	16QAM	QPSK/16QAM/64QAM
Antenna Gain	29dBi	32dBi
Max Transmission Speed	80Mbps	600Mbps/1Gbps

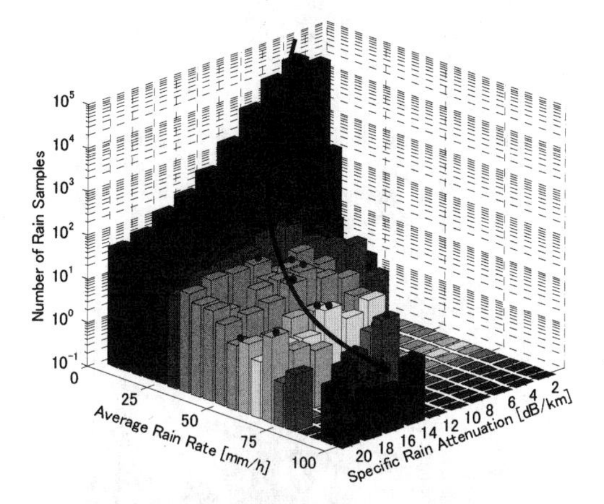

Fig. 3: Number of average rain rate between 2 points of link classified according to rain attenuation of 38 GHz using three years' 1-min rain rate data (Mar. 2010 - May 2012) for links 493, 595 and 1020 m.

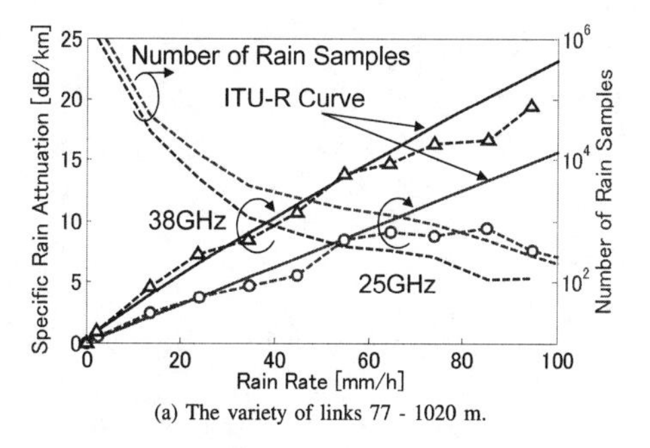

(a) The variety of links 77 - 1020 m.

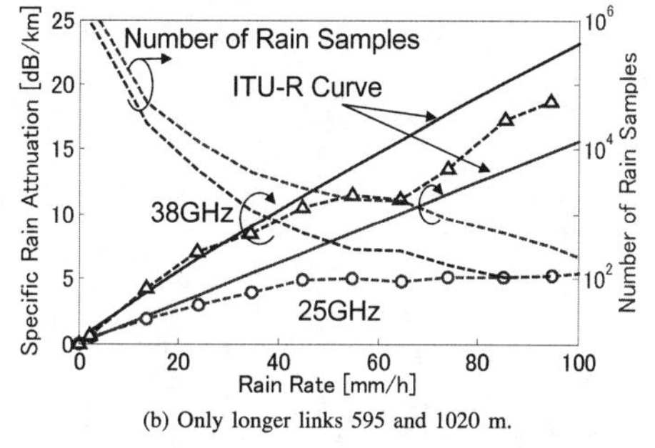

(b) Only longer links 595 and 1020 m.

Fig. 4: Specific rain attenuation of 25 GHz (data of Apr. 2009 - May 2012), 38 GHz (data of Mar. 2010 - May 2012) as a function of 1-min rain rate for the variety of links 77 - 1020 m (a), for only the longer links 595 and 1020 m (b).

III. Spatial Variability of Rainfall

Fig. 3 shows the distribution of number of samples as function of the rain rate averaged between 2 end points of each link with 493, 595 and 1020 m, and the specific rain attenuation [dB/km] in 38 GHz. Here the specific rain attenuation is grouped with 2 dB/km steps (e.g. 1-3, 3-5, 5-7...) and plotted at (e.g. 2, 4, 6...) on the x-axis while the y-axis shows the rain rate which grouped with 5 mm/h steps (e.g. 0-5, 5-10, 10-15...) and plotted at (e.g. 2.5, 7.5, 12.5...). It can be observed that the number of samples rapidly decreases when the rain rate and rain attenuation increase. The peak for each rain rate is indicated by the black dots and a likelihood relation between rain rate and the specific rain attenuation is indicated by the solid line; the rain attenuation has strong correlation with the rain rate. It is noted however that for the strong specific rain attenuation, the rain rate averaged over the 2 end points, is not always large but is spread widely from 0 mm/h. This result suggests that in the strong rain and large rain attenuation, rainfall may be localized and spatially non-uniform, which may not be identified by the conventional statistical analysis and will be focused in the later section of this paper.

In the millimeter-wave band, the attenuation due to rain increases as the frequency goes up. This is also verified as shown in Fig. 4. The experimental specific rain attenuation is calculated from Rx level (data of Apr. 2009 - May 2012 for 25 GHz and data of Mar. 2010 - May 2012 for 38 GHz) for three links in two groups. Fig. 4 (a) is for the variety of links 77 - 2020 m while Fig. 4 (b) is for only the longer links 595 and 1020 m. For each rain rate, the rain attenuation is plotted as function of the link distance and then the specific rain attenuation [dB/km] is calculated as the slope of linear approximation in the sense of least mean squares, where the shortest link 77 m is used in both for calibrating out the attenuation by radome. The specific rain attenuation thus obtained at the particular rain rate is compared in Fig. 4 with the ITU-R curve [7]. From the figure, it is observed that the attenuation for 38 GHz link is larger than that of 25 GHz link. Moreover, the experimental data in Fig. 4 (a) follows the ITU-R curve in case of rain rate below 60 mm/h, while in Fig. 4 (b) it departs from ITU-R curve in case of rain rate above 30 mm/h and is smaller than ITU-R curve, though in higher rain rate, the accuracy of statistics may be degraded due to smaller number of sample data. This suggests the possibility that the rainfall is not uniformly distributed along the whole path in the strong rain, or put differently, the longer the path is the more variability the rainfall is. This localized feature is the unique

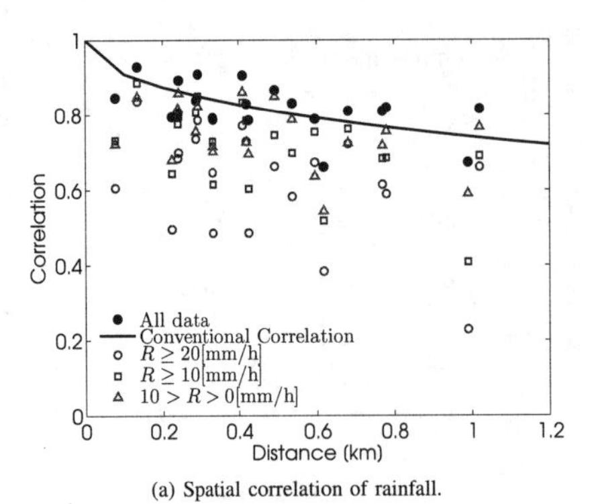

(a) Spatial correlation of rainfall.

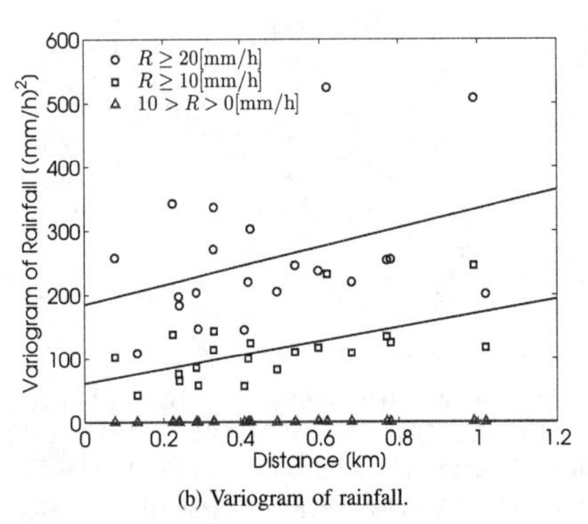

(b) Variogram of rainfall.

Fig. 5: Spatial correlation and variogram of rainfall classified according to rainfall intensity (1-min rain rate data of Apr. 2009 - May 2012).

aspect of strong rainfall and large rain attenuation, which should be focused for the millimeter-wave network consisting of the links shorter than 1 km. Most of the propagation studies in the past have focused on the weak rainfall which is critical for the long distance radio links utilizing lower frequencies [2]

In order to quantify the spatial variability of rainfall, a classical tool, namely variogram, is considered in this paper. The detailed analysis will be explained. Variogram is a key tool used in geostatistics to investigate and quantify the spatial of a random function and it is also a plot of average variance between points vs. distance between those points [9]. It is expressed as

$$\gamma(\boldsymbol{d}) = 0.5E\left(\left[z(\boldsymbol{d}+\boldsymbol{x}) - z(\boldsymbol{x})\right]^2\right), \quad (1)$$

where $\boldsymbol{x}$ is a position vector and $\boldsymbol{d}$ is a distance separation vector. E denotes the expectation and $z(\boldsymbol{x})$ is random function. The classical widely used sample variogram is given as follows [10]

$$\gamma(\boldsymbol{d}) = \frac{1}{2N}\sum_{x_2-x_1\approx d}\left(\left[z(x_2) - z(x_1)\right]^2\right), \quad (2)$$

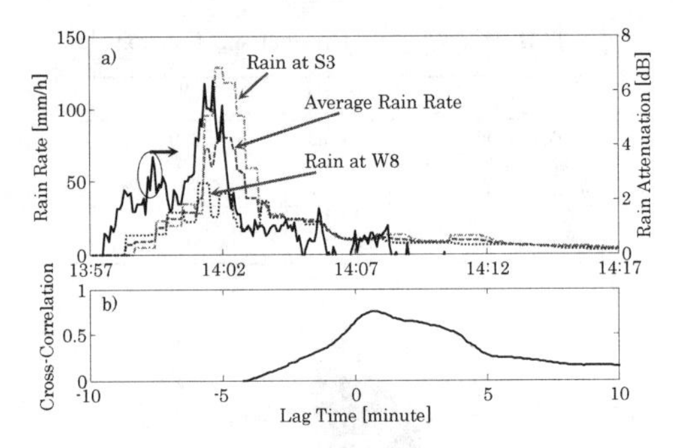

Fig. 6: Rain, rain attenuation (a) and correlation between average rain rate and rain attenuation (b) in heavy rainfall event (peak 1-min average rain rate of 85 mm/h) in 1-min time interval for link 7.

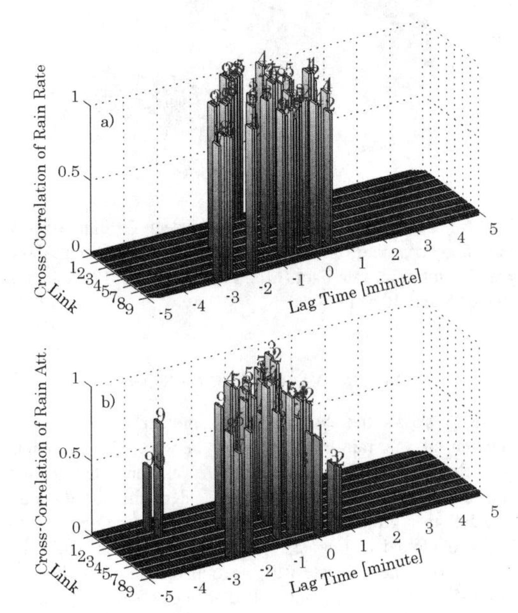

Fig. 7: Peak cross-correlation values between the average rain rate of one link and other links (a), rain attenuation of one link and other links (b) in the heavy rainfall event (peak 1-min average rain rate of 85 mm/h) for whole network.

where N denotes the number of pairs (x_1, x_2) separated by a distance equal to $\boldsymbol{d}$. If data are spatially correlated, variance generally increases with distance and if data are spatially uncorrelated, it will form a straight line.

Fig. 5 shows the spatial correlation (a) [8] and variogram of rainfall (b) classified according to rainfall intensity. For conventional correlation, $\rho = \exp(-0.3\sqrt{d})$ is used. Because the scale of the network is small, a linear variogram model is used [3]. From Fig. 5 (a), correlation of all rainfall data shows reasonable agreement with conventional formula. Also the correlation coefficient decreases with the increase in link distance. It is observed that the heavy rainfall tends to decrease correlation of rainfall. Fig. 5 (b) clearly shows that the variability of rainfall which was analyzed by using variogram correspond to the correlation coefficient. Variogram of rainfall increases with the increase in distance. The variability observed over the network is larger for heavy rainfall than weak rainfall, and this trend becomes much more clearly than that of correlation coefficient. It is noted that the variogram of less than 10 mm/h is very small which is related to large number of sample data.

Fig. 6 shows rain, rain attenuation (a) and correlation between average rain rate and rain attenuation (b) in heavy rainfall event (peak 1-min average rain rate of 85 mm/h) in 1-min time interval for link 7. Fig. 7 shows the peak cross-correlation values between the average rain rate of one link and other links (a), rain attenuation of one link and other links (b) in the heavy rainfall event (peak 1-min average rain rate of 85 mm/h) for whole network. The localized behavior of rainfall and attenuation in space together with their delays in time suggest the effectiveness of diversity even in such a small area, provided the rain is heavy.

IV. CONCLUTION

In this paper, the spatial variability of rainfall and rain attenuation were evaluated by using variogram. The statistical analysis results showed that the rain attenuation at both high rain rates and long distances was affected by the localized behaviors of rain affects. Also it has been suggested that the spatial variability of rainfall and rain attenuation are a good parameter for diversity.

ACKNOWLEDGEMENT

This work was conducted in part as “the Research and Development for Expansion of Radio Wave Resources” under the contract of the Ministry of Internal Affairs and Communications.

REFERENCES

[1] J. Joel, B. Alexis, “Quantification of the small-scale spatial structure of the raindrop size distribution from a network of disdrometers,” J. Appl. Meteor. Climatol., vol. 51, pp. 941-953, 2012.

[2] S. Nomoto, K. Nakama, and Y. Kishi, “Route diversity effect of mesh network taking account of space correlation of rainfall,” IEICE Tech. Rep., CS2001-50, pp. 49-56, June 2001.

[3] T. Manabe and T. Yoshida, “A study on rain attenuation characteristics of quasi millimeter waves,” IEICE Trans. Commun., vol. J78-B-II, no. 1, pp. 11-20, Jan. 1995.

[4] K. Ono and Y. Karasawa, “Consideration on characteristics of spatial correlation coefficient of nth power of 1-minute rainfall rate for accuracy improvement of rain attenuation probability estimation method,” IEICE Trans. Commun., vol. J89-B, no. 10, pp. 1998-2011, Oct. 2006 (in Japanese).

[5] H. V. Le, T. Hirano, J. Hirokawa, M. Ando, “Link correlation property in Tokyo Tech millimeter-wave model network,” 2013 IEEE International Symposium on Antennas and Propagation (AP-S 2013), Lake Buena Vista, Florida, USA, July 2013.

[6] H. V. Le, T. Hirano, J. Hirokawa, M. Ando, “Localized rain effects observed in Tokyo Tech millimeter-wave model network,” 2013 International Symposium on Electromagnetic Theory (EMTS 2013), Hiroshima, Japan, May 2013.

[7] Rec. ITU-R P.838-3, “Specific attenuation model for rain for use in prediction methods,” ITU-R Recommendations and Reports, ITU, Mar. 2005.

[8] J. L. Rodgers and W. A. Nicewander, “Thirteen ways to look at the correlation coefficient,” The American Statistician, vol. 42, pp. 59-66, 1988.

[9] G. Matheron, “The theory of regionalized variables and its applications,” Elcole national supelrieure des mines, 1971.

[10] J.-P. Chiles and P. Delfiner, “Geostatistics: modeling spatial uncertainty,” 2th ed. Wiley, 2012.

A Nyström-Based Esprit Algorithm for DOA Estimation of Coherent Signals

Yuanming Guo, Wei Li, Yanyan Zuo, Junyuan Shen
School of Electrical and Information Engineering of Harbin
Institute of Technology Shenzhen Graduate School, Shenzhen, China 518055
guoym2010@163.com
li.wei@hitsz.edu.cn
Juny_shen@hotmail.com

***Abstract*—Sample covariance matrix (SCM) and its eigenvalue decomposition (EVD) are needed in the conventional subspace based algorithm. Under normal circumstances, the computation burden is really heavy, especially for the estimation of the coherent signals which also needs to use the spatial smooth technology. In this paper, we firstly use the Nyström-Based method to obtain the signal subspace, which results in reducing the computational complexity greatly. To further reduction of the computational complexity under the completely coherent condition, we use an improved SVD method to calculate the equivalent forward and backward covariance matrices. Then we use the ESPRIT method to estimate coherent signals' direction of arrival (DOA), and we also compare the performance among the Nyström-Based algorithm, the SCM-EVD method and the classical space smooth algorithm. The simulation results show that this new method achieves almost the same accuracy with SCM-EVD based method and classical spatial smooth method with a much less computation burden.**

Keywords - Nyström; ESPRIT; coherent; DOA estimation

I. INTRODUCTION

Subspace based algorithm has been widely used for DOA estimation. It is well known that the subspace based method which depends on the decomposition of the observation space into signal subspace and noise subspace can provide high resolution DOA estimation with good accuracy. While the classical subspace based methods involve the estimation of the covariance matrix and eigen-decomposition, such as the MUSIC (Multiple Signal Classification)[1][2] and ESPRIT (Estimation Signal Parameter via Rotational Invariance Techniques) [3][4]. As a result, the classical subspace based methods are rather computationally expensive, especially for the case that the model's order number in those matrices is large. Recently, the method based on the Nyström Algorithm [5]-[7] simplifies the computation complexity, which use the block matrix method to find the signal subspace without the computation of SCM and its EVD. Coherent signals environment is common in reality, such as multi-path effect. The coherent signals make the rank of array spatial covariance matrix a loss and causes ineffectiveness in some super-resolution subspace algorithms such as MUSIC and ESPRIT. To deal with the coherent signals, expecially completely coherent signals [4], we need to do de-coherent processing to get the full rank sample covariance matrix. There are many methods to realize. For instance, forward spatial smoothing algorithm [3], backward spatial smoothing method. The improved SVD(singular value de-composition) is used for coherent signals to estimate the DOA [4], but it needs to do the spatial smoothing processing to calculate the covariance matrix. In this paper, we account for the complete coherent case, combining the Nyström-Based Algorithm with the improved SVD algorithm. In this way. it can greatly reduce the computational complexity. In addition, we have also compared it with the classical spatial smooth method and the SCM-EVD method[4]. The simulation results show that the method presented by this paper achieves almost the same accuracy with the SCM-EVD method and a little suboptimal to the classical spatial smooth methed.but it has less computation burden.

II. PROBLEM FORMULATION

A. Data Model

Let us consider an uniform linear array (ULA) composed of M isotropic sensors. Assume that P $(P<M)$ narrow-band signals impinge upon the ULA from distinct direction $\theta_1, \theta_2 \cdots \theta_P$, The $M\times 1$ output vector of the array, which is corrupted by additive noise, at the k th snapshot can be expressed as

$$\mathbf{x}(k)=\sum_{i=1}^{P}\mathbf{a}(\theta_i)*s_i(k)+\mathbf{n}(k) \quad k=0,\cdots,N-1 \tag{1}$$

Where $s_i(k)$ is the scalar complex waveform referred to as the i th signal, $\mathbf{n}(k)\in\mathbb{C}^{M\times 1}$ is the complex noise vector. N and P denote the number of snapshots and the number of signals respectively. $\mathbf{a}(\theta_i)$ is the steering vector of the array toward direction θ_i and take the following form

$$\mathbf{a}(\theta_i)=\frac{1}{\sqrt{M}}\left[1,e^{j\varphi_i},\cdots,e^{j(M-1)\varphi_i}\right]^T \tag{2}$$

where $\varphi_i=(2\pi d\sin\theta_i)/\lambda$ in which $\theta_i\in(-\pi/2,\pi/2)$, d is inter-element spacing and λ is the wavelength. In matrix form, the received data becomes

This work was supported by the National Natural Science Foundation of China granted No.61102157 and the National Basic Research Program of China (973 Program) granted No.2013CB329003.

978-7-5641-4279-7

$$\mathbf{x}(k)=\mathbf{A}(\theta)\mathbf{s}(k)+\mathbf{n}(k) \quad k=0,1\cdots,N-1 \tag{3}$$

Where

$$\mathbf{A}(\theta)=\left[\mathbf{a}(\theta_1),\mathbf{a}(\theta_2),\cdots,\mathbf{a}(\theta_p)\right] \tag{4}$$

$$\mathbf{s}(k)=\left[s_1(k),s_2(k),\cdots,s_p(k)\right] \tag{5}$$

are the $M\times P$ steering matrix and the $P\times 1$ complex signal vector, respectively. Furthermore, the background noise uncorrelated with the signal is modeled as a zero-mean Gaussian complex random process , which is both stationary spatially and temporally.

Then the sample covariance matrix can be calculated as

$$\mathbf{R_X}=E[\mathbf{XX}^H] \tag{6}$$

Where $\mathbf{X}=\left[\mathbf{x}(1),\mathbf{x}(2),\cdots,\mathbf{x}(N)\right]$ is a $M\times N$ data matrix, and the superscript $(\bullet)^T$ and $(\bullet)^H$ denote the transpose and conjugate transpose, respectively.

B. Nyström Method For Matrix Approximation

As we know the $\mathbf{X}$ is a $M\times N$, we can divide the data into following form

$$\mathbf{X}=\begin{bmatrix}\mathbf{X}_1\\ \mathbf{X}_2\end{bmatrix} \tag{7}$$

$\mathbf{X}_1$ is a $P\times N$ matrix. then the covariance matrix can be partitioned as

$$\mathbf{R_X}=\begin{bmatrix}\mathbf{R}_{11} & \mathbf{R}_{12}\\ \mathbf{R}_{21} & \mathbf{R}_{22}\end{bmatrix} \tag{8}$$

where $\mathbf{R}_{11}=E\left[\mathbf{X}_1\mathbf{X}_1^{\mathbf{H}}\right],\mathbf{R}_{12}=E\left[\mathbf{X}_1\mathbf{X}_2^{\mathbf{H}}\right],\mathbf{R}_{21}=E\left[\mathbf{X}_2\mathbf{X}_1^{\mathbf{H}}\right],\mathbf{R}_{22}=E\left[\mathbf{X}_2\mathbf{X}_2^{\mathbf{H}}\right]$. Since the covariance matrix is a symmetric matrix, we approximate the covariance matrix by Nyström Method[6][7]. Then we obtain

$$\mathbf{R}_{\mathbf{NCE}}=\begin{bmatrix}\mathbf{R}_{11} & \mathbf{R}_{12}\\ \mathbf{R}_{12}^{\mathbf{H}} & \mathbf{R}_{12}^{\mathbf{H}}\mathbf{R}_{11}^{-1}\mathbf{R}_{12}\end{bmatrix} \tag{9}$$

the matrix $\mathbf{R}_{\mathbf{NCE}}$ is the approximation of the covariance $\mathbf{R_X}$,and we can use the $\mathbf{R}_{\mathbf{NCE}}$ to estimate the signal subspace. Here we give the result directly

$$\mathbf{R}_{\mathbf{NCE}}=\mathbf{U_S}\mathbf{\Lambda}\mathbf{U_S^H} \tag{10}$$

Where

$$\mathbf{U_S}=\mathbf{F}\mathbf{U_F}\mathbf{\Lambda_F^{-1/2}} \tag{11}$$

and

$$\mathbf{F}=\begin{bmatrix}\mathbf{R}_{11}\\ \mathbf{R}_{12}^{\mathbf{H}}\end{bmatrix}\mathbf{R}_{11}^{-1/2} \tag{12}$$

$$\mathbf{F^H F}=\mathbf{U_F}\mathbf{\Lambda_F}\mathbf{U_F^H} \tag{13}$$

The Nyström-based method avoids the calculation of the SCM and its SVD. In particular, it only needs to compute $\mathbf{R}_{11}$ and $\mathbf{R}_{12}$. since the estimation accuracy and computation complicity has been evaluated in [7]. The computational complexity of SCM-SVD method is $o\left(M^3\right)+o\left(M^2N\right)$, but the computational complexity of the proposed Nystrom method is $o\left(P^2N\right)+o\left((M-P)NP\right)+o\left(MP^2\right)$. Usually the number of sensors and the number of signals is very small. So by this method we can reduce much complexity, here we give out the MUSIC spectrum based on this method for DOA estimation Fig 1.

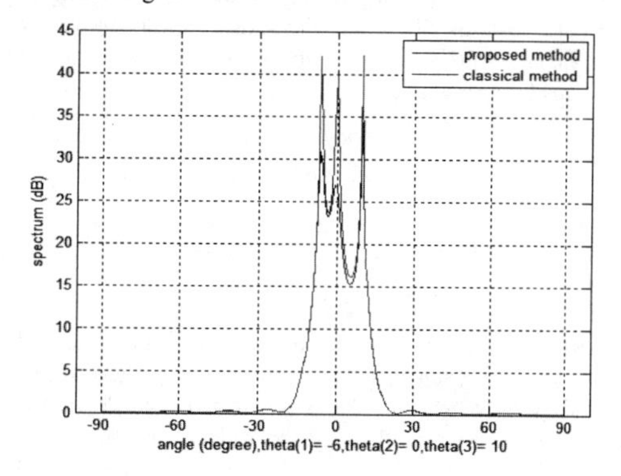

Fig.1 MUSIC based on the Nyström Method and the classical method

In the simulation the SNR(signal to noise ratio) is 20dB, from the Fig 1 we see that Nyström-based method almost has the same performance as the classical SCM-EVD based method.

III. Completely Coherent Signals Model and Improved SVD Algorithm

When the sources are completely coherent, namely the sources differ only in complex constant, we can express it as

$$s_i(t)=\alpha_i s_1(t) \quad i=2,3\cdots P \tag{14}$$

The rank of sources covariance matrix is one. After the eigenvalue decomposition of $\mathbf{R_X}$, the dimension of the signal subspace is less than the rank of the array manifold $\mathbf{A}(\boldsymbol{\theta})$, which leads to the consequence that the steering vector is no longer orthogonal to the noise subspace and makes failure of subspace algorithm. Recently a method based on the eigenvector of the maximum eigenvalue [4] has been represented. We assume the noise is temporal white and the noise covariance matrix $\mathbf{R_N}$ is full-rank. As the steering vectors span the signal subspace, so we have linear representation form as:

$$\mathbf{R_N}\mathbf{e_k}=\sum_{n=1}^{P}\alpha_k(n)\mathbf{a}(\theta_n) \tag{15}$$

Where $\mathbf{e_k}\ (1\le k\le K)$ is an eigenvector of the received signal covariance matrix (the first K eigenvectors corresponding to first K eigenvalue in decreasing order), $\alpha_k(n)$ is a linear combination factor. When the noise covariance matrix is an identity matrix, the above equation (15) is simplified as:

$$\mathbf{e}_k = \sum_{n=1}^{P} \alpha_k(n)\mathbf{a}(\theta_n), \quad 1 \le k \le K \tag{16}$$

For completely coherent case, namely $K=1$, the above equation is reduced to:

$$\mathbf{e}_1 = \sum_{n=1}^{P} \alpha_1(n)\mathbf{a}(\theta_n) \tag{17}$$

It indicates that the largest eigenvector of the largest eigenvalue contains all the signal information. So the vector $\mathbf{e}_1$ can be used to reconstruct a equivalence covariance matrix $\mathbf{Y}$, it can be constructed as:

$$\mathbf{Y}_f = \begin{bmatrix} e_{11} & e_{12} & \cdots & e_{1g} \\ e_{12} & e_{13} & \cdots & e_{1,g+1} \\ \vdots & \vdots & \ddots & \vdots \\ e_{1m} & e_{1m+1} & \cdots & e_{1M} \end{bmatrix} \tag{18}$$

Where $m = M - g + 1, m > P, g > P$.

As we known, the performance of the forward and backward method is more better than the only forward method, which can also be used to solve the coherent situation. So in this paper we also use the similar way, the backward matrix is :

$$\mathbf{Y}_b = \mathbf{J}_m \mathbf{Y}^* \mathbf{J}_g \tag{19}$$

Where $\mathbf{J}_m$ (m -dimension) and $\mathbf{J}_g$ (g -dimension) are anti-diagonal eye matrices, $(\bullet)^*$ denotes the conjugate operation, then we obtain the equivalence covariance matrix:

$$\mathbf{Y} = \frac{1}{2}[\mathbf{Y}_f + \mathbf{Y}_b] \tag{20}$$

Then we do SVD of the factorization of $\mathbf{Y}$ to obtain the signal subspace. And we can use the ESPRIT algorithm to estimate the DOA of the sources as follows:

$$\mathbf{Y}^H\mathbf{Y} = \mathbf{U}\mathbf{\Lambda}\mathbf{U} \tag{21}$$

Where

$$\mathbf{U}_1 = [eye(g-1), zeros((g-1)\times 1)] * \mathbf{U}_s \tag{22}$$

$$\mathbf{U}_2 = [zeros(g-1)\times 1, eye(g-1))] * \mathbf{U}_s \tag{23}$$

$\mathbf{U}_s$ is the P largest eigenvector of $\mathbf{U}$. we can get the DOA (θ_k) of the signals by solving following equation

$$\mathbf{\Phi} = \left(\mathbf{U}_1^{\dagger}\right)\mathbf{U}_2 = \mathbf{H}^{-1}\mathbf{\Psi}\mathbf{H} \tag{24}$$

$$\mathbf{\Psi} = diag(e^{j\varphi_1}, e^{j\varphi_2} \cdots, e^{j\varphi_P}) \tag{25}$$

$$\varphi_k = (2\pi d / \lambda)\sin\theta_k \tag{26}$$

Where $(\bullet)^{\dagger}$ denotes the pseudo-inverse,and $\mathbf{H}$ is a full rank matrix . In this way, when we do the de-coherent of the signals we don't need to do the spatial processes, which is more computation costly.

IV. SIMULATION RESULTS

In this section, the performance of the proposed method is evaluated by computer simulation. For comparison purpose , we also do the simulation of the spatial method ,which is based on the spatial smooth technology, firstly ,we calculate the covariance matrix ,then do its EVD decomposition ,after this, we apply ESPRIT algorithm to estimate the DOA. Another method ,which is called SCM-EVD method[4] is also performed there. For simplicity, the array herein is assumed to be an ULA with nine isotropic sensors whose spacing equals half-wavelength. Suppose that there are three completely coherent signals with equal power impinging upon the ULA and the true DOAs are $\{-20^\circ, 0^\circ, 40^\circ\}$. We assumption the background noise is a zero mean, white Gauss stationary process. In Fig.2 the snapshot is 200, and the number of Monte Carlo trials is 500, RMSE is calculated an follows:

$$RMSE = \sqrt{\frac{1}{T}\sum_{i=1}^{T}\left(\hat{\theta}_i - \theta\right)^2} \tag{27}$$

Where T is the number of trials, and $\hat{\theta}_i$ is the estimated angle, θ_i is the real angle

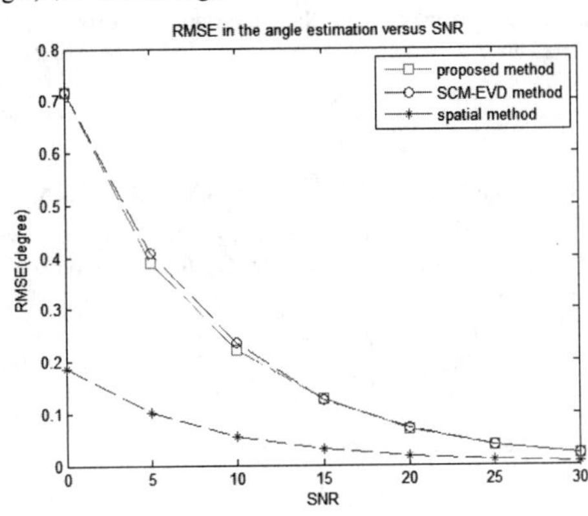

Fig. 2 RMSE in the angle estimation versus SNR of the three methods

we can observe that the RMSE of the proposed method is very closed to the SCM-EVD method, the curves of the two are almost coincide to each other, but both of them are a little biger than the spatial method. What's more, they all have the same trend, as the SNR increase the RMSE of the three methods are both decreased. To some extend, we can see that the proposed method achieved the computation simplicity with the sacrifice of its performance, compared with the SCM-EVD method, we have the same performace but with much less computation burden.

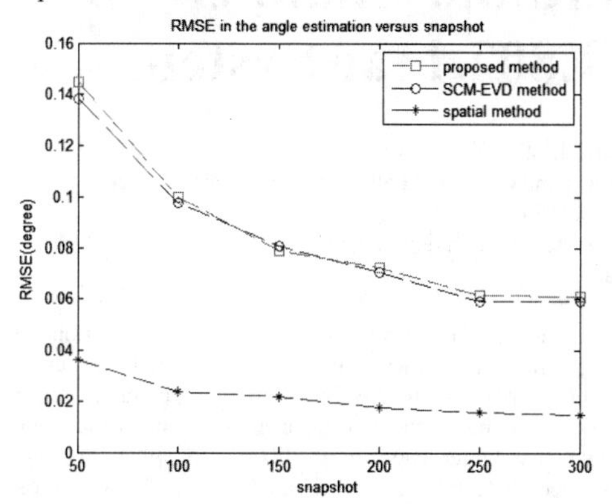

Fig. 3 RMSE in the angle estimation versus snapshot of the three methods

In Fig.3 the SNR is 20dB and the number of Monte Carlo trials is 500. From Fig.3 we can obtain the same conclusion as from the Fig.2. The reason why the proposed method have poor performance than the spatial method is that the Nyström-Based Esprit Algorithm noly use the $\mathbf{R}_{11}$ and $\mathbf{R}_{12}$ to calculate the signal subspace, while the spatial method have used all the sample data to calculate the signal subspace, and in this way its more closer to the real situation.

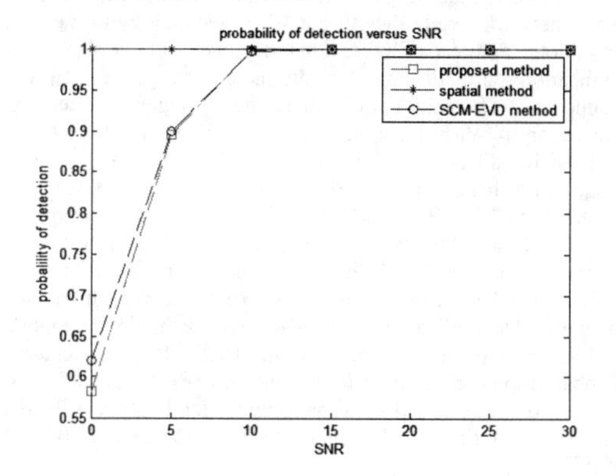

Fig. 4 probability of detection the three methods versus SNR

In Fig.4 the snapshot is 200 and the number of Monte Carlo trials is 500. From Fig.4 we can obtain the same conclusion as from Fig.2.and Fig.3. the probability of detection of the proposed method and the SCM-EVD method almost coincide. The detection probability of the three increase with the SNR, When the SNR is exceed 10 dB, the detection pabability is almost one.

V. CONCLUSION

In this paper, we use the Nyström-Based ESPRIT Algorithm for DOA Estimation of completely coherent signals. The total computation complexity of the proposed method is $o\left((M-P)NP+NP^2+MP^2\right)+o\left(m^2g+mg^2+g^3\right)+o\left(P^3\right)$. While the total computation complexity of the SCM-EVD method is $o\left(MN^2\right)+o\left(M^3\right)+o\left(m^2g+mg^2+g^3\right)+o\left(P^3\right)$, while the spatial method is $o\left(P-1)N(M-P+1)^2\right)+o\left((M-P+1)^3\right)+o\left(P^3\right)$. Generally, the snapshot is very large, but the sensor number and the number of signals are not large. Then we can see that the proposed method can greatly reduce the complexity. From the simulation results, we can obtain that the proposed method has strong de-correlation ability and high estimation accuracy. In addition, the performance of the proposed method almost achieves the same performance with the SCM-EVD method and the classical spatial smooth method ,but wih less computation burden.

REFERENCES.

[1] Pillai, S.U.; Kwon, B.H., "Performance analysis of MUSIC-type high resolution estimators for direction finding in correlated and coherent scenes," Acoustics, Speech and Signal Processing, IEEE Transactions on , vol.37, no.8, pp.1176,1189, Aug 1989

[2] Huang L, Wu S, Zhang L. A novel MUSIC algorithm for direction-of-arrival estimation without the estimate of covariance matrix and its eigendecomposition[C]//Vehicular Technology Conference, 2005. VTC 2005-Spring. 2005 IEEE 61st. IEEE, 2005, 1: 16-19.

[3] Shan T J, Wax M, Kailath T. On spatial smoothing for direction-of-arrival estimation of coherent signals[J]. Acoustics, Speech and Signal Processing, IEEE Transactions on, 1985, 33(4): 806-811.

[4] Zhi-Jin Z, Yang W, Chun-Yun X. DOA Estimation of Coherent Signals Based on Improved SVD Algorithm[C]//Instrumentation, Measurement, Computer, Communication and Control (IMCCC), 2012 Second International Conference on. IEEE, 2012: 524-528.

[5] Arcolano N, Wolfe P J. Estimating principal components of large covariance matrices using the Nyström method[C]//Acoustics, Speech and Signal Processing (ICASSP), 2011 IEEE International Conference on. IEEE, 2011: 3784-3787.

[6] Fowlkes C, Belongie S, Chung F, et al. Spectral grouping using the Nystrom method[J]. Pattern Analysis and Machine Intelligence, IEEE Transactions on, 2004, 26(2): 214-225.

[7] Qian C, Huang L. A Low-Complexity Nyström-Based Algorithm for Array Subspace Estimation[C]//Instrumentation, Measurement, Computer, Communication and Control (IMCCC), 2012 Second International Conference on. IEEE, 2012: 112-114.

[8] Pesavento M, Gershman A B. Maximum-likelihood direction-of-arrival estimation in the presence of unknown nonuniform noise[J]. Signal Processing, IEEE Transactions on, 2001, 49(7): 1310-1324.

Impact of Reconfiguring Inclination Angle of Client's Antenna on Radio Channel Characteristics of IEEE802.11ac System

Hassan El-Sallabi[1], Mohamed Abdallah[2] and Khalid Qaraqe[3]
hassan.el_sallabi@qatar.tamu.edu[1], mohamed.abdallah@qatar.tamu.edu[2] and khalid.qaraqe@qatar.tamu.edu[3]
Wireless Research Laboratory,
Electrical Engineering Department, Texas A&M University at Qatar
Doha, Qatar

***Abstract* — This work presents the impact of reconfigurable antenna on radio channel characteristics of IEEE 802.11ac system, which is the next evolution of Wi-Fi systems. Reconfigurable antennas (RA) can be used to change radio channel characteristics in favor of receiver design. The results show that RA can be used to reduce power of some of the delay clusters, which lead to minimizing excess delays of these delay clusters by pushing their power to a level lower than noise floor. This is reflected in reducing the rms delay spread and increasing the coherence time of the radio channel, which has a direct impact on receiver design and operation.**

Keywords—radio channel; reconfiguable antenna; 5G Wi-Fi

I. INTRODUCTION

The successful design of any wireless communications system depends heavily on tested channel models. Channel models are supposed to describe radio channel characteristics (RCC) to some accurate extent. The RCC vary according to propagation environment such as indoor versus outdoor. The outdoor RCC vary for urban, suburban and rural, which also differ for forest area and open area as well as for line of sight and non-line of sight RF propagation [1]. The indoor RCC differ for different building categories such hotels, shopping malls, hospitals, airports, residential areas, office buildings as well as ground floor and high floors in high rise buildings. In addition to impact of the environment on RCC, there are other communication system parameters that affect the RCC. These parameters include operating frequency, bandwidth, antennas and their characteristics at the two ends of the communication link. These communication system parameters make the transmitted electromagnetic (EM) waves to interact with environments according to particular physics laws. The higher frequency of EM wave results in larger loss in both free space propagation and diffraction loss [2],[3]. The wide system bandwidth allows higher delay resolution at the receiver end. This makes multipath characteristics be different with bandwidth and operating frequency. Therefore, there is a need for channel characterization of parameters new merging Wi-Fi system under these considerations.

In this work, we investigate the impact of reconfigurable inclination angle of client's antenna on radio channel characteristics of IEEE 802.11ac system. The inclination angle changes the three dimensional pattern for both vertical and horizontal polarization. We show how it can be used to reduce powers of some of the multipath components, which would change their interaction in a different manner that cause controlled different channel characteristics. We also present how the RA can be used to change the root mean squared delay spread as well as the coherence time of the tested radio channel.

II. 5G WI-FI AND RECONFIGURABLE ANTENNA

A. IEEE 802.11ac

Wireless access to the Internet is becoming the default scenario. The IEEE released in 1997 first Wi-Fi standard and adopted IEEE 802.11a, b, g and n versions. The 802.11n version operates at 2.4- and 5 GHz and allows up to four data streams at a time. The increasing demand for multimedia streaming over Wi-Fi networks made the current Wi-Fi technology to meet its limitations and capabilities. This demand and similar other applications are driving for dramatic changes to current technology. The initial problem is the throughput. The next generation of Wi-Fi technology (aka 5G Wi-Fi) of IEEE 80.11 is the IEEE 802.11ac. This system works at 5 GH frequency range with bandwidth of 80 MHz and optionally can be expanded to 160 MHz. The 5 GHz band is less used compared to 2.4 GHz and thus would experience less interference. This increase in bandwidth is the biggest factor in increasing the throughput. This makes it possible to deliver very high data rates of video applications to handset. This technology supports multi-user, multi-input, multi-output (MU-MIMO) scenarios compared to single-user MIMO found in IEEE 802.11n. It also utilizes the 256 quadrature amplitude modulation (QAM), which is four times higher than the 64 QAM of the 802.11n [4],[5].

B. Reconfigurable Antenna

Other frontier that may result in significant improvement in performance of wireless communications system is the use of reconfigurable antenna (RA). It can be used in a manner to change RCC in favor of designing wireless system [6]. The RA, per control, can change its field pattern, frequency and polarization. The key part in gaining benefit from RA is based on understanding the interplay between superposition of

978-7-5641-4279-7

complex signals of multipath components of radio channel, three dimensional antenna patterns, and velocity of mobile terminal. This interplay would affect channel characteristics in delay, direction and Doppler domains, which will have impact on their corresponding correlation parameters such coherence spectra (i.e., frequency correlation), spatial correlation and coherence time, respectively. The RF agility of RA can be used to change some of these channel correlation properties such as the coherence time and power weighted multipath dispersion metrics.

III. Radio Channel

Indoor environments where the IEEE 802.11ac is supposed to work can be categorized according to radio propagation in to different categories. The RF propagation in airports is different from RF propagation in office buildings and hotels, shopping malls, parking garage, residential buildings, etc. The RF propagation is generally confined inside the building and penetrating the building to outdoor after experiencing indoor to outdoor loss, which depends on electrical properties of the wall. The dimension of this indoor environment would have direct impact in signal multipath dispersion in different domains. In order to characterize the radio channel in different domains, the rays are described in multi-dimensions in terms of delays, angle of arrivals and angle of departure [7]. These different rays would experience different antenna gains and losses as a function of angular information, azimuthal and co-elevation departure and arrival directions.

In this work we consider cubically shaped indoor environment where multipath components include line of sight and multiple reflections. Multiple specular reflection take place via EM signal bouncing between every opposite surfaces (walls or ceiling and floor) or combination between them. This particular shape of indoor environment includes different propagation scenarios such as a corridor, office, lecture hall, convention center, etc. The core principle of the adopted and developed model in its essence is similar to that presented in [8]. The inputs to the model include environment parameters such as ceiling height, distance of AP antenna to reflecting walls and their electrical properties in addition to communication system parameters. The system parameters include operating frequency, system bandwidth, polarization of AP antenna, antenna field pattern and antenna height at transmitter and receiver. The geometrical configuration of the communication link ends with scatterers' spatial distribution has key factor in channel characterization. In this physical model, each ray is determined by its parameters defined by its delay, azimuth-co-elevation angle of arrival, and azimuth-co-elevation angle of departure. The complex amplitude of each ray is computed with electromagnetic formulations for free space loss and interaction loss due to the interaction EM signal with the scatterers in the environment. The interaction loss depends on the interaction, wave-front and geometrical properties of impinging rays and physical properties of reflecting surfaces. The reflected EM signal is related to the incident wave via a reflection coefficient defined by polarimetric matrix. Different coefficients can be used to capture the interaction losses that depend on the transmit waveform type: plane wave, cylindrical wave or spherical wave. The most commonly used reflection coefficient is the Fresnel reflection coefficient, which is valid for an infinite boundary between two mediums. The computed loss due to this propagation mechanism depends on polarization, frequency, and electrical properties of reflecting surface in terms of permittivity and conductivity of each media. The received signal is obtained as a sum of multi-ray components as vector superposition of the N individual rays, which can be represented as follows:

$$h(t;\lambda,\tau,\phi,\theta,\varphi,\vartheta,\mathbf{V}) = \sum_{n=1}^{N} A_n \delta(t-\tau_n) e^{-jk(r_n - \mathbf{V}\cdot\mathbf{\Psi}_n t)}$$

where $\mathbf{V}$ is the velocity vector of the client station, which is assumed as the receiver in this notation, and defined by $\mathbf{V} = v_x\vec{x} + v_y\vec{y} + v_z\vec{z}$, λ denotes to wavelength of the operating frequency, k is wave number given as $k = \frac{2\pi}{\lambda}$, and $\mathbf{\Psi}_{Rx}$ is the arrival direction vector defined for ray n as

$$\mathbf{\Psi}_n = \cos(\phi_n)\sin(\theta_n)\vec{x} + \sin(\phi_n)\sin(\theta_n)\vec{y} + \cos(\theta_n)\vec{z}$$

where ϕ_n is the horizontal arrival angle relative to the x-axis of ray n, and θ_n is the elevation arrival angle relative to z-axis of ray n, r_n is the path length of ray n

$$r_n = \sum_{p=1}^{P_n} d_{n,p}.$$

where parameter $d_{n,p}$ denotes the distance traversed by the specular wave between the (p-1) and p-th boundary intersections and the complex amplitude A_n is defined as

$$A_n = \frac{\lambda}{4\pi r_n}\sqrt{G_{tx}(\varphi_n,\vartheta_n)G_{rx}(\phi_n,\theta_n)}\prod_{p=1}^{P_n}\Gamma_p e^{-jkd_{n,p}}$$

Γ_p stands for the surface reflection coefficient for the p-th wave-interface intersection, while the term $\frac{\lambda}{4\pi r_n}$ is the free space path loss that accounts for the wave spreading loss, and $G_{tx}(\varphi_n,\vartheta_n)$ and $G_{rx}(\phi_n,\theta_n)$ are the transmitter and receiver antenna gain, respectively. In this work, we assume that client station antenna has reconfigurability in its inclination angle. For half-wavelength dipole antenna, the reconfigurable inclination angle antenna gain pattern model for vertical polarization can be written as [9]

$$G_V(\theta,\phi) = 1.64\,(\cos\theta\cos\phi\sin\alpha - \sin\theta\cos\alpha)^2\frac{\cos^2(\pi\xi/2)}{(1-\xi^2)^2}$$

where $\xi = \sin\theta\cos\phi\sin\alpha + \cos\phi\cos\alpha$ and the angle α is the inclination angle of the antenna element from z-axis in the vertical zx-plane.

IV. Numerical Results and Analysis

The accuracy of ray theory increases with increasing operating frequency. Thus, modeling radio channel of IEEE 802.11ac Wi-Fi system with ray theory is more accurate than that of

IEEE 802.11n Wi-Fi system operating at 2.4 GHz. The simulation environment of this work is built to study the impact of reconfigurability of inclination angle of receiver antenna at the client station of 5G Wi-Fi IEEE802.11ac system. The parameters of the communication system are given by IEEE standard where operating frequency is 5 GHz and bandwidth is 80 MHz. The dimensions of the indoor environment are selected to represent a banquet hall with height of 10 m, width of 15 m and length of 50 m. It is assumed that the AP antenna is placed on ceiling and antenna height of client station is 1.7 m. The client is assumed to move with speed of 3 km/hr, which is defined as the pedestrian speed in 3GPP standard [10]. Maximum reflection order per surface is selected to be six. More reflection orders take place for rays bounce between multiple walls. Reflecting surfaces have relative permittivity of 5 and conductivity of 0.02. The simulated temporal range is for one second for every spatial location. The temporal sampling rate is 26000 samples/sec at every spatial location. The simulated travelled route is 2 m till 10 m horizontal distance from AP with spatial resolution of 2.5 cm. Polarization of AP antenna is vertical. The results shown below are for two selected inclination angles 0° and -30°. The corresponding antenna patterns are shown in Figure 1. It can well be seen that the donut shape has significantly changed with -30° inclination angle, which has direct impact on RCC as shown in Figure 2 and Figure 3. Figure 2 shows the spatial variant power delay profile of the channel along the measured route. The spread of excess delays and multiple delay clusters can be well seen in Figure 2a for normal scenario of conventional dipole antenna. The impact of inclination angle -30° is shown in Figure 2b. It can be observed that the excess delay has significantly been reduced and power of many delay clusters have been pushed down below noise floor. The noise floor for 80 MHz is -95 dBm. Figure 2 shows samples of power delay profiles beginning and end of measured route. The difference in distance clearly shows different delay clusters. The long delay multipath components, which may cause severe frequency selective fading and inter-symbol interference (ISI), have been pushed down noise floor with reconfiguring antenna's inclination angle. This will lead to a significant simplification in receiver design and higher data rate transmission. This reduction in powers of delay clusters to below noise floor is reflected in rms delay spread, which is directly related to frequency selectivity of the channel and has direct relation to ISI and complexity of receiver design. The rms delay spread is a measure of power weighted dispersion of multipath components based on power delay profile. If the rms delay spread is reduced by reconfiguring antenna's inclination angle, so that the delay spread is smaller than the symbol period, then the channel become a flat fading channel. The impact of antenna's inclination angle on rms delay spread is shown in Figure 4. The inclination angle -30° made significant reduction in rms delay spread compared to conventional dipole antenna (0° degree inclination angle). The calculations formula of rms delay spread is presented in [1], [11]. Figure 5 shows clear impact of reconfiguring inclination angle on channel coherence time. The channel coherence time characterizes the rate at which the channel changes. The radio channel coherence time is extended with -30° inclination angle. This makes the radio channel to be coherent for longer time, which makes the receiver to see similar RCC for this prolonged time. This results in less adaptively rate in receiver operations. The coherence time is computed from temporal channel correlation, which is presented in [11]. We considered the correlation value of 0.5 is the threshold value.

V. Conclusion

Reconfigurable antenna at client station may have significant improvement in the new evolution of Wi-Fi approved by IEEE standard under IEEE 802.11ac. The results of this work show that reconfiguring inclination angle can make significant changes in RCC that are directly related to performance of wireless communications system. It has significantly reduced the rms delay spread and increase the coherence time when the inclination angle was -30° relative to the corresponding RCC when the antenna was the convention half-wave dipole antenna.

Acknowledgment

This publication was made possible by NPRP grants #: NPRP 5-653-2-268 from the Qatar National Research Fund (a member of Qatar Foundation). The statements made herein are solely the responsibility of the authors.

References

[1] H. L. Bertoni, Radio Propagation for Modern Wireless Systems. Englewood Cliffs, NJ: Prentice Hall, 2000.

[2] A. Ishimaru, Electromagnetic Wave Propagation, Radiation and Scattering. Englewood Cliffs, NJ: Prentice-Hall, 1991

[3] J. A. Kong, Electromagnetic Wave Theory. Cambridge: EMW Pulishing, 2005.

[4] G. Goth "Next-Generation Wi-Fi: As Fast as We'll Need?" *IEEE Internet Computing*, vol 16, no. 6, 2012, pp. 13 – 16.

[5] L. Garber "Wi-Fi Race into Faster Future," Computer, vol. 45, no. 3, 2012, pp. 13-16.

[6] A. Grau, J. Romeu, M.-J. Lee, S. Blanch, L. Jofre, and F. De Flaviis, "A Dual-Linearly-Polarized MEMS-Reconfigurable Antenna for Narrow-band MIMO Communication Systems," *IEEE Trans. on Antennas and Proagat.* vol 58, no. 1, 2010, pp. 4-17.

[7] A. Molisch *Wireless Communications, chapter: Wideband and Directional Channel Characterization*, pp. 101 – 123, 2011.

[8] W. Q. Malik, C. J. Stevens, and D. J. Edwards, "Spatio-temporal ultrawideband indoor propagation modelling by reduced complexity geometric optics," *IET Commun.*, vol. 1, no. 4, pp. 751-759, 2007.

[9] T. Taga "Analysis for Mean Effective Gain of Mobile Antennas in Land Mobile Radio Environments," *IEEE Trans. Veh. Technol.* vol. 39, no. 2, pp. 117 – 131, 1990.

[10] http://www.3gpp.org/specifications

[11] B. H. Fleury "First- and Second-Order Characterization of Direction Dispersion and Space Selectivity in the Radio Channel," *IEEE Trans. on Information Theory*, vol. 46, no. 6, pp. 2027 – 2044, 2000.

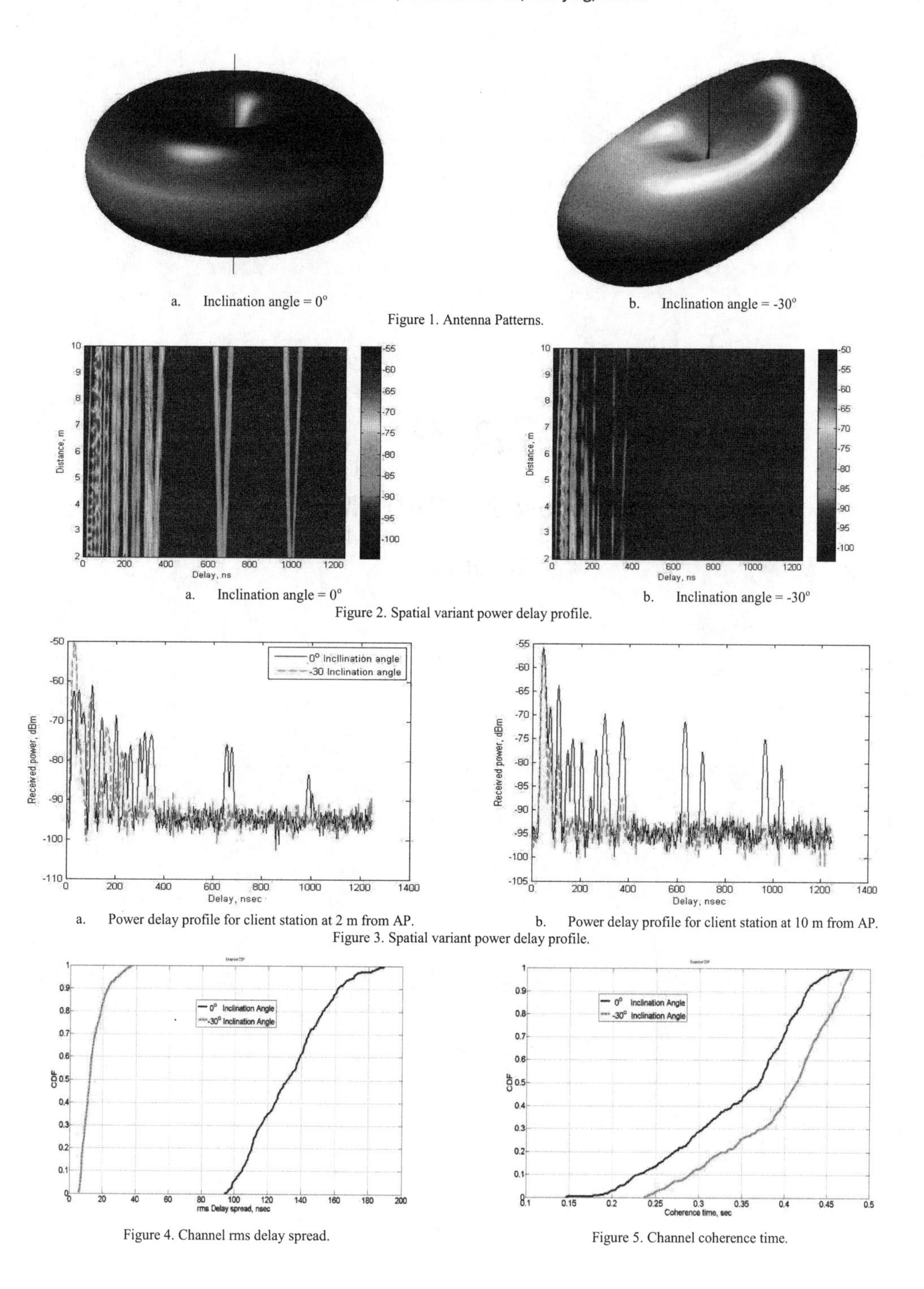

a. Inclination angle = 0°

b. Inclination angle = -30°

Figure 1. Antenna Patterns.

a. Inclination angle = 0°

b. Inclination angle = -30°

Figure 2. Spatial variant power delay profile.

a. Power delay profile for client station at 2 m from AP.

b. Power delay profile for client station at 10 m from AP.

Figure 3. Spatial variant power delay profile.

Figure 4. Channel rms delay spread.

Figure 5. Channel coherence time.

TA-P

October 24 (THU) AM

Room E

Low Frequency Characteristics of Electric Wire Antenna onboard Scientific Spacecraft

Tomohiko Imachi[1], Ryoichi Higashi[2], Mitsunori Ozaki[3], Satoshi Yagitani[3]
[1]Information Media Center, Kanazawa University
Kakuma-machi, Kanazawa, Ishikawa pref., 920-1192, Japan, imachi@imc.kanazawa-u.ac.jp
[2]Ishiakwa National College of Technology
[3]Graduate School of Natural Science and Technology, Kanazawa University

Abstract **For satellite observation of the electric fields of waves, a dipole antenna using a wire (wire antenna) is often used. For calibration of a wire antenna, the effective length is a problem. In our present research, we clarified that the characteristics of the effective length depends on the frequency and the structure of the wire antenna, such as the wire insulator and the shape of tips by holding "rheometry" experiments[1]. In this paper we describe about "DPS method" on which we calculated the characteristic of the effective length theoretically.**

I. Introduction

Observation of plasma waves in space is an important task for scientific satellites. The electromagnetic waveform detected by an electromagnetic sensor on the satellite is transformed to a voltage by the sensor, is converted to digital data by A-D conversion during the observation equipment, and then transmitted to the earth by telemetry. Hence, in order to accurately determine the magnitude of the original electromagnetic field, the calibration of the observation equipment and the sensor must be accurate. Most of the transfer function needed for this purpose is acquired in ground tests prior to launching of the satellite. However, the effective length (h_{eff}) of the electric field sensor cannot be accurately obtained in the ground test. If the electric field is E (V/m) and the antenna output voltage is V (V), the effective length is an extremely important parameter providing the ratio of the two:

$$V = h_{eff} \cdot E \tag{1}$$

For observation of the electric field component of the wave, a dipole antenna using a wire (the wire antenna) is often used. Since sensitivity must be assured at frequencies of several kilohertz and below for plasma wave observation, the total length of the wire is extremely large, ranging from tens of meters to 100 m. Since the surrounding environment at the time of observation is plasma medium, it is difficult to acquire the needed characteristics in a ground test.

II. Rheometry Experiment

In order to study the transition of the effective lengths of the wire antennas, the authors performed a rheometry experiment [1]. The experimental setup is shown in Figure. 1.

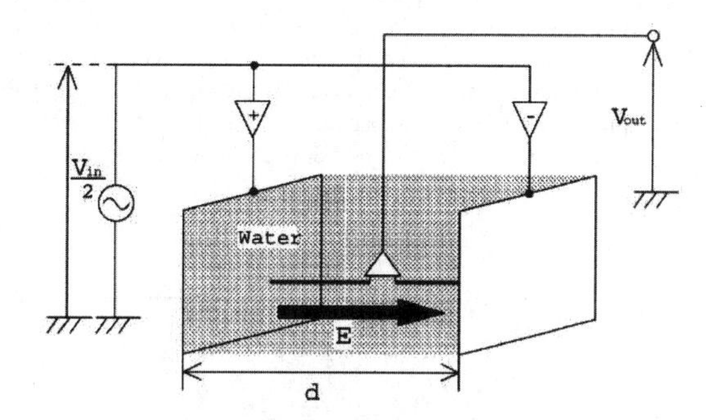

Figure 1. Setup for a "rheometry" experiment..

In this experiment, a signal is applied to two electrodes placed in parallel in water so that a quasi-static electric field is generated between the electrodes. This field is considered as the electric field component of an electromagnetic wave and is received by an antenna whose output voltage is measured. If the distance between the electrodes is d and the amplitude of the signal is V_{in}, then the amplitude of the generated electric field is

$$E = V_{in}/d \tag{2}$$

Therefore, if the output voltage is V, the effective length becomes

$$h_{eff} = \frac{V}{E} = \frac{Vd}{V_{in}} \tag{3}$$

In free space, the impedance between the elements of a dipole antenna is purely capacitive and is extremely large at low frequencies, however, by adding conductivity in parallel to capacitance when the antenna is immersed in water, measurement is made possible. In contrast, the electric lines of force in water cannot escape into the air by passing through the water surface, because the permittivity of water is about 80 times that in air. As a result, an extremely uniform electric field can be formed in water.

Since this experiment models the electric field component of an electromagnetic wave by a quasi-static electric field, the measurable frequencies are limited. The wave length must be

978-7-5641-4279-7

much longer than the antenna length where the phase does not vary for a location shift of about the antenna length. On the other hand, if the frequency is too low, no electric field may be formed as the water is polarized. In the experiments reported in this paper, measurements were performed at frequencies of 10 Hz to 100 kHz for an antenna with 0.3 m length.

For the experiment, the three types of antennas shown in Figure. 2 were prepared and measurements of variations of the characteristics due to differences in structure were performed: (1) The antenna without insulation, so that the entire length of the wires (copper) was exposed to the water, (2) the antenna with its sides insulated so that only the cross sections of the two ends were exposed to the water, and (3) the antenna with aluminum spheres attached on both ends of the wires with their sides insulated. The wire diameter was 1.3 mm. A differential amplifier with input impedance consisting of a parallel connection of 10 MΩ and 2 pF between the wires was inserted.

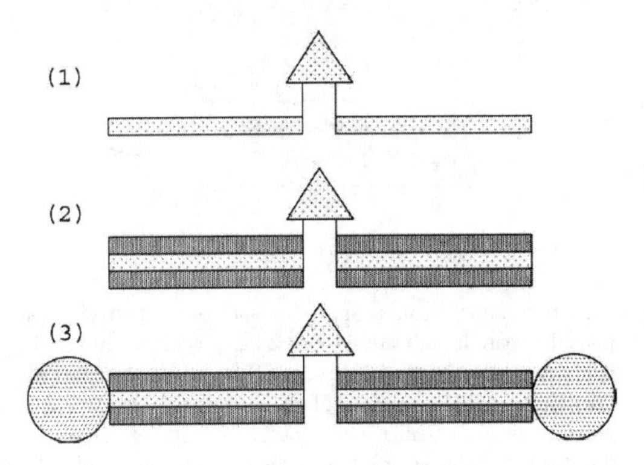

Figure 2. Antennas applied to the experiment: (1) without isolations; (2) insulated cylindrically and exposed to the water at both tips; (3) with sphere probes at both tips.

Figure 3 shows the experimental results. The horizontal axis of the figure represents frequency over the range of 10 Hz to 100 kHz. The vertical axis represents h_{eff}/L where L is antenna length. If the value is 1, the effective length and the wire length are equal. "Naked" denotes the case without insulation, "Cylindrical" denotes insulation only over the sides, and "With Probe" denotes the structure with spherical probes.

As shown in Figure. 3, three entirely different characteristics are obtained for the three antennas. There is no observable phase difference between the input waveform and the output waveform. Below we present a theoretical analysis of the results based on an equivalent circuit.

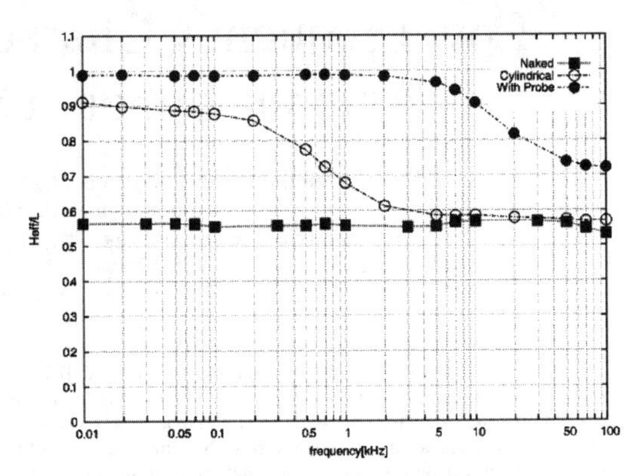

Figure 3. Experimentally measured effective lengths.

III. DPS METHOD

The circuit elements located on the wire are considered to be those in Figure. 4. The insulating coat is considered as a capacitance C_I between the wire core and the water and is connected in series to the grounding resistor R_I of the wire through the water. Grounding resistances exist from the wire cross section and the probe surface and are designated as R_M and R_P.

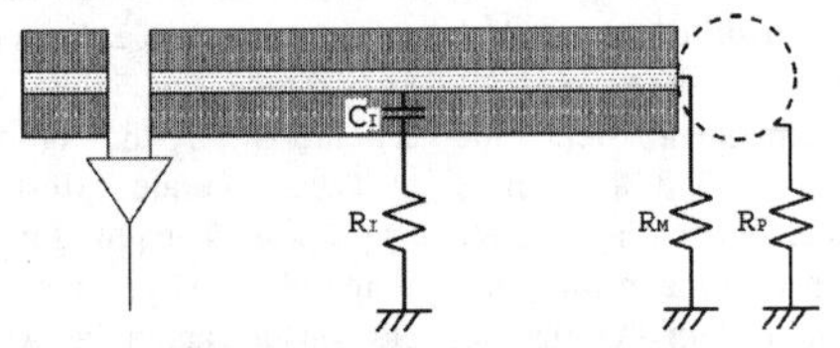

Figure 4. Equivalent of impedance of a wire.

When a wire antenna is placed parallel to a uniform static electric field, a potential difference arises as the location is moved away from the center, where the potential is 0, and a differential amplifier is installed. The potential becomes the maximum at the end. This phenomenon can be expressed by an equivalent circuit in which many voltage sources are connected in parallel through impedances as shown in Figure. 5. Then, the voltage of each source is $(L/2)EV$ if the voltage is 0 V at the center and the length of the wire on one side is $L/2$ m. The voltage variation along the wire is linear. Hence, if the center is the 0th and the end is the n-th, the voltage V_k at the k-th supply is

$$V_k = \frac{k}{n} \cdot \frac{L}{2} E \tag{4}$$

Then the potential of the wire core is V in Figure. 5, which becomes the input voltage to one side of the differential amplifier. In the following discussions, the value of this V is evaluated.

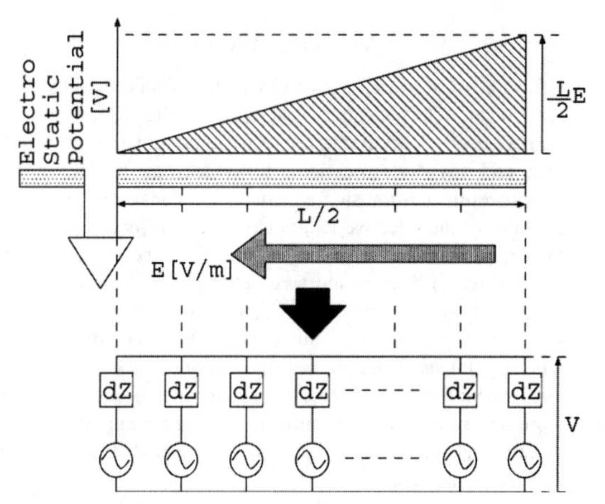

Figure 5. Basic concept of DPS method

For example, let us consider the case in which coatings exist only on the sides of the wire. If the impedance of the coating is Z_I and the impedance of the end surfaces where the conductors are exposed is Z_M, then the equivalent circuit is as shown in Figure. 6.

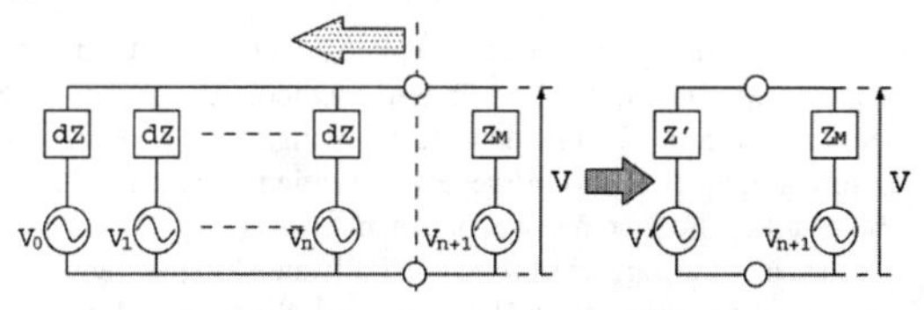

Figure 6. An equivalent circuit for the case of a cylindrical insulator.

Here,

$$V' = \frac{\sum_{k=0}^{n} V_k}{n+1} \tag{5}$$

$$Z' = \frac{dZ}{n+1} \tag{6}$$

$$dZ = (n+1)Z_I \tag{7}$$

When the output voltage is derived by means of the Ho-Thévenin theorem, we have

$$V = V_{n+1} + \frac{Z_M}{\frac{dZ}{n+1} + Z_M}\left(\frac{\sum_{k=0}^{n} V_k}{n+1} - V_{n+1}\right)$$

$$= \frac{\frac{dZ}{n}V_{n+1} + \frac{Z_M}{n}\sum_{k=0}^{n} V_k}{Z_M + \frac{Z_M + dZ}{n}} \tag{8}$$

Where

$$V_k = \frac{k}{n}\cdot\frac{L}{2}E \tag{9}$$

$$V_{n+1} = \frac{L}{2}E \tag{10}$$

Therefore,

$$V = \frac{\frac{dZ}{n}\cdot\frac{L}{2}E + \frac{Z_M}{n}\sum_{k=0}^{n}\frac{k}{n}\cdot\frac{L}{2}E}{Z_M + \frac{Z_M + dZ}{n}}$$

$$= \frac{\frac{dZ}{n} + \frac{Z_M}{2} + \frac{Z_M}{2n}}{Z_M + \frac{Z_M}{n} + \frac{dZ}{n}}\cdot\frac{L}{2}E \tag{11}$$

If n goes to infinity,

$$\frac{Z_M}{n} \to 0 \tag{12}$$

$$\frac{dZ}{n} = \frac{(n+1)Z_I}{n} \to Z_I \tag{13}$$

Hence,

$$V|_{n\to\infty} = \left(1 - \frac{1}{2}\frac{Z_M}{Z_I + Z_M}\right)\frac{L}{2}E \tag{14}$$

As described above, by regarding the electric field to be "distributed potential source", we can calculate the output voltage of a wire antenna with complex structure. We call this "DPS (Distributed Potential Source) " method.

IV. Applying DPS Method to the Experiment Result

Let up apply DPS method to the rheometry experiment. In the case of naked wire, dZ is dominated by the grounding resistance of the sides of the wire. If the output voltage V is calculated in Figure. 5, we obtain

$$V = \frac{\sum_{k=0}^{n} V_k}{n+1} \tag{15}$$

Therefore, the output voltage is

$$V = \frac{1}{n+1}\cdot\sum_{k=0}^{n}\frac{k}{n}\cdot\frac{L}{2}E$$

$$= \frac{L}{4}E \tag{16}$$

The output voltage of the differential amplifier is twice, namely, $L/2 \cdot$ E, and the effective length is

$$h_{eff} = \frac{L}{2} \tag{17}$$

On the other hand, when the impedance of the exposed metal portions is only the grounding resistance R_M and the impedance of the insulating parts on the sides is a series connection $R_I + 1 / j\omega C_I$ of the capacitance of the coating C_I and the grounding resistance R_I; then the output voltage is the following according to Eq. (14)

$$V = \left(\frac{1 + j\omega C_I (R_I + \frac{R_M}{2})}{1 + j\omega C_I (R_I + R_M)} \right) \frac{L}{2} E \tag{18}$$

Since there is little phase difference, only the amplitude characteristics are evaluated. The absolute value of V is

$$|V| = \sqrt{\frac{1 + \omega^2 C_I^2 (R_I + \frac{R_M}{2})^2}{1 + \omega^2 C_I^2 (R_I + R_M)^2}} \frac{L}{2} E \tag{19}$$

The output voltage of the differential amplifier is twice the above and hence the amplitude characteristics of the frequency dependence of the effective length are given by

$$h_{eff} = \sqrt{\frac{1 + \omega^2 C_I^2 (R_I + \frac{R_M}{2})^2}{1 + \omega^2 C_I^2 (R_I + R_M)^2}} \cdot L \tag{20}$$

When probes are attached to both ends, the equation for deriving the effective length is Eq. (20), with R_M replaced by the grounding resistance R_P, which has a smaller value. Therefore, the effective length is

$$h_{eff} = \sqrt{\frac{1 + \omega^2 C_I^2 (R_I + \frac{R_P}{2})^2}{1 + \omega^2 C_I^2 (R_I + R_P)^2}} \cdot L \tag{21}$$

Figure 6 shows the result of DPS method calculation according to Eq. (17), (20) and (21) applying the actual physical parameters of the experiment environment. The solid lines show the calculation results, and the symbols shows the measurement results.

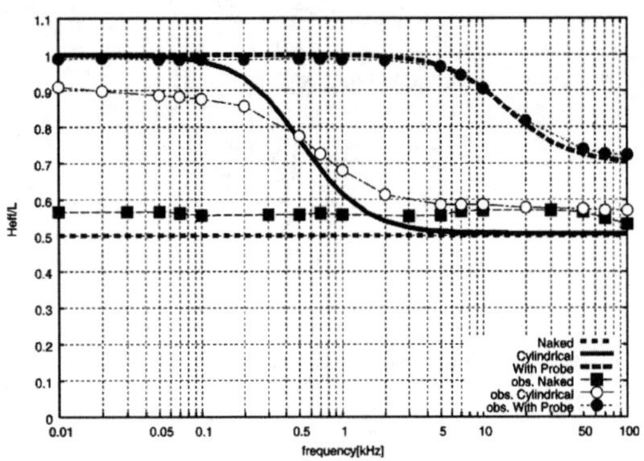

Figure 6. The result of theoretical calculation by using DPS method.

When the numerical values and the measured values are compared in each case, the frequencies at which the effective length starts decreasing at all cases agree well.

V. CURRENT WORK

Currently, we are working on an advanced rheometry experiment. The advanced factors are the following: (1) improvement of accuracy, (2) clarification of the applicable limit and (3) diversification of the antenna model. In the current result, the transit frequencies are consistent, however, the value of the effective length have some difference, between the experiment and DPS calculation. We try to clarify the cause of the difference and take better result by improvement of accuracy of the experiment. To clarify the applicable limit, we use the computer simulation. By reproducing the experimental environment on the computer, we identify the requirement for the experiment, such as the tank size, the water depth and so on. For diversification of the antenna model, we make models of antenna that have complicated structure, such as Hockey Pack antenna.

In parallel, we make a detail analysis about the experiment result by using computer simulations. By simulating the detail structure of potential distribution around the antenna models, we try to clarify the mechanism of the effective length determination, comparing to DPS calculation.

VI. CONCLUSION

In this paper, we made explain about the rheometry experiment and DPS calculation method. By using DPS method, we can calculate the effective length of a wire antenna at low frequencies, and the result is clarified to be consistent to the result of the rheometry experiment.

Currently, we are working on an advanced experiment, and the computer simulation. We will show the result of them in near future.

As a future work, we try to apply DPS method calculation to the wire antenna aboard actual scientific spacecraft.

REFERENCES

[1] T. Imachi, et al., "Rheometry Experiment for a Wire Antenna Aboard Spacecraft at Low Frequencies", Vol. J89-B, No. 4, April 2006, pp. 552–559.

Design of Broad Beam Circular-Polarized Microstrip Antenna

D.P. Fan Z.X. Wang B. Huang W. Zhang
State Key Lab of Millimeter Waves
Southeast University,Nanjing,210096,P.R.China
windyfan43@163.com

Abstract-In this paper, broad beam circular-polarized microstrip antennas are studied. Using a corner-truncated structure with conducting post[1], a patch antenna with a height of 0.1λ only is designed and exhibits a 3dB gain beam width of over 160°, but the axial ratio is poor. Then a stacked squared-ring microstrip antenna is designed using hybrid perturbation method in order to increase 3dB AR beam width. Simulation results for operating at 2.06 GHz are presented and discussed.

I. INTRODUCTION

Microstrip antenna has been widely used in mobile terminals and satellite communication systems for its lightweight, low profile, low cost and easy to conform. However, the above application systems require that the antenna can provide uniform response over approximately the entire upper hemisphere and high gain at low-angle [2]. But the 3dB gain beam width of conventional microstrip antenna is only 70°~110° and gain at low-angle is about -7dB~-3dB. Some methods are presented to solve the problem, such as using high permittivity substrate, employing a pyramidal configuration or operating at higher order modes. But these methods only meet the requirement in some degree.

In this paper, a microstrip antenna example operating at 2.06GHz is presented. The designed patch antenna has a height of 0.1λ only and exhibits a 3dB gain beam width of more than 160° [3]. To increase the 3dB gain beam width, some techniques are employed [4]. The patch presented is enclosed partially by some conducting posts to increase the 3dB gain beam width to more than 130°. Corner-truncated method is adopted to realize circular-polarization (CP) and the 3dB Axial Ratio(AR) beam width can reach about 80°. According to the cavity theory, size of the corner can be determined. To improve circular-polarization characteristic further, a stacked squared-ring microstrip antenna designed using hybrid perturbation[5] method is adopted to increase 3dB AR beam width. Simulation results for operating at 2.06 GHz are presented and discussed.

II. ANTENNA CONFIGURATION

Antenna A in Fig.1 shows the geometry of the normal square-patch antenna which is mounted on a plane. The patch is printed on a square Arlon AD255a(tm) substrate of thickness 40mil, relative permittivity 2.55 and size 100mm×100mm, and is probe fed at a point which is 14.3mm from the patch center. The side length L of the patch is 55.6mm. Moreover, an air layer with a thickness of 1.5mm is embedded between the ground and substrate layer.

To study the effects of the conducting posts and the truncated corner, two other prototypes in Fig.1 are implemented and investigated. The first one, denoted by antenna B, consists of truncated corner but no posts exist. The side length L of the patch is 55.6mm and the length ΔL of the truncated corner is 7.0mm. The second one, antenna C, is constructed by adding posts to antenna B. The distance between two adjacent posts is 10mm and the height of the posts is 14.5 mm . Antenna parameters, such as the VSWR , gain beam width, 3dB axial-ratio beam width are simulated.

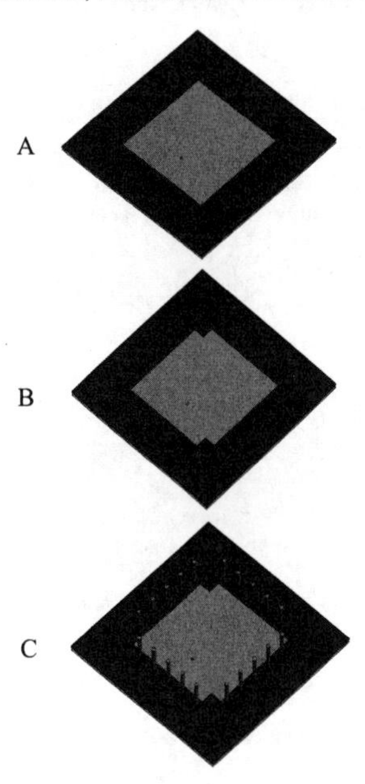

Figure 1. Configuration of proposed antennas(A:normal patch antenna; B:corner-truncated CP patch antenna; C: final broad beam circularly-polarized patch antenna)

978-7-5641-4279-7

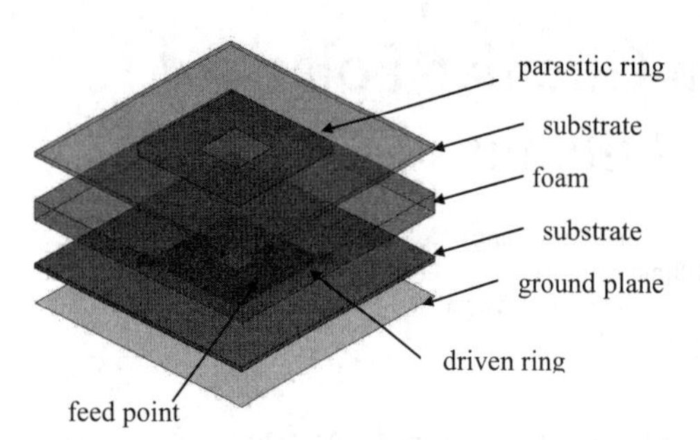

Figure 2. Configuration of hybrid perturbation CP antenna

We know, a CP patch antenna can be achieved by truncating its diagonal corners (a negative perturbation) as shown in Fig.1. However, the 3dB AR beam width is not satisfied for some communication systems. Fig.2 shows a unique hybrid perturbation stacked squared-ring microstrip antenna to obtain good polarization characteristic, where both positive and negative perturbation are used. The substrate[6], in the case, adopts Arlon 255a(tm),which has a dielectric constant of 2.55, thickness of 40mil. A foam layer is embedded between the two substrate. To decrease antenna's size for a given frequency, the regular patches are perforated at the center to form ring antennas. Only the driven ring is probe fed at a point which is 11.3mm from the center.

III. DISCUSSIONS & RESULTS

Fig .3 shows the return loss of antenna C. It is shown that antenna C can provide a impedance bandwidth(VSWR<2) of over 170MHz. The radiation pattern is presented in Fig 4, and

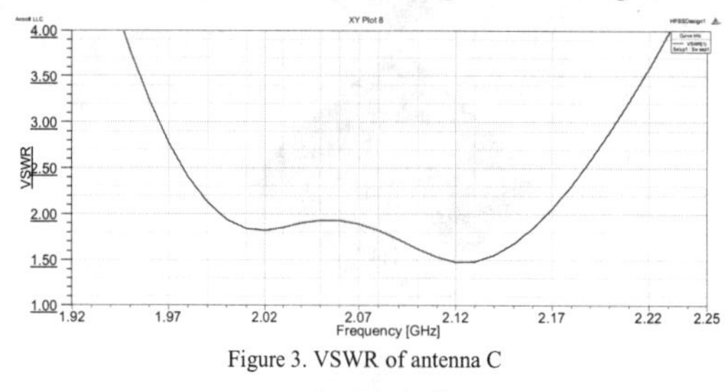

Figure 3. VSWR of antenna C

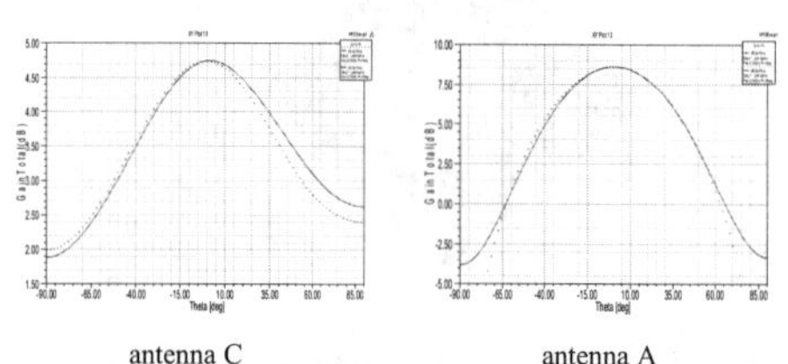

antenna C antenna A

Figure 4. gain beam width of antenna A and antenna C at 2.06GHz in both E and H plane

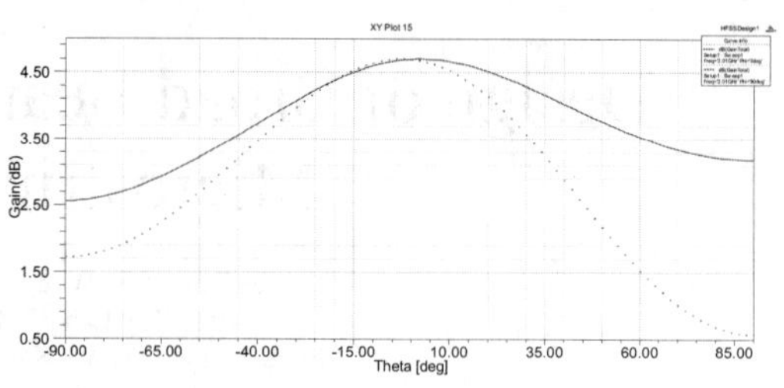

Figure 5. gain beam width of antenna C at 2.01GHz in both E and H plane

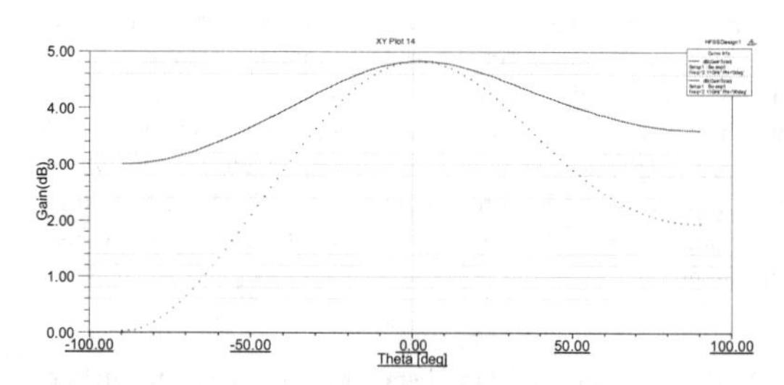

Figure 6. gain beam width of antenna C at 2.11GHz in both E and H plane

antenna C's 3dB gain beam width is over 160° in both E plane and H plane, contrarily, antenna A can only provide a gain beam width of about 80° . To understand antenna C's radiation characteristic at different frequency, we choose 2.01GHz and 2.11GHz as investigation points. Fig.5 and Fig.6 illustrate the radiation pattern respectively. Obviously, the 3dB gain beam width are both over 130° .

The circular-polarization technique we adopt in antenna C can provide a 3dB AR beam width of about -40° ~40° in both E plane and H plane as is shown in Fig.7. In order to obtain a better circular-polarization characteristic, hybrid perturbation method is adopted. The 3dB AR beam width presented in Fig.8 is about 140° ,which is improved greatly compared to antenna C.

IV. CONCLUSION

A corner-truncated patch antenna with conducting posts has been concisely studied for operating at 2.06GHz. Employing

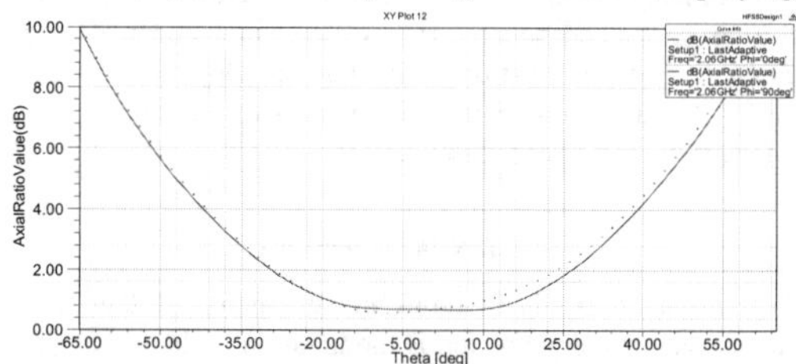

Figure 7. 3dB AR beam width of antenna C

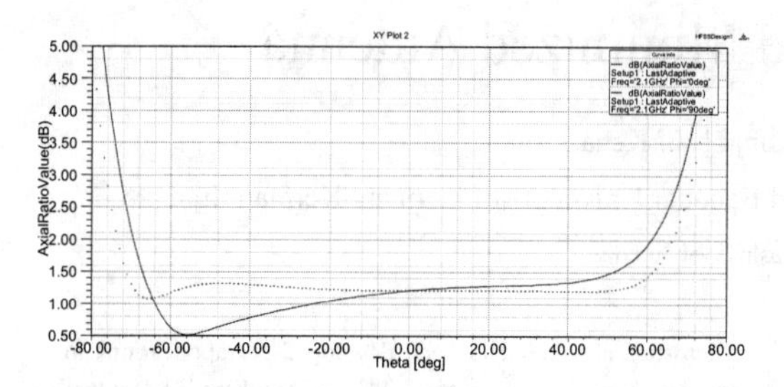

Figure 8. 3dB AR beam width of hybrid perturbation antenna

this technique, the antenna can achieve a wide 3dB gain beam width of up to 160° and the 3dB AR beam width is about 80°. To obtain a better circular-polarization, hybrid perturbation method is adopted. The 3dB AR beam width can reach about 140° ,which is improved greatly compared to the former.

ACKNOWLEDGMENT

This paper is supported by NSFC 61071046, NSFC 61101020.

Joint supported by Aeronautical Science Foundation and RF simulation Key Lab of Avionics System.

REFERENCES

[1] H.H. Heo, J.Y. Jang, J.M. Woo, 'Miniaturization of Microstrip Antenna with Folded Structure', *Journal of Korea Electromagnetic Engineering Society*, vol. 16, No. 5,pp. 526~533, 2005.05.

[2] Laubner,TL,and Schilling, R: 'Microstrip antenna with improved low angle performance', *US Patent Publication* No. 1049520 A1, 2002.

[3] Tang,C.L, Chiou, J.Y., and Wong, K.L.: 'Beam width enhancement of a circularly polarized microstrip antenna mounted on a three-dimensional ground structure', *Microw. Opt. Technol.* Lett., 2002,32, (1), pp. 149–153.

[4] Nakano, H., Shimada, S., Yamauchi, J., and Miyata, M.: 'A circularly polarized patch antenna enclosed by a folded conducting wall'. *IEEE Conf. on Wireless Communication Technology,* Honolulu, Hawaii, 2003,pp. 134–135.

[5] Tesshirogi,T.,Tanaka,M., and Chujo,W.:'Wide and circularly polarized array antenna with sequential rotations and phase shift of elements'. *Proc. ISAP*, Kyoto, Japan,1985,pp.117-120.

[6] Bafrooei, P.M., and Shafai, L.: 'Characteristics of single- and double microstrip square ring antennas', *IEEE Trans. Antennas Propag*, 1999, 47, (10), pp. 1633–1639 .

Ultra Wide Band and Minimized Antenna

Mohammad Alibakhshi Kenari

Electrical Engineering Department of Shahid Bahonar University of Kerman, Kerman, Iran

Naeem.alibakhshi@yahoo.com

***Abstract* — In this paper, a novel ultra-wideband (UWB) miniature antenna based on the composite right-left handed transmission line (CRLH-TL) structure with enhancement gain is proposed and investigated. With CRLH metamaterial (MTM) technology embedded, the proposed UWB and miniature antenna is presented with best in bandwidth, size, efficiency and radiation patterns. To realize characteristics of the antenna, the printed Π -shaped gaps into the rectangular radiation patches are used. This antenna is constructed of the two unit cells, also presented antenna is designed from 2.25 GHz to 4.7 GHz which corresponding to 70.5% bandwidth. The overall size of the presented antenna is 10.8mm×6.9mm×0.8mm or $0.09\lambda_0\times 0.05\lambda_0 \times 0.006\lambda_0$ at the operating frequency f = 2.5 GHz (where λ_0 is free space wavelength). The radiation peak gain and the maximum efficiency which occurs at 4.6GHz, are 3.96dBi and 63.6%, respectively.**

***Index Terms* — Composite right/left-handed Transmission Line (CRLH-TLs), antenna, wireless.**

1. Introduction

In recent years, with development of broadband and minimizing technology for high resolution and high data transmission rates and foot print area reduction in modern communication systems, there is increasing demand for small low-cost antenna with unidirectional radiation patterns, dispersive and broad band characteristics. The printed antennas have received great attention in broadband applications due to their advantages of compact, planar, low cost, light weight, broadband, compatibility and easy integration with other microstrip circuits. Applications in present-day mobile communication systems usually require smaller antenna size in order to meet the miniaturization requirements of mobile units. Thus, size reduction and bandwidth enhancement are becoming major design considerations for practical applications of microstrip antennas [1].

Metamaterial (MTM) [2], have recently been extensively discussed and studied for special properties. Metamaterials (MTMs) are manmade composite materials, engineered to produce desired electromagnetic propagation behavior not found in natural media [2], [3]. Those unusual properties were used to improved performances of antennas and circuits. Microstrip antennas had been developed for applications in present communication systems [4], [5], but there is a fact that the size reduction levels remain unsatisfactory to the electromagnetic community. Several techniques were suggested to reduce antenna size [6], however, such techniques usually suffer from increasing the design complexity. The occurrence of metamaterial may be a solution for this challenge [7], [8]. In this work, we using of the metamterial technology and the simple techniques for foot print area reduction, enhancement bandwidth and improvement gain of the antenna, which consist of employing of the printed planar mushroom structure based on CRLH-TL and suitable structural parameters. Various implementations of metamaterial structures have been reported and demonstrated [2]. In this paper a metamaterial CRLH antenna with two unit cells which each unit cell embrace of two printed Π -shaped gaps capacitors and the spiral inductor accompanying a metallic via connected to ground plane is presented. The printed Π -shaped structure exhibit wide bandwidth, miniature and improvement gain property which useful for UWB and miniaturized antennas.

This paper is organized in the following way. A UWB and miniature antenna prototype with high gain and efficiency employing the proposed concept will be depicted in Section 2. Followed by section 3 where various performance including dimension, impedance bandwidth and radiation patterns characteristics of the recommended antenna are demonstrated. Further discussion and conclusion are raised at last.

2. Theory of the Proposed Antenna

As discussed in [2], [9], several implementations can be used to realize the CRLH-TL unit cell including surface mount technology (SMT) chip components and distributed lines. However, lumped elements are not appropriate in antenna design because of their lossy characteristics and discrete values. We using printed planar technique for our antenna design, since printed planar structures are good candidate for antenna design because of their advantages

978-7-5641-4279-7

which include foot print area reduction, loss less and non-discrete values [13], [14]. A novel UWB and miniature antenna with improvement gain based on CRLH-TL presented in here, which consists of two unit cells while each unit cell will be designed by two rectangular radiation patches with printed Π-shaped gaps into patches, and the spiral inductor accompanying metallic via connected to the ground plane. Fig. 1 shows geometry of the proposed antenna and Fig. 2 display equivalent circuit model of each cell as CRLH unit cell. In this structure, port 1 is excited with input signal and port 2 is matched with 20Ω load impedance. The antenna structure is based on a composite right-left handed (CRLH) transmission line (TL) model used as a periodic structure. Because the lowest mode of operation is a LH mode, the propagation constant approaches negative infinity at the cutoff frequency, and reduce its magnitude as frequency is increased. Making use of this phenomenon, an electrically large but physically small antenna can be developed.

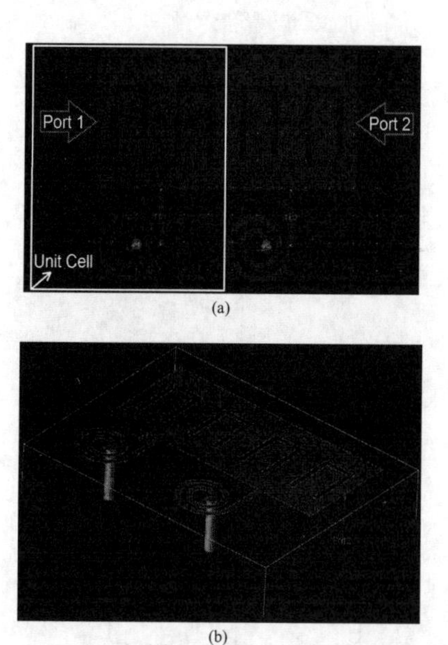

(a)

(b)

Fig.1. Configuration of the presented UWB miniature antenna composed of the two unit cells based on CRLH MTM-TL. a) Top view, b) Isometric view.

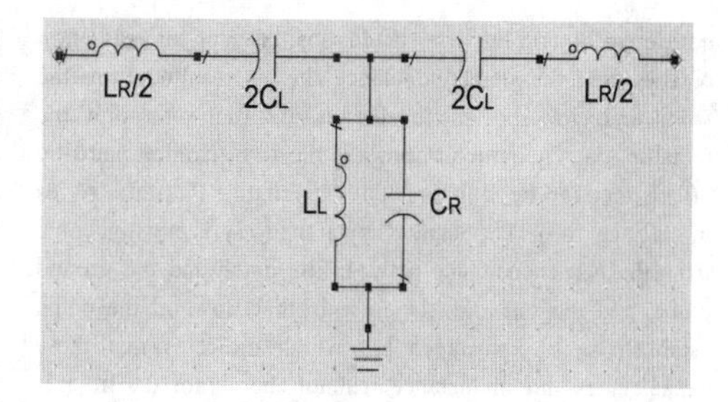

Fig.2. Proposed Antenna: Equivalent circuit model of the CRLH MTM antenna for one unit cell.

By means of the Π-shaped gaps and spiral inductors with shorting via-hole connecting to ground plane, the series capacitance (C_L) and shunt inductance (L_L) can be easily implemented in a compact fashion. The host TL possess the right-handed parasitic effects that can be seen as shunt capacitance (C_R) and series inductance (L_R).

In this paper, we employing of metamaterial (MTM) technology and the printed planar approach that results to foot print area reduction of the proposed antenna. Overall size of this antenna is $0.09\lambda_0 \times 0.05\lambda_0 \times 0.006\lambda_0$ at the operating frequency f = 2.5 GHz where λ_0 is the free space wavelength and also with choosing smaller distance between printed Π-shaped gaps edges, we will be obtained wide bandwidth from 2.25 GHz to 4.7 GHz which corresponding to 2.45 GHz bandwidth. Furthermore, with acceptable selecting of the number unit cells (*N*) constructing antenna structure and structural parameters of the spiral inductors such as number of turns (*N*), inner radius measured to the center of the conductor (R_i), conductor width (*W*) and conductor spacing (*S*) we will be achieved good radiation performances. The gain and efficiency of the proposed antenna are changed from 0.4dBi to 3.96dBi and 19.5% to 63.6%, respectively, into frequency band 2.25 - 4.7 GHz, that shown very good radiation characteristics. Therefore, the MTM antenna designed is miniature and ultra-wideband with high gain and efficiency. The proposed antenna based on CRLH-TL made very small size and wide bandwidth to support today's multi-band modern wireless applications.

Fig. 1 shows configuration of the recommended antenna constructed of the two unit cells based on CRLH-TL structure that was designed on a FR_4 substrate, with a dielectric constant of 4.6, a thickness of 0.8mm and Tan δ=0.001. This mushroom type unit cell consisted of 5.4 mm×6.9 mm or $0.045\lambda_0 \times 0.05\lambda_0$ patch, printed on top of the Substrate which in each unit cell, the series capacitance (C_L) is developed by two the printed Π-shaped gaps into radiation patches, and the shunt inductance (L_L) is resulted from the spiral inductor shorted to the ground plane through the metallic via. The structure possess the right-handed parasitic effects that can be seen as shunt capacitance (C_R) and series inductance (L_R). The shunt capacitance C_R is mostly come from the gap capacitance between the patch and the ground plane, and the unavoidable currents that flow on the patch establish series inductance L_R, which indicates that these capacitance and inductance cannot be ignored. In this structure, port 1 is excited with input signal and port 2 is matched to 20Ω load impedance, as illustrated in Fig.1. The proposed design keeps the overall size of the unit cell compact while aims at reducing the ohmic loss to improve gain and radiation efficiency. This antenna can support all cellular frequency bands from 2.25 GHz to 4.7 GHz, using single or multiple feed designs, which eliminates the need for antenna switches. All of these attributes make the proposed antenna is well suitable for electromagnetic requirements such as millimeter waves and emerging wireless applications [10], [11].

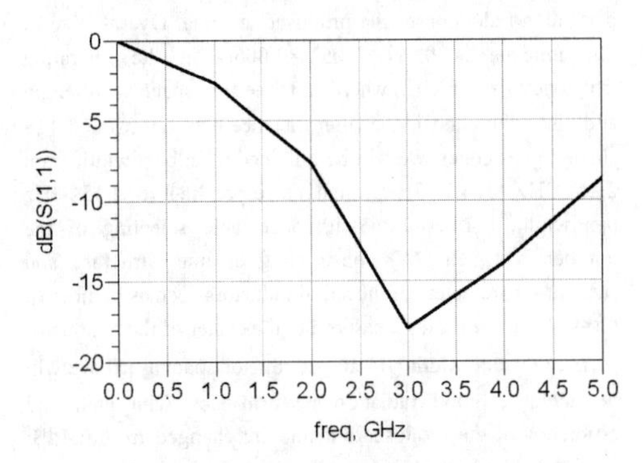

Fig.3. Simulated reflection coefficient S_{11}.

3. Simulation Results and Discussion

The proposed metamaterial antenna is designed as a CRLH antenna where the substrate has dielectric constant $\varepsilon_r = 4.6$, thickness h = 0.8 mm and Tan δ=0.001. UWB and miniature recommended antenna is simulated by using the full-wave simulator (ADS). The simulated reflection coefficient (S_{11} parameter) displayed in Fig. 3, and simulated radiation gain patterns in 2.3, 3.4 and 4.6 GHz are plotted in Fig. 4. The radiation patterns are unidirectional characteristics. The radiation gains at 2.3, 3.4, 4.6 GHz are 0.4, 2.8, and 3.96dBi, respectively. The radiation efficiency is 19.5% at 2.3 GHz, 47.8% at 3.4 GHz, and 63.6% at 4.6 GHz. To validate the design procedure the proposed antenna was compared with some of the antennas and their dimension and radiation characteristics were summarized in Table 1.

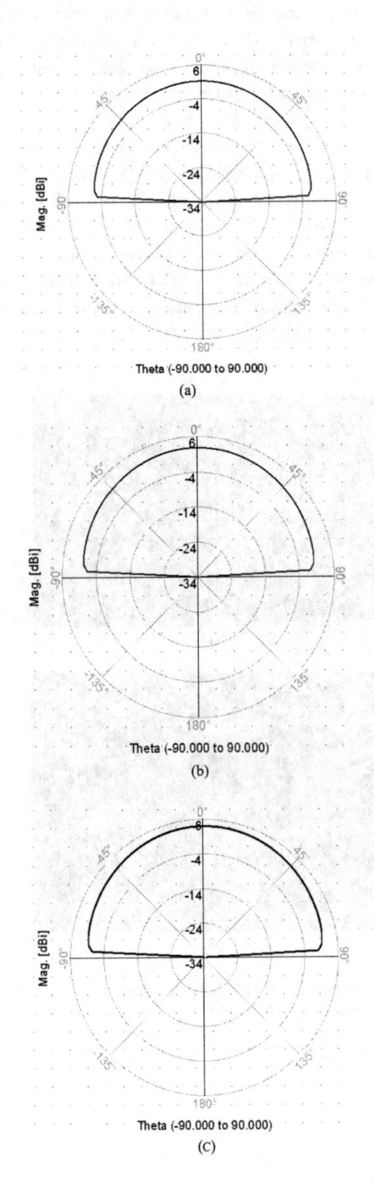

Fig.4. The Radiation Patterns (Gains) of the proposed antenna in elevation ($\Phi = 0$ degree). a) 2.3 GHz, b) 3.4 GHz, and c) 4.6 GHz.

The two unit cells of the UWB and miniature antenna is designed from 2.25 GHz to 4.7 GHz and this antenna exhibit good matching between this frequency band for 20Ω impedance port. The physical length, width and height of the suggested antenna are 10.8 mm, 6.9 mm and 0.8 mm ($0.09\lambda_0 \times 0.05\lambda_0 \times 0.006\lambda_0$), respectively. The gain and the radiation efficiency of this antenna are varies from 0.4dBi to 3.96dBi and from 19.5% to 63.6%, respectively, into the frequency range 2.25 GHz to 4.7 GHz.

Table 1. Dimension and Radiation Characteristics of the Some of the Antennas in Comparison to the Proposed Antennas.

Parameters	**[11]**	**[12]**	**Proposed Antenna**
Gain	0.45 dBi	0.6 dBi	3.96 dBi
Bandwidth	0.8-2.5 GHz	1-2 GHz	2.25-4.7 GHz
Efficiency	53.6%	26%	63.6%
Dimension	$0.4\lambda_0 \times 0.03\lambda_0 \times 0.03\lambda_0$	$0.07\lambda_0 \times 0.07\lambda_0 \times 0.03\lambda_0$	$0.09\lambda_0 \times 0.05\lambda_0 \times 0.006\lambda_0$

4. Conclusion

In this paper, we introduced a new concept of antenna size reduction with broad bandwidth accompanying enhancement gain based on a metamaterial design methodology. A practical UWB, miniature and high gain antenna with a simple feed structure and planar circuit integration possibilities has been demonstrated. Overall size of the recommended antenna is 10.8mm× 6.9 mm× 0.8 mm or $0.09\lambda_0 \times 0.05\lambda_0 \times 0.006\lambda_0$ at the operating frequency f=2.5 GHz where λ_0 is free space wavelength. A return loss below -10dB from 2.25 GHz - 4.7 GHz was obtained which corresponding to 70.5% bandwidth. The peak gain and the maximum efficiency of the proposed antenna which occurs at f = 4.6 GHz, are 3.96dBi and 63.6%, respectively. This antenna has the advantages of ultra-wideband, compact size, high gain, unidirectional radiation patterns, low cost and simple implementation. The recommended antenna can be used for millimeter wave applications, mobile handsets and wireless communication implementations.

Acknowledgement

The author would like to express his sincere thanks to Research Institute for ICT of Iran (Contract number 6987/500/T).

References

[1] C. J. Lee, K. M. H. Leong, and T. Itoh, "Broadband Small Antenna for Portable Wireless Application," in proc. International Workshop on Antenna Technology: Small Antennas and Novel Metamaterials, pp. 10- 13, 2008.

[2] C. Caloz and T. Itoh, Electromagnetic Metamaterials: Transmission Line Theory and Microwave Applications, The Engineering Approach, New York, John Wiley & Sons, 2005.

[3] R. A. Shelby, D. R. Smith, and S. Schultz, "Experimental Verifcation of a Negative Index of Refraction," Science,292, 55 14, 2001, pp. 77-79.

[4] W. L. Stutzman, Antenna Theory and Design, Second Edition: J. Wiley & Sons, 1997.

[5] Aparna Sankarasubramaniam, "Design guidelines for tunable coplanar and microstrip patch antennas," Microwave Conference, European, 2007.

[6] K. L. Wong, Planar Antennas for Wireless Communications, Wiley-Interscience, 2003.

[7] Andrea Alù, "Subwavelength, compact, resonant patch antennas loaded with metamaterials," IEEE Transaction on Antennas and Propagation, vol. 55, no. 1, 2007.

[8] Ourir, A, "Phase-varying metamaterial for compact steerable directive antenna," Electronics Letters, vol.43, pp. 493-494, 2007.

[9] A. Lai, C. Caloz, and T. Itoh, "Composite right/left-handed transmission line metamaterials," IEEE Microwave., Mag., vol. 5, no. 3, pp. 34–50, sept. 2004.

[10] C. J. Lee, M. Achour, and A. Gummalla, "Compact Metamaterial High Isolation MIMO Antenna Subsystem," in proc. Asia Pacifc Microwave Conference, pp. 1-4, 2008.

[11] Y. Li, Z. Zhang, J. Zheng and Z. Feng, "Compact heptaband reconfigurable loop antenna for mobile handset," IEEE Antennas and Wireless Propagation Letters, vol. 10, pp. 1162- 1165, 2011.

[12] C. J. Lee, K. M. K. H. Leong, and T. Itoh, "Composite right/left-handed transmission line based compact resonant antennas for RF module integration," IEEE Trans. Antennas and Propagation., vol. 54, no. 8, pp. 2283-2291, Aug. 2006.

[13] Mohammad Alibakhshi-Kenari, Masoud Movahhedi and Hadi Naderian, "A New Miniature Ultra Wide Band Planar Microstrip Antenna Based on the Metamaterial Transmission Line" *2012 IEEE Asia-Pacific Conference on Applied Electromagnetics (APACE 2012), December 11-13, 2012, Melaka, Malaysia.*

[14] Mohammad Alibakhshi-Kenari, Masoud Movahhedi and Ahmad Hakimi, "Compact and Ultra Wide Band Planar Antenna Based on the Composite Right/Left-Handed Transmission Line Accompanying Improvement" *First Iranian Conference on Electromagnetic Engineering (ICEME 2012), Tehran, Iran, December 2012.*

Miniaturized Dual-Polarized Ultra-Wideband Tapered Slot Antenna

Fuguo Zhu, Steven Gao
University of Kent, Canterbury, UK

Anthony TS Ho, Tim WC Brown
University of Surrey, Guildford UK

Jianzhou Li, Gao Wei, Jiadong Xu
Northwestern Polytechnical University, Xi'an, China

***Abstract*-A compact dual-polarized tapered slot antenna is proposed for ultra-wideband (UWB) through-wall imaging applications. The dual-polarized antenna is obtained by shifting one single-polarized element orthogonally into the other one. The miniaturization of the antenna is realized by employing a folded balun comprising a folded slot on the right side of the radiator and a stepped microstrip line is used for improving the impedance matching. In addition, the fabrication complexity of the antenna is reduced due to the use of a folded balun. The antenna occupies a small size of 30 × 30 × 16 mm^3. Both simulated and measured results confirm that it can achieve a wide impedance bandwidth from 3.1 to over 12 GHz. Both ports have similar radiation characteristics with an isolation of 20 dB and the gain response is varying from 2.5 to 6.8 dBi.**

Keywords: UWB antenna, dual-polarized antenna, tapered slot antenna, compact size, folded balun

I. INTRODUCTION

The allocation of the frequency range from 3.1 to 10.6 GHz has increasingly attracted much interest in various UWB applications including short-range, high-data-rate communication [1] and high-resolution imaging radar and sensor [2-3] due to its large bandwidth. The tapered slot antenna [4-5] is a promising candidate for UWB applications as it has many attractive characteristics: wide bandwidth, directional radiation patterns, low profile and easy fabrication.

Compared to single-polarized UWB antennas, the performance of imaging systems can be significantly improved by exploiting dual-polarized UWB antennas [6-7]. Moreover, the dual-polarized tapered slot antenna can be obtained by shifting two single-polarized elements orthogonally into each other [8]. In order to reduce the volume, a compact dual-polarized UWB tapered slot antenna of 35×35×53 mm^3 in [9] is realized by dielectric loading. It is also noticed that, the metallization on the backside of one element has to be interrupted to accommodate the other one, thus leads to increasing the fabrication complexity.

In this paper, a compact dual-polarized UWB tapered slot antenna with a folded balun is proposed. The size of the antenna is significantly reduced by employing the folded balun on one side of the element. In addition, the fabrication complexity of the prototype is eased due to the folded balun. To prove the concept, a prototype is developed and measured in the frequency domain. Both simulated and measured results are presented and discussed. The rest of the paper is organized as follows: the configuration and design of the single-polarized tapered slot antenna are demonstrated in Section II, Section III illustrates the design of the dual-polarized tapered slot antenna. A conclusion is drawn in Section IV.

II. SINGLE-POLARIZED TAPERED SLOT ANTENNA

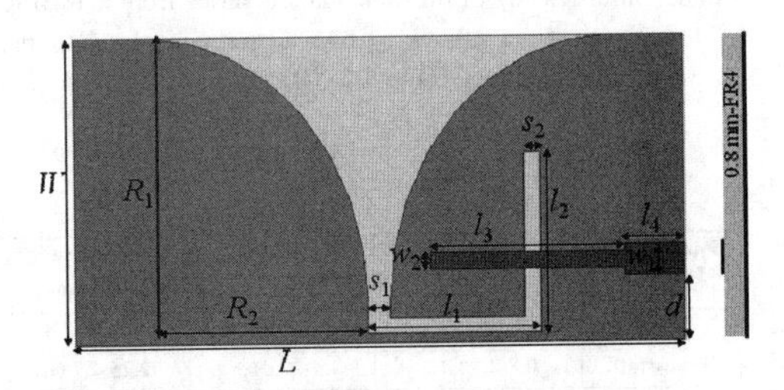

Figure 1. Geometry of the proposed single-polarized tapered slot antenna.

Figure 1 presents the geometry of the proposed single-polarized tapered slot antenna, which is printed on an FR4 substrate with a thickness of 0.8 mm and a relative permittivity of 4.55. The elliptical tapered slot is embedded in a rectangular plane on the bottom layer of the substrate and has major and minor radii of R_1 and R_2 respectively. The microstrip-line-to-slotline transition consists of a folded slot and a stepped microstrip line with distinct widths on the top layer of the substrate. The impedance matching of the antenna can be improved due to the stepped mcirostrip line. It is worthwhile to mention that, the right part of the radiator also behaves as a ground plane for the microstrip line. As the folded slot is inserted in the radiator and doesn't require additional space, the overall size of the antenna is miniaturized. Wideband performance can be achieved by selecting proper dimensions of the antennas. The optimized values of the dimensions are as follows: W = 15 mm, L = 30 mm, R_1 = 14.55 mm, R_2 = 12.1 mm, l_1 = 8.4 mm, l_2 = 8.8 mm, l_3 = 9.5 mm, l_4 = 3 mm, s_1 = 1.1 mm, s_2 = 0.75 mm, w_1 = 1.52 mm and w_2 = 0.75 mm.

Since the microstrip-line-to-slotline plays an important role in the impedance matching of the antenna, four parameters l_1, l_2, l_3, and l_4 are selected to study its effect. The parametric analysis is implemented while holding the remaining parameters with the optimized values. The effects on the reflection coefficient by the microstrip line and folded slot are illustrated in Figures 2 to 5. As shown in Figure 2, a wide frequency range for $|S_{11}| \leq -10$ dB is from 3.05 to over 12 GHz when the length of the horizontal slot l_1 is 8.4 mm. When it's too short, poor impedance matching near 4 and 11 GHz can be observed, while large l_1 leads to limited bandwidth. The result

978-7-5641-4279-7

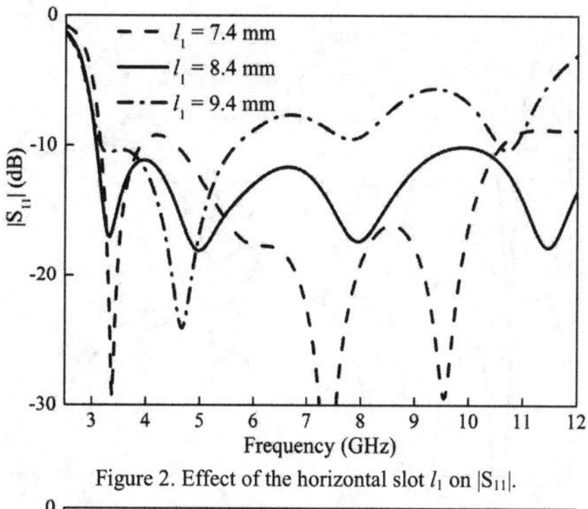

Figure 2. Effect of the horizontal slot l_1 on $|S_{11}|$.

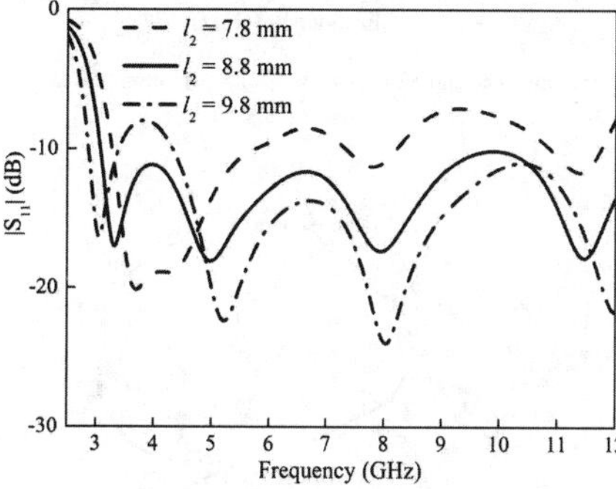

Figure 3. Effect of the vertical slot l_2 on $|S_{11}|$.

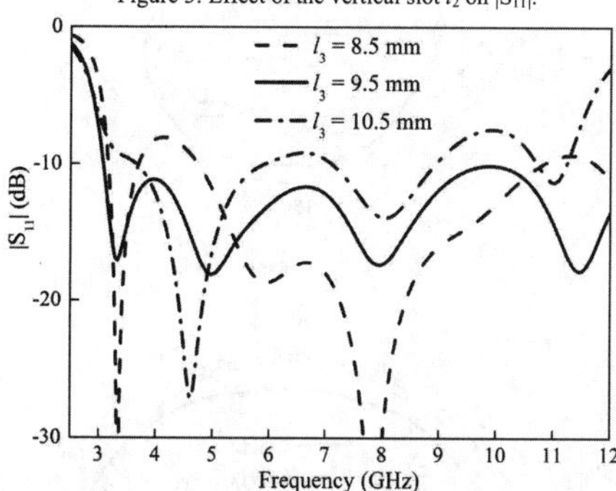

Figure 4. Effect of the narrow microstrip line l_3 on $|S_{11}|$.

in Figure 3 indicates that short vertical slot causes poor impedance matching in the upper band whereas impedance mismatching in the lower band is observed when l_2 is larger than 8.8 mm. It is also noticed that the lowest operating frequency for $|S_{11}| \leq$ -10 dB is ranging from 3.28 to 2.78 GHz when l_2 is increasing from 7.8 to 9.8 mm. The results in Figures 4-5 demonstrate that the effect of the microstrip line is similar to that of the horizontal slot. Limited bandwidth is obtained owing to the long microstrip line while short microstrip line will lead to poor impedance matching near 4 GHz. Interestingly, $|S_{11}|$ is larger than -10 dB near 11 GHz

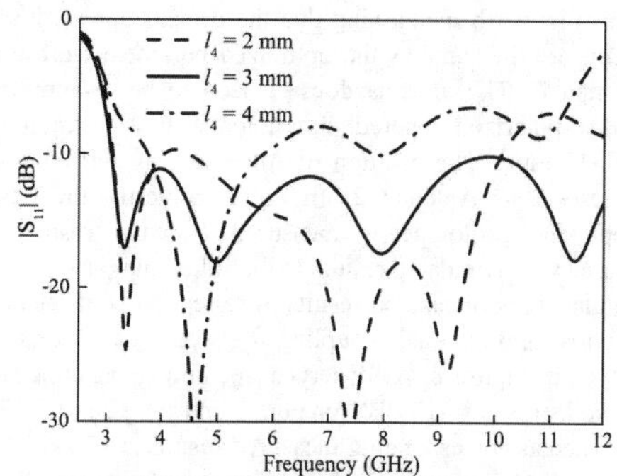

Figure 5. Effect of the wide microstrip line l_4 on $|S_{11}|$.

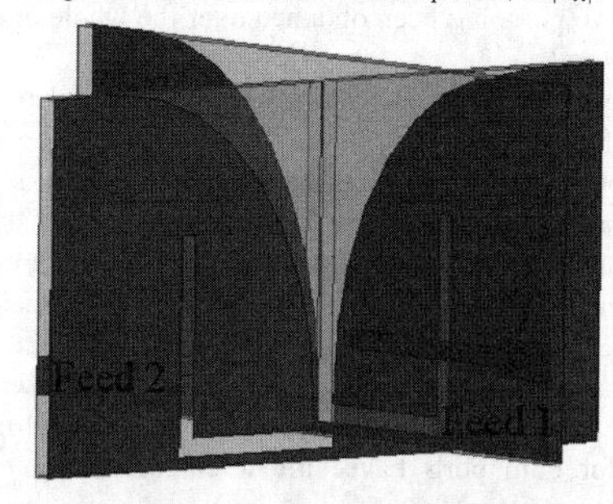

Figure 6. Geometry of the proposed dual-polarized tapered slot antenna.

Figure 7. Photo of the fabricated prototype and a coin.

when the length of the wide mcirsotrip line is less than 3 mm. It can be summarized that, good impedance matching within a wide frequency band can be achieved by adjusting the dimensions of the microstrip-line-to-slotline.

III. DUAL-POLARIZED TAPERED SLOT ANTENNA

The configuration of the proposed dual-polarized tapered slot antenna is shown in Figure 6. The dual-polarized tapered slot antenna can be easily obtained by shifting one element (Antenna 1) orthogonally into the other one (Antenna 2). As observed, the metallization of Antenna 2 is not interrupted as Antenna 1 can be easily inserted into Antenna 2 by cutting a slot in the substrate of Antenna 2. The photos of the fabricated prototype and a coin are shown in Figure 7. The compactness of the antenna can be clearly observed when comparing it with a coin. It is worth mentioning that the dimensions of the two elements are the same as the optimized parameters presented in Section II. The antenna doesn't need to be re-optimized. The dual-polarized tapered slot antenna has a volume of 30×30×16 mm^3. The position of Antenna 1 is shifted 1 mm with respect to Antenna 2. In order to verify the design concept, the prototype is measured and the results are compared with simulated results in the following part.

Simulated and measured results of reflection coefficient for two ports and mutual coupling between two ports are displayed in Figure 8. As observed, the simulated impedance bandwidths ($|S_{11}| \leq -10$ dB) for port 1 and port 2 are 3.05-12 GHz, whereas corresponding measured results are 3.3-12 GHz and 3.2-12 GHz, respectively. A high isolation of 20 dB between two ports has been obtained over the whole operating frequency range.

The measured radiation patterns in two principal planes at different frequencies (3, 6, 9 and 12 GHz) for two ports are shown in Figures 9 and 10, respectively. The results in Figure 9 show that, the pattern in the *E*-plane is shifted off the end-fire direction at 3 GHz and omni-directional radiation in the *H*-plane can be observed at 3 GHz. At other frequencies, the patterns focus on the end-fire direction and have higher front-to-back ratio compared to 3 GHz. Figure 10 presents the measured radiation patterns for port 2. It is observed that the patterns for both ports have similar characteristics as they have the same dimensions.

Figure 11 presents the simulated and measured gain response for port 1 and port 2. As observed, the gain variations of the two ports have the same trend across the frequency band. The measured gain is found to be from 2.5 to 6.8 dBi.

IV. Conclusion

A compact dual-polarized tapered slot antenna with a size of 30 × 30 × 16 mm^3 has been presented in this paper. The volume of the antenna is significantly reduced owing to the use of the folded balun which consists of a folded slot and a stepped microstrip line. The antenna is excited by the microstrip line through a slot. In addition, it can help to ease the fabrication complexity of the prototype. The proposed antenna can achieve a wide frequency band from 3.1 to over 12 GHz with the antenna gain varying from 2.5 to 6.8 dBi for both ports. The measured mutual coupling between two ports is below -20 dB across the operating frequency range.

Acknowledgment

Authors wish to thank Prof. Raed A. Abd-Alhameed and Dr. Chan H. See at University of Bradford, UK for their help in the antenna measurement.

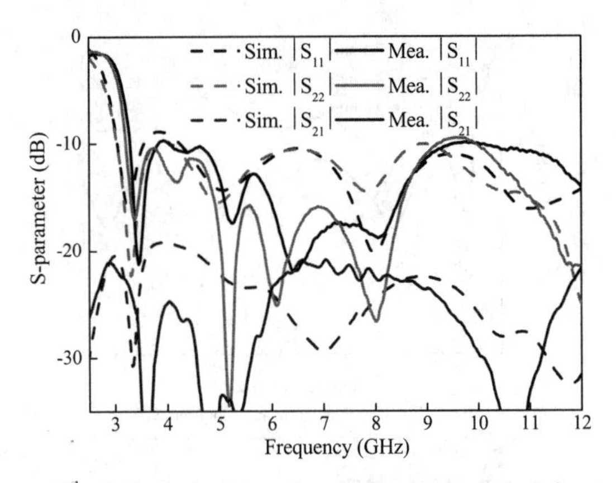

Figure 8. Simulated and measured results of S-parameters for the dual-polarized tapered slot antenna.

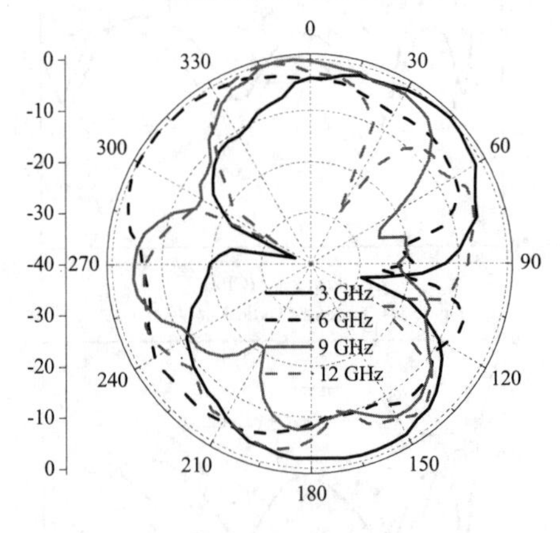

(a) *E*-plane

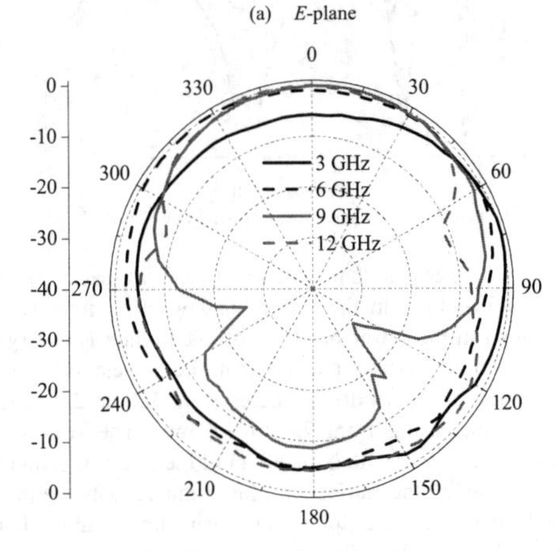

(a) *H*-plane

Figure 9. Measured radiation patterns at different frequencies for Port 1.

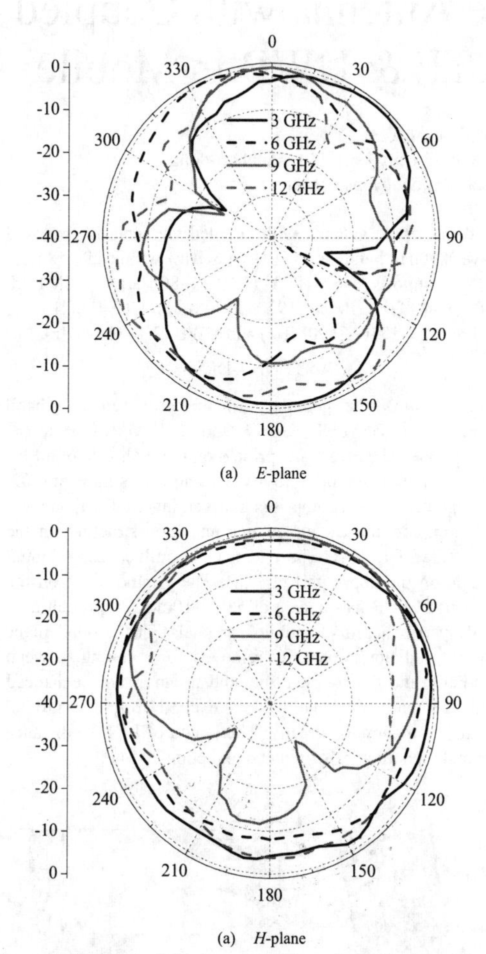

(a) *E*-plane

(a) *H*-plane

Figure 10. Measured radiation patterns at different frequencies for port 2.

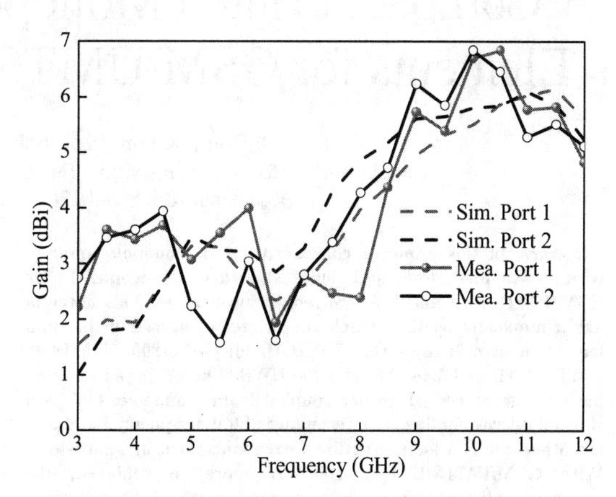

Figure 11. Simulated and measured gain response for port 1 and port 2.

REFERENCES

[1] I. J. G. Zuazola, J. M. H. Elmirghani, and J. C. Batchleor, "High-speed ultra-wide band in-car wireless channel measurements," *IET Commun.*, vol. 3, no. 7, pp. 1115-1123, 2009

[2] Y. Yang, and A. E. Fathy, "Development and implementation of a real-time see-through-wall radar system based on FPGA," *IEEE Trans. Geosci. Remote Sens.*, vol. 47, no. 5, pp. 1270-1280, 2009.

[3] J. Li, Z. F. Zeng, J. Sun, and F. Liu, "Through-wall detection of human being's movement by UWB radar," *IEEE Geosci. Remote Sens. Lett.*, vol. 9, no. 6, pp. 1079-1083, 2012.

[4] J. Shin, and D. H. Schaubert, "A parameter study of stripline-fed Vivaldi notch-antenna arrays," *IEEE Trans. Antennas Propagat.*, vol. 47, no. 5, pp. 879-886, 1999.

[5] P. J. Gibson, "The Vivaldi aerial," *Proc. 9th European Microw. Conf.*, pp. 101-105, 1979.

[6] X. Li, G. Adamiuk, M. Janson, and T. Zwick, "Polarization diversity in ultra-wideband imaging systems," *2010 IEEE International Conference on Ultra-Wideband (ICUWB)*, vol. 1, pp. 1-4, 2010.

[7] L. Zwirello, G. Adamiuk, W. Wiesbeck, and T. Zwick, "Measurement verification of dual-orthogonal polarized UWB monopulse radar system," *2010 IEEE International Conference on Ultra-Wideband (ICUWB)*, vol. 2, pp. 1-4, 2010.

[8] G. Adamiuk, T. Zwick, and W. Wiesbeck, "Dual-orthogonal polarized Vivaldi antenna for ultra-wideband applications," *17th International Conference on Microwaves, Radar and Wireless Communications*, pp. 1-4, 2008.

[9] G. Adamiuk, T. Zwick, and W. Wiesbeck, "Compact, dual-polarized UWB- antenna, embedded in a dielectric," *IEEE Trans. Antennas Propagat.*, vol. 58, no. 2, pp. 279-286, 2010.

A Compact Printed Monopole Antenna with Coupled Elements for GSM/UMTS/LTE & UWB in Mobile

OhBoum Kwon, Woojoong Kim, YoungJoong Yoon
Department of Electrical & Electronic Engineering, Yonsei University 134 Shinchon-dong,
Seodaemun-Gu, Seoul 120-749, Korea, yjyoon@yonsei.ac.kr

***Abstract*- In this paper, a compact printed monopole antenna with a coupled line and loop structure is proposed for GSM/UMTS/LTE and UWB in mobile application. This antenna has a monopole feeding which couples to a microstrip line and loop structures. It supports 1.7~2.7GHz for GSM1800, GSM1900, UMTS, LTE and 3.1~10.6 GHz for UWB. The proposed antenna has two structures which are coupled from monopole. One is a shorted microstrip line for a resonance of low frequency band and the other one is a loop structure for enhancement of bandwidth. With GSM/UMTS/LTE and UWB operation achieved, the proposed antenna only occupies a small space of 29 × 8 mm2 physical size. CST simulation was utilized in the design stage. The antenna was constructed on a substrate, FR-4, with the thickness of 0.8mm and relative permittivity of 4.3. The size of the substrate is 58 × 110 mm2.**

I. INTRODUCTION

Recently, wireless charging has become a hot topic in the mobile phone industry and it began to be applied to several models. Wireless charging will be generalized within a few years. USB port will be only used for Data Communication with PC. Thus, in addition to that the need is reduced, in terms of design and size, the need for wireless USB will become increasing. UWB that is described in wireless USB standard is expected to be applied to the mobile [1]. There are attempt to include UWB in USB dongles and mobile [2], [3].

For UWB, after approval of the FCC in the frequency band of 3.1~10.6GHz, many studies have been made on its application. However, to be applied to a mobile phone, more specialized environments must be considered. Already, mobile has supported several frequency bands like CDMA, GSM, UMTS and LTE. Furthermore, it must also support diversity and MiMO according to the communication method and additional services like NFC, GPS, WiFi and Wireless charging etc. Normally, LTE smartphone has 6 antennas or more. Therefore, adding antenna just for UWB is very difficult due to the reasons of a space and a number of antennas. It is the work that is meaningful for this reasons, it is to support as possible as small UWB and frequency of existing as an antenna. Recently, lots of researches have been carried out to provide small antennas for covering both UWB and existing frequency band. Various type of monopole antennas, such as an octagonal-shaped slot fed by a beveled and stepped rectangular patch with size 25 × 28mm2 [4] and ellipse-shaped monopole with size 30 × 30mm2 [5] and various type of slot antenna, such as a printed open and slot antenna with size 60 × 18.5mm2 [6], were proposed. However, they are still large to be applied to the mobile phone. In this paper, the compact printed monopole antenna with coupled elements is proposed to have 8.9GHz bandwidth(1.7~ 10.6GHz) within 29 × 8mm2 size. It supports GSM1800 (1710~1880MHz), GSM1900 (1850~1990MHz), UMTS (1920~2170MHz), LTE Band1, 2, 3, 4, 5, 9 and 25 (1710~2690MHz) and UWB (3.1~10.6GHz).

II. ANTENNA DESIGN

Fig. 1 shows the geometry of the proposed multiband antenna for the GSM/UMTS/LTE and UWB operation in the mobile phone. The antenna is printed on the both side of a FR4 substrate which has a permittivity 4.3 and a loss tangent 0.02. The proposed antenna comprises a driven inverted-L monopole on the top side and a coupled line and loop structure on the bottom side of a PCB. The antenna is confined 29 × 8mm2 rectangle on the upper part of PCB. The ground plane on the lower part of PCB has an area of 58 × 110mm2, representing a typical system board of mobile terminals. The ground plane below the radiator is partially removed for coupling between the radiator and the coupled line, and it can also be enhanced impedance matching. The feed is excited from the ground plane to the monopole directly. The length of L is the distance between loop structure's right end and coupled line.

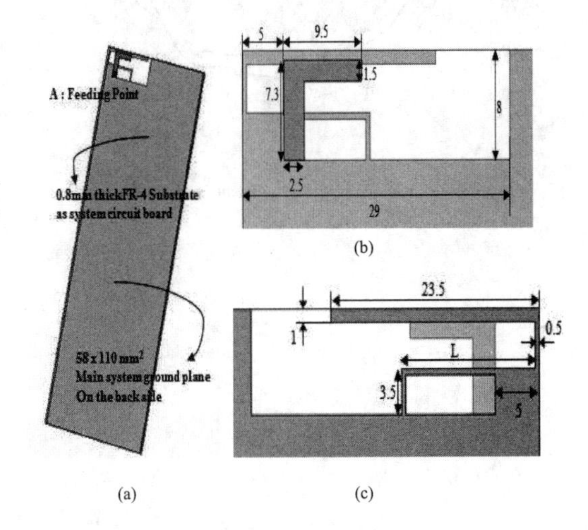

Fig. 1 Geometry of the proposed antenna
(a) Perspective view (b) Top view (c) Bottom view

978-7-5641-4279-7

III. Design Result And Discussion

Fig. 2 shows the simulated return loss of the proposed antenna. It shows ultra-wide operated bandwidth about 8.9GHz. With 3:1 VSWR (under -6dB) which is widely used for practical internal mobile application antenna. In case of UWB, with 3:1 VSWR definition is used for portable antenna [7]. Therefore, the return loss result of proposed antenna covers the desired low band 1.7~2.7GHz and UWB 3.1~10.GHz.

Fig. 3(c) shows the simulated return loss of the Ant1 (the printed monopole antenna) and the Ant2 (the printed monopole antenna and coupled line). Referring to Fig. 3(a) and (b), the Ant1 is single monopole type that has length 14.8mm only (about λ/4 at 4.2GHz). The Ant2 is capacitive coupled antenna with short strip on the bottom side of PCB. It is seen that the impedance level of the resonance at about 4.2GHz is effectively decreased owing to the use of the coupled line. The coupled line makes two resonances at about 3GHz and 10GHz. However, it cannot cover two desired band.

Fig. 4 shows the simulated return loss, compared with the Ant2, Ant3 and Ant4. The Ant3 is that the 1st loop structure is added to the Ant2. The loop structure is configured to be connected to the ground so along the left side of the monopole of the Ant2 on the bottom side of PCB. The Ant4 is that the 2nd loop structure is located along the right side of the monopole. Referring to Fig. 4 (c), it can be seen that the characteristics of the low band is widened by the addition of a loop structure. In Fig 4(d), each loop structure is large effects on the impedance matching of the low band. It improves the impedance matching for frequencies over the lower band and decreases the real part of impedance over the high band (8~10.6GHz).

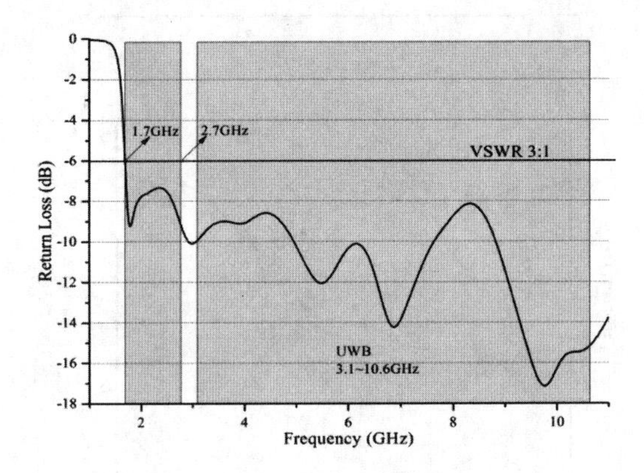

Fig. 2 Return Loss of Proposed Antenna

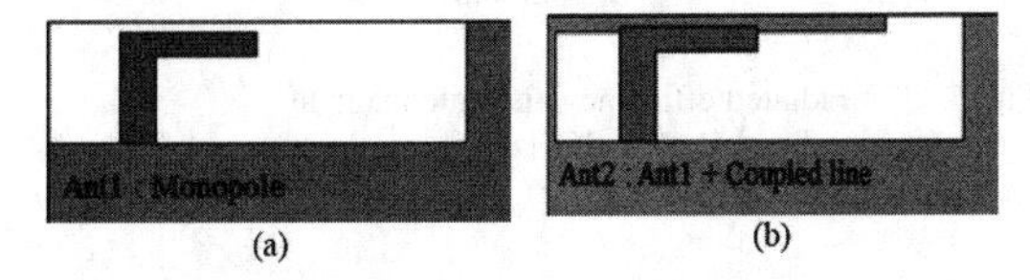

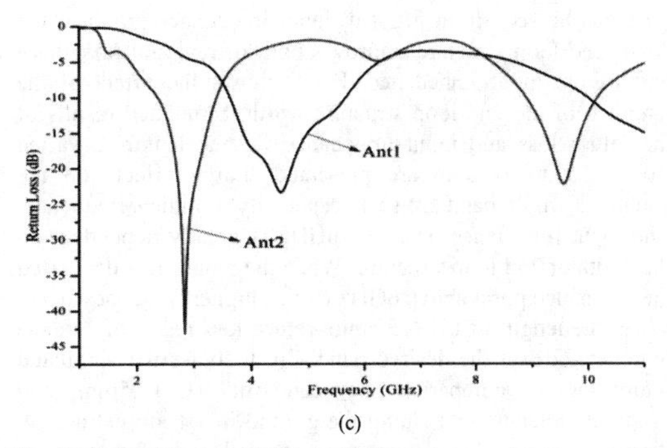

Fig. 3 (a) Ant1 Structure (b) Ant2 Structure
(c) Simulated Return Loss of Ant1 & Ant2

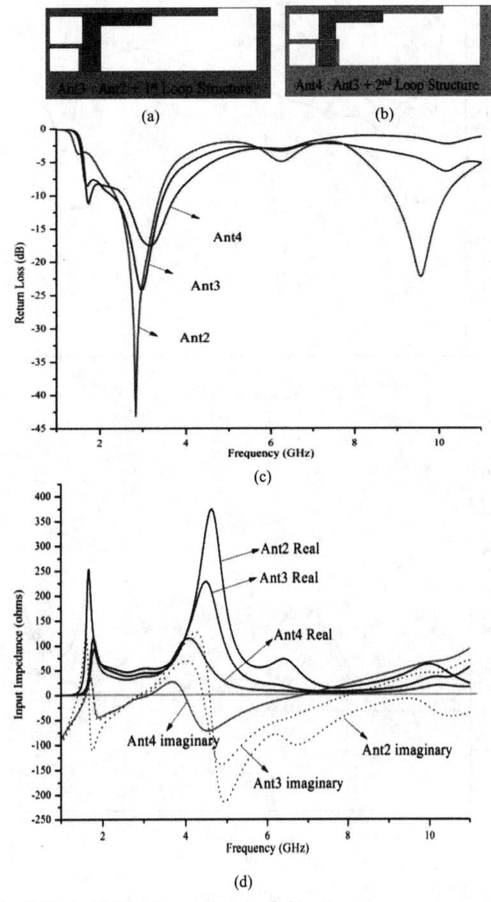

Fig. 4 (a) Ant3 Structure (b) Ant3 Structure
(c) Simulated Return Loss of Ant2, Ant3 & Ant4
(d) Input Impedance for the Ant2, Ant3 & Ant4

As can be seen from Fig 4(d) input impedance graph of the Ant4, 2nd loop structure improves both of real and imaginary part of the input impedance. Fig 5 shows the effect of the length L of the 2nd loop structure width. Simulated results of the return loss and input impedance for the length L varied from 12.5 to 18 mm are presented. Large effects on the antenna's whole band are seen, especially at higher frequency. The input impedance of above 6GHz is greatly dependent on the width of 2nd loop structure. When the length L is decreased, the resonance point above 6GHz moves higher frequency band. When the length of L is 15.5mm, return loss result of antenna is under -6dB at the desired band. Fig 6 shows the simulated return loss of comparison between Ant4 (L=15.5mm) and proposed antenna with filling the ground in 1st loop structure. Proposed antenna has a more good return loss performance on low frequency band (under 2.5GHz).

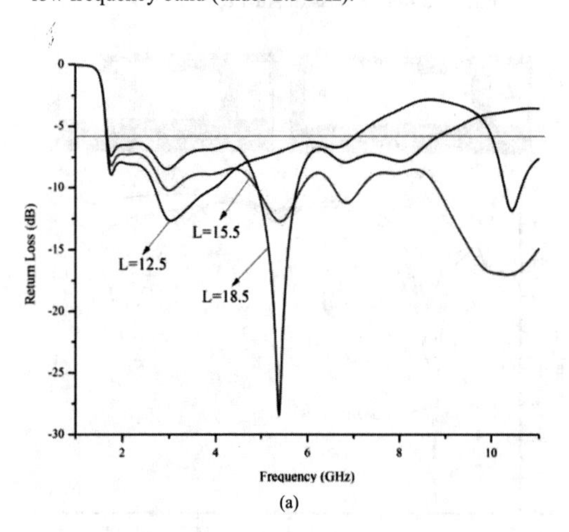

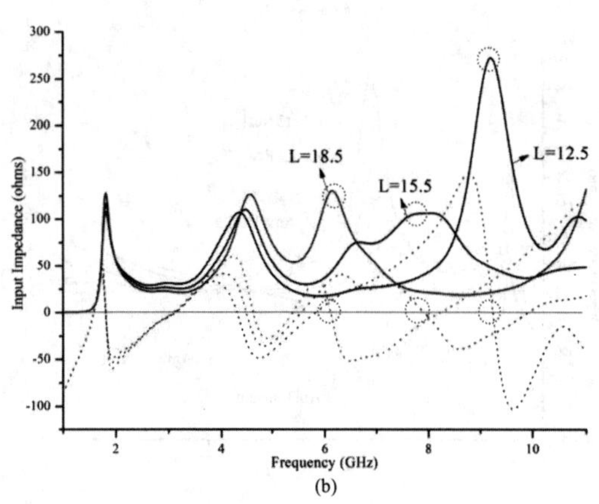

Fig. 5 Effect of 2nd loop structure length L
(a) Simulated Return loss (b) Input impedance

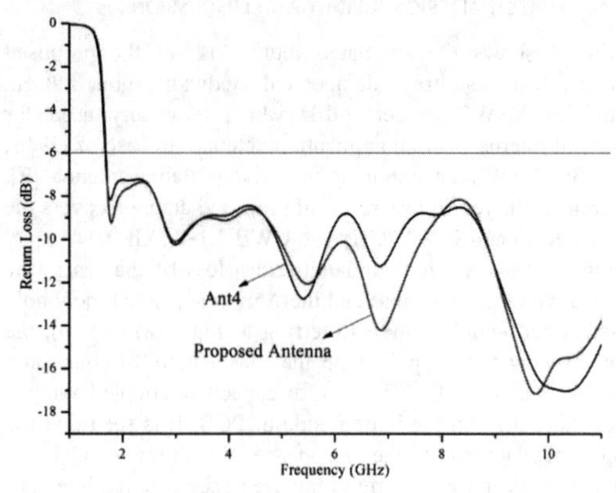

Fig. 6 Comparison between ant4 (L=15.5mm) and proposed antenna

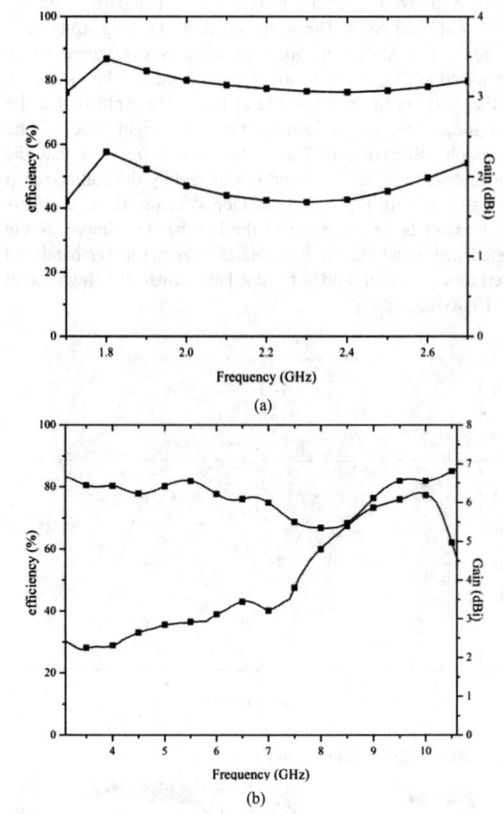

Fig. 7 The radiated efficiency and antenna gain
(a) Low band (1.7~ 2.7GHz)
(b) UWB (3.1~ 10.6GHz)

The simulated radiation efficiency and antenna gain are shown in Fig 7. The simulated radiation efficiency over the lower band (1.7~2.7GHz) and UWB is, respectively, about 76 ~ 87% and 66.7 ~ 85.8%. The simulated gain is about 1.68 ~ 2.3dBi and 2.2 ~ 6.25dBi over the lower band and UWB. Good radiation characteristics are generally obtained for the proposed antenna.

IV. Conclusion

This paper presents a new compact printed monopole antenna with a coupled line and loop structure for mobile. Using a coupled line and loop structure to improve the desired impedance matching bandwidths, which can generate one wide operating band about 1.7~10.6GHz to cover GSM1800, GSM1900, easily printed on the non-ground region of the system board of the mobile at low cost and occupies a very compact region of 28 × 8 mm2 only. It has a wide bandwidth performance and good radiation efficiency higher than 70% over the full operating bandwidth, the proposed antenna is attractive for mobile industry.

ACKNOWLEDGMENT

This work was supported by Defense Acquisition Program Administration and Agency for Defense Development under the contract UD130007DD.

References

[1] G. Eason, B. Noble, and I.N. Sneddon, "On certain integrals of Lipschitz-Hankel type involving products of Bessel functions," *Phil. Trans. Roy. Soc. London,* vol. A247, pp. 529-551, April 1955.

[2] D.D. Krishna, M. Gopicrishna, C.K. Aanaadan, P. Mohanan, and k.Vasudevan, "Ultra-wideband slot antenna for wireless USB dongle applicaions" Electron Lett 44, 2008, pp1057-1058

[3] S.-W.Su, J.-H.Chou, and K.L.Wong, "Internal ultrawideband monopole antenna for wireless USB dongle applications:, IEEE Trans Antennas Propaf 55 pp1180-1183

[4] M. Bod, H.R. Hassani, and M.M. Samadi Taheri, "Compact UWB Printed Slot Antenna With Extra Bluetooth, GSM, and GPS Bands" *IEEE Antennas And Wireless Propagation Letters, Vol. 11, pp 531-534*

[5] Guihong Li, Huiqing Zhai, Tong Li, Xiaoyan Ma, and Changhong Lian, "Design of a Compact UWB Antenna Integrated with GSM/WCDMA/WLAN Bands" *Progress In Electromagnetics Research, Vol. 136 , 2013 pp 409-419.*

[6] Wei-hua Zong, Xiao-yun Qu, Yong-xin Guo and Ming-Xin Shao "An Ultra-Wideband Antenna for Mobile Handset Applications" *Advanded Materials Research Vols. 383-390 2012,pp4457-4460.*

[7] Ian Oppermann, Matti Hamalainen and Jari linatti "UWB theory and applications" *Wiley, p132*

A Compact Triple-band Monopole Antenna for WLAN/WIMAX Application

Hui-Fen Huang, Shao-Fang Zhang
South China University of Technology
Guangzhou, 510641 China

***Abstract*-A compact triple-band monopole antenna is proposed to cover the frequency band for WLAN/WIMAX application in this paper. The triple-band monopole antenna proposed in this letter has the advantage of compact size and relatively wide bandwidth. The proposed antenna consists of a modified E- shaped radiation patch and slotted ground plane. It has a compact size of only $15\times30\times1.2$ mm^3. The compactness of the triple-band antenna is achieved by adding slots on the ground plane. The triple-band antenna has three operation bands, ranging from 2.4 GHz to 2.493 GHz, 3.01 GHz to 3.84 GHz and 4.89 GHz to 6 GHz, which can apply for 2.4/5.2/5.8 GHz WLAN bands (2.4-2.484 GHz, 5.15-5.35 Hz, and 5.725-5.825 GHz) and 3.5/5.5 GHz WIMAX (3.3-3.79 GHz and 5.25-5.85 GHz) band. The operating mechanism and the effect for parameters on the antenna performance are studied. Experimental results show that the antenna gives omnidirectional radiation patterns and good antenna gains in the operating bands.**

I. INTRODUCTIN

Wireless communication systems have experienced enormous growth over the last decade. Wireless local area network (WLAN) and the Worldwide Interoperability for Microwave Access (WiMAX) technology have extensively been used in commercial, medical, and industrial applications. The WLAN has a lower frequency band 2.4–2.484 GHz for the 802.11b/g standards, and two higher frequency bands 5.15–5.35 GHz and 5.725–5.825 GHz for the 802.11a standard. The WIMAX has a lower frequency band 3.3-3.79 GHz and a higher frequency band 5.25-5.85 GHz for the 802.16 standard. The multiband antenna must have simple structure, small size, light weight, low cost, and easily integrated with RF circuits. So far, numerous antennas have been proposed to reduce size, enhance bandwidth for WLAN application [1-14]. Different shapes of monopole antenna used to apply for WLAN application, such as G-shaped, F-shaped, 9-shaped, T-shaped and L-shaped [10]–[13]. However, the sizes of most antennas are large. For WLAN and WIMAX application, the smaller size of antenna for 2.4/3.5/5.2/5.8 GHz wireless applications is 22×29 mm^2 [14], which is still larger than our proposed antenna of only size 15×30 mm^2.

In this paper, a compact dual-band monopole antenna is proposed to cover the frequency band for 2.4/3.5/5/2/5.8 GHz for WLAN and WIMAX application. The designed antenna has wider bandwidth for WIMAX application compared to antenna [6]. In the mean time, it has a compact size

Fig.1 Geometry of the p

of $15\times30\times2$ mm^3, which is smaller than the antenna [14]. The triple-band antenna has three operation bands, ranging from 2.40 GHz to 2.493 GHz, 3.01 GHz to 3.84GHz and 4.89 GHz to 6 GHz which can apply for 2.4/5.2/5.8 GHz WLAN bands (2.4-2.484 GHz, 5.15-5.35 GHz, and 5.725-5.825 GHz) and 3.5/5.5 GHz WIMAX (3.3-3.79 GHz and 5.25-5.85GHz) band. The operating mechanism and the effect for parameters on the antenna performance are studied. The triple-band antenna is fabricated and experimental results are presented. Experimental results show that the antenna gives omnidirectional radiation patterns and good antenna gains over the operating bands. The design of antenna and operating mechanism are shown in Section II. Experimental results are shown in Section III. The conclusion is in Section IV.

II. ANTENNA DESIGN

A. Antenna Configuration

The geometry of the proposed antenna is shown in Fig. 1 and the parameters are given in Table I. The antenna is constructed on the FR-4 substrate with dielectric constant of 4.4, loss tangent 0.02, thickness of 1.2 mm. The antenna has a compact size of 15×30 mm^2. The antenna consists of a modified E-shaped radiation patch on the top side of the substrate and slots on the ground plane. The point A is used to excite the resonant modes at 5.55 GHz, and the point B and the point C on the slotted ground excite the resonant modes for 3.3 GHz and 2.45 GHz, respectively.

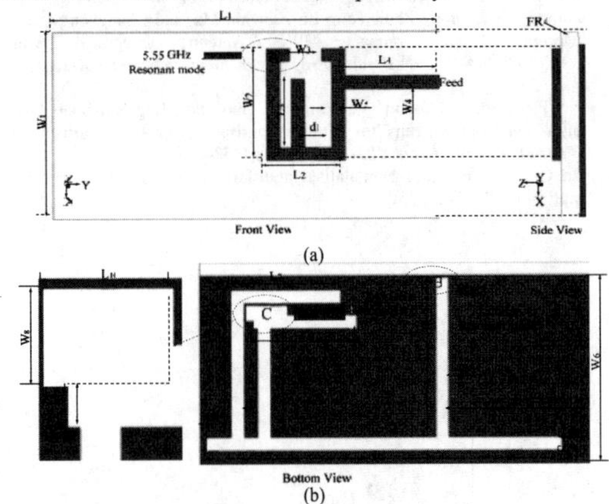

Fig.1 Geometry of the proposed antenna (a) Front view and Side View (b) Bottom View

978-7-5641-4279-7

TABLE I

PARAMETERS OF THE PROPOSED ANTENNA (UNITS: MM)

W_1	L_1	W_2	L_2	W_3	L_3	d_1
15	30	9	6	2.5	5.5	2.1
W_4	L_4	W_5	d_2	d_3	d_4	d_5
1.2	7.5	1	2.4	1	0.3	0.6
d_6	d_7	d_8	d_9	W_6	W_7	W_8
0.4	11	2.3	1.1	14	1	1.2
L_5	L_6	L_7	L_8	L_9	L_{10}	L_{11}
27.2	10.9	8.4	8.1	7.9	3.2	11.8

B. *Current Distribution analysis of the antenna*

To further examining the mechanism of the proposed antenna, the surface current distributions on both the radiator and the ground is studied. Fig. 2 shows the current distributions for the three resonant frequencies at 5.55, 3.3, and 2.45 GHz. At 5.55 GHz, the current is mainly distributed along the point A of modified E-shaped patch, as is shown in Fig.2 (a). At 3.3 GHz, the current is mainly distributed along the point B of slot L_{11}, as is shown in Fig. 2 (b). At 2.45 GHz, the current was mainly distributed along the point A of slots L_9 and L_{10}, as is shown in Fig. 2 (b).

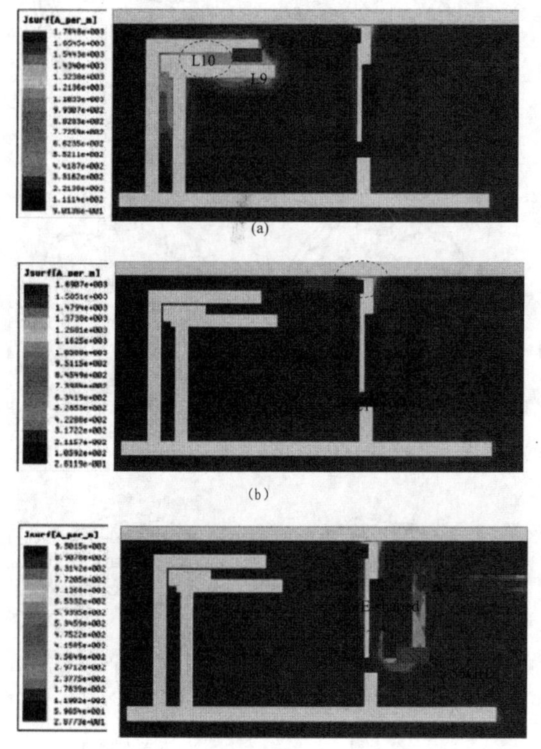

Fig.2 Simulated current distribution at (a) 2.45 GHz (b) 3.3 GHz (c) 5.55 GHz.

C. *Simulated and measured Reflection Coefficient*

The compact triple-band monopole antenna for WLAN / WI-MAX application is simulated by High-Frequency System Simulation software. The prototype of the proposed antenna is fabricated and measured. Fig. 3 is the prototype of the antenna. The simulated and measured S11 for the antenna is shown in Fig. 4.The S11 of the compact antenna is measured by using an Advantest R3770 network analyzer. It is observed that the proposed triple-band antenna has three operation bands, ranging from 2.4 GHz to 2.493 GHz, 3.01 GHz to 3.84 GHz and 4.89 GHz to 6 GHz. The discrepancy between the measured and simulated results is due to fabrication and measurement deviation.

D. *Parametric Analysis*

The parameter effects on the impedance are shown in Fig.5 (a)-(c). First, for modified E-shaped patch, the influence of L_3 is shown in Fig. 5 (a). L_3 for E-shaped patch only affects the higher frequency at 5.55 GHz. The resonant mode moves towards lower frequency with the increasing of L_3. Second, for the slots, Fig. 5 (b) presents the effect for the slot L_9.The lower frequency at 2.45 GHz moves down with L_9 increasing, but the resonant modes at higher frequency don't change. Third, theinfluence for the slot L_{11} is also observed in Fig. 5 (c). With L_{11} decreasing, the band of the higher frequency 3.3 GHz and 5.55GHz go narrower. It is observed that the resonant mode at 2.45GHz is affected by slot L_9, the resonant mode at 3.3 GHz is affected by slots L_5 and L_{11}, and the resonant mode at 5.55GHz is affected by the branch L_3 of the modified E-shaped patch and slots L_5, L_9, and L_{11}.

Fig.3 The fabricated antenna.

Fig.4 Simulated and measured S11 of the proposed antenna

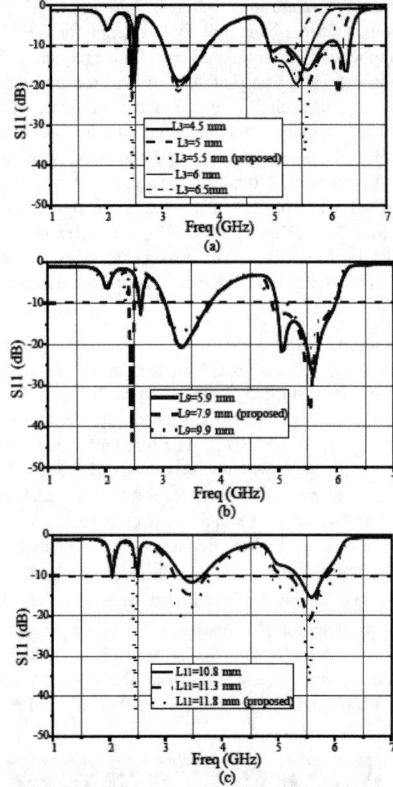

Fig.5 Parameters studies of the antenna (a) L_3 (b) L_9 (c) L_{11}

III. EXPERIMENTAL RESULTS

The radiation characteristics such as radiation pattern and peak gain across the three operating bands for the proposed antenna have been measured. The simulated and measured radiation patterns including the vertical (E_θ) and horizontal (E_ψ) polarization in the x-z plane and y-z plane for the antenna at the three resonant modes of 2.45, 3.3, and 5.55 GHz are shown in Fig. 6 (a)–(c), respectively. First, the radition patterns are nearly omnidirectional in the x-z plane which is similar to monopole-like radiation pattern and nearly dumbbell in the y-z plane. The radiation patterns are quite stable throughout the WLAN and WiMAX bands. Second, the average peak gain for the triple band at 2.45, 3.3, and 5.55 GHz are about -8.45, 1.83, and 1.12 dBi, and the average radiation efficiency of the compact dual-band monopole antenna is almost 85%. Considering the small size of the antenna, the efficiency and gain values are acceptable.

IV. CONCLUSION

A compact triple-band monopole antenna for WLAN/ WIMAX application is proposed, analyzed and measured. The triple-band monopole antenna proposed in this letter has the advantage of compact size 15×30×1.2 mm^3 and relatively wide bandwidth. The compactness of the triple-band antenna is achieved by adding slots on the ground plane. The operating mechanism and the effect for parameters on the antenna performance are studied. By controlling the parameters of the antenna, the operation bands can be changed. The triple- band antenna is fabricated, and experimental results are presented. Experimental results show that the compact triple-band antenna gives omnidirectional radiation patterns and good antenna gains over the operating bands.

ACKNOWLEDGMENT

This work is supported by the National Natural Science-Foundation of China under Grant 61071056.

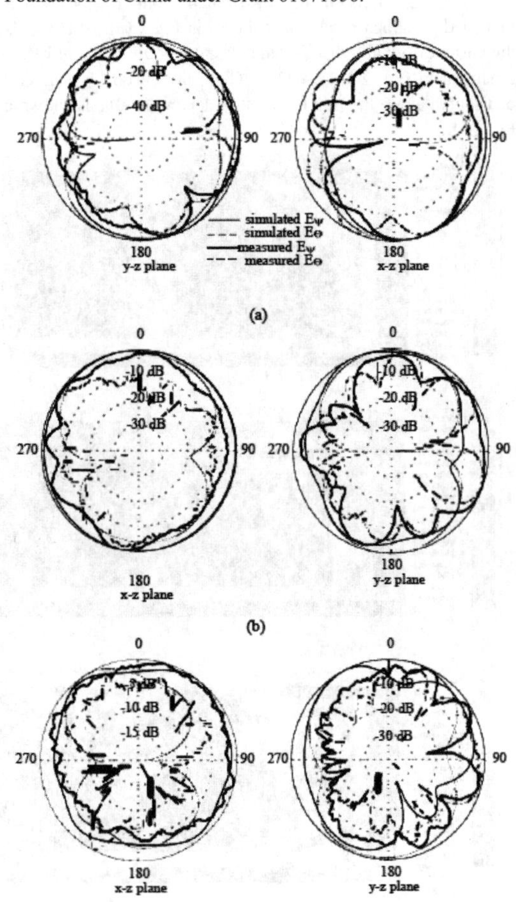

Fig.6 Simulated and measured radiation patterns of the antenna (a) 2.45 GHz (b) 3.3 GHz (c) 5.55 GHz.

REFERENCES

[1] Y. J. Wu, B. H. Sun, J. F. Li, and Q. Z. Liu "Triple-band omni-directional antenna for WLAN application," Progress Electromagn. Res.,vol.76, pp. 477–484, 2007.

[2] X. L. Quan, R. L. Li, Y. H. Cui, and M. M. Tentzeris, "Analysis and Design of a Compact Dual-Band Directional Antenna," IEEE Antennas Wireless Propag. Lett. ,vol. 11, pp. 547-550,2012

[3] P. L. Shu, and Q. Y. Feng, "Design of a compact quad- band hybrid antenna for Compass/WIMAX/WLAN application," Progress Electromagn. Res.,vol.138, pp. 585–598,2013.

[4] J. R. Panda and R. S. Kshetrimayum, "A printed 2.4 GHz/5.8 GHz dual-band monopole antenna with a protruding stub in the ground plane for WLAN and RFID," Progress Electromagn. Res.,vol.117, pp. 425–434, 2011.

[5] C. S. Chuang, B. S. Yang, "Compact dual broadband planar slot antenna for wireless LAN application," IEEE International Conference Electrical and Electronics Engineering (ELECO), PP. 223-225, 2011.

[6] W. C. Liu, C. M. Wu, and Y. Dai, "Design of triple- frequency microstrip-fed monopole antenna using defected ground structure," IEEE Trans. Antennas Propag., vol. 59, no.7, pp. 2457-2463, Jul. 2011.

[7] X. L. Sun, L. Liu, S. W. Cheung, and T. I. Yuk, "Dual- band antenna with compact radiator for 2.4/5.2/5.8 GHz WLAN applications," IEEE Trans. Antennas Propag., vol. 60, no. 12, pp. 5924-5931, Dec. 2012.

[8] H. F. Abutarboush, H. Nasif, R. Nilavalan and S. W. Cheung, "Multiband and wideband monopole antennafor GSM900 and other wireless applications," IEEE Antennas Wireless Propag. Lett. , vol. 11, pp. 539-542, 2012.

[9] D. Yu, Z. H. Zhang, W. L. Liu, B. Yang, "Dual-band miniaturized printed antenna for WLAN applications," IEEE International Conference on Microwave and Millimeter Wave Technology (ICMMT), vol. 3, pp. 1-4, 2012.

[10] W. C. Liu, "Optimal design of dualband CPW-feed G- shaped monopole antenna for WLAN application Application," Progress Electromagn. Res.,vol.74, pp. 21–38, 2007.

[11] S. H. Huang. Yeh and K. L. Wong, "Dual-band F-shaped monopole antenna for 2.4 /5.2 GHz WLAN application," IEEE Antennas and Propagation Society International Symposium (APSURSI), vol.4, pp. 72-75, July. 2002.

[12] A. S. R. Saladi, J. R. Panda and R. S. Kshetrimayum, "A Compact printed 9-shaped dual-band monopole antenna for WLAN and RFID applications," IEEE International Conference on Computing Communication and Networking Technologies (ICCCNT), pp. 1-4, Jul. 2010.

[13] E. Debono, A. Muscat and C. J. Debono, "Dual frequency 2.4 GHz T-shaped and 5.2 GHz L-shaped nonopole antenna for WLAN applications," IEEE International Conference on Microwaves, Radar & Wireless Communications, pp. 1015-1018, May. 2006.

[14] A. Mehdipour, A. Sebak, W. Trueman and T. A. Denidni, "Compact multiband planar antenna for 2.4/3.5/5.2/5.8-GHz wireless applications," IEEE Antennas Wireless Propag. Lett. , vol. 11, pp. 144-177,Jan. 20

A Hexa-band Coupled-fed PIFA Antenna for 4G Mobile Phone Application

Zhenglan Xie, Wenbin Lin*,
Institute of Electromagnetic of Southwest Jiaotong University,Chengdu,610031,China

Guangli Yang*
Motorola Solutions Inc., Holtsville, NY 11742

***Abstract*-A coupled-fed microstrip antenna for 4G mobile system is presented. The PIFA antenna operates at 704-960 MHz and 1710-2170 to cover frequency bands of LTE 700/850/900/1800/1900/2100(Tx/Rx). An L-shaped coupled-fed structure, an inductor and a shorting line to ground are compactly designed together to create dual resonances at both low and high bands to achieve wide bandwidth and better impedance matching. The overall dimension of mobile system is 58×119×1.2mm^3 and the antenna size occupies area of 58×16mm^2.**

***Index Terms*-PIFA (planar inverted-F antenna); coupled-fed; LTE(Long Term Evolution); multiband**

I. INTRODUCTION

With the vast development of science and technology and people's increased demands on communication, today's mobile terminals need to operate in more than one standard, such as WAN, WLAN, GPS, BT and RFID [1-5]. Particularly, WAN is more important to support multimode and multiband for voice and data communications. The new generation mobile terminal typically needs to support 2G, 3G and 4G network, including LTE700(704-787MHz), GSM850(824-894MHz), GSM900(880-960MHz), GSM1800(1710-1880MHz), GSM1900(1850-1990MHz) and UMTS(1920-2170MHz). How to design a compact and multiband antenna is always a challenge.

In this paper, a coupled-fed [6-8] PIFA antenna is proposed. The PIFA consists of a ground plane, a top patch, a feed wire, a shorting line, an inductor and a plastic housing. The coupled-fed method and an inductor are used to improve matching and increase bandwidth. The shorting line can add one resonant mode. Section □ describes the details of each part of the proposed antenna design. Section III analyze and discuss the simulation results, and lastly a conclusion is summarized in section □.

II. ANTENNA DESIGN

The configuration of the proposed PIFA antenna for mobile phone application is given in Figure 1(a). The substrate is put in the center of the 1 mm thick plastic box with relative permittivity of 3, loss tangent of 0.02, and dimension of 58×119×11.2mm^3. In this paper, the PIFA antenna is printed on a 1.2 mm thick FR4 material substrate (size 58×103 mm^2). The relative permittivity of FR4 is 4.4 and dielectric loss tangent is 0.02. Figure 1(b) shows the ground plane, which is under the substrate.

Figure 1(c) shows the detailed dimensions of the antenna pattern in its planar structure. For proof of concept study, the antenna material setting is simplified as PEC. The antenna is printed on the non-ground portion mostly and occupies a wide radiating plate of size 16×58 mm^2. The L-shaped coupled-fed is used to replace direct-fed to increase the resonant bandwidth significantly. Point A of L-shaped feeding strip is the antenna's feeding point. The antenna and ground plane are connected by shorting point B. In addition, to ensure enough electrical length and bandwidth, a chip inductor with size of 1.7×2 mm^2 and inductance of 3.2 nH is proposed at point C.

Fig. 1(a). Geometry of the proposed antenna.

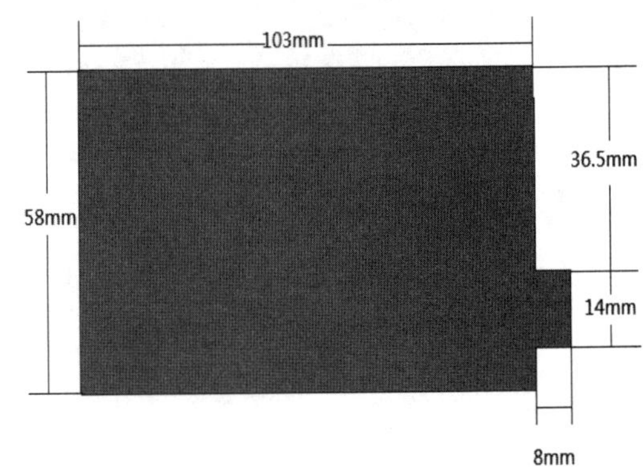

Fig. 1(b). Geometry of the ground plane.

* Corresponding authors: Wenbin Lin (wl@swjtu.edu.cn) and Guangli Yang (guangli.yangs@gmail.com)
978-7-5641-4279-7

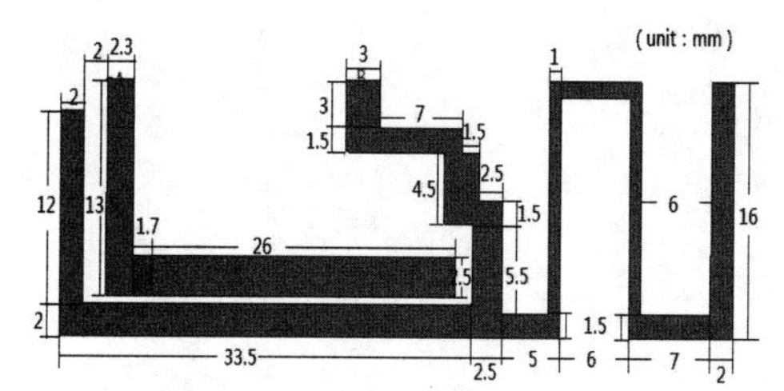

Fig.1(c). Detailed dimensions of the antenna pattern.

III. RESULTS AND DISCUSSION

The proposed antenna for multiband phone application was studied and simulated by software. Figure 2(a) gives the simulated return loss (3:1 VSWR) of L-shaped coupled-fed method. Obviously, the operating bandwidth and impedance matching of direct-fed is quite poor. The lower and higher band are formed by two adjacent resonant modes respectively where low band covers bandwidth from 683 MHz to 977MHz and high band covers from 1697 MHz to 2379MHz. This can be explained by the following: For the low band, the first resonance is from feeding point A coupling to bottom arm toward the meander line end, similarly like a traditional inverted L antenna. The second resonance is from coupled structure between feeding point A and ground point B, this is evidenced by the frequency shift observed in experiment when changing coupling gap and inductor value. Properly tuning the two resonances closer can enhance low band bandwidth reported in Fig.2 (a). For the high band, the first resonance is from the third resonance of long inverted L meander line arm, the second resonance is from coupling feed A connected arm itself.

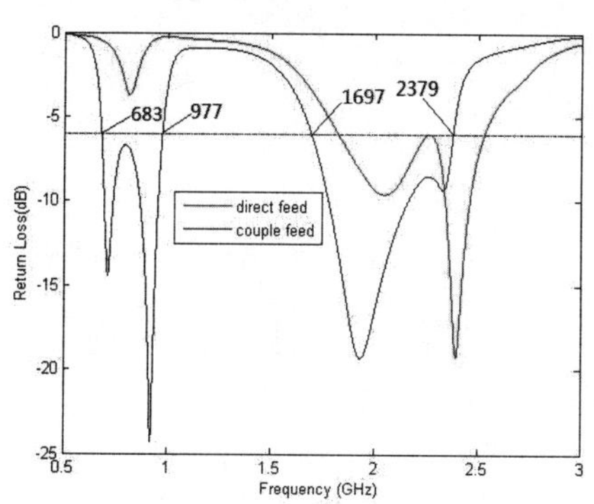

Fig.2(a). Simulated return loss of the proposed antenna.

Figure 2(b) shows comparisons of return loss with different inductance, the inductance varies from 2.2nH, 2.7 nH, to 3.2nH. As we can see, the inductor can shift the resonant frequency because it increases the effective electrical length. When the inductance is higher, the effective electrical length is longer, therefore the resonant frequency is shifted lower.

To better understand the antenna principle, Figure 3 shows the surface current distribution simulated on the proposed antenna's top radiator at f1=0.7GHz and f2=1.9GHz. The red part indicates a strong field and the blue part indicates a weak field. As we can see, the strongest field of low frequency is at grounding line whereas the strongest field of high frequency is distributed on feeding line, which further proves our explanations of this antenna.

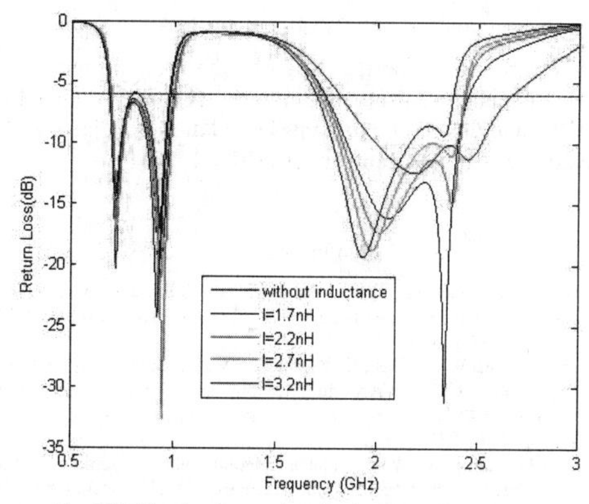

Fig. 2(b). Simulated return loss for the proposed antenna.

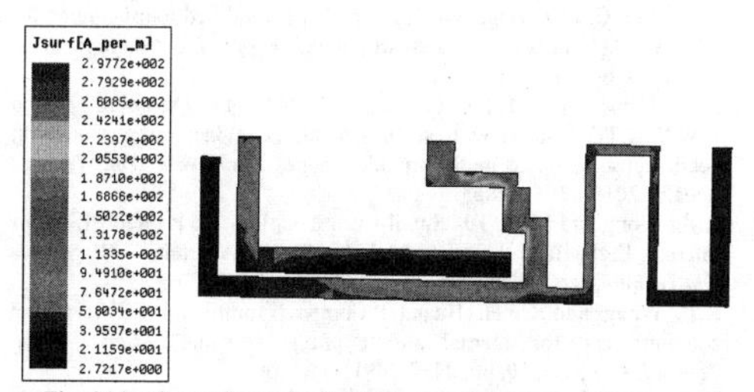

Fig .3(a). The surface current distribution simulated on the proposed antenna's top radiator at f1=0.7GHz.

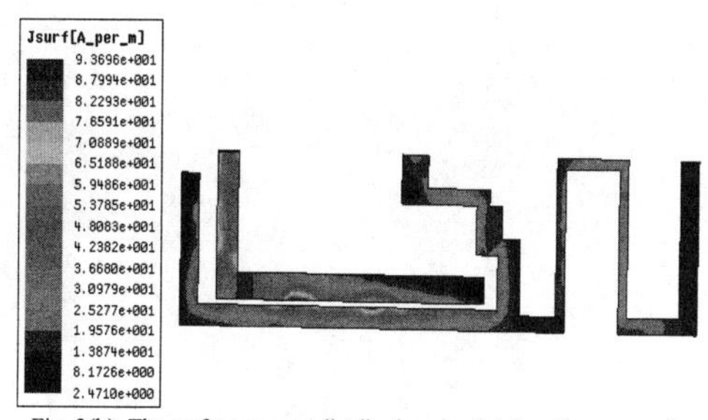

Fig. 3(b). The surface current distribution simulated on the proposed antenna's top radiator at f1=1.9GHz

IV. CONCLUSIONS

In this paper, a compact L-shaped coupled-fed PIFA antenna is proposed. An inductor, a shorting line and coupled-fed are designed together to create dual resonances at both low and high bands and better impedance matching. The antenna covers very wide bandwidth from 683 MHz to 977MHz for low band and 1697 MHz to 2379MHz for high band good for potential LTE700, GSM850, GSM900, GSM1800, GSM1900, UMTS applications. The future work will focus on antenna prototype and efficiency measurement and report results shortly.

ACKNOWLEDGEMENT

This work was supported in part by the Program for New Century Excellent Talents in University (Grant No. NCET-10-0702) and the Ph.D. Programs Foundation of Ministry of Education of China (Grant No. 20110184110016).

REFERENCES

[1] C.H. Chang and K.L. Wong, "Printed λ/8-PIFA for penta-band WWAN operation in the mobile phone," *IEEE Trans Antennas Propagat.*, vol. 57, pp. 1373-1381, 2009.

[2] K.L. Wong, W. Y. Chen, C. Y. Wu, and W. Y. Li, "Small-size internal eight-band LTE/WWAN mobile phone antenna with internal distributed LC matching circuit," *Microwave Opt. Technol. Lett.*, vol. 52, pp. 2244–2250, Oct. 2010.

[3] C.T.Lee and K. L. Wong, "Planar monopole with a coupling feed and an inductive shorting strip for LTE/GSM/UMTS operation in the mobile phone," *IEEE Trans. Antennas Propag.*, vol. 58, no. 7, pp. 2479–2483, Jul. 2010.

[4] C.W.Chiu, C. H. Chang, and Y. J. Chi, "A meandered loop antenna for LTE/WWAN operations in a smart phone," *Progr. Electromagn. Rese. C*, vol. 16, pp. 147–160, 2010.

[5] K.L. Wong, M.F. Tu, C.Y. Wu, and W.Y. Li, On-board 7-band WWAN/LTE antenna with small size and compact integration with nearby ground plane in the mobile phone, *Microwave Opt Technol Lett*52 (2010), 2847–2853.

[6] K.-L. Wong and M.-F. Tu "Small-sized Coupled-Fed Printed PIFA For Internal Eight-Band LTE/GSM/UMTS Phone Antenna," *Microwave Opt.Technol.Lett*,vol.52,pp.2123-2128,September 2010.

[7] K.L. Wong and C. H. Huang, "Compact multiband PIFA with a coupling feed for internal mobile phone antenna," *Microw. Opt. Technol. Lett.*, vol. 50, pp. 2487-2491, Oct. 2008.

[8] L.Lu and J. C. Coetzee, "A modified dual-band microstrip monopole antenna," Microw. Opt. *Technol. Lett.*, vol. 48, pp. 1401-1403, Apr. 2006.

A Single Feed Circularly Polarized RFID Reader Antenna with Fractal Boundary

Ming Hui Cao[1], Zhuo Li[2]
College of Electronic and Information Engineering, Nanjing University of Aeronautics and Astronautics; Nanjing ,China
[1]cmhhgd123@163.com
[2]lizhuo@nuaa.edu.cn

Abstract- ***A single feed circularly polarized RFID reader antenna with Minkowski fractal boundary is proposed. A 2-itration Minkowski patch is adopted for antenna compactness and four rectangular slots are etched on the ground to improve the antenna performance. The simulation result shows that this antenna behaves good characteristics in the design band, which has a -10dB impedance bandwidth of 52MHz(906-958MHz),3dB (Axial Ratio)AR bandwidth of 12MHz(917-929MHz). This design shows that the antenna structure is simple, easy for manufacture and integration.***

Index terms – circularly polarized, fractal, RFID , reader antenna , slot

I. INTRODUCTION

In recent years, radio frequency identification (RFID) technique is widely used in the logistic, distribution market, merchant flow tacking and other areas. The RFID reader antenna, which is the critical component in the RFID reader system, has been thoroughly investigated. Generally, the tag antennas are arbitrarily oriented. So it's advisable that a circularly polarized reader antenna is needed for reliable detection. Microstrip patch antenna have many advantages with low profile, low cost, easy fabrication and easy to realize dual-frequency and circular polarization [1]. These properties make compact microstrip very popular and attractive for RFID and wireless communication system.

The "fractal" concept was first proposed by B. Mandelbrot in 1975 [2]. Most fractal objects have self-similar shapes and space filling ability. This property can achieve antenna miniaturization and band improvement. Fractal theory has been widely used in antenna design. Monopole antenna and fractal loop antenna have achieved remarkable progress in reducing antenna's size [3] [4]. A wideband Minkowski fractal dielectric resonator antenna is proposed. Parametric study is carried out to investigate the antenna design [5]. By replacing each side of square patch with Koch curve of 2nd stage and having fractal slot of same indentation angle but scaled down in the centre of the radiation patch, the proposed antenna get 1.2% 3dB AR bandwidth[6],but this antenna have a big slot dimensions sensibility, this would increase the difficulty of fabrication.

In this paper, we propose a compact, circularly polarized mircostrip RFID reader antenna based on Minkowski structure. The antenna is fed by coaxial probe along the patch diagonal axes. The antenna can achieve circularly polarized radiation by trimming resonance length. Because of folding boundary of fractal, the gain and radiant efficiency of dominant mode decrease. For compensation of the antenna radiation pattern, four identical rectangular slots are etched on the ground for lower the dielectric constant of the antenna equivalently [7].

II. DESIGN OF ANTENNAWIHT MINKOWSKI FRACTAL BOUNDARY

The generation of Minkowski fractal patch

The construction of many ideal fractal shapes is usually carried out on initiator by applying an infinite number of times (iterations). The iteration determine the inner structure of the fractal graph. The initiator and iteration of the Minkowski fractal curve are depicted in Fig. 1. Suppose that the initiator length is L, displace the middle one-third of each straight segment by the fold line. Indentation factor ρ is defined here as the ratio of indentation width to the indentation length of the fold line. Changing the indentation factor causes a shift in the resonant frequencies, we can get different 1st Minkowski fractal curve.

The Minkowski fractal patch can be obtained by replacing each side of square patch by Minkowski fractal curve of 2nd stage. The generation procedure for the patch with Minkowski

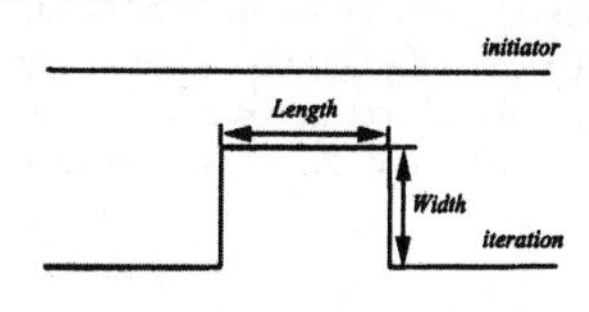

Fig.1. initiator and iteration of the Minkowski fractal curve

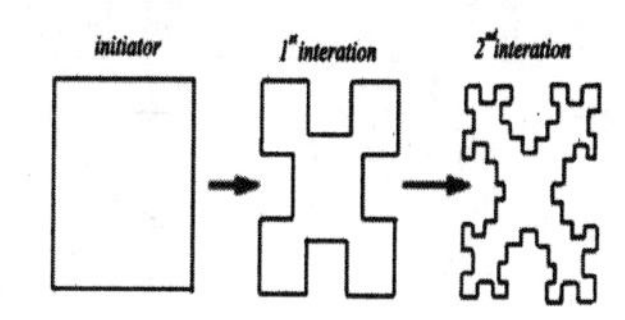

Fig.2 . Generation procedure for the patch with Minkowski fractal boundary

978-7-5641-4279-7

fractal boundary is depicted in Fig. 2. The N order patch antenna with ideal Minkowski boundary can be achieved by N order iteration, but it is impossible to get an infinite fractal structure along with increasing the difficulty of fabrication and appearance of more high order mode which lead to decrease of the antenna radiant efficiency. This paper discussed the compact microstrip antenna based second order Minkowski fractal boundary.

The structure of Minkowski fractal patch antenna

The configuration of the proposed antenna is depicted in Fig.3. The antenna is designed on a FR4 substrate (thickness h=5mm, dielectric constant=4.4 and loss tangent=0.02). The mode TM_{01} and TM_{10} are excited when the microstrip antenna is fed by a coaxial probe along the diagonal line. The two modes will have the same frequency, amplitude and phase if the length of the patch equals the width. The two orthogonal modes, which have the same amplitude and phase difference of 90 degrees, are obtained by trimming resonance length. On the basis of microstrip fundamental theory, the simulation results are studied by using the commercial FEM solver Ansoft HFSS. The coaxial-feed location is on the diagonal with a coordinate of (x0=y0=11mm). The length(A) and the width(B) of the rectangular patch are 69.5mm and 72mm respectively. The ground-plane area of the antenna(100mm×100mm), which is selected based on suitability for common RFID reader applications. Fig.4. shows the reflection coefficient characteristics when the indentation factor ρ is varied. It can be seen that with an increase of ρ the frequency keeps decreasing, which is due to the bend of the current 's electrical length, hence a decrease in the patch dimension. But it also causes the drop of the antenna's gain and radiant efficiency. On synthesizing this contradictory condition, we choice ρ =0.35. To further compensate the gain and radiant efficiency, four rectangular slots with each of size (15mm×8mm) are etched on the ground plane and they are symmetrically located on the y-axis and x-axis. This configuration is exactly symmetrical with little influence on the antenna radiation pattern. However, the slots change the dielectric constant of the substrate to a certain extent, which lead to the increasing of the antenna radiation efficiency.

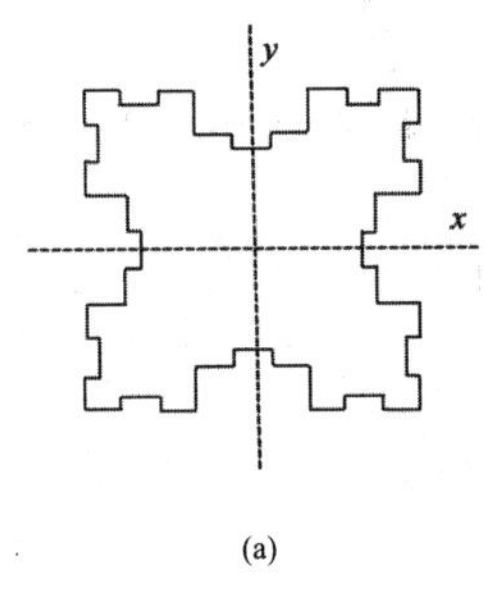

(a)

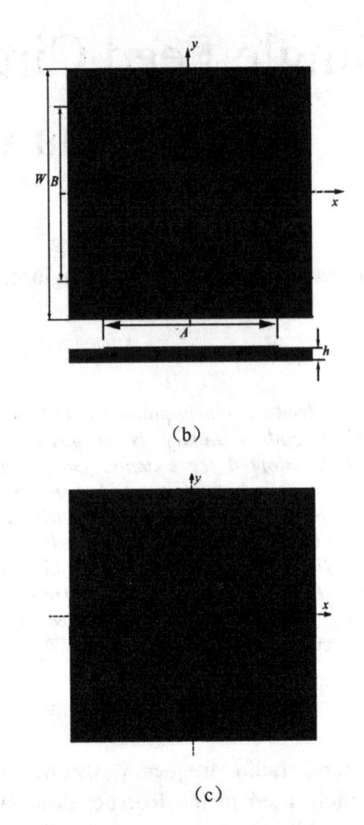

(b)

(c)

Fig.3. geometry of proposed antenna (a)the patch element(ρ =0.35) (b) top view and the side view（P is the feed point） (c) view from the groud-plane side

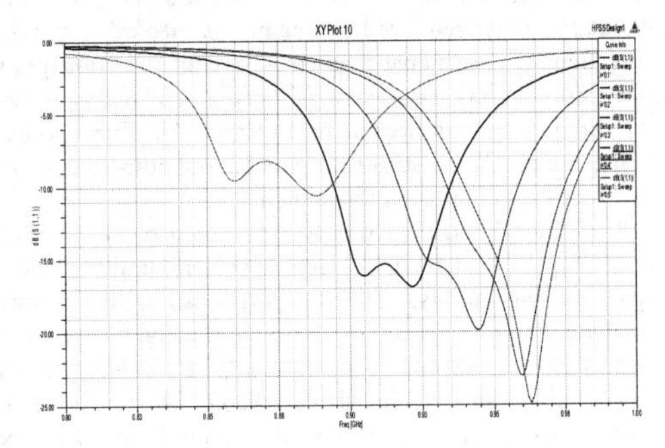

Fig.4.Simulated S11(dB) for different indentation factor

III. NUMERICAL RESULTS

With the ground slots, the impedance bandwidth, for 10 dB return loss, is 52MHz，ranging from 906MHz to 958MHz , as plotted in Fig. 5. It can also be observed that the antenna has broader bandwidth than that without the ground slots. The axial ratio and gain of the propose antenna with and without ground slots are shown in Figs.5 and 6. It is shown that the 3dB bandwidth with ground slots is 12MHz (from 917MHz to 929MHz) with minimum axial ratio of 0.21 dB. And the peak

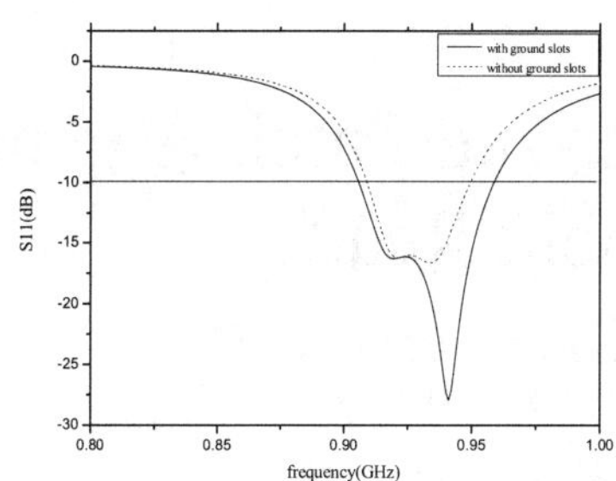

Fig. 5.Return loss of the proposed antenna with and without ground slots

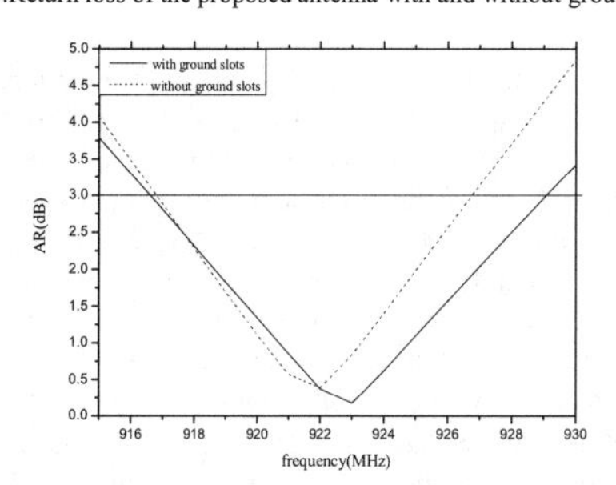

Fig.6.Axial-ratio of the proposed antenna with and without ground slots

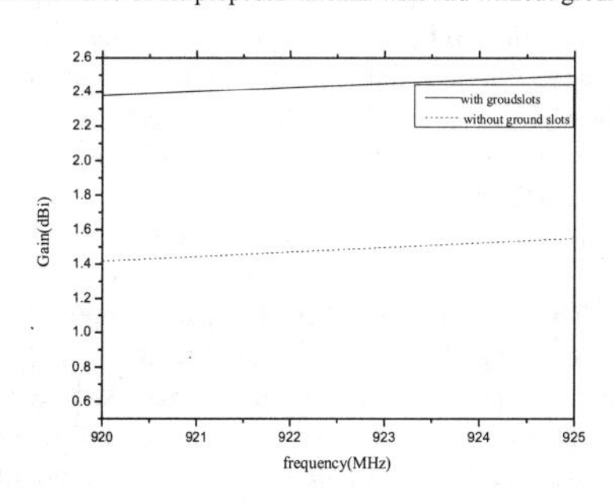

Fig.7. Gain of the antenna with and without ground slots

gain of this condition is from 2.38dBi to 2.49dBi in the UHF band (920 - 925MHz), which is compared to the status of that without ground slots. Approximately 1 dB increase in the peak gain and better AR behavior are obtained.

IV. CONCLUSION

A compact circularly polarized patch antenna with Minkowski fractal boundary is presented for UHF RFID applications. By etching four rectangular ground slots, the impedance bandwidth, 3dB AR bandwidth and the gain of proposed design have some degree of enhancement. Final simulated results show that this antenna has good performance in the band of 920-925MHz(the UHF RFID band in China). And this design has the characteristic of simplicity, low cost, easy to fabricate and useful for integration and array designs.

REFERENCES

[1] I.J.Bahl and P.Bhartia,"MICROSTRIP ANTENNAS",Artech Honse , 1980, pp.2-3.

[2] Mandelbrot Benoit. B, The fractal geometry of nature [M]. San Francisco: Freeman, 1983.

[3] Ahmed M. A. Salama,Kaydar andM. Quboa ,"A New Fractal Loop Antenna For Passive UHF RFID Tags Applications ,"*ICTTA 3rd International Conference* , Damascus ,7-11 April,2008,PP.1-6.

[4] Andrey S. Andrenko," Conformal Fractal Loop Antennas for RFID Tag Applications ", *18th ICEC International Conference* ,12-14 Oct, 2005 , PP.1-6.

[5] ayantan Dhar, Rowdra Ghatak, Bhaskar Gupta and D.R. Poddar," A Wideband Minkowski Fractal Dielectric Resonator Antenna , " *IEEE Trans ,Antennas and Propagation*, Vol.61 , Issue. 6, pp. 2895 – 2903 , june,2013

[6] P. Nageswara Rao and N. V. S. N. Sarma,"A Single Feed Circularly Polarized Fractal Shaped Microstrip Antenna with Fractal Slot,"*Progress In Electromagnetics Research Symposium*, China, March 24-28, 2008,pp.195-197.

[7] Debatosh Guha, Manotosh Biswas, and Yahia M. M. Antar, "Microstrip Patch Antenna With Defected Ground Structure for Cross Polarization Suppression"*IEEE Trans. Antennas and Wireless Propagation* , Vol.4, pp.455-458,feb.27,2005.

Novel Compact Circularly Polarized Patch Antenna for UHF RFID Handheld Reader

M. J. Chang[1], H. M. Chen[1], Y. F. Lin[1], Wilson W. C. Chiu[2]
[1]Institute of Photonics and Communications, Kaohsiung University of Applied Sciences,
Kaohsiung 807, Taiwan
[2]Wirop Industrial CO., LTD., Kaohsiung 806, Taiwan
Email: hmchen@kuas.edu.tw

Abstract- **A compact X-shaped slotted square-patch antenna is proposed for circularly polarized (CP) radiation. A cross-strip is embedded along the X-shaped slot for a novel proximity-fed technique to generate CP radiation in the UHF band. In the proposed design, two pairs of T-shaped slots are etched orthogonally on the square patch, connected to the center of the X-shaped slot for CP radiation and antenna size reduction. Proper adjustment of the length and coupling gap of the cross-strip will excite two orthogonal modes with 90^0 phase difference for good CP radiation. Simulated and measured results indicate that the proposed structure can achieve circular polarization. A measured impedance bandwidth (VSWR$\leq$2) of about 3.0% (909-937 MHz) and a 3-dB axial-ratio (AR) bandwidth of about 1.3% (917-929 MHz) were obtained.**

I. INTRODUCTION

Radio frequency identification (RFID) system in the ultra-high frequency (UHF) band has gained much interest in several service industries, manufacturing companies and goods flow systems [1]. An RFID system generally consists of a reader and a tag. In North America and Taiwan, these UHF RFID systems operate between 920-928 MHz, and between 865-867 MHz in Europe. A reader with an antenna sends a radio frequency signal to a tag and receives a backscattered signal from the tag. The RFID reader antenna is one of the important components in RFID systems and has been designed with CP operation in mind, because circularly polarized antennas can reduce the loss caused by the multi-path effects between the reader and the tag antenna. RFID handheld readers require a light-weight CP antenna with a low profile and small size.

A typical technique for the generation of CP radiation is to produce two orthogonal linearly polarized modes on the radiating element, with a 90^0 phase difference. Single-fed circularly polarized annular-ring, square and circular patch antennas with symmetrical or asymmetrical perturbation elements have been reported [2]. Using perturbation cuts or strips to differentiate suitably the two orthogonal modes at resonant frequency enables the antenna to radiate easily CP waveforms.

In recent years, small CP patch antennas attracted a lot of research interest, since they are extensively used in UHF RFID systems. Different methods for the single-feed patch antenna have been published in the literature [3-7]. In [3], the antenna demonstrated an arrow-shaped slot-coupled CP technique using different lengths of the cross-strip embedded at the center of the patch for RFID handheld reader application. The overall antenna volume is $0.46\lambda_0 \times 0.31\lambda_0 \times 0.005\lambda_0$ at 925 MHz. Coaxial-fed cross-shaped slotted patch antenna was also proposed for CP radiation and size reduction, while the overall antenna volume is $0.27\lambda_0 \times 0.27\lambda_0 \times 0.013\lambda_0$ at 910 MHz [4]. A single-layer square-ring patch antenna with a dual-feed Wilkinson power divider for CP radiation in UHF band is presented, but it provided a relatively larger CP bandwidth, and has an overall antenna volume of $0.29\lambda_0 \times 0.29\lambda_0 \times 0.005\lambda_0$ at 925 MHz [5]. As is well known, using high dielectric substrate is another effective method for achieving size reduction, but this antenna still operates at its half wavelength mode [6]. A small, single-layer, crossed-dipole antenna for CP radiation with antenna volume of $0.19\lambda_0 \times 0.19\lambda_0 \times 0.005\lambda_0$ at 925 MHz is proposed in [7]. To our knowledge, this antenna has the smallest volume for handheld RFID reader application.

In this letter, a novel and compact square patch antenna with an X-shaped slot has been proposed for achieving CP radiation. The excitation mechanism of the proposed antenna is the radiating patches along the perimeter of an X-shaped slot, which was proximity fed from the cross-strip embedded in the X-shaped slot. However, studies on the single-layer proximity-fed square patch antenna with circular polarization have been scarce. The antenna operates at its fundamental orthogonal modes for the UHF band (902-928 MHz). The two near-degenerated resonant modes for circular polarization are generated by etching unequal cross-strips on the diagonal of the patch.

II. ANTENNA DESIGN

The configuration of the proposed antenna is shown in Fig. 1, in which a square patch with the same side length of 60 mm is printed on the upper side of single FR4 substrate, with thickness of 0.8 mm and relative permittivity of 4.4. A same FR4 substrate is chosen as the ground plane, and the height of the air substrate is selected as 13.4 mm. An X-shaped slot of 57.9×7.5 mm^2 is etched along the diagonal of the patch radiator. A cross-strip with width of 1.5 mm is embedded in the X-shaped slot as a feeding line, and has tuning stubs of different lengths (L_a = 17.5 mm and L_b = 22.5 mm). The single coaxial probe is located at the center of the patch and connected to the cross-strip. The tuning stubs can also improve the phase difference of the proposed antenna. The

978-7-5641-4279-7

slotted patch is electromagnetically coupled from the cross-strip through a gap distance of 0.75 mm. At the same time, the cross-strip feed line is employed to excite directly the patch along the perimeter of the X-shaped slot. Therefore, CP radiation from the patch can be enabled. In addition, a pair of T-shaped slots of unequal length are etched on the square patch and connected to the X-shaped slot at an angle of 45^0. With the T-shaped slots of unequal length (S_a and S_b), the fundamental resonant modes of the slotted square patch can be split into two orthogonal resonant modes with equal amplitudes and a 90^0 phase difference for CP radiation requirements, as well as for good impedance matching.

To illustrate the radiation mechanism of the proposed antenna, two resonant paths of surface current on the slotted patch are depicted in Fig. 2. As clearly seen, the two current paths are inherently formed to be orthogonal. Varying the length of each T-shaped slot provides an adjustable resonant mode of the proposed antenna. In this study, S_x and S_y are selected to be 17.9 and 18.9 mm, respectively. For Fig. 2(a), the resonant length of the radiating square patch is about 154.8 mm, which operates at its first mode and corresponds to about 0.473 λ_0 of 916 MHz. Similarly, the resonant length of the second mode in Fig. 2(b) is about 152.8 mm, corresponding to about 0.473 λ_0 at 930 MHz. The proposed antenna with optimal dimensions given in Figure 1 is fabricated and tested. Note that the antenna configuration will radiate a left-hand CP (LHCP) wave.

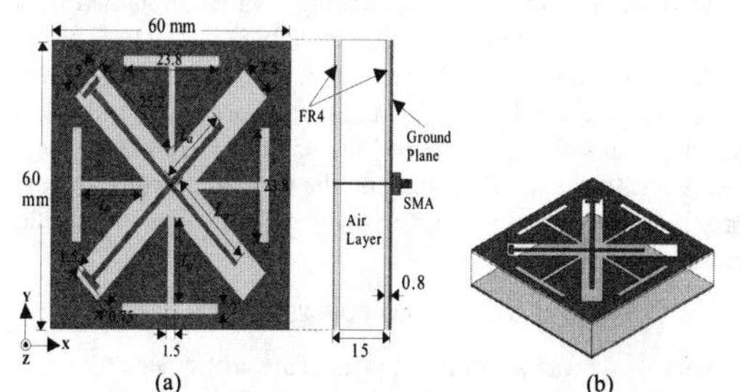

(a) (b)

Figure 1. Geometry of the circularly polarized X-shaped slotted patch antenna. (a) top and side view; (b) 3D structure.

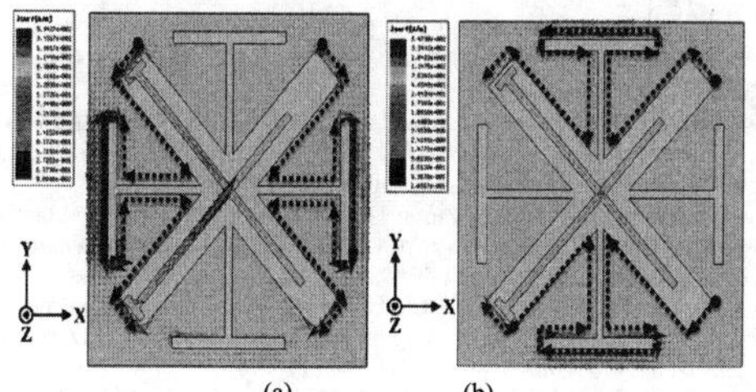

(a) (b)

Figure 2. Simulated path of the surface current on the slotted patch. (a) at 916 MHz, (b) at 930 MHz.

III. EXPERIMENTAL RESULTS AND DISCUSSION

The proposed CP antenna is designed to operate at the center frequency of about 923 MHz in the UHF band for RFID readers. The return loss is measured using an Agilent N5230A vector network analyzer, and axial ratio and radiation patterns are evaluated in an anechoic chamber with an NSI-AMS2000 antenna measurement system. Fig. 3 shows the simulated and measured return loss of the proposed antenna. The measured impedance bandwidth for 10 dB return loss is 3.03%, ranging from 917 to 929 MHz, and agrees well with the HFSS-simulated results (917-929 MHz). In addition, the operational principle of this CP antenna is derived from the fact that the generated mode can be separated into two orthogonal modes (916 and 930 MHz) of equal amplitude and 90^0 phase difference. The two resonant modes, 916 and 930 MHz, are also observed in Fig. 2. The measured and simulated axial ratio in the boresight direction, versus frequency, is presented in Fig. 4. The measured 3-dB axial-ratio CP bandwidth is about 12 MHz, from 917 to 929 MHz or 1.3% around the center frequency of 923 MHz, and 6 MHz from 922 to 928 MHz or approximately 0.65%, with respect to the center frequency of 925 MHz for simulation. Note that the measured minimum axial-ratio is about 0.3 dB at 923MHz, indicating that the circular polarization generated is very pure. Fig. 5 presents the simulated amplitude ratio and phase difference in the broesight direction. As seen in these results, the amplitude ratio (E_θ/E_ϕ) and phase difference for the 3-dB AR bandwidth are close to -1~0 dB and 70^0~110^0, respectively, for producing a good LHCP radiation. The measured and simulated radiation pattern at 923 MHz is plotted in Fig. 6. The measured and simulated 3-dB beamwidths are about 60^0 (-30^0~30^0) in both x-z and y-z planes. The measured gain was obtained using the gain transfer method with the standard gain horn antenna used as a reference. The obtained peak gain is from 3.5 to 4 dBic and the efficiency is about 40% over the UHF band (920-928 MHz), and the gain is similar to those in previous studies [3-7].

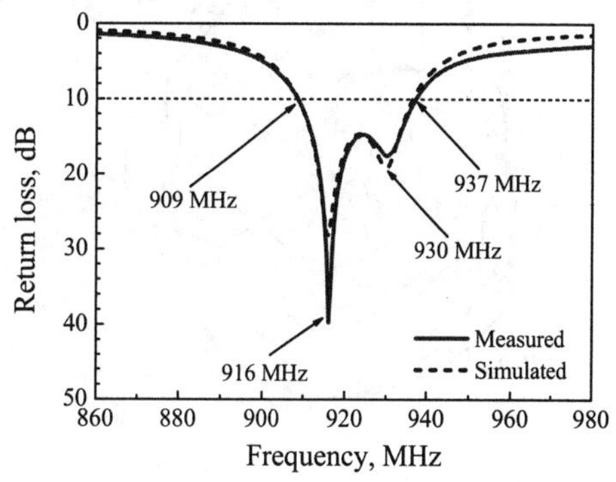

Figure 3. Measured and simulated return loss of the proposed antenna.

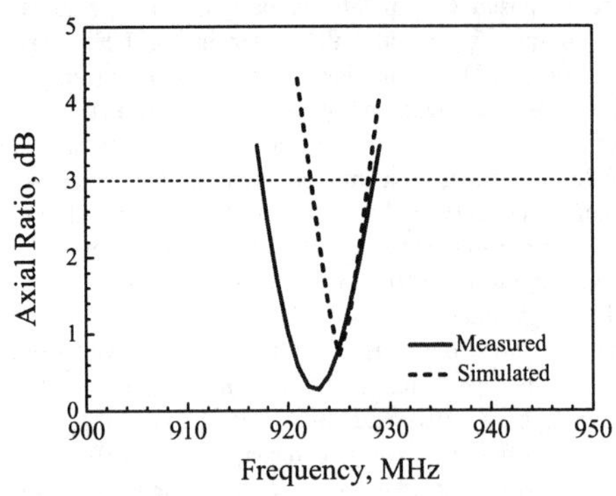

Figure 4. Measured and simulated axial ratio of the proposed antenna.

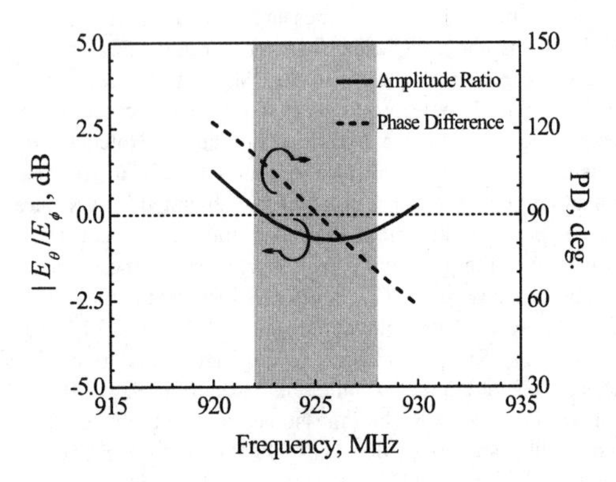

Figure 5. Simulated amplitude ratio and phase difference.

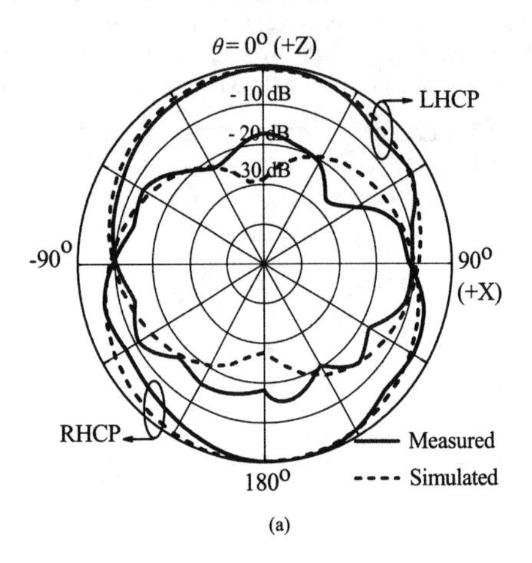

(a)

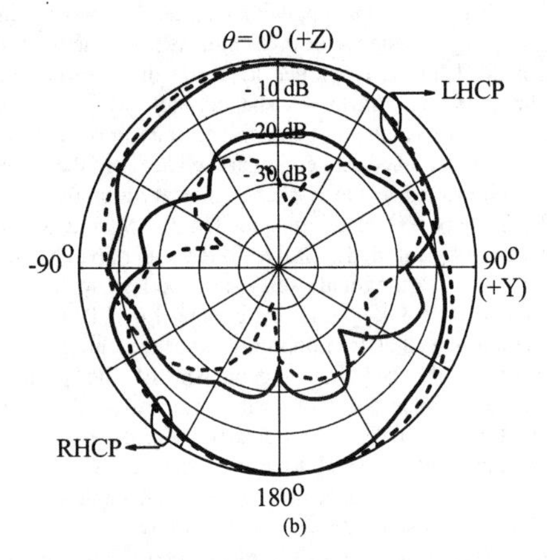

(b)

Figure 6. Measured and simulated normalized CP radiation patterns at 923 MHz. (a) x-z plane; (b) y-z plane.

IV. CONCLUSION

A new feed coupling mechanism of circularly polarized slotted patch antenna with cross-strips of unequal length for RFID readers is designed and evaluated. By applying the proximity-fed technique to this square patch, good CP bandwidth and impedance matching can be obtained. The overall antenna volume is $0.19\lambda_0 \times 0.19\lambda_0 \times 0.046\lambda_0$ at 923 MHz. Since the proposed antenna uses a simple design, with lower weight, smaller volume and lower cost, it can thus serve as the transmitting antenna in handheld RFID readers. However, this is the first study on the CP patch antenna with a proximity-fed at the same layer, which merits further exploration.

ACKNOWLEDGMENT

This work was supported by the Taiwan Science Council under Contract NSC100-2221-E-151-054-MY3.

REFERENCES

1. K. Finkenzeller, *RFID Handbook*, 2nd Ed, New York: Wiley, 2004.
2. K. L. Wong, *Compact and Broadband Microstrip Antennas,* Chapter 5, New York: Wiley, 2002
3. Y. F. Lin, H. M. Chen, S. C. Pan, Y. C. Kao, and C. Y. Lin, "Adjustable axial ratio of single-layer circularly polarized patch antenna for portable RFID reader, " *Electron. Lett.*, vol. 45, pp. 290-291, 2009.
4. Nasimuddin, Z. N. Chen, and Z. Qind, "A compact circularly polarized cross-shaped slotted microstrip antenna," *IEEE Trans. Antennas and Propagat.*, vol. 60, pp. 1584-1588, 2011.
5. Y. F. Lin, H. M. Chen, F. H. Chu, and S. C. Pan, "Bidirectional radiated circularly polarized square-ring antenna for portable RFID reader," *Electron. Lett.*, vol. 44, pp. 1383-1384, 2008.
6. X. Tang, H. Wang, Y. Long, Q. Xue, and K. L. Lau,"Circularly polarized shorted patch antenna on high permittivity substrate with wideband," *IEEE Trans. Antennas and Propagat.*, vol. 66, pp. 1588-1592, 2011.
7. Y. F. Lin, Y. K. Wang, H. M. Chen, and Z. Z. Yang, "Circularly polarized crossed dipole antenna with phase delay lines for RFID handheld reader," *IEEE Trans. Antennas and Propagat.*, vol. 66, pp. 1221-1227, 2011.

Direction-of-Arrival Estimation for Closely Coupled Dipoles Using Embedded Pattern Diversity

Yanhui Liu[1]*, Xiaoping Xiong[1], Shulin Chen[1], Qing Huo Liu[1,2], Kun Liao[1], Jinfeng Zhu[1],
[1] Department of Electronic Science, Xiamen University, Fujian 361005, China
[2] Department of Electrical Engineering, Duke University, Durham, NC 27708, USA
*Email: yanhuiliu@xmu.edu.cn (Y. Liu)

***Abstract*-direction-of-arrival (DOA) estimation for very closely spaced dipoles (no larger than 0.1 wavelength) is considered. In contrast to reducing the mutual coupling effect in conventional DOA methods, we demonstrate in this work that the mutual coupling can produce amplitude and phase difference of embedded element patterns, which can be utilized to greatly improve DOA estimation performance by incorporating the pattern diversity into the estimation algorithm. Simulation results show that two coupled dipoles achieve much higher DOA estimation accuracy than the ones without mutual coupling (for example, with the basic multiple signal classification (MUSIC) algorithm, the two coupled dipoles can achieve the root-mean-squared error (RMSE) of 1° within 120° arriving angle range for the spacing of 0.1 wavelength and RMSE of 2° within 90° range for only 0.02 wavelength, at moderately high SNR and sampling condition)**

I. Introduction

Accurate DOA estimation is significant for many civil and military applications and has received much attention over the past years [1]-[6]. Usually, the DOA estimation accuracy depends on the maximum size of antenna array, and compact or electrically small arrays have very poor performance mainly due to the following reasons: a) the signals received by different elements only have little phase difference or very small time delay, and the phase difference is hard to measured accurately due to both noise contamination and non-ideal channel response of receiver; b) strong mutual coupling existing between adjacent elements causes the element pattern deviate from the ideal steering vector used in conventional DOA estimation algorithms. Up to now, there have been many state-of-the-art methods designed to mitigate the degradation of DOA estimation due to element mutual coupling. These methods include, for example, self calibration method based on mutual coupling model [2], mutual coupling compensation using electromotive force (EMF) method [3] and receiving mutual impedance [4]. Especially, the receiving mutual impedance method is a new concept of characterizing the mutual impedance which may be more appropriate for the receiving antenna array, since it has better performance of compensating the mutual coupling effect [4].

In this work, the DOA estimation in presence of strong mutual coupling for very closely spaced dipoles is considered. It is shown, for example in [7], that the electromagnetic coupling is very useful for enhancing the element pattern diversity. Here, we demonstrate that the amplitude and phase difference of embedded element patterns can be used to greatly improve DOA estimation accuracy by simply modifying the steering vector in the MUSIC algorithm.

II. DOA Estimation with Embedded Pattern Diversity

A. Embedded Patterns of Coupled Dipoles

Consider two coupled dipole antennas with spacing of d , as shown in Fig. 1. The embedded pattern is defined as the radiation pattern of one antenna in the situation where all the other antenna elements in the array are matched with appropriate loading impedances. Here, we assume that 50Ω load is used. To obtain the embedded pattern under element electromagnetic coupling, we solve this problem by using finite element method (FEM). In the simulation, we set the dipole length equal to 470mm, the diameter 10mm and the spacing 20mm or 100mm (the two spacing cases correspond to 0.02λ and 0.1λ at the frequency of 300MHz).

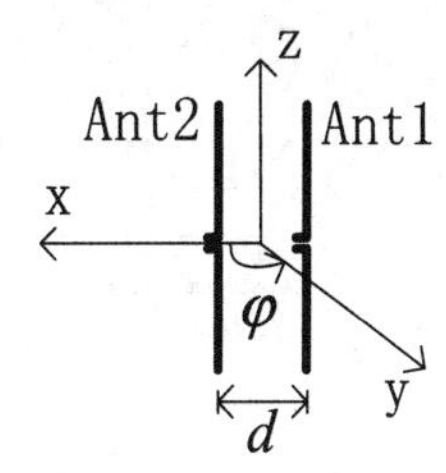

Figure 1. Configuration of two coupled dipoles

Fig. 2 (a) and (b) show the amplitude difference of embedded pattern for the cases of $d=0.02\lambda$ and 0.1λ , respectively. It is seen that amplitude difference exists between the two patterns, especially for the case of $d=0.1\lambda$. In particular, the parasitic element plays the role of reflector for the case of $d=0.1\lambda$ but director for the case of $d=0.02\lambda$. Fig. 3 shows phase difference of embedded pattern. It is seen that compared with the ideal phase difference of two isolated dipoles which is only due to the optical path difference, the phase difference of coupled dipoles can be enlarged by 2 and 3 times for $d=0.1\lambda$ and $d=0.02\lambda$. This means that the coupled dipoles are equivalent to the case of two isolated ones with larger spacing, and therefore more accurate DOA estimation can be achieved from the responses of them. It is

978-7-5641-4279-7

worth noting that the amplitude and phase difference of embedded pattern for the transmitting array are exactly equal to the difference of received voltage response between different antenna excited by a plane wave. This can be verified by both theory and electromagnetic simulation. Hence, the pattern difference can be incorporated into the DOA algorithm with modification of array steering vectors.

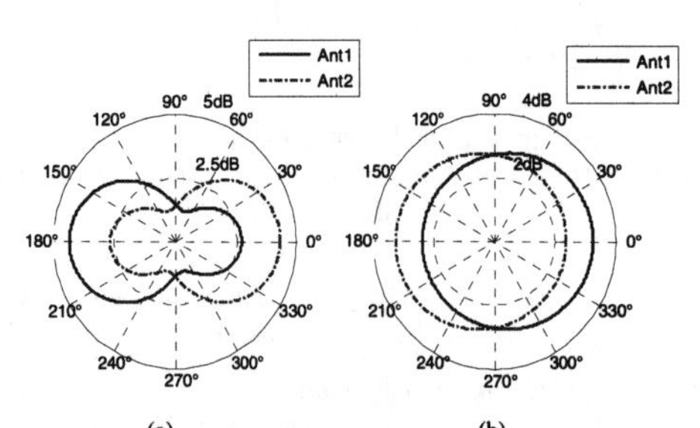

Figure 2. Element gain patterns of two coupled dipoles: (a) d=0.1λ, and (b) d=0.02λ (they are 100 mm and 20mm at 300MHz, respectively).

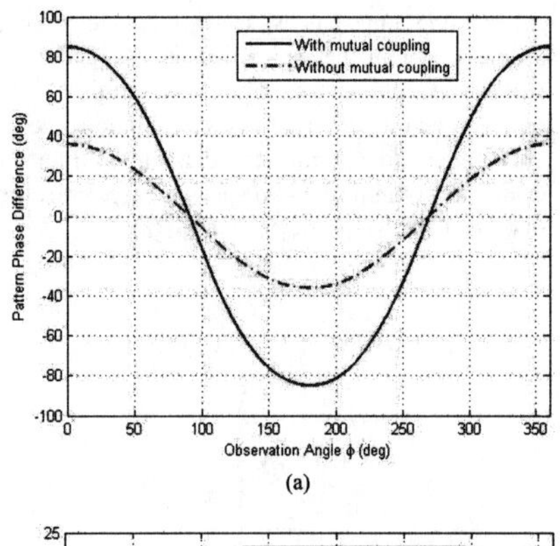

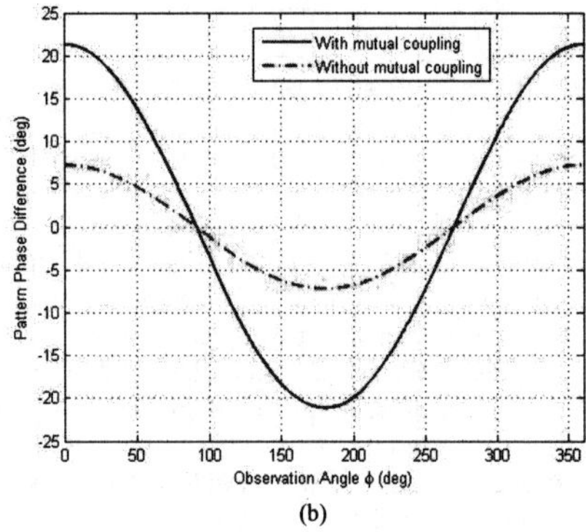

Figure 3. Pattern phase difference of two isolated/coupled dipoles: (a) d=0.1λ, and (b) d=0.02λ (they are 100 mm and 20mm at 300MHz, respectively).

B. *DOA Estimation with Pattern Diversity*

MUSIC algorithm is one of popular DOA estimation methods which are based on the fact of the orthogonality between the array steering vector and noise subspace [1]. This method can be applied to the array with a non-ideal steering vector only if the real steering vector is already known. The MUSIC spectrum can be formulated into the following equation,

$$P_{music}(\varphi) = \frac{1}{\left|\mathbf{a}(\varphi)^H \mathbf{E}_N \mathbf{E}_N^H \mathbf{a}(\varphi)\right|}$$

where $\mathbf{E}_N$ is noise subspace obtained from eigenvalue decomposition, and $\mathbf{a}(\varphi) = [a_1(\varphi) \;\; a_2(\varphi)]^T$ is a vector consisting of complex embedded patterns. The DOA of a plane wave can be estimated by maximizing the above spectrum.

III. SIMULATION RESULTS

In the following simulation, we always set $SNR = 15$ dB , $f_s = 3f_0 = 900$ MHz . And set the data length equal to $K = 300$. Assume that a plane wave arrives from $45°$. Fig. 4 compares the MUSIC spectrum of two coupled/isolated dipoles for $d = 0.02\lambda$ and 0.1λ, as are shown in Fig. 4 (a) and (b), respectively. Ten trials are used for roughly checking the statistic performance. It is clear that the coupled dipoles with embedded pattern diversity obtain much more concentrative spectrum around the real arriving angle for the both spacing cases, which corresponds to higher estimation accuracy, compared with the case of isolated dipoles without mutual coupling effect.

Fig. 5 compares the performance of coupled and isolated dipoles for different arriving angle cases in terms of root-mean-squared error (RMSE) of DOA estimation. As can be seen, the coupled dipoles achieve much lower RMSE than the isolated dipoles for both $d = 0.02\lambda$ and 0.1λ . In particular, the RMSE of coupled dipoles is less than 1° for the incident angle $\varphi_0 \in [30°,150°]$ at $d = 0.1\lambda$, and less than 2° for the impinging angle $\varphi_0 \in [50°,130°]$ at $d = 0.02\lambda$. This means that high DOA estimation accuracy can be achieved by very closely spaced dipoles if the mutual coupling effect on the embedded patterns is appropriately included into the direction finding algorithm.

IV. CONCLUSIONS

Embedded patterns of very closely spaced dipoles are obtained by electromagnetic simulation. The embedded pattern diversity is utilized to greatly improve the DOA estimation accuracy by incorporating the difference of pattern amplitude and phase into the MUSIC algorithm. Simulation results show that two coupled dipoles can achieve much higher DOA estimation accuracy than the ones without mutual coupling. At moderately high SNR condition, the root-mean-squared error can reach only 1° and 2° within a large range of arriving angle for the spacing of 0.1λ and 0.02λ, respectively.

ACKNOWLEDGMENT

This work was supported in part by the Fundamental Research Funds for the Central Universities under Grant 2012121036, in part by Specialized Research Fund for the Doctoral Program of Higher Education under Grant 20120121120027, and in part by the Natural Science Foundation of Fujian Province of China under Grant 2013J01252.

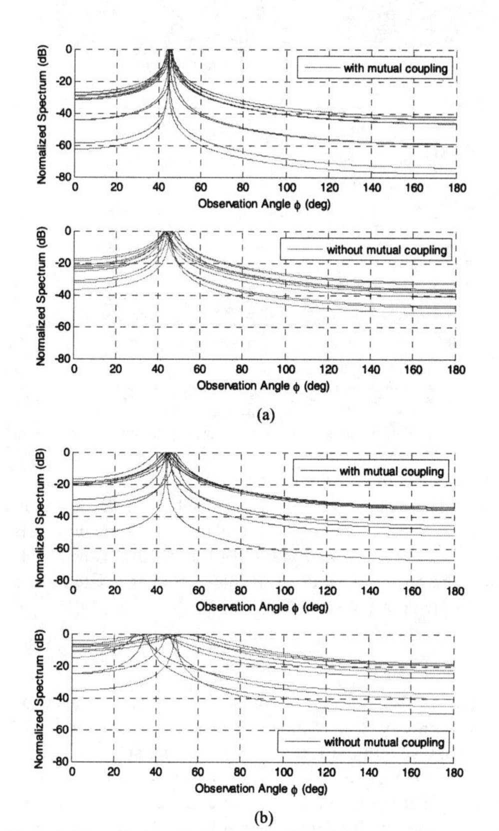

Figure 4. Normalized MUSIC spectrum of two isolated/coupled dipoles for a plane wave arriving from 45°: (a) d=0.1λ, and (b) d=0.02λ. Ten trials are performed.

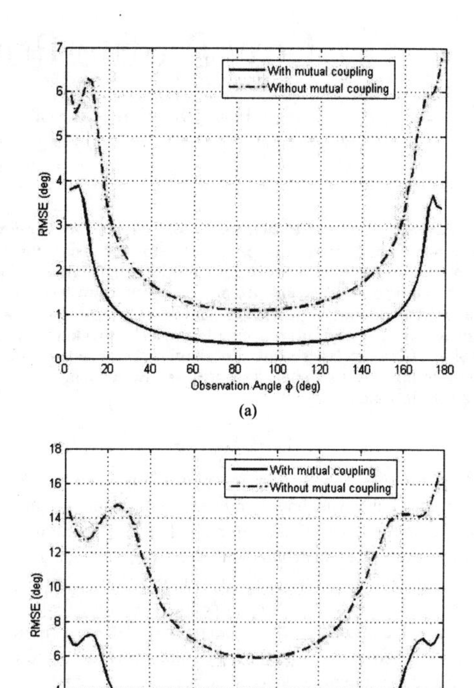

Figure 5. Root-mean-squared error (RMSE) of DOA estimation for two isolated/coupled dioples: (a) d=0.1λ, and (b) d=0.02λ.

REFERENCES

[1] R. O. Schmidt, "Multiple emitter location and signal parameter estimation," IEEE Trans. Antennas Propag. , vol. 34, no. 3, pp. 276-280, 1986.

[2] B. Friedlander, and A. J. Weiss, "Direction finding in the presence of mutual coupling, IEEE Trans. Antennas Propag., vol. 39, no. 3, pp. 273-284, Mar. 1991.

[3] Y. Wu and Z. Nie, "New mutual coupling compensation method and its application in DOA estimation," Front. Electr. Electron. Eng. China,vol. 4, no. 1, pp. 47-51, 2009.

[4] Y. T. Yu, H. S. Lui, C. H. Niow, and Hon Tat Hui, "Improved DOA estimations using the receiving mutual impedances for mutual coupling compensation: an experimental study," IEEE Trans. Wireless Commun. vol. 10, no. 7, pp. 2228-2233, Jul. 2011.

[5] H. S. Lui, and H. T. Hui, "Direction-of-arrival estimation of closely spaced emitters using compact arrays," International Journal of Antennas and Propagation, vol. 2013. Online: http://dx.doi.org/10.1155/2013/104848.

[6] B. K. Lau, J. B. Andersen, "Direction-of-arrival estimation for closely coupled arrays with impedance matching," 2007 6th International Conference on Information, Communications & Signal Processing (ICICS2007).

[7] A. Khaleghi, "Diversity techniques with parallel dipole antennas: radiation pattern analysis," Progress In Electromagnetics Research, PIER 64, pp. 23–42, 2006

Low Profile Printed Dipole Array

Huang Jingjian, Xie Shaoyi, Wu Weiwei and Yuan Naichang
College of Electronic Science and Engineering, National University of Defense Technology,
Changsha 410073, China

***Abstract*-A low profile printed dipole array for wireless application is presented. The proposed array by using folded feeding technique lowers the height of the whole array. The dipoles and integrated baluns are printed on one substrate. The whole feeding network is printed on another substrate. These two substrates are perpendicular. A circular array of this kind with a diameter of 260mm is designed. The simulated and tested results show that this novel structure can achieve low sidelobe and low cross polarization.**

I. INTRODUCTION

Many wireless communication applications need low cost, low profile, and wideband antennas. Integrated balun-fed printed dipole antennas [1,2] can meet these requirements. To produce a unidirectional radiation, it entails a planar ground which is placed at a distance about below the dipole arms [3], where is the wavelength of the center operating frequency (10GHz). Traditionally, the feeding network and printed dipole linear array are printed on the same substrate board [4]. So the feeding network increases the height of the array.

In this letter, a folded feeding printed dipole is proposed. By using this folded feeding dipole and a planar feeding network, the height of a planar array is limited to about . The simulated and tested results show that this circular array achieves good performances. Its sidelobe levels is lower than -20dB and cross polarization is lower than -25dB. It is a good candidate for the wireless communication systems.

II. ANTENNA ELEMENT DESIGN

Fig.1 shows the configuration of a folded feeding printed dipole. The dipole and the integrated balun are printed on each side of a 0.5mm thick substrate (so-called 'substrate-I') with a dielectric constant of . The feeding network with a grounded plane is printed on another substrate (named 'substrate-II') with a 0.5mm thickness and a dielectric constant of 2.2. There is a slot on the substrate-II, so substrate-I can be vertically inserted into the slot. Each integrated balun and the microstrip line of the feeding network is perpendicularly soldered together to form the folded feeding microstrip line.

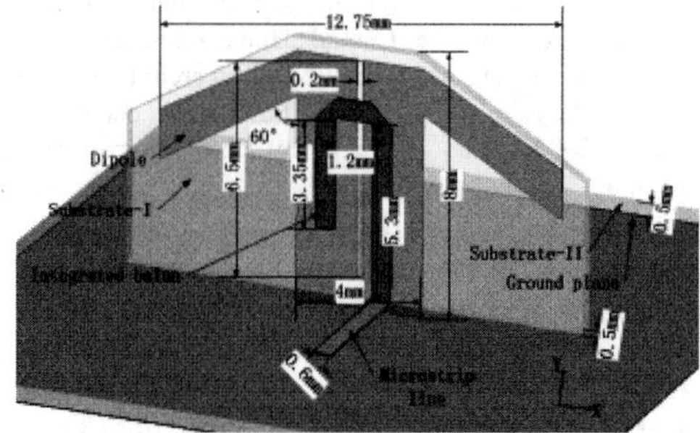

Fig.1 Configuration of antenna element.

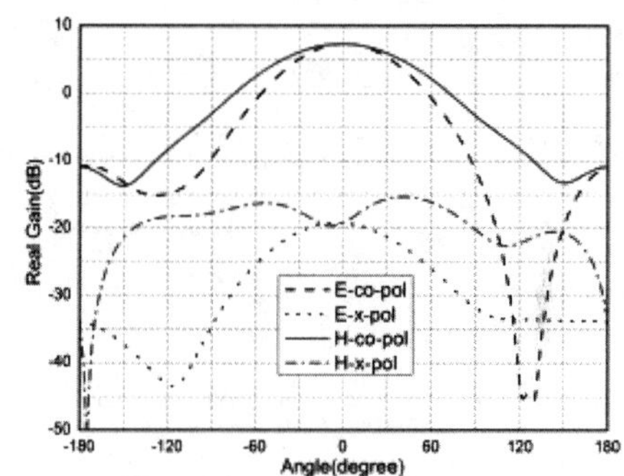

Fig.2 Radiation pattern of an antenna element.

Fig.2 shows the far-field patterns of an antenna element at 10GHz. The realized gain of the element is 7.3dB. The cross polarization of the H-plane and E-plane are all below -23dB. And the cross polarization of the H-plane is 5dB lower than E-plane. The results show that this folded feeding printed dipole can achieve good performance.

III. ARRAY DESIGN

Based on the element designed above, a two dimensional circular array with a 260mm diameter is designed. Fig.3 shows the structure of the two dimensional array. Fig.3a shows the top view of the array. It clearly exhibits the feeding networks. Each element in the sub-array and each sub-array are both series fed by the microstrip lines. The feeding ends of the sub-arrays are combined at point A. The height of the array is the same as the single element. So using such structure can greatly lower the array profile. Fig.3b shows the perspective view of the array. The elements in the north half

978-7-5641-4279-7

are mirror image of the south half. So the total E-field along the Y direction is cancelled. There forth, in the main radiation direction, the H-plane cross-polarization generated by the Y-direction E field is greatly suppressed.

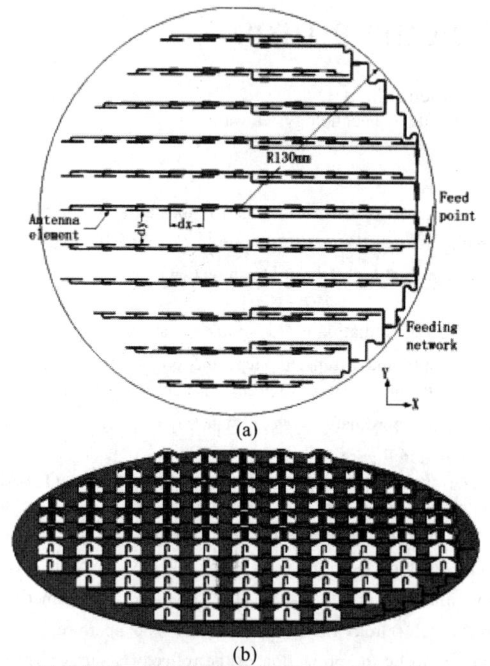

Fig.3 Configuration of the folded feeding printed dipole Array. (a) Top view, (dx=dy=21.7mm). (b) Perspective view.

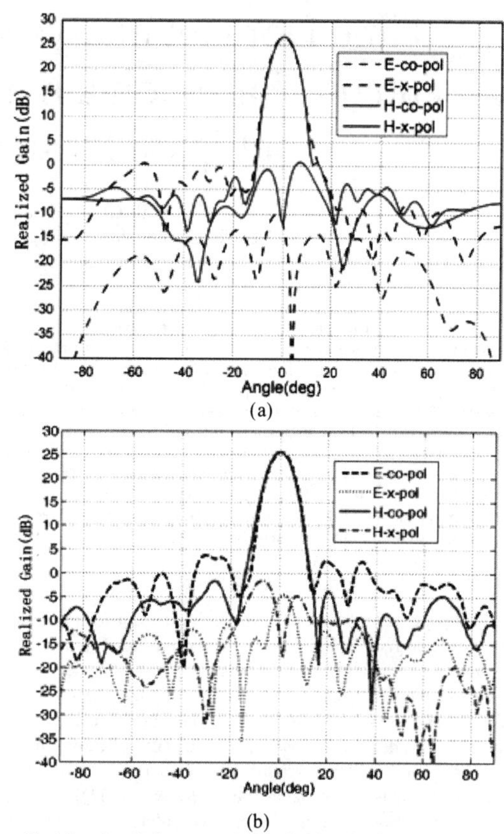

Fig.4 Array's radiation pattern. (a) Simulated results, (b) Tested results.

The realized gain patterns are shown in Fig.4. Fig.4a shows simulated results. The sidelobe level of the E-plane main polarization pattern is below -25dB. Due to the asymmetry feeding networks, the right sidelobe level is about 8dB lower than the left sidelobe. E-plane achieves good cross-polarization performance which is below -35dB. The sidelobe level of the H-plane main polarization pattern is below -25dB too. But the H-plane cross-polarization level is about 10dB higher than the E-plane's, owing to Y-direction E-field generated between the integrated balun and the printed dipole antenna. As we can see the H-plane cross polarization pattern has a null in the main beam direction.

Fig.4b shows tested results. The tested gain is about 1dB lower than the simulated results, owing to the substrate loss and the ohmic loss of the feeding network. The main polarization sidelobe levels are all below -20dB, which can meet the engineer application demands. And the tested cross-polarization levels are all below -25dB.

IV. CONCLUSION

This paper presents a new low profile printed dipole array. This array is suitable for wireless system applications. Novel folded feeding structure greatly lowers the array profile by height. The simulated and tested results show that this array can achieve good performance of low sidelobe (<-20dB) and low cross polarization (<-25dB).

REFERENCES

[1] Brian Edward, Daniel Rees. 'A broadband printed dipole with integrated balun' Microwave Journal, 1987, 30(5): 339-344.

[2] D.Jaisson, 'Fast design of a printed dipole antenna with an integrated balun,' Proc. Inst. Elect. Eng. Microw. Antennas Propag., 2006, 153(4):398–394.

[3] J R Bayard, M E Cooley, D H Schaubert. 'Analysis of infinite arrays of printed dipoles on dielectric sheets perpendicular to a ground plane', IEEE Trans. Antennas and Propagat. 1991, AP-39(12): 1722-1732.

[4] Yong Cai, Y. Jay Guo, and Pei-Yuan Qin. 'Frequency Switchable Printed Yagi-Uda Dipole Sub-Array for Base Station Antennas', IEEE Trans. Antennas and Propagat. 2012, AP-60(3): 1639-1642.

A Time Domain analysis for a hyperbolic reflector antenna based on a mathematic continuation of ellipsoidal surface curvatures

Shih-Chung Tuan [1] and Hsi-Tseng Chou[2]
1. Dept. of Communication. Eng., Oriental Institute of Technology, Taiwan
2. Dept. of Communication Eng., Yuan Ze University, Taiwan

Abstract- **A transient analysis of a hyperbolic reflector antenna is performed based on a mathematic continuation of surface curvatures of an ellipsoidal reflector antenna with the analysis previously performed in [1]. This work makes the time-domain (TD) analysis useful for the design of reflector antennas since both ellipsoidal and hyperbolic reflectors are widely used as sub-reflectors for dual-reflector antenna system. In particular, this work interprets the scattering phenomena in terms of reflected and diffracted fields with respect to the frameworks of geometrical theory of diffractions, and allows one to analyze the reflector based on the local scattering mechanism.**

I. INTRODUCTION

This paper extends previous works on the transient analysis of electromagnetic (EM) fields scattering from a perfectly conducting ellipsoidal reflector in [1], and presents the corresponding analysis for the case of a hyperbolic reflector. This work is motivated by the fact that both ellipsoidal and hyperbolic reflectors are widely used as sub-reflectors in the dual-reflector antenna systems [2]. The transient analysis presented in this paper makes the time-domain (TD) analysis of sub-reflector antennas more complete for practical applications. In particular, the analysis is performed based on a mathematic continuation of surface curvatures of an ellipsoidal reflector [1], and as a result, the analysis of such ellipsoidal reflectors can be applied straightforwardly. Furthermore, the analysis successfully decomposes the scattering fields into components of reflection and edge diffractions, and allows one to interpret the scattering phenomena within the frameworks of geometrical theory of diffraction (GTD) [3] and its wave propagation mechanisms. Examples are presented to demonstrate the wave scattering phenomena.

II. ANALYSIS DESCRIPTION

The hyperbolic reflector, illustrated in Figure 1(a), is described by a part of the following surface

$$\frac{(z+c)^2}{a^2}-\frac{x^2+y^2}{b^2}=1, \tag{1}$$

and truncated at $z = z_a$, where a and b ($a>b$) are related to the radii of its two principal axes with $c=\sqrt{a^2+b^2}$. The two focuses, F_1 and F_2, are located at $z=-c$ and $z=c$, where F_2 was selected as the origin of a global coordinate system I the following analysis while the feed is located at F_1 for a focus feeding. Given an arbitrary point, Q_s , on the hyperbolic surface, it is straightforward to show that

$$|F_1Q_s-Q_sF_2|=2a\ . \tag{2}$$

The feed's radiation is a $\hat{x}_f$ -polarized spherical wave with a cosine-taper and a transient step function, and is described by (2)-(3) in [1]. Here (r_f,θ_f,ϕ_f) is defined in the feed's spherical coordinate system with $\hat{z}_f=+\hat{z}$ and $\hat{x}_f=\hat{x}$ as illustrated in Figure 1(a).

The analysis employs a TD aperture integration (TD-AI) technique, where the scattering fields are found by the radiation from a set of equivalent currents, $(\bar{J}_a,\bar{M}_a)$, defined on an aperture, S_a , in the front of the reflector. They are found from the aperture fields, $(\bar{E}_a,\bar{H}_a)$, obtained by geometrical optics (GO) from the incident fields. The aperture, S_a , is selected to be the projection of the reflector's surface onto a sphere's surface, S_r , which is centered at F_2 with a radius r_1, and makes $(\bar{E}_a,\bar{H}_a)$ have a uniform distribution of propagation time delay. Figure 1 (a) illustrate the equivalent aperture configurations of an ellipsoidal reflector described in [1], and the hyperbolic reflector of interest, respectively, where in both cases the shapes of equivalent apertures are identical and may share the similar procedure in the TD scattering analysis. As a result, the radiation integral of scattering field in TD can be expressed in (6) of [1].

III. ANALYTIC DEVELOPMENT FOR SCATTERED FIELDS

The scattering field is found by using the continuation of surface curvature from an ellipsoidal surface to hyperbolic reflectors in [1].

(A) The Continuation of Surface Curvature

The hyperbolic surface in (1) can be analogously described as

$$\frac{(z+c)^2}{a^2}+\frac{x^2+y^2}{b_{eq}^2}=1 \tag{3}$$

where $b_{eq}=jb$ is a complex radii of surface curvature, and makes the surface an ellipsoidal one. The trigonometric

978-7-5641-4279-7

representation of an arbitrary position vector, $\bar{r}' = (x', y', z')$ is described by

$$\bar{r}' = (b_{eq} \sin\theta_{eq} \cos\phi, b_{eq} \sin\theta_{eq} \sin\phi, -c + a\cos\theta_{eq}) \qquad (4)$$

where $\theta_{eq} = j\theta$ with (θ,ϕ) being illustrated in Figure 1. Based on this curvature continuation into a complex space, the solution in [1] can be extended for the analysis of hyperbolic reflector.

(B) Contribution Contours

The scattering field at t is contributed from the equivalent currents on an equal time-delay contour, C_δ, formed by the intersection of S_a and S_R with S_R being a sphere centered at the field point, $\bar{r}_o$, with a radius, $R \geq 0$. The condition of this contour is $R + r_1 - 2a = vt$, where r_1 is selected to make S_a close to the hyperbolic reflector as shown in Figure 1. In particular, C_δ shares the properties as that in [1], and will not be repeated for brevity. However, it is circular with its center, Q_c, located on the straight line going through F_2 and $\bar{r}_o$, the reflection path of feed's radiation to arrive $\bar{r}_o$. The scattering field vanishes as C_δ vanishes. It is noted that when C_δ is a closed circle, then the scattering field corresponds to the reflected fields. Otherwise, it is a partial circle with ends at the boundary of S_a, and will consist of reflection and edge diffractions. The mathematic expression of C_δ is identical to (9) in [1] by replacing $R = vt - (2a - r_1)$ in [1] with $R = vt - r_1 + 2a$ here.

(C) Resulting Solution

The scattering field at $\bar{r}_o$ due to the illumination of a TD step-function feed radiation can be expressed in a closed-form as

$$\bar{E}_s(\bar{r}_o,t) = \left(\bar{E}_r(\bar{r}_o,t) \cdot T(\phi_e - \phi_b) + \bar{E}_e(\bar{r}_o,t)\big|_{\phi_p'=\phi_e} - \bar{E}_e(\bar{r}_o,t)\big|_{\phi_p'=\phi_b} \right) \Pi(t) \qquad (5)$$

where $T(\varphi) = \varphi/(2\pi)$ and $\Pi(t) \equiv u(t-t_1) - u(t-t_2)$ with $u(t)$ is a Heaviside step function. In (5), t_1 and t_2 are the starting and ending time that the contributing contour, C_δ, overlaps with the reflecting surface. Also $\phi_{b,e}$ are the two angular intersection points between C_δ and the edge of reflecting surface, spanned from the center of C_δ. When the entire C_δ overlaps on the reflecting surface, $\phi_e = \phi_b + 2\pi$. The reflected and edge diffracted components are given by

$$\vec{E}_{r,e}(\bar{r}_0,t) = \frac{r_1 V_0}{4\pi z_p} \left\{ \begin{array}{l} (1+\hat{R}(t)\cdot\hat{r}_1)\bar{\Omega}_1^{r,e} - \hat{z}_p \left(\dfrac{z_p^2}{R^2(t)} + \dfrac{(R(t)-r_1) z_p z_c}{R^2(t) r_1} \right) (\hat{z}_p \cdot \bar{\Omega}_1^{r,e}) \\ + \dfrac{(R(t)-r_1) z_p \rho_t}{R^2(t) r_1} \left[(\hat{z}_p \cdot \bar{\Omega}_c^{r,e}) \hat{x}_p + (\hat{z}_p \cdot \bar{\Omega}_s^{r,e}) \hat{y}_p \right] \end{array} \right\} \qquad (6)$$

The sign change due to the difference of surface's unit normal direction has been considered in (6). The parameters not defined in this letter are identical to these in [1] with the new parameter mapping in (3) and (4) used.

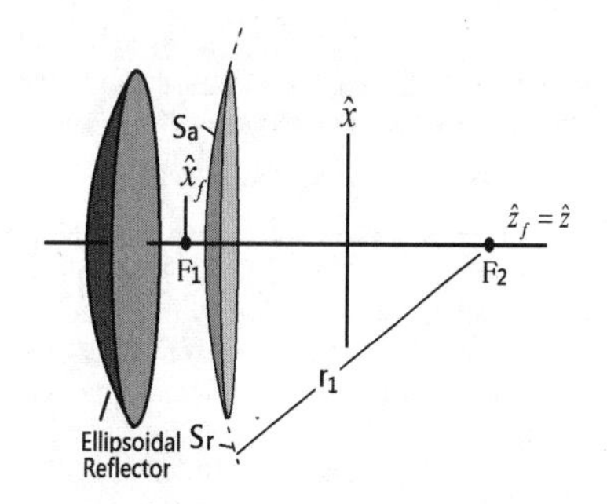

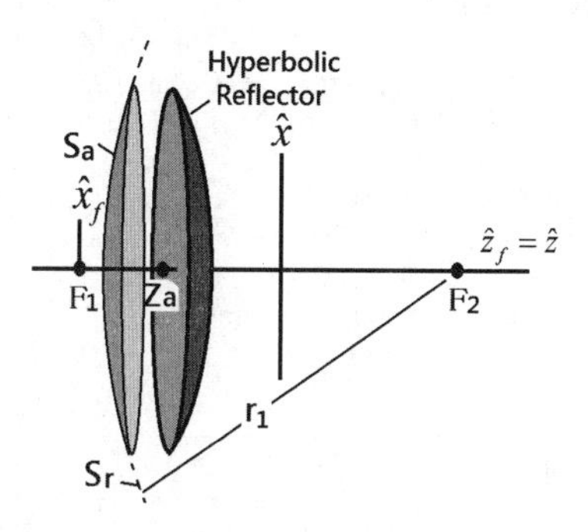

(a) Reflectors and their apertures

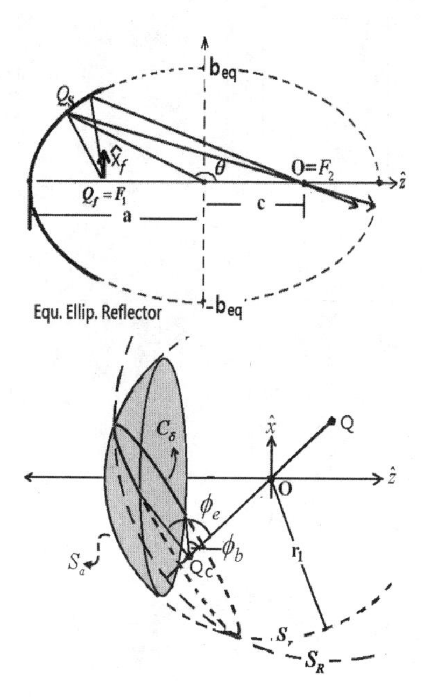

(b) Geometrical parameters

Fig.1: A rotationally symmetric hyperbolic and a near-field focused reflector which is taken from a part of hyperbolic.

IV. NUMERICAL EXAMPLES

The examples consider a reflector size with the parameters in (1) given by $a = 4\text{m}$, $b = 3\text{m}$ and $c = 5\text{m}$. The reflector in Figure 1 is rotationally symmetric with a radius $r_a = 1\text{m}$. Figure 2 (a) shows the TD response on the z-axis with z_p being the distance measured from F_2 toward the –z direction. In contrast to the radiation of an elliptical reflector, this reflector radiates defocusing fields. The time duration becomes shorter at the observation of farer distance. Figure 2(b) shows the TD response at $r_o = 15\text{m}$ with various observation angles. In this case, the field at the axis corresponds to the main beam while that at the far away angles are the sidelobes. Figure 3 (a) and (b) show the cross-polarization and curvature effects in comparison with a parabolic reflector antenna with a focal length, 1m, where the observation is at a same distance to the reflector surface. In this case the parameters of hyperbolic reflector is same to those in Figure 2(b) with the observation at $r_o = 15\text{m}$ on the -z-axis.

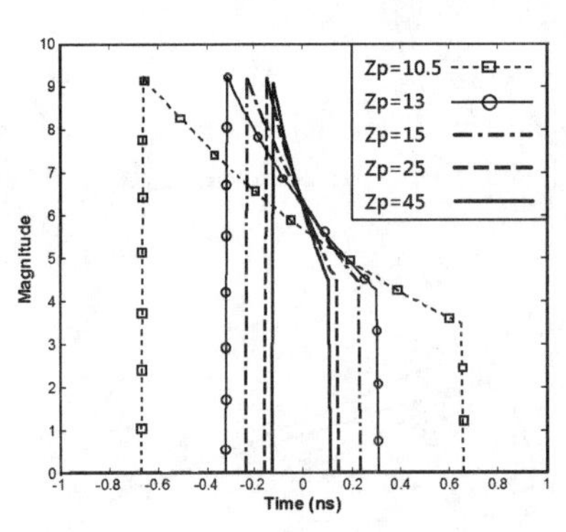

(a) Observation on the axis

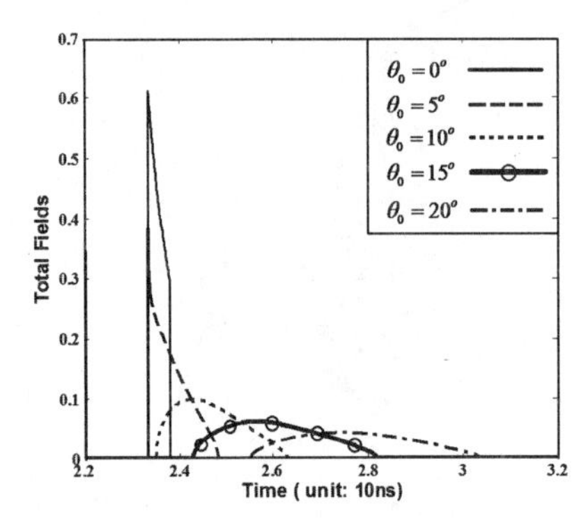

(b) Observation at $r_o = 15\text{m}$

Fig. 2: TD field response of scattering fields.

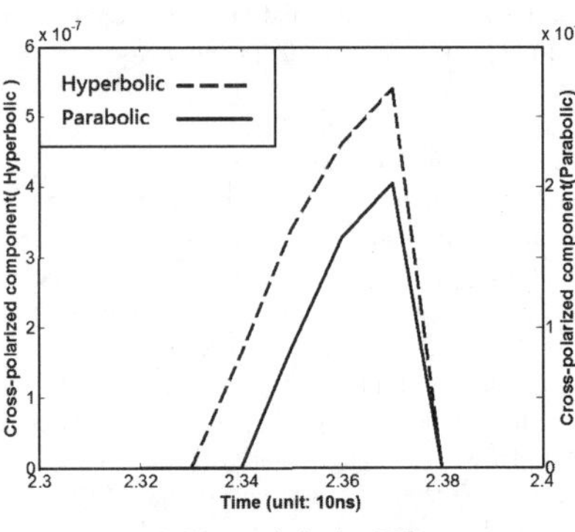

(a) Cross-polarization Effects

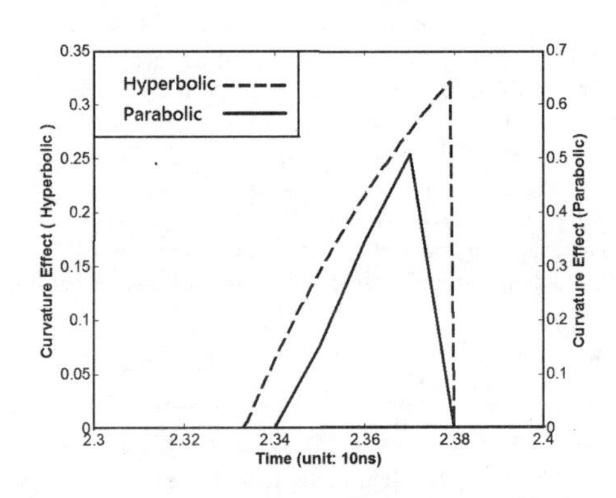

(b) Curvature effects

Fig. 3: Cross-polarization and curvature effects in comparison with a parabolic reflector.

V. CONCLUSIONS

This paper presents the TD analysis of a hyperbolic reflector based on the mathematic continuation of an ellipsoidal reflector antenna. The formulation has been developed with numerical results presented to validate the results.

REFERENCES

[1] S.-C. Tuan, H.-T. Chou, K.-Y. Lu and H.-H. Chou, "Analytic Transient Analysis of Radiation From Ellipsoidal Reflector Antennas for Impulse-

Radiating Antennas Applications," Antennas and Propagation, IEEE Transactions on , vol.60, no.1, pp.328-339, Jan. 2012
[2] Houshamand, B.; Rahmat-Samii, Y.; Duan, D.W.; , "Time response of single and dual reflector antennas," Antennas and Propagation Society International Symposium, 1992. AP-S. 1992 Digest. Held in Conjuction with: URSI Radio Science Meeting and Nuclear EMP Meeting., IEEE , vol., no., pp.1161-1164 vol.2, 18-25 Jul 1992
[3] Rousseau, P.R.; Pathak, P.H.; , "Time-domain uniform geometrical theory of diffraction for a curved wedge," Antennas and Propagation, IEEE Transactions on , vol.43, no.12, pp.1375-1382, Dec 1995

A Beam Steerable Plane Dielectric Lens Antenna

Yingsong Zhang[1,2], *Member, IEEE.* and Wei Hong[1], *Fellow, IEEE.* Yan Zhang[1], *Member, IEEE.*
1, State Key Laboratory of Millimeter Waves, Department of Radio Engineering,
Southeast University, Nanjing 210096. P. R. China
2, Institute of Communication Engineering, PLA University of Science and Technology, Nanjing 210007
yszhang@emfield.org, weihong@seu.edu.cn

Abstract- **A planar dielectric lens is designed, and the performance is simulated with HFSS. The performances of plane layers dielectric lens antenna are analyzed when microstrip patch antenna is adopted as the feeding antenna. The influence of the distance from the lens to feeding antenna is discussed. The deviation from the central axis and the unparallel between the feeding antenna and the lens towards the lens antenna are investigated. A beam steerable plane dielectric lens antenna is achieved. The maximum gain of the lens antenna is about 17.43 dB which is about 9.07 higher than that of the patch antenna and the feeding antenna is allocated flexible.**

I. INTRODUCTION

The conventional Luneburg lens is sphere symmetrical structure. The dielectric is radial, which decreases from centre of the sphere to the edge of the sphere. That makes a high consistency between the multi-beams[1],[2]. But the symmetrical distribution of the dielectric leads to that it is difficult to fabricate. So some transformative Luneburg lens have been attached importance by the researchers recently. A sliced spherical Luneburg lens (SLL) is proposed [3] easy to manufacture. The idea was to drill radial holes in a homogeneous dielectric substrate in such a way that their radii change with position in the substrate in order to produce an approximation of the continuous Luneburg distribution. Afterwards all the substrates are cascaded into Luneburg lens. But the air slot between layers impacts on the lens performance. Using in a polymer-jetting rapid prototyping technique the gradient permittivity is realized by the filling ratio of a polymer [4]. Recently a nanostructure Luneburg lens is investigated. The lens used electron beam lithography on silicon on insulator wafer with Hydrogen silsesquioxane as the resist material and mask for reactive ion etching of the silicon structure layer [5]. The designs showed high-directivity, low FSLL, and steering capabilities outperforming other antenna systems in the literature [6].

In this paper, a plane dielectric lens antenna is proposed and investigated. The lens is easy to fabricate and possess good performance such as: high directivity, the flexible location for the feeding antenna, and insensitive to the declination of the feeding antenna.

TABLE I
PARAMETERS OF THE PLANE LENS

Layers	Relative Dielectric Constant	Radius（mm）
1	12	17
2	7.4	25.7
3	6.8	34
4	4.8	38
5	6.2	44
6	2	49

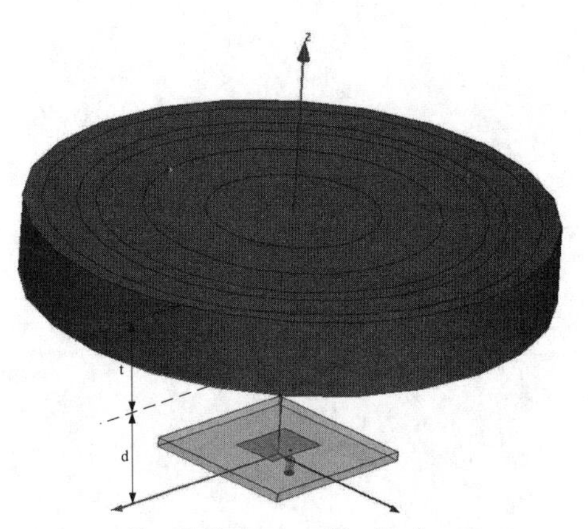

Figure 1. The structure of the plane lens antenna

II. DESIGN OF THE LENS ANTENNA

A. *Structure of the Lens*

Fig. 1 shows the structure of the plane lens antenna. Taking into account manufacturing limitation, the lens is discretized into 6 layers. The thickness of lens is t. A patch antenna operating at 10 GHz with dimensions 11.9mm×9mm is used as feeding antenna, and is designed on the dielectric substrate with a thickness of 1.6 mm and a relative permittivity of 2.2. The patch antenna emits a spherical-like wave, which is transformed in the lens, and a high directive beam achieves. The distance from feeding antenna to the lens is d.

The size of lens is about $3.2\lambda_0$.

The relative dielectric constant and radius of each layer are listed in TABLE I.

B. *Performances*

Fig. 2 gives S11 of the patch antenna and lens antenna, when the distance between patch antenna and lens is λ_0. The bandwidth of the lens antenna is wider than patch antenna, but reflection creases because of the discretizing for the lens makes that every dielectric layer is weakly reflecting the incident wave.

978-7-5641-4279-7

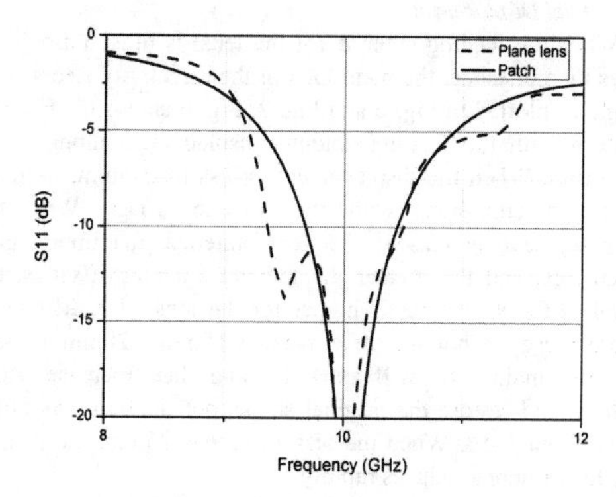

Figure 2. S11 of the patch antenna and lens antenna.

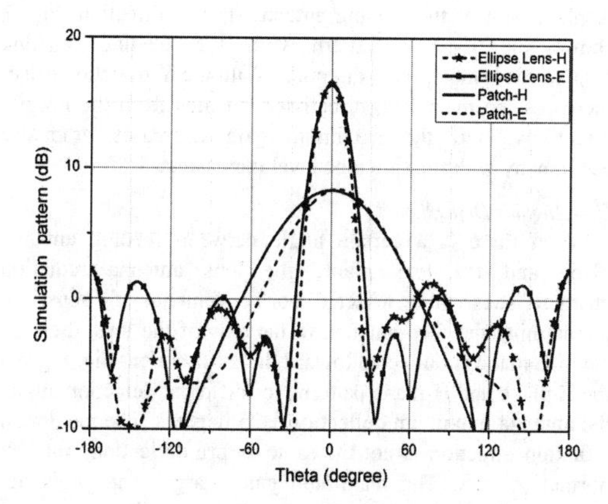

Figure 3. Comparison of the E-plane and H-plane pattern between patch antenna and lens antenna.

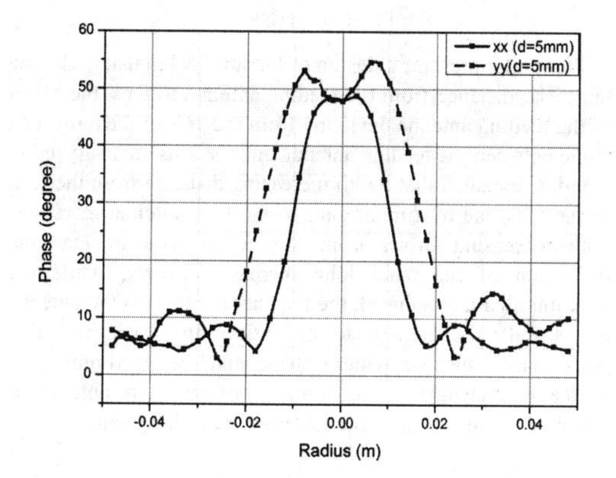

Figure 4. The phase distribution on XX=5mm and YY=5mm cross the lens.

Fig. 3 represents the E-plane and H-plane pattern of the patch antenna and the lens antenna for comparison. The maximum gain of the patch antenna is about 8.36 dB. With electromagnetic wave converging when breaking through the lens, the gain of the lens antenna is strengthened. The maximum gain of the lens antenna is about 17.43 dB which is about 9.07 dB higher than that of the patch antenna accompanying with a narrower beam width for E-plane and H-plane. In TABLE II the performances comparison of the two antennas are listed.

TABLE II
PERFORMANCES OF THE PLANE LENS ANTENNA AND THE PATCH ANTENNA

Performance	Patch antenna with lens	Patch antenna
Gain(dB)	17.43	8.36
E-HPBW	22°	78°
H-HPBW	22°	54°
E-FSLL(dB)	19.11	-
H-FSLL(dB)	16.7	-

The phase velocity is variation when the dielectric constant is different in different dielectric lens location. The higher of the dielectric constant is, the slower of the phase velocity is. Thus the wave front of the lens is approximately a plane front which makes the energy propagates ahead on an equal footing. This is the primary cause for the lens antenna with high gain.

In Fig. 4 the phase distribution on line XX and YY at d=5mm across the dielectric lens is descript. For the inconformity of the radiation pattern of the patch antenna on xoz plane and yoz plane, the phase distribution is also different. In the x direction for a radius of about 17mm, the phase fluctuation is about 48 degrees, but from -6mm to 6mm there is hardly no fluctuation. In y direction the phase fluctuation radius is slightly larger than that of x direction, for a radius of about 20mm, the phase fluctuation is about 50 degrees, but from -10mm to 10mm the phase fluctuation is about 6 degrees. In this range of the planar lens there is only a single high dielectric. It is not enough for the phase adjustment, which can be resolved to increase the number of hierarchical lens, but the more layers , the design industry will be more complicated. Outside the above range, the phase fluctuation is about 10 degrees which is so small to be ignored.

III. BEAM STEERABLE PERFORMANCE

A. *The Lens Focal Length*

Focal length of the lens greatly affect its performance. Figure 5 gives the maximum gain as the distance increases between the feeding and the dielectric lens. With the distance increasing, the maximum gain increases until f = 35mm. In the range from 20mm to 45mm the focal length, the maximum gain of the lens antenna is greater than 17 dB. With the increasing of the focal length exceeding 35mm, the gain decreases for the lens being small to cover the 3 dB beamwidth of E-plane and H-plane of the patch antenna.

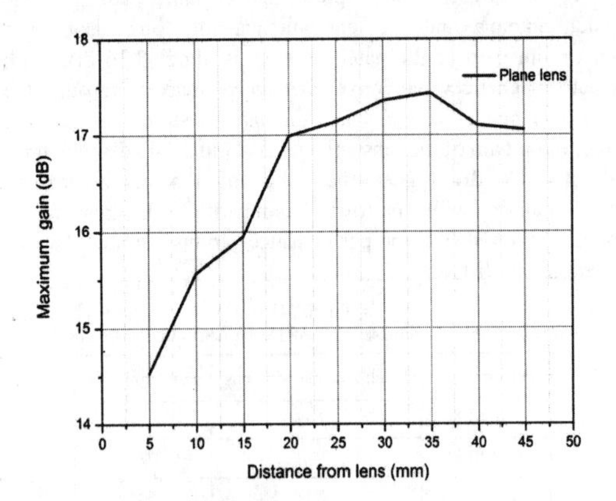

Figure 5. The maximum gain of the lens antenna when the distance from lens to feeding antenna is increases.

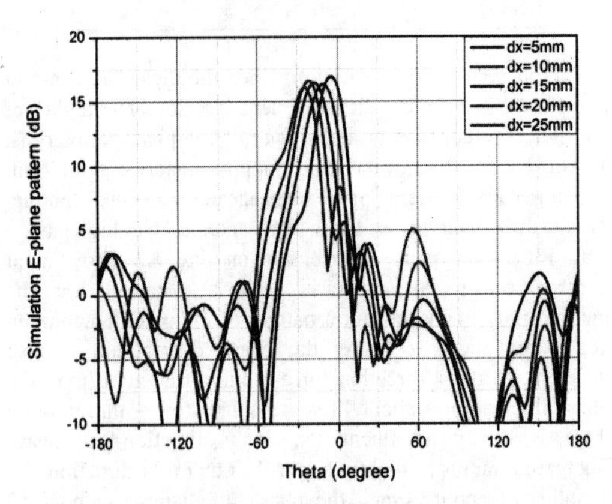

Figure 6. The E-plane pattern of the lens antenna when the feeding antenna is moved along x direction.

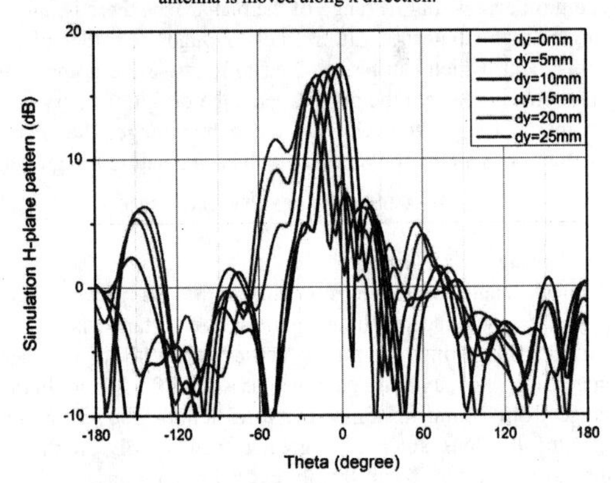

Figure 7. The H-plane pattern of the lens antenna when the feeding antenna is moved along y direction.

B. Axial Displacement

When the feeding antenna (or the lens) is moved along the axes by a distance, the main lobe of the directivity steers at an angle as plotted in Fig. 6 and Fig. 7. Fig. 6 shows the E-plane pattern with the feeding antenna displacement along the x direction. When the displacement is less than 5mm, its main beam is consistent with that of no offset. With the displacement increases, the lens antenna maximum gain decreases, and the greater the feeding antenna offset is, the level of the side lobes is higher for the lens edge diffracting more energy. When the offset reaches 15mm ~ 20mm the lens antenna main lobe will expand , and then increase offset pattern and restore the original shape, but the gain has fallen only about 1 dB. When the offset exceeds 25mm, the gain of the lens antenna reduces rapidly.

Since three-dimensional distribution of the feeding antenna is not rotational symmetry, its beam shift is not the same for displacement of the feeding antenna in x, y direction. Fig. 7 shows the H-plane pattern with the feeding antenna displacement along the y direction. With the Y-direction offset increases the main lobe directivity deviates from the normal directivity, and the maximum gain decreases gradually accompany with the side lobe level increasing.

C. Angular Deflection

When there is a certain angle between feeding antenna plane and the lens plane, the lens antenna radiation characteristics will be affected. For convenience of discussion, maintaining the feed antenna to the axis of the lens, the feed antenna plane is deflected toward the x- direction. Fig. 8 gives the E-plane and H-plane patterns for different defection angle. the antenna maximum deflection ± 25 degrees. The maximum radiation direction is not the same as pre-deflection, but still normal to lens. The maximum gain margin change is not obvious. Feeding antenna along the x- deflection angles caused only a H-patterns side lobe level changes. Deflecting from negative to positive the side lobe level rises gradually.

IV. Conclusion

This paper presents a design of hierarchical planar dielectric lens. The distance from the feeding antenna to lens, the offset of the feeding antenna deviating from the axis of the lens, feed angle between the feeding antenna and the lens are analyzed in detail. Concluded that , with increasing distance from the feed to the lens, the maximum gain of the lens antenna increases, with increasing offset from the center axis of lens the deflection of the main lobe increases largely, while the maximum gain is reduced, the feed angle between the antenna and the lens changing do not affect the basic radiation performance such as pattern shape and the maximum gain basically unchanged. The structure of the lens antenna is simple, easy to process. and the lens obtains high gain.

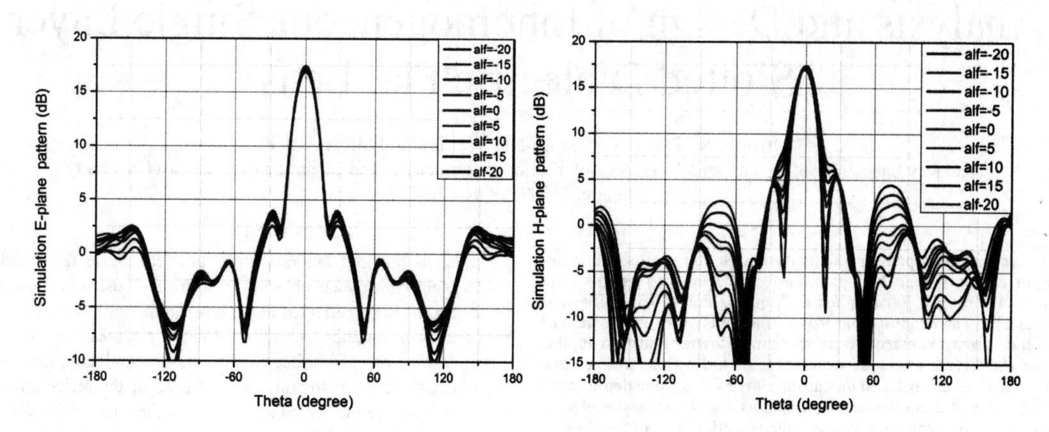

(a) E-plane pattern (b) H-plane pattern

Figure 8. The pattern of the lens antenna when the feed antenna plane is deflected toward the x- axis.

References

[1] SChoenlinner B. , Wu X. . Wide-scan Spherical-Lens Antennas for Automotive Radars. *IEEE TranS. on MTT* 2002, VOL 50, Num. 9, pages 2166-2175.

[2] B. Fuchs, Olivier L. , S. Palud, L. Le Coq, M. Himdi, Design Optimization of Multishell Luneburg Lenses. *IEEE TRANSACTIONS ON ANTENNAS AND PROPAGATION*, VOL. 55, NO. 2, FEB. 2007. Pp: 283-289.

[6] A. Demetriadou and Y. Hao, A Grounded Slim Luneburg Lens Antenna Based on Transformation Electromagnetics. *IEEE ANTENNAS AND WIRELESS PROPAGATION LETTERS*, VOL. 10, 2011. PP: 1590-1593.

[3] Rondincau S. , Himdi. M. , Sorieux J. . A sliced spherical Luneburg lens. *IEEE Antennas and Wireless Propagation Letters*, 2003, Vol. 2, Issue 1, pp: 163-166.

[4] M. Liang, W. R. Ng, K. H. Chang, M. E. Gehm and H. Xin, An X-Band Luneburg Lens Antenna Fabricated by Rapid Prototyping Technology.

[5] Satoshi Takahashi, Chih-hao Chang, Se Young Yang, Design and Fabrication of Dielectric Nano-structured Luneburg Lens in Optical Frequencies. *2010 International Conference on Optical MEMS & Nanophotonics.* PP: 179-180.

Analysis and Design of Inhomogeneous Single Layer Slotted Dielectric Flat Lens

Mustafa K. Taher Al-Nuaimi and Wei Hong, Fellow, IEEE
State Key Laboratory of Millimeter Waves, School of Information Science and Engineering, Southeast University
Nanjing, 210096, P. R. China

***Abstract*-This paper presents the design and results of low cost, light weight, planar, high gain, no feed blockage and single layer slotted dielectric flat lens for high gain antenna applications at 73.5GHz. The proposed flat lens is completely dielectric where no metal used to construct it so the ohmic losses, mutual coupling and the surface waves are expected to be lower. The unit cell that forming the lens is based on cutting slots having same depth in the both sides of a commercial non magnetic low loss host dielectric substrate to produce a phase compensation from $0^o \rightarrow 360^o$ by controlling the width of the cut slot with losses less than -1dB. A 21 cell × 21 cell lens that is occupied an area of 42 mm × 42 mm and illuminated by a pyramidal horn antenna is designed and simulated. A peak gain of 28dBi is achieved at the design frequency with almost flat gain from 71GHz to 76GHz, beamwidth of 5.1^o and aperture efficiency of 47.41%. The radiation characteristics of the lens are calculated using Finite integral technique (CST Microwave Studio) and finite element method (HFSS).**

I. Introduction

In modern wireless communication systems such as radar (Military and civil), satellite systems and deep space exploration systems high gain antennas are highly required. Parabolic reflector antenna is the most used high gain antenna in many applications such as point to point communication systems [1]. Although these parabolic reflector antennas are efficient high gain radiators but parabolic reflectors are too heavy, bulky and can not be installed easily. To overcome these drawbacks of conventional curved parabolic reflector, reflectarrays were first introduced in 1963 by Berry and Malech [2]. Reflectarray antennas have been used in many applications when a high gain is required and for many years considered to be the future high gain alterative antenna candidate because of its important characteristics such as light weight, low side lobe levels, ease of fabrication, narrow beam width and ease of installation [3].

Reflectarray is combining the best features of phased arrays and the reflector antennas where its passive and operate principally on the basis of its geometrical shape. Reflectarray consist of number of repeated unit cells in both *x*-axis and *y*-axis printed on a grounded dielectric slab in order to make the reflectarray having a flat aperture and thus reduce the volume and the mass of the reflector [4-6]. These unit cells which are composed the flat reflectarray is usually illuminated by a horn antenna as a feed source . The unit cells is designed in such away to correct the phase delay of the incident EM-waves that are radiated by horn antenna with spherical phase front and reflect them with planar phase front in front of the reflectarray and direct the main reflected beam toward the desired direction. However at millimeter and sub-millimeter regimes the ohmic losses, excitation of surface waves and mutual coupling is sever which increases the amount of losses of the reflectarray used for these frequency bands. Further more, using the feed source in front of the reflectarray will increase the blockage of the reflected waves. Although that shifting the feed antenna by a certain degree to reduce the blockage is a good technique but sometimes results in high side lobe level and it might destroy the symmetry of the radiation pattern.

Quite recently, there is an increasing interest in the design of planar dielectric lenses (also called Transmit-array and discrete lens) for millimeter and sub-millimeter waves where the lenses become adequate in terms of weight and size for many applications in this region of EM spectrum. Besides, very low loss dielectric materials are available, and present-day numerically controlled machines enable low-cost fabrication of quite sophisticated lenses made with very good tolerances [7,8]. In this paper a low cost, light weight, high gain, no feed blockage and having flat geometry dielectric lens is presented for 73.5GHz. The proposed lens is completely dielectric where no metallic layers used to construct it so the mutual coupling and excitation of surface waves is expected to be lower. Using only single dielectric layer to build the proposed lens make it a good candidate for integration with other planar devices.

II. Principle of Operation of the Lens

The operation principle of dielectric flat lens type structures presented in this paper can simply be explained as follows: when the lens in the transmitting mode the lens will collimating the incoming divergent EM waves with a spherical wave front on one side, processing it by prevent it from spreading in undesired directions and then retransmitted a narrow shaped beam of EM wave of enhanced directivity with planar wave front on the other side after correcting the phase of the incoming EM waves just like the parabolic reflectors. On the other hand, in the receiving mode the lens will focus the incoming EM waves with planar phase front onto the feeding point. In other words, the lens is resembles to a two set of receiving and transmitting antennas connected to each other via phase shifters. These phase shifters are used to correct the phase delay of the incoming EM waves and controlling the direction of the transmitted beam.

978-7-5641-4279-7

III. Proposed Lens Design and Results

The design of the unit cell that forming the proposed dielectric lens is based on phase correction technique. The proposed unit cell is achieved by cutting slots having the same depth on both sides of a dielectric slab. Cutting slot in the host dielectric substrate will increase the amount of free space that having dielectric constant of ε_{r1} and make it along with the host dielectric substrate that having dielectric constant of ε_{r2} compose a new martial having dielectric constant of ε_{eff} as depicted in Fig.1 .

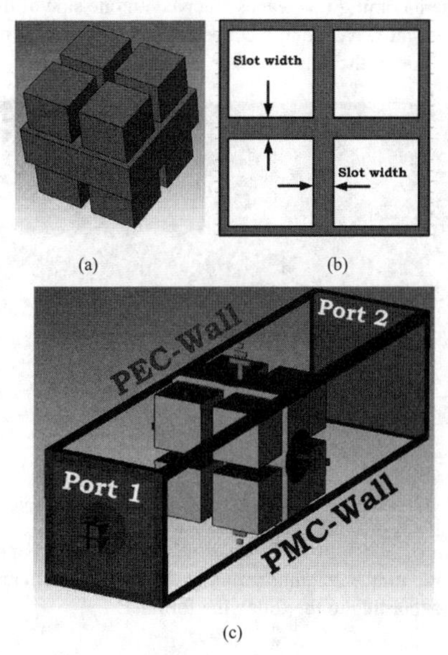

Figure 1. Geometry of the proposed unit cell (a) 3D view, (b) top view and (c) Simulation setup in CST.

The dielectric substrate used is RO6010 that having dielectric constant of 10.2 and dissipation factor of 0.0023 and thickness of 3.175 mm . Choosing the unit cell is an important point in the design of such lens.

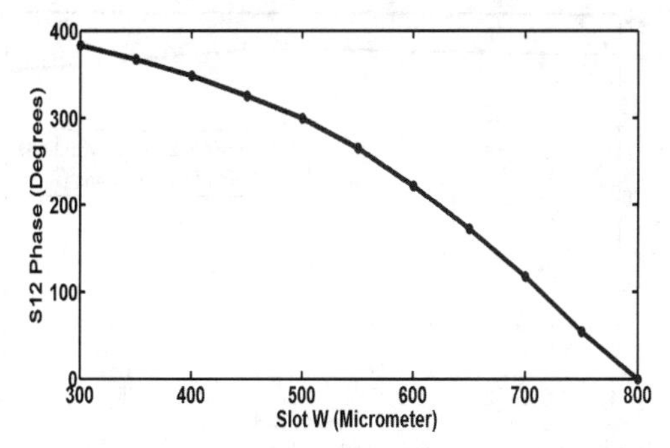

Figure 2. Transmitted wave (S_{12}) phase versus slot width.

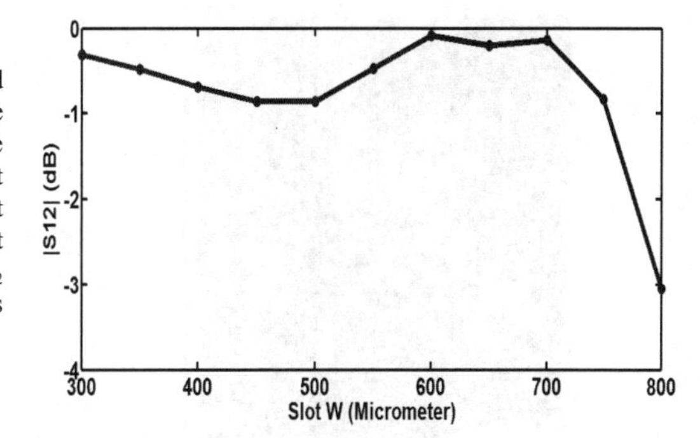

Figure 3. Transmitted wave (S_{12}) magnitude versus slot width.

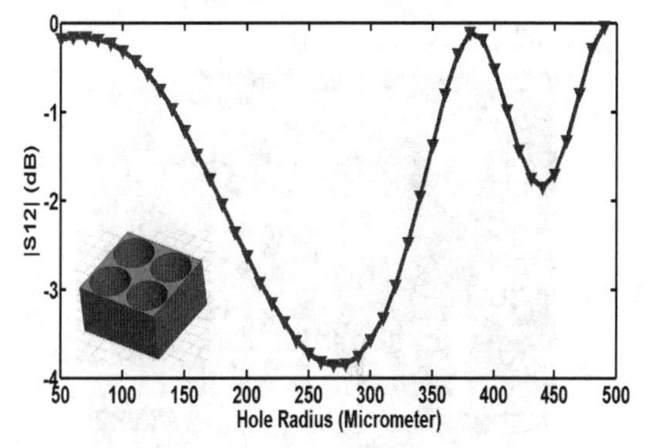

Figure 4. Transmitted wave (S_{12}) magnitude for perforated unit cell.

After detailed analysis it's found that the unit cell size will highly affect the resulted side lobe level. In this paper the unit cell size is set to 2mm × 2mm (the distance between any two adjacent unit cell centers). First the phase diagram which is represents the transmitted wave phase (and magnitude) as a function to the slot width is calculated and presented in Fig.2. The CST simulation setup in Fig. 1 (c) is used to compute the phase diagram. A single unit cell of the proposed lens is placed at the center of TEM-waveguide. The side walls of the TEM-waveguide are formed by two E-walls boundary condition and two H-walls boundary condition and the other open ends are terminated to two waveguide ports and then the phase of the transmitted signal is recorded at the surface of the unit cell.

It's clear that the proposed unit cell can correct the phase from $0^{o}\rightarrow 360^{o}$. Further more, the magnitude of the transmitted signal versus the slot width is computed too. As depicted in Fig. 3 the transmitted losses is less than -1dB for the whole range of slot width expect for slot width from 0.78 to 0.8 mm and compared to perforated type unit cell that having the same size, the proposed unit cell exhibit lower losses as in Fig.4. So, The performance of the proposed lens is expected to be better than perforated type lens having the same dimensions. Furthermore, in order to obtain a almost flat gain a horn antenna with almost flat gain over 71-76GHz band is used to illuminate the lens.

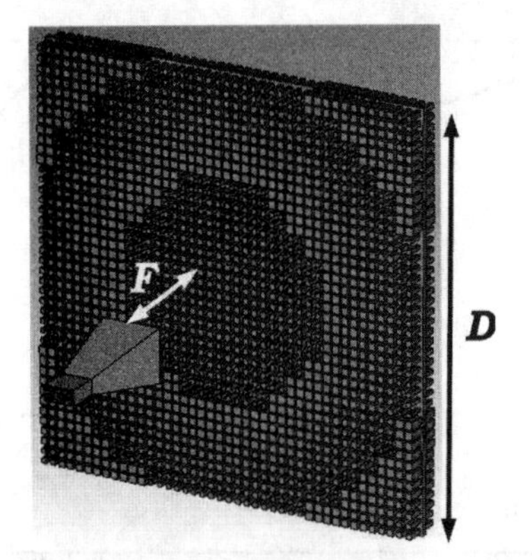

Figure 5. 3D view of the proposed lens with the feeder.

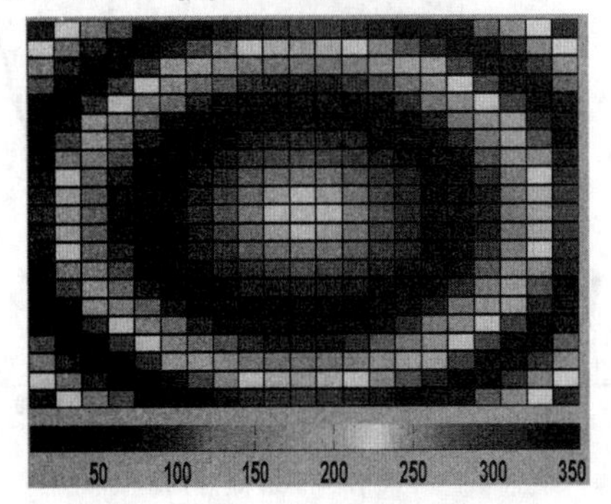

Figure 6. The required phase correction (in degrees) at each unit cell.

A flat dielectric lens consist of 21 cells × 21 cells and occupied an area of 42mm × 42mm and having a thickness of 3.175mm is designed and simulated. The layout configuration of the proposed center fed lens is presented in Fig. 5. The lens is assumed to be along *x*-axis and *y*-axis, then required phase correction $\Phi(x_i, y_i)$ at each unit cell to collimate the beam in the desired direction is calculated by the fowling formula:

$$\Phi(x_i, y_i) = k_o\left[d_i - \sin(\theta_o)(x_i\cos(\varphi_o) + y_i\sin(\varphi_o))\right] \tag{1}$$

$$d_i = \sqrt{(x_i - x_f)^2 + (y_i - y_f)^2 + (z_f)^2} \tag{2}$$

Where $k_o = (2\pi/\lambda_o)$ is the free space propagation constant, d_i is the distance from the unit cell positioned at (x_i, y_i, z_i) to the feed antenna positioned at (x_f, y_f, z_f). The values of the required phase correction at each unit cell were obtained using MATLAB® code and the results are presented in Fig. 6. The proposed lens is illuminated by an E-band linearly polarized pyramidal horn antenna of size W × H × L is 6 mm × 5.6mm × 13.6 mm. The dimensions of the horn antenna are chosen to get almost a flat gain over the frequency range 71-76GHz. Determining the phase center point (the point at which the EM waves propagates spherically outwards) of the horn antenna is very important and for the horn antenna used in this paper its found to be at (0, 0, 13.2) mm. Then the lens along with the horn antenna was simulated using the time domain transient solver of the full wave CST Microwave studio package. In order to shrink the simulation volume and the required time for simulation, symmetry boundary conditions are used. A computer having 16-GB of RAM and 8-CPUs is used. The simulation time is around 14 hours. The design of the proposed lens is based on the phase transformation and as depicted in Fig. 7 the incoming EM waves received on one side of the lens with spherical wave front is converted to planar wave front on the other side or the lens.

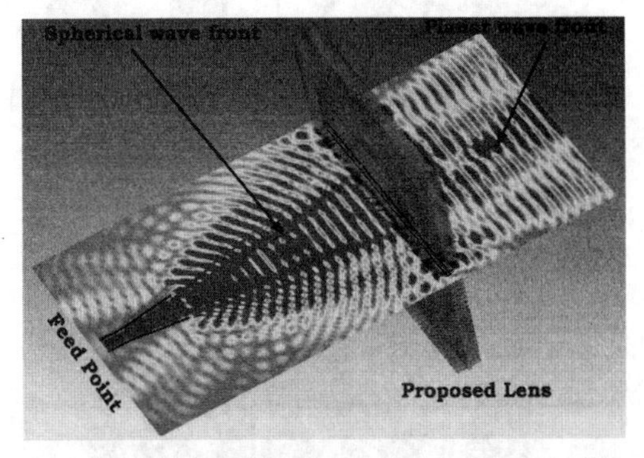

Figure 7. 3D view of the phase transformation concept of the proposed lens.

A gain of 28dBi at 73.5GHz is achieved. The lens physical area of 42 mm × 42 mm provides an aperture efficiency of 47.41% according to the following formula.

$$\eta_{app} = \frac{G \times \lambda^2}{4\pi \times A_{physical}} \tag{3}$$

Figure 8 shows the achieved gain over the 71-76GHz band for the horn antenna with/without the proposed lens.

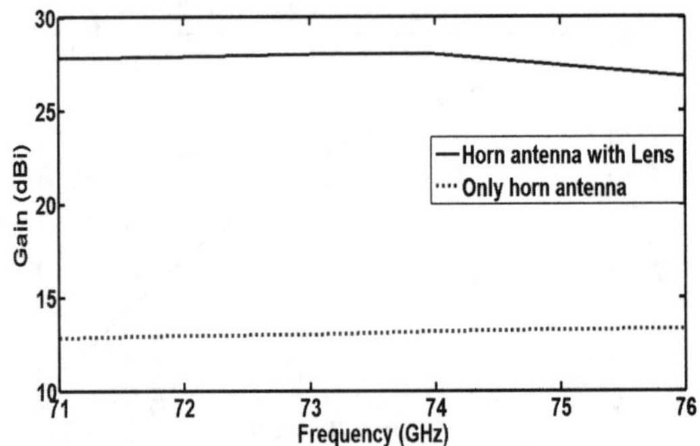

Figure 8. Achieved gain with and without the lens.

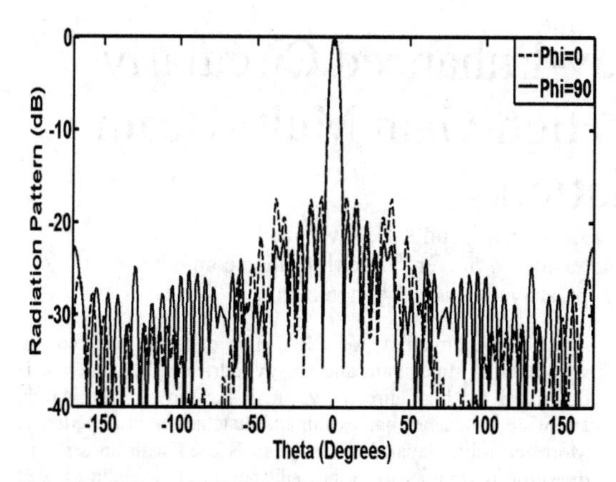

Figure 9. Gain Pattern for F/D=1 at 74GHz.

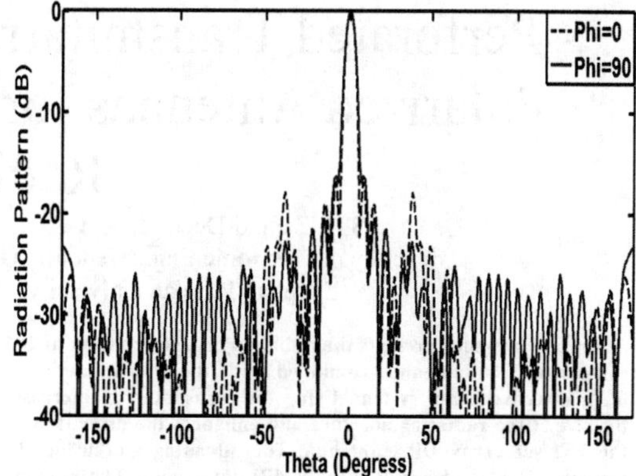

Figure 10. Gain Pattern for F/D=0.9 at 74GHz.

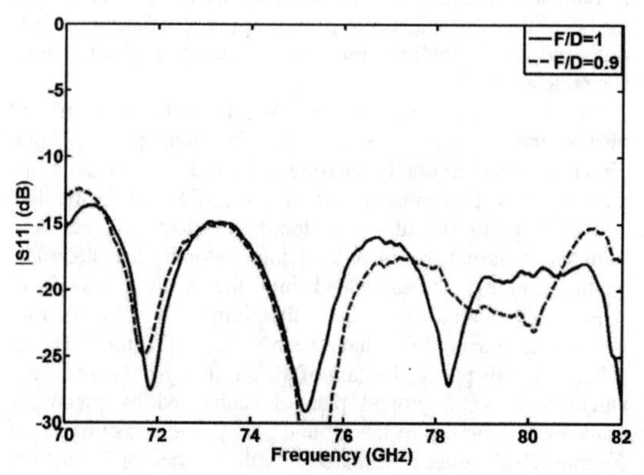

Figure 11. Return loss of the horn antenna with the proposed lens.

The computed radiation pattern plot for $\Phi=0^{o}$ plane and $\Phi=90^{o}$ plane is presented in Fig. 9 and Fig. 10. Focal length-to-diameter ratio (F/D) of the lens is adjusted carefully and optimized in order to have the highest possible peak gain with lower side lobe levels and the radiation gain pattern for F/D=1 and F/D=0.9 cases are presented here. As depicted in Fig. 9 and Fig. 10 the transmitted beam is broad side with peak gain of 27.7 dBi (F/D=1) and 28 dBi (F/D=0.9) and the main transmitted beam have 3-dB beamwidth of 5.1^{o} and SLL around -13.7 dB. Its important here to point out that the standard gain horn antenna that have been used have maximum gain of 13.2 dBi, compared with 28 dBi gain that achieved using the proposed lens it means that a 14.8 dBi gain improvement is the result of using the proposed lens as in Fig.8. The return loss at the horn antenna port within the design frequency band is presented in Fig. 11 where $|S_{11}|$ <-10dB for the whole frequency band for the two cases F/D=1 and F/D=0.9.

IV. Conclusion

In this paper the design of light weight, high gain, low cost, no feed blockage and single layer slotted dielectric flat lens for high gain antenna applications is presented. The unit cell consists of only dielectric material having slots of same depth on both sides which makes the fabrication process easier. A lens consists of 21cell × 21 cell and occupied an area of 42mm × 42mm and having a thickness of 3.175mm is designed and simulated. The design of the unit cell and the whole lens was verified using CST and HFSS software. A gain of 28dBi at 73.5GHz is achieved with 3-dB beam width of 5.1^{o}, SLL of -13.7 and aperture efficiency of 47.41% with a highly shaped radiation pattern in both E-plane and H-plane. The results show that the proposed lens could be a good alternative solution for the conventional dielectric lens (hyperboloid or spherical shape lens which are difficult to machine and a bit heavy with some complexity in integrating it with other devices) due to its simple configuration and high radiation performances.

References

[1] W. L. Stutzman and G. A. Thiele, "Antenna Theory and Design", 3rd edition, John Wiley & Sons: New York, 2012.

[2] D. G Berry and R. G. Malech, "The Reflectarray Antenna," IEEE Trans. Antennas Propagat., Vol. 11, Nov. 1963.

[3] D. M. Pozar, S. D. Targonski, and H. D. Syrigos, "Design of millimeter Wave Microstrip Reflect Arrays," IEEE Trans. Antennas Propagat., Vol. 45, No. 2, Feb. 1997.

[4] W. Hu, M. Arrebola, R. Cahill, J. A. Encinar, V. Fusco, H. Gamble, Y. Alvarez and F. Las-Heras, " 94GHz Dual-Reflector Antenna With Reflectarray Sub-Reflector", IEEE Trans. Antennas Propag., vol. 57, Oct. 2009.

[5] W. Menzel, M. Al-Tikriti and R. Leberer, "A 76 GHz Multiple-Beam Planar Reflector Antenna", EuMC, Milano, Italy, Sept. 2002..

[6] J. Huang and J. A. Encinar, Reflectarray Antennas, John Wiley & Sons Inc., Hoboken, NJ, 2007.

[7] J. Thornton and K. C. Huang, "Modern Lens Antennas For Communications Engineering", John Wiley & Sons, Inc., Hoboken, New Jersey , 2013.

[8] A. Petosa, N. Gagnon, A. Ittipiboon, "Effects Of Fresnel Lens Thickness On Aperture Efficiency", ANTEM 2004, Ottawa Canada, pp. 175–178, July 20–23 2004.

Perforated Transmitarray-Enhanced Circularly Polarized Antennas for High-Gain Multi-Beam Radiation

S.H. Zainud-Deen[1], S.M. Gaber[2], H.A. Malhat[1*], and K.H. Awadalla[1]
[1]Faculty of Electronic Eng., Menoufia University, Egypt, *er_honida1@yahoo.com
[2]Egyptian Russian University, Egypt, shaymaa.gaber@yahoo.com

Abstract- **The paper presents the radiation characterization of 2×2 sub-array DRA antennas combined with the transmitarray to form directive beam system. Using the transmitarray increases the size of the radiating aperture and enhances the directivity of the 2×2 sub-array DRA antenna. The advantages of using the transmitarray is observed to be 6 dBi extra gain. The antenna array design is optimized by full-wave simulation for high gain radiation up to 18.8 dB. The transmitarray designed at frequency of 10 GHz. The unit cell element in the transmitarray achieves 360 degrees of phase agility with less than 3 dB of variation in transmission magnitude in the tuning range. Single-feed dual-beam perforated transmitarray antenna designs will be investigated for covering two regions at the same time. The transmitarray introduces dual beam at ± 45 degrees. The antenna introduces maximum gain of 12.5 dB at the dual direction with circular polarization characteristics.**

I. INTRODUCTION

High gain antennas are desired in various applications, such as satellite communications, and wireless broadcasting. Recently, the gain and radiation patterns of planar printed antenna can be improved by covering the antenna with a superstrate at a specific distance in free space [1]. Various configurations of superstrates were used to improve the radiation performance of the antenna, such as dielectric slabs [2], electromagnetic bandgap (EBG) structures [3], highly-reflective surfaces [4], and the most recently artificial magnetic superstrates [5]. In these entire configurations, the substrate-superstrate spacing and the thickness of the superstrate are optimized for antenna gain enhancement. High permeability superstrate is required to achieve very high radiation from the patch antenna. Artificial magnetic structures (AMS) can be used to provide such superstrate [5]. DRAs have attracted broad attention in many applications due to their attractive features such as high radiation efficiency, wide bandwidth, light weight, and small size [6]. One drawback with DRAs is that they have low directivity. In 1968, the idea of left-handed metamaterial (LHM) with simultaneously negative permittivity and permeability was conceived by Veselago [7]. Recently, they have started to be used as superstrates for planar antennas, working in the transmission bands of the LHM [8,9] with the goal of improving the radiation performance of antennas. A high-directivity DRA with metamaterial is presented in [10]. The metamaterial was positioned over a cylindrical DRA mounted on curved ground plane to improve its directivity. Three different types of metamaterials as superstrates, viz., S-shape, split ring resonator, and cubic high dielectric resonator are discussed. The metamaterial structure is used as a lens to improve the directivity of DRA. The distance between the metamaterial superstrate and the cylindrical DRA antenna is optimized for high directivity, small half-power beamwidth, and wide impedance bandwidth and matched input impedance. More recently, planar dielectric slab is used with an array of dielectric resonators to further enhance the array gain as well as reducing the number of the array elements by 75% in [11]. In [12], the lens-enhanced phased array configuration was proposed and studied for implementing high-directivity steerable antennas.

A transmitarray antenna combining the features of lens and phased array antenna is a paradigm for implementing high-directivity reconfigurable apertures. Like an optical lens, the elements of a transmitarray produce specific phase shift that, when properly tuned, can focus incident waves. The transmitarrays can be integrated into radomes for electronic beam scanning, or embedded into the walls or roofs of structures to create large high-directivity apertures. Various transmitarray approaches have been described in the literature [13-15]. In this paper, the gain of the circularly-polarized DRA antennas above a ground plane is enhanced by placing a transmitarray on top of the ground plane. The transmitarray is composed of a dielectric material with a lattice of holes. The separation between the holes and the number of the holes are optimized to maximize the transmission through the structure. The performance of the structure is validated by simulation using the finite element method (FEM) [16] and the finite integration technique (FIT) [17].

II. NUMERICAL RESULTS

Transmitarray using perforated dielectric material is shown in Fig.1a. The transmitarray is constructed from one piece of perforated dielectric material sheet. The perforations result in changing the effective dielectric constant of the dielectric material. The simplicity of the structure makes it practical in terms of cost, space, and ease of fabrication. The dielectric sheet had a thickness of 14 mm and a relative permittivity of ε_r =12. The transmitarray is composed of 9 × 9 unit cell elements and is covered an area of 13.5×13.5 cm^2 in x-y plane. Single elliptical DRA element fed the transmitarray is shown in Fig.1b. The configuration of the proposed unit cell is shown in Fig.2a. A square cell with length L =15 mm, and substrate thickness h = 14 mm with ε_r =12 is used. The cell has four circular holes of equal diameters. The number of the holes in the cell element is optimized to maximize the transmission coefficient through the structure. The FIT method was used to

978-7-5641-4279-7

simulate the unit cell element of this transmitarray antenna. The unit cell is simulated in an infinite periodic array. A circularly polarized normal incident plane wave is illuminated the two-dimensional infinite array of similar unit cells. The variations of the transmission magnitude and phase versus the hole radius are illustrated in Fig.3. The results are compared with that calculated using the FEM method at 10 GHz to validate them. Good agreement between the two approaches is obtained. Simulation results are showing from 0° to 345° of phase variation with less than -4 dB of variation in transmission magnitude throughout the tuning range.

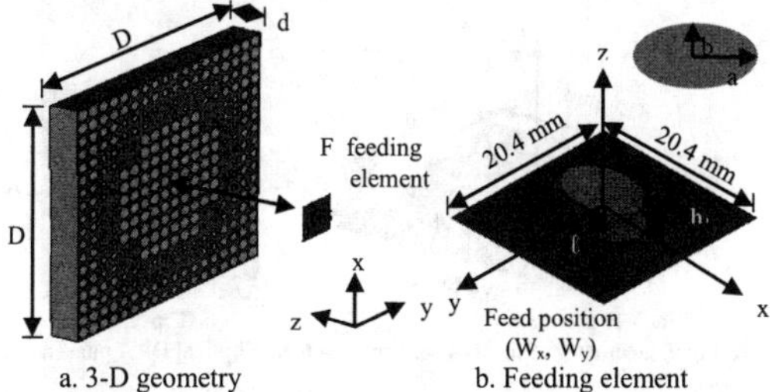

a. 3-D geometry b. Feeding element

Fig.1. Geometry of the 9×9 perforated transmitarray fed by single elliptic dielectric resonator antenna excited by a single probe.

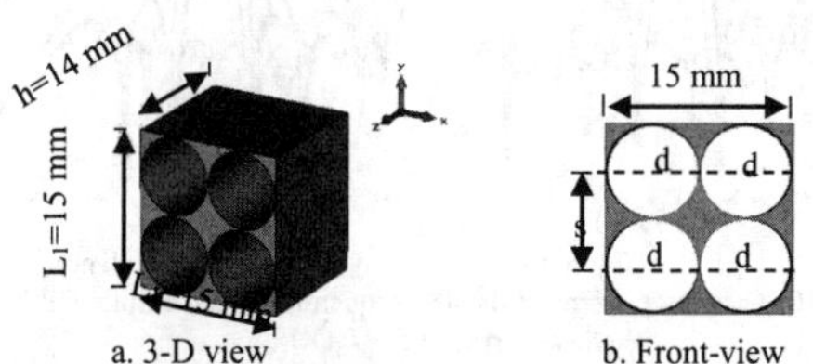

a. 3-D view b. Front-view

Fig.2. The detailed dimensions of the unit-cell and the waveguide simulator arrangement.

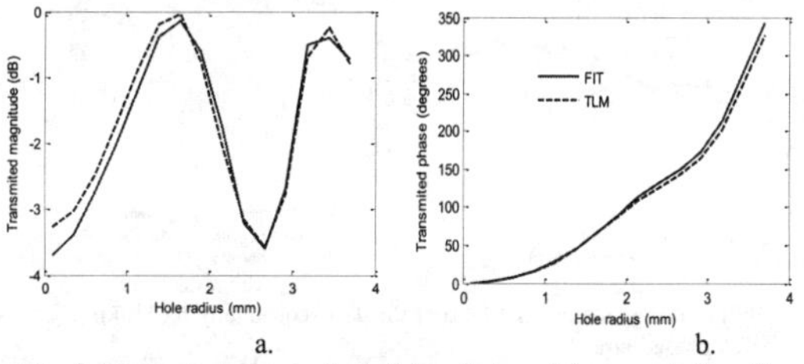

a. b.

Fig. 3. Transmission magnitude (a) and phase (b) versus hole radius.

In this paper, two numerical examples are given. In the first one, single elliptical DRA element fed the transmitarray is considered. Figure 1b shows the geometry of the structure. The semi-major axis a=5.25 mm, the semi-minor axis b=3.5 mm, h_l= 3.25 mm, ε_r= 12, ℓ= 2.78 mm and the feed probe radius of 0.25 mm. The probe is located (1.85 mm, 1.85 mm). The focal length–to-diameter ratio, F/D, is optimized for lower side lobe levels and highest transmitarray gain. Five transmitarrays are simulated using FIT for F/D = 0.5, 0.6, 0.7, 0.9, and 1. The results are summarized in Table I. As the total structures of antennas are very large, simulation with FEM method requires a huge amount of memory which makes it difficult. Consequently, the antennas were modeled and simulated by FIT method. The transmitarray is illuminated by the fields radiated from elliptical DRA antenna. The E-plane and H–plane patterns for the transmitarray are shown in Fig. 4 for F/D=0.6 and f=10 GHz. The output beam in this case points to the boresight direction, and directivity is 18.7 dBi that is 13.1 dBi higher than the value found for antenna alone (5.6 dBi). The transmitarray gain and axial ratio variations versus frequency are shown in Fig.5. The 1-dB gain variation bandwidth is 0.5 GHz. The band of circular polarization is not changed as the electromagnetic wave passes through the transmitarray.

Table I. Comparison between the radiation characteristics of 9×9 perforated transmitarray fed by single elliptic DRA at different ratios of F/D.

Array position F/D	Gain in dB at f= 10 GHz	1-dB gain variation BW	3-dB Axial ratio BW
1	16.2	0.5	zero
0.9	17.6	1.4	0.2
0.7	18.4	0.7	0.1
0.6	18.7	0.5	0.1
0.5	17.7	0.7	0.1

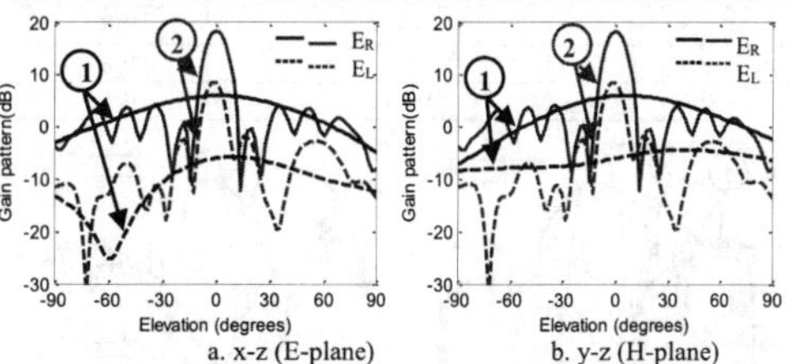

a. x-z (E-plane) b. y-z (H-plane)

Fig.4. The gain pattern and of the 9×9 perforated transmitarray fed by single elliptic DRA at F/D=0.6 at 10 GHz.

① Single DRA ② Transmitarray fed by Single DRA

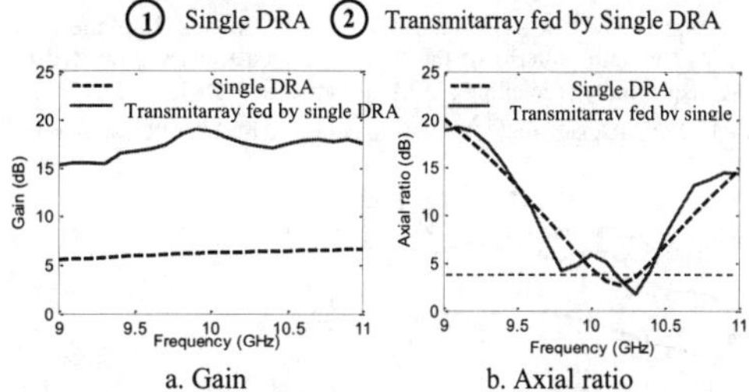

a. Gain b. Axial ratio

Fig.5. The gain and axial ratio of the 9×9 perforated transmitarray at F/D=0.6.

In the last one, four elliptical DRA elements fed the transmitarray is considered. Fig.6a shows the geometry of four elliptical DRA elements in such 2×2 sub-array improve the gain of the antenna. Each element is fed uniformly in power from orthogonal feed points F_1 and F_2. The elements in one diagonal are 90° out of phase and rotated 90° in orientation relative to the elements in the other diagonal as shown in Fig.6b. The results of circular polarization components are summarized in Table II for F/D = 0.5, 0.6, 0.7, 0.9, and 1. The E-plane and H–plane patterns for the transmitarray are shown in Fig.7 for F/D=1. Narrow beamwidth with HPBW of 13 degrees is obtained in both the E-plane and H–plane. The axial ratio versus frequency is shown in Fig.8. The sub-array provides an extremely broadband axial ratio bandwidth due to the orthogonality of the array elements in phase and orientation.

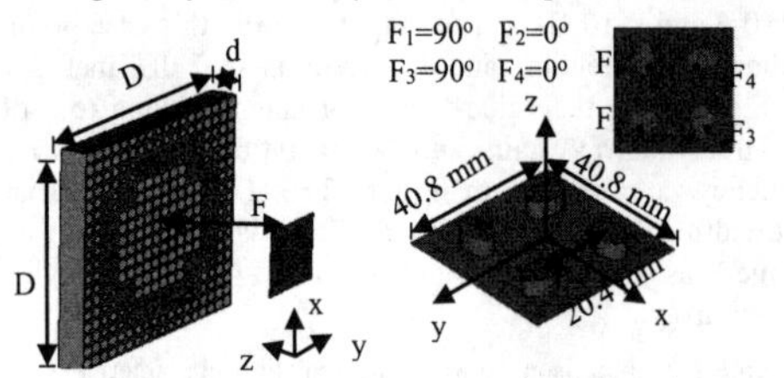

Fig.6. Geometry of the 9×9 perforated transmitarray fed by 2×2 sequentially feed elliptic DRA sub-array.

Table II. Comparison between the radiation characteristics of 9×9 perforated transmitarray fed by 2×2 sequentially feed elliptic DRA sub-array at different ratios of F/D.

Array position at F/D	Gain in dB at f= 10 GHz	1-dB gain variation BW
1	19	0.4
0.9	18.6	1.2
0.8	18	0.8
0.7	18.7	0.8
0.6	16.8	0.5

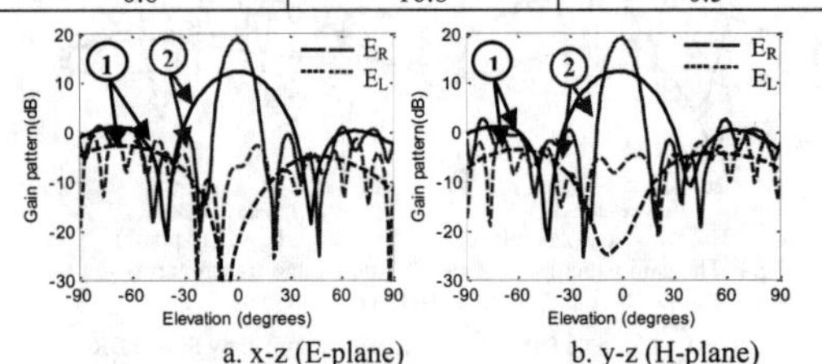

Fig.7. The gain patterns of the 9×9 perforated transmitarray fed by 2×2 sequentially feed elliptic DRA sub-array at F/D=1 and f= 10 Ghz.

① 2×2 DRA sub-array ② Transmitarray fed by 2×2 DRA sub-array

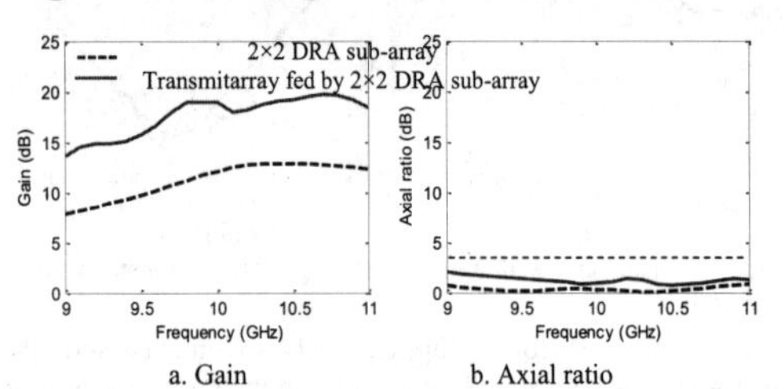

Fig.8. The gain and axial ratio of the 9×9 perforated transmitarray fed by 2×2 sequentially feed elliptic DRA sub-array at F/D=1.

To account for the number of DRA elements required for increasing the DRA gain and keeping on the circular polarization bandwidth 4×4 sequentially fed DRA phased array is investigated. Figure 9 shows a complete structure of the 4×4 sequentially fed DRA phased array in which the 2×2 DRA sub-array is employed and is repeated in sequential feeding manner. The E-plane and H–plane patterns for 4×4 sequentially fed DRA phased array compared with that of the transmitarray fed by 2×2 DRA sub-array are shown in Fig.10. The transmitarray introduces more directive beam with HPBW of 13 degree compared with 18 degrees for the 4×4 sequentially fed DRA phased array. The gain and axial-ratio variation versus frequency for 4×4 sequentially fed DRA phased array compared with that of the transmitarray fed by 2×2 DRA sub-array are shown in Fig.11. Approximately the same behaviors of the gain and axial ratio variation versus frequency are obtained with the transmitarray with reduction of the DRA elements to about 75%.

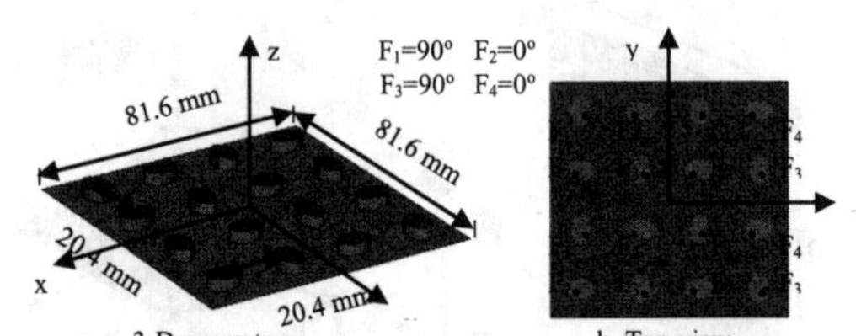

Fig.9. Geometry of the 4×4 sequentially feed elliptical DRA phased array.

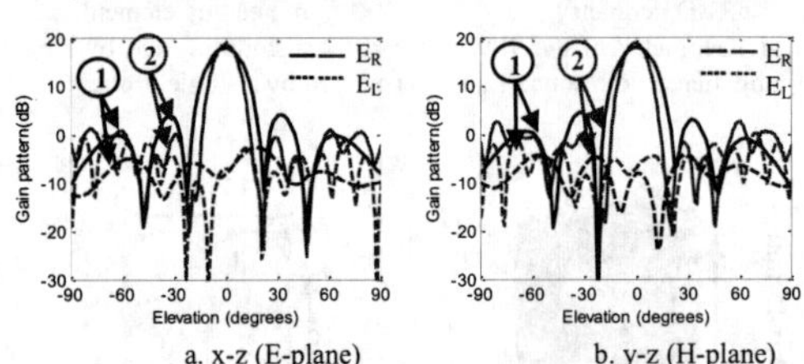

Fig.10. The gain patterns of the 4×4 sequentially feed elliptic DRA phased-array at f= 10 GHz.

① 4×4 DRA phased-array ② Transmitarray fed by 2×2 DRA sub-array

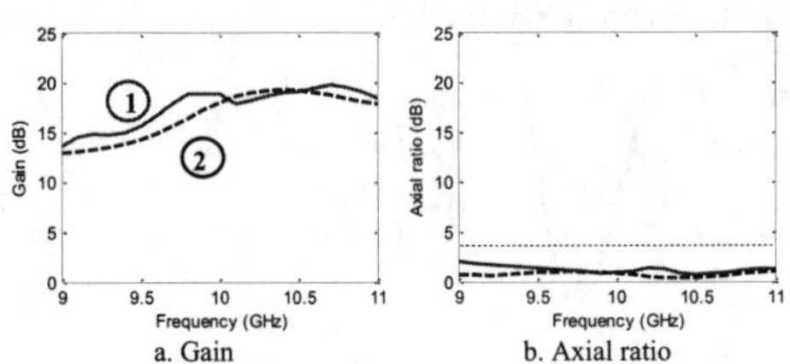

Fig.11. The gain and axial ratio of the 4×4 sequentially feed elliptic DRA phased-array.

A. Dual-beam transmitarray

Many applications, as crash avoidance systems, imaging systems, base stations for wireless communication systems, etc., require antennas capable of supporting several radio links at the same time. Dielectric lens antennas with multiple feeds are used in [18]. In this section, two transmitarray structures each one with single-feed dual-beam will be investigated for covering two regions at the same time. The first designed transmitarray has main beam at $\theta_o = \pm 25^o$ in x-z plane (H-plane) as shown in Fig.12a and the second transmitarray has main beam at +20° and -45° in x-z plane (H-plane) as shown in Fig.12b. Each transmitarray is designed at 10 GHz and is fed by 2×2 DRA sub-array circularly polarized DRA with F/D ratio is set to 1. The cell phase compensation's is used to compensate the transmitted wave's different paths to be coherent in phase at certain plane. The gain patterns in x-z plane of the two dual beam transmitarray are shown in Fig.13. The first transmitarray introduces dual beam at ± 25° in x-z plane with maximum gain of 12.3 dB while the second transmitarray introduces dual beam at + 20° and -45° in x-z plane with maximum gain of 13.1 dB. The gain and axial ratio variations versus frequency are shown in Fig.14.

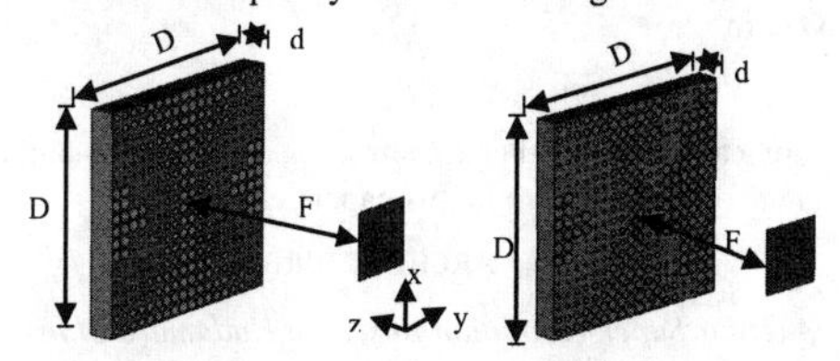

a. Dual beam at ±25° b. Dual beam at +20°, -45°
Fig.12. Geometry of the dual beam 9×9 perforated transmitarray fed by 2×2 sequentially feed elliptic DRA sub-array.

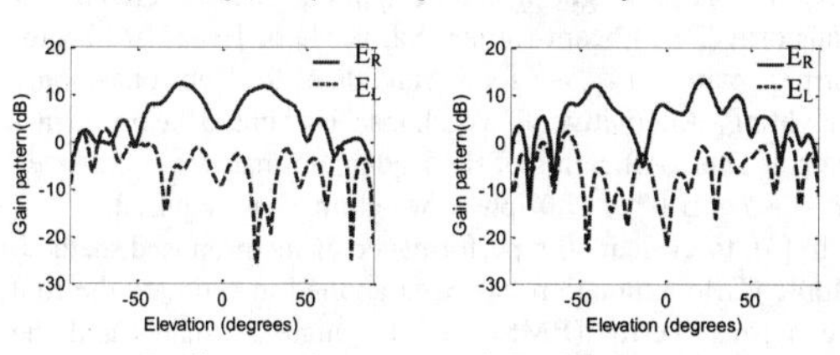

a. Dual beam at ±25° b. Dual beam at +20°, -45°
Fig.13. The gain pattern of the dual beam 9×9 perforated transmitarray fed by 2×2 sequentially feed elliptic DRA sub-array.

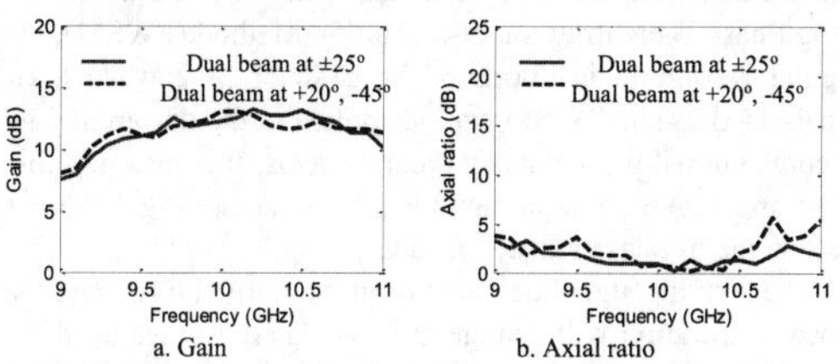

a. Gain b. Axial ratio
Fig.14. The gain and axial ratio variation versus frequency for the dual beam 9×9 perforated transmitarray fed by 2×2 sequentially feed elliptic DRA sub-array.

III. CONCLUSIONS

In this paper, the radiation characteristics of DRA antenna characteristics combined with the transmitarray to form high directive beam systems are investigated. The transmitarray is constructed from one piece of perforated dielectric material sheet. The transmitarray is composed of 9x9 unit cell elements and is covering an area of 13.5x13.5 cm^2. Two numerical examples are considered. Single elliptical DRA element fed the transmitarray is investigated in the first example. The directivity is 18.7 dBi that is 13 dBi higher than the value found for antenna alone (5.6 dBi). The band of circular polarization is not changed as the electromagnetic wave passes through the transmitarray. In the second example, four elliptical DRA element fed the transmitarray is considered. Narrow beamwidth with HPBW of 13 degrees is obtained in both the E-plane and H-plane. For comparison 4x4 sequentially fed DRA phased array is considered. Approximately the same behaviors of the gain and axial ratio variations versus frequency are obtained with the transmitarry fed by 2x2 DRA sub-array with reduction of the DRA elements to about 75%. Also, two transmitarray structures each one with single-feed dual-beam is investigated for covering two regions at the same time. The first designed transmitarray has main beam at θ_o = ± 25° in x-z plane and in the second transmitarray has main beam at +20° and -45° in the same plane.

REFERENCES

[1] D. Jackson and N. Alexopoulos, "Gain enhancement methods for printed circuit antennas," *IEEE Trans. Antennas Propagat.*, vol. 33, no. 9, pp.976–987, Sep. 1985.

[2] H. Vettikalladi, O. Lafond, and M. Himdi, "High-efficient and high-gain superstrate antenna for 60-ghz indoor communication," *IEEE Antenna Wireless Propagat. Lett.*, vol. 8, pp. 1422–1425, 2009.

[3] Y. J. Lee, J. Yeo, R. Mittra, and W. S. Park, "Application of electromagnetic bandgap (EBG) superstrates with controllable defects for a class of patch antennas as spatial angular filters," *IEEE Trans. Antennas Propagat.*,vol. 53, no. 1, pp. 224–235, Jan. 2005.

[4] A. Foroozesh and L. Shafai, "Investigation into the effects of the patch type fss superstrate on the high-gain cavity resonance antenna design," *IEEE Trans. Antennas Propagat.*, vol. 58, no. 2, pp. 258–270, Feb. 2010.

[5] H. Attia, L. Yousefi, M.M. Bait-Suwailam, M.S. Boybay, and O.M. Ramahi, "Enhanced-gain microstrip antenna using engineered magnetic superstrates," *IEEE Antenna Wireless Propagat. Lett.*, vol. 8, pp. 1198-1201, 2009.

[6] K. M. Luk and K. W. Leung, Dielectric Resonator Antennas: Research Studies Press LTD Baldock, Hertfordshire, England, 2003.

[7] V. G. Veselago, "The electrodynamics of substances with simultaneously negative values of e and fl," Sov. Phys. Usp., vol. 10, pp. 509-514, 1968.

[8] K. Jaewon and A. Gopinath, "Application of cubic high dielectric resonator metamaterial to antennas," in IEEE Antennas and Propagation Society International Symposium, Honolulu, HI, pp. 2349-2352, USA, 2007.

[9] B.-I. Wu, W. Wang, J. Pacheco, X. Chen, T. M. Grzegorczyk, and J. A. Kong, "A study of using metamaterials as antenna substrate to enhance gain," Progress In Electromagnetics Research B, PIER B, vol. 51, pp. 295-328, 2005.

[10] S.H. Zainud-Deen, Mourad S. Ibrahim, and A.Z. Botros, "A High-Directive Dielectric Resonator Antenna Over Curved Ground Plane Using Metamaterial," 28th National Radio Science Conference (NRSC 2011), National Telecommunication Institute, April, 2011, Egypt.

[11] Ahmed A. Kishk, "DRA-array with 75% reduction in elements number," Radio and Wireless Symposium (RWS), Austin, TX, USA, pp.70-72, January 2013.

[12] A. Abbaspour-Tamijani, L. Zhang, and H.K. Pan, "Enhancing the directivity of phased array antennas using lens-arrays," Progress In Electromagnetics Research M, PIER M, vol. 29, pp.41-64, 2013.

[13] J. Y. Lau, and S. V. Hum," Analysis and characterization of a multipole reconfigurable transmitarray element," IEEE Trans. Antennas Propag.,vol. 59, no. 1, January 2011.

[14] C. G. M. Ryan, M. R. Chaharmir, J. Shaker, J. R. Bray, Y. M. M. Antar, and A. Ittipiboon," A wideband transmitarray using dual-resonant double square rings," IEEE Trans. Antennas Propagat., vol. 58, no. 5, pp.1486-1493, May.2010.

[15] S.H. Zainud-Deen, Sh. M. Gaber, and K.H. Awadalla, "Transmitarray using perforated dielectric material for wideband applications," Progress in Electromagnetics Research M, PIER M, vol.24, pp.1-13, 2012.

[16] A.C. Polycarpou, Introduction to the Finite Element Method in Electromagnetics, Morgan and Claypool, USA, 2006.

[17] M. Clements, and T. Weiland, "Discrete electromagnetism with the finite integration technique," Progress in Electromagnetics Research, PIER, vol.32, pp.65-87, 2001.

[18] K.K. Chan, G.A. Morin and S.K. Roa, "EHF multiple beam dielectric lens antenna," in IEEE Antennas and Propagation Society International Symposium, Newport Beach, CA, vol.1, pp.674-677, USA, 1995.

Research on a Novel Millimeter-wave Linear MEMS Array Antenna Based on Hadamard Matrix

Shu-yuan Shi, Yong-sheng Dai
Nanjing University of Science & Technology
No.200 Xiaolingwei, Nanjing, China
shuyuan_shi90@163.com

***Abstract*-In this paper, a new angular super-resolution technique called Phase Weighting Super-resolution Method (PWSM) based on Hadamard matrix is proposed, which is used in conventional phased array radar for improving angle resolution. The phase shifting necessary to steer the main beam of the aperture coupled micro-strip antenna is achieved by 0-π MEMS phase shifters, which are integrated in a hybrid fashion between the antenna and feed network. The results of the measured pattern are consistent with the theoretical calculation results.**

I. Introduction

It is well known that the angle resolution is very important in some radar applications. In order to improve the angular resolution, we use an antenna with a large diameter, which leads to an increase in cost, volume, and wind resistance. Due to this reason angular super-resolution has gained a great concern. It can be applied to either element-space or beam-space observations. In the element space-estimations, there have appeared many Fourier transform-based algorithms beyond the conventional methods. Individual complex response can be obtained through the element-space estimation. In this paper, PWSM is a combination of the phase weighting method and the non-linear spectral estimated algorithm, and to realize the angular super-resolution for the conventional phased array radar, beam-space data are observed in [3] by using a Hadamard matrix as the beam-former matrix.

We design a double-layer aperture-coupled micro-strip antenna. Micro-strip antenna has quite a few advantages such as a lower profile, a smaller size and a lighter weight and so on. Double-layer aperture-coupled antenna with thick substrate can achieve a wideband. To obtain a large gain and achieve a specific directivity, an antenna array is often applied. Our antenna array utilizes an 8-element linear array with combination of the feed structure.

Recently, RF Micro-electro-mechanical system (MEMS) switch has drawn much attention due to its rapid isolation and connection, low insertion loss, low turn-on voltage, high reliability, as well as good performance of the reconfiguration in microelectronics, which is better than the PIN diode and MESFET switch. Based on MEMS switch, MEMS phase shifter, switch filter and switch array can be made and widely used in radar, satellite communications, wireless communications and other systems, e.g. in the application of the antenna arrays shown in this paper.

II. Architecture

A. Angular Super-resolution Based on Hadamard Matrix

In most of the conventional phased array radar, only the phase corresponding to each antenna element can be adjusted, i.e., only phase weighting can be applied. The discrete Fourier transform (DFT) beam-former matrix [1] or Hadamard beam-former matrix [4] is good candidate for the phase-only weighting. Alternatively, we choose Hadamard beam-former matrix. The coefficients of the Hadamard matrix are either +1 or –1, so only 0° or 180° phase weightings are required.

In [3], to evaluate the performance of the proposed method, Monte Carlo simulation has been applied to estimate the root mean square error (RMSE) of the angle estimates and the spatial resolution SNR threshold in cases of either nonfluctuating targets or fluctuating targets. For an X band conventional phased array antenna with 139 elements, by using Phase Weighting Super-resolution Method (PWSM), the angular resolution is improved by a factor of 2 when SNR equals 15 dB. The PWSM can be applied in the design of new or conventional phased array radar systems. It is also possible to be applied in some communication systems. Fig. 1 shows the scheme of phased array antenna.

If the arriving signal deviates from the normal direction, the phase distribution will change quickly. Therefore we need the space intensive sampling. However, in the phased antenna array, it is easy to align scanning beam to the target. We use a space sparse sampling to reduce the number of phase weight statuses and accelerate the real-time processing speed. For example, after aligning the transmitting and receiving beam to the target of interest, we choose the appropriate value of M to form Hadamard matrix and realize the super-angle resolution. In this way, the number of phase weight statuses is reduced to half of the element number of the antenna array.

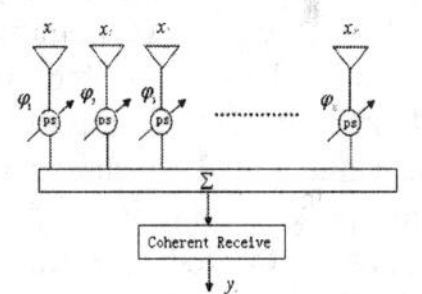

Fig.1. Scheme of phased array antenna.

978-7-5641-4279-7

B. *The Design of Double-layer Aperture-coupled Micro-strip Antenna Array*

Based on the requirement and the recent situation in relevant fields, we proposed the issue: millimeter-wave MEMS 0-π phased one-dimension 1×8 super-resolution array antenna, whose feed network, phase scanning, antenna array are respectively realized by micro-strip antenna array, micro-strip feeder, millimeter-wave MEMS and drive circuit. In this way we can use existing technique and qualification to realize super-resolution character by using millimeter MEMS vary the phase of 0-π. The radiate implement utilizes micro-strip antenna unit and RT/5880 substrate, whose permittivity is 2.2. The input port utilizes W28 waveguide. The guide line is: center frequency at 35GHz, bandwidth larger than 5%, the largest gain better than 13dB when all units of array are at the same phase. One difficulty is the solution of the mutual coupling of the antenna array and the feed structure.

The three-dimensional analysis of aperture coupled micro-strip antenna is shown in Fig.2. It has a rectangle patch of size $a \times b$. The micro-strip patch fed by the micro-strip line through the aperture or the crack in common ground is also shown in Fig. 1. The aperture is with center of (x_0, y_0) and size of $L_a \times W_a$. The width of micro-strip line is W. The thickness of substrate is t.

In our design, we choose the double-layer media made of RT/5880 substrate with the thickness of 0.254 mm. There is an air gap of 0.5 mm thickness between the double media, which can be used to widen the wideband of our antenna. With the aid of IE3D, we get the patch length is 2.7mm and the width is 3.4mm. The coupling gap length is 2.4 mm and the width is 3.4 mm. The micro-strip feed is 0.73 mm beyond the center of gap, whose width is 0.78 mm. While debugging, we notice that coupling resistance reduces significantly as the width of patch increase, which indicates the gap coupling declines. The resonant frequency is sensitive to the length of patch. Generally gap length should not be too much long, otherwise back radiation will increase, but coupling cannot be ensured if it is too short. So, we must compromise while designing. The VSWR and the gain are shown in Fig. 3.

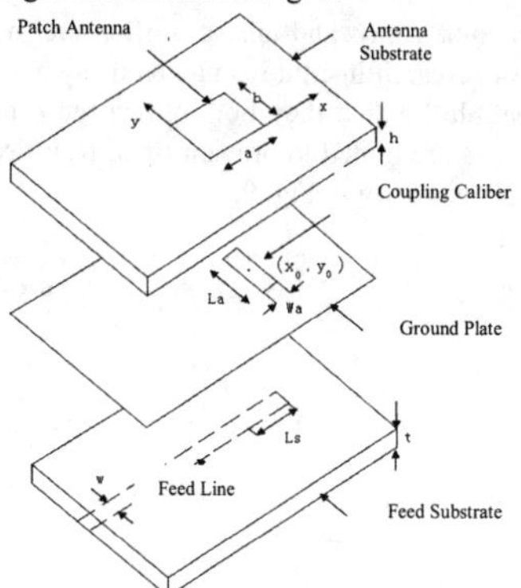

Fig. 2. Antenna model analysis.

To obtain a larger gain and achieve a specific directivity, the simplest way is a linear array. Due to the simple and compact feed network the transmission loss of series-fed array [5] is small. The combination of multiple series feed array can easily meet the beam required.

The antenna array is fed by one divide eight power divided network, in which the distance from feed source to every unit is all uniform. The distance between antenna unites is 0.8 λ_0 .The 1×8 coupling antenna array simulation block diagram is shown in Fig. 4. The simulation results of the antenna array are shown in Fig. 5.

We use a waveguide to micro-strip probe device to transform waveguide to micro-strip, which occupied a small space, the waveguide and micro-strip are orthogonal to each other. Too big waveguide hatch will arouse seriously leak and influence the pattern. While it is hard to be fixed and completely separate the waveguide from the micro-strip with too small waveguide patch, which will lead to short-circuit. Through the optimization, we get that the patch length is 1.6mm, the width is 0.66mm, and the distance from probe to jumper is 1.35mm. The feeding structure is shown in Fig. 6.

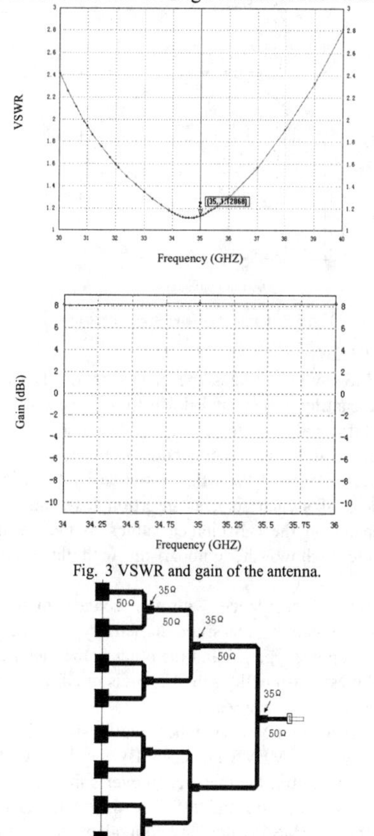

Fig. 3 VSWR and gain of the antenna.

Fig. 4 Simulation block diagram of coupling antenna array.

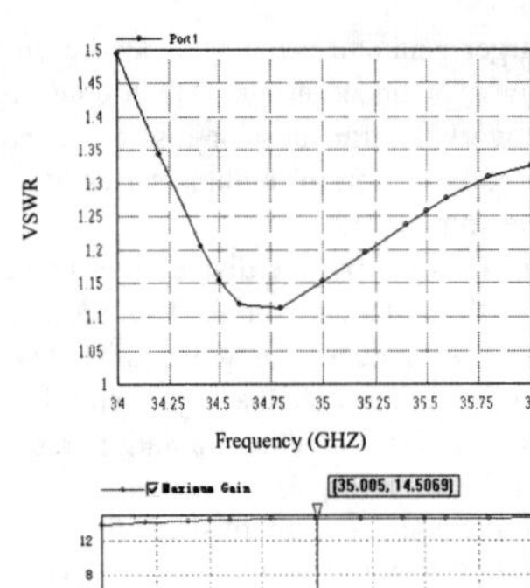

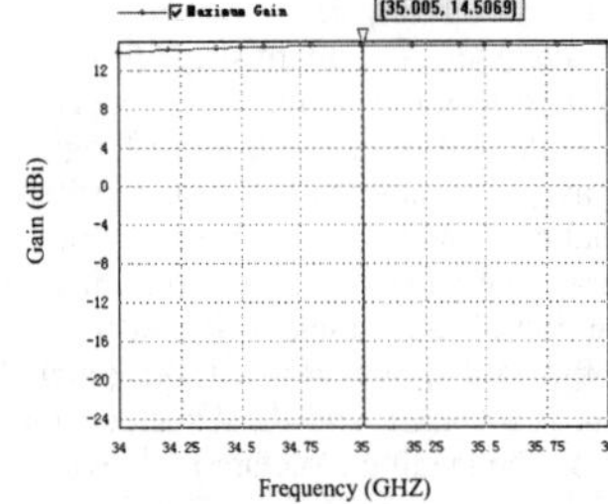

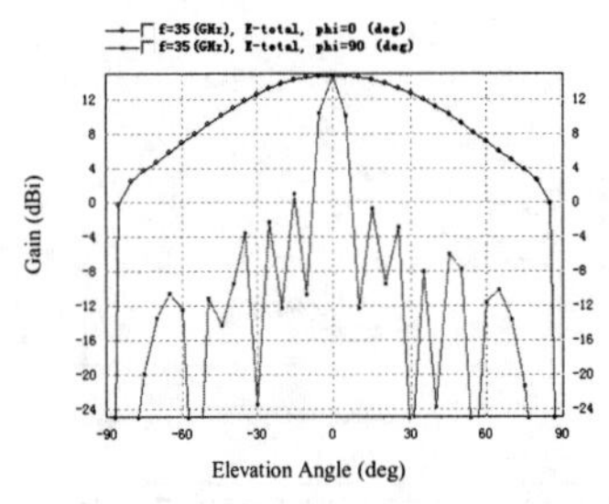

Fig. 5 Performance of the array antenna.

C. *MESFET Switches*

RF MEMS switch utilizes the electrostatic adsorption to control the cantilever's up and down to achieve the switch's on and off, and eliminates the PN junction in the semiconductor and metal-semiconductor junction in the RF device. We can control the MEMS switch by DC bias. The performances of switch such as insertion loss and isolation degree depend on the capacitance values of the on and off states. Table I shows the comparison with three types of switches.

Generally, in accordance with the contact manner, RF MEMS switch can be classified in two types: one is the capacitive coupling type [2] and the other is the metal contact type [6]. We establish both of their models in HFSS to discuss their principles and structures.

Metal contact MEMS switches work similarly to the capacitive coupling MEMS switches. By the driving of static electric we can control the metal cantilever's movements, thus realizing the circuit's on and off. The difference is a metal contact type MEMS switch is equivalent to a variable capacitance connected in series in the circuit.

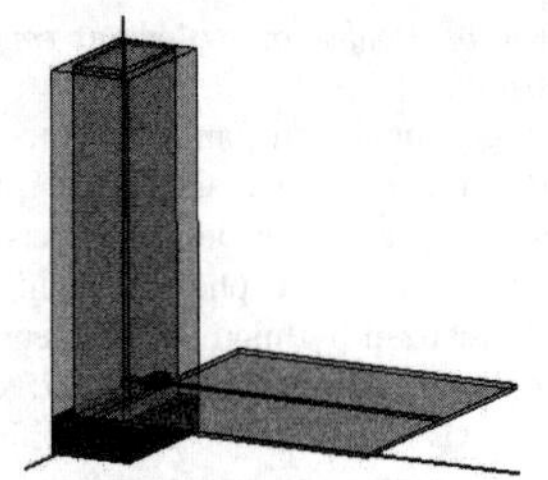

Fig. 6. Feed structure.

TABLE I
PROPERTIES SUMMARY OF THREE TYPES OF SWITCHING DEVICES

Type of switching devices	MEMS switches	PIN diodes	GaAs MESFET switches
Insertion loss	Excellent	Good	Good
Isolation degree	Excellent	Good	Good
Power capacity	Excellent	Good	Bad
Driving power	Excellent	Bad	Excellent
Switch speed	Bad	Good	Excellent
Cost	Good	Good	Bad

D. *The Switched-line MEMS Phase Shifter*

The phase shifter above is no longer applicable when the frequency rises to the millimeter wave band. The application of MEMS technology to the millimeter-wave phase shifter can greatly reduce size and weight, especially for aerospace weapons.

According to the analysis above, we need the phase shifter to realize the super angle resolution. We choose metal-contact cantilever MEMS series switch to control the 0-π phase shifter. As is shown in Fig. 7, the phase shifting delta phi is 179.88° to ensure that the signal into the antenna array is of equal amplitude and inverted.

The photo of physical metal-contact cantilever MEMS serious switch is present in Fig. 8 (a). The switched-line MEMS phase shifter is present in Fig. 8 (b).

III. MEASUREMENT

The antenna array and phase shifter are both made by Nanjing 14 research institute. Antenna array and phase shifter are compatible for that they both utilize the same media. All metal patches are gilded to prevent them from oxidation. The final sample is shown in Fig. 9.

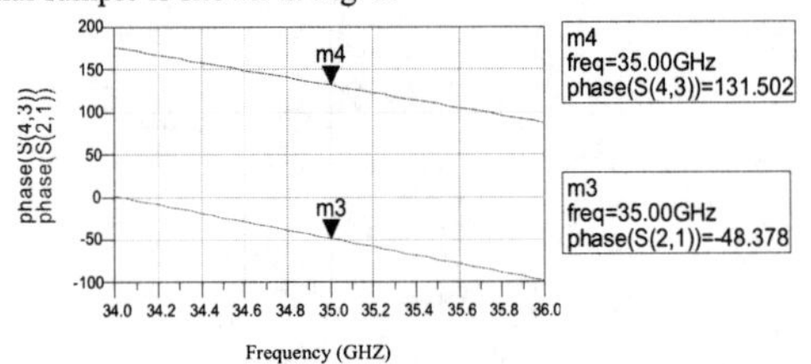

Fig. 7. Phase shifting of phase shifter.

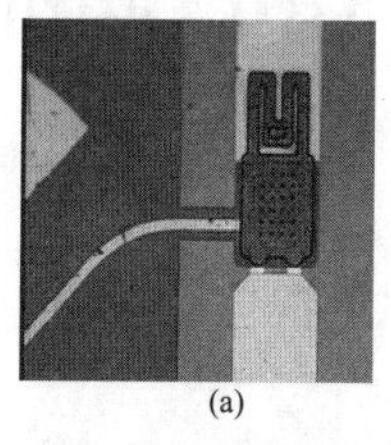

(a)

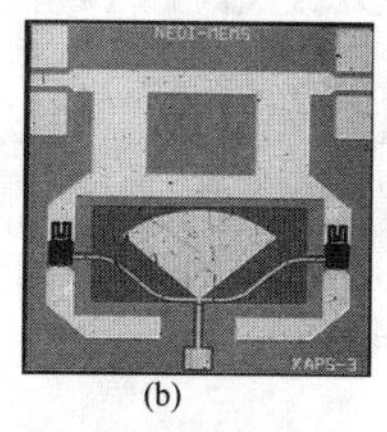

(b)

Fig. 8. Physical metal-contact cantilever MEMS switch and phase shifter. (a) MEMS switch. (b) MEMS phase shifter.

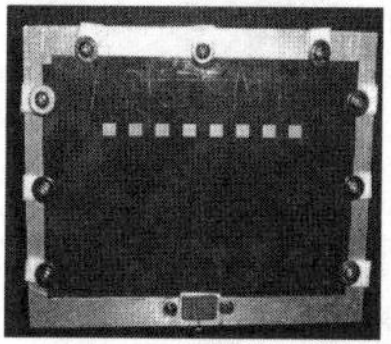

Fig. 9. Sample of antenna.

Simulate (dotted line) and measured (solid line) VSWR is shown in Fig. 10 (a). The simulated and measured gain of antenna array is shown in Fig. 10 (b).When eight antennas are all at the same shift phase of 0°, the measured pattern is shown in Fig. 10 (c). When shift phase of eight antennas are alternate with 0°and 180°, the measured pattern is shown in Fig. 10 (d).When the first four antennas are at the shift phase of 0° and the other antennas are at the shift phase of 180°, the measured pattern is shown in Fig. 10 (e). It can be seen that the results of the measured pattern are consistent with the theoretical calculation results. The join of the phase shifter in antenna array results in the small distance between the feed network and itself.

We chose a phase shifter online testing, i.e., by putting a phase shifter into a uniform antenna array and measuring a received signal's amplitude and phase distribution, we can estimate the amplitude and phase distribution of the caliber of the antenna according to the inverse change of Walsh-Hadamard matrix, then judge the working condition of the phase shifter. In millimeter wave band, especially when the phase shifter and the antenna array integrated together, a separate test of phase shifter is very difficult, so the method above is a valid and reliable one.

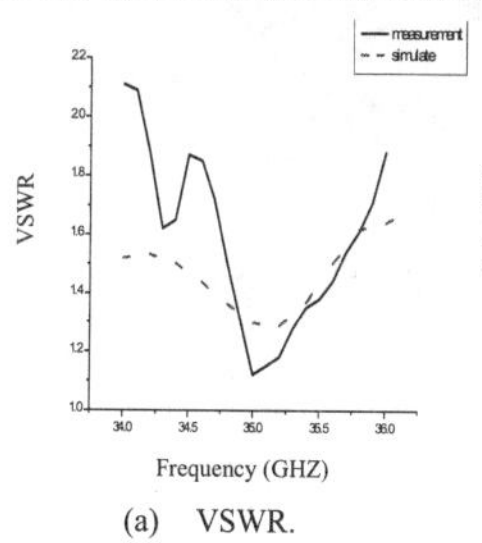

(a) VSWR.

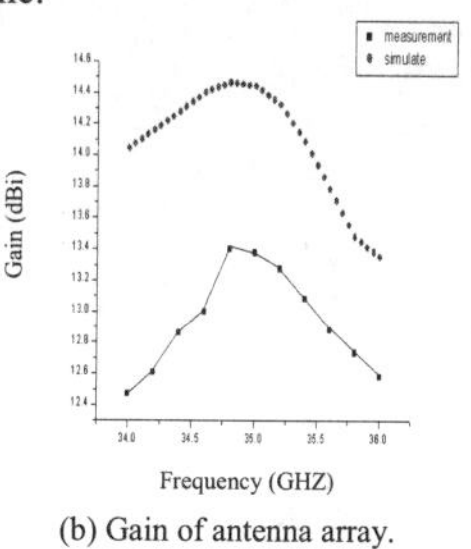

(b) Gain of antenna array.

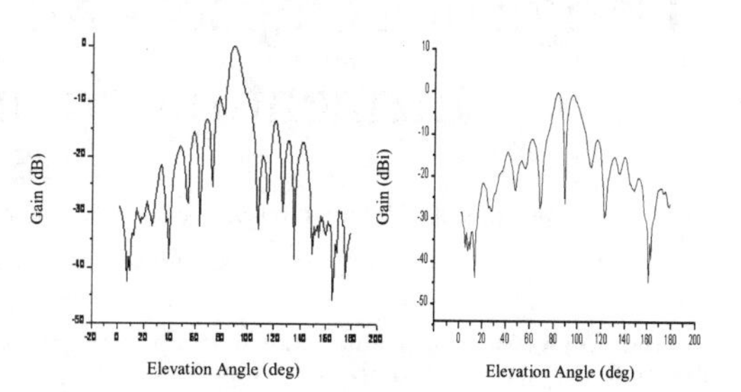

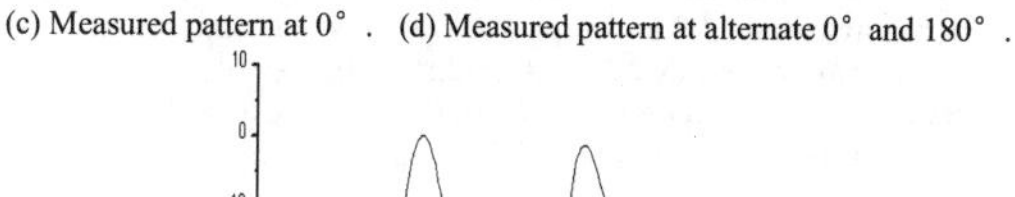

(c) Measured pattern at 0° . (d) Measured pattern at alternate 0° and 180° .

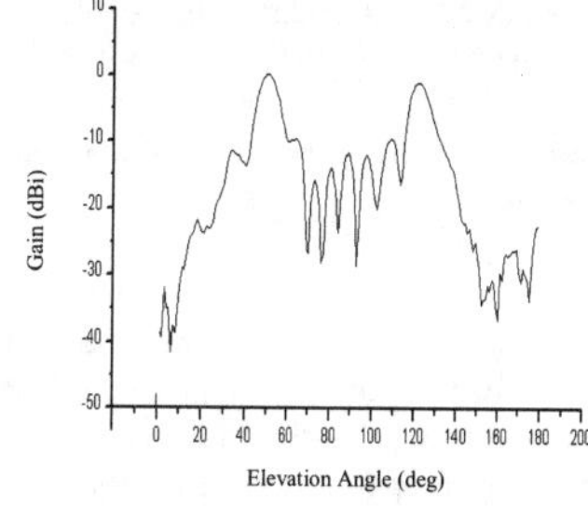

(e) Measured pattern at the shift phase of 0° in the first four antennas and 180° in the other antennas.

Fig. 10. Performance of measured pattern.

IV. CONCLUSIONS

In this work an 8×1 MEMS antenna array with phase shifter for 34.2GHz-36GHz has been designed. The VSWR of the antenna is better than 2. The non-inverting gain is better than 13 dB and the horizontal beam-width is 5.8°-6.2°.

ACKNOWLEDGMENT

The authors are grateful to Nanjing University of Science & Technology.

REFERENCES

[1] Zoltowski, M. D., Kautz, G. M., and Silverstein, S. D. 1993 Beamspace Root-MUSIC.*IEEE Transactions on Signal Processing*, 41, 1993, pp. 344-364.

[2] C. Goldsmith, T. H. Lin, B. Powers, W. R. Wu, Norvell B, "Micromechanical membrane switches for microwave applications" *Tech. Digest, IEEE Microwave Theory and Techniques Sypm*, 1995, pp.91-94.

[3] W. X. Sheng, D. G. Fang, "Angle super-resolution for phased antenna by phase weighting", *IEEE Transactions on Afrospace and Electronic Systems*, 2001, VOL. 37, NO. 4.

[4] Fang, D. G., Sun, J. T., Wang, Y., and Sheng, W. X. 1995. "Realization of radar cross-range resolution by array phase weighting," *Chinese Journal of Radio Science*, 10, 1995, pp. 1-3.

[5] J. X. Yin, K. C. Liu, "Calculation of series-feed micro-strip antenna array's gain and pattern," *National University of Defense Technology*, Vol. 22, pp. 43-46.

[6] J. J. Yao, M. F. Chang, "A surface micromachined miniature switch for telecommunications applications with signal frequencies from DC up to 4 GHz, " *Tech. Digest, 8th Int. Conf. on Solid-State Sensors and Actuators*. 1995. pp. 384-387.

Design of Compact 4X4 X-band Butler with Lump Element Based on IPD Technology

Lu Ning, Liguo Sun
Department of Information Science and Technology, University of Science and Technology of China
Email: ninglu@mail.ustc.edu.com

Abstract—A X-band 4×4 Butler Matrix in IPD (Integrated Passive Device) technology is presented. Coupler and the phase shifter constituted by lumped circuits are utilized for Butler miniaturization. The compact chip area of implemented IPD Butler only occupies 2.12×2.07 mm^2 including all test pads. The result show the isolation of >12dB, the average loss of 3dB and the phase errors are within 5° at 9.5-10.5GHz.

***Index Terms*—Lumped parameter, Butler matrix, IPD technology, X band**

I. INTRODUCTION

In modern communication systems, multi-beam and beam-scanning antennas are crucial. Butler matrices are important components in beam-forming antenna systems. A lot of ways has been proposed to build up butler matrix. Micro-strip line and suspended strip-line were used to get butler respectively in[1]-[3]. SIW (substrate integrated waveguide) also has been used in the implementation of the butler[4]. In the above traditional methods, good performance but large chip area appeared.

Miniaturization methods also have been reported. Butler are realized by the distribution parameter in 0.18um CMOS and IPD respectively[5][6]. Sizes of them are in millimeter, while traditional one is in decimeter. The size reduced greatly. However, the distribution parameter circuit has advantage in high frequency but not in low frequency because of the size. The lump parameter circuit is used in low frequency instead of the distribution parameter circuit. In this paper it is put forward that the lumped parameter circuit in IPD technology is used to achieve miniaturization of the butler under circumstances of relatively low frequency.

In the paper, a X-band 4-ways Butler Matrix composed by the lumped parameter in IPD technology is demonstrated. As far as we know, this is the first demonstration of using lumped parameter in IPD for butler.

II. DESIGN OF X BAND BULTER MATRIX

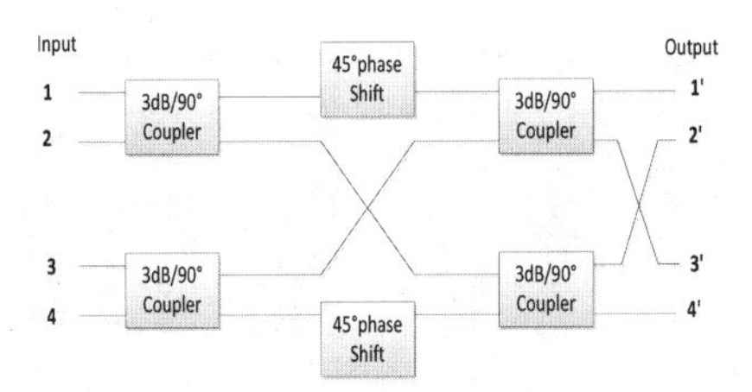

Figure1. The structure of 4×4 Butler

The general structure of 4×4 Butler is shown in Figure1. It consists of 3dB quadrature coupler and 45° phase shift. 4×4 Butler matrix is a circuit in which the input signal is equally divided among output ports, the adjacent phase shift between the output ports is equal and the value of the phase shift depends on the selected input port[1]. Table I presents retardation of each adjacent output signal when the signal from the four inputs, respectively.

TABLE I

OUTPUT PHASE DIFFERENCE

Input port	1	2	3	4
Retardation of adjacent output	-45°	135°	-135°	45°

A. Miniaturized Coupler by Lumped Parameter

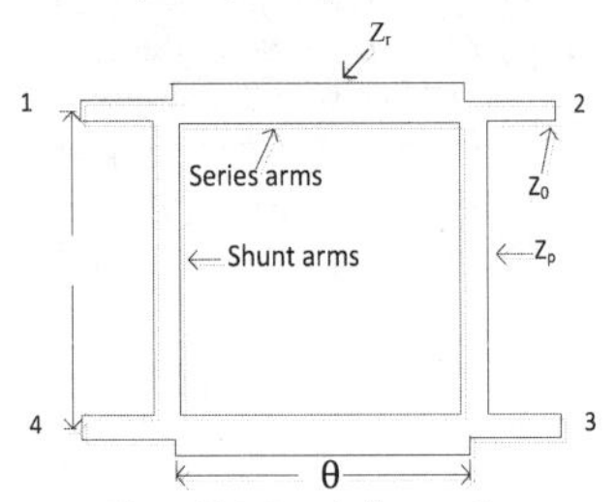

Figure 2(a). Branch line coupler

978-7-5641-4279-7

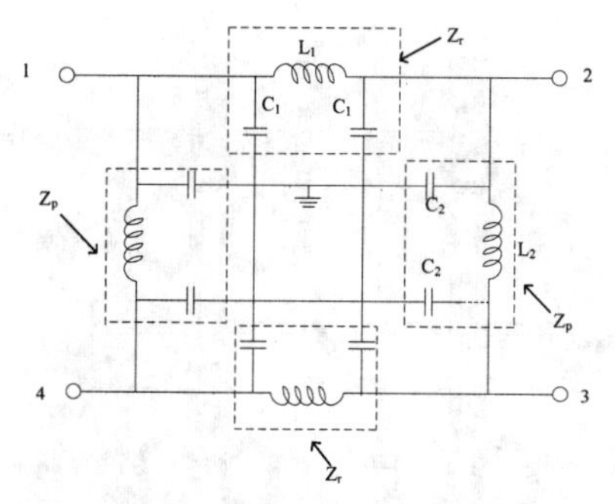

Figure 2(b). Coupler constituted of lumped parameter

Figure.2 (a) is the structure of branch line coupler and figure.2 (b) is the structure of lumped parameter coupler which is transformed by branch line coupler. Each transmission line can be equivalent to a π network. The conversion process is shown in following formula.

ABCD matrix of the transmission line[7],

$$\begin{pmatrix} A & B \\ C & D \end{pmatrix} = \begin{bmatrix} \cos\theta & jZ_r \sin\theta \\ j\dfrac{1}{Z_r}\sin\theta & \cos\theta \end{bmatrix} \tag{1}$$

ABCD matrix of the π network can be obtained by the matrix multiplying of three lumped elements.

$$\begin{pmatrix} A & B \\ C & D \end{pmatrix} = \begin{bmatrix} 1-\omega^2 L_1 C_1 & j\omega L_1 \\ j\omega C_1\left(2-\omega^2 L_1 C_1\right) & 1-\omega^2 L_1 C_1 \end{bmatrix} \tag{2}$$

By combining (1) and (2), (3) can be obtained

$$L_1 = \frac{Z_r \sin\theta}{\omega} \tag{3a}$$

$$C_1 = \frac{1}{\omega Z_r}\sqrt{\frac{1-\cos\theta}{1+\cos\theta}} \tag{3b}$$

$$L_2 = \frac{Z_p \sin\theta}{\omega} \tag{3c}$$

$$C_2 = \frac{1}{\omega Z_p}\sqrt{\frac{1-\cos\theta}{1+\cos\theta}} \tag{3d}$$

For traditional quadrature branch line coupler, θ=90°, $Z_r = \frac{Z_o}{\sqrt{2}}, Z_p = \sqrt{2}Z_r$ and Z_0=50Ω. The initial value is easy to calculate with above formula.

B. Phase Shift by Lumped Parameter

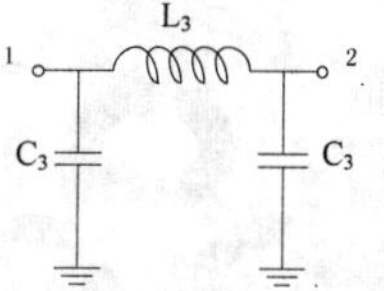

Figure3. The structure of 45°phase shift

Figure3 shows the 45° phase shift constituted of inductance and capacitance.

Due to the simple structure, circuits can be optimized by ADS software to get desired results.

TABLE II

THE ORIGINAL VALUE OF EACH PART COMPONENT

L_1	C_1	L_2	C_2	L_3	C_3
0.5627nH	0.4502pF	0.7985nH	0.3183pF	0.4204nH	0.1685pF

The original value of coupler and phase shift can be seen in Table II which comes from a series of calculation. The circuit simulation shows good performance with these original values. In the EM simulation component value will be further changed and optimized due to parasitic effects in the layout.

III. IPD TECHNOLOGY DESCRIPTION AND APPLICATION

A. Description of IPD

The IPD mentioned in this article is copper process based on high-resistivity silicon with ε_r=9.6. The structure of IPD is shown in Figure4. There are three metal layers (MCAP, M1 and M2) and two dielectric layers (PI-1and PI-2). M1 and MCAP are aluminum, M2 is copper and inductor is mainly produced in M2. The bottom and top plate of capacitors are respectively MACP and M1[8]. In order to reduce the area of inductors, M1 and M2 are combined to make it. M1 is right below M2.

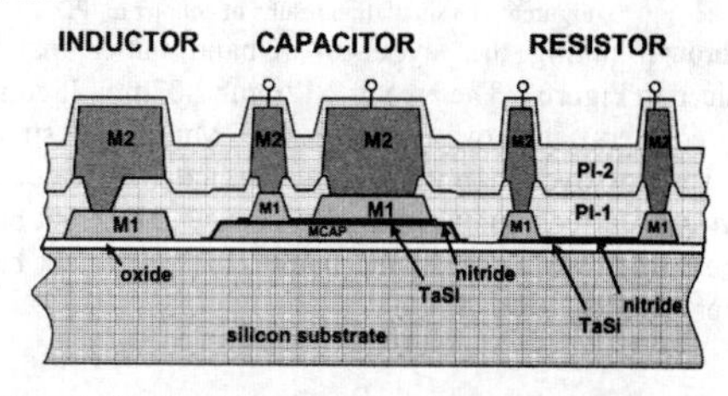

Figure4. Thin film Integrated Passive Device (IPD) structure [8]

B. Butler in IPD

After optimizing the circuit by ADS software, the layout of coupler can be seen in figure5 and the EM simulation result is in figure6. About 0.7dB insertion loss is produced because of dielectric loss and conductor loss. The phase difference between the port2 and port3 is 90° .

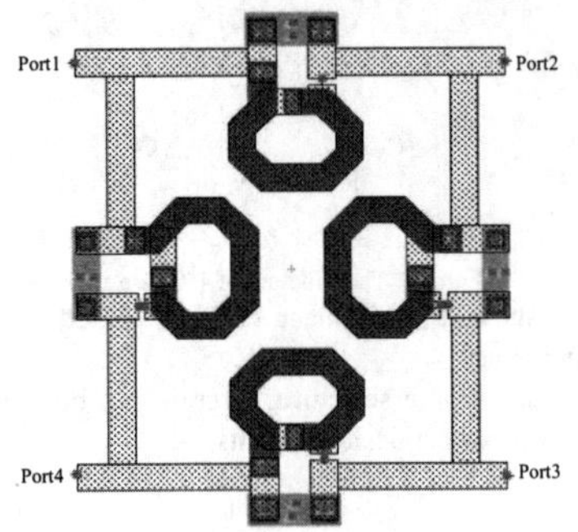

Figure5. The layout of coupler in IPD

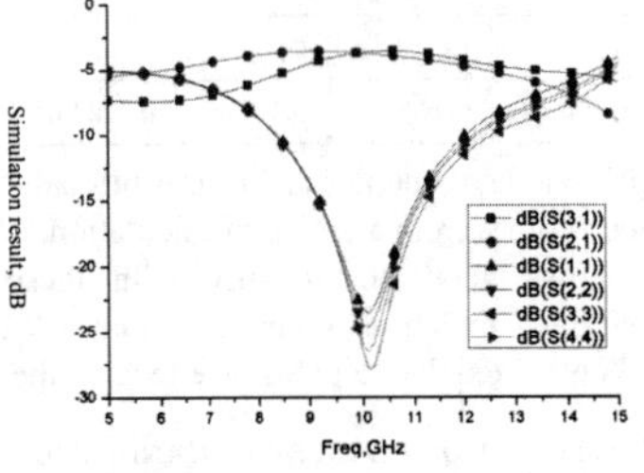

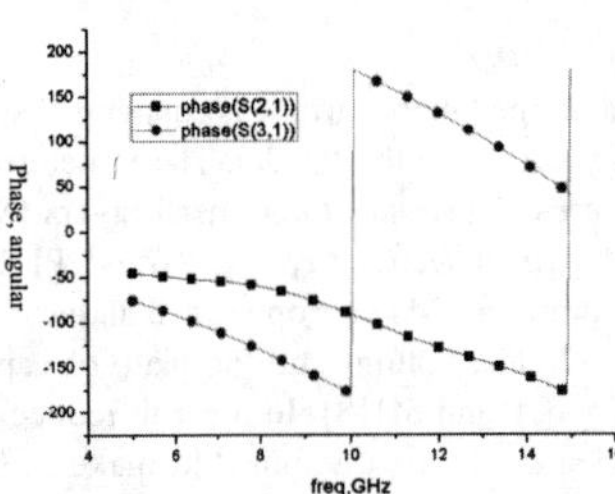

Figure6. The simulation results of coupler in IPD

Through tuning the layout of X-band butler matrix is obtained in Figure7. The area is 2.12mm×2.07mm. It contains four couplers and two phase-shifts. Multilayer structure achieves crossover simply and make the butler more compact. There are fourteen bumps in the IPD layout. The eight bumps on both sides are for input and output, the rest of six bumps are for electrical ground.

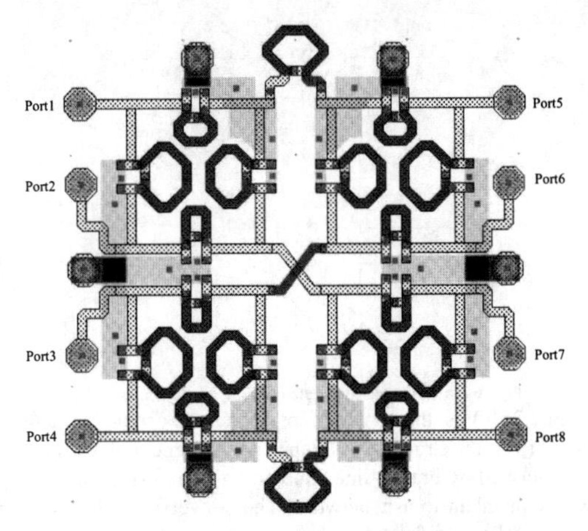

Figure7. The layout of desired Butler Matrix in IPD

EM simulation results for the layout is shown in Figure 8, in which, as an example, the port 1 is excited along with 45° phase shift between adjacent output ports. The overall phase error is less than 5° and the insertion loss is about 3dB within 9.5-10.5GHz.

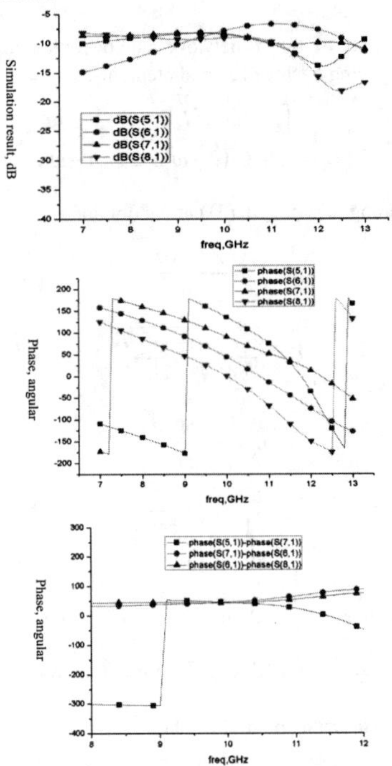

Figure8. Simulation results of desired Butler

IV. Validation of Results by IPD filter

In order to verify the accuracy of the ADS simulation

results in butler design, an IPD filter with the center frequency of 2.45GHz is presented in figure9. The filter was fabricated and measured. The photograph of filter is presented in figure10. The good agreement between ADS EM simulation and measurement results for the filter is shown in Figure11. Because the simulation setup in the filter design is same as that in our current butler design it indirectly verify the EM simulation of butler design in the paper.

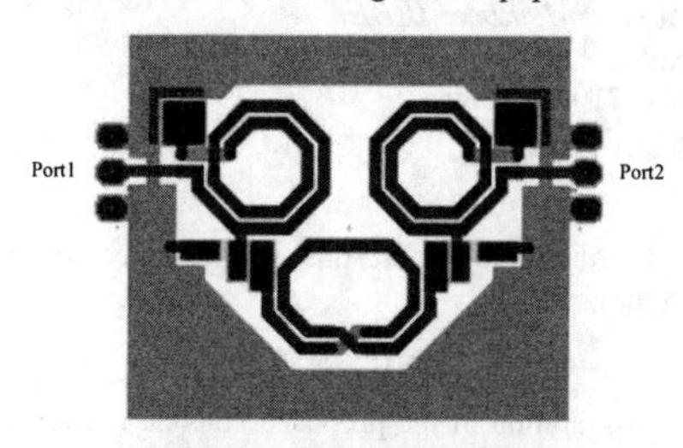

Figure9. The layout of filter in IPD

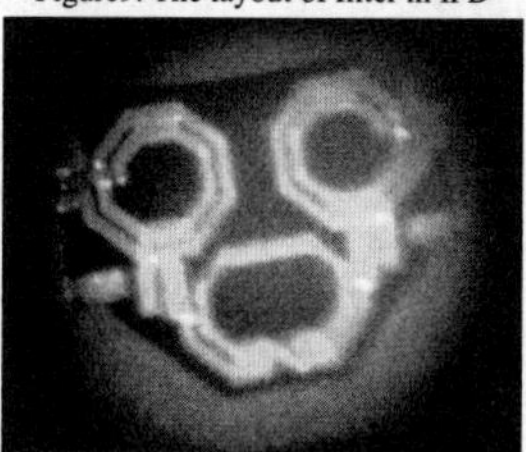

Figure10.Photograph of the IPD filter

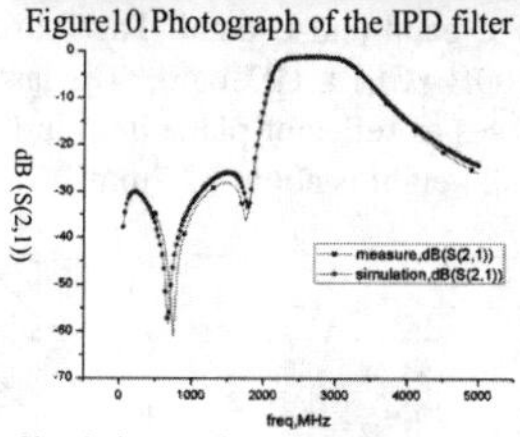

Figure11. Simulation result compared with measured result

V. CONCLUSION

A lumped butler matrix is presented in the paper. The butler is achieved with passive lumped elements in IPD technology. The proposed butler matrix is much smaller than a classic design based on transmission lines so that big size reduction has been achieved. The insertion loss is not as good as distributed one and therefore further study should be carried out.

ACKNOWLEDGMENT

The authors wish to acknowledge the assistance and support of the ISAP organizing committee.

REFERENCES

[1] Krzysztof Wincza, Slawomir Gruszczynski and Krzysztof Sachse, *"Ultrabroadband 4×4 Butler Matrix with the Use of Multisection Coupled-line Directional Couplers and Phase Shifters"*,Microwaves, Radar and Remote Sensing Symposium, pp.118-122, August 2011

[2] Th.Bechteler,B.Mayer, R.Weigel, *"A New High-temperature Superconducting Double-Hybrid Coupler With Wide Bandwidth"*, IEEE MTT-S Digest, pp.311-314, 1997

[3] M. Bona, L. Manholm, J. P. Starski,B. Svensson" *Low-Loss Compact Butler Matrix for a Microstrip Antenna",* IEEE trans. on microwave theory and techniques, vol.50, no.9, September 2002, pp.2069-2075

[4] Tarek Djerafi, Ke Wu, *"A Low-Cost Wideband 77-GHz PlanarButler Matrix in SIW Technology",* IEEE trans. on antennas and propagation , vol. 60, no. 10, October 2012,pp.4949-4954

[5] Ting-Yueh Chin, Sheng-Fuh Chang, Chia-Chan Chang, and Jen-Chieh Wu, *" A 24-GHz CMOS Butler Matrix MMIC for Multi-Beam Smart Antenna Systems",* IEEE Radio Frequency Integrated Circuits Symposium, pp.633-636, 2008

[6] Ibrahim Haroun, Ta-Yeh Lin, Da-Chiang Chang, Calvin Plett, *"A Compact 24-26 GHz IPD-Based 4x4 Butler Matrix for Beam Forming Antenna Systems"*, APMC 2012, Kaohsiung, Taiwan, pp.965-967

[7] Bahl,I.J., *"Lumped Elements for RF and Microwave Circuits",* ARTECH HOUSE, 2003

[8] Hyun-Tai Kim, Kai Liu, Robert C. Frye, Yong-Taek Lee, Gwang Kim and Billy Ahn, *"Design of Compact Power Divider Using Integrated Passive Device (IPD) Technology,"*2009 Electronic Components and Technology Conference

Design of a Dual-band Shared-Aperture Antenna Based on Frequency Selective Surface

Bo Yan, Yuanming Zhang, Long Li*, *Senior Member, IEEE*
School of Electronic Engineering, Xidian University
No.2 South Taibai Road, Xi'an, Shaanxi 710071, China
*Corresponding email: lilong@mail.xidian.edu.cn

Abstract **- A new design of a dual-band shared-aperture antenna based on frequency selective surface (FSS) for synthetic aperture radar (SAR) applications is presented. Bow-tie patches are used as the radiating elements both at L-band and C-band. The centre frequency of the L-band antenna is 1.3 GHz and the centre frequency of the C-band antenna is 5.2 GHz. The simulated impedance bandwidth ($S_{11} \leq -10dB$) of the antenna reaches 16.2% for L- band and 26.8% for C-band. The simulated cross-polarization levels are all less than -48 dB for both L-band and C-band. The work confirms the practicability of the dual-band shared-aperture antenna based on frequency selective surface (FSS) design, and good bandwidth and cross-polarization performance of the antenna has been achieved.**

I. Introduction

The radar system with a variety of frequency bands can effectively improve the versatility of the radar, especially when combining a high frequency and low frequency, the radar system can provide more information about the imaging region. Synthetic aperture radar SAR not only can be used to obtain the image of a large area on the ground, but also has a strong penetrating power. Now SAR is widely used in resource exploration, major disaster estimates, land surveying, and other fields, and SAR has unique advantages in the military.

Various solutions for shared-aperture dual-band dual-polarized microstrip antenna arrays have been proposed in the last two decades. The typical configurations of array may be roughly classified into two types: 1) perforated structure, including perforated patches/patches [1], [2], which a C/X and L/X shared aperture array designs with the configuration of perforated patches and patches are presented, such as ring/patch [3], [4], cross-patch/patch [5], [6]; 2) interlaced layout, such as interlaced patch with slot/dipole [7-9]. An example is the DBDP array of L/C-band interlaced slots with patches by R. Pokuls et.al [7]. Another example is an S/X-band array composed of interlaced dipoles with patches which is proposed by S. S. Zhong, et.al [9].

In this paper, a new dual-band shared-aperture antenna based on frequency selective surface (FSS) is presented which operates at L- and C-bands. The configuration and design of shared-aperture antenna are described in Section II, and the simulated results of the antenna are discussed in Section III. In Section IV, the simulated results are presented to verify the design approach.

II. Antenna Design

The configuration of the proposed antenna is shown in Fig. 1. The antenna consists of five layers. Starting from the top of the structure, the first layer is the L-band antenna which embedded in twelve frequency selective surface units. This kind of frequency selective surface unit produces the property of band-pass for C-band antenna. The second layer is the ROHACELL HF ($\varepsilon_r = 1.07$) foam, whose thickness is about 43.5mm. The third layer is the C-band antenna. To minimize the volume of antennas, the C-band antenna and the L-band antenna is placed orthogonal to share the aperture of the L-band antenna. To obtain a wide bandwidth, bow-tie patches are selected for L-band and C-band. The fourth layer is about 14.4mm thick ROHACELL HF foam. The last layer is ground plane which serves as reflector plane both in L-band and in C-band. The overall height is about 57.7mm.

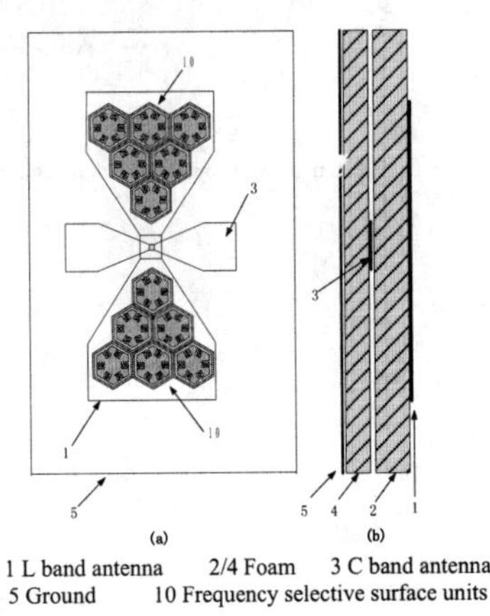

Fig. 1.configuration of the proposed antenna (a) top view, (b) side view

As shown in Fig. 2, L-band antenna is presented, and the antenna consists only of a single bow-tie radiating patch. Due to the larger size of the L-band antenna and placed on the C-band antenna, this means that L-band serves as a reflector plane for C-band antenna. To reduce the effect of the L-band

978-7-5641-4279-7

antenna for C-band antenna, FSS units which produce the property of band-pass for C-band antenna are embedded in the L-band antenna. Each of the frequency selective surface unit includes a regular hexagonal metal ring and a regular hexagonal metal patch with six double-F type aperture. And there is a regular hexagonal aperture between the regular hexagonal metal patch and the regular hexagonal metal ring. The details of the L-band antenna dimensions are given in Table 1.

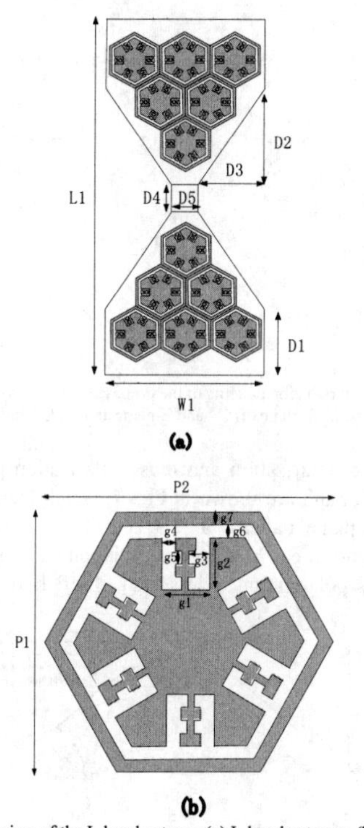

Fig. 2. Top view of the L-band antenna (a) L-band antenna, (b) FSS unit cell

TABLE I
DIMENSIONS OF THE ANTENNA UNIT(MM)

L1	W1	D1	D2	D3
73	31	12	22	13
D4	D5	P1	P2	g1
5	5	9.5	11	1.2
g2	g3	g4/g5	g6/g7	
2	0.5	0.2	0.2	

The simulated transmission and reflection of the FSS unit are shown in Fig. 3.To meet the requirements of bandwidth, bow-tie patch is used for the C-band antenna, as shown in Figure 4. The key dimensions of the antenna are as follows: L2=18.2mm, W2=9mm, S1=3.1mm, S2=5mm, S3=3.5mm, S4=2mm, S5=2mm.

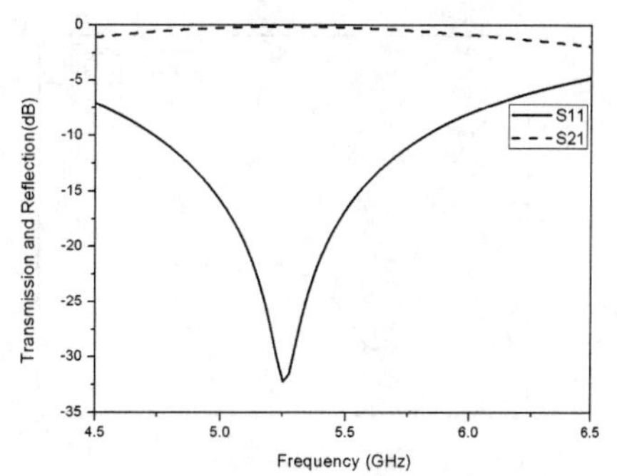

Fig. 3. Transmission and Reflection of the FSS unit

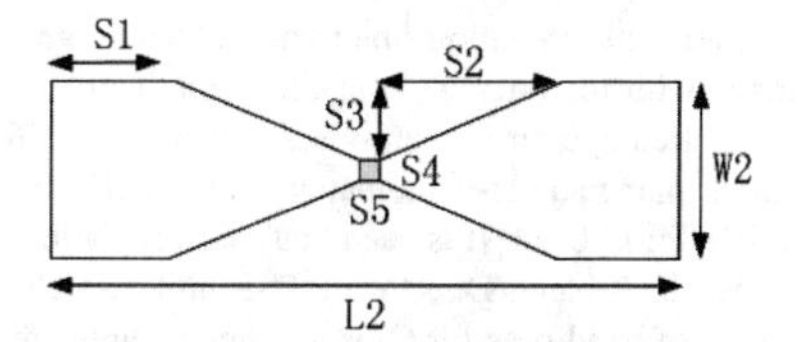

Fig. 4. Top view of the C-band antenna

III. SIMULATED RESULTS

In this section, the simulation and design of the proposed antenna was done by using HFSS based on the finite-element method (FEM), and the performance of the array presented. The simulated S-parameters of the L-band and C-band in comparison with the ones without FSS units embedded in L-band of shared-aperture antenna are shown in Figure 5a and b, respectively. The simulated frequencies of the L-band and C-band are slightly shifted down compared with the ones without FSS units. The simulated relative bandwidth that the reflection coefficient is better than -10 dB is 16.2% in L-band. For the C-band antenna, there is a 26.8% relative bandwidth within the operating band. The reflection coefficient of the C-band antenna is improved by embedding the FSS units in the L-band antenna.

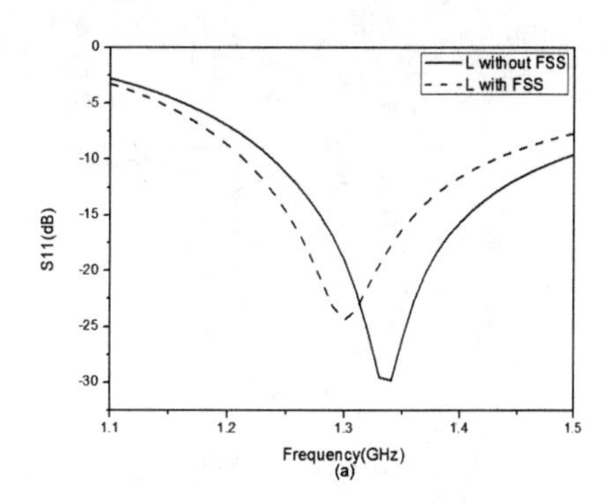

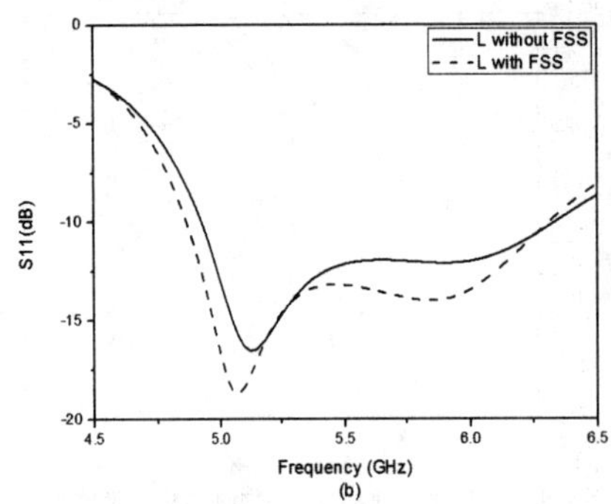

Fig. 5.Simulated S11 of the proposed antenna (a) L-band, (b) C-band

The simulated radiation patterns of the antenna in comparison with the ones without FSS units embedded in L-band of shared-aperture antenna are given in Fig. 6, which shows the E-plane and H-plane patterns at 1.3GHz for L-band and 5.2GHz for C-band. It is seen that good radiation patterns are obtained in L-band. Due to the FSS units which produce the property of band-pass for C-band antenna embedded in the L-band antenna, the front and back patterns of C-band antenna are improved. The gain increases from 8.1 to 9.3 dB in C-band. This paper focuses on a design to integrate L-band antenna and C-band antenna (or antennas) in a compact planar aperture. This means that some of C-band antennas are placed under the L-band radiation patch. These C-band antennas are obtained worse radiation pattern in H-plane pattern. In this situation, FSS units embedded in the L-band antenna make a great impact for C-band antenna, as shown in Fig. 6.

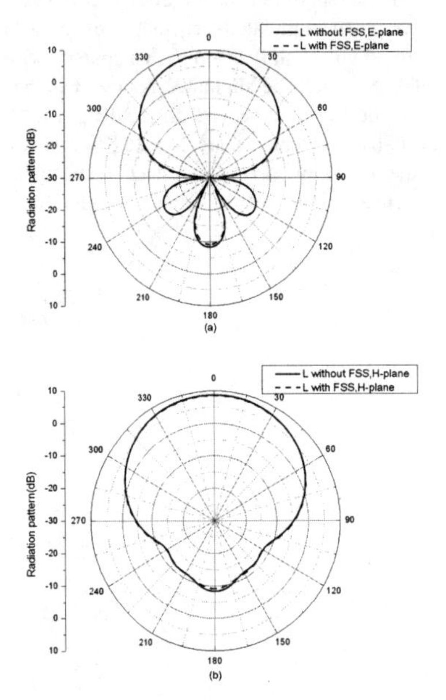

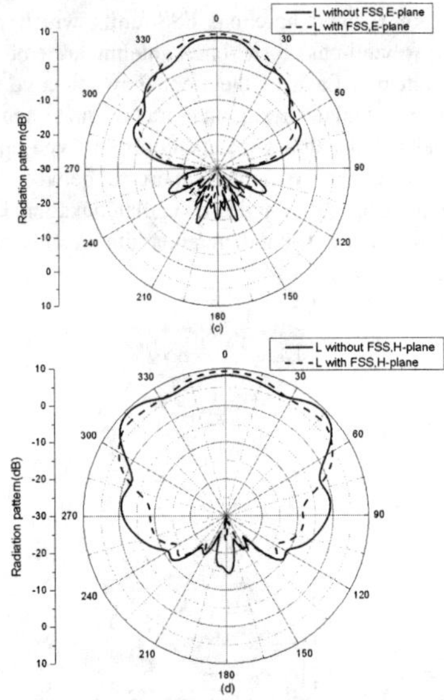

Fig. 6. Simulated radiation patterns of the proposed antenna (a) L-band E-plane, (b) L-band H-plane,(c) C-band E-plane and (d) C-band H-plane

Both the co-polarization and cross-polarization patterns of L-band and C-band are shown in Fig. 7, which include the E-plane and H-plane patterns at 1.3 GHz for L-band and 5.2 GHz for C-band. It can be seen that radiation patterns are good and the cross polarization is less than -48dB both in L-band and C-band.

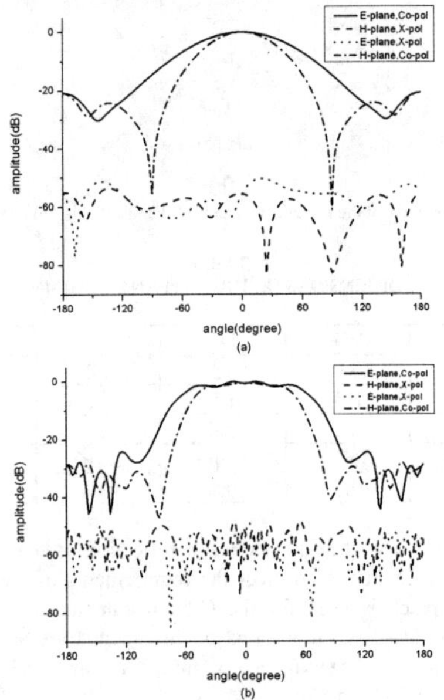

Fig. 7. Normalized simulated patterns of the antenna (a) L-band, (b) C-band

IV. Conclusion

This paper presents a kind of dual-band shared-aperture antenna with wideband and low cross-polarization. The bow-tie patches are selected for L- and C-band design and frequency selective surface units which produce the property of band-pass for C-band antenna are embedded in the L-band antenna. The simulated results show that the bandwidths of $S_{11} \leq -10dB$ for the L-band and C-band are 16.2% and 26.8%, respectively. The level of cross-polarization patterns is lower than -48dB both in L-band and C-band, and the antenna achieved good patterns over the whole frequency bands.

Acknowledgements

This work is supported by National Natural Science Foundation of China under Contract No. 61072017, and supported partly by the Program for New Century Excellent Talents in University of China, by Natural Science Basic Research Plan in Shaanxi Province of China (No. 2013JZ019), National Key Laboratory Foundation, Fundamental Research Funds for the Central Universities (K5051202051, K5051302025).

References

[1] L. L. Shafai, W. A. Chamma, M. Barakat, P. C. Strickland, and G. Séguin, "Dual-band dual-polarized perforated microstrip antennas for SAR applications," *IEEE Trans. Antennas Propag.*, vol. 48, no. 1, pp. 58-66, Jan. 2000.

[2] D. M. Pozar and S. D. Targonski, "A shared-aperture dual-band dual – polarized microstrip array," *IEEE Trans. Antennas Propag.*, vol. 49, no. 2, pp. 150-157, Feb. 2001.

[3] C. Mangenot and J. Lorenzo, "Dual band dual polarized radiating subarray for synthetic aperture radar," in *Proc. IEEE Antennas Propag. Soc. Int. Symp.*, vol. 3, pp. 1640-1643, 1999.

[4] S. H. Hsu, Y. J. Ren, and K. Chang, "A dual-polarized planar-array antenna for S-band and X-band airborne applications," *IEEE Trans. Antennas Propag.*, vol. 51, no. 4, pp. 70-78, Aug. 2009.

[5] C. Salvador, L. Borselli, A. Falciani, and S. Maci, "Dual frequency planar antenna at S and X bands," *Electron. Lett.*, vol. 31, no. 20, pp.1706-1707, Sep. 1995.

[6] A. Vallecchi, G. B. Gentili, and M. Calamia, "Dual-band dual polarization microstrip antenna," in *Proc. IEEE Antennas and Propagation Society Int. Symp.*, vol. 4, pp. 134-137, 2003.

[7] R. Pokuls, J. Uher and D. M. Pozar, "Dual-frequency and dual polarization microstrip antennas for SAR applications", *IEEE Trans. Antennas Propag.*, vol. 46, no. 9, pp.1289-1296, 1998.

[8] X. Qu, S. S. Zhong and Y. M. Zhang, "Dual-band dual-polarized microstrip antenna array for SAR applications," *Electronics Letters*, vol.42, no. 24, pp. 1376-1377, 2006.

[9] S. S. Zhong, X. Qu, Y. M. Zhang and X. L. Liang, "Shared-aperture S/X dual-band dual-polarized microstrip antenna array", *Chinese Journal of Radio Science*, vol. 23, no. 2, pp.305-309, 2008. .

Research on Side-lobe Radiation Characteristic of Printed Dipole Array

Hailiang Xiong, Dongan Song, Qi Zhang, Mingliang Huang, Kai Zhang
Science and Technology on Electromagnetic Compatibility Laboratory
China Ship Development and Design Center
Wuhan, China
xhl19881227@163.com

***Abstract*-In the paper, the radiation characteristic of printed dipole array along side-lobe direction is researched. The electric field intensity values in side-lobe area are calculated using FEKO under different situation. It is also analyzed how metal plate and scanning angle influence electric field intensity values in side-lobe region. Besides, through the peak values and valley values of electric field intensity along side-lobe direction, the possibility of using peak value and valley value identify the number of side lobe is discussed.**

I. INTRODUCTION

A large number of array antennas are equipped in very limited platform, so the near-field high-power radiation is the potential electromagnetic hazards to open-air electronics equipment, fuel and staff [1]. In order to control and resolve the electromagnetic hazards, the designer would like to originally predict the distribution of the near-field radiation and potential hazards of array antenna, which is a new challenge for electromagnetic compatibility design of platform.

Research on the radiation characteristics of antenna array in literatures mainly focus on the main-lobe radiation characteristics, but few on the side-lobe radiation characteristics. Therefore, it will research side-lobe radiation characteristics of printed dipole array and the coupling between printed dipole array and surrounding metal plate. The radiation characteristic of printed dipole array along side-lobe direction has been researched by simulation method, electric field intensity values in side-lobe area are calculated under different situations. It is also analyzed how metal plate and scanning angle influence side-lobe area. Besides, through the peak values and valley values of electric field intensity along side-lobe direction, the possibility of using peak value and valley value identify the number of side lobe is discussed.

II. MODELING

The radiating elements of array antenna are generally organized and fed in the light of some rules. The most common forms of the array are rectangular array and triangular array, and triangular array has the advantages of saving elements and relevant components, curbing grating lobe near the main lobe [2]. In addition, the triangular array could be considered as the sum of the two rectangular arrays, so the basic theory of rectangular array can be valid while triangular array is analyzed.

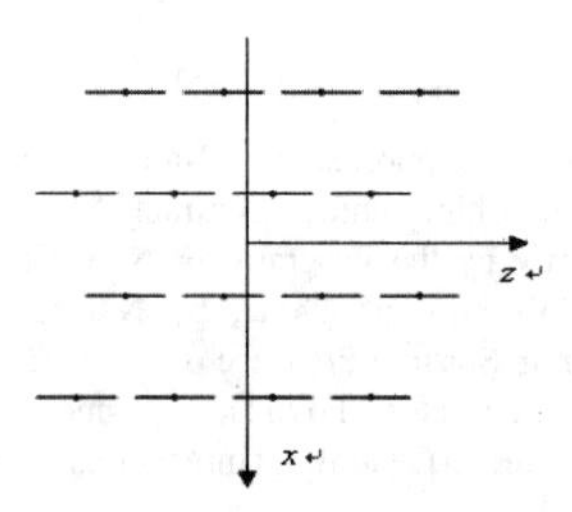

Figure 1. The whole dipole array sketch map

In view of above reasons, in this paper, the triangular array has been chosen as the object making use of EM simulation software FEKO which is based on MOM. As shown in figure.1, the printed dipole array is working in S band. The whole array includes 16 elements which are organized as triangular array with 4 rows and 4 columns, each element is a printed dipole antenna. As shown in Figure 2, the radiating elements distance along X and Z axis are respectively $dx = \lambda$ and $dz = \lambda / 2$.

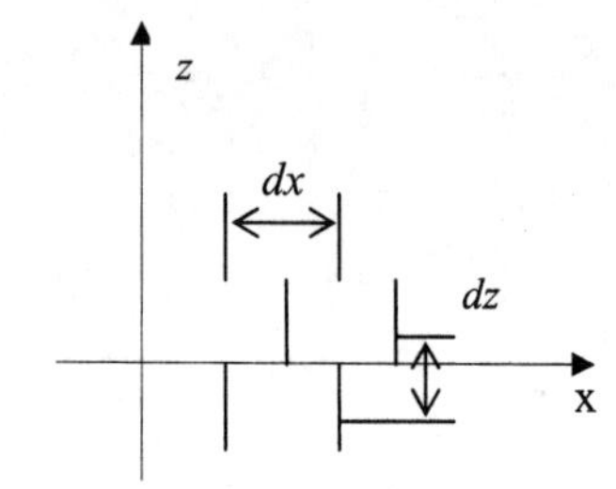

Figure 2. Triangular array model

It is also shown in the figure.3 that the elevation plane is YOZ plane and the azimuth surface is XOZ plane. The normal direction of the printed dipole array is Y direction and the tangential direction is X direction. Each printed dipole antenna is vertically placed on the 1×1 m perfect conducting plane and the printed dipole array is placed on the center of the plane. The ground plane (XOY surface) is infinite perfect

978-7-5641-4279-7

conducting plane. The height between the center of printed dipole array and the ground plane is 0.9m. What's more, the height between the ground plane and the measuring point is h.

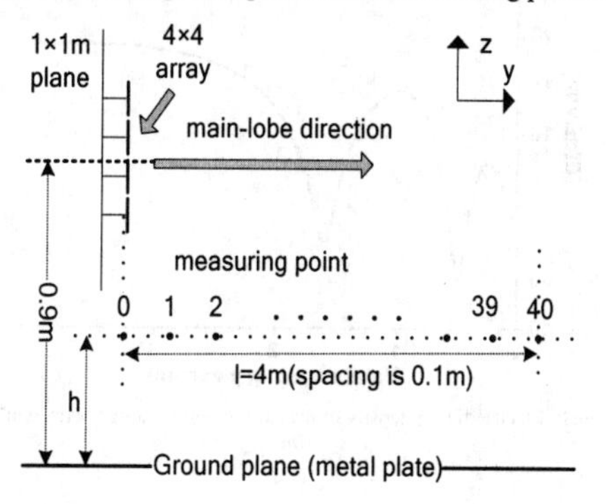

Figure 3. Measuring point of side-lobe radiation field sketch map

As shown in Figure.3, in order to analyze the side-lobe radiation characteristic of printed dipole array, we prefer to radiation characteristics just below h of the main-lobe direction as the measuring point. The first measuring point just below the printed dipole array and the spacing along the horizontal direction is 0.1m.

In fact that printed dipole antenna in S band is made of metal and media, so if calculating the antenna based on MOM will generate a large number of unknowns and can't finished. In order to reduce the number of unknowns, we make an equivalent model which is only consisted of metal, as shown in the figure 4.

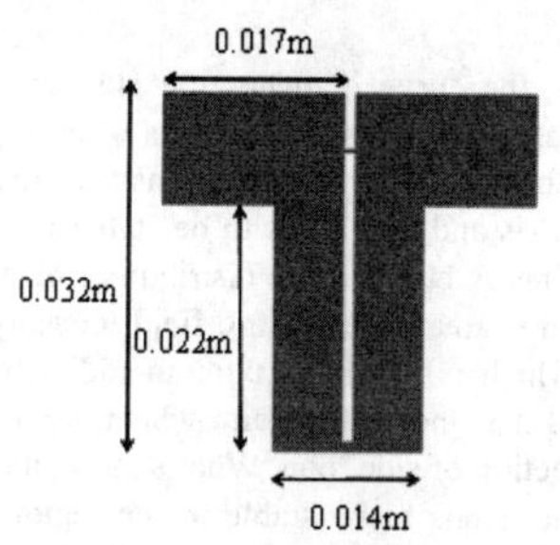

Figure 4. Equivalent model of printed dipole antenna

Based on the theory of planar array, the beam of planar array can be controlled in two-dimensional space. In the spherical coordinate system, the space direction of half spherical space is according to the two coordinate of θ and φ. According to the simplified method proposed by Von Aulock, the phase of the mnth element (x_m, z_n) can be given as follows on the condition that the main beam is available [3].

$$\Psi_{mn} = \frac{2\pi}{\lambda}(x_m \sin\theta\cos\varphi + z_m \cos\theta) \qquad (1)$$

Where x_m is the coordinate of antenna along X axis and z_n is the coordinate along Z axis. According to formula (1), it is very easy to obtain the phase of each element, where θ is the elevation angle and ϕ is the azimuth angle.

III. SIMULATION RESULT ANALYSIS

The electric field intensity values in side-lobe area are calculated under following different situations, the radiation characteristic of print dipole array along side-lobe direction is researched by the simulation results.

1) Analyze how the horizontal metal plate influences the side-lobe radiation characteristic of printed dipole array.

2) Analyze how the scanning angle influences the side- lobe radiation characteristic of printed dipole array.

3) Through the peak values and valley values of electric field intensity along side-lobe direction, the possibility of using peak value and valley value identify the number of side lobe is discussed.

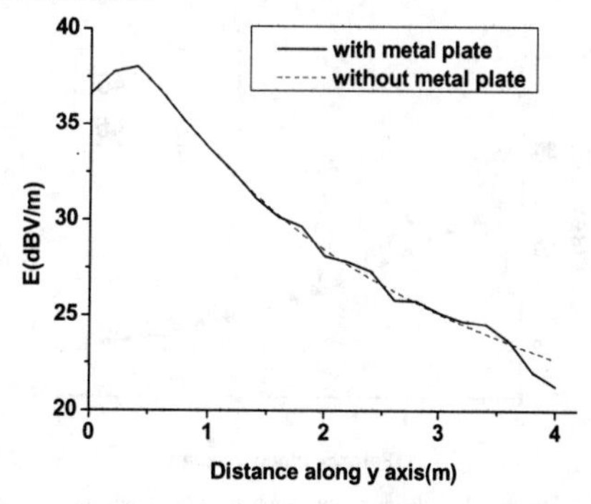

Figure 5. Comparison of the simulation result with the situation of existing metal plate and without metal plate, while h=0.8m

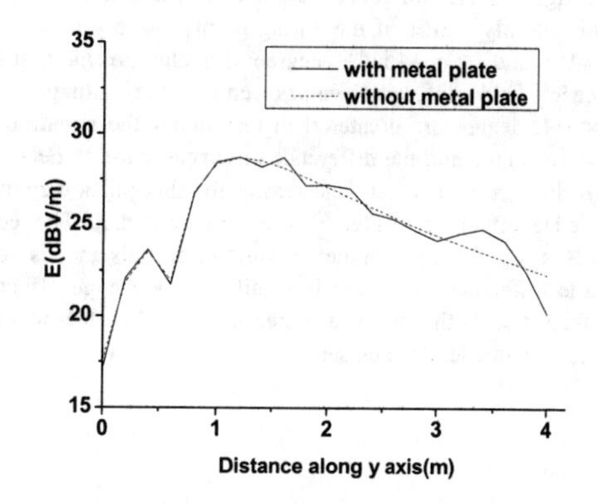

Figure 6. Comparison of the simulation result with the situation of existing metal plate and without metal plate, while h=0.6m

In order to analyze how the horizontal metal plate affects the side-lobe radiation characteristic of printed dipole array, figure 5 and 6 provide the electric field intensity values of the measuring points along y axis when h=0.8m and h=0.6m respectively. The solid line indicates the situation of existing metal plate and the other dotted line indicates the situation without metal plate. From the curves, it is clear to find that the electric field intensity values under the situation of existing metal plate and without metal plate agree well when the distance along y axis is less than 1.3m. However, the electric field intensity values under the situation of existing metal plate is oscillated in both sides of that without metal plate when the distance is greater than 1.3m. What's more, the oscillation is more obvious with the distance increasing. That is due to the mirror effect caused by metal plate [4]. Because the printed dipole antennas in model are perpendicular to the ground metal plate, the current amplitude in printed dipole and its mirror are equal and the phase is the same. The superposition of phase is changed with the distance along y axis, then the electric field intensity values under the situation of existing metal plate is oscillated.

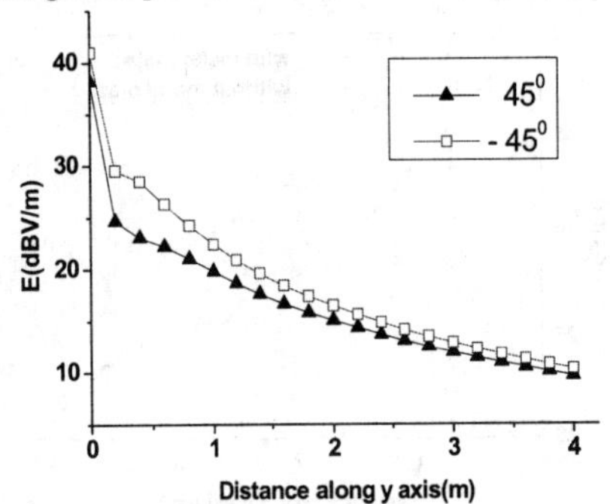

Figure 7. Simulation result vary with the azimuth angle

In figure.7, the two curves respectively indicate the electric field intensity values of measuring points along y axis when $\theta=90$ and $\phi=-45,45$ degree. It is clear to find that the electric field intensity values under the situation of $\phi=-45$ degree are greater than that under the situation of $\phi=45$ degree and the difference of curves starts to decrease with distance increasing. The reason for this phenomenon is the edge effect of printed dipole array antenna. The edge effect is significant when the measuring point distance is near, so the difference of curves is significant; the edge effect is weakened with the distance increasing and the electric field intensity value tends to be same.

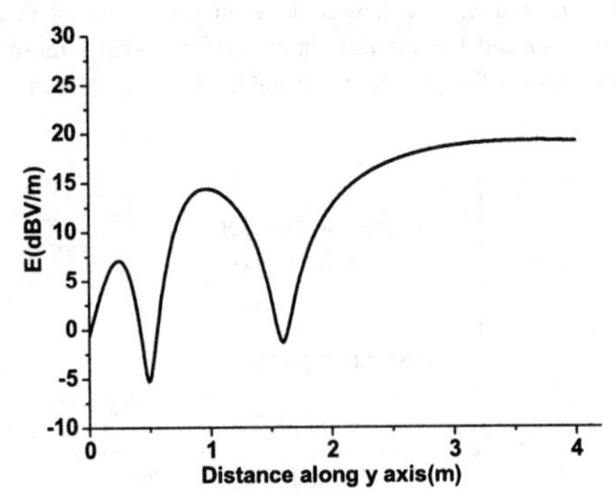

Figure 8. Electric field intensity of measuring points along y axis, while h=0 m

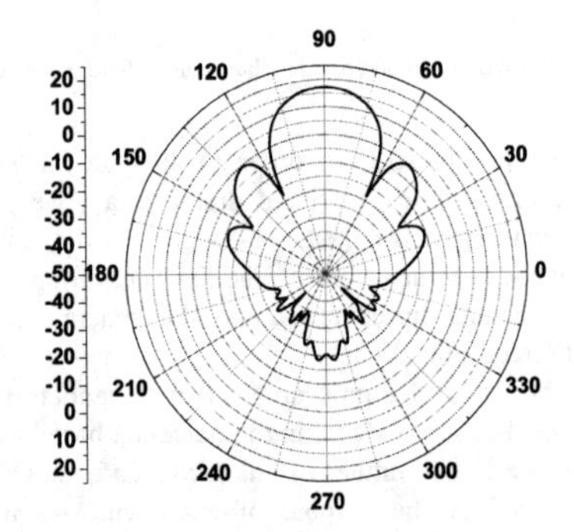

Figure 9. The far-field pattern

In figure.8, the curve indicates the electric field intensity values of measuring points along y axis when h=0m. It can be clearly find that the electric field intensity value is oscillated along the Y axis and that tends to be stable with the distance increasing. This is because the distribution of main lobe and side lobe in this area. The electric field intensity value is the peak point which is on the maximum radiation direction of side lobe and it is the valley point which is on the minimum radiation direction of side lobe. What's more, the electric field intensity value tends to be stable in the region which is the area of main lobe. From the peak and valley points alternating occur, we can identify the number of side lobe and main lobe in this region. There are two peak points in the region, so the number of side lobe in the region is two. There is just a main lobe in this region because of the one and only stable region, which is verified by the far-field pattern, as shown in figure 9.

IV. CONCLUSION

According to above simulation results, the impact from metal plate should be taken into account when side-lobe radiation characteristic is to be researched. Due to the mirror effect, electric field intensity values under the situation of existing metal plate is oscillated in both sides of that without metal plate with the distance increasing. The scanning angle is the other factor which influences the side-lobe radiation characteristic, and the difference will be decrease with the distance increasing. Finally, the simulating results of radiation characteristic in this region do not correspond with the desired result, which is because that the side-lobe effect cannot be ignored, hence the number of side lobe and main lobe in this region can be identified.

REFERENCES

[1] Guangyi Zhang. Principles of Phased Array Radar. Beijing: National Defence Industrial Press, 1994.

[2] Dongan Song. Prediction analysis of near field radiation for Shipborne Phased Array Radar Antenna. Ship science and technology, Vol.2, No.20, pp.46-48, 1998.

[3] Zhengyue Xue, Weiming Li, Wu Ren. Analysis and synthesis of array antenna. Beijing: Beihang University Press, 2011.

[4] W. D. Burnside, N. Wang, E. Pelton. Near Field Pattern Analysis of Airborne Antennas. IEEE Trans. A P, Vol.5, No.28, pp.318-327, 1980.

Experimental Study of a Ka-Band Waveguide-Fed Longitudinal Slot Phased Array

L. Qiu, H. S. Zhang, L. F. Ye, Y. Liu, S. L. Chai
IEEE Conference Publishing
College of Electronic Science and Engineering, National Univ. of Defense Technology
Changsha, 410073 China

***Abstract*-A Ka-band waveguide-fed longitudinal slot array designed as a phased array to achieve a $\pm 8^{\circ}$ E plane beam scan is presented. A three-layer structure, which is easy to fabricate, has been applied to produce the array. The measured results show that, the array possesses a high gain of more than 34.5dBi in the entire scan range and shows an aperture efficiency of more than 54%. The measured E plane side lobe level is below -16.5dB while the beam scans in the angle range.**

I. Introduction

Lots of works about electronically scanned slot arrays had been published during the past decades [1-4]. Most of them had applied edge slot arrays as their scan elements [1, 2, 4], since narrow element spaces could be applied to avoid the grating lobes in a wide scan angle range. However, the operation bands of the edge slot phased arrays were usually below the K-band. Because in higher frequency bands, such as the millimeter wave (MMW) band, high fabrication accuracy of individual waveguides and the edge slots on them will be needed, which can hardly be achieved and will be extremely expensive.

As well known, the waveguide-fed longitudinal slot arrays had got precise and efficient design method [5] which can sufficiently consider the mutual coupling effect, and had already been widely applied to design high gain and high efficient slot array antennas in MMW band [6, 7]. Single layer structures were usually applied to construct the arrays, since it could be fabricated by die-casting technique with enough fabrication accuracy needed in the MMW band and at very low cost [8].

However, only a few of waveguide-fed longitudinal slot arrays have been applied in phased arrays, because wide scan angles will not be achievable due to the scan blindness and the grating lobes caused by the big element spaces and serious E plane mutual coupling factors. In [9], a large Ka-band longitudinal slot array was designed and analyzed for digital beam-forming application. The beam-forming was achieved in the H plane to avoid the serious E plane coupling. However, there were no experimental results of the scanned beam patterns or beam-forming patterns in the article. In [10], a method was proposed for analyzing the performance of the phased longitudinal slot arrays. The E plane scan pattern of a phased longitudinal slot array in K-band was calculated, yet no experimental results were provided.

This paper presents a Ka-band waveguide-fed longitudinal slot array designed as a phased array to achieve a $\pm 8^{\circ}$ E plane beam scan. The experimental results show that the array maintains a high gain performance of more than 34.5dBi in the required scan range. Good radiation pattern performances have been achieved in the entire scan range.

II. Design Of The Slot Array

A. Slot Element

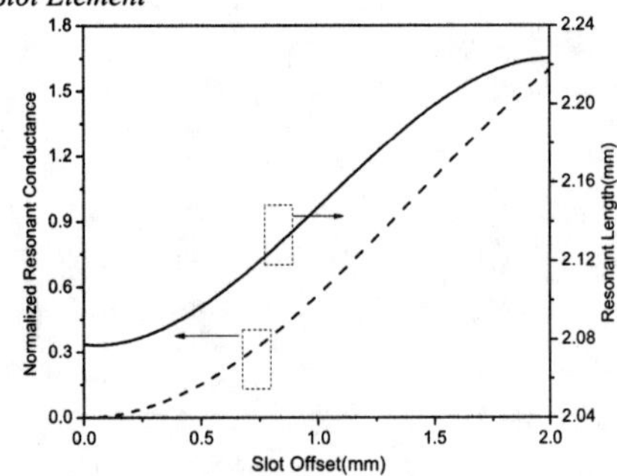

Figure 1. Normalized resonant conductance curve and resonant length curve versus slot offset at the center frequency f_0.

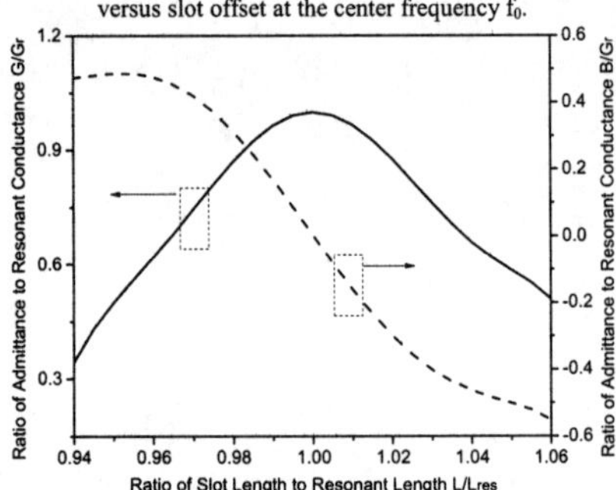

Figure 2. Normalized self-admittance components of the shunt slot at f_0.

The width of the waveguide which contains the radiating longitudinal shunt slots is 5.5mm, and the height is 2.5mm. The width of the radiating slots is set to be 0.6mm, and the shim that contains the slots has a thickness of 0.5mm. The normalized resonant conductance curve and the resonant length curve versus the slot offset at the center frequency f_0 have been calculated by using the method of moments (MoM)

978-7-5641-4279-7

[11], and are shown in Fig. 1. The ratio of admittance to resonant conductance curve versus ratio of slot length to resonant length at f_0, which is needed in the design process, is also calculated and presented in Fig. 2.

B. Structure Of One Linear Array

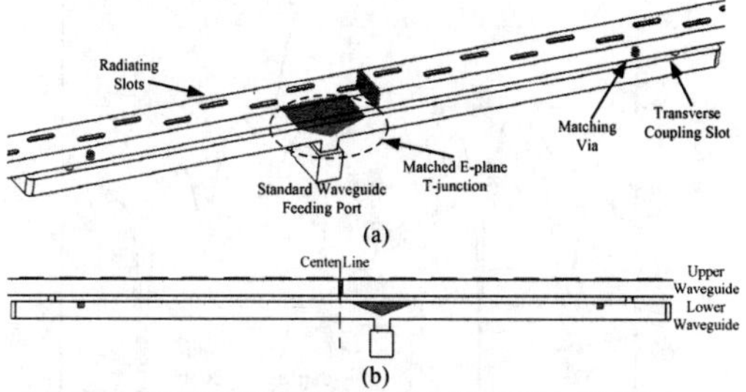

Figure 3. Structure of one linear slot array, (a) 3D view; (b) side view.

The planar array contains 16 linear arrays, each of which contains 42 radiating slots and has an individual feeding port. The structure of one linear slot array is shown in Fig. 3. It should be noted that, only the air parts of the waveguide and the slot have been shown in the figure to clearly depict the structure. In order to meet a 1% bandwidth requirement, the linear array is equally divided into two sub-arrays, which are separated by a metal plate. The energy will be received by the radiating slots, then coupled to the lower feeding waveguide through the coupling slots, and finally synthesized together by using a matched E plane T-junction to the feeding port thus the receiver. The waveguides in this paper have set to be smaller than the standard BJ320 waveguides. Thus step transmission sections have been applied to transmit the waveguides to the standard ones. To leave enough space for the standard waveguide ports and maintain an opposite phase transmission to the two sub-arrays, the waveguide port is set to have a $\lambda_g/2$ offset from the center line, where λ_g denotes the waveguide wavelength of the feeding waveguide at f_0.

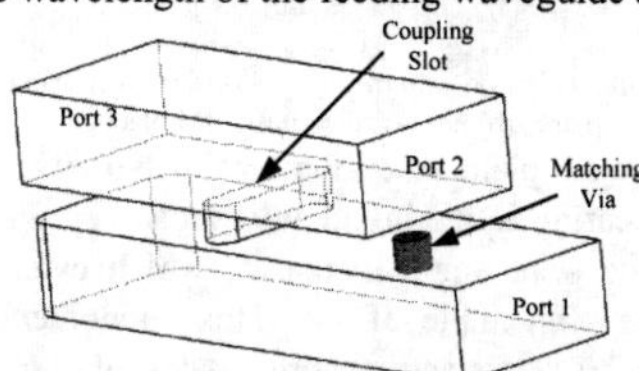

Figure 4. Broad wall transverse coupling slot.

The broad wall transverse coupling slot is depicted in Fig. 4. The thickness of the common wall of the waveguides is 0.5mm. The slot has a width of 0.5mm, a length of 3.4mm, and is $\lambda_g/2$ away from the waveguide shorting end. A metal matching via has been placed 4.5mm before the coupling slot to obtain a good impedance match. The via has a diameter of 1mm, a height of 0.8mm and a offset of 0.7mm from the center line of the waveguide.

The matched E plane T-junction is depicted in Fig. 5. A metal roof-top structure has been added to obtain a good impedance match.

The structures of the broad wall transverse coupling slot and the matched E plane T-junction have been simulated by applying Ansoft's High Frequency Structure Simulator (HFSS). Both of their simulated return losses are below -20dB in the 1% operation bandwidth around f_0.

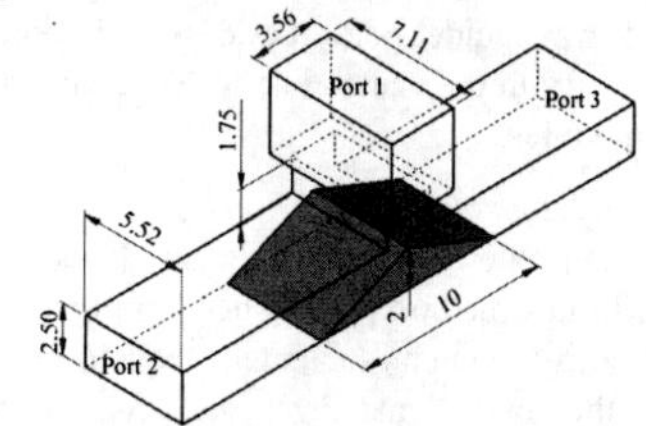

Figure 5. Matched E plane T-junction.

C. Structure Of The Planar Array

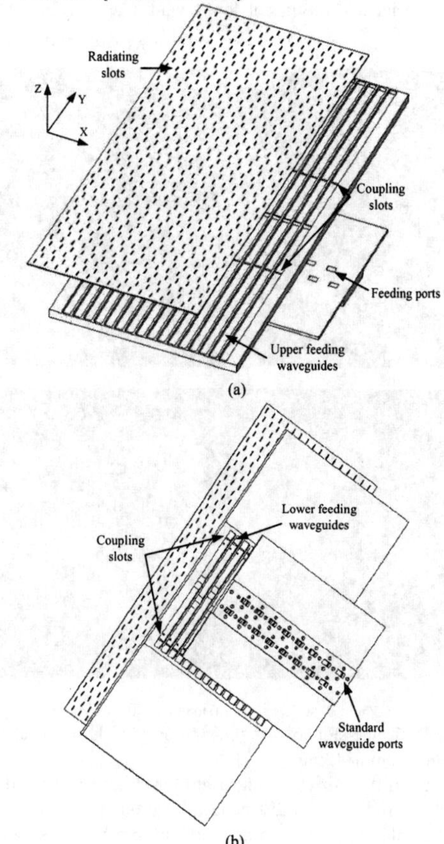

Figure 6. Structure of the planar slot array, (a) top view; (b) bottom view.

The planar array is shown in Fig. 6. It mainly includes three parts: the shim part that contains the radiating slots, the waveguide part and the plate that contains the feeding waveguide ports. At last, the three parts are welded together to construct the final array. This three layers structure can be fabricated by die-casting technique with enough fabrication accuracy and at very low cost, thus suitable for mass production. The element space at the X direction is $0.81\lambda_0$, which is decided by the waveguide width and the thickness of the metals wall which construct the waveguides. Noted that, the standard waveguide ports besides each other have an opposite offset from the center line to leave enough space for the receive modules.

D. Slot Array Design Method

By using the property curves of the single slot in Fig.1 and Fig. 2, the Elliott's method [5] has been applied to design the planar slot array, which accurately considers the mutual coupling of the total planar array and insures the far field pattern performance. The array distribution has set to be a Taylor distribution, with a -30dB side lobe level in H plane (YZ plane) and a -25dB in side lobe level E plane (XZ plane).

III. Experimental Investigation

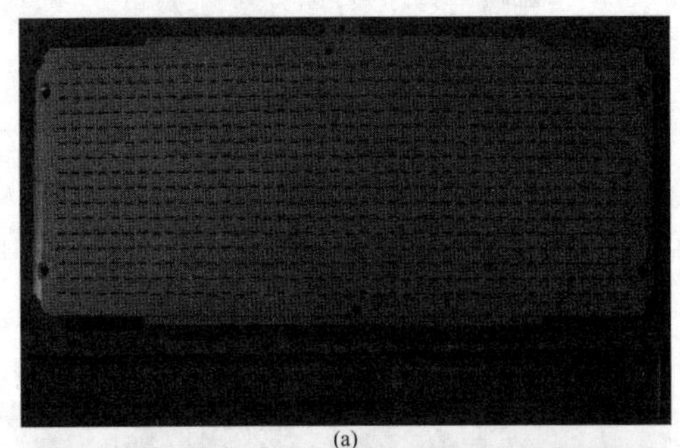

(a)

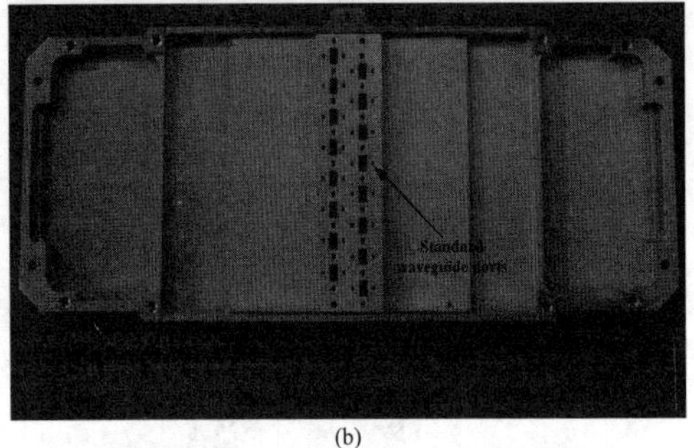

(b)

Figure 7. Photo of the slot array, (a) top view; (b) bottom view.

The fabricated slot array is shown in Fig. 7. Its size is about 275mm×110mm×13mm.

The return loss of every linear array has been measured to be below -10dB in a 1.2% bandwidth around f_0, under the condition that all the feeding ports of other arrays being matched.

The none-scanned antenna gain at f_0 is measured to be 35dBi, which represents that the slot array has gained an aperture efficiency of about 60% at the Ka-band.

The measured E plane radiation patterns are shown in Fig. 8(a). It can be observed that, the array only experiences a slight gain drop as it scans at the E plane, which is at most -0.5dB as it scans to -8°. This means that the array possesses a high gain of more than 34.5dBi and an aperture efficiency of more than 54% in the entire scan range.

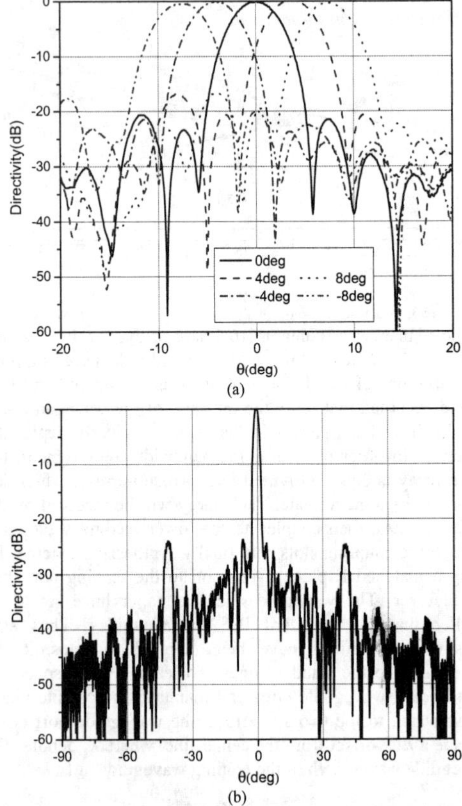

Figure 8. Measured radiation patterns of the slot array, (a) scan patterns at E plane; (b) none-scan pattern at H plane.

The measured E plane side lobe level is below -21dB as the main beam steering at the broad side. However, the side lobe rises up as the scan angle increases, which even rises to -16.5dB at the scan angle of -8°. This phenomenon can be explained by that, the active admittances of the slots vary when the mutual coupling experienced by the slot elements change as the beam scanned, so the radiation amplitude and the phase of the elements change accordingly.

The measured H plane pattern as the array is steering at the broad side is shown in Fig. 8(b). It can be observed that, the measured H plane side lobe level is below -24dB as the main beam steering at the broad side. However, two side lobes at the angle of ±45° rise up to about -24dB. From the theory in [12], it can be deduced that, there must be at least 0.03mm misplacement at the X direction between the shim which contains the radiating slots and the waveguide plate, thus a periodical amplitude distribution error occurs along the Y direction, which causes the ±45° side lobe. This can be overcome through constructing the radiating slots and the waveguides on one plate or applying a more restrict locating method with less misplacement error.

IV. CONCLUSION

A Ka-band waveguide-fed longitudinal slot phased array which achieves a ±8° E plane scan range has been presented in this paper. A three-layer structure has been applied to produce the array, which makes the fabrication process accurate and suitable for mass production. The measured results have shown that, the array possesses an extremely high gain of more than 34.5dBi in the entire scan range and shows an aperture efficiency of more than 54%. The E plane side lobe level has been measured to be below -16.5dB in the entire scan range. And the H plane side lobe level has been measured to be below -24dB while the main beam steering at the broad side. The reasons for the side lobe rising of the E plane scanned patterns and the ±45° side lobe of the H plane pattern have been discussed.

REFERENCES

[1] J. L. Hilburn, R. A. Kinney, R. W. Emmett, and F. H. Prestwood, "Frequency-scanned X-band waveguide array," IEEE Trans. Antennas Propag., pp. 506-509, Jul. 1972.

[2] J. L. Harrison and B. D. Cullen, "A large edge-slot array for satellite communications," in Antennas and Propagation Society International Symposium, vol. 11, Aug. 1973, pp. 336-339.

[3] A. G. Derneryd and T. C. Lorentzon, "Design of a phase/frequency scanned array antenna with non-resonant slotted ridge waveguide elements," in Antennas and Propagation Society International Symposium, vol. 3, Jun. 1991, pp. 1728-1731.

[4] R. R. Kinsey, "An edge-slotted waveguide array with dual-plane monopulse," IEEE Trans. Antennas Propag., vol. 47, no. 3, pp. 474-481, Mar. 1999.

[5] R. S. Elliott, "An improved design procedure for small arrays of shunt slots," IEEE Trans. Antennas Propag., vol. 31, no. 1, pp. 48-53, Jan. 1983.

[6] M. Ando and J. Hirokawa, "High-gain and high-efficiency single-layer slotted waveguide arrays in 60 GHz band," in 10th International Conference on Antennas and Propagation, vol. 1, Apr. 1997, pp. 464-468.

[7] Y. Kimura, M. Takahashi and M. Haneishi, "76 GHz alternating-phase fed single-layer slotted waveguide arrays with maximum gain and sidelobe suppression," in Antennas and Propagation Society International Symposium, vol. 3, Jun. 2005, pp. 1042-1045.

[8] S. Park, Y. Tsunemitsu, J. Hirokawa, and M. Ando, "Center feed single layer slotted waveguide array," IEEE Trans. Antennas Propag., vol. 54, no. 5, pp. 1474-1480, May 2006.

[9] S. R. Rengarajan, M. S. Zawadzki and R. E. Hodges, "Design, analysis, and development of a large Ka-band slot array for digital beam-forming application," IEEE Trans. Antennas Propag., vol. 57, no. 10, pp. 3103-3109, Oct. 2009.

[10] R. V. Gatti, R. Sorrentino and M. Dionigi, "Fast and accurate analysis of scanning slotted waveguide arrays," in Microwave Conference, 32th European, 2002, pp. 1-4.

[11] R. F. Harrington, Field computation by moment methods, IEEE Press, New York, 1993.

[12] I. P. Theron and J. H. Cloete, "On slotted waveguide antenna design at Ka-band," in South African Symposium on Communications and Signal Processing, Sep. 1998, pp. 425-426.

Wide-Angle Impedance Matching of Phased-Array Antenna Using Overlapped Subarrays

Run-Liang Xia, Shi-Wei Qu, Ming-Yao Xia, and Zai-Ping Nie
Department of Microwave Engineering, University of Electronic Science and Technology of China (UESTC)
Chengdu, 611731, China
rlxia2012@gmail.com

***Abstract*-For phased array antennas, the active impedance of the radiating elements changes considerably with scan angle. Therefore it is ordinarily possible to match the active impedance at only one scan angle. In this paper, a microstrip patch phased array using overlapped subarrays is designed and simulated. It is found that by using overlapped subarrays good impedance matching over a wide range of scan angle can be achieved. Simulated results show that the investigated phased array has much better scan performance compared with conventional microstrip patch phased arrays.**

I. Introduction

One of the most challenging design aspects of phased array is the wide impedance matching technique. An unusual issue exists in phased array design is the active impedance of element varies considerably with scan angle. The inherent mutual coupling among the elements is the main cause that leads to the impedance of each element in a way changes with the scan angle [1]. Consequently, the elements can be impedance matched at only one scan angle by ordinary matching techniques. The mismatch at other scan angles may seriously reduces the gain realized by the antenna and deteriorates other specifications e.g. radiation pattern, polarization purity, and amplifier stability [2], [3].

Wide-angle impedance matching (WAIM) technique is used to compensate for the change of reflection coefficient with scan angle [4]. There have been various studies on the WAIM technique by a number of researchers. In [5], Hannan *et al.* put a connecting circuit between neighboring elements which introduces a signal, variable with generator phasing, i.e., with scan angle, into each array element. This approach can effectively compensate the variation of element active scan impedance. In [4], Magill and Wheeler employed a planar thin sheet with high dielectric constant located in front of, and parallel to the array aperture to compensate the varied reactance due to the array scan so that wide-angle impedance matching can be realized. In [6], Sajuyigbe *et al.* investigated the use of metamaterials in achieving wide angle impedance matching. The anisotropic properties of a metamaterial layer could provide more degrees of freedom than an isotropic dielectric layer, thus a wider range of scan angle can be achieved. Recently Awida *et al.* proposed a class of microstrip patch phased arrays backed by substrate-integrated cavities which perform wider scan angle than that of conventional counterpart [7].

According to previous studies, overlapped subarrays are often employed to reduce the side lobe level in a large array. For moderate scan angles, overlapped subarrays allow for the side lobe levels of down to -20dB [8]. Recently Bhattacharyya studied the wide-angle performance of phased arrays with overlapped subarrays and numerical results prove that overlapped subarrays have wider scan performance than conventional subarrays with and without WAIM layer [9]. His research provides a new approach to wide-angle impedance matching in phased arrays.

In this paper, an E-plane microstrip patch phased array using overlapped subarrays is developed. In Section II, the phased array configuration with overlapped subarrays is illustrated. In Section III, a unit cell of the investigated array is first designed utilizing the periodic boundary condition. A simple feeding networks for the best impedance matching performance over a wide range of scan angle in the E plane is simulated, following the results in [9]. In Section IV, a 16 patch elements array is built by assembling 16 well designed unit cell elements mentioned above. Simulated results show that the designed phased array has a wide scan performance of better than 60°. Finally, Section V concludes the paper.

II. Phased Array Configuration

Geometry of the designed phased array is shown in Fig. 1, including top view, side view, and feeding network. Patch elements are etched on the top side of a substrate layer (antenna layer) with a relative permittivity of 2.2 and thickness h_1 = 3 mm. The patch element spacing is 20 mm, i.e., $0.5\lambda_0$ at the operating frequency 7.5 GHz. The feeding network in Fig. 1(c) is printed on the other substrate (feeding layer) with a relative permittivity of 2.2 and thickness h_2 = 0.5 mm, which is isolated from the antenna layer by a common aluminum ground plane of thickness 2 mm. A simple two-way power divider is adopted as the feeding network and the characteristic impedances of the input and output ports are 50 and 100Ω, respectively. Each radiating element is shared by two adjacent power dividers as depicted in Fig. 1(c). By adjusting the length of the 100-Ω transmission line at the output ports, a good impedance matching can be obtained. The two-way power divider has two arms of different lengths thus can provide branch currents with phase difference which is the main cause leads to the improvement according to [9].

978-7-5641-4279-7

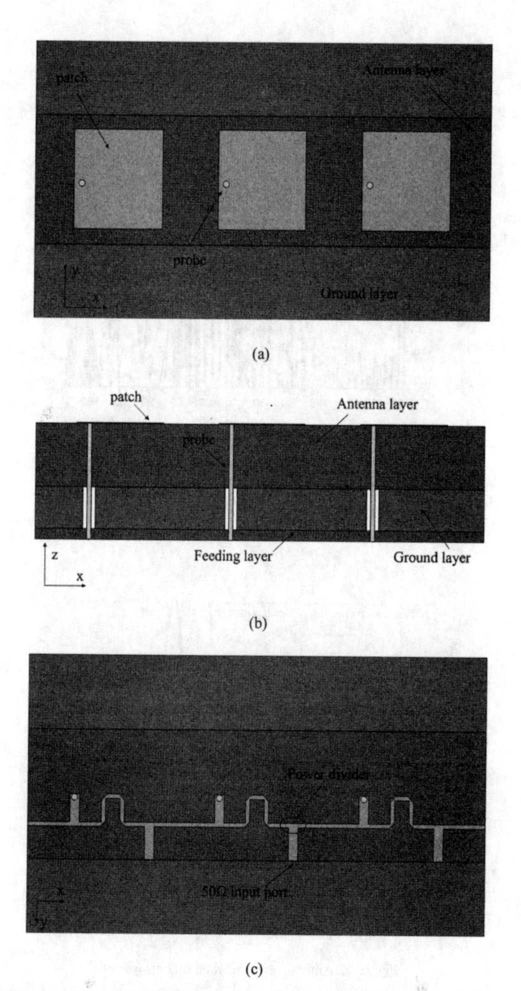

Figure 1. Geometry of the E-plane overlapped array. (a) Top view (b) Side view (c) Feeding network.

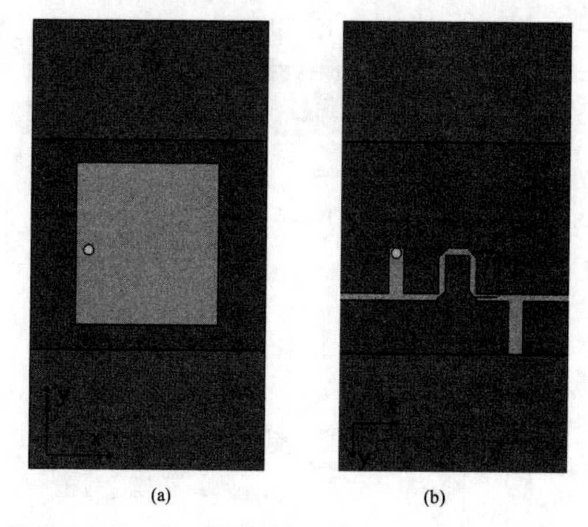

Figure 2. (a) Top view. Cell size=40.4mm × 20mm. patch size=12.2mm × 13mm. probe location = 5 mm from patch center. probe diameter=0.84 mm. (b) Feeding layer.

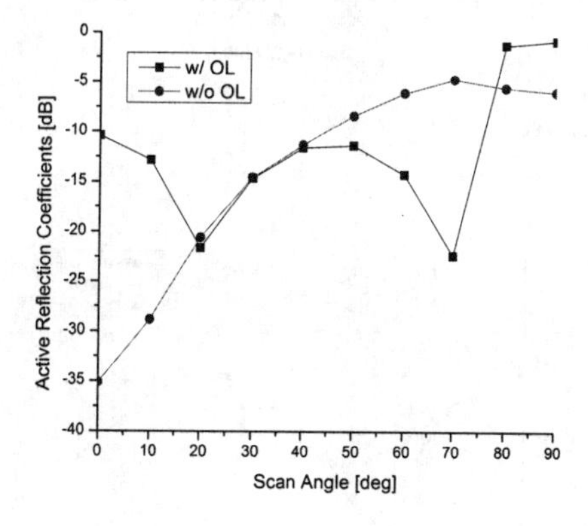

Figure 3. Simulated active reflection coefficient of a unit cell

III. ARRAY ELEMENT DESIGN

Impedance matching performance of the investigated phased array has been investigated using HFSS, commercial software based on the finite element method (FEM). First in our numerical study that the phased array is assumed to be infinite and in order to simulate an array unit cell the periodic boundary condition is utilized in HFSS. The geometry of one unit cell is shown in Fig. 2. Then the active reflection coefficients of the investigated phased array are calculated versus the scan angles. The length l of the transmission line at the output port of the divider is adjusted to obtain the best active reflection coefficients as the array is scanned from 0° to 60°. The scan range is defined in this work by a value in which the reflection coefficient is below -10dB. Fig. 3 shows the simulated active reflection coefficients versus scan angles as l varies. For comparison, that of the conventional microstrip array is also shown in Fig. 3, as a reference.

As seen from the figure, the conventional microstrip phased array presents a scan range of 44° and a weak scan blindness at about 70°. Unlike the referenced microstrip arrays, the E-plane scan performance of the investigated phased array gets much broader, and its scan range in the case of l=4.05 mm is about 75°. At boresight, overlapped subarrays technique slightly worsens the active reflection coefficient but it is still less than -10dB. This is a tolerable sacrifice for applications.

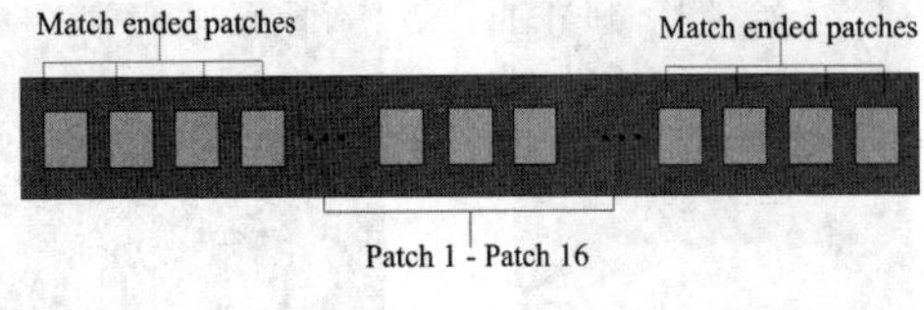

Figure 4. A 16 patch elements array

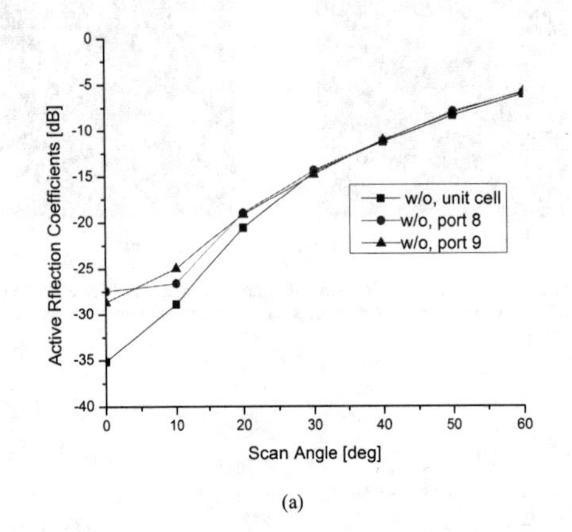

(a)

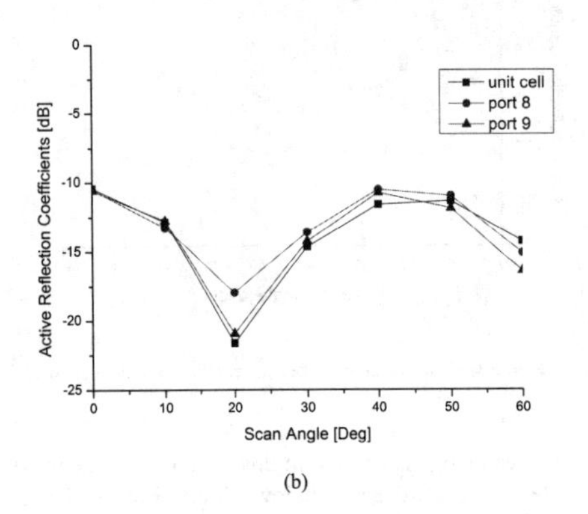

(b)

Figure 5. Simulated active reflection coefficients of the 16 elements array in case of (a) the conventional microstrip phased array (b) the investigated phased array with overlapped subarrays.

IV. SCAN PERFORMANCE OF THE 16 ELEMENTS ARRAY

In Section III, the unit cell has been well designed in an infinite phased array environment. In this section a finite microstrip phased array is built by assembling 16 unit cell elements and another 4 elements match-ended in each side of the array (see Fig. 4). The simulated active reflection coefficients versus scan angles from Port 8 and Port 9 corresponding to Patch 8 and Patch 9 in the middle of the array is presented in Fig. 5. The reason of choosing Ports 8 and 9 is that the electromagnetic environment in the middle of the array is the most similar to an infinite array. A conventional array without overlapped subarrays, as a reference, is also built for comparison.

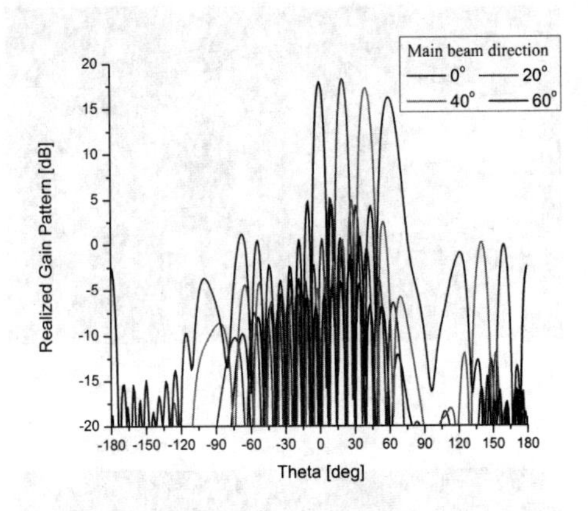

Figure 6. Pattern scanning of the investigated phased array.

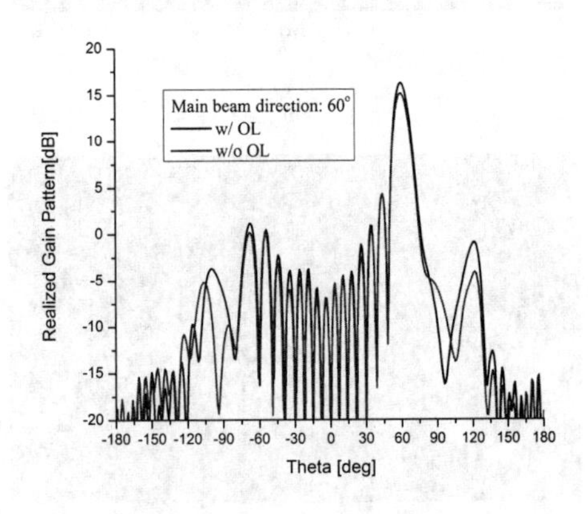

Figure 7. Antenna pattern when scanning to 60°

TABLE I
REALIZED GAIN COMPARISON OF THE TWO KIND OF PHASED ARRAY

Scan angle		0	20	40	60
Realized Gain[dB]	w/ OL	17.94	18.27	17.27	16.22
	w/o OL	18.34	18.39	17.52	15.11

The active refection coefficients versus scan angles of the referenced array is shown in Fig. 5(a) which agrees well with the simulated result in Fig. 3 and the scan range is still about

44°. In Fig. 5(b), the impedance matching performance of the 16 patch elements array using overlapped subarray is given. A good impedance matching is obtained as the array is scanned from the boresight to 60°. The realized gain pattern of the investigated array is also shown in Fig. 6. As seen from the figure, the investigated array can scan its main beam from 0° to 60° with a gain fluctuation less than 2.1 dB and the maximum gain of 18.27 dB occurs at the scanning angle of 20°. The overlapping results in about 1 dB gain improvement at 60° scan angle, as shown in Table I compared to the referenced array at 60°.

V. CONCLUSION

This paper addressed the use of overlapped subarrays to improve the impedance matching performance over wide scan angle. Simulated results show that the scan range of the investigated E-plane arrays is much wider than the conventional microstrip phased arrays with equal dimensions.

REFERENCES

[1] H. A. Wheeler, "Simple relations derived from a phased-array antenna made of an infinite current sheet," *IEEE Trans. Antennas Propag.*, vol. 13, pp. 506-514, Jul. 1965.

[2] P. W. Hannan, "The element-gain paradox for a phased-array antenna," *IEEE Trans. Antennas Propag.*, vol. 12, pp. 423-433, Jul. 1964.

[3] J. Blass and S .J. Rabinowitz, "Mutual coupling in two-dimensional arrays," 1957 *IRE WECON Conv. Rev.*, pt. 1, pp. 134-150, 1957.

[4] E.G. Magill and H. A. Wheeler, "Wide-angle impedance matching of a planar array antenna by a dielectric sheet," *IEEE Trans. Antennas Propag.*, vol. 14, pp. 49-53, Jan. 1966.

[5] P. W. Hannan, D. S. Lerner and G. H. Knittel, "Impedance matching a phased-array antenna by a connecting circuits ," *IEEE Trans. Antennas Propag.*, vol. 13, pp. 28-34, Jan. 1965.

[6] S. Sajuyigbe, M. Ross, P. Geren, S. A. Cummer, M. H. Tanielian and D.R. Smith, "Wide angle impedance metamaterials for waveguide-fed phased-array antennas," *IET Microw, Antennas Propag.*, vol. 4, pp. 1063-1072, 2010.

[7] M. H. Awida, Aladin. H. Kamel, and A. E. Fathy, "Analysis and design of wide-scan angle aide-band phased arrays of substrate-integrated cavity-backed patches ," *IEEE Trans. Antennas Propag.*, vol. 61, pp. 3034-3041, Jun. 2013.

[8] R. J. Mailoux, "A low sidelobe partially overlapped constrained feed network for time delayed subarrays," *IEEE Trans. Antennas Propag.*, vol. 49, pp. 39-40, Feb. 2001.

[9] A. K. Bhattacharyya, "Floquet modal based analysis of overlapped and interlaced subarrays," *IEEE Trans. Antennas Propag.*, vol. 60, pp. 1814-1820, Apr. 2012.

Design of the Automatic Test System of Active T/R module

Han Liu[1] Xin Zheng[1] Zhipeng Zhou[2] Qiang Zhang [2] Yiyuan Zheng[3]
(1.Nanjing Research Institute of Electronics Technology, 210039 China; 2. Science Technology on Antenna and Microwave Laboratory (STAML), Nanjing 210039 China; 3.Nanjing University of Science and Technology, Nanjing 210094, China)

Abstract- Active array radar (DAR) is a kind of phased radars, which utilizes beam-forming technique both in receiving and transmitting. Active T/R module is a key technique in the next generation of T/R modules for active array radar. In order to realize the automatic test of the active T/R modules, the advanced modern automatic test system (ATS) is built based on GPIB bus. The principle of the ATS for the active T/R module is introduced. The composition and structure of the ATS is described and the designing method of the hardware and software is discussed. The measurement results of the ATS verify that the design of the ATS is reasonable, because its performance is stable and the efficiency of the active T/R module test increases by using it.

I. INTRODUCTION

Active array radar technology has been an important development for the Radar industry [1, 2]. The active array radar has dramatically increased the operational capability of modern Radars compared to the conventional phased array radars [3, 4]. The Beam Forming technology can not only improve the performance of phased array radars, but also can extend their function, so it has been increasingly used in modern radar system [5, 6].

Active T/R module is a core part of the active array radar. The performance index of it will have a direct impact on that of the active array radar. The performances tests of the active T/R module are complex and hard work. In order to meet the measurement requirement of radar system, new advanced automatic test system (ATS) for active T/R modules are needed.

II. THE PRINCIPLE AND FUNCTION INDEX OF THE ACTIVE T/R MODULE

A typical active T/R module is composed of phase shifter, TR switch, power amplifier, circulator, LNA etc [7-9]. A schematic presentation of the active T/R module is shown in Fig1.

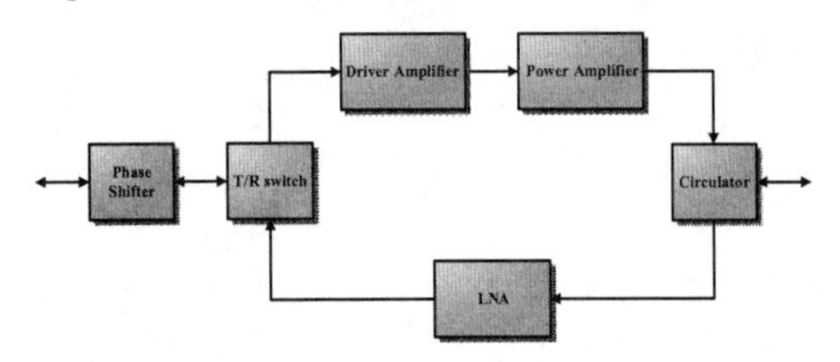

Figure1.A schematic representation of the active T/R module

In the transmitter mode, the signal is generated by the microwave signal source. The signal is amplified by the driver amplifier and power amplifier. The transmitted signals are transferred to the antenna though circulator. In the receiver mode, the received signal is received though the circulator and amplified by a low noise amplifier (LNA). The Phase Shifter in each channel translates the control signal to phase variation. The measurements of the active T/R module in transmitter mode are as follows:

Output Power
Efficiency
Spurious
Harmonic
Droop
Rise Time
Fall Time

III. HARDWARE DESGIN OF AUTOMATIC TESTING SYSTEM

Automatic testing systems of active T/R module can automatically do system testing, data recording and data processing for active T/R module with the least manual participation, and output result in the appropriate mode. The development of it is based on the technologies of computer and test bus. The computer brought the automation of test technology, and the GPIB bus is bringing the test system into network era. A schematic presentation of the ATS setup used in this work is shown in Fig2.

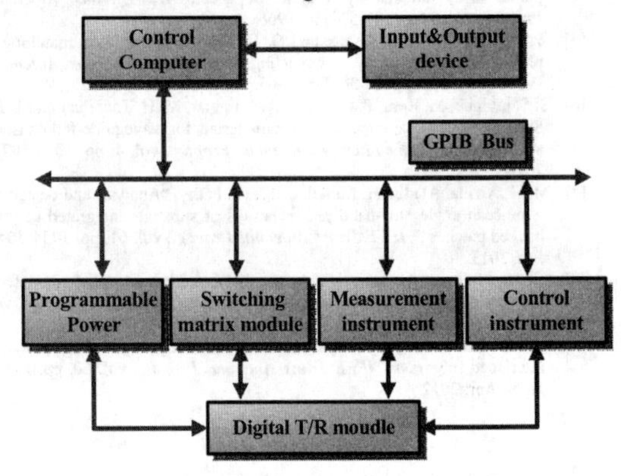

Figure2.hardware architecture of ATS of active T/R modules

The ATS of active T/R modules is constituted of programmable power, switching matrix module, measurement instrument, control instrument and other peripheral devices. All of devices are connected with each other by GPIB bus. The programmable power can supply multiplexing programmable control power motivations, and give the measurements a practical work power environment. The test channel and working mode is selected by control computer though the switching matrix module and control instrument. The measurement instrument modules include the following devices, which are random curve generator, lowpass filter, highpass filter, spectrum analyzer, power

978-7-5641-4279-7

analyzer, vector signal generator. The filter and spectrum analyzer can realize the measurement of harmonic and spurious. Random curve generator is used to supply the mandatory actuating signals and self-check signals of the system such as square wave, TTL clock and so on. The power analyzer is used to test output power, droop, risetime and falltime. The vector signal generator can supply the frequency actuating signals and some other special actuating signals.

IV. SOFTWARE DESGIN OF AUTOMATIC TESTING SYSTEM

The system software architecture is the key factor that affects the usability of general test system and the operability of test system development. The modularization and hierarchy design is adopted in software architecture that accomplishes the transplantation and expansibility of the system. Fig. 3 shows the basic software architecture. The system module is reducible, transplantable and exchangeable with users and provides relevant service through interface.

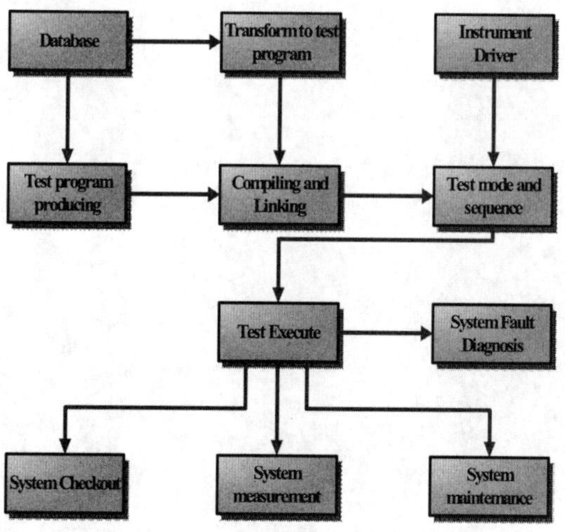

Figure3.software architecture of ATS of active T/R modules

V. TEST RESULT

Once the ATS had been built, the measurement of active T/R modules could be made. Take one channel of active T/R module for example. The test results are as follows:

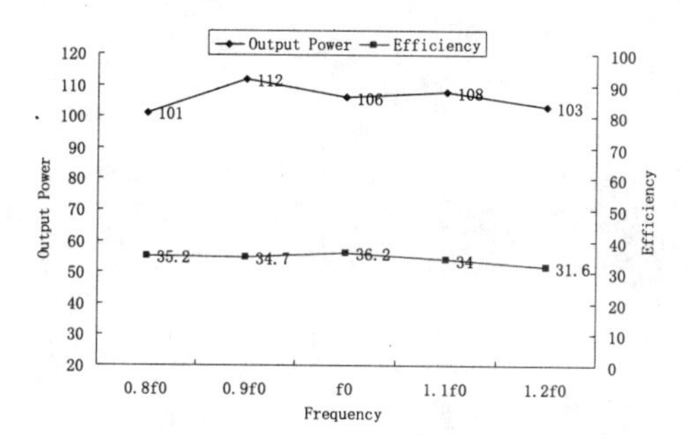

Figure4.Output Power & Efficiency of ATS of active T/R modules

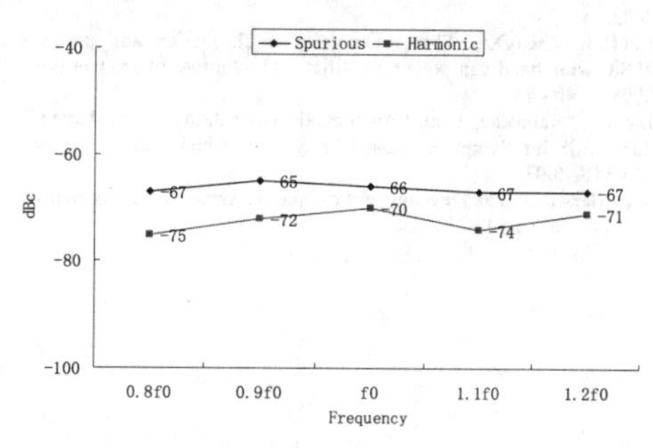

Figure5. Spurious & Harmonic of ATS of active T/R modules

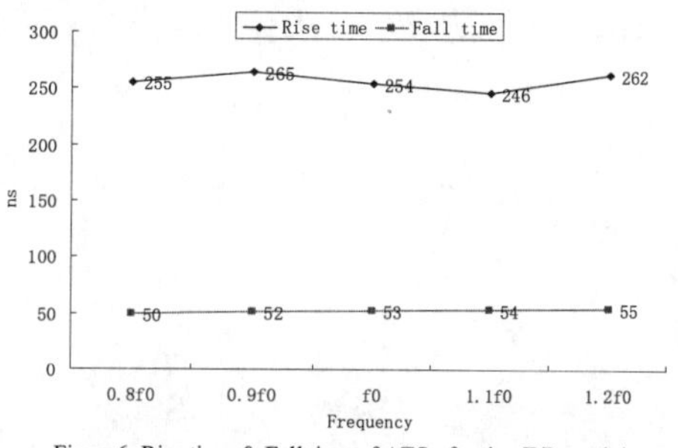

Figure6. Rise time & Fall time of ATS of active T/R modules

The picture of the ATS of active T/R modules is shown as figure 7.

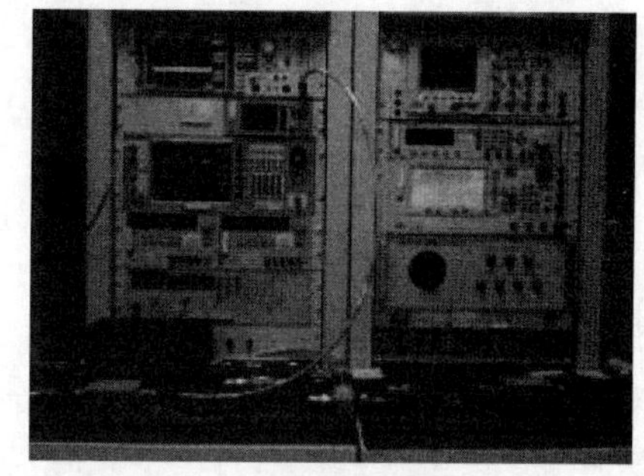

Figure7. The picture of ATS of active T/R modules

VI. CONCLUSIONS

We have demonstrated an automatic test system (ATS) that can be used to measure the active T/R module. The system can perform automatic test and fault diagnosis of the active T/R module. The hardware and software architecture is researched and described based on the system. The results showed the ATS can meet application requirements and it increases the efficiency of the active T/R module test.

REFERENCES

[1] YU Zhen-kun, ZHENG Xin . "The application analysis of SiC power device in radar transmitter" ,journal of microwaves. Vol 23,pp.61-65, 2003.

[2] Liu Han, ZhengXin, Shang Jian-gang et al. Design and practice of SiC wide band gap power amplifier〔J〕 Journal of microwaves. 2008, 5: 40 - 44

[3] Daniel J. Rabideau, et al, "An S-band active Array Radar Tested," 2003 IEEE Int. Symp. On Phased Array System and Technology pp. 113-118, 2003.

[4] WU Man-qing, The Development of Active Array Radar. Journal of CAEIT, 2006, pp.11-16.

[5] Wang Yan, Wu Manqing. "The Development of DBF Phased Array Radar System," 2001 CIE Int. Conference on Radar Proceeding,2001, 61-64

[6] R. Ney, J.J. Berthelier . "Electronic Active Beamforming Implementation For Radars," MST 10 Workshop on Technical and Scientific Aspect of MST,Session 5,2003..

[7] Lu Jiaguo Wu Manqin Jin Xueming, et al. Active phased-array antenna based on DDS[J]. Acta Electronica Sinica, 2003,31(2) :199-202.

[8] Chen Tian, Chen Zhuming ,Zhou Peng, et al. Design and implement of wide band active T/R module[J]. Modern Radar, 2008,30(3):89-92.

[9] Inder Bahl Prakash Bhartia. "Microwave Solid State Circuit Design "ZHENG Xin translate publishing house of Electronics Industry. 2006.

Application of GaN High Power Chips in T/R Modules

CAN LIN [1], HAN LIU [1], YIYUAN ZHENG [2]

(1. Nanjing Research Institute of Electronics Technology, Nanjing 210039, China; 2. Nanjing University of Science and Technology, Nanjing 210094, China)

***Abstract*-Digital transmit/receive modules (digital T/R modules) play a vital role in active phased array antennas for electronic warfare applications. High power amplifier (HPA) chain is the key component in the next generation of T/R modules for the future S band phased array antenna. Today GaAs MMIC HPA is widely used in most of T/R modules. However, the drawbacks of GaAs MMIC HPA such as low efficiency and limited output power level (usually in the range of 5W to 15W) have made it not to be a competent candidate of power chips. In this paper, the GaN power chip with better characteristics is introduced and analyzed. Compared with GaAs MMIC HPA, it has remarkable advantages such as high power, high-gain, high efficiency, wide operating bandwidth, and etc. An S band 85W GaN power chips amplifier is developed and investigated. The performance tests of the amplifier have verified that the performance of T/R modules can be improved by using the GaN power chips. The improvements of the T/R modules system in output power, efficiency, reliability, size and weight have illustrated that the GaN power chips is a prime candidate for warfare systems and electronic warfare application.**

I. INTRODUCTION

Microwave power chips which used in T/R modules are one of the most important components of active phased array antenna. It determines whether the T/R modules can meet the performance of warfare systems. Thus the development level of T/R modules technology to some extent depends on the development level of power devices. The main power device which commonly used in S-band solid-state T/R modules is mainly GaAs MMIC. Its typical output power level is in range of 5W to 15W [1-3]. Modern warfare systems for military applications place new requirements on RF power amplifiers due to the desire to reduce system size, weight, and improve system reliability. High output power, high power density, wide broadband and high efficiency microwave power devices are needed [4-6]. Power device requirements for T/R modules of power electronics are at a point that the present GaAs MMIC cannot handle. In order to meet the requirement of warfare systems, new semiconductor materials for power device applications are needed. Wide band gap semiconductors technologies like gallium nitride (GaN) are used to create new type T/R module systems to support the next generation active phased array antenna. The improved RF power device is made possible due to the much improved material properties: wide band gap, high electric breakdown field and high saturated electron drift velocity. With the GaN RF power chips offering particularly high performance, it showed that this will ultimately result in reduced circuit complexity, improved power density, gain, efficiency, and higher reliability. In particular, the abilities of warfare systems detection will benefit from the development of this technology.

Compared with GaAs MMIC, GaN power chips offer high power capabilities in S band, high power density, high gain, and high efficiency and thus, they would be used more and more widely in modern warfare systems. In order to fully investigate the application analysis of GaN power chips in T/R module systems, it is crucial to design RF power amplifiers intended for T/R modules applications with GaN power chips. In this paper, An S band 85W GaN power chips amplifier for warfare applications was developed. The performance test of the GaN power chips amplifier was carried out. Additional comparisons of the test results were made to investigate the application analysis of GaN wide bandgap power chips in T/R module systems.

II. ADVANTAGE OF GAN WIDE BANDGAP SEMICONDUCTOR

TABLE I

PHYSICAL CHARACTERISTICS OF GAAS AND GAN WIDE BANDGAP SEMICONDUCTOR

material	Bandgap Eg (eV)	Electric Breakdown Field,Ec (kV/cm)	Thermal Conductivity (W/cmk)	Saturated Electron DriftVelocity ($\times 10^7 cm/s$)
GaN	3.45	2000	1.3	2.2
GaAs	1.13	400	0.46	1

GaN wide band gap semiconductor materials have superior electrical characteristics compared with GaAs semiconductor materials. GaN wide bandgap semiconductor has many properties ideal for electronic devices for high temperature, and high power and radiation hard applications. GaN have bandgap and electric field values which are significantly higher than GaAs. Generally, wide bandgap and high electric breakdown field values is desirable. Semiconductors with wider bandgaps can operate at higher temperatures. Higher electric breakdown field results in power devices with higher breakdown voltages. As shown in Table 1[7-8], the theoretical breakdown voltage of a GaN semiconductor material is 5 times more than that of GaAs semiconductor materials. The bandgap in the GaN semiconductor is significantly higher than GaAs semiconductor. This permits high breakdown voltages and high operating temperatures to be developed. With the advantages of the material, the width of the drift region can be reduced, which can reduced power device size and improve power density of power device. GaN wideband semiconductor offers an unmatched combination of electronic and physical properties which enable the fabrication of new classes of power devices for applications ranging from L-band, S-band and X-band to radiation-hard and nuclear environment. In comparison with GaAs, GaN is

978-7-5641-4279-7

also physically rugged and chemically inert which is an advantage is for semiconductor power devices required to operate in harsh environments.

III. 85W GaN HIGH POWER CHIPS AMPLIFIER DESIGN

A. The principle of GaN power amplifier

In this paper, the design and realization of GaN high power chips amplifier are researched and analyzed, which covers from fo to 1.6fo GHz frequency range in S band. The output power of power chips amplifier is 85W. The amplifier uses 4 GaN power chips with 8mm gate width. The basic specification for the GaN power amplifier is shown in Table II.

TABLE II
SPECIFICATIONS FOR THE GaN POWER CHIPS AMPLIFIER

Frequency Band	Frequency Range	Power Gain	Output Power	Drain Efficiency	Duty Cycles
S	fo to 1.6fo GHz	> 11 dB	> 85W	>45%	10%

The power amplifier was consisted of wideband divide/combine circuit and match network. The block schematic diagram of the GaN power chips amplifier is shown in Fig.1.

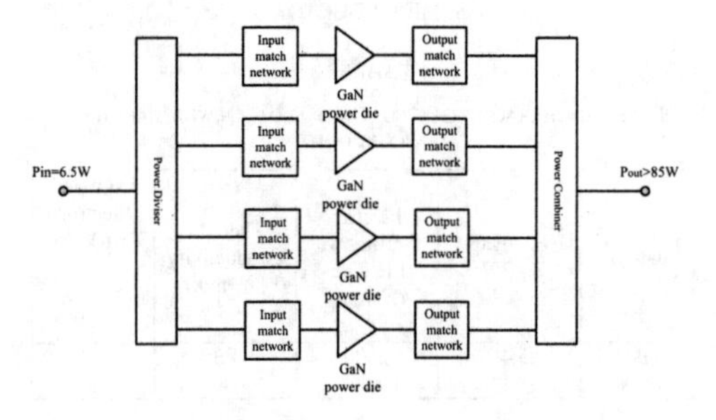

Fig 1. Block schematic diagram of the GaN power chips amplifier

B. Design of match network

The output impedance match was designed to achieve the optimum load impedances for the dies. These impedances were obtained from load-pull simulations taking the bond wires into consideration. The input impedance match was designed to conjugate match. The hybrid circuit incorporating microstrip technology is based on ceramic substrates with 10 mil thickness. The necessary bonds interconnect between the transistors and the microstrip circuit is analyzed with a 3D electromagnetic field simulator. Fig 2 show the block schematic diagram of the match circuit

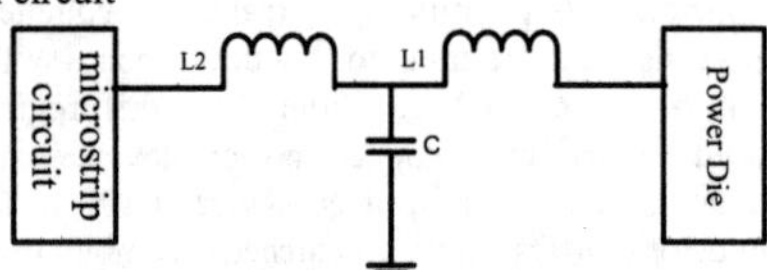

Fig 2. Block schematic diagram of the match network

C. Design of wideband division and combination

In the chip, a two branches divider and combiner is designed with the chebyshev broadband match method [9], the schematic diagram was shown as Fig. 3. The computer-aided design of the matching network using Agilent ADS tools was accomplished by an optimization. If isolation is reduced, the power chips are unstable. So the divider and combiner is designed that all port is isolated with the isolating resistor.

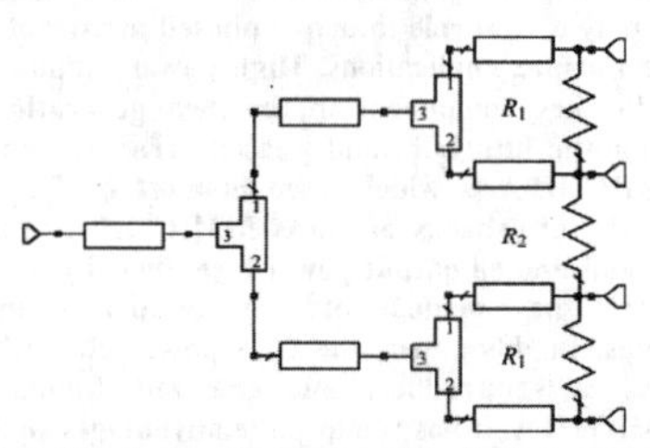

Fig 3. Block schematic diagram of the divider/combiner

IV. PERFORMANCE TEST OF THE GaN POWER HIGH CHIPS AMPLIFIER

Fig4 shows the view of the GaN high power chips amplifier. The size of the amplifier is 16mm × 16mm.

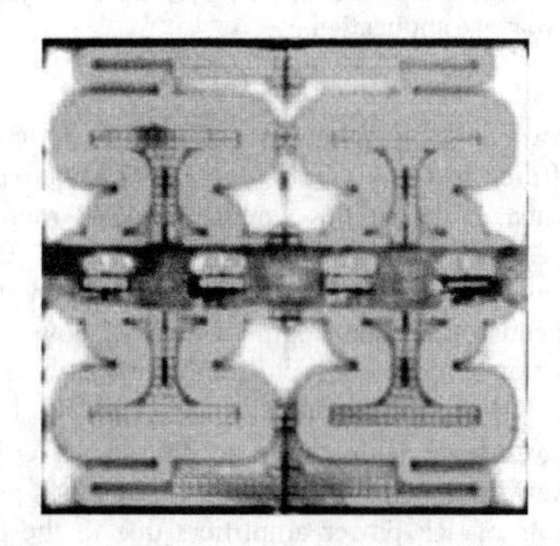

Fig 4. The photograph of the GaN high power chips amplifier

The GaN high power chips amplifier was measured under various pulsed conditions intended to replicate typical Warfare operating conditions. For all measurement both the RF and the drain supply were pulsed. The amplifier were measured with 38dBm (pulsed) input power with a duty cycles of 10%. Additional measurements were also made to investigate the application analysis of GaN wide bandgap power chips in T/R modules. Output Power &efficiency vs. Frequency for GaN high power chips amplifier is shown in Fig 5. Power gain &spurious vs. Frequency for GaN high power chips amplifier is shown in Fig 6.

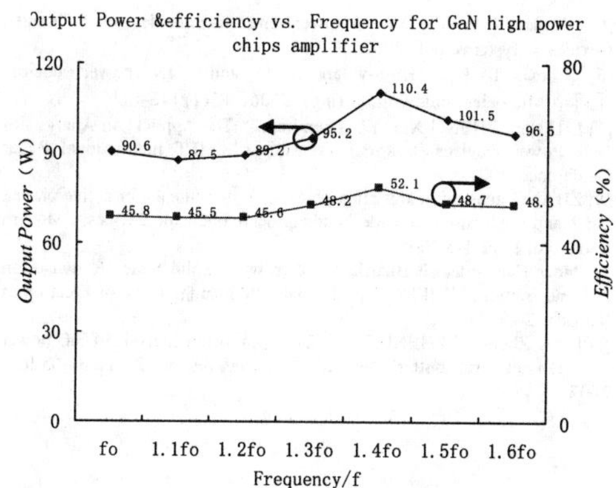

Fig 5.Output power &Efficiency vs. Frequency for GaN high power chips amplifier

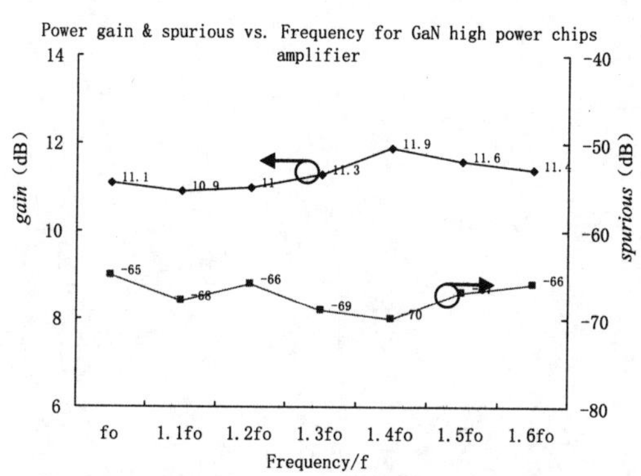

Fig 6.Power gain &Spurious vs. Frequency for GaN high power chips amplifier

V. DATA ANALYSIS

A. Output power

GaN chips offer important advantages for high power applications due to the GaN wide band-gap. GaN-based dies offer power densities in the few W/mm range. Typically GaAs MMIC has demonstrated high out-power of 5-15 W. A principal reason for this was the lack of bang-gap. Compared with GaAs MMIC GaN chips supplies higher out-power for T/R modules. Besides this, The GaN power chips support long plus length with high duty cycle waveform which GaAs power amplifier can not handle. This results in an improvement of the warfare systems detection range.

This would make it possible in the future to upgrade existing phased arrays antenna by replacing the GaAs power device with GaN power device having high power in a limited space. This reduces the volume of the T/R modules and provides either an improvement in search volume or an increase in track range.

B. Gain and efficiency

Modern warfare systems for military applications place new requirements on T/R modules due to the desire to reduce system size, weight, and cost. A major shift in T/R module specifications focuses more on efficiency to reduce DC power requirements. Lower component power dissipation can also improve phased arrays warfare systems reliability. GaAs MMIC typical gain is 6-8 dB in one stage and efficiency is 30%. As Fig5 shows, the GaN chips amplifier has higher efficiency and gain compared with GaAs power amplifier. The total Efficiency of warfare T/R module is expressed in (1) as follows [10]:

$$\eta_M = \eta_c L_c (1 - \frac{1}{G}) \qquad (1)$$

Where η_c is the efficiency of the amplifier

L_c is the efficiency of the combine networ'.

G is the Gain of the amplifier

Compared with the GaAs power amplifier, for the same L_c, the total efficiency of T/R module can be improved with the GaN power chips due to its high efficiency and gain. The higher efficiency and gain reduces DC power requirements and simplifies cooling. This is an important advantage, since cost and weight of cooling systems is a significant fraction of the weight and cost of a T/R module.

C. Reliability

GaN technology offers much higher impedance of the chips. This enables lower complexity and lower cost impedance matching in T/R modules. With the application of GaN power devices with simple matching due to high impendence, reliability of T/R modules can be improved. GaN devices can easily operate at 32 V and potentially up to 42 Volts. The high voltage feature eliminates or at least reduces the need for voltage conversion. Furthermore, the wide bandgap offers a rugged and reliable technology capable of high drain-source breakdown voltage. In this paper, the drain-source breakdown voltage of the GaN power chips is 100V. However, the operate voltage of the GaAs power device is just only 8V. The higher drain-source breakdown voltage results in higher reliability, which T/R modules system applications can benefit from this advantage.

VI. CONCLUSION

In this paper, An S band 85W high power amplifier for T/R modules applications was developed with the GaN power chips. The performance test of the GaN power amplifier was carried out. Compared with GaAs MMIC, it showed that the output power, efficiency, and reliability of the phased arrays warfare systems can be improved with the application of GaN wide bandgap semiconductor power chips. The size and weight of the T/R Modules system can be reduced with the application of GaN power Chips.

REFERENCES

[1] ZHENG Xin. "Research on the application of novel semiconductor power device in modern radar". Semiconductor Technology. vol 34, pp. 828-832. 2009

[2] Adolph M, Hackenberg U, Reber R, et al. "High-precision temperature drift compensated T/R-Module for satellite based SAR applications". European Microwave Conference. Paris: IEEE Press, 2005.

[3] YU Zhenkun, LIU Dengbao. "Application research on broadband amplifier with GaN power chips in S band". Journal of Microwaves. vol 27, pp.68-71, Apr.2011

[4]Milligan J W, Sheppard S, Pribble W, et al. SiC and GaNwide bandgap device technology overview. IEEE Radar Conference. Boston MA, 2007. 960-964

[5] Millan, J. Wide band-gap power semiconductor devices . Circuits, Devices & Systems, IET. 2007: 372~379

[6] Chow T P . High-voltage SiC and GaN power devices [J] . Microelectronic Engineering, 2006, 83(1):112-122

[7] LIU Han, ZHENG Xin, YU Zhen-kun. " The Application Analysis of GaN Power Devices in Radar Transmitter". IET International Radar Conference 2009.

[8] ZHANG guangyi Wang bingru. "Some requirements for active phased radar and application of wide bandgap semiconductor devices", Modern Radar vol.27,pp.1-5 2005

[9] Inder Bahl Prakash Bhartia " Microwave Solid State Ciruit Design (Second Edition)" ZHENG Xin translate Publishing House of Electronics Industry, 2006

[10] YU Zhenkun, ZHENG Xin. "The application analysis of SiC power device in radar transmitter", Journal of Microwaves. vol.23, pp.61-65,June 2007

Compact Multi-band Circularly Polarized Antenna for GNSS Applications

J.X. Li, H.Y. Shi, H. Li, K. Feng, and A.X. Zhang
School of Electronic and Information Engineering, Xi'an Jiaotong University
No.28, Xianning West Road, Xi'an, Shaanxi, 710049, P.R. China

***Abstract*-A novel compact multi-band stacked patch antenna with circularly polarization is proposed for GNSS applications. The antenna has been designed to operate at the satellite navigation frequency bands including GPS L1, GLONASS L1, BDS-1 L, BDS-1 S, and BDS-2 B3. The proposed antenna comprises of four layer stub-loaded circular microstrip patches, all of which are dual probe-fed. The feed network which includes four broadband baluns for the antenna is also designed. The antenna has been designed and fabricated. Details of the design considerations are described. Both simulated and measured results are presented and discussed.**

I. INTRODUCTION

Global navigation satellite systems (GNSS) can provides reliable positioning, navigation, and timing services for users on a continuous basis in all weather, day and night, anywhere on or near the Earth which has four or more visible GNSS satellites. Therefore, nowadays satellite navigation systems have been intensively used both in civilian and military areas. At present, GNSS includes GPS (USA), GLONASS (Russia), Galileo (Europe), BDS-1/-2 (China), etc. The American GPS and the Russian GLONASS are already operable worldwide. In 2012, it has been announced that Chinese BDS began to provide service for Asia-Pacific region (55°E-180°E, 55°S-55°N) [1]. In the future, satellite navigation receivers will provide multi-mode operation using different satellite navigation systems to improve positioning accuracy and reliability. So it is practically desirable to design terminal antenna for multi-mode satellite navigation systems.

Recently, GNSS antenna has attracted great research attention and kinds of GNSS antennas have been reported, including microstrip antennas, quadrifilar helix antennas, and many other type antennas [2-7]. However, most of GNSS antennas developed can only cover a single band [2-4], or operate in a single mode [5- 6]. The antenna proposed in [7] is able to operate at GPS, GLONASS, Galileo, and BDS-2 bands. However, the bands for BDS-1 which can provide both navigation and communication services has not been covered.

In this paper, a multi-band circularly polarized (CP) antenna is proposed. The proposed antenna has right hand circularly polarized (RHCP) radiation patterns for GPS L1 (1575 ± 5 MHz), GLONASS L1 (1602 ± 8 MHz), BDS-1 S (2492 ± 5 MHz), and BDS-2 B3 (1268 ± 10 MHz) bands, while left hand circularly polarized (LHCP) radiation patterns for BDS-1 L (1616 ± 5 MHz) band. In the proposed antenna, due to its advantage of low profile, easy fabrication, and easy integration with passive and active devices, four layer stub-loaded circular microstrip patches are stacked to achieve multi-band operation. Details of the design considerations and experimental results will be presented and discussed in the following sections.

II. ANTENNA DESIGN

The geometry of the proposed multi-band GNSS antenna is presented in Fig. 1. The four stub-loaded circular patches overlap each other without any air gaps, which usually result in an increase of antenna thickness, fabrication complexity and cost [8]. From top to bottom, the patches are deigned to operate at BDS-1 S, BDS-1 L, BDS-2 B3, and GPS L1 and GLONASS L1 successively. The top and bottom substrates have thickness of h_1 and relative permittivity of ε_{r1}, while the second and third substrates have thickness of h_2 and relative permittivity of ε_{r2}. Stubs are loaded for each patch for the convenience of tuning the impedance matching. The commercial electromagnetic simulation software Ansoft HFSS is used to design and optimize the geometric parameters of the proposed antenna. In the end, the optimal antenna parameters are determined as follows: h_1 = 3.2 mm, h_2 = 2.5 mm, r_{s1} = 17.0 mm, r_{s2} = 22.5 mm, r_{s3} = 30.0 mm, r_{s4} = 35.0 mm, r_{p1} = 12.5 mm, r_{p2} = 18.0 mm, r_{p3} = 26.0 mm, r_{p4} = 31.5 mm, ε_{r1} = 6.15, ε_{r2} = 10.0, d_{f1} = 3.5 mm, d_{f2} = 7.5 mm, d_{f3} = 12.5 mm, d_{f4} = 17.8 mm. A ladder-type cylindrical station and a screw were employed to fix and support the microstrip patches.

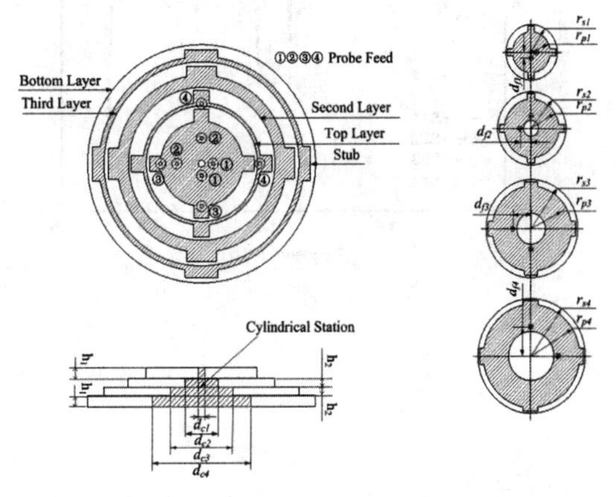

Figure 1. Geometry of the proposed GNSS antenna

978-7-5641-4279-7

The proposed structure of the feed network is shown in Fig. 2. It includes four broadband baluns, each of which comprises of a cascade of 3-dB Wilkinson power divider and a 90° broadband phase shifter. The feed network is printed on substrate with thickness of h_1 = 1.0 mm and relative permittivity of ε_{r2} = 2.65. To decrease the coupling between the patches and the feed network, the baluns are printed on the opposite side of the ground plane of the patches. Certain amounts of grounding holes are adopted to enhance the amplitude balance and phase orthogonality of the outputs for each balun.

The return loss parameters of the proposed antenna with and without the feed work are simulated, which are respectively shown in Fig. 3. It can be seen that good performance of return loss has been obtained. The return loss is less than -15 dB over all the operating frequency bands, which attributes to the adoption of the broadband baluns of the feed network.

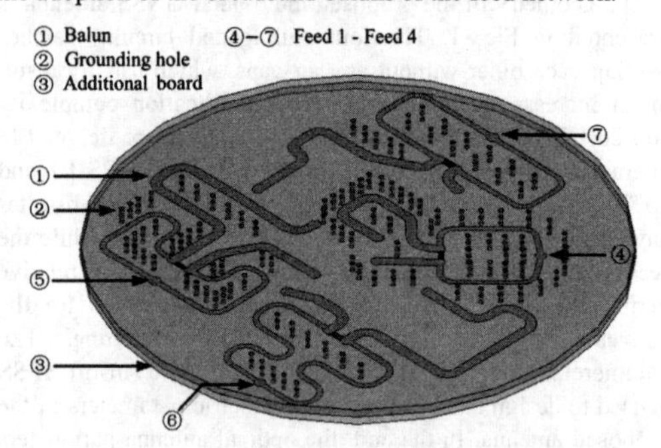

Figure 2. Structure of the proposed feed network

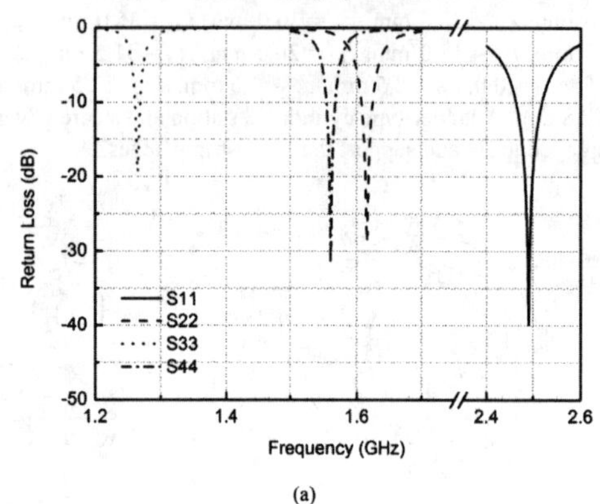

(a)

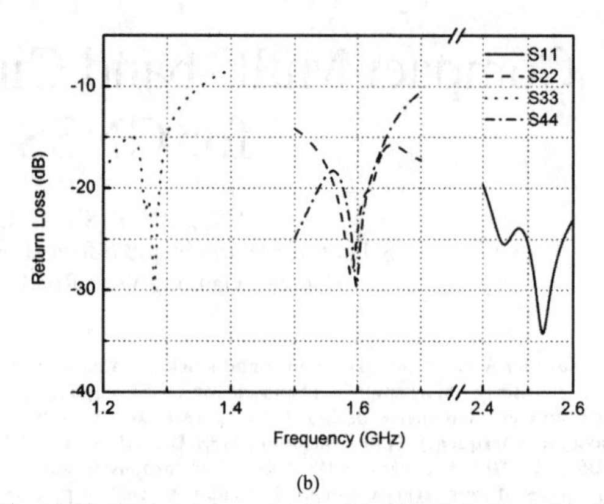

(b)

Figure 3. Simulated return loss parameters: (a) without feed network, (b) with feed network

III. Results and Discussions

The proposed GNSS antenna integrated with the feed network is fabricated as shown in Fig. 4. Fig. 5 shows the measured return loss parameters. As seen, the measured return loss demonstrates high performance of less than -15 dB (corresponding to VSWR < 1.5) for all the operating bands. Compared with Fig. 3(b), it can be observed that the measured results agree well with the simulated ones, and good impedance matching performance has been achieved.

Fig. 6 presents the normalized measured far-field radiation patterns respectively at 1268 MHz, 1575 MHz, 1616 MHz and 2492 MHz. It shows that good RHCP performance is achieved for 1268 MHz, 1575 MHz and 2492 MHz, and good LHCP performance is achieved for 1616 MHz. The CP isolation is above 12 dB for all designed bands. The measured peak gain is 0.38 dBi, 2.93 dBi, 2.62 dBi, and 3.56 dBi for 1268 MHz, 1575 MHz, 1616 MHz and 2492 MHz. The axial ratio in the broadside direction also have been measured, which is 2.1 dB at 1268 MHz, 1.8 dB at 1575 MHz, 1.9 dB at 1616 MHz and 2.3 dB at 2492 MHz.

Figure 4. The fabricated multi-band GNSS antenna

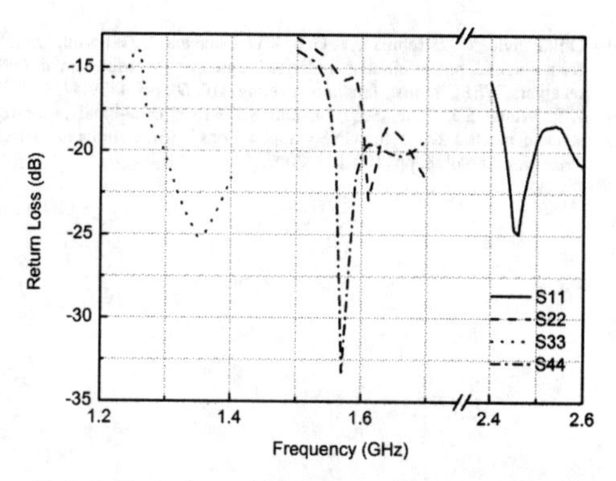

Figure 5. Measured return loss parameters of the fabricated antenna

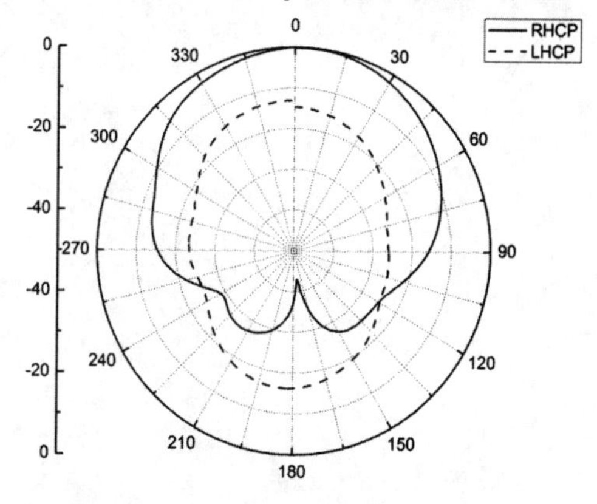

(a)

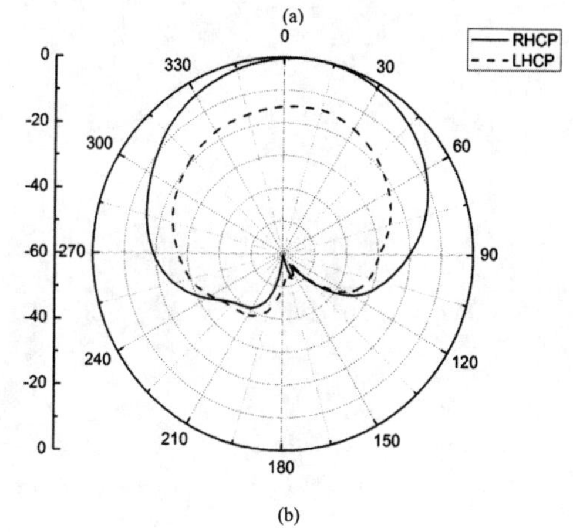

(b)

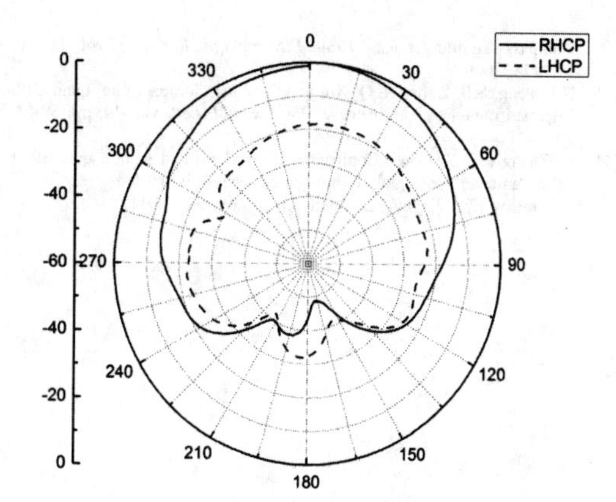

(c)

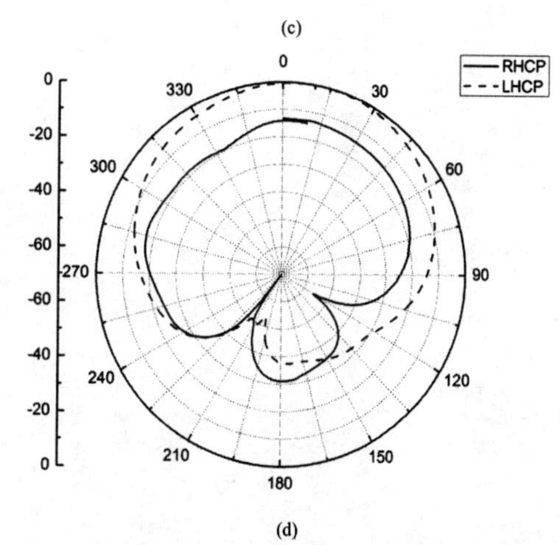

(d)

Figure 6. Measured normalized radiation patterns at: (a) 1268 MHz, (b) 1575 MHz, (c) 2492 MHz, and (d) 1616 MHz

IV. CONCLUSION

In this paper, a novel compact multi-band stacked patch antenna for GNSS applications is proposed. The antenna can achieve good impedance matching performance and circularly polarized radiation patterns within GPS L1, GLONASS L1, BDS-1 L, BDS-1 S, and BDS-2 B3 bands. A prototype has been developed and measured to verify the concept. Both the simulated and the measured results demonstrate the high performance of the proposed antenna.

REFERENCES

[1] China Satellite Navigation Office, "BeiDou Navigation Satellite System Signal in Space Interface Control Document Open Service Signal B1I." Available: www.beidou.gov.cn.

[2] L. Boccia, G. Amendola, and G. Di Massa, "A shorted elliptical patch antenna for GPS applications," *IEEE Antennas Wirel. Propag. Lett.* vol. 2, pp. 6-8, 2003.

[3] J.Y. Deng, Y.Z. Yin, Y.H. Huang, J. Ma, and Q.Z. Liu, "Compact circularly polarized microstrip antenna with wide beamwidth for

Compass satellite service," *Prog. Electromagn. Res. Lett.* vol. 11, pp. 113-118, 2009.

[4] X.F. Peng, S.S. Zhong, S.Q. Xu, and Q. Wu, "Compact dual-band GPS Microstrip antenna," *Microwave Opt. Technol. Lett.* vol. 44, pp. 58-61, 2005.

[5] L. Zheng and S. Gao, "Compact dual-band printed square quadrifilar helix antenna for global navigation satellite system receivers," *Microwave Opt. Technol. Lett.* vol. 53, pp. 993-997, 2011.

[6] O.P. Falade, M.U. Rehman, Y. Gao, X.D. Chen and C.G. Parini, "Single feed stacked patch circular polarized antenna for triple band GPS receivers," *IEEE Trans. Antennas. Propag.* vol. 60, pp. 4479-4484, 2012.

[7] Z.B. Wang, S.J. Fang, S.Q. Fu, and S.W. Lv, "Dual-band probe-fed stacked patch antenna for GNSS applications," *IEEE Antennas Wirel. Propag. Lett.* vol. 8, pp. 100-103, 2009.

Simulation of Two Compact Antipodal Vivaldi Antennas With Radiation Characteristics Enhancement

Qiancheng Ying, Wenbin Dou
State Key Laboratory of Millimeter Waves, Southeast University, Nanjing 210096, People's Republic of China
Email: seuyqc@163.com; njdouwb@163.com

***Abstract*—In this paper, two compact antipodal Vivaldi antennas (AVA) with improved Radiation Characteristics are presented. U-typed tapered slot edge (U-TSE) AVA and comb-shaped AVA with U-TSE are simulated and optimized. The -10-dB reflection coefficient bandwidth of comb-shaped AVA spans from 2.3 GHz to 18 GHz (7.83:1) and covers 15.7 GHz, while the bandwidth of conventional AVA which has the same dimensions spans from 3.3 GHz to 18 GHz (5.45:1) and covers 14.7 GHz. The U-TSE structure has the capacity to extend the low-end bandwidth limitation and improve the radiation characteristics in the lower frequencies. Compared with the conventional AVA, this modification can miniaturize the electrical size of the AVA antenna about 30%. The results regarding return loss, far field pattern, antenna gain and axial ratio are illustrated.**

I. INTRODUCTION

With the development of modern communication technology in both military and civil applications, the demand for compact, smart, wideband, and multifunctional antennas increases. The tapered slot antenna (TSA), which is still one of the most widely used wideband antennas, has a history of more than 30 years. The first TSA was introduced by Gibson [1] with exponential profile in 1979, which is also known as the ETSA or Vivaldi antenna. Vivaldi antenna is widely investigated and applied in satellite communication, remote sensing, and radio telescopes due to its broad bandwidth, low cross-polarization, and high directivity [2-4]. However, the operating bandwidth of conventional Vivaldi antenna is usually limited. First, the high-end working band is restricted by the transition structure from the microstrip to slotline. In the low-end working band, on the other hand, the bandwidth is limited by the width of the antenna patch.

In order to overcome the limitation, Gazit proposed the antipodal Vivaldi antenna (AVA) [5] in 1988. The antipodal Vivaldi antenna separates the patches with the substrate and centers them. The antenna is fed by a standard microstrip transmission line. A microstrip-to-symmetric-double-sided-slotline transition that has extremely wide operating band is employed instead of the conventional microstrip-to-slotline transition. In recent years, although many researches [6-8] on AVA have been down, how to decrease low-end cutoff frequency limited by the antenna aperture while improving the antenna radiation characteristics is still a problem. In [9], the tapered slot edge (TSE) structure is first proposed to reduce the low-end cutoff frequency while keeping the dimensions of the antenna unchanged.

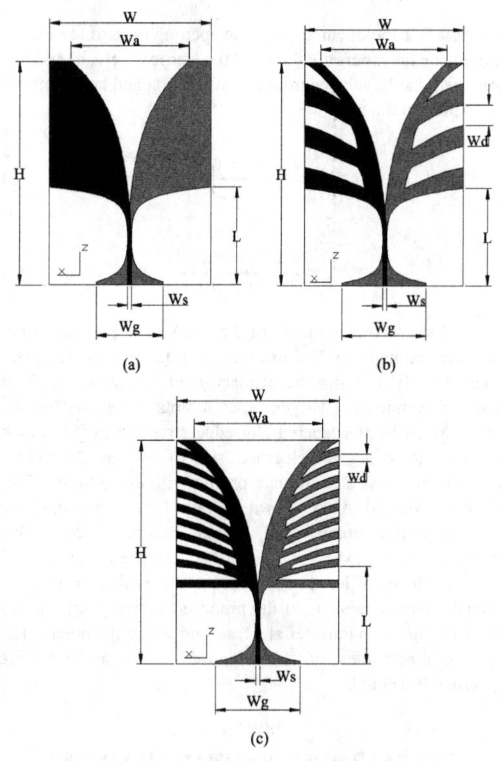

Fig. 1. Proposed evolution of the AVA configuration. (a) Conventional AVA. (b) U-TSE AVA. (c) Comb-shaped AVA with U-TSE.

In this paper, U-type tapered slot edge (U-TSE) structure is proposed. Novel U-TSE AVA and comb-shaped AVA with U-TSE are designed. By using U-TSE, the low-end cutoff frequency of the antenna shows significant reduction, indicating that the electrical size of the antenna can be miniaturized. Also, U-TSE structure can improve the antenna's radiation characteristics, such as antenna gain and

978-7-5641-4279-7

axial ratio. The Ansoft simulation software high-frequency structure simulator (HFSS) [10] is used to optimize the design. Good return loss and radiation pattern characteristics are obtained in the frequency band of interest.

II. Antenna Design and Configuration

Two proposed antipodal Vivaldi antennas fed by a 50Ω microstrip transmission line are illustrated in Fig. 1(b) and (c), which are printed on the FR4 substrate of thickness 0.6 mm, relatively permittivity 4.4 and loss tangent 0.02. A 50Ω microstrip feed line has a metal strip of width Ws=1.18 mm. The inner and outer edge tapered curves of the antenna are defined as

$$x = c_1 \cdot e^{Rz} + c_2 \tag{1}$$

Where R is the opening rate. The opening rates of the inner and outer edge tapers are 0.06 and 0.4, respectively. c_1 and c_2 are determined by the coordinates of the first and last points of the exponential curve.

$$c_1 = \frac{x_2 - x_1}{e^{Rz_2} - e^{Rz_1}} \tag{2}$$

$$c_2 = \frac{x_1 e^{Rz_2} - x_2 e^{Rz_1}}{e^{Rz_2} - e^{Rz_1}} \tag{3}$$

Fig. 1 shows the evolution of the AVA configuration. First, the conventional AVA illustrated in Fig. 1(a) is designed. Then, Fig. 1(b) shows the primary modified structure. Two pairs of symmetrical U-type tapered slots are etched on the fins inspired by the tapered slot edge structure in [9], which was used to reduce the electrical size of antenna. As the U-TSE structure has a coordinate profile with the radiation fins, the comb-shaped AVA shown in Fig. 1(c) is proposed by increasing the number of U-type tapered slots. This modification is expected to further lengthen the overall effective electrical length and improve the performance of the antenna. The parameters of the proposed antennas are studied by changing one parameter at a time and fixing the others. The optimal dimensions of the proposed AVA antennas are specified in Table I.

TABLE I

THE OPTIMAL DIMENSIONS OF THE PROPOSED AVA ANTENNAS

	Conventional AVA	U-TSE AVA	Comb-shaped AVA
W(mm)	48	48	48
H(mm)	64	64	64
Wa(mm)	35	38	38
L(mm)	28	28	28
Wg(mm)	20	25	25
Wd(mm)	-	6	2

III. Results and Discussion

A. Return Loss

Fig. 2 illustrates the return loss curves of conventional AVA, U-TSE AVA, and comb-shaped AVA, respectively.

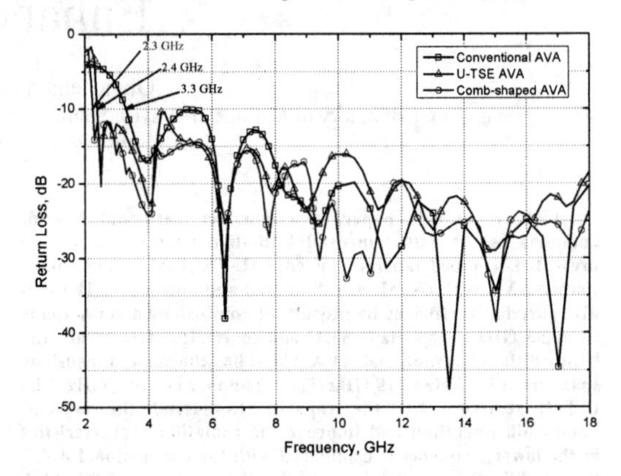

Fig. 2. Simulated return loss of the proposed AVA antennas

As shown in the figure, the low-end cutoff frequency of the conventional AVA for $S_{11} \leq -10$ dB is 3.3 GHz, while the U-TSE AVA extends it to 2.4 GHz. The comb-shaped AVA further extends the low-end cutoff frequency to 2.3 GHz and has ultra-wideband performance spanning from 2.3 to more than 18 GHz. Additional resonant points can be observed around the lower limitation of working band in both U-TSE and comb-shaped AVA. The dimensions of the conventional AVA are 48×64 mm^2, which is approximately $0.53\lambda \times 0.7\lambda$, where λ is the wavelength of 3.3 GHz. Also, the dimensions of the comb-shaped AVA are 48 × 64 mm^2, which is approximately $0.37\lambda \times 0.49\lambda$, where λ is the wavelength of 2.3 GHz. It is demonstrated that the U-TSE structure has the capacity to miniaturize the electrical size of the antenna. Although the low-end cutoff frequencies of U-TSE AVA and comb-shaped AVA are almost the same, the return loss of comb-shaped AVA is better than U-TSE AVA in the frequency band of interest, especially in the higher frequencies.

B. Current Distributions

In order to further understand the behavior of the U-TSE structure, especially in the lower frequencies, current distribution patterns of conventional AVA, U-TSE AVA, and comb-shaped AVA at 2.4 GHz are given in Fig. 3.

As the figure reveals, the surface current of the conventional AVA in region A is very small at 2.4 GHz. It is demonstrated that most of the input energy is reflected and can not be radiated. Thus, the return loss of the conventional AVA shown in Fig. 2 is very bad from 2.3 GHz to 3.3 GHz. On the other hand, by etching the U-type tapered slots, significant currents are observed in both region B and region C, indicating that the effective length of the current path on the antenna is lengthened through the modification. Besides, compared with U-TSE AVA, there are more U-type tapered slots etched on the fins of comb-shaped AVA. The effective length of the current path should be further lengthened. This is the reason why the comb-shaped AVA has the lowest cutoff frequency and the best return loss.

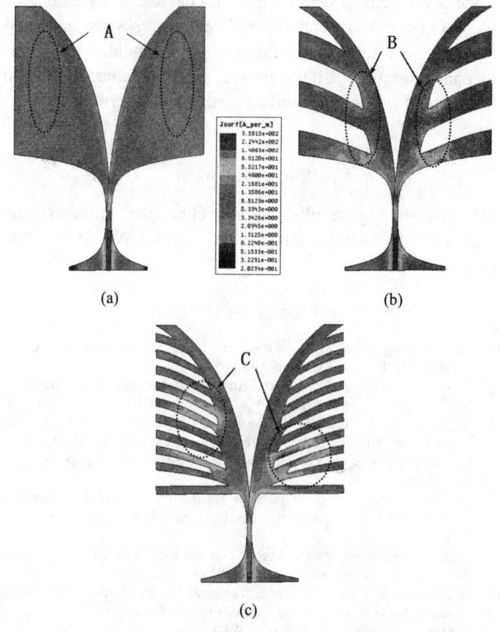

Fig. 3. Surface current distribution of the proposed antennas.(a) Conventional AVA. (b) U-TSE AVA. (c) Comb-shaped AVA.

C. *Radiation Patterns*

Fig. 4 illustrates simulated far-field E-plane (x-z plane) and H-plane (y-z plane) radiation patterns of the proposed AVA antennas at 4.4, 8.4, 13.5 and 17.5 GHz, including the co-polarization and cross-polarization.

As the figure shows, the proposed AVA antennas have endfire characteristics with the main lobe in the axial direction of the tapered slotline (z-direction in Fig. 1). At all four frequency points, the radiation patterns of the comb-shaped AVA show significant improvement in side lobe level and front to back ratio compared with the conventional one. The cross polarization levels are very small for all the simulated patterns and a co-pol/cross-pol ratio of better than 20 dB is observed in the direction of z-axis. Besides, the proposed antennas have symmetrical H-plane radiation pattern, while the E-plane has the slight asymmetry.

D. *Antenna Gain and Axial Ratio*

The realized gain variation with frequency of conventional AVA, U-TSE AVA, and comb-shaped AVA in the direction

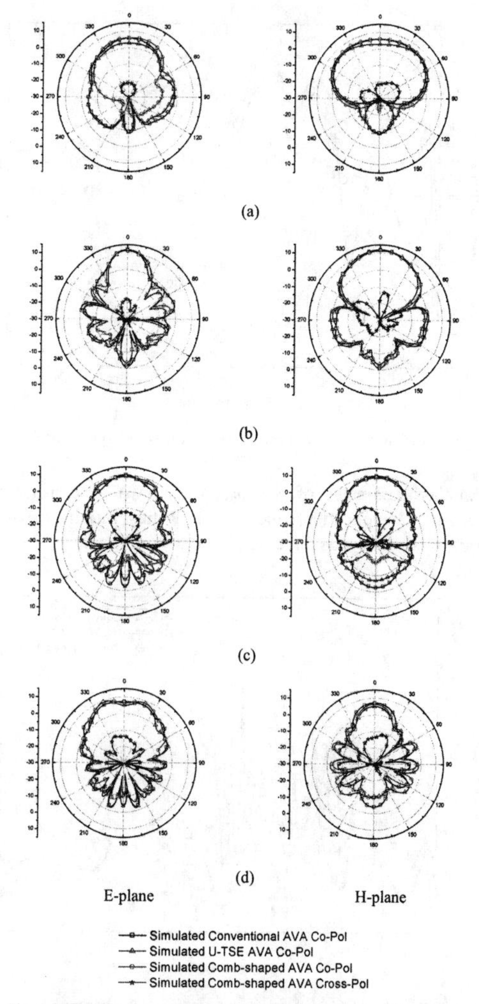

Fig. 4. E-plane (x-z) and H-plane (y-z) radiation patterns of the proposed AVA antennas at (a) 4.4, (b) 8.4, (c) 13.5, and (d) 17.5 GHz.

of z-axis is illustrated in Fig. 5.

As shown in the figure, in the band from 2.3 GHz to 10 GHz, especially from 2.3 GHz to 6 GHz, the realized gain curves of both U-TSE AVA and comb-shaped AVA show great improvement. The minimum gain of comb-shaped AVA is at the frequency 5.7 GHz with a value about 4.4 dB and maximum gain at 8.8 GHz with a value about 11.5 dB. However, in the band after 10 GHz, the realized gain becomes worse. It is indicated that the dielectric loss is larger in the high frequencies compared with the conventional one. The reason is possibly that the longer effective length of current path on the U-TSE and comb-shaped AVA antennas causes more loss in the high frequencies.

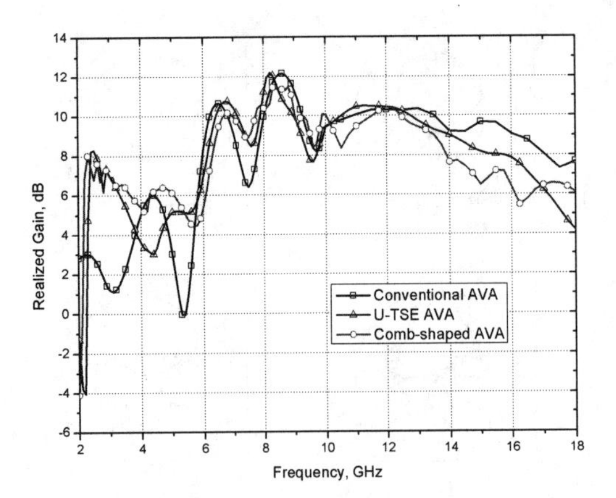

Fig. 5. Variation of the realized gain with frequency of the proposed AVA antennas.

Another parameter of the proposed AVA antennas for polarization application, the Axial Ratio variation with frequency in the z-axis direction, is shown in Fig. 7.

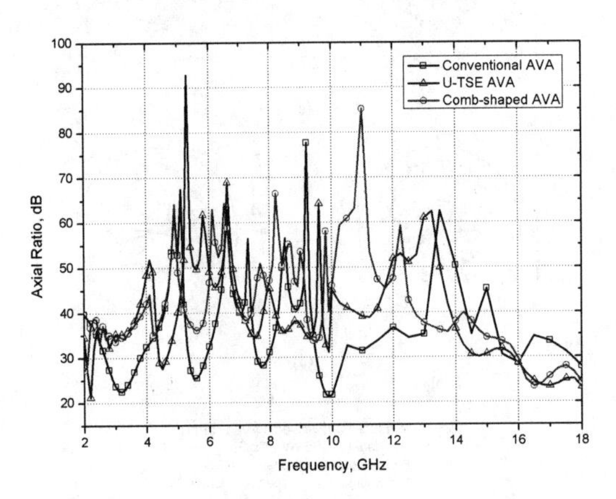

Fig. 7. Variation of the axial ratio with frequency of the proposed AVA antennas.

As the figure reveals, in the band from 2.3 GHz to 14 GHz, the axial ratio (AR) curves of both U-TSE AVA and comb-shaped AVA show significant improvement. The minimum AR of comb-shaped AVA is at the frequency 2.5 GHz with a value about 33.1 dB. It is demonstrated that U-TSE structure can help enhance the co-pol/cross-pol ratio, indicating excellent polarization purity. However, in the higher frequencies, the axial Ratio decreases because the effective thickness of the dielectric substrate increases.

IV. CONCLUSION

In this paper, two novel antipodal Vivaldi antennas with U-type tapered slot edge modification are proposed. With return loss better than -10 dB, the impedance bandwidth of U-TSE AVA spans from 2.4 GHz to 18 GHz, while the bandwidth of comb-shaped AVA is over the frequency range of 2.3 GHz to 18 GHz (7.83:1). The best low-end cutoff frequency for $S_{11} \leq$ -10 dB is extended to 2.3 GHz from the conventional AVA 3.3 GHz. This modification can miniaturize the electrical size of the antenna about 30%. Moreover, U-TSE AVA and comb-shaped AVA behave better radiation characteristics compared with the conventional one. Simulated results show that two modified AVA antennas could be a good candidate for ultra-wideband communication systems. The implementation and measurement of the proposed antenna can be carried out in the future.

ACKNOWLEDGMENT

The authors gratefully acknowledge the students and teachers of State key laboratory of Millimeter Waves, for their helpful comments.

REFERENCES

[1] P. J. Gibson, "The Vivaldi aerial," in *Proc. 9th Eur. Microw. Conf.*, Brighton, U. K., Jun. 1979, pp. 101-105.

[2] R. Janaswamy, D. H. Schaubert, "Analysis of the Tapered slot antenna," *IEEE Trans. Antennas Propag.*, vol. 35, no. 9, pp. 1058-1065, Sept 1987.

[3] K. S. Yngvesson, T. L. Korzeniowski, Y. Kim, and E. L. Kollberg, "The tapered slot antenna-A new integrated element for millimeter-wave application," *IEEE Trans. Microw. Theory Tech.*, vol. 37, pp. 365-374, Feb. 1989.

[4] H. Oraizi and S. Jam, "Optimum Design of Tapered Slot Antenna Profile," *IEEE Trans. Antennas Propag.*, vol. 51, no. 8, pp. 1987-1995, 2003.

[5] E. Gazit, "Improved design of the Vialdi antenna," *Proc. Inst. Elect. Eng. H*, vol 135, no. 2, pp. 89-92, Apr. 1988.

[6] A. Z. Hood, T. Karacolak, and E. Topsakal, "A Small Antipodal Vivaldi Antenna for Ultrawide-band Applications," *IEEE Antennas Wireless Propag. Lett.*, vol. 7, pp. 656-660, 2008.

[7] X. D. Zhuge, A. Yarovoy, Leo P. Ligthart, "Circularly Tapered Antipodal Vivaldi Antenna for Array-Based Ultra-wideband Near-Field Imaging," *Proc. 6th Eur Radar Conf.*, Rome, Italy, Sept. 2009, pp. 250-253.

[8] Y. X. Che, K. Li, X. Y. Hou, and W. M. Tian, "Simulation of A Small Sized Antipodal Vivaldi Antenna for UWB Applications," *ICUWB.*, Nanjing, China, Sept. 2010, pp. 1-3.

[9] P. Fei, Y.C. Jiao, W. Hu, and F. S. Zhang, "A Miniaturized Antipodal Vivaldi Antenna With Improved Radiation Characteristics," *IEEE Antennas Wireless Propag. Lett.*, vol. 10, pp. 127-130, 2011.

[10] Ansoft Corporation, Ansoft High Frequency Structure Simulation (HFSS), Ver. 13, Ansoft Corporation, Pittsburgh, PA, USA, 2010.

Design of a Compact UWB Band-notched Antenna with Modified Ground Plane

Tong Li, Huiqing Zhai, Jianhui Bao and Changhong Liang
School of Electronic Engineering, Xidian University
Xi'an, 710071, China

***Abstract*- A compact ultra-wideband (UWB) antenna with dual band-notched characteristics is presented. The proposed antenna consists of a folded fork-shaped radiator and a modified ground plane. The modified ground plane with trapezoid shaped slots on its top edge effectively increases the bandwidth of the proposed antenna. To avoid potential interferences, the arms of the forked-shaped radiator are folded to form a pair of $\lambda_g/4$-coupled lines, and a $\lambda_g/2$ semicircular slot is embedded on the radiator. The antenna, having an entire size as small as 30 × 25 mm², achieves an operation frequency band from 3GHz to 17GHz with good rejection to the Worldwide Interoperability for Microwave Access (WiMAX) and the wireless local area network (WLAN) bands in both simulated and measured results.**

I. INTRODUCTION

Ultra-wideband technology has been widely used in wireless communication owing to its high data transmission rate, large bandwidth, and short-range characteristics. As a key component of UWB system, UWB antenna has gained increasing attentions. Various planar UWB antennas with wide impedance bandwidth and good radiation characteristics have been proposed and investigated. However, most of them only focus on the band of 3.1 – 10.6 GHz [1]-[4]. On the other hand, their sizes are relatively large, which cannot meet the demand of the consumer electronics applications. Although literature [5] proposed a fairly compact UWB antenna with a size of 11 × 20.5 mm², the -10dB impedance bandwidth is only from 3.1 to 5 GHz. An improved design of a planar elliptical dipole antenna for UWB applications has also been developed recently [6]. By using elliptical slots on the dipole arms, the antenna has achieved an operating bandwidth of 94.4%. However, the antenna does not process a physically compact size, having dimensions of 106 × 85 mm². Besides the above mentioned challenge, UWB antenna also needs to solve electromagnetic interference (EMI) problem since there exist several undesired radio signals within UWB bandwidth, such as WiMAX (3.3-3.6 GHz) and WLAN (5.15-5.825 GHz).

Based on these points, a novel UWB planar monopole antenna with compact size, wider bandwidth and dual notched bands is presented in this letter. The antenna has a modified ground plane which effectively increases the operation bandwidth. To achieve dual band-notched function without increasing the dimensions of the antenna, the arms of the forked-shaped radiator are folded to form a pair of $\lambda_g/4$-coupled lines, and a $\lambda_g/2$ semicircular slot is embedded on the radiator.

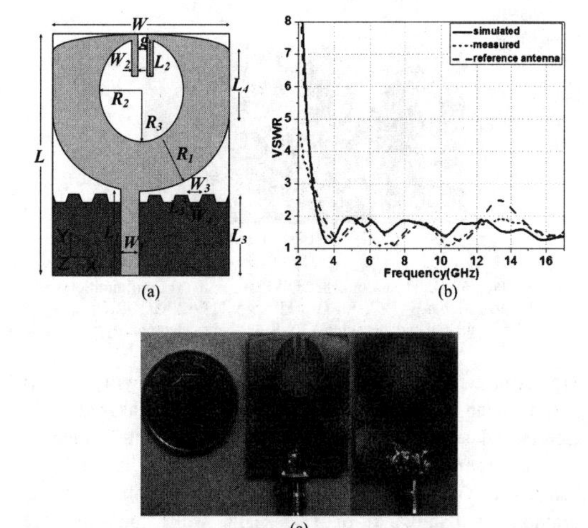

Figure 1. Proposed compact UWB antenna. (a) Configuration with the dimensions: $W = 25$, $L = 30$, $H = 1.6$, $W_1 = 2.56$, $L_1 = 10.65$, $W_2 = 1$, $g= 1$, $L_2 = 4.8$, $L_3 = 11$, $L_4 = 6.85$, $R_1 = 12.5$, $R_2 = 6$, $R_3 = 7$, $W_3 = 2.2$, $W_4 = 1.6$, $L_5 = 1$. (b) Simulated and measured VSWR. (c) Fabricated prototype.

II. ANTENNA CONFIGURATION AND DESIGN

A. *Compact UWB Antenna Design*

Fig. 1(a) shows the geometry and dimensions of the proposed compact UWB antenna. The proposed radiator has rounded edge to broaden the bandwidth and to produce smooth transitions from one resonant mode to another. By folding the arms of the forked-shaped radiator, the current path is enlarged, thus the antenna demonstrates a fairly compact dimensions, 30×25 mm² in physical size. To enhance the impedance bandwidth, the ground plane is modified to a symmetrical sawtooth shape by cutting trapezoid shaped slots on its top edge. This antenna is printed on the FR4 substrate with thickness of 1.6 mm, relative dielectric constant ε_r =4.4, and loss tangent of 0.02. The antenna is fed by a 50Ω microstrip line and connected to a SMA connector, as shown in Fig. 1(c). The antenna was measured with an Agilent N5230A vector network analyzer. Both simulated and measured VSWRs of the proposed UWB antenna are shown in

978-7-5641-4279-7

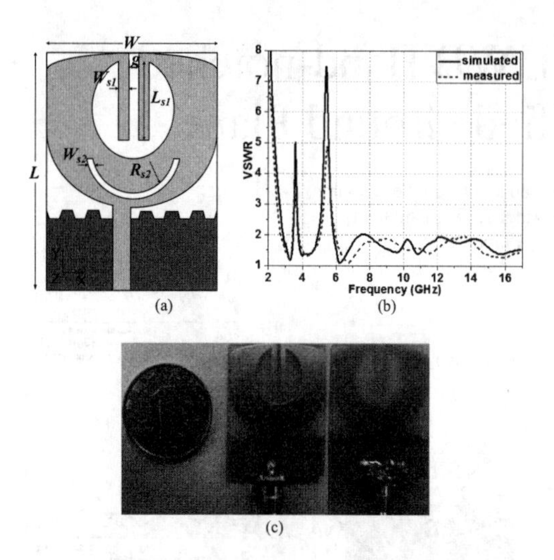

(a) (b)

(c)

Figure 2. Proposed dual band-notched UWB antenna. (a) Configuration with the dimensions: Ws1 = 1.3, Ls1 = 11.5, g = 1, Ws2 = 0.8, Rs2 = 6. (b) Simulated and measured VSWR. (c) Fabricated prototype.

Fig. 1(b), where the primitive UWB antenna with regular ground plane is also given as a reference. The measured result clearly indicates that the proposed antenna with sawtooth shaped ground plane covers an ultra wide impendence bandwidth (VSWR < 2) of 3.1 to more than 17 GHz, providing a bandwidth of 13.9 GHz, while the reference antenna with regular ground plane only provides an impendence bandwidth of 3.1 to 12.3GHz. Compared to the reference antenna, the proposed UWB antenna can enhance the bandwidth by 51% (4.7GHz).

B. Dual Band-Notched UWB Antenna Design

In order to mitigate EMI problems, we developed a dual band-notched UWB antenna, as shown in Fig. 2(a). The folded arms form a pair of coupled lines. When the length of the lines equal to $\lambda_g/4$, a notched band is created. By simply adjusting the length and gap width of the coupled lines, its notch frequency can be tuned to 3.5 GHz. To reject interference of WLAN band, a $\lambda_g/2$ semicircular slot is arranged in the middle of the patch close to the feeding strip. By adjusting the radius and gap width of the slot, its notch frequency can also be tuned easily.

The antenna was fabricated on FR4 substrate(Fig.2(c)). Measured and simulated VSWR values are compared in Fig. 2(b). Good agreement is observed. As can be seen, the proposed UWB antenna successfully exhibits dual notched bands of 3.3–3.63 GHz and 5.1–5.83 GHz, while it still maintains good impedance matching at the rest of the UWB band. It should be noted that the lowest operation frequency shifts to 2.7 GHz for the folded arms enlarge the current path.

To provide guidance for the design of notching structures, the effect of varying notching structures' parameters on

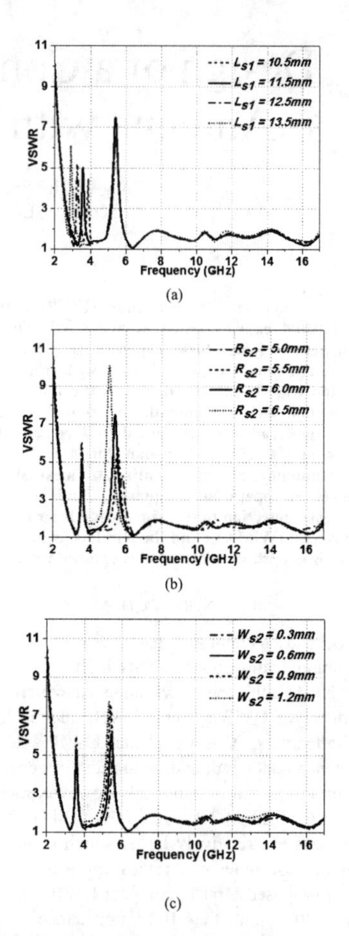

(c)

Figure 3. Effect of varying notching structures' parameters on notched bands. (a) Length L_{s1} of the coupled lines. (b) Radius R_{s2} and (c) width W_{s2} of the semicircular slot.

notched bands are analyzed. As shown in Fig. 3, the simulated lower notch band shifts to higher frequency with the decrease of length L_{s1} of the coupled lines, and the upper notch frequency increases as radius R_{s2} and width W_{s2} of the semicircular slot decrease. It should be noted that the variation of one notched band almost has no effect to the other, which means that these two notched bands can be controlled independently.

To explain the generation of the notched bands, surface current distributions are analyzed. As shown in Fig. 4(a), at the notch frequency of 3.5 GHz, most current concentrates along the folded arms of the radiator. While the current is maximum at one end, it is minimum at the other end, which is in agreement with the $\lambda_g/4$ dimension of arms. It can also be seen that the directions of the current vectors in the arms are opposite to that in the radiating patch, resulting in a notched

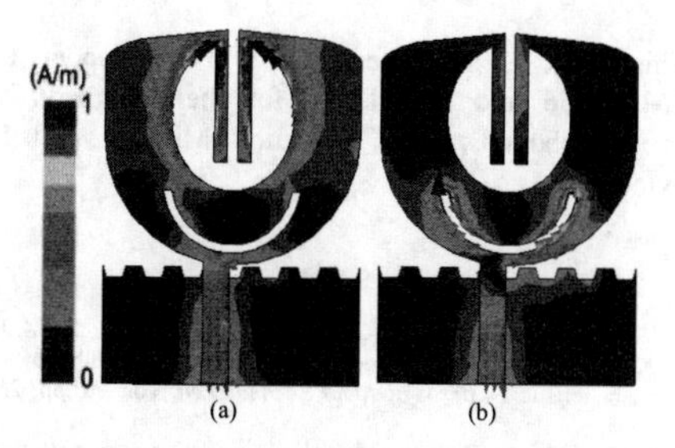

Figure 4. Surface current distributions at (a) 3.5GHz and (b) 5.5 GHz.

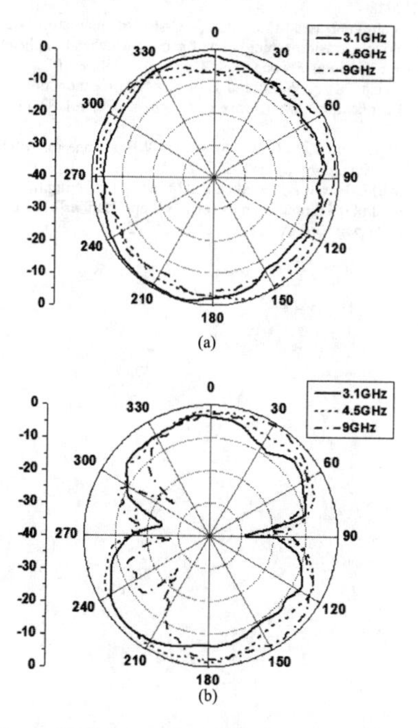

Figure 5. Measured radiation patterns at (a) xz-plane and (b) yz-plane

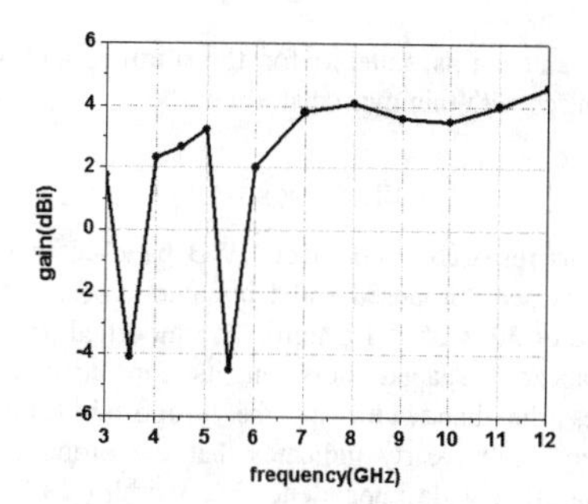

Figure 6. Measured gains of the proposed antenna

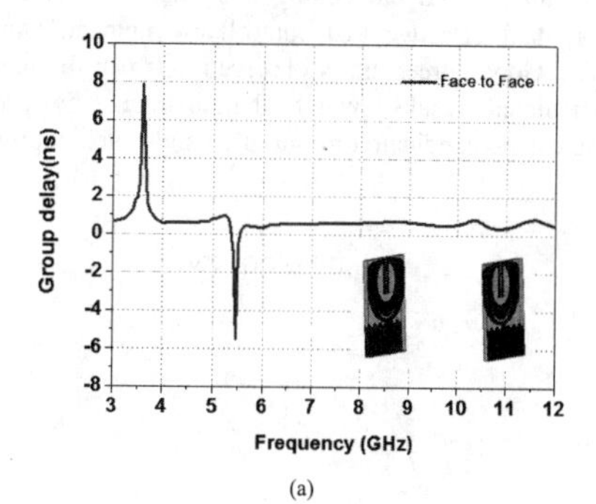

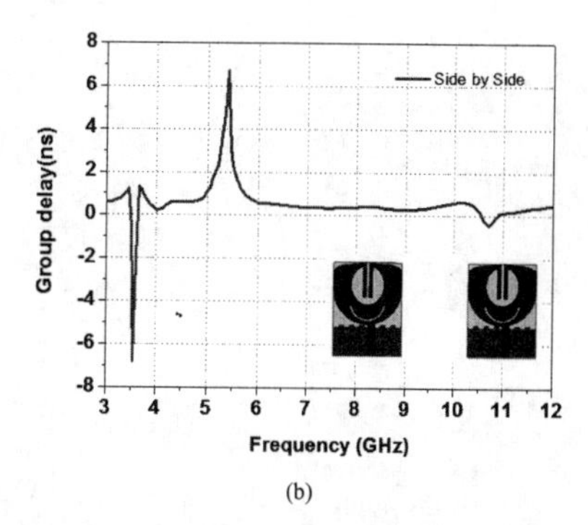

Figure 7. Simulated group delay of the proposed antenna for both (a) face-to-face and (b) side-by-side configurations.

band at 3.5 GHz. Similarly, as shown in Fig. 4(b), most current concentrates around the edge of the semicircular slot and flow in opposite directions (out of phase) between interior and exterior of the slot, causing a notched band at 5.5 GHz. The current distribution in Fig. 4(b) also clearly indicates that the length of the slot is about $\lambda_g/2$. It should be noted that, at the desired frequency, only the corresponding notching structure is active while the other is inactive, confirming the independence of the notched bands.

The measured radiation patterns of the proposed antenna at the passband (among dual notch bands) frequencies of 3.1, 4.5 and 9GHz in xz-plane (H-plane) and yz-plane (E-plane) are shown in Fig. 5. It can be observed that the antenna displays a good omnidirectional radiation pattern in xz-plane (H-plane), even at high frequencies.

Fig. 6 illustrates the measured maximum gain of the dual band-notched UWB antenna. Only several frequencies have been chosen for measurement. It is observed that gains are generally flat over the operating band of the antenna while significantly decrease at each notched band due to the frequency-rejected function.

Fig. 7 presents the simulated group delay of the proposed antenna for both face-to-face and side-by-side configurations. As can be seen, group delay is almost smooth in the entire UWB range except at the rejected bands, which demonstrates that the antenna is suitable for transmitting and receiving UWB pulses with minimum distortion.

III. CONCLUSION

In this research, a compact UWB band-notched antenna design has been proposed and fabricated. The antenna has a total size of 30 × 25 × 1.6 mm^3. The modified ground plane with trapezoid shaped slots on its top edge effectively increases the bandwidth of the proposed antenna. The measured result clearly indicates that the proposed antenna covers an ultra wide impendence bandwidth (VSWR < 2) of 13.9 GHz (3.1 to 17 GHz). By folding the arms of the forked-shaped radiator and embedding a semicircular slot on the radiator, dual effective and controllable notched bands are achieved. Good agreement is observed between the measured and simulated results, which demonstrates the proposed antenna a good candidate for multiple band-notched problem.

ACKNOWLEDGMENT

This work is supported by the NSFC under Contract No.61101066 and Foundation for the Returned Overseas Chinese Scholars, State Education Ministry and Shaanxi Province.

REFERENCES

[1] Y. L. Zhao, Y. C. Jiao, G. Zhao, L. Zhang, Y. Song and Z. B. Wong, "Compact planar monopole UWB antenna with band-notched characteristic", *Microwave Opt Technol Lett*, vol. 50, pp. 2656-2658, 2008.

[2] W. F. Chen, C. C. Tsai and C.Y. Huang, "Compact wide-slot antenna for ultra-wideband communications", *Electron. Lett.*, vol. 44, pp. 892 – 893, 2008.

[3] M. Koohestani and M. Golpour, "U-shaped microstrip patch antenna with novel parasitic tuning stubs for ultra wideband applications", *IET Microw. Antennas Propag.*, vol. 4, pp. 938 – 946, 2010.

[4] J. X. Xiao, M. F. Wang and G. J. Li, "A ring monopole antenna for UWB application", *Microwave Opt Technol Lett*, vol. 52, pp. 179 – 182, 2010.

[5] T. S. P See and Z. N. Chen, "A small UWB antenna for wireless USB", *ICUWB, Singapore*, pp. 198–203, 2007.

[6] H. Nazli, E. Bicak, B. Turetken and M. Sezgin, "An improved design of planar elliptical dipole antenna for UWB applications", *IEEE Antennas Wirel. Propag. Lett.*, vol. 9, pp. 264 – 267, 2010.

A Compact Ultra Wideband Antenna with Triple-Sense Circular Polarization

Guihong Li, Huiqing Zhai, Tong Li, Long Li, Changhong Liang
School of Electronic Engineering, Xidian University
Xi'an, 710071, China, guihonger@sina.com

***Abstract*-In this letter, a new compact antenna design with triple-sense circular polarization (CP) is presented for ultra-wideband (UWB) communication applications. The UWB property is obtained by utilizing a compact complementary planar monopole. In order to achieve multiband CP waves on base of UWB operation, the measurement has been taken such as etching a T-shaped slit in the antenna ground and adding a rectangular parasitic stub to the antenna radiator. The antenna has been fabricated and measured. Good agreement is achieved between the simulation and measurement, which shows that the presented antenna covers the UWB operation from 3.1GHz to 11.8GHz with good CP waves at Worldwide Interoperability for Microwave Access (WiMAX) (3.63-3.77 GHz bands) for left hand CP (LHCP), the wireless local area network band (WLAN) (4.95-5.18 GHz bands) for right hand CP (RHCP), and X band (9.55-9.97 GHz bands) for left hand CP (LHCP).**

I. INTRODUCTION

In recent decades, circularly polarized (CP) antennas have received much attention for finding applications in satellite positioning, radio frequency identification (RFID) and sensor systems with the development of modern wireless communication. To be resistant to the depolarization effects, bad weather conditions and multi path propagation, it is imperative for the antennas to have CP characteristics, which is insensitive to the respective orientations on the useful frequency operation when applied to launch and receive electromagnetic waves. CP waves can be generated when two degenerate modes of equal amplitude and phase difference of 90° are excited. Right-hand circular polarization (RHCP) and left-hand circular polarization (LHCP) can be defined by a 90° phase lead and lag, respectively. In [1]-[6], various techniques were proposed for multiband CP patch antenna designs. For example, eight rectangular perturbations with unequal length in [1] or two asymmetric pentagonal slots in [2] were introduced to excite a pair of degenerate modes for CP radiation. By loading a pair of L-shaped stubs outside the truncated patch, an outer mode and an inner mode were excited respectively [3]. In [4], the dual-band circular polarization was provided by dual resonant elements. In [5], a novel broadband monopole antenna with dual band CP radiation was presented using a ground plane embedded with an inverted-L slit. Furthermore, a triple band CP antenna was proposed in [6] by utilizing two ring slots and an L-shaped feed line.

Since the Federal Communications Commission (FCC) specifies the unlicensed 3.1-10.6 GHz band for ultra-wideband (UWB) commercial usages in 2002 [7], the increasing demands have stimulated many researches into UWB technology because of its low cost, low power consumption, high data transmission rates, resistance to severe multipath and jamming, etc. Meanwhile, antenna designs for UWB applications are facing many challenges including the miniaturized design, impedance matching, radiation stability, and electromagnetic interference (EMI) problems [8]-[10]. As is known, the printed monopoles are widely utilized to design the UWB antennas due to their attractive features such as simple structure, low profile, light weight, wide impedance bandwidth, and omnidirectional radiation patterns. However, the radiation patterns of the printed monopole antennas are always linearly polarized while is difficult to radiate CP radiation Thus, a monopole antenna design which can generate both the LP and CP waves is an interesting issue that is worthy of being researched.

In this letter, a new monopole antenna design with triple-sense CP radiation property is proposed for UWB application. The research is based on a circular-shaped monopole antenna which covers UWB operation. By employing a T-shaped slit in the ground and a rectangular parasitic stub on the edge of the radiator, triple-sense CP radiation at the centre frequency of 3.7, 5.1, and 9.7 GHz are generated, which belong to Worldwide Interoperability for Microwave Access (WiMAX), wireless local area network (WLAN), and X bands. The employment of T-shaped slit and the rectangular parasitic stub to the basic circular-shaped monopole can obtain the triple-sense CP radiation while keeping the UWB property. Since a 50-Ohm feed line connected to an impedance transformer is adopted, the matching of the proposed antenna is improved. Meanwhile, the introduced T-shaped slit and the rectangular parasitic stub hasn't occupied the more space, leading to a miniaturized design. Furthermore, the antenna's performances are investigated according to the key parameters, from which we can find the antenna's controllable characteristics.

II. ANTENNA DESIGN

Fig.1 (a) shows the geometry of the presented UWB antenna with CP radiation in WiMAX/WLAN/X wireless communication design. The photograph of the fabricated antenna is displayed in Fig.1 (b). The antenna is fabricated on a dielectric substrate with relative permittivity of 4.4 and loss tangent of 0.024. The antenna is designed with the overall size of 27mm×27mm×1mm, where the size of the ground plane is 27mm×9mm and the radius of the circular radiation patch is 8.5mm. The detailed design dimensions are listed in Tab. 1. In

978-7-5641-4279-7

this letter, a 50-Ohm feed line of width 2mm and length 3mm is terminated with SMA connector to improve the impedance matching of the antenna. Compared with traditional UWB antenna, the presented antenna obtains triple-sense CP waves by embedding a T-shaped slit in the ground plane, and adding a rectangular parasitic stub in the circular radiation patch. Through the antenna's performance analysis, the introduction of the T-shaped slit and rectangular parasitic stub can redistribute the magnetic currents in the slot so that two orthogonal resonant modes with equal amplitude and a 90° phase difference can be excited for CP operation.

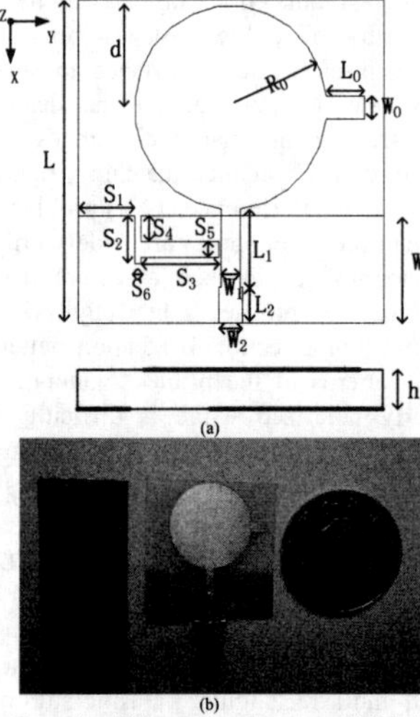

Figure 1. (a) Configuration of the presented antenna (b) Fabricated prototype.

To investigate the operation of resonant modes, surface current distributions on the antenna at 3.7, 5.1, and 9.7 GHz are presented in Fig. 2. In Fig. 2(a), maximum currents are localized mainly in the ground to produce resonant modes at 3.7 GHz. At 5.1 GHz, the ground and patch both contribute to the resonance from Fig. 2(b). In Fig. 2(c), the maximum current is distributed mainly along the rectangular parasitic stub in the circular patch to excite the resonant mode at 9.7 GHz. Fig. 2 shows that the three resonant modes of the antenna are influenced by the ground embedded with T-shaped slit and the patch connected with a parasitic stub. In other words, the T-shaped slit and the rectangular parasitic stub can be used to excite triple CP radiation.

TABLE I
DIMENSIONS OF THE OPTIMIZED ANTENNA DESIGN

Parameters	(mm)	Parameters	(mm)
L	27	S_1	5
L_0	3.46	S_2	4
L_1	6.5	S_3	6.9
L_2	3	S_4	2.2
W	9	S_5	1.3
W_0	2	S_6	0.6
W_1	1.6	R_0	8.5
W_2	2	d	9

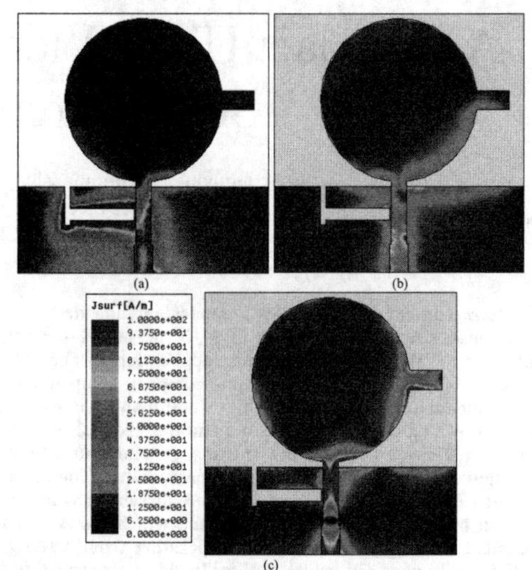

Figure 2 Current distributions on the antenna. (a) at 3.7GHz, (b) at 5.1GHz, (c) at 9.7GHz.

The two CP characteristics at lower frequencies are determined by the size of the T-shaped slit, while the CP wave at upper frequency is controlled by the size of the rectangular stub, which will be discussed in the following section. In this design, the upper CP frequency is approximated by

$$f_0 = c / 4L_0\sqrt{\varepsilon_r} \tag{1}$$

where c is velocity of light, L_0 is the length of the rectangular stub, and ε_r is the relative dielectric constant of the substrate. Given a desired CP resonance frequency, one can use Eq. (1) to define the initial total size of the stub for an initial design. Here, the stub in the radiator is designed with the side width of 2mm and length of 3.46mm. The measurement is carried out by using Agilent E8357A vector network analyzer. The measured simulated S11 results of the proposed UWB antennas with triple-band CP radiation are shown in Fig.3, which have good agreement with simulated result carried out by the full-wave electromagnetic software Ansoft High Frequency Structure Simulator (HFSS) [11]. According to the measured results, it can be found that the proposed compact antenna can effectively cover the impedance bandwidths of 3.1-11.8GHz, which can well satisfy the UWB band.

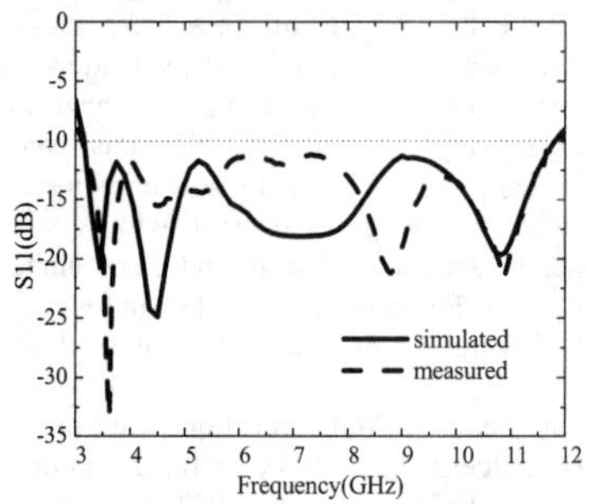

Figure 2 Measured and simulated return losses.

Fig. 4 depicts illustrates the measured and simulated AR against frequency in the +Z direction at the desired CP operation bands. It can be seen that the presented antenna can achieve CP radiations at WiMAX band from 3.63 to 3.77 GHz, WLAN band from 4.95 to 5.18 GHz, and X band from 9.55 to 9.97 GHz. The 3-dB AR bandwidths at WiMAX band, WLAN band, and X-band are 3.78%, 4.54%, and 4.3%, respectively. The reflection coefficients in these bands are all below -10 dB. The measured result is in good agreement with numerical prediction. The variation of the measured and simulated results is possibly caused by the rough welded joint of the SMA connector, the leakage current of the coaxial cable, and anechoic chamber measurement error.

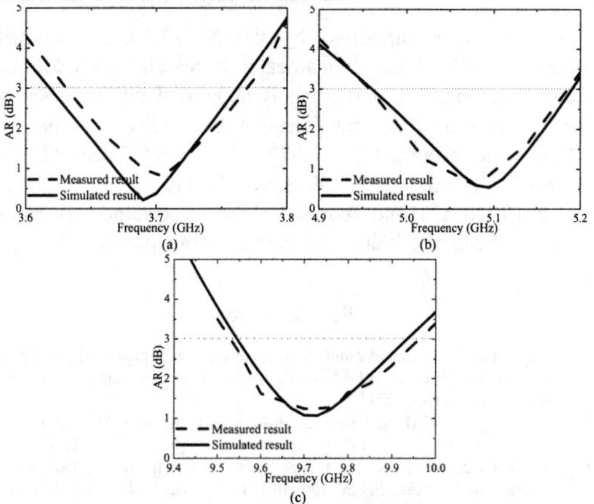

Figure 4 Measured and simulated AR at CP operating bands. (a) WIMAX band, (b) WLAN band, (c) X band.

Fig. 5-7 display the measured normalized radiation patterns at 3.7, 5.1, and 9.7 GHz in the x-z and y-z planes, respectively. The measured radiation patterns of the antenna include RHCP and LHCP components. The figures show that reasonably good LHCP radiations at 3.7 and 9.7 GHz along with good RHCP radiation at 5.1 GHz are realized according to the comparison of RHCP and LHCP components. It is noted that the radiation patterns are not omnidirectional because the proposed antennas is not symmetrical and the radiation patterns are influenced by the employed slit and stub. The measured peak gain is 1.8 dBic at 3.7 GHz, 2.5 dBic at 5.1 GHz, and 3.8 dBic at 9.7 GHz.

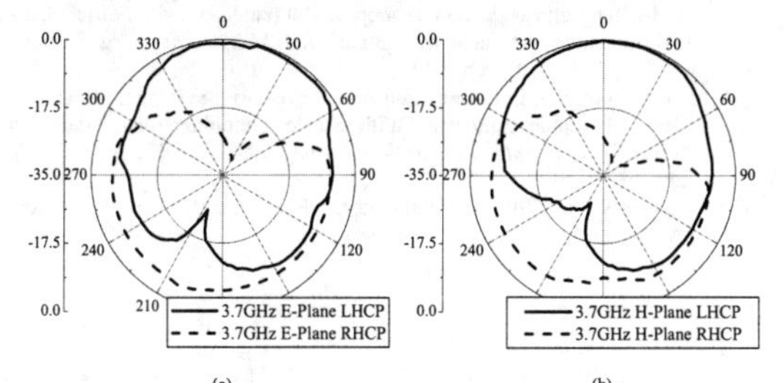

Figure 5 Measured radiation pattern at 3.7 GHz. (a) on XZ-plane, (b) on YZ-plane.

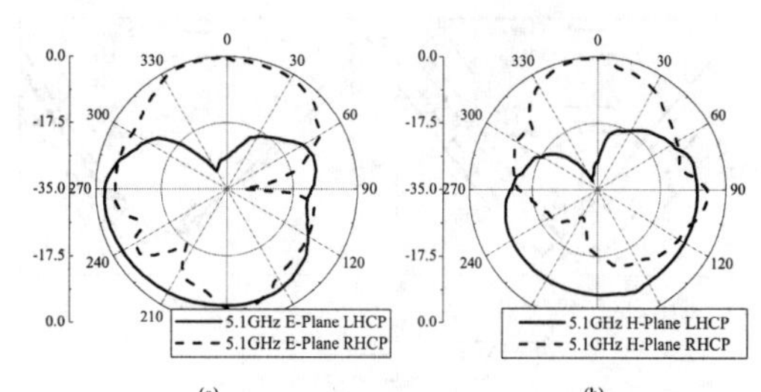

Figure 6 Measured radiation pattern at 5.1 GHz. (a) on XZ-plane, (b) on YZ-plane.

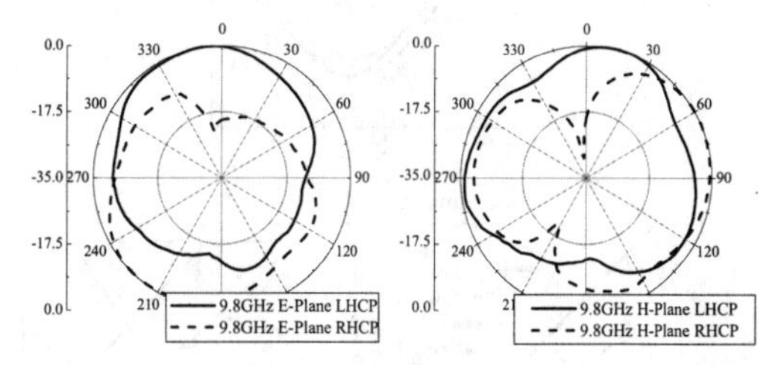

Figure 7 Measured normalized radiation patterns at 9.7 GHz. (a) on XZ-plane, (b) on YZ-plane.

III. PARAMETRIC ANALYSIS

For further investigation, the vital parameters of the antenna are discussed to find their impacts on the AR characteristics in the triple CP bands. The lengths of L-shaped slit arms S_2, S_3 and the parasitic stub length L_0 are especially examined to study their influences on the AR property. All the parameters keep their initial values unless stated. To note, the S11 curve is always less than -10 dB in the UWB band, which is insensitive to the parameter variation.

Fig. 8 shows the effects of various dimensions of S_2 on the triple CP bands. In Fig. 8, it is observed that the AR frequency center of CP operations shifts downwards in WiMAX band, whereas AR results are affected slightly in WLAN/X band when S_2 increases. Fig. 9 shows the effects of various S_3 on the triple CP bands. It can be seen that the AR frequency center of CP operations tends to decrease at WiMAX/WLAN band, but moves little in X band with the increase of S_3. The influence of L_0 on AR results is shown in Fig. 9. The value of L_0 is decreased from 3.66 to 3.26 mm progressively. It is observed that the AR results get worse in X band and have nearly no change in WiMAX/WLAN bands.

From Fig. 8-10, it can be seen that the desired CP bands are mainly controlled by the added T-shaped slit and parasitic stub. Specifically, the WiMAX band is mainly controlled by the parameter S_2 and S_3, WLAN band is mainly controlled by S_3 and L_0, and X band depend mainly on L_0. The above result can verify the current analysis in the upper section. Therefore, we can easily change their dimensions for the desired CP bands without the need to redesign the whole antenna.

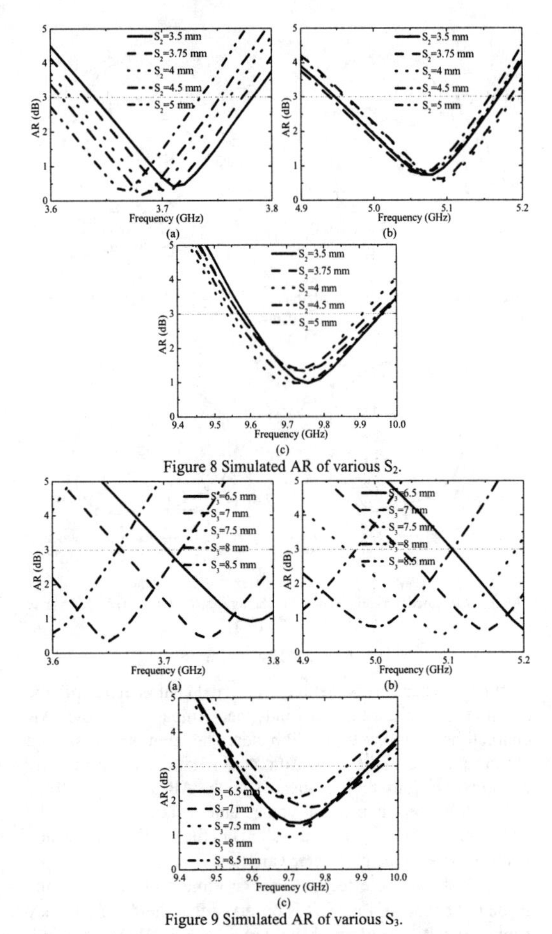

Figure 8 Simulated AR of various S_2.

Figure 9 Simulated AR of various S_3.

Figure 10 Simulated AR of various L_0.

IV. CONCLUSION

A triple-sense circularly polarized UWB antenna is presented in this research, which is obtained by embedding a T-shaped slot element in the ground plane and adding a rectangular stub element in the monopole structure. The measured and simulated values for the antenna designs verify their predicted performance characteristics, including small size, easy fabrication, UWB impedance bandwidth, triple-sense CP characteristics. Parametric analysis shows that the above characteristics can be modified by adjusting antenna sizes to provide CP radiation in other useful bands.

ACKNOWLEDGMENT

This work is supported by the NSFC under Contract No.61101066 and the Fundamental Research Funds for the Central Universities, and partially supported by the Program for New Century Excellent Talents in University of China, the NSFC under Contract No.61072017, Natural Science Basic Research Plan in Shaanxi Province of China (No.2010JQ8013), and Foundation for the Returned Overseas Chinese Scholars, State Education Ministry and Shaanxi Province.

REFERENCES

[1] S. A. Rezaeieh, "Dual band dual sense circularly polarized monopole antenna for GPS and WLAN applications," Electron. Lett., vol. 47, pp. 1212-1214, October 2011.

[2] Y. Sung, "Dual-Band Circularly Polarized Pentagonal Slot Antenna," *IEEE Antennas Wireless Propag. Lett.*, vol. 10, pp. 259-261, April 2011.

[3] C. H. Chen and E. K. N. Yung, "A Novel Unidirectional Dual-Band Circularly-Polarized Patch Antenna," *IEEE Trans. Antennas Propag.*, vol. 59, pp. 3052-3057, August 2011.

[4] C. Deng, Y. Li, Z. Zhang, G. Pan, and Z. Feng, "Dual-Band Circularly Polarized Rotated Patch Antenna With a Parasitic Circular Patch Loading," *IEEE Antennas Wireless Propag. Lett.*, vol. 12, pp. 492-495, April 2013.

[5] C. F. Jou, J. W. Wu and C. J. Wang, "Novel broadband monopole antennas with dual-band circular polarization, " *IEEE Trans. Antennas Propag.*, vol. 57, pp. 1027–1034, April 2009.

[6] L. Wang, Y. X. Guo, and W. Sheng, "Tri-Band Circularly Polarized Annular Slot Antenna for GPS and CNSS Applications," *IEEE Early Access Articles.*

[7] First Report and Order FCC, Federal communications commission revision of part 15 of the commission's rules regarding ultra-wideband transmission systems, FCC, 2002.

[8] G. Teni, N. Zhang, J. Qiu, and P. Zhang, "Research on a Novel Miniaturized Antipodal Vivaldi Antenna With Improved Radiation, " *IEEE Antennas Wireless Propag. Lett.*, vol. 12, pp. 417-420, April 2013.

[9] E. A. Akbari, M. N. Azarmanesh, and S. Soltani, "Design of miniaturised band-notch ultra-wideband monopole-slot antenna by modified half-mode substrate-integrated waveguide," *IET Microw. Antennas Propag.*, vol. 7, pp. 26–34, March 2013.

[10] F. Fereidoony, S. Chamaani, and S. A. Mirtaheri, "Systematic Design of UWB Monopole Antennas With Stable Omnidirectional Radiation Pattern, " *IEEE Antennas Wireless Propag. Lett.*, vol. 11, pp. 752-755, July 2012.

[11] High Frequency Structure Simulator (HFSS). Ansoft Corp., Pittsburgh, PA.

Design of a Broadband Cavity-Backed Multislot Antenna

Jing-yu Yang[1], Yan-ming Liu[1], Wen-jun Lu[1,2*] and Hong-bo Zhu[1]
1. Jiangsu Key Laboratory of Wireless Communications, Nanjing University of Posts & Telecommunications, Nanjing, China
2. State Key Laboratory of Millimeter Waves, Nanjing, China
*wjlu@njupt.edu.cn

***Abstract*-A novel cavity-backed microstrip-fed multislot antenna (CBMSA) for wideband application is proposed. The antenna is simply fed by a microstrip line terminated with a wide rectangular stub. By incorporating multi-slot configuration and backed cavity, unidirectional beam can be obtained within a broad bandwidth. The return loss, radiation pattern and gain over the antenna's operational bandwidth have been numerically and experimentally studied. It is shown that the impedance bandwidth of the antenna is from 5 to 10.3 GHz, which is about 69% (for $|S_{11}|$ lower than -10 dB). Good polarization purity with cross-polarization level lower than 18 dB and front-to-back ratio better than 12 dB are obtained. Hence, the antenna can be used in different broadband communication or identification systems.**

***Index Term*-Cavity-backed, multislot, slot antenna, microstrip-fed**

I. INTRODUCTION

Slot antennas have been studied since the 1940s [1]. As one kind of low-profile and cost-effective antennas, microstrip-fed narrow slot antennas have been initially studied theoretically and rigorously by employing full-wave numerical approaches [2]. However, the inherent narrow impedance bandwidth of such antennas restricts their applications. By using different feed structures [3-4] or modifying the shapes and numbers of the slots [5-6], various wideband slot antennas have been developed. In terms of these methods, impedance bandwidths varying from 30% to over 100% can be achieved.

With the improvement of the bandwidth and their intrinsic advantages, the slot antennas can be applied in broadband wireless systems. However, some applications, such as indoor base station or identification systems, require the antenna to have a unidirectional beam, and the aforementioned omnidirectional slot antennas should be modified to unidirectional ones. Usually, a slot antenna can be backed by a mental cavity to achieve unidirectional radiation. The cavity height of these cavity-backed slot (CBS) antennas is approximately one or three-quarter guide-wave lengths at the resonant frequency so that impedance matching is substantially preserved. Cavity-backed microstrip-fed slot antennas have been initially studied in [7]. Nevertheless, the measured bandwidth of this antenna is less than 10%. In order to increase the bandwidth of a CBS antenna, a bandwidth enhancement design with a via-hole above the slot has been proposed [8]. By introducing an additional resonance at high frequency band, a broadband design is obtained. However, the location of the via-hole is quite difficult to be determined. Another wideband slot antenna with a mesh cavity has been proposed [9] more recently. As is shown, the antenna has wide bandwidth, stable radiation pattern and low cross-polarization level. However, the antenna has a complicated structure and additional complexity may be introduced to its design, analysis and fabrication. To design a CBS antenna while retaining its simple structure and pretty good broadband performance is still an important and challenging task.

The objective of this paper is to propose a novel, broadband cavity-backed microstrip-fed multislot antenna with a simple structure. Two different narrow, rectangular slots are combined as one radiator and the antenna is fed by microstrip line terminated with a wide, rectangular patch. A metallic cavity is placed under the feed line to obtain unidirectional radiation pattern. The antenna is studied numerically first, and then, prototypes are fabricated and experimentally studied. Simulated and measured results are compared with each other and discussed in detail.

II. ANTENNA DESCRIPTION

The geometry of the proposed CBS antenna is shown in Fig. 1. It is seen that the cavity-backed slot antenna proposed here is based on the multislot antenna in [6]. The objective of the proposed design is to obtain unidirectional beam, wide bandwidth using a simple structure. In the proposed design, two different narrow, rectangular slot radiators are etched on the conductor ground plane. In order to obtain broadband impedance matching as well as simple feed structure, a fatter, open-circuit tuning stub is terminated at the end of the microstrip feed line. A metallic cavity with a height of 15mm is arranged at the side of the microstrip line so as to suppress the back-lobe. It is seen that proposed antenna has a simple structure. The characteristic impedance of the feed line is 50Ohm.

The antenna is simulated by using *Zeland's* IE3D based on method of moments (MoM). After a few simulation, an initial antenna with parameters as shown in Table 1 is designed on a dielectric substrate with low relative permittivity ε_r=2.2, loss tanδ of 0.0008 and thickness h=0.8mm.

This work was supported in part by the National Natural Science Foundation of China (61001079), Program for New Century Excellent Talents in University (NCET) of Ministry of Education, Basic Research Program of Jiangsu Province (BK2011027) and Research Project of State Key Laboratory of Millimeter Waves (K201202).

978-7-5641-4279-7

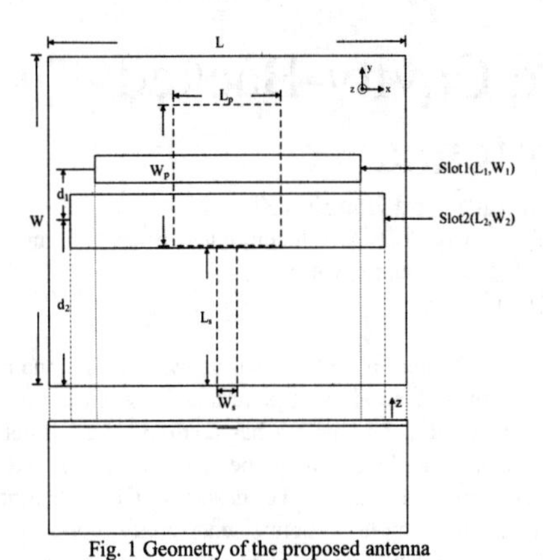

Fig. 1 Geometry of the proposed antenna

TABLE 1
PARAMETERS OF THE INITIAL ANTENNA

Length of cavity	L	52mm
Wide of cavity	W	46mm
Depth of cavity	H	15mm
Length of solt1	L_1	39mm
Wide of slot1	W_1	4mm
Length of slot2	L_2	46mm
Wide of slot2	W_2	7.5mm
Stub length	L_p	16.4mm
Stub width	W_p	19.7mm
Feed line length	L_s	2.8mm
Feed line width	W_s	29.3mm
Distance from slot1 to slot2	d_1	7.25mm
Distance from slot2 to feed point	d_2	22.57mm

Sensitive dimension parameters will be studied first so that an optimal antenna can be obtained. In the numerical simulation, only one parameter is varied, whereas the others are kept constant. All simulations are performed by employing *Zeland's* IE3D.

As described in Fig. 1, a microstrip line width step exists at the junction of the microstrip feed line and the tuning stub. Such a discontinuity will affect the results in the slot coupled structure. The relative position between the junction and the lower edge of slot2 can be defined as a sensitive parameter named α, which is given by

$$\alpha = d_2 - W_2/2 - L_s \quad (1)$$

The effect of varying α on the input performance is plotted in Fig. 2(a). It can be seen that the frequency of all resonances are quite sensitive to α. Even α varies within a relatively small range, the return loss of the antenna will change greatly. It is seen that an optimal value of α is 0.3mm.

The lengths of the slots also affect the antenna's performance significantly. Fig. 2(b) shows the effect of the length of slot1 on the antenna's return loss. It can be seen that the lowest and the highest resonances are not sensitive to the value of L_1. However, it is seen that the coupling between the two resonances at 8-9GHz band can be flexibly controlled by modifying L_1.

Fig. 2(c) demonstrates the tendency of the antenna's return loss when varying the length of slot2. It shows that the lowest resonant frequency is also not sensitive to L_2. Meanwhile, three other resonances at higher frequency band are sensitive to L_2. As L_2 increases, the second resonant frequency is increases, and the third and fourth resonant frequencies decreases. It is seen that the variations of L_1 and L_2 have opposite effect on the coupling between two resonances at 8-9GHz. Thus, L_1 and L_2 can be simultaneously used to finely tune the return loss at high frequency band. From Fig. 2(b) and (c), an optimal combination should be L_1=39mm and L_2=46mm.

The study of the influence of three most sensitive parameters on the antenna's return loss has been done. These results are useful in optimizing the dimension parameters of the fabricated prototypes.

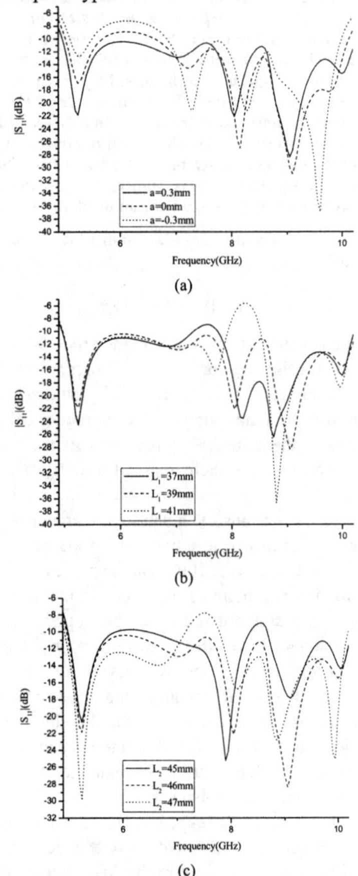

Fig. 2 $|S_{11}|$ versus parameters: (a) α (b) L_1 and (c) L_2

III. NUMERICAL AND EXPERIMENTAL RESULTS

Based on numerical results discussed in Section□, prototypes are fabricated on a dielectric substrate with low relative permittivity ε_r=2.2, loss tanδ of 0.0008 and thickness h=0.8mm. The photograph of a fabricated prototype is shown in Fig. 3. The fabricated antenna is measured with *Agilent's*

8720ET vector network analyzer (VNA) first and then measured in an anechoic chamber by using *NSI's* NSI-800F-10x far-field antenna measurement system.

The simulated and measured return loss of the proposed antenna in Table 1 are plotted and compared with each other in Fig. 4. It is observed that the measurement result is in good agreement with the simulation one. The measured 10 dB return loss bandwidth is about 5.3 GHz (5.0-10.3 GHz) which is about 69%. The slight deviations between the measured and simulated results may be due to the non-ideal materials and fabrication errors. The cavity is copper with the thickness of 0.5 mm in reality, while in the simulation, it is assumed to be made from zero-thickness perfect electrical conductor (PEC).

The radiation patterns of the antenna at 6GHz, 7GHz and 9.7GHz in both E-plane (z-y plane) and H-plane (z-x plane), are measured by using NSI-800F-10x far-field antenna measurement system in an anechoic chamber and displayed in Fig. 5. The difference between the simulated and measured results is mainly introduced by the asymmetric factors introduced in the fabrication process, as well as the antenna mounting and aligning errors. It is observed that relatively stable unidirectional radiation patterns are achieved within the operational frequency band. The cross-polarization level in the H-plane is about 18 dB below the co-polarization one. The measured front-to-back ratio (f/b) ranges between 12 and 20 dB within the whole frequency range.

Moreover, the peak gain of the antenna is also measured in a chamber, as plotted in Fig. 6. The gain of the proposed antenna is from 4.5 to 8.6 dBi which is relatively constant over the whole frequency band. It also can be observed the measured gain is lower than the simulated at the whole frequency band. The difference between the measured and simulated results is caused by material loss and measurement errors, as well as the undesired reflections in the cavity excited by the parasitic cavity modes.

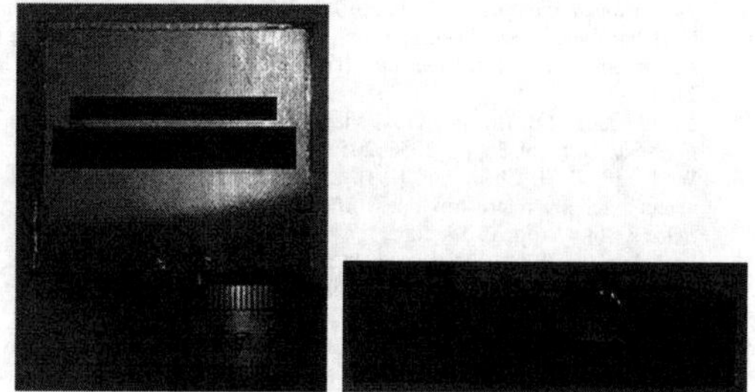

Fig. 3 Photograph of fabricated prototypes

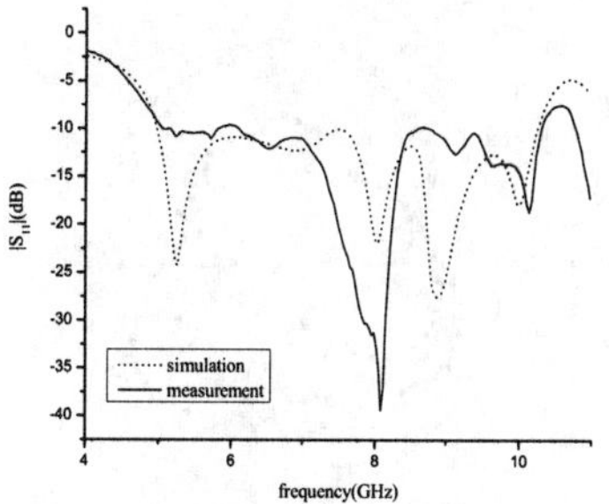

Fig. 4 Measured and simulated return loss

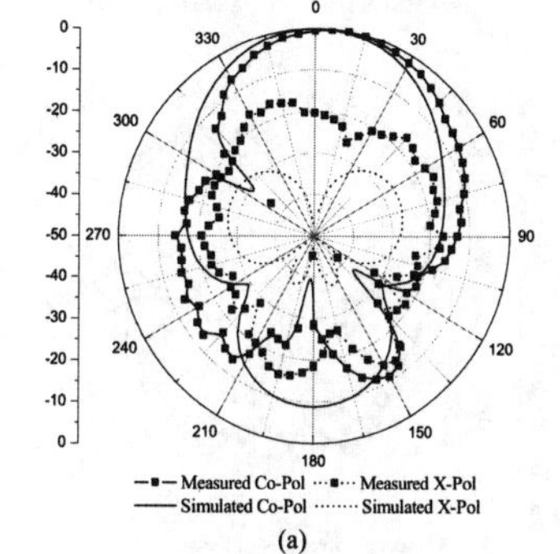

(a)

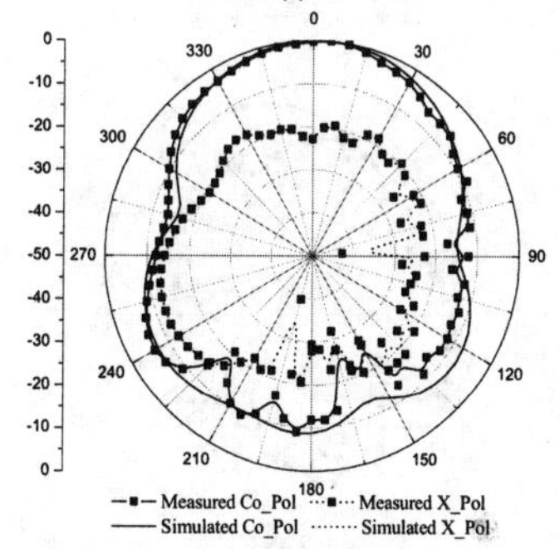

(b)

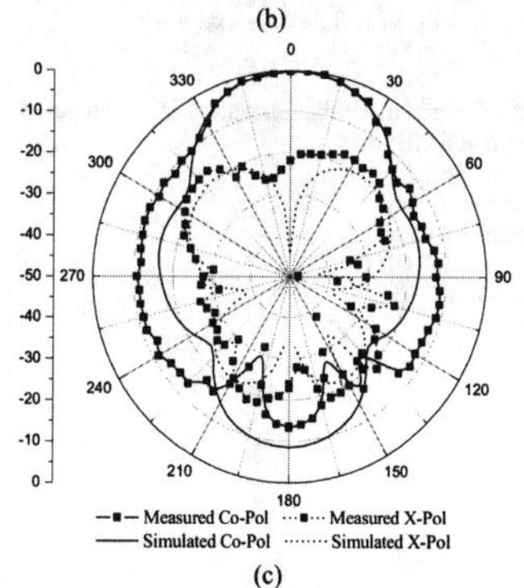

(c)

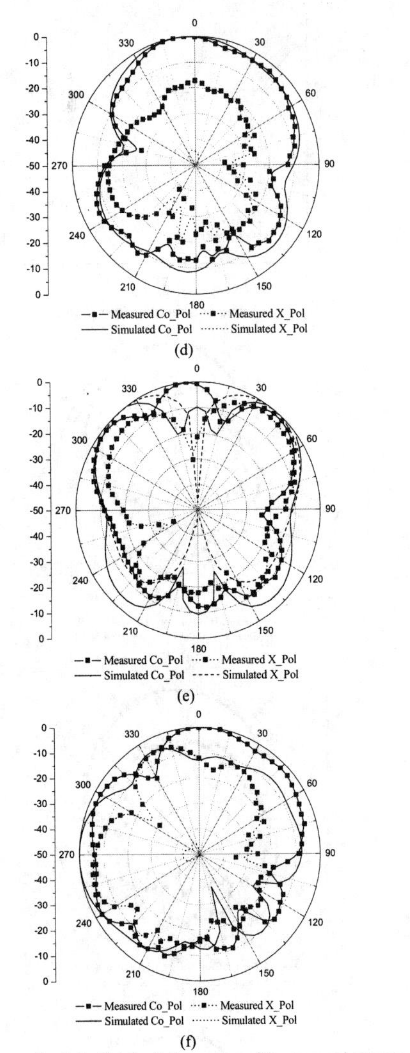

Fig. 5 Measured and simulated radiation pattern of the proposed antenna
(a) H-plane pattern at 6 GHz
(b) E-plane pattern at 6 GHz
(c) H-plane pattern at 7 GHz
(d) E-plane pattern at 7 GHz
(e) H-plane pattern at 9.7 GHz
(f) E-plane pattern at 9.7 GHz

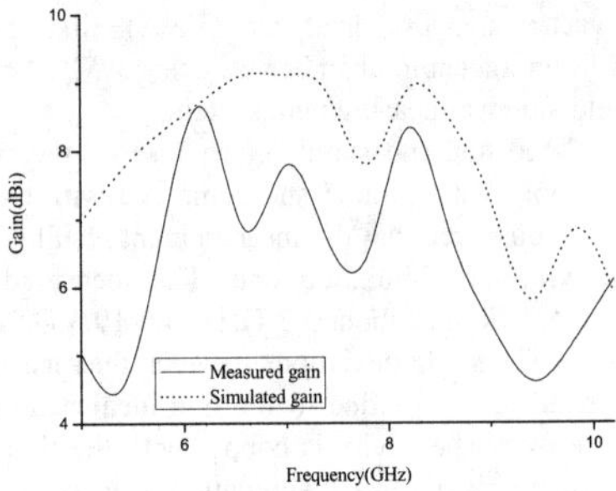

Fig. 6 Measured and simulated gain of the antenna

IV. CONCLUSION

A simple broadband cavity-backed slot antenna has been presented and experimentally studied. Multislot configuration introduces multiple resonances with close frequencies and an operation bandwidth of more than 69% can be achieved. Stable unidirectional beam with good polarization purity is obtained across the bandwidth, with front-to-back ratio better than 12 dB, cross-polarization level lower than 18 dB and maximum gain of 8.6 dB. The antenna has a simple structure and can be easily designed and fabricated. The operation bandwidth of the antenna covers the C and X band, so it can be applied to 5GHz-WLAN (i.e., 5.15-5.35 GHz), 5GHz-RFID (i.e., 5.725-5.875 GHz) and European standard UWB systems (i.e., 6-8.5 GHz). On the other hand, the antenna is also suitable for indoor base station applications after scaling-in-large to L and S band.

REFERENCES

[1] H. G. Booker, "Slot aerials and their relation to complementary wire aerials", *J IEE*, vol. 26, pp. 620–626, 1946.

[2] D. M. Pozar, "Reciprocity method of analysis for printed slots and slot coupled microstrip antennas," *IEEE Trans Antennas Propagat*, vol. 34, pp. 1439–1446, 1986.

[3] Y. Yoshimura, "A microstrip line slot antenna," *IEEE Trans. Microwave Theory and Techniques*, vol. 20, pp. 760-762, 1972.

[4] N. Behada and K. Sarabandi, "A multiresonant single-element wideband slot antenna," *IEEE Antenna and Wireless Propagat Letters*, pp. 5–8, 2004.

[5] Y. W. Jang, "Broadband cross-shaped microstrip-fed slot antenna," *Elec-tron Lett*, vol. 36, pp. 2056–2057, 2000.

[6] W. J. Lui, C. H. Cheng and H. B. Zhu, "A novel broadband multislot antenna fed by microstrip line," *Microwave and Optical Technology Letters*, vol. 45, pp. 55-57, 2005.

[7] S. Tokumaru and H. Takahashi, "Cavity backed two slots antennas," *Antennas and Propagation Society International Symposium*, vol. 14, pp. 383-386, 1976.

[8] S. Yun, D. Y. Kim and S. Nam, "Bandwidth enhancement of cavity-backed slot antenna using a via-hole above the slot," *IEEE Antenna and Wireless Propagat Letters*, vol. 1, pp. 98-101, 2012.

[9] C. R. Medeiros, E. B. Lima, J. R. Costa and C. A. Fernandes, "Wideband slot antenna for WLAN access points," *IEEE Antennas and Wireless Propagation Letters*, vol. 9, pp. 79-82, 2010.

Development of a Compact Planar Multiband MIMO Antenna for 4G/LTE/WLAN Mobile Phone Standards

Foez Ahmed*, Ronglin Li†, *Senior Member, IEEE*, and Ying Feng*, *Member, IEEE*
*School of Automation Science and Engineering
†School of Electronic and Information Engineering
South China University of Technology, Guangzhou, China-510640
Email: foez28@ru.ac.bd, lirl@scut.edu.cn, yfeng@scut.edu.cn

***Abstract*-A compact planar multiband MIMO antenna system is developed. The MIMO antenna consists of two symmetric antenna elements, each of which comprises three effective radiators: a driven monopole, a meandered S- and F-shaped strip and all are printed on the front side of a thin low-cost substrate. At VSWR ≤ 2.75, the proposed MIMO antenna operates in the frequency range of LTE-1, LTE-2, LTE-3, LTE-7, LTE-40, and WLAN 2.4 GHz bands. The experimental results verify the simulations. Higher isolation (> 18 dB) is achieved. The received signals satisfy the condition $P_i \approx P_j$. The measured correlation coefficient is lower than 0.01 and radiation patterns of each antenna unit can cover complementary space region that confirms both the spatial and pattern diversity.**

Keywords – MIMO antenna, Broadband, LTE, Handsets

I. Introduction

Doubtless that the modern wireless communication system demands reliable transmission with higher - data rate, channel capacity and spectral efficiency in a rich scattering environment. But the channel capacity of a conventional single-input single-output (SISO) communication system is limited according to the Shannon's theorem [1]. The Long Term Evolution (LTE) standard for mobile broadband inclusion of Multiple Input Multiple Output (MIMO) transmission technique confirms the improved performance in terms of coverage, spectral efficiency and high data rate without additional power requirements [2][3]. However, the multiband MIMO antenna design for mobile handset is more challenging in many aspects unlike other applications due to its limited space and compactness.

Over the last decade, a significant research has been carried out to design broadband MIMO antenna for mobile handsets. In [4], a printed diversity monopole antenna is presented for WiFi/WiMAX (2.4-4.2 GHz) applications. The achieved isolation is greater than 17 dB. The author of [5] has also studied MIMO antenna for WLAN (2400-2484 MHz) and WiMAX (2500-2690 MHz) applications. The proposed antenna has total system size of $100\times50\times0.8\text{mm}^3$ and isolation is roughly 15 dB. In [6], another wideband MIMO antenna has been presented to cover a wideband frequency of 2.4-6.55 GHz with high isolation (>18 dB). A very recent work, proposed by [7] covers a wide band frequency range of 1700-2900 MHz. The total system size is about of $55\times99\times1.6\text{ mm}^3$ and isolation is better than 15 dB. Though the antenna has overall better performance but the structure is not fit for farther 3/4-element antenna array systems into the limited volume of mobile handset without compromising system's size. However, in this paper a planar compact wideband MIMO antenna has been presented for future wireless applications of LTE#1 (1920-2170 MHz), LTE#2 (1850-1990 MHz), LTE#3 (1710-1880 MHz), LTE#7 (2500-2690 MHz), LTE#40 (2300-2400 MHz), and WLAN (2400-2484 MHz) bands. The total systems size is $60\times110\times0.8\text{mm}^3$ that is a typical handset's size for practical usage as well as comparable with the IPhone4 mobile handset.

II. MIMO Antenna Design and Measurements

A. Antenna Structure

The optimized layout with dimensional details of the proposed multiband MIMO antenna is illustrated in Fig. 1. A single antenna unit comprises of a driven monopole (orange colored), an S-shaped strip (blue colored) and F-shaped strip (red colored). The S-strip is short circuited to the ground plane though Via (H). Both of the S- and F-shaped strips are capacitively coupled to each other as well as electromagnetically coupled to the driven monopole. The driven monopole is fed with a 50Ω microstrip line whereas other radiators are coupled-fed. The antenna arrays are printed on the front side of a low cost FR-4 substrate. On the back side of the substrate a system ground plane of size $60\times95\text{mm}^2$ is printed. Due to minimizing mutual coupling effects, a protruded T-shaped metal strip with rectangular cutting slots, and two symmetric ground slots have been introduced. The antenna was analyzed using Ansoft HFSS v13 and optimized values are listed in Table I.

A prototype of proposed MIMO antenna is viewed in Fig. 2. The antenna was fabricated on a FR-4 substrate (ε_r=4.6, thickness = 0.8mm, tan(δ) = 0.02) of dimension $110\text{x}60\text{mm}^2$. The measured VSWR of proposed MIMO antenna is compared with the simulation results in Fig. 3. At VSWR ≤ 2.75, the achieved bandwidths are around 27.4% (1.67-2.2 MHz) and 25.6% (2.28-2.95 GHz).

978-7-5641-4279-7

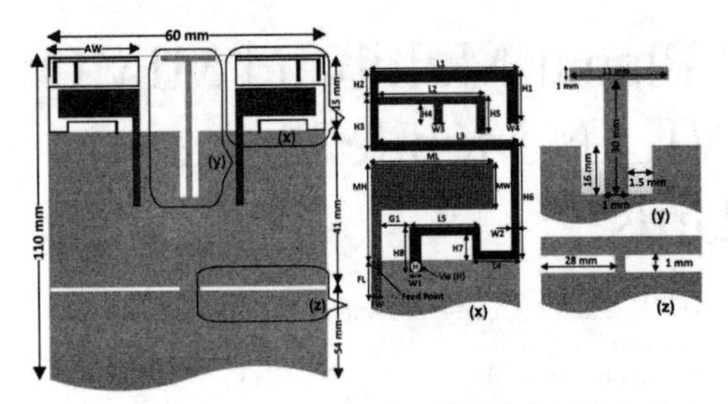

(a) Full Structure (b) Optimized Values

Figure 1. Optimized Geometry details of the proposed 2×2 MIMO antenna

TABLE I

OPTIMIZED PARAMETERS & VALUES (UNIT: MILLIMETERS)

Parameter	Value	Parameter	Value	Parameter	Value
MH	8.5	ML	16.5	MW	5.5
FL	17.5	FW	1.45	G1	0.6
H1	4	H2	0.8	H3	5.2
H4	4.15	H5	4.4	H6	9.25
H7	2.25	H8	2.75	L1	18.5
L2	17.25	L3	18.5	L4	4
L5	10.25	W1	0.5	W2	0.25
W3	0.75	W4	0.04	AW	18.5

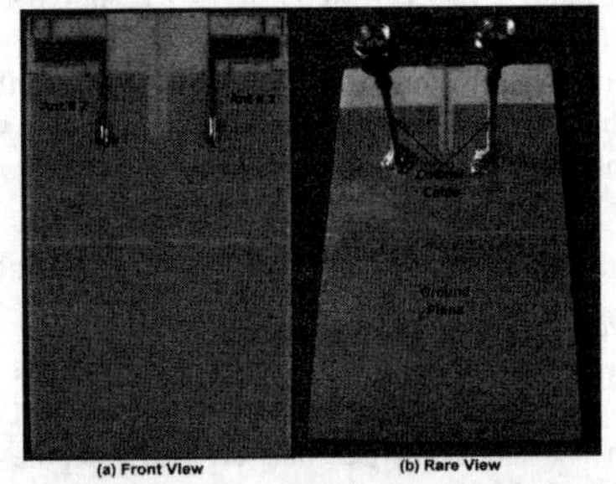

Figure 2. Prototype of the proposed multiband MIMO antenna

B. *Operating Principle and Parameter Study*

The fundamental operating principles are analyzed with the aid of simulated return loss shown in Fig. 4. The corresponding dimension of Ref. 1, Ref. 2, and Ref. 3 are fixed as the optimized values listed in Table I. In case of driven monopole only (Ref. 1), a wide resonant mode is observed at around 2.8 GHz. The length of the driven monopole is about 25mm up to feed point (A) that can excite a quarter-wavelength ($\lambda/4$) resonant mode. The length of the S-shaped strip is around 75mm that can excite a half-wavelength ($\lambda/2$) resonant mode. Farther by adding F-shaped coupled strip (Ref. 3), a 3rd resonant mode is formed at around 1.83 MHz and the impedance of the 2nd resonant mode is improved. The length of F-shaped strip itself is about 20mm, but with the partial S-shaped strip the total length is about 67mm that generates a half-wavelength ($\lambda/2$) resonant mode at 1.83 GHz. The reason behind of exciting ($\lambda/2$) resonant mode is that both the S- and F-shaped strips are electromagnetically coupled with the driven monopole.

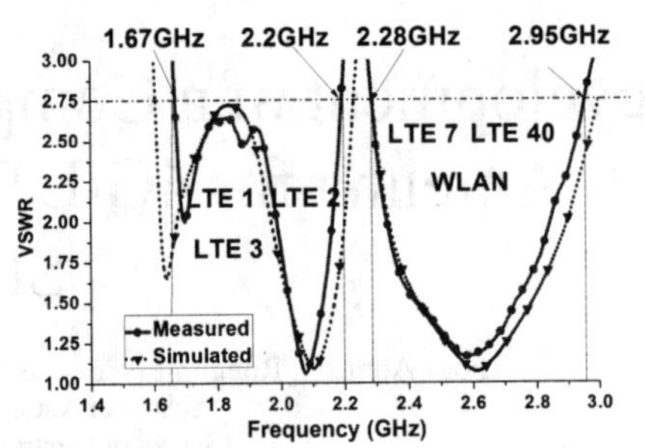

Figure 3. Measured and Simulated results for VSWR characteristics

Unlike direct feeding, couple-fed generally decreases the input impedance and hence lengthens the electrical length of the radiator [8]. Moreover, F-strip is farther capacitively coupled with the S-shaped strip and hence the inductive reactance is compensated by the coupling capacitance introduced by the small gap between the S- and F-shaped strip which extends the length of F-strip electrically and hence excites the 3rd resonant mode [6].

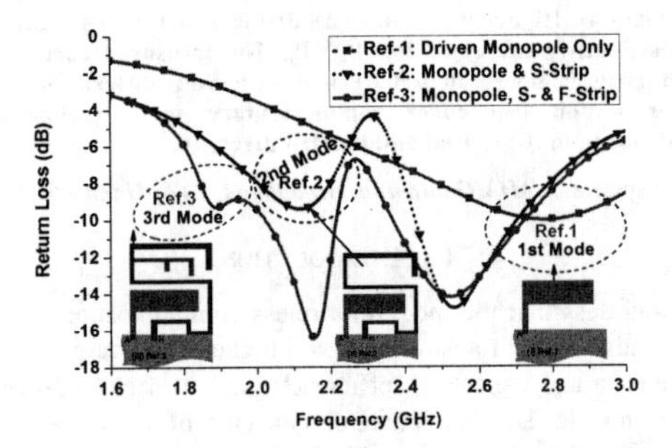

Figure 4. Simulated Return Loss for the case of (i) Driven Monopole only (Ref. 1), (ii) S-shaped Strip with Monopole (Ref. 2), and (iii) F-shaped Strip with S-strip & Monopole (Ref. 3)

Fig. 5 shows the simulated return loss with varying length of H1 while other values are fixed as listed in Table I. As H1 (hence the length of S-strip) is decreased, the 2nd resonant frequency is shifted towards the higher frequency or vice-versa without having any effect on other two resonant modes. In Fig. 6, it is evident that the 3rd resonant frequency is shifted towards the higher frequency as H5 (hence the length of F-strip) is decreased or vice-versa with having minor effect on other two resonant modes. Therefore, it is inferred that radiation characteristic within the 2nd and 3rd resonant modes over the operating bands are controlled and contributed by the S- and F-shaped strip respectively.

C. *Isolation and Decoupling Technique Analysis*

As a decoupling technique, two symmetric rectangular slots and a protruded T-shaped strip with rectangular slots have been studied into the ground plane. In principle, its function is to provide anti-phase coupling currents to eliminate the original

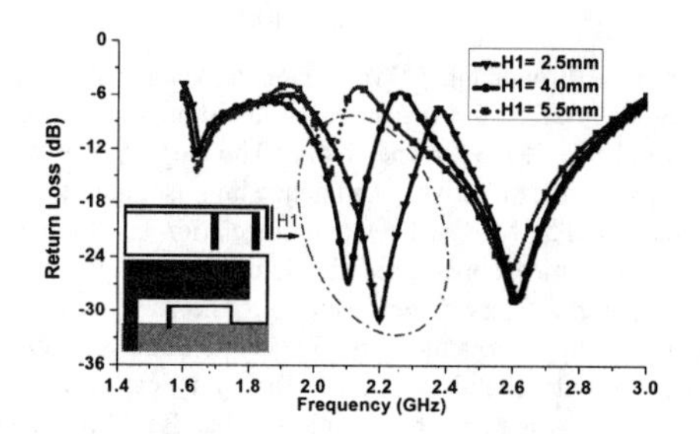

Figure 5. Effect of the length (H1) of S-Strip on the return loss of the proposed MIMO antenna, where H1= 4mm is an optimized value

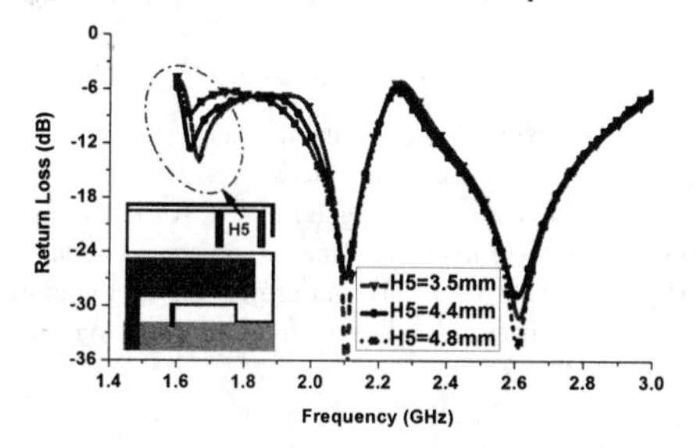

Figure 6. Effect of the length (H5) of F-Strip on the return loss of the proposed MIMO antenna, where H5= 4.4mm is an optimized value

coupling currents. Whereas the cutting slots reduce the mutual coupling effects caused by distributed ground surface currents. The ground slots act as a band stop filter and also make slow the surface currents flow over the finite ground. Surface current distribution shown in Fig. 7 demonstrates the effectiveness of the applied techniques as well. However, the measured isolation curves for the proposed MIMO antenna are compared with the simulation results and plotted in Fig. 8. It is observed that the isolation is greater than 18 dB over the operating bands.

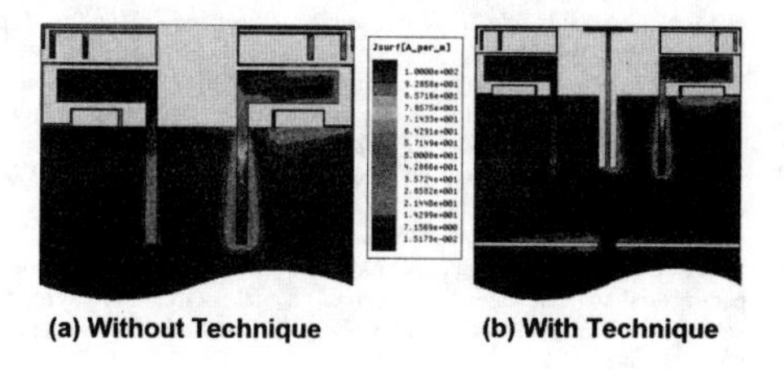

Figure 7. Surface current distribution of 2×2 MIMO antenna at 2GHz (a) Without decoupling techniques and (b) With decoupling techniques

III. Antenna Performance Evaluation

A. Envelope Correlation Coefficient (ρe)

In the case of isotropic/uniform signal propagation environment, the envelope correlation coefficient (ρ_{ij}) can be

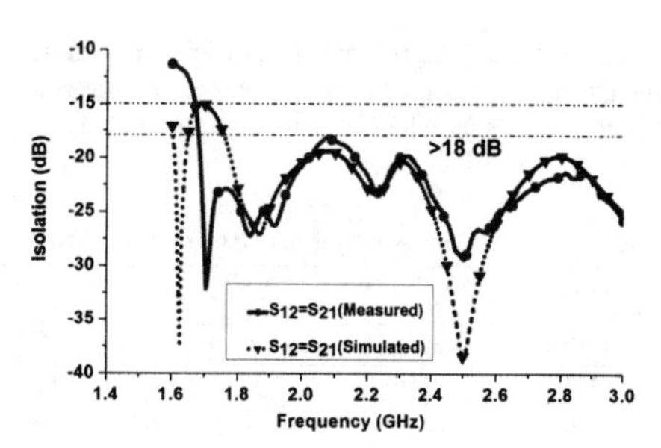

Figure 8. Isolation comparison of the proposed multiband MIMO antenna

computed by Eq. (1) [9],

$$\rho_{ij}^{es} = \left| \frac{\left|S_{ii}^{*}S_{ij} + S_{ij}^{*}S_{jj}\right|}{\sqrt{\left(1-\left|S_{ii}\right|^{2}-\left|S_{ji}\right|^{2}\right)\times\left(1-\left|S_{jj}\right|^{2}-\left|S_{ij}\right|^{2}\right)}} \right|^{2} \tag{1}$$

The simulated and computed envelope correlation coefficients obtained from measured S-parameter values are plotted in Fig. 9. It is observed that in any of two antenna elements combinations the measured ρ_e the proposed MIMO antenna are always below 0.01, which leads to a perfect performance in terms of diversity.

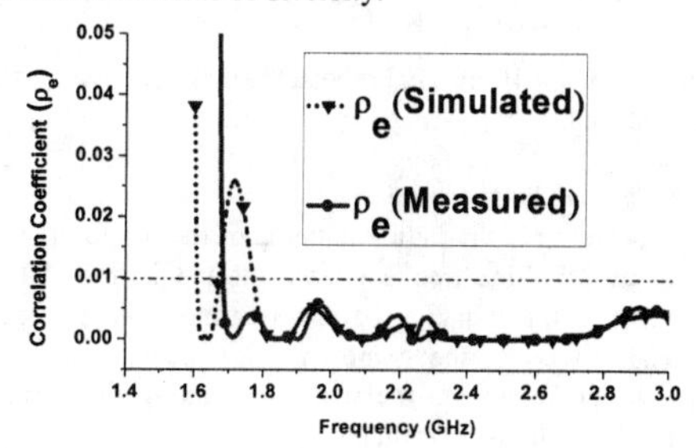

Figure 9. Correlation of the proposed MIMO antenna

B. Mean Effective Gain (MEG) and Diversity Gain (DG)

In the case of mobile wireless environment for a series of assumptions [10], the MEGs can be calculated by Eq. (2). The MEGs obtained from measured radiation data are listed in Table II. It is found that the MEG for each antenna element is almost identical. In worst case situation, the maximum ratio of the MEG between any of the two antenna units is not more than 1.7 dB. Therefore, the received signals satisfy the conditions $P_i \approx P_j$ ($|MEG_i/MEG_j| < 3$ dB) [11].

$$MEG_i = \frac{1}{2\pi} \times \left[\int_0^{2\pi} \left[\frac{\Gamma}{1+\Gamma} G_{\theta i}\left(\frac{\pi}{2},\varphi\right) + \frac{1}{1+\Gamma} G_{\varphi i}\left(\frac{\pi}{2},\varphi\right) \right] d\varphi \right] \tag{2}$$

Where $G_{\theta i}(\Omega)$ and $G_{\varphi i}(\Omega)$ are the power gain patterns of *i*th antenna, and Γ represents the cross polarization discrimination; assuming Γ= 0 dB in indoor fading environment [11].

TABLE II
MEAN EFFECTIVE GAIN AND DIVERSITY GAIN FOR THE PROTOTYPE

Center Frequency (GHz)	Mean Effective Gain in dB (Indoor, Γ = 0 dB)		K (dB)	DG, 1% (dB)
	Ant. #1	Ant. #2		
1.80	-17.50	-17.46	0.04	9.95
1.92	-22.49	24.18	1.69	8.30
2.10	-09.83	-08.52	1.31	8.68
2.35	-08.78	-08.33	0.44	9.56
2.45	-09.42	-08.73	0.69	9.31
2.60	-11.54	-10.73	0.81	9.19

Under the selected combining scheme and at 99% link reliability, the computed diversity gain (DG) by using Eq. (3) [12], which are obtained from the measured data, are presented in Table II. The high diversity gains are obtained over the operating bands.

$$DG = DG_0 \cdot \left(\sqrt{1-\rho_{ij}^{e}}\right) \cdot K \quad (3)$$

$$K = \min\left(\frac{MEG_1}{MEG_2}, \frac{MEG_2}{MEG_1}\right) \quad (4)$$

Where, DG_0 = 10 dB [12] represents the ideal diversity gain of the antenna.

C. *Radiation Characteristics*

The measured 2D radiation patterns of the proposed MIMO antenna at 2.5 GHz frequency are plotted in Fig. 10. It is observed that the radiation pattern is almost omnidirectional and tends to cover the complementary space region. This approach can overcome the multipath fading problems and enhance the systems performance.

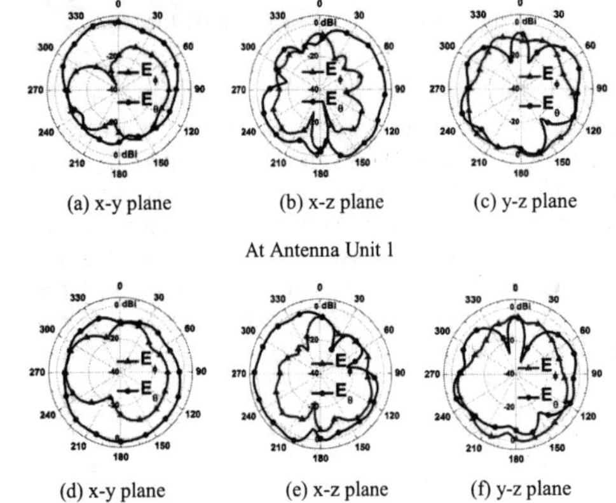

(a) x-y plane (b) x-z plane (c) y-z plane

At Antenna Unit 1

(d) x-y plane (e) x-z plane (f) y-z plane

At Antenna Unit 2

Figure 10. Measured 2D radiation patterns at 2.5 GHz

IV. CONCLUSION

A printed broadband MIMO antenna system is presented. At VSWR ≤ 2.75, the measured bandwidths are around 530 MHz and 670 MHz, respectively. The applied decoupling techniques work effectively and hence high isolation (> 18 dB) is achieved. The MEGs, DGs and correlation coefficients are computed from the measured far-field radiation patterns and ensuring high diversity performance. The proposed MIMO antenna occupies total size of $60\times110\times0.8mm^3$, which is a typical mobile handset's size for practical usage and comparable with the IPhone4 mobile handset. Moreover, a footprint of each single antenna unit is only $15\times18.5mm^2$, which makes it fit for antenna array.

ACKNOWLEDGMENT

The author would like to thank the Key Laboratory of Autonomous Systems and Networked Control, SCUT, Ministry of Education, Guangzhou. The work was partially supported by the Funds for Natural Science Foundation of China under Grant 61074097 and Project on the Integration of Industry, Education and Research of Guangdong Province (2012B091100039).

REFERENCES

[1] C. E. Shannon, "A mathematical theory of communication," *Bell Syst. Tech. J.*, vol. 27, pp. 379-423, Jul. 1948.

[2] E. Dahlman, S. Parkvall, J. Skold, and P. Beming, *3G Evolution: HSPA and LTE for Mobile Broadband*, 2nd ed., Academic Press, Oxford, 2007.

[3] G. J. Foschini and M. J. Gans, "On limits of wireless communications in a fading environment when using multiple antennas," *Wireless Pres. Commun.*, vol. 6, no. 3, pp. 331-335, 1998.

[4] S. H. Chan, R. A. Abd-Alhameed, Z. Z. Abidin, N. J. McEwan, and P. S. Excell, "Wideband printed MIMO/Diversity monopole antenna for WiFi/WiMAX Applications," *IEEE Trans. Antennas Propag.*, vol. 60, no. 4, pp. 3936-3939, Apr. 2012.

[5] C.-C. Hsu, K. –H. Lin, and H.-L. Su, "Implementation of Broadband Isolator Using Metamaterial-Inspired Resonators and a T-Shaped Branch for MIMO Antennas," *IEEE Trans. Antennas Propag.*, vol. 59, no. 10, pp. 3936-3939, Oct. 2011.

[6] J. Li, Q. Chu, and T. Huang, "A compact wideband MIMO antenna with two novel bent slits," *IEEE Trans. Antennas Propag.*, vol. 60, no. 2, pp. 482-489, Feb. 2012.

[7] D. Shen, T. Guo, F. Kuang, X. Zhang, and D. Wu, " A novel wideband printed diversity antenna for mobile handsets," in *Proc. IEEE Vehicular Technology Conference (75th VTC Spring)*,pp. 1-5, May. 2012.

[8] C. H. Chang and K. Wong, "Printed λ/8-PIFA for Penta-Band WWAN Operation in the Mobile Phone," *IEEE Trans. Antennas Propag.*, vol. 57, no. 5, pp. 1337-1381, May. 2009.

[9] H. Paul, "The significance of radiation efficiencies when using s-parameters to calculate the received signal correlation form two antennas," *IEEE Antennas Wireless Propag. Lett.*, vol. 4, no. 1, pp. 97-99, Jun. 2005.

[10] S. C. K. Ko and R. D. Murch, "Compact integrated diversity antenna for wireless communications," *IEEE Trans. Antennas Propag.* Vol. 49, pp. 954-960, Jun. 2001.

[11] R. G. Vaughan and J. B. Anderson, "Antenna diversity in mobile communications," *IEEE Trans. Veh. Technol.*, vol. 36, pp. 147-172, Nov. 1987.

[12] Y. Gao, X. D. Chen, and Z. N. Ying, "Design and performance investigation of a dual-element PIFA array at 2.5 GHz for MIMO terminal," *IEEE Trans. Antennas Propag.*, vol. 55, no. 12, pp. 3433-3441, Jun. 2007.

A Conical Quadrifilar Helix Antenna for GNSS Applications

Baiquan Ning, Juan Lei, Yongchao Cao, and Liang Dong

National Key Laboratory of Antenna and Microwave Technology, Xidian University
Xi'an, 710071, China

***Abstract*-A broadband conical quadrifilar helix antenna is presented in this paper. The broadband characteristic is achieved by tapering the helices, utilizing conical geometry and adjusting the dimensions of the cone-shape support. The bandwidth, defined for a VSWR<2, is about 25% across 1155-1327MHz and 1447-1617MHz, covering all of the frequencies of Global Navigation Satellite System including GPS, GLONASS, Galileo and Compass. The antenna is fed by a wideband four-output-ports feed network with equal magnitude and consistent 90° phase shift. The final antenna obtains excellent circular polarisation, a good gain (greater than 4.5dB at boresight) and good axial-ratio performance over a wide angular range for Global Navigation Satellite System applications.**

Key Words: Conical Quadrifilar Helix Antenna, Global Navigation Satellite System (GNSS), wideband feed network.

I. INTRODUCTION

Global Navigation Satellite System (GNSS) will in effect be fully deployed and operational in a few years [1], with its spectra spreading densely across 1164–1300 and 1559–1610 MHz, therefore, a broadband or dualband antenna is required for its applications. Moreover, hemispherical radiation patterns with good circular polarization in the bandwidth are required for global positioning systems [2]. In these systems, circular polarization antennas are widely used for signals transmission due to their insensitivity to ionospheric polarization rotation.

A prime candidate for these applications is the resonant quadrifilar helical antenna (QHA) [3] and the printed quadrifilar helical antenna (PQHA) [4] due to their salient features, such as: low cost, light weight, hemispherical coverage and good circular polarization. A typical quadrifilar helix antenna consists of four helices equally spaced and wrapped around a common cylindrical structure. At one end of that structure, each helix is connected to a feed network or an equal-power 4-way combiner with the following phase relationship: 0, -90, -180, -270 degrees.

To cover all the GNSS frequencies, these antennas require broadband properties. However, the bandwidth of a conventional PQHA operating under resonant modes, is typically 5-8% while the tapered PQHA (TPQHA) is 14% in L-band [5], both of which are insufficient for GNSS applications. In the past decades, several techniques have been described in the literature for dual-band behavior [6]-[8]. Other techniques have been proposed to broaden the bandwidth of the QHA: for instance, using a conical geometry [9] obtains a 18.5% bandwidth. The FPQHA presented in [10] considerably increases the QHA bandwidth, as 30% has been achieved. But unfortunately GPS-L5 and Galileo-E5a (1166.22-1186.68MHz) are out of the band.

In this paper, the helices of the proposed antenna are tapered and wrapped around a conical structure of relative permittivity $\varepsilon_r = 2.2$ to obtain a Conical PQHA (CPQHA). By adjusting the dimensions of the cone, the equivalent impedance of the antenna will be changed and matched. Finally, the bandwidth of the proposed antenna is about 25%, covering all the GNSS frequencies. In the design process, commercial simulation software HFSS has been used [11].

The remainder of this paper is organized as follows. Section II describes the antenna configuration and the feed network. Measurement results are given in Section III. Section IV presents conclusions.

II. CONFIGURATION

A. Antenna Configuration

The planar unwrapped antenna is illustrated in Figure 1. The four arms of the antenna were printed onto a thin dielectric substrate of relative permittivity $\varepsilon_r = 2.2$ and of thickness h=0.15mm, wrapped around a conical support, and mounted on the ground plane of the feed network which will be presented in Part B.

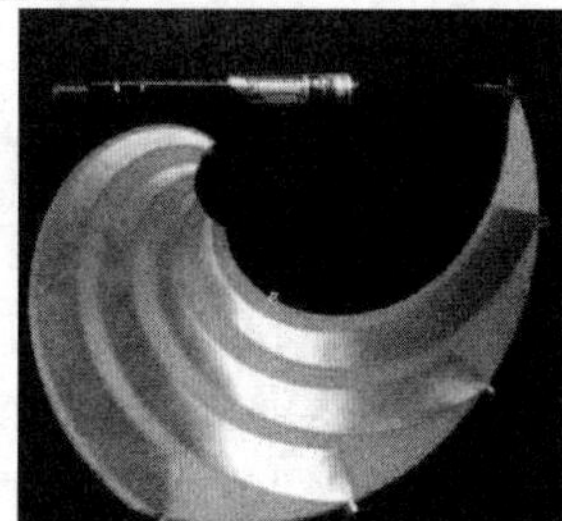

Figure 1. Picture of the unwrapped helix.

As presented in [5] and [9], tapering technique and conical geometry can enhance the bandwidth of the QHA. In this paper, these two techniques are utilized to broaden the bandwidth. One of the helical arms, starting from $+x$ axis in the x-o-y plane, can be described by the following equations:

Equations of the lower line:

978-7-5641-4279-7

$$\begin{cases} x(\varphi) = (r_1 - \dfrac{h\tan\theta}{2n\pi}\varphi)\cos(-\varphi) \\ y(\varphi) = (r_1 - \dfrac{h\tan\theta}{2n\pi}\varphi)\sin(-\varphi) \quad 0 \le \varphi \le 2n\pi \\ z(\varphi) = \dfrac{h}{2n\pi}\varphi \end{cases} \tag{1}$$

Equations of the upper line:

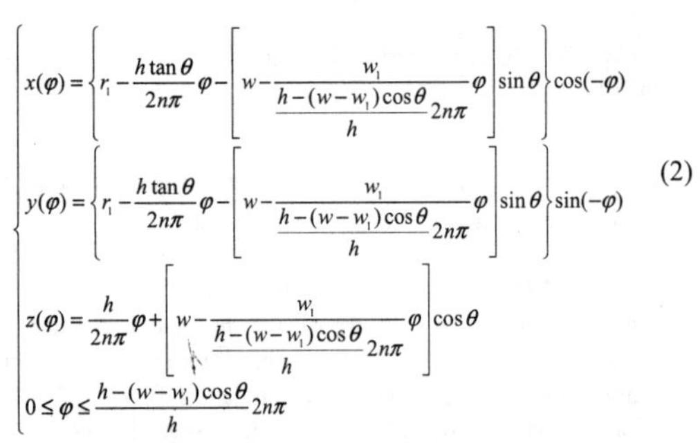

$$\begin{cases} x(\varphi) = \left\{ r_1 - \dfrac{h\tan\theta}{2n\pi}\varphi - \left[w - \dfrac{w_1}{\dfrac{h-(w-w_1)\cos\theta}{h}2n\pi}\varphi \right]\sin\theta \right\}\cos(-\varphi) \\ y(\varphi) = \left\{ r_1 - \dfrac{h\tan\theta}{2n\pi}\varphi - \left[w - \dfrac{w_1}{\dfrac{h-(w-w_1)\cos\theta}{h}2n\pi}\varphi \right]\sin\theta \right\}\sin(-\varphi) \\ z(\varphi) = \dfrac{h}{2n\pi}\varphi + \left[w - \dfrac{w_1}{\dfrac{h-(w-w_1)\cos\theta}{h}2n\pi}\varphi \right]\cos\theta \\ 0 \le \varphi \le \dfrac{h-(w-w_1)\cos\theta}{h}2n\pi \end{cases} \tag{2}$$

where n is the number of turns, r_1 is the initial radius, w is the initial width of the arms and h is the height of the support which consists of two parts as shown in Figure 2(b) and (c).

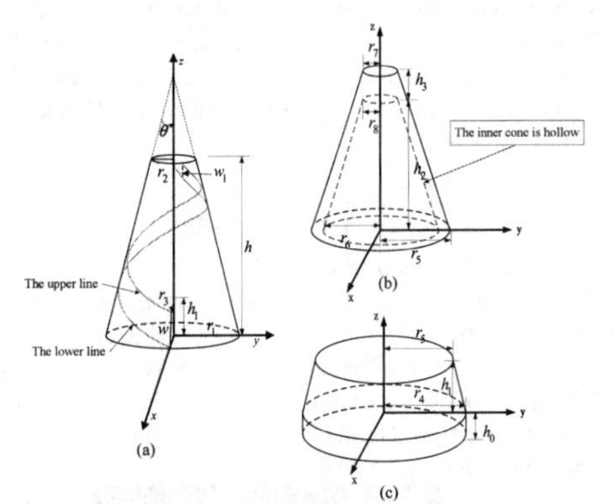

Figure 2. (a) One arm wrapped around the cone, (b) The upper part of the support, and (c) The lower part of the support.

The geometrical parameters of the final antenna are listed in Table I. Figure 3 is the simulated VSWR of the antenna. It is observed that the voltage standing-wave ratio (VSWR) is lower than 2 across 1155-1327 MHz and 1447-1617 MHz, covering all the GNSS frequencies.

TABLE I
GEOMETRICAL PARAMETERS

r_1(mm)	36.65	h_0(mm)	2
r_2(mm)	5.15	h_1(mm)	7
r_3(mm)	30.79	h_2(mm)	65
r_4(mm)	36.5	h_3(mm)	8
r_5(mm)	33.74	w(mm)	16
r_6(mm)	30.24	w_1(mm)	8.5
r_7(mm)	5	θ(°)	21.49
r_8(mm)	5	n	1
h(mm)	80		

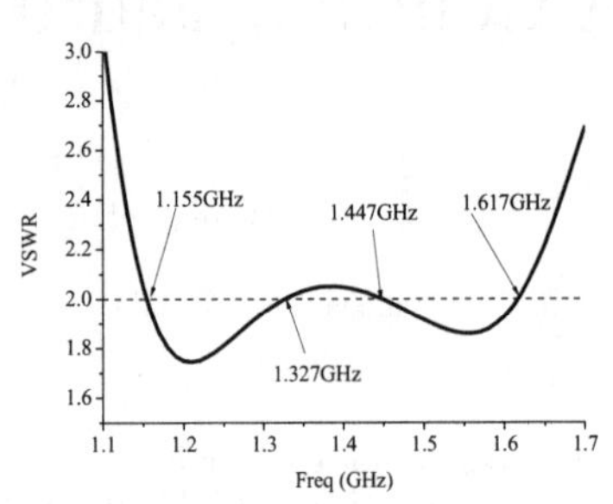

Figure 3. The simulated VSWR of the proposed antenna.

B. Feed Network Configuration

The feeding system has to provide constant magnitudes and 90° phase differences to the four helices of the CPQHA over the operating frequency band. However, the conventional directional coupler and rat-race coupler generate a narrow bandwidth of 90° phase shift, which are not adequate for the proposed antenna. To feed the proposed antenna, a wideband feed network composed of three Wilkinson dividers and three broadband phase shifters [12] has been designed. According to [12], the wideband 90° phase shift was achieved when $z_{m1} = 61.9\Omega$, $z_{s1} = 125.6\Omega$ while $z_0 = 50\Omega$; the condition for 180° phase shift is $z_{m2} = 80.8\Omega$, $z_{s2} = 62.8\Omega$. The designed feed network is centered at 1.4GHz, and achieved by adjusting the length and width of every segment of the microstrip line on a PC board with relative permittivity of 2.65. The simulation and measurement results of the feed network will be described in Section III. Figure 4 is the photograph of the designed feed network.

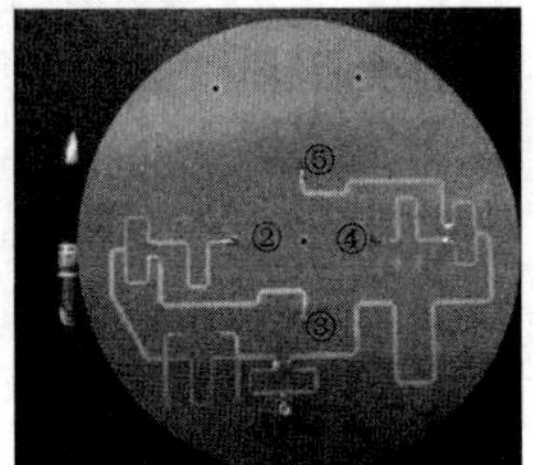

Figure 4. Photograph of the designed feed network.

III. Experimental Results

Measured VSWR, magnitude response and phase difference of the designed feed network were obtained using the Agilent vector network analyzer as shown in Figures 5, 6 and 7. A frequency shift mainly due to the machining error can be seen from Figures 5 and 6. In addition, it is observed that the maximum difference between measured magnitude and that of simulated is about 0.3 dB in the band of 1.15-1.65 GHz. The measured phases, with respect to that of Port 2, of the output ports at several frequencies are listed in Table II. One can see that the phase shift unbalance is less than 4°.

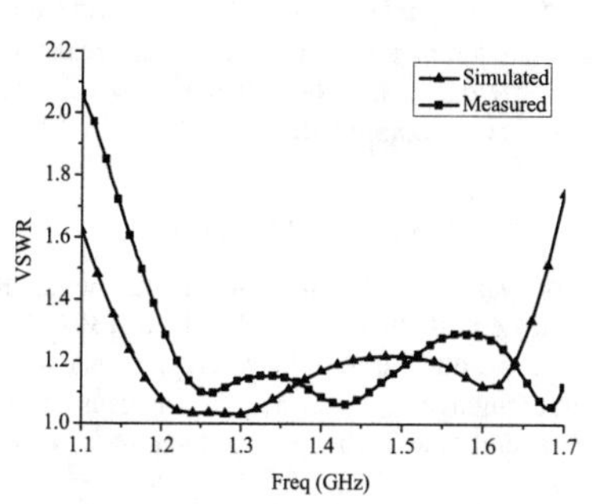

Figure 5. The VSWR against frequency of the designed feed network.

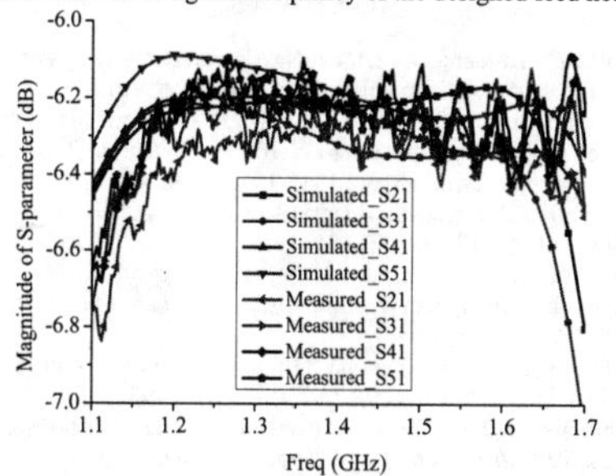

Figure 6.The magnitude response of the designed feed network.

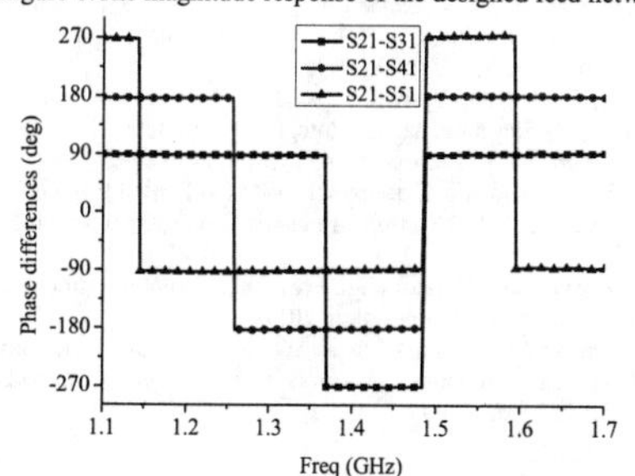

Figure 7. The measured phase difference of the designed feed network.

TABLE II
THE MEASURED PHASES AT SEVERAL FREQUENCIES

Freq(MHz) / Port	1176	1561	1575	1602
2	0	0	0	0
3	88.437	89.678	91.015	90.150
4	176.074	180.298	180.804	180.282
5	92.634	273.679	273.683	86.159

The far-field radiation patterns, measured in an anechoic chamber in magnitude and phase for linear polarizations and combined to give the circular polarization, are shown in Figure 8 at frequencies of 1176, 1561, 1575 and 1602 MHz.

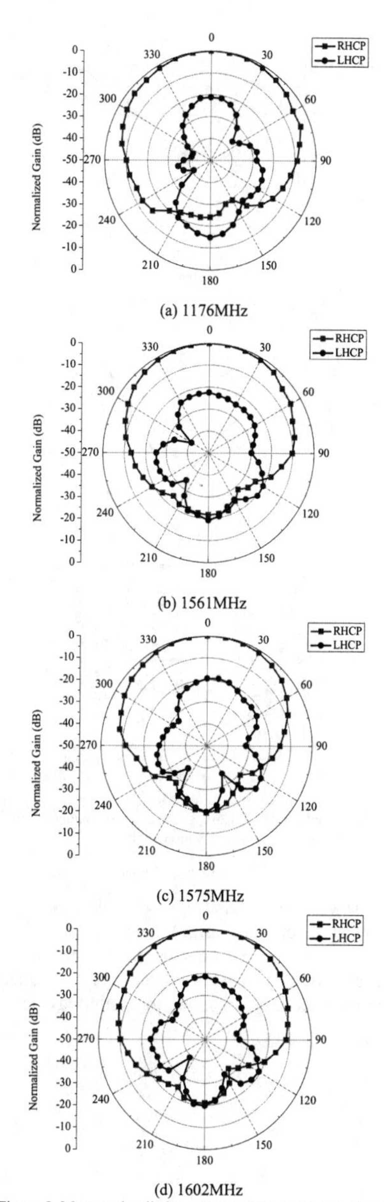

Figure 8. Measured radiation pattern of the proposed antenna.

Figure 9(a) and (b) depict the dual broadband characteristics of the RHCP gain and axial ratio. The measured RHCP realized gain is greater than 4.5 dB in the interesting GNSS bands. Moreover, the axial ratio is less than 3 dB from -80° to 80° in the elevation plane, showing that the antenna has a broad pattern coverage.

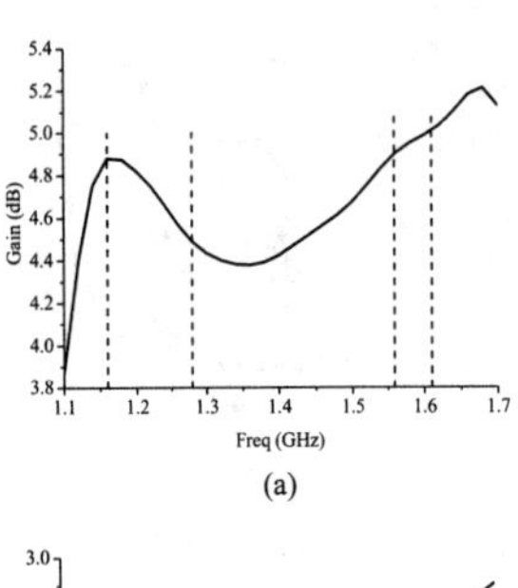

(a)

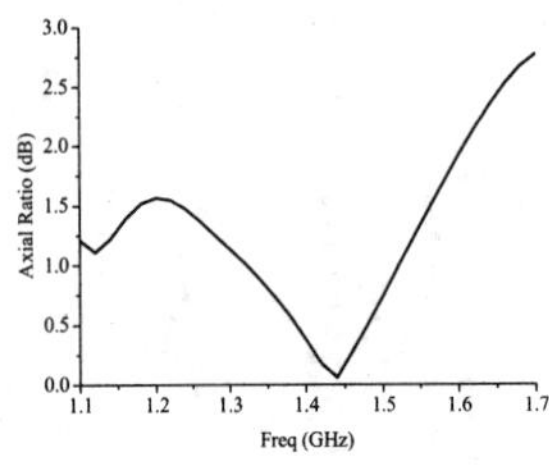

(b)

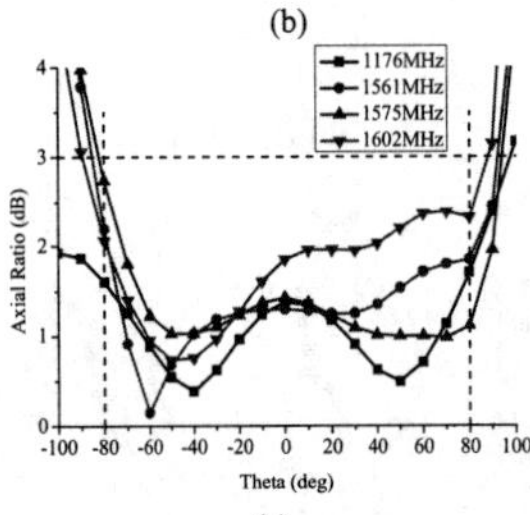

(c)

Figure 9. Measured (a) RHCP Gain, (b) AR versus frequency, and (c) AR versus elevation angle.

IV. CONCLUSION

In this paper, a broadband conical PQHA is reported. By tapering the arms, utilizing conical geometry and adjusting the dimensions of the cone-shape support, the antenna obtains a dualband behavior and covers all the GNSS frequencies. A wideband feed network is applied to excite high-performance RHCP radiation. The axial ratio is less than 3 dB from -80° to 80° in the elevation plane and the RHCP realized gain is greater than 4.5 dB at boresight which make the antenna a good choice for GNSS applications.

ACKNOWLEDGMENT

This work was supported by " the Fundamental Research Funds for the Central Universities (No. K5051302034)".

The author would like to acknowledge excellent tutor Juan Lei for her encouragement and helpful criticism and guidance; good partners and others who gave helps in the course of study. Best wishes!

REFERENCES

[1]. J. H. Wang, "Antennas for Global Navigation Satellite System (GNSS)," *Proceedings of the IEEE,* vol.100, pp.2349-2355, 2012.

[2]. J. M. Tranquilla and S. R. Best, "A study of the quadrifilar helix antenna for global positioning system (GPS) applications," *IEEE Trans. Antennas Propag.,* vol. 38, pp. 1545-1550, 1990.

[3]. C. C. Kilgus, "Resonant quadrafilar helix," *IEEE Trans. Antennas Propag.,* vol. AP-17, pp. 349-351, 1969.

[4]. A. Sharaiha, C. Terret, and J. P. Blot, "Printed quadrifilar resonant helix antenna with integrated feeding network," *Electron. Lett.,* vol. 33, pp.256-257, 1997.

[5]. J. C. Louvigne and A. Sharaiha, "Broadband tapered printed quadrifilar helical antenna, " *Electron. Lett.,* vol. 37,pp.932-933,2001.

[6]. A. Sharaiha and C. Terret, "Overlapping quadrifilar resonant helix antenna, " *Electron. Lett.,* vol. 26, pp. 1090-1092, 1990.

[7]. D. Lamensdorf, M. Smolinski, and E. Rosario, "Dual band quadrifilar helix antenna," *IEEE Antennas Propag. Symp.,* vol. 3, pp. 488-491,2002.

[8]. M. Ohgren and S. Johansson, "Dual frequency quadrifilar helix antenna, " U.S. Patent 6 421 028, Jul. 16, 2002.

[9]. S. Yang, S. H. Tan, Y. B. Gan, and C. W. See, "Broadband conical printed quadrifilar antenna with integrated feed network," *Microw. Opt. Technol. Lett.,* vol. 35, no. 6, pp. 491 - 493, Dec. 2002.

[10]. Yoann Letestu and Ala Sharaiha, "Broadband Folded Printed Quadrifilar Helical Antenna," *IEEE Trans. Antennas Propag.*, vol.54, pp.1600-1604, 2006.

[11]. Ansoft Corporation, Ansoft High Frequency Structure Simulator (HFSS) Version 13.0, Ansoft Corporation, 2011.

[12]. S. Y. Eom and H. K. Park, "New switched-network phase shifter with broadband characteristics," *Microw. Opt. Technol. Lett.,* vol. 38, no. 4,pp. 255–257, Aug. 2003.

Design of a Dual Frequency and Dual Circularly Polarized Microstrip Antenna Array with Light Weight and Small Size

YU Tongbin[1], LI Hongbin[2], ZHONG Xinjian[1], ZHU Weigang[1], YANG Tao[1]
1. Institute of Communications Engineering, PLA University of Science and Technology, Nanjing 210007, China;
2. The First Engineers Research Institute of the General Armaments Department, Wuxi 214035, China

***Abstract*-A novel dual frequency and dual circularly polarized microstrip antenna array is introduced in this paper. By placing two 2 × 2 microstrip antenna arrays on the same broad with unique fashion and taking out the dielectric substrate around the element, the small size antenna array with light weight was designed and manufactured. The array is designed and optimized with the help of HFSS10 software. The return loss, radiate pattern, gain and the axial ratio plots of the array are measured, which agree well with the simulated results.**

I. INTRODUCTION

Microstrip antenna (MPA) has been widely used in the mobile and satellite communications, due to its light weight, low profile and circularly polarized performance [1-3]. Recently, more researchers are interested in the design of dual-polarized patch antenna. The reason is that it can provide polarization diversity to reduce the multipath fading of the received signals or furnish frequency reuse to double the capacities of the mobile communication systems [4]. However, these antennas usually has a complex configuration, such as stacked fabric [5] or complex feeding structure [4][6], which brings much difficulty to the fabrication and production of the antenna.

In this paper, a novel dual frequency and dual circularly polarized microstrip antenna array of S band is presented, which is constituted of two irrelated 2 × 2 microstrip antenna array that works at different frequency and different polarization. And by laying the two array across to each other, all the patches are designed on one substrate which size is 20cm×20cm. By taking out the dielectric substrate around the element, the weight of the array is just 72% of that has no substrate cut. With the help of HFSS10 software, the lower band array has a simulated 1dB gain bandwidth of 3.5% from 1.94 to 2.01GHz, and a maximal gain of 12.19dB. The upper band array has a simulated 1dB gain bandwidth of 3.9% from 2.14 to 2.1925GHz, and a maximal gain of 11.4dB. Across the gain bandwidth, all axial ratios of the array are less than 2dB. Compared with the measured results, preferable agreement has been obtained.

II. ELEMENT DEGIGN

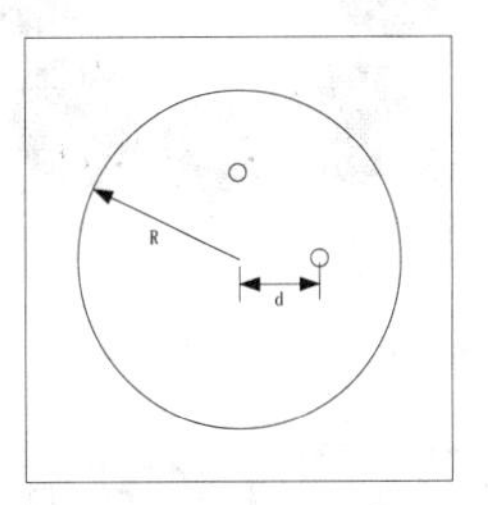

Figure 1. The configuration of element

The element shown in Figure1 is designed with a single layer to remain low profile. Circular patch is chosen for convenience embedding the two arrays, which radius is R. Two feeds with 90° phase difference are placed along the orthogonal directions with a spacing d away from the patch center, which realizes circular polarization. The relative permittivity and the height of the substrate are 2.2 and 3mm partly. The radius of the patch that works at 2GHz is 22.8cm while the radius of the patch that works at 2GHz is 20.5cm. The substrate between the patch and ground has a thickness of 3mm and relative permittivity of 3.5. The position of the feeding point d directly affects the return loss S11 and the coupling degree S21. Via optimizing the position of the feeding point d, it can be found that the excellence position d of the antenna that works at 2GHz is 7cm, while that works at 2.19GHz is 5cm.

III. ARRAY DESIGN

Basing on the circular patch discussed above, the dual frequency and dual circularly polarized microstrip antenna array is designed, which is shown in Figure 2. The radiating elements shown in Figure 2(a) consists of 8 circular patches, in which 4 patches work at 2GHz and the other 4 patches work at 2.19GHz, and they are interveined one another to reduce the horizontal size. Because the Wilkinson power divider has advantages such as small output power difference and high isolation between the two ports [7]. Here Wilkinson power divider is chosen to form feeding network, as shown in Figure 2(b), in which two irrelated feeding networks are intersected to each other . The outside length of the ground plane is 20cm. The space between the patches that work at 2 GHz is denoted to be d_1 which is 98mm(0.65λ, λ is the free space wavelength), while that work at 2.19 GHz is denoted to be d_2 which is 84.8mm(0.62λ). As seen in figure 2(a), the substrate is designed in ring form, thus the weight of the array is reduced

978-7-5641-4279-7

by 28% compared with that filled completely with the dielectric substrate.

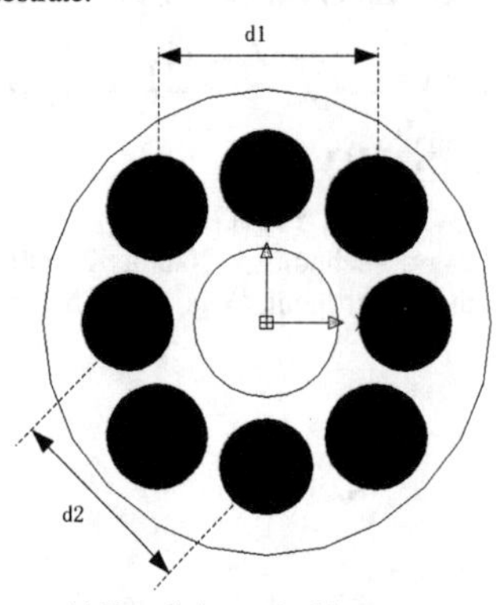

(a) The radiating patch of the array

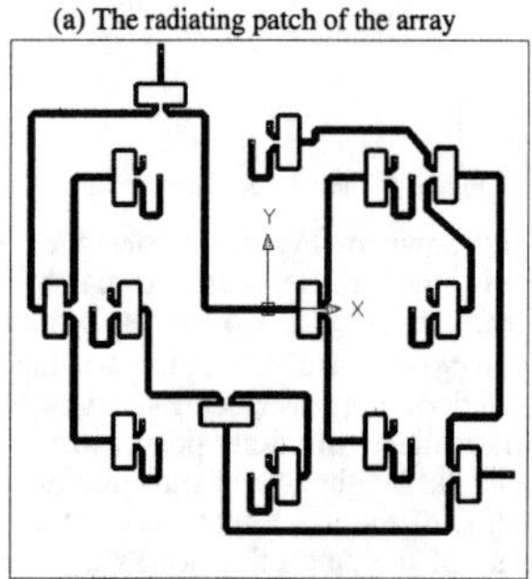

(b) The feeding network of the array

Figure 2 The structure of the array

IV. MEASUREMENTS

The corresponding mirostrip antenna array has been fabricated. Because the array has two irrelated sub-arrays, Measurements are implemented one another. Figure3(a) shows the simulated and measured return loss characteristics (S_{11}) of the array that works at 2GHz, where the simulated bandwidth defined by S_{11} less than -10dB is about 25% from 1.7 to 2.19GHz. Figure3(b) shows the simulated and measured return loss characteristics (S_{11}) of the array that works at 2.19GHz, where the simulated bandwidth defined by S_{11} less than -10dB is about 22.7% from 1.9 to 2.4GHz.

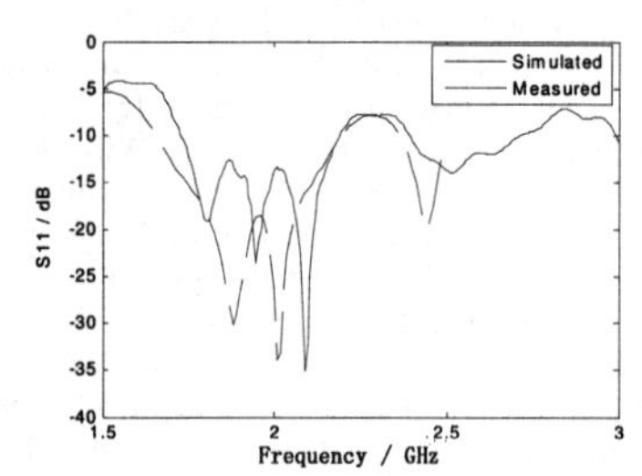

(a) The simulated and measured S_{11} of the lower band array

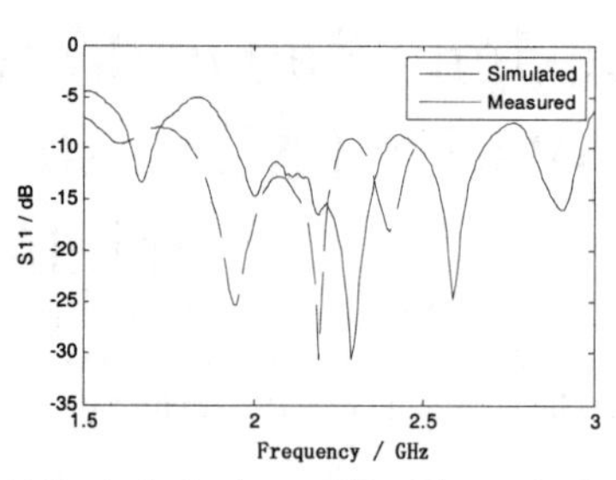

(a) The simulated and measured S_{11} of the upper band array

Figure 3 The simulated and measured S_{11}

To verify the simulated results of the gain and the axial radio (AR), the antenna array is measured in the anechoic chamber. Figure 4 shows the simulated plots and measured points of the gain and AR against frequency. It can be seen that, the maximum gain of the array works at 2 GHz is about 12dB, which is similar to the simulated result. But the shift of the center frequency can be detected and the maximal gain appears at 1.96GHz instead of 2GHz. The difference is probably owing to the fabrication error and the tolerance of the relative permittivity of the substrate. Similarly the array works at 2.19GHz has difference of the center frequency, which maximum gain appears at 2.16GHz instead of 2.19GHz. From the measured points of the AR, during the 1-dB gain bandwidth the measured AR is less than 3dB, which agrees well with the simulated results.

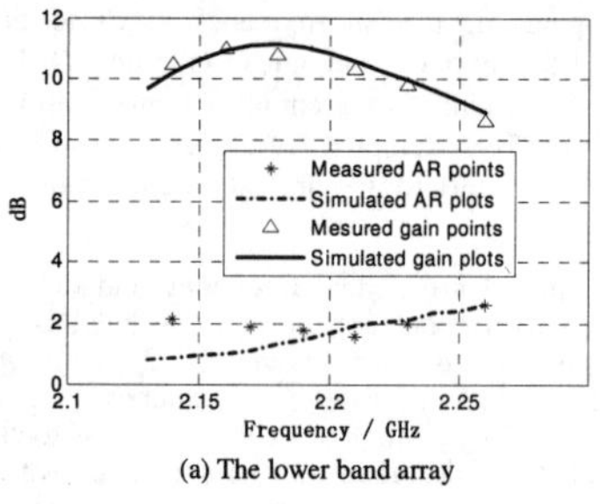

(a) The lower band array

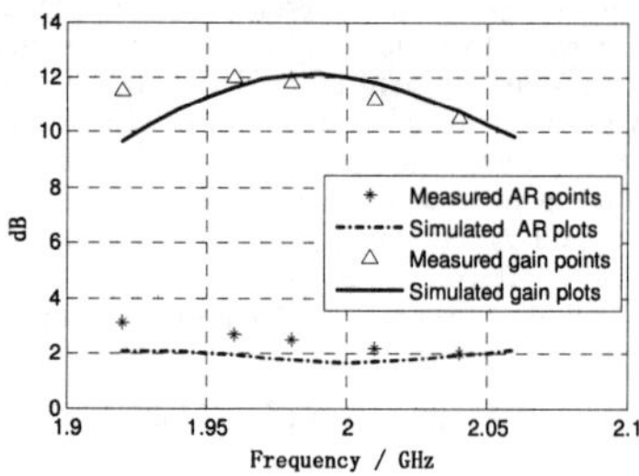

(a) The upper band array

Figure 4 The simulated plots and measured points of the gain and AR against frequency

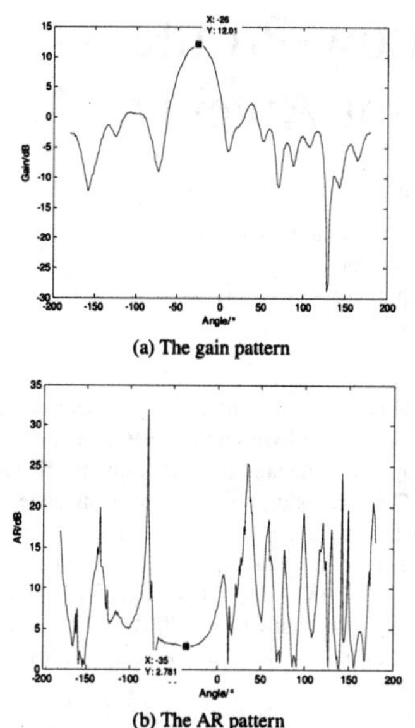

(a) The gain pattern

(b) The AR pattern

Figure 5 The measured far-field radiation patterns at 1.96GHz

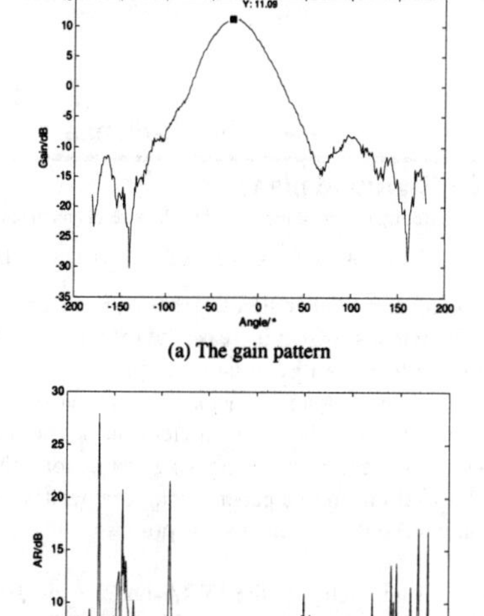

(a) The gain pattern

(b) The AR pattern

Figure 6 The measured far-field radiation patterns at 2.16GHz

Figure 5 and Figure 6 gives the measured far-field radiation patterns at the frequency 1.96GHz and 2.16GHz. It can be seen that the maximum gain appears at -28° not at 0° ,the reason is simply that the initialized position of the rotating platform is not at the 0° direction. Still good symmetric patterns have been obtained. Figure 7 shows the fabricated antenna.

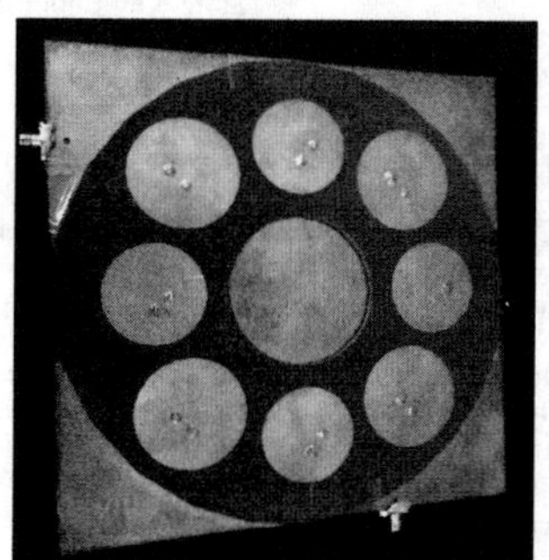

Figure 7 The fabricated antenna

V. CONCLUSION

A novel dual frequency and dual circularly polarized microstrip antenna array is presented in this paper. By laying the two array across to each other, the horizontal size of the antenna array is reduced. And by taking out the dielectric substrate around the element, the weight of the array is reduced by 28%. The designed array was fabricated and measured in this paper, which shows preferable performance.

References

[1] F. S. Chang, K. L. Wong and T. W. Chiou,' Low-cost broadband circularly polarized patch antenna', IEEE Trans. Antennas Propagat., vol. 51,no.10, Oct.2003: 3006-3009 .

[2] N. C. Karmakar, ' Investigations into a cavity-backed circular-patch antenna', IEEE Trans. Antennas Propagat., vol. 50,no.12, Dec. 2002: 1706-1715 .

[3] K. L. Lau and K. M. Luk, 'A Novel Wide-Band Circularly Polarized Patch Antenna Based on L-Probe and Aperture-Coupling Techniques', IEEE Trans. Antennas Propagat., vol. 53,no.1, Jan.2005: 577-580 .

[4] K. L. Lau, K. M. Luk.'A Wideband Dual-Polarized L-Probe Stacked Patch Antenna Array', IEEE Trans. Antennas Propagat.,vol. 6,no.8, Sep. 2007: 529-532 .

[5] J. M. Laheurte, 'Dual-Frequency Circularly Polarized Antennas Based on Stacked Monofilar Square Spirals', IEEE Antennas Propag., vol. 51,no.3, Mar. 2003: 488-492.

[6] E.Aloni, R.Kastner, 'Analysis of a dual circularly polarized microstrip antenna fed by crossed slots' IEEE Trans. Antennas Propagat., vol. 42,no.8, Aug. 1994:1053-1058 .

[7] K. L. Wong and T. W. Chiou.' Broad-band single-patch circularly polarized microstrip antenna with dual capacitively coupled feeds', IEEE Trans. Antennas Propagat., vol.49,no.1, Jan. 2001: 41-44.

Wideband High-Gain Low-Profile 1D Fabry-Perot Resonator Antenna

Y. Ge[1,2], C. Wang[1], and X. Zeng[1]

[1]College of Information Science and Engineering, Huaqiao University
Xiamen, Fujian Province 361021, China
[2]State Key Laboratory of Millimeter Waves, Southeast University
NANJING, 210096, CHINA

***Abstract*-Wideband high-gain low-profile Fabry-Perot resonator antennas (FPRAs), composed of a single superstrate and a PEC ground, are proposed in this paper. Two thin dielectric slabs are used to construct the superstrate that exhibits increasing reflection phase at the designated frequency band, making the Fabry-Perot cavity resonate at a wider band and hence leading to a wideband low-profile FPRA. The design example validates the design principle. By truncating the size of the superstrate to about 1.3λx1.3λ, the 3dB-gain directivity bandwidth of the antenna can be extended from 12.6% to 34.8%, with a peak gain of 15.7 dBi.**

I. INTRODUCTION

Fabry-Perot (FP) resonator, originally used in optical region, was first applied to enhance the gain of radiators in 1956 [1]. Afterwards, dielectric Fabry-Perot resonator antennas (FPRAs) were well studied [2]. In the recent decade, progress in photonic band-gap (PBG) and electromagnetic band-gap (EBG) materials inspired the study on FPRAs. High-gain, wideband and multiband directive FPRAs were developed [3-6] using various EBG materials in the microwave region.

One of the crucial properties of FPRAs concerned by researches is the bandwidth, due to the inherent narrowband resonant cavity. Several methods have been developed to overcome the problem. An active reconfigurable FPRA is designed to obtain an operating bandwidth of about 13.5% [4]. An FPRA with dual-resonator configuration and patch array feeding can achieve a bandwidth of 13.2% [5]. A tapered FSS can be designed to form a wideband FPRA. Theoretically, the increasing reflection phase with frequency of the EBG can be utilized to design wideband FPRAs. This principle has been utilized in practice to develop wide FPRAs. Recently, a single dielectric layer with dual dipole arrays etched on the two sides is developed to form wideband FPRAs [6], due to its increasing reflection phase in the operating frequency band.

To the best of the author's knowledge, most EBG or partially reflective surface (PRS) structures forming FPRAs published in literatures are multiple-layer, which will result in the increase of the antenna size and side-lobes at higher frequencies. Another problem, as mentioned in [6], the radiation pattern will deteriorate at higher frequencies, due to out-of phase at the surface of the EBG/PRS structure, though the antenna gain is still within the 3-dB gain bandwidth.

In this paper, a wideband high-gain low-profile 1D FPRA is proposed. The increasing reflection phase is generated by two stacked dielectric layers with no air gap. To overcome the deteriorated radiation pattern, the PRS is truncated to have a size of about 1.3λx1.3λ. This will further enhance the effective 3-dB gain bandwidth of FPRAs. Example shows that a bandwidth of about 34.8% and a peak gain of 15.7 dBi can be achieved.

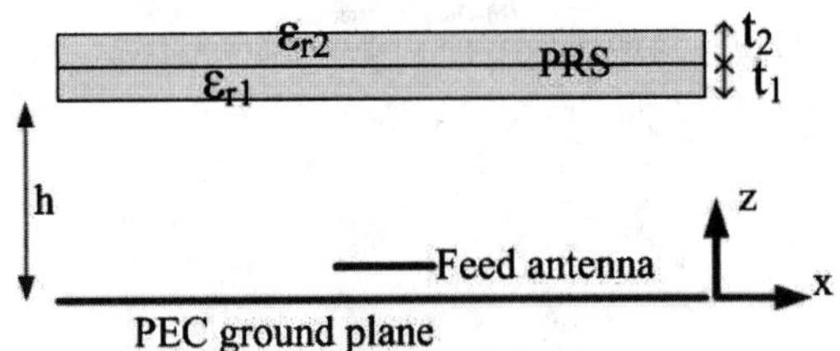

Figure 1. Configuration of the 1D Fabry-Perot resonator antenna.

II. 1D PRS FOR WIDEBAND FABRY-PEROT CAVITY

Most wideband Fabry-Perot resonator antennas have multiple superstrates, which increase the antenna size. In [6], a single superstrate and a PEC round are applied for a wideband FPRA, making the antenna structure simple and low-profile. In the superstrate, periodic resonant elements are printed on its two sides to generate the increasing reflection phase for wideband operation. In this paper, a single superstrate without loading any resonant elements is proposed for wideband FPRAs.

The general geometry of the FPRA under consideration is shown in Figure 1. The structure is ideally infinite in the *x-y* plane. The superstrate is composed of two dielectric slabs and there is no air gap between the two slabs. Each dielectric layer is assumed lossless, homogeneous, and isotropic. A horizontal Hertzian dipole feeds the Fabry-Perot cavity, and the separation area between the PEC and the superstrate is considered free space. By following the same derivation procedure described in [1], the total far-field pattern of the FPRA is

978-7-5641-4279-7

$$E = f(\alpha)\sqrt{\frac{1-\Gamma^2(\theta)}{1-2\Gamma(\theta)\cos\left(\varphi-\pi-\frac{4\pi}{\lambda}\cos\theta\right)+\Gamma^2(\theta)}} \quad (1)$$

where Γ, φ, and θ are reflection coefficient, reflection phase and orientation angle, respectively. $f(\alpha)$ is the radiation pattern of the horizontal Hertzian dipole. The reflection coefficient, reflection phase of the superstrate can be calculated by using the cascade ABCD network, as detailed in [7]. The far-field pattern can be approximately calculated by

$$E = \sqrt{\frac{1-\Gamma^2(\theta)}{1-2\Gamma(\theta)\cos\left(\varphi-\pi-\frac{4\pi}{\lambda}\cos\theta\right)+\Gamma^2(\theta)}} \quad (2)$$

This analytical model is based on the ray-tracing method and accurate in calculating the far-field pattern of the FPRA.

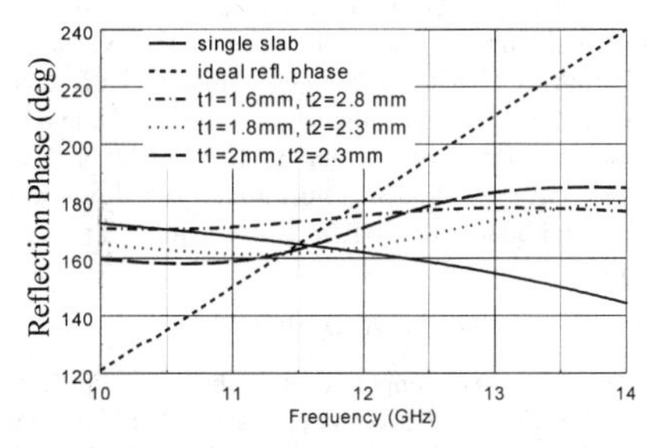

Figure 2. Reflection phases from the single dielectric slab and various dual dielectric slabs.

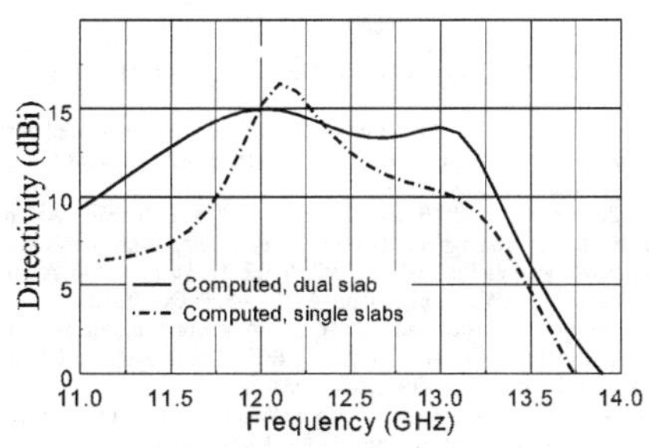

Figure 3. Directivities vs frequency.

It is well known that the reflection phase of a single dielectric slab decreases with frequency, as can be shown in Figure 2. The superstrates made out of dual slabs with no air gap, as shown in Figure 1, were investigated and the results show that increasing phase with frequency can be obtained. Parameters of such superstrate examples presented here are ε_{r1}=25 and ε_{r2}=4.4. Other parameters like cavity height h and the thickness of the two slabs t_1 and t_2 are adjustable for better performance. Figure 2 gives some results of the investigation. The reflection phases of the proposed superstrate with three sets of values of t_1 and t_2 are plotted. It can be seen that all three results have the increasing phase within a frequency range, demonstrating the potential feasibility of wideband Fabry-Perot resonator antennas. The ideal increasing phase needed to maintain the resonance at the corresponding Fabry-Perot cavity is also plotted in Figure 2, for reference.

The superstrate with t_1=*1.8 mm* and t_2=*2.3 mm* is applied to construct an FPRA to verify the design principle. Formula (2) is employed to calculate the directivity and the radiation patterns, which are shown in Figure 3 and Figure 4 respectively. For comparison, the directivity obtained from the FPRA based on a single-slab superstrate is also plotted in Fig. 3. It can be seen that the antenna with dual-slab superstrate exhibits a much wider 3-dB directivity bandwidth. The radiation patterns shown in Figure 4 have a proper main lobe and low side lobes at 12 GHz and 13 GHz, demonstrating the effectiveness of the design of wideband high-gain low-profile FPRAs, together with the results in Figure 3.

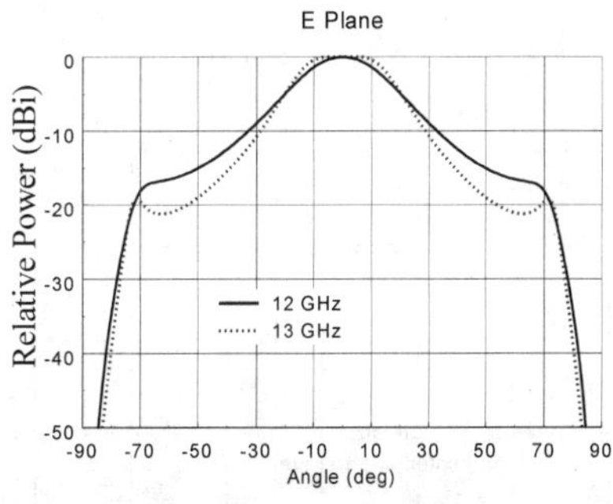

Figure 4. Far-field patterns in E plane of the proposed FPRA at 12 GHz and 13 GHz.

Figure 5. Prototype of wideband high-gain low-profile 1D FPRA.

III. 1D Wideband Fabry-Perot Resonator Antenna

The structure of the final design of the wideband high-gain low-profile FPRA is shown in Figure 5. It is composed of a dual-slab superstrate and a PEC ground and fed by a WR75 rectangular waveguide. The design parameters of the dual-slab superstrate are ε_{r1}=*25*, ε_{r2}=*4.4*, t_1=*1.8 mm* and t_2=*2.6 mm*. The cavity height h is *14 mm*. The sizes of the superstrate and the ground are *80×80 mm²* and *90×90 mm²*, respectively. The aperture size of the feeding WR75 waveguide is *19×9.5 mm²*. To improve the impedance matching of the feeding waveguide, thin metallic irises are added at the radiating aperture of the waveguide. Therefore, the aperture size on the ground surface is *12×5 mm²*. Ansoft HFSS is employed to simulate the designed FPRA.

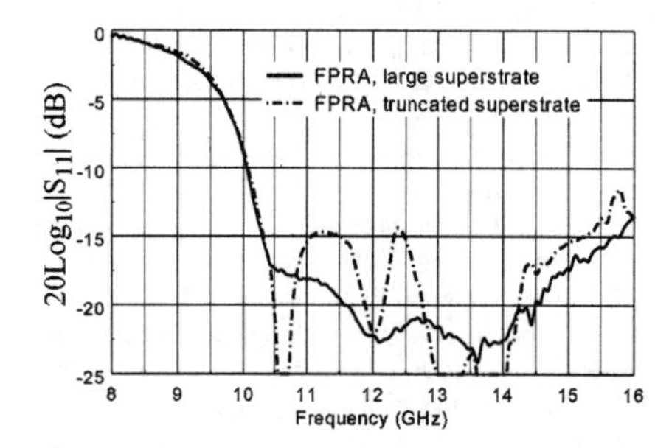

Figure 6. Reflection coefficient.

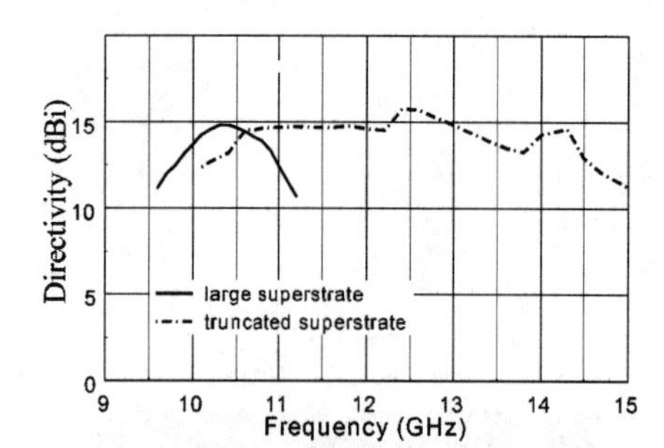

Figure 7 Directivities vs frequency.

The computed reflection coefficient is shown in Figure 6. It can be seen that good impedance matching is obtained from 10.1 GHz to over 16 GHz. The calculated directivity is shown in Figure 7. The peak directivity is about 15 dBi and the 3-dB directivity bandwidth is about 12.6%. To further investigate the FPRA, the size of the dual-slab superstrate is truncated to be about 1.3λ×1.3λ. The simulation results, also plotted in Figures 6, and 7, respectively, show that the similar impedance matching can be obtained, while the peak directivity and the 3-dB directivity bandwidth become 15.7 dBi and 34.8%, showing a much improved bandwidth performance with a much smaller size and without compromising any other performance. The computed radiation patterns of the truncated FPRA at 11 GHz, 12 GHz, 13 GHz and 14.3 GHz are plotted in Figure 8. All patterns at the four frequencies are similar except that higher sidelobe level is obtained at higher frequencies.

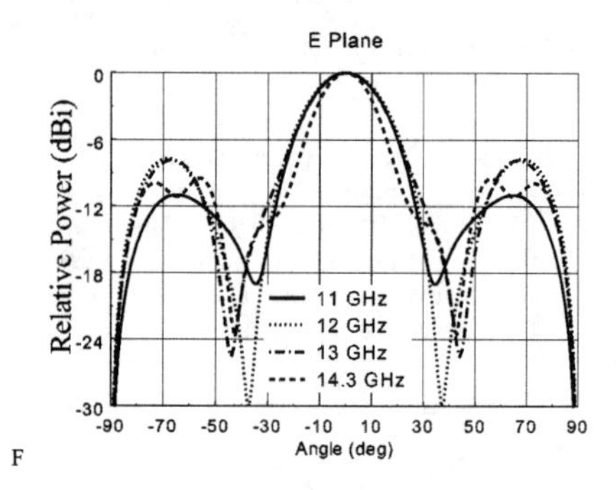

F Hz.

IV. CONCLUSION

A superstrate composed of two thin dielectric slabs with no air gap has been successfully designed to give increasing reflection phase within a designated frequency band. This superstrate is then applied to construct Fabry-Perot resonator antennas. Simulations show that the antenna achieves a 3-dB gain bandwidth of 12.6%., with a peak directivity of 15 dBi. By truncating the size of the superstrate to 1.3λ×1.3λ and then constructing an FPRA, a bandwidth of 34.8% is obtained, with a peak directivity of 15.7 dBi. Thus, a compact, low-profile, wideband, high-gain Fabry-Perot resonator antenna is successfully developed.

ACKNOWLEDGMENT

This research was supported by the start-up grants (11BS301) of Huaqiao University, Xiamen 361021, China, Natural Science Foundation of Fujian Province (2012J01276), China, and Open Research Program (K201212) from State Key Laboratory of Millimeter Waves, Nanjing 210096, China.

REFERENCES

[1] G. V. Trentini, "Partially Reflecting Sheet Array", IRE Trans. Antennas Propag., vol. 4, pp. 666-671, Oct. 1956.

[2] D. R. Jackson, and N. Alexopoulos, "Gain Enhancement Methods for Printed Circuits Antennas", *IEEE Trans. Antennas Propagat.*, vol. 33, no. 9, pp. 976 - 987, Sept. 1985.

[3] B. A. Zeb, Y. Ge, K. P. Esselle, Z. Sun, and M. E. Tobar, "A Simple Dual-Band Electromagnetic Band Gap Resonator Antenna based on Inverted Reflection Phase Gradient", IEEE Transactions on Antennas and Propagation, vol. 64, no. 10, pp. 4522 – 4529, Oct. 2012.

[4] A. R. Weily, T. S. Bird and Y. J. Guo, "A Reconfigurable High-Gain Partially Reflecting Surface Antenna", IEEE Trans. Antennas Propag., vol. 56, no. 11, pp. 3382-3390, Nov. 2008

[4] P. Feresidis, J. C. Vardaxoglou, "A broadband high-gain resonant cavity antenna with single feed", in Proc. EuCAP 2006, Nice, France, 2006

[5] L. Moustafa and B. Jecko, "EBG Structure with Wide Defect Band for Broadband Cavity Antenna Applications", IEEE Antennas and Wireless Propag. Lett., vol.7, pp. 693-696, Nov. 2008.

[6] Y. Ge, K P. Esselle, and T. S. Bird, "The Use of Simple Thin Partially Reflective Surfaces with Positive Reflection Phase Gradients to Design Wideband, Low-Profile EBG Resonator Antennas ", IEEE Trans. Antennas Propag., vol. 60, pp. 743-750, Feb. 2012.

[7] H. A. Macleod, Thin-film optical filters, 2nd ed., Adam Hilger Ltd., Bristol, 1986.

Bandwidth Improvement Through Slot Design on RLSA Performance

I.M. Ibrahim[1,2], T.A.Rahman[2], M.I.Sabran[2]

[1]Faculty of Electronics and Computer Engineering, Universiti Teknikal Malaysia Melaka, Malaysia
[2]Wireless Communication Centre, Universiti Teknologi Malaysia, Skudai, Johor, Malaysia

***Abstract*-Radial Line Slot Array Antenna is a slot planar antenna. The slot has been arranged radially to create a concentration of wave into a single point. The body of this antenna normally a metal with hollow cavity and slots. However, recent development on RLSA has utilized the FR4 as a substrate material. The copper is attached to the substrate and the slot is not hollow as compare to the conventional approach. This research is studying the effect of slot with dielectric substrate to the Radial Line Slot Array Antenna (RLSA). From the measurement, it is found that the slot without substrate provide a better return loss as compare to slot with dielectric. It is also improve the bandwidth of the RLSA.**

I. Introduction

RLSA antenna for point to point microwave link based on 802.11a standard currently a popular candidate for this application due to its capability of carrying high speed signal [1]. RLSA prototypes has been designed and developed at the frequency range of 5725 – 5875 MHz by few researchers [2-4]. The classic design was using an air gap as a separator between radiation surface and ground plane [5]. Then, the polypropelene has been used as a slow wave element in the RLSA structure. This material normally give 2.3 dielectric value [6-7].

Recent development on RLSA has utilized FR4 board as a part of the antenna structure [7-11]. The copper is attached to the substrate and the slot is not hollow as compare to the conventional way. This research is studying the effect of slot with dielectric substrate to the Radial Line Slot Array Antenna (RLSA)

II. ANTENNA STRUCTURE

In this research, the FR4 board with air gap distance to the ground plane has been introduced as shown in Figure 1. The thickness of overall cavity is 9.6mm where the thickness of open air gap is 8mm. A 50Ω single coaxial probe coated with Teflon is used to feed the signal into the cavity. The aluminum plate is used as a platform to hold the antenna and also become a ground plane. The FR4 with 1.6mm thickness with 5.4 permittivity value is used as a first layer substrate to the radiating surface. Figure 2 shows the front view of the antenna.

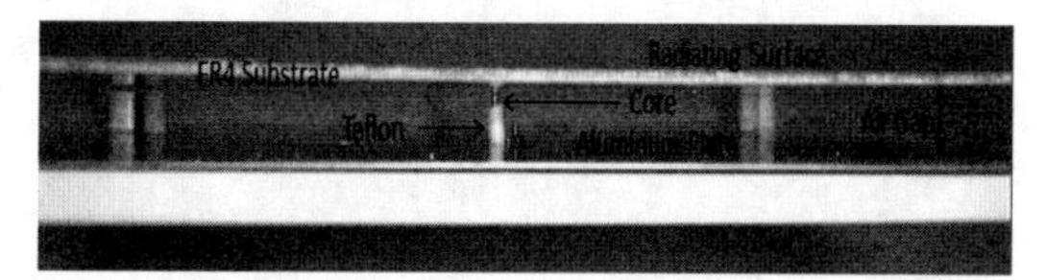

Figure 1: Fabricated open ended Air Gap RLSA Structure from side view

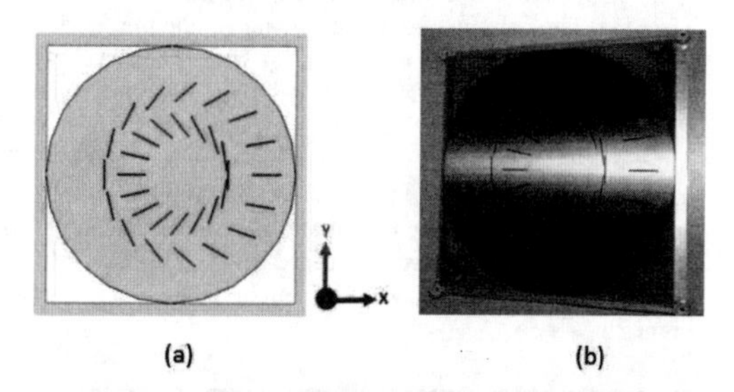

Figure 2: The slots arrangement on the surface of Air Gap RLSA (a) simulated, (b) fabricated

III. Antenna Development

The designed antenna has been developed as shown in Figure 1 and Figure 2(b). The slot has been constructed as shown in Figure 3 base on FR4 board. The arrow showed the wave movement to the air. However, the wave has to pass through the dielectric substrate that might provide the resistant value to the circuit.

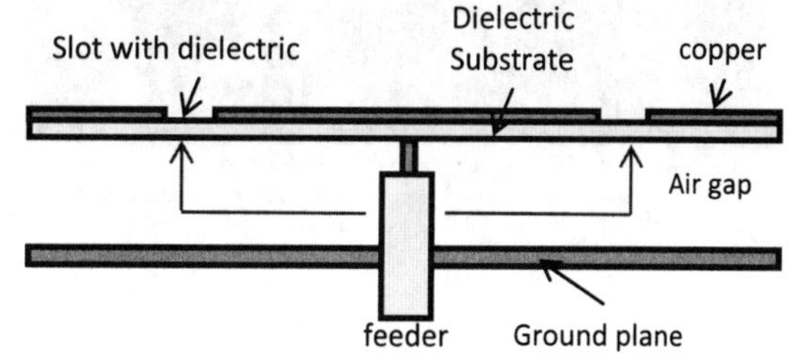

Figure 3: Slot antenna with dielectric substrate

978-7-5641-4279-7

Another approach to maximize the wave excitation to the air is to provide the hollow structure on the slot. The slot hole has been constructed as shown in Figure 4. It is expected that the wave will be fully excite to the air because the resistant element has been remove.

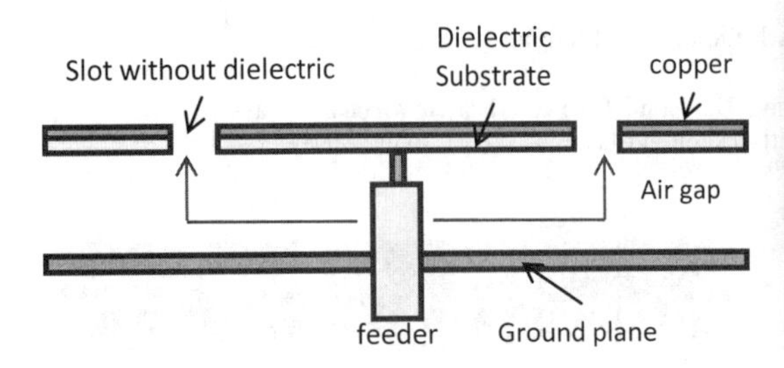

Figure 4: Slot antenna without dielectric substrate

The hollow slot has been constructed using DMF Milling Machine. The computerized machine has made the fabrication process become more easier. The FR4 board with hollow slot shown in Figure 5. A 2mm drill has been used to create the slots. The machine read the CATIA drawing file from it computer. The drawing originally drawed using CST softeware before converted to CATIA drawing file. Using this high end machine, an accurate slot dimention was succesfully constructed. The hollow slot board then installed into the antenna platform and become a complete antenna as shown in Figure 6.

Figure 5: Slot antenna without dielectric substrate fabricated using DMF Milling Machine

Figure 6: RLSA antenna with slot without dielectric substrate

IV. Result and Discussion

The concept of slot with and without dielectric was simulated using CST software. The simulated results show a smooth graph on slot with dielectric but not the slot without dielectric. The slot with dielectric performed better on reflection coefficient. The return loss of antenna slot with dielectric is from 5.3-6.9 GHz while the return loss antenna slot without dielectric are from 5.1-6.3GHz. The return loss at 5.8GHz antenna slot with dielectric is 16dB while antenna slot without dielectric is 12dB. However, the return loss on antenna slot without dielectric seems close to real environment. The researchers decided to proceed to fabrication and measured the real performance of the antenna.

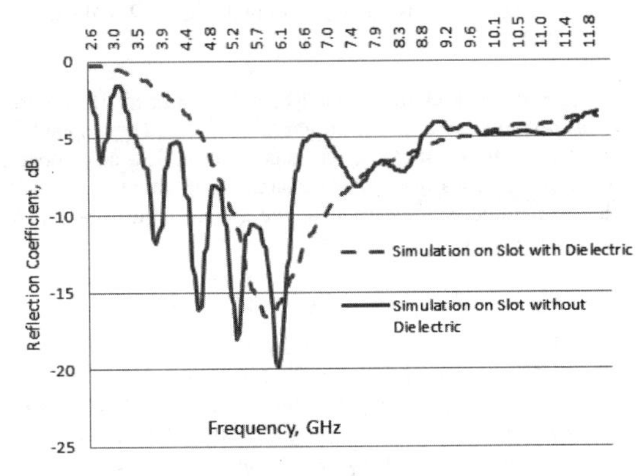

Figure 7: Simulated reflection coefficient of RLSA antenna slot with and without dielectric substrate

The fabricated antenna without slot has been measured it performances. A comparison on simulated and fabricated result for antenna slot without dielectric has been performed and presented in Figure 8. The fabricated results show a better performance. The return loss of the antenna is from 4.2-6.6 GHz. The 44% bandwidth recorded. It is 23% better than simulation. The return loss at 5.8 GHz for fabricated result is 24dB which is 12 dB better than simulated result.

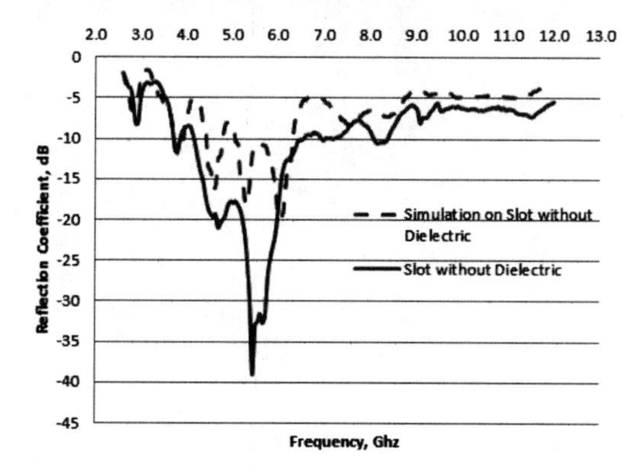

Figure 8: Simulated vs measured reflection coefficient of RLSA antenna slot without dielectric substrate

The antenna slot with and without dielectric are measured their performance and presented in Figure 9. The reflection coefficient result on antenna slot without dielectric demonstrate better performance as compare to antenna slot with dielectric.

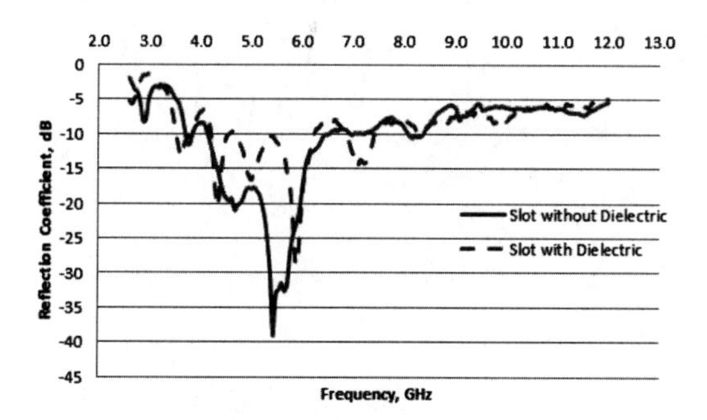

Figure 9: Measured reflection coefficient of fabricated RLSA antenna slot with and without dielectric substrate

Antenna slot without dielectric performed return loss from 4.2-6.6 GHz while antenna slot with dielectric performed from 4.2-6.2GHz. The 6% bandwidth improvement recorded for te antenna slot without dielectric. The return loss on that region show the antenna slot without dielectric is more dominant while antenna with dielectric demonstrated a few band notch and fluctuation. The return loss at 5.8 GHz for antenna without slot is 24dB which is only 2dB better than antenna slot with dielectric

The overall results are compressed on Table 1. It is show that the simulation performance show the antenna with dielectric slot perform better than antenna without dielectric slot. However, the measured performance show antenna without dielectric slot performed better.

Table 1: Comparison on simulated vs measured performances of RLSA antenna withy design slot with and without dielectric substrate

Design	Slot with Dielectric		Slot Without Dielectric	
Parameter Measured	Bandwidth	Return Loss at 5.8GHz	Bandwidth	Return Loss at 5.8GHz
Simulation	5.3-6.9 GHz, (26%)	16dB	5.1-6.3 GHz, (21%)	12dB
Measurement	4.2-6.2 GHz, (38%)	22dB	4.2-6.6 GHz, (44%)	24dB

V. CONCLUSION

It is show that the antenna without slot performed a better bandwidth and return loss. The return loss on the perform region also very dominant. The dielectric substrate has contributed a resistance element to the circuit. It has created resonant and band notch to the result. When this resistance element been removed from the circuit, it is show the wave has been smoothly travel through the slots. In overall, this research concluded that the slot design without dielectric has improve the bandwidth and return loss of the RLSA antenna.

VI. Acknwoledgement

The authors would like to acknowledge and express sincere appreciation to Universiti Teknologi Malaysia and Ministry of Higher Education (MOHE) for financing the research project. Appreciation also goes to Universiti Teknikal Malaysia Melaka and Ministry of Higher Education Malaysia for funding the author's scholarship.

The authors also would like to acknowledge Advance Manufacturing Centre, Universiti Teknikal Malaysia Melaka and Radiation Laboratory, Universiti Malaysia Perlis for the equipment assistance on this research.

REFFERENCES

[1] T. S. Lim, A. R. Tharek, W. A. Wan Khairuddin, and A. Hasnain, "Prototypes development for reflection canceling slot design of radial line slot array (RLSA) antenna for direct broadcast satellite reception," in *Applied Electromagnetics, 2003. APACE 2003. Asia-Pacific Conference on*, 2003, pp. 34-37.

[2] M. I. Imran and A. R. Tharek, "Radial line slot antenna development for outdoor point to point application at 5.8GHz band," in *RF and Microwave Conference, 2004. RFM 2004. Proceedings*, 2004, pp. 103-105.

[3] M. I. Imran, A. Riduan, A. R. Tharek, and A. Hasnain, "Beam squinted Radial Line Slot Array antenna (RLSA) design for point-to-point WLAN application," in *Applied Electromagnetics, 2007. APACE 2007. Asia-Pacific Conference on*, 2007, pp. 1-4.

[4] M. I. Imran, A. R. Tharek, and A. Hasnain, "An optimization of Beam Squinted Radial Line Slot Array Antenna design at 5.8 GHz," in *RF and Microwave Conference, 2008. RFM 2008. IEEE International*, 2008, pp. 139-142.

[5] M. Unno, J. Hirokawa, and M. Ando, "A Double-Layer Dipole-Array Polarizer with a Low-Sidelobe Radial Line Slot Antenna," in *Microwave Conference, 2007. APMC 2007. Asia-Pacific*, 2007, pp. 1-4.

[6] J. I. Herranz-Herruzo, A. Valero-Nogueira, E. Alfonso-Alos, and D. Sanchez-Escuderos, "New topologies of radial-line slot-dipole array antennas," in *Antennas and Propagation, 2006. EuCAP 2006. First European Conference on*, 2006, pp. 1-5.

[7] M. R. ul Islam and T. A. Rahman, "Novel and simple design of multi layer radial line slot array (RLSA) antenna using FR-4 substrate," in *Electromagnetic Compatibility and 19th International Zurich Symposium on Electromagnetic Compatibility, 2008. APEMC 2008. Asia-Pacific Symposium on*, 2008, pp. 843-846.

[8] M. F. Jamlos, O. A. Aziz, T. A. Rahman, M. R. Kamarudin, P. Saad, M. T. Ali, and M. N. Md Tan, "A Reconfigurable Radial Line Slot Array (RLSA) Antenna for Beam Shape and Broadside Application," *Journal of Electromagnetic Waves and Applications,* vol. 24, pp. 1171-1182, 2010/01/01 2010.

[9] M. F. Jamlos, O. A. Aziz, T. A. Rahman, M. R. Kamarudin, P. Saad, M. T. Ali, and M. N. Md Tan, "A Beam Steering Radial Line Slot Array (RLSA) Antenna with Reconfigurable Operating Frequency," *Journal of Electromagnetic Waves and Applications,* vol. 24, pp. 1079-1088, 2010/01/01 2010.

[10] I.M. Ibrahim, Tharek A. R, M.I. Sabran" Wide Band Open Ended Air Gap RLSA Antenna at 5.8GHz Frequency Band", presented in 2012 IEEE International Conference on Wireless Information Technology and System (ICWITS2012) Hawaii, United States, 11-16 November 2012.

[11] I.M. Ibrahim, Tharek A. R, P. Teddy,U. Kesavan" Wide Band Open Ended Air Gap RLSA Antenna at 26GHz Frequency Band", Accepted at PIERS Conference in Taipei, Taiwan, March 2013

A Broadband Center-Fed Circular Microstrip Monopolar Patch Antenna with U-Slots

Yue Kang, Juhua Liu, Zhixi Liang, and Yunliang Long
Department of Electronics & Communication Engineering
Sun Yat-Sen University, Guangzhou, 510006, P. R. of China
Kangyue945@126.com

***Abstract*-A novel center-fed circular microstrip patch antenna is presented. The proposed antenna consists of a circular ground plane and a circular patch antenna. A set of concentric U-shape slots are etched on the patch to improve the operating bandwidth. With the proposed structure, a broad corresponding impedance bandwidth of 15.4% is obtained from 2.3 GHz to 2.72 GHz. A monopole like radiation pattern is obtained and the average gain is about 6dBi. Details of the proposed antenna including reflection coefficient, radiation patterns are presented and discussed.**

***Index Terms*— circular patch antenna (CPA), microstrip antenna, U-Slot**

I. INTRODUCTION

Monopole antennas are widely used in the wireless communication systems since they radiate an omnidirectional pattern in the horizontal plane. Typically, monopole antennas are in the form of a straight rod and placed vertically with a height of λ/4 [1, 2]. Due to low cost and easy to manufacture, planar monopole antennas have been achieved. These planar monopole antennas are printed on a substrate for mobile phone, laptop and other applications [3-5]. However, in these designs, a relatively large height is still needed. To obtain a monopole like radiation pattern, researches have been made investigations on the circular microstrip patch antenna [6-10]. In Reference [6], a circular microstrip antenna equivalent to simple monopole was designed. The antenna has a very narrow band width of about 1.5 % for a profile of 0.0152λ. To reduce the size of the circular patch, slots [7] and shorting via [8] are added to the antenna. Apart from miniaturization, multiband performance has also been achieved [9]. However, the narrow bandwidth is still limited for wireless communication application.

Recently, center-fed circular microstrip patch antennas excited in the TM01 or TM02 mode have been proposed [10, 11]. These designs have the advantages of low profile and wide operating band. Reference [10] achieved a 12.8% corresponding bandwidth, but with a large radius of about 0.72λ. Reference [11] improved the corresponding bandwidth to 18%, but too many shorting pins are difficult to manufacture. Adding U-slot is an efficient and convenient way to improve bandwidth for microstrip patch antenna. U-slot on the patch introduces an additional resonant mode near the fundamental mode resonance frequency. Reference [12] made a conclusion that disk patch loaded with U-slot exhibited broad bandwidth. But the antenna in reference [12], fed by an off-center coaxial probe, excited a broadside radiation.

In this paper, we introduce a novel center-fed circular microstrip patch antenna with U-slots for bandwidth improvement. To maintain the omnidirectional radiation pattern of the circular patch antenna, a set of concentric U-shape slots were etched on the patch appropriately. With the proposed structure, an extra resonant mode is excited near the fundamental resonant frequency. The achieved corresponding bandwidth is improved to 15.4%. The parameters of the U-slot are studied with the simulated results. Details of the proposed antenna including reflection coefficient, radiation patterns are presented and discussed.

II. ANTENNA GEOMETRY

Fig. 1 shows the geometry of the center-fed circular microstrip patch antenna with U-slots, which is printed on a substrate with a thickness of 10 mm (0.084 λ) and a dielectric constant of 2.33. The radius of the ground plane (R_g) is 90 mm (0.756 λ) and the radius of the patch (R_p) is 48 mm (0.403λ). A 50Ω coaxial connector is attached through the center of the ground.

On the top of the patch, a set of N U-slots opening to outwards had been concentrically etched around the patch. The inner edge of each u-slot is at a distance (a) to from the

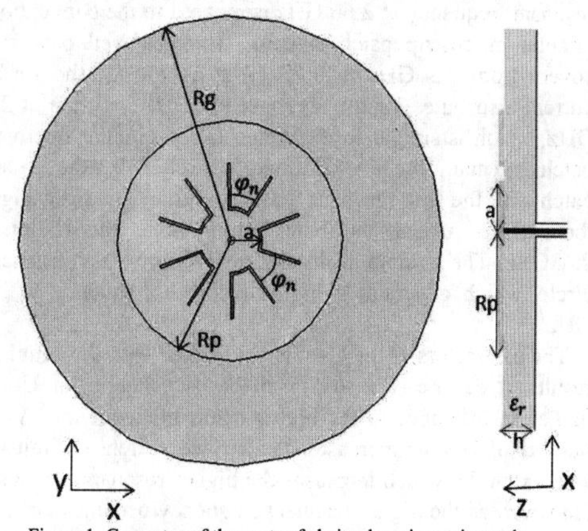

Figure 1. Geometry of the center-fed circular microstrip patch antenna with U-slot.

978-7-5641-4279-7

TABLE I
PARAMETERS OF THE PROPOSED ANTENNA

Parameter	Value
N	5
R_p	48 mm
a	12.5 mm
R_g	90 mm
L	19 mm
w	1 mm
h	10 mm
ε_r	2.33

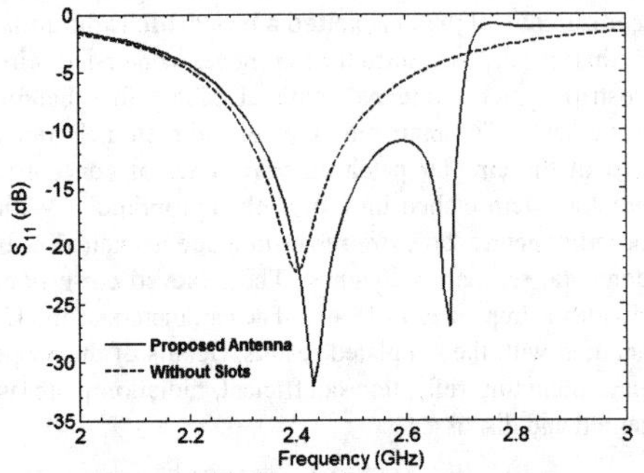

Figure 2. Reflection coefficients of the proposed antenna and without slots

center of the circular patch. Other parameters of U-slots are: length of the slot (L), the number of U-slot N, base corner angle ($\varphi_n = \pi/N$) and width of the slot (w). Through optimization and adjustment based on simulated result, the final antenna parameters are shown in Table I.

III. Results and Discussion

As shown in Fig. 2, the proposed antenna generates an extra resonant frequency at 2.66 GHz compared to the conventional circular microstrip patch antenna. The achieved bandwidth covers from 2.3 GHz to 2.72 GHz. In Fig. 3, the surface current distributes almost invariant in the Ø-direction at 2.44 GHz, which is similar to the conventional circular microstrip patch antenna. The TM02 mode is excited by the circular patch with the first resonant frequency (2.44 GHz). In Fig. 4, the surface current distributes mainly near the U-slots at 2.68GHz. The U-slots limits the surface current in a smaller circle, which generates a higher resonant frequency at 2.68 GHz.

The parameters of the U-slots are studied with the simulated results. It can be seen in Fig. 5 that the number of U-slots makes an influence to the higher resonant frequency. As the number of U-slots increases, the surface current is limit in a smaller circle, which increases the higher resonant frequency. However, as the higher resonant frequency becomes far away from the fundamental resonant frequency, the impedance match gets worse, that because the higher resonant frequency

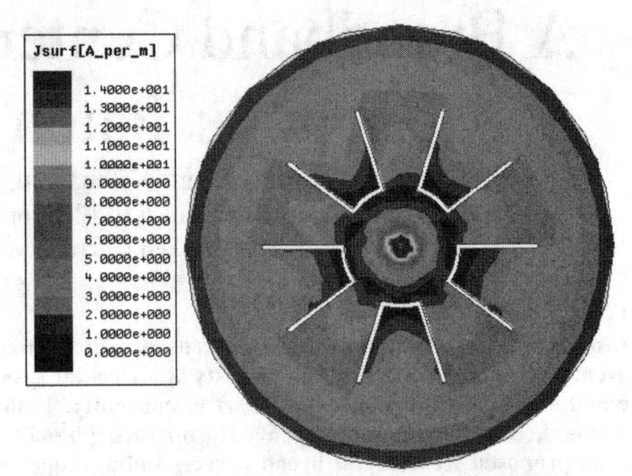

Figure 3. Simulated current distributions at 2.44GHz.

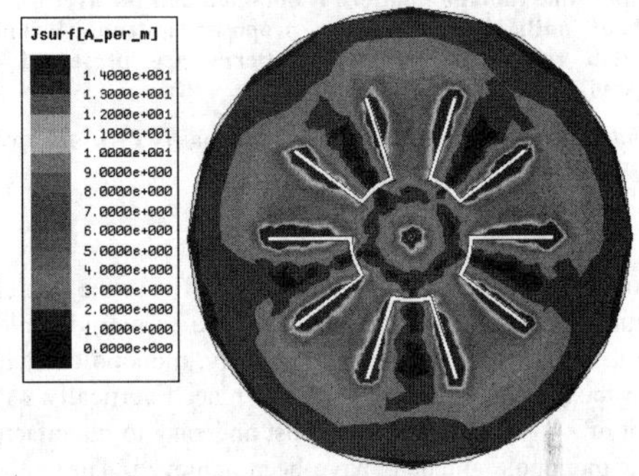

Figure 4. Simulated current distributions at 2.68GHz.

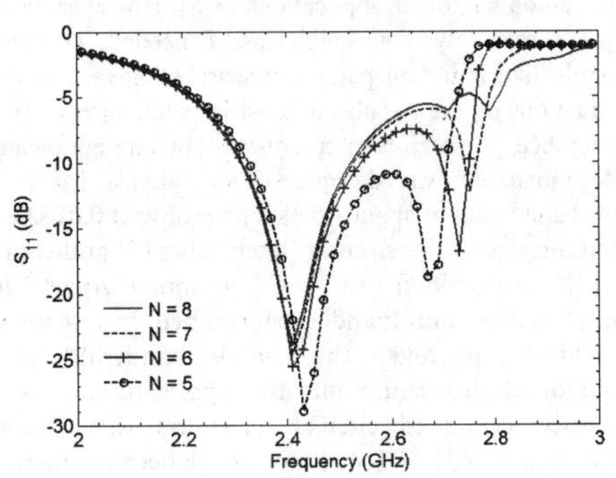

Figure 5. Simulated reflection coefficients for different number of the U-slots, other parameters are the same in Table I.

is excited by the fundamental mode. In Fig. 6, the higher resonant frequency falls, when the distance (a) between the U-slots and the center of the patch in opposite side increases. As the distance (a) increase, the limitation of the U-slots becomes weaker and the surface current can distribute in a larger area. The same effect can be seen in Fig. 7. As the length (L) of the U-slot increase, the higher resonant frequency falls. In general, the parameters of the U-slots control the limitation of the surface current in the second resonant mode.

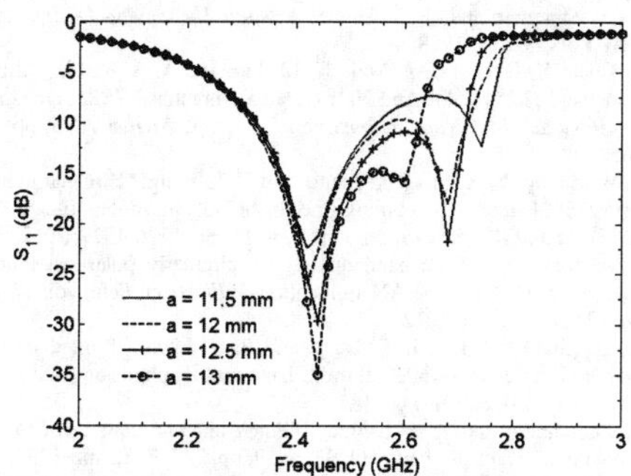

Figure 6. Simulated reflection coefficients for different distance of the U-slots, other parameters are the same in Table I.

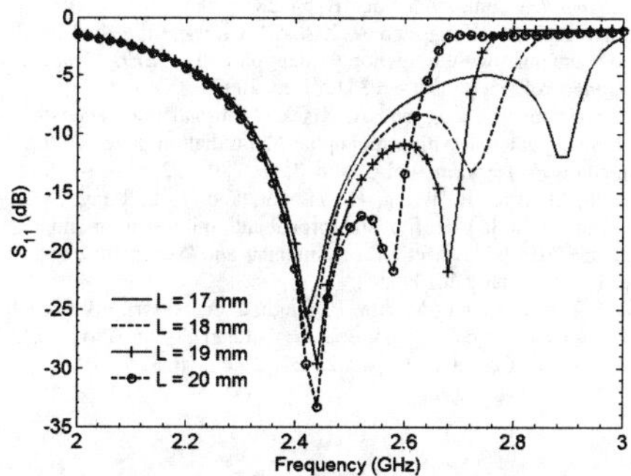

Figure 7. Simulated reflection coefficients for different length of the U-slots, other parameters are the same in Table I.

Fig. 8 shows the simulated radiation patterns at 2.44 GHz and 2.68 GHz for the proposed antenna. The simulated radiation patterns of the proposed at two different frequencies are identical to the conventional monopole antenna. A omnidirectional radiation pattern is achieved in the H-plane. The patterns at these two frequencies are similar. The Simulated results show that the cross polarization is 20 dB below the co-polarization level. When the frequency increases, the cross polarization becomes stronger. As shown in Fig. 9, the gain has a relatively stable value about 6 dBi within the bandwidth from 2.3 GHz to 2.72GHz. The maximum gain is achieved at 2.64 GHz with 7dBi.

IV. Conclusion

A new microsrtip circular antenna fed at center is presented. This antenna has a monopole like radiation pattern over the whole operating band. A set of U-slots are cut inside at appropriate position the patch concentrically, which enhance the bandwidth. The parameters of U-slots have been studied with the simulated results. It obtains a wide impedance corresponding bandwidth of 16.3%, from 2.3 to 2.72 GHz.

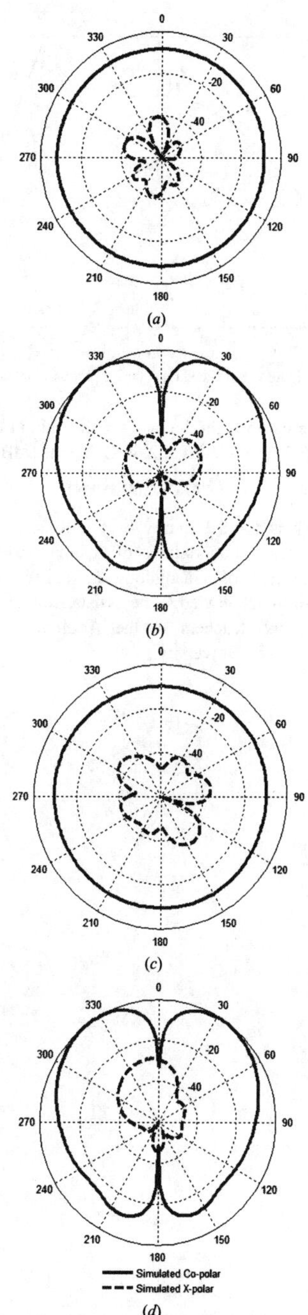

Figure 8. Simulated H-plane radiation patterns of the proposed antenna at: (*a*) 2.44 GHz; (*c*) 2.68 GHz. Simulated E-plane patterns of the proposed antenna at: (*b*) 2.44GHz; (*d*) 2.68 GHz.

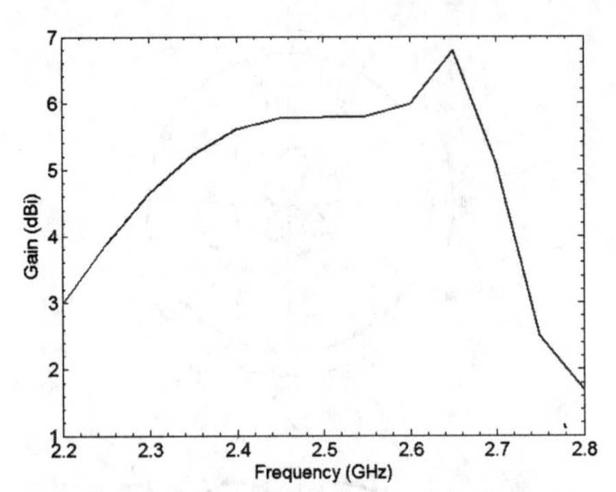

Figure 9. Simulated Gain for the proposed antenna.

The average gain within the bandwidth is about 6 dBi and the maximum gain is achieved at 2.64 GHz with 7dBi.

ACKNOWLEDGMENT

This work is funded jointly by the Research Program of Natural Science Foundation of China (61172026) and Research Project of Guangdong (2012B091100050). The authors wish to acknowledge the assistance and support of all the fellows and teachers in the Applied Electromagnetic Lab, Sun Yat-sen University.

REFERENCES

[1] W. L. Stutzman and G. A. Thiele, *Antenna Theory and Design*, 2nd ed. New York: Wiley, 1998.

[2] R. Chair, K. F. Lee, C. L. Mak, K. M. Luk and A. A. Kishk, "Miniature Wideband Half U-Slot And Half E Patch Antennas," *IEEE Transactions on Antennas And Propagations*, vol. 52, no. 8, August 2005, pp. 2645-2652..

[3] Z. W. Zhong, Y. X. Li, Z. X. Liang, and Y. L. Long, "Biplanar monopole with DSPSL feed and coupling line for broadband mobile phone," *IEEE Antennas and Wireless Propa. Lett.*, vol. 11, pp. 1326-1329, 2012.

[4] S. A. Rezaeieh, "Dual band dual sense circularly polarized monopole antenna for GPS and WLAN applications," *Electron. Lett.*, vol. 47, no. 12, pp. 1212-1214, Oct. 2012.

[5] D. B. Lin, H. P. Lin, I. T. Tang, and P. S. Chen, "Printed inverted-F monopole antenna for internal multi-band mobile phone antenna," in Proc. *VTC*, May. 15-18, 2011, pp. 1-5.

[6] L. Economou. and R. J. Langley, "Patch antenna equivalent to simple monopole," *Electron. Lett.*, vol. 33, no. 9, pp. 727–728, Apr. 1997.

[7] C. B. Ravipati, "Compact circular microstrip antenna for conical pattern," in Proc. *IEEE AP-S Int. Symp.* Dig., Jun. 2004, vol. 4, pp. 3820–3823.

[8] C. B. Ravipati, D. R. Jackson, and H. Xu, " Center-fed microstrip antennaswith shorting vias for miniaturization, " in Proc. *IEEE AP-S Int.Symp. Dig.*, Jul. 2005, vol. 3B, pp. 281 - 284.

[9] A. Al-Zoubi, F. Yang, and A. Kishk, "A low-profile dual band surface wave antenna with a monopole-like pattern," *IEEE Trans. Antennas Propag.*, vol. 55, pp. 3404 - 3412, Dec. 2007.

[10] A. Al-Zoubi, F. Yang, and A. Kishk, "A broadband center-fed circular patch-ring antenna with a monopole like radiation pattern," *IEEE Trans. Antennas Propagation*, vol. 57, no. 3, pp. 789–792, Mar. 2009.

[11] J. Liu, Q. Xue, H. Wong, H. W. Lai, and Y. L. Long, "Design and analysis of a low-profile and broadband microstrip monopolar patch antenna," IEEE Transactions on Antennas and Propagation, vol. 61, no. 1, pp. 11-18, January 2013.

[12] N. P. Yadav, Anurag Mishra, P. Singh, J. A. Ansari, "A Broadband U-slot loaded circular disk patch Antenna," in Proc. ELECTRO International Conference, Dec. 22-24, 2009 , pp. 317- 319.

Design of a Compact Multi-Band Circularly-Polarized Microstrip Antenna

YangJie, Lu Chunlan, Shen Juhong
CCE, PLAUST
Nanjing 210007, China

***Abstract*-A multi-band circularly-polarized microstrip antenna is designed in this paper. The antenna is made up of two patches. Both the patches are placed on one substrate. The centre patch with four T-shaped slits is aimed to obtain dual-band operation. The outer patch with two square slits can work at another band. The antenna is excited with two feeding points to satisfy the port isolation. The antenna is compact and easy to fabricate. Simulated and measured results show that the antenna can generate circular polarized in three bands and can satisfy satellite communication.**

***Key words*-microstrip antenna, multi-band, circularly-polarized.**

I. INTRODUCTION

With the development of RF technology, satellite communication systems have been widely used in recent years [1]. For these systems, circular polarized antennas are needed due to their insensitivity to ionospheric polarization rotation[2]. In order to increase communication capacity and simplify the system structure, the antenna needs to receive and radiate signals at the same time. So, multi-band circularly-polarized antennas become more and more important. As satellite terminal antennas, small size and high isolation are needed. Microstrip antenna has been widely used due to its small size, lightweight, low profile, low cost and easy to conform. There are many literatures[3-5] in dual-band circularly-polarized microstrip antenna, but seldom in three band circularly-polarized microstirp antenna.

In literature[6], four T-shaped slits and four L-shaped slot are used to generate dual-band operation. In this paper, a three-band circularly-polarized is designed. The antenna only has one layer. And the patches are feed by two probes independently. The antenna is compact and easy to fabricate. The simulated and measured results show that the antenna has good electrical properties.

II. ANTENNA STRUCTURE

The structure of the proposed antenna is shown in Fig. 1. The antenna has only one layer. But the patches are made up of two parts. The centre part is a square patch with four T-shaped slits inserted at the radiating edges of the patch. This patch can generate dual-band operation, and works at 1.3GHz and 3GHz. The other part surrounding the centre patch is a square ring with two square slits, and works at 1.98GHz. The patches are placed on substrate with a dielectric constant of $\varepsilon_r = 4.4$ with thickness of h=1.5mm. The antenna has two feeding points. The centre patch and the outer patch are feed by probes separately. So, high isolation can be achieved.

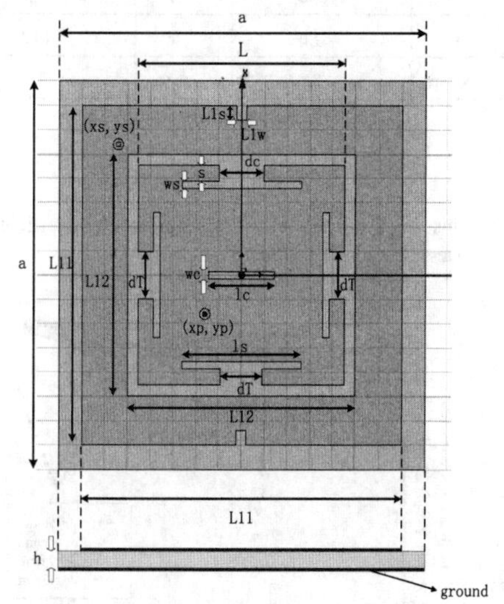

Figure 1. Configuration of the proposed antenna

The centre patch is a square patch with four T-shaped slits. So it can be regarded as a regular patch and the size can be calculated approximately by the following equations[7] :

$$f = \frac{c}{2w_e\sqrt{\varepsilon_e}} = \frac{c}{2(w+2\Delta l)\sqrt{\varepsilon_e}},$$

where f is the resonant frequency,

$$\varepsilon_e = \frac{\varepsilon_r+1}{2} + \frac{\varepsilon_r-1}{2}\left(1+\frac{10h}{w}\right)^{-1/2},$$

$$\Delta l = 0.412h\frac{(\varepsilon_e+0.3)(w/h+0.264)}{(\varepsilon_e-0.258)(w/h+0.8)}$$

$$w_e = w + 2\Delta l.$$

The size calculated by the equations is not the exact and final characteristics. The proposed antenna has been simulated

978-7-5641-4279-7

using HFSS software. The final physical characteristic of the antenna is given in Table 1.

PARAMETERS			
	VALUE(mm)		VALUE(mm)
L11	69	L12	49.5
L1w	2	L1s	3.1
a	85	h	1.6
L	45.2	ls	26
dc	9	dT	9.8
ws	1.6	s	3.2
wc	1.6	lc	14.2
xp,yp	8,8	xs,ys	27,27

Table 1: Physical characteristics of the antenna

III. PARAMETRIC STUDY

In this section, the effect on reflection coefficient of parameters dc and dT is studied.

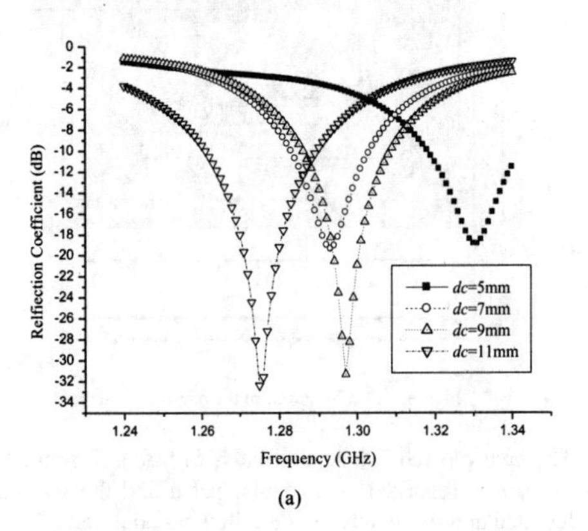

(a)

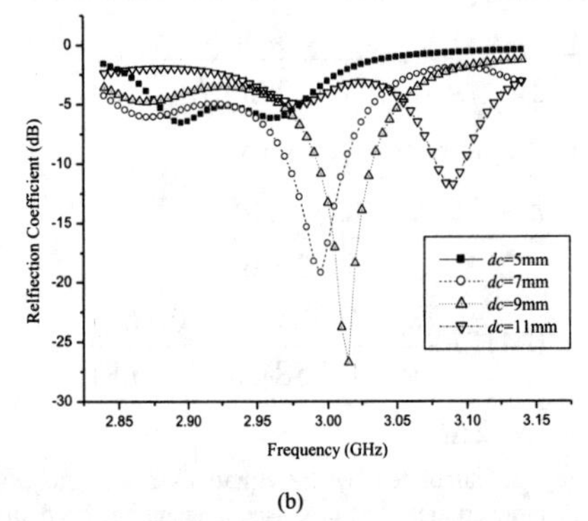

(b)

Figure 2. Simulated refletion coefficient versus dc of the antenna

In Fig.2, the reflection coefficient versus frequency for different values of dc is presented when dT is 9.8mm. In low frequency, the increase of "dc" causes a decrease of the reflection coefficient. While in high frequency, the increase of "dc" causes an increase of the reflection coefficient. When the antenna works at 1.3GHz and 3.0GHz, dc is around 9mm.

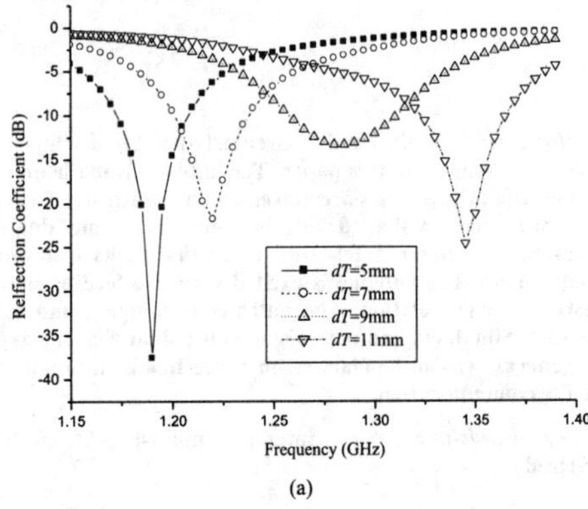

(a)

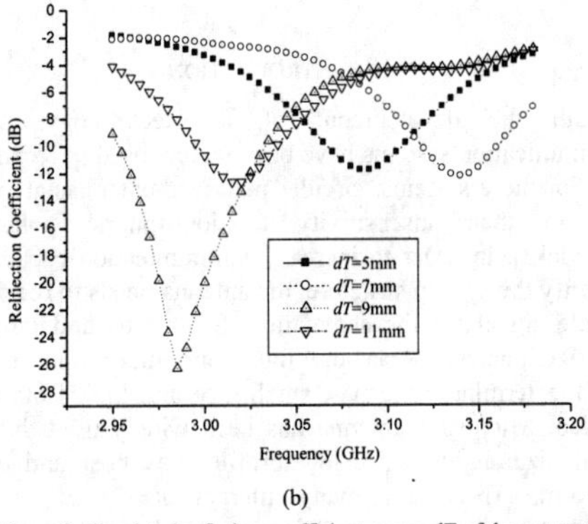

(b)

Figure 3. Simulated refletion coefficient versus dT of the antenna

In Fig.3, the reflection coefficient versus frequency for different values of dT is presented when dc is 9mm. We can note that in low frequency, the increase of "dT" causes a increase of the reflection coefficient. But in high frequency, there is no regularity with the change of "dT". When the value of dT is from 9mm to 11mm, the antenna can works at 1.3GHz and 3.0GHz. From further optimization, the value of dT is decided to 9.8mm.

IV. EXPERIMENTAL RESULT

Simulation is based on the software of HFSS. The antenna is fabricated and measured at last. The photograph of the antenna is shown in Fig. 4.

Figure 4. Photograph of the antenna

The simulated and measured results of reflection coefficient versus frequency are shown in Fig. 5. When the reflection coefficient is lower than -10dB, the simulated bandwidths of the antenna are 1.278GHz-1.315GHz, 1.965GHz-1.995GHz and 2.968GHz-3.032GHz. However, the measured results of reflection coefficient are 1.267GHz-1.283GHz, 1.972GHz-1.998GHz and 2.955GHz-2.985GHz. The difference between measured and simulated results in low frequency and high frequency is bigger. The measured bands are lower than the simulated. The reason is that there are many slits on the patch, and high machining precision is requirement. There are errors in fabricating.

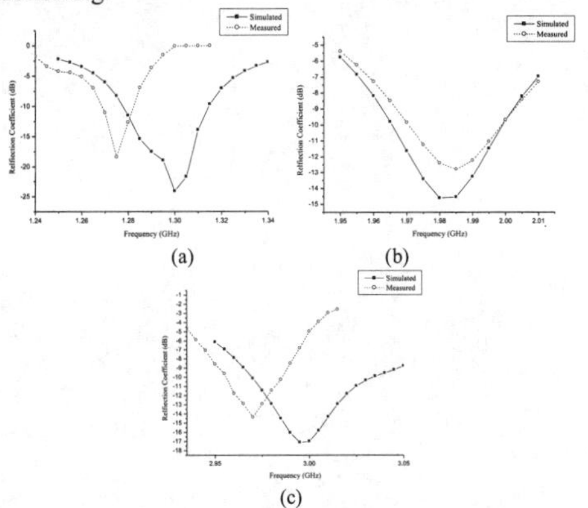

Figure 5. Simulated and measured reflection coefficient versus frequency of the antenna

Fig. 6 gives out the simulated and measured results of axial ratio versus frequency of the antenna. From Fig. 6, we can conclude that when the axial ratio is lower than 3dB, the simulated bandwidths of axial ratio are 1.297GHz-1.308GHz, 1.967GHz-1.994GHz and 2.955GHz-3.035GHz. The measured bandwidths of axial ratio are 1.277GHz-1.283GHz, 1.975GHz-1.993GHz and 2.942GHz-2.977GHz. When the antenna works at 1.98GHz, the bandwidth of axial ratio is narrower than that at 1.3GHz and 3GHz.

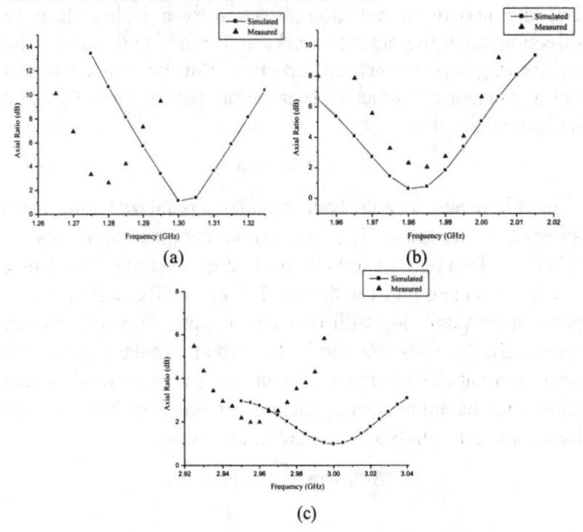

Figure 6. Simulated and measured axial ratio versus frequency of the antenna

From Fig. 5 and Fig. 6, we can obtain that the bandwidth of the outer patch, including impendence bandwidth and axial ratio bandwidth， is not as wide as the centre patch. This is because that the bandwidth of the square ring is narrow, and not wider than the square patch.

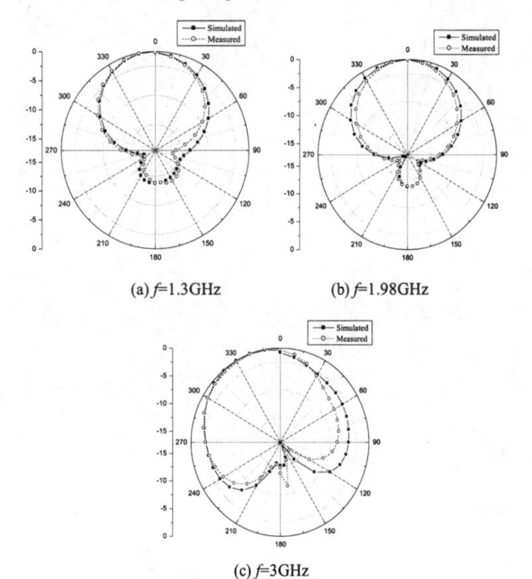

Figure 7. Simulated and measured normalized radiation patterns of the proposed antenna at xoz plane

The simulated and measured normalized radiation patterns at xoz plane are shown in Fig. 7. When the antenna works at 1.98GHz, the maximum radiation direction is +z direction. But the maximum radiation direction is a little offset +z direction when the antenna works at 1.3GHz and 3GHz. The square ring is a symmetrical structure. But the centre patch is not a symmetrical structure because the parameters of dc and dT are not equal.

V. CONCLUSION

In this paper, a multi-band circularly-polarized microstrip antenna is designed. The antenna is compact and easy to fabricate. Two patches are placed on one substrate. The centre patch is a square patch with four T-shaped slits, and the outer patch is a square ring with two square slots. The antenna can work 1.3GHz, 1.98GHz and 3GHz. Good circularly polarized operation can be generated. Simulated and measured results show that the antenna can generate circular polarized in three bands and can satisfy satellite communication.

ACKNOWLEDGMENT

This work is supported by National Natural Science Foundation of China (No. 61271103).

REFERENCES

[1] J. V. Evans. "Satellite systems for personal communications", IEEE Antennas Propag. Mag., vol.39, pp. 7-20, 1997.

[2] Yoann Letestu and Ala Sharaiha. "Size reduced multi-band printed quadrifilar helical antenna", IEEE Trans. Antennas Propag., vol. 59, pp. 3138-3143, 2011.

[3] Xiulong Bao, M. J. Ammann. "Dual-Frequency Dual-Sense Circularly-Polarized Slot Antenna Fed by Microstrip Line", IEEE Transactions. Antennas Propagation, vol. 56, pp. 645-649, 2008.

[4] K. P. Yang and K. L. Wong. "Dual-band circularly-polarized square microstrip antenna", IEEE Trans. Antennas Propagation, vol.49, pp. 377-382, 2001.

[5] X. L. Bao and M. J. Aman. "Compact annular-ring embedded circular patch antenna with a cross-slot ground plane for circular polarization", Electronics Letters, vol. 42, pp. 192-193. 2006

[6] Behzad Naimian, Mohammad H. Neshati. "A New Novel Dual-Band Circularly-Polarized Microstrip Antenna", Electrical Engineering (ICEE), 2011 19th Iranian Conference on, 2011.

[7] Zhang Jun, Liu Ke-cheng, Zhang Xiang-yi and He Chong-jun. "Microstrip antenna theory and engineering", 1988.

A Pattern Reconfigurable Quasi-Yagi Antenna with Compact Size

Pengkai Li[1], Zhenhai Shao[1], Yujian Cheng[2], Quan Wang[1], Long Li[1] and Yongmao Huang[1]
[1]Greating-UESTC Joint Experiment Engineering Center,
[2]Fundamental Science on Extreme High Frequency Laboratory,
University of Electronic Science and Technology of China,
Chengdu, 611731, P. R. China

Abstract- **A compact pattern reconfigurable quasi-Yagi antenna is presented. It consists of a driven element, which is simply realized by a microstrip monopole and only one parasitic element with four PIN diodes integrated on the arm. The effective electrical length of the parasitic element is controlled by changing the biasing voltages. This configuration allows two end-fire patterns with relatively high gain and opposite beam direction and an omnidirectional pattern to be maintained across a wide bandwidth. A commonly available impedance bandwidth of 12.65% is achieved corresponding to the same center operating frequency of 2.45 GHz for the three modes. As this antenna has reconfigurable pattern and relatively small size, it is very suitable for the indoor wireless communication systems.**

I. Introduction

Pattern diversity antenna has gained extreme attention in indoor wireless communication. Although most wireless local area network (WLAN) applications utilize omnidirectional antennas, directional and quasi end-fire antennas, such as Yagi antenna, have been employed to suppress unwanted radio frequency (RF) emissions as well as unwanted interference in other directions [1]. In addition, reconfigurable antenna patterns provide pattern diversity that could be used to provide dynamic radiation coverage and mitigate multi-path fading [2], [3]. The Yagi-Uda antenna is one of the most popular end-fire antennas which can be designed to achieve a medium gain with relatively low cross-polarization levels. A traditional Yagi-Uda antenna consists of a driven element, a reflector, and one or more directors. Many reconfigurable Yagi and quasi-Yagi antenna designs have been reported in the literature [4]-[8].

In this paper, a compact pattern reconfigurable quasi-Yagi antenna is proposed. The structure of the proposed antenna is simple and planar. It consists of a driven element, which is simply realized by a microstrip monopole and only one parasitic element with four PIN diodes integrated on the arm. The effective electrical length of the parasitic element is adjusted by changing the "on" and "off" state of the four PIN diodes to achieve three different radiation patterns, two end-fire patterns and an omnidirectional pattern over the same operating frequency. A well-designed microstrip feed monopole-like driven element is introduced to match the different impedance of the three operation modes [9]. The design procedure with integrated PIN diode switches and biasing circuitry is presented.

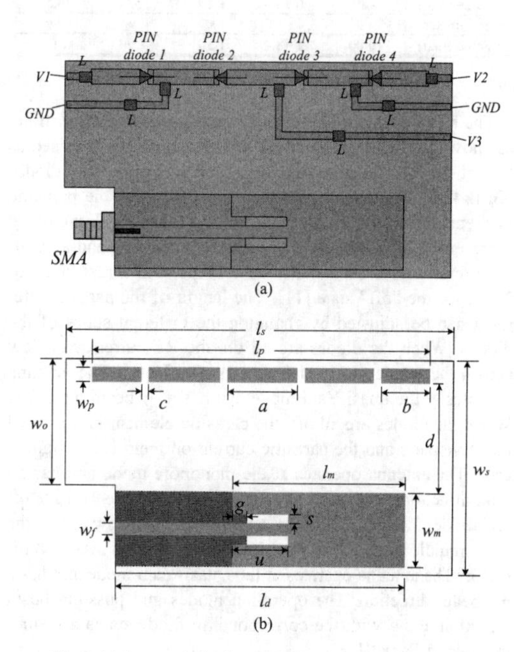

Fig. 1. (a) Antenna design layout. (b) Geometry and dimensions of the antenna

II. Antenna Design

A schematic layout and dimensions of the proposed antenna are shown in Fig. 1. The parameters and dimensions of the antenna are given in Table I. The antenna metallization is etched on both sides of a 0.8mm-thick FR4 substrate with a dielectric constant of 4.4, and total dimensions of 55.09 × 29.00mm^2. The top metallization is composed of a microstrip fed driven monopole element and a metal strip parasitic element divided into five parts by four PIN diodes. The bottom side metallization is a truncated ground plane with the same width w_m as monopole element. The printed monopole functions as a driven element with broadband characteristics fed by a microstrip line. The microstrip line has signal width w_f of 1.6 mm to achieve the 50 Ω characteristic impedance feed line.

978-7-5641-4279-7

The driven monopole element is broadband designed by cutting two slits on the feeding edge and well optimized to match the different impedance of the three operation modes. The proposed antenna has a very simple feeding structure without the need of any external balun, transition, tapered or truncated structures.

TABLE I
DIMENSIONS OF THE PROPOSED RECONFIGURABLE QUASI-YAGI ANTENNA

Parameter	l_s	w_s	l_d	l_p	w_p
Value (mm)	55.09	29.00	41.09	48.00	2.00
Parameter	a	b	c	d	l_m
Value (mm)	9.80	6.90	1.20	15.00	24.57
Parameter	w_m	g	s	u	w_f
Value (mm)	10.00	1.48	1.30	7.90	1.60

The bias line is printed on the upper surface of the substrate, as shown in Fig. 1(a). Eight 330nH inductors (L) are used as RF choke to separate the RF signal. Four PIN diodes (GMP4201, Microsemi) are used as switches in the parasitic element. According to the PIN diode datasheet [10], the diode represents a resistance of 2.3 Ω for the "on" state and a parallel circuit with a capacitance of 0.18 pF and a resistance of 30kΩ for the "off" state [11]. The length of the parasitic element can be adjusted by changing the different states of the diodes. When the diodes are all on, the RF current can flow across the whole arm which acts as a reflector. The antenna operates at the quasi-Yagi mode and has a $-y$ beam direction. When the diodes are all off, the parasitic element is separated into five parts and the parasitic current on them is not significant. The antenna operates at the monopole mode and has an omnidirectional radiation pattern. When the diode 1, 4 are off and diodes 2, 3 are on, the RF current can only flow across the three middle parts of the parasitic element which acts as a director. The antenna operates at the Quasi-Yagi mode and has a $+y$ beam direction. The operation modes and possible beam direction along with the corresponding diode states are summarized in Table II.

TABLE II
OPERATION STATES OF THE PROPOSED RECONFIGURABLE QUASI-YAGI ANTENNA

	PIN 1, 4	PIN 2, 3	Mode	Beam Direction
State 1	ON	ON	Quasi-Yagi	$-y$
State 2	OFF	OFF	Monopole	Omnidirectional
State 3	OFF	ON	Quasi-Yagi	$+y$

III. SIMULATED RESULTS AND DISCUSSION

Optimization of the antenna was carried out using HFSS 14. The simulated S_{11} for the three modes are given in Figs. 2. According to the simulation results, the 10-dB impedance bandwidth commonly available to all modes is 310 MHz from 2.35 to 2.66 GHz, which is 12.65% with respect to the designed frequency at 2.45GHz. As can be seen from Fig.2, the bandwidth of state 1 or state 3 is narrower than state 2. This is due to the reactance introduced by the parasitic element, which acts as a reflector or a director. At state 2, when all the diodes are off, the influence of the parasitic element is not notable, thus the antenna bandwidth is mainly depending on the driven monopole, which is broadband designed to overcome the impedance variety of different operating modes.

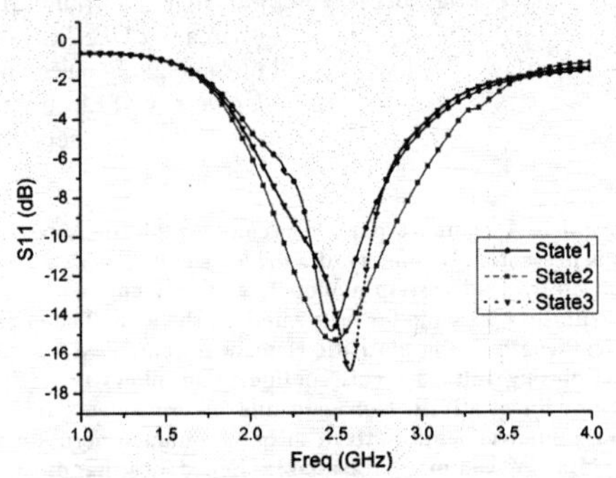

Fig. 2. Simulated S_{11} for the three different operating modes.

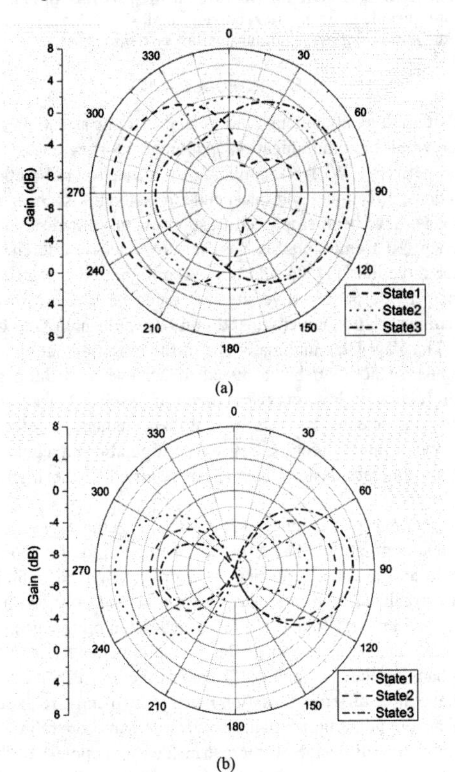

Fig. 3. Simulated radiation patterns for the three different operating modes. (a) y-z plane and (b) x-y plane.

The simulated radiation patterns (*y-z* and *x-y* plane) of the proposed antenna for three operating modes at 2.45 GHz are shown in Fig. 3. The simulated peak gains for the $-y$, omnidirectional and $+y$ modes are 5.53, 2.9 and 5.17 dBi, respectively. The front-to-back ratio is 6.3 dB for $-y$ mode and 5.6 dB for $+y$ mode. The maximum ripple level of *H*-plane for state 2 is 2.2 dB. Since the ripple level is quite low, this antenna at state2 has nearly omnidirectional *H*-plane pattern. To sum up, the proposed antenna has significant characteristics of pattern reconfiguration.

IV. Conclusion

A pattern reconfigurable quasi-Yagi antenna using integrated pin diodes is presented. The effective electrical length of the parasitic element is adjusted without changing the driver monopole when the operating frequency is constant by varying the biasing voltages. This configuration allows two endfire patterns with relatively high gain and opposite beam direction and an omnidirectional pattern to be maintained across a wide bandwidth. The impedance bandwidth are 12.65% at state 1, 25.51% at state 2 and 16.33% at state 3 with the same center operating frequency of 2.45 GHz. The simulated peak gain for the $-y$, omnidirectional and $+y$ modes is 5.5, 2.9 and 5.2 dBi, respectively. The pattern reconfigurable quasi-Yagi antenna has the advantages of compact and easy to manufacture, which makes it a good candidate for adaptive beam switching in indoor wireless applications.

References

[1] Z. Shi, R. P. Zheng, J. Ding, and C. J. Guo, "A Novel Pattern-Reconfigurable Antenna Using Switched Printed Elements," *IEEE Antennas Wireless Propag. Lett.*, vol. 11, pp. 1100–1103, 2012.

[2] O. N. Alrabadi, J. P. Carrier, and A. Kalis, "MIMO Transmission Using a Single RF Source: Theory and Antenna Design," *IEEE Trans. Antennas Propag.*, vol. 60, no. 2, pp. 654–664, Feb. 2012.

[3] Y. Y. Bai, S. Q. Xiao, M. C. Tang, Z. F. Ding, and B. Z. Wang, "Wide-Angle Scanning Phased Array With Pattern Reconfigurable Elements," *IEEE Trans. Antennas Propag.*, vol. 59, no. 11, pp. 4071–4076, Nov. 2011.

[4] Y. Cai, Y. J. Guo, and A. R. Weily, "A Frequency-Reconfigurable Quasi-Yagi Dipole Antenna," *IEEE Antennas Wireless Propag. Lett.*, vol. 9, pp. 883–886, 2010.

[5] H. A. Majid, M. K. A. Rahim, M. R. Hamid and M. F. Ismail, "Frequency and Pattern Reconfigurable Yagi Antenna," *Journal of Electromagnetic Waves and Applications*, vol. 26, no. 2-3, 379-389, 2012.

[6] J. W. K, S. Pyo, T. H. Lee, and Y. S. Kim, "Switchable Printed Yagi-Uda Antenna with Pattern Reconfiguration," *ETRI Journal*, vol. 31, no. 3, Jun. 2009.

[7] B. H. Sun, S. G. Zhou, Y. F. Wei, and Q. Z. Liu, "Modified Two-element Yagi-Uda Antenna with Tunable Beams," *Progress In Electromagnetics Research*, PIER 100, 175-187, 2010.

[8] S. Lim and H. Ling, "Design of electrically small, pattern reconfigurable Yagi antenna," *Electron. Lett.*, vol. 43, no. 24,1326-1327, Nov. 2007.

[9] H. D. Lu, L. M. Si and Y. Liu, "Compact planar microstrip-fed quasi-Yagi antenna," *Electron. Lett.*, vol. 48, no. 3, 140-141, Feb. 2012.

[10] GMP4201 datasheet. Available online at http://www.microsemi.com

[11] A. Edalati, and T. A. Denidni, "Frequency Selective Surfaces for Beam-Switching Applications," *IEEE Trans. Antennas Propag.*, vol. 61, no. 1, pp. 195–200, Jan. 2013.

TP-1(A)

October 24 (THU) PM

Room A

Body-central Antennas

Multi-Functional Small Antennas for Health Monitoring Systems

C. H. Lin, K. Ito, M. Takahashi, and K. Saito
Chiba University
1-33 Yayoi-cho, Inage-ku
Chiba 263-8522, Japan

***Abstract*-Body-centric wireless communications (BCWCs) have become an active area of research due to their wide range of applications. Therefore, as an interface between the transceiver and the propagation environment, antennas in these systems need to be carefully designed. Moreover, in this paper, we proposed two multi-functional antennas for on-body (10 MHz) and off-body (2.45 GHz ISM band) communications in health monitoring system. In on-body communications, the received voltage and electric-field distribution are analyzed. In addition, in off-body communications, the reflection coefficients and radiation patterns are discussed.**

I. INTRODUCTION

Recently, body-centric wireless communications (BCWCs) have become a very active area of research because of numerous applications, such as personal healthcare, smart home, personal entertainment, and identification systems [1]. Especially, many researchers considered health monitoring systems as the biggest potential application with all kinds of wearable wireless devices.

In health monitoring systems, at least two important scenarios are required: on-body communications (collecting body signal data) and off-body communications (exchanging data with external equipment). For on-body communications, relatively low frequency bands (tens of MHz) are more suitable [2-3], while for the off-body communications, 2.45 GHz ISM band (2.40-2.48 GHz) is a good candidate [4]. Fig. 1 shows the proposed health monitoring system. The transmitters (sensors) and data collector (dual-mode antenna or switchable antenna) are mounted on the human body. The body signal information is transmitted to the data controller by the sensors at low frequency (10 MHz). The data controller also sends the information to an external device by ISM band at 2.45 GHz. In this way, the health monitoring system for home care or telemedicine system can be achieved. Thus, a multi-functional small antenna is the key component for health monitoring systems. In this paper, we propose two antennas named dual-mode antenna [5] and switchable antenna with on-body and off-body functions for different needs in medical applications. In dual-mode antenna, both on-body and off-body modes are operated simultaneously. In switchable antenna, a simple switching circuit is integrated in the antenna for on-body and off-body communications. Therefore, both on-body and off-body modes can be switched, it is suitable to be used in home care system.

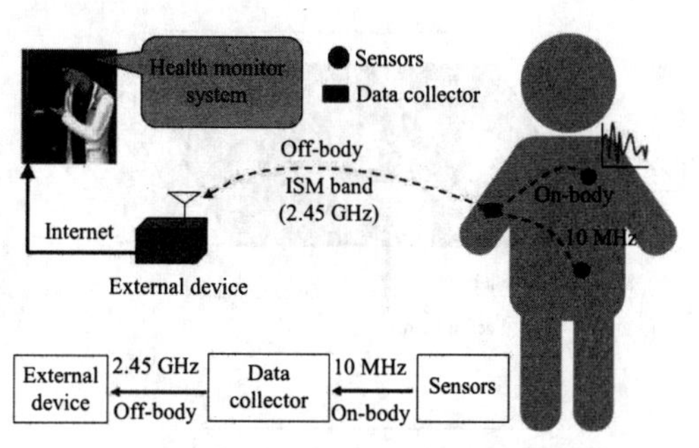

Figure 1. Proposed medical application.

II. ANTENNA DESIGN AND PHANTOMS

A. Dual-Mode Antenna Design

Fig. 2 presents the dual-mode antenna [5]. It consists of a feeding pin, a signal electrode and an L-shaped slit embedded in the ground plane. Therefore, the proposed antenna is similar to a pair of metallic electrodes. The area of the ground plane and the electrode are 30 × 36.5 and 30 × 10 mm^2, respectively. The height of the signal electrode is 4 mm and the feeding pin is located in the center of the signal electrode. An L-shaped slit with 2 mm in width is embedded in the ground plane for 2.45 GHz ISM band.

B. Switchable Antenna Design

Fig. 3 (a) presents the structure of the proposed switchable planar inverted-F antenna. This antenna includes a ground plane, a radiator, a feeding pin and a switching circuit board, which mainly includes a pin diode and replaces the conventional shorting plate. The radiator and ground plane have the same dimension of 30 × 12 mm^2. The height of the antenna is 8 mm. In the lower frequency band, the characteristic of the proposed antenna is equivalent to an electrode operating at 10 MHz while the pin diode located on the circuit board is in the off state. In the higher frequency band (2.45 GHz ISM band), the antenna operates as a planar inverted-F antenna, while the pin diode is in the on state. As shown in Fig. 3 (b), the area of the switching circuit board is 10 mm × 8 mm. In our design, the pin diode HVU131 from Renesas Tech. [6] was selected; thus, a resistor of 200 Ω is used to provide the bias for the diode. In addition, an inductor of 33 nH plays the RF choke and a capacitor of 100 pF is used to block the DC voltage.

978-7-5641-4279-7

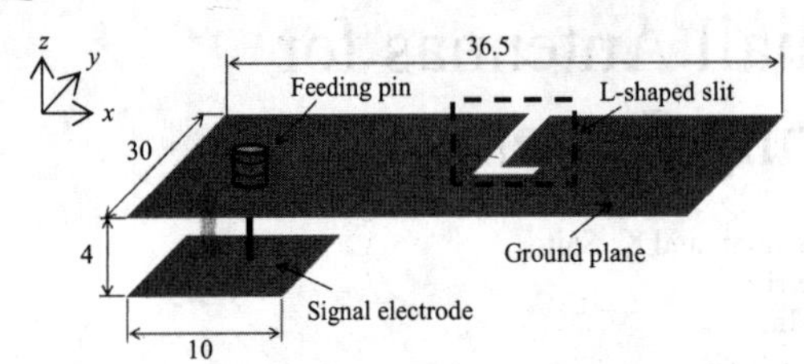

Figure 2. Structure of dual-mode antenna [5] (unit: mm).

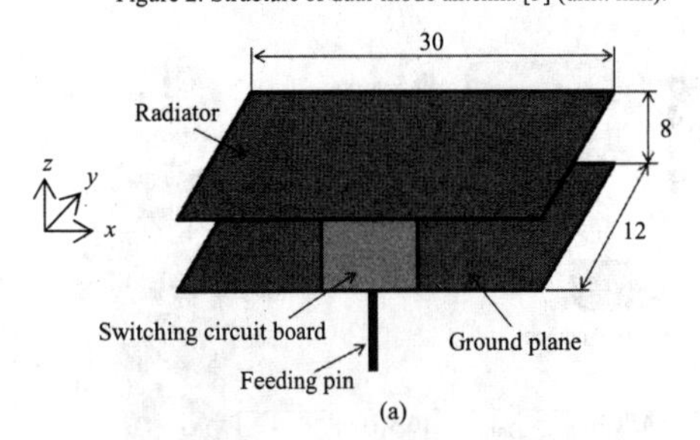

(a)

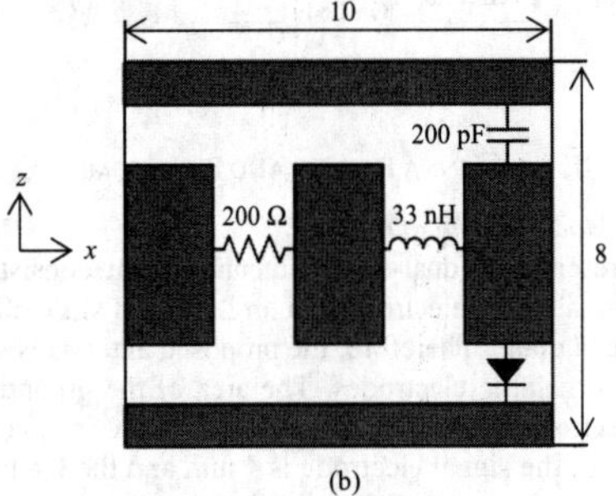

(b)

Figure 3. Structure of switchable planar inverted-F antenna: (a) 3D view and (b) switching circuit board (unit: mm).

C. *Phantoms*

Since the proposed antennas are applied in BCWCs, we discuss the results when the antenna is placed close to two kinds of two-thirds muscle equivalent arm phantoms: 10 MHz (ε_r = 106.73, σ = 0.42 S/m) and 2.45 GHz (ε_r = 35.2, σ = 1.16 S/m) for on-body and off-body communications. The ingredients of the two phantoms are shown in Table I.

TABLE I
COMPOSITION OF THE TWO PHANTOMS

	10 MHz	2.45 GHz
Material	Amount [g]	Amount [g]
Water	1291.2	3375
Agar	54	104.6
Sodium chloride	5.52	7
Polyethylene powder	0	1012.6
TX-151	0	30.1
Sodium salt	1.32	2

III. RESULTS

A. *Dual-Mode Antenna*

Fig. 4 presents the simulated and measured reflection coefficients in 2.45 GHz ISM band when the proposed dual-mode antenna is on the arm phantom (50 mm×50 mm ×500 mm). The measured bandwidth is 300 MHz (2.25-2.55 GHz) and the measured result is close to the simulated one. The simulated and measured radiation patterns at 2.45 GHz in *xz* and *yz* planes are shown in Fig. 5. From the results, the radiation patterns are relatively omni-directional and are of no deep nulls in the half-sphere above the arm phantom. Therefore, the proposed antenna is a good candidate for off-body communications.

As follow, we used the whole human models to discuss the received voltage at 10 MHz for on-body mode. As shown in Fig. 6, the calculation includes simple and high-resolution human models. We fixed the transmitter on the waist and located the two receivers on the right and left chest and two receivers on the right and left wrists. The height of the two human body models are 1750 mm and there is a 30 mm gap between the two feet and the earth ground. From the simulated results in Fig. 7, due to the shorter distance between the transmitter and the receiver, RX 3 obtained higher received voltage than the other receivers. In addition, the received voltage with the simple human phantom is higher than that with high-resolution human model. It is because the surface of the simple human phantom is flat and the signal electrode can attach to the surface of human model very well.

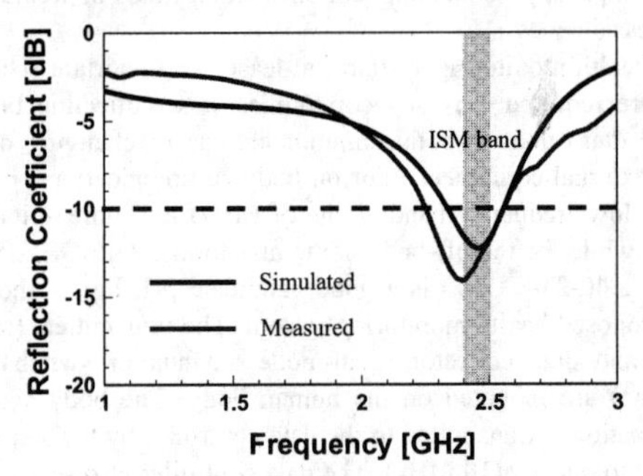

Figure 4. Simulated and measured reflection coefficients for dual-node antenna [5].

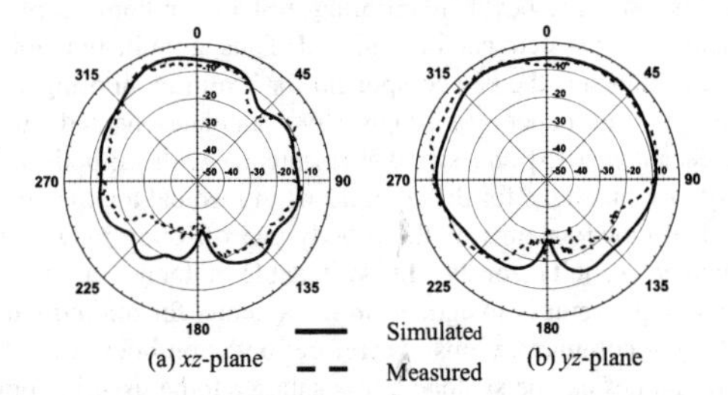

(a) *xz*-plane (b) *yz*-plane

Figure 5. Simulated and measured radiation patterns for dual-mode antenna at 2.45 GHz [5] (unit: dBi).

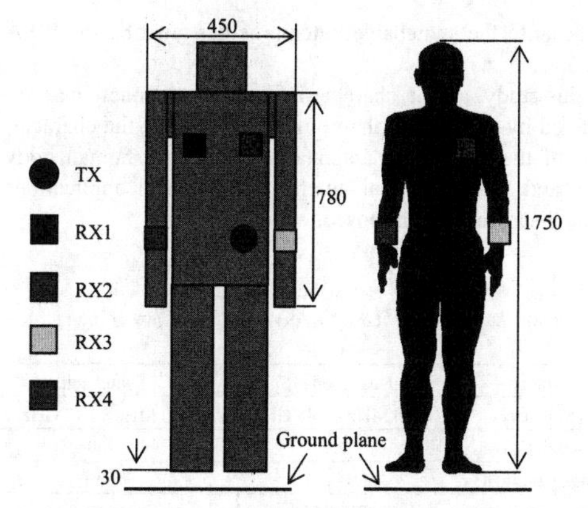

Figure 6. Simple and the high-resolution human models (unit: mm).

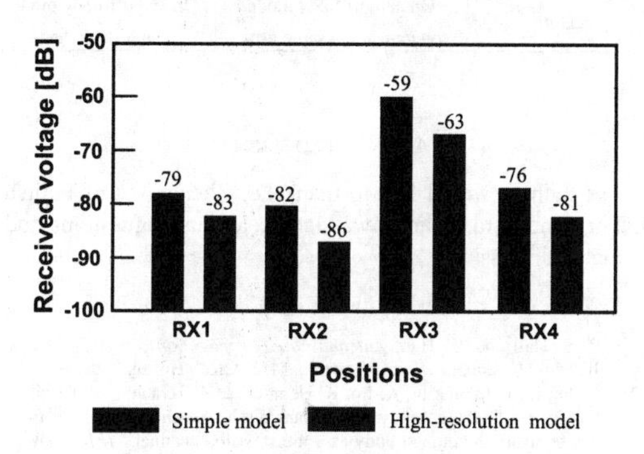

Figure 7. Received voltage for dual-mode antenna at 10 MHz on the human models. The received voltage is normalized by the input voltage of the transmitter.

B. Switchable Antenna

Fig. 8 presents the simulated and measured reflection coefficients of the switchable antenna at 2.45 GHz with the same arm phantom while the diode was in the on state (3V is supplied by a battery box). The measured bandwidth was 350 MHz (2.30-2.65 GHz) and the measured result was close to the simulated one. Fig. 9 shows the simulated and measured radiation patterns at ISM band 2.45 GHz in the *xz* and *yz* planes. The results suggest that the radiation patterns for the ISM band are relatively omni-directional without deep nulls in the half-sphere above the arm phantom; the weaker radiation toward the phantom is due to absorption by the phantom. Therefore, the proposed antenna is shown to be a good candidate for off-body communications.

In this section, we will evaluate the performance of the proposed switchable antenna for on-body communications while the diode was in the off state at 10 MHz. Fig. 10 presents the simulation setup and received voltage results, the proposed antenna is installed at all the transmitter and receivers; all are attached to the chest phantom (525 mm × 370 mm × 70 mm). From the result, the received voltage between the transmitter and the receiver decreases with the increasing distance. It is because the mutual coupling decreased as the increase of the distance between the transmitter and the receiver. Fig. 11 illustrates the electric-field distribution at 10 MHz on the surface of the chest phantom. From the result, the intensity of the electric-field distribution decayed with the increasing distance between the transmitter and the receiver.

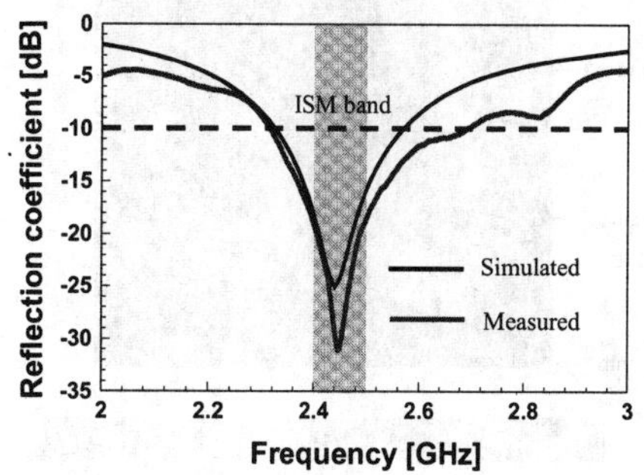

Figure 8. Simulated and measured reflection coefficients for switchable antenna.

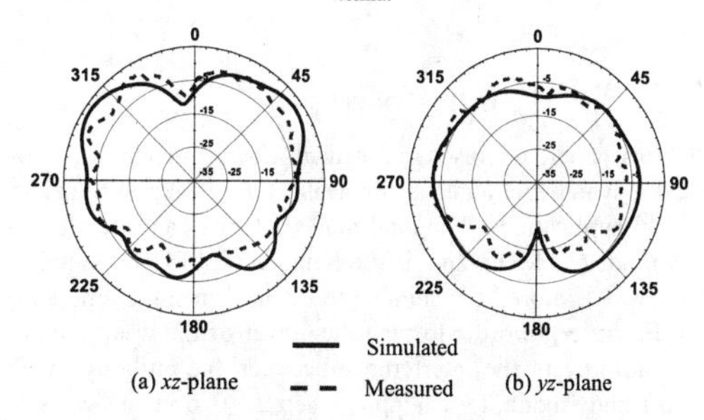

Figure 9. Simulated and measured radiation patterns for switchable antenna at 2.45 GHz (unit: dBi).

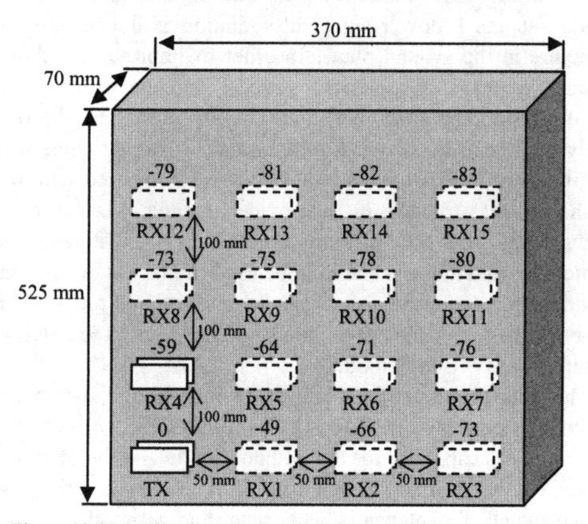

Figure 10. Normalized received voltage for switchable antenna at 10 MHz on the chest phantom (unit: dB).

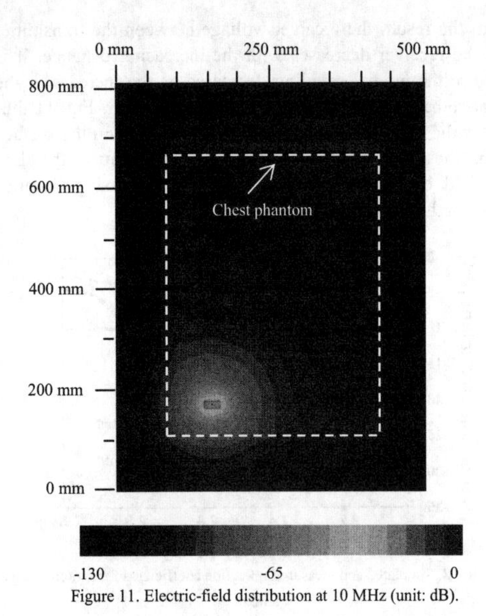

Figure 11. Electric-field distribution at 10 MHz (unit: dB).

IV. Conclusion

In this paper, we discussed a dual-mode antenna and proposed a switchable antenna for health monitoring systems in medical applications. The dual-mode antenna can operate on-body mode (10 MHz) and off-body mode (2.45 GHz) simultaneously. Therefore, it is suitable to be used in some emergencies. However, in order to save the power of the wearable device and prevent the interference between the on-body mode and off-body mode, for example, when the two modes work at the same time, the interference may be caused by the leakage from the off-body mode. Besides, the antenna gain in dual-mode antenna is lower since the radiation is due to the slit structure in the ground plane. In order to improve the above issues, we proposed a planar inverted-F antenna with switching function. Therefore, both the on-body mode and the off-body mode can be switched and the gain is higher since it is equivalent to a PIFA antenna at 2.45 GHz compared with the L-shaped slit structure in dual-mode antenna. Since the on body mode and the off-body mode can be switched, the switchable antenna is suitable to be used in home care system. For example, the body signals can be detected and recorded in on-body mode for one or two hours and transmitted to external equipment in off-body mode.

The comparison between dual-mode antenna and switchable antenna is presented in Table II. From this table, (a) both the two antennas can work for the on-body mode and the off-body mode, (b) the performance such as the bandwidth and the gain of the switchable antenna is better than the dual-mode antenna. It is because the switchable antenna is designed by the PIFA structure.

In this study, all the characteristics of the two antennas are discussed by the human phantoms. In the future, the characteristics of the switchable antenna on the whole human body model and more practical antennas in medical applications will be discussed and proposed.

TABLE II
Comparison between Dual-Mode Antenna [5] and Switchable Antenna

Parameter	Dual-mode [5]	Switchable
Frequency	10 MHz / 2.45 GHz	10 MHz / 2.45 GHz
Bandwidth [MHz]	320	400
Average gain [dBi]	-8	-5
Size [mm^3]	30×36.5×4	30×12×8
Feature	On and off-body mode operate at the same time	On and off-body mode can be switched

Acknowledgment

The authors would like to thank Dr. Zhengyi Li of Fujitsu Laboratories Ltd., Kanagawa, Japan, for his suggestions and valuable discussions.

References

[1] P. S. Hall and Y. Hao, *Antennas and Propagation for Body-Centric Wireless Communications*. Norwood, MA: Artech House, 2006.

[2] K. Fujii, M. Takahashi, K. Ito, K. Hachisuka, Y. Terauchi, Y. Kishi, K. Sasaki and K. Itao, "Study on the transmission mechanism for wearable device using the human body as a transmission channel," *IEICE Trans. Commun.*, vol.E88-B, no. 6, pp.2401-2410, Jun. 2005.

[3] T. G. Zimmerman, "Personal area networks: Near-field intrabody communication," *IBM Syst. J.*, vol. 35, no. 3/4, pp. 609-917, 1996.

[4] A. Tronquo, H. Rogier, C. Hertleer and L. Van Langenhove, "Robust planar textile antenna for wireless body LANs operating in 2.45 GHz ISM band," *Electron. Lett.*, vol. 42, issue 3, pp. 142–143, Feb. 2006.

[5] C. H. Lin, Z. Li, K. Ito, M. Takahashi and K. Saito, "Dual-mode antenna for on-/off body communications (10 MHz / 2.45 GHz)," *Electron. Lett.*, vol.48, issue 22, pp. 1383–1385, Oct. 2012.

[6] Renesas Electronics, http://www.renesas.com

Performance of An Implanted Tag Antenna in Human Body

#H. Y. Lin, M. Takahashi, K. Saito, K. Ito
Graduate School of Engineering, Chiba University
1-33, Yayoi-cho, Inage-ku, Chiba-shi,
Chiba 263-8522, Japan

***Abstract*- RFID system is a growing technology for various applications such as logistics management and automatic object identification, and medicine management. In recent years, it is suggested to combine with in-body wireless communications to reduce medical error and improve quality of live (QOL) of patient. In this paper, we designed the tag antenna which integrates with an integrated circuit (IC) chip of 9.3 – j55.2 Ω. The antenna is embedded into the three-layered human arm phantom and the performance is simulated by finite-difference time-domain (FDTD) method. In addition, by use of the handy reader, the maximum read range of 1.3 cm of the proposed antenna can be reached, that is approaching the theoretical value of 1.7 cm by the link budget.**

I. Introduction

RFID system is a growing technology and useful for various applications such as logistics management and automatic object identification, and medicine management. In recent years, it is suggested to combine with in-body wireless communications to reduce medical error and improve quality of live (QOL) of patient [1] in the hospital.

In addition, the in-body technology is being used in medical application, where cardiac pacemaker is implanted in the human body to provide the electrical impulse to the human heart in asystole caused a ventricular contraction. Besides, microchip implant use identifying integrated circuits for tagging and tracking of animal [2]. Today, in-body wireless communication is suggested to be applied to deliver the information of human body such as temperature, blood pressure, and cardiac rate [3] in the hospital. Therefore, if RFID system combines with in-body wireless communication that will be a great benefit to the medical application. However, the human body is a complex environment that is not attractive for wireless signals because of the mutual influence between the human body and the implanted antennas. Moreover, antenna muse be small and thin because the implanted device is embedded in the human body. Previous study describes that the antenna performance is attenuated by the loss of the human body [4].

Nowadays, many reports focusing on the low frequency (LF, 30–300KHz) and the high frequency (HF, 3–30 MHz) [3] have been presented, because the attenuation from the human body is not strong. However, these systems are limited by drawbacks of short communication range and low data rates; moreover, the design of small size antenna in a limited implantable environment is a challenge. The medical implanted communication service (MICS) has been suggested to be used for in-body wireless communication. However, the physical antenna size is still large for embedding antenna into human body. Therefore, the industry-science-medical (ISM) band of 2.4 GHz is a suitable candidate due to the high data rates and reduction of the antenna size. Some references regarding these issues have been published; for instance, The planar inverted-F antennas (PIFA)-like structure on a substrate with high dielectric loss and a cavity slot antenna design have been proposed [5]–[7]. Moreover, The relationship between the RF transmission and the human body for link budget was proposed in references [8] and [9].

In this report, we proposed an implanted tag antenna which combines with a IC chip (μ-chip). In order to match the conjugate impedance of 9.3 – j55.2 Ω, the loop structure is adopted in the antenna design. Moreover, the proposed antenna is coated by a glass coating of 16.75 mm × 5 mm × 2 mm to avoid touching human tissue and keep antenna working for a long time. The antenna is embedded into the three-layered arm phantom and the antenna performance is simulated by FDTD method. The maximum read range of the proposed antenna is measured by the handy reader.

II. Antenna Structure and Human Arm Model

Fig. 1 (a) illustrates the configuration of the proposed antenna which is designed to match the conjugate impedance of 9.3 – j55.2 Ω by the adopting loop structure. The antenna size is 15.75 mm (L) × 4 mm (W). The glass coating (ε_r = 5.0) is introduced to cover the proposed antenna for reducing influence from the human body. Therefore, the whole size of the proposed antenna will become 16.75 mm × 5 mm × 2 mm.

Figs. 2 shows the human arm phantom which is used to represent a realistic human arm. The phantom of 150 mm × 60 mm × 60 mm is composed of a skin (ε_r = 38.0, σ = 1.5 S/m), a fat (ε_r = 5.3 σ = 0.1 S/m) and a muscle (ε_r = 52.7, σ = 1.7 S/m) at 2.45 GHz [10]. The thicknesses of each tissue in the three-layered phantom are 2, 4, and 54 mm, respectively. Moreover, the proposed antenna is embedded into the fat of the phantom with a depth of 3 mm from surface of skin, since the loss of fat is less than at the skin and muscle layers.

978-7-5641-4279-7

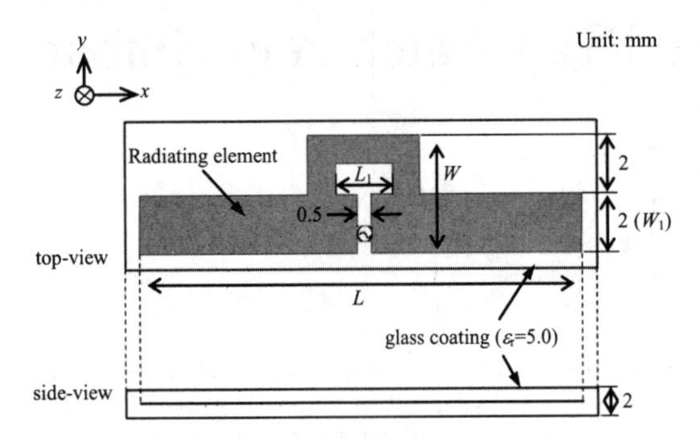

Figure 1. Antenna structure with glass coating.

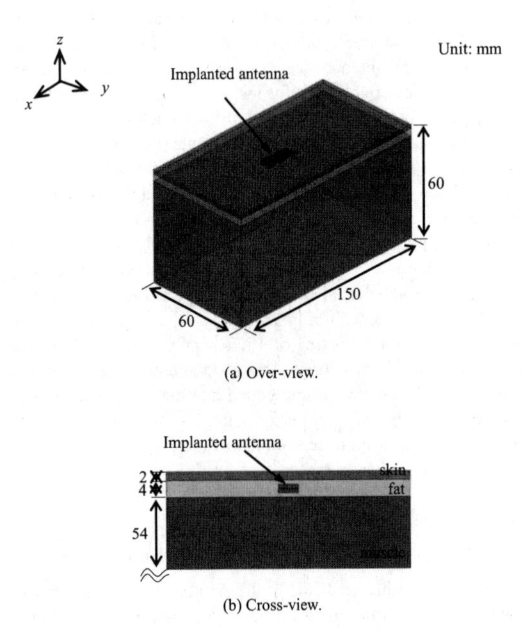

(a) Over-view.

(b) Cross-view.

Figure 2. Human arm phantom.

III. Result and Discussion

Since the loop structure is added for improving the impedance matching, the effect on antenna performance is investigated by varying the parameters of loop structure L, L_1 and W_1 (position of the feeding point). The simulated results are shown in Figs. 3 (a)–(f). In Figs. 3 (a) and (b), it is found that when the L is changed from 13.75 mm to 17.75 mm, the resistance is changed from 5.2 Ω to 16.3 Ω at the desired frequency of 2.45 GHz but the reactance does not changed. From the Figs. 3 (c) and (d), simulated result describes that varying parameter of L_1 not only causes resistance to be increased but also changes the reactance. Thus, when L_1 equals to 2.0 mm, the good impedance matching can be obtained. From the results of Fig. 3 (e) and (f), the input impedance can be fin tuned to achieve good impedance matching by varying the parameter W_1 from 0.0 mm to 2.0 mm. Therefore, when these parameters of the L, L_1 and W_1 are set to be 15.75 mm, 2 mm and 1.5 mm, respectively, the input impedance of the proposed antenna can easily match the conjugate impedance of $9.3 - j55.2$ Ω.

Fig. 4 shows the simulated and measured input impedances. As the result, it is found that the measured result of $9.5 + j55.1$ Ω which is approaching to the IC chip of $9.3 - j55.2$ Ω. Moreover, it also shows a good agreement with simulated result.

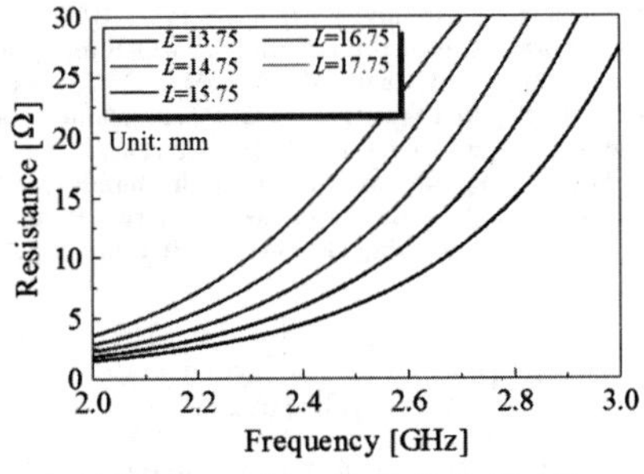

(a) Resistance of parameter L. (L_1=2 , W_1=1.5)

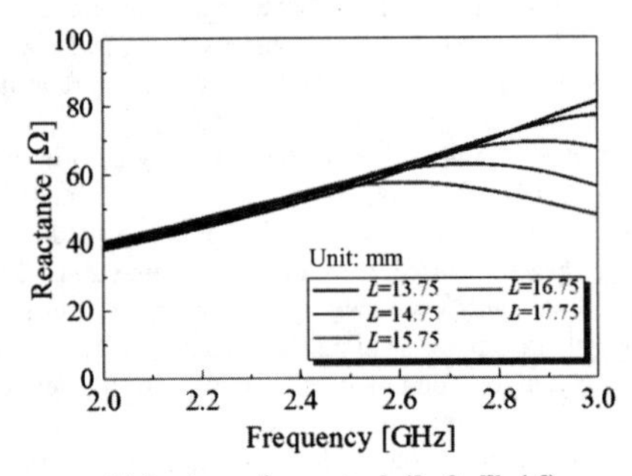

(b) Reactance of parameter L. (L_1=2 , W_1=1.5)

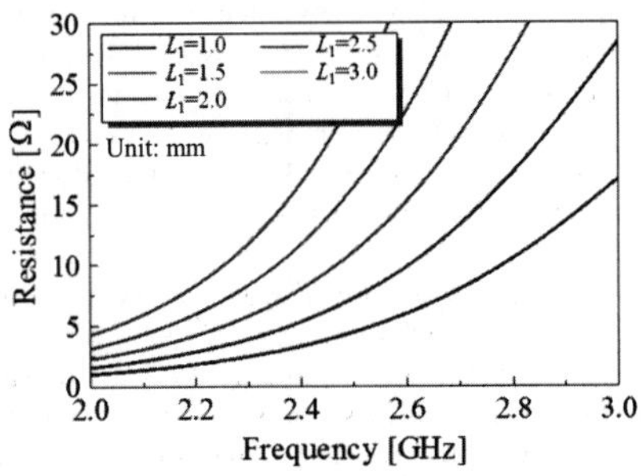

(c) Resistance of parameter L_1. (L=15.75 , W_1=1.5)

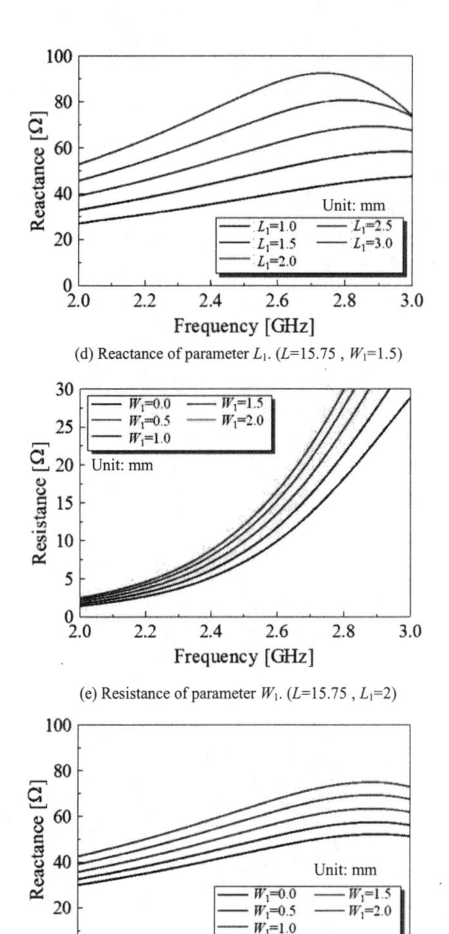

(d) Reactance of parameter L_1. (L=15.75 , W_1=1.5)

(e) Resistance of parameter W_1. (L=15.75 , L_1=2)

(f) Reactance of parameter W_1. (L=15.75 , L_1=2)

Figure 3. Impedance characteristics of the proposed antenna.

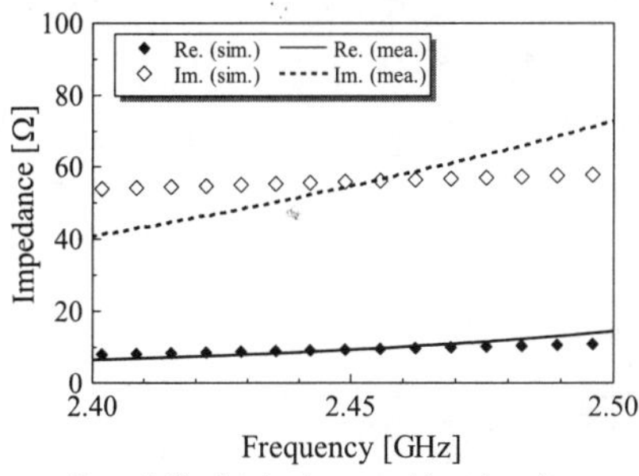

Figure 4. Simulated and measured input impedances.

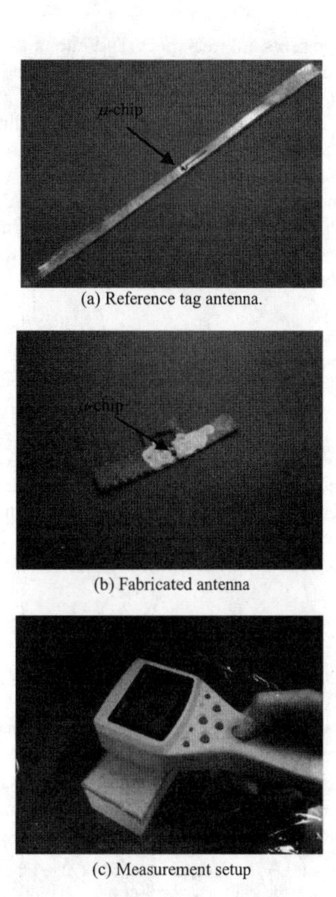

(a) Reference tag antenna.

(b) Fabricated antenna

(c) Measurement setup

Figure 5. Reference antenna, fabricated antenna and the measurement setup.

IV. LINK BUDGET AND MEASUREMENT

Figs. 5 show the reference antenna, fabricated prototype antenna and a handy reader. In order to validate the design, we measured the maximum read rage of the proposed antenna by a handy reader. Before the measurement, we can calculate the theoretical value of the maximum read range by the friis transmission formulas:

$$r = \frac{\lambda}{4\pi}\sqrt{\frac{P_t G_t G_r}{P_r}} \tag{1}$$

In measurement setup of Fig. 5 (c), the reader is an R001M (Sekonic Company) with the output power of 10 mW/MHz. Due to the Bandwidth of 20 MHz, the transmission power P_t by the reader is 200 mW. The minimum threshold power P_r is 2.2 mW [11]. G_r is the gain of the tag antenna (-15.7 dBi). G_r is the gain of the reader. The gain of the reader can be calculated as follows. Fig 5 (a) shows the reference tag antenna [12] which is designed to combine with a μ-chip. The antenna gain of 2.3 dBi presents in free-space and the maximum reader range of 14 cm can be detected by a handy reader under the measurement. Therefore, by Eq. (1), the gain of the handy reader of 1.25 dBi can be obtained. According to above discussion, the theoretical value of the maximum read range r of an implanted tag antenna is 1.7 cm. From the measured result, the maximum read range of the proposed antenna is approximately 1.3 cm between the proposed antenna and a handy reader. Moreover, it is very approaching the theoretical value of 1.7 cm.

V. Conclusions

An implanted tag antenna is proposed and realized for in-body wireless communication. The result shows that the antenna can match the conjugate impedance of $9.3 - j55.2\ \Omega$ very well, even though antenna is embedded in human arm. Moreover, the measured result confirmed the reliability of the simulated result. In addition, we also confirmed that the measured maximum read range of tag antenna is 1.3 cm which approaches the theoretical value of 1.7 cm.

References

[1] P. S. Hall, Y. Hao, Antennas and propagation for body-centric wireless communications, Artech House, Boston, MA, 2006.

[2] Positive ID web site : http://www.postiveidcorp.com/

[3] A. Sani, M.Rajab, R. Foster, and Y. Hao, "Antennas and Propagation of Implanted RFIDs for Pervasive Healthcare Applications," *Proceedings of the IEEE*, Vol. 98, No. 9, pp. 1648-1655, Sep. 2010.

[4] H. Y. Lin, M. Takahashi, K. Saito, and K. Ito, "Performances of Implantable Folded Dipole Antenna for In-Body Wireless Communication," *IEEE Trans. Antennas Propag.*, vol.61, no.3, pp.1363-1370, Mar. 2013.

[5] Merli, F.; Bolomey, L.; Zurcher, J.-F.; Meurville, E.; Skrivervik, A.K., "Versatility and tunability of an implantable antenna for telemedicine," Antennas and Propagation (EUCAP), page(s): 2487-2491, Apr. 2011

[6] W. Xia, K. Saito, M. Takahashi, and K. Ito, "Performances of an Implanted Cavity Slot Antenna Embedded in the Human Arm," *IEEE Trans. Antennas Propag.*, vol. 57, no. 4, pp. 894–899, Apr. 2009.

[7] K. Kawaski, M. Takahashi, K. Saito, and K. ITO, "Design of planar antenna for small implantable devices," *International Symposium on Antennas and Propagation* (ISAP 2012), pp.1264-1267, Nagoya, Japan, Oct. 2012.

[8] W. G. Scanlon, J. B. Burns and N. E. Evans, "Radiowave Propagation from a Tissue-Implanted Source at 418 MHz and 916.5 MHz," *IEEE Trans. Bio. Eng.*, vol. 47, no. 4, pp. 527–534, Apr. 2000.

[9] T. S. P. See, X. Qing, Z. N. Chen, C. K. Goh, and T. M. Chiam, "RF Transmission In/Through the Human Body at 915 MHz," *Antennas and Propagation Society International symposium (SPSURSI)*, pp. 1–4, Apr. 2010.

[10] Dielectric Properties of Body Tissues (IFAC), http://niremf.ifac.cnr.it/tissprop/

[11] M. USAMI, et al, "The μ-chip: An Ultra-Small 2.45 GHz RFID Chip for Ubiquitous Recognition Applications," *IEICE Trans. Electron.*, Vol. B86-C, No. 4, pp. 521-528, 2003.

[12] Daisuke OCHI, Masaharu TAKAHASHI, Koichi ITO, Kouichi UESAKA, and Aya OHMAE, "Performances of the wrist band type RFID antenna," *Proceedings of the 2006 IEICE general conference*, p.71, Kanazawa, Sep. 2006.

Design of Low Profile On-body Directional Antenna

Juneseok Lee, and Jaehoon Choi
Department of Electronics and Computer Engineering, Hanyang University 17 Haengdang-Dong,
Seongdong-Gu, Seoul, 133-791, Korea

Email : juneseok.lee@gmail.com, choijh@hanyang.ac.kr (corresponding author)

***Abstract*-This paper proposes a low profile on-body directional antenna, which suits for on-body system. The antenna generates surface wave along the body surface in the industrial, scientific, and medical (ISM) band. The proposed antenna has a size of 65 mm × 65 mm × 2mm. On the human body equivalent phantom with the proposed antenna, simulated S-parameters show that the impedance bandwidth is lower than - 10 dB. The proposed antenna is a good candidate for on-body applications.**

Index Terms —ISM band, On-body antenna, Wireless Body Area Network (WBAN), On body communication, Low profile antenna

I. INTRODUCTION

People's interest on medical applications of microwave has been rapidly increased recently[1]. Brain neurostimulators, bladder pressure sensors, and pacemakers are good examples. The critical biomedical data obtained from medical devices need to be communicated in the form of microwave with monitoring devices. In order to achieve good communication link between on-body medical device and monitoring device, a directional antenna is necessary[2]. The low profile and mobility are key requirements for such directional antenna. Previously proposed on-body directional antenna has relatively thick thickness (around 8mm) that is inconvenient to install on a human body[2]. In this paper, we propose a low profile on-body directional antenna utilizing stacked guiding patches. The proposed antenna has low profile and generates electric field along the surface of a body to form a surface wave.

II. ANTENNA DESIGN

Figure 1(a) shows the side view of the proposed low profile on-body directional antenna. The proposed antenna consists of stacked guiding patches, a feed layer, and a ground plane. The antenna occupies the volume of 65 mm×65 mm×2 mm, which can be easily installed on a human body. Figure 1(b) shows the top view of the proposed antenna which is fabricated with Taconic CER-10 substrates (εr = 10.1). The ground plane of the proposed antenna has the size of 65 mm×65 mm. The feed layer is printed on the top side of a 0.4 mm thickness Taconic CER-10 substrates with the size of 43 mm×43 mm. The stacked guiding patches are printed on each corner of four Taconic CER-10 substrates (0.4 mm thickness) with different sizes. The dimensions of guiding patches are 9.5 mm×9.5mm, 11.5 mm×11.5 mm, 13.5 mm×13.5 mm , and 15.5 mm×15.5 mm, respectively. The guiding patches are stacked as shown in Figure 1(b).

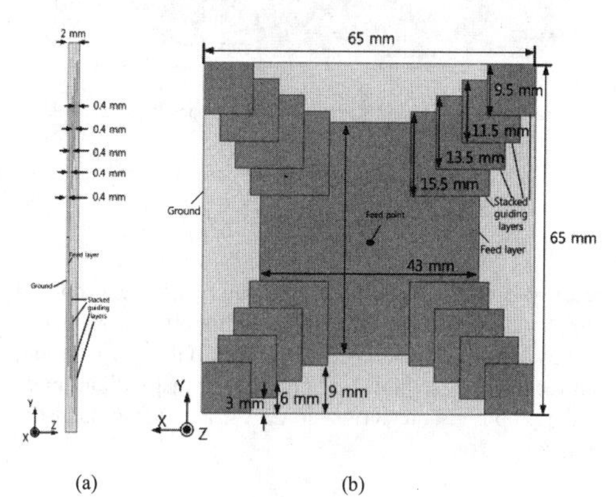

Figure 1. Geometry of the proposed antenna:
(a) side view, (b) top view

III. SIMULATION SETUP

Figure 2 shows the simulation set-up. Three proposed antennas are located 1 mm above the sphere shaped body tissue (ε r = 52.5, σ = 1.78 S/m) having the radius of 100 mm. Figure 3 shows E-field distribution near the surface of body tissue.

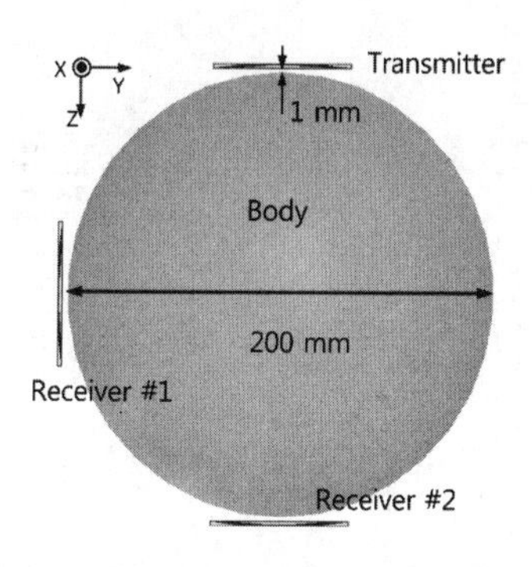

Figure 2. Simulation set-up of the proposed antennas with body tissue

978-7-5641-4279-7

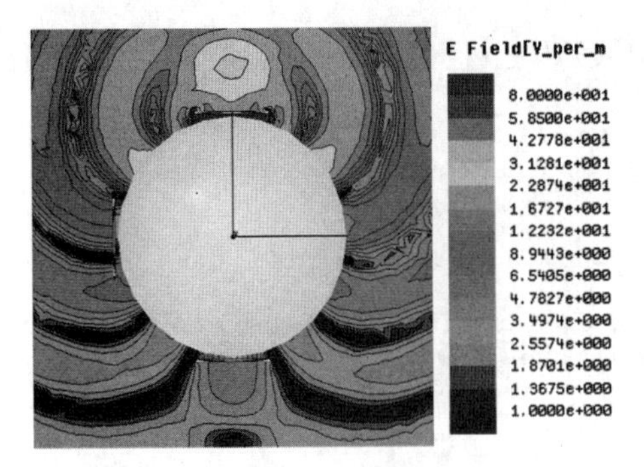

Figure 3. Simulated E- field distribution of the proposed antennas with body tissue (2.465GHz)

IV. Simulation Result

Figure 4 shows simulated S-parameters of the proposed antenna with and without body tissue. The proposed antenna with body has S11 around -25 dB at 2.46 GHz. In addition, S21 of receiver #2 is about 3dB higher then that without body since the proposed antenna generates surface wave along the body tissue.

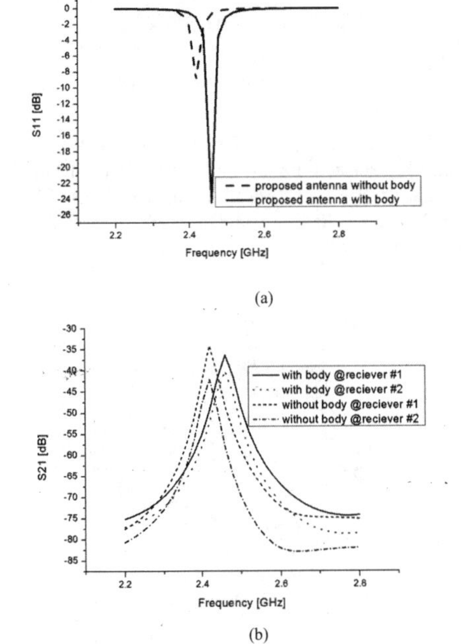

Figure 4. Simulated S11 and S21 characteristic of the proposed antenna

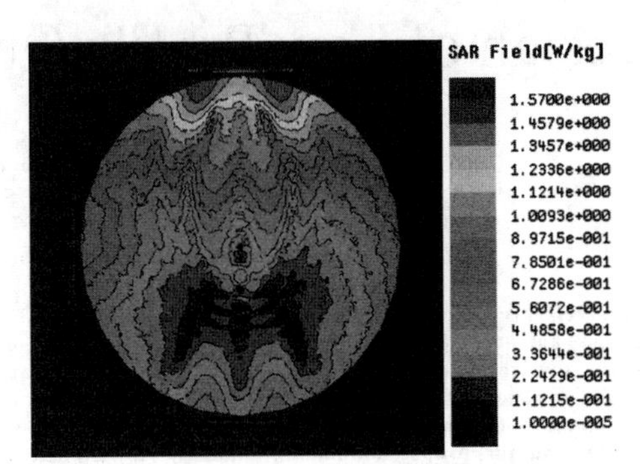

Figure 5. Simulated SAR distribution of the proposed antenna (2.465 GHz)

The simulated SAR distribution is shown in Figure 5. The SAR was simulated with ANSYS HFSS 14.0. The input power to the proposed antenna was 250mW,which is usually used for SAR measurement of mobile application devices, is used to measure SAR. Hot spots are observed around edge of the proposed antenna where the stacked guiding patches are located. The maximum SAR value was 1.57 W/kg at 2.465 GHz. Although a high input power of 250mW was delivered to the antenna, the maximum value of the measured SAR was still below the regulated SAR limitation (1.6W/kg) of the American International Journal of Antennas and Propagation 5 National Standards Institute (ANSI/IEEE) for short-distance biotelemetry [4].

V. Conclusion

This paper proposed a low profile on-body directional antenna for medical devices. The proposed antenna operates in the ISM band. The antenna is modeled with stacked structure for low profile and used guiding patches to generate E-field along the desired direction. In addition, the E-field distribution of the antenna is advantageous for communication between on-body medical devices in the ISM band. The simulated maximum SAR value was low enough to conform to the SAR limitation of the ANSI. Consequently, the proposed antenna shows good performances as a suitable candidate for on-body system owing to the low profile structure and E-field distribution.

Acknowledgment

This research was funded by the MSIP(Ministry of Science, ICT & Future Planning), Korea in the ICT R&D Program 2013

REFERENCES

[1] P. S. Hall andY.Hao, Antennas and Propagation for Body-Centric Wireless Communications, ArtechHouse, Norwood,Mass, USA, 2006.

[2] G. A. Conway and W. G. Scanlon, "Antennas for Over-Body-Surface Communication at 2.45 GHz," IEEE Trans. on Antennas and Propagation, vol. 57, pp. 844-855, April 2009.

[3] R. Garg, P. Bhartia, I. Bahl, and A. Ittipiboon, Microstrip antenna design handbook, Artech House, Canton, MA, 2001.

[4] "IEEE Standard for Safety Levels with Respect to Human Exposure," IEEE Standard C95. 1-1999, 1999.

K-factor Dependent Multipath Characterization for BAN-OTA Testing Using a Fading Emulator

Kun Li Kazuhiro Honda Koichi Ogawa
Graduate School of Engineering, Toyama University
3190 Gofuku, Toyama-shi, Toyama, 930-8555 Japan
m1271031@eng.u-toyama.ac.jp

***Abstract*—This paper presents a new methodology for BAN-OTA Testing using a developed fading emulator with a dynamic phantom. The key to the success of the proposed apparatus is to provide the fading emulator with a proper K-factor that can represent an actual propagation environment under consideration. Firstly, a configuration of the developed fading emulator used for BAN-OTA Testing is shown. Based on the configuration, an analytical investigation of the channel modeling considering K-factor dependent multipath characterization is presented. For getting the knowledge of K-factor, an experiment has been carried out by moving the phantom with different locations of dipole antennas. Finally, a method of calibration by setting the attenuators in fading emulator is also introduced and used in a preliminary experiment for confirming the validity of the proposed channel model. The experimental results show that a variety of K-factors can be controlled in a wide range, which indicates that a realistic BAN radio wave propagation environment can be realized in BAN-OTA Testing using the developed fading emulator.**

Keywords- BAN-OTA Testing, Fading emulator, Rice channel model, K-factor

I. INTRODUCTION

OTA (Over-the-Air) Testing is a method that can evaluate the performance of mobile-device more accurately. An OTA Testing to assess a handset MIMO antenna in multipath environments has been developed [1]. However, in Body Area Network (BAN) systems, the OTA technique has not been included in the previous studies. As a next step for future mobile systems, we are aiming at developing an OTA methodology using a fading emulator for the evaluation of BAN systems in this paper.

There are two major differences between an OTA for cellular MIMO and an OTA for BAN radios. The first difference is that in BAN radios we must consider dynamic channel variations, commonly referred to as shadowing, caused by the motion of an operator, such as the arm swinging while walking. The second difference is that since a BAN sensor module may be attached to the head, as shown in Fig. 1, where there is a strong direct wave coming from a sensor module to an access point attached to the waist. This situation creates a Rice propagation environment in cooperation with reflected fields from surrounding objects. To solve the first subject, we have developed the arm-swinging dynamic phantom that can simulate a natural walking style of humans [2]. Using the phantom we can provide a fading emulator with the fading–shadowing combined effects. As for the second difficulty, to get the knowledge of effects caused by the direct path in indoor environments, such as a hospital, is an indispensable subject in the evaluation of BAN radio systems, and this is the core subject of this paper.

Firstly, a configuration of the developed fading emulator used for BAN-OTA Testing is shown. Based on the configuration, an analytical investigation of the channel modeling considering K-factor dependent multipath characterization is presented. For getting the knowledge of K-factor, an experiment has been carried out by moving the phantom with different locations of dipole antennas. Finally, a method of calibration by setting the attenuators in fading emulator is also introduced and used in a preliminary experiment for confirming the validity of the proposed channel model. The experimental results show that a variety of K-factors can be controlled in a wide range, which indicates that a realistic BAN radio wave propagation environment can be realized in BAN-OTA Testing using the developed fading emulator.

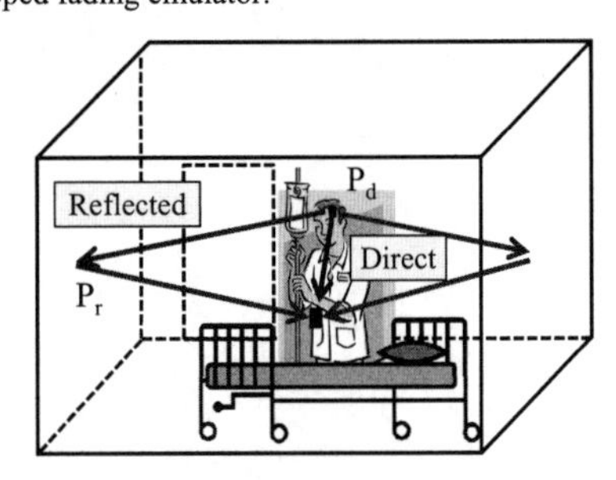

Fig. 1 Typical use scene of BAN radios

978-7-5641-4279-7

II. Configuration of BAN Fading Emulator

A new fading emulator used for BAN-OTA Testing is shown in Fig. 2. The arm-swing dynamic phantom is located at the center of the fading emulator. The phantom can swing both arms to emulate the shadowing effects during scatterers comprised of dipole antennas surrounding the phantom create fading signals at the Doppler frequency of walking. A dipole antenna attached to the head creates direct waves towards a dipole antenna simulating an access point located at waist. The circuit assembly of fading emulator consists of a DA converter, an amplifier, a power combiner, 8 phase shifters, and 8 attenuators. With the control voltage changing from 0 to 10 volts, the phase shifter can change the range of 360 degrees, and the amplitude can change the range of 34 dB, which is sufficient to create the radio waves in different phases and amplitudes as we need in BAN-OTA Testing.

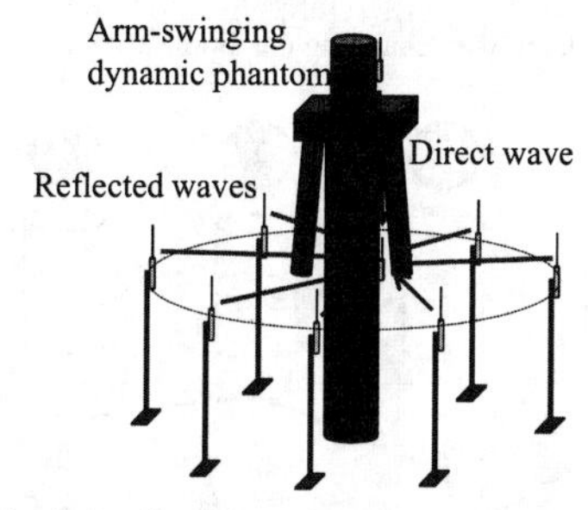

Fig. 2 Configuration of the BAN fading emulator

III. Analytical Investigation of Channel Modeling

Based on the configuration of the BAN fading emulator, an analytical investigation of the channel modeling considering K-factor dependent multipath characterization is presented in this section.

As shown in Fig. 3(a), the average power of reflected waves is obtained as follows:

$$P_r = \frac{1}{S}\sum_{s=1}^{S}|h_s|^2 \tag{1}$$

where h_s indicates the channel response of reflected fields represented as:

$$h_s = \sum_{k=1}^{K} h_k \exp\left\{j\frac{2\pi d}{\lambda}\cos(\phi_k - \phi_v)\right\} = \sum_{k=1}^{K} h_{kr} \tag{2}$$

Then, the channel response of direct wave can be expressed as:

$$h_d = \sqrt{K_c P_r} \tag{3}$$

where the value K_c is defined as the ratio of P_d and P_r, usually known as the Rice factor or K-factor.

As shown in Fig. 3(b), the signal response of each path can be obtained as:

$$h_{kc} = h_d + h_{kr} \tag{4}$$

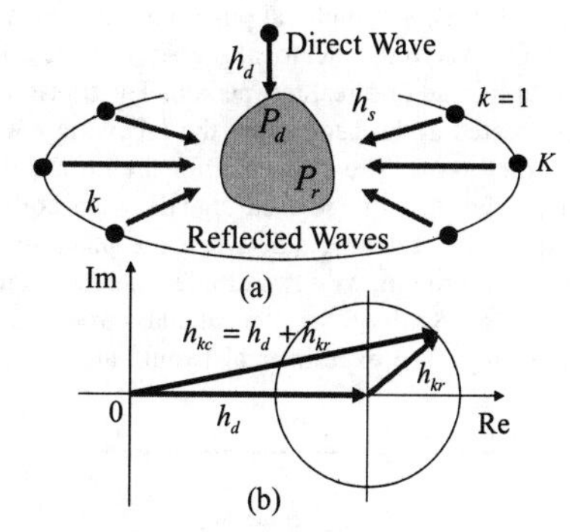

Fig. 3 Channel modeling for BAN-OTA Testing

Therefore, the overall channel response in a Rice environment is expressed as follows:

$$h_{s,c} = \sum_{k=1}^{K} h_{kc} = \sum_{k=1}^{K}(h_d + h_{kr})$$

$$= \sum_{k=1}^{K}\left(\sqrt{K_c P_r} + h_k \exp\left\{j\frac{2\pi d}{\lambda}\cos(\phi_k - \phi_v)\right\}\right) \tag{5}$$

Using the formulas mentioned above, an analysis can be advanced by calculating the CDF characteristic with different values of K-factor in a Rice propagation environment. As shown in Fig. 4, the CDF curves in different values of K factors as shown by the blue curves agree well with their theoretical curves as shown by the pink curves, indicating the validity of the proposed analytical channel model in BAN-OTA Testing.

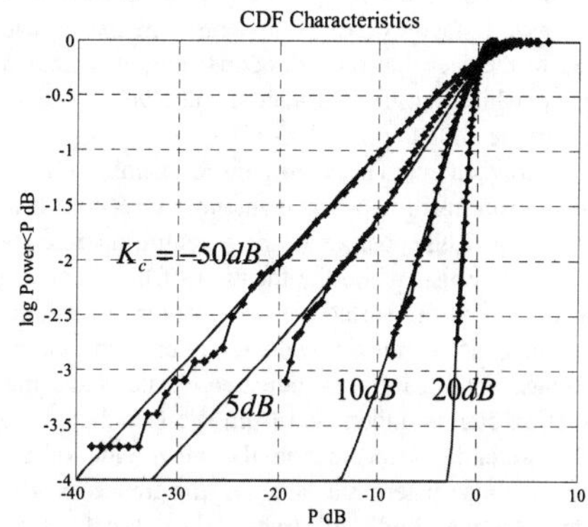

Fig. 4 CDF characteristic with different values of K factor in a Rice propagation environment

IV. Measurement of K-Factor

In order to get the knowledge of the actual value of K-factor in BAN situations indoors, a preliminary experiment has been carried out and analytical results are shown for the effects caused by various conditions in BAN situations in this section.

As shown in Fig. 5, a cylindrical phantom with salt water inside, which is close to the electrical property of the human body, was used instead of a real test person. The transmitting antenna was located at the head while the receiving antenna was located at the waist. The frequency for the measurement was 950MHz. The data of received signals were collected and analyzed on the receiving side when the phantom was moved in a distance of 6m, with two dipole antennas attached to the body in an 8m-by-6.5m typical class room in the Toyama University. The experimental results are shown as follows.

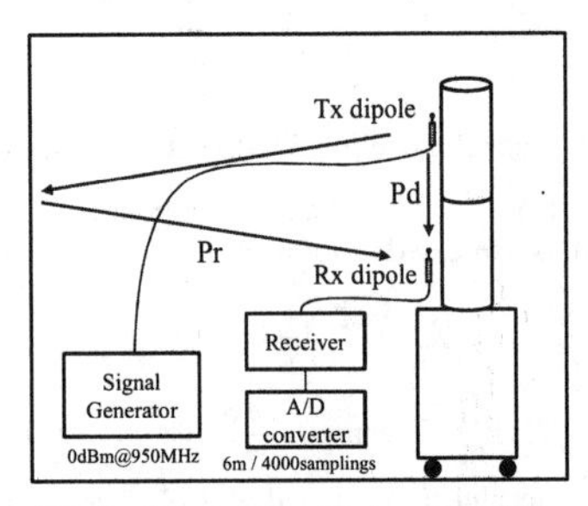

Fig. 5 Measurement setup of K-factor

Fig. 6 shows the cumulative distribution function (CDF) of the received signals. The theoretical curve for the Rayleigh response is also included as the black line. As can be seen, when we use the vertical dipole antenna attached at the waist, the CDF curve approaches the Rayleigh theoretical curve. On the other hand, when we use the parallel dipole antenna, it is found that a large K-factor of 13 dB appears. The reason can be attributed to the fact that the orthogonal alignment has a bad directivity, which causes the direct signal level to reduce to a large extent, leading to a small K-factor.

We have conducted another experiment to examine effects of the relative positions of two sleeve antennas on K-factor in more detail. In Fig. 7, the position of the transmitting antenna is fixed at the left waist, while the location of the receiving antenna varies around the surface of the phantom in a same horizontal plane at 45-degree intervals. It can be seen that the K-factor changes significantly, as indicated by the black line. This observation can be understood from the fact that when the receiving antenna is located on the other side of the transmitting antenna, where the angular difference is 135 degrees, the human body obstructs the direct wave considerably, resulting in a small K-factor.

Moreover, the blue line in the same figure shows the value of path loss between the two dipole antennas calculated by the method of moments. It can be seen that the K-factor decreases as the value of path loss increases. Furthermore, the amount of the variations for the K-factor and path loss coincides approximately with each other, indicating that the K-factor changes mainly due to the variations in the direct wave between the two dipole antennas, rather than the variations in the scattered waves from surrounding objects, such as walls and furniture.

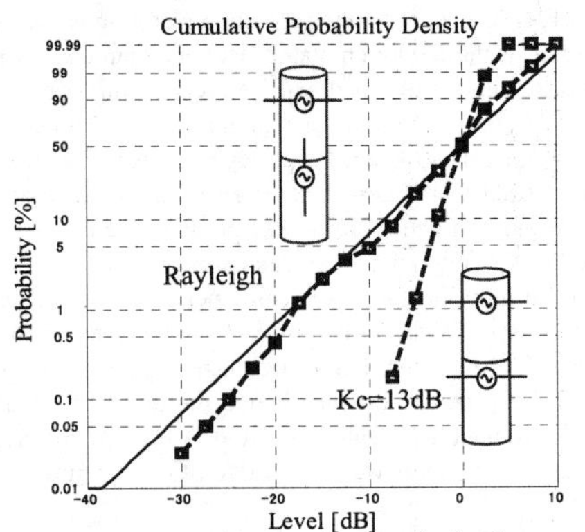

Fig. 6 CDF with changing the dipole alignment

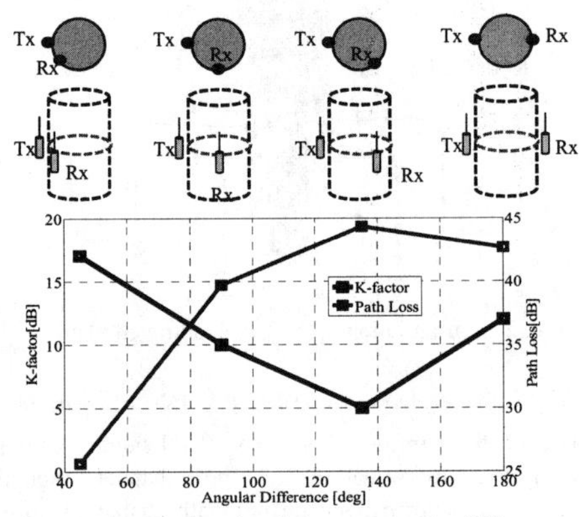

Fig. 7 K-factor and path loss vs. angular difference

V. Calibration in Fading Emulator

Once K-factor in a specified environment is determined as described above, we can calibrate the fading emulator by setting the value of attenuators using Eq. (6) and (7), as follows.

If Pr > Pd, then

$$ATT1 = 0\ (dB),\ ATT2 = E[Pr]\ (dBm) - Pd\ (dBm) + Kc(dB) \quad (6)$$

If Pr < Pd, then

$$ATT1 = Pd\ (dBm) - E[Pr]\ (dBm),\ ATT2 = Kc(dB) \quad (7)$$

where Pd denotes the power of direct wave in dBm. E[Pr] signifies the expectation or average of the scattered signals Pr. The value Kc in Eq. (6) and (7) is defined as the ratio of Pd and Pr, usually known as the Rice factor or K-factor. Knowing the K-factor estimated by the experiment mentioned in Sec. IV, the attenuators in the fading emulator can be set to adjust the power ratio to create a realistic radio propagation environment as we expect.

Then, the method of calibration can be adopted to realize the Rice channel model by a preliminary experiment using the developed fading emulator. As can be seen in the left side in Fig. 8, a sleeve antenna in 950MHz located at the center is used to receive the combined signals, where another sleeve antenna beside it is used to create a strong direct radio wave. Furthermore, 7 scattering units with radio wave absorbers behind them can create Rayleigh fading signals by varying the phases and amplitudes in the control circuit of fading emulator, as shown in the right side in Fig. 8.

Fig. 9 shows the instantaneous response of the combined outcome of the direct and multipath signals of a dipole antenna with a walking distance of 50 wavelengths at 950 MHz in different values of K-factor calibrated by the fading emulator. As can be seen, with the value of K-factor increasing, the deep nulls disappear due to an increase in the power ratio of direct and reflected waves controlled by the fading emulator.

Fig. 10 shows the cumulative distribution function (CDF) of signals created using the channel model mentioned in Sec. III. It can be seen that the CDF curves in different values of K-factor coincide with their theoretical curves, which corresponds to the instantaneous response shown in Fig. 9 and the CDF characteristic shown in Fig. 4, indicating that the Rice channel model in the indoor environment can be realized by adjusting the K-factor using the developed fading emulator in BAN-OTA Testing.

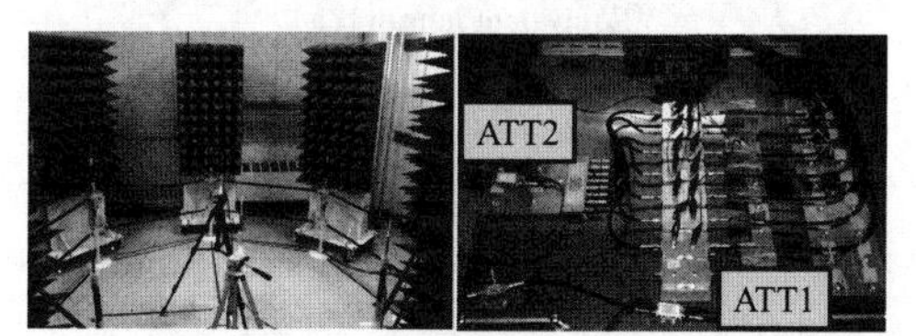

Fig. 8 Multipath measurement with the calibration of different K values using fading emulator

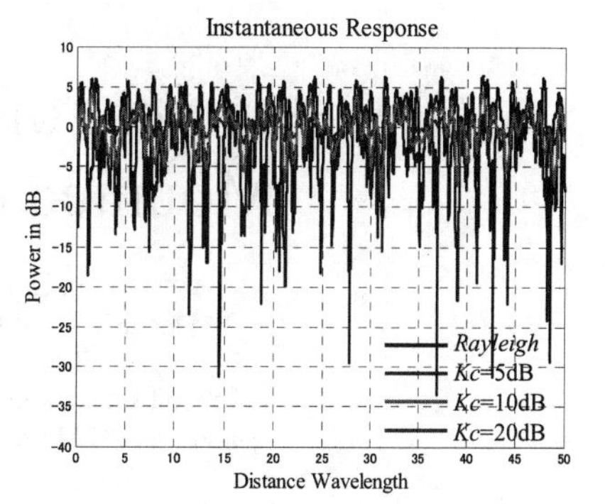

Fig. 9 Instantaneous response of fading signals in different K values

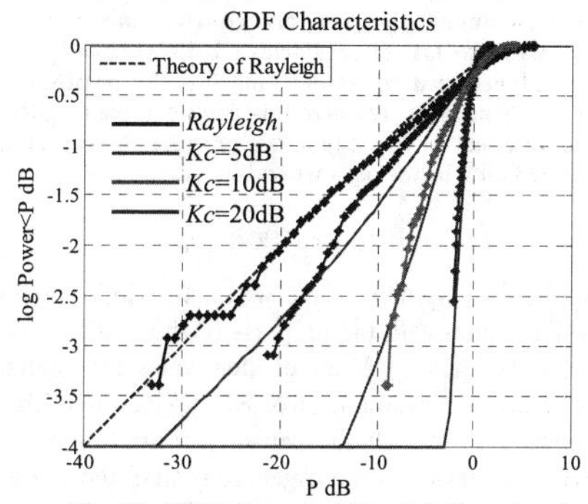

Fig. 10 CDF characteristic of fading signals in different K values

VI. CONCLUSION

This paper presents a new methodology for BAN-OTA testing using a fading emulator with a dynamic phantom. The key to the success of the proposed apparatus is to provide the fading emulator with a proper K-factor that can represent an actual propagation environment under consideration. Moreover, an analytical investigation of the Rice channel model in BAN-OTA Testing considering K-factor in the multipath characterization has been proposed. In order to confirm the validity of the channel model, a preliminary experiment has been carried out, which indicates that the Rice channel model with a variety of K-factors can be realized by adjusting the K-factor using the developed fading emulator. As a future plan, we will use the dynamic phantom which can swing both the arms to create the shadowing effects for testing the realistic propagation environment considering both on-body and off-body situations in BAN-OTA Testing using the developed fading emulator. The final target is to realize a BAN-OTA apparatus that can make communication quality assessments of commercially available BAN radios. This will be addressed in future studies.

ACKNOWLEDGMENT

This work was supported by JSPS KAKENHI Grant Number 25420363.

REFERENCE

[1] T. Sakata, A. Yamamoto, K. Ogawa, and J. Takada: "MIMO Channel Capacity Measurement in the Presence of Spatial Clusters Using a Fading Emulator," PIMRC2009 Intl. Symp. Digest (Tokyo, Japan), Session D1 no. 2, Sep. 2009.

[2] N. Yamamoto, N. Shirakata, D. Kobayashi, K. Honda and K. Ogawa, "BAN Radio Link Characterization Using an Arm-Swinging Dynamic Phantom Replicating Human Walking Motion," IEEE Transactions on Antennas and Propagation, Vol. 61, No. 8, pp.***-***, Aug. 2013. (to be published).

Development of VHF-band Antenna Mounted on the Helmet

Yuma Ono, Yoshinobu Okano
Department of Information Engineering
Tokyo City University
1-28-1, Tamazutsumi, Setagaya-Ku, Tokyo, Japan
E-mail: g1281511@tcu.ac.jp, y-okano@tcu.ac.jp

***Abstract-* Recently, the VHF band attracts attention for mobile use. For example, around 150 MHz is used in fire fighting and police radio et al. The microwave in VHF-band is especially useful for the communication network construction under the environment with a lot of obstacles. However, a present transceiver's antenna that sticks out long has the possibility to obstruct the user's activity. To solve this problem, we decided to use wearable antenna. In this paper, we propose the inverted-L antenna mounted on a helmet as a wearable antenna.**

I. INTRODUCTION

Recently, the development of the wearable wireless device is being paid attention with the progress of the human centric communication technology. Many of them show the tendency to concentrate on the development of RF device intended for high frequency uses though the capacity restriction for the wearable wireless device is more generous than the portable wireless device [1]-[3]. On the other hand, the wireless system that uses a lower frequency band is more advantageous than the system that uses the ISM band concerning the propagation loss or multiple reflection influence. Especially, the VHF-band wireless system used for the police, fire fighting or a military communication is robust for interception with barriers. However, it is thought that the development of the wearable antenna that can be used in this frequency band (for instance, 146-156MHz in Japan) is a considerably challenging problem. For this problem, the development of the antenna that was able to use the wireless system hands-free and was able to exclude the communication interception with system user's body was set to the goal. Additionally, the design for the antenna mounting method that is able to suppress the microwave exposure to users than a present VHF-band transceiver was also included.

For the above-mentioned requirements, when the antenna is mounted on the metallic helmet often used in the police, fire fighting or the army, etc., antenna's radiation and input property are described in this report. Concretely, the antenna radiation and the input property when PIFA is mounted on a metallic helmet are shown in Chapter 2. The antenna feature when PIFA is transformed to improve its performance is also shown. In next Chapter 3, to emphasize bandwidth than PIFA, the performance of helmet antenna that adopts PILA (Planar Inverse L-shape Antenna) is described. Moreover, to emphasize bandwidth additionally, the antenna performance of PILA that adds a parasitic element is also described.

II. BASIC STRUCTURE OF HELMET ANTENNA BASED ON A PIFA

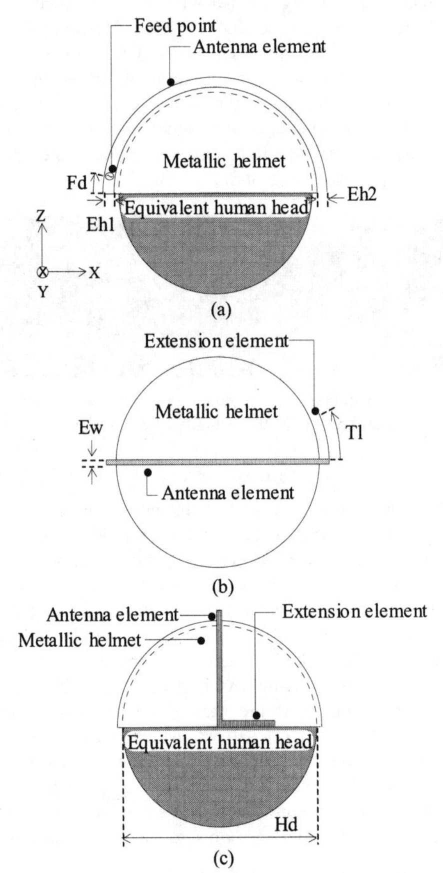

Figure 1. Basic structure of helmet antenna based on a PIFA. (a) Side view of antenna. (b) Bird's eye view of antenna. (c) Antenna back side.

978-7-5641-4279-7

Figure 1 shows the antenna configuration when PIFA is adopted as the element to evaluate the possibility of the helmet antenna that can be used in VHF-band. The each parts of antenna dimension are shown in Table I collectively.

TABLE I THE EACH PARTS OF ANTENNA DIMENSION. UNIT [mm]

Fb	5	Ew	5
Ehl	10	Tl	75
Eh2	10	Hd	200

In the helmet antenna shown in Fig. 1, PIFA of a round element is composed on a metallic hemispherical shell as a ground plane.

Figure 2 shows the result of simulating the antenna input property in the helmet antenna loading on the equivalent human head. Because the helmet antenna is used near the human head, it is necessary to simulate the human head effect for the antenna design. In the analysis FDTD method is used [4]. The cell size is set to 2 mm in X, Y and Z directions. The time step is 3.85ps based on the Curant condition. As the absorbing boundary condition, an 8-layered Berenger's PML is used. To evaluate the worst case with the antenna input property, the equivalent human head is modeled with the homogeneous brain tissue (relative permittivity εr =60.19, conductivity σ =0.479 S/m [5]).

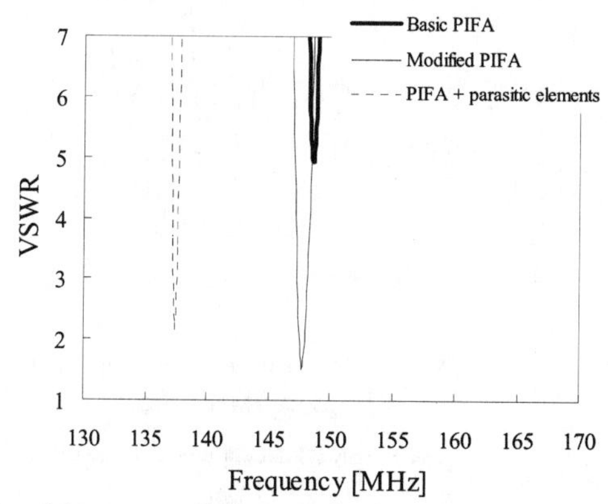

Figure 2. Input property of helmet antenna based on a PIFA

The antenna resonates around 150 MHz by adding extension element like the tail because the gap of the element and ground plane is free space. However, antenna bandwidth is so narrow to use for mobile communication system. Also, VSWR is not satisfied to use. On the other hand, it is proven to be able to construct the antenna that can be used in VHF-band with loading the serviceable size element on the helmet. Hence, the above-mentioned antenna is modified for the practical improvement as follows.

(1) The space of the element and ground plane is filled with dielectric spacer.
(2) The edge of the element of PIFA is expanded (Eh1 is maintained, and only Eh2 is expanded).
(3) Parasitic elements are loaded to PIFA.

Figure 3 shows configuration of the antenna which adopts the modification of (1) and (2). In addition, the antenna configuration that adopts the modification until (3) is shown in Figure 4.

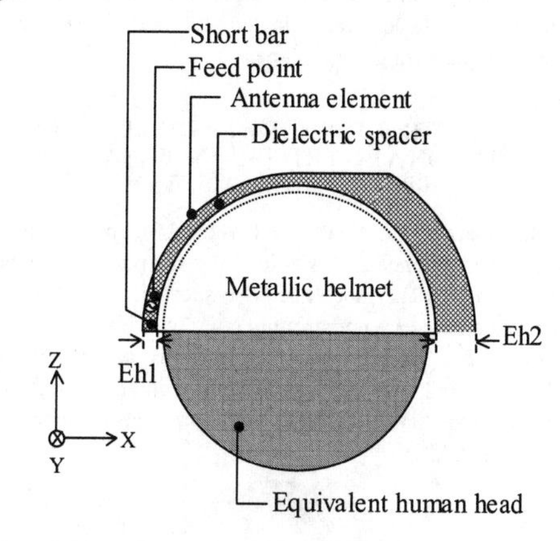

Figure 3. The configuration of modified PIFA with dielectric spacer.

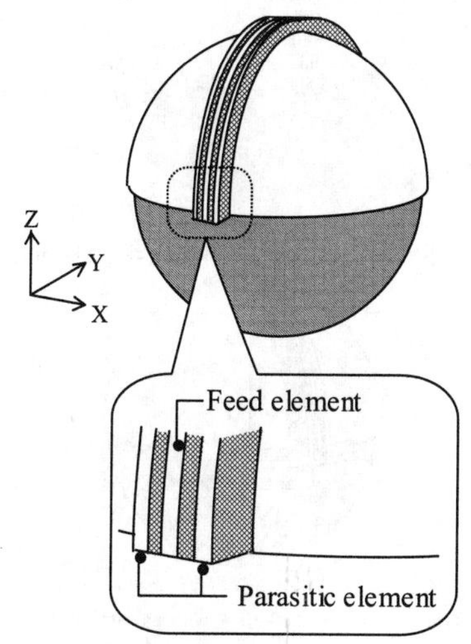

Figure 4. The configuration of modified PIFA with parasitic elements.

Figure 2 shows VSWR as the function of the frequency for the modified PIFA that expanded the PIFA element edge (only Eh2 is expanded in Figure 3). The impedance matching is achieved by the loadings of dielectric spacer (εr =4.0, conductivity σ =0.0005 S/m) around 150MHz though there is no extension element. Moreover, because the height of PIFA on opposite side of feed point is increased, bandwidth and VSWR are improved. However, it turned out not effective of the bandwidth emphasis though the effect of the resonance frequency reduction appeared to a further loadings of parasitic elements (see Figure 2). In the case of PIFA, the radiation element is coupled with the ground plane strongly. Accordingly, it is assumed that an electromagnetic coupling between the radiation element and parasitic elements becomes unremarkable. In other words, the parasitic elements effect is assumed to be demonstrated in the antenna which does not strongly couple with ground plane.

III. BASIC STRUCTURE OF WEARABLE ANTENNA BASED ON A INVERTED-L ANTENNA WITH PARASITIC ELEMENTS

In this chapter, PILA (Planar Inverse L-shape Antenna) is adopted as an antenna that is able to use the parasitic elements effect. Because PILA does not have short bar like PIFA, the radiation elements are not strongly coupled with ground plane. Therefore, it is expected that the effect of parasitic elements be demonstrated. Moreover, the improvement technique of bandwidth and VSWR for PIFA in the previous chapter can be applied to PILA.

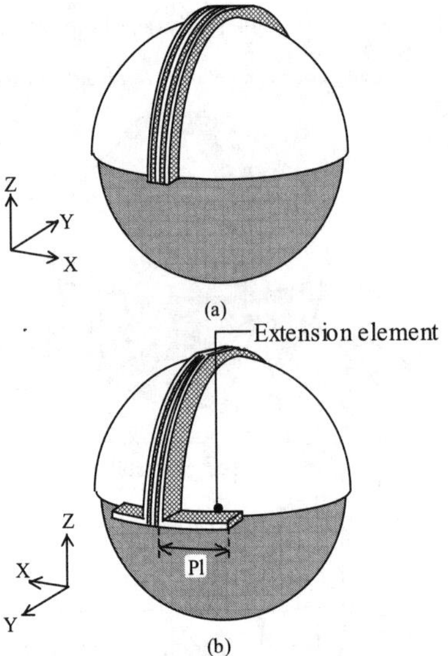

Figure 5. The configuration of PILA with parasitic elements. (a) Front view of PILA with parasitic elements. (b) Antenna back view.

Figure 5 shows the outline of PIFA with parasitic elements. The structure of PILA is almost same as the antenna shown in Fig. 3. However, there is no short bar other than the feed point. Also, the extended element for parasitic elements has been added as shown in Figure 5(b). The parasitic elements are short-circuited to ground plane at the same position of feed point. As for PIFA, the miniaturization is easy, and the extension element for parasitic elements is unnecessary though antenna bandwidth is narrow. On the other hand, the influence of ground plane decreases in PILA, and coupling between parasitic elements and the radiation element comes to stand out though the antenna miniaturization is difficult.

Figure 6 shows the input property of the antenna shown in Figure 5. Because coupling between the parasitic and the radiation elements strengthened, bandwidth has been greatly expanded. Furthermore, the resonance frequency can be controlled only by extending parasitic elements with the radiation element fixed. The resonance frequency transition by extension parasitic element's length ('Pl' in Figure 6) as the parameter is also shown in Figure 6. This is thought to be a feature profitable to design the antenna. As the result of Figure 6, the effect on the VSWR improvement is low though parameter 'Pl' is effective for the resonance frequency control. It is attempted to move the parasitic elements short-circuited position from the bottom to the upper side in Figure 5(a) as VSWR improvement technique. The result is shown in Figure 6 as explanatory note of 'Pl=11cm+short offset'. It is found that the antenna input property is rapidly improved by this technique.

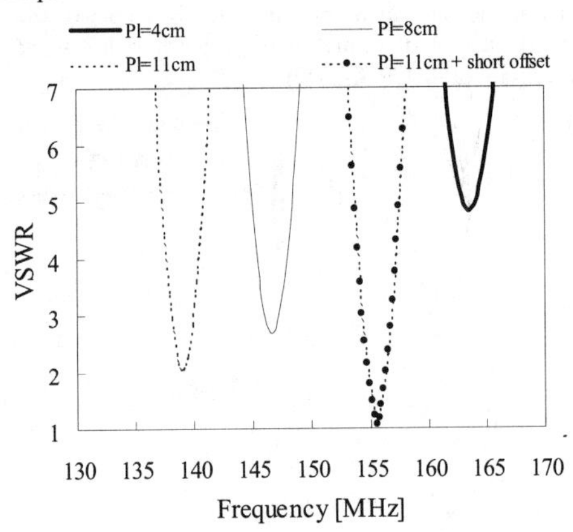

Figure 6. Input property of PILA with parasitic element.

IV. COMPARISON BETWEEN THE PRESENT TRANSEIVER ANTENNA AND PROPOSAL ANTENNA

In this chapter, the comparison result concerning the radiation performance of a proposal helmet antenna and a present transceiver is described. The radiation gain and the radiation efficiency in VHF-band of both antennas are compared for the radiation performance evaluation.

An actual situation that uses present transceiver near the human torso and the numerical model to analyze those situations are shown in Fig. 7. The dimension of human torso and the transceiver model is noticed to Table 2. Transceiver is composed of metal box and normal mode helical antenna. The gap between transceiver and human torso model is set to 10mm. The mixture tissue such as the fat, muscles and bones are adopted as an averaged electric property of human torso. They roughly correspond to 2/3 of muscular tissue's electric properties (relative permittivity εr =41.78, conductivity σ =0.45 S/m [5]).

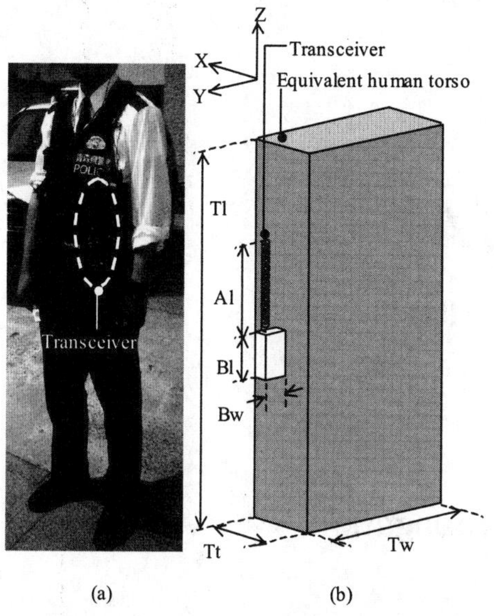

Figure 7. The present transceiver's uses situation and the numerical model configuration. (a) An actual uses situation of present transceiver by policeman. (b) The canonical model configuration for numerical analysis.

TABLE II THE DIMENSION OF HUMAN TORSO AND THE TRANSCEIVER MODEL.. UNIT [mm]

Tl	700	Al	205
Tt	200	Bl	95
Tw	270	Bw	60

Figure 8 shows the simulated comparison results about the radiation properties of proposal helmet antenna and present transceiver. Figure 8(a) shows the vertical-polarization radiation pattern comparison result between the normal mode helical antenna and helmet antenna on X-Y plane. The transceiver's radiation in opposite direction toward torso is about 6dB superior compared with the helmet antenna. However, there is deflection of 14.5dB in the transceiver's radiation though the radiation deflection by helmet antenna is settled to about 8dB on X-Y plane. On the other hand, the comparison of the horizontal-polarization radiation pattern in Z-X plane with the proposal antenna and transceiver is shown in Figure 8(b). The proposal antenna's radiation is more greatly than transceiver's superior in Z-X plane. The maximum radiation gain of proposal antenna is about -4dBi, and this is about 4dB more superior to the transceiver's maximum gain. Moreover, the helmet antenna's radiation efficiency is around 74 %, though transceiver's radiation efficiency is 28 %. Accordingly, it is thought that the proposal antenna has the advantage for present transceiver in the radiation property.

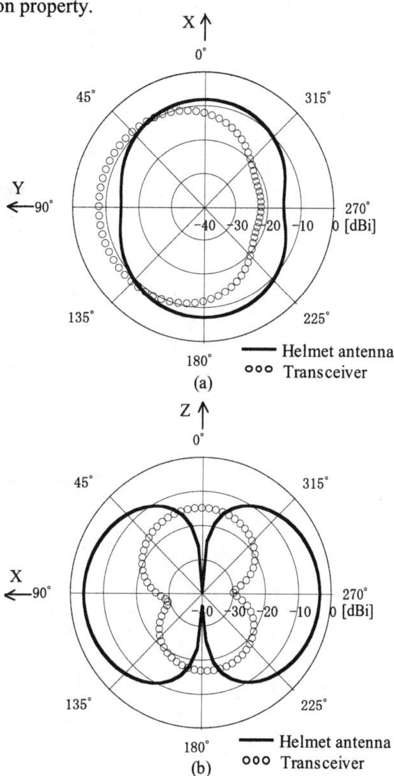

Figure 8. Radiation property comparison between helmet antenna and present transceiver. (a) Vertical-polarization radiation pattern on X-Y plane @155 MHz. (b) Horizontal-polarization radiation pattern in Z-X plane @155 MHz.

REFERENCES

[1] Y. Ouyang, W. J. Chappell, "High frequency propertes of electro-textiles for wearable applications," IEEE Trans. on AP, vol.56, no.2, pp.381-389, Feb. 2008.
[2] E. Jovanov, "Wireless thecnorgy and system intergration in body area networks for m-health applications," in Proc. of IEEE-EMBS 27th Annu. Int. Conf. 2005, pp.7158-7160.
[3] P. Salonen, Y. Rahmat-Samii, H. Hurme and M. Kuvikoski, "Effect of conductive material on wearable antenna performance: A case study of WLAN antenna, " in Proc., IEEE Antennas Propag. Soc. Int. Symp., 2004, vol.1, pp.455-458.
[4] K. S. Kunz and R. J. Luebbers, "The finite difference time domain method for electromagnetic" CRC Press, 1993.
[5] URL: http://transition.fcc.gov/oet/rfsafety/dielectric.html

TP-2(A)

October 24 (THU) PM

Room A

Body-central Propagation

Signal Propagation Analysis for Near-Field Intra-Body Communication Systems

[1]Kohei Nagata, [1]Tomonori Nakamura, [1]Mami Nozawa, [1]Yuich Kado, [1]Hitoshi Shimasaki [2]Mitsuru Shinagawa,
[1]Department of Electronics Kyoto Institute of Technology Sakyo-ku Kyoto, Japan
[2]Faculty of Science and Engineering, Hosei University, Koganei-shi, Tokyo, Japan

***Abstract*- We studied MHz-band communication between a TRX on the surface of the human body (wearable TRX) and a TRX embedded in the environment (embedded TRX). In this work, we propose a medical application to take advantage of MHz-band characteristics on the human body by embedding a TRX into the human body (in-body TRX) and examined the feasibility of in- and on-body communication. We also fabricated an electrically isolated probe (E/O-O/E probe) because conventional probes overestimate the received signal due to additional capacitance. The E/O-O/E probe eliminates this additional capacitance and can measure the received signal precisely. We measured the path loss around a phantom equivalent to a human and found that the proposed application is very useful in terms of engineering output power and received sensitivity.**

I. INTRODUCTION

Wireless body area networks around the human body are expected to play an important role in various areas of application [1]. Several recent studies [2]–[6] have focused on body-channel communication (BCC), which is communication between TRXs on the surface of the human body (wearable TRXs) that uses the human body as a communication path. We previously proposed technology for human-area networking based on near field coupling communication (NFCC) using the megahertz band that consisted of wearable TRXs and a transceiver embedded in floors, PCs, and equipment (embedded TRXs) to expand the areas to which BCC can be applied. NFCC can provide more versatile systems than those with BCC and enables intuitive communication because people can touch and connect. There are many advantages to MHz-band communication, including short communication distance, low power consumption, and less absorption of energy for the human body. These characteristics are attractive for various medical applications such as the hospital scenario outlined in Fig. 1. Using wearable TRXs that contain personal information enables medical professionals to know the physical location of patients and the condition of their health. There are also potential applications to security and convenience, e.g., a scenario in which only people who have wearable TRXs can unlock doors by just stepping up to them and walking into an examination room. A TRX embedded in the human body (in-body TRX) collects medical information and communicates with the wearable TRX on the body so that information can be monitored at all times. Moreover, healthcare and medical applications over networks can be combined by using NFCC between an embedded TRX connected to the network and the wearable TRX. When NFCC is applied, medical information leaks can be prevented and the influence on medical equipment and on humans can be limited. It can be used long-term without having to replace batteries.

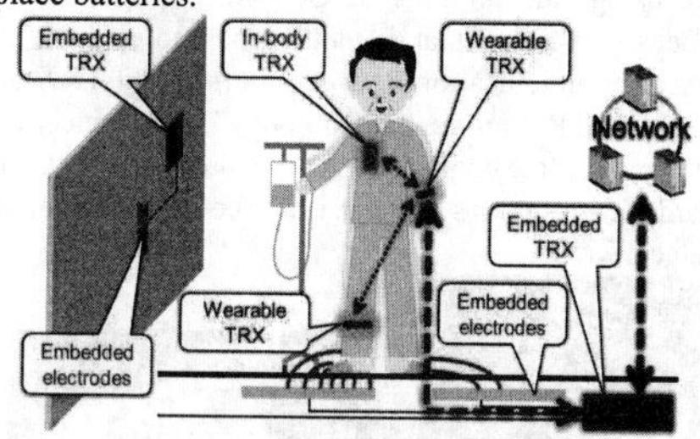

Figure 1. Model of scenario for NFCC application.

II. COMMUNICATION PATH

Figure 2 shows the basic communication model of NFCC between a wearable TRX and an embedded TRX. The signal loop consists of two paths: a forward path and a return path. The forward path runs from the body-side electrode of the wearable TRX through the human body to the upper electrode of the embedded TRX, while the return path runs from the lower electrode of the embedded TRX through the air to the outside electrode of the wearable TRX. When we measure the signal intensity around the human body with the measurement instrument to determine the signal characteristics, an additional path emerges in the equivalent circuit. This makes the return path seem bigger than it is and leads to measurement error.

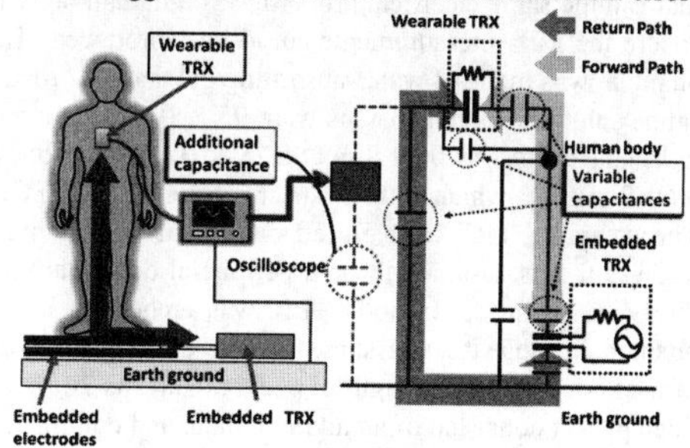

Figure 2. Communication circuit model

III. TRANSCEIVER CONFIGURATION

We fabricated three TRXs —in-body, wearable, and embedded—that could be installed into various objects in

978-7-5641-4279-7

our surroundings, such as doors, floors, and equipment. Figure 3 shows photographs of the configurations for the wearable and in-body TRXs. The prototype used a 6.75-MHz carrier frequency and binary phase-shift keying, achieving a transmission rate of 420 kbps. The wearable TRX was mainly composed of a multi-chip module (MCM), a power circuit, and a coin battery. It could operate for approximately one year on a single CR3032 button lithium-ion battery and incorporated EEPROM that could be used to store user data for applications. The wearable TRX had an asymmetric parallel plate 78 × 49 mm electrode and a horseshoe-shaped electrode to increase space for the button battery. We used two different output voltages: an 8-V "Normal V_out" and a 16-V "Double V_out". We measured the packet error rate (PER) using the in-body TRX with two kinds of electrodes, "3*4" types and "card" types, to evaluate the quality of communication between the in-body TRX and the wearable TRX on the human body. The 3*4 electrode was a parallel plate composed of a 3 × 4 cm electrode and the card electrode was the same as the one used in the wearable TRX.

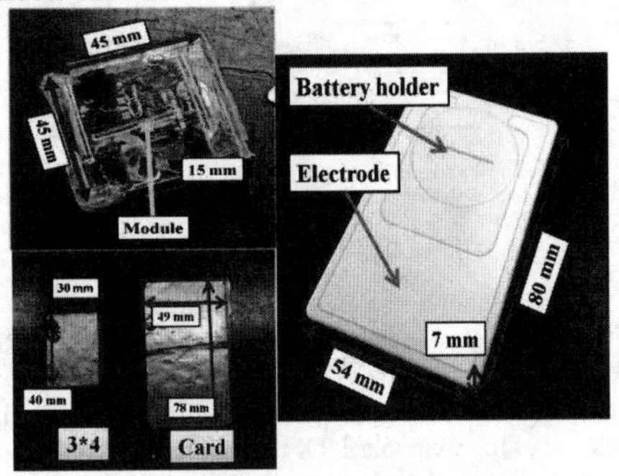

Figure 3. In-body, Wearable TRX configuration and electrodes

IV. Medical Application

A. *Experimental system*

A schematic of the configuration for the system to measure the PER is shown in Fig. 4. We used a phantom that had the same electrical properties as a human body to ensure the PER measurements could be reproduced. The phantom was made of water-absorbing gel that absorbed a saline solution. Its dimensions were 95 × 19 × 19 cm and its conductivity was 0.59 S/m at 6.75 MHz (human body: 0.60 S/m) [7], which is the carrier frequency of NFCC. The wearable TRX was placed on top of the phantom because it was assumed to be a peripheral device around the body, while the in-body TRX was embedded in the phantom because it was assumed to be a device embedded in the body. The total length of packet data was 22 bytes. Each packet consisted of an address, data, and commands. Ten thousand packets were sent at 1-ms intervals. After all the packets had been sent, we confirmed the number that had been successfully received. We also calculated PER. We measured PER between the in-body and wearable TRXs. When the in-body TRX communicates with the wearable TRX in actual use, there are interspaces between the human body and the wearable TRX such as those resulting from clothes and air space. Polystyrene boards (0–80 mm) were placed between the phantom and the wearable TRX to reproduce this situation and to vary the distance between them. We reproduced the depth at which the in-body TRX was embedded in the human body by varying the embedding depth in the phantom.

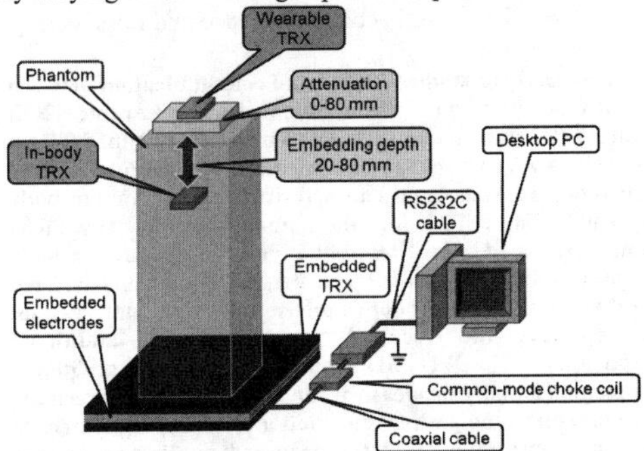

Figure 4. Schematic of PER measurements

B. *Results*

Figure 5 plots the PER for NFCC from the in-body TRX to the wearable TRX for each electrode. We used double V_out. The solid lines represent the approximate lines for 3*4 electrode and the dotted lines represent those for card electrode. The parameters (20–80 mm) represent embedded depth from the in-body TRX to the surface of the phantom. More stable communication was established when we used a larger electrode with greater distances between the TRXs.

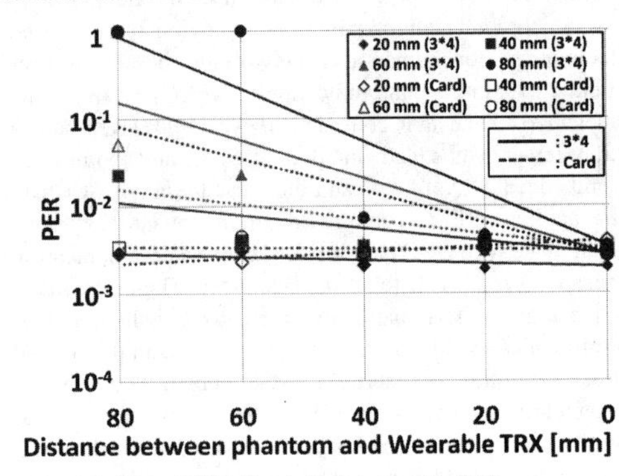

Figure 5. PER from In-body to wearable TRX

Figure 6 plots the PER for communication from the wearable to in-body TRX as parameters of the depth (20–80 mm) for the 3*4 electrode. We only used double V_out TRX because our previous study indicated that communication using a normal V_out TRX was not established in this direction. The PER sharply degraded at a specific distance between the surface and the wearable TRX. For the card electrode, PER was less than 10^{-4}. This means that the signal intensity received by the in-body TRX is a critical factor for PER.

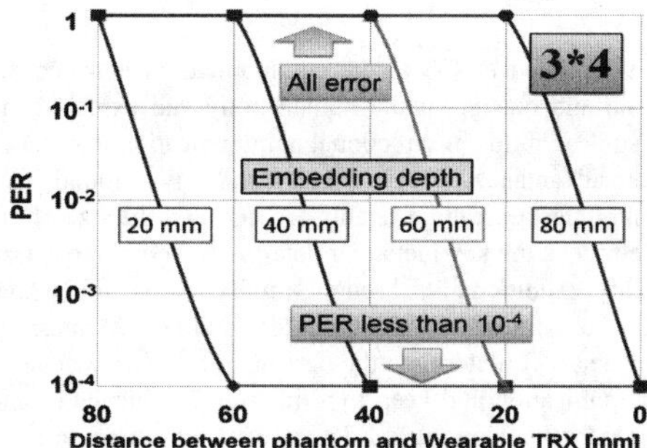

Figure 6. PER from wearable to In-body TRX with 3*4 electrodes

Figure 7 shows the results of the simulation for the electric field distribution around the phantom. The output voltages of the in-body and the wearable TRX were 4.4 V and their signal frequency was 6.75 MHz. The size and the electrical properties of a box placed at the center were the same as those of the phantom. Fig. 7 (a) shows the electrical field generated by the in-body TRX and Fig. 7 (b) shows the electrical field generated by the wearable TRX. As seen in the figures, the range of communication from the wearable to the in-body TRX is narrower than that of backward communication. These results coincide with those we obtained from measurements.

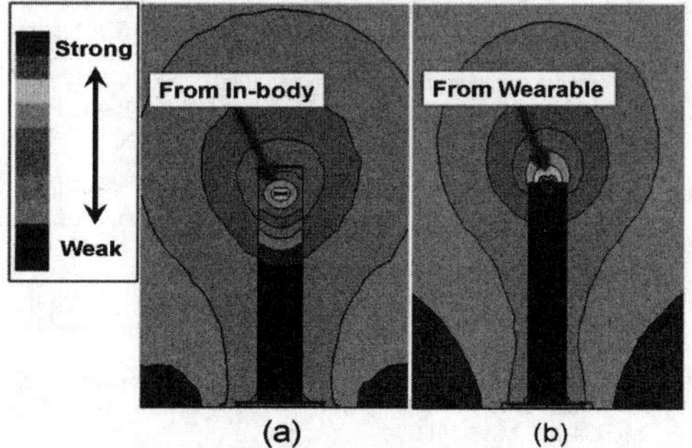

Figure. 7 (a) Electrical distribution from In-body TRX and (b) Electrical distribution from Wearable TRX

V. E/O-O/E

A. E/O-O/E characteristics

Measuring the signal intensity on the human body is very important to determine the output of the TX and the sensitivity of the RX. As discussed in chapter Ⅱ, an additional path appears when we directly measure the signal with a measuring instrument, and we incorrectly measured the signal intensity due to the above. We need a tool to precisely measure the signal intensity on the communication path and to design a stable communication network, so we fabricated the electrically isolated probe and tested it to determine the effectiveness.

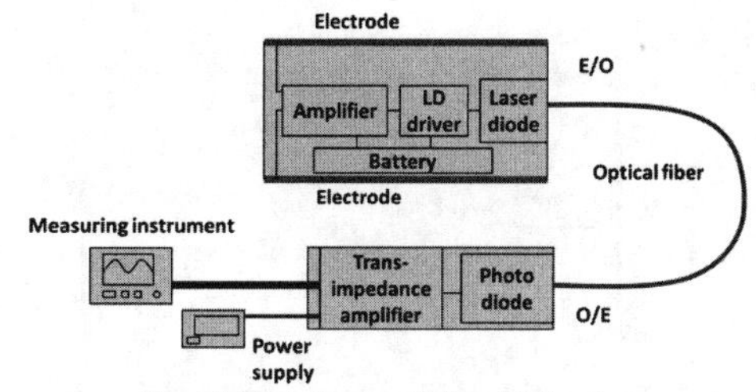

Figure 8. Block diagram of E/O and O/E probe

We measured signal voltages that the probe received when the ground electrode was directly connected to the measuring instrument and when it was not connected. The experimental system is shown in Fig. 9. As the measuring instrument, we chose a battery-operated spectrum analyzer that was independent of the earth grounding. We used two 350-mm-square copper plates as electrodes for the embedded TRX. The distance between the plates was 10 mm. We entered a sine wave of 6.75 MHz by using a signal generator, and the ground electrode of the E/O probe was connected to the spectrum analyzer with a lead line. The ratio between applied and received voltages and changes due to the height of the spectrum analyzer are plotted in Fig. 10. The closer the spectrum analyzer was to the earth ground, the larger the received signals were, as was measured with a conventional electric probe. In contrast, the measured values using the E/O and O/E probe were constant and were not affected by the position of the measuring instrument. In addition, the measured values were much smaller than those obtained with the conventional electric probe because the additional capacitance was excluded from the signal loop by an electrical isolation method. These results revealed that the E/O and O/E probe solved the problem of overestimating the measuring signals and built a repeatable measurement system with the E/O and O/E probe.

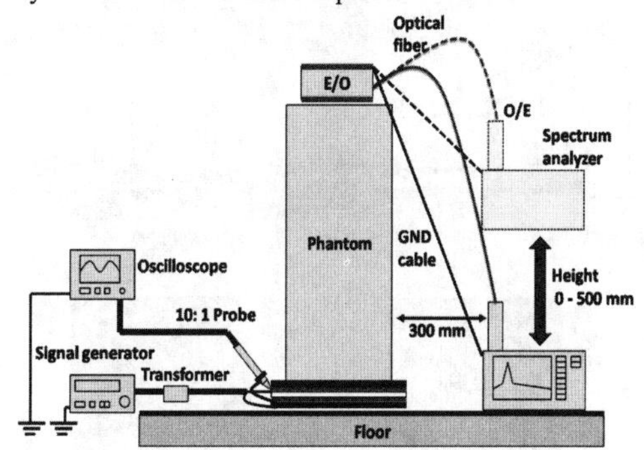

Figure 9. Experimental system to evaluate effect of conventional electric probe on measured value.

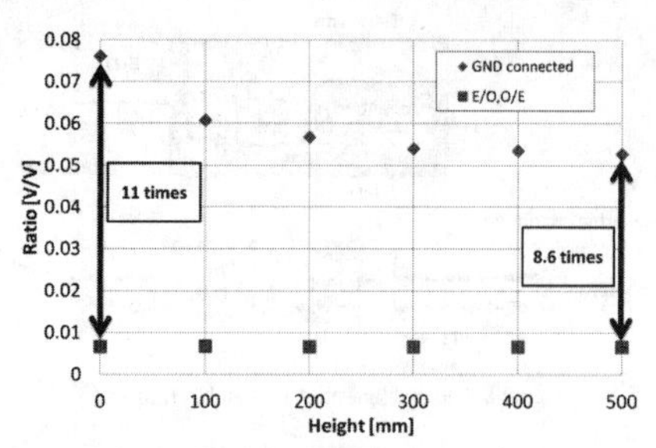

Figure 10. Ratio of applied voltages to received voltages as function of spectrum analyzer height.

B. *Signal distribution around human body*

Figure 11 shows a schematic of the experimental system. We measured the received voltage of the surrounding phantom by using the E/O and electric probes attached to a pole made of PVC pipe placed on wooden rails. This probe was always kept vertical to the ground. We also measured the difference in the potential of the embedded electrodes and provide a graph indicating that the vertical and horizontal attenuation depends on the distance from the phantom (Fig. 12). It is obvious that measuring signals with the electric probe overestimated the signal intensity. As a result, we can precisely confirm the path loss distribution with the E/O probe around the phantom. These results are extremely useful for appropriately determining the receiver sensitivity and transmitter power in order to achieve good communication quality and prevent interference problems in hospital.

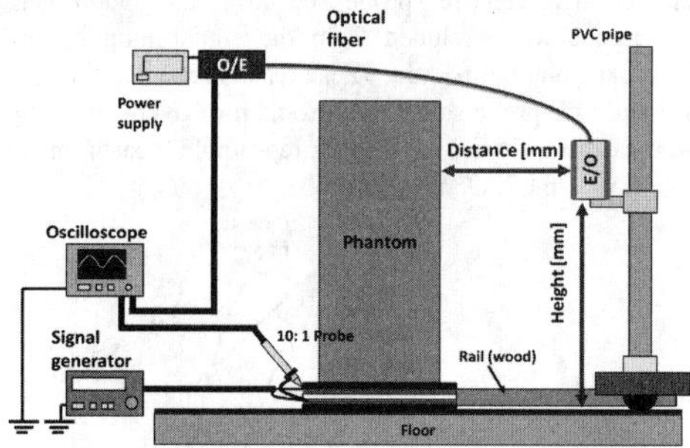

Figure 11. Measuring system for path loss around phantom.

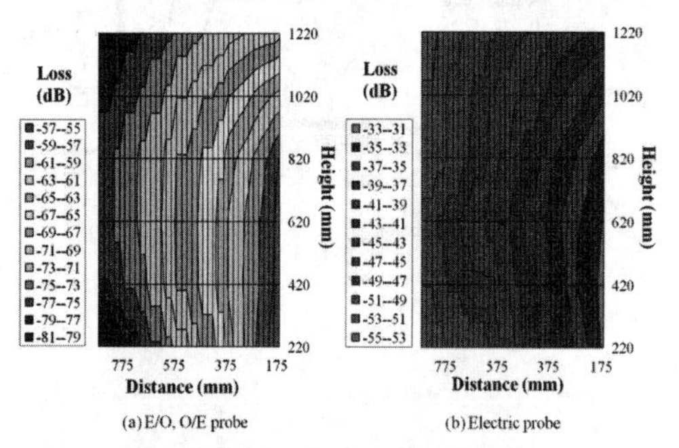

Figure 12. Path loss distribution around phantom.

VI. Conclusion

We applied NFCC to the communication between the inside and outside of the human body and explored the feasibility of this bi-directional communication in order to take advantage of the MHz band. We found that communication can be established and that the size of the electrode is the key factor for determining communication quality. Additionally, through simulation, we found that signal characteristics become altered when TX transmits the signal. To determine the best output power to achieve communication and keep to a minimum influence on the human body or peripheral devices, we fabricated an E/O-O/E probe that eliminates such influence through the measuring instrument. Results showed that it could achieve signal measurement on the human body precisely. In the future, our target is to fabricate the tool for use not only on the body but in the body as well.

Acknowledgment

Part of this work was supported by a Grant-in-Aid for Scientific Research (A) 23246073 from the Ministry of Education, Culture, Sports, Science and Technology of Japan.

References

[1] N. Cho, J. Yoo, S. J. Song, J. Lee, S. Jeon, and H. J. Yoo,"The Human Body Characteristics as a Signal Transmission Medium for Intrabody Communication," *IEEE Trans. Microwave Theory and Techniques,* Vol. 55, pp. 1080-1086, May 2007.

[2] A. Fazzi, S. Ouzonov, and J. v. d. Homberg,"A 2.75mW Wideband Correlation-Based Transceiver for Body-Coupled Communication," *IEEE ISSCC,* pp. 204-205, Feb 2009.

[3] J. Bae, H. Cho, K. Song, H. Lee, and H. J. Yoo, "The Signal Transmission Mechanism on the Surface of Human Body for Body Channel Communication," *IEEE Trans. Microwave Theory and Techniques,* Vol. 60, pp. 582-593, March 2012.

[4] T. G. Zimmerman, "Personal Area Networks: Near-field intrabody communication," *IBM Syst. J.* , Vol. 35, no. 3--4, pp. 609--617, 1996.

[5] Y. Kado and M. Shinagawa, "AC Electric Field Communication for Human-Area Networking," *IEICE Trans. Electron.*, Vol.E93-C, pp. 234-243, MAR 2011

[6] A. Fazzi, S. Ouzonov, and J. V. D. Homberg, "A 2.75 mW Wideband Correlation-Based Transceiver for Body-Coupled Communication," IEEE ISSCC, pp. 204–205, Feb. 2009.

[7] S.Gabriel, R.W.Lau and C.Gabriel, "The dielectric properties of biological tissues: II .Measurements in the frequency rage 10 Hz to 20 GHz", Phys. Med. Biol. Vol.41, pp.2251-2269 ,1996

[8] Y. Kado, T. Kobase, T. Yanagawa, T. Kusunoki, M. Takahashi, R. Nagai, O. Hiromitsu, A. Hataya, H. Shimasaki, and M. Shinagawa : "Human-Area Networking Technology Based on Near-Field Coupling Transceiver" *2012 IEEE Radio & Wireless Sym. (RWS 2012),* pp. 119 - 122, Santa Clara, California, USA, Jan. 2012.

[9] K. Nagata, Y. Kado : "Transmission Characteristics between In-body and On-body Transceivers Using MHz-band Near-field Coupling Technology" *7th International Symposium on Medical Information and Communication Technology,* pp 71-75, Tokyo, Japan, Mar, 2013

[10] M. Nozawa, T. Nakamura, H. Simasaki, Y. Kado and M. Shinagawa, "Signal Measurement System Using Electrically Isolated Probe for MHz-Band Near-Field Coupling Communication", *IEEE International Instrumentation and Measurement Technology Conference 2013, pp. 37 – 40,* Minneapolis, MN, USA, May. 2013.

Numerical investigation on a Body-Centric Scenario at W Band

K. Ali, A. Brizzi, A. Pellegrini, Y. Hao
School of Electronic Engineering and Computer Science
Queen Mary, University of London
London, UK
khaleda.ali; alessio.brizzi; alice.pellegrini; yang.hao@eecs.qmul.ac.uk

***Abstract*- Numerical analysis has been presented for a short-range on-body channel at 94 GHz. Finite Difference Time Domain technique has been adopted to investigate the scenario. Since at higher frequency FDTD becomes computationally expensive, a parallel version of the method has been implemented in an in-house software. Path loss values have been calculated for head to shoulder link. Results are compared with measurements and commercially available software adopting a ray-based approach. A further analysis is provided for both parallel and perpendicular polarization of the source.**

Index Terms— FDTD; Ray-tracing, BAN, Path Loss, V band

I. INTRODUCTION

In recent years, interest on body centric wireless communication systems has expanded significantly. Various systems operating at microwave frequencies ranging from 401 MHz of MICS to 10 GHz of UWB region [1] have earned popularity in the research community. However these portions of the spectrum have become saturated due to a strong overlap with the frequencies dedicated to mobile communication systems. As a consequence, new on-body systems might struggle to acquire licenses. Moving up to higher frequencies seem to provide a plausible solution [2], specifically V and W bands, which jointly cover from 60 GHz to 110 GHz. Small wavelengths, enabling miniature antennas with higher data rate and greater bandwidth make these spectrums potentially more lucrative. Adoption of this band may accelerates the development of many innovative, personalized and integrated electronic applications, suitable from health care to defense, from lifestyle to fitness. Since at these frequencies, losses increase significantly at higher distances, they also offer the additional advantage of an easier confinement of the energy around the body, with reduced problems in terms of interference, safety and security.

It is worth mentioning that at V and W band frequencies the human body is large compared to the wavelength and acts as a highly lossy medium for the propagating signal. Therefore even for shorter link channels a correct characterization becomes crucial. A good amount of groundwork has already been performed in terms of experimentation for analyzing the channel characteristics at mm-wave frequencies [3], [4]. On the other hand, numerical approaches give the flexibility to consider the propagation scenario closely and overcome many issues arisen during of measurement campaign. Generally, ray-tracing based techniques, hybridized along with Finite Difference Time Domain (FDTD), Finite Element Method (FEM) and Method of Moments (MoM) have been proposed to study outdoor propagation and radiation problems dealing with large objects [5], [6]. Preliminary studies involving the implementations of ray-based techniques in body-centric scenarios have been presented [7], [8]. However, Finite Difference Time Domain method is one of the most versatile full wave numerical approaches for bio electromagnetic simulations. Despite the fact that at higher frequencies FDTD becomes highly expensive in terms of computational burden it can still be efficiently employed for short-range communication.

In this paper, firstly the analysis has been presented for a head-to-shoulder link at 94 GHz in terms of path loss values. Initially this link has been investigated with the aid of in-house software implementing a full wave Parallel FDTD (P-FDTD) method. Subsequently the results have been compared with commercial software RemCom XGTDv2.5, which is based on Geometrical Optics (GO) and Uniform Theory of Diffraction (UTD), and measured data. In addition, both parallel and perpendicular polarization of the source has been considered for the above mentioned link in PFDTD domain.

II. SIMULATION SETUP

In order to characterize an on body channel at W band, different experimental techniques can be adopted. However measurement campaigns include a lot of artifacts with a limitation in repeatability. Numerical approaches, on the other hand, provide considerable flexibility in investigating the propagation scenario. More specifically, full wave numerical techniques are highly useful for channel characterization. In this paper FDTD, a highly reliable full wave method has been adopted. Since at the investigated frequency human body is large compared to the wavelength, to obtain a good tradeoff between accuracy and computational effort only half of the investigated structure has been taken into consideration in the FDTD simulation environment. The numerical model used in the full wave simulation domain is presented in Fig. 1 (a). In fact, due to the high losses in human tissues at 94 GHz, the

978-7-5641-4279-7

contribution of the other half of the phantom can be considered negligible. This model has been generated statistically from the MRI (Magnetic Resonance Imaging) scanned images of various human subjects representing different body shapes and proportions [9]. A high frequency based ray based method is compared along with the measurement results as in Fig.1 (b).

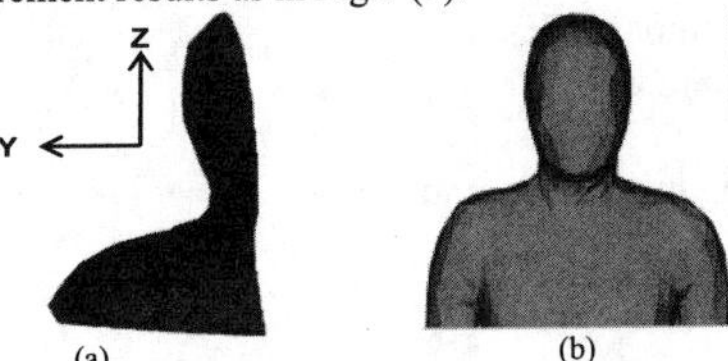

Figure 1. Numerical model of the human body(a)Model used in Ray Tracing (b)Model used for FDTD simulation

The models are considered to be homogeneous with the dielectric properties as that of dry skin at 94GHz.This are due to the fact that at this frequency penetration depth is significantly lower and the transmitted signal diminishes within a few millimeters from the top of the skin. Therefore, according to [10], a dielectric permittivity, ε_r of 5.97 and an electric conductivity σ of 38.19 S/m have been assigned to the model.

The Parallel FDTD Simulation

In the simulation domain, a properly voxelized version of the geometry shown in Fig. 1 (a) has been imported and further separated in several subdomains. A parallel version of FDTD has been adopted by analyzing each subsection by different processors. The whole volume is constituted by $603\times543\times1010$ cells which discretize a domain of about $20\times20\times30$ cm^3. The cell size has been chosen equal to 0.3 mm. A hard source has been imposed in order to polarize the electric field along the x axis and y axis for exciting parallel polarization and perpendicular polarization respectively. A continuous sinusoidal function at 94 GHz is applied to excite the transmitter. The source is positioned at a distance of 1cm from the head.

The Ray-based approximation using XGTD

After importing the structure shown in Fig. 1 (b), in XGTD, a set of receivers have been defined near the shoulder. The transmitting antenna has been placed at 1cm far from the head, above the ear, as mentioned above. Two open-ended standard rectangular waveguides WR-10 (Fig. 2c) have been used both as transmitter and receiver. The patterns on the E-plane and H-plane, shown in Fig. 2(a), (b), (d) and € respectively, have been obtained by simulating the waveguide in proximity of a digital phantom representing the properties of human skin at 94 GHz. in order to correctly take into account the modification of the radiation pattern due to the proximity of the head and the shoulder.. The WR-10 at the transmitter end is positioned such that the E-plane is perpendicular to head. On the receiver side the rectangular waveguide provides an E-plane parallel to the shoulder. In order to correctly take into account the modification of the radiation pattern due to the proximity of the head, then imported and interpolated in XGTD.

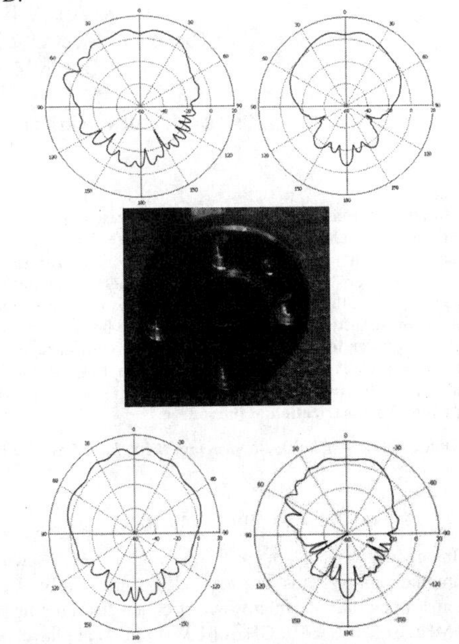

Figure 2. Radiation pattern in E-plane for perpendicular polarization(a) and on H-plane for same polarization(b)WR-10 used in measurements (c), 3D pattern (b), Radiation Pattern on E-plane for parallel polarization (d) and on H-plane (e)

III. EXPERIMENTATION SETUP

An experimental procedure has been carried out for a real human subject with regard to numerical simulation. Results have been compared in terms of path loss for two of the cases. The measurement has been performed as shown in Fig. 5. Signal has been generated by a Continuous Wave (CW) generator at 10.4 GHz. It acts as an input for the frequency multiplier. A 20 dB directional coupler has been connected at the output of multiplier in order to obtain a reference signal. Both the reference and the received signals at 94 GHz are down-converted to 20MHz by using a 9th harmonic mixer and the signal generated by a Local Oscillator (LO).

On the receiving end, a mechanical scan remotely controls the flanged waveguide WR-10 and enables it to move over a vertical plane with a precision of 0.1 mm. In the end, path loss values are obtained from the VNA by using the reference signal and the received one.

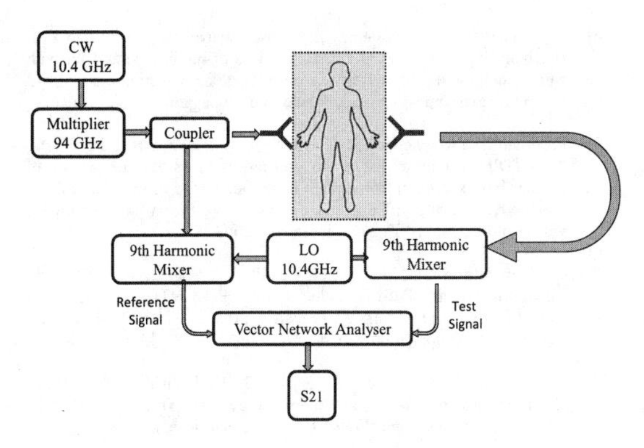

Figure 3. Setup for measurement

IV. RESULTS

The path loss distribution obtained from FDTD simulation is presented in longitudinal view (Figure-4(a) and 4(c)) and cross sectional view (Figure-4(b) and 4(d)) for perpendicular and parallel polarization.

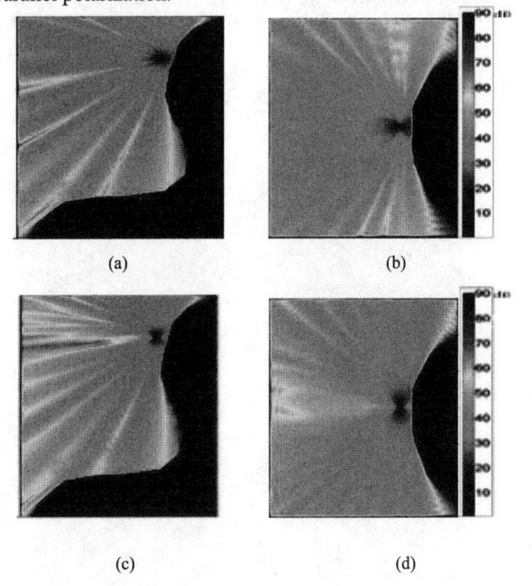

Figure 4. Path loss distribution[dB] for parallel polarization in longitudinal view(a) and in crosssectional view(b)Path loss distribution[dB] for perpendicular polarization in longitudinal view(c) and in crosssectional view(d)

It is observed from figure 4 that for both the cases parallel and perpendicular polarizations signal penetration inside body is negligible.

TABLE I. COMPARISON BETWEEN PATH LOSS VALUES ACHIEVED FOR THE HEAD-SHOULDER LINK WITH FDTD MEASUREMENT AND RAY TRACING BASED TECHNIQUE, D=10MM

Distance (cm)	*FDTD Simulated Result[dB]*	*Measurement result [dB]*	*RayTracing Simulation(dB)*
24	47.03	36.5	52.5
26	49.4	39	46.8
28	52.8	40.8	41.0

It is observed from the above mentioned table that for all three cases of source to receiver distance path loss obtained from FDTD is 10dB higher than the measurement result. This occurs due to the fact that the gain of the rectangular waveguide is more than the point source adopted in FDTD domain. However the constant increase in path loss from full wave simulation agrees with the fact that FDTD is providing stable solution.

Finally, the effect of different polarizations at 94 GHz has been observed. Path gains have been monitored from FDTD along a curve (along the blue line inset of figure 5) near shoulder in the source plane. Path gain distribution with the change of distances is plotted in figure 5 in comparison with free space path loss values.

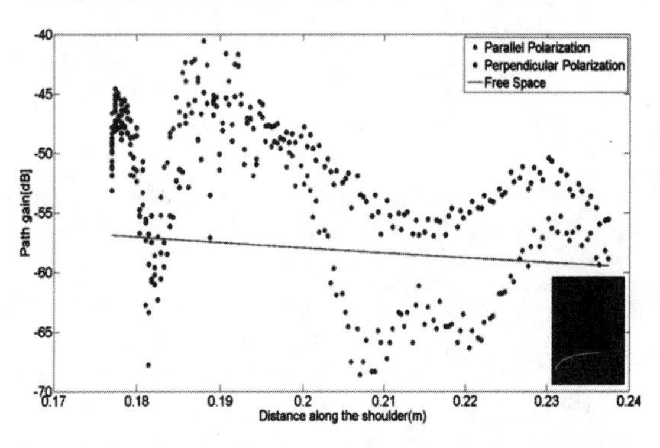

Figure 5. Receiving locations for calculating path gain(a) Path gain distribution for paralleland perpendicular polarization(b)

Figure 5 depicts that along the receiving locations near the shoulder, perpendicular polarization offers better link with higher path gain compared to the parallel polarization. As the receiving probe moves toward the edge of the shoulder, path gain values for both polarizations approximate the free space path gain.

V. CONCLUSIONS

A numerical study has been performed for an on body channel at 94GHz.Parallel FDTD technique has been applied to analyze the head–shoulder link in terms of path loss.

Results are compared with measured values and commercially available software XGTD suitable for high frequency operation. It has been observed that although the antenna gains are different in measurement and FDTD domain, path loss values follow a similar trend with a constant 10dB difference between the two, whereas the ray based algorithm fails to maintain the trend. This is due to the fact that ray based approach does not consider guided wave propagation and a more detailed model might be useful. A further analysis has been performed in FDTD with two different polarizations of transmitters: parallel and perpendicular polarization. It is observed that for head to shoulder link perpendicular polarization allows stronger link than parallel polarization. However to the edge of the shoulder, path loss tends to approximate free space losses.

REFERENCES

[1] P. S. Hall, Y. Hao, Antennas and Propagation for Body-Centric Wireless Communications, Boston, MA: Artech House, 2006.

[2] Report and Order of the Federal Communications Commission, FCC 03-248, November 2003.

[3] I P. Usai, A. Pellegrini, A. Brizzi, L. Zhang, A. Monorchio, Y. Hao, "Analysis of On-Body Propagation at W band by using Ray Tracing Model And Measurements", 2012 Antennas and Propagation Society Symposium, Chicago, USA.

[4] Y. Nechayev, C. Constantinou, S. Swaisaenyakorn, O. Rakibet, J. Batchelor, P. Hall, C. Parini, J. Hunt, "Use of motion capture for path gain modelling of millimetre-wave on-body communication links," 2012 International Symposium on Antennas and Propagation, Nagoya, JP

[5] P.Bernardi, M. Cavagnaro, R. Cicchetti, S. Pisa, E. Piuzzi and O. Testa, "A UTD/FDTD investigation on procedures to assess compliance of cellular base-station antennas with human-exposure limits in a realistic urban environment," IEEE Trans. on microvawe theory and techniques, vol. 51, no. 12, pp. 2409-2417, December 2003.

[6] R. Fernández-Recio, L. García-Castillo, I. Gómez-Revuelto and M.Salazar-Palma, "Fully coupled hybrid FEM-UTD method using NURBS for the analysis of radiation problems," IEEE Trans. on Antennas and Propagation, vol. 56, no. 3, pp. 774-783, March 2008.

[7] S. Alipour, F. Parvaresh, H. Gajari, D. F. Kimball, "Propagation characteristics for a 60 GHz Wireless Body Area Network (WBAN)," 2010 Military Communications Conference, San Jose, CA.

[8] A. Pellegrini, A. Brizzi, L. Zhang, Y. Hao, "Body-centric Wireless Communications at 94GHz", 2012 Symposium on Antennas and Propagation, Nagoya, JP.

[9] S. L. Lee, K. Ali, A. Brizzi, J. Keegan, Y. Hao, and G. Z. Yang, "A Whole Body Statistical Shape Model for Radio Frequency Simulation," 33rd Annual International Conference of the IEEE EMBS, Boston, Massachusetts,USA, August 30 - September 3, 2011.

[10] D. Andreuccetti, Fossi R., C. Petrucci, "Calculation of the dielectric properties of body tissues", http://niremf.ifac.cnr.it/tissprop

A Wearable Repeater Relay System for Interactive Real-time Wireless Capsule Endoscopy

S. Agneessens, T. Castel, P. Van Torre, E. Tanghe, G. Vermeeren, W. Joseph and H. Rogier
Department of Information Technology, IMEC-INTEC, Ghent University,
St. Pietersnieuwstraat 41, 9000 Ghent, Belgium

***Abstract*-Real-time wireless capsule endoscopy offers more flexibility and more precise screening options over commercially available passive endoscopy systems by allowing physicians to steer endoscopy capsules in real time. Yet, this requires reliable uninterrupted high frame-rate video streaming. In this contribution, we present a wearable repeater relay system that overcomes the impairments of the in-to-out body propagation channel and reliably relays implant data to a remote access point. The system consists of a set of wearable textile repeater nodes, exploiting receive diversity to provide a sufficiently large instantaneous carrier-to-noise ratio for live video streaming. Each wearable node, combining a dedicated receive antenna capturing the implant signal, an amplifier and an off-body transmit antenna, is fully implemented in textile materials, such that the comfort of the patient is not disturbed by the relay system. After outlining the design steps for the wearable relay node, in particular demonstrating stable robust antenna characteristics for the textile receive antenna oriented towards the body, we experimentally verify that a 6th-order diversity system provides the best compromise between user comfort and signal quality.**

I. INTRODUCTION

In the last decade, smart textile interactive fabric systems have shown great potential in health-care applications [1], enabling to remotely monitor patients in an unobtrusive and comfortable manner both in a hospital and home environment. Research on body-centric communication focused in particular on on-body [2], off-body [3] and in-body [4] wireless links, but only to a lesser extent on the in-to-out body wireless propagation channel. Yet, wireless communication with implants has recently gained increased attention thanks to the advent of Wireless Capsule Endoscopy (WCE) [5] and the advances made in the field of implanted antennas [6-8]. Setting up high data-rate wireless communication links between implants and remote base stations is hampered by the large attenuation of high-frequency signals inside the body, such that a trade-off must be made between link quality and available communication bandwidth. The lower frequency bands, such as the 402-405MHz Medical Implant Communication Services (MICS) band [9], offer low path loss at the expense of narrow communication bandwidth and large or inefficient antennas, whereas the higher frequency bands, such as the 2.45 GHz Industrial, Scientific and Medical (ISM) band [10] and the 3.4–4.8 GHz low-UWB band, provide large available bandwidth, achievable with smaller components, at the cost of excessive channel attenuation. Actively controlled capsules, however, require large bandwidths to support higher frame rates to provide real-time video feedback to the physician [11]. This enables real-time steering of the capsule's movement to interactively focus on diagnostically important features. Therefore, for these systems, the 2.45GHz ISM band might be the best option, provided the implant makes use of the maximum allowed transmit power of 20dBm and additional measures are taken to ensure the quality and reliability of the wireless in-to-out body communication channel.

In this paper, we present the design, implementation and experimental validation of a wearable repeater relay system for interactive real-time wireless capsule endoscopy. The textile multi-antenna relay system provides the diversity gain required to leverage the signal quality and reliability to sustain high-data communication between the implanted capsule and a remote access point. In the meanwhile, the comfort of the patient is guaranteed by fabricating each wearable repeater node fully based on textile materials, as described in Section II. The receive textile antenna of each repeater node, pointing towards the body to capture the implant's signals, exhibits robust and stable antenna characteristics when deployed on different parts of the body. The received signal is then amplified and relayed to a remote access point by means of a textile off-body antenna. Section III outlines how these nodes are distributed over a patient's body to achieve maximum diversity gain. In addition, the experimental setup is described that implements a two-step validation procedure of the system. The results of this experimental verification are detailed in Section IV. In the first phase, it is shown that a single wearable repeater indeed provides a significant increase in signal-to-noise ratio of the signal received at the remote access point. In the second phase, it is demonstrated that a 6th-order diversity system provides the best compromise between patient comfort, signal quality and link robustness.

II. TEXTILE REPEATER NODE

A. System overview

A general schematic overview of the textile repeater node is shown in Fig. 1, representing the repeater positioned on a human torso, consisting of a textile receive antenna RX_{rep}, capturing the signals emitted by the implant, an analog amplifier and a transmit antenna TX_{rep} that establishes the off-body wireless link with a remote access point. The receive RX_{rep} antenna's boresight, pointing along the y-axis, is

978-7-5641-4279-7

directed towards the body, whereas the direction of main gain of the transmit antenna TX_{rep} is oriented along the negative y-axis. For the latter antenna TX_{rep}, we make use of the dual-polarized textile antenna described in [8], yielding a vertical linear polarization and 8.2 dBi gain when fed at one single terminal (with the other terminal, corresponding to horizontal polarization, correctly terminated by 50Ω). Although in this contribution only one single antenna terminal is used, the availability of a dual-polarized antenna offers the flexibility to later extend the functionality of the node by implementing transmit diversity at the transmit antenna TX_{rep} or by connecting a second receive antenna RX_{rep} to the other antenna terminal. An amplifier, with 43.8 dB gain at 2.45GHz, is used to amplify the signal received by RX_{rep}, which is then retransmitted by TX_{rep}. The textile ground planes provide sufficient shielding between RX_{rep} and TX_{rep}, as the isolation between the RX_{rep} and TX_{rep} antennas was measured to be 55 dB. More details about the system are found in [12].

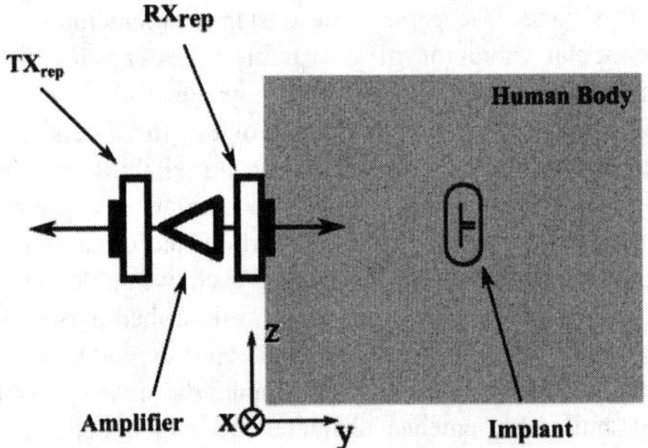

Figure 1. Schematic overview of the textile repeater node.

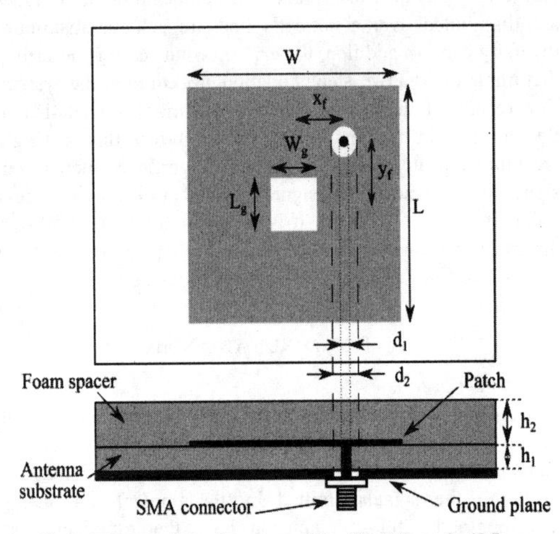

Figure 2. Repeater's receive antenna RX_{rep} (W=40.9mm, L=48.7mm, Wg=8.8mm, Lg=13.2mm, Xf=7.8mm, Yf=18.5mm, h1=3.94mm, h2=7.92mm, d1=1.3mm, d2=5.5mm)

B. *Textile receive antenna*

The wireless node's textile receive antenna RX_{rep}, shown in Fig. 2 with its dimensions specified in the caption of the figure, was specifically designed to exhibit optimal stable performance when deploying the antenna with its direction of main gain towards the body. To ensure robust 50Ω impedance matching when positioned at different locations on the body, a foam spacer with thickness 7.92 mm was placed on top of the antenna plane. Experimental verification shown in Fig. 3 indeed demonstrates that the return loss exceeds 10dB in the complete ISM band when positioning the antenna on the stomach back and sections of the body of an average test person.

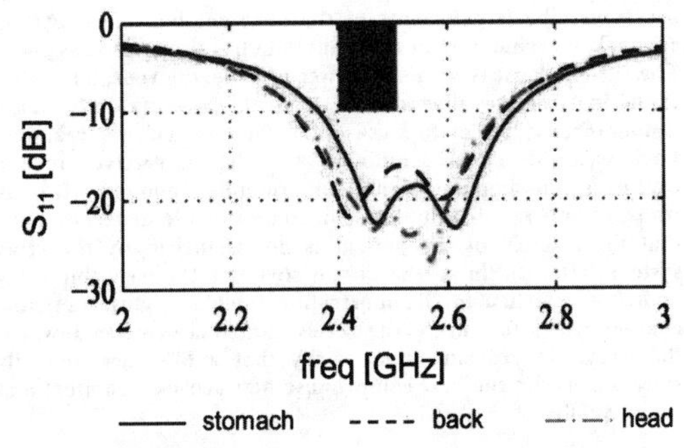

Figure 3. Reflection coefficient of receive antenna RX_{rep}, when deployed on stomach, back and head sections of an average person's body.

III. Multi-Antenna Diversity System

In this section, we describe the practical deployment of the system and illustrate how reproducible validation of the system's figures of merit is performed.

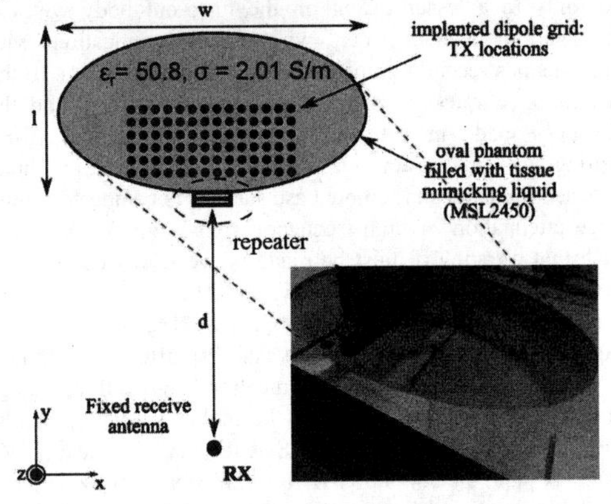

Figure 4. Wearable repeater on a standardized phantom, with w = 0.8 m, l = 0.5 m and h = 0.12 m, setting up a link between an implanted dipole antenna and a remote access point at d = 2m.

A. Single-node standardized experimental setup

Fig. 4 shows the flat phantom filled with tissue mimicking fluid MSL2450, used to emulate the torso of a patient and to yield reproducible measurements following the IEC 62209 standard. The wearable repeater is deployed on the side-wall of the phantom. This setup enables us to assess the data link quality between an implanted transmit antenna, being a half-wavelength dipole insulated by polytetrafluorethylene (ε_r =2.07) and resonating at 2.457 GHz, placed at various depths in the human body, and a remote receiver at d = 2m.

B. Multi-node standardized experimental setup

To exploit spatial receive diversity by improving the signal quality, by means of array gain, and the channel robustness, by means of diversity, we deploy multiple textile repeater nodes, distributed over the body for optimal coverage as shown in Fig. 5. Eight antennas are placed on the side and the back of the patient, such that they may be comfortably integrated into a garment. We concentrate on the improvement provided by the multi-antenna over a single-antenna setup by directly monitoring the signals at the output terminal of the receive antennas RX_{rep} at the different nodes using a Signalion wireless testbed. Again, a reproducible experimental setup is obtained by deploying the different wireless nodes on the wireless phantom with muscle-simulating fluid, as described in Section III.A. In the next section, the link improvement is studied as a function of the number of receive antennas. A more detailed description of the setup and its experimental validation is found in [13].

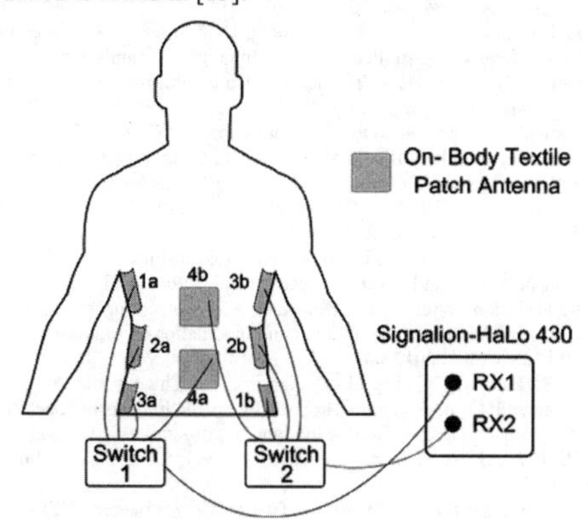

Figure 5. Multi-antenna repeater system, with 6 side (1a, 2a, 3a, 1b, 2b, 3b) and 2 back (4a, 4b) antennas, deployed on a patient, with the receive antenna outputs connected to the Signalion-HaLo 430 wireless testbed.

IV. EXPERIMENTAL VALIDATION

A. Single-node setup

To experimentally verify the benefits of the wearable repeater system, we compare the received signal level at the remote access point with and without the wearable repeater, for different implant (represented by the vertically placed dipole antenna, Section III.A) positions, varying from -10 cm to 10 cm along the x-axis with respect to the repeater's receiver placed at the origin, and depths, ranging from 1 cm to 10 cm (y-axis). The transmit power at the implant is set to 10 mW, which corresponds to about half the allowed SAR limit of 2 W/kg averaged over 10 g of tissue [IEC 62209-1]. The received SNR levels without repeater and with a repeater are displayed in Figs. 6a and 6b, respectively. The received SNR without repeater, averaged over the scanning grid, equals 8 dB, which only allows for low data rate communications. The SNR values achieved at remote terminal are significantly higher when using a repeater, with an average SNR of 33.0 dB.

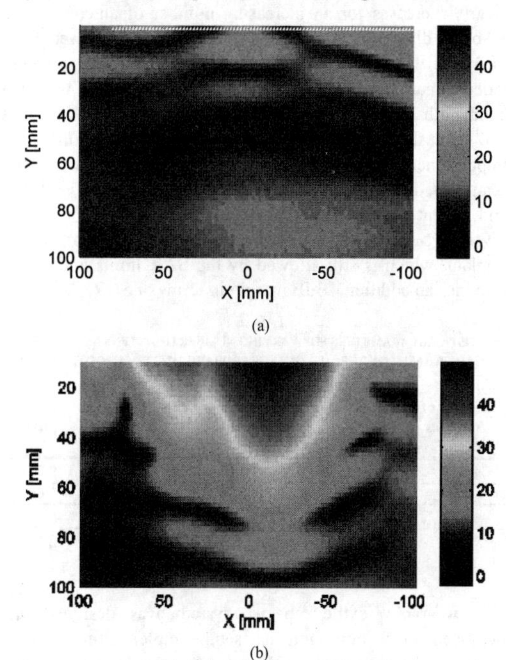

Figure 6. Received SNR on RX, (a) without repeater, (b) with wearable repeater acting as a relay (Y-axis = TX depth in body; X-axis = position relative to RX_{rep} antenna's center).

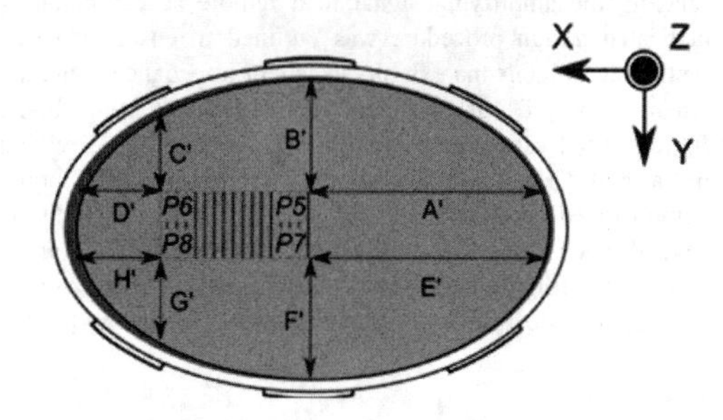

Figure 7. Scan area for the evaluation of the multi-node setup

B. *Multi-node setup*

We now consider a scan area deep inside the body phantom, between the points P5(192mm,105mm), P6(385mm,105mm), P7(192mm,150mm) and P8(385mm,150mm), as shown on Fig. 7. In Table I, the received signal-to-noise (SNR) ratios are compared for different orders of diversity. For dual-order diversity, antennas 4a and 4b are used, antennas 1a, 1b, 3a, 3b realize four-order diversity and all side antennas implement 6th-order diversity. We notice that making use of the antennas positioned on the back of the patient yields the highest signal levels. However, in Table 1 the most relevant figure in terms of links reliability is the minimum achieved SNR, which clearly increases for an increasing number of antennas. The 8th-order diversity setup guarantees a minimal received SNR of 10dB, which is required for highly reliable live wireless video streaming at a minimal bitrate of 3.5Mbit/s within a bandwidth of 1MHz. Yet, including the bottom antennas 4a and 4b in the 8th-order diversity scheme only yields marginal improvements in diversity gain, as they only partly cover the scan area. Therefore, a better tradeoff between link reliability and patient comfort might be obtained by using the 6th-order diversity scheme with a transmit power level of 20mW at the implant, which is still allowed by the SAR limits and which provides an additional 3dB margin in terms of SNR.

TABLE I
STATISTICAL PARAMETERS FOR VARYING NUMBER OF ANTENNAS, WHERE THE GAIN IS BASED ON THE 10% OUTAGE PROBABILITY LEVELS OF THE CDF

Number of Antennas	Max. (dB)	Min. (dB)	Mean (dB)	Median (dB)	Gain (dB)
8	27.99	10.85	16.89	17.32	9.25
6	11.83	8.74	10.25	10.16	7.65
4	10.97	6.87	8.69	8.56	6.15
2	27.92	3.34	14.40	16.34	2.45

V. CONCLUSION

A wearable textile repeater system was designed and deployed in a multi-antenna setup implementing receive diversity. Each repeater node consists of a receive antenna pointing towards to body to capture the signals of an implant, an analog amplifier and an off-body textile transmit antenna relaying the amplifying signal to a remote access point. A dedicated design procedure was outlined to ensure that the textile receive antenna exhibits excellent impedance matching when deployed at different locations on the body. Experimental validation by means of a standard setup relying on a body phantom filled with muscle-mimicking liquid demonstrated a significant increase in SNR. In addition, several nodes were deployed in a multi-antenna diversity reception configuration. It was shown that 6th-order diversity provides the ideal tradeoff between patient comfort and a link reliability that is sufficient to implement high frame-rate video transmission.

ACKNOWLEDGMENT

The authors thank Sioen Industries for their support and the Belgian Science Policy Office (IUAP Program) and the FWO-V (Research Foundation - Flanders) for their financial support.

REFERENCES

[1] Godara, Balwant; Nikita, Konstantina S (Eds.), *Wireless Mobile Communication and Healthcare*, Third International Conference, MobiHealth 2012, Paris, France, November 21-23, 2012, Revised Selected Papers, ISBN 978-3-642-37892-8.

[2] I. Khan, P. S. Hall, A. A. Serra, A. R. Guraliuc, and P. Nepa, "Diversity performance analysis for on-body communication channels at 2.45 GHz," *IEEE Transactions on Antennas and Propagation*, vol. 57, no. 4, pp. 956–963, Apr. 2009.

[3] P. Van Torre, L. Vallozzi, C. Hertleer, H. Rogier, M. Moeneclaey, J. Verhaevert, "Indoor off-body wireless MIMO communication with dual-polarized textile antennas," *IEEE Trans. on Antennas and Propagation*, vol. 59, no. 2, pp.631-642, Feb. 2011.

[4] D. Kurup, W. Joseph, G. Vermeeren and L. Martens, "In-body path loss model for homogeneous human tissues," *IEEE Trans. on Electromagnetic Compatibility*, vol. 54, no. 3, pp. 556 - 564, June 2012.

[5] G. Iddan, G. Meron, A. Glukhovsky and P. Swain, "Wireless Capsule Endoscopy," *Nature*, vol. 405, no. 6785, p. 417, May 2000.

[6] A. Kiourti and K. Nikita, "Miniature Scalp-Implantable Antennas for Telemetry in the MICS and ISM Bands: Design, Safety Considerations and Link Budget Analysis," *IEEE Trans. on Antennas and Propagation*, vol. 60, no. 8, pp. 3568-3575, Aug. 2012.

[7] M. Scarpello, D. Kurup, H. Rogier, D. V. Ginste, F. Axisa, J. Vanfleteren, W. Joseph, L. Martens and G. Vermeeren, "Design of an Implantable Slot Dipole Conformal Flexible Antenna for Biomedical Applications," *IEEE Trans. on Antennas and Propagation*, vol. 59, no. 10, pp. 3556-3564, Oct. 2011.

[8 F.-J. Huang, C.-M. Lee, C.-L. Chang, L.-K. Chen, T.-C. Yo and C.-H. Luo, "Rectenna Application of Miniaturized Implantable Antenna Design for Triple-Band Biotelemetry Communication," *IEEE Trans. on Antennas and Propagation*, vol. 59, no. 7, pp. 2646-2653, Jul. 2011.

[9] Federal Communications Commission, "§ 95.627 MedRadio transmitters in the 401-406 MHz band," in *Code of Federal Regulations, Title 47-Telecommunications, Chapter 1-FCC, Subchapter D-Safety and Special Radio Services, Part 95-Personal Radio Services, Subpart E-Technical Regulations*, 2012.

[10] European Telecommunications Standards Institute, "Electromagnetic compatibility and Radio spectrum Matters (ERM); Wideband transmission systems; Data transmission equipment operating in the 2,4 GHz ISM band and using wide band modulation techniques," ETSI EN 300 328 V1.8.1, 2012-04.

[11] G.-S. Lien, C.-W. Liu, J.-A. Jiang, C.-L. Chuang and M.-T. Teng, "Magnetic Control System Targeted for apsule Endoscopic operations in the Stomach-Design, Fabrication, and in vitro and ex vivo Evaluations," *IEEE Trans. on Biomedical Engineering*, vol. 59, no. 7, pp. 2068-2078, July 2012.

[12] S. Agneessens, P. Van Torre, E. Tanghe, G. Vermeeren, W. Joseph and H. Rogier, "On-body wearable repeater as a data link relay for in-body wireless implants," *IEEE Antennas and Wireless Propagation Letters*, vol. 11, pp. 1714 - 1717, 2012.

[13] T. Castel, P. Van Torre, E. Tanghe, S. Agneessens, G. Vermeeren, W. Joseph, and H. Rogier, "Improved Reception of In-Body Signals by Means of a Wearable Multi-Antenna System" , *Int. Journal on Antennas and Propagation*, paper no. 328375, 8 pages, 2013.

Phase Characterization of 1-200 MHz RF Signal Coupling with Human Body

Nannan Zhang, Zedong Nie * and Lei Wang, Member, IEEE

Shenzhen Institutes of Advanced Technology
and Key Lab for Health Informatics, Chinese Academy of Sciences

***Abstract*-To continuously monitor health information using wireless sensors placed on a person is a promising application. Human body communication (HBC) is proposed as a low power, security and light-weight communication technology that could be applied in aforementioned applications. In this work, the characterization of HBC propagation delay and phase deviation were investigated for the first time. *In-situ* experiments were performed in our lab in the frequency bands:1MHz-200MHz. Four HBC propagation channels, i.e., from left arm to right arm, from left arm to right leg, from right leg to left leg and from left arm to left leg were investigated. Group delay was measured from 13 volunteers, and then the propagation delay and phase deviation were estimated by statistical analysis. The propagation delay values are almost equal for four communication channels in each sub-band:1-100MHz and 100-200MHz. It is concluded that propagation delay is independent with propagation channel. However, from 1MHz to 100MHz the delay is generally longer than that in the frequency band 100-200MHz. i.e., in channel 1, the mean delay is 17.06 ns in frequency band 1-100MHz, while the mean delay is 15.23 ns in frequency band 100-200MHz. Lognormal model is found to be the best fitting distribution for the normalized phase deviation according the maximum likelihood estimation (MLE) and Akaike information criterion (AIC).**

***Keywords*-human body communication (HBC), propagation delay, phase deviation, statistical analysis**

I. INTRODUCTION

The increasing use of real-time healthcare monitoring without constraining the activities of the user demands new designs of portable and wearable biomedical sensors, as well as convenient communication technology for transmitting distributed information from sensors to base in further processing[1]. Majority of body area network (BAN) researches attempted to apply existing wireless communication technologies, such as ultra wideband techniques [2] [3],Bluetooth , Zigbee [4], and other industrial scientific medical (ISM) band-based communication protocols. Recently, a novel near field communication technology-human body communication (HBC) which uses the human body as propagation medium, shows its great potential for BAN applications[5]. Until now, many researches have shown that HBC has several advantages over wireless body area network (WBAN) schemes[6, 7]. The attenuation of body channel is much lower than that of the air channel because most of the signal from the transmitter is confined to the body. Little interference also is an advantage over human body, due to its near-field-coupling operation.

Human body communication channel modeling is important to the understanding of the communication mechanism and to the transceiver and system design. Many studies have been made on the development of the HBC, they were mainly focused on the following issues:

- Electromagnetic field distributions around human body and equivalent circuit models [8-10].
- HBC propagation path gain, fade duration and fade depth with different HBC system design elements, such as: the different electrodes size, the different frequency bands, the channel is dynamic or not[11, 12].

However, the detailed phase characterization of HBC is not well understood. Phase distortion has a great influence for digital communication which would cause inter symbol interference (ISI) and increase bit error ratio (BER). In this work, we have investigated the phase characterization of 1-200 MHz RF signal coupling with human body.

The paper is organized in four sections. Experimental setup is introduced in Section II. Section III describes the results of propagation delay, normalized phase deviation and its fitting distribution. Finally, section IV draws the conclusions.

II. EXPERIMENTAL SETUP

A. Group Delay

Group delay is a measure of the time delay of the amplitude envelopes of a signal through a device under test (DUT), and is a function of frequency for each component. Group delay could be calculated by differentiating the DUT's phase response versus frequency, which is also a useful measure of phase distortion.

The function for group delay T_g is:

$$T_g = -\frac{1}{2\pi}\frac{d\phi(f)}{df} \tag{1}$$

Where $\phi(f)$ is the phase response, f is the frequency of signal. Group delay variation means that signals consisting of multiple frequency components will suffer distortion because these components are not delayed by the same amount of time. In this experiment, the propagation delay is calculated by averaging group delay value at each measured point while

* Zedong Nie is with the Shenzhen Institutes of Advanced Technology, Chinese Academy of Science, Shenzhen, China. E-mail:zd.nie@ siat.ac.cn

978-7-5641-4279-7

phase distortion is indicated by the phase deviation from average delay[13].

B. Experimental Configuration

In HBC, the frequency band is suggested to be below 100MHz. When the carrier frequency is above 100MHz, human body acts as an antenna and the communication is no longer limited to the human body [14]. In order to better understand the HBC phase characteristics, the frequency band 1-200MHz is set in this paper, and in the following analysis, the frequency band 1-200MHz is divided into two sub bands:1-100MHz and 100-200MHz.

The Agilent E5061A Vector Network Analyzer (VNA) was adopted to measure the group delay. Just as Fig. 1(a) shows, the volunteer stood in front of the VNA and were tied with a pair of electrodes by bandages. Fig. 1(b) shows the different electrode positions placed on the body and the measurement channels. The four HBC propagation channels are from left arm to right arm, from left arm to right leg, from right leg to left leg and from left arm to left leg respectively. Fig. 1(c) illustrates a 4cm×4cm copper electrode which is used to couple the signal to the body.

There were 13 volunteers (8 male and 5 female) selected, they were requested to keep standing still in whole experiment. The body weight of the volunteers varied from 40 to 70 kg, and the body height varied from 150 to 175 cm with the average age being 24 years old.

III. RESULTS

A. The Results of Propagation Delay

Fig. 2 shows the mean value of propagation delay and its standard deviation of two measured frequency bands: 1-100MHz and 100-200MHz, Table I illustrates the detailed value.

From 1-100 MHz, the delay values of four channels are basically equal. For example, in four measured channels, channel 2 has the maximum propagation delay 18.37 ns and channel 4 has the minimum propagation delay 16.86 ns. The delay values of channel 1 and channel 3 are 17.06 ns and 17.74 ns respectively. The maximum value of propagation delay is only larger 1.56 ns than the minimum one in four channels.

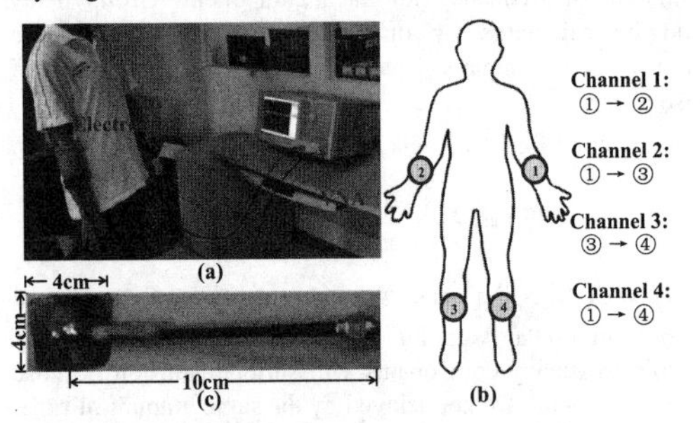

Figure 1. (a) Experimental scenario setup; (b)Measurement configuration, channel 1:left arm to right arm ;channel 2:left arm to left leg; channel 3:left leg to right leg ; channel 4： left arm to right leg;(c) TX and RX signal electrode.

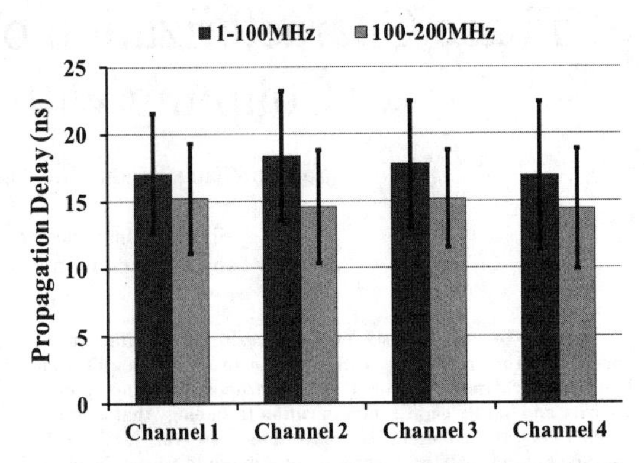

Figure 2. The propagation delay value in the frequency bands: 1-100MHz and 100-200MHz against channel for all subjects. The error bar represents the standard deviation from the mean

TABLE I

MEAN DELAY AND STANDARD DEVIATION AT EACH CHANNEL

	Mean(ns)		Standard Deviation(ns)	
	1-100MHz	100-200MHz	1-100MHz	100-200MHz
Channel 1	17.06	15.23	4.47	4.09
Channel 2	18.37	14.56	4.79	4.19
Channel 3	17.74	15.14	4.71	3.61
Channel 4	16.86	14.39	5.51	4.45

From 100-200 MHz, the delay values are almost equal. For instance, channel 1 has 15.23 ns propagation delay and channel 3 has 15.14 ns group delay. Channel 2 has 14.56 ns delay value and channel 4 has 14.39 ns propagation delay. It was obvious observed that maximum delay value is larger 0.84 ns than the minimum of four channels in frequency band 100-200MHz.

Based on the analysis above, the delay values are almost equal in the frequency band 100-200MHz in different propagation channels. In frequency band 1-100MHz, the propagation delays in four channels are also equal. It could be concluded that the propagation delay is independent on propagation channel. However, as the Fig. 2 shows, the signal propagation delay values in the frequency band 1-100MHz are greater than these in the frequency band 100-200MHz.

B. The Result of Phase Distortion

Linear phase delay is represented by a flat group delay response. The deviation of constant group delay indicates phase distortion of the measured channel at each frequency point. Phase distortion has a great influence for digital communication which would cause inter symbol interference (ISI) and increase bit error ratio (BER). The result of phase distortion provides reference for transceiver designs.

Fig. 3 depicts the normalized distortion of phase against frequency from 1 MHz to 200MHz. It is illustrated that the phase deviation value is lower and varies slightly in the frequency band 100-200MHZ. The lowest value of normalized phase deviation is 2.53 ns at 150 MHz frequency and the largest value is 7.69 ns at 130MHz frequency in the frequency band 100- 200MHz. In the contrary, from 1 MHz to 100MHz, the phase deviation varies largely between 3.41 ns to 12.99 ns.

C. *The Probability Density Function of Phase Distortion*

To model the phase deviation accurately ,three well-known distribution models for normalized phase deviation of four channel such as： Lognormal, Gamma, Weibull distributions are adopted to find the besting fitting distribution. We have obtained distribution parameters by using the maximum likelihood estimation (MLE) first, then calculate AIC value according Akaike information criterion (AIC) [15].

The AIC is defined as follow:

$$AIC = 2k - 2\ln(L) \quad (2)$$

Where k is the number of parameters in the statistical model, and L is the maximized value of the likelihood function for the estimated model.

Table II summarizes the best fitting distribution functions and the estimated parameters by AIC and MLE, respectively. It is observed that Lognormal is the preferred model which has the minimum AIC value. Fig. 4 indicates the probability density function (PDF) of normalized phase deviation and fitting model. All parameters were calculated on a 95% confidence interval (CI).

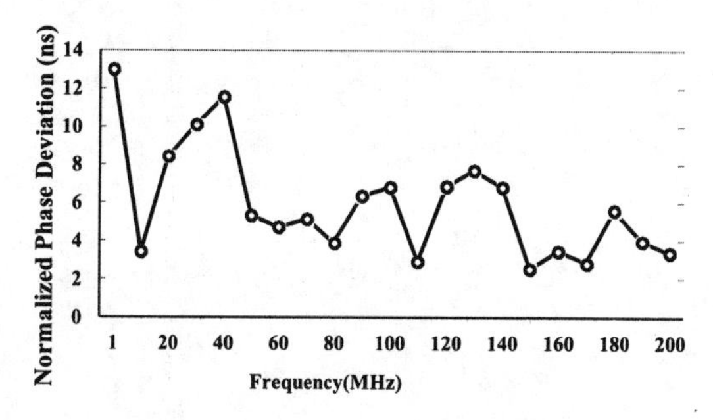

Figure 3. Normalized distortion of phase against frequency

TABLE II

STATISTIC PARAMETER AND CANDIDATED MODEL

Model	Log likelihood	AIC
Lognormal	-712.559	1429.118
Gamma	-720.262	1444.524
Weibull	-711.421	1468.788

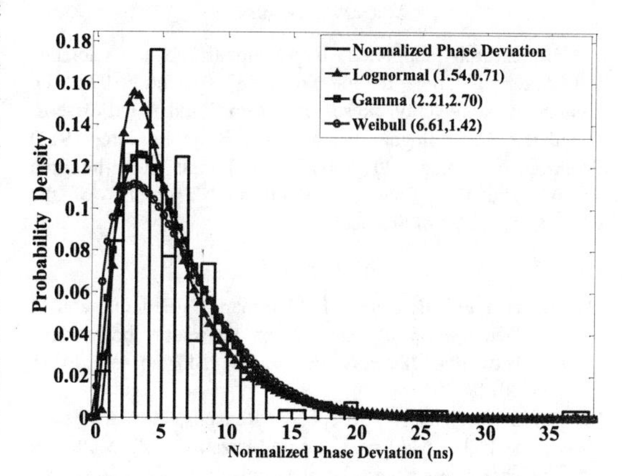

Figure 4. Empirical PDF of the normalized phase deviation for four channels and the fitting model

IV. CONCLUSION

In this paper, the phase characteristic of 1-200MHz RF signal coupling with human body is characterized *in situ*. We acquired group delay in the frequency band 1-200MHz with 13 subjects. The statistical results show that the phase characteristic is quiet different between two measured frequency bands: 1-100MHz and 100-200MHz.

From the general statistics view and considering the possible accidental error, the propagation delay is equal approximately in each frequency sub-band 1-200MHz and 100-200MHz. It could be concluded that the propagation delay is independent on propagation channel in that two sub-bands. However, the signal propagation delay in the frequency band 1-100MHz is generally longer than these in the frequency band 100-200MHz.We infer that the delay difference between two sub-bands may be caused by the different transmission mechanism:

1) Between1-100MHz, most signal is coupled with body and human body is treated as a special kind of transmission medium
2) Between 100-200MHz, the signal mainly couple with the air around body. The human body acts as an antenna and the communication is no longer limited to the human body.

Furthermore, we have given the profile of normalized phase deviation against frequency. The profile implies that the frequency band 100-200MHz has more optimal phase-frequency characteristic because the lower normalized phase deviation in this band. By using the MLE and AIC criterion, Lognormal distribution is found to be the best fitting distribution for the phase deviation.

ACKNOWLEDGMENT

The research supported by National Natural Science Foundation of China (Grant Nos.60932001 and 61072031), Guangdong Innovation Research Team Fund for Low-cost Healthcare Technologies, the National Basic Research (973) Program of China (2010CB732606) ,the "One-hundred Talent" and the "Low-cost Healthcare" Programs of the Chinese Academy of Sciences.

REFERENCES

[1] B. Latré, B. Braem, I. Moerman, C. Blondia, and P. Demeester, "A survey on wireless body area networks," *Wireless Networks,* vol. 17, no. 1, pp. 1-18, 2011.

[2] L. Roelens, W. Joseph, E. Reusens, G. Vermeeren, and L. Martens, "Characterization of scattering parameters near a flat phantom for wireless body area networks," *Electromagnetic Compatibility, IEEE Transactions on,* vol. 50, no. 1, pp. 185-193, 2008.

[3] Y. P. Zhang and Q. Li, "Performance of UWB impulse radio with planar monopoles over on-human-body propagation channel for wireless body area networks," *Antennas and Propagation, IEEE Transactions on,* vol. 55, no. 10, pp. 2907-2914, 2007.

[4] E. Monton, J. Hernandez, J. Blasco, T. Herve, J. Micallef, I. Grech, A. Brincat, and V. Traver, "Body area network for wireless patient monitoring," *IET communications,* vol. 2, no. 2, pp. 215-222, 2008.

[5] T. G. Zimmerman, "Personal area networks: near-field intrabody communication," *IBM systems Journal,* vol. 35, no. 3.4, pp. 609-617, 1996.

[6] M. Seyedi, B. Kibret, D. T. Lai, and M. Faulkner, "A Survey on Intrabody Communications for Body Area Network Applications," 2013.

[7] M. S. Wegmueller, M. Oberle, N. Felber, N. Kuster, and W. Fichtner, "Signal transmission by galvanic coupling through the human body," *Instrumentation and Measurement, IEEE Transactions on,* vol. 59, no. 4, pp. 963-969, 2010.

[8] R. Xu, H. Zhu, and J. Yuan, "Electric-field intrabody communication channel modeling with finite-element method," *Biomedical Engineering, IEEE Transactions on,* vol. 58, no. 3, pp. 705-712, 2011.

[9] K. Fujii, M. Takahashi, and K. Ito, "Electric field distributions of wearable devices using the human body as a transmission channel," *Antennas and Propagation, IEEE Transactions on,* vol. 55, no. 7, pp. 2080-2087, 2007.

[10] N. Haga, K. Saito, M. Takahashi, and K. Ito, "Equivalent Circuit of Intrabody Communication?? Channels Inducing Conduction Currents inside the?? Human Body," 2013.

[11] D. B. Smith, L. W. Hanlen, J. A. Zhang, D. Miniutti, D. Rodda, and B. Gilbert, "First-and second-order statistical characterizations of the dynamic body area propagation channel of various bandwidths," *annals of telecommunications-annales des télécommunications,* vol. 66, no. 3-4, pp. 187-203, 2011.

[12] Z. Nie, J. Ma, Z. Li, H. Chen, and L. Wang, "Dynamic propagation channel characterization and modeling for human body communication," *Sensors,* vol. 12, no. 12, pp. 17569-17587, 2012.

[13] A. Agilent, "1287-1: Understanding the Fundamental Principles of Vector Network Analysis," ed: Agilent Technologies, 2002.

[14] H. Baldus, S. Corroy, A. Fazzi, K. Klabunde, and T. Schenk, "Human-centric connectivity enabled by body-coupled communications," *Communications Magazine, IEEE,* vol. 47, no. 6, pp. 172-178, 2009.

[15] K. P. Burnham and D. R. Anderson, *Model selection and multi-model inference: a practical information-theoretic approach*: Springer, 2002.

Electromagnetic Wave Propagation of Wireless Capsule Antennas in the Human Body

Zhao Wang, Enggee Lim, Meng Zhang, Jingchen Wang, Tammam Tillo and Jinhui Chen
Department of Electronic and Electrical Engineering, Xian Jiaotong-Liverpool University,
No. 111, Ren'ai Road, Suzhou,
215123, P. R. China

***Abstract*- Wireless Capsule Endoscopy (WCE) uses an ingestible small capsule-shaped device to detect various diseases within the digestive system. It is superior to traditional endoscopy as WCE lacks the limitations of traditional wired diagnostic tools, such as the cable discomfort and the inability to examine highly convoluted sections of the small intestine. However, a number of obstacles still need to be overcome to improve the clinical applications. This paper attempts to investigate the performance of a WCE system by studying its electromagnetic (EM) wave propagation through the human body, which allows the capsule's positioning information to be obtained. The WCE transmission channel model is constructed to evaluate signal attenuations and to determine the capsule position.**

I. INTRODUCTION

The use of endoscopes to examine the body's internal organs dates back to the 19th century [1], where a Mainz scientist developed the 'Lichtleiter' to examine human bladder and bowel with candle light. Later, various types of endoscopes were developed to examine the body's internal organs in greater detail. Timely detection and diagnosis are extremely important since the majority of gastrointestinal (GI) cancers are curable if caught early. Surgical treatments through the use of endoscopies were developed into two branches: gastroscopy for examining the stomach and colonoscopy for the colon. Each branch developed rapidly in the last two decades, eventually culminating in the birth of capsule endoscopy. Compared to earlier techniques, capsule endoscopy is non-invasive and hence more comfortable to patients. It can examine deeper GI tracts in the human body inaccessible with existing wired endoscopes.

Wireless Capsule Endoscopy (WCE) is a technique in which a small capsule-shaped device containing a video camera, LED lights, a power source and a wireless transmitter, is ingested in order to detect various diseases within the digestive system (e.g. in duodenum, jejunum, ileum, etc.). There are many different types of WCE systems and they are mostly developed and manufactured by Olympus, Intromedic and Given Imaging. Although this technique develops very fast, there are still some drawbacks limiting the application of WCE. First of all, the collected physiological data, such as the GI tract images, are insufficient for clinical diagnosis without the presence of capsule positioning data. Secondly, most capsules are powered by an internal battery that restricts the capsule miniaturization. Lastly, current systems do not have continuous communication due to random orientations of the capsule [2].

This paper investigates the performance of a WCE communication system by studying its EM wave propagation. This investigation serves to determine signal attenuation and capsule position. There are two main reasons for this proposed research. The first one is that some EM energy is absorbed by the organs when waves are transmitted through the human body, which could lead to large signal distortions. In addition, the human body is a frequency dispersive system with frequency dependent dielectric properties (i.e. permittivity and conductivity) [3] that influence the electrical and magnetic properties of the signal transmission channel. The parameters change when wide-band signals are applied to the system, which require human body models to simulate the signal transmission with frequency dependent permittivity and conductivity. Furthermore, positioning of the capsule in the human body can be achieved by studying the EM wave propagation of the system, enabling tracking of position and orientation of the capsule without adding additional sensors. This allows more capsule space to be allocated for other components.

For the above reasons, this project works through three aspects. Firstly, the level of signal attenuation in the transmission system is investigated. Secondly, the EM wave propagation properties of WCE with different transmission distances when the transmitting and receiving antennas are in the same work plane (z=0 plane) are evaluated through simulation. Lastly, the second step is repeated but with varying work planes to simulate cases where the transmitting and receiving antennas are not in the same plane ($z\neq0$ planes).

II. WCE COMMUNICATION SYSTEM MODEL

EM wave propagation of the WCE transmission channel is studied by examining signal distortions, extracting the WCE's location information and determining the capsule position. Therefore, a WCE communication system is built with a transmitter, a receiver and a communication channel in the following subsections.

A. *EM wave propagation environment*

The abdominal environment is highly complex and the small intestines, which lie in close proximity, greatly influence the performance of the capsule. Therefore, the inhomogeneous human body module is simplified to a homogeneous body

978-7-5641-4279-7

model which uses muscle material whose relative permittivity equals 56 and conductivity equals 0.83 S/m [4]. Based on previous studies, the shape of the body model, cylinder or elliptical cylinder, does not influence the results much. In this paper, a cylinder with the radius of 100 mm is used as the human trunk model.

B. *Transmitting and Receiving Antennas*

To implement the communication system, suitable transmitting (TX) and receiving (RX) antennas are selected to operate in the human body environment. The WCE antenna should be less sensitive to human tissue influences as the EM wave transmits in the body. Lossy dielectric material absorbs a number of waves and thus attenuates the receiving signal, causing strong negative effects on the EM wave propagation. A much wider bandwidth is required to enable transmission of high resolution images and large amounts of data. The detection of transmitted signal is preferred to be independent of the transmitter's position and hence, the transmitting antenna should have an omnidirectional radiation pattern.

Two sets of transmitting and receiving antennas are selected for the study of the communication system: a pair of spiral antennas designed by Yoon [4] and a pair of planar antennas proposed by our group [5].

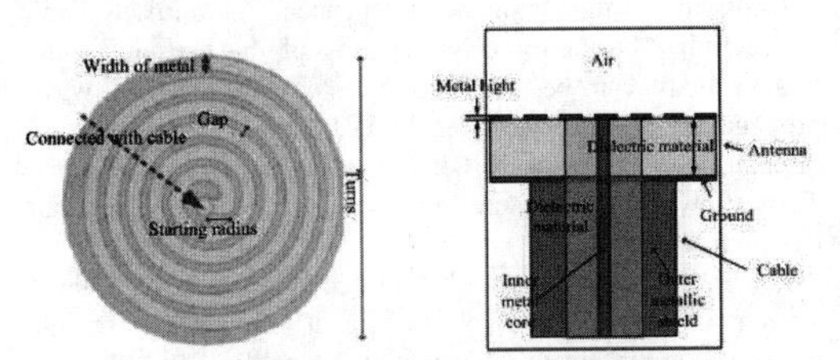

Figure 1. Top and side views of the spiral antenna [4]

The top and side views of the spiral antenna are illustrated in Figure 1. By adjusting the parameters, the antenna can be tuned to work at 403 MHz with 85 MHz bandwidth. In section III, the spiral antennas are used as TX and RX to investigate the signal transmission in the WCE system.

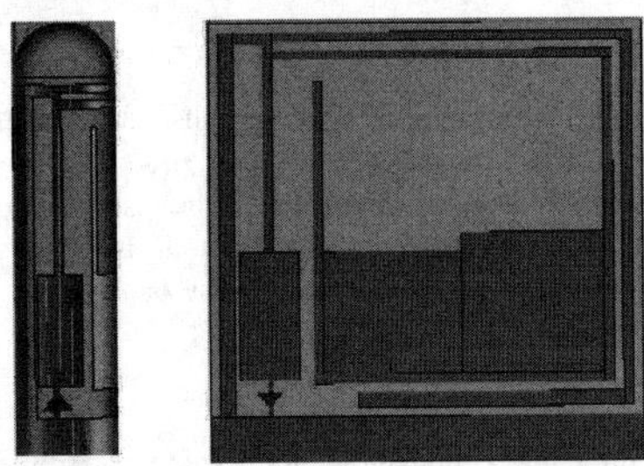

Figure 2. Rolled and planar microstrip line antenna [5]

A carefully designed planar microstrip antenna (as shown in Figure 2) can also be used in the WCE system. It is rolled up and attached to the surface of the capsule shell to work as the transmitter, as shown in the left of Figure 2. When working as receiver, the planar structure (in the right of Figure 2) is used. The center operating frequencies of these two antennas are 410 MHz and the bandwidths are more than 180MHz. It is used in section III.C to compare with spiral antennas.

C. *The two-port network for WCE communication system*

The communication system including the transmitter (TX), the intermediate material and the receiver (RX) can be considered as a two port network as shown in Figure 3.

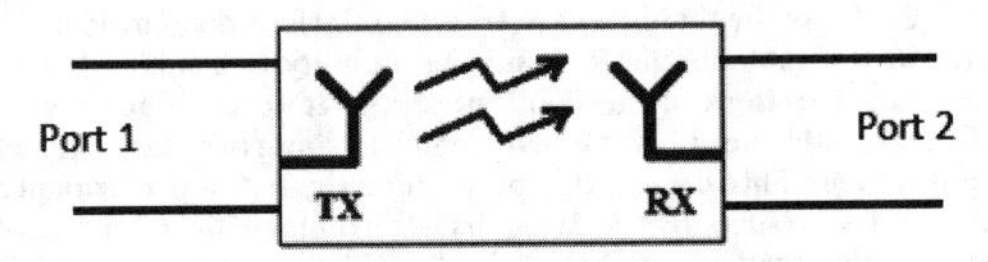

Figure 3. The two-port network of the WCE system

The terminal of the TX's cable is set as the port 1, whilst port 2 is at the end of RX's cable. Therefore, S-parameters can be used to analyze this system. S11 is the return loss used to determine the channel bandwidth. S21 the forward transmission parameter is used to evaluate the signal transmission between TX and RX.

III. EM WAVE PROPAGATION EVALUATIONS

A. *Relative angle position between TX and RX (z=0 plane)*

Antennas are not symmetrical in general; therefore the influence of the antenna radiation pattern (RP) should also be taken into consideration. To test the system, one direction of the transmitter with the appropriate radiation pattern is chosen. Excitation signals are supplied into the transmitting antenna while signals at the receiving antenna are compared to the input signal to check for attenuations. The receiver is placed in z=0 plane at different angles, surrounding the transmitter and separated by 45 degrees each (illustrated in Figure 4).

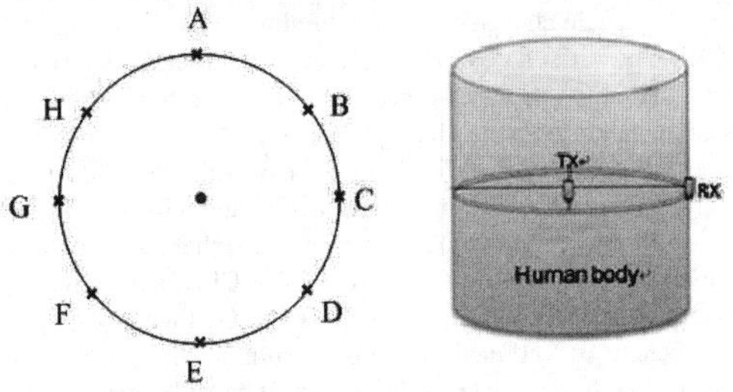

Figure 4. Testing points for evaluating the RP effects

The calculated forward transmission parameters S21 at 403 MHz are plotted in below of Figure 5, and compared with the transmitter's directivity calculated at different angles in the examined plane (as plotted in above of Figure 5).

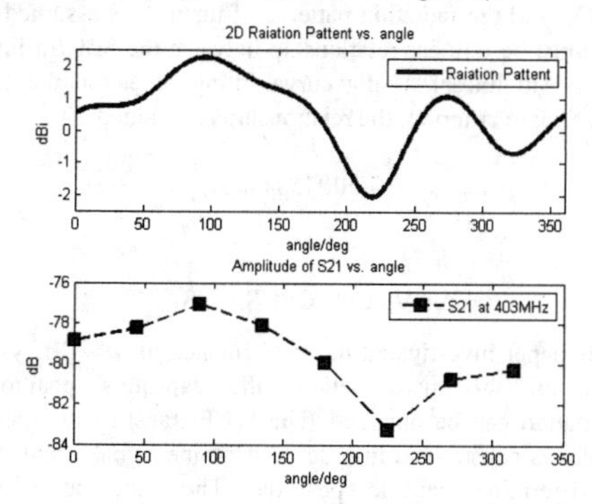

Figure 5. Radiation pattern's 2D plot vs. S21 at 403MHz

According to the results shown in Figure 5, the S21 is almost proportional to the radiation pattern of tested antenna plane when the receiving antenna is placed at different positions. The unstable forward voltage gain influences the accuracy of localization results. If the capsule is rotated around the center of the outer shell, errors are introduced to the localization calculations. This influence needs to be taken into consideration while performing the localization estimation.

B. *Relative Distance between TX and RX (z=0 plane)*

At this stage, different offsets between transmitter and receiver are applied to this system to collect the simulation results of S21. To perform capsule localization, signal transmission distances are swept from 0 to 160 mm with 20 mm steps. With the position of RX fixed, TX is moved in the human body model to obtain the EM wave propagation properties of WCE with different offsets as shown in Figure 6.

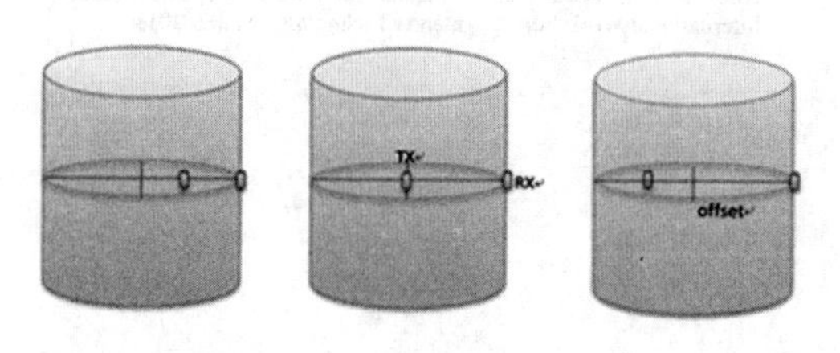

Figure 6. Layouts of TX and RX (offsets: 50, 100, and 150 mm)

The simulation results of S21 at different frequencies are shown in Figure 7.

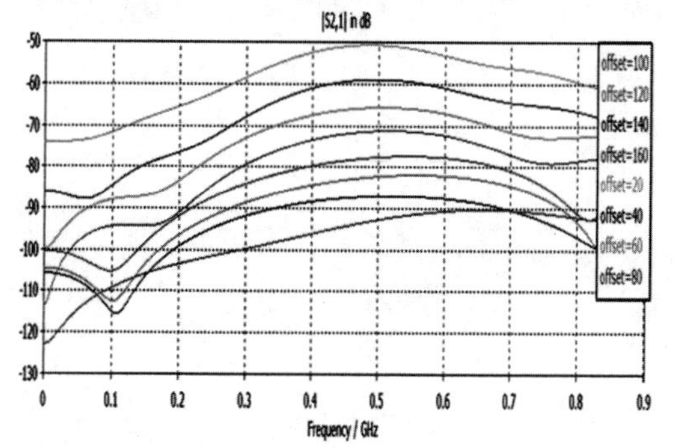

Figure 7. S21 (in dB) vs. various offsets between TX and RX

It can be observed that within 300 MHz to 500 MHz, the S21 results are regular and do not have sudden changes. Hence, these groups of data are used to fit the channel transfer function and used in the evaluation of the communication system. Curve fitting is performed based on the Least Mean Square criterion, and the fitted transfer function is a function of frequency (*f*) and offset (*r*) as follows:

$$S21(f,r)(\mathrm{dB}) = 20\log_{10}(c_1(f)r^{-2} + c_0(f)) \qquad (1)$$

where, *r* is offset in mm, and *f* is frequency in Hz. $c_1(f)$ and $c_0(f)$ are the frequency dependent coefficients:

$$c_1(f) = -18.72f^2 + 20.31f - 3.699$$

$$c_2(f) = 0.0008371f^2 - 0.0008211f + 0.0001555$$

S21 calculated from the transfer function and S21 obtained from simulation are compared in Figure 8.

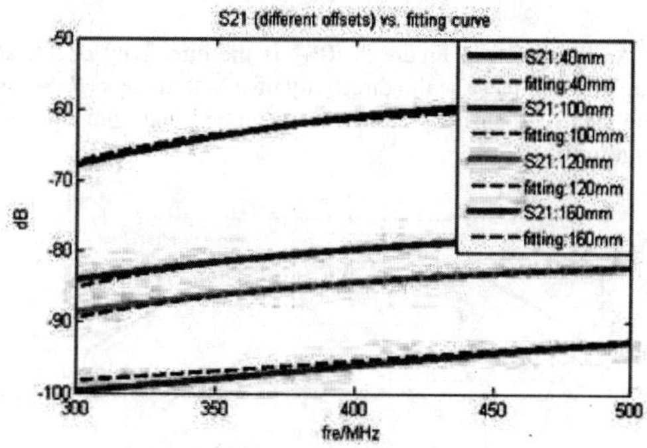

Figure 8. Fitted transfer function compare with simulated S21 (offset: 40, 100, 120 and 160mm)

Figure 8 reveals the compared S21 results calculated from the transfer function and the results achieved by simulation using Gaussian signal as the excitation signals. Based on the transfer function, the position of TX can be estimated according to the received signal.

C. *Relative Position between TX and RX (z≠0 plane)*

Since the position of the capsule endoscope in the gastrointestinal tract is changing, knowing the transmission characteristics of the wireless signal in the same plane is not sufficient. The step in Section III.B will be repeated but with the TX and RX in different planes so that the location and quantity of the receiver and transmitter can be determined.

Three-dimensional positioning and tracking of the capsule endoscope is therefore possible.

Moving TX with RX fixed is equivalent to moving RX with TX. In this study, for the simplicity of modeling, the TX is fixed at the center of the body model. As RX moves along the z-axis, the radio propagation properties of WCE with different signal transmission distances when the transmitting and receiving antennas in the different plane (z≠0 plane) can be obtained. The forward transmission coefficient S21 at specific frequency can be obtained from the simulation, which is related to the radiation patterns of TX, RX and the distance between them.

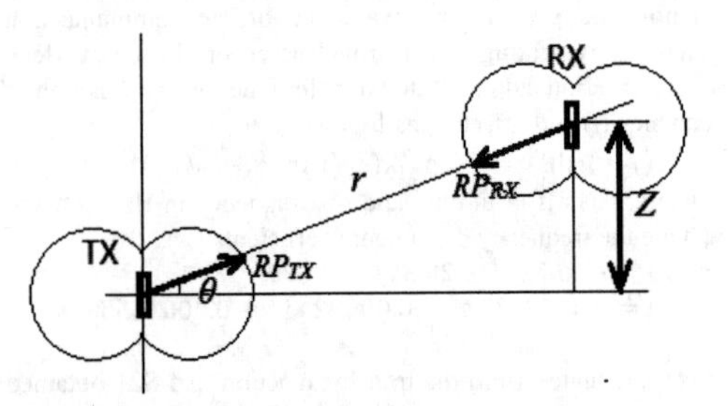

Figure 9. The relationship between RP_{TX} and RP_{RX}

As illustrated in Figure 9, RP_{TX} is the directivity of TX at angle θ, and RP_{RX} is the directivity of RX at angle $\pi+\theta$, S_{RP} is the sum of them. S_{RP} and S21 are plotted and compared in Figure 10.

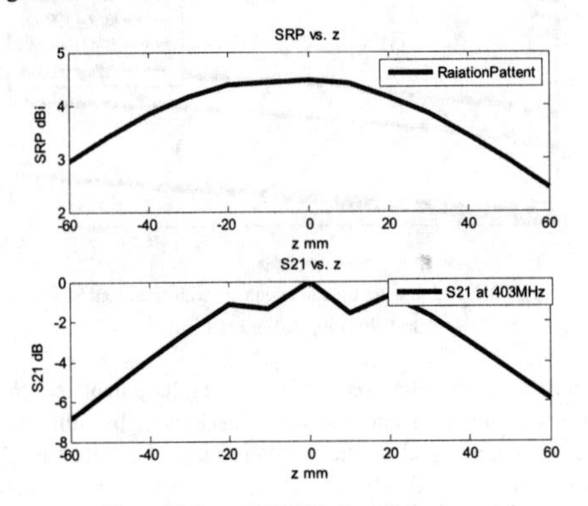

Figure 10. S_{RP} and S21 VS z for spiral antenna pair

Both SRP and S21 have peak value at z = 0 mm, and reducing with the increasing z. Irregular values of S21 are observed at z = ±10 mm, which is probably due to the ground plane shielding effect of the spiral antenna. By replacing the spiral antenna with the microstrip antennas, the irregular drops at z = ±10 mm disappear as shown in Figure 11.

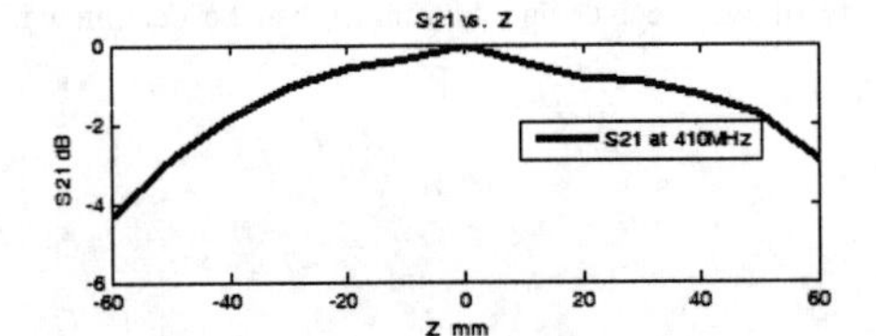

Figure 11. S21 VS z for microstrip antenna pair

On the basis of the Friis formula, the total loss between the transmitter and receiver is related to the distance between TX and RX, and the radiation patterns of them. It is assumed that there must be a linear relationship between the S21 (in linear) and S_{RP} (in linear). Using curve fitting based on the Least Mean Square criterion, the relationship is obtained:

$$S_{21(linear)} = 0.0001S_{RP(linear)} - \frac{0.0056}{r}$$

IV. CONCLUSIONS

This paper investigated the performance of a WCE system. Based on this investigation, the capsule's positioning information can be obtained. The WCE transmission channel model was constructed in order to examine signal attenuations and determine capsule position. The outcome of this investigation will be useful for researchers to carry out further research in locating the WCE position within the human body.

ACKNOWLEDGMENT

This work is partially supported by the Natural Science Foundation of Jiangsu province (No. BK2011352) and Suzhou Science and Technology Bureau (No. SYG201211).

REFERENCES

[1] Endoscopy (2011), From Wikipedia, the free encyclopedia: http://en.wikipedia.org/wiki/Endoscopy.

[2] Z. Fireman, and Y. Kopelman, "New frontiers in capsule endoscopy". J Gastroenterol Hepato, pp. 1174–1177, 2007.

[3] "The Finite Integration Technique' (2011), CST Computer Simulation Technology AG", available at: http://www.cst.com/Content/Products/MWS/FIT.aspx.

[4] S. I. Kwak, K. Chang, and Y. J. Yoon. "The helical antenna for the capsule endoscope system". in Proc. IEEE Antennas Propag. Symp. vol. 2B, pp. 804–807, Jul. 2005.

[5] J. C. Wang, E. G. Lim, Z. Wang, Y. Huang, T. Tillo, M. Zhang and R. Alrawashdeh, "UWB Planar Antennas for Wireless Capsule Endoscopy". International Workshop on Antenna Technology, March 2013.

TP-1(B)

October 24 (THU) PM

Room B

SIW Antennas & Devices

Substrate Integrated Waveguide Antenna Arrays for High-Performance 60 GHz Radar and Radio Systems

Ajay Babu Guntupalli and Ke Wu

Poly-Grames Research Center, Electrical Engineering department
Ecole Polytechnique (University of Montreal)
Montreal, Canada, H3T 1J4.

***Abstract*- This article presents and discusses some of the recent developments on 60 GHz antenna arrays in substrate integrated waveguide (SIW) technology. They include planar linearly polarized (LP) antenna, three dimensional (3-D) high gain LP antenna, multi-layer single circularly polarized (CP) antenna, and 3-D dual circularly polarized antenna (DCP) over 60 GHz frequency range. Single LP antenna is realized by loading anti-podal linearly tapered slot antenna (ALTSA) with dielectric rod antenna. The dielectric rod antenna is used as a radiating element to construct 4 × 4 antenna array. Single right-hand circularly polarized (RHCP) antenna is constructed by using SIW as feeding network and CP patch as radiating element. 2 × 2 RHCP array makes use of an aperture coupling method to excite each antenna element. In the final prototype, DCP antenna is constructed by using multi-layer E-plane coupler and rod radiating element. All the proposed antenna arrays are experimentally validated and compared with the simulated counterparts.**

I. INTRODUCTION

Millimeter-wave wave (or mm-wave) antenna arrays with single linear and dual circular polarization characteristics are widely being used for numerous applications including wireless data communication, radar sensors, passive imaging, energy harvesting, and cognitive radio systems. The simplified block diagram of a radar transceiver using single LP antenna and CP antenna is shown in Fig. 1a, b. Radar with CP antenna can detect chest and heart displacement irrespective of the patient position [1]-[3].The sensor can always detect amplitude variations even polarization of patient is not matched with the transmitting antenna. Another application of CP antenna in energy harvesting techniques, where the rectified output voltage would become constant when compared with the scenario of an LP antenna. To satisfy front-end requirements of these systems, two different techniques have been proposed to obtain circular polarization at mm-wave frequency.

Among different types of feeding mechanism, waveguide is an excellent candidate to implement low-loss feeding networks and high gain antenna arrays over mm-wave frequency range [4]-[5]. Those waveguide-based antennas have been exhibiting excellent radiation characteristics, but they are not easy to

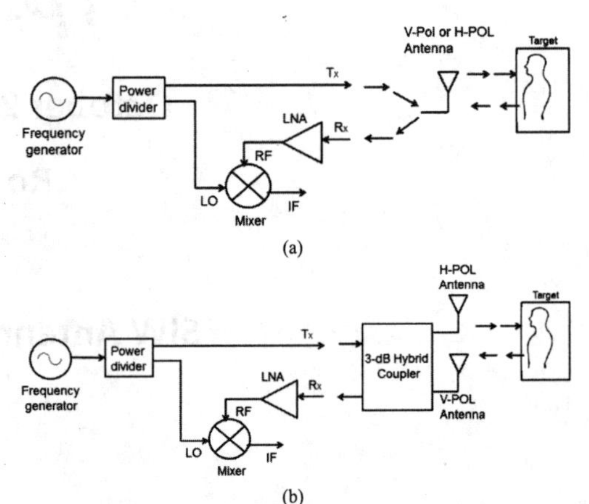

Figure.1. General simplified block diagram of a radar sensor using dual circular polarization antenna to detect human heart beat and respiration.

integrate with active components on a single substrate. The microstrip-fed patch array proposed in [6] is easy to characterize and to integrate on a single substrate. However, antenna efficiency is lower than 50% at the 60 GHz frequency range. Low temperature co-fired ceramic (LTCC) technology was used to implement high efficiency antenna arrays in [7]-[11]. The antenna efficiency was shown better than 90%, but the fabrication cost would be high.

At mm-wave frequency, SIW (substrate integrated waveguide) is an oustanding candidate to implement low loss and low cost feeding mechanism. SIW-fed antenna is able to yield high radiation efficiency and broadband impedance behavior. At 60 GHz, SIW-fed antenna arrays with linear polarization were proposed in [12]-[16] and circular polarization in [17] and [18]. In this work, frequency range from 57 GHz to 64 GHz is chosen to design a number of high performance antenna arrays with linear and dual circular polarizations. The proposed CP antenna can be integrated with the front end of the systems proposed in [19] and [20].

978-7-5641-4279-7

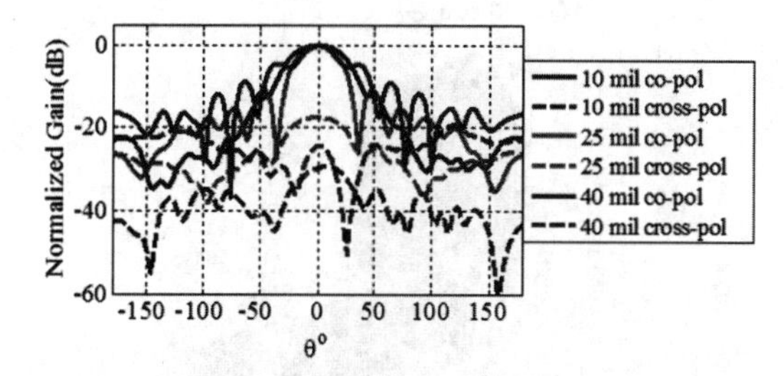

Figure. 2. E-plane co-pol and cross-pol radiation patterns as a function of substrate thickness.

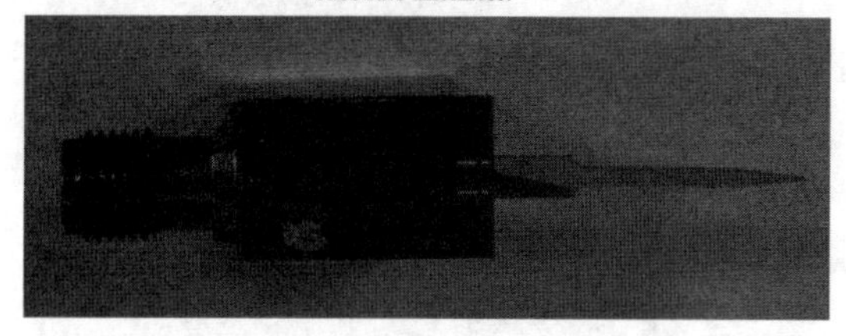

Figure. 3. ALTSA is loaded by a tapered dielectric rod.

II. PLANAR DIELECTRIC ROD ANTENNA

Among the substrate integrate circuits (SICs) family, substrate integrated image guide (SIIG) have been used in [21] and integrated non-radaitive dielectric waveguide (SINRD) have been used in [22] to feed a dielectric rod antenna. These two families of SICs require an additional transition to interface with the other parts of circuit. The third and well-known family member of SICs, is SIW that has also been used for the design of feeding networks, e.g., in [23] and to develop high-gain dielectric rod antenna arrays. In this work, ALTSA antenna is loaded with planar dielectric rod antenna to amplify the gain of the combination [18].

Antenna cross-pol and side-lobe level as a function of the substrate thickness is given in Fig. 2. As the substrate thickness increases, cross pol level and side lobe levels also increase. As the thickness increases, fields are loosely bounded within the dielectric substrate hence the polarization purity is also reduced. For 10 mil thickness, the cross-pol value is less than -29.35 dB and the worst side lobe value is about -26 dB. Experimental antenna prototype is shown in Fig. 3. To obtain a better matching condition, a grounded coplanar waveguide (GCPW) to SIW transition is designed and used to measure the cascaded section of ALTSA and rod antenna. The transition behavior is similar to the operating principle of a diploe antenna. Input impedance and radiation pattern are measured by using the V-connector. Thus, the final architecture of the proposed antenna is a series combination of GCPW transition, linearly tapered slot antenna and dielectric rod antenna.

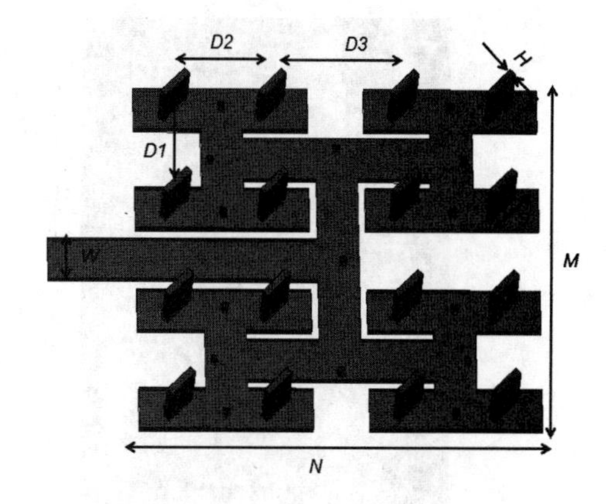

Figure. 4. Three dimensional view of the feed network where input port on the *XY*-plane and 16 output ports are located on the vertically placed SIW lines.

III. HIGH GAIN ANTENNA ARRAY

In this work, a 45° LP planar array utilizing 16 radiating elements is proposed for the 60 GHz frequency systems. Array occupies the total volume of $2.8\lambda \times 3.2\lambda \times 5.1\lambda$ with an average peak gain of 17.5 dBi over 8.3% of pattern bandwidth. Dielectric rod antenna gain is a function of rod length. SIW-fed rod antenna discussed in the previous section is selected as a radiating element to implement high efficiency antenna array. The feed network of 4 × 4 rod antenna array is shown in Fig. 4. The planar power divider and bend used at each intersection are optimized to have better impedance match over all the bandwidth. Sixteen antenna elements and feed network are integrated together to realize the final prototype. Antenna performance characteristics are measured by using a Southwest microwave end-launch connector. Simulated and measured input impedances (with $|S11|<-8$ dB) are matched over a bandwidth from 57 GHz to 64 GHz.

Far-field radiation pattern is measured in an anechoic chamber with 45° polarization set at transmitting horn antenna. Antenna gain is measured as 16.5 dB, 17 dB and 17.5 dB at 59 GHz, 60 GHz and 62 GHz frequency. Side lobe levels in both E-plane and H-plane are lower than –10 dB and cross-polarization is lower than -18 dB from the maximum gain value of 18 dBi. The thickness of 10 mil is chosen to improve the cross-polarization level of rod antenna. Nevertheless, the mechanical strength is reduced and also the rod antenna end-points are slightly misaligned along *Z*-axis of orientation.

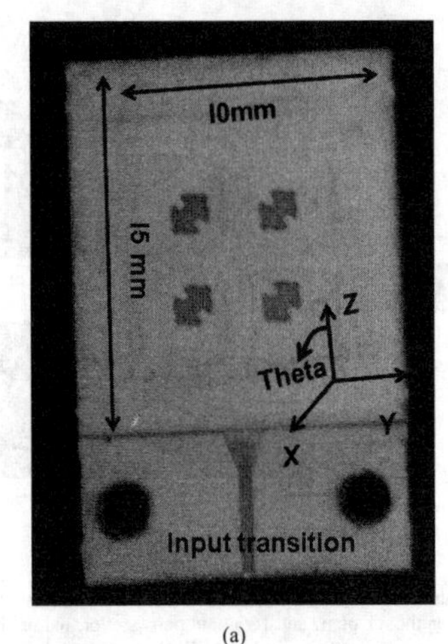

(a)

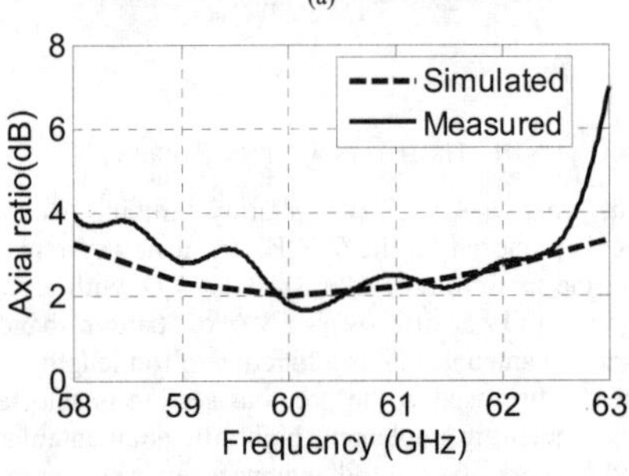

(b)

Figure. 5. Fabricated prototype of the Single RHCP array.

IV. SINGLE CP ANTENNA ARRAY

A parallel type of feeding topology is chosen to remove the beam squint effect, which naturally occurs in a series feeding topology. An SIW power divider and its physical parameters obtained through modeling are given in Fig. 3a. Magnitude of the input reflection coefficient is less than – 15 dB from 57 GHz to 64 GHz. The power is divided equally among the four output ports.

The final array construction in 3 layers is shown in Fig. 5a. The SIW feed network is integrated on layer 1.The slots provide required excitation coefficients for an array. Layer 2 works as an air gap between feeding layer 1 and antenna layer 3. Mutual coupling between antenna elements is reduced by placing an array of vias in a rectangular cavity under each antenna element. Layer 3 is a radiating layer, where the circularly polarized patch antennas are integrated on an ultrathin substrate. The feed network printed on a high dielectric permittivity allows a freedom to choose the spacing between antenna elements. The optimum spacing between antennas sums up the individual element patterns to contribute to the maximum total gain. Simulated and measured axial ratios as a function of frequency are compared in Fig. 5b. Measured AR value is less than 3.5 dB from 58 GHz to 62 GHz.

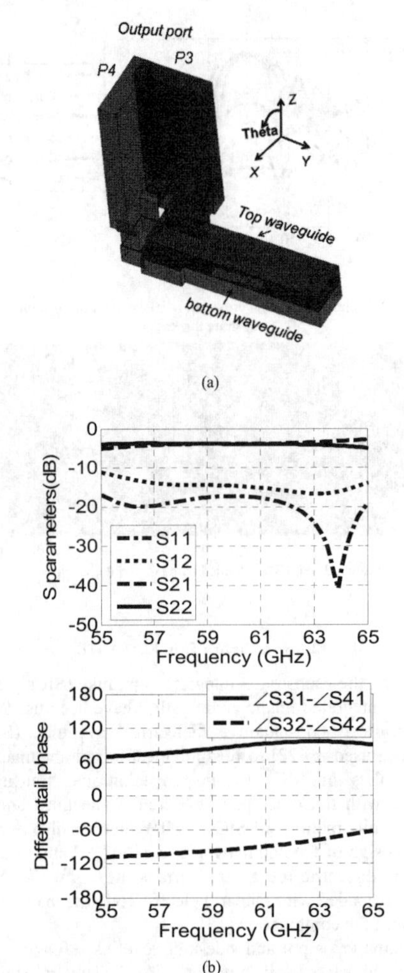

(a)

(b)

Figure. 6. CP antenna feed network (a) architecture (b) amplitude and phase performance.

V. DUAL CIRCULARLY POLARIZED ANTENNA

In this work, E-plane coupler is used as a feeding topology to supply equal amplitude for two LP components with 90° phase shift. The proposed DCP antenna is working in left-hand circularly polarized (LHCP) mode and right-hand circularly polarized wave (RHCP) mode for port 1 and port 2, respectively. Four port feeding network shown in Fig. 6a is used to obtain a DCP radiation at the 60 GHz frequency range. Input ports *P1*, *P2* are located on the horizontal plane and output ports *P3*, *P4* are located on the vertical plane. Four SIW transmission lines are arranged in a three dimensional (3-D) configuration, where vertical and horizontal waveguides are connected through an SIW vertical interconnect. Coupler is used to feed with equal power division between *P3* and *P4*. A differential phase shift of +90° is realized for excitation at *P1* and -90° for excitation at *P2*. Hence, the condition for CP operation is satisfied in the proposed feed network. The output port lengths are unequal for the two orthogonally polarized array ports. To retain 90° phase shift, an unequal width and unequal length phase shifter is integrated between output ports of the coupler. As shown in Fig. 6a, the antenna feed network is designed in the SIW technology and later used to feed an SIW fed rod antenna. Output ports *P3* and *P4* are loaded with two SIW fed dielectric rod antennas to realize the final prototype. Simulated amplitude and phase performance as a function of frequency is plotted in Fig. 6b. All interconnects are matched from input to output ports, so all the field is coupled to the output ports. For port *P1*, a differential phase shift between output ports varies from 77° to 91° and amplitude coefficient varies between -3.7 dB and -4.1dB over the frequency range from 57 GHz to 65 GHz. Measured half-power beam width (HPBW) in both the pattern cuts is 39° with side lobe levels are lower than -10 dB for both input ports *P1* and *P2*.

VI. CONCLUSION

In this work, two linearly polarized antennas and two circularly polarized antennas have been proposed and experimentally validated in SIW technology. Feed network and antenna are designed to share the SIW transmission line. Antenna efficiency greater than 70% is obtained for all the proposed techniques. The proposed antennas can be used for radio and radar front-ends that require single linear or single circular or even dual circular polarization at mm-wave frequency.

ACKNOWLEDGMENT

The authors wish to thank T. Antonescu for his help in the fabrication of the prototypes and M. Thibault for his help in measuring the radiation pattern in our anechoic chamber.

REFERENCES

[1] T.Y.J Kao,Y. Yan, T.M. Shen, A.Y.K Chen, and J.Lin, "Design and Analysis of a 60-GHz CMOS Doppler Micro-Radar System-in-Package for Vital-Sign and Vibration Detection," *IEEE Transactions on Microwave Theory and Techniques*, vol.61, no.4, pp.1649,1659, April 2013

[2] T.M. Shen, T.J. Kao, T.Y. Huang, J. Tu, J. Lin, R.B. Wu, "Antenna Design of 60-GHz Micro-Radar System-In-Package for Noncontact Vital Sign Detection," *IEEE Antennas and Wireless Propagation Letters*, vol.11, pp.1702-1705, 2012.

[3] K.K.M. Chan, A.E. Tan, and K. Rambabu, "Circularly Polarized Ultra-Wideband Radar System for Vital Signs Monitoring," *IEEE Transactions on Microwave Theory and Techniques*, vol.61, no.5, pp.2069-2075, May 2013.

[4] T. Tomura, Y. Miura, Z. Miao; J. Hirokawa, and M. Ando, "A 45° Linearly Polarized Hollow-Waveguide Corporate-Feed Slot Array Antenna in the 60-GHz Band," *IEEE Transactions on Antennas and Propagation*, vol.60, no.8, pp.3640-3646, Aug. 2012.

[5] D. Kim; M. Zhang; J. Hirokawa, and M. Ando, "Design of dual-polarization waveguide slot array antenna using diffusion bonding of laminated thin plates for the 60 GHz-band," 2012 *IEEE Antennas and Propagation Society International Symposium* (APSURSI), , vol., no., pp.1-2, 8-14 July 2012.

[6] N. Chahat, M. Zhadobov, S.A. Muhammad, L. Le Coq, R. Sauleau., "60-GHz Textile Antenna Array for Body-Centric Communications," *IEEE Transactions on Antennas and Propagation*, vol.61, no.4, pp.1816-1824, April 2013.

[7] S-B Yeap, X. Qing, M. Sun, and Z. N. Chen, "140-GHz 2×2 SIW horn array on LTCC," *IEEE Asia-Pacific Conference on Antennas and Propagation* (APCAP), 2012, vol., pp.279-280, 27-29 Aug. 2012.

[8] M. Sun, Y.X. Guo, M.F. Karim, and L.C. Ong, "Linearly polarized and circularly polarized Arrays in LTCC Technology for 60GHz Radios," 2010 *IEEE Antennas and Propagation Society International Symposium* (APSURSI), pp.1-4, 11-17 July 2010.

[9] C. R. Liu, Y. X. Guo, X. Y. Bao and S. Q. Xiao "60-GHz LTCC integrated circularly polarized helical antenna array",*IEEE Trans.Antennas Propag.*, vol. 60, no. 3, pp.1329 -1335 2012.

[10] Y. Li, Z. N. Chen, X. Qing, Z. Zhang, J. Xu, and Z. Feng, "axial ratio bandwidth enhancement of 60-GHz substrate integrated waveguide-Fed circularly polarized LTCC antenna Array," *IEEE Trans. Antennas Propag*, vol.60, no.10, pp.4619-4626, Oct. 2012.

[11] L. Wang, Y.-X. Guo, and W.X. Sheng, "wideband high-gain 60-GHz LTCC L-Probe patch antenna array with a soft surface," *IEEE Trans. Antennas Propag*, vol.61, no.4, pp.1802-1809, April 2013.

[12] S. Mei, Q. Xianming and C. Zhi Ning, "60-GHz antipodal Fermi antenna on PCB," *Proceedings of the 5th European Conference on Antennas and Propagation* (EUCAP), vol., no., pp.3109-3112, 11-15 April 2011.

[13] X. P. Chen, K. Wu, L. Han, and F. H. , "Low-cost high gain planar antenna array for 60-GHz band applications," *IEEE Trans. Antennas Propag.*, vol. 58, no. 6, pp. 2126–2129, Jun. 2010.

[14] M. Ohira, A. Miura and M. Ueba "60-GHz wideband substrate-integrated-waveguide slot array using closely spaced elements for planar multisector antenna", *IEEE Trans. Antennas Propag.*, vol. 58, no. 3, pp.993 -998 2009.

[15] Y. Zhang, Z.N. Chen, X. Qing, and W. Hong, "Wideband Millimeter-Wave Substrate Integrated Waveguide Slotted Narrow-Wall Fed Cavity Antennas," *IEEE Transactions on Antennas and Propagation*, vol.59, no.5, pp.1488-1496, May 2011.

[16] O. Kramer, T. Djerafi, and K. Wu, "Very small footprint 60 GHz stacked Yagi antenna array," *IEEE Trans. Antennas Propag.*, vol. 59,no. 9, pp. 3204–3210, Sep. 2011.

[17] Q. H. Lai, C. Fumeaux, W. Hong, and R. Vahldieck, "60 GHz aperture-coupled dielectric resonator antennas fed by a half-mode substrate integrated waveguide," *IEEE Trans. Antennas Propag.*, vol. 58, no. 6, pp. 1856–1864, Jun. 2010.

[18] A.B. Guntupalli, and K. Wu, "Polarization-Agile millimeter wave antenna arrays" *Proceedings of Asia-Pacific Microwave Conference, APMC 2012*, Kaohsiung, Taiwan, pp. 148-150, Dec. 4-7,2012.

[19] N. Athanasopoulos, D. Makris, and K.Voudouris, "Millimeter-wave passive front-end based on Substrate Integrated Waveguide technology," *2012 Antennas and Propagation Conference (LAPC)*, Loughborough, vol., no., pp.1-5, 12-13 Nov. 2012.

[20] L. Chioukh, H. Boutayeb, K. Wu, and D. Deslandes, "Monitoring vital signs using remote harmonic radar concept," *European Radar Conference (EuRAD)*, 2011, vol., no., pp.381-384, 12-14 Oct. 2011.

[21] A. Patrovsky and K. Wu, "94-GHz planar dielectric rod antenna with substrate integrated image guide (SIIG) feeding," *IEEE Antennas Wireless Propag. Lett.* , vol. 5, pp. 435–437, 2006.

[22] N. Ghassemi and K. Wu, "Planar Dielectric Rod Antenna for Gigabyte Chip-to-Chip Communication," *IEEE Transactions on Antennas and Propagation*, vol.60, no.10, pp.4924-4928, Oct. 2012.

[23] R. Kazemi, A.E Fathy and R.A Sadeghzadeh, "Dielectric Rod Antenna Array with Substrate Integrated Waveguide Planar Feed Network for Wideband Applications," *IEEE Transactions on Antennas and Propagation*, vol.60, no.3, pp.1312-1319, March 2012.

A New E-plane Bend for SIW Circuits and Antennas Using Gapwave Technology

Jian Yang*, Ali Razavi Parizi†,

*Dept. of Signals and Systems, Chalmers University of Technology, S-41296 Gothenburg, Sweden
Email: jian.yang@chalmers.se
†Dept. of electrical engineering, Ferdowsi University of Mashhad, Iran
Email: alirazavi_parizi@yahoo.com

***Abstract*—The SIW (substrate integrated waveguide) technology makes use of metal vias in a dielectric substrate, electrically connecting two parallel metal plates, to make a waveguide. The main advantages of SIW are simple geometry, low manufacture cost and integratability with MMIC (monolithic microwave integrated circuit) or other circuits. It is often required to have E-plane bend components in the whole SIW circuits or antenna systems for the integration, for example, in multilayer configurations. However, it is difficult to make an E-plane bend by using only SIW technology. We present a new solution to E-plane bend for SIW circuits and antennas by combining the SIW technology and the so-called gap waveguide (or gapwave) technology in the paper, with the latter also realized in PCB (printed circuit board) technology, and therefore keeping its above-mentioned advantages.**

***Index Terms*—SIW technology, gap waveguide technology, E-plane bend, PCB technology**

I. Introduction

SIW (substrate integrated waveguide) technology is a new transmission line technology at frequencies of mm-waves and sub mm-waves [1]–[3], as shown in Fig. 1. It makes use of metal vias in a dielectric substrate, electrically connecting two parallel metal plates, to make a waveguide. The main advantages of the SIW technology are simple geometry, low manufacture cost and integratability with MMIC (monolithic microwave integrated circuit) or other circuits.

Since the first introduction with the name of post-wall waveguide [1] and laminated waveguide [4], the SIW technology has been applied to many different microwave components, for examples, H-plane bends [5], filters [2], directional couplers [6], oscillators [6], power amplifiers [7], slot array and leaky antennas [8], and circulators [9].

However, from the technical literature, it seems that the E-plane bend with SIW technology does not exist. An E-plane bend associated with applications of SIW is often an important component, especially in multilayer configurations, like the SWE antenna in [10]. In the present paper, we propose a solution to E-plane bend suitable for use in SIW circuit and antenna systems, by employing the gapwave technology.

II. Gapwave Technology

Fig. 2 (the cross section of a long parallel-plate waveguide) shows the principle of gapwave technology [11]–[13]: no wave

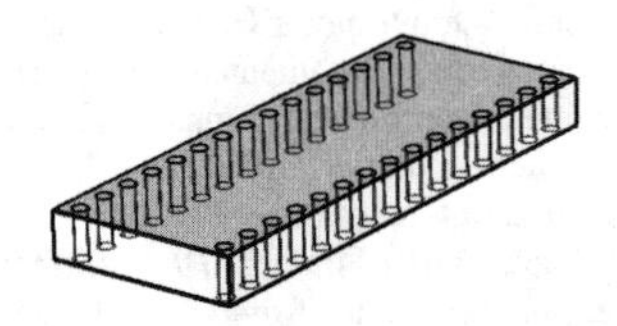

Fig. 1. Geometry of SIW (substrate integrated waveguide)

can propagate between a PEC (perfect electric conductor) plate and a PMC (perfect magnetic conductor) plate when the distance h between the two plates is smaller than a quarter wavelength, and thus waves can propagate only along the waveguide formed by the two parallel PEC plates, i.e. the range of the blue E-fielld vectors drawn in Fig. 2. Note that there is no leakage into the PEC/PMC stop region even if there are no side metal walls physically. We can refer to this stop range as an invisible EM (electromagnetic) wall.

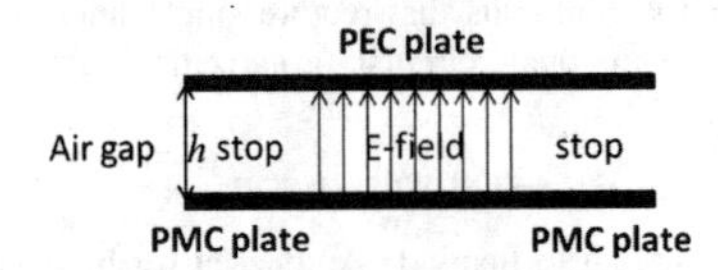

Fig. 2. Principle of Gapwave technology

Good metals, such as copper, are very close to PEC, and PMC (not existing in nature) can be realized by a mushroom structure in PCB (printed circuit board) or metal posts on a metal plate, easily with an octave (2:1) bandwidth; see Fig. 3.

Gapwave technology has been also applied to make different microwave components for mm-waves and sub-mm-waves, such as filters [14], power divider [15], hybrid ring coupler [16], [17], packaging [18], [19], etc.

III. SIW E-plane Bend using Gapwave Technology

The proposed SIW E-plane bend and its geometrical parameters are illustrated in Fig. 4. In this structure, two substrate integrated waveguides (29×0.787 mm^2, where the width of 29 mm is the same as that of WG15 waveguide and the thickness of 0.787 mm is a standard thickness of Rogers substrate) are

978-7-5641-4279-7

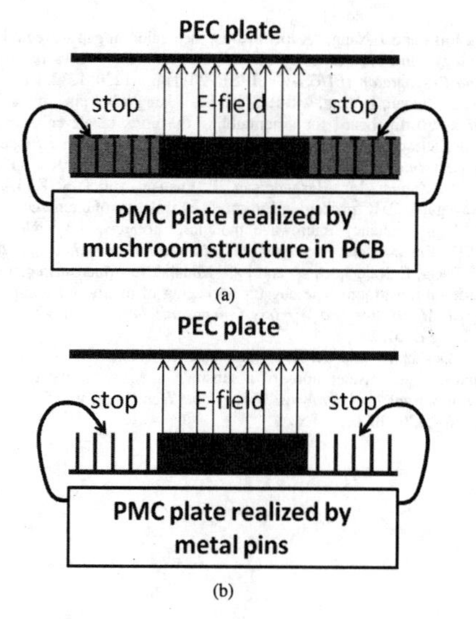

Fig. 3. Realization of gapwave technology by (a) mushroom structure in PCB (printed circuit board) and (b)metal posts.

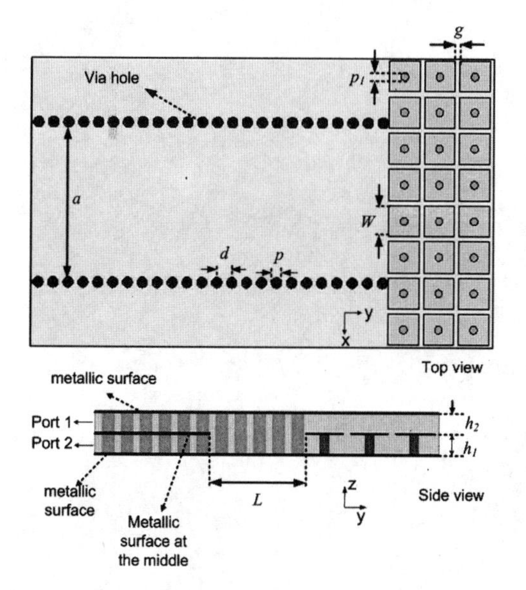

Fig. 4. Configuration of the proposed SIW E-plane bend with mushroom structure of gapwave technology.

connected to each other and loaded with a E-plane bend by a mushroom structure in PCB and an gap filled with the same dielectric (Rogers 5880) above the mushrooms.

In a certain frequency range, the mushroom structure, together with the top smooth plate with a small gap, creates a

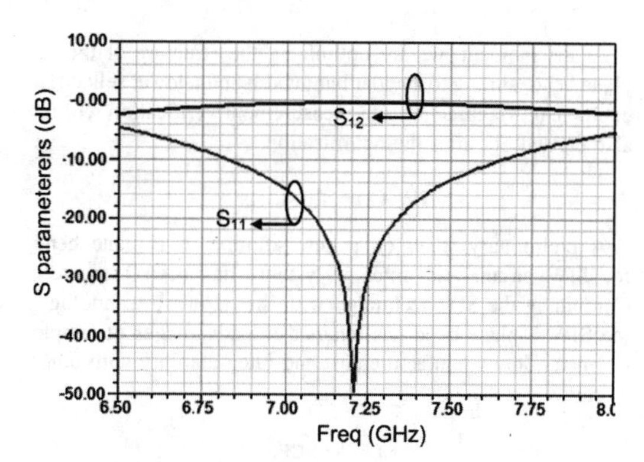

Fig. 5. Simulated S-parameters of the E-plane bend.

stopband, at which the existing mode in SIW is not able to propagate through the gap. As a result, the propagating mode in SIW tends to reflect back and continue its way through the E-plane bend. By proper choice of L, the minimum reflection and the maximum transmission through the E-plane bend can be achieved.

The dimensions of the mushroom structure are obtained from the one in [20], with a stopband of 5.3–11.3 GHz. Moreover, the values of a, p and d are determined according to the design rules coming from [8] and [21]. The number of mushroom columns should be both large enough in order to stop the propagation of SIW mode in the substrate gap and compact in size. As shown in Fig. 4, a structure of 3 columns is used in this preliminary study. The spacing between vias in SIW should be small enough so that the leakage through the SIW side walls become negligible and the width a of SIW is chosen to have the same width of WG15 standard waveguide with an operating frequency range of 7–10 GHz.

Based on the above discussion, the structure has been designed with the dimensions listed in Table I, where their definitions are illustrated in Fig. 4. The Rogers 5880 with relative permittivity of $\varepsilon_r = 2.2$ and tangent loss of $tan\delta = 0.0009$ is used as the substrate for both the SIW part and the mushroom gapwave part. The designed structure is simulated using HFSS and the simulation results are illustrated in Fig. 5, where the reflection coefficient and the insertion loss of the designed structure are depicted. It is observed that over the frequency range of 6.8–7.6 GHz (11.7%), the reflection coefficient is below -10 dB and the insertion loss is between 0.2–0.7 dB.

TABLE I
DIMENSIONS IN MILLIMETER OF THE DESIGNED STRUCTURE DEPICTED IN FIG. 4

Parameter	Value	Parameter	Value
a	29 mm	W	6 mm
p	3 mm	p_1	1.5 mm
d	2 mm	g	1.0 mm
h_1	0.787 mm	L	15 mm
h_2	0.787 mm		

At the moment, we are optimizing the structure of the E-plane bend, aiming to a wider bandwidth and a lower reflection coefficient. Manufacture and measurement verification will be also carried out after the optimization.

IV. CONCLUSIONS

A preliminary study on a new solution to E-plane bend for SIW circuit and antenna systems has been presented. Combining the SIW technolgy with the gapwave technology, the SIW E-plane bend has retained the advantages of simple geometry, low manufacture cost and integratability with other circuits.

REFERENCES

[1] J. Hirokawa and M. Ando, "Single-layer feed waveguide consisting of posts for plane tem wave excitation in parallel plates," *IEEE Transactions on Antennas and Propagation*, vol. 46, no. 5, pp. 625–630, 1998.

[2] D. Deslandes and K. Wu, "Single-substrate integration technique of planar circuits and waveguide filters," *IEEE Transactions on Microwave Theory and Techniques*, vol. 51, no. 2, pp. 593–596, 2003.

[3] M. Bozzi, A. Georgiadis, and K. Wu, "Review of substrate-integrated waveguide circuits and antennas," *IET Microwaves, Antennas & Propagation*, vol. 5, no. 8, pp. 909–920, 2011.

[4] H. Uchimura, T. Takenoshita, and M. Fujii, "Development of a laminated waveguide," *IEEE Transactions on Microwave Theory and Techniques*, vol. 46, no. 12, pp. 2438–2443, 1998.

[5] L. Geng, W. Che, and K. Deng, "Propagation characteristics of H-plane folded substrate integrated waveguide bends," *Microwave and Optical Technology Letters*, vol. 49, no. 12, pp. 2977–2981, 2007.

[6] J.-X. Chen, W. Hong, Z.-C. Hao, H. Li, and K. Wu, "Development of a low cost microwave mixer using a broad-band substrate integrated waveguide (siw) coupler," *Microwave and Wireless Components Letters, IEEE*, vol. 16, no. 2, pp. 84–86, 2006.

[7] H. Jin and G. Wen, "A novel four-way ka-band spatial power combiner based on hmsiw," *IEEE Microwave and Wireless Components Letters*, vol. 18, no. 8, pp. 515–517, 2008.

[8] L. Yan, W. Hong, G. Hua, J. Chen, K. Wu, and T. J. Cui, "Simulation and experiment on siw slot array antennas," *IEEE Microwave and Wireless Components Letters*, vol. 14, no. 9, pp. 446–448, 2004.

[9] W. D'Orazio and K. Wu, "Substrate-integrated-waveguide circulators suitable for millimeter-wave integration," *IEEE Transactions on Microwave Theory and Techniques*, vol. 54, no. 10, pp. 3675–3680, 2006.

[10] J. Yang, "The SWE gapwave antenna – a new wideband thin planar antenna for 60GHz communications," in *2013 Proceedings of the Fourth European Conference on Antennas and Propagation (EuCAP)*, 2013, pp. 1–5.

[11] P.-S. Kildal, E. Alfonso, A. Valero-Nogueira, and E. Rajo-Iglesias, "Local metamaterial-based waveguides in gaps between parallel metal plates," *IEEE Antennas and Wireless Propagation Letters*, vol. 8, pp. 84–87, 2009.

[12] P.-S. Kildal, A. U. Zaman, E. Rajo-Iglesias, E. Alfonso, and A. Valero-Nogueira, "Design and experimental verification of ridge gap waveguide in bed of nails for parallel-plate mode suppression," *IET Microwaves, Antennas & Propagation*, vol. 5, no. 3, pp. 262–270, 2011.

[13] H. Raza, J. Yang, P.-S. Kildal, and E. Alfonso, "Resemblance between gap waveguides and hollow waveguides," *accepted for publication in IET Microwaves, Antennas & Propagation*, 2013.

[14] A. U. Zaman, P.-S. Kildal, and A. A. Kishk, "Narrow-band microwave filter using high-Q groove gap waveguide resonators with manufacturing flexibility and no sidewalls," *IEEE Transactions on Components, Packaging and Manufacturing Technology*, vol. 2, no. 11, pp. 1882–1889, 2012.

[15] H. Raza and J. Yang, "Compact UWB power divider packaged by using gap-waveguide technology," in *2012 6th European Conference on Antennas and Propagation (EUCAP)*. IEEE, 2012, pp. 2938–2942.

[16] J. Yang and H. Raza, "Empirical formulas for designing gap-waveguide hybrid ring coupler," *Microwave and Optical Technology Letters*, vol. 55, no. 8, pp. 1917–1920, 2013.

[17] H. Raza and J. Yang, "A low loss rat race balun in gap waveguide technology," in *Proceedings of the 5th European Conference on Antennas and Propagation (EUCAP)*. IEEE, 2011, pp. 1230–1232.

[18] A. U. Zaman, J. Yang, and P.-S. Kildal, "Using lid of pins for packaging of microstrip board for descrambling the ports of eleven antenna for radio telescope applications," in *2010 IEEE Antennas and Propagation Society International Symposium (APSURSI)*. IEEE, 2010, pp. 1–4.

[19] A. U. Zaman, M. Alexanderson, T. Vukusic, and P.-S. Kildal, "Gap waveguide PMC packaging for improved isolation of circuit components in high frequency microwave modules," *accepted for publication in IEEE Components, Packaging and Manufacturing Technology*, 2013.

[20] E. Pucci, E. Rajo-Iglesias, and P.-S. Kildal, "New microstrip gap waveguide on mushroom-type ebg for packaging of microwave components," *IEEE Microwave and Wireless Components Letters*, vol. 22, no. 3, pp. 129–131, 2012.

[21] D. Deslandes and K. Wu, "Accurate modeling, wave mechanisms, and design considerations of a substrate integrated waveguide," *IEEE Transactions on Microwave Theory and Techniques*, vol. 54, no. 6, pp. 2516–2526, 2006.

Simplified Wavelength Calculations for Fast and Slow Wave Metamaterial Ridged Waveguides and their Application to Array Antenna Design

[1,2] Hideki KIRINO and [2] Koichi OGAWA
1 Panasonic Healthcare Co., Ltd. 247 Fukutake-kou, Saijo-shi, Ehime, 793-8510 Japan
2 Toyama University 3190 Gofuku, Toyama-shi, Toyama, 930-8555 Japan
E-mail: kirino.hideki@jp.panasonic.com, ogawa@eng.u-toyama.ac.jp

***Abstract*- Simple equations for calculating the wavelengths of the Fast and Slow waves in combined-mode metamaterial ridged waveguides are derived. The transverse resonance method is used in analyzing the fast wave and a distributed transmission line with series inductive reactance is used for the slow wave. The equations are used to show how the wavelength characteristics can be controlled and to find the angles of undesired lobes of array antennas in actual examples.**

I. INTRODUCTION

Metamaterial Ridged Waveguides (MRW) [1][2] and GAP waveguides [3][4] have the advantages of ease of manufacture and the reliability of non-contacting metal structures compared to ordinary hollow waveguides. In this paper, we use an analytical method to derive simple equations for the characteristics of MRWs.

The most significant parameter in designing a waveguide is the wavelength. Longer wavelengths than those in free space are preferable for electronic circuits in order to apply a uniform voltage over the terminals, and shorter ones than those in free space are preferable for feeding the lines of array antennas in order to suppress undesired lobes. For these flexible requirements, the fast and slow wave combined mode MRWs presented in the literature [5][6] provide adequate solutions.

On the other hand, fast and slow wave combined mode MRWs have complicated structures compared to ordinary waveguides, and require an EM-simulator for circuit design. However, in practice, it is hard to justify the time taken and cost of EM-simulation for calculating the wavelengths. For this reason, we propose a simple analysis and equations for calculating the wavelengths of fast and slow wave combined mode MRWs, which reduce the calculation time and improve design efficiency. Using these equations, the variation in wavelength with the physical size of the structure can be evaluated much faster than with EM-simulators. As an example to demonstrate their use, at each end of sections II and III, we apply the equations to an actual design in order to change the angles of undesired lobes.

In addition, we think this paper is beneficial for new researchers of MRWs, in that it provides an understanding of the physical phenomena involved in the propagation modes in MRWs with complex structures.

II. FAST WAVE

A. *Analysis of the fast wave*

Fig. 1 shows the structure and parameters of a fast and slow wave combined mode MRW. In the metamaterial region, the metal rods are $\lambda/4$ in height. The upper and lower plates are not in contact with each other with the gap between them selected to be $\lambda/8$ (the essential condition is that this is less than $\lambda/4$) in order to prevent parallel plate mode propagation and to confine the electro-magnetic energy to the ridge.

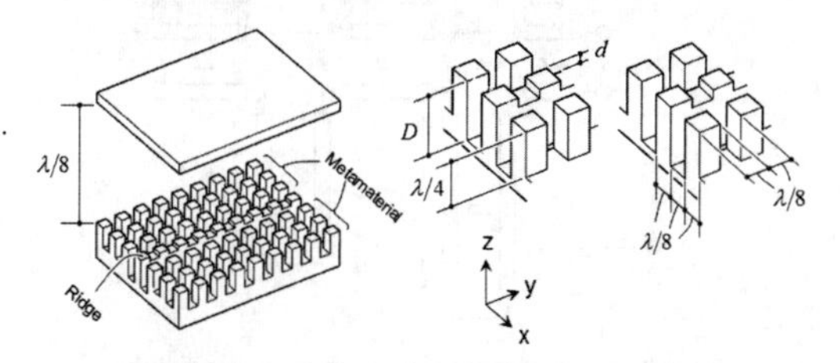

Figure 1. Structure and parameters of the fast and slow wave combined mode MRW.

The transverse resonance method is a well known conventional method for analysing the fast wave. As shown in the literature [5][6], the wavelength of the fast wave λ_f can be described by the following equation, where ϕ is the phase of the reflection from the EMB (Equivalent Magnetic Boundary) of a TEM wave injected from the centre of the ridge in the transverse direction (x-direction in Fig. 1), and where the reflection coefficient at the EMB is 1.

$$\frac{\lambda_f}{\lambda} = \sqrt{\frac{1}{1-(2\pi/\phi)^2}} \tag{1}$$

where λ is the wavelength in free space.

If the characteristic impedance at the centre of the ridge is Z_o, and the input impedance in the transverse direction from the edge of the ridge is Z_t, then ϕ can be described by the following equation.

$$\phi = Ang\left(\frac{Z_t/Z_o - 1}{Z_t/Z_o + 1}\right) \tag{2}$$

978-7-5641-4279-7

Note that an equivalent circuit for the transverse resonance from the centre of ridge can be constructed, as shown in Fig. 2. There are parasitic reactances at the corners of the ridge as shown in Fig. 3, which can be calculated by equations obtained from the literature [7], with the results shown in Table 1. We assume the position of EMB is as shown in Fig. 2 and therefore Z_m is the impedance of an open-ended-line, $\lambda/8$ in length. Note that Z_t is the impedance containing Z_m, a short-ended-line of length D (see Fig. 1) and a line of length $\lambda/16$, therefore Z_t can be written as follows.

$$Z_t/Z_o = \left(Z + jZ_o \tan\frac{\pi}{8}\right) \Big/ \left(Z_o + jZ \tan\frac{\pi}{8}\right) \tag{3}$$

$$Z = jZ_o\left(-\cot\frac{\pi}{4} + \tan 2\pi\frac{D}{\lambda}\right) \tag{4}$$

These equations mean that the wavelength ratio of fast wave to TEM wave, λ_f/λ, can be calculated from D/λ. It is necessary to include parasitic reactances, such as shown in Fig. 3, and select angles of $-720<\phi<-360$ before substituting into equation 1, because the waveguide boundary is at EMB.

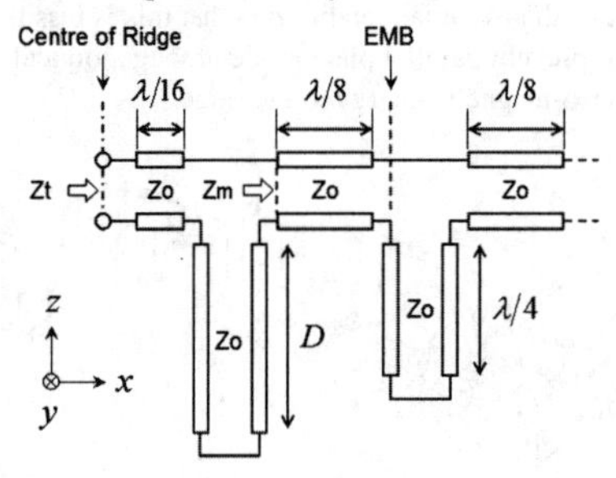

Figure 2. Equivalent circuit in transverse resonance direction.

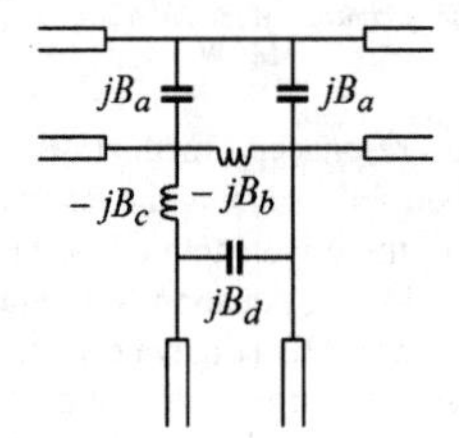

Figure 3. Parasitic reactance at a corner.

TABLE I
Parasitic reactance values.

B_a	B_b	B_c	B_d
$\frac{0.333}{Z_o}$	$\frac{0.109}{Z_o}$	$\frac{1.273}{Z_o}$	$\frac{0.196}{Z_o}$

B. Analytical result

Fig. 4 shows the wavelength ratio λ_f/λ as a function of trench depth D, calculated using the analytical method presented in this paper. This is compared with the results from an EM-simulator. The values are shown on the left ordinate, where λ_{fo} is the wavelength for $D=4\lambda/16$. As can be seen in Fig. 4, the two plots are similar, both decreasing as D increases. The difference between the results is less than 5% at $D=6\lambda/16$, which means the analytical method presented in this paper is sufficiently accurate for approximating the wavelength in a MRW.

The curve produced by the analytical method can be represented by the following 3rd order polynomial.

$$\frac{\lambda_f}{\lambda_{fo}} = -16.79\left(\frac{D}{\lambda}\right)^3 + 21.40\left(\frac{D}{\lambda}\right)^2 - 9.95\left(\frac{D}{\lambda}\right) + 2.41 \tag{5}$$

Using equation 5, the variation in wavelength ratio can be obtained much more quickly and easily than with an EM-simulator.

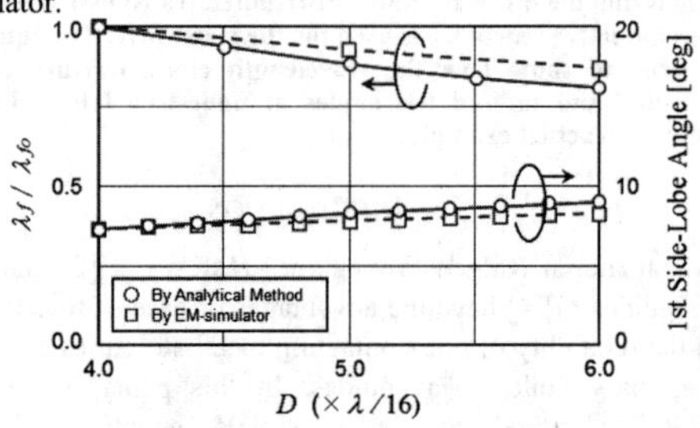

Figure 4. Wavelength of the fast wave and the angle of the 1st side lobe of a 10 element array with a resonant feed line calculated using equation 5.

C. Design example

A design example using the equations derived above is given in this section. One of the applications that adopts the fast and slow wave combined mode MRW is the feed line of an array antenna, as mentioned at the beginning of the paper. There are two types of feed line for array antennas, one is a travelling wave type and the other is a resonance type. The resonance type was selected because its waveguide structure is the same over the whole circuit for narrow bandwidth systems [1][2]. For automotive radar applications, as described in [1][2], the performance of undesired lobes is very important for evaluating reflected waves. Therefore, in our work, the variation in the angle of the 1st side lobe was calculated for the wavelengths obtained using equation 5 and the EM-simulator. The results are shown in Fig. 4 with the 1st side lobe angle given on the right ordinate. Fig. 5 shows the side-lobe characteristics as a function of the depth D for a 10 element liner array antenna. In Figs. 4 and 5, the value of λ_{fo} for $D=4\lambda/16$ is taken from simulation; the values of λ_f are calculated from equation 5.

As can be seen in Fig. 4, for $4\lambda/16 < D < 6\lambda/16$, the angle of the 1st side lobe from equation 5 and the EM-simulator are in good agreement, showing that there is no significant difference between the analytical method and EM-simulation.

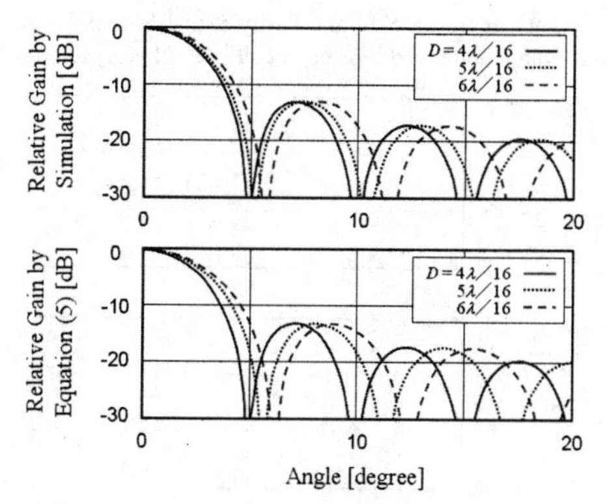

Figure 5. Side-lobe characteristics as a function of the depth D for 10 elements liner array antenna.

III. Slow Wave

A. Analysis of the slow wave

In general a slow wave effect can be realized by having a periodic stepped structure to increase the equivalent length of the waveguide. This is equivalent to inserting inductive impedances in series with the waveguide. Fig. 6 shows cross sections of the MRW with and without periodic steps and their equivalent circuits. As shown in Fig. 6b, the periodic steps with depth d ($d<\lambda/4$: see Fig. 1) work as short-ended-lines with impedance Z_s. In series with the waveguide these produce a waveguide with a longer equivalent length.

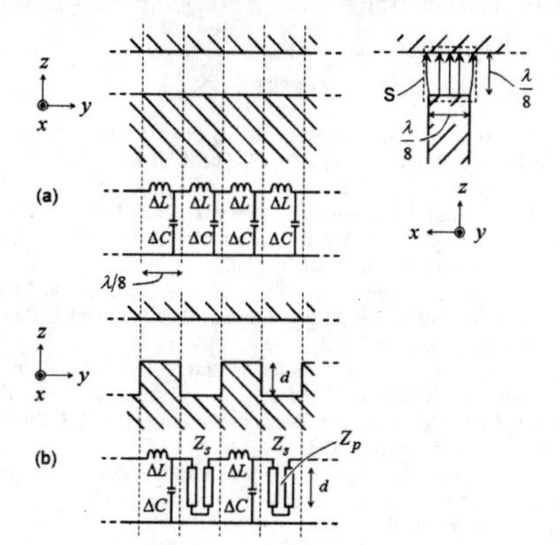

Figure 6. Cross sections of the MRW with and without periodic steps and their equivalent circuits.

Analysis of the slow wave proceeds as follows. The equivalent circuit for each $\lambda/8$ section has a small series inductance, ΔL, and a small shunt capacitance, ΔC. With periodic steps ($d\neq0$), alternate $\lambda/8$ sections are replaced by Z_s, as shown in Fig. 6b. Therefore the ratio of the wavelengths of the combined fast and slow wave mode ($d\neq0$) to the fast wave ($d=0$), λ_{fs}/λ_f, is expressed as

$$\frac{\lambda_{fs}}{\lambda_f}=\sqrt{\frac{\Delta L\Delta C}{\left(\frac{\Delta L}{2}+\frac{Z_s}{\omega}\right)\frac{\Delta C}{2}}}=2\sqrt{\frac{\Delta L}{\Delta L+\frac{Z_s\lambda}{\pi v_c}}} \tag{6}$$

where v_c is the velocity of light.

Equation 6 means that λ_{fs}/λ_f can be obtained from ΔL and Z_s. ΔL is obtained as follows. In the MRW structure, it can be assumed that most of the electric field is confined to the region S, shown on the upper-right hand side of Fig. 6. Therefore, ΔC is equal to the ideal parallel plate capacitance of a $\lambda/8$ cube, given by

$$\Delta C=\varepsilon\frac{(\lambda/8)^2}{\lambda/8}=\varepsilon\frac{\lambda}{8} \tag{7}$$

where ε is the permittivity of air.

Define the characteristic impedance of the ridge for $d=0$ as Z_f

$$Z_f=\sqrt{\frac{\Delta L}{\Delta C}} \tag{8}$$

Since the height and width of region S are equal ($\lambda/8$), the characteristic impedance of the ridge for a TEM wave propagating in region S is 120π. However, a fast wave (TE wave) propagates on the ridge, so the magnetic field normal to the axis of the waveguide, H', can be given in terms of the wavelength ratio of the fast wave to that in free space, λ_f/λ, by

$$H'=H\frac{\lambda}{\lambda_f} \tag{9}$$

This means that Z_f can expressed as

$$Z_f=\frac{E}{H'}=\frac{E}{H}\frac{\lambda_f}{\lambda}=120\pi\frac{\lambda_f}{\lambda} \tag{10}$$

From equations 7, 8 and 10

$$\Delta L=\varepsilon\frac{\lambda}{8}\left(120\pi\frac{\lambda_f}{\lambda}\right)^2 \tag{11}$$

Z_s is obtained as follows. The periodic step is equivalent to a short-ended-line with a cross section of $\lambda/8$ square and no waveguide wall in the $-z$ direction, as shown in Fig. 6. Therefore, a TEM wave propagates in it. Its characteristic impedance, Z_p, (see Fig. 6) is 120π, so that the impedance of the periodic step, Z_s, is

$$Z_s=Z_p\tan\frac{2\pi d}{\lambda}=120\pi\tan\frac{2\pi d}{\lambda} \tag{12}$$

With ΔL and Z_s the wavelength ratio of combined mode of the fast and slow wave to the fast wave, λ_{fs}/λ_f, can be calculated from equation 6. The parasitic reactances at the corners, with the same values given in Table 1, also need to be included.

B. *Analytical result*

Fig. 7 shows the wavelength ratio λ_{fs}/λ_f as a function of step depth *d*, calculated using the analytical method presented in this paper and this is compared with the results from an EM-simulator. The values are shown on the left ordinate, where λ_{fso} is the wavelength for $d=\lambda/16$. The value of λ_f in equation 9 was calculated with $D=\lambda/4$. As can be seen immediately from Fig. 7, the EM-simulations didn't converge at higher values of *d*, and therefore no results were obtained for $d>2\lambda/16$, which sometimes occurs for low loss structures in time domain simulators. This has a serious effect on the design efficiency. On the other hand, the analytical method presented here not only gives results which are in accordance with the simulated results for $d<2\lambda/16$, but also calculates the wavelength ratio for $d>2\lambda/16$, showing that the wavelength ratio is reduced by around 50% over the whole range.

The curve produced by the analytical method can be represented by the following 3rd order polynomial.

$$\frac{\lambda_{fs}}{\lambda_{fso}} = -124.93\left(\frac{d}{\lambda}\right)^3 + 43.99\left(\frac{d}{\lambda}\right)^2 - 8.56\left(\frac{d}{\lambda}\right) + 1.39 \quad (13)$$

Using equation 13, the variation in wavelength ratio can be obtained much more quickly and easily than with an EM-simulator.

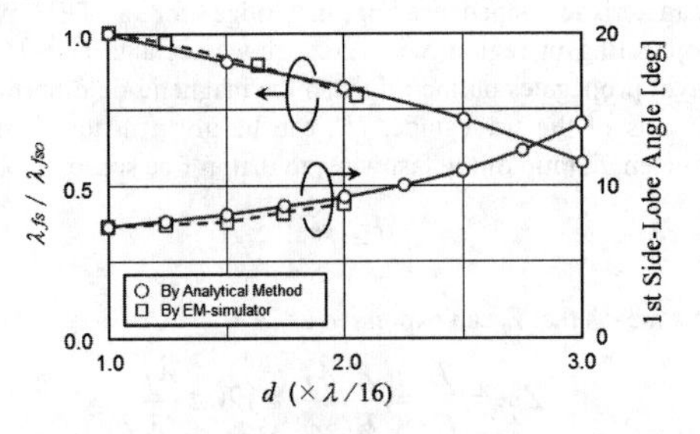

Figure 7. Wavelength of the fast and slow wave combined mode, and the 1st side lobe angle of a 10 element array with a resonant feed line calculated using equation 13.

C. *Design example*

As was done for the fast wave, the variation in the angle of the 1st side lobe was calculated for the wavelengths obtained using equation 13 and the EM-simulator. The results are shown in Fig. 7, with the values shown on the right ordinate. Fig. 8 shows the side-lobe characteristics as a function of the depth *d* for a 10 element liner array antenna. In Figs. 7 and 8, the value of λ_{fso} for $d=\lambda/16$ is taken from simulation; the values of λ_{fs} are calculated from equation 13.

As shown in Fig. 7, the EM-simulator cannot evaluate the 1st side lobe angle for $d>2\lambda/16$ because the simulations didn't convergence, whereas the analytical method can evaluate this, and shows that the 1st side lobe angle changes from 7° to 14° when the depth *d* varies from $\lambda/16$ to $3\lambda/16$.

The above study demonstrates that the analytical method presented in this paper can accurately calculate wavelength changes not only for the fast wave but also for the combined mode much more quickly and easily than EM-simulation and also evaluate undesired lobe characteristics of array antennas.

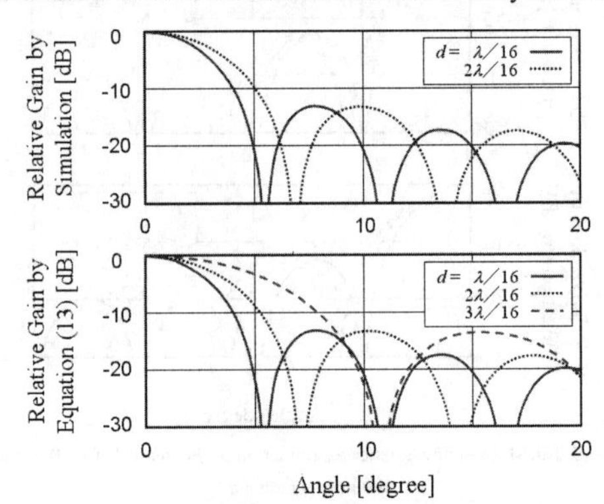

Figure 8. Side-lobe characteristics as a function of the depth *d* for a 10 element linear array antenna.

IV. Conclusion

Simple equations for the wavelengths of Fast and Slow wave combined-mode MRWs were proposed. The transverse resonance method was used to analyze the fast wave and a distributed transmission line with series inductive reactance was used for the slow wave.

Whereas the EM-simulations didn't converge for $d>2\lambda/16$, the analytical equations covered the range $\lambda/16<d<3\lambda/16$ and revealed the characteristics of the undesirable lobes of array antennas in actual examples.

References

[1] H. Kirino and K. Ogawa, "A 76GHz Phased Array Antenna Using a Waffle-iron Ridge Waveguide, " *EuCAP2010*, C32P2-2, Barcelona, Spain, Apr. 2010.

[2] H. Kirino and K. Ogawa, "A 76GHz Multi-Layered Phased Array Antenna Using a Non-Metal Contact Metamaterial Waveguide," *IEEE Trans. Antennas Propag.*, vol.60, No.2, pp. 840-853, Feb. 2012.

[3] P.-S. Kildal, E. Alfonso, A. Valero and E. Rajo, "Local Metamaterial-Based Waveguides in Gaps Between Parallel Metal Plates," *IEEE Antennas Propag. Lett.*, vol. 8, Sept. 2009.

[4] E. Pucci, A. U. Zaman, E. Rajo-Iglesias, P. –S. Kildal and A. Kishk, "Losses in Ridge Gap Waveguide Compared with Rectangular Waveguides and Microstrip Transmission Lines," *EuCAP2010*, C32P1-5, Barcelona, Spain, Apr. 2010.

[5] H. Kirino and K. Ogawa, "Metamaterial Ridged Waveguides with Wavelength Control for Array Antenna Applications," *ISAP Intl. Symp.* 4E2-1, Digest, 2012.

[6] H. Kirino and K. Ogawa, "A Fast and Slow Wave Combined-Mode Metamaterial Ridged Waveguide for Array Antenna Applications, " *EuCAP2013*, CA02a.5, Gothenburg, Sweden, Apr. 2013.

[7] N. Marcuvitz, "Waveguide Handbook," McGraw-Hill, pp. 336-338, 1951.

A Novel SIW Slot Antenna Array Based on Broadband Power Divider

Dongfang Guan, Zuping Qian, *Member, IEEE*, Yingsong Zhang, *Member, IEEE*, and Yang Cai

College of Communications Engineering, PLA University of Science and Technology, Nanjing 210007
gdfguandongfang@163.com

***Abstract*- A novel substrate integrated waveguide (SIW) slot antenna array is proposed in this paper. The array, based on SIW scheme, consists of one compact SIW 4-way power divider and 4 radiating SIWs each supporting 4 radiating slots. The 4-way power divider uses the self-compensating phase shift technique, which can make the feeding network operate over a broadband. Comparing with conventional SIW slot antenna arrays, this array has simple structure and broadband bandwidth. The antenna array is fabricated and the measured results are in agreement with the simulated ones. A relative bandwidth over 10% is achieved with return loss below -10 dB.**

I. Introduction

Waveguide longitudinal slot array antennas are widely applied in radar and communication systems featuring high gain, high efficiency, low cross-polarization levels and great capability of accurate control of the radiation patterns.

However, classical rectangular waveguide is costly, heavy, and bulky. In order to eliminate the mentioned drawbacks and make integration with planner circuits possible, substrate integrated waveguide (SIW) as a preferred choice over the classical rectangular waveguide has been proposed and design [1]. Actually, SIW has been proposed to design many high-quality microwave and millimeter wave components because of its advantages of low profile, low insertion loss, low interference and easiness of integrating with planar circuits. Many types of SIW slot antenna arrays have been extremely investigated in recent years [2]-[4]. The feeding networks of SIW slot antenna arrays usually consist of multilevel T-junctions or Y-junctions. As using parallel feeding and multilevel structure, with the increase of output branches, these feeding networks become more and more complex to design. To overcome this drawback and design feeding networks with compact and simple structure is an important issue for SIW slot antenna arrays. In [2], a compact SIW 12-way power divider is used as feeding network for 12 radiating SIWs. This power divider is compact because it is divided into 12 ways directly. However, it only can keep in phase in a narrow bandwidth, so the operation bandwidth of the whole SIW array antennas has been limited. In [5], an alternate phase SIW power divider is proposed. Alternate phase technique and in series structure are used to make the power divider compact. Since SIW is a dispersive guided-wave structure, the effective bandwidth is also narrow.

In this paper, a novel 4-way power divider is proposed. The power divider uses alternate phase technique to make structure compact. Besides, self-compensating phase shift technique is used to make the feeding network operate over a broadband. Based on this broadband feeding network, a 4×4 SIW slot antenna array is design. Comparing with conventional SIW slot antenna arrays, this array has simple structure and broadband bandwidth.

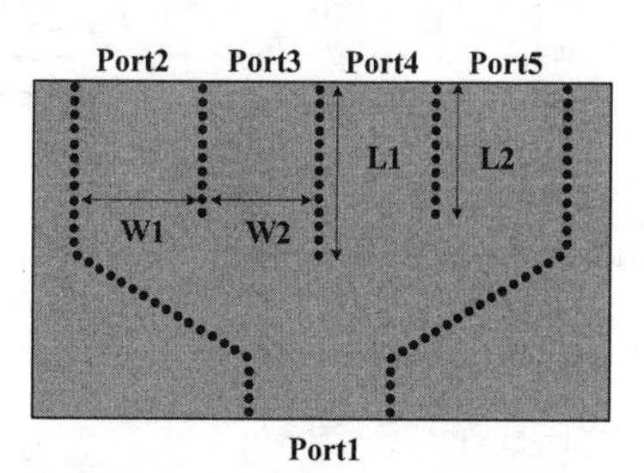

Fig. 1 Geometry of the proposed 4-way power divider.

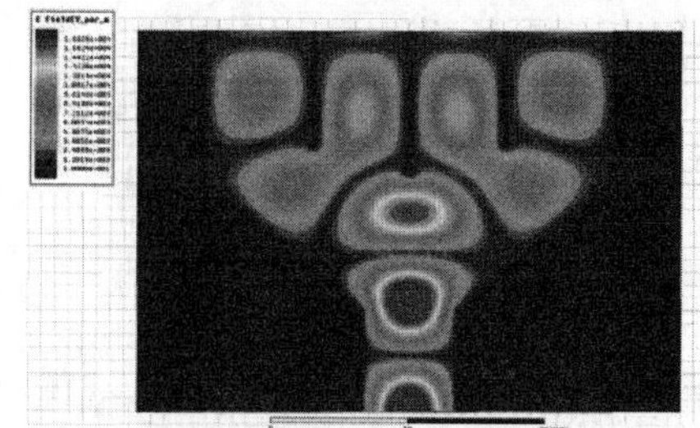

Fig. 2. E-field distribution of the proposed 4-way power divider.

II. Design Of The Antenna

The geometry of the proposed 4-way power divider is shown in Fig. 1. Using tapered transition without multilevel structure, the input port splits into 4-way output ports directly, hence the power divider becomes compact. Besides, this power divider does not need metalized via holes to adjust the magnitude and phase of output ports. Through adjusting the parameters of **L1** and **L2**, power equality can be achieved. The E-field distribution is shown in Fig. 2. The adjacent output ports, port2 and port3 (accordingly, port4 and port5), are spaced by a half guided wavelength, so there exists an alternating-phase of 180^0. Since SIW is a dispersive guided-

978-7-5641-4279-7

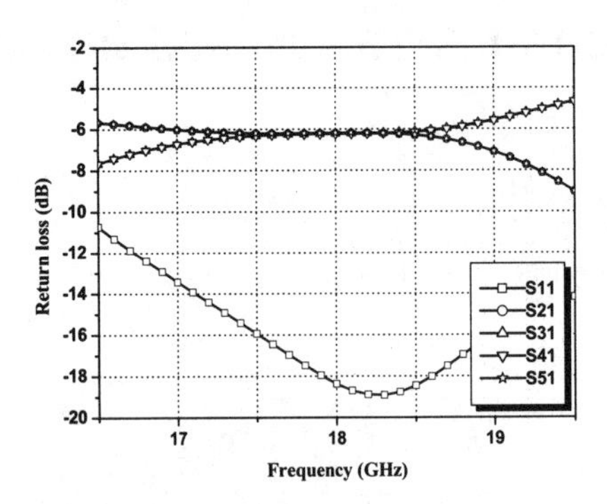

Fig. 3. Simulated S-parameter of the network.

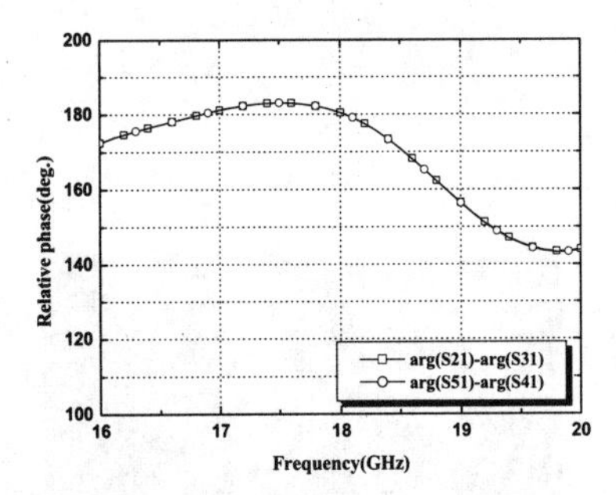

Fig. 4. Simulated phase difference of the network.

wave structure, the phase difference can only keep 180^0 in narrow bandwidth.

In order to broaden the bandwidth of the feeding network, self-compensating phase shift technique is used. In [6], a concept and mechanism of phase compensation combining delay line and equal-length unequal-width phaser is proposed, which can make the phase shift almost constant over a very wide band. As shown in Fig. 1, the length and width of SIW output branches is different. Though tuning the parameters of **L1, L2, W1, W2**, the phase difference between port2 and port3 (accordingly, port4 and port5) can keep 180^0 in a wide bandwidth.

Fig. 3 depicts the characteristic of S parameters after optimization. The power equality for S_{n1} (n=2, 3, 4, 5) is -6 dB ±0.5 dB in 17.1-18.9 GHz. The return loss is better than -10 dB over the frequency band from 16 to 20 GHz. The results of phase difference between port2 and port3 (accordingly, port4 and port5) are shown in Fig 4. The phase difference is 178^0 $\pm 5^0$ from 16.1 to 18.4 GHz covering a 12% relative bandwidth.

Based on this broadband feeding network, a 4×4 SIW slot antenna array is design. The geometry of the proposed antenna array is shown in Fig. 5. The feed consists of a 50 Ω microstrip line with a tapered transition to provide impedance matching between the microstrip and SIW. The 4-way SIW power divider is to feed 4 linear SIW slot arrays, and each of them carries 4 radiation slots etched on the broad wall of SIW. The SIW structure is terminated with a short-circuit quarter guided wavelength beyond the centre of the last radiation slot. In order to allocate the slots at the standing wave peaks and excite all the slots with the phase condition, the slots in a linear array are placed half a guided wavelength at the required centre frequency and the adjacent slots have the opposite offset with respect to the SIW centre line. The slots lengths are half a wavelength in the free space to ensure good radiation. The width of radiation slot should be much smaller

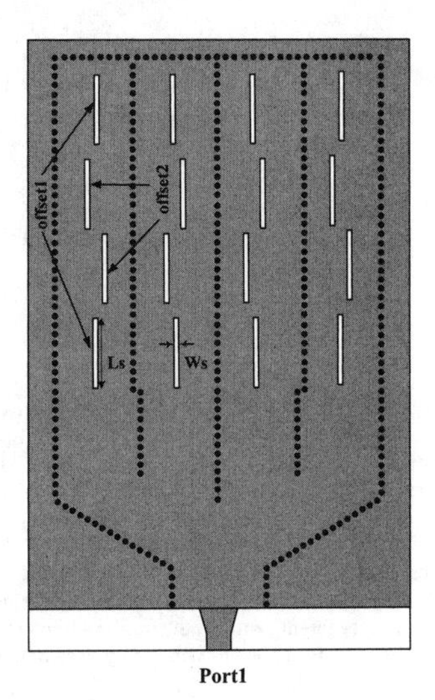

Fig. 5 Geometry of the antenna array.

TABLE I
PARAMETERS OF THE STRUCTURE

L1	L2	W1	W2
11.4mm	8.5mm	8mm	7mm
Ls	Ws	offset1	offset2
7.2mm	0.22mm	0.13mm	0.4mm

Fig. 6. Photograph of the antenna array.

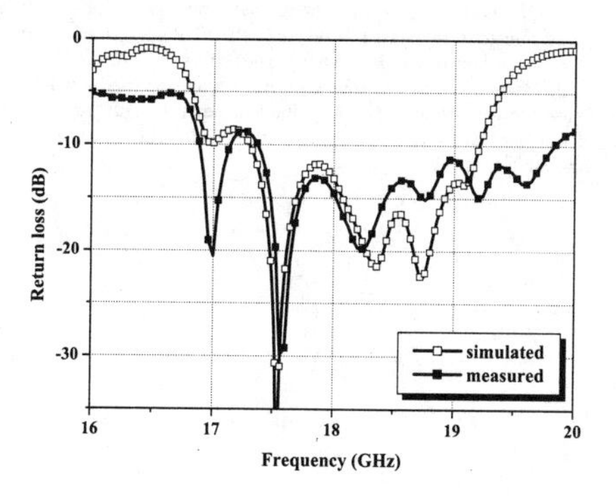

Fig. 7. The simulated and measured return loss of the antenna array.

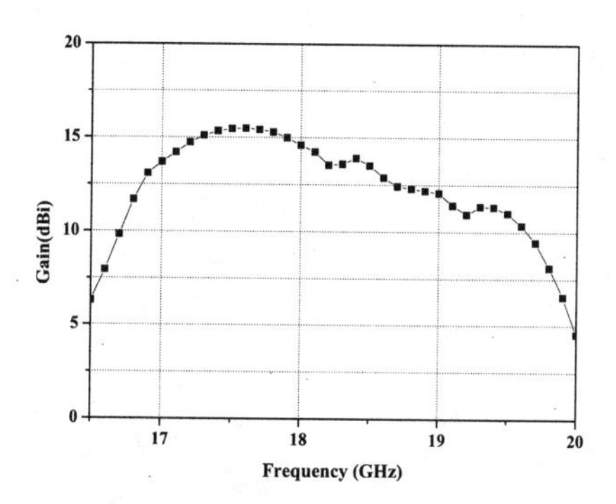

Fig. 8. The simulated gain of the antenna array.

than the slot length, usually between one tenth and one twentieth of the slot length. Because the radiating SIWs are excited with alternating-phase of 180^0, in order to keep 4 linear SIW slot antenna arrays in phase, the offset of slots following port2 and port3 (accordingly, port4 and port5) need to be set opposite.

Simulated and Experiment Results

After optimization with Ansoft HFSS, the geometry parameters for the array are listed in Table I. A Rogers-Duroid 5880 high-frequency substrate with thickness h=0.5mm, ε_r=2.2, tan δ=0.0009 is used in our experiments. According to the optimized simulation parameters, we fabricate and measure the antenna array. Fig. 6 shows photograph of our proposed array.

Fig. 7 depicts simulated and measured return loss of the antenna array. The measured results follow the trend of the simulated ones well. The simulated return loss below -10 dB is from 17.3 GHz to 19.2 GHz, while the measured result is from 17.4 GHz to 19.8 GHz. The measured impedance

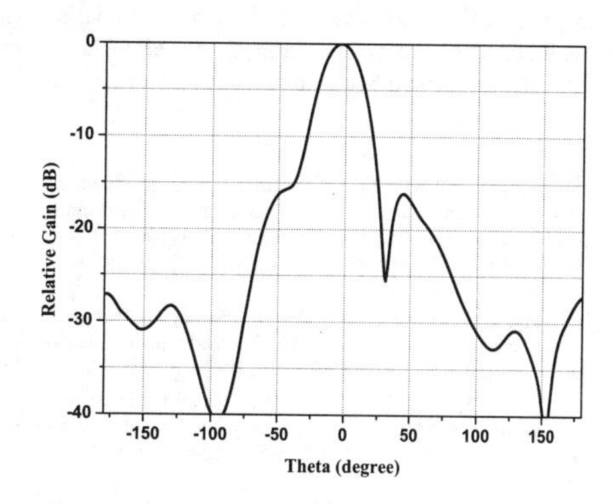

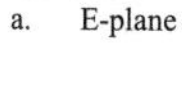

a. E-plane

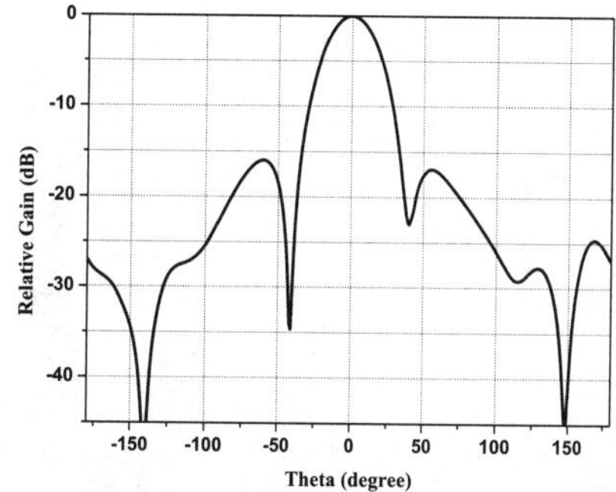

b. H-plane

Fig. 9 The simulated normalized radiation patterns of the array at 17.5 GHz

bandwidth reaches up to 13%, which is two times more than that of traditional SIW slot antenna array.

Fig. 8 shows the simulated gain of the proposed array at the boresight direction. We can observe that the simulated gain is above 12.5dBi from 16.8 GHz to 18.7 GHz and the peak gain can reach up to 15.5 dBi at 17.5 GHz. The normalized simulated radiation patterns in E-plane and H-plane of the array at 17.5 GHz are plotted in Fig. 9. Their 3 dB beamwidths in E-plane and H-plane are about 29^0 and 33^0. The sidelobe levels (SLLs) are below -16 dB in the E-plane and below -17 dB in the H-plane.

Conclusion

A novel substrate integrated waveguide (SIW) slot antenna array is proposed. The array consists of a broadband SIW 4-way power divider and 4 radiating SIWs each supporting 4 radiating slots. The 4-way power divider uses the self-compensating phase shift technique, which can make the feeding network operate over a broadband. Comparing with conventional SIW slot antenna arrays, this array has simple structure and broadband bandwidth.

REFERENCES

[1] K. Wu, D. Deslandes, and Y. Cassivi, "The substrate integrated circuits—a new concept for high-frequency electronics and optoelectronics," in *Proc. 6th Int. Conf. Telecommunications Modern Satellite, Cable Broadcasting Service (TELSIKS'03)*, vol. 1, Oct. 1-3, 2003, pp. P-III-P-X.

[2] Xiao-Ping Chen, K. Wu, Liang Han, and Fanfan He, "Low-Cost High Gain Planar Antenna Array for 60-GHz Band Applications," in *IEEE Trans. Antennas Propag.*, vol. 58, No. 6, Jun. 2010, pp. 2126-2129.

[3] J. F. Xu, W. Hong, P. Chen, and K. Wu, "Design and Implementation of Low Sidelobe Substrate Integrated Waveguide Longitudinal Slot Array Antennas," in *IET Microw. Antennas and Propag.*, 2009, vol. 3, pp. 790-797.

[4] Y. J. Cheng, Wei Hong, and Ke Wu, "94 GHz Substrate Integrated Monopulse Antenna ArraY," in *IEEE Trans. Antennas Propag.*, vol. 60, No. 1, Jan. 2012, pp. 121-129.

[5] Bing Liu, Wei Hong, Zhang-cheng Hao, and Ji-xin Chen, "Alternate Phase Substrate Integrated Waveguide (SIW) Power Divider," in *Acta Electronica Sinica*, vol. 35, no.6, pp. 1061-1064, Jun. 2007.

[6] Y. J. Cheng, W. Hong, and K. Wu, "Broadband Self-Compensating Phase Shifter Combining Delay Line and Equal-Length Unequal-Width Phaser," in *IEEE Trans. Microw. Theory Tech.*, vol. 58, no.1, pp. 203-209, January. 2010.

Novel Antipodal Linearly Tapered Slot Antenna Using GCPW-to-SIW Transition for Passive Millimeter-Wave Focal Plane Array Imaging

Wen Wang[#1], *Student Member, IEEE*, Xuetian Wang[#3], Wei Wang[#4] and Aly E. Fathy[*2], *Fellow, IEEE*
[#] School of Information and Electronic, Beijing Institute of Technology
Beijing 100081, China
[1] wangwenbit@gmail.com
[*] Department of Electrical Engineering and Computer Science, University of Tennessee
Knoxville, TN 37996 USA
[2] fathy@eecs.utk.edu

***Abstract*- In this work, a novel antipodal linearly tapered slot antenna (ALTSA) using grounded coplanar waveguide (GCPW)-to-substrate integrated waveguide (SIW) transition is proposed and demonstrated. The antenna is well designed for using as the feed antenna in the focal plane array (FPA) imaging system. The S11 of the proposed antenna is below -13dB on Ka-band, meanwhile, the E-plane side lobe is almost 15dB down and 40dB down after reflecting by a parabolic antenna when placing along a line on its focal plane, which show us its high availability for passive millimeter-wave (PMMW) focal plane array imaging system.**

Index Terms- Passive millimeter-wave; Focal plane array; Antipodal linearly tapered slot antenna; Grounded coplanar waveguide; Substrate integrated waveguide

I. INTRODUCTION

Passive millimeter-wave (PMMW) imaging method is based on the passive detection of naturally occurring millimeter-wave radiation from a targeted scene. Upon receiving the thermal radiation of an object in contrast to its background, the PMMW imaging would have the ability to produce images during day or night times; in clear weather or in low-visibility conditions--such as haze, fog, clouds, smoke, or sandstorms; and even through clothing [1]. Such capabilities make PMMW imaging of great importance in many civilian and military applications, such as atmospheric research, all-weather landing guidance, non-destructive security check, and also car collision avoidance [2].

A highly popular method for PMMW imaging is the Focal Plane Array (FPA) Imaging where the receiving units are placed along a line on the focal plane of a reflector antenna, resulting in a multi-beam coverage sight. Because of the relatively higher performance and lower cost of the feed units and their associated receivers; the FPA imaging system has become the main method in PMMW imaging. The packing density, the noise figure, and the data handling speed performance of the receivers are main factors determine the spatial sampling rate of the millimeter-wave focal plane array (MFPA) imaging systems [2]. Undeniably, the performances of the feed antenna, such as, better side-lobe level, better S11 and narrower main beam are extremely needed for better accuracy of the imaging.

In this paper, a novel antipodal linearly tapered slot antenna (ALTSA) with grounded coplanar waveguide (GCPW)-to-substrate integrated waveguide (SIW) transition is proposed and its performance is fully analyzed, simulated and proved by measurement results.

II. PLANAR ANTIPODAL LINEARLY TAPERED SLOT ANTENNA

The ALTSA features high-gain, low-cost, low-side lobe, ease of fabrication, light-weight and compact-size as well as a better performance than the conventional feed elements, i.e. the horn antennas.

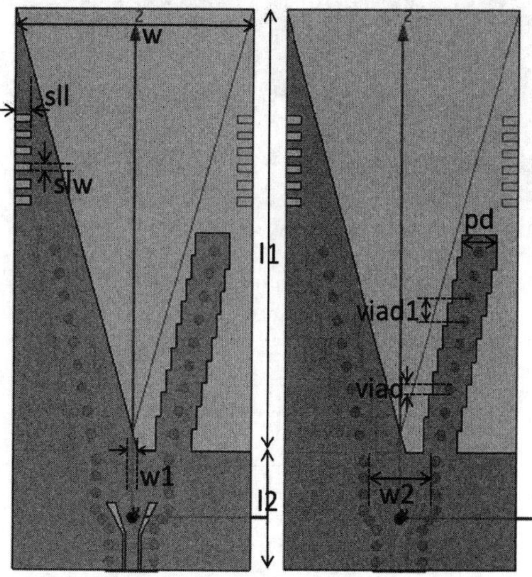

Fig. 1. Modeling of ALTSA (front and ground) in HFSS

978-7-5641-4279-7

The developed ALTSA design is a suitable feed antenna for the reflector for the PMMW imaging system, and covers the Ka-band (32-38GHz). The proposed antenna structure (Fig. 1) is designed using a RT\Duroid6002 substrate with a 2.98 relative permittivity, a dielectric loss tangent of 0.0012, and a 0.508 mm thickness, which is relatively thick for lower SIW conductors' loss.

The double-sided-metallization of the substrate is flared linearly in opposite directions with an overlap width (*w1*) to form tapered slots to reduce mismatch losses. The parameters *w1*, *w* and *l1* are optimized to achieve wide band performance. Cylindrical vias with rectangular patch on one side are added to the antenna structure to increase gain especially at the high frequency end, while reduce side lobe level particularly in E-plane [3]. Additional periodic slits (*sll**slw*) are placed on top of the antenna's tapered structure to effectively increase the gain, and decrease the side lobe and cross-polarization levels. The photograph of fabricated antenna with connector is shown in Fig. 2.

Fig. 2. Photograph of fabricated antenna (front and back with connector)

All optimized antenna dimensions are listed in Table I.

TABLE I
DIMENSIONS OF ALTSA (UNIT:MM)

Symbol	Value	Symbol	Value
l1	27.1	tcsl	2
w	15	tcsw	0.55
w1	0.7	tcsa	62°
l2	7.2	cw	0.8
w2	4	cs	0.15
viad	0.6	cl	2.1
viad2	1	slw	0.5
viad1	1.4	sll	1
pd	2.2	viad3	0.6

III. DESIGN OF GCPW-TO-SIW TRANSITION

For low cost and easy fabrication, compactness, light weight and low loss performance, substrate integrated waveguide technology is utilized, where waveguide side-walls are emulated using rows of metallic vias placed in double-sided substrate for the top and bottom waveguide walls. SIW technology makes it possible to realize the waveguide in a substrate and provides an elegant way to integrate the waveguide with microwave and millimeter wave planar circuits using the conventional low-cost PCB processing technologies [4]. The parameters *w2*, *viad* and *viad2* are calculated by [5] and optimized to make sure that SIW can be replaced by a dielectric filled rectangular waveguide perfectly and good matched to ALTSA.

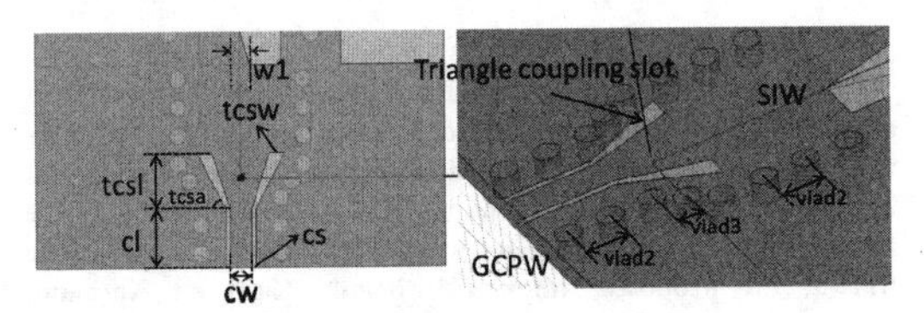

Fig. 3. GCPW to SIW transition (2D and 3D)

It is really a big challenge on how to connect the SIW to coaxial connector. SIW components are typically fed by coplanar waveguide [6], or microstrip-to-SIW transitions [7], here we use the former. In this work, we propose a simple and compact transition (Fig. 3) between GCPW and SIW which can be easily realized on the substrate of SIW circuit with PCB technologies. The sidewalls of the SIW are tapered along the triangle-shaped coupling slot, in such a way that the direction of the electric field on the coupling slot is always perpendicular to the SIW sidewalls. The tapered coupling slot also serves as an impedance transformer to transform any arbitrary impedance line in SIW to the CPWG port impedance. The parameters *viad3*, *tcsa*, *tscl* and *tcsw* are optimized for a smooth transition and perfect match. The optimized length of the coupling slot (*tcsl*) is almost quarter-wavelength of the center frequency in order to achieve a wider bandwidth. Additionally, a 2.92mm connector is utilized in the implement of this transition, while *cl*, *cw* and *cs* are decided by size of the connector and calculated by [8]. All optimized parameters are shown in Table I.

IV. FOCAL PLANE ARRAY

After perfect design of single antenna, we can use antennas in the focal plane array imaging system. In our PMMW imaging system, the reflector we choose for focal plane array is parabolic antenna, which is proved to be suitable to get a multi-beam coverage sight. In our PMMW-FPA imaging system, each antenna is individually fed and independently connected to a receiver for better spatial sampling rate and imaging, while the whole array is placed linearly along its E-plane

To evaluate the overall performance, we simulated the overall system including parabolic antenna that is fed by the focal plane linear array at 35GHz. In our simulation, we choose 8 antennas for a small and compact system, as well as, a parabolic antenna with a 500 mm focal distance, while its aperture diameter is also 500 mm, which are designed for portable far-distance-imaging system. Fig. 4 shows the model of parabolic antenna with a linear 8x1 array which is placed along a line on the focal plane of the reflector. The off-set angle and distance of each single antenna is calculated and optimized for perfect reflection.

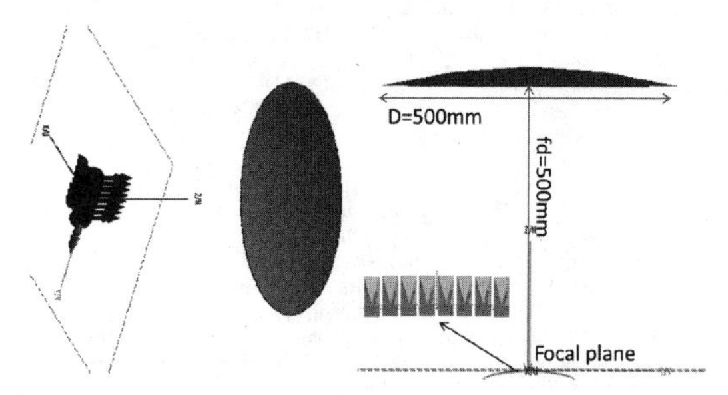

Fig. 4. Modeling of parabolic reflector antenna with a linear 8x1 array in FEKO(3D and 2D)

V. RESULTS

Fig. 5 illustrates simulation and measurement results of the input return loss S11 (32-38GHz) of ALTSA. The S11 of the proposed antenna is below -13dB in a really wide band from 32GHz to 38GHz, especially below -17dB at 35GHz.

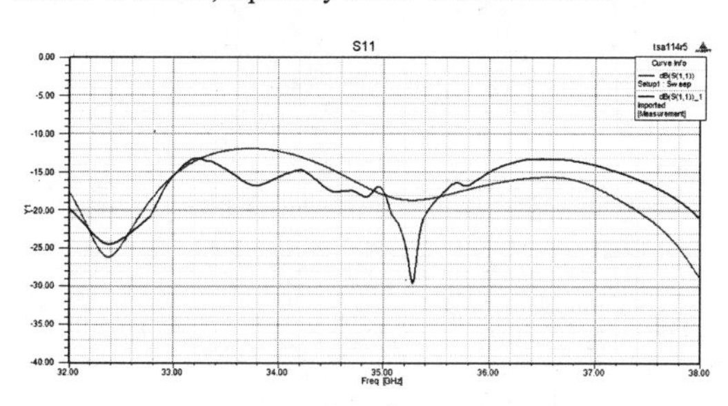

Fig. 5. The S11 (32-38GHz) results of ALTSA (Red: Simulation, Blue: Measurement)

Fig. 6(a, b, c) show simulation and measurement results of the E-plane/H-plane normalized radiation patterns at 33, 35, 37GHz separately for the ALTSA antenna. It is essential to have low side lobe levels in the E plane to minimize the reception of unwanted noise; given that we arrange the elements in the focal plane along the E- plane. The asymmetric of the side lobe may be resulted by the coupling between the connector and the antenna. Absorbing materials are used to cover the connected coaxial cable during the measurement, which maybe the reason that the measured radiation at the back is lower than simulated one.

The measured gain of main beam is about 14.5dB, meanwhile, the E-plane side lobe is almost 15dB down, which indicates the ALTSA high potential for high performance in the MFPA imaging system. The measurement results are really close and even better than the simulation results.

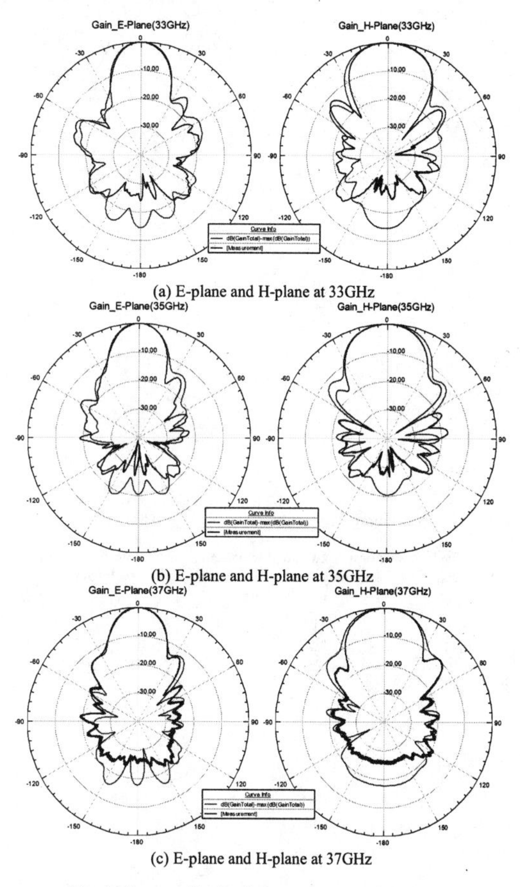

Fig. 6. The normalized radiation pattern results of ALTSA (Red: Simulation, Blue: Measurement)

The normalized radiation pattern at 35GHz of a single ALTSA element feeding the parabolic antenna has a side lobe in the E-plane of almost 40 dB down with a 1.3° HPBW (Fig. 7(a)), also the shoulders in the radiation pattern of ALSTA are disappeared after reflector, which illustrates our design is really promising. For the 2x1 array, the side lobe level in the E-plane is almost 37 dB down with the 3° HPBW (Fig. 7(b)), while 4x1 array with almost 40 dB down side lobe level and 7°

HPBW (Fig. 7(c)), lastly, Fig. 7(d) shows almost 38 dB down side lobe level with the 14.5° HPBW for 8x1 array. The high-gain, flat and narrow main beam made by the reflected multi-beam coverage is becoming larger while the number of the ALTSA elements is increasing, which are meaningful for large range and far distance imaging.

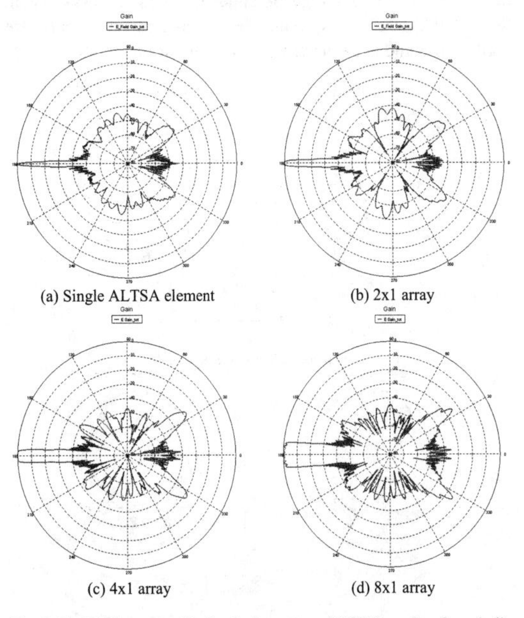

(a) Single ALTSA element (b) 2x1 array

(c) 4x1 array (d) 8x1 array

Fig. 7. The E-Plane normalized radiation pattern (35GHz) results of parabolic antenna with linear antenna array placed along a line on focal plane

All of these above theoretical and measurement results are very encouraging and indicates their high availability for PMMW focal plane array imaging.

ACKNOWLEDGMENT

The author--Wen Wang is now a visiting student at The University of Tennessee at Knoxville, which is sponsored by China Scholarship Council.

REFERENCES

[1] Yujiri L., Shoucri M., Moffa P., "Passive millimeter wave imaging," *Microwave Magazine, IEEE*, vol. 4, no. 3, pp.39-50, Sept. 2003.

[2] Wen Wang, Xuetian Wang, Lili Fang, "LTSA with microstrip-slotline transition for MFPA imaging systems," *Microwave Technology & Computational Electromagnetics (ICMTCE), 2011 IEEE International Conference on*, pp.230-232, May 2011.

[3] Ghassemi, N., Wu, K., "Planar High-Gain Dielectric-Loaded Antipodal Linearly Tapered Slot Antenna for E- and W-Band Gigabyte Point-to-Point Wireless Services," *Antennas and Propagation, IEEE Transactions on*, no.99, pp.1.

[4] S. Lin, S. Yang, A. E. Fathy, and A. Elsherbini, "Development of a novel UWB vivaldi antenna array using SIW technology," *Progress In Electromagnetics Research*, Vol. 90, 369-384, 2009.

[5] Deslandes, D., "Design equations for tapered microstrip-to-Substrate Integrated Waveguide transitions," *Microwave Symposium Digest (MTT), 2010 IEEE MTT-S International*, pp.1,May 2010.

[6] S. Yang, "Antennas and arrays for mobile platforms - Direct broadcast satellite and wireless communication," PHD dissertation, University of Tennessee, Knoxville, USA, 2008.

[7] Miralles, E., Esteban, H., Bachiller, C., Belenguer, A., Boria, V.E., "Improvement for the design equations for tapered Microstrip-to-Substrate Integrated Waveguide transitions," *Electromagnetics in Advanced Applications (ICEAA), 2011 International Conference on*, pp.652-655, Sept. 2011.

[8] M. A. Calculator. "http://wcalc.sourceforge.net/cgi-bin/coplanar.cgi," *Microwave Theory and Techniques, IEEE Transactions on*, vol. 55, no. 9, pp. 1880–1886, Sept. 2007.

TP-2(B)

October 24 (THU) PM

Room B

Integrated MMW Antennas

The Substrate and Ground Plane Size Effect on Radiation Pattern of 60-GHz LTCC Patch Antenna Array

Lei Wang [1,3], Yong-Xin Guo [1,2], and Wen Wu [3]

[1]National University of Singapore, Singapore 117583
[2]National University of Singapore (Suzhou) Research Institute, Suzhou, Jiangsu Province, China, 215123
[3]Ministerial Key Laboratory of JGMT, Nanjing University of Science and Technology, Nanjing, China
Email: eleguoyx@nus.edu.

***Abstract*-The influence of finite substrate and ground plane size on the radiation pattern performance of L-probe feed thick patch antenna-in-package array in low-temperature co-fired ceramic (LTCC) for 60-GHz wireless communications is investigated and compared when the antenna array is with and without the soft surface structure, which is used to reduce surface wave effect and improve the antenna performance. It is shown that the effect of the substrate and ground plane size effect on the radiation patterns of the array is significant and the shape of the main lobe will be distorted obviously when the antenna array is without the soft surface structure. On the other hand, the antenna radiation performance will not be affected too much by the substrate and ground plane size when the soft surface structure is added into the antenna array to reduce surface wave effect.**

I. INTRODUCTION

Recently, there exists a surge of interest in wide unlicensed frequency band around 60 GHz for wireless short-range communications. The low-temperature co-fired ceramic (LTCC) multilayer technology based antenna-in-package (AiP) solutions has become a hot topic for 60-GHz applications [1]-[9]. Many types of antennas, such as a dipole, a slot, a patch, a helical, etc. [2]-[5], as the elements used in the design of 60-GHz LTCC arrays have been investigated. The patch antenna, because of its geometrical simplicity and other attractive features, has been widely used in the designs of the 60-GHz LTCC antenna arrays [6]–[8]. In [9], we proposed a wideband LTCC L-probe fed thick patch antenna for 60-Ghz applications and a soft surface was added to reduce the surface wave. For mobile terminals applications, the antennas or arrays need be integrated into the system mother board with a large substrate and ground plane size. Thus, it is necessary to study the effect of the substrate and ground plane size on 60-GHz LTCC antenna array characteristics.

In this paper, we investigate the effect of the substrate and ground plane size on the radiation performance of the 60-GHz L-probe fed thick patch antenna array in LTCC technology with and without the soft surface structure. It is shown that the effect of the substrate and ground plane size effect on the radiation patterns of the array is significant and the shape of the main lobe will be distorted obviously when the antenna array is without the soft surface structure. On the other hand, the antenna radiation performance will not be affected too much by the substrate and ground plane size when the soft surface structure is added into the antenna array.

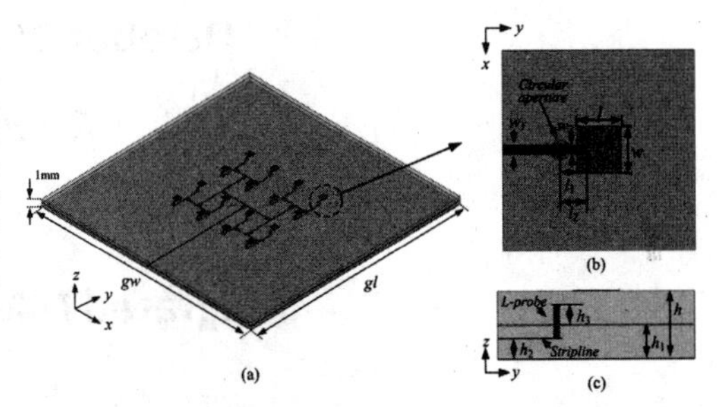

Fig. 1. Geometry of the 60-GHz LTCC based L-probe fed patch antenna array in [9], (a) 3D view of the array, (b) Top View of the single element, (c) Side View of the single element.

TABLE I DETAILED L-PROBE PATCH ANTENNA DIMENSIONS

Parameters	Dimensions (mm)	Parameters	Dimensions (mm)
w	0.7	l	0.7
w_1	0.15	l_1	0.25
w_2	0.1	l_2	0.4
h	1	h_1	0.5
h_2	0.3	h_3	0.3

II. ANTENNA GEOMETRY

The 60-GHz LTCC based L-probe fed patch antenna array which was proposed in [9] as shown in Fig.1. The multilayer LTCC substrate used is Ferro A6-M with dielectric permittivity ε_r=5.9 and loss tangent tanδ=0.001. The detailed dimensions of the array element are shown in Table I. The total thickness of the proposed array is 1 mm. The distance between the top rectangle patch and the top ground of the stripline feeding network is 0.5mm, which is about 10% free space wavelength (λ_0) at 60 GHz. The size of the substrate and ground is $gw \times gl$. In this work, the element spacing of 3.7 mm ($0.75 \times \lambda_0$) was chosen.

978-7-5641-4279-7

III. Effects of Substrate and Ground Plane Size on the Radiation Pattern of the Array Located in the Center of the Mother Board

A. Array Without the Soft-surface Structure

The effects of the substrate and ground plane size on the radiation pattern of the L-probe fed thick patch antenna array without any soft surface structure are investigated. The comparison of simulated radiation pattern in XoZ-plane and YoZ-plane of the array with different substrate and ground plane size at 60 GHz are shown in Fig. 2.

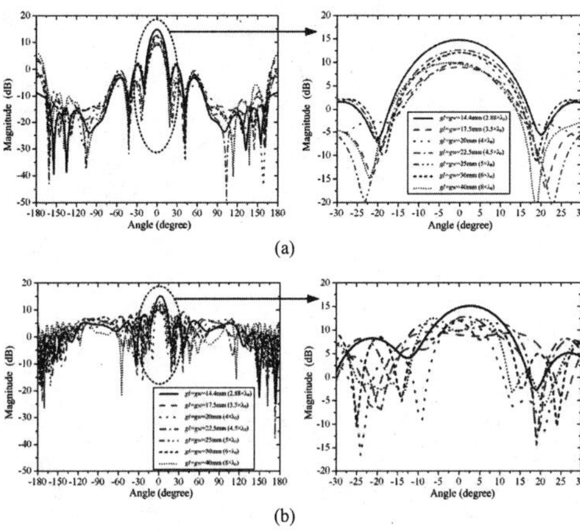

Fig. 2. The radiation pattern of the LTCC based L-probe fed patch antenna array for different substrate and ground plane size at 60 GHz, (a) XoZ-plane, (b) YoZ-plane.

As shown in Fig 2, with the substrate and ground plane size increasing, the shape of the main-lobe radiation pattern in XoZ-plane (H-plane) will not change too much. However, the main lobe shape of radiation pattern in YoZ-plane (E-plane) is changed obviously with the substrate and ground plane size. The E-plane radiation pattern is changed more significantly than H-plane radiation pattern. The distortion of the patch antenna array's radiation pattern with the substrate and ground size is due to the surface waves in the antenna array.

B. Array With the Soft-surface Structure

To reduce the surface wave effect and improve the radiation patterns performance of the array located on the large size of the substrate and ground, seven strip-shaped soft surface structures constituted of metal strips on top layer and via fences are loaded in the array. The top metal strip width of the proposed soft-surface structure is 1.4mm. The via fences are composed of a row of vias. Following the fabrication process requirement, the diameter of each via is 0.1mm, and the distance between the centers of two adjacent vias is 0.25mm. Fig. 3 shows the layout of the proposed array with soft surface structures on the large substrate and ground size.

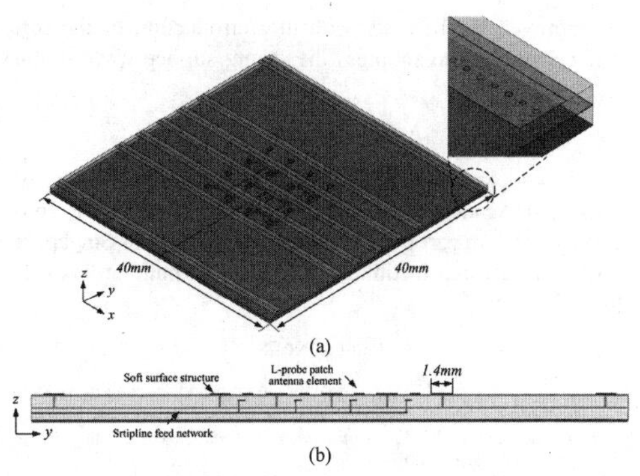

Fig. 3. Geometry of the LTCC based L-probe feed patch antenna array with soft-surface structures on the large substrate and ground size, (a) 3D view, (b) Side View.

Fig. 4 compares the simulated radiation patterns of the proposed arrays with or without the soft surface structure on the large substrate and ground size at 60GHz. For the array without soft surface structure, when the substrate and ground size is 40mm × 40mm ($8\lambda_0 \times 8\lambda_0$), the radiation pattern is distorted significantly. There is an obvious notch appeared at the broad direction (*z*-axis) in YoZ-plane (E-plane). Compared with the array without soft surface structure, it is very clearly seen form Fig. 4(b) that the radiation patterns performance of the array with soft surface structure is improved obviously, especially the radiation pattern in YoZ-plane. Therefore, for the array with the soft surface structure, the distortion of main lobe can be ignored; and good radiation pattern and high gain performances of the array on the large substrate and ground size can be achieved.

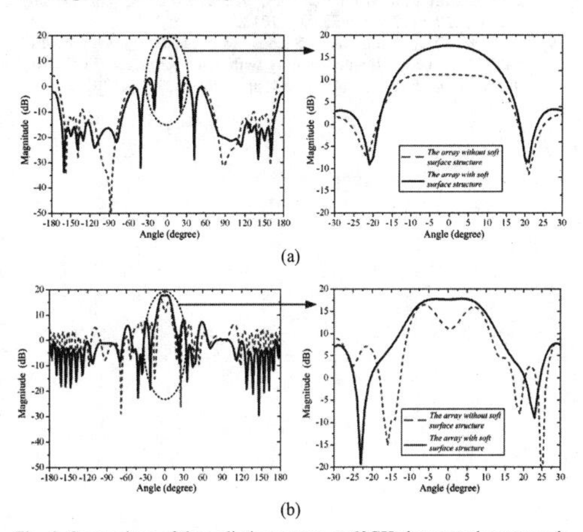

Fig. 4. Comparison of the radiation pattern at 60GHz between the proposed arrays without or with the soft surface structure on a large substrate and ground sizee, (a) XoZ-plane, (b) YoZ-plane.

IV. Conclusion

In this paper, the effect of the substrate and ground size on the radiation pattern of the 60-GHz L-probe fed thick patch antenna array in LTCC technology has been investigated. It has shown that the radiation pattern of the antenna array can be improved significantly with the introduction of the soft surface structure to suppress the strong surface wave in the array.

ACKNOWLEDGMENT

This work was supported in part by Singapore Ministry of Education Academic Research Fund Tier 1 project R-263-000-667-112, in part by the National University of Singapore (Suzhou) Research Institute under the grant number NUSRI-R-2012-N-010.

REFERENCES

[1] Y. P. Zhang and D. Liu, "Antenna-on-chip and antenna-in-package solutions to highly integrated millimeter-wave devices for wireless communications," *IEEE Trans. Antennas Propag.*, vol. 57, no. 10, pp. 2830–2841, Oct. 2009.

[2] H. Chu, Y. X. Guo, and Z. Wang, "60-GHz LTCC wideband vertical off-center dipole antenna and arrays," *IEEE Trans. Antennas and Propag.*, vol. 61, no. 1, pp. 153–161, Jan. 2013.

[3] J. Xu , Z. N. Chen, X. Qing, and W. Hong, "Bandwidth enhancement for a 60 GHz substrate integrated waveguide fed cavity array antenna on LTCC," *IEEE Trans. Antennas Propag.*, vol. 59, no. 3, pp. 826-832, Mar. 2012.

[4] M. Sun, Y. Q. Zhang, Y. -X. Guo, M. F. K., L.C. Ong, and M. S. Leong, "Integration of circular polarized array and LNA in LTCC as a 60-GHz active receiving antenna," *IEEE Trans. Antennas Propag.*, vol. 59, no.8, pp. 3083-3089, Aug. 2011.

[5] C. Liu, Y. -X. Guo, X. Bao, and S. Q. Xiao, "60-GHz LTCC integrated circularly polarized helical antenna array," *IEEE Trans. Antennas Propag.*, vol. 60, no. 3, pp. 1329-1335, Mar. 2012.

[6] Y. Li ,Z. N. Chen, X. Qing, Z. Zhang, J. Xu, and Z. Feng, "Axial ratio bandwidth enhancement of 60-GHz substrate integrated waveguide-fed circularly polarized LTCC antenna array," *IEEE Trans. Antennas and Propag.*, vol.60, no. 10, pp. 4619-4626, Oct. 2012.

[7] H. Sun, Y.-X. Guo, and Z. Wang, "60-GHz circularly polarized U-slot patch antenna array on LTCC," *IEEE Trans. Antennas Propag.*, vol. 61, no. 1, pp. 430-435,Jan. 2013.

[8] S. B. Yeap, Z. N. Chen, and X. Qing, "Gain-enhanced 60-GHz LTCC antenna array with open air cavities," *IEEE Trans. Antennas Propag.*, vol.59, no. 9, pp. 3470-3473, Sep. 2011.

[9] L. Wang, Y.X. Guo, W.X. Sheng, "Wideband high-gain 60-GHz LTCC L-probe patch antenna array with a soft surface," *IEEE Trans. Antennas and Propag.*, vol.61, no.4, pp. 1802-1809, Apr. 2013.

Ultra-broadband Tapered Slot Terahertz Antennas on Thin Polymeric Substrate

Masami Inoue[#], Masayuki Hodono[#], Shogo Horiguchi[*],
Masayuki Fujita[*], and Tadao Nagatsuma[*]

[#] Functional Design Technology Center, Nitto Denko Corp., Suita, Osaka 565-0871, Japan
[*] Graduate School of Engineering Science, Osaka University, Toyonaka, Osaka 560-8531, Japan

Abstract— **We developed a planar tapered slot antenna on a thin polymeric substrate with low dielectric constant for terahertz-wave applications. The efficiency and the gain of this antenna are not degraded even for higher frequencies. Due to the ultra-broadband features, we demonstrate multiband terahertz wireless error-free communications at 1.5 Gbits/s for both 120 and 300 GHz bands.**

I. INTRODUCTION

Research that explores terahertz (THz) waves at frequencies from 100 GHz to 10 THz has recently attracted much attention, especially since THz waves are suited to such novel applications as spectroscopic sensing, non-destructive imaging, and ultra-broadband wireless communications [1-5]. Planar antennas fabricated on semiconductor substrates have attracted great interest for THz applications [6-16], because planar structures offer great potential for integration with other planar devices. Here, as a planar antenna, we focus on a tapered slot antenna, which is a kind of traveling-wave antenna, that can achieve ultra-broadband operation since it has no resonance at a specific frequency. However, serious practical problems exist: the antenna gain decreases, and the antenna pattern diverges due to unnecessary substrate modes that are excited when the substrate thickness is increased against the wavelength of interest [16-19]. At THz-wave frequencies, the wavelength becomes less than one millimeter, and then this problem becomes remarkable.

In this paper, we present our recent progress in the development of ultra-broadband planar antennas fabricated on a polymeric substrate. We investigate both the theoretical and experimental performances of tapered slot antennas at sub-THz frequencies. We also describe wireless transmission experiments using a receiver module integrated with an antenna that can be operated at both 120 and 300 GHz.

II. ANTENNA STRUCTURE AND SIMULATIONS

Figure 1 shows a schematic of the tapered slot antenna studied in this work. It consists of a substrate, a copper (Cu) pattern, and is nickel (Ni)-gold (Au) plated. THz waves are detected or radiated at the wider edge of the tapered slot. Such semiconductor substrates as indium phosphide (InP) have so far been used in transmitter and receiver modules [11-16]. However, the high relative dielectric constant ε_r of these substrates, typically 12, deteriorates the radiation efficiency and the antenna directivity, because the electromagnetic-wave radiation is attracted to the semiconductor substrate when the frequency increases [16, 19]. To overcome this problem, we proposed a tapered slot antenna on a polyimide film with a ε_r as low as 3 [20, 21].

To compare the semiconductor and polymeric substrates, we changed the dielectric constant and the thickness in the simulation. Table I summarizes the material properties of the substrates. The simulations were performed by a finite-element method considering the dielectric loss tangent (tanδ) and conductivity σ of the metal layers.

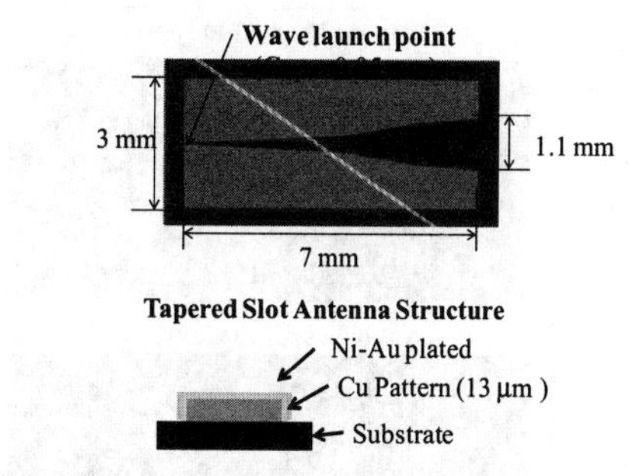

Fig. 1 Schematic of a tapered slot antenna. Conductivities of Cu, Ni, and Au are set to 5.80×10^7 S/m, 1.45×10^7 S/m, and 4.10×10^7 S/m, respectively.

TABLE I
Material characteristics of substrate

Material	Dielectric constant (ε_r)	tanδ	Thickness [μm]
InP	12.4	0.007 (@1GHz)	100
Polyimide	3.2	0.0105 (@1MHz)	25

The simulated electric-field distributions at 300 GHz are shown in Fig. 2. The antenna on a 25-μm-thick polyimide substrate produces a good radiation level. On the other hand, when a 100-μm-thick InP substrate is used, the THz wave is confined to the substrate due to its high dielectric constant. Although the thickness of the InP substrate must be reduced to overcome this problem, handling such a fragile InP substrate is very difficult. Thus, using low dielectric constant and thin

978-7-5641-4279-7

substrates is essential for tapered slot antennas in THz applications.

Figure 3 shows the antenna pattern of the polyimide and InP substrates for an E-plane at 300 GHz. The maximum gains of the polyimide and InP antennas are estimated to be 12.4 and 6.0 dBi, respectively. The polyimide antenna only radiates toward the tapered slot direction. On the other hand, the antenna InP pattern has many side lobes due to the excitation of the higher order modes inside the high dielectric constant substrate, which is much thicker than the wavelength.

Next we calculated the radiation efficiency for various frequencies (Fig. 4). Radiation efficiency η is defined by

$$\eta = P_{\text{rad}} / (P_{\text{rad}} + P_{\text{loss}}) \quad (1),$$

where P_{rad} is the total radiated power and P_{loss} is the dielectric and conductive losses. The efficiency of the polyimide antenna is almost constant and higher than that of the InP one from 100 to 500 GHz. This is because the polyimide substrate hardly affects the propagation of the THz waves. On the other hand, the THz wave is strongly confined by the InP substrate and suffers material losses.

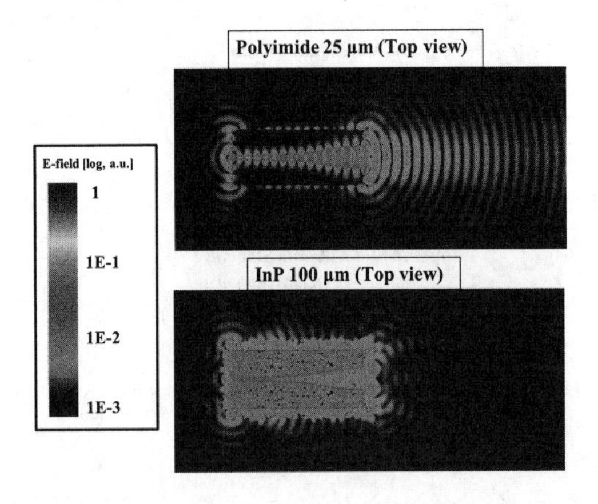

Fig.2 Simulated electric-field intensity distributions at 300 GHz (top view). Upper and bottom figures show 25-μm-thick polyimide and 100-μm-thick InP, respectively.

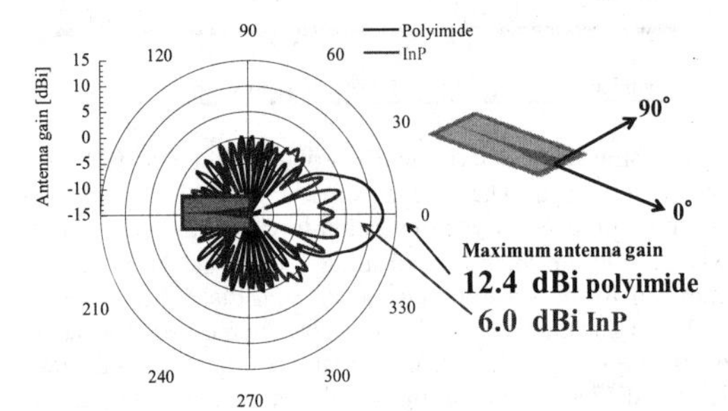

Fig.3 Simulated E-plane antenna patterns for polyimide and InP.

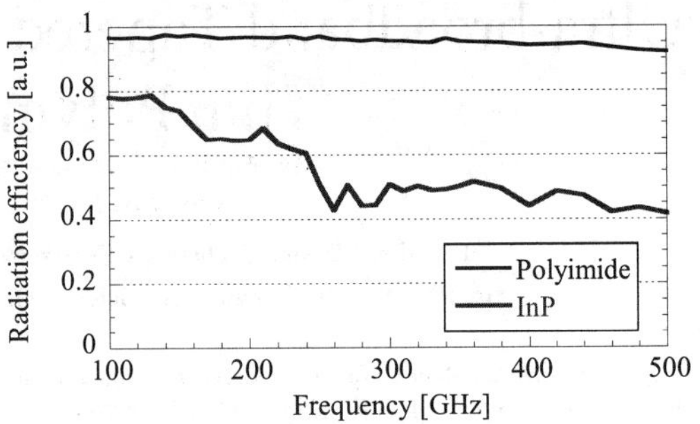

Fig. 4 Radiation efficiency vs. frequency for polyimide and InP.

III. Antenna Pattern Measurements

For our experiments, we fabricated a receiver module that consists of a SMA connector, a rigid board, and an antenna on a 25-μm-thick polyimide film layer (Fig. 5). The rigid board is designed to allow air space under the antenna to decrease the effective dielectric constant. A Schottky barrier diode (SBD) chip [22] was mounted on the antenna by flip-chip bonding.

We evaluated the antenna properties by measuring the directional characteristics with a photonics-based transmitter (Fig. 6). In the transmitter, a sinusoidally intensity-modulated optical signal at 120 or 300 GHz is generated by two sets of wavelength-tunable lasers with a wavelength difference of 0.96 or 2.4 nm that corresponds to 120 or 300 GHz, respectively. The optical signal is modulated with an optical intensity modulator at 100 MHz by a signal generator. The optical signal is converted into a THz signal by an ultrafast photodiode [23]. In the receiver module, the detected THz signal is demodulated by the envelope detection, and the demodulated signal is amplified by a pre-amplifier that has a bandwidth of up to 1.5 GHz. The transmission distance was set to 90 mm. Fig. 7 shows the antenna patterns. Our experimental results agree well with the designed ones. Consequently, the polyimide antenna demonstrated a single lobe in the direction of a tapered slot at both the 120- and 300-GHz bands.

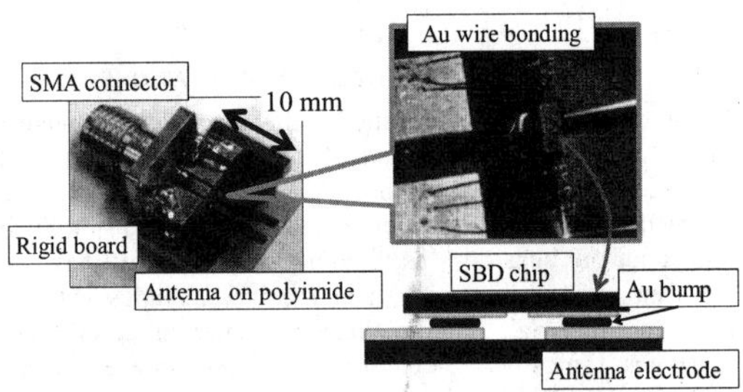

Fig. 5 Receiver module with a tapered slot antenna on polymeric substrate.

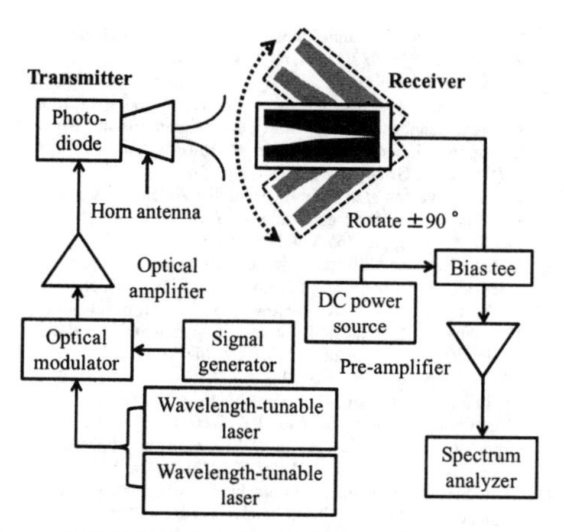

Fig.6 Block diagram of antenna pattern measurements.

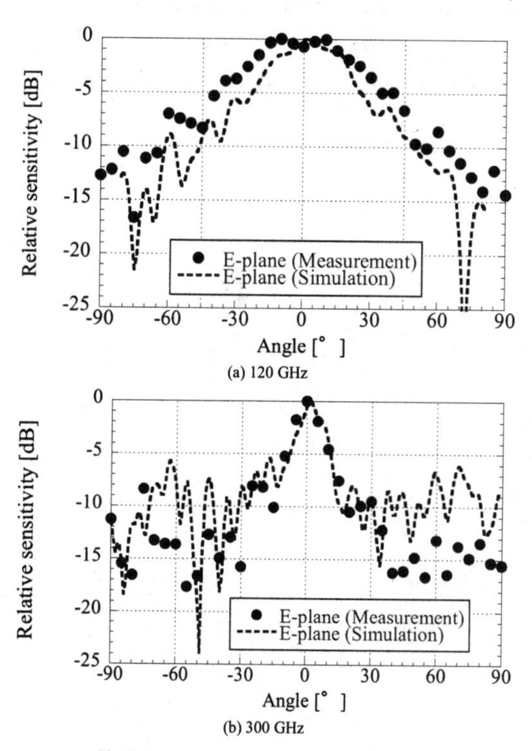

Fig. 7 Antenna patterns: (a) 120 and (b) 300 GHz.

IV. THz Wireless Communication

We performed experiments on THz communication for both 120- and 300-GHz bands to demonstrate the ultra-broad property of the antennas. Compared with the set-up in Fig. 6, the optical signal was ON-OFF modulated with an optical intensity modulator, which was driven by the pulse pattern generator. The demodulated signal was reshaped by a limiting amplifier. In addition, we introduced dielectric lenses to collimate the THz waves.

Figure 8 shows the relationship between the photocurrent of the transmitters, which is proportional to the square root of the transmitted power, and the bit-error-rate (BER) for the 120- and 300-GHz carrier frequencies. This result indicates that the transmission power required for error-free transmission for 300 GHz is larger than that for 120 GHz. This is due to the frequency-dependent sensitivity of the SBD chip used in our experiment; the 3-dB bandwidth is around 110 GHz.

Figure 9 depicts eye diagrams at 1.5 Gbit/s for the 120- and 300-GHz bands. Error-free transmission was confirmed not only by the BER tester (BER < 10^{-11}) but also by transmitting uncompressed high-definition television (HDTV) video data (Fig. 10).

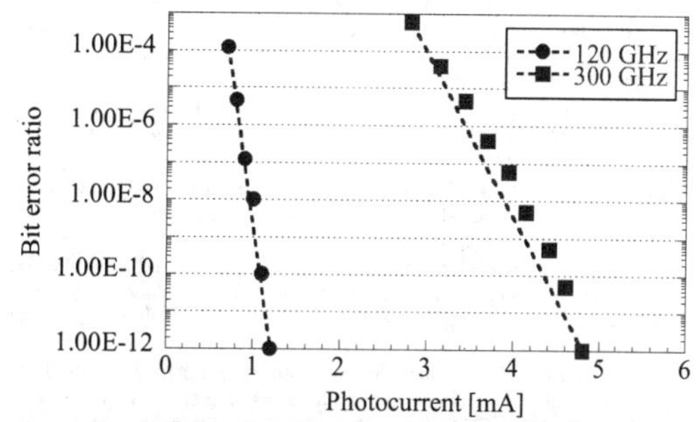

Fig. 8 BER vs. photocurrent for 120 and 300 GHz.

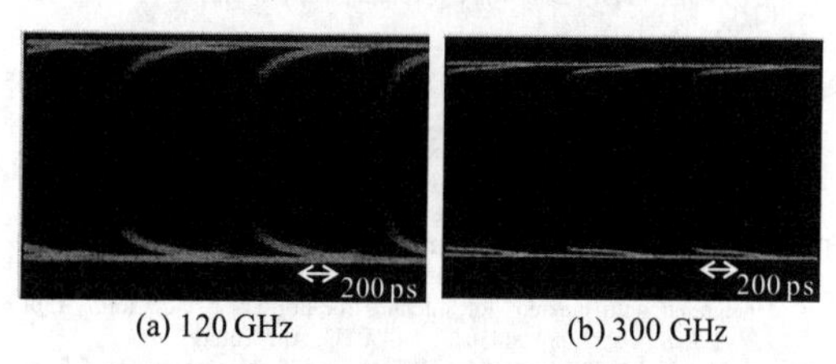

Fig. 9 Measured eye-diagrams at 1.5 Gbit/s: (a) 120 and (b) 300 GHz.

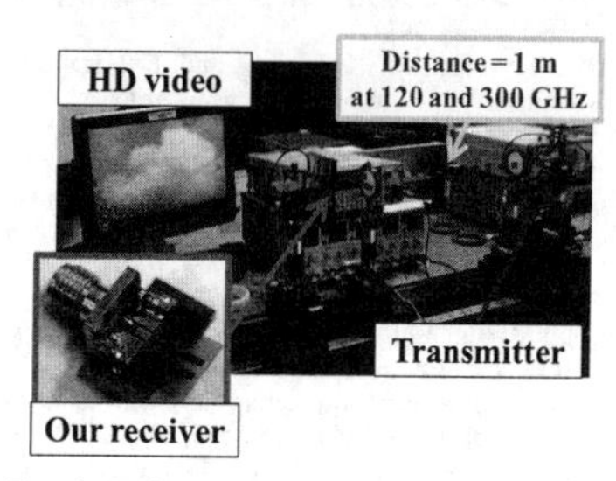

Fig. 10 Experiment for transmitting uncompressed HDTV video data.

V. Conclusion

We have developed an ultra-broadband THz antenna using a polymeric substrate. Due to the low dielectric constant of the polymeric substrate, we have successfully demonstrated multiband THz receivers covering frequencies from 100 to 300 GHz bands in antenna pattern measurements and wireless transmission experiments.

Acknowledgments

The authors thank Yusuke Minamikata and Daiki Tsuji for their experimental support. This work was supported in part by the JST-ANR WITH program and the Strategic Information and Communications R&D Promotion Programme (SCOPE) by the Ministry of Internal Affairs and Communications, Japan.

References

[1] M. Tonouchi, "Terahertz technologies: present and future," Nat. Photonics, vol. 1, no. 2, pp. 97-105, Feb. (2007).

[2] T. Nagatsuma, "Terahertz technologies: present and future," IEICE Electron. Express, vol. 8, no. 14, pp. 1127-1142, Jul. 2011.

[3] H.-J. Song and T. Nagatsuma, "Present and future of terahertz communications," *IEEE Trans. THz Sci. Technol.*, vol. 1, no. 1, pp. 256–263, Sep. 2011.

[4] T. Kleine-Ostmann and T. Nagatsuma, "A review on terahertz communications research," *J. Infrared, Millim. Terahertz Waves*, vol. 32, no. 2, pp. 143–171, Feb. 2011.

[5] J. Federici and L. Moeller, "Review of terahertz and subterahertz wireless communications," *J. Appl. Phys.*, vol. 107, no. 11, p. 111101-1–111101-22, Jun. 2010.

[6] N. Orihashi, S. Hattori, and M. Asada, "Millimeter and submillimeter oscillators using resonant tunneling diodes with stacked-layer slot antennas," *Jpn. J. Appl. Phys.*, vol. 43, no. 10A, pp. L1309- L1311, Oct. 2004.

[7] R. Piesiewicz, M. N. Islam, M. Koch, and T. Kürner, "Towards short-range terahertz communication systems: basic considerations", in *18th International Conference on Applied Electromagnetics and Communications, Dubrovnik, Croatia*, Techn. Digest. p. 153, Oct. 2005.

[8] M. N. Islam and M. Koch, "Terahertz patch antenna arrays for indoor communications," *Int. Conference on Next-Generation Wireless Systems 2006*, (Dhaka, Bangladesh, 2006).

[9] M. Asada, S. Suzuki, and N. Kishimoto, "Resonant tunneling diodes for sub-terahertz and terahertz oscillators," *Jpn. J. Appl. Phys.*, vol. 47, no. 6, pp. 4375-4384, Jun. 2008

[10] K. Urayama, S. Aoki, S. Suzuki, M. Asada, H. Sugiyama, and H. Yokoyama, "Sub-terahertz resonant tunneling diode oscillators integrated with tapered slot antennas for horizontal radiation," *Appl. Phys. Exp.*, vol. 2, pp. 044501-1-044501-3, Apr. 2009.

[11] T. Mukai, M. Kawamura, T. Takada, and T. Nagatsuma, "1.5-Gbps wireless transmission using resonant tunneling diodes at 300 GHz," *Tech. Dig. Optical Terahertz Science and Technology 2011 Meeting*, MF42, Santa Barbara, 2011.

[12] T. Shiode, T. Mukai, M. Kawamura, and T. Nagatsuma, "Giga-bit wireless communication at 300 GHz using resonant tunneling diode detector," in *Proc. Asia-Pacific Microw. Conf. (APMC2011)*, pp. 1122-1125, Dec. 2011.

[13] T. Shiode, M. Kawamura, T. Mukai, and T. Nagatsuma, "Resonant-tunneling diode transceiver for 300 GHz-Band Wireless Link," in *Proc. Asia-Pacific Microw. Photon. 2012 (APMP2012)*, WC-1, Kyoto, Apr. 2012.

[14] K. Ishigaki, M. Shiraishi, S. Suzuki, M. Asada, N. Nishiyama, and S. Arai, "Direct intensity modulation and wireless data transmission characteristics of terahertz-oscillating resonant tunnelling diodes," *Electron. Lett.*, vol. 48, no. 10, pp. 582-583, May. 2012.

[15] A. Kaku, T. Shiode, T. Mukai, K. Tsuruda, M. Fujita, and T. Nagatsuma, "Characterization of resonant tunneling diodes as receivers for terahertz communications," *Int. Sympo. Frontiers in THz Technology (FTT2012)*, Nara, no. Pos1.51, Nov., 2012.

[16] A. Kaku, T. Shiode, T. Mukai, K. Tsuruda, M. Fujita, and T. Nagatsuma, "3-Gbit/s error-free terahertz communication using resonant tunneling diode detectors integrated with MgO hyper-hemispherical lens," *Asia-Pacific Microwave Photonics Conference (APMP2013)*, Gwangju, no. MD-1, Apr. 2013.

[17] K. S. Yngvesson *et al*, "The Tapered Slot Antenna - A new integrated element for mm wave applications," *IEEE Trans.Microwave Theory and Tech.*, 37, 2, pp. 365-374, Feb. 1989.

[18] G. M. Rebeiz, "Millimeter-wave and terahertz integrated circuit antennas," *Proc. IEEE*, vol. 80, no. 11, pp.1748-1770, Nov. 1992.

[19] K. S. Yngvesson, T. L. Korzeniowski, Y. S. Kim, E. L.Kollberg, and J. F. Johansson, "The tapered slot antenna-A new integrated element for millimeter-wave applications," *IEEE Trans. Microw. Theory Tech.*, vol. 37, no. 2, pp. 365–374, Feb. 1989.

[20] M. Inoue, M. Hodono, S. Horiguchi, K. Arakawa, M. Fujita and T. Nagatsuma, "Ultra-broadband Receiver for Multiband Terahertz Communications," *International Workshop on Optical Terahertz Science and Technology 2013 (OTST 2013)*, Kyoto, F2B-2, Apr., 2013.

[21] M. Inoue, M. Hodono, S. Horiguchi, K. Arakawa, M. Fujita, and T. Nagatsuma, "Ultra-broadband receivers using polymeric substrate for multiband terahertz communications," *2013 International Symposium on Electromagnetic Theory (EMTS2013)*, Hiroshima, May, 2013.

[22] L. Liu, J. L. Hesler, X. Haiyong, A.W. Lichtenberger, and R.M. Weikle, "A broadband quasi-optical terahertz detector utilizing a zero bias schottky diode," *IEEE Microw. Wireless Comp. Lett.*, vol. 20, no. 9, pp. 504-506, Sept. 2010.

[23] T. Nagatsuma, H. Ito, and T. Ishibashi, "High-power RF photodiodes and their applications," *Laser Photon. Rev.*, vol. 3, pp. 123–137, no. 1-2, Feb. 2009.

A D-Band Packaged Antenna on Low Temperature Co-Fired Ceramics for Wire-Bond Connection with an Indium Phosphide Power Meter

Bing Zhang, Li Wei and Herbert Zirath
Microwave Electronics Laboratory, Department of Microtechnology and Nanoscience MC2, Chalmers University of Technology, SE-41296, Gothenburg, Sweden
bing.zhang@chalmers.se

***Abstract*-A D-band packaged grid array antenna (GAA) on Ferro A6M low temperature co-fired ceramic (LTCC, ε_r = 5.74, tanδ = 0.0023 @ 145 GHz) is presented. It is designed for wire-bond connection with an indium phosphide (InP) power meter as a demonstration of a D-band packaged radio. Dimensions of the GAA (x, y, z) = (12, 12, 0.5) mm^3. Simulated by Ansoft HFSS, the antenna's impedance bandwidth is 139 - 149.3 GHz, maximum gain of 20.9 dBi @ 147.8 GHz with 3-dB gain bandwidth 139.5 – 150.8 GHz, vertical beams of 10° beamwidth in the broadside are detected 139 – 151 GHz on both E- and H-planes. The 25 um diameter co-planar bonding wires that bridge the InP chip and LTCC antenna substrate over a 250 um separation are studied, showing acceptable insertion loss and limited bandwidth.**

I. INTRODUCTION

The D-band radiometer is attractive for real-time imaging, in which the system's angular resolution is directly related with the radiation beamwidth of the antenna [1]. Silicon lens are widely adopted for this application because of the focused beams over a wide spectrum [2]. However, the relatively high fabrication cost of the silicon lens prevents its wide-spread industrial application which requires a cost-effective mass production capability. Moreover the protruding contour of a silicon lens can result in a bulky package that also contradicts the need for compact front-ends.

We investigate the antenna for D-band real-time imaging applications from the perspective of electromagnetics instead of the optical view. For the small wavelength of millimeter-wave (mmWave) spectrum, a single chip radio becomes possible in which most of the antennas fall into categories of antenna-on-chip (AoC) and antenna-in-package (AiP) [3]. In this paper, grid array antenna (GAA) is adopted on Ferro A6M low temperature co-fired ceramic (LTCC) as a D-band AiP prototype [4-7]. Simulated by Ansoft HFSS, the antenna shows comparable performance with the silicon lens [1], [2]. A cavity on the back of the GAA is designed to host an indium phosphide (InP) D-band power meter [8]. The co-planar boding wires bridging the InP power meter and the GAA are studied, showing acceptable insertion loss and narrow bandwidth.

II. DESIGN OF THE D-BAND GAA

The GAA has dimensions of 12 mm × 12 mm × 0.5 mm in Fig. 1. It consists of four LTCC (in green) and five metallic layers (in brown). All the metallic layers are made of 0.01 mm thick gold (conductivity = 2.5 × 10^8 s/m). The radiating array (rad) is composed of 16 subarrays, each including 8 loops, with separation of 0.05 mm between the adjacent subarrays. The lengths of the short and long sides of each loop are 0.51 mm and 1.02 mm respectively. The widths of the loop short and long sides are governed by impedance, transmission and radiation properties [9]. The width of the short side of each loop is made equal to that of the long side which is 0.13 mm. The LTCC substrate layer for the radiating array (sub$_1$) is 0.2 mm thick. The metallic ground plane (gnd$_1$) is for the radiating elements and the stripline feeding network (sfn). In the LTCC process, it is difficult to achieve smooth finishing surface with lumped resistors that are sandwiched between two substrates. To overcome the problem, the ideal Wilkinson power splitters are replaced by T-junctions in sfn which eliminate the necessity of the lumped resistor to dissipate the odd mode. The ceramic substrates (sub$_2$ and sub$_3$) for the sfn have an equal thickness of 0.1 mm. The stripline feeding network has one input port and sixteen output ports. The sixteen output ports are connected to the sixteen subarrays respectively by vias of length 0.3 mm through sixteen circular openings on the gnd$_1$. The fencing vias of length 0.2 mm are located around each feeding via to connect the first and second ground planes. The fencing vias are designed for smooth transition between the sfn and rad. The sfn supports a transverse electromagnetic (TEM) mode whose equal phase front is perpendicular to gnd$_1$ and gnd$_1$, however equal phase front of the TEM mode on the feeding vias is parallel with gnd$_1$ and gnd$_2$. As expected, there will be diffraction, reflection as well as radiation when the input signal of the GAA is propagating from the sfn to the rad. Fences of more vias can ensure smoother field transition between the vertical and horizontal TEM modes, stronger suppression of the parallel-plate mode, as well as maintain an equal potential between ground planes gnd$_1$ and gnd$_2$. The 0.1 mm bottom ceramic layer (sub$_4$) provides the substrate for the ground-signal-ground (GSG) testing pad (tp) or the power, signal and ground traces to be wire-bonded with InP power meter. A cavity of dimensions (x, y, z) = (0.35, 2.5, 1) mm^3 is realized on sub$_4$ to host the InP power meter.

978-7-5641-4279-7

Simulated by Ansoft HFSS, the GAA shows impedance bandwidth 139 - 149.3 GHz in Fig. 1. In Fig. 2, the maximum gain of 20.9 dBi appears at 147.8 GHz with 3-dB gain bandwidth 139.5 – 150.8 GHz. Radiation patterns in Fig. 3 show vertical beams on both E- and H-planes from 139 – 151 GHz. The beamwidths are 8° @ 139 GHz with -10 dB side lobe level (SLL), 10° @ 145 GHz with -16.2 SLL and 10° @ 151 GHz with -20 dB SLL. Considering the high gain and broad impedance bandwidth of the GAA, it features a capable candidate for D-band imaging applications.

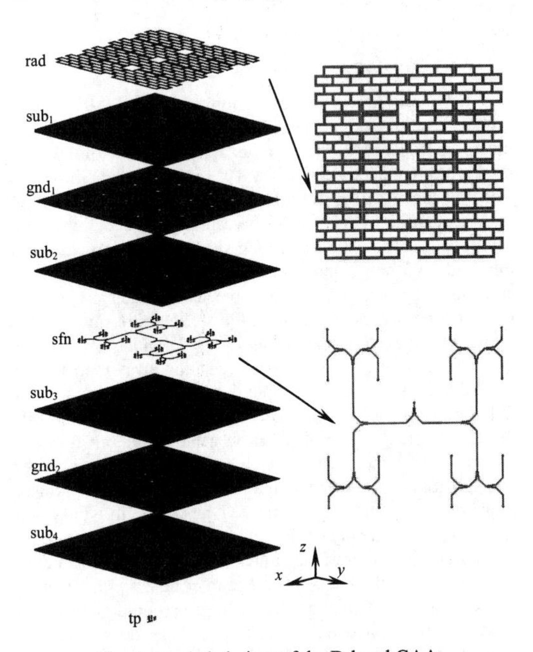

Fig. 1. Exploded view of the D-band GAA.

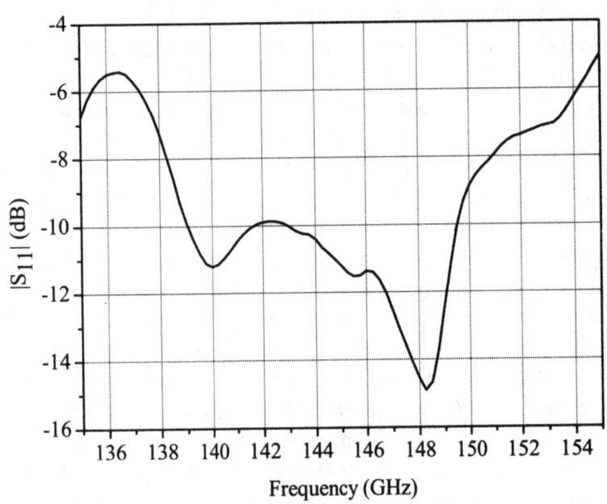

Fig. 2. Simulated $|S_{11}|$ of the D-band GAA.

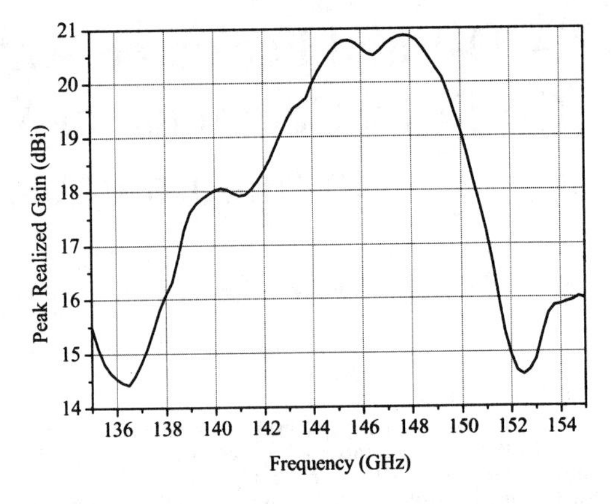

Fig. 3. Simulated peak realized gain of the D-band GAA.

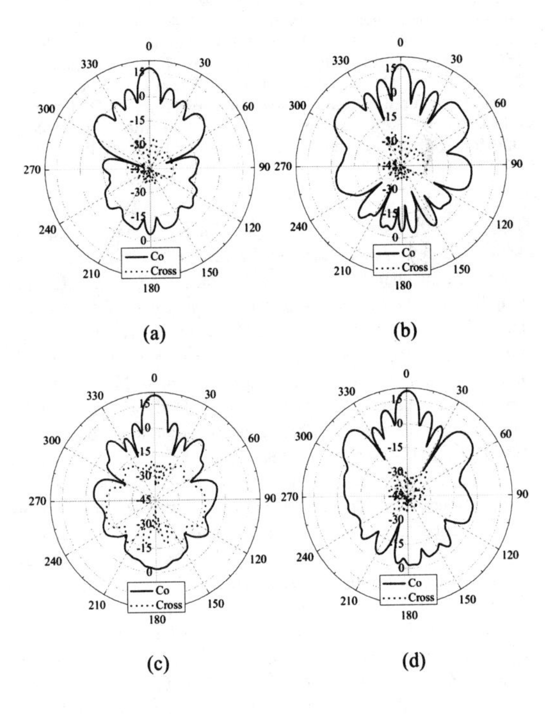

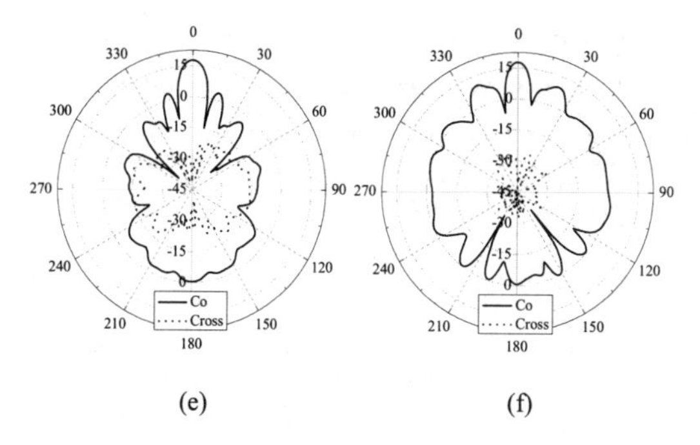

(e) (f)

Fig. 4. Simulated radiation patterns of the D-band GAA: (a) 139 GHz E-plane, (b) 139 GHz H-plane, (c) 145 GHz E-plane, (d) 145 GHz H-plane, (e) 151 GHz E-plane and (f) 151 GHz H-plane.

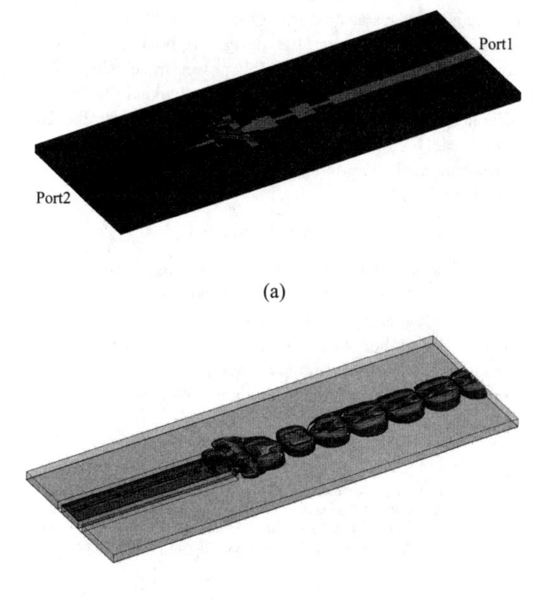

(a)

(b)

Fig. 5. Wire-bond connection between the InP power meter and the LTCC GAA: (a) HFSS model and (b) E-field distribution at 146 GHz.

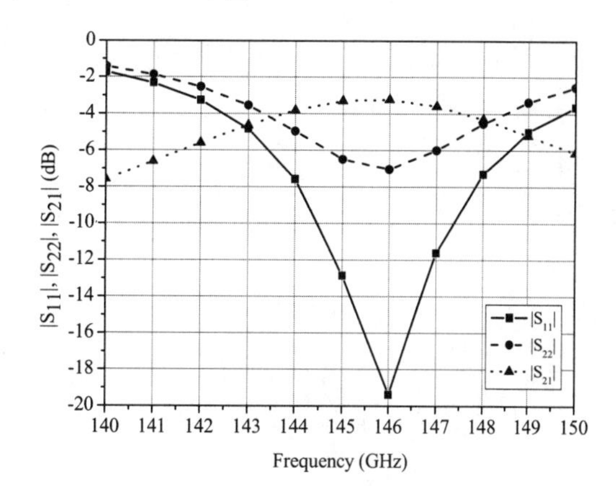

Fig. 6. Simulated frequency response of the wire-bond connection between the InP power meter and LTCC GAA.

III. Study of the Co-Planar Bonding Wires

In this work, for the easy implementation and robustness against chip thermal expansion (or contraction), Au co-planar bonding wires of 25 um diameter are adopted to interconnect an InP power meter with the GAA in Fig. 1. They bridge over a separation of 250 um with dual-wire connections between the GSG pad on the benzocyclobutene (BCB, $\varepsilon_r = 2.7$) layer of 5 um thickness over the 100 um InP substrate and the other GSG pad on the Ferro A6M LTCC substrate of 100 um thickness as shown in Fig. 5. Port 2 is located at the end of the BCB substrate while Port 1 is located at the end of the LTCC substrate, on which matching stubs are designed for smoother transition.

At low frequencies, bonding wires are modeled as multi-stage low-pass filters with fixed cut-off frequency. The influential parasitic inductance is usually compensated by properly designed series capacitor [10]. As frequency increases, the length of bonding wire reaches significant fractions of wavelength and exhibits transmission line properties. As a result, the characteristic impedance and effective dielectric constant become dominant factors instead of the parasitic inductance. Former research focused on the specific condition of bonding wires over a uniform substrate [11]. In this paper, co-planar bonding wires overpass two different substrates will bring about drastic discontinuity of the characteristic impedance, which results in relatively high insertion loss and limited bandwidth. Simulated by Ansoft HFSS, it is shown in Fig. 6 that after de-embedding, the minimum insertion loss is 3.2 dB @ 146 GHz which is comparable with [12], the 3-dB pass band is 141-150 GHz. Since the GAA is going to be used as a receiving antenna, though the 3.2 dB insertion loss is acceptable for power consideration it can double the noise level and decrease the sensitivity of the following InP power meter. The input impedance at Port 1 ($|S_{11}| < -10$ dB) is 144.5 – 147.5 GHz, which is not broad enough for imaging applications. Future work will be involved in the optimization of the bonding wires for lower insertion loss and broader bandwidth.

IV. Conclusions

In this paper, a GAA is designed on Ferro A6M LTCC substrate to be wire-bonded with a D-band InP power meter. Simulated by Ansoft HFSS, it shows comparable performance with silicon lens in terms of gain and beamwidth, while having advantages as conformal profile and the elimination of extra package. However, the design of bonding wires that connect the LTCC GAA and InP power meter shows large insertion loss as well as limited bandwidth for the discontinuity of the characteristic impedance over the BCB band the LTCC substrates. Further effort will be taken into the optimization of the bonding wires for low insertion loss and broad bandwidth.

ACKNOWLEDGMENT

The author would like to thank Vessen Vassilev and Vedran Furtula for the discussion on the InP power meter, as well as Bertil Hansson and Per-Åke Nilsson for the assistance in chip dicing. The author would also like to thank Zhongxia He, Niklas Wadefalk and Marcus Gavell for the discussion on bonding wires, and Mattias Ferndahl for probe testing of the antenna.

REFERENCES

[1] Y. Yan, Y. B. Karandikar, S. E. Gunnarsson, B. M. Motlagh, S. Cherednichenko, I. Kallfass, A. Leuther, and H. Zirath, "Monolithically integrated 200-GH double-slot antenna and resistive mixers in a GaAs-mHEMT MMIC process," *IEEE Trans. Microw. Theory Tech.*, vol. 59, no. 10, pp. 2494-2503, Oct. 2011.

[2] D. F. Filipovic, S. S. Gearhart and G. M. Rebeiz, "Double-slot antennas on extended hemispherical and elliptical silicon dielectric lens," *IEEE Trans. Microw. Theory Tech.*, vol. 41, no. 10, pp. 1738-1749, Oct. 1993.

[3] Y. P. Zhang and D. Liu, "Antenna-on-chip and antenna-in-package solutions to highly integrated millimeter-wave devices for wireless communications," *IEEE Trans. Antennas Propag.*, vol. 57, no. 10, pp. 2830-2841, Oct. 2009.

[4] J. D. Kraus, "A backward angle-fire array antenna," *IEEE Trans. Antennas Propag.*, vol. 12, pp. 48-50, Jan. 1964.

[5] H. Nakano, T. Kawano, H. Mimaki, and J. Yamauchi, "A fast MoM calculation technique using sinusoidal basis and testing functions for a wire on a dielectric substrate and its application to meander loop and grid array antennas," *IEEE Trans. Antennas Propag.*, vol. 53, no. 10, pp. 3300-3307, Oct. 2005.

[6] B. Zhang, D. Titz, F. Ferrero, C. Luxey, and Y. P. Zhang, "Integration of quadruple linearly-polarized microstrip grid array antennas for 60-GHz antenna-in-package applications," *IEEE Trans. Comp. Packag. Manuf. Technol.*, in press.

[7] S. Beer, C. Rusch, B. Göttel, H. Gulan, and T. Zwick, "D-band grid-array antenna integrated in the lid of a surface-mountable chip-pakcage," *EuCAP2013*, Gothenburg, Sweden, Apr. 8-12, 2013, pp. 1273-1277.

[8] V. Vassilev, H. Zirath, V. Furtula, Y. Karandikar and K. Eriksson, "140-220 GHz imaging front-end based on 250 nm InP/InGaAs/InP DHBT process," in *Proc. Defence Security+Sensing Technical Program SPIE 2013*, Baltimore, Maryland, Apr. 29 – May 3, 2013.

[9] B. Zhang and Y. P. Zhang, "Analysis and synthesis of millimeter-wave microstrip grid array antennas," *IEEE Antennas Propag. Mag.*, vol. 53, no.6, pp. 42-55, Dec. 2011.

[10] Y. P. Zhang, M. Sun, K. M. Chua, L. L. Wai, and D. Liu, "Antenna-in-package design for wirebond interconnection to highly integrated 60-GHz radios," *IEEE Trans. Antennas. Propag.*, vol. 57, no. 10, pp. 2842-2852, Oct. 2009.

[11] K. W. Goossen, "On the design of coplanar bond wires as transmission lines," *IEEE Microw. Guid. Wave Lett.*, vol. 9, no. 12, pp. 511-513, Dec. 1999.

[12] T. Krems, W. Haydl, H. Massler, and J. Rudiger, "Millimeter-wave performance of chip interconnections using wire bonding and flip chip," in *IEEE MTT-S Int. Microwave Symp. Dig.*, vol. 1, San Francisco, CA, Jun. 1996, pp. 247- 250.

Circuit Model and Analysis of Antenna-in-Package

Li Li, Wenmei Zhang, *Member, IEEE*

***Abstract*—This letter presents a circuit model of the antenna-in-package (AiP) employing the via holes to connect the antenna ground and the system ground. A closed-form expression for the inductance of ground via is provided in case that two ground vias are arranged under the non-radiating edge of the AiP. The new expression takes account of the radius and length of the vias, the relative positions of the vias and the distances between the vias and the feed point of the antenna. Then, the influences of the vias on the performance of the AiP are analyzed using the model. It is found that the bandwidth of the AiP can be improved effectively by properly arranging two via holes under the non-radiating edge of antenna. The simulated results agree with the modeled results.**

***Index Terms*—Antenna-in-package, equivalent circuit, RF transceiver, via holes**

I. Introduction

HIGHER levels of integration are driven by technological advancements in electrical performance requirements. The antenna-in-package (AiP) provides an advanced solution to the system architect. Lots of research about AiP were been done [1-6]. In AiPs, the multiple via holes are used to connect the antenna ground with the system ground. The influences of the number and positions of the ground vias on the performance of the antenna and package were studied in [7-8]. In this letter, the circuit model of the investigated AiP is presented. A closed-form expression for the inductance of the ground via arranged under the non-radiating edge of the AiP is deduced. The new formula takes account of not only the radius and length of the vias, the relative positions of the vias, but also the distances between the via holes and the feed point of the antenna. Based on this circuit model, the influences of the positions of the vias on the AiP are studied. The results indicate the bandwidth of the investigated AiP can be effectively improved by arranging two via holes properly.

II. The Circuit Model of the AiP

This work was supported by the National Science Foundation of China (61172045, 61271160), the Natural Science Foundation of Shanxi province (2010021015-1) and Program for the Top Young Academic Leaders of Higher Learning Institutions of Shanxi

Li Li, Wenmei Zhang are with the College of Physics and Electronics Engineering, Shanxi University, Shanxi, 030006, China. (e-mail: zhangwm@sxu.edu.cn).

978-7-5641-4279-7

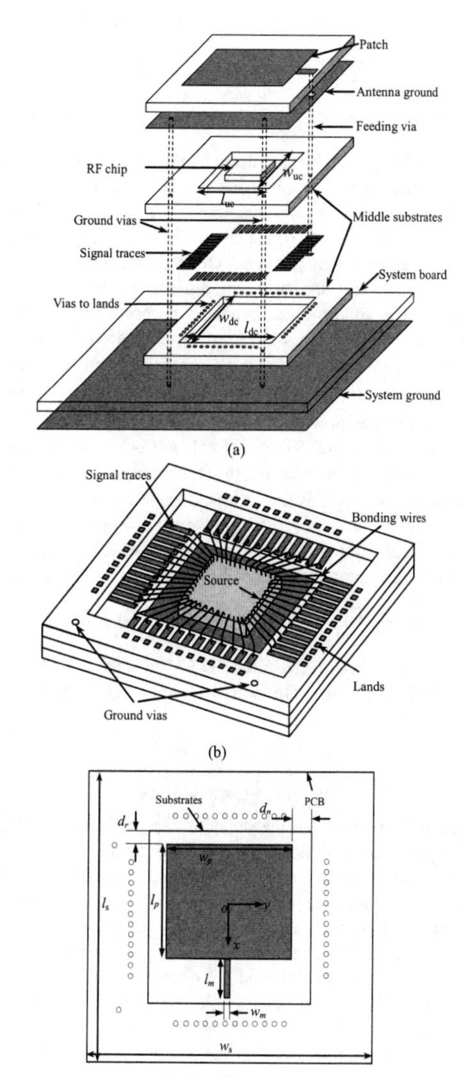

Fig. 1. Architecture of the investigated AiP: (a) explored view, (b) bottom view, (c) top view.

Fig. 1 (a) shows the explored view of the investigated AiP. It consists of three substrate layers and three metallization layers.

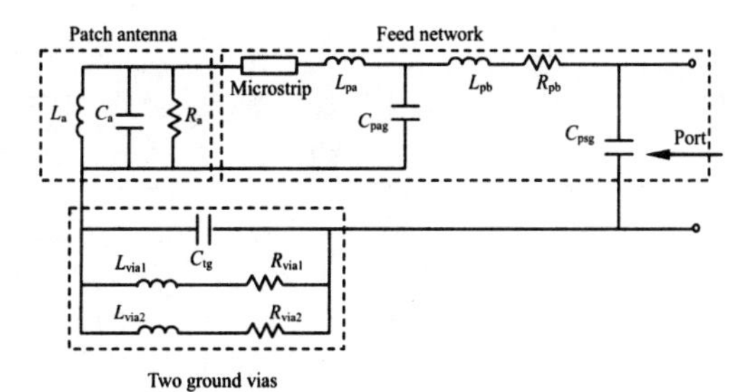

Fig. 2. Equivalent circuit model of the AiP.

TABLE I
PARAMETERS IN THE CIRCUIT MODEL

R_a(Ohm)	L_a(nH)	C_a(pF)	L_{pa}(nH)	R_{pb}(Ohm)
130.5	0.225	4.95	0.44	0.2
L_{pb}(nH)	C_{pag}(pF)	C_{psg}(pF)	C_{tg}(pF)	
0.87	0.38	0.1	3.6	

The top metallization layer is the radiator of the antenna, while the second and third metallization layer is used for the antenna ground and the signal traces, respectively. Under the antenna ground, a stepped cavity is formed in the second and third substrate layers. The RF chip can be packaged in the cavity then connected to the system board. Two vias called ground vias are used to connect the antenna ground with the system ground. Fig. 1 (b) shows the bottom view of the AiP. The RF chip is attached facing downwords to the antenna ground plane with conductive adhesive. Through the bond wires the chip is connected to the inner ends of the signal traces, then connected to the lands through the vias. Finally, the whole packaged module can be connected with the system board by the lands. The dimensions of AiP are l_p=13.5 mm, w_p=15 mm, l_m=3.5 mm, w_m=0.5 mm, l_s=w_s=20 mm, d_n =2 mm, d_r=2.5 mm, l_{uc}=w_{uc}=10 mm, and l_{dc}=w_{dc}=14 mm respectively. The size of the system board is 40 × 40 mm^2. All substrate layers are with the thickness of 0.8 mm, relative dielectric constant of 4.4 and loss tangent of 0.02.

The equivalent circuit model of the investigated AiP is shown in Fig. 2 and it comprises three subcircuits which represent the patch antenna, the feed network and two ground vias, respectively. For the microstrip patch antenna, it is usually modeled as a simple parallel resonant *RLC* circuit according to the cavity theory. The *RLC* values are given as follows [6],

$$R_a = \frac{Q_{total} H}{\pi f \varepsilon_{dyn} W L_{eff}} \cos^2\left(\frac{\pi X_{eff}}{L_{eff}}\right) \quad (1)$$

$$L_a = \frac{R_a}{2\pi f_r Q_{total}} \quad (2)$$

$$C_a = \frac{Q_{total}}{2\pi f_r R_a} \quad (3)$$

The feed network consists of the microstrip line and the feeding via. The feeding via above the antenna ground plane is represented by an inductance L_{pa}. L_{pb} and R_{pb} are the inductance

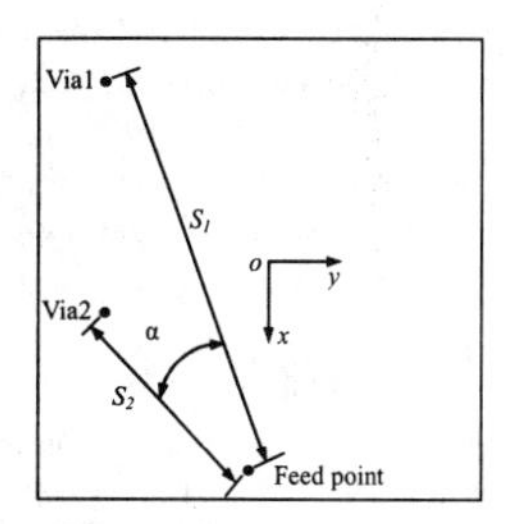

Fig. 3. Relative positions of the feed point and two vias.

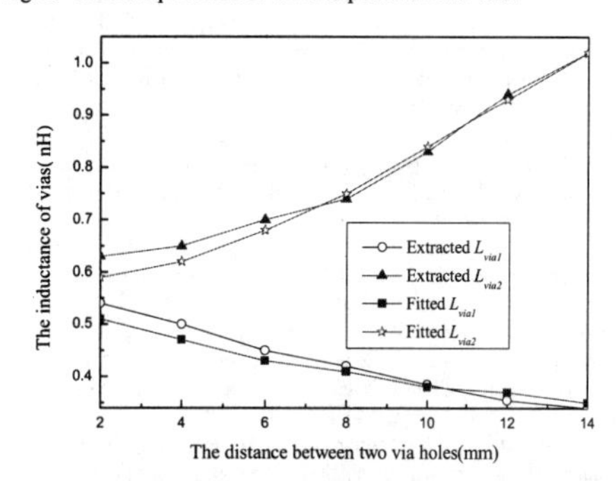

Fig. 4. Inductances of two ground vias when the first via hole is placed at (-8, -8) and the second via hole shifts along the positive direction of x axis.

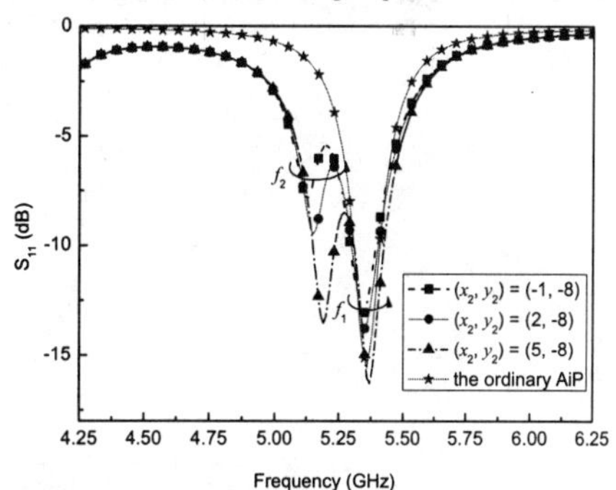

Fig. 5. S parameters calculated by the circuit model.

and the resistance of the feeding via below the antenna ground [6]. C_{pag} and C_{psg} are the capacitances between the feeding probe and two ground planes, respectively.

The via holes connect two ground planes and each one offers a route for the current to the system ground, so it is represented with a series branch of an inductance L_{viai} and a resistance R_{viai}. L_{viai} depends on the positions of two vias because the portion of the current on the antenna ground plane crowds down to each via hole and it contributes to the magnetic flux wrapping it. Further, L_{viai} is related to the distances between the via holes and the feed point, the separation of two ground planes and so on. R_{viai} represents the loss and is highly dependent on the fabrication process. They can be extracted from 3D EM full-wave field solver. C_{tg} represents the capacitance between

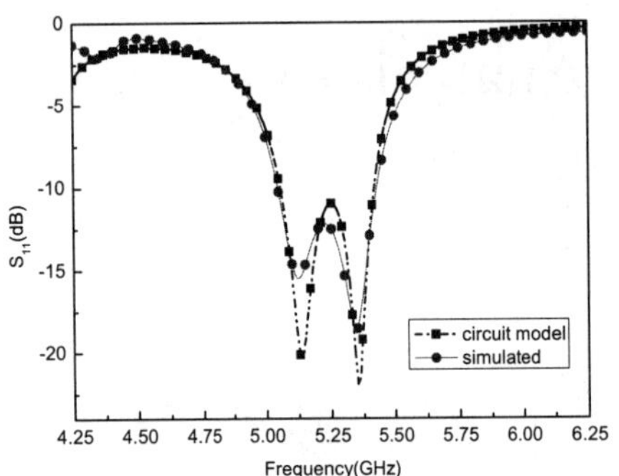

Fig. 6. Input impedances for the AiP placed two vias at (-8, -8) and (6, -8).

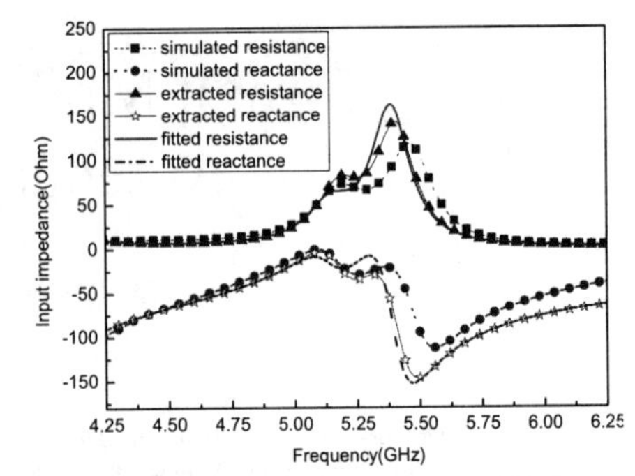

Fig. 7. S_{11} for the AiP placing two via holes at (-8, -8) and (6, -8).

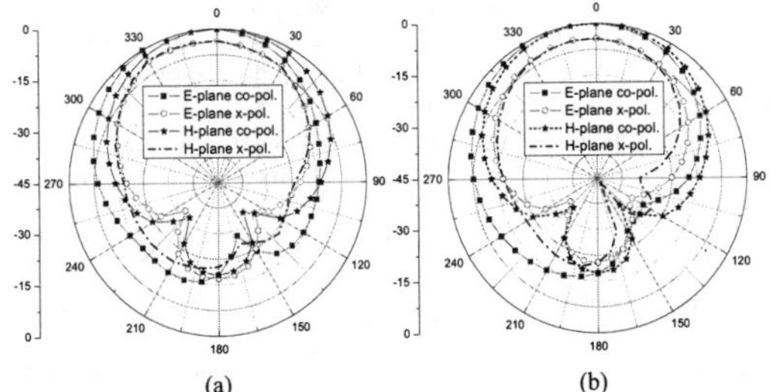

Fig. 8. Simulated radiation patterns at two operating frequencies: (a) 5.14 GHz, (b) 5.35 GHz.

two ground planes and is calculated by the method of moments. The calculated parameters in Fig. 2 are shown in Table I.

When two via holes are arranged under the non-radiating edge, as is shown in Fig. 3, the closed-form expression for L_{viai} can be obtained by the nonlinear curve fitting. Firstly, extract the parameters for the different dimensions from 3D EM full-wave field solver. Then, find the appropriate function to fit the parameters. An expression for the inductance of a cylinder via, $h \cdot \ln(h/r)$, is selected as the basis of the function, where h and r are the length and radius of the via. Considering the effect of the relative positions between two vias and the feed point, an item with regard to (S_j/S_i) (S_i, S_j (i=1, j=2 or i=2, j=1) are the distances between two vias and the feed point) is added. Also, a natural exponential function of α (the angle between two lines connecting the vias and the feed point) is supplemented to introduce the influences of the relative positions of the vias. The final expression for L_{viai} can be got as,

$$L_{viai} = \frac{Exp(-0.4\alpha/\pi)\mu h}{4\pi}[\ln(\frac{h}{r}) + 1.65(\frac{S_j}{S_i})^{0.85}\ln\frac{S_j}{S_i}] \quad (4)$$

Fig. 4 shows L_{viai} extracted from 3D EM full-wave field solver and calculated by the expression (4). It indicates the expression (4) can fit L_{viai} well.

III. Influence of Vias

The influences of the via holes on the performance of the investigated AiP will be analyzed in this section.

Fig. 5 shows the S parameters calculated by the circuit model when the first via hole is placed at (-8, -8) and the second via hole shifts along the x-axis. It indicates for the ordinary AiP with one ground via at (-8, -8), there is a main resonant frequency at f_1 = 5.23 GHz and f_1 is determined by the patch of the antenna. When another via hole is added, a new resonant frequency f_2 around 5 GHz occurs and f_2 is related with the patch and the positions of the vias. By properly adjusting the positions of the ground vias, f_2 can be close to f_1 that the bandwidth of the AiP can be improved.

According to the above analysis, an AiP operating at the center frequency of 5.2 GHz is realized. In order to obtain a wider bandwidth, two resonant frequencies of 5.14 GHz and 5.35 GHz are expected to appear. In this case, two vias are arranged at (-8, -8) and (6, -8).

IV. Validation of Circuit Model

Fig. 6 presents the modeled, and simulated S_{11} of the AiP with two via holes at (-8, -8) and (6, -8). The simulated resonant frequencies are 5.12 GHz and 5.34 GHz, respectively. The simulated bandwidth is 370 MHz. Fig. 7 shows the input impedances of the AiP. It is obvious that there are two resonant frequencies around 5.2 GHz. The impedances calculated by the closed-form expression agree with the simulated. Also, it can be seen that the circuit model is valid around the operation frequency (from 4.5GHz to 6.25GHz).

The simulated radiation patterns at two resonant frequencies are also plotted in Fig. 8. It is noted that the same polarization planes and similar radiation characteristics are obtained at two frequencies. Due to the influences of the ground vias on the current distribution on the patch, the cross-polar levels are higher. Moreover the simulated peak gains are 3.2 dBi and 4.1 dBi, respectively.

V. Conclusion

In this paper, an equivalent circuit model of the AiP with two via holes connecting the antenna and the system ground planes is proposed. When two ground vias are arranged under the non-radiating edge of the AiP, a closed-form expression for the inductance of the ground via is further provided as a function of the via radius and length, the via relative positions and the distances between the via holes and the feed point. Based on the

model, the influences of the positions of the via holes on the performance of the AiP are investigated. It is found that the positions of the via holes will affect the performance of the AiP. Moreover, the bandwidth of the investigated AiP can be effectively improved by arranging two via holes properly. The modeled results agree with the simulated results.

REFERENCES

[1] Y. P. Zhang, "Integrated of microstrip antenna on ceramic ball grid array package," *Electron. Lett.*, vol. 38, no. 1, pp. 1307-1308, 2002.

[2] Y. P. Zhang, M. Sun, and W. Lin, "Novel Antenna-in-Package Design in LTCC for Single-Chip RF Transceivers," *IEEE Trans. Antennas Propag.*, vol. 56, pp. 2079–2088, Jul. 2008.

[3] G. Felic and S. Skafidas, "Flip-chip interconnection effects on 60 GHz microstrip antenna performance," IEEE Antennas Wireless Propag. Lett., vol. 8, pp. 283–286, Feb. 2009.

[4] R. Suga, H. Nakano, Y. Hirachi, J. Hirokawa, and M. Ando, "Cost effective 60 GHz antenna package with end-fire radiation for wireless file-transfer system," IEEE Trans. Microw. Theory Tech., vol. 58, no. 12, pp.3989–3995, Dec. 2010.

[5] W. Wang and Y. P. Zhang, "0.18-m CMOS push-pull power amplifier with antenna in IC package," *IEEE Microw. Wireless Compon. Lett.*, vol. 14, pp. 13–15, Jan. 2004.

[6] J. J. Wang, Y. P. Zhang, C. W. Lu, and K. M. Chua, "Circuit model of microstrip patch antenna on ceramic land grid array package for antenna-chip codesign of highly integrated RF transceivers," *IEEE Trans. Antennas Propag.*, vol. 53, pp. 3877–3883, Dec. 2005.

[7] S. H. Wi et al., "Package-level integrated LTCC antenna for RF Package Application", IEEE Trans. Adv. Packag., vol. 30, no.1, Feb. 2007, pp. 132–140.

[8] S. H. Wi et al., "Package-level integrated antennas based on LTCC technology", *IEEE Trans. Antennas Propag.*, vol. 54, Aug. 2006, pp. 2190–2197.

Design, Simulation and Measurement of a 120GHz On-Chip Antenna in 45nm CMOS for High-Speed Short-Range Wireless Connectors

Noël Deferm and Patrick Reynaert

KU Leuven ESAT/MICAS, Kasteelpark Arenberg 10, 3001 Leuven, Belgium

***Abstract*—In this paper an on-chip antenna solution for high-speed millimeter wave wireless integrated transceivers in CMOS is presented. Accurate design and analysis of the complete 3D antenna structure has lead to an efficiency as high as 69% and an input bandwidth of 48GHz. A peak antenna gain of 4.4dBi is also achieved. The bondwire dipole is fabricated on a 45nm CMOS chip as part of a fully integrated 120GHz wireless communication front end. Measurements of the E-plane and H-plane radiation pattern are in good agreement with the simulation results.**

***Index Terms*—Millimeter-wave, on-chip antenna, CMOS.**

I. Introduction

Recent developments in the CMOS process technology enable the design of fully integrated mm-wave data transmitters for short range communication. The scaling of the CMOS transistors has lead to an increase of the devices ft and fmax, which allows the implementation of integrated mm-wave systems. These high frequencies enable the use of a high available bandwidth.

One of the most important building blocks in integrated wireless transceivers is the antenna as it provides the interface to the transmission channel. The performance of the antenna can have a large impact on the bandwidth and SNR of the transceiver. Therefore, accurate antenna design and analysis is essential in the development of a mm-wave wireless transceiver.

As the wavelength of the transmitted carrier is scaled down to the order of millimeters, antenne dimensions become comparable to chip dimensions. This leads to the possibility of integrating antennas in the same package, or even on the same die as the integrated circuits. Of course, integration of an antenna in a standard CMOS technology imposes restrictions on the geometrical properties of the antenna, making the design challenging.

Fully integrated planar antennas were already presented in the past [1]. Despite their high factor of integration, their performance suffers from the geometrical and physical technology limitations. Due to the large difference in permittivity between the air and the silicon substrate, the largest part of the radiated power gets trapped in the silicon chip in the form of surface waves. Due to the lossy substrate, part of the surface wave power is dissipated in the silicon. The finite dimensions of the chip result in radiation of the remaining surface wave power at the edges of the chip, which disturbes the radiation pattern of the antenna. To deal with the problem of surface waves, several techniques are already discussed in literature. One technique is to apply back etching of the silicon chip to prevent the generation of surface waves (figure 1(a)) [2]. Unfortunately, this technique results in a reduction of the mechanical stability of the chip, which is undesirable. Moreover, post processing steps are required to create the back cavity in the substrate. This has to be avoided for a high volume target application. Another technique, which does not incorporate post processing of the silicon chip itself, is to add a silicon lens at the backside of the chip to convert the surface wave power to useful radiated power (figure 1(b)) [3]. The problem with this technique is the assembly of the complicated structure with the dielectric lens on which the chip has to be mounted. Several materials have to be assembled and aligned accurately, which is a time consuming and costly post processing step.

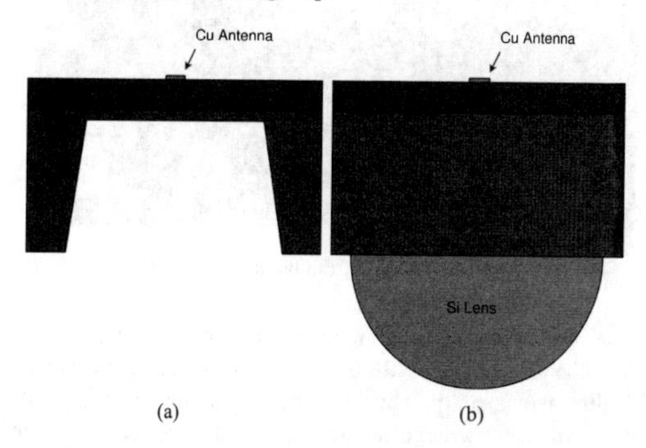

Fig. 1: Integrated antenna with silicon backetching (a) and silicon lens (b).

The solution discussed in this paper, places a reflector between the antenna and the substrate, rather than underneath the substrate. This prevents the entrapment of the radiated power in the substrate as surface waves. The only way to accomplish this when the radiating element is designed in the top metal layers, is to design the reflector shield in the lower metals of the CMOS metalstack. Due to the geometrical limitations of a standard CMOS technology, the distance between the antenna and the reflector is greatly reduced to a couple of micrometers. The small antenna-reflector spacing results in a vast reduction of the radiation resistance. The overall antenna resistance will therefore be dominated by the ohmic losses in the copper structure, resulting in an even lower radiation efficiency [3].

An elegant technique to increase the antenna-reflector spacing is to implement a full three dimensional structure by means of bond wires [4]. In this way, both the antenna efficiency and directivity can be increased. Although this solution is not a completely integrated solution, the antenna can still be included in the same package without any additional processing steps. Also, the impact of the variation on the antenna dimensions and position are rather limited, making this a very attractive solution for integrated mm-wave wireless communication systems and mm-wave laboratory chip measurement setups.

978-7-5641-4279-7

II. Antenna Design and Analysis

In this section, the design and analysis of the presented 120GHz bondwire dipole is discussed. It is a complete 3D structure as part of the radiating elements are lifted 500μm with respect to the substrate and reflector. Figure 2 shows the 3D view of the bondwire dipole.

Fig. 2: 3D view of the bondwire dipole.

The bondwire antenna is mounted on a 45nm CMOS silicon chip. The chip itself will be mounted on an FR4 carrier. The dimensions of the chip and the interaction with the FR4 carrier will have a large impact on the antenna performance, so they have to be taken into account in the simulation of the antenna. To make a fair comparison between different antenna types, typical dimensions were chosen for the chip on which a complete CMOS transceiver is integrated. The dimensions of the FR4 carrier were chosen as large as possible while maintaining an acceptable simulation time. The complete structure together with the antenne is simulated in Ansoft HFSS. In figures 3 and 4 the front and top view of the silicon chip mounted on the FR4 carrier with copper ground plane are respectively shown. Intuitively one can see that this ground plane will behave as a reflector for electromagnetic waves with a wavelength significantly smaller than the ground plane dimensions. This will result in a shift of the main lobe of the radiation pattern along the z-axis.

Figure 5 shows the 3D radiation pattern of the bondwire dipole. The peak gain of 4.4dBi is pointing in the positive z-direction. There are also 2 large side lobes pointing in the x-direction under an angle of 70° compared to the z-axis. Peak gain of the side lobes is around 3.4dBi. The high directivity of the antenna results in a narrow main lobe. The radiation efficiency of this bondwire dipole is 69%.

The antenna is resonant at 30GHz with an input impedance of approximately 50Ω. The input impedance at 120GHz is 69-22jΩ. Figure 6 shows the input impedance over frequency. Input bandwidth under conjugate match at 120GHz ranges from 107GHz up to 155GHz. Apart from the input bandwidth of the antenna, the 3dB antenna gain bandwidth also plays an important role when the target application is data communication. The 3dB gain bandwidth of the antenna is depicted in figure 7 and is ranging from 107 up to 140GHz. The large antenna

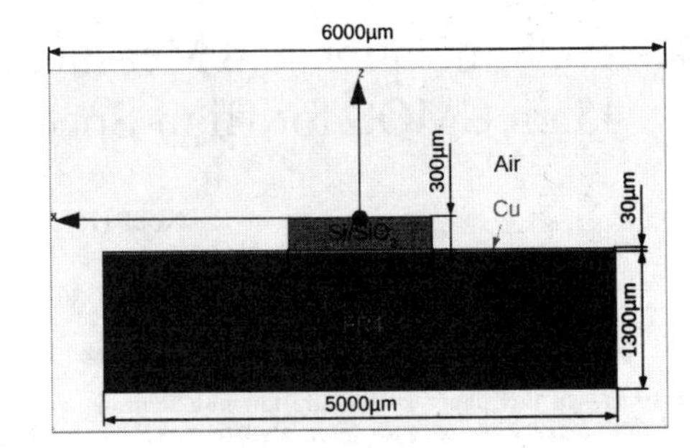

Fig. 3: Front view of the silicon chip on PCB carrier with reflector.

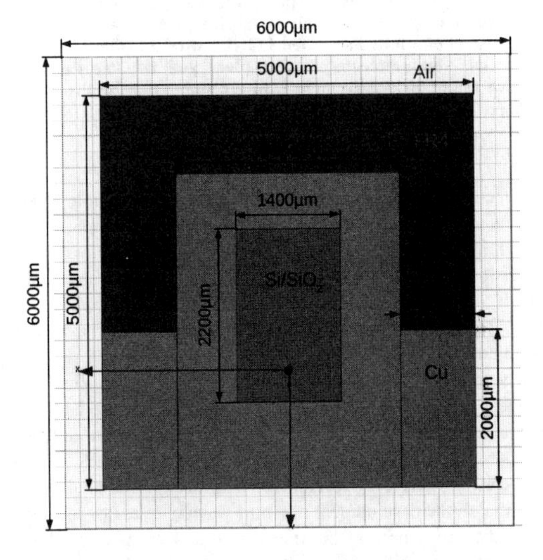

Fig. 4: Top view of the silicon chip on PCB carrier with reflector.

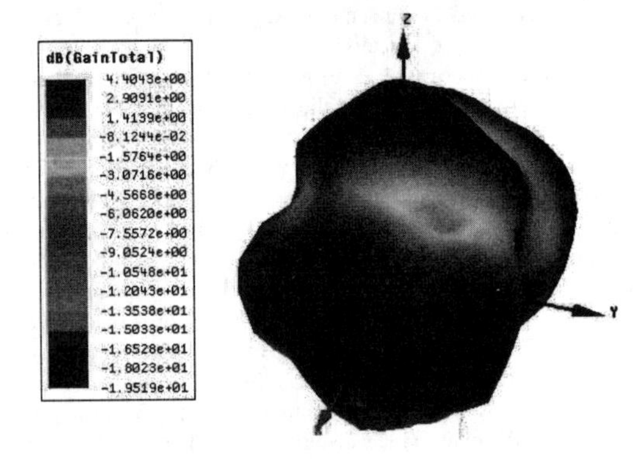

Fig. 5: 3D radiation pattern of the bondwire dipole.

gain bandwidth leads to very reliable operation when integrated together with the power amplifier. Ideally, the center frequency of the PA should be located near the center of the antenna gain bandwidth. Uncertainties in the fabrication of the antenna can lead to a shift of the bandwidth. Thanks to the large gain bandwidth, variations of the antenna center frequency and PA center frequency due to variations in the fabrication process can be tolerated.

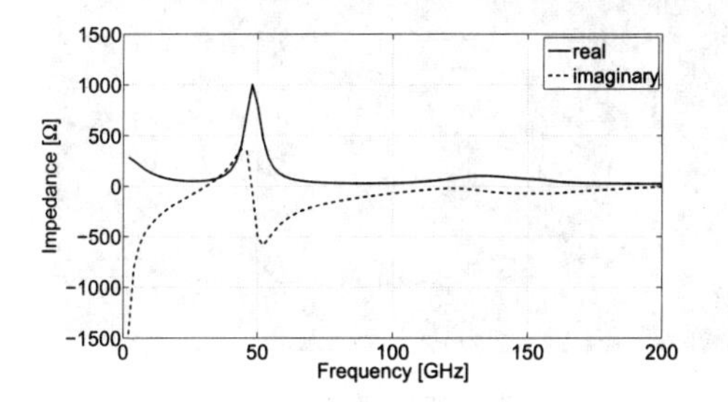

Fig. 6: Antenna input impedance as a function of frequency.

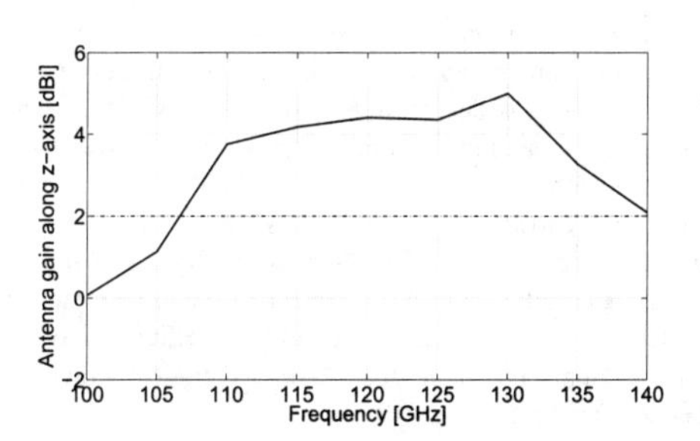

Fig. 7: Antenna gain as a function of frequency.

The main antenna characteristics are highly depending on the shape and dimensions of the antenna. Therefore, changes in the shape or dimensions as a result of variations in the fabrication process will have an impact on the electromagnetic properties of the antenna. Manufacturing the bondwire dipole antenna is a semi-automatic process of which the bondwire positions in the XY-plane can be controlled with an accuracy of one to several micrometers. The height of the bondwire however, cannot be controlled with the same accuracy. To have an idea of the sensitivity of the antenna to height variations, simulations were performed in which the height of the bondwire antenna was swept. When a height variation of 100μm around the optimal value of 500μm is considered, the antenna gain in the z-direction only varies between 3.7 and 4.4dBi. For the same height variation, the radiation efficiency varies between 60% and 69%. In figure 8 the variation of the antenna gain in the z-direction and the radiation efficiency are shown for a height sweep from 200μm up to 950μm.

The antenna behaves like a double wave dipole of which the frequency of first resonance is approximately 30GHz. When a height variation from 400μm to 600μm is considered, the resonance frequency varies between 29GHz and 34GHz. The

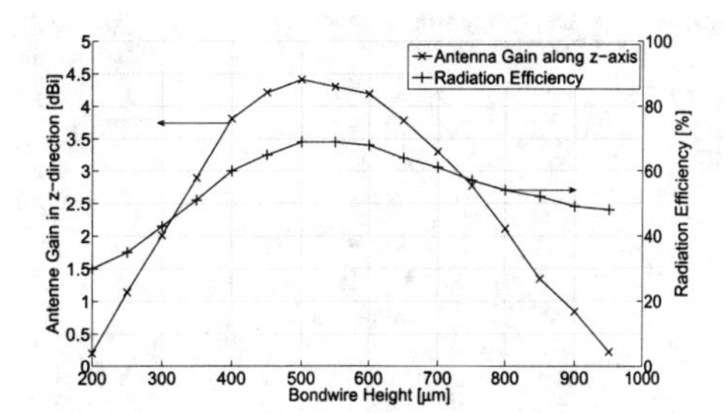

Fig. 8: Simulated variation of the antenna gain and efficiency with respect to the bondwire height.

frequency variation over the complete height sweep is shown in figure 9. Thanks to its large bandwidth, frequency shifts resulting from fabrication process variations can be tolerated up to a value of 20GHz.

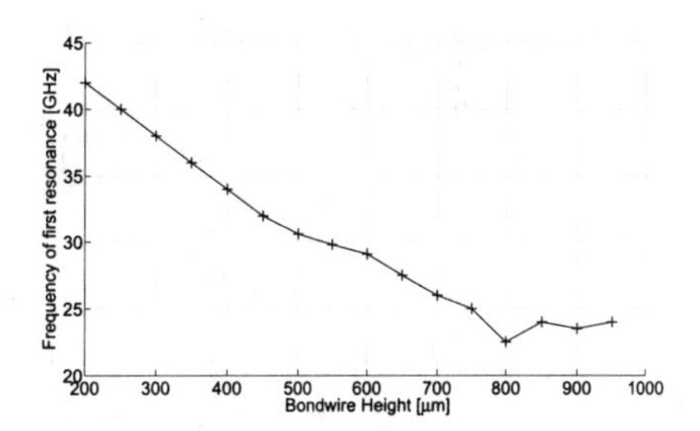

Fig. 9: Simulated variation of the antenna resonance frequency with respect to the bondwire height.

III. Measurements

The bondwire dipole antenna is part of a complete 120GHz transmitter with an on-chip 120GHz LO generator (figure 10) [5]. By integrating the mm-wave signal generation circuits on the same chip as the antenna, the use of probes can be omitted. Without the on-chip signal generator, a large probe (typically in the order of a couple of cm) is needed to drive the antenna. The presence of this probe would have a severe impact on the antenna radiation pattern and efficiency, which impedes the characterization of the stand-alone antenna. To measure the antenna, a continuous wave signal is generated on-chip and applied to the input of the antenna. Measurements show that optimal performance of the transmitter is achieved for a carrier frequency of 114.3GHz, so the analysis of the measured antenna radiation pattern will be carried out at this frequency.

To measure the antenna properties, the on-chip generated and radiated 114.3GHz signal has to be captured and converted down. Figure 11 shows the measurement setup to characterize the bondwire antenna. An F-band SGH antenna is used to receive the transmitted 114.3GHz signal. A wideband mixer, driven by a 59.65GHz external signal, is used to convert the received signal down to an IF of 5GHz. This IF signal is subsequently analyzed with a spectrum analyzer.

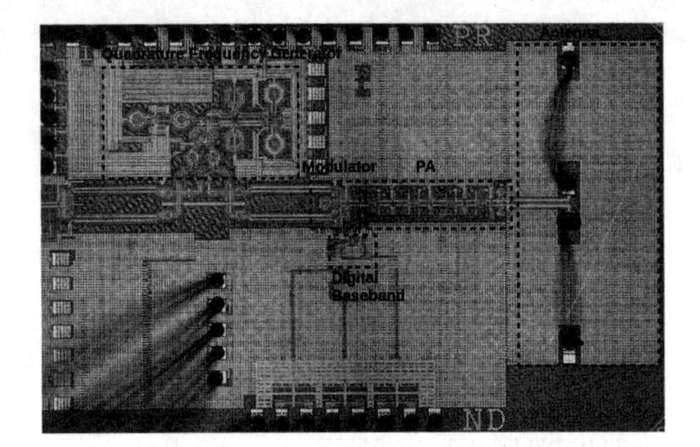

Fig. 10: Chip photograph of the 45nm CMOS 120GHz transmitter with bondwire antenna.

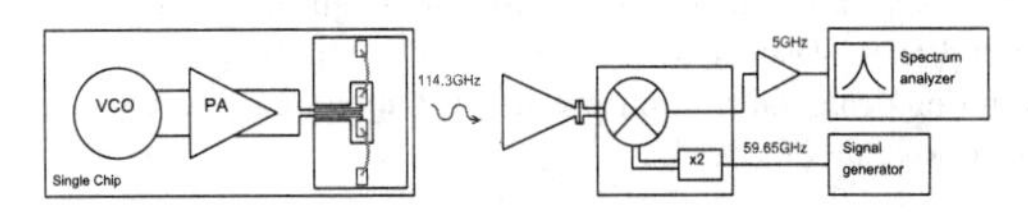

Fig. 11: Block diagram of the radiation pattern measurement setup.

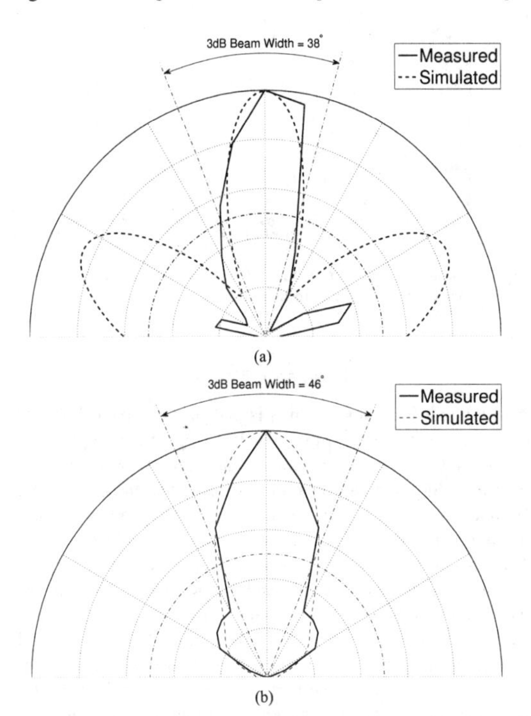

Fig. 12: Measured and simulated normalized E-plane (a) and H-plane (b) radiation pattern of the bondwire dipole at 114.3GHz.

To measure the E-plane and H-plane radiation pattern of the antenna, the chip is respectively rotated around the y-axis and x-axis according to figure 2. Figures 12(a) and 12(b) respectively show the E-plane and H-plane radiation patterns of the antenna. To compare the measured and simulated patterns, normalized values are plotted on the graphs. Good agreement is achieved for both main lobes. Measurements show an electrical field main lobe beam width of 38° and a magnetic field main lobe beam width of 46°. In figure 13 a picture is shown of the golden bondwire dipole, manufactured on a 45nm standard CMOS silicon substrate.

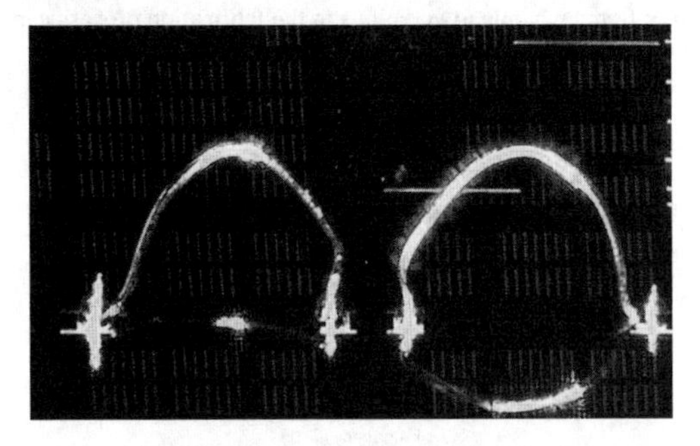

Fig. 13: Photograph of the bondwire dipole.

IV. Conclusion

In this paper, an on-chip antenna solution was presented for high-speed integrated mm-wave CMOS transceivers. The bondwire dipole was designed as part of a 120GHz integrated transmitter, fabricated in a 45nm low power CMOS technology. Accurate design and analysis of the 120GHz 3D antenna structure and carrier in Ansoft HFSS was carried out. The antenna occupies an area of 1100μm by 600μm. The radiating elements are lifted 500μm with respect to the silicon substrate and reflector which has lead to a radiation efficiency of 69% and an antenna gain of 4.4dBi. Also, an input bandwidth of 48GHz was achieved.

Acknowledgment

This research is partly supported by the ERC Advanced Grant 227680, the CHIPS K.U.Leuven Program Financing (Circuit design for smart and high-performance electronic systems). Furthermore, the authors would like to thank NXP research Eindhoven to support this work. The authors also thank the Institute for the Promotion of Innovation through Science and Technology in Flanders (IWT Vlaanderen).

References

[1] T. Al-Attar and T. Lee, "Monolithic integrated millimeter-wave IMPATT transmitter in standard CMOS technology," *Microwave Theory and Techniques, IEEE Transactions on*, vol. 53, no. 11, pp. 3557–3561, Nov. 2005.

[2] M. Nezhad Ahamdi, G. Rafi, and S. Safavi-Naeini, "An Efficient Integrated Antenna Structure on Low-Resistivity Silicon Substrate," in *Antennas and Propagation Society International Symposium 2006, IEEE*, July 2006, pp. 4353–4356.

[3] A. Babakhani, X. Guan, A. Komijani, A. Natarajan, and A. Hajimiri, "A 77-GHz Phased-Array Transceiver With On-Chip Antennas in Silicon: Receiver and Antennas," *Solid-State Circuits, IEEE Journal of*, vol. 41, no. 12, pp. 2795–2806, Dec. 2006.

[4] U. Johannsen, A. Smolders, J. Leiss, and U. Gollor, "Bond-wires: Readily available integrated millimeter-wave antennas," in *Microwave Conference (EuMC), 2012 42nd European*, 2012, pp. 197–200.

[5] N. Deferm, W. Volkaerts, J. Osorio, A. de Graauw, M. Steyaert, and P. Reynaert, "A 120GHz Fully Integrated 10Gb/s Wireless Transmitter with On-Chip Antenna in 45nm Low Power CMOS," in *ESSCIRC (ESSCIRC), 2013 Proceedings of the*, 2013, p. To be published.

Author Index